Book of Abstracts of the 60th Annual Meeting of the European Association for Animal Production

EAAP - European Federation of Animal Science

The European Association for Animal Production wishes to express its appreciation to the
Ministero delle Politiche Agricole e Forestali (Italy) and the
Associazione Italiana Allevatori (Italy)
for their valuable support of its activities.

Book of Abstracts of the 60th Annual Meeting of the European Association for Animal Production

Barcelona, Spain, August 24th - 27th, 2009

ISBN 978-90-8686-121-7
ISSN 1382-6077

First published, 2009

Sponsorships

Below the different types of Sponsorships

1. Meeting sponsor – From 3000 euro up
- Acknowledgements in the book of abstracts with contact address and logo.
- One page allowance in the final programme booklet of Barcelona.
- Advertising/information material inserted in the bags of delegates.
- Advertising/information material on a stand display.
- Acknowledgement in the EAAP Newsletter with possibility of a one page of publicity.
- Possibility to add session and speaker support (at additional cost to be negotiated).

2. Session sponsor – from 2000 euro up
- Acknowledgements in the book of abstracts with contact address and logo.
- One page allowance in the final programme booklet of Barcelona.
- Advertising/ information material in the delegate bag.
- ppt at beginning of session to acknowledge support and recognition by session chair.
- Acknowledgement in the EAAP Newsletter.

3. Speaker sponsor from 1000 euro up
- Half page allowance in the final programme booklet of Barcelona.
- Advertising / information material in the delegate bag.
- Recognition by speaker of the support at session.
- Acknowledgement in the EAAP Newsletter.

4. Registration Sponsor (equivalent to a full registration fee of the Annual Meeting)
- Acknowledgements in the book of abstracts with contact address and logo.
- Advertising/information material in the delegate bag.

Contact and further information
If you are interested to become a sponsor of the "EAAP Program Foundation" or want to have further information, please contact the Treasurer/Secretary Andrea Rosati:
rosati@eaap.org / eaap@eaap.org
Fax: +39 06 86329263
Phone +39 06 44202639

Acknowledgements

CRV Holding BV
P.O.Box 454
6800 AL ARNHEM
The Netherlands
crv@crvholding.com

IMV Technologies
10, rue Georges Clemenceau
61302 L'Aigle Cedex
France
www.imv-technologies.com

European Association for Animal Production (EAAP)

President: Kris Sejrsen
Secretary General: Andrea Rosati
Address: Via G.Tomassetti 3, A/I
I-00161 Rome, Italy
Phone: +39 06 4420 2639
Fax: +39 06 8632 9263
E-mail: eaap@eaap.org
Web: www.eaap.org

Organising Secretariat

President
Carlos Escribano — Director General de Recursos Agrícolas y Ganaderos, Ministerio de Medio Ambiente, Medio Rural y Marino; Madrid

Vice President
Isabel García Sanz — Subdirectora General de Medios de Producción Ganaderos, Ministerio de Medio Ambiente, Medio Rural y Marino; Madrid

Executive Secretary
Gerardo Caja — Departament de Ciència Animal i dels Aliments, Universitat Autònoma de Barcelona, Bellaterra, Barcelona

Members
Daniel Babot — Departament de Producció Animal, Universitat de Lleida, Lleida
Antoni Cambredó — Expoaviga, Fira de Barcelona, Barcelona
Montserrat Castellanos — Subdirección General de Medios de Producción Ganaderos, Ministerio de Medio Ambiente y Medio Rural y Marino, Madrid
Jose Antonio Fernández — Federación Española de Asociaciones de Ganado Selecto, Madrid
Carmen Garrido — Subdirección General de Medios de Producción Ganaderos, Ministerio de Medio Ambiente y Medio Rural y Marino, Madrid
Jose María Gómez-Nieves — Subdirección General de Ordenación y Buenas Prácticas Ganaderas, Ministerio de Medio Ambiente y Medio Rural y Marino, Madrid
Eduardo González — Subdirección General de Comercio Exterior y otras Producciones, Ministerio de Medio Ambiente y Medio Rural y Marino, Madrid
Joan Guim — Col·legi Oficial d'Enginyers Agrònoms de Catalunya, Barcelona
Francesc Monné — Col·legi Oficial de Veterinaris de Barcelona, Barcelona
Jesús Piedrafita — Departament de Ciència Animal i dels Aliments, Universitat Autònoma de Barcelona, Bellaterra
Josefina Plaixats — Institució Catalana d'Estudis Agraris, Barcelona; Departament de Ciència Animal i dels Aliments, Universitat Autònoma de Barcelona, Bellaterra
Joaquim Porcar — Subdirecció de Ramaderia, Departament d'Agricultura, Alimentració i Acció Rural, Generalitat de Catalunya, Barcelona
Francesc Puchal — Academia de Ciències Veterinàries de Catalunya, Barcelona
Joan Tibau — Institut de Recerca i Tecnologia Agroalimentària, Monells, Girona
Eduardo Torres — Departament d'Agricultura, Alimentació i Acció Rural, Generalitat de Catalunya, Barcelona
Andrea Urdampilleta — Expoaviga, Fira de Barcelona, Barcelona
Isabel Vázquez — Instituto Nacional de Investigación y Tecnología Agraria y Alimentaria, Madrid
Representante of ESAB — Escola Superior d'Agricultura de Barcelona, Universitat Politècnica de Catalunya, Barcelona

61st EAAP Annual Meeting of the European Association for Animal Production

Heraklion, Crete Island, Greece 23-27 August, 2010.
www.eaap2010.org

Organizing Committee of the 61st EAAP Meeting 2010

President
Prof. George Zervas — Rector of the Agricultural University of Athens and President of the Hellenic Society of Animal Production

Vice-President
Mr. George Zacharopoulos — General Director of Animal Production, Dept Ministry of Rural Development and Food

Executive Secretary
Prof. Kostas Fegeros — Dept of Animal Sciences and Aquaculture of the Agricultural University of Athens

Members
Prof. Stelios Deligeorgis — Dept of Animal Sciences and Aquaculture, Agricultural University of Athens

Prof. Andreas Georgoudis — Dept of Animal Sciences, Aristotle University of Thessaloniki

Assoc. Prof. Eleftheria Panopoulou — Dept of Animal Sciences and Aquaculture, Agricultural University of Athens

Assist. Prof. John Hadjigeorgiou — Dept of Animal Sciences and Aquaculture, Agricultural University of Athens

Dr. George Papadomichelakis — Dept of Animal Sciences and Aquaculture, Agricultural University of Athens

Dr. Christina Ligda — Nacional Agricultural Research Institution, Thessaloniki

Dr. Eleni Tsiplakou — Dept of Animal Sciences and Aquaculture, Agricultural University of Athens

Dr. Katerina Tsolakidi — Dept of Animal Production, Ministry of Rural Development and Food

Official Congress Organizing Bureau

Erasmus Conferences Tours & Travel Sa
1, Kolofontos & Evridikis str. - 161 21 Athens, Greece
Tel: +30 210.7257693 - Fax: +30 210 7257532
E-mail: info@eaap2010.org - Web-site: www.erasmus.gr

37th ICAR Session

Riga, Latvia, 1-4 June, 2010
Contact: ICAR Secretariat, Rome, Italy
Email: icar@eaap.org

S A B R E

CUTTING EDGE GENOMICS FOR SUSTAINABLE ANIMAL BREEDING

SABRE Conference @ EAAP

Food Quality and Safety

4th SABRE Conference integrated with 60th EAAP Annual Meeting

SABRE previously organised 3 conferences to disseminate the plans and results of the project: *"Sustainable Animal Breeding"* (September 2006), *"Genomics for Animal Health"* (June 2007 in collaboration with EADGENE - European Animal Disease Genomics Network of Excellence) and *"Welfare and Quality Genomics"* (September, 2008).

We would like to thank the organising committees of EAAP for the opportunity to integrate the 4th SABRE Conference into the 60th EAAP Annual Meeting. We are very pleased that over 30 papers which stated ***"These results are obtained through the EC-funded FP6 project SABRE."*** have been accepted by the EAAP committees to be presented in the poster halls and lecture theatres. These papers will be presented throughout various sessions, in an effort to reach the audiences who are interested in the subjects dealt within the SABRE project.

The following pages will provide you with a first insight into the project. For further details we invite you to attend the SABRE presentations or to visit www.sabre-eu.eu.

Chris Warkup, SABRE Coordinator
Toine Roozen, SABRE Operations Manager

Cutting Edge Genomics for Sustainable Animal Breeding

Animal breeders have made considerable progress in recent decades in improving the economic efficiency of food production (this is one of the reasons the real price of food has fallen), but in recent years animal breeding has become more complex with breeders needing to broaden their breeding objectives. Nowadays breeders want to improve a wide range of traits, such as product quality, welfare related fitness traits and disease resistance. Many of these traits are difficult or expensive to measure and this is where the science of genomics is valuable. Through research such as the SABRE project, scientists are beginning to unravel which genes and which variants of these genes are important to explaining the genetic component of these new selection traits. 'Sustainable Animal BREeding' is the main focus of the SABRE Integrated Research Project.

The European Integrated Research Project "SABRE" (Cutting Edge Genomics for Sustainable Animal BREeding) is an innovative four-year, €23 million pan-European project which utilises the latest techniques in genetic science to develop more economically and environmentally sustainable production systems for cattle, pigs and chickens. The 'headline' objectives of the project are:

- To provide fundamental knowledge on the genomics and epigenetics relating to livestock
- To provide understanding of biological systems central to sustainability
- To identify genes and markers allowing focused breeding for sustainability goals
- To demonstrate the effectiveness of genomics for sustainable breeding
- To disseminate existing knowledge and new results to the user community
- To develop skills and training to best capitalise on new genomics knowledge.

Thirty three leading animal breeding research groups and businesses have joined forces in the project which commenced in April 2006 and has been made possible by a €13.9 million grant under Thematic Priority 5, "Food Quality and Safety", of the 6th Research Framework Programme of the European Union (FP6). The SABRE work programme, involving almost 200 scientists in 14 countries, is divided into 13 Work Packages. These harness key areas of emerging genomic and epigenetic science to generate new knowledge and apply it in practical breeding improvement strategies throughout Europe.

SABRE will provide the fundamental knowledge of the genomics and epigenetics of animal health, food safety and food quality traits of livestock species, together with the strategies to deliver such technologies for use in selection. This will enable producers to move animal breeding and production towards more sustainable, environmentally and welfare friendly, low-input systems, that deliver safe and high quality foods in line with consumer expectations and European Policy.

Our overall strategy is to combine the power of gene mapping technologies, gene expression studies in target tissues and modern bioinformatics tools with available and expanded genome sequences, to determine the origin of genetic variation in key traits in important livestock species. These new breeding strategies will help industry improve animal health and welfare, adopt lower chemical and energy inputs, reduce livestock waste and pollution, produce safer and better quality foods whilst maintaining biodiversity and economic sustainability. The project will undoubtedly have a lasting positive impact on the EU animal breeding industry, scientific community, farmers and consumers long after its completion in 2010.

SABRE is designed to provide a range of new breeding strategies to improve animal health and welfare; reduce chemical and energy inputs; minimise livestock waste and pollution; and, maximise food safety and quality. The mammary gland, the digestive system and fertility are the focus of separate basic research packages, with more applied research aimed at enhancing eggshell quality for food safety; improving animal behaviour linked to welfare; and eliminating boar taint in pig meat. Three Work Packages address underpinning science, these are; numerical genomics, epigenetics and genomics with bioinformatics.

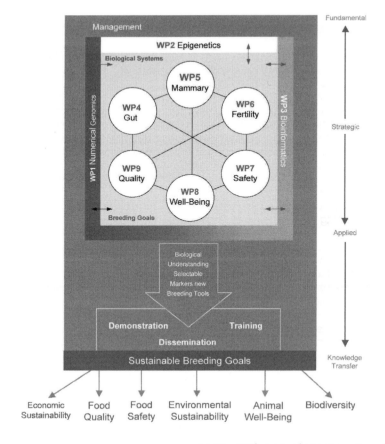

Finally, we have activity on demonstration of genome-wide selection in dairy cattle, and coordinated activity on training/mobility and dissemination of project outcomes.

SABRE Co-ordinator: Chris Warkup
SABRE Operations Manager: Toine Roozen
Genesis Faraday Partnership
Roslin BioCentre, Roslin, EH25 9PS, UK
t: +44 (0)131 527 4358
f: +44 (0)131 527 4335
info@sabre-eu.eu
www.sabre-eu.eu

SABRE Participants

No.	Participant	Country
1	Genesis Faraday Partnership	UK
2	Institut National de la Recherche Agronomique	F
3	ASG Lelystad	NL
4	The Roslin Institute and R(D)SVS, University of Edinburgh	UK
5	University of Aarhus	DK
6	Wageningen University	NL
7	Argentix Ltd	UK
8	Cordoba University	E
9	Parco Tecnologico Padano	I
10	Agricultural Research Organization, The Volcani Center	IL
11	MTT Agrifood Research Finland	FIN
12	Genus International plc	UK
13	Institute of Animal Genetics, Nutrition and Housing, University of Berne	CH
14	CNRS-UPR	F
15	Research Institute for the Biology of Farm Animals, FBN-Dummerstorf	D
16	Norwegian University of Agricultural Sciences	N
17	University of Bonn	D
18	Institut De Recerca I Tecnologia Agroalimentaries	E
19	Lohmann Tierzucht	D
20	University of Copenhagen	DK
21	University of Glasgow	UK
22	University of Munich	D
23	Cogent Breeding Ltd	UK
24	Genome Research Limited	UK
25	Institute for Pig Genetics	NL
26	BioBest Ltd	UK
27	Scottish Agricultural College	UK
28	Institute for Animal Health	UK
29	University of Medical Sciences Poznan	PL
30	JiangXi Agricultural University	CN
31	Zhejiang University	CN
32	China Agricultural University	CN
33	Universidade Federal De Viscosa	BR

Enter the Zulvacuum
You can't be more prepared
www.Bluetongue.eu

Serotypes 1, 4, 8, 1+8

Protected by Zulvac

FORT DODGE

 "la Caixa"

Caja de Ahorros y Pensiones de Barcelona, "la Caixa" was established in 1990 as a result of the merger of Caja de Pensiones para la Vejez y de Ahorros de Cataluña y Baleares, founded in 1904, and Caja de Ahorros y Monte de Piedad de Barcelona, founded in 1844, and it is therefore the lawful universal successor to the personality of those institutions, namely their nature, aims, rights and obligations.

It was established as a non-profit financial institution providing beneficent, community services with a private board of trustees, and is separate from any company or entity.

The corporate purpose of "la Caixa" is to encourage all authorized forms of saving, to carry out beneficent work for the wellbeing of the community and invest the related funds in safe and profitable assets of general interest.

The basic purposes set forth in its bylaws are:

- To encourage savings as the individual economic expression of collective interests.
- To encourage providence generally as the expression of an individual and collective interest.
- To provide financial services and services for the wellbeing of the community.
- To finance and support beneficent and/or community activities.
- To develop the Institution so as to enable it to fulfill its purposes in the most appropriate manner.

The Mission and Vision of "la Caixa," which are its strategic keys, have been shaped by its origins, corporate purpose and basic aims.

The Mission of "la Caixa"

To encourage saving and investment by offering the best and most comprehensive range of financial services to the greatest number of customers and make a decisive contribution to society to provide flexible tailored coverage of basic financial and social needs.

The Vision of "la Caixa"

Leading financial group in the Spanish market generating value for society, its customers and employees.

The strategic activity of "la Caixa" is underpinned by its identifying values, guidelines and characteristic convictions. The most important of all of the Institution's values are: quality, decentralization, responsibility, innovation, efficiency and safety.

AgroCaixa, developing rural areas

We at "la Caixa" show our strong commitment to arable and livestock farmers by assisting with the paperwork involved in the day-to-day running of a business and providing financial solutions that are specific to the arable and livestock industries and, by extension, to the rural world. In short, we stand alongside you to provide you with everything you need to achieve **optimum yield from your farm** and so assure your future.

AgroCaixa

This specialisation in financial products and services has led to our setting up **AgroCaixa**, aimed at farmers. Along with our network of branches in rural and agricultural areas, AgroCaixa provides the solutions that your farm or business deserves.

Sponsors

www.marm.es

Generalitat de Catalunya
**Departament d'Agricultura,
Alimentació i Acció Rural**

www20.gencat.cat/portal/site/DAR

www.expoaviga.com

www.lacaixa.es

CZ Veterinaria, S.A.

www.czveterinaria.com

www.fortdodge.com

www.bcn.es

www.ancce.es

www.illumina.com

www.agronoms.org

www.colvet.es

www.veterinaris.cat

www.asas.org

www.adsa.org

www.fruitsponent.com

www.invac.org

www.feagas.es

www.ancri.org

www.mercabarna.es

www.cag.es

www.cesfac.com

www.brunadelspirineus.org

www.eurocai.com

www.jamondehuelva.com

www.riojawine.com

www.gencat.cat

D.O. Catalunya

www.uab.es

www.irta.es

www.udl.es

Contributors

I. G. P. Cordero de Extremadura – Corderex (www.corderex.com)

Xata Roxa – Ternera Roxa (www.viaganadera.com/prodycar)

I. G. P. Ternera Asturiana (www.terneraasturiana.org)

I. G. P. Carne de Ávila (www.carnedeavila.org)

I. G. P. Ternera de Navarra (www.terneradenavarra.com)

Asociación Nacional de Criadores de Vacuno de Raza Retinta (www.retinta.es)

D. O. Jamón de Teruel (www.jamondeteruel.com)

Fundación Jamón Serrano (www.fundacionserrano.org)

D. O. Dehesa de Extremadura (www.dehesa-extremadura.com)

D. O. Jamón Los Pedroches (www.jamondolospedroches.com)

D. O. Queso de Mahón (www.quesomahonmenorca.com)

D. O. Queixo Tetilla (www.queixotetilla.org)

D. O. Queso Cabrales (www.fundacioncabrales.com)

D. O. Queso de Murcia y Queso de Murcia al vino (www.quesosdemurcia.com)

D. O. Queso Nata de Cantabria, Quesucos de Liébana y Picón Bejes-Tresvino

(www.mapa.es/es/alimentacion/pags/Denominacion/queso_lacteo)

D. O. Jerez-Xérèz-Sherry (www.sherry.org)

D. O. Valdepeñas (www.dovaldepenas.es)

Grupo Damm (www.damm.es)

Novartis (www.novartis.com)

Universitat Politècnica de Catalunya (www.upc.es)

Fundació Sagrada Familia (www.sagradafamilia.cat)

Institució Catalana d'Estudis Agraris (www.iecat.net)

Scientific Programme EAAP 2009

Monday 24 August 8.30 – 12.30	Monday 24 August 14.00 – 18.00	Tuesday 25 August 8.30 – 12.30	Tuesday 25 August 14.00 – 18.00
Session 1 Local breeds: what future? 1. Selection Chair: H. Simianer (DE)	Session 9 Local breeds: what future? 2. Farming systems and products Chair: I. Casasús (ES)	Plenary session Leroy Fellowship Award Lecture W. Martin-Rosset (FR)	Session 18 Practical implementation of marker-assisted selection in pig and poultry breeding Chair: P. Knap (DE)
Session 2 Symposium on protein metabolism and N excretion in dairy cattle Chair: A. van Vuuren (NL) and C. Thomas (UK)	Session 10 EAAP-ASAS-ADSA Growth and Development Symposium: Methods and animal models used to study physiological aspects of postnatal growth and development in farm animals Chair: M. Vestergaard (DK)	Conference: EU Policy on supporting animal research J. M. Silva (EU – DG Research)	Session 19 Horse genetics with special emphasis on defects and longevity Chair: S. Janssens (BE)
Session 3 The impact of competition between food, feed and fuel on livestock industry Chair: A. Van der Zijpp (NL)	Session 5 Nutrition and product quality Chair: J. Aguilera (ES)	Session: Impact of Darwin's theories on livestock breeding	Session 20 Are organic farming systems sustainable? Chair: J. Hermansen (DK)
Session 4 Impact of global market on cattle breeding programs and practices Chair: J. Juga (FIN) and G. Thaller (DE)	Session 11 Future of non-production traits for breeding and management of beef and dairy husbandry Chair: M. Coffey (UK)		Session 21 What contributes to low still birth and succesfull herd replacement in cattle? Chair: F. Hocquette (FR)
Session 7 Dog breeding and behaviour Chair: E. Strandberg (SE)	Session 12 Indicators and analysis of risk factors in livestock welfare Chair: N. Bareille (FR)		Session 22 Promising applications of Nutrigenomics in animal science Chair: A.J. Connolly (IE)
Session 8 Sustainability in livestock production gains from improved coordination of livestock research	Session 13 Practical strategies and tools for the genetic management of farm animal populations Chair: M.A. Toro (ES)		Session 23 Animal transportation (welfare, handling, risk assessment, economics) Chair: G. Giovagnoli (IT)
	Session 14 Genetics of meat animals free communications Chair: J.S. Szyda (PL)		Session 24 Selection in harsh environments: methods and results and Sheep and goat genetics free communications Chair: L. Dempfle (DE)
	Session 15 Dairy genetics free communications Chair: G. Schopen (NL)	Session 17 Writing and Presenting Scientific Papers Chair: B. Malmfors (SE)	
	Session 16 FABRE-TP Technology platform in animal breeding: updating of the EC strategic research agenda Chair: A.M. Neeteson (NL)	Tuesday 25 August 18.00 – 19.30 Poster session Poster viewing and discussion with authors	

Wednesday 26 August 8.30 – 12.30	Wednesday 26 August 13.30 – 18.00	Thursday 27 August 8.30 – 12.30	Thursday 27 August 14.00 – 18.00
Session 25 Sheep and goat reproduction Chair: L. Bodin (FR)	S33-S41 Commission future programme and elections meetings (14.00 – 15.00)	Session 42 Health issues and immunocompetence in pig production Chair: S. Chadd (UK)	Session 50a Biosecurity and free range systems Chair: S. Bastian (FR) Session 50b Emerging zoonotic diseases Chair: C. Fourichon (FR)
Session 26 Physiology and genetics of stress and behaviour Chair: X. Manteca (ES)	Followed by Free communications on (15.00 – 18.00)	Session 44 The role of livestock farming in rural development Chair: K. Eilers (NL)	
Session 27 Methods to assess livestock farming systems dynamics: adaptive strategies to changing socio-economic environment Chair: S. Ingrand (FR)	Session 33 Animal Genetics (methodology) Chair: V. Ducrocq (FR) Session 34 Livestock Farming Systems Chair: V. Matlova (CZ)	Session 45 Molecular tools for disease resistance Chair: A. Sánchez (ES)	Session 51 Management of pig feeding: health, environment and social implications Chair: D. Torrallardona (ES)
Session 28 Genomic selection Chair: E. Strandberg (SE)	Session 35 Animal Management and health Chair: C. Fourichon (FR)	Session 47 Feed additives to improve diet utilisation Chair: G. van Duinkerken (NL)	Session 52 Animal Nutrition Free Communications Chair: J.E. Lindberg (SE)
Session 29 Reproduction technologies in horses Chair: E. Palmer (FR)	Session 36 Animal Nutrition Chair: J.E. Lindberg (SE)	Session 48 Cattle Network: Sustainable production in Mediterranean countries Chair: K. Osoro (ES)	Session 53 Genetics Free Communications (methodology) Chair: V. Ducroq (FR)
Session 30 Management and processing for high quality forages Chair: L. Bailoni (IT)	Session 37 Animal Physiology Chair: J. Rátky (HU) Session 38 Cattle Production Chair: G. Keane (IE)	Session 49a Sheep and goat organic farming and product marketing (8.30 – 10.15) Chair: M. Gauly (DE) Session 49b	Session 54 Cattle Production Free Communications Chair: M. Klopcic (SLO)
Session 31 Ways of valorization of local animal products (10:30 – 12:30)	Session 39 Horse Production (management and nutrition) Chair: M.J. Fradinho (PT)	Free communications on sheep and goat feeding and breeding (10.45 – 12.30) Chair: M. Gauly (DE)	Session 55 Incorporation of ethical considerations in professions in livestock industry Chair: M. Marie (FR)
Session 32 Animal fibre science Chair: C. Renieri (IT)	Session 40 Pig Production Chair: P. Knap (DE) Session 41 Sheep and Goat Production Chair: M. Schneeberger (CH)	Session 43 Symposium on "Advances and use of electronic identification" Chair: J. Robles (ES), O.K. Hansen (DK), B. Besbes (FAO) and J.W. Oltjen (USA) Session 46 Horse Network: Performance Horses. Open questions and challenges for future research and practice Chair: N. Miraglia (IT) and W. Martin-Rosset (FR)	

Commission on Animal Genetics

Dr Ducrocq	President	INRA
	France	Vincent.ducrocq@dga.jouy.inra.fr
Prof. Dr Simianer	Vice-President	University of Goettingen
	Germany	simianer@genetics-network.de
Dr Gandini	Vice-President	University of Milan
	Italy	gustavo.gandini@unimi.it
Dr Strandberg	Secretary	SLU
	Sweden	Erling.Strandberg@hgen.slu.se
Dr Szyda	Secretary	Agricultural University of Wroclaw
	Poland	szyda@karnet.ar.wroc.pl
Alfred de Vries	Industry rep.	CRV

Commission on Animal Nutrition

Dr Lindberg	President	Swedish University of Agriculture
	Sweden	jan-eric.lindberg@huv.slu.se
Dr Bailoni	Vice-President	University of Padova
	Italy	Lucia.bailoni@unipd.it
Dr Connolly	Vice-President/	
	Industry rep.	Alltech
	Ireland	aconnolly@alltech.com
Dr Moreira	Secretary	University of the Azores
	Portugal	ocmoreira@netcabo.pt
Dr Cenkvàri	Secretary	Szent Istvan University
	Hungary	Czenkvari.Eva@aotk.szie.hu

Commission on Animal Management & Health

Dr Fourichon	President	Veterinary School – INRA
	France	Fourichon@vet-nantes.fr
Dr Spoolder	Vice-President	ASG-WUR
	Netherlands	Hans.spoolder@wur.nl
Dr Geers	Vice-President	Zootechnical Centre - K.U.Leuven
	Belgium	Rony.geers@agr.kuleuven.ac.be
Dr Edwards	Secretary	University of Newcastle Upon Tyne
	United Kingdom	Sandra.edwards@ncl.ac.uk

Commission on Animal Physiology

Dr Vestergaard	President	Aarhus University
	Denmark	Mogens.Vestergaard@agrsci.dk
Dr M. Kuran	Vice-President	Gaziosmanpasa University
	Turkey	mkuran@gop.edu.tr
Dr Royal	Vice-President	University Liverpool
	UK	mdroyal@liverpool.ac.uk
Dr Bruckmaier	Vice-President	University of Bern
	Switzereland	Rupert.bruckmaier@physio.unibe.ch
Dr Quesnel	Secretary	INRA Saint Gilles
	France	Helene.quesnel@rennes.inra.fr
Dr Scollan	Secretary	Institute of Biological, Environmental and rural sciences
	UK	ngs@aber.ac.uk

Commission on Livestock Farming Systems

Dr Bernués Jal	President	CITA
	Spain	abernues@aragon.es
Dr Hermansen	Vice-President	DIAS
	Denmark	john.hermansen@agrsci.dk
Dr Leroyer	Vice president/ industry rep.	ITAB
	France	Joannie.leroyer@itab.asso.fr
Dr Matlova	Vice-President	Res. Institute for Animal Production
	Czech Republic	matlova.vera@vuzv.cz
Dr Peters	Vice-President	Humboldt-University Berlin
	Germany	k.peters@agrar.hu-berlin.de
Dr Eiler	Secretary	Wageningen University
	Netherlands	karen.eilers@wur.nl
Dr Ingrand	Secretary	INRA/SAD
	France	ingrand@clermond.inra.fr

Commission on Cattle Production

Dr Kuipers	President	Wageningen UR
	Netherlands	abele.kuipers@wur.nl
Dr Thaller	Vice-President	Animal Breeding and Husbandry
	Germany	Georg.Thaller@tierzucht.uni-kiel.de
Dr Keane	Vice-President	TEAGASC
	Ireland	gkeane@grange.teagasc.ie
Dr Coffey	Vice president/ Industry rep.	SAC, Scotland
	UK	Karin.hendry@sac.ac.uk
Dr Hocquette	Secretary	INRA
	France	hocquet@clermont.inra.fr
Dr Klopcic	Secretary	University of Ljublijana
	Slovenia	Marija.Klopcic@bfro.uni-lj.si

Commission on Sheep and Goat Production

Dr Schneeberger	President Switzerland	ETH Zentrum markus.schneeberger@inw.agrl.ethz.ch
Dr Ringdorfer	Vice-President Austria	LFZ Raumberg-Gumpenstein ferdinand.ringdorfer@raumberg-gumpenstein.at
Dr Bodin	Vice President France	INRA-SAGA Loys.bodin@toulouse.inra.fr
Dr Papachristoforou	Vice President Cyprus	Agricultural Research Institute Chr.Papachristoforou@arinet.ari.gov.cy
Prof. Gauly	Secretary Germany	University Göttingen mgauly@gwdg.de
Dr Milerski	Secretary/ Industry rep. Czech Republic	Research Institute of Animal Science m.milerski@seznam.cz
Dr Sagastizabal	Secretary/ Industry rep.	NEIKER-Tecnalia eugarte@neiker.net

Commission on Pig Production

Dr Knap	President Germany	PIC International Group pieter.knap@pic.com
Dr Chadd	Vice-President UK	Royal Agric. College steve.chadd@royagcol.ac.uk
Dr Torrallardona	Vice-President Spain	IRTA David.Torrallardona@irta.es
Dr Pescovicova	Secretary Slovak Republic	Research Institute of Animal Production peskovic@vuzv.sk
Dr Manteca	Secretary Spain	Universitat Autònoma de Barcelona xavier.manteca@uab.es

Commission on Horse Production

Dr Miraglia	President Italy	Molise University miraglia@unimol.it
Dr Burger	Vice president Switzerland	Clinic Swiss National Stud Dominique.burger@mbox.haras.admin.ch
Dr Janssen	Vice president Belgium	BIOSYST Steven.janssens@biw.kuleuven.be
Dr Lewczuk	Vice president Poland	IGABPAS d.lewczuk@ighz.pl
Dr Coenen	Vice president Germany	University of Leipzig coenen@vetmed.uni-leipzig.de
Dr Palmer	Vice president/ Industry rep. France	CRYOZOOTECH ericpalmer@cryozootech.com
Dr Holgersson	Secretary Sweden	Swedish University of Agriculture Anna-Lena.holgersson@hipp.slu.se
Dr Hausberger	Secretary France	CNRS University Martine.hausberger@univ-rennes1.fr

Session 01. Local breeds: what future? 1. Selection

Date: 24 August '09; 08:30 - 12:30 hours
Chairperson: Simianer (DE)

Session 02. Symposium on protein metabolism and N excretion in dairy cattle

Date: 24 August '09; 08:30 - 12:30 hours
Chairperson: A. van Vuuren (NL) and C. Thomas (UK)

Session 03. The impact of competition between food, feed and fuel on livestock industry

Date: 24 August '09; 08:30 - 12:30 hours
Chairperson: Van Der Zijpp (NL)

Session 04. Impact of global market on cattle breeding programs and practices

Date: 24 August '09; 08:30 - 12:30 hours
Chairperson: J. Juga (FIN) and G. Thaller (DE)

Session 05. Nutrition and product quality

Date: 24 August '09; 14:00 - 18:00 hours
Chairperson: Aguilera (ES)

Session 07. Dog breeding and behaviour

Date: 24 August '09; 08:30 - 12:30 hours
Chairperson: Strandberg (SE)

Session 09. Local breeds: what future? 2. Farming systems and products

Date: 24 August '09; 14:00 - 18:00 hours
Chairperson: Casasus (ES)

Session 10. EAAP-ASAS-ADSA Growth and Development Symposium

Date: 24 August '09; 14:00 - 18:00 hours
Chairperson: Vestergaard (DK)

Session 11. Future of non-production traits for breeding and management of beef and dairy husbandry

Date: 24 August '09; 14:00 - 18:00 hours
Chairperson: Coffey (UK)

Genetic relationships between metabolic diseases and fertility in Fleckvieh dual
purpose cattle 15 110
Koeck, A., Fuerst-Waltl, B., Fuerst, C. and Egger-Danner, C.

Session 12. Indicators and analysis of risk factors in livestock welfare

Date: 24 August '09; 14:00 - 18:00 hours
Chairperson: Bareille (FR)

Session 13. Practical strategies and tools for the genetic management of farm animal populations

Date: 24 August '09; 14:00 - 18:00 hours
Chairperson: Toro (ES)

Mating animals by minimising the covariance between ancestral contributions generates more genetic gain without increasing rate of inbreeding in breeding schemes with optimum-contribution selection
Henryon, M., Sørensen, A.C. and Berg, P.

Session 14. Genetics of meat animals free communications

Date: 24 August '09; 14:00 - 18:00 hours
Chairperson: Szyda (PL)

Session 15. Dairy genetics free communications

Date: 24 August '09; 14:00 - 18:00 hours
Chairperson: Schopen (NL)

Theatre   **Session 15 no. Page**

Session 18. Practical implementation of marker-assisted selection in pig and poultry breeding

Date: 25 August '09; 14:00 - 18:00 hours
Chairperson: Knap (DE)

Session 19. Horse genetics with special emphasis on defects and longevity

Date: 25 August '09; 14:00 - 18:00 hours
Chairperson: Janssens (BE)

Session 20. Are organic farming systems sustainable?

Date: 25 August '09; 14:00 - 18:00 hours
Chairperson: Hermansen (DK)

Session 21. What contributes to low still birth and succesfull herd replacement in cattle?

Date: 25 August '09; 14:00 - 18:00 hours
Chairperson: Hocquette (FR)

Session 22. Promising applications of Nutrigenomics in animal science

Date: 25 August '09; 14:00 - 18:00 hours
Chairperson: Connolly (IE)

Session 23. Animal transportation (welfare, handling, risk assessment, economics)

Date: 25 August '09; 14:00 - 18:00 hours
Chairperson: Giovagnoli (IT)

Session 24. Selection in harsh environments: methods and results and sheep and goat genetics free communications

Date: 25 August '09; 14:00 - 18:00 hours
Chairperson: Dempfle (DE)

Session 25. Sheep and Goat reproduction

Date: 26 August '09; 08:30 - 12:30 hours
Chairperson: Bodin (FR)

Session 26. Physiology and genetics of stress and behaviour

Date: 26 August '09; 08:30 - 12:30 hours
Chairperson: Manteca (ES)

Gene expression and ontological analyses reveal a key role of LINEs in endurance horse race-induced stress conditions

Cappelli, K., Capomaccio, S., Galla, G., Barcaccia, G., Felicetti, M., Silvestrelli, M. and Verini-Supplizi, A.

Session 27. Methods to assess livestock farming systems dynamics: adaptive strategies to changing socio-economic environment

Date: 26 August '09; 08:30 - 12:30 hours
Chairperson: Ingrand (FR)

Session 28. Genomic selection

Date: 26 August '09; 08:30 - 12:30 hours
Chairperson: Strandberg (SE)

Session 29. Reproduction technologies in horses

Date: 26 August '09; 08:30 - 12:30 hours
Chairperson: Palmer (FR)

Session 30. Management and processing for high quality forages

Date: 26 August '09; 08:30 - 12:30 hours
Chairperson: Bailoni (IT)

Session 32. Animal fibre science

Date: 26 August '09; 08:30 - 12:30 hours
Chairperson: Renieri (IT)

Session 33. Animal genetics (methodology) free communications

Date: 26 August '09; 14:00 - 18:00 hours
Chairperson: Ducrocq (FR)

Session 34. Livestock farming systems free communications

Date: 26 August '09; 14:00 - 18:00 hours
Chairperson: Matlova (CZ)

Session 35. Animal Management and Health Free Communications

Date: 26 August '09; 14:00 - 18:00 hours
Chairperson: Fourichon (FR)

Session 36. Animal Nutrition Free Communications

Date: 26 August '09; 14:00 - 18:00 hours
Chairperson: Lindberg (SE)

Session 37. Animal Physiology Free Communications

Date: 26 August '09; 14:00 - 18:00 hours
Chairperson: Ratky (HU)

Theatre   **Session 37 no.** **Page**

Session 38. Cattle Production Free Communications

Date: 26 August '09; 14:00 - 18:00 hours
Chairperson: Keane (IE)

Session 39. Horse Production (management and nutrition)

Date: 26 August '09; 14:00 - 18:00 hours
Chairperson: Fradinho (PT)

Session 40. Pig Production Free Communications

Date: 26 August '09; 14:00 - 18:00 hours
Chairperson: Knap (DE)

Session 41. Sheep and Goat Production Free Communications

Date: 26 August '09; 14:00 - 18:00 hours
Chairperson: Schneeberger (CH)

Session 42. Health issues and immunocompetence in pig production

Date: 27 August '09; 08:30 - 12:30 hours
Chairperson: Chadd (UK)

Session 43. Symposium on 'Advances and use of electronic identification'

Date: 27 August '09; 08:30 - 18:00 hours
Chairperson: J. Robles (ES), O.K. Hansen (DK), B. Besbes (FAO) and J.W. Oltjen (USA)

Session 44. The role of livestock farming in rural development

Date: 27 August '09; 08:30 - 12:30 hours
Chairperson: Eilers (NL)

Theatre   **Session 44 no.**   **Page**

Session 45. Molecular tools for disease resistance

Date: 27 August '09; 08:30 - 12:30 hours
Chairperson: Sanchez (ES)

Session 46. Performance Horses. Open questions and challenges for future research and practice

Date: 27 August '09; 08:30 - 18:00 hours
Chairperson: N. Miraglia (IT) and W. Martin-Rosset (FR)

Session 47. Feed additives to improve diet utilisation

Date: 27 August '09; 08:30 - 12:30 hours
Chairperson: Van Duinkerken (NL)

The effectiveness of a recombinant cellulase used to supplement a barley-based feed for free-range broilers is limited by crop beta-glucanase activity and barley endoglucanases 6 516
Ponte, P.I.P., Guerreiro, C.I.P.D., Crespo, J.P., Crespo, D.G., Ferreira, L.M.A. and Fontes, C.M.G.A.

Session 48. Cattle Network: Sustainable production in Mediterranean countries

Date: 27 August '09; 08:30 - 12:30 hours
Chairperson: Bonnet (FR)

Session 49a. Sheep and goat organic farming and product marketing

Date: 27 August '09; 08:30 - 10:15 hours
Chairperson: Gauly (DE)

Session 49b. Free communications on sheep and goat feeding and breeding

Date: 27 August '09; 10:45 - 12:30 hours
Chairperson: Gauly (DE)

Session 50a. Biosecurity and free range systems

Date: 27 August '09; 14:00 - 16:15 hours
Chairperson: Bastian (FR)

Session 50b. Emerging zoonotic diseases

Date: 27 August '09; 16:30 - 18:00 hours
Chairperson: Fourichon (FR)

Session 51. Management of pig feeding: health, environment and social implications

Date: 27 August '09; 14:00 - 18:00 hours
Chairperson: Torrallardona (ES)

Poster   **Session 51 no. Page**

Session 52. Animal Nutrition Free Communications

Date: 27 August '09; 14:00 - 18:00 hours
Chairperson: Lindberg (SE)

Theatre   **Session 52 no. Page**

Session 53. Animal genetics (methodology) free communications

Date: 27 August '09; 14:00 - 18:00 hours
Chairperson: Ducrocq (FR)

Session 54. Cattle Production Free Communications

Date: 27 August '09; 14:00 - 18:00 hours
Chairperson: Klopcic (SLO)

Session 55. Incorporation of ethical considerations in professions in livestock industry

Date: 27 August '09; 14:00 - 18:00 hours
Chairperson: Marie (FR)

Session 01

Why and how to select the Iberian pig breed?

Silió, L. and García-Casco, J., INIA, Mejora Genética Animal, Carretera A Coruña km 7.5, 28040 Madrid, Spain; garcia.juan@inia.es

The production of Iberian pigs increased along the last two decades providing the raw material of meat products of high sensorial quality. Most of the 3×10^6 animals slaughtered per year are crossbred Duroc x Iberian pigs reared in conventional farms, but purebred Iberian pigs fattened with acorns and pasture are the source of the most expensive dry-cured products. Genetic improvement of the low reproductive performance of Iberian sows is necessary because maternal Iberian origin is mandatory for labelling Iberian products. Joint selection of dam lines for litter size and mothering ability using multi trait models is an advisable approach. Estimated genetic parameters for number of live born piglets (NBA) and litter weight at 21days (LW) showed changes in the genetic basis of these traits between first and later parities: $h^2=0.15$ (NBA1), 0.12 (NBA2+), 0.22 (LW1-2), 0.15 (LW3+), and genetic correlations (r_g) from 0.44 to 0.84. Moreover, selection of Iberian sire lines is required for elite production based on purebred pigs. The four main productive traits are daily growth along the final fattening, percentages on carcass weight of trimmed hams, forelegs and loins. Moderate to high h^2 values (0.37-0.48) have been estimated for these traits, being positive the genetic correlations between the percentages of diverse premium cuts (0.36-0.69). Selection for this breeding goal -based on records of these traits from relatives or descendents- may be effective but delayed by a long generation interval. Estimated values of r_g between intramuscular fat content (IMF) and daily growth (-0.05) or percentage of forelegs (0.07) were not significant, but significant r_g values were found between IMF and percentages of hams (-0.19) and loins (-0.23). This genetic antagonism points risks of deterioration of sensorial meat quality due to selection in a medium-time horizon. Molecular genetic tests -mainly based on genes related to feed intake and lipid metabolism- can be useful by reducing the generation interval or by increasing the accuracy of traits of expensive recording.

Session 01

Is there room for selection for meat quality in local breeds? A simulation study

Gourdine, J.L.[1], Sørensen, A.C.[2], De Greef, K.[3] and Rydhmer, L.[4], [1]INRA, UR143, Department of Animal Genetics, Domaine Duclos, 97170 Petit Bourg, Guadeloupe, France, [2]Aarhus University, Department of Genetics and Biotechnology, P.O. Box 50, 8830 Tjele, Denmark, [3]WUR, Animal Sciences Group, Box 65, 8200 AB Lelystad, Netherlands, [4]SLU, Department of Animal Breeding and Genetics, P.O. Box 7023, 75007 Uppsala, Sweden; Lotta.Rydhmer@hgen.slu.se

The present simulation study was conducted within the EU-supported project Q-Porkchains. Except for large populations, like the Iberian pig, there is no well defined breeding program with specified traits to be improved in local pig breeds. Due to small population size, focus is on management of genetic diversity. However, in presence of pedigree information, optimal contribution selection (OCS) can be applied to control inbreeding rate and to avoid low performance in valuable traits. The aim of this study is to assess to what extent a breeding program to improve a meat quality trait (MQT) in a small local breed population is feasible. Simulations are performed by applying the stochastic simulation program ADAM to 16 scenarios, different with regard to 1) population size (2,000 to 8,000 reproducers); 2) type of mating (AI vs. natural); 3) selection rules (random vs. OCS) and 4) heritability of MQT ($h^2=0.2$ vs. 0.4). It is assumed that the local breed is reared in an extensive system (24 sows/herd) for a high meat quality market and that MQT is measured at the slaughterhouse. As age at slaughter is assumed to be higher than pubertal age, selection is based on pedigree information. The simulations (still running) will result in quantitative examples of sustainable breeding schemes for local pig breeds. We expect that OCS can help us achieve genetic progress at the same rate of inbreeding as obtained using random selection.

Genetic progress and inbreeding control in simulated cattle breeds with low population numbers

Gandini, G.[1], Del Corvo, M.[2], Spagnoli, E.[2], Jansen, G.[3] and Stella, A.[4], [1]Università degli Studi di Milano, Department VSA, Via Celoria 10, 20133 Milano, Italy, [2]IBBA-CNR, Via Einstein, Loc. Cascina Codazza, Lodi, Italy, [3]Dekoppel Consulting, Chiaverano, 10010, Italy, [4]CERSA-PTP, Via Einstein, Loc. Cascina Codazza, Lodi, Italy; gustavo.gandini@unimi.it

Most local breeds do not benefit from modern breeding techniques. Selection programmes capable of increasing their genetic ability for productivity, and consequently profitability for the farmer, would be beneficial. However breeding goals should respect the conservation value of the breed and selection schemes should take into account maintenance of genetic variation, adaptation to the farming environment and cost feasibility. Genetic progress in a milk trait, and alternatively in one milk and one meat trait, was analysed in simulated demographically structured dairy and dual purpose cattle populations of 500 to 4,000 cows. A scheme considering only selection of dams of sires was chosen when simulating dairy cattle because its adaptability to a wide range of farming contexts, including low input breeding systems. With dual purpose cattle, selection of dams of sires on a milk trait and selection of sires on pedigree index after performance test were simulated. In the simulations inbreeding rates were controlled by computing optimal genetic contributions given a penalty to average additive genetic relationship among future breeding animals. The effect of including in the model male and female genetic contributions of previous selection decisions was also evaluated. Average genetic responses of about 0.2 standard deviations per year were observed at an inbreeding rate of approximately one percent per generation.

Marker assisted EBV using a dense SNP markers map: application to the Normande and Montbéliarde breeds

Guillaume, F.[1,2], Fritz, S.[3], Ducrocq, V.[2], Croiseau, P.[2] and Boichard, D.[2], [1]Institut de l Elevage, 149 rue de Bercy, 75595 Paris, France, [2]INRA, UMR1313-GABI, Domaine de Vilvert, 78352 Jouy-en-josas, France, [3]UNCEIA, 149 rue de Bercy, 75595 Paris, France; francois.guillaume@jouy.inra.fr

Marker assisted selection based on linkage equilibrium has been applied in France since 2001, in the Holstein, Normande and Montbéliarde breeds. This programme led to the confirmation of numerous QTL on a large population (respectively, on 20,000, 6,000 and 8,000 individuals). In 2008, the use of a 54K SNP chip allowed the inclusion of linkage disequilibrium in the evaluation model, and noticeable increases in accuracy were demonstrated in the Holstein breed. The present study investigates the benefits of SNP genotypes observed in the Normande and Montbéliarde breed. Respectively 661 and 601 sires of these two breeds were genotyped. Among them, 152 and 144 young sires did not have any phenotype (Daughter Yield Deviations - DYD) in 2004. Based on phenotypes available in 2004, polygenic EBV and MA-EBV were computed for the young sires, and compared to their 2008 DYD. With the current MA-EBV model (a two-step approach using known QTL with a haplotype-based model), gains in accuracy between polygenic and MA-EBV ranged from + 0.29 for milk yield to +0.14 for somatic cell count in Normande breed and +0.28 for protein percentage to +0.17 for fertility in Montbéliarde breed. These results are currently being compared to Holstein results and to genomic selection approaches. Key factors for an efficient MA-EBV implementation in local breed populations will be discussed.

Deterministic and stochastic prediction of the evolution of QTL genotype frequencies with overlapping generations and finite population size

Ytournel, F. and Simianer, H., Georg-August University, Goettingen, Department of Animal Sciences, Animal Breeding and Genetics Group, Albrecht-Thaer-Weg 3, 37075 Goettingen, Germany; fytourn@gwdg.de

The evolution of the molecular and statistical methods provides the opportunity of a more precise knowledge of the positions of Quantitative Trait Loci (QTL) and even the discovery of some causal mutations. Deterministic methods have been proposed to include them in the optimization of breeding programs. We chose here a method focusing on the optimization of the genetic progress over the generations and accounting for the linkage disequilibrium between the QTL and the polygene. It supposes both an infinite population size and a constant polygenic variance. We assumed that two QTL were known and included in the selection process which lasted over ten generations. We looked at the influence of a finite population size (100 and 10,000 individuals) and overlapping generations on the realized development of genotype frequencies in simulated populations. Using simulations also gave us an insight into the variability of the genotype frequencies. Both small population sizes and overlapping generations delayed the fixation of the most favourable genotype compared to the deterministic expectation. The factor 'overlapping generations' showed the largest effect. The standard deviations of these frequencies were also higher for small population sizes and the highest values were reached later in populations with overlapping generations. This was due to the remaining unfavourable alleles in the population and to the reduction of the polygenic variance. These results show the necessity of accounting for the real population parameters to estimate the efficiency of a selection program.

On the importance of diffusion management for local breeds selection

Labatut, J., Girard, N., Astruc, J.M. and Bibé, B., INRA AGIR, SAGA, Institut elevage, BP 52627, 31326 Castanet Tolosan Cedex, France; julie.labatut@toulouse.inra.fr

Researches on local breeds have mainly focused on the production of genetic gain, while its diffusion used to be taken for granted, or considered as of less importance as the State was subsidizing official breeding schemes. However, diffusion and sustainability of small local breeding schemes are threatened by current changes in breeding activities and organizations - diversification of breeding objectives, liberalization of public policies, decrease in public support. Local breeds are particularly concerned, as they may be threatened by more competitive and widespread ones. Indeed, the management of the diffusion dimension of breeding activities gets a greater importance. Thus, there is a need for a better understanding of the market of genetic gain and the strategies of its participants. To investigate this question, we study with quantitative and qualitative data the way of the genetic market works in the case of local dairy sheep breeds in the Western-Pyrenees. In this area, the use of artificial insemination outside nucleus flocks is weak. The diffusion is mainly based on the exchanges of live breeding animals, but the number and substance of the exchanges are unknown. We analyze two types of markets set up: the official sale of breeding animals, organized by the breeding center; the parallel market of rams' exchanges by mutual agreement between farmers. We find several paradoxical results: the more expensive animals are sold outside of the breeding schemes; the breeding center does not find enough buyers for its rams, while there is a shortage of rams in the region; outside the breeding schemes, the parallel market of ram's is dominant. We also identify that the diversity of prices on the market can not be explained only according to scientific evaluation of animals. We show the existence of a secondhand market of rams. In conclusion, we argue that when the use of artificial insemination is quite limited, organisms of selection have to give more importance on the market of rams.

Breeding goals for local sheep breeds in Austria

Baumung, R. and Fuerst-Waltl, B., University of Natural Resources and Applied Life Sciences Vienna, Dep. Sust. Agric. Syst., Div. Livestock Sciences, Gregor-Mendel-Strasse 33, 1180 Vienna, Austria; roswitha. baumung@boku.ac.at

A great variety of different autochthonous sheep breeds exists in the Eastern part of the Alps. The main breed in Austria is Tyrol Mountain (Tiroler Bergschaf). It is an a-seasonal extensive breed, which is kept for meat production but also for landscape management in alpine regions. Mountain sheep are adapted to the harsh alpine conditions. The actual population size (herd book ewes) is 12,300. The current breeding goal focuses on conformation and type traits. To ensure competitiveness in the future a new breeding programme based on an economic breeding goal must be established. A total merit index with relative economic values of approximately 85% for functional and 15% for fattening traits is recommended. However, an analysis of auction prices showed that conformation traits, especially type score representing the breed characteristics in Mountain sheep, influence the price at auction in ewes and rams. Therefore it seems justified to discuss the breeders' desire to include the conformation complex in a future total merit index according to the derived economic weights. However, for most other autochthonous local breeds in Austria breeding goals do not focus on classical economic aspects but rather on maintenance of genetic variability, where matings between related animals are avoided and a balanced contribution of breeding rams is aimed for. These breeds are: Alpines Steinschaf, Montafoner Steinschaf, Tiroler Steinschaf, Krainer Steinschaf, Waldschaf, Kärntner Brillenschaf, Zackelschaf and Braunes Bergschaf. All of them are considered being endangered due to their small actual population size, the largest population consists of about 2,200 individuals, the smallest of less than 50. The long-term survival of those breeds will depend on the effectiveness of the actual breeding programmes to preserve genetic variability allowing adaptation to environmental changes and the willingness of farmers to keep those animals even with lower or no federal subsidies.

User specific breeding goals in dairy goat breeding

Herold, P.[1,2], Wenzler, J.-G.[3], Jaudas, U.[2], Kanz, C.[1] and Valle Zárate, A.[1], [1]Institute of Animal Production in the Tropics and Subtropics, Universität Hohenheim, Garbenstrasse 17, 70593 Stuttgart, Germany, [2]Goat breeders association of Baden-Wuerttemberg e.V., Heinrich-Baumann-Strasse 1-3, 70190 Stuttgart, Germany, [3]Rural municipalitites, Agriculture (Animal breeding), Heinrich-Baumann-Strasse 1-3, 70190 Stuttgart, Germany; pherold@uni-hohenheim.de

In Germany, goats have been traditionally kept for backyard milk production for home-consumption. Commercial orientation developed only during the last 20 years. Today, goat keepers can be mainly divided to 'breeders', keeping goats for hobby and presentation on breeding shows, and 'farmers', keeping goats for dairy and goat cheese production. A case study in southern Germany focuses on breeding goal development for the two different user groups. All breeders presenting animals on goat shows in 2008 (n=30) and 17 farmers (70% of goat farmers in the region) were addressed in personal interviews. Results on breeding goals were ranked according to their average score. Twenty-seven breeders keep goats for hobby, 3 as sideline activity. Twelve farmers keep goats as main activity, 5 as sideline activity. Average goat number of breeders is 13, of farmers 122. All breeders are members of the breeders association, all animals are registered in the herdbook and does are under milk recording. Only half of the farmers are members of the breeders association and 24% of farmers do milk recording. Breeders ranked milk yield, fertility and longevity highest. Farmers ranked longevity highest, followed by forage feed efficiency, in third place character and milk yield. From the users' point of view a breeding program has to be focused on milk yield and on sustainability traits like longevity and forage feed efficiency. In feedback seminars with farmers the low interest in milk yield is discussed.

The role and organization of cryopreservation for local cattle breeds in France, Finland, Italy and the Netherlands

Duclos, D.[1], Avon, L.[1], Maki-Tanila, A.[2], Pizzi, F.[3], Woelders, H.[4] and Hiemstra, S.J.[4], [1]Institut de l Elevage, 149 rue de Bercy, 75 595 Paris Cedex 12, France, [2]MTT Agrifood Research Finland, Biotechnology and Food Research, 31600 Jokioinen, Finland, [3]University of Milano, Institute of Biology and Biotechnology, Dept for food security, Via Trantacoste 2, 20122 Milan, Italy, [4]Centre for Genetic Resources, the Netherlands (CGN), Wageningen UR, P.O. Box 65, 8200 AB, Lelystad, Netherlands; sipkejoost.hiemstra@wur.nl

Within the framework of the EU funded Action EURECA (AGRI GEN RES 870/2004) cryopreservation activities related to local cattle breeds were studied in detail in four countries: Finland, France, Italy and the Netherlands. For each country, historic developments, existing structures and institutions involved in cryopreservation as well as their role, strategies and protocols related to local cattle breeds were investigated. Moreover, we made a complete inventory of semen and embryo collections, available at commercial AI organizations, with breed associations or individual breeders or in national cryoreserves. The inventory included bull variation, number of doses per bull and sanitary and legal status of the material. For each local breed we determined the availability of semen for short term breeding objectives and for long term conservation of within breed genetic variation. General objectives of cryopreservation activities are comparable between countries. However, countries have specific strategies and policies, and roles of particular national stakeholders are country-specific. One example of identified differences is the relationship between long term cryoreserves and commercial collection and use of semen (by AI centres). We carried out a SWOT (strengths, weaknesses, opportunities, threats) analysis for the situation in each country, we identified risks related to cryopreservation activities, and suggested options for further development at national and European level.

A unified approach for monitoring of genetic resources in German cattle breeds

Köhn, F., Edel, C., Emmerling, R. and Götz, K.-U., Bavarian State Research Centre for Agriculture, Institute of Animal Breeding, Prof.-Dürrwaechter-Platz 1, 85586 Poing-Grub, Germany; friederike.koehn@lfl.bayern.de

In accordance with the European order for monitoring genetic resources in agriculture a routine monitoring system for German cattle breeds has to be developed. Our task is the development of a monitoring system for 22 cattle breeds in Germany. The results for Gelbvieh and Hinterwälder are presented for illustration. For each breed the unified procedure consists of a data preparation step, descriptive analyses of census data and the calculation of genetic resource parameters. Here effective population size based on the increase of average inbreeding coefficients (N_eI) and coancestries (N_eC), the pedigree completeness index and the effective number of founders (f_e) and ancestors (f_a) were calculated. The census size of the Gelbvieh population is constantly decreasing over the last decade from 5,000 to 1,500 milk recorded cows per birth year. The calculated N_eI of 150 does not reflect the constant replacement of Gelbvieh genes by Fleckvieh genes. By restricting the reference population to animals with deep breed specific pedigrees significant lower N_eI of 120 was calculated. N_eC for the same restricted reference population was substantially lower (60) indicating that non-random mating with respect to inbreeding was applied during recent years. Low drift related measures like f_e (57) and f_a (45) suggest that the random loss of breed specific genes is a major problem in the Gelbvieh population. In contrast, the census size of the Hinterwälder population shows increasing number of cows per birth year from 120 in 1995 to 250 in 2005. N_eI over all animals was 92. When analysing the more breed specific pedigree the calculated N_eI was 65 and N_eC was 70 suggesting only small deviations from random mating. Relatively high values for f_e and f_a (90 and 71, respectively) are well in accordance with a growing population and a breeding scheme that is mainly based on natural service.

How to maintain declining Dutch local cattle breeds?

De Haas, Y.[1], Hiemstra, S.J.[2], Bohte-Wilhelmus, D.[1], Windig, J.J.[1], Hoving, A.H.[1,2] and Maurice-Van Eijndhoven, M.H.T.[1], [1]Animal Sciences Group of Wageningen UR, Animal Breeding and Genomics Centre, P.O. Box 65, 8200 AB Lelystad, Netherlands, [2]Centre for Genetic Resources, the Netherlands (CGN) of Wageningen UR, P.O. Box 16, 6700 AA Wageningen, Netherlands; Yvette.deHaas@wur.nl

Conservation of local breeds should not be limited to endangered breeds, but also directed at non endangered but declining breeds. A detailed assessment was made of the Groningen White Headed (GWH) cattle, based on (1) demographic and genetic analyses on pedigree data, and (2) interviews with GWH farmers. The demographic analysis showed that the number of purebred GWH calves born per year decreased from 3,000 in 1980 to 500 nowadays. Average inbreeding coefficient in the purebred GWH population increased in the last decades to 3% in 2005. Rate of inbreeding between 1980 and 2005 was 0.45% per generation, based on a generation interval of 5.5 years. Average mean kinship was 3.9% for GWH in 2005. Similar genetic results were found in a previous study on Meuse-Rhine-Yssel (MRY) and Dutch Friesian (FH) cattle. Based on outcomes of 23 interviewed GWH farmers an analysis of the strengths, weaknesses, opportunities and threats (SWOT) was made. According to the farmers, the strengths of the GWH cattle are their strong legs, good fertility, easy calvings and efficient feed intake, while low milk production and limited availability of AI-bulls can be considered as weaknesses of the breed. Similarly, for MRY and FH cattle functional traits are strong, while milk production and opportunities for breeding and selection are weak. Preservation of local cattle breeds requires a strong breed organization, where breeding programs and structures have to be discussed. When limited number of purebred bulls are available through AI, genetic gain and accuracies of breeding values need attention. Possible options to cope with this are a 'cold sire system' to avoid excessive use of individual sires, or a 'fundament based' breeding system, including limited exchange of semen between herds.

Combining local breeds in conservation schemes?

Bennewitz, J.[1], Simianer, H.[2] and Meuwissen, T.H.E.[3], [1]Institute of Animal Husbandry and Breeding, University of Hohenheim, Garbenstrasse 17, 70599 Stuttgart, Germany, [2]Institute of Animal Breeding and Genetics, University of Göttingen, Albrecht-Thaer-Weg, 37075 Göttingen, Germany, [3]Institute of Animal and Aquacultural Sciences, University of Aas, Dröbackveien, 1432 Aas, Norway; j.bennewitz@uni-hohenheim.de

The genetic diversity of livestock species can be found within and between breeds. The diversity is threatened by extinction of breeds and by genetic drift; the need to conserve genetic diversity by conservation schemes is widely accepted. In order to maintain the between breed diversity, usually breeds are kept separately in live conservation schemes. However, in some cases it might be very difficult or even impossible to conserve a highly endangered local breed in a closed population. If this breed is important for diversity, it might be beneficial to merge it with one or more breeds in order to conserve a part of the diversity that is contributed by this breed. The present study introduces a general framework that enables us to decide when it is beneficial to form a synthetic breed that includes highly endangered breeds in order to maximize conserved diversity and when to keep the breeds separate. Expected future diversities were estimated using a kinship based diversity measure together with extinction probabilities of the breeds. Using a small hypothetical data set, the pattern of diversity and its two components, the within and the between breed diversity, was analyzed in detail when forming a synthetic breed. The suggested approach was applied to a data set of 13 central European red and yellow cattle breeds. The results suggested forming a synthetic breed by combining a non-endangered breed with one of the two highly endangered and local breeds, which would result in an increase in conserved diversity.

Importance of the local breeds within the context of Spain

Luque-Cuesta, M. and Fernández, J.A., Federación Española de Asociaciones de Ganado Selecto (FEAGAS), Technical Services, c/Castelló, 45, 28001, Spain; manuel_luque@feagas.es

The Spanish territory enjoys a great diversity which can be explained by several factors: Its large surface and unique position allow it to receive the important influence of the Cantabrian and Mediterranean Seas on one hand, and the Atlantic Ocean on the other. All these elements have resulted in numerous ecosystems with a rich variety of livestock breeds. The lack of attention given to local breeds throughout the past decades has led to many of them being in danger of extinction; some of them have even disappeared already. At present, the situation has begun to change, in spite of the difficulties that affect the sector. Thankfully, there is a better knowledge and more appreciation of the potential of Spanish local breeds. Consequently, top quality products are able to be produced under natural conditions. Better results with some Spanish local breeds have been achieved due to the implementation of the new strategies. Currently, the success obtained with some of these breeds constitutes a path which can be followed and applied to the rest of them to improve their production and condition.

Using mid-infrared analysis to investigate the milk fat composition of different dairy cattle breeds

Maurice - Van Eijndhoven, M.H.T.[1,2], Rutten, M.J.M.[1] and Calus, M.P.L.[1,2], [1]Wageningen University, Animal Breeding and Genomics Centre, Marijkeweg 40, 6709 PG Wageningen, Netherlands, [2]Wageningen University and Research Centre, Animal Sciences Group, Animal Breeding and Genomics Centre, Edelhertweg 15, 8219 PH Lelystad, Netherlands; myrthe.maurice-vaneijndhoven@wur.nl

Nowadays there is an increasing interest in possibilities to modify milk composition, including genetic selection, because of its relation to human health. The genetic background of fatty acid composition in bovine milk, therefore, is a major topic in several studies. Heritabilities have been estimated for a whole range of individual milk fatty acids, mainly within the Holstein Friesian (HF) cattle breed. The regularly used Gas-Chromatography (GC) analysis to explore the fatty acid composition in milk is, however, expensive. Recently, Mid Infrared spectrometry (MIR) was recognised as a cheaper and high throughput analysing method. Calibration equations were developed to translate and interpret the MIR profiles, using GC analysis as a golden standard on a large dataset of HF milk samples. From the perspective of conservation of genetic diversity an important question is to what extent different breeds add to the genetic variation in milk composition (or quality). Predictive ability of the earlier developed (HF based) calibration equations to analyse the fatty acid composition in other breeds needs to be validated. In this study in total 160 milk samples will be analysed by both MIR and GC for validation. These milk samples originate from five different breeds: Meuse-Rhine-Yssel, Dutch Friesian, Groningen White Headed, Jersey, and Montbéliarde. The cows were sampled on in total 13 farms throughout the Netherlands. Approval of the calibration equations enables valuation of these breeds with respect to milk fat composition.

Influence of coat color trends on genetic variability of the Spanish Purebred horse population

Bartolomé, E.[1], Gutiérrez, J.P.[2], Molina, A.[3], Cervantes, I.[2], Goyache, F.[4] and Valera, M.[1], [1]University of Seville, Ctra.Utrera,1, 41013Seville, Spain, [2]UCM, Avda.PuertadelHierro,s/n, 28040Madrid, Spain, [3]University of Cordoba, Ctra.Madrid-Cordoba,396a, 14071Cordoba, Spain, [4]SERIDA, CmnClaveles,604, 33203Gijón, Spain; v92bamee@gmail.com

The Spanish Purebred (SPB) is the most important horse breed in Spain. The most frequent coat colors in this breed are grey and bay, but minority coat colors are getting more popular, increasing their frequency. Possible effects of balancing colors subpopulations on the genetic diversity of this population were measured on coancestry analysis. A method to standardize between- and within-subpopulation mean coancestries was developed to account for these different population sizes. Data included 166,264 horses registered on the SPB StudBook. Animals born in the last 11 years (1996-2006) were selected as 'reference population'. Animals were grouped following the coat colors in 8 subpopulations: grey (64,836 animals), bay (33,633), black (9,414), chestnut (1,243), buckskin (433), roan (107), isabela (57) and white (37). Contributions to total genetic diversity were first assessed on the actual subpopulations and later compared with two scenarios with equal subpopulations size: mean population size (20,783) and a simulated minimum population size (100). Wright parameters showed not any well coat color differentiated subpopulation (FIS=0.0199, FST=0.0019 and FIT=0.0217). Ancestor analysis revealed a very similar origin of the different groups except for 6 ancestors that were only present in one of the groups probably being responsible of the correspondent color. Standardization showed that balancing subpopulation size of each coat color would contribute to preserve the genetic diversity of the breed. The adjusting methodology developed here seems to be useful for the study of the genetic structure of populations with unbalanced size.

Morphofunctional characterisation in the Menorca horse using geometric morphometrics methods

Cervantes, I.[1], Gómez, M.D.[2], Molina, A.[2] and Valera, M.[3], [1]University Complutense of Madrid, Avda. Puerda de Hierro s/n, 28040, Madrid, Spain, [2]University of Córdoba, CU Rabanales, 14071, Córdoba, Spain, [3]University of Sevilla, Ctra Utrera km1, 41013, Sevilla, Spain; icervantes@vet.ucm.es

The Menorca Horse is an endangered breed located in the Balearic Island. The maintenance of this black riding horse in the autochthonous environment is ensured because of its strong relationship with the cultural events. This cultural link is signed with a specific dressage style called 'Menorca Dressage', which is regulated by the Balearic Equestrian Federation. Their uses in sport events have increased in the last years; therefore, breeders are interested on the development of objective methods to select the animals from a morpho-functional point of view. Nowadays, a breeding program is being developed in order to maintain the genetic variability and to select horses for conformation and sportive objectives. The aim of this study was to describe the morpho-type in this breed analysing the size (by morphological variables) and the shape (by geometric variables using Geometric Morphometrics Method -GMM-). GMM eliminates size factors and focus on the shape of individuals. Our data consisted of 37 body measurements and 8 angles from 158 Menorca horses (82 males and 76 females) with two different breeding goals, 'morphological show' and 'dressage aptitude'. A MANOVA analysis was performed using 3 fixed effects in the model: gender (male and female), age class (≤4, 5-8, and ≥9 years old) and the breeding goal for both the direct measurement and the geometric variables. The gender and the age were statistically significant for both criteria. The analyses of morphofunctional characteristics using morphological traits did not reveal the existence of significant differences between animals bred for different purposes. Whereas for geometric analysis the significant differences between morphological show and dressage aptitude were found. These results might be valuable for the breeding program to improve the morphofunctional aptitude of this breed.

Build up a synthetic reference map of bovine muscle proteins

Roncada, P.[1], Gaviraghi, A.[2], Deriu, F.[2], Greppi, G.F.[3] and Bonizzi, L.[2], [1]Istituto Sperimentale Italiano L. Spallanzani, Laboratorio di Proteomica, c/o Università degli Studi di Milano, Facoltà di Medicina Veterinaria, via Celoria 10, 20133 Milano, Italy, [2]Università degli Studi di Milano, DIPAv, sezione di diagnostica sperimentale e di laboratorio, via Celoria 10, 20133 milano, Italy, [3]Università degli Studi di Sassari, Dipartimento di Scienze Zootecniche, via E. De Nicola 9, 07100 Sassari, Italy; paola.roncada@unimi.it

Variability of meat quality is a major concern for industry and consumers. Many factors can affect final meat quality, such as animal welfare, breeding, feeding and transport conditions, slaughtering conditions, electrical stimulation, and chilling conditions. Two-dimensional gel electrophoresis (2-DE) applied to meat science is a powerful tool for biomarker discovery and to understand factors related to both inter- or intraspecific meat quality. Aim of this work is to evaluate protein expression profiles in different bovine species towards the building up of an informative synthetic map of bovine muscle proteins. A synthetic gel consists of a representative set of spots generated from several registered gel images. This is a useful tool to obtain a representative profile of the proteome of bovine muscle in meat sciences. Besides, analysis of expression profiles could be a key to find protein markers for meat quality, and give deep understanding of characteristics of more suitable genotypes for meat and milk production. Work supported by SELMOL Project, S.U.O. ISILS P.R.

Exploring growth functions to describe changes in weight in the fattening period in Avileña Negra Ibérica beef calves

Pérez-Quintero, G.[1], Carabaño, M.J.[2], Piles, M.[3] and Díaz, C.[2], [1]Universidad Centroccidental Lisandro Alvarado, Area de Genética Animal. Decanato de Ciencias Veterinarias, Nucleo, Cabudare, Venezuela, [2]INIA, Depto Mejora Genética Animal, Apdo. 8111, 28080 Madrid, Spain, [3]IRTA, Unidad de Cunicultura, 68140, Caldes de Montbui, Spain; cdiaz@inia.es

Weights of Avileña Negra-Ibérica calves are routinely recorded in commercial feedlots to evaluate the genetic potential for growth in this breed. The aim was to study the suitability of several non-linear functions to describe growth patterns of animals during fattening. After edits, 28,166 data of body weight in an age range of 138 to 689 d corresponding to 6,447 calves were used. Individual's trajectories were adjusted to either a Brody, Gompertz, Logistic or Von Bertalanffy functions using a Bayesian hierarchical model. Each parameter of the functions was assumed to be influenced by the environmental factors of feedlot-year, herd of origin, starting age-season of fattening and maternal environmental effects, in addition to the animal's additive genetic component. Error variance was assumed to be constant throughout the whole growth trajectory. Gibb sampling and Metropolis-Hastings algorithms were used to draw samples from known and unknown parameter distributions, respectively. Both, model predictive ability and goodness of fit at different points were evaluated for each growth curve. The model fitting a logistic curve showed a better performance. Heritabilities varied between 0.15 to 0.52 for a, b and k parameters. Genetic correlations among parameters ranged from low to moderate. According to the magnitude of the genetic parameters, growth curve could be modified by selection to accommodate the production system requirements, but selection to increase the slope of the growth curve would probably modify adult weight.

Performance and effective population size of genetic resources in Czech Republic – Czech gold brindled hen

Gardiánová, I., Šebková, N. and Vaníčková, M., Czech University of Life Sciences Prague, Department of Animal Science and Ethology, Kamýcká 129, 16521 Prague 6 - Suchdol, Czech Republic; sebkova@af.czu.cz

Czech hen springs from peasant hen. It was the aboriginal domestic breed, which was kept on a large scale within Bohemia as well as within Moravia up the half of 19th century. Since the half of 19th century domestic hen were cross-bred with imported breeds from abroad. Around 1913 poultry regeneration was done by Škoda and Sedlák from the rest of peasant hen found within Bohemia and Bohemian and Moravian highlands. Škoda created two strains, one named Komorovicky and the second one Czechsicendorf. In 1924 these strains were recognized as a breed. The Czechsicendorf strain became Czech partridge hen and the Komorovicky strain became the Czech brindled gold hen. In the 1936 was this breed added amongst agricultural significant breeds for its high level of utility. Its breeding and other improvement became supported by the country. The performance of this breed is in average 160.28 eggs per year, a average egg weight is 57.1 g and hatchability 68.38%. Average effective population size were in range 59.84-102.69 among the years 2000-2008, with the average effective population size 78.74. Therefore this breed has been included in endangered breeds. The total number of fowl in each year from 2000 to 2008 is 299 cocks and 276 hens.

Characteristics of Karayaka sheep in Turkey

Ulutas, Z.[1], Sen, U.[1], Aksoy, Y.[1], Sirin, E.[1] and Kuran, M.[2], [1]Gaziosmanpasa University, Faculty of Agriculture, Department of Animal Science, Gaziosmanpasa Universitesi Ziraat Fakultesi Zootekni Bolomu, 60240 Tokat, Turkey, [2]Ondokuz Mayis University, Faculty of Agriculture, Department of Animal Science, Ondokuz Mayıs Üniversitesi Merkez Kampüsü - Kurupelit, 55139 Samsun, Turkey; zulutas@gop.edu.tr

Karayaka sheep is one of the native breeds of Turkey raised coastline of Black Sea Region with the number of 883,000 head. Karayaka sheep are well suited to the harsh climate, poor pasture and severe conditions that are the characteristics of the hills and uplands. They are a carpet-wool breed kept also for meat production. The color of body of Karayaka Sheep is white, and there might be black and brown plaque in head, ear, leg and body and occasionally black or brown animals are seen. Karayaka sheep has long thin tail, and also has got a piece of wool on head called 'Hotoz'. The rams are usually horned and the ewes are usually polled. A project was initiated in 2005 to investigate production traits of Karayaka sheep in Agricultural farm of Gaziosmanpasa University, covered the period of 2005-2008. Live weight of mature Karayaka rams and ewes are ranged 65-90 kg, and 40-60 kg respectively. Birht weight, 56 day of age and 140 day of age of Karayaka lambs were 4.16 kg, 16.47 kg and 30.63 kg respectively. On the other hand, twins' ratio, milk yield of lactation, lactation period, and total greasy wool yield were 8-29%, 60-90 kg, 100-160 day and 2-3.5 kg, respectively. These results showed that there is wide variation of Karayaka sheep characteristics and these could be improved by using new breeding tools. Preliminary studies also showed that Karayaka has got high quality meat due to mosaic dispersion pattern of fat in among muscle fiber. Therefore, Karayaka sheep needs to be taken under the breeding programs in order to improve growth and carcasses characteristics.

Creation of a germplasm bank of 'Gochu Astur-Celta' pig: sperm characterization

Hidalgo, C.O.[1], Rodríguez, A.[1], De La Fuente, J.[2], Merino, M.J.[1], Fernández, A.[1], Benito, J.M.[1], Carbajo, M.[3] and Tamargo, C.[1], [1]SERIDA, Camino Claveles 604, 33202 Gijón, Spain, [2]INIA, Carret. Coruña, 28040 Madrid, Spain, [3]Facultad Veterinaria, Campus Vegazana, 24007 León, Spain; cohidalgo@serida.org

The Gochu Astur-Celta pig is an endangered autochthonous breed from the north of Spain and its semen has never been described before. For the conservation of its genetic biodiversity and long-term survival, sperm parameters must be studied to establish a germplasm bank. Semen was collected by the gloved-hand technique from six boars (aged 13-24 months), twice a week (N=109). Sperm-rich ejaculate fractions were evaluated for volume (V), concentration (C), morphological abnormalities of sperm head (HA), midpiece (MA), tail (TA) and cytoplasmic droplets (CD), functional integrity of sperm membranes (hypoosmotic swelling test) and acrosome integrity rate (NAR). For freezing, semen was extended (1:1, v/v) with BTS, cooled to 17 ºC, centrifuged and pellets were re-extended with lactose-egg yolk (LEY, 20%, v:v egg yolk) extender. Ejaculates were cooled to 4 ºC, and resuspended with LEY-Glycerol-Orvus ES Paste (9% glycerol, 1.5% Equex STM) extender to a final concentration of 1000×10^6 cells/mL, before packaging into 0.5 ml straws and freezing for storage in liquid nitrogen. Total motility rate (TM) were assessed after collection, at 4 ºC and after thawing. Data are expressed as means ± standard error. Fresh semen characteristics were: V = 82.5±4.0 ml; C = 560.7×10^6±22.4 spz/ml and TM = 85.0%±1.0. Percentage of HA was 1.2%±0.1; MA 0.6%±0.1; TA 2.8%±0.3; CD 2.8%±0.4; NAR 98.9±0.2 and membrane integrity 89.1%±0.6. After refrigeration, the % of TM was 71.4±1.2, and the post-thawing survival rate was 32.8±1.4. However complementary studies are needed to ensure that banks are correctly created, our results indicate the possibility of collecting sperm that survive freezing/thawing procedures with satisfactory quality to use it as fresh and for its cryopreservation. Work performed in collaboration with ACGA. Supported by FEDER.

Use of an endangered dual purpose cattle for a quality beef production scheme

Henning, M.D., Ehling, C. and Koehler, P., Institute of Farm Animal Genetics (FLI), Breeding and Genetic Resources, Hoelty Strasse 10, 31535 Neustadt, Germany; martina.henning@fli.bund.de

The old German dual purpose Black and White cattle (DSN) which is well known as the ancestry of the Holstein population is not competitive in milk production and therefore endangered. As an alternative concept of *in situ* conservation 30 cows were chosen for a cross breeding beef production scheme with Limousin (Lim) and Angus (Ang). All cows in the programme should produce one female offspring for remount. Afterwards they raise a Lim x DSN or Ang x DSN calf in a low input system. Cows and calves, young steers and heifers are grazing in spring, summer and fall. During the winter period they are kept indoors in a loose housing system on straw bedding, and fed in two intensity groups (one group grass silage only, the other group grass and corn silage and concentrates). At 20 month of age animals are slaughtered and carcasses graded. A rib sample (9th to 11th rib) is taken and analysed for intramuscular fat content and other quality criteria. 91 animals have been evaluated as yet. Castrated males (525.2 kg) and females (474.4 kg) differ in live weight by appr. 50 kg at the end of the trial, the grass fed group (both sexes) weighing 492.5 kg compared to 507.2 kg. Fat coverage of the carcass is higher in heifers than steers, intramuscular fat content, determined in the M. long. dorsi, is also significantly influenced by feeding intensity and sex (0.5% in each case). These differences do not effect cooking loss and tenderness. There is a tendency for better tenderness in Ang x DSN compared to Lim x DSN, but quality was all in all above average. A comparison with pure DSN offspring will follow. Several marketing schemes for products from local breeds do exist already, and could also support beef from DSN cattle.

Genetic parameters for calving interval in three Portuguese autochthonous breeds of cattle
Carolino, N.[1,2], Gama, L.T.[1,3], Sousa, C.O.[1], Santos-Silva, M.F.[1], Bressan, M.C.[1] and Carolino, M.I.[1], [1]URGRMA-INRB, I.P., Fonte Boa, 2005-048 Vale de Santarém, Portugal, [2]Escola Universitária Vasco da Gama, Estrada da Conraria, 3040-714 Coimbra, Portugal, [3]Faculdade de Medicina Veterinária, Av. Univ. Técnica, 1300-477 Lisboa, Portugal; carolinonuno@sapo.pt

Calf output produced annually in a beef herd is fundamental for Portuguese breeds of cattle, particularly in autochthonous breeds, which are kept under extensive range conditions and are often used as dam lines in crossbreeding programs. Therefore, improving reproductive performance is one of the major priorities for these breeds. Data on calving intervals in the Alentejana (n=80,746), Barrosã (n=69,385) and Mertolenga (n=73,462) breeds of cattle, and corresponding pedigree information, were obtained from the breed associations responsible for their management. The data set covered a period of about 40 years, and was analysed separately for each breed, with the purpose of estimating genetic parameters by Restricted Maximum Likelihood, using the MTDFREML package. The univariate Animal Model included the fixed effects of herd-year, month, calf sex and age at calving, and the random, additive genetic and permanent environmental effects. Mean calving interval was 456.1±145.5, 444.7±108.8 and 438.8±135.1 days in Alentejana, Barrosã and Mertolenga, respectively. The estimated heritability in the same breeds was 0.03±0.004, 0.09±0.008 and 0.07±0.006, and the proportional contribution of permanent environmental effects was 0.05±0.004, 0.02±0.008 and 0.04±0.004, respectively. Even though the estimated heritability for calving interval tends to be low, the variability observed for this trait is quite large, indicating that it is feasible to select for reproductive performance in the breeds studied. Furthermore, selection response can be improved by expanding the use of artificial insemination, which will also provide better genetic connectedness among herds. Acknowledgements: The authors express their appreciation to the breed associations which have made their data records available for this study.

BLUESEL: an INTERREG France-Wallonie-Vlaanderen project aiming at the conservation and the use of the genetic heritage of the dual-purpose Blue Breeds in Belgium and Northern France
Colinet, F.G., Gembloux Agricultural University, Passage des Déportés 2, 5030 Gembloux, Belgium; colinet.f@fsagx.ac.be

Dual-purpose Belgian Blue and North Blue cattle are mainly located on both sides of the border between France and Belgium. Even if these Belgian and French Blue breeds are related because of their common ancestors in the former Mid and High Belgium cattle, these breeds diverged slightly under differentiated selection objectives in both countries. Within the BLUESEL project, a first aim consists to create a working group cross-border which will develop common guidelines for selection of bull dams and elite-matings for this dual-purpose Blue Breeds. This working group will create and help to conserve a common pool of bulls available for breeding in both countries. The project will also develop tools to harmonize the collection of phenotypic data (milk production and morphology). A joint genetic evaluation for production traits will be developed, adapted to the specifities of these breeds and integrating data provided by both countries. Others objectives of BLUESEL are the implementation of a technical and economical guidance of these farms and the improvement of profitability of farms through advice on management and on improvement of livestock. The valorisation of these breeds through the development of new and specific products (e.g. cheese products) is another objective. In summary, the whole project should contribute maintaining biodiversity in this cross-border region through conservation and use of animals naturally adapted. The BLUESEL project was launched in July 2008. It is conceived for four years and is supported by the European Union, the Walloon Region, the Nord-Pas-de Calais Region and the General Council of Department of Nord.

Conservation and improvement of native livestock breeds in Portugal

Gama, L.T.[1,2], Afonso, F.P.[3] and Carolino, N.[1,4], [1]URGRMA-INRB, I.P., Fonte Boa, 2005-048 Vale de Santarém, Portugal, [2]FMV, Av. Univ. Técnica, 1300-477 Lisboa, Portugal, [3]DGV, R. Ant. Serpa, 1050-027 Lisboa, Portugal, [4]EUVG, Estrada da Conraria, 3040-714 Coimbra, Portugal; ltgama1@yahoo.com

Portugal currently has 45 native breeds of livestock recognized, including 15 cattle, 15 sheep, 5 goat, 3 swine, 3 equine, 1 donkey and 3 poultry breeds. Due to their low census, the majority of these breeds is considered to be endangered. The importance of native livestock breeds in sustainable rural development has been recognized in Portugal for a long time, and measures have been taken over the years to recover the more endangered breeds and prevent further losses of genetic diversity. These measures have included financial support to farmers maintaining breeds considered at risk of being abandoned, according to EU regulations, and have been successful in slowing down the decline in numbers observed in many native breeds in the second half of the 20th century. Furthermore, efforts have been made to establish coordinated *ex situ* conservation programs through a national animal germplasm bank, which now covers nearly all livestock species and breeds. Recent studies have shown that losses in within-breed genetic diversity represent a serious threat in many native breeds, which have effective population sizes far below the minimum recommended for maintenance of genetic variability, indicating that concerns with within-breed genetic erosion should be taken into account, in addition to the prevention of breed extinction. Policy measures have recently been adopted, whereby breed associations have an approved conservation or genetic improvement program, which includes support to carry out activities such as herdbook registration, performance recording, genetic and demographic characterization, artificial insemination, germplasm banks, paternity testing, breed promotion and genetic evaluation. All native breeds are covered by these measures, with emphasis in a specific breed either on conservation measures or on the implementation of a selection program.

Influence of genetic markers on carcass and meat quality traits in Nellore cattle

Rezende, F.M., Ferraz, J.B.S., Mourão, G.B., Eler, J.P. and Meirelles, F.V., FZEA - University of Sao Paulo, Basic Sciences - GMAB, Rua Duque de Caxias Norte, 225, 13635-900 Pirassununga, SP, Brazil; frezende@usp.br

Data on 1,889 Nellore cattle, reared under pasture conditions in southwestern Brazil, and measured by ultra-sound for carcass traits and 674 bulls finished in a feedlot for 90 to 120 days and slaughtered at age from 21 to 29 months were analyzed to verify the association with genetic markers (Single nucleotide polymorphism or SNP), to evaluate the use of those markers as auxiliary criteria for selection in Nellore, the most important breed in the Brazilian beef herd. Carcass traits measured by ultra-sound were ribeye area (REA_US), backfat (BF_US) and fat depth at rump (FD_P8) and, measured after slaughter, were hot carcass weight (HCW), ribeye area (REA), backfat (BF). Meat quality traits measured after 7, 14 and 21 days of ageing were weep loss (WL7, WL14 and WL21), shrink loss (SL7, SL14 and SL21) and tenderness (TEND7, TEND 14 and TEND 21). Total lipids and cholesterol content in 100 g of samples aged for 7 days, were, also, measured and included on the analysis. The genotypes of DNA markers were carried out in laboratories licensed by a private company using theirs micro-array panels. Allele substitution effects were estimated in single or multi-polymorphism analysis. Additive and dominance effects were also estimated. Many DNA polymorphisms analyzed showed to be fixed or the frequencies for one of the alleles were too high, more than 99%. In those cases, analysis could not be performed. However, for many others polymorphisms there was observed variability on allele frequencies what make possible to do the association analysis. All traits analyzed were influenced by, at least, four polymorphisms with statistically significant ($P \leq 0.05$) or suggestive ($0.05 < P \leq 0.20$) effects, thus DNA polymorphisms can be used as additional and auxiliary criteria on selection process of carcass and meat quality traits in Nellore cattle.

Four-trait joint estimation of variance components in Nellore cattle

Ferraz, J.B.S.[1], Eler, J.P.[1], Pedrosa, V.B.[1], Balieiro, J.C.C.[1], Mattos, E.C.[1] and Groeneveld, E.[2], [1]FZEA - University of Sao Paulo, Basic Sciences - GMAB, Av Duque Caxias Norte 225, 13635900, Brazil, [2]FAL, Animal Breeding, Höltystraße 10, Neustadt, D-31535, Germany; jbferraz@usp.br

The objective of this research was to estimate (co)variance components and genetic parameters jointly for 4 traits in a population of Nellore cattle reared in Brazil, considering management group at weaning as a random effect to increase the size of contemporary groups for post weaning traits. Traits analyzed were weaning weight (WW, kg, N=103,554), post weaning gain (PWG, kg, N=81,908), scrotum circumference (SC, cm, N=39,960) and muscle visual score (MUS, N=72,787). The full animal model had 161,865 animals in A^{-1} and the models of analysis considered, for each trait, age of dam (linear and quadratic), age at measurement (linear), Julian date (linear and quadratic) and contemporary group, as well random effects of animal (direct and maternal, this not for MUS), permanent environment (PE, not for MUS) and management group at weaning (MGW, not for WW). The program used for the estimation was VCE 6.0 (Groeneveld, 2008). Direct heritability estimates (standard error) were, for WW, 0.22 (0.006), 0.24 (0.006) for PWG, 0.402 (0.007) for SC and 0.236 (0.005) for MUS. Maternal heritability estimates were 0.086 (0.003, WW), 0.036 (0.002, PWG) and 0.090 (0.003, SC). Correlation between direct and maternal genetic effects were low for majority of trait combinations, but around 0.5 between PWG and maternal WW, and MUS and maternal WW and -0.60 (0.026) between direct and maternal PWG and -0.71 (0.024) between maternal WW and PWG. Ratios over phenotypic variance were, for MGW, 0.14 (0.005) for PWG, 0.04 (0.003) for SC and 0.04 (0.002) for MUS> Ratios for PE were 0.12 (0.003), as concerned to WW, 0.04 (0.002), for PWG and 0.012 (0.002) for SC. Genetic correlations between direct effects were all positive and medium, but between PWG and MUS, that correlation was 0.70 (0.018). Estimates of (co)variance components were all within reasonable interval, when compared to two-trait estimates.

Analysis of inbreeding in the Swedish Gotland pony using pedigree information and microsatellite markers

Eriksson, S., Näsholm, A., Andersson, L. and Mikko, S., Swedish University of Agricultural Sciences, Dept. of Animal Breeding and Genetics, P.O. Box 7023, SE-750 07 Uppsala, Sweden; Sofia.Mikko@hgen.slu.se

The Swedish indigenous breed Gotland pony is considered endangered by the Swedish Agricultural Board. The breed was near extinction in the beginning of the 20th century, but the number of horses has since then increased, and in 2007 there were approximately 600 breeding females and 110 breeding males in Sweden. The ponies are rather small, hardy and long-lived and are used mainly for pleasure riding and pony trotting races. Traditionally, they were bred in a free-range system in the forests on the island of Gotland. Today a herd of about 50 mares are still kept under similar conditions. The majority of Gotland ponies are however bred in the mainland of Sweden. The aim of this study was to describe the population structure and investigate the inbreeding situation using both pedigree data and microsatellite marker information. Pedigree information was available since 1900 and comprised in total 14 941 individuals. Molecular data from microsatellites was available for 343 ponies. The generation interval was on average 10.4 years. Pedigree completeness was generally very high for ponies born after 1940. The average inbreeding coefficient for foals born 2000-2009 was over 11%. The rate of inbreeding was however not alarming, and the effective population size was estimated at about 60 animals. The results from the pedigree data will be compared with results from the molecular data.

Analysis of the relationship among indigenous Cokanski Tsigai rams semen quality, body condition and thickness of subcutaneous fat

Oláh, J., Harangi, S., Fazekas, G., Kusza, S., Pécsi, A., Kovács, A. and Jávor, A., University Debrecen, Institute of Animal Science, Debrecen, Böszörményi str 138, 4032, Hungary; harangis@agr.unideb.hu

Correlations between the quantity and quality of ejaculated semen, body condition and thickness of subcutaneous fat of indigenous Cokanski Tsigai rams were investigated on the Experimental Farm of the University Debrecen, Institute of Animal Science in the main breeding season. The volume of fresh ejaculate, semen density and live cell proportion were recorded. The body condition was determined by the Kilkenny method, and the body weight and the thickness of subcutaneous backfat were measured using ultasonic equipment. One-compound analysis of variance was used for statistical analysis. Cathegories were determined for all above mentioned factors and rams were cathegorized. It was confirmed that condition, body weight and fat thickness have effect on semen quality. There were no significant differences in the live cell proportion between the different fatty groups, but yes between the ejaculate quantity and density. There were significant differences between the ejaculate volume and density of groups of thin and medium fat layer, and between the groups of thin and thick fat layer groups, but not between the medium and thick fat layer groups.This study will be continued on more rams of more breeds because the quality and quantity of semen has main role in the sheep breeding.

Future of Dalmatian Turkey: traditional local form of poultry in Croatia

Ekert Kabalin, A.[1], Menčik, S.[1], Štoković, I.[1], Horvath, Š.[2], Grgas, A.[3], Balenović, T.[1], Sušić, V.[1], Karadjole, I.[1], Ostović, M.[1], Pavičić, Ž.[1], Marković, D.[4], Marguš, D.[5] and Balenović, M.[6], [1]University of Zagreb Faculty of Veterinary Medicine, Heinzelova 55, 10000 Zagreb, Croatia, [2]outer collaborator, 10000 Zagreb, Croatia, [3]Croatian Agricultural Extension Institute, Fra Andrije Kačića Miošića 9/III, 10000 Zagreb, Croatia, [4]State Institute for Nature Protection, Trg Mažuranića 5, 10100 Zagreb, Croatia, [5]National Park Krka, Trg Ivana Pavla II, 22001 Šibenik, Croatia, [6]Croatian Veterinary Institute, Poultry Center, Heinzelova 55, 10000 Zagreb, Croatia; istokovic@vef.hr

Since 1994 Croatia has been involved into FAO project for preservation of rare animal breeds. Today, in the National Register of Autochthonous Breeds of Croatia are included only two breeds of poultry: Hen Hrvatica and Zagorje Turkey. But process of identification and characterisation of autochthonous breeds and local forms is still continuing. A local, archaic form of turkey traditionally has been reared on the area of Dalmatian hinterland. During history its spreading on wild territory or islands was mostly limited by mountains. This local form has quite specific phenotypic characteristics that mostly maintained during hundred of years, as a result of extensive production system in small flocks where turkeys are keeping and feeding outdoor the most part of the year. At the beginning of 2009 project for identification of majority of flocks as well as their morphological and physiological characterisation has beginning. For that reason we try to establish average phenotypic traits of breeding animals in parental flocks on few family farms. That include: determination of feather colour, body mass and average body and head measures (body length, length of sternum, length of shank, body width, depth of chest, distance from sternum to pubic bone, head width and length and beak length). Those will present the first steps toward preservation of this local form for further generations.

The role of coat colour varieties in the preservation of the Hungarian Grey cattle
Radácsi, A., Posta, J., Béri, B. and Bodó, I., University of Debrecen, Institute of Animal Science, Böszörményi 138., 4032 Debrecen, Hungary; postaj@agr.unideb.hu

When preserving genetic resources one of the most important tasks is to maintain the typical characteristics of the breed in order to avoid loosing the available genetic variability. Therefore, traits without economic value at the moment should also be conserved. The phenotypic and genotypic qualities of the Hungarian Grey cattle are subject to several research projects, however, many relationships remained unclear. Aim of our paper was to survey the different coat colour varieties characteristic to the Hungarian Grey cattle. For objective measurement of coat colour the Minolta Chromameter CR-410 was applied. The great variability of coat colours is characteristic to both the calves and the adult animals. Three coat colour varieties (reddish, light reddish and dark reddish) of new-born calves were separated and their ratio was determined in the observed population. The coat colour of adult Hungarian Grey animals ranges from silvery to dark crane and the colour of the bulls are more diversified. Results of our survey showed that the ratio of grey-coloured animals was the highest (45.58%) in the observed population. More than quarter of the population (26.19%) was crane-coloured, while the ratio of silvery and light silvery animals were 20.69% and 7.54%, respectively. Results of statistical analyses confirmed that colour variables (L^*, a^* and b^*) measured on all three measurement areas (neck-shoulder, side, thigh-croup) are important for separating coat colour varieties. The L^* values (lightness) proved to be the strongest discriminant factors. Analyses of factors influencing coat colour confirmed that the variability of coat colour is associated with the age and the sex of animals and is affected by the season of measurements (winter and summer coat).

Genealogical control of bovine breed Rubia Gallega through 17 DNA microsatellite genetic markers
Moreno, A.[1], Viana, J.L.[2], Lopez, M.[2], Sanchez, L.[3] and Iglesias, A.[4], [1]Asociación Nacional Criadores Ganado Vacuno Raza Rubia Gallega, Ramón Montenegro 18, 27002 Lugo, Spain, [2]Laboratorio Xenética Molecular Xenética Fontao, Fontao-Esperante Apdo 128, 27080 Lugo, Spain, [3]Spin-off Deinal. Facultad Veterinaria Lugo. USC, Campus Universitario Lugo, 27002 Lugo, Spain, [4]Facultad Veterinaria Lugo. USC., Anatomia y Producción Animal, Campus Universitario Lugo, 27002 Lugo, Spain; secretarioejecutivo@acruga.com

The Rubia Gallega breed is the most characteristic cattle racial biotype of Spanish northwest, representing the best example of the agricultural Galician profile and constituting an identity signal ofGalicia as its landscape, customs or language. The census of registered pure-bred animals in the Herd Book managed by the National Association of Rubia Gallega Beef Cattle Breeders (ACRUGA) amounts to more than 45,000 animals of which 60% are used as parents (26,846 females older than 24 months and 359 males older than 14 months). The Molecular Genetic Lab of Xenética Fontao SA. has developed a work protocol based on high polymorphic DNA markers analysis such as microsatellites, chosen among the proposed lists of ISAG (International Society for Animal Genetics). In this study we present the results of the analysis of 17 markers, and other 13 additional markers used when necessary to solve complex cases, which are therefore applicable in studies of molecular characterization and analysis of pedigree and traceability. This markers constitute a more precise and unambiguous manner to identify individual animals. The analysis of these markers could be used as a very reliable tool for parentage assignment to authenticate pedigrees, to study the population dynamics and the inbreeding degree in our herds, and to design the most appropriate crosses to avoid, as far as possible, future inbreeding, as well as to ensure the traceability of its products. Currently, we have developed 18,685 genotyping analyses on 47% of the parents censed in ACRUGA that represents 100% of the males and 50% of the females.

Study of the myostatin gene and its relationship with the double-muscled phenotype in Rubia Gallega cattle breed

Moreno, A.[1], Viana, J.L.[2], Sanchez, L.[3] and Iglesias, A.[4], [1]Asociación Nacional Criadores Ganado Vacuno Selecto Rubia Gallega, Ramón MOntenegro 18, 27002 Lugo, Spain, [2]Laboratorio Genética Molecular. Xenética Fontao, Fontao-Esperante. Apdo 128, 27080 Lugo, Spain, [3]Spin-Off Deinal, Facultad Veterinaria USC, 27002 Lugo, Spain, [4]Departamento Anatomía y Producción Animal, Facultad Veterinaria USC, 27002 Lugo, Spain; secretarioejecutivo@acruga.com

Bovine muscle hypertrophy (HM) or double-muscled phenotype is characterized by an increase in muscular mass associated to a reduction in fat and connective tissue, sometimes accompanied by undesirable effects such as finer bones, defective postures, macroglosy, calving difficulties, poor mothering skills and decreased milk production. Muscular hypertrophy is a character with a recessive inheritance model whose genetic cause is due to several mutations in the Myostatin gene for nt419 (del7ins10), Q204X, E226X, nt821 (del11), C313Y and E291X; also, homozygosis in D182N potentiates its appearance. In this work we present the results of the analysis of these 7 mutations of Myostatin gene studied in a group of animals of the Rubia Gallega breed of both sexes and with the least possible inbreeding. An striking result when analysing the mutation nt821 (del11) was its much more common appearance in adult males than in females (84% vs. 44%), probably because of the farmer use of females without double-muscled phenotype while this is a preferred phenotype in males. This genetic analysis in individual animals, specially in those used as parents, would be very useful as an additional information to improve the selection of the Rubia Gallega breed in optimizing this character through its rational use in both purebred and crossbred animals.

Preliminary results of a genome scan of Marchigiana cattle for carcass yield using a Illumina 54,000 SNP panel

Valentini, A.[1], Pariset, L.[1], Bongiorni, S.[1], D'andrea, M.[2], Pilla, F.[2], Guarcini, R.[3], Filippini, F.[3], Williams, J.L.[4], Ajmone Marsan, P.[5] and Nardone, A.[1], [1]UNITUS, via de lellis, 01100 Viterbo, Italy, [2]UNIMOL, via de Sanctis, 86100 Campobasso, Italy, [3]ANABIC, via viscioloso, 06132 San Martino in colle (PG), Italy, [4]PTP, via Einstein, 26900 Lodi, Italy, [5]UNICATT, via E. Parmense, 29100 Piacenza, Italy; alessio@unitus.it

Carcass yield (CY) is an important characteristic for producers and processing industries. A genome scan for CY recorded on 228 male individuals was carried out using the 54,000 Illumina SNP panel. Assuming codominance, a linear model was used to estimate the effect of each SNP on the trait. A total of 16 SNP were detected with statistical significance level of $P<0.01$ (not corrected for multiple hypotheses). A comparison of the chromosomal locations of these SNP with Quantitative Trait Loci reported in the literature revealed that 12 of the SNP (75%) occurred close to QTL regions affecting CY. Only two adjacent SNPs were found both significant. These map on chromosome 7 in the same chromosomal region as the gene coding for lysyl oxidase (LOX), which has been implicated in several meat characteristics. The results here presented are considered as preliminary because of the small number of samples and the simple model used for the analyses. More appropriate analyses models are currently under development.

Genetic markers influence on growth traits in Nellore beef cattle

Ferraz, J.B.S., Rezende, F.M., Meirelles, F.V., Balieiro, J.C.C., Eler, J.P. and Mattos, E.C., FZEA - University of Sao Paulo, Basic Sciences - GMAB, Rua Duque de Caxias Norte, 225, 13635-900 Pirassununga, SP, Brazil; jbferraz@usp.br

Data on growth traits of 3,844 Nellore cattle, reared under pasture conditions on two different farms in southwestern Brazil, were analyzed to verify their association with genetic markers (DNA Single nucleotide polymorphism or SNP), with the objective of detecting association of those markers with economically relevant growth traits and the possible use as auxiliary tools for selection in that breed that influences 80% of the 200 million heads herd of Brazil. Traits considered were birth weight (BW), weaning weight (WW), yearling weight, measured at 18 mo (YW), post weaning weight gain (WG345), visual scores for carcass conformation (CONF), finishing (PREC) and muscle content (MUSC). The genotypes of DNA markers were carried out in laboratories licensed by a private company using theirs micro-array panels. Allele substitution effects were estimated in single or multi-polymorphism analysis. Additive and dominance effects were also estimated. Many DNA polymorphisms analyzed showed to be fixed or the frequencies for one of the alleles were too high, more than 99%. In those cases, association analysis could not be performed. However, for many others polymorphisms there were observed variability on allele frequencies, what made possible to carry out the association analysis. All traits analyzed were influenced by, at least, four polymorphisms with statistically significant ($P \leq 0.05$) or suggestive ($0.05 < P \leq 0.20$) effects. Those results indicate that DNA polymorphisms can be used as additional and auxiliary criteria on selection processes for growth traits in Nellore cattle, raised in pastures. As individual allele substitution effects explain only a small part of the phenotype, the results of this study suggest that the effect of genetic markers should be considered together.

Differentiation of Andalusian autochthonous bovine breeds: morphometric study through discriminate analysis

Gonzalez, A.[1], Herrera, M.[1], Luque, M.[2], Gutierrez-Estrada, J.C.[3] and Rodero, E.[1], [1]Universidad de Cordoba, Unidad de Etnologia. Dpto. Produc. Animal, Campus Universitario de Rabanales, 14071. Cordoba, Spain, [2]Federacion de Ganado Selecto (FEAGAS), Calle Castelló, 45, 2 Izq., 28001. Madrid, Spain, [3]Universidad de Huelva, Ciencias Agroforestales, Avenidad 3 de marzo, s/n, 21071. Huelva, Spain; manuel_luque@feagas.es

Four Andalusian autochthonous bovine breeds were studied (Berrenda en Colorado-BC, Berrenda en Negro-BN,Cárdena Andaluza-CA and Negra Andaluza-NA)which, in spite of their better adaptation capacity to extensive production systems,are in danger of extinction, mainly,because of the crossbreeding among them, and, in other cases,because of the crossbreeding with other breeds greatly specialized to produce meat. Six zoometric traits were analyzed (height at withers-AC, height at rump-AP, width at thorax-DB, front pelvic width-AEA, rear pelvic width-AG and rump length-LG on an initial sample of 518 females (BC:179, BN:214,CA:48 and NA:77).A differentiation study was carried out through of the classic discriminate and the heuristic methods(artificial neuronal networks of the perceptron multilayer-MLP and probalistics-PNN). The forward stepwise method was used for the classic discriminante analysis. Over the validation phase, the discriminate analysis provided an average success rate of 53.23%,which was broken by the MLP networks as well as the PNN(67.49%). In the three models,the Berrenda en Colorado and Berrenda en Negro Cattle breeds had the lower success rate.

Y chromosome diversity in Portuguese sheep breeds

Santos-Silva, M.F.[1], Sousa, C.O.[1], Carolino, I.[1], Santos, I.[2] and Gama, L.T.[1], [1]INRB, URGRMA, Fonte Boa, 2005-048 Vale de Santarém, Portugal, [2]Universidade do Algarve, Campus de Gambelas, 8005-139 FARO, Portugal; santossilva.fatima@gmail.com

Y Chromosome (Cry) has a pseudoautosomal region (PAR), that recombines with X and a Y-specific (MSY) that does not undergo recombination. Analysis of genetic variation at MSY can give important information on the origin, and breed development of today's breeds. Studies of genetic variation on Cry in sheep are scarce, and in Portuguese breeds they are inexistent, that we know. This work aims to evaluate Cry diversity at 15 Portuguese sheep breeds, Algarvia (AL), Badana (BA), Bordaleira de Entre Douro e Minho (EM), Campaniça (CA), Churra da Terra Quente (TQ), Galega Bragançana (GB), Galega Mirandesa (GM), Merino Branco (MB), Merino da Beira Baixa (BB), Merino Preto (MP), Mondegueira (MO), Saloia (SA), Serra da Estrela (SE), Merino Precoce (MPR) and Assaf (AS), assessed by a multi-allelic microsatelite, SRYM18, and a bi-allelic SNP, oY1, and define haplotypes for these breeds. Two fragments containing SRYM18 and oY1 loci were amplified by PCR, in 15 males per breed. SRMY18 detection was performed on an ABI 310. Fragment containing oY1, was digested by an endonuclease and products visualised by gel electrophoresis. SRYM18 shows at least three alleles, 140, 142 and 144 with different frequencies: 142 (84.9%) is largely predominant in all populations, 144 (5,8%) appear at EM, MP, MB, GB and GM and 140 (1.8%) appear at GM and MPR. oY1 has a predominant allele (a) in 90.2% of males analyzed, while the second allele (g) represents only 5.3, distributed by EM, MP. Combination of SRYM8 and oY1 at Portuguese breeds originates at least four haplotypes from ovis aries, H6, H5, H8 and H4, and/or one H1 from Ovis Dali. Our results show low levels of polymorphism at the Cry in Portuguese breeds, mostly at EM, GB, GM, MB, MP and MPR populations, what is in agreement with most studies in different species. We acknowledge doctors J. Meadows and J. Kijas for their valuable help in the realization of this work.

The Iberian pig genetic improvement scheme

Diéguez, E., Ureta, P., Álvarez, F., Barandiarán, M. and Garcia Casco, J.M., Iberian Pig Breeders Association, San Francisco 51, 1 dcha., 06300 Zafra, Spain; edieguez@aeceriber.es

The Genetic Improvement Scheme of the Iberian pigs was approved by the Spanish Agricultural Ministry in December of 1992, five years after of the foundation of the herd book. From the beginning, the management and application of both breed tools is responsibility of the Iberian Pig Breeders Association (AECERIBER), with the technical support of the Animal Breeding Department of INIA (Madrid). The Scheme includes two selection indexes: a) Piglet Index, focussed to farmers who main activity is to send piglets or young breeding animals and b) Finishing Cycle Index, more suitable for farmers that produce pigs to final slaughter weight at 160 kg. This presentation makes references to this index. The traits included in the genetic-economic index are: - Daily growth during the last feeding period called 'Montanera' (from 100 to160 kg). - Trimmed hams and forelegs weights. - Loins weight (fat free). Since 1999 the intramuscular fat percentage (IFP), measured through NIR in a sample of the longissimus muscle is also included in the index, in order to keep this favourable characteristic of the Iberian pigs under control. The animals tested are castrate males, come from several collaborator herds and are controlled - since five months of age to slaughter - at the same farm with an extensive management system. Breeding values are obtained by a traditional multivariate BLUP-animal model after the current REML estimates of genetic and environmental parameters. At present the data base has around 6,000 slaughter pigs, born from 500 boars and 2,500 sows, in 20 parities (year-season effect), proceeding from 55 herds and 80 slaughter series. AECERIBER publish every year the main results of the genetic evaluations in a Stud Boar Catalogue. Heritabilities and correlations results indicate that selection could be effective for improvement carcass trait but the negative genetic correlations between IFP and hams and loins advised to be cautious to avoid a deterioration of the meat suitability for dry-curing.

Estimation of genetic and phenotypic parameters for meat and carcass traits in Nellore bulls

Rezende, F.M.[1], Ferraz, J.B.S.[1], Groeneveld, E.[2], Mourão, G.B.[1], Oliveira, P.S.[1], Bonin, M.N.[1] and Eler, J.P.[1], [1]University of Sao Paulo, Basic Sciences, Av Duque Caxias Norte 225, 13635900, Brazil, [2]FAL, Animal Breeding, Höltystraße 10, D31535, Germany; frezende@usp.br

Data on hot carcass weight (HCW), rib eye area (REA), backfat (BF), shear force (SF), total lipids (LIP) and cholesterol (CHOL) of 656 Nellore bulls were used to estimate genetic parameters. The full relationship matrix had 4,734 animals. Estimation of (co)variance components was performed by REML, using VCE 6.0 software in two three-trait analysis. The animal model included the fixed effects of contemporary groups, the effects of analysis date (LIP and CHOL), and, as covariates, age of animal at slaughter, backfat (LIP, CHOL and SF), pH measured 24 hours after slaughter (SF) and temperature of samples (SF). Random effects of direct additive genetics and residual were also considered. Descriptive statistics described HCW with an average of 290.21, with a range from 225.5 to 393.0 kg. Average for REA was 73.30 and a range from 56.0 to 101.0 cm^2. The measures of BF varied from 1.0 to 15.0 mm, with an average of 4.38 mm. SF with an average of 5.93, with a range from 1.82 to 9.99 kg. Average for LIP was 2.18 and a range from 0.96 to 4.52 g/100 g of meat. The measures of CHOL varied from 28.76 to 83.95, with an average of 56.28 mg/100 g of meat. Phenotypic correlations estimates were 0.35 (HCW x REA), 0.05 (HCW x BF), -0.13 (REA x BF), 0.004 (SF x LIP), 0.01 (SF x CHOL) and 0.23 (LIP x CHOL). Estimates of heritability and their standard errors for HCW, REA, BF, SF, LIP and CHOL were 0.38 (0.106), 0.35 (0.088), 0.52 (0.117), 0.18 (0.120), 0.23 (0.114) and 0.002 (0.011), respectively. Genetic correlations estimates and their standard errors were -0.07 (0.185, HCW x REA), 0.36 (0.178, HCW x BF), -0.40 (0.150, REA x BF), -0.32 (0.089, SF x LIP), -0.77 (0.069, SF x CHOL) and -0.35 (0.070, LIP x CHOL). The results of this research indicate that selection can be effective to increase HCW, REA and BF and can also promote a moderate genetic gain for SF and LIP, but almost no gain for CHOL.

Kinship breeding in theory and practise

Nauta, W.J.[1], Baars, T.[2] and Cazemier, C.H.[3], [1]Louis Bolk Institute, Animal production, Hoofdstraat 24, 3972-LA, Netherlands, [2]University of Kassel, BD-farming, Nordbahnhofstrasse 1a, D-37213 Witzenhausen, Germany, [3]Association of Dutch Frisian Cattle breeding, Dokkumlaan 19, 6835 JW Arnhem, Netherlands; w.nauta@louisbolk.nl

Kinship breeding is a farm based breeding system and can be used as a basis of organic breeding. This system is used by a group of Dutch farmers that breed the native Dutch Friesian cow breed. There are 15 different breeding farms in the Netherlands and 3 in France. These breeding farms breed their cows mainly with bulls that are bred on the farm. In a total breeding population of about 800 cows at these farms, 47 breeding bulls from 50 different cow families were used for breeding in 2007. With this number of bulls genetic variation is kept at a high level, which is important for the survival of the breed of which in total 1500 animals are milked in the Netherlands and a few thousand in Germany, Ireland and Great Brittain. Baars and Endendyk described the kin breeding system at farm level and it is clear that not all breeding farms practice kinship breeding in the same way. The biggest bottle neck is the need to use 4 to 5 new breeding bulls per year. However it is important that the breeding farms use enough bulls to keep their inbreeding trend at the farm low and they should not use to many bulls from other farms to develop different lines within the breed. To find out about how farmers have managed to use this system over many decades, data of pedigrees and production and fertility records are collected from the Dutch and analysed. Primarily results show that inbreeding is used at farms but did not have impact on longevity of cows. Further investigations will show us how farmers deal with the kinship breeding system and the results can help organic farmers to set up a similar system to meet the principles of organic farming, also for selective breeding.

Potential of milk production of Iranian Water buffaloes

Sanjabi, M.R.[1], Naderfrad, H.R.[2], Moeini, M.M.[3], Lavaf, A.[4] and Ahadi, A.H.[1], [1]Iranian Research Organization for Science &Technology(IROST), Animal Science, No.71,Forsat St, Enghelab Av, 15815-3538-Tehran, Iran, [2]Deputy minister of jihad e Keshavarzi, Water Buffalo, Cross of taleghani and valie asr Av, 15815-Tehran, Iran, [3]Razi University, Animal Science, Taghebostan, University St, Kermanshah, Iran, [4]Azad University,Karaj Branch, Animal Science, Mehr shar Bolvard, Karaj, Iran; msanjabii@gmail.com

Milk production and fat percentage of 65,534 individual milk records of Iranian water buffaloes has been studied on 473 herds in 6 provinces. The data were analyzed by SAS software in GLM procedure and heritability and Breeding Value has been calculated by DFRML Procedure. The average milk yield per lactation, days of lactation, fat percentage and LSM of fat percentage were 1,513 kg, 202 Days,5.04% and 6.77 respectively. The effects of year and season of calving were significant on milk production($P \leq 0.05$). The estimated heritability of milk was 0.16. The LSM of average milk production of in the provinces of Gilan, Mazandaran, E.Azarbaijan, W.Azarbijan, Khuzistan and Ardabil were 1,452, 1,586, 1,382, 1,183, 2,135 and 1,189 kg respectively. The results indicated that the potential of milk production of Iranian water Buffaloes are economics and acceptable especially on those farms who are using proper diet formulation and using concentrate. The top five highest breeding value bulls have been introduced to A.I Station of Uromia.

Genetic analysis of meat quality traits in two commercial cuts in *Avileña negra* iberica

López De Maturana, E.[1], Carabaño, M.J.[1], G Cachán, M.D.[2] and Diaz, C.[1], [1]INIA, Dpto. Mejora genética, Apdo. 8111, 28080 Madrid, Spain, [2]ITACyL, Estación Tecnológica de la Carne, Apdo. 58, 37770 Guijuelo, Spain; ldematurana.eva@inia.es

The purpose of this study was to analyze genetically the meat quality characteristics of two commercial cuts with largely different functions and monetary value, shin and sirloin, as well as to determine the existence of residual variance heterogeneity. Samples of those commercial cuts were obtained from 400 calves of Avileña-Negra Ibérica breed, fed in six fattening places controlled by the Breed Association. The following characteristics were determined: intramuscular fat, cholesterol, collagen, protein, and dry matter content; colorimeter readings (L, a*, b*); Warner Bratzler shear force; thawing and cooking losses; and pH. Each trait was analyzed using univariate animal repeatability models considering both homogeneous and heterogeneous residual variances regarding the commercial cut, and Bayesian methods via a Markov chain Monte Carlo implementation were used. Models were compared in terms of goodness of fit and predictive ability using two different criteria: the Bayes Factor and a checking function for the cross-validation predictive densities of the data. Results revealed the difference of quality between both commercial cuts: sirloin showed higher intramuscular fat and dry matter content, higher luminosity, b* value and cooking and thawing losses than shin. Meanwhile, shin contained higher cholesterol, protein and collagen, and showed higher color a* and pH values. Further, the existence of heterogenous residual variances was assessed for all traits, except for pH, b* value and cook loss. Estimates of residual variances tended to be larger for traits measured in shin, except for intramuscular fat, dry matter and thawing losses. The Bayes Factor pointed at the heteroscedastic model as best, except for the thawing losses and pH. Both models showed similar predictive ability except for L parameter. Heritability estimates were low for all traits (0.02-0.17), but relevant genetic variability was detected.

Monitoring of ovarian activity in female of local breed cattle of difficult handling through the levels of progesterone in serum and faeces

De Argüello Díaz, S., Fernández Irizar, J., Ortiz Gutiérrez, A. and Chomón Gallo, N., CENSYRA Torrelavega (Cantabria), Sierrapando S/N, 39300 Torrelavega (Cantabria), Spain; dearguello@cantabria.org

To determine the ovarian activity in female of indomitable character, using a non-invasive method, and minimizing physical contact with such animals. Therefore, through knowledge of follicular development by detecting the levels of progesterone in samples of faeces, it will be able to improve the performance of reproductive treatment programs and to undertake the collection or embryo transfer. Establishment of the correlation profiles of plasma progesterone and fecal samples (validation of the technique): Sampling of faeces and blood of 2 females of Monchina breed for 2 sexual cycles. Fecal samples (50 g) to be frozen. Blood samples were centrifuged (30 m at 4 °C at 3,500 rpm.) and frozen. Validation of serum: direct technique or without removal of the hormone Validation of fecal samples: method of Isobe. Determination of analytical progesterone levels by RIA competition technique. Periodic sampling of faeces and blood every other day from 6 sexually active females, Monchina breed. Conservation as above. Analysis of pairs of data for the concentration of progesterone in serum and faeces through the SAS statistical package to assess the possible correlation between them. According to data analysis, there is a statistically significance ($P<0.0001$) existing positive correlation between progesterone levels in faeces and serum (0.56) Correlation of progesterone levels in faeces at 48 - 72 hours of its appearance in serum. The concentration of progesterone in faeces reflects the activity of the corpus luteum in a similar way as in the serum samples. Therefore, ovarian activity in sexually active females can be monitored through the determination of the hormone in faeces.

Innovative and practical management approaches to reduce nitrogen excretion by ruminants: REDNEX

Van Vuuren, A.[1] and Thomas, C.[2], [1]ASG Animal Production, Wageningen UR, P.O. Box 65, 8200 AB Lelystad, Netherlands, [2]EAAP, Via G Tomassetti, Rome, Italy; ad.vanvuuren@wur.nl

Dairying is an important sector of EU agriculture, but intensification has been accompanied by an increase in N surplus. This has a negative impact on groundwater (pollution with nitrates), surface water (eutrophication) and on the atmosphere (de-nitrification and ammonia volatilisation). The EU seeks to stimulate measures that improve management of nutrients, waste and water as a start to move to management practices beyond 'usual good-farming practice'. The objective of REDNEX is to develop innovative and practical management approaches for dairy cows that reduce N excretion into the environment through the optimization of rumen function, an improved understanding and prediction of dietary N utilization for milk production and excretion in urine and faeces. Novel tools for monitoring these processes and predicting the consequences in terms of N losses on–farm will be developed. At the centre of the project is a detailed mathematical model of N utilization by the cow which will act to integrate results from previous work and from new research carried out in the project. This interlinked research aims to improve the supply of amino acids to be absorbed relative to the quantity and quality of amino acids and carbohydrates in feed allowing a reduction in N intake. Research to understand amino acid absorption, intermediary utilization and the processes involved in the transfer of urea N from blood to the gastro-intestinal tract will further underpin model development and indicate strategies to reduce N losses. To predict N losses on-farm and the impact on profitability, a harmonised applied model will be derived from the mechanistic model and will be supported by tools to better describe feeds and biomarkers to indicate N status. Impact of the research will be enabled by dissemination and knowledge interaction using a participatory approach to include the views of stakeholders and recognition of the need to provide support to EU neighbours.

The impact of EU directives on N-management in dairy farming

Aarts, H.F.M., Wageningen-UR, Postbus 616, 6700AP Wageningen, Netherlands; frans.aarts@wur.nl

Dairy farming is an intensive form of livestock farming. Most farms are situated in rather densely populated areas with high land prices, forcing a high milk production per unit farm area. Thus huge amounts of animal manure are available, which is a difficult to manage and for the farmer a 'risky' fertilizer. To produce high quality forage, desirable for dairy cows, farmers prefer to apply high amounts of mineral fertilizer and accept high grazing and harvesting losses. They focus their attention on cattle performance and disregard an efficient utilization of 'home made' manure and forage, instead increasing purchases of feeds and fertilizers. This process has been strengthened by the withdrawal of EU governments from farm advice. Public advisory services were privatized so farmers have to pay for independent advice. At present, representatives of companies selling fertilizers and feeds are by far the main advisors of dairy farmers. Losses of ammonia, nitrous oxides and nitrates from the N-cycle, as a consequence of an inadequate utilization of manure and forage, forced EU to formulate directives to reduce these to acceptable levels. Member states have to implement directives as Action Plans, but can take specific conditions or desires into account. The best way to meet the targets of EU directives is to restore the disturbed cycling of N. Results of commercial and experimental farms show that on most dairy farms the required progress can be realized in a profitable way. Reductions in purchases overcompensate costs of investments and additional work. However, pilot farms as illustrative examples should be available to inspire farmers and to give them confidence, access to relevant information should be improved and governments should include rewards for (checkable) excellent nutrient management, as elements of their Action Plans. Impacts of past, existing and forthcoming environmental legislation on the cycling of N and on farm income will be illustrated with figures of commercial farms, pilot farms and an experimental farm on leaching sensitive soils in the Netherlands.

Methods and systems used for protein evaluation for ruminants

Hvelplund, T. and Weisbjerg, M.R., Aarhus University, Animal Health, Welfare and Nutrition, Tjele, P.O. Box 50, 8830, Denmark; Torben.Hvelplund@agrsci.dk

Reduction of nitrogen excretion by ruminants has been in focus during the last decades and to help this development a variety of protein evaluation systems have been introduced in different countries both within and outside EU. In a new EU project the aim is to further improve nitrogen utilization in ruminants. For this an initial inventory on methods and systems is performed. Information on different systems, their practical use on farm level, feed characteristics used as input for the systems and methods used for obtaining the input parameters will be reported. Major differences in systems and their impact on protein evaluation and utilization will be discussed.

The control of voluntary feed intake and diet selection in cattle with particular reference to nitrogen
Forbes, J.M., University of Leeds, Leeds, LS2 9JT, United Kingdom; j.m.forbes@leeds.ac.uk

While the control of food intake in ruminants has focussed on energy and bulk it is clear that nutrients play in important role. Of these, protein, nitrogen (N) and amino acids (AAs) are of particular importance. There are requirements for essential AAs and a deficiency, excess or imbalance of one or more limits voluntary feed intake and choices. Not only does the rumen microflora play a vital role in the N economy of the animal, it also breaks down fibre thereby participating in the 'physical limitation' of food intake. There are optimum rates of supply of N and AAs (as well as for the other feed resources such as ME and NDF) for a given animal and inextricable links between N-containing compounds, energy and bulk in the control of food intake and diet selection. Commonly energy (metabolisable energy, ME) and bulk (neutral detergent fibre, NDF) are emphasised and the daily intake is either that needed to meet the ME 'requirements' or that limited by food bulk and the physical capacity of the rumen. This principle of 'first limiting factor' is physiologically unsatisfactory, as is the concept of fixed limits or set-points. 'Minimal Total Discomfort' (MTD) allows several factors affecting intake and choice to be incorporated simultaneously in a semi-quantitative model of food intake and selection. The optimum daily intakes of each feed resource (ME, N, protein, NDF.) are specified for the animal; the feed(s) as their concentrations of these resources; a rate(s) of intake of the feed(s) is chosen arbitrarily and the difference between the optimal intake and the current intake of each resource is calculated and used to estimate the total 'discomfort'. This process is repeated for many rates of intake and the intake(s) for which total discomfort is minimum is the predicted intake for that combination of animal and feed(s). The presentation will show how discomforts due to excesses or deficiencies of resources can be calculated and will provide examples of how MTD is constructed and some predictions.

Urea flux from blood to gut: a physiological entity under metabolic control?
Kristensen, N.B.[1] and Reynolds, C.K.[2], [1]Aarhus University, Dept. Animal Health, Welfare and Nutrition, Blichers Allé, DK-8830 Tjele, Denmark, [2]The University of Reading, Dept. Agriculture, Early Gate, Reading, RG6 6AR, Berkshire, United Kingdom; nbk@agrsci.dk

The N efficiency of dairy cattle expressed as milk N output/dietary N input at zero body N balance is relatively low, on average 27% for Danish dairy farms. N efficiency can be increased when reducing N supply to the cows; however, this increase in efficiency is often at the expense of a decreasing milk production. The inability of dairy cattle to capture a larger proportion of dietary N in milk is associated with excretion of N in feces and urine as: un-digested dietary N and microbial products of hindgut fermentation, purine derivatives of rumen bacterial and endogenous origin, xenobiotics excreted in urine (e.g. hippuric acid), endogenous metabolites excreted in urine (e.g. creatinine), and urea excreted in urine. One of the obvious targets for improving efficiency of the dairy industry is minimizing urea excretion in urine by maximizing utilization of endogenous urea as N source for ruminal fermentation. However, the exact nature of the transport proteins assumed to facilitate urea flux to the rumen across gut epithelia is still unresolved and so are the signalling pathways involved in regulation of epithelial urea flux. In fact, it remains to be proven whether or not epithelial urea flux is regulated by the cow or just an intrinsic function of the epithelium responding to epithelial mass and blood flow without actually changing the expression of transporter protein in the epithelium. The renal handling of urea is of critical importance to the recycling of urea as renal urea reabsorption conserves urea in the extra cellular pool. Work in progress within the EU project, REDNEX, will address fundamental questions related to inter-organ urea fluxes in dairy cows with the overall aim of developing nutritional strategies that improve N efficiency of dairy cattle.

Challenges to modelling intermediary nitrogen metabolism

France, J.[1], Kebreab, E.[2], Lopez, S.[3], Hanigan, M.D.[4], Crompton, L.A.[5], Bannink, A.[6] and Dijkstra, J.[6], [1]University of Guelph, Animal & Poultry Science, Guelph, Canada, [2]University of Manitoba, Department of Animal Science, Winnipeg, R3T 2N2, Canada, [3]Universidad de Leon, Departamento de Produccion Animal, 24007 Leon, Spain, [4]Virginia Tech University, Dairy Science Department, Blacksburg, VA 24061, USA, [5]University of Reading, School of Agriculture, Policy and Development, Reading RG6 6AR, United Kingdom, [6]Wageningen UR, Animal Sciences Group, Wageningen, Netherlands; jfrance@uoguelph.ca

The variation in intermediary nitrogen metabolism in gut wall, liver and mammary gland of dairy cows is described. Isotopic data show that various intra- and inter-organ flows are more pronounced than suggested by simple net fluxes alone. Further, the differing behaviour of individual amino acids (AA) in gut, liver and mammary gland is highlighted. Current protein evaluation systems cannot deal with this variation and many assume fixed efficiencies of utilization of metabolisable or absorbed protein. Representation of the principal individual AA in at least liver, gut wall and mammary gland is required in future and both kinetic and process-based simulation models need to be constructed at the organ/tissue level. The interplay between the two types of modelling is demonstrated. Simple kinetic models of AA and protein metabolism in gut, liver and mammary gland which resolve *in vivo* tracer data are described. These provide 'building blocks' for process-based simulation models. A review of current process-based simulation models for gut, liver and mammary gland utilizing dynamic, non-linear differential equations based on the rate:state formalism is presented, together with a review of information required to produce them. To be a truly mechanistic representation of actual biology, such a model should address each key AA individually, uptake via primary transporters should be described bi-directionally as observed experimentally, and the requirements for essential proteins of organ/tissue origin addressed. Such a representation should also include interactions among individual AA and between supply of AA and supply of other nutrients. A conceptual generic framework for such a model is advanced, and a scheme for integrating process-based organ models into a model of whole-body metabolism is suggested.

Impact of competition between biofuels, food and feed and its impact on the livestock industry: North American perspective

Radcliffe, J.S., Purdue University, Animal Sciences, 125 S. Russell Street, West Lafayette, IN 47907-2042, USA; jradclif@purdue.edu

As fossil fuel stocks continue to decrease, and prices continue to increase, there is increasing pressure to find alternative energy sources. Biofuels are attractive, because they are renewable, and can be sold as a 'green' technology. However, the implications of using biomass, previously used as food or feed, are long reaching and in general not well understood by the average person. Globally, we are facing an energy shortage, and the challenge both scientifically and ethically is to find new sources of energy and to prioritize how that energy should be used. The U.S. is on pace to have 35% of its corn crop being used to produce ethanol in the next few years. Historically, the number one use of corn has been for livestock feed. This increased demand for corn for ethanol production has resulted in increased corn prices, decreased use of corn by livestock, and increased use of ethanol by-products (primarily wet and dry distillers grains with solubles) in livestock feeds. Distillers dried grains with solubles (DDGS) have similar energy content to corn, but are higher in protein and mineral content. Therefore, if formulating on an energy basis, nutrient excretion will increase. In addition, the fatty acid profile of DDGS can alter the fatty acid profile of the carcass unfavorably. Ethically, there are debates over the use of corn to produce fuel instead of being used for food or feed, and many opponents or corn ethanol production promote cellulosic ethanol production. However, this would inevitably lead to decreased corn acreage to make room for the production of cellulosic biomass. In addition, the practical logistics of moving and handling adequate volumes of cellulosic biomass are daunting. In summary, there are no easy answers to the current energy crisis. Livestock is merely one energy user, and will likely continue to be asked and/or expected to utilize the by-products of biofuel production.

Session 03 Theatre 2

The impact of biofuel production from cereals on the European pig and poultry sectors

Lynch, P.B., Teagasc, Pig Production Development Unit, Moorepark Research Centre, Fermoy, Co. Cork, Ireland; Brendan.Lynch@teagasc.ie

Biofuel production from cereals is at an early stage in Europe compared with the US. Nevertheless, competition for cereals from the biofuel industry will continue to stimulate the price of cereals to the European animal feed industry. Removal of a large portion of the starch from cereals leaves by-products that are well utilised by ruminants but of lesser value in pig and poultry diets on account of their relatively higher indigestible fractions. The availability of a large tonnage of biofuel by-products, European and especially imported, will have a major effect on the European feed ingredient market in the medium term. The process of EU authorisation for GM varieties already commercialised in the Americas and low tolerance for admixture of GM and non-GM varieties are for the moment major barriers to the importation of greater tonnages of these cereal by-products into Europe.

Session 03 Theatre 3

Biofuel: consequences for feed formulation

Van Der Aar, P.J. and Doppenberg, J., Schothorst Feed Research B.V., Meerkoetenweg 26, 8218 NA Lelystad, Netherlands; pvdaar@schothorst.nl

The production of biofuels from vegetable sources implies that residues of this process will be available as feedstuff for livestock. Biofuels can be divided into bioethanol and biodiesel, both having specific co-products. The value of DDGS, and residues from the biodiesel (meal, expeller and glycerol) will be discussed. DDGS is characterized as fibre and protein rich, containing high amounts of NSP's.The nutritional value of especially DDGS is highly variable. This variation is due to the type of grain used and the production process and processing conditions. The substrate used for the fermentation has a relative predictable effect on the nutritional value. However, factories may change their substrate. The largest variation is due to the production process. Fractionation before fermentation, the amount of solubles added to the DDG&S and the drying process exert an affect on either composition or digestibility. Too high temp during he drying process reduces the rate of fermentation in the rumen, lysine content and its digestibility in the intestine. For non-ruminants this is a very negative effect, however for ruminants a moderate heat treatment can be positive as the bypass values increase and the intestinal digestion is only slightly decreased. These byproducts will replace SBM in the diets. However they are relative to SBM low in lysine/other amino acids. Therefore these diets may have more pure lysine. Besides the variation in nutritional value DDG&S may require specific monitoring of mycotoxins whereas rapeseed meal require monitoring regarding glucosinolates. The economic value of the byproducts alters with alterations of raw materials in the commodity markets. The economic value of DDGS is highest for ruminants and lowest for gestating sows. In small amounts maize DDG&S can be an interesting feedstuff for laying hens in diets in which the use of rape is limited. For an optimal use of co-products of the biofuel production by the feed industry rapid methods to quantify the the nutritional value is essential.

Low carbon farming: sustainability indicators from the dairy stewardship alliance
Matthews, A., University of Vermont, Center for Sustainable Agriculture, 106 High Point Center, Colchester, Vermont 05466, USA; allen.matthews@uvm.edu

Which sustainable practices contribute to reducing the carbon foot print of our dairy farms? To be truly sustainable, these practices must enhance the natural environment and herd health, support profitability and improve the quality of life for farmers and their communities. The Dairy Stewardship Alliance 'low carbon farming' module measures improvements in farming practices related to reduced GHG emissions and increased soil carbon. We are developing a network of pilot projects of innovative approaches to decrease atmospheric concentrations of greenhouse gases by increasing carbon sequestration and/or by reducing greenhouse gas emissions from agricultural operations. Low Carbon Farming is examining how GHG offsets can create financial opportunities for dairy farmers and create product value for their co-ops. The Dairy Stewardship Alliance's research on sustainability indicators is a collaborative effort of the University of Vermont, Ben & Jerry's, St. Albans Coop and Vermont's Agency of Agriculture. This self-assessment of sustainability indicators for dairy farmers promotes a broader use of sustainable agriculture practices. We are involved in an industry wide interest in identifying ways to reduce GHG emissions and the carbon footprint throughout the production and distribution system. Direct support is provided for farmers to develop a better understanding of their production practices, explore alternatives and implement changes to improve the sustainability of their farm operations. Farmers complete a self assessment of sustainability indicators, receive summary reports and identify sustainable practices to implement. 75% of participating farms improve sustainable practices to meet Environmental Quality Improvement certification requirements. Of the 500 co-op dairy farms involved, 10% will participate in the Dairy Stewardship Alliance's research on sustainability indicators and implement changes in their production practices related to low carbon farming.

Biofuels, feed and food security
Van Der Zijpp, A.J., Wageningen University, Animal Production Systems Group, Department Animal Sciences, P.O.Box 338, 6700AH Wageningen, Netherlands; akke.vanderzijpp@wur.nl

With diminishing resources of fossil fuels (resulting in higher prices before the financial credit crisis) the search for alternative sources of energy (biofuel and biodiesel) has been directed to food/feed crops like maize and sugarcane. A rising demand of meat, milk and eggs of increasingly affluent urban populations has created a rising demand for feeds. An increasing world population needs more food. Crops, livestock and energy agricultural systems have to address a complex set of issues simultaneously to reach optimal solutions. Sofar policy makers have nationally addressed single, often geo-political issues, which unfortunately have had global effects like rising prices. Decisions on national and global land use for fuel, food and feed functions will have to adequately manage environmental impact (water, nutrients, climate change, biodiversity), food security and quality of the diet of poor (high percentage of income spend on food, low animal protein consumption) and rich consumers (overconsumption of animal proteins and energy leading to obesity, hart disease and diabetes), socio-economic equity of income of poor (subsistence farmers, risk averse, less organised and informed) and rich producers (market driven, organised, access to credit). Market distorsions can be the result of national policies not accounting for the global effects of national actions. These may appear as food price increases or loss of employment both contributing to social inequity. System comparisons will be presented to increase understanding of complex optimisations of fuel, food and feed production and their trade offs.

Future animal improvement programs applied to global populations
Vanraden, P.M., Animal Improvement Programs Laboratory, USDA, Building 5 BARC-West, Beltsville, MD, 20705, USA; Paul.VanRaden@ars.usda.gov

Breeding programs evolved gradually from within-herd phenotypic selection to local and regional cooperatives to national evaluations and now international evaluations. In the future, breeders may apply reproductive, computational, and genomic methods to global populations as easily as with national populations now. Countries could merge phenotypes for standard traits such as production, SCS, and longevity across borders to reduce evaluation efforts within each country and to simplify across border marketing. Larger farms collect much automatic data, but might not provide it for use in evaluations unless paid to do so. Phenotypes for new or less heritable traits will become a limiting factor as the supply of genotypes rapidly expands and the price of genotyping decreases. Individual country data sets for traits recorded only recently such as heifer fertility may be too small for reliable genomic predictions, whereas a combined international file could give good results. Dairy cattle breeders exchange traditional breeding values worldwide via Interbull, and methods are now needed to exchange either genomic evaluations or genotypes. Goals are to adapt multi-trait across country evaluation (MACE) in the short term and to merge genotypes in the long term. Swine and poultry breeding companies may find that more open exchange such as in dairy cattle leads to more rapid progress in the genomic era. Separate breeding companies can each pay to test their own animals, but shared investment in genotyping of reference populations can result in larger returns. Genotyped young animals are rapidly replacing progeny tested bulls and phenotyped cows as sources of breeding stock. A new market could also develop for genotyped frozen embryos. Marker subsets may be selected to provide, for example, 40% of the benefit of the full set for only 10% of the cost, allowing wide application of low density chips. The global population of animal breeders will develop and apply many other new tools to improve the global population of animals.

Practical cattle breeding in the future: commercialised or cooperative, across borderlines between countries and organisations
Bo, N., VikingGenetics, Ebeltoftvej 16, DK 8960 Randers SO, Denmark; nbo@vikinggenetics.com

Due to the development of new technologies the dairy cattle breeding industry is facing many changes. Selection of bulls will be changed from progeny testing scheme to genomic selection of young animals with a dramatic drop in the number of progeny tested bulls. The breeding goal will be efficient milk- and beef production from sound, healthy animals with respect to animal welfare and ethics. This paradigm will lead to further globalization of the dairy cattle industry. The effectiveness of the breeding programme will still depend of intensive registration of phenotypic data on farm and common use of data through central databases. Traditional cooperatives like A.I. centres and cattle breeding associations need to collaborate or merge within countries and across borderlines to secure continuous development of breeding programmes and resulting genetic progress in the population.

Country profiles regarding the use of imported dairy bulls

Dürr, J.W. and Jakobsen, J.H., SLU, Interbull Centre, Dept. Animal Breeding and Genetics, Box 7023, 750 07 Uppsala, Sweden; joao.durr@hgen.slu.se

Interbull currently provides international genetic evaluation services for 29 countries, divided in 73 populations of 6 different dairy breeds and involving 38 different traits. Most of the international trading of dairy genetics refers to animals that have been evaluated at Interbull in order to be marketed in several countries. It is safe to assume that the multi-country pedigree file at Interbull is one of the most complete sources of information in global animal breeding of any species. Therefore, an analysis of both importing and exporting countries profiles from Interbull pedigree is proposed. A total of 180,617 bulls with official national evaluations were included and among those 37,256 were identified as foreign bulls by the reporting country (same bull may have been reported by different importing countries). This represents that 20.6% of dairy bulls worldwide are imported. Cluster analysis was applied in data from six dairy breeds to group countries according to two criteria: ratio between number of imported bull by country of origin and total number of bulls reported, and ratio between number of imported bulls by country of origin and total number of imported bulls. Methodology allowed clear distinction between heavy, medium and small importers and also indicated the country profiles regarding the preferred origin of the imports. Red dairy breeds practice the lowest amount of trading (7.8%), while Holstein is the breed with the largest proportion of imported bulls (24.9). US are consistently the largest exporters of dairy genetics across all but Simmental and the red breeds, followed by the Netherlands, Canada, Germany and France. Heavy importers (more than 80% of imported bulls in at least one breed) were Belgium, Canada, Germany, Ireland, Italy, the Netherlands, New Zealand and Slovakia. Group D created by the cluster analysis showed that importing country profile is influenced by region and production system.

Breeding for a global dairy market using genomic selection

De Roos, A.P.W.[1], Schrooten, C.[1], Veerkamp, R.F.[2] and Van Arendonk, J.A.M.[3], [1]CRV, P.O. Box 454, 6800 AL Arnhem, Netherlands, [2]Wageningen University and Research Centre, Animal Breeding and Genomics Centre, P.O. Box 65, 8200 AB Lelystad, Netherlands, [3]Wageningen University and Research Centre, Animal Breeding and Genomics Centre, P.O. Box 338, 6700 AH Wageningen, Netherlands; sander.de.roos@crv4all.com

Global dairy producers may prefer different bulls as a result of different breeding objectives or genotype by environment interaction. With traditional progeny testing, daughter information introduced extensive variation in EBVs so superior bulls for various markets could be identified and commercialised. With the use of genomic selection, however, much more variation among bulls can already be observed directly after birth, and bulls may be commercialised before progeny testing. Consequently, global breeding organisations must actively breed for multiple breeding objectives. The aim of this study was to optimise a dairy cattle breeding program that uses genomic selection to breed bulls for multiple market segments. A closed nucleus breeding program was simulated in which 2,000 calves, generated from 200 dams and 40 sires, were born and genotyped annually, out of which 1,600 were culled directly after birth. The remaining 200 heifers and 200 bulls received a phenotype, based on own or progeny performance when they were 3 or 5 years old, respectively. Breeding values were estimated for two traits, A and B, using BLUP to combine phenotypes, pedigree, and marker information. Marker information for both traits was included as a simulated phenotype for a correlated trait with a heritability of 1. Three scenarios were compared: (1) selection based on the average of the EBVs for trait A and B, (2) dividing the breeding program in one half devoted to trait A and one half devoted to trait B, and (3) an intermediate scenario where animals that ranked high for either trait A or trait B were selected. Breeding programs were compared by the number of top-ranked bulls for trait A and B, while varying the correlation between A and B and the accuracy of the marker information.

Utilizing international gene pool, a Nordic experience

Stålhammar, H., VikingGenetics, Box 64, 532 21 Skara, Sweden; hans.stalhammar@vikinggenetics.com

In the breeding programmes of Nordic countries health traits have been included for several decades for both Holstein as well as the red breeds. For the Holstein, international sires of sons have also been used. A problem has been that several of the traits in the Nordic breeding profile, were not included in other countries set of EBVs and no international proofs were available for these traits. The effect of this was enlarged due to the unfavourable genetic correlation between production traits and some of the health traits. When national EBV for functional traits first were calculated in exporting countries, this information was considered in the selection of bulls to use as sire of sons. Later international proofs from Interbull became available for production, conformation and also functional traits. This new information was valuable and utilized in the sourcing of candidates from the international gene pool. Results is presented as genetic trend for total merit index (NTM), milk production, daughter fertility and mastitis resistance for age groups of Holstein test bulls for the years 1990-2004. The gain in NTM has been large for the period and is increasing over time. The composition of the progress has changed over time to be in better accordance to the breeding objective. During the first part of the period the genetic trend for milk production was large, but it has decreased during the latter part. Initially the trend for female fertility was negative, but this has changed during the last years. The difference between the age groups' EBVs for mastitis resistance is smaller than for the other traits.

Global vs local dairy cattle breeding: beyond GxE

Madalena, F.E., Federal University of Minas Gerais, Av. Torino 270 B. Bandeirantes, 31340-700 Belo Horizonte-MG, Brazil; iprociencia@terra.com.br

Conventional global genetic evaluation relies on high genetic correlations (r_g) among performances in different countries/regions. Published estimates of r_g between milk yield in temperate, semen exporter, and tropical, semen importer countries, have been high (0.7-1), but restricted to purebred Bos taurus breeds (mostly Holsteins), kept in high input systems at high altitudes or otherwise attenuated climatic direct and indirect impacts on performance. However, low r_g estimates (0.1-0.4) were reported in the few papers in hot climates. Besides the GxE, low correlations between pure and crossbred performance might be expected on theoretical grounds, a hypothesis that has not yet been properly tested. Selective recording and parentage errors, common in many developing countries, would also reduce the correlations. Thus, the effectiveness of selection in temperate countries needs still to be checked under the prevailing tropical production circumstances of crossbred progeny, hot climate and low inputs. Although global dairy cattle breeding might conceivably make use of the vast information available in the developed countries, its effectiveness in tropical countries depends on the selection programme adopted, which is likely to have different desired goals in both environments. Economic dairying in tropical countries usually is based on low quality forages (e.g. grasses, sugar cane) which nonetheless provide low cost nutrients because of their high yield. Thus, high milk yield per animal is less important relative to other traits than in the developed countries systems, and fertility, resistance to parasites and diseases, heat tolerance and low live weight are traits of major importance. In addition, the practice of milking with the stimulus of the calf, proven economical in most tropical production circumstances, might conceivably also require different genetics than in artificial calf rearing.

Challenges and opportunities for global dairy cattle breeding: a Canadian perspective

Miglior, F.[1,2] and Chesnais, J.[3], [1]Agriculture and Agri-Food Canada, Dairy and Swine Research and Development Centre, 2000, College Street, P.O. Box 90 - STN Lennoxville, J1M 1Z3 Sherbrooke, Quebec, Canada, [2]Canadian Dairy Network, 660 Speedvale Avenue West, Suite 102, N1K 1E5 Guelph, Ontario, Canada, [3]Semex Alliance, 130 Stone Road West, N1G 3Z2 Guelph, Ontario, Canada; miglior@cdn.ca

Around 6,000 Holstein bulls are newly proven each year worldwide, and 80% of those newly proven bulls are sampled in 9 countries. The major semen exporters are from US, Canada and the Netherlands, but other countries have been increasing their share of the global market. In 2008, Canada has exported $71M of dairy semen. The advent of genomic selection will provide new opportunities and challenges in the global dairy semen market. The market will partly shift from proven sires to young genotyped bulls, provided one can confirm over the next few months that the genetic level and accuracy of evaluation of these young bulls are as high as expected. There is a global race to be among the first to offer this new 'product'. However, it becomes a priority to be cautious and take all the steps necessary to offer on the market something that is consistent and reliable. Because of the large number of tested bulls, genomics will be first applied in the Holstein breed. The other dairy breeds, like Jersey, Ayrshire and Brown Swiss will run the risk to fall behind, if genomic selection is attempted only at the national level. Global cooperation among countries for those breed may be the best alternative to test and adopt this new technology. Another might be the use of SNP haplotypes to mark a large number of QTL, an approach that would require a smaller 'training set' of bulls than genomic selection per se. Finally, genome wide selection may make it easier for multinational AI organizations to offer different groups of young bulls for different selection objectives corresponding to various local markets.

Monitoring sustainability of international dairy breeds

Philipsson, J., Forabosco, F. and Jakobsen, J.H., Interbull Centre, Dept of Animal Breeding and Genetics, SLU, Box 7023, S-750 07 Uppsala, Sweden; Jan.Philipsson@hgen.slu.se

Characteristics of a sustainable breeding program are broad breeding objectives, managing inbreeding rates and continuous genetic improvement to keep populations competitive. We have monitored some measures of sustainability of the six dairy breeds Brown Swiss, Guernsey, Holstein, Jersey, Red Dairy Cattle and Simmental currently involved in the Interbull evaluations. Globally, selection within dairy breeds has for several decades almost entirely been practiced for production and conformation traits. Due to the negative genetic correlations between production and many functional traits such narrow selection has detrimental effects on some functional traits. These traits have therefore, in one country after the other, been included in national breeding objectives and genetic evaluations. This resulted in a demand for the same traits to be evaluated internationally. The Interbull service portfolio now counts seven trait groups including 38 sub-traits. At each Interbull evaluation a joint pedigree file is formed per breed. In the current study, the rate of inbreeding was computed for bulls with a pedigree completeness of at least 80%. Among Holstein bulls 46% had a pedigree completeness of more than 80%. The average inbreeding coefficient of Holstein bulls increased from 0.011 in 1980 to 0.035 for bulls born in 1998, i.e. close to a rate of 1% per generation, but has then decreased to 0.028 in 2006. Genetic improvement can be monitored globally by computing genetic trends of the traits for all bulls in the Interbull evaluations on different country scales. In general, large positive genetic trends have been seen for production traits of all six dairy breeds, whereas genetic trends for somatic cell count and longevity have been rather flat. However, negative genetic trends have been observed for female fertility in some breeds, especially in Holstein. In summary, data and pedigree information received by Interbull at each evaluation can be utilized to monitor sustainability of international dairy breeds at the global level.

Brown Swiss x Holstein crossbreds compared to pure Holsteins for production in first two lactations
Bloettner, S., Wensch-Dorendorf, M. and Swalve, H.H., Institute of Agricultural and Nutritional Sciences, Group Animal Breeding, Adam-Kuckhoff-Str. 35, 06108 Halle, Germany; hermann.swalve@landw.uni-halle.de

Brown Swiss x Holstein crossbreds (BSH, n=55) were compared to pure Holsteins (HOL, n=50) during the first two lactations for milk, fat and protein production at the experimental station in Iden, Germany. Cows calved from September 2005 to November 2007 and were housed in the same environment under the same management conditions. All animals originated from a designed experiment. A minimum of 5 tested cows per weekly test day was required. The actual model for genetic evaluations in Germany was used, fixed regressions were based on the Wilmink function. Coefficients were estimated simultaneously for each breed group. Random regressions were third order Legendre Polynomials. Variables in the model for statistical analysis were estimated with SAS (proc mixed). The fixed effects of breed group, age of first calving and test day were included. In addition, three intervals for days in milk (DIM) I1 = 5-100 DIM, I2 = 101-200 DIM and I3 = 201–300 DIM were defined. No significant differences ($P<0.05$) were found for breed group effects for production in first and second lactation. BSH (1.70) tended ($P<0.10$) to have lower SCS than HOL (2.02) in first lactation. The first 21 DIM in I1 in first lactation showed significant differences ($P<0.05$) in milk yield. No significant differences ($P>0.05$) between BSH and HOL were found in I1 (-2.00 kg), I2 (-0.54 kg) and I3 (0.70 kg). Differences between BSH and HOL tended to be smaller from first to second lactation. BSH had more capacity in production at the end of each lactation. However, BSH had lower milk production level in early (I1) and mid (I2) lactation.

Genotype x environment interaction in the Greek Holstein population
Tsiokos, D.[1], Chatziplis, D.[2] and Georgoudis, A.[1], [1]Aristotle University of Thessaloniki, Dept. of Animal Production, Faculty of Agriculture, Thessaloniki, 54006, Greece, [2]Technological Educational Institute of Thessaloniki, Dept. of Animal Production, School of Agricultural Technology, P.O.Box 141, Thessaloniki, Greece; dtsiokos@agro.auth.gr

Genetic improvement of dairy cattle, in the Greek dairy industry, is based mainly on the use of imported semen of progeny tested bulls or direct importation of 7 month pregnant heifers. As there are major differences from other European Countries in the production system, i.e. herd management and feeding system, the study of the magnitude of genotype x environment interactions (GxE) on milk yield in the Greek dairy industry is of significant importance. The data set for this study was provided by the Holstein Association of Greece covering a time period from 1999 to 2008, and comprised of 14,200 completed lactations from 39 dairy farms. The sires under study originated from 12 countries, in alphabetical order these were Belgium, Canada, Czech Republic, Denmark, France, Germany, Hungary, Italy, Luxemburg, Netherlands, New Zealand and USA. Breeding values, for the bulls under study, were estimated in Greece using an animal model. The fixed factors included were herd, herd production group, year-season of calving, number of lactation. Repeated milk records of 7,335 cows having at least the first three consecutive lactations were used in the analysis. The heritability of 305 days milk production was 0.23. The determination of any possible GxE interaction was based on the ranking correlation of the bulls breeding values as estimated from the responsible organizations in each of the sire's origin with the corresponding breeding values estimated from the Greek dairy population. The correlation coefficients were between 0.03 and 0.627 depending on the country of the bull's origin. However, none of the correlations were significant. Moreover, the further analysis using genetic groups indicate possible genotype x environment interaction regardless of the origin of the bull and of the management level of the dairy farm.

Organic or inorganic selenium in a diet for fattening bulls: effects on the selenium content in meat and on meat quality

Robaye, V.[1], Dotrippe, O.[1], Hornick, J.L.[1], Paeffgen, S.[2], Istasse, L.[1] and Dufrasne, I.[1], [1]Liege University, Nutrition Unit, Bd de Colonster 20, 4000 Liege, Belgium, [2]YARA, Bygdøy allé 2, P.O. Box 2464, Solli, 0202 Oslo, Norway; listasse@ulg.ac.be

In animals dietary selenium (Se) is offered on an inorganic (selenite) or on an organic form (Se enriched yeasts or Se incorporated in the feedstuffs). The aim of the present work was to compare two forms of Se (inorganic as selenite vs organic as feedstuffs) in a concentrate diet for finishing fattening bulls. Twelve Belgian Blue double muscled bulls were divided in three groups. The bulls from the control group were offered a diet based on sugar beet pulp (46%) and supplemented with spelt (13%), barley (24%) and soja bean meal (13%). In the inorganic Se group, the composition was similar except that sodium selenite was included in the mineral mixture. In the organic Se group, the incorporated barley and spelt were grown with Se enriched fertilizers spread at a rate of 4 g Se as selenate/ha at the 2nd and 3rd nitrogen applications. Furthermore, linseed meal with a high Se content owing to the Canadian origin of the linseeds was substituted to soja bean meal and included at a rate of 15,4%. The Se contents in the barley and spelt were 187 and 203 µg/kg DM compared to 36 and 57 µg/kg DM for the control cereals. It was 878 and 12 µg/kg DM for the linseed meal and soja bean meal. The Se content was 80 µg/kg DM in the control ration and was 206.1 and 194.7 µg/kg DM with the organic and the inorganic rations respectively. The organic Se diet significantly increased the total Se content in the Longissimus thoracis muscle (505 vs 279 µg/kg DM) while the increase was rather small with selenite. Except for the drip losses, there were no significant differences on any of the parameters related to meat characteristics. The lack of effects of both forms of Se could be associated to its role as cofactor in selenoproteins involved in biochemical reactions in live animals and therefore not anymore some days after slaughter.

Biomarker development for meat quality in pork production chains for Spanish dry cured ham using proteomics technology

Te Pas, M.F.W.[1], Keuning, E.[1], Kruijt, L.[1], Hortos, M.[2], Diestre, A.[3], Evans, G.J.[3], Hoving-Bolink, A.H.[1] and Hoekman, A.J.W.[1], [1]ASG-WUR, ABGC, P.O. Box65, 8200AB Lelystad, Netherlands, [2]IRTA, Finca Camps i Armet, Monells, Spain, [3]PIC, Kingston, Oxfordshire, United Kingdom; marinus.tepas@wur.nl

Improvement of the meat quality of pork is a major aim in pork production. Special products such as Spanish dry cured ham are produced from lines bred and managed especially for this aim. Establishing a standard high quality of the hams would increase profitability of this sector. Since the proteome of muscle tissue composes a major component of meat and changes in the proteome are related to meat quality we investigated the relationship between proteome profiles and meat quality traits. The longissimus muscle was sampled in 2 batches of pigs bred with a terminal Duroc boar that were managed with the aim of increasing intramuscular fat content for Spanish dry cured ham production. Each batch (N=70) consisted of pigs from 2 independent producers. The determined meat quality parameters included: drip loss, pH at 24 h, and IMF (measured in the gluteus medius). Furthermore several carcass parameters were measured including lean meat percentage and loin34FOM. Meat quality parameters did not differ between batches or producers. The proteome profiles of all samples were investigated with SELDI-TOF-MS technology using 3 different array types. Statistics were performed using the software provided with the equipment and with the Biomarker Patterns software. The results were analyzed for each batch independently and for the 2 batches taken together. To exclude false positives, only peaks that associated in all 3 analyses were regarded positive. We show statistically significant associations between a number of peaks (indicated by the mass of the protein) and the meat quality traits mentioned above. The peaks may be composed of one or more proteins that could be used as biomarkers. Presently, experiments are ongoing to identify the proteins of the peaks that have the best associations. Biomarker tests will be developed based on these proteins.

Comparative proteomic profiling of two muscles from five divergent pig breeds using SELDI-TOF proteomics technology

Mach, N.[1], Keuning, E.[1], Kruijt, L.[1], Hortós, M.[2], Arnau, J.[2] and Te Pas, M.F.W.[1], [1]Wageningen University and Research Centre-Animal Science Group, Breeding and Genomics, Edelhertweg, 15, 8219PH, Lelystad, Netherlands, [2]IRTA (Institut de Recerca i Tecnologia Agroalimentàries), Tecnologia de Processos, Finca Camps i Amet, 17121, Monells, Spain; nuria.mach@wur.nl

The aim of this study was to identify biomarkers for breed and muscle type through a proteome differential analysis. Samples from the semimembranosus muscle (SM) and longissimus muscle (LM) were collected 24 h post-mortem from 79 male pigs (174±6 d) of the Landrace, Duroc, Large White, Pietrain and Belgian Landrace breeds, following a 2 x 5 factorial design. The Surface-enhanced laser-desorption/ionization time-of-flight mass spectrometry analysis was performed using the Protein Chip System Series 4000 equipment. The CM10 ProteinChip® array type and binding buffer pH5 containing 0.1% Triton was chosen. Data was normalized and corrected for background. Peak intensity was log-transformed. The number of peaks (36.33±0.87) was not affected by breed and muscle type. However, the average intensity of peaks was affected ($P<0.001$) by muscle type in 5 out of 36 detected m/z values. A large proportion was higher expressed ($P<0.001$) in LM compared with SM. In fact, m/z 8,126, and 8,485 kDa proved very good biomarkers for muscle type condition. The average intensity of 8,126 kDa peak was 2.34-times greater ($P<0.001$) in SM compared with LM, whereas the average intensity of 8,485 kDa peak was 3.94-times greater in LM than in SM. Furthermore, for the 14,847 kDa peak, hierarchical cluster analysis perfectly discriminates the muscle type ($P<0.001$). In fact, the intensity of 14,847 kDa peak greater than 1.17 was 2% and 100% for LM and GM, respectively. No differences were detected in the average intensity of peaks between breeds. Present results demonstrate a difference in composition between muscles and show that proteomics is a useful tool to investigate the molecular bases for physiological differences in muscles between pig breeds.

Extruded linseed fed with or without Sel-Plex® to dairy cows: effect on Se and Omega 3 milk levels, milk stability and production parameters

Andrieu, S.[1], Warren, H.[1] and Nollet, L.[2], [1]Alltech Biotechnology Centre, Sarney Road, Dunboyne, Ireland, [2]Alltech Belgium, Gentsesteenweg 190/1, Deinze, Belgium; sandrieu@alltech.com

Feeding extruded linseed to dairy cows increases omega 3 levels in milk but might reduce milk oxidative stability. Thus high levels of Vit E are added but Vit E efficacy is dependent on dietary selenium supply. Three groups of dairy cows were made based on previous milk production, parity and DIM. The 3 treatments consisted of control feed (C); Control feed + extruded linseed at 2 kg/cow/day (L); Control feed + extruded linseed at 2 kg/cow/day + 6 mg/cow/day organic Se as Sel-Plex (SP). Number of cows per group was 10, 18 and 21, parity was 3.3, 2.8 and 2.9, while initial DIM were 110, 128 and 141 for C, L and SP groups. Control feed contained 1.5 mg/cow/day Se as sodium selenite. Extruded linseed contained 600 mg Vit. E/kg. During the 8 weeks trial, performance data were recorded from monthly official milk control. Milk samples were taken at week 8 for Se analysis, fatty acid profile and milk stability (resp. 5, 10 and 10 per treatment). Milk yield decreased over the 8 weeks by 6; 7.5 and 3.1% for C, L and SP. Milk fat level didn't differ between L and SP but both had lower values than C. No effect was seen on milk protein levels. Feeding Sel-Plex® increased Se in milk (respectively 15.6[a]; 12.7[a] and 41.2[b] mg/ml for C, L and SP; $P<0.001$). Fat oxidation measurement at 24 h was increased in L treatment vs C while SP was similar to C (resp. 0.565; 0.505; 0.498 Units absorbance). L treatment only resulted in a significant reduction on SFA and increase in MUFA in comparison to control. The C18:1t11+t10 level was however improved significantly for SP compared to C, while L didn't differ from C. L treatment increased significantly C18:2 n-6 and C18:3 n-3 levels. SP supplementation didn't influence directly these levels but lowered significantly the C18:2 n-6/C18:3 n-3 ratio when compared to L treatment (respectively 1.59[a]; 1.13[b]; 1.03[c] for C, L and SP treatments; $P<0.05$)

The effect of dietary fatty acid level on the fatty acid composition of the adipose and lean tissues of immunocastrated male pigs
Pauly, C.[1], Spring, P.[1] and Bee, G.[2], [1]SCA, Länggasse 85, 3052 Zollikofen, Switzerland, [2]ALP, Rte de la Tioleyre 4, 1725 Posieux, Switzerland; giuseppe.bee@alp.admin.ch

At a given supply of unsaturated dietary fatty acids (FA) the degree of unsaturation of adipose tissue (AT) lipids increases with decreasing carcass fat deposition. Because carcasses of immunocastrated male pigs (IC) fed standard grower-finisher diets are leaner than those of barrows one can expect higher levels of unsaturated FA in the AT, which then might negatively affect its oxidative stability and firmness. The goal of this study was to determine the effects of castration method and dietary FA supply on the lipid composition of the AT and intramuscular fat (IMF). Forty-eight Swiss Large White male pigs were blocked by litter and assigned by BW to 4 experimental groups: barrows were fed a grower-finisher diet with a PUFA-MUFA-Index [PMI=1.3 × MUFA (g/MJ DE)+PUFA (g/MJ DE)] of 1.7 (C), IC pigs were fed a grower-finisher diet with a PMI of 1.7 (IC17), 1.5 (IC15) or 1.3 (IC13). All pigs had ad libitum access to the diets from weaning to 107 kg BW. Compared with C, IC17 tended ($P<0.10$) to grow slower but consumed less ($P<0.01$) feed, were more ($P<0.001$) feed efficient and their carcasses were leaner ($P<0.01$). The lower saturated fatty acid (SFA) concentration of the AT ($P=0.12$) and IMF ($P<0.05$) in IC17 than C was compensated by the higher PUFA level (AT: $P<0.01$; IMF: $P<0.05$). The PMI-level had no ($P_l>0.05$) effect on growth performance and carcass composition of IC17, IC15 and IC13 whereas total feed intake linearly decreased ($P_l<0.05$) with decreasing PMI level. In the AT, but not the IMF, the SFA ($P_l<0.01$) and MUFA ($P_l<0.10$) concentration linearly increased and that of PUFA linearly decreased ($P_l<0.001$) with decreasing dietary PMI level. These results confirmed that the lower lipid deposition in carcasses of IC markedly increased the degree of unsaturation of AT and IMF lipids. To insure a comparable processing quality of the AT like that found for barrows the dietary MUFA and PUFA supply for IC must be restricted.

The effect of castration method and dietary raw potato starch supply on sensory quality of pork
Pauly, C.[1], Spring, P.[1], Ampuero Kragten, S.[2] and Bee, G.[2], [1]SCA, Länggasse 85, 3052 Zollikofen, Switzerland, [2]ALP, Rte de la Tioleyre 4, 1725 Posieux, Switzerland; giuseppe.bee@alp.admin.ch

The most common method to avoid boar taint, which is an off-odour and –flavour of pork, is surgical castration. The aim of the study was to assess by trained panellists the effects of feeding raw potato starch (RPS) and applying different castration methods on boar odour, boar flavour, tenderness and juiciness of loin and neck chops from Swiss Large White pigs. In experiment 1 and 2, 18 pigs from 6 litters (1 barrow, 1 entire male [EM] and 1 EM fed additionally RPS the last week prior to slaughter) and 33 pigs from 11 litters (1 barrow, 1 immunocastrate (IC) and 1 EM), respectively, were selected based on the androstenone (A: <1.9ppm) and skatole (S: <0.24ppm) levels in the adipose tissue. In both experiments, neck chops had a stronger boar odour and flavour, were more tender and juicier than loin chops ($P<0.05$ for each). In experiment 1, A levels in the adipose tissue of barrows were lower ($P<0.05$) than in EM with and without RPS supply. On contrary, S levels were lower ($P<0.05$) in barrows and EM fed RPS than in EM without RPS supply. Nevertheless pork from both EM groups had stronger ($P<0.01$) boar odour and flavour than barrows, probably due to the higher A tissue concentration. The pork from EM without RPS was more ($P<0.05$) tender than the pork from barrows and EM fed RPS. Although A but not S level was higher ($P<0.05$) in the adipose tissue of EM compared with barrows and IC, boar odour and flavour intensities did not ($P>0.05$) differ between groups. However, pork from IC and EM was less ($P<0.05$) juicier than from barrows. In conclusion, the discrepancy between the well recognized boar taint compounds A and S and the sensory perception indicate that other factors affected the panellist's perception for boar taint.

The effect of the inclusion grape seed extract and *Cistus ladanifer* in lambs diet fed with dehydrated lucerne supplemented with oil. 2. Intramuscular fatty acid composition

Jerónimo, E.[1], Alves, S.P.[1], Dentinho, M.T.P.[1], Santos-Silva, J.[1] and Bessa, R.J.B.[2], [1]INRB, Unidade Produção Animal L-INIA, Fonte Boa, 2005-048 Vale Santarém, Portugal, [2]Faculdade Medicina Veterinária, UTL, DPASA, Polo Universitario do Alto da Ajuda, 1300-477 Lisboa, Portugal; rjbbessa@fmv.utl.pt

Thirty-six ram lambs were used to evaluate the effect of diet's supplementation with oil and dietary sources rich in polyphenols, on fatty acid (FA) composition of L. dorsi muscle. Animals were randomly assigned to 12 groups of 3 lambs each, and two groups were assigned to each diet. The six pelleted diets were: Control diet (C) - 90% dehydrated lucerne and 10% of wheat bran; Control diet with 6% of oil blend - sunflower oil and linseed oil (1:2) - (CO); control with 2.5% of grape seed extract (CG); control with 2.5% of grape seed extract and 6% of oil blend (CGO); control with 25% of *Cistus ladanifer* shrub (CC); control with 25% of Cistus ladanifer and 6% of oil blend (CCO). Lambs were slaughtered after 6 weeks of trial. Intramuscular fat (IMF) were extracted and FA methyl esters prepared and analysed by gas chromatography using a 100m CP-Sil 88 capillary column. IMF content was higher in CCO group than the other groups (31 vs 21 mg/g fresh muscle). The major FA were 18:1cis-9 and 16:0, that increased with oil supplementation (26 vs 31% and 19% vs 23% of total FA respectively), followed by 18:0 (15% of total FA). Oil supplementation increased 18:1trans-11 and 18:2cis-9,trans-11, however in diet with *Cistus ladanifer* induced a greater increase of these FA, suggesting that *Cistus ladanifer* determines changes in ruminal ecology and in biohydrogenation intermediates. Oil supplementation increased 18:2n-6, except in diet with *Cistus ladanifer*, and decreased 20:4n-6. The proportion of 18:3n-3 and 20:5n-3 increased with oil, while that 22:5n-3 and 22:6n-3 were not affected. The n-6/n-3 ratio was higher in CC (3.5) that in other diets without oil (2.8) and decreased with oil supplementation (1.9). The P/S ratio increased with oil supplementation (0.18 vs 0.31).

Utility of plasma carotenoid profile to discriminate between pasture-diet and dry unifeed-diet in goats

Alcalde, M.J.[1], Stinco, C.[2], Meléndez, A.J.[2], León, M.[3] and Vicario, I.M.[2], [1]Univ Seville, Ctra Utrera Km1, 41013 Seville, Spain, [2]Univ Seville, C/ Profesor García González 2, 41012 Seville, Spain, [3]CSIC, Av/ Padre García Tejero, 41012 Seville, Spain; aldea@us.es

The type of feeding affects the carotenoid profile of ruminants' plasma. Fresh pasture has been reported to be rich in carotenoid pigments vs dry unifeed. The aim was to establish the carotenoid profile of goats´s plasma corresponding to animals grazed pastures (P), in comparison with animals fed in confinement (C), receiving dry-unifeed. Blood samples corresponding to 10 animals of each type of feeding were taken from the jugular vein. Plasma was obtained by centrifugation and stored at -80 °C until analysis. Carotenoids were analyzed by High Performance Liquid Chromatography coupled with ultraviolet diode array detector (DAD) by reverse phase. The main carotenoids identified in the plasma of the animals eating fresh grass were b-carotene and lutein while in the intensive group of animals the main carotenoids identify were z-carotene by far and lutein. This fact could be relevant for the trazability of the goats feeding.

Treated linseed increases ALA in lambs

Kronberg, S.L.[1], Scholljegerdes, E.J.[1] and Murphy, E.J.[2], [1]USDA-ARS, P.O. Box 459, 58554 Mandan, ND, USA, [2]Univ. of ND, 3902 15th Ave., 58201 Grand Forks, ND, USA; scott.kronberg@ars.usda.gov

Red meat can supply more n-3 fatty acids to people if a safe technique is found to protect α-linolenic acid (ALA) in feed from hydrogenation by ruminal microbes. Twenty-four lambs were randomly divided into 4 groups. All were fed a basal ration of lucernepellets at 08:30 and 16:30 h. Each lamb was fed 567 g/feeding of lucerne for 1st 51 days, then the 1,630 h amount was increased to 667 g/lamb. Lambs were penned together except during morning feedings and had free-choice access to water and salt. One group (LI) received 136 g/day of non-treated linseed, a 2nd group (T1) received 136 g/day of treated linseed, a 3rd group (T2) received 136 g/day of a 2nd treated linseed, and a 4th group (CT) received a mix of maize and soya meal with similar levels of CP and DE as other treatments. For morning feedings on the 1st 10 days, all lambs were individually fed the maize-soya mix and lucerne. On 10th day, a jugular blood plasma sample was collected for each lamb. For morning feedings of next 14 days, lambs were fed their treatment rations and lucerne then a second plasma sample was collected. This feeding regime continued for 66 days then lambs were slaughtered and muscle collected. Plasma total lipid extraction and fatty acid analysis was done as previously described by authors. Treatment means were compared using a mixed model with animal as random factor. Initial plasma levels of ALA were similar ($P=0.67$) for all groups. Change from initial to post-treatment plasma levels of ALA differed ($P=0.001$). Ingestion of treated linseed resulted in higher ALA levels ($P=0.005$) than ingestion of the LI treatment. There was no difference ($P=0.64$) in change of ALA levels for T1 and T2. Muscle phospholipid ALA levels were 5.9, 7.8, 8.3, and 3.1 mole%, respectively, for LI, T1, T2, and CT and were different ($P<0.001$). Levels of ALA were different ($P<0.001$) for LI, T1, and T2; however, there was no difference in ALA levels between T1 and T2 ($P=0.85$). Linseed treatments raised n-3 levels in lamb blood and muscle.

Effect of using forages in lamb fattening on the profile and content of fatty acids in roast leg meat

Borys, B.[1], Borys, A.[2], Grzeskiewicz, S.[2] and Grzeskowiak, G.[2], [1]National Research Institute of Animal Production, Experimental Station Koluda Wielka, Parkowa st. 1, 88-160 Janikowo, Poland, [2]Meat and Fat Research Institute, Jubilerska st. 4, 04-190 Warszawa, Poland; bronislaw.borys@onet.eu

The effect of using forages and breed origin of lambs on the profile and content of fatty acids in roast leg meat (RLM) was determined. The experiment consisted of two replications and involved 36 Koluda sheep (KS) lamb-rams and Ile de France ´ KS (IF´KS) hybrids that were fattened intensively to 32-37 kg body weight. Three groups of lambs were fed *ad libitum* with a concentrate mixture and different roughage supplements: grass hay in group C, field forage fed in F, and pasture grazing (4 h/day) in P. The intake of basic nutrients in the groups was similar, with a higher PUFA intake in groups F and P compared to C by 20.2 and 14.0%, respectively. The use of forages resulted in non-significant differences in RLM fat content (C – 53.3; F – 50.2; P – 55.9 g/kg) and cholesterol content (0.63; 0.67; 0.66 g/kg, respectively). There were no significant differences in the percentage of SFA, MUFA, PUFA and CLA in MPU fat. Differences in RLM fat content caused greater differences in the absolute content of PUFA, which was 9.1% lower in group F and 9.4% higher in group P compared to C. There was a tendency towards more favourable parameters of RLM health quality in the groups receiving forages, especially in pasture grazed lambs. Breed origin of the lambs only had a clear effect on PUFA content and health parameters calculated on the basis of these acids; the meat of KS lambs compared to IF´KS lambs contained more PUFA (4.02 vs. 3.68 g/kg) and CLA (0.34 vs. 0.30 g/kg) and had a higher PUFA:SFA ratio (0.184 vs. 0.166). In Overall, the present study showed that the use of field forage supplement or pasture grazing had little effect on modifying the health-promoting quality of the meat of intensively fattened lambs.

Effect of the diet, fatness degree and ageing on meat sensory quality from Pirenaica calves

Panea, B.[1], Albertí, P.[1], Sañudo, C.[2], Olleta, J.L.[2] and Campo, M.M.[2], [1]Centro de Investigación y Tecnología Agroalimentaria de Aragón, Avenida de Montañana, 930, 50059 Zaragoza, Spain, [2]Universidad de Zaragoza, Miguel Servet, 177, 50013 Zaragoza, Spain; bpanea@aragon.es

It is often argued that beef meat has a composition in fatty acids undesirable to human health because of its high content in saturated fatty acids and because of its ratio ω 6 /ω 3 is higher than nutritionist recommendations, if reared intensively. Therefore, in the last years, polyunsaturated fatty acids (PUFA) have been included in the feed for cattle. Nevertheless, since PUFA are susceptible to oxidation, is necessary to add an antioxidant in the feed formulation. The aim of the present study was to study the effect of feeding cattle with a feed rich in ω 3 PUFA plus antioxidants on meat sensory quality in calves from Pirenaica breed. Forty-eight young bulls were used in a factorial design including 3 diets (control, 5% linseed seed or 5% linseed seed plus 200 IU vitamin E/kg), two fatness degrees at slaughter (3 or 4 mm of subcutaneous fat) and two ageing periods (2 or 14 days) under vacuum conditions. Longissimus thoracis was sliced into 2cm-thick steaks which were tasted by a 9-member- trained panel. Diet had not effect on meat sensory quality, whereas both fatness degree and ageing time had a significant effect. The higher the fatness level, the higher the intensity of rancid odour, the tenderness, the juiciness, the beef odour intensity and the acid flavour intensity and the lower the fibrousness. Throughout ageing tenderness, beef flavour, acid flavour and rancid flavour increased and fibrousness decreased. It would be recommended long ageing periods since they increased positives attributes such as tenderness, juiciness and beef flavour. Nevertheless, since rancid flavour also increased with ageing, a balance within positive and negative attributes should be assured.

Carcass and meat quality from yearling bulls managed under organic or conventional systems

Oliván, M., Sierra, V., Castro, P., Martínez, A., Celaya, R. and Osoro, K., SERIDA, Apdo 13, 33300 Villaviciosa (Asturias), Spain; mcolivan@serida.org

One of the major goals of organic farming is the production of high quality food. The aim of this study, carried out along two years (2006 and 2007) was to compare carcass and meat quality from yearling bulls reared under organic or conventional systems. A total of 35 animals from 'Asturiana de los Valles' breed were slaughtered at 19 to 20 months of age, coming from two different finishing treatments: organic (pasture + concentrate), being concentrate 40% of total intake, and conventional indoors fattening with concentrate meal and barley straw ad libitum. Animals from organic production showed lower slaughtering weight (580 vs 626 kg; $P=0.064$) and produced leaner carcasses, as assessed by carcass fatness score (2.6 vs 5.4 in 1-15 scale; $P<0.001$) and total fat content (7.4 vs 11.9%; $P<0.001$). Feeding treatment did not affect carcass conformation. There was a significant effect of the year on live weight at slaughter (554 and 649 kg in 2006 and 2007, respectively; $P<0.001$) and carcass weight (309 vs 367 kg; $P<0.001$), mainly due to the lower live weight gains and age of animals slaughtered in 2006, which probably influenced carcass fat colour, that showed higher lightness and lower yelowness in the first year of study. Carcass fat colour was also affected by feeding treatment, with higher b* values (yelowness) in organic carcasses (13.8 vs 8.1; $P<0.001$) due to storage of pigments consumed from green forage. With respect to meat quality, chemical composition was significantly affected by feeding treatment. Organic meat had lower intramuscular fat (1.4 vs 2.5%; $P<0.001$) and higher moisture (75.1 vs 74.3%; $P<0.001$) and protein (22.5 vs 22.1%; $P<0.001$) contents than conventionally-produced meat. Feeding also affected pigment content and meat colour, organic meat showing lower myoglobin content (4.5 vs 5.4 mg/kg; $P<0.001$) and a* redness index, probably due to the above mentioned lower slaughtering weight, although in general meat from organic production was darker than that produced conventionally.

Effects of feeding expired bakery products to pigs on production and quality of meat
Inoué, T., Ishida, M. and Kobayashi, J., Miyagi University, School of Food Agricultural and Environmental Sciences, Hatatate 2-2-1, Sendai, 982-0215, Japan; inoueta@myu.ac.jp

While almost all grains for livestock feeds in Japan are imported from overseas, use of residues from the food processing industries has been encouraged in order to improve self-sufficiency and to reduce environmental load of waste disposal. Unlike many European countries, the Japanese consumers favour well marbled meat and for this reason, feeding carbohydrate at a high level for the finishing period is especially important for increased intramuscular fat which has a major role for tenderness and flavour of the pork when they are fed to pigs. Expired wheat products such as bread or other bakery products are often available to local pig producers, otherwise subjected to disposal. We fed pigs with expired bakery products with chocolate fillings and examined its effects on growth performance and meat quality. Fifteen ternary crossbred pigs (LWD) were divided into three groups to receive diets in which a commercial diet has been replaced with chocolate rolls by 0% (control), 15% (CR15) and 30% (CR30). The pigs were slaughtered at about 110 kg of body weight and meat production performances were recorded. Loins (longissimus thoracis) from each animal were examined for meat quality. The group on CR30 diet had a higher daily gain ($P<0.05$) but a lower dressing rate ($P<0.05$) compared with the control group. While loin weight and back fat thickness were similar among the groups, total lipid content in the loin increased with levels of chocolate rolls in the diet. There were no major differences in fatty acid among the groups but release of free amino acids which are related to sweetness (glutamic acid and aspartic acid) and to savoury (serine, glycine and threonine) tended to increase with levels of chocolate rolls after four-day aging. There was no major difference among the groups in preferences assessed by a sensory panel. The mean urinary excretion of pigs on CR30, however, was more than the double of that of the control group ($P<0.05$) due possibly to theobromine contained in the chocolate.

Consequences of including high levels of sorghum-based distillers dried grains with solubles (DDGS) on nursery and finishing pig diets
Cerisuelo, A.[1], Bonet, J.[2], Coma, J.[2] and Lainez, M.[1], [1]Instituto Valenciano de Investigaciones Agrarias (IVIA), Centro de Investigación y Tecnología Animal (CITA), Pol. La Esperanza, 12400 Segorbe, Spain, [2]Vall Companys Group, Pol. Industrial El Segre, 25191 Lleida, Spain; cerisuelo_alb@gva.es

A study was conducted to evaluate the consequences of feeding distillers dried grains with solubles (DDGS) from sorghum grains to pigs on growth performance, body composition and carcass yield. Chemical composition of DDGS was analysed. A total of 204 animals of 13.9±2.60 kg body weight (BW) were allocated in 2 treatments. Two different diets were formulated in an iso-nutrient basis and fed to the animals during 28 days: a conventional diet (C) and a diet containing 15% DDGS (DDGS). At 38.9±6.33 kg BW, 194 pigs coming from the nursery phase continued in the same treatment (C or DDGS) during the finishing phase. DDGS inclusion level was increased to 30% in the period from 40 to 80kg BW and 35% from 80 kg to slaughter. At the end of the study, backfat thickness (BF) and loin depth (LD) were measured at P2 using ultrasounds. DDGS contained 90.2% dry matter, 4.93% ash, 33.7% crude protein, 10.8% crude fat, 26.50% neutral detergent fibre and 8.80% acid detergent fibre. Amino acid composition was determined and lysine to crude protein ratio was 2.08%. Regarding pig performance, during the nursery period no significant differences were found on average daily gain, but daily feed intake and final weight in DDGS group were slightly lower than in C group ($P<0.10$). During the finishing period, the inclusion of 30 to 35% DDGS did not lead to differences on growth performance. Pigs fed sorghum DDGS showed higher BF than pigs fed C diet ($P<0.001$); no differences were found on LD. At slaughter, carcass weight and carcass yield were similar between treatments. Thus, inclusion of 15% DDGS from sorghum grains in nursery diets might reduce feed intake and performance. However, inclusion rates up to 30-35% of DDGS during the finishing phase do not affect growth performance if an acceptable quality of DDGS is used, but might lead to fatter carcasses.

Growth performance and carcass characteristics of Holstein bulls reared exclusively on grass or finished with ground maize

Rosa, H.J.D.[1], Rego, O.A.[1], Evangelho, L.R.[1], Silva, C.C.G.[1], Borba, A.E.S.[1] and Bessa, R.J.B.[2], [1]CITA-A, Departamento de Ciências Agrárias, Universidade dos Açores, Angra Heroismo, 9701-851, Portugal, [2]Faculdade de Medicina Veterinária, Alto da Ajuda, Lisboa, 1300-477, Portugal; hrosa@uac.pt

Holstein bulls reared exclusively on grass (traditional beef production system of Azores) has better fat (lower n-6/n-3 ratio - 1.46 vs 7.92 – and more CLA - 0.59% vs 0.33%) than bulls fed concentrate but carcasses are very lean and present very poor conformation (60% graded 1 for fatness and 13 n% graded P for conformation). The objective of this study was to investigate the effects of supplementing previously grass reared Holstein bulls with two amounts of ground maize for a short finishing period on growth rate and carcass quality parameters. Thirty three Holstein bulls (15.4±1.9 (SD) months and 388±51.5 (SD) kg weight) previously reared exclusively on grass were finished for 85 days exclusively on grass (P) or supplemented with 4 (PM4) or 8 kg (PM8) ground maize. PM4 and PM8 bulls grew, 17 and 36% faster than bulls fed pasture only (ADG was 1,228, 1,441 and 1,667 g day^{-1}; $P<0.0001$). PM8 carcasses were 14% heavier than P ($P<0.05$). Dressing percentage was improved by the supplementation with 8 kg maize/head day $^{-1}$ (49.2 vs 46.2%; $P<0.0001$). Maize supplementation dramatically increased fat cover and fat content of carcasses. Fat score increased by 45% in PM8 carcasses and subcutaneous fat dept increased by 285% and 255% with the supplementation of 4 and 8 kg ($P<0.01$). Intramuscular fat of Longissimus dorsi increased from 1.84 to 2.96% (61%) with 4 kg maize and to 3.24% (76%) with 8 kg ($P<0.05$). Carcass conformation improved with 8 kg maize supplementation (carcass grading - 3.1 versus 4.4; $P<0.05$). The supplementation promoted the thickness of the carcasses mainly when 8 kg/head day^{-1} were supplied (1.88 versus 2.05 kg/cm; $P<0.05$). No significant effect was detected for carcass length, area of muscle Longissimus dorsi, proportion of saleable meat and proportion of joints of different quality categories.

Fatty acid composition of intramuscular fat of Holstein bulls fed exclusively on grass or finished with ground maize

Silva, C.C.G.[1], Rosa, H.J.D.[1], Rego, O.A.[1], Prates, J.A.M.[2] and Bessa, R.J.B.[2], [1]CITAA University of Azores, Angra Heroismo, 9701-851, Portugal, [2]Faculdade de Medicina Veterinária, Alto da Ajuda, Lisboa, 1300-477, Portugal; celia@uac.pt

In the traditional beef production system of the Azores Holstein bulls are reared exclusively on grass. In this system bulls produces beef with very healthy fat (n-6/n-3 ratio=1.46; CLA=59% in total fatty acids) but carcasses have bad conformation and lacks fat, therefore needing improvement e. g. through a finishing period with cereal. The aim of this study was to evaluate the extent to which the high level of linoleic (18:2n-6) and low level of α-linolenic (18:3n-3) acids of maize, provided during a short period immediately before slaughter, deteriorates the favorable fatty acid profile of total and both neutral and polar fractions of intramuscular fat of cattle previously fed exclusively on pasture. Thirty three Holstein bulls (15.4±1.9 (SD) months and 388±51.5 (SD) kg weight) reared exclusively on grass were finished for 85 days on grass (P) or supplemented with 4 (PM4) or 8 (PM8) kg ground maize/head day $^{-1}$. Evaluations were made on intramuscular fat of muscle longissimus dorsi. Intramuscular fat increased from 1.84 to 2.96% (61%) with 4kg maize and to 3.24% (76%) with 8kg ($P<0.05$). The n-6:n-3 PUFA ratio of total fat increased ($P<0.001$) linearly with inclusion of maize in diet (1.6, 2.4 and 3.5 respectively for P, P4 and P8) while the ratio PUFA:SFA was not affected by treatments (0.21, 0.22 and 0.22). Concentrations of EPA, DPA and DHA decreased by 30% in P4 and 60% in P8 ($P<0.05$). Maize in diets had no effect on the concentration of either total CLA or c9,t11 CLA in total fat or in any fat fraction. The intake of maize resulted in a decreased ($P<0.01$) concentration of linolenic FA in total and both fat fractions accompanied by an increase in concentration of linoleic FA ($P<0.01$).

Increasing possibilities of lycopene in poultry meat and egg yolk

Vitina, I.I., Krastina, V., Jemeljanovs, A., Miculis, J., Cerina, S. and Konosonoka, I.H., Latvia University of Agriculture, Research Institute of Biotechnology and Veterinary Medicine Sigra, Instituta street No 1, Sigulda LV 2150, Latvia; sigra@lis.lv

Lycopene as carotenoids antioxidant, plays important role in mitigating the damaging effects of oxidative stress on cells in human organism and in improving human health. The goal of our investigations was to evaluate the influence of antioxidants (carotenoids, vitamin E, Se) containing additives in poultry ration on contents of lycopene and amount of total carotenoids in meat and eggs yolk. In commercial produced eggs yolk of cross Lohmann Brown laying hens contained lycopene 0.60-0.90 mg/kg that is 4.49-8.78% from total carotenoids amount in eggs yolk (10.24-16.35 mg/kg). Content of lycopene in commercial produced broiler chicken meat of cross ROSS-308 is average 0.12-0.22mg/kg amount of total carotenoids is average 0.42-0.54 mg/kg. Enriching poultry ration with additives of natural carotenoids (by 1.8%), increased level of lycopene in broiler chicken meat by 0.12-0.15 mg/kg and in egg yolk by 0.26-0.83 mg/kg in comparison with commercial production. Amounts of total carotenoids in egg yolk increased by 2.20-4.86 mg/kg in comparison with commercial egg. The transfer level of amount of lycopene from feed to egg yolk is in average 20.64%. Feeding feed with additives of antioxidants containing complex increased content of lycopene in broiler chicken meat by 0.15-0.20 mg/kg and in egg yolk by 0.77-1.33 mg/kg in comparison with commercial production. Amount of total carotenoids in egg yolk increased by 7.21-15.91 mg/kg. Content of lycopene average in egg yolk composed 7.17-7.38% of total amount carotenoids in yolk. By using feed with additives of antioxidants complexes lycopene transfer level from feed to egg was 20.05-20.88%, from feed to meat was 18.34-20.15%. Obtained broiler chicken meat and egg were characterized with higher amount of lycopene and these products are more healthy for human diet.

Effects of dietary *Ferula elaeochytris* root powder on laying performance, egg quality and plasma metabolites of hens

Filik, G. and Kutlu, H.R., University Of Çukurova, Animal Science, Balcali / Adana, 01330, Turkey; gfilik@hotmail.com

The present study was conducted to evaluate whether dietary ferula (*Ferula elaeochytris*) root powder would affect laying performance, egg quality, egg cholesterol level and plasma glucose, cholesterol,triglycerides and calcium levels of Nick Brown layers. Thirty four weeks old laying hens were divided into 4 treatment groups of similar mean weight, comprising 18 birds each. The birds were fed standard layer diet having 177.6 g crude protein, 36 g calcium and 2,800 Kcal ME in each kg. Treatment birds were fed on their own diets made by 0, 2, 4 or 8 g dietary Ferula root powder as supplement to each kg standard layer diet for 8 weeks. Birds housed in individual cages were employed a 16:8 hours light:dark photoperiod. Feed and water were given ad libitum. Laying performance was determined by recording feed intake, egg weight, egg production daily; egg quality, egg cholesterol weekly and blood parameters (glucose, cholesterol, triglyceride and calcium level) biweekly. The results showed that dietary supplemental ferula root powder did not have any significant ($P>0.05$) effects on body weight, feed intake, egg yield (hen-day production), egg mass, egg weight and feed conversion ratio,. However, Ferula root powder improved egg shape index ($P<0.05$), reduced plasma glucose concentration ($P<0.05$) and decreased egg yolk cholesterol level ($P<0.01$) without affecting plasma cholesterol, triglyceride and calcium levels ($P>0.05$) without causing any toxicity incidence and symptoms in layers with respect to health status, voluntarily feed intake and egg production, suggesting that ferula root powder, at least, could be used in layer diet to produce egg with lower cholesterol content and better shape index.

Effect of different fattening methods and feeding sunflower seeds on the composition of carcass and beef in crossbred angus growing fattening bulls

Hajda, Z.[1], Lehel, L.[1], Várhegyi, J.[1], Kanyar, R.[2], Várhegyi, J.N.É.[1] and Szabó, F.[3], [1]Research Institute for Animal Breeding and Nutrition, Ruminant Nutrition, Gesztenyés út 1., 2053 Herceghalom, Hungary, [2]Hubertus Bt., Nimród út 1., 8646 Balatonfenyves, Hungary, [3]University of Pannonia, Georgikon Faculty of Agriculture, Department of Animal Science and Husbandry, Deák Ferenc utca 16., 8360 Keszthely, Hungary; hajda.zoltan@atk.hu

The aim of the experiment was to study the effect of the restricted feeding in the first period and feeding sunflower seed at the end of fattening on the composition of carcass and meat. Four homogenous (45-45 animals) groups of growing fattening bulls were chosen after weaning. Two experimental groups received high energy density rations from the beginning of fattening until slaughter. In the case of the other two groups the energy supply was restricted up to 400-450 kg BW, followed by intensive feeding. Two of the four groups received sunflower seed with high linoleic acid content after reaching the 500-550 kg BW. The results showed that apply of the restricted feeding period compared to continuous energy supply did not influence significantly on the rate of meat, bone and tallow in the carcass but there was a tendency that fat content decreased and protein content increased in the meat. The linoleic acid supplementation significantly decreased the rate of lean ($P<0.01$) and increased the rate of separated tallow ($P<0.05$). The protein content of longissimus dorsi and semitendinosus muscles decreased ($P<0.05$) and the fat content increased ($P<0.01$). From the aspect of human healthcare the use of sunflower seed, with high linoleic acid content, has a positive effect on the fatty acid composition of meat and tallow, the proportion of unsaturated fatty acid and conjugated linoleic acid has grown and the rate of omega-6/omega-3 fatty acid narrowed.

Eating quality of bacon from swine feed with fish oil

Kayan, A.[1], Jaturasitha, S.[2], Soonthornneth, A.[2] and Phongpiachan, P.[2], [1]Institute of Animal Science, Animal Breeding and Husbandry Group, Endenicher Allee 15, D-53115, Bonn, Germany, [2]Chiang Mai University, Animal Science, Chiang Mai, 50200, Thailand; akay@itw.uni-bonn.de

The 96 crossbred pigs were randomly selected from 480 pigs for more in-depth studies of pork quality. The focus of study was put on the efficiency to enrich tissue with n-3 fatty acids and the expression of adverse side-effects on pork quality. This study was conducted in CRD experimental design. They were assigned to a control (0% of fish oil) and diet with 2% of fish oil. The average weight of finishing pigs were began at 60 kg until slaughter weight 90, 100 and 110 kg, respectively. The result showed that feed had no effect on sensory evaluation but fish oil group had less triglyceride ($P<0.01$), tended to have lower cholesterol ($P>0.05$) and higher TBA value than control group ($P<0.001$). Furthermore, Fish oil group had higher EPA and DHA which were essential fatty acid than control group. The n-6:n-3 fatty acid ratio of fish oil group was lower than control group ($P<0.001$). In term of sex, sensory evaluation showed no significant difference between groups ($P>0.05$). Bacon from gilts had more cholesterol and TBA value, less MUFA and PUFA than bacon from barrows. Slaughter weight effect at 90 kg had the highest TBA value ($P<0.05$) but the lowest in total n-3 fatty acid level. In conclusion supplement 2% fish oil began at 60 kg of live weight which selected barrow and slaughter weight at 110 kg was suitable to produce smoked bacon.

Relation between mesophilic and psychrotrophic aerobe sporulating microorganisms in raw cow's milk

Foltys, V. and Kirchnerová, K., Animal Production Researche Centre Nitra, ABPQI, Hlohovecká 2, 951 41 Lužianky, Slovakia (Slovak Republic); foltys@scpv.sk

We studied the occurrence of mesophilic and psychrotrophic aerobe sporulating microorganisms (MPAS) in raw cow's milk and their relations to microflora in milk. We took 294 samples of raw cow's milk from 14 farms during one year. Briefly the method for MPAS assessment is to inactivate the milk sample by heating it to 80-82 °C for 30 minutes. Mesophilic aerobe sporulates are incubated at 30 °C for 3 days –, and psychrotrophic aerobe sporulates at 7 °C for 10 days. Results of studied microbiological parameters characterize the sampled milk as complying with requirements of the Directive 852/2004 EC. MPAS count was within the span 2.5–340 CFU/ml. The average value of MPAS was 59.4 CFU/ml, with variation coefficient 93.1%. Counts up to 50 CFU/ml were in 55.4% samples, the value was not higher than 100 in 85%, and in 3.1% of the samples the MPAS count was higher than 200. MPAS do not show correlation with any of the studied microbiological parameters; marked influences of season were not observed either. On the basis of obtained results, it is possible to support the proposal of an initial limit of maximum 200 CFU/ml for the introduction of a MPAS parameter. MPAS count found in the same dishes at incubation for mesophilic and subsequently strictly psychrophilic microorganisms was 56.9 CFU/ml on average. This represents 95.8% of total CFU sums of individual dishes at two temperatures. The correlation coefficient of these two types of results, r=0.99, gives evidence of close dependence expressed by the linear regression equation. Use of two incubation temperatures, one after another with an identical set of dishes, enables us to exclude overestimation of results due to sporulates able to grow at both incubation temperatures.

Yea-Sacc®1026 for high-producing dairy cows: commercial response in a Dutch herd

Andrieu, S.[1], Warren, H.[1], Fitie, A.[2], Fievez, V.[3] and Nollet, L.[4], [1]Alltech Biotechnology Centre, Sarney Road, Dunboyne, Ireland, [2]Feed Innovation Services, Generaal Foulkesweg 72, 6703 BW Wageningen, Netherlands, [3]University of Ghent, Sint-Pietersnieuwstraat 25, 9000 Ghent, Belgium, [4]Alltech Belgium, Gentsesteenweg 190/1, 9800 Deinze, Belgium; sandrieu@alltech.com

The objective of this trial was to evaluate the effect of Yea-Sacc®1026 (Saccharomyces Cerevisiae, CBS 493.94) at a dose of 10 g/cow/day on performance and milk fatty acid pattern in a dairy herd fed a grass based diet. Cows between 60 and 230 DIM at the start of the trial were used to produce 2 equal groups of 22 cows based on DIM, parity, milk production, milk composition and SCC from the previous 4 months before the trial. The 2 groups received the same diet but YS group was supplemented with Yea-Sacc®1026. The ration consisted of grass silage, concentrates, extruded linseed, potatoes, hay, minerals and vitamins. Initial DMI, parity, milk production were respectively 120 vs 118; 2.68 vs 2.59; 29.6 vs 29.5 kg/d for Control (C) and YS groups. Cows were milked using a milking robot; official milk control took place at day 35, 64 and 91 during the trial. The fatty acid (FA) profile of 20 milk samples/treatment (20 individual cows selected per treatment) was controlled during week 12. YS group had a significantly higher milk production than C group: 28.8 vs 28.4 (NS) at day 35; 29.2 vs 27.6 (P<0.01) at day 64 and 27.7 vs 26.0 at day 91 (P<0.01). Milk composition didn't differ significantly between treatments however a tendency (P=0.07) to a higher C18:3 n-3 content (g/kg milk) for the YS group was noticed. Also, the total amount of saturated FA decreased in favor of the level of mono- and polyunsaturated FA in the YS group. It can be concluded that Yea- Sacc®1026 at an inclusion at 10 g/cow/day in a grass based diet resulted in a similar milk response to that noted on more typical starch-based diets, i.e. an increase in milk production of about 1.5 kg/cow/day.

The structure of buffalo milk of a German herd

Schafberg, R.[1], Thiele, M.[2] and Swalve, H.H.[1], [1]Institute of Agricultural and Nutritional Sciences, Group Animal Breeding, Adam-Kuckhoff-Str. 35, 06108 Halle, Germany, [2]Landgut Chursdorf GbR, Landgutweg 25, 09322 Penig/OT Chursdorf, Germany; renate.schafberg@landw.uni-halle.de

Fat droplets are generated in the glandular epithelium of the mammary gland by apocrine secretion. Thus, every fat globule is covered by a cell membrane. Native milk fat globules (MFG) diameter in Friesian cows range from 0.1 to 17 μm, the mean diameter is around 3.6 μm. To study MFG size in buffalo raw milk, 630 samples were obtained from a German buffalo herd with about 20 milking cows (Chursdorf, Saxony). Sampling was done jointly with the monthly routine milk recording for three years (60 cows, 1-5 lactation, 5.9 kg milk yield). Average fat percentage was 8.1%, protein 4.7%, the volume-surface average diameter of MFG was 5.0 μm. This diameters ranged from 0.8 (own methodological lower threshold) to 30.4 μm. MFG diameters were influenced by individual variation as well as effects like stage of lactation or the month of sampling. In contrast to Holstein cows our results showed a negative correlation between MFG sizes and fat or protein content. The increase in size of the buffalo milk MFG may be a result of an enhanced cell metabolism of the buffalo mammary gland which is accompanied by the ability to produce more MFG by secretion. The size of MFG affects the physical properties of the emulsion and hence influences characteristics of the cheese manufacturing process like storage potential, gel structure, or firmness.

The study of ensiling residues of Iran fruit and vegetable in autumn

Karkoodi, K.[1], Fazaeli, F.[2] and Ebrahimnezhad, Y.[3], [1]Islamic Azad University, Saveh Branch, Animal Science, Saveh, Iran, [2]State Animal Science Institute, Karaj, Iran, [3]Islamic Azad University, Shabestar Branch, Animal Science, Shabestar, Iran; kkarkoodi@yahoo.com

The objective of the present study was to study ensiling residues of Iran fruit and vegetable in autumn. Samples of fruit and vegetables residues were collected during 3 months of autumn every other week per month. After separation of exogenous materials, all samples were dried and prepared to be ensiled. In a 3 x 4 factorial completely randomized experiment with 3 replicates, silage characteristics of fruit and vegetable residues were studied, while treatments were: 25, 30, 35 and 40% of DM (a mixture of fruit and vegetable residues and wheat straw) each with 0, 2 and 4% of sugar beet molasses respectively. The apparent evaluation showed that silages were scored from 5.3-16.68 (based on the maximum of 20), that were significantly different.The pH values were from 4.22-5.33 and DM content between 25.2-42% that were significantly varied among the treatments. There were also significant differences between the treatments for the content of OM, Ash, Total Nitrogen, ammonia Nitrogen, ammonia Nitrogen ratio to Total Nitrogen,Total VFA and *in vitro* digestibility. In general, the silage contained 30% of DM and 4 percent of molasses showed to be superior treatment for ensiling of fruit and vegetables residues.In this treatment the means of DM, OM, Ash, Total Nitrogen, Amonia Nitrogen, Ammonia Nitrogen ratio to Total Nitrogen, DMD, OMD and DOMD were 29.68, 78.07, 21.93, 1.06, 0.07, 6.67, 62.92, 68.85, 59.46% (DM basis) respectively and the mean of pH and score were 4.27 and 15.80 respectively and the mean of Total VFA was 72.80 mMol/100g (DM basis).

The effects of different dietary energy and protein levels at fixed slaughter weight on performance and carcass characteristics of Arabi fattening lambs

Dabiri, N.[1], Haydari, K.[2] and Fayazi, J.[2], [1]Islamic Azad University, Karaj Branch, Animal Science, 31485-313, Karaj, Iran, [2]Ramin Agricultural and Natural Resources Unversity, Animal Science, Mollasani, Ahvaz, Iran; Naj_dabiri@yahoo.com

Forty eight Arabi fattening lambs were randomly allocated to six dietary treatments in a 2×3 factorial experiment. The treatments included low (E_L=2.53 Mcal/kg$_{DM}$ ME), medium (E_M=2.71 Mcal/kg$_{DM}$ ME) and high (E_H=2.89 Mcal/kg$_{DM}$ ME) levels of dietary energy in combination with low (P_L=16% cp) and high (P_H=18% cp) levels of dietary protein. The body weight (BW), average daily gain (ADG), average daily feed (ADF) and feed conversion ratio (FCR) of lambs were measured two weeks interval until the end of experiment. Carcass components were recorded at the end of trial. Total ADG of fattening lambs fed the diets containing P_H and E_H levels were significantly higher ($P<0.05$) than lambs fed the other diets. Final BW of lambs fed the diets containing P_H and E_H level were greater than other groups. Total FCR of fattening lambs fed the diets containing P_H and E_H level were no significantly lower than lambs fed other diets. ADF amountsof the lambs fed diets containing P_L were not significantly differ than lambs fed the diet containing P_H, while were greater for the lambs fed the diet containing E_L level than did other groups. Carcass characteristics were not affected by changes in dietary protein and energy levels, except the lambs fed diet containing P_H and E_H levels had greater dressing (%) and hot carcass weight. Significant interactions between protein and energy levels in the ration were noted for final BW, FCR and ADG. According greater performance of lambs fed diets containing 18% protein and high energy level, it is concluded that, using this treatment would be useful for performance and carcass characteristics of fattening Arabi lambs.

Olive cake in laying hen diets for modification of yolk lipids

Abd Elsamee, L.D. and Hashish, S.M., National research center, Animal production, 36 Iran street, 12311-Dokki-Cairo, Egypt; samiahashish@hotmail.com

The possibility of improving egg yolk lipids of laying hens by olive cake (OC) feeding was investigated. Forty-two, 54-week-old, Lohman laying hens were fed for 12 weeks on 3 diets formulated to contain 0 (control), 28.5 or 57 g OC/kg diet, providing 0, 3.8 or 7.5 g olive oil/kg diet respectively. Inclusion of OC in hen diets at 28.5 or 57g/kg decreased ($P<0.001$) plasma cholesterol and triglycerides concentration compared to the control, without affecting plasma high-density lipoproteins. Olive cake feeding at 28.5 or 57 g/kg diet decreased ($P<0.001$) yolk concentration of total lipids, triglycerides. cholesterol, low-density lipoproteins and phospholipids. Olive cake at 28.5 or 57 g / kg diet decreased yolk concentration of total saturated fatty acids by 37.3 and 38.3% respectively. Total monosaturated fatty acids was decreased by 30.1% with OC feeding at 28.5g /kg of the diet, while it was increased by 17.3% with the 57 g/kg dietary OC. Olive cake feeding at 28.5 or 57 g.kg diet increased yolk concentration of total polyunsaturated fatty acids (2.8 and 2.6 fold, respectively), n-6 polyunsaturated fatty acids (2.7 and 2.5 fold, respectively), and n-3 polyunsaturated fatty acids (3.1 and 3.0 fold, respectively), resulting in 10.4 and 13.1% decreases in the ratio of n-6: n-3 polyunsaturated fatty acids of egg yolk respectively, compared to the control.

Effect of calcium soap supplementation without or with rumen-protcted methionine and lysine in lactating buffaloes' ration on milk production and composition

Kholif, A. and Ebeid, H., National Research Center, Dairy Science Department, El-Behoss St., Dokki, Giza, 12622 Dokki, Egypt; hossam_ebaid@yahoo.com

This investigation was carried out to study the effect of calcium soap fatty acid as energy concentration in diets without or with rumen-protected amino acids (methionine and lysine) on lactating Egyptian buffaloes performance. Four lactating buffaloes were used in single 4 x 4 Latin square experiment. All animals were subjected to four different experimental diets; 1) control diet, consisted of concentrate feed mixture, berseem clover and rice straw without supplementation, 2) control diet supplemented with 400 g protected fat source (calcium soap, Magnapac®); 3) control diet supplemented with rumen-protected amino acids (15g Methionine + 40 g Lysine); and 4) control diet supplemented with (400 g Magnapac® + 15g Methionine + 40g Lysine). Results obtained indicated that actual milk yield not significantly increased, but the amount of 7% fat corrected milk (FCM) increased significantly ($P<0.05$) in response to protected fat and protected amino acid supplementations. Milk composition also not significantly changed. Short and medium-chain fatty acids in milk decreased with treatments compared to control, while long chain fatty acids were improved in all treatment than control. Also, essential amino acids and limiting (Methionine and Lysine) in milk improved with treatments than control.

Can canthaxanthin affect on egg yolk lipid oxidation?

Esfahani-Mashhour, M., Moravej, H., Mehrabani-Yeganeh, H. and Razavi, S.H., Tehran University, Animal Science, Karaj, Tehran, 3158777871, Iran; hmoraveg@ut.ac.ir

A $2 \times 3 \times 4$ factorial experiment was planned to study the influence of storage temperature (4 °C, 24 °C) and dietary doses of canthaxanthin (CX) (0, 4, 8 mg/kg of feed) and storage time (d 1, d 15, d 30, d 45) on eggs lipid oxidation. 90 Hy-Line laying hens, 68 weeks old, were used in this study. The birds were allotted to 3 dietary treatments (0, 4 and 8ppm canthaxanthin) with each treatment replicated 3 times with 10 hens per replicate. At the end of excremental period collected eggs were allotted to 2 groups and stored in separated cabinet with different temperature (4 °C, 24 °C) to be analyzed for TBARS value in egg yolk. Collected eggs were cracked; yolks were separated, homogenized and prepared for sampling. Laying rate, egg weight, daily feed intake, and feed efficiency were not influenced by dietary treatments of canthaxanthin but Egg yolk color was affect by Canthaxanthin supplementation ($P<0.05$). Analysis of variance for main effects of Canthaxanthin supplementation levels, storage temperature, storage time and their interactions showed significantly difference for amount of TBARS value ($P<0.05$).

Nutritional value of barley malt rootlets in growing lamb rations

Bampidis, V.A.[1], Christodoulou, V.[2], Hučko, B.[3], Nistor, E.[4], Skapetas, V.[1] and Nitas, D.[1], [1]Alexander Technological Educational Institute, Department of Animal Production, P.O. Box 141, 57400 Thessaloniki, Greece, [2]National Agricultural Research Foundation, Animal Research Institute, Giannitsa, 58100 Giannitsa, Greece, [3]Czech University of Life Sciences, Department of Microbiology, Nutrition and Dietetics, Prague, 16521 Prague, Czech Republic, [4]Banat's University of Agricultural Sciences and Veterinary Medicine – Timisoara, Department of Agricultural Biology, Timişoara, 300645 Timişoara, Romania; bampidis@ap.teithe.gr

In an experiment with 72 male growing Florina (Pelagonia) lambs, effects of replacing soybean meal, wheat bran, alfalfa meal, and sugar beet pulp with barley malt rootlets (BMR), wheat grain, and corn gluten meal on performance and carcass characteristics were determined. In the 8 week experiment, lambs were allocated to one of four dietary treatments (BMR0, BMR100, BMR200, and BMR300) of 18 lambs each. Lambs had an initial body weight (BW) of 14.3 ± 1.9 kg, and were fed one of four isonitrogenous (crude protein 187 g/kg, dry matter – DM basis) and isoenergetic (net energy for gain 8.22 MJ/kg, DM basis) concentrate mixtures ad libitum and alfalfa hay (0.18 kg/lamb/day, DM basis). The BMR was added to the concentrate mixture at inclusion levels (as mixed basis) of 0, 100, 200, and 300 kg/t for treatments BMR0, BMR100, BMR200, and BMR300, respectively. No differences ($P>0.05$) occurred among BMR treatments in final BW (26.4 kg), BW gain (217 g/day), DM intake (0.84 kg/day), and feed conversion ratio (3.89 kg DM intake/kg BW gain). Moreover, carcass characteristics were not affected ($P>0.05$) with increased BMR feeding. Barley malt rootlet supplementation, at levels up to 300 kg/t, in isonitrogenous and iso (net energy) energetic diets for growing lambs did not affect their performance and carcass characteristics.

Effect of fattening lambs with sunflower cake and linseeds with or without vitamin E on fatty acid profile of intramuscular fat

Borys, B.[1], Kaczor, U.[2] and Pustkowiak, H.[2], [1]National Research Institute of Animal Production, Experimental Station Koluda Wielka, Parkowa st. 1, 88-160 Janikowo, Poland, [2]University of Agriculture, Mickiewicza st. 24/28, 30-050 Krokow, Poland; bronislaw.borys@onet.eu

The effect of fattening lambs with sunflower cake and linseeds and vitamin E supplementation on fatty acid profile and content of m. longissimus lumborum (MLL) fat was investigated. Subjects were 18 Koluda sheep (KS) lamb-rams and Ile de France ´ KS (IF´KS) hybrids, which were fattened intensively to 32-37 kg body weight with concentrates and grass hay supplement. The control group (C) was fed a diet based on cereal components (>50%) and rapeseed meal (RM; 20%). The experimental groups received sunflower cake and linseeds instead of RM: 23.5 and 5%, respectively (group SCL), and additional supplements of vitamin E (group SCL+E). The experimental diets contained 11.0% more fibre and over twice as much fat (in DM: C – 4.31%, SCL – 8.75% on average). Lambs from SCL groups consumed less C18:1 c11 (by 63.4% on average) and more C18:0, C18:1 c9, C18:2 and C18:3 (by 395.6; 58.2, 124.1 and 336.4%, respectively). Feeding sunflower cake and linseeds as well as vitamin E supplements had a clear effect on fatty acid composition and content of MLL in the lambs. With a similar total SFA content, the fat of SCL and SCL+E lambs contained less MUFA (39.6 and 42.2% vs. 46.9%) and more PUFA (18.8 and 15.2% vs. 10.7%) compared to the fat of C lambs. In the group supplemented with vitamin E (SCL+E), changes in the FA profile were smaller than in the SCL group, with a clear decrease in MLL fat content in relation to groups C and SCL (2.25% vs. 2.82 and 3.05%, respectively). Overall, considerable differences were found in the absolute content of most fatty acids between SCL and C groups, and in some acids also between SCL and SCL+E groups. Breed origin had no significant effect on FA profile and content of MLL. The study was preliminary in nature and will be continued.

The optimum intake of row and soaked chicken pea in broiler chicken
Moeini, M., Souri, M. and Haedari, M., Razi University, Animal Science, Kermanshah, 67155, Iran;
msouri@yahoo.com

Chicken pea could be used as a cheap protein source for poultry nutrition in west of Iran. In this study it has been tried to determine the optimum intake of Chicken pea (Desi variety) in broilers diet. Total 147 day old chicks broilers (Cobb breed) were chosen in a completely randomized design in seven treatment groups: 15%, 25% and 35% raw pea; 15%, 25% and 35% soaked pea in water for 48 hours and control group (without pea). Each treatment had three replications with 7 observations in each treatment. The rations were standard and balanced for protein and energy. Feed intake, weight gain and feed conversion ratio (FCR) were measured weekly during six weeks experiment (from day 7 to day 49). At the end of experiment 9 chicks from each treatment selected randomly for the percentage of leg, chest and fat percentage of ventricle area. Statistical analyses were performed using Duncan's mean test. The results showed that the mean daily gain and FCR had no significant difference among treatments, but the mean weight gain was higher in treatment T6 (35% soaked pea) compared with other treatments ($P<0.05$). There was no significant difference between treatments for ventricle area, the percentage of leg, fat and chest. It can be concluded that up to 25% of row split pea or at the level of 35% when treated in water for 48 hours could be used as a good protein and energy source in broiler feed.

Effect of dietary extruded chickpea supplementation on meat quality of broiler turkeys
Christodoulou, V.[1], Bampidis, V.A.[2], Labrinea, E.[3], Ambrosiadis, J.[4], Arkoudelos, J.[3] and Hučko, B.[5],
[1]National Agricultural Research Foundation, Animal Research Institute, Giannitsa, 58100 Giannitsa,
Greece, [2]Alexander Technological Educational Institute, Department of Animal Production, P.O. Box
141, 57400 Thessaloniki, Greece, [3]National Agricultural Research Foundation, Institute of Technology of
Agricultural Products, Athens, 14123 Athens, Greece, [4]Aristotle University, School of Veterinary Medicine,
Thessaloniki, 54006 Thessaloniki, Greece, [5]Czech University of Life Sciences, Department of Microbiology,
Nutrition and Dietetics, Prague, 16521 Prague, Czech Republic; vchristodoulou.arig@nagref.gr

In an experiment with 200 day old broiler turkeys (B.U.T. 9), the effect of partial and total replacement of soybean meal with chickpeas (*Cicer arietinum* L.) on meat quality was determined. In the 84 day experiment, turkeys were allocated to five dietary treatments being: ECKP0, ECKP200, ECKP400, ECKP600 and ECKP800 of 40 birds each (five subgroups of 8 birds in each treatment), and received a diet *ad libitum*. The diet for ECKP0 treatment had no chickpeas (control), while those for treatments ECKP200, ECKP400 ECKP600 and ECKP800 included 200, 400, 600 and 800 kg/t of wet extruded (at 120 °C for 20 sec) chickpeas, respectively. At the end of the experiment, 5 turkeys, randomly selected from each treatment (1 from each subgroup), were fasted for 18 h (water was allowed), weighed, and euthanized. After dressing, samples of the right breast muscles (m. *pectoralis superficialis and m. pectoralis profundus*), and the right leg muscles, of the carcass of all birds were removed for chemical composition analysis, color evaluation, sensory evaluation, and fatty acid (FA) and cholesterol analysis. Results showed that extruded chickpeas used as an alternative protein source to replace soybean meal in broiler turkey diets, at inclusion levels up to 800 kg/t, did not affect meat quality.

Effect of dietary attapulgite clay on performance and blood parameters of lactating Holstein cows

Bampidis, V.A.[1], Christodoulou, V.[2] and Theophillou, N.[3], [1]Alexander Technological Educational Institute, Department of Animal Production, P.O. Box 141, 57400 Thessaloniki, Greece, [2]National Agricultural Research Foundation, Animal Research Institute, Giannitsa, 58100 Giannitsa, Greece, [3]Geohellas S.A., P. Faliro, 17564 Athens, Greece; vchristodoulou.arig@nagref.gr

Sixteen lactating Holstein cows were used in an experiment to determine effects of dietary attapulgite clay (AC) supplementation on productivity and milk composition, and blood parameters. Attapulgite and saponite, that form AC, are complex hydrated magnesium aluminum silicates. In the experiment, which started on week 12 postpartum, cows were allocated, after equal distribution relative to milk yield and lactation number (i.e., 2 or 3), into 2 treatments being AC0 and AC10 of 8 cows each. For a period of 12 weeks (i.e., weeks 12-24 postpartum), cows were fed one of two isonitrogenous (crude protein 178 g/kg, dry matter – DM basis) and isoenergetic (net energy for lactation 7.85 MJ/kg, DM basis) concentrates (12.3 kg DM/cow/day), alfalfa hay (5.4 kg DM/cow/day), corn silage (3.3 kg DM/cow/day) and wheat straw (1.8 kg DM/cow/day, DM basis). The AC was added to the concentrate mixture at inclusion levels (as mixed basis) of 0 and 10 kg/t for treatments AC0 and AC10, respectively. During the experiment, there were no differences between AC0 and AC10 treatments ($P>0.05$) in milk fat (31.7 g/kg), protein (33.6 g/kg), lactose (48.9 g/kg) or ash (6.7 g/kg) contents. Average milk yield (25.2 kg/day), yields of components and somatic cell counts were not affected ($P>0.05$) with increased AC feeding, except for protein yield that increased (0.82 vs. 0.86 kg/day, $P<0.001$) and colony forming units that decreased (59.3 vs. 41.2 ×1000/ml, $P<0.001$). Moreover, no differences in blood parameters occurred ($P>0.05$) between treatments. Dietary AC supplementation, at levels up to 10 kg/t, in isonitrogenous and iso (net energy) energetic diets for lactating cows did not affect their performances.

Effect of alfalfa hay supplementation on grazing time and milk parameters in a rationed dairy sheep grazing system

Arranz, J.[1], Amores, G.[2], Virto, M.[2], Barrón, L.J.R.[2], Beltrán De Heredia, I.[1], Abilleira, E.[2], Ruiz De Gordoa, J.C.[2], Nájera, A.[2], Ruiz, R.[1], Albisu, M.[2], Pérez-Elortondo, F.J.[2], De Renobales, M.[2] and Mandaluniz, N.[1], [1]NEIKER, P.O. Box 46, Vitoria, Spain, [2]University of Basque Country, P.O. Box 450, Vitoria, Spain; nmandaluniz@neiker.net

Dairy sheep systems in the Basque Country are based on pasture use by part-time grazing (PTG) during spring. In this period forage resources are managed by matching grazing and indoor supplementation with production requirements. The aim of this study was to evaluate the effect of alfalfa hay supplementation on the quality (% fat and protein) and yield of standardized dairy milk (DMYs). Moreover, grazing time was monitored. The experiment was conducted over 4 weeks with multiparious Latxa dairy ewes. Sheep were blocked into homogeneous groups of 12, and each group was randomly assigned to 3 different alfalfa hay rates (AR): 1) Low: 300g DM/day, 2) Intermediate: 600g DM/day and 3) High: 900g DM/day. Each ewe received 500g DM concentrate/day at milking and ewes accessed pasture 4 hours/day. Data were analysed considering as fixed effects AR, week, their interaction and initial values of each parameter as covariate. Results showed that ewes receiving the low and intermediate alfalfa rates spent significantly more time in grazing activity (228, 224 and 209 min/day, respectively) and produced the same amount of DMYs (1.21, 1.18 and 1.21 l/day, respectively) and quality (fat: 6.16, 6.46 and 6.29%; and crude protein: 4.96, 5.12 and 4.97%, respectively) as the group of high rate. These results show the possibility to manage grazing in a PTG system by reducing indoor feeding to increase the use of locally available resources, without compromising milk yield and quality. In conclusion, no productive advantage is achieved when offering more than 600g alfalfa/day and this management could be interesting in areas with enough pasture availability due to its positive effects on the milk conjugated linoleic acid isomer content and costs reduction.

Chemical composition in raw and cooked lamb

Sañudo, C.[1], Campo, M.M.[1], Cerra, Y.[2], Muela, E.[1], Olleta, J.L.[1], Pérez, P.[3], Robles, J.[4] and Oliván, A.[4], [1]Universidad de Zaragoza, Producción Animal y Ciencia de los Alimentos, Miguel Servet, 177, 50.013 Zaragoza, Spain, [2]INTA, Análisis químico, Málaga 5, 44.002 Teruel, Spain, [3]Consejo Regular Ternasco de Aragón, Técnico, Edificio Centro Origen, 50.014 Zaragoza, Spain, [4]Pastores Grupo Cooperativo, Técnico, Edificio Pastores, 50.014 Zaragoza, Spain; csanudo@unizar.es

Consumers are currently concern about their health and about the possible benefits of a healthy diet. On the other hand, red meat is considered as an unhealthy product. However, not many studies have demonstrated the relationship between diet (in the meat case) and health. The aim of this study was to analyse basic chemical composition on raw meat and after three selected cooking recipes. Sixty deboned legs without knuckle, from thirty lamb carcasses (IGP Ternasco de Aragón) from animals reared intensively and slaughtered at 3 months of age, were analysed. The left side was used raw and the right side after cooking in three different procedures (10 legs each): grilled with salt and olive oil, roasted with a traditional recipe, and boiled with almonds, olive oil, wine and salt. Meat was cleaned prior to homogenate. Fat, protein, humidity and ashes were analysed by ISO procedures. Raw vs cooked samples and cooking method were analysed using a GLM test. Cooking vs raw and cooking method itself had a significant effect on chemical composition, especially on humidity and protein content. Humidity was lower in boiled and roasted meat (58.4%) than in grilled meat (62.6%) or raw meat (71.5%). Protein content was higher in boiled and roasted recipes (25.6 and 25.7%, respectively) than in grilled (21.8%) or raw meat (17.6%). Fat content was lower in raw meat (9.6%) that in roast, grill or boiled meats (11.1%, 12.1% and 14.8%, respectively). Ashes varied between 3.0% (raw) and 4.8% (grilled). In conclusion, roast seems to be the healthiest cooking method (more protein, less fat) on this cut (leg), although other cuts and cooking procedures would need to be analysed.

Marbling IA: software for fat content estimation on digital image of cutted beef

Riha, J.[1], Homola, M.[2] and Hegedusova, Z.[1], [1]Research Institute for Cattle Breeding, s.r.o., Vyzkumniku 267, 788 13 Vikyrovice, Czech Republic, [2]Agroresearch Rapotin, s.r.o., Vyzkumniku 267, 788 13 Vikyrovice, Czech Republic; jan.riha@vuchs.cz

Marbling IA v1.0 is software solution for estimation of fat content and marbling of beef by usage image analysis and machine learning methods. Software is designed for usage in research and experimental laboratories and for end-user (consumer) too. It can be used for routine, fast and cheap analysis of digital image of cutted beef from digital camera or scanner. Input for this software is an digital image of cutted beef and model which classifies pixels (each pixel and its 8-neighbourhood and some neighbourhood measures as average colour information etc.) to classes 'fat' or 'loan meat'. Classification models are upgradeable and based on machine learning algorithms (J48, WEKA 3-4-6). We calibrated the model with 50 digital photos of cutted beef with following classification parameters. Correctly classified instances -99.982%, Kappa statistic -0.999, Mean absolute error -0.002,Root mean squared error -0.013, precision for loan meat class -1.000, precision for fat class -0.999. After classification of pixels of whole image, cluster analysis is performed. For each cluster of 'fat' their area, relative area, circularity and perimeter are calculated. For whole image algorithm calculates whole area, loan meat content, fat content (absolute and relative), cluster count, average cluster area (absolute and relative), average cluster perimeter, average circularity. With this parameters marbling of beef can be estimated after calibration. User interfaces is created in Borland Delphi 2005 with usage of containers implementation for real-time implementation. User can load images, perform digital image filters and analysis of fat content. Also outputs and export functions to MS Excel are included in this interface. Software is available on the company website: www.vuchs.cz/marblingIA. Authorized software Marbling IA was created with support of research projects of MŠMT ČR – 2B08037 and INGO LA330.

Effect of diet supplementation on visual appearance of lamb
Muela, E., Sañudo, C., Campo, M.M. and Beltran, J.A., University of Zaragoza, Animal Production and Food Science Department, Miguel Servet 177, 50013-Zaragoza, Spain; ericamola@hotmail.com

Meat appearance plays an important role on consumer acceptance and purchase, where colour is one of the main factors. To preserve colour stability, diet supplementation with antioxidant substances is a main strategy. The objective of this study was to evaluate if flavonoids supplementation on diet had a benefit on colour acceptability in lamb. Sixty lambs of Rasa Aragonesa breed were divided into 6 batches, depending on diet supplementation (DS): 3 batches with vitamin E (100, 200 or 300 ppm, respectively), 1 batch with vitamin E and LEBEN® (natural flavonoids) (100/100 ppm), 1 batch with LEBEN® (150 ppm), and a control group without supplementation. Animals were reared intensively for 1 month with the DS. Carcasses (cold carcass weight: 11-13 kg) were aged for 48 h. From each carcass right side, 3 chops of the thoracic segment were packed in a commercial modified atmosphere (O_2:CO_2:N_2), and the trays were exposed in a commercial expositor, under permanent light exposition and at refrigeration temperature (2-4 °C). Throughout 12 d-display, 15 consumers evaluated colour acceptability (from 1 -very bad- to 8 -very good-) and purchase intention (Yes/No). Until 2 d-display, there were no statistical differences between DS. At 3 d-display, there were no statistical differences between vitamin E+LEBEN® (with the lower acceptability in almost all the display period) and 150 ppm LEBEN® batches, but there were between vitamin E+LEBEN® batch and the others. Control, 100 ppm vitamin E, and 100-100 vitamin E-LEBEN® batches were purchased by consumers until 6 d-display, whereas 150 ppm LEBEN® batch had 1 d of extent at purchase intention compared with them, which may be related to LEBEN® dose. At every time, 200 and 300 ppm vitamin E batches showed the highest acceptabilities (purchase intention until 8 or 9 d-display, respectively), and they differed significantly from the others at any display time after 2 d. Further investigation on flavonoids dose is necessary.

Application of agricultural and factory byproducts to growing poultry ration with 45% replacement of imported feed ingredients
Tobioka, H., Miyagi, A., Shinozaki, K. and Tashiro, M., Tokai University, School of Agriculture, Minamiaso-mura, Aso-gun, Kumamoto, 869-1404, Japan; hisaya.tobioka@agri.u-tokai.ac.jp

Agricultural and factory byproducts such as sweet potato, tofu cake and confectionary byproduct have been examined on the nutritive values with ducks. Presently focus was made on poultry ration. The feed which included byproduct ingredients more than 45% on dry matter was prepared, ensiled and fed to broiler. Growth performance, edible meat yield and organ weight were investigated. After mixing of ungraded sweet potato, tofu cake, confectionary byproduct, soy sauce cake and fish silage, the byproduct feed, that is, eco-feed (Eco) was prepared and ensilaged more than a month. The target nutritive values were 17% crude protein and 12MJ ME. Thirty cross-bred chickens were divided into reference (Ref) and Eco groups with 3 replicates of 5 birds per cage. After preliminary feeding of 3 weeks, growth performance was evaluated for 5 weeks and slaughtered at age of 56 day. Edible meat and organs were weighed and relative weight of respective organs to body weight was calculated. Dry matter intake and daily weight gain for Eco group were 169 g/d and 76 g which were 14% and 17% larger than those of Ref group, respectively. Feed conversion ratio for Eco group showed 13% improvement compared to that of Ref group. The rate of weight of edible tissue to carcass was 67% with 1.6% increase than that of Ref group. The percentage of weight of heart and liver relative to body weight showed the higher tendency for Ref group, however, the reverse tendency was observed for the digestive organs. With this result, Eco-feed prepared with a half of ingredients from byproducts is applicable to broiler feeding without any adverse effects.

Ageing time, diet and breed effects on the meat color and pH of Spanish dairy goat kids

Alcalde, M.J.[1], Ripoll, G.[2], Horcada, A.[1], Sañudo, C.[3] and Panea, B.[2], [1]Univ Seville, Ctra Utrera Km1, 41013 Seville, Spain, [2]CITA, Av Montañana 930, 50059 Zaragoza, Spain, [3]Univ Zaragoza, Miguel Servet, 177, 50013 Zaragoza, Spain; aldea@us.es

The influence of breed (Murciano-Granadina vs Malagueña), ageing time (24 vs 72 h) and diet (dam's milk vs milk replacer) effects on the colour and pH of 60 male kids (4.6±0.1kg cold carcass weight) were evaluated. The variables were measured in the Longissimus dorsi lumborum after 1 h blooming. The average pH was 5.83±0.01, L* was 52.3±0.6, a* was 2.6±0.2, b* was 11.7±0.3, C* was 76.7±1.1 and h* was 12.2±0.3. b* was significantly different in all the effects and also it was significant the interaction between ageing time*breed. pH was affected by diet and ageing time. h* was influenced by ageing time and breed and the interaction ageing time*breed. L* had differences between the ageing time and interactions between breed*diet and breed*ageing time. And finally a* and C* only were influenced by breed and the interaction between diet and breed. The yellow index (b*) was the variable that explains the most part of the model variability (R^2=0.9). The ageing time was the effects with the highest discrimination power (98.3% of accuracy) while the diet effect only discriminate 60% and breed effect the 75.7%.

Consumer acceptability of pork enriched with CLA

Font I Furnols, M.[1], Reixach, J.[2], Oliver, M.A.[1], Gispert, M.[1], Francàs, C.[1] and Realini, C.[1], [1]IRTA, Granja Camps i Armet, 17121 Monells, Spain, [2]Grup Batallé, Avda. Segadors s/n, 17421 Riudarenes, Spain; maria. font@irta.es

Entire male and female pigs ((LandrancexDuroc) x Pietrain) fed Control (C) or CLA (1.0%) diets during 38 d and slaughtered at 115 kg live weight were used to evaluate consumer acceptability of pork enriched with CLA. Two hundred consumers selected by their age and sex according to the Spanish population distribution participated in the study carried out in Barcelona. Loins from 18 pigs fed CLA (CLA levels in intramuscular fat were 0.38% c9t11 and 0.16% t10c12) and 18 pigs fed C were evaluated in 18 sessions of 8-15 consumers. Each consumer evaluated the overall acceptability, tenderness, juiciness and taste of two samples, one from each diet, using a 9 point scale (1 'dislike extremely', 8 'like extremely). The 1.5 cm slices of longissimus lumborum were cooked in a convection oven until samples reached an internal temperature of 72 °C. Pork slices were cut into 1.5x1.5x4 cm samples, each one wrapped with aluminium foil, coded, kept warm and served. Session effect was corrected; diet and sex were considered fixed effects and consumer random effect. Tukey test was applied. No interaction was found between diet and sex effects and there were no sex differences (P>0.05) in pork acceptability. Consumers evaluated all pork attributes significantly better for meat from animals fed C than CLA (6.26 vs. 5.87 overall acceptability; 6.15 vs. 5.62 tenderness; 5.78 vs. 5.27 juiciness; 6.13 vs. 5.74 taste). Pork from animals fed CLA was rated lower acceptability scores by consumers than meat from animals fed C.

Fatty acid profile in small ruminants of Spanish breeds
Horcada, A.[1], Sañudo, C.[2], Polvillo, O.[1], Campo, M.M.[2] and Alcalde, M.J.[1], [1]Universidad de Sevilla, Agricultural and Forestry Science, University of Sevilla, EUITA, 41.013 Sevilla, Spain, [2]Universidad de Zaragoza, Animal Production and Food Science, Miguel Servet, 177, 50.013, Spain; csanudo@unizar.es

Fatty acid composition of kidney knob (KK) and intramuscular (IM) fat of Longissimus dorsi was studied on 40 Spanish goat kids and 10 lambs. Moncaina (5.31+0.26 kg live weight, LW), Blanca Celtiberica (6.55+0.68LW), Negra Serrana (5.83+1.13LW), Murciano Granadina (4.37+0.43LW) goat kids and Churra (5.66 + 0.57 LW) lambs were reared in the origin farms. Animals were fed exclusively with natural milk and slaughtered on the day of weaning. After slaughtering, KK fat and Longissimus dorsi were obtained. Total fatty acid was extracted by Aldai *et al.* (2006). Separation and quantification of the fatty acid methyl esters (FAMEs) was carried out by gas chromatography. FAMEs were separated on a 100-m Supelco SPTM-2560 fused silica capillary column with an internal diameter of 0.25mm and a film thickness of 0.20mm. FAMEs were identified by comparing their retention times with those of an authenticated individual standard (Sigma Chemical Co. Ltd., Poole, UK). Fatty acids were expressed as a percentage of total fatty acids identified and grouped as follows: saturated (SFA), monounsaturated (MUFA) and polyunsaturated (PUFA) and total CLA. Profile fatty acid in goat kids and lambs are in agreement with Arsenos *et al.* (2006) and Cañeque *et al.* (2005). C18:1, C16:0 and C18:0 were the main fatty acids making up in IM and KK fat depots as corresponding ruminants. Significant differences in fatty acid composition from KK between kids and lambs were observed. SFA was lower in lambs than goat kids ($P<0.001$). Higher MUFA ($P<0.01$) and PUFA ($P<0.001$) were observed in lambs. CLA was higher in lambs than in kids. No significant differences in IM fatty acids between goat kids and lambs were observed. This study showed that in earlier fat deposits, such as KK, differences in fatty acid composition are observed between small ruminant species.

Terpene content in ewe's milk: feeding management
Abilleira, E.[1], Barrón, L.J.R.[1], Arranz, J.[2], Virto, M.[1], Nájera, A.I.[1], Beltrán De Heredia, I.[2], Albisu, M.[1], Pérez-Elortondo, F.J.[1], Ruiz, R.[2], Ruiz De Gordoa, J.C.[1], Mandaluniz, N.[2] and De Renobales, M.[1], [1]Univ. Basque Country, P.O. Box 450, Vitoria, Spain, [2]NEIKER, P.O. Box 46, Vitoria, Spain; mertxe.derenobales@ehu.es

The aim of this work was to study the effect of the feeding management, consisting of different fresh pasture and alfalfa hay levels, on the terpene content of ewe's milk fat. The experiment was conducted in spring with sheep of the latxa breed and it lasted 4 weeks plus 1 more week for adaptation. Ewes were got into 4 homogeneous groups of 12. First 3 groups differed only in the amount of alfalfa hay supplement (300, 600 and 900 g/d). All sheep received 500 g/d of concentrate and were given access to cultivated pasture during 4 hours. Fresh pasture intake was determined indirectly by monitoring real grazing time. Along with these 3 groups, a fourth control group of the same characteristics was kept indoors consuming 600 g/d of alfalfa hay, 500 g DM/d of concentrate and 1000 g/d of grass hay instead of the fresh pasture. Bulk milk samples from each group were taken once a week. A total of 20 samples were analyzed by means of HS-SPME-GC/MS. Terpene concentrations were calculated using 1,3,5-triisopropylbenzene (TIPB) as internal standard. A mixed model of repeated-measures anova was performed. Feeding management type was used as fixed factor whereas the experimental week was the repeated-measures factor. Monoterpenes and sesquiterpenes were significantly higher in the low alfalfa supply group which was the group that grazed significantly more time. No significant differences were found between the other 3 groups suggesting that, among other factors, a certain minimun intake of fresh pasture is needed to see a significant increase of these compounds in ewe's milk.

Effect of freezing method on lamb acceptability
Muela, E., Sañudo, C., Campo, M.M. and Beltran, J.A., University of Zaragoza, Animal Production and Food Science Department, Miguel Servet 177, 50013-Zaragoza, Spain; ericamola@hotmail.com

Despite the numerous advantages of freezing, consumer's perception of frozen meat is not good. The objective of this study was to evaluate the influence of freezing method on lamb meat acceptability. Ninety out of 100 lamb carcasses of Rasa Aragonesa breed (11-13 kg cold carcass weight) were divided into 3 batches, depending on freezing method (FM): air blast freezer (-30 °C, 30 h), freezing tunnel (-40 °C, 18 min), or nitrogen chamber (-75 °C, 15 min). After 48h in refrigeration, the lumbar segment of the left side was cut into chops and randomly allocated to one of the three FM. Chops were covered with a retractile oxygen-permeable film and then frozen. They were stored at -20 °C (6 months maximum). Fresh meat samples (control lot, 10 carcasses) followed the same procedure but they were aged in a refrigerator for 72h before analysis. In 5 sessions performed over controlled conditions, 100 consumers evaluated overall, tenderness, and flavour acceptability with an 8-point scale. 24h before each session, samples were thawed in a refrigerator. Chops were cooked wrapped in aluminium foil at 200 °C on a double-grill hotplate, until 70 °C of internal muscle temperature. Then, the Longissimus dorsi muscle from each chop was cut in portions wrapped individually in aluminium foil and assigned a random code. Samples were kept warm until they were served in plates one at a time in random order among consumers. It was found that sensory quality was significantly affected by FM but acceptability differences were not found between fresh and thawed meat. Cluster division confirmed this fact, except for a group (39% of consumers), which distinguished fresh meat. The 24% of the population preferred thawed vs. fresh meat.

Cows' udder microorganisms: raw milk contaminants
Konosonoka, I.H., Jemeljanovs, A. and Ikauniece, D., Latvia University of Agriculture, Research Institute of Biotechnology and Veterinary Medicine Sigra, Instituta street No 1, Sigulda LV 2150, Latvia; sigra@lis.lv

The mammary system of the cow is very important for milk production and its quality. Bacteria present inside of the mammary gland and colonize teat end on the one hand may cause cow's udder infections, and on the other hand may contaminate raw milk with pathogens and opportunistic pathogens accordingly lowering quality of raw milk. The objective of the current study was to investigate raw milk contaminants from cows' udder. Aseptically taken cows' milk samples from four dairy herds were bacteriologically examined at the Scientific Laboratory of Biochemistry and Microbiology of the Research Institute 'Sigra'. Complex and selective culture media were used for the isolation and differentiation of bacteria. Appropriate biochemical tests and Becton Dickinson BBL Crystal gram-positive and gram-negative kits were used for identification of bacteria to species level. In total, 631 samples were analysed. Acquired results showed that 5,3% (n=357) of healthy cows' milk samples contained food born pathogens Staphylococcus aureus, 5,5% (n=274) *Listeria monocytogenes*, and 5,1% (n=274) *Salmonella* spp. 6,7% (n=357) of milk samples contained bacteria from the family Enterobacteriaceae int.al. *Escherichia coli, Klebsiella pneumoniae, Pseudomonas aeruginosa, Enterobacter aerogenes*. Our investigations showed that 72,8% (n=357) of raw milk samples contained different coagulase negative staphylococci (CNS) species. The most often isolated CNS species were *Staphylococcus haemolyticus* (31,3%; n=196), *Staphylococcus simulans* (18,2%; n=196) and *Staphylococcus xylosus* (14,5%; n=196). Results of investigations showed that 31,7% (n=101) of aseptically taken milk samples contained moulds. Our investigations let us conclude that different milk contaminating microorganisms from cows' udder and teat end get into raw milk and decrease milk quality and safety for consumers.

Innovative veal production
Sterna, V., Jemeljanovs, A., Konosonoka, I.H., Jansons, I., Lujane, B., Cerina, S. and Strazdina, V., Latvia University of Agriculture, Research Institute of Biotechnology and Veterinary Medicine Sigra, Instituta street No 1, Sigulda LV 2150, Latvia; sigra@lis.lv

It was concluded that supplementing basic feed of dairy cows with carotenoids gave good results on cows health, milk production and composition. Carotenoids in feed or food act as antioxidant and improve helth and body condition. Therefor was investigated possibility to increase content of carotenoids, amount of w-6 and w-3, to decrease content of cholesterol in meat by supplementing feed with carotenoids and linseed oil. Two groups of beef calves was complected (Holsteins Black and White). In the experimental group basic feed (milk, concentrate, fodder) was supplemented with cocentrate of carotenoids mixed in linseed oil (linseed oil composed 1% of feed amount). The chemical analyses were done: dry matter was determined by a drying method, fat content by Sochlet method, protein content by Kjeldahl, cholesterol content by Blur. The fatty acid composition was analysed by the gas-chromatography. The studies were carried out in the 2008. Aim of investigation was to produce meat with functional properties, improved quality and chemical composition. Results of investigation demonstranted that supplement contributed development of calves. After sixth month trial average livingmass of experimental groups calf was greater (254 kg) than control groups calf (234 kg). Results of biochemical testing demonstrated higher protein 21.6%, fat 2.43% and carotinoids 1.45 mg/kg contents compared with veal of control group (respectively 20.06%, 1.05% and 0.67 mg/kg). Experimental groups vial has higher content of w-6 16,27% and w-3 8,13% and ratio was more optimal (w-6/w-3=2) than veal of control group respectively 12.97%, 6.4% and w-6/w-3=2.4. It was concluded that supplementing feed of calves with carotenoids in linseed oil ensure producing veal with functional properties improving quality of meat.

Concentrate level and slaughter body weight effects on growth performances and carcass quality of barbarin lambs
Majdoub-Mathlouthi, L., Saïd, B. and Kraiem, K., Institut Supérieur Agronomique Chott Mariem, Production animale, Chott Mariem-Sousse, 4042, Tunisia; lmajdoub@lycos.com

The objective is to evaluate the effects of concentrate level (CL) and slaughter body weight on growth performances and carcass quality in barbarin lambs. 24 male Lambs receiving a hay-based diet were allowed into two groups. The high concentrate level group (HCL) received 600 g of concentrate and the low level concentrate group (LCL) received 300 g of concentrate. Six lambs of each group were slaughtered at a body weight of 35 kg (SW1). The others were slaughtered at a body weight of 42 kg (SW2).Carcass and different tissues (heart, lungs, liver, digestive tract, kidney and testis) were weighted. After 24 hours, refrigerated carcasses were separated into 7 parts and each part was dissected into meat and bone. Data was analyzed using the GLM procedure. Concentrate level, slaughter weight and their interactions were considered. Average daily gain (ADG) was higher for HCL group (121.7 vs 77.1 g for HCL and LCL groups respectively, $P<0.001$). Carcass weight and hot dressing percentage were also higher for HCL group (47.8 vs 44.9% for HCL and LCL groups respectively). Viscera percentage was not affected by CL. Proportions of the different parts were affected by CL, especially for leg and loin proportion. Proportion of leg was lower (26.6 vs 32.5%) and that of loin was higher (10.3 vs 5.7%) for HCL. Meat proportion in the carcass was lower for HCL group (78.6 vs 82.9% for HCL and LCL groups, respectively). The carcass weight and dressing percentage were higher for SW2. Kidney fat and tail proportions were higher for SW2 ($P<0.001$). No clear effect of SW on parts and meat proportions in the carcass appeared. In conclusion, increasing concentrate level improves growth performances, carcass weight and dressing percentage without affecting carcass adiposity. However, it affected negatively meat proportion in the carcass. Late slaughtering improves carcass weight and dressing percentage. It increases the carcass adiposity, especially the tail proportion.

Meat quality traits in chicken supplemented with different sources and concentrations of selenium

Ramos, E.M.[1], Bertechini, A.G.[1], Bressan, M.C.[2], Rodrigues, E.C.[1], Rossato, L.V.[1], Botega, L.M.G.[1], Gomes, F.A.[1], Ramos, A.L.S.[1], Silva, R.A.G.[1] and Gama, L.T.[2], [1]UFLA, C.p. 37 -Lavras-MG, 37200 000, Brazil, [2]INRB, Fonte Boa-Santarém, 2005-048, Portugal; mcbressan1@hotmail.com

An experiment was conducted to evaluate the effect of diet supplementation with selenium on meat quality characteristics of chicken breast. Two levels of selenium (0.15 and 0.30mg/kg) from organic (trademark A and B) or inorganic (sodium selenite) sources were used and compared with a control treatment (no selenium supplementation). A total of 1440 one-day old male chicks from the Cobb-500 line was distributed in 48 plots, with 7 (supplemented) and 6 (control) replicates, such that each experimental unit was composed of 30 birds. The diets were formulated according to the stage of development of the birds. At 45 days of age, a random sample of one chicken was taken from each plot, transported and slaughtered in a commercial abattoir. After cooling, the carcasses were deboned and the breasts were frozen (-35 °C) and stored (-18 °C) for 4 weeks. The pH was measured with a potentiometer in 6 points per breast. The color was determined on the breast surface, CIELAB system, in 6 points. The shear force was determined in 6 subsamples from the pectoralis major muscle. Supplementation with different sources and levels of selenium did not affect ($P>0.05$) the pH, L*, a*, b*, C* (chroma), h* (hue angle), cooking loss or shear force. The overall means for pH (5.90±0.09), cooking loss (26.57±3.61%), shear force (2.40±0.43kg), L* (42.97±3.46), a* (4.24±0.94), b* (4.79±1.44), C* (6.49±1.31) and h* (47.85±9.78) found in this work, are within the range usually described for chicken breast, from animals raised under intensive system. The supplementation with inorganic or organic selenium at 0.3 and 0.5mg/kg levels does not affect the meat quality of chicken breast in any of the attributes evaluated.

Changes in milk production and milk fatty acid composition of cows switched from pasture to a maize silage based-total mixed ration

Rego, O.A.[1], Rosa, H.J.D.[1], Cabrita, A.R.J.[2], Borba, A.R.[3], Fonseca, A.J.M.[2] and Bessa, R.J.B.[4], [1]CITA-A, Universidade Açores, Angra Heroismo, 9701-851, Portugal, [2]ICBAS, Universidade do Porto, Vairão, 4485-661, Portugal, [3]SRAF, Angra Heroismo, 9700, Portugal, [4]Faculdade Medicina Veterinária, Lisboa, 1300-477, Portugal; orego@uac.pt

Eight lactating Holstein cows (562±50 kg BW; 24.3±3.8 kg daily milk yield; 179±76 DIM) were used in a 52-d experiment to study the changes in fatty acid composition after transition from and to pasture. Experiment was divided into 3 periods. In the first 10 d cows grazed a pasture supplemented with 5 kg d^{-1} of concentrate (P1). In the next 21 d, cows fed a TMR (60% corn silage and 40% concentrate) (P2). In the last 21 d cows were turned out to pasture and supplemented with 5 kg of concentrate (P3). Milk samples were collected on day 10 of P1, and on days 2, 4, 7, 15 and 21 of P2 and P3. TMR feeding significantly increased DM intake and milk production and decreased milk fat content. Treatments had no effect on solids production and milk protein content. TMR significantly increased the concentration of saturated short and medium chain FA (6:0 to 16:0) and decreased the concentration of branched chain FA, MUFA and PUFA, excepting 18:2 n-6 and 18:2 t10-c12, which increased. Mean milk fat concentration were 0.77, 0.49 and 0.73% for linolenic acid, 2.63, 1.85 and 3.09 for vaccenic acid and 1.71, 0.85 and 1.58 for rumenic acid respectively in P1, P2 and P3. Concentration of rumenic acid was 1.7% in P1, decreased gradually until day 21 on P2 (TMR diet) when reached a minimum of 0.44%. After turnout to pasture (P3) its concentration increased gradually until the 7th day and stabilized thereafter until day 21 when reached the maximum of 2.16%. Therefore, rumenic acid increased 5 fold from last day of TMR to the last day of grazing P3. The concentration of vaccenic acid decreased from 2.63% in P1 to a minimum of 1.32% on day 14 of P2 and after turnout to pasture (P3) increased to a maximum of 3.82% on day 7, reaching a plateau thereafter.

Studies on the effects of L-carnitine on performance and hematology value in broiler chickens

Taraz, Z.[1] and Dastar, B.[2], [1]Gonbad High Eudaction Center, Animal Science, Gonbad, 49718557781, Iran, [2]Gorgan Agricultral Science & Natural Resources University, Animal Science, 49718557766, Iran; z_taraz@yahoo.com

This study was conducted to investgate the effect in L-carnitine of diets with different levels of protein on the performance and hematology value of ross 360 broiler chickens were kept for 42 days.The experiment consisted of six dietary treatments and four replicates per treatment with15 broiler chickens per replicate in a completly randomized design. Dietary treatments consisted of: 1) Basal diet; 2) Basal diet with 125 mg/kg L-carnitine; 3) Basal diet with 250 mg/kg L-carnitine; 4) low protein diet without L-carnitine; 5) low protein diet with 125 mg/kg L-carnitine; and 6) low protein diet with 250 mg/kg L-carnitine. The results showed that during the experimental period, brids were fed with sufficient quantity of protein had higher weight gain as compared to those were fed low protein diet ($P<0.05$). Reduction of dietary protein level, caused significant increase in food intake and food conversion ratio ($P<0.05$). Also L-carnitine caused significant increase in ratio of the weight of breast and the weight of cookable chiken and significant decrease in the obdominal fat ($P<0.05$). Dietary protein and L-carnitine supplementation didn't have any significant effects on the blood factors.

Effect of raw soybeans on chemical composition, fatty acid profile and CLA of sheep milk fat

Eleftheriadis, J.[1], Nitas, D.[2], Petridou, A.[3], Karalazos, V.[4], Mougios, V.[3], Marmaryan, G.[2], Michas, V.[2], Nita, S.[5] and Karalazos, A.[1], [1]AUTH, School of Agriculture, University Campus, 54124, Thessaloniki, Greece, [2]A.T.E.I.Th., Dept of Animal Production, P.O BOX 141GR, 57400, Thessaloniki, Greece, [3]AUTH, Dept of Physical Education and Sport Science, University Campus, 54124, Thessaloniki, Greece, [4]University of Thessaly, Dept of Ichthyology and Aquatic Environment, Fytokou Str, 38446, Volos, Greece, [5]Ministry of Education, Panormo, 74057, Rethymno, Greece; dnitas@ap.teithe.gr

Raw soybeans (SBs) were fed to 24 ewes to study the influence on milk yield, milk fatty acid composition and conjugated linoleic acid (CLA) content. Ewes were fed one of three diets A, B and C. The control diet containing soybean meal, maize grain, barley grain, alfalfa hay and wheat straw compared with diets B and C containing 10 and 20% SBs, respectively. Measurements were made during the last 2 days of each of the 6 periods of the experiment. The replacement of a part of soybean meal and maize grain by either 10 or 20% SBs in the diet, did not affect the initial or final body weight of ewes, the daily milk yield and composition and the somatic cell contents (SCC). The proportions of short and medium-chain fatty acids 6:0, 8:0, 10:0, 12:0, 14:0 and 16:0 decreased and the proportions of long-chain unsaturated fatty acids 18:1n9t, 18:1n9c, 18:2n6t, 18:2n6c, 18:3n3 and 20:5n3 (EPA) were increased in milk fat when SBs were fed compared with the proportions of fatty acids when the control diet was fed. The proportion of CLAs was not affected by the inclusion of 10 or 20% SBs in the diet. The atherogenicity index (AI) and the thrombogenicity index (TI) in milk fat of ewes decreased a mean of 15.9% and 26.5% by the inclusion of 10 or 20% SBs in the diet of ewes, respectively. The results demonstrate that unsaturated fatty acids content of sheep milk can be substantially increased by the inclusion of soybeans in the diet at 10 or 20%, without any negative effects on animal performance.

Effects of different carbohydrates provided through drinking water during feed withdrawal on metabolite concentrations of broilers

Kop, C. and Ocak, N., Ondokuz Mayis University, Agricultural Faculty, Animal Science Department, 55139, Samsun, Turkey; canankop@gmail.com

Blood metabolites such as glucose, triglyceride, uric acid, uric acid nitrogen and total lipid concentrates indicate a negative energy balance of broilers related to feed withdrawal. Therefore, these metabolites were used to assess the effect of glucose, saccarose or corn starch provided through drinking water during feed withdrawal on metabolic processes. Treatments consisted of: full-fed control broilers fed the standard broiler diet and water for the full 10 h (C); fasted broilers receiving only water (FW); 3 g glucose (G), saccarose (S) or corn starch (CS) L^{-1} supplemented water for the 10 h. A total of 200 broilers (Ross 308) were allocated randomly to five treatments consisted of 4 replicates at 43 days, and total 40 birds (20 males and 20 females) were slaughtered to determine the plasma metabolite concentrations. Carbohydrates provided through drinking water did not influence plasma cholesterol and total lipid concentrates. Full-fed broilers had higher glucose, uric acid nitrogen and uric acid concentrations than S birds ($P<0.05$). Plasma triglyceride concentrate of C birds was higher than that of other treatment birds ($P<0.01$). These results show that the last day of the broiler's life is associated with a negative energy balance and stress, which are related directly with meat quality, and the impact of different carbohydrates provided during feed withdrawal on metabolite concentrations was not at the same direction and stability.

Rapid and sensitive methods to assure food safety and quality

Caprita, A. and Caprita, R., University of Agricultural Sciences and Veterinary Medicine, Calea Aradului 119, 300645 Timisoara, Romania; rodi.caprita@gmail.com

Real time methods for monitoring microorganisms are essential for implementation of HACCP, determine contamination, or the food quality. We tested two methods for rapid monitoring and detection of microorganisms in milk: the ATP and the electrical conductivity method. Milk samples were analyzed for CFU, SCC and RLU. For determination of total ATP by the luciferin/luciferase enzymatic reaction we used the luminometer Betz Bioscan Monitor RHS 055. CFU was determined by counting the colonies grown on nutrient agar plates. The number of viable microbial cells was determined using the BIO-KOBE (Japan) colony counter. The electrical conductance of the milk samples was measured with conductivity meter type OK-102/1 (Radelkis). Since ATP is found in all living cells its detection is indicative of living material being present. High RLU values indicate either a subclinical mastitis or a bacterial contamination. We observed a high correlation in raw milk between Relative Light Units (RLU) and the sum of somatic cell count (SCC) and CFU (r=0.9390). The electrical conductivity method is based on the measurement of changes in electrical impedance of a medium or a reaction solution resulting from the growth of bacteria. During the process of microbial metabolism, large molecules (carbohydrates, fats and proteins) are broken down into smaller and more highly charged components (fatty acids, amino acids and other organic acids). The highly charged molecules cause a change in the media's electrical conductivity/resistance. These changes indicate the presence of microorganisms in the original sample. We observed a positive correlation between milk ATP and electrical conductivity (r=0.9522). ATP bioluminescence and electrical conductivity assays can be applied to bring a rapid microbial test, allowing more effective management of the freshness/quality balance. The bioluminescence assay of ATP and the electrical conductivity assay are very rapid, non-polluting methods.

The biochemical indices of the milk and meat in goats with and without digestive strongilate invasion
Birgele, E., Keidane, D. and Ilgaža, A., Preclinica institute, Faculty of Veterinary Medicine, LUA, Helmana-8, LV-3002, Latvia; edite.birgele@llu.lv

The objective of this study was to determine is there and how digestive strongilate invasion in goats influence the biochemical indices of milk and meat. The total amount of 30 goats in the age between two and three years were used in this study. All animals were kept and feeded in the same circumstances, like in biological farm. Based on coprological results all animals were divident in two groups - 15 goats with digestive strongilates invasion and 15 goats without strongilate invasion. In the milk sample we determine total fat (%), urea (mmol/l), replaceable and non-replaceable amino acids (%). We established that invasion of digestive strongilate do influence some biochemical indices of milk. We established the tendency of non-replaceable amino acids like hystidine, treonine, valine, metionine, lizine, isoleucine and fenylalanine to increase in the milk of goats with digestive strongilates ($P<0.05$). Replaceable amino acids did not change significantly. Interestingly that the level of fat and urea in milk was lower in non-invaded animals that in invasted goats. In the biochemical finding of meat we established statistically significant higer total protein level in non-invaded animal meat ($P<0.05$). That approves also the perceptual relation between some of the amino acids. In the meat from the goats without digestive strongilates there is a significantly higher non-replaceable amino acids like phenylalanine, valine ($P<0.01$), and histidine, isoleucine ($P<0.05$). The amount of replaceable amino acids did not changed in goats meat.

The effect of the inclusion grape seed extract and *Cistus ladanifer* in lambs diet fed with dehydrated Lucerne supplemented with oil. 1 - growth performance, carcass composition and meat quality
Santos-Silva, J.[1], Dentinho, M.T.[1], Jerónimo, E.[1] and Bessa, R.[2], [1]INRB, UPA, Fonte Boa, Vale de Santarém, Portugal, [2]FMV, Alto da Ajuda, Lisboa, Portugal; josesantossilva.ezn@mail.telepac.pt

Supplementation with lipids is very effective in the modulation of ruminant's fat, increasing the proportion of some bioactive fatty acids without compromise animal's growth or product quality. This approach results in more unsaturated meat fat, increasing the risk of oxidation during storage. In this trial, thirty-six ram lambs were used, to evaluate the effects of supplementation with oil and polyphenols, grape seed extract and *Cistus ladanifer*, a Mediterranean shrub, on growth, carcass composition and meat quality. Animals were assigned to 12 groups of 3 lambs each and two groups were submitted to each diet. The six experimental pelleted diets were: Control diet (C) - 90% dehydrated lucerne and 10% of wheat bran; Control diet with 6% of oil blend - sunflower oil and linseed oil (1:2) - (CO); control with 2.5% of grape seed extract (CG); control with 2.5% of grape seed extract and 6% of oil blend (CGO); control with 25% of Cistus ladanifer (CC); control with 25% of *Cistus ladanifer* and 6% of oil blend (CCO). Trial last for 6 weeks and intake was controlled daily and weight weekly. Lambs were slaughtered at Fonte Boa abattoir, carcass composition was estimated and meat quality was assessed at the L. dorsi muscle. A consumer's panel was used to verify how *Cistus ladanifer* and grape seed extract inclusion in the diets affected meat sensory traits. Dry matter intake was independent of treatments (1616 g/day). Oil supplementation resulted in higher growth rate (293 vs 264 g/day), dressing percentage (44 vs 41%) and kidney knob channel fat (KKCF) (2.5 vs 1.8%). Cistus ladanifer inclusion decreased carcass muscle (54 vs 59%) and increased dressing percentage (45 vs 42%) and KKCF (2.6 vs 1.9%). Meat color, parameters were not affected by treatments. Panelists could not distinguish meat from lambs fed with polyphenol supplemented diets.

Conjugated linoleic acid (CLA) content of camel milk produced under semi-extensive management conditions

Ayadi, M.[1] and Casals, R.[2], [1]Institut Supérieur de Biologie Appliquée de Médénine, SCQ, Km 22.5 Route el Djorf, Médénine, 4119, Tunisia, [2]Universitat Autònoma de Barcelona, G2R, Facultat de Veterinària, 08193 Bellaterra, Spain; ramon.casals@uab.cat

Camel milk is an important nutrition source for inhabitants in arid and semiarid areas, but its fatty acids (FA) profile, particularly conjugated linoleic acid (CLA), is not well known. The major CLA isomer in milk fat of ruminants, the cis-9, trans-11 CLA, has anticarcinogenic properties. This isomer is primarily a product of endogenous synthesis in the mammary gland by the desaturation of vaccenic acid (VA, trans-11 C18:1). Both CLA and VA are considered components of functional foods. To our knowledge CLA and VA contents in camel milk have not been reported previously. Therefore, the main objective of this work was to evaluate the CLA and VA contents in camel milk under semi-extensive management conditions. Four primiparous and 13 multiparous Tunisian Maghrebi dairy dromedaries (Camelus Dromedarius) at the beginning of lactation (31 ± 11 DIM) were hand–milked, and milk samples were analyzed by gas chromatography to asses their FA profile. Camels grazed in a halophyte pasture (6% CP) in the Southeast of Tunisia and received a daily supplement of olive cake (1 kg.), wheat bran (0.5 kg), and barley (0.5 kg). Milk yield was greater in multiparous than in primiparous camels (3.37 ± 0.46 vs. 1.00 ± 0.18 L/d, $P<0.001$), but the milk FA profile was similar regardless of lactation number. Main FA of camel milk were: C18:1 (oleic, 30.5% of total FA), C16:0 (palmitic, 25.2%), C18:0 (stearic, 17.2%), C14:0 (myristic, 6.4%), and C18:2 (linoleic, 3.3%). The cis-9, trans-11 CLA and VA contents were 1.1% and 2.8% of total FA, respectively, and significantly correlated ($R^2=0.74$; $P<0.05$). These values are similar to those observed in ruminants receiving diets with moderate quality pastures or not supplemented with sources of linoleic acid. In conclusion, camel milk produced under semi-extensive conditions is a source of CLA and VA, which reinforces its health benefits for people in arid areas.

The effect of feeding potassium iodide on performance and iodine excretion in Holstein dairy cows

Norouzian, M.A. and Valizadeh, R., Ferdowsi University of Mashhad, P.O. Box: 91775-1163, Iran; manorouzian@ymail.com

Sixteen Holstein dairy cows with daily milk yield of 32.9 ± 2.4 kg allocated to 4 treatments in a complete randomized design with 4 replications to evaluate the effect of iodine supplementation on performance of dairy cow and iodine excretions especially in milk. The treatments were 1) basal diet, 2, 3 and 4, the basal diet plus 2.5, 5 and 7.5 mg/kg diet DM Potassium Iodide, respectively. There were no significant difference between treatments for dry matter intake, milk yield and compositions and the milk production efficiency. Iodine contents in blood, urine, raw and pasteurized milk were significantly ($P<0.01$) affected by the iodine supplementation. Blood T3 and T4 concentrations were not significantly affected by the treatments. No adverse effect of iodine supplementation on performance and health of dairy cow were detected in this study. It was concluded that iodine supplementation above of NRC recommendation (0.5 mg/kg diet DM) led to a desirable level of iodine in the milk ready for human consumption without adverse effects on dairy cows performance and health. This finding could be an excellent recommendation for the area with iodine deficiency mainly for children's.

Effect of season on muscle characteristics of camel calves

Abdelhadi, O.M.A.[1], Babiker, S.A.[2], Bauchart, D.[3], Picard, B.[3], Jurie, C.[3], Faye, B.[4] and Hocquette, J.F.[3], [1]University of Kordofan, Dept. of Animal Prod., P.O. Box:716, Khartoum, Sudan, [2]University of Kordofan, Dept. Meat Prod., P.O. Box:716, Khartoum, Sudan, [3]INRA, UR1213, Herbivore Research Unit, Theix, 63122 Saint-Genes Champanelle, France, Metropolitan, [4]CIRAD, Headquarter, Montpellier, France, Metropolitan; abusin911@yahoo.com

Thirty 2-3 year-old camel calves fattened by local farmers in Sudan were used for this study. Ten calves were slaughtered according to different seasons (winter, summer and autumn) to examine the muscle characteristics according to season. Muscle samples were taken from Longissimus thoracis (LT) at the 5th thoracic vertebra. The overall chemical composition of LT showed mean values of 3.1, 17.2, 2.7 and 1.6% for DM, crude proteins, total lipids and fatty acids and ash, respectively. Unlike for CP and ash ($P<0.001$), no significant differences were found between seasons in DM and total lipids. Unlike in cattle, the results obtained from electrophoresis test indicated the presence of two muscle fibers only. The mean percentages were: type I 66.6% and type IIa 33.4%. Higher proportions ($P<0.001$) were observed in winter for type I (85.2%) and in autumn for type IIa (47.1%). Positive correlation coefficient was observed (0.80) between the proportion of fibers type IIa and isocitrate dehydrogenase (ICDH) enzyme activity. Enzymes mitochondrial (ICDH and COX) activities as well as phosphofructokinase activity were higher ($P<0.001$) during autumn season compared to summer and winter. In conclusion, muscle fiber characteristics in camel (except intramuscular fat content) are highly regulated by seasonal factors.

Carcass traits and carcass composition of roe deer (*Capreolus capreolus* L. 1758)

Popovic, Z. and Bogdanovic, V., Faculty of Agriculture, University of belgrade, Nemanjina 6, 11080 Zemun-Belgrade, Serbia; vlbogd@agrif.bg.ac.rs

Roe deer (*Capreolus capreolus* L. 1758) is one of the most important species of wild ruminants in Europe. Primarily, the value of roe deer is represented by value of animal trophy. On the other hand, venison of this species becomes more and more interested for additional income. However, research on carcass characteristics of roe deer are seldom carried out since the farming of roe deer is not so common. In order to determine carcass traits and carcass composition data of 36 roe deer (25 male and 11 female) shot in one open hunting ground in central Serbia were used. Analysed traits were body weight after shooting, hot carcass weight, hot dressing percentage, leg weight, back weight, shoulder weight, neck weight and thorax weight. Average body weight after shooting was 26.4 kg and 26.12 for male and female, while hot carcass weight was 15.95 kg and 15.20 kg for male and female, respectively. Hot dressing percentage was 60.44% and 58.14% for male and female, respectively. The weights of the most important part of carcass were: 5.22 kg and 5.18 kg for male leg weight and female leg weight, respectively; 1.86 kg and 1.64 kg for male back weight and back leg weight, respectively; 2.43 kg and 2.29 kg for male shoulder weight and female shoulder weight, respectively; 1.19 kg and 0.94 kg for male neck weight and female neck weight, respectively; and 1.77 kg and 1.62 kg for male thorax weight and female thorax weight, respectively. Sex had significant effect on hot dressing percentage ($P<0.01$), back weight ($P<0.01$) and neck weight ($P<0.005$), while for other analysed traits sex of animals had no effect.

Effects of increasing prepartum dietary protein level using poultry by-product meal on productive performance and health of multiparous Holstein dairy cows

Hossein Yazdi, M.[1], Amanlou, H.[2], Kafilzadeh, F.[1] and Mahjoubi, E.[2], [1]Razi University, Animal Science, Kermanshah, 213, Iran, [2]Zanjan University, Animal Science, Zanjan 45195, Iran; e_mahjoubi133@yahoo.com

The aim of this study was to compare the effects of two levels of crude protein (CP) using poultry by-product (PBPM) meal fed during late gestation on the performance, blood metabolites, and colostrum composition of Holstein cows. Sixteen multiparous cows 26±6 d before expected calving were assigned randomly to two treatments containing 1) 14% crude protein [3.4% PBPM] 2) 16% crude protein [7.5% PBPM]. The cow's BCS was 3.56±0.5 on average, at the beginning of the trial. Yields of milk, protein, lactose, fat, and SNF were not affected by prepartum dietary CP level. Colostrum composition (fat%, CP% and Total solids%), blood metabolites (Ca, Glucose, Total protein, Albumin, Globulin, Urea N and Cholesterol), and metabolic diseases incidence were not influenced by prepartum dietary CP level. There was no significant difference between treatments in body weight and BCS changes. As expected, blood urea N before calving was higher in the cows fed 16% CP diets ($P<0.002$). Serum cholesterol during prepartum ($P<0.03$) and postpartum ($P<0.01$) periods was significantly different between dietary treatment groups. In general, although postpartum glucose level increased in cows which received 16% CP in the diet, it seems that no other obvious advantages over feeding the 14% CP diet are apparent. So feeding this last diet to close up cows is recommended.

The effect of calcium level on egg quality characteristics

Englmaierova, M., Tumova, E., Charvatova, V., Klesalova, L. and Heindl, J.,; englmaierova@af.czu.cz

The aim of the study was to determine the effect of calcium content in feed mixture on eggshell and egg content quality characteristics. The experiment was realized with ISA Brown hens from the 20th to 60th week of age and housed in conventional cages. The laying hens were fed by commercial feed mixture with different calcium level. The control group had 3.5% of calcium in diet and in the experimental group was used mixture with lower share of calcium (3.0%). The egg quality determination was carried out by the devices from firm TSS England. The findings showed that calcium content had significant effect on yolk quality characterized by yolk index ($P<0.05$). The higher values of yolk index were recorded in hens fed by feed mixture with higher content of calcium (45.7%) compared to experimental group (45.2%). But no influence of calcium level was observed for albumen quality characteristics. Calcium content in feed mixture influenced especially eggshell quality. There were find out significant differences in shell thickness ($P<0.05$), shell strength ($P<0.001$) and shell percentage ($P<0.05$). The values of shell thickness (0.370 versus 0.361 mm), strength (4893 vs. 4651 g.cm^{-2}) and percentage (12.3 versus 12.2%) grew with increasing of calcium share in diet.

Determination of metabolizable energy and organic matter digestibility of two food industrial by-products using gas production technique
Mirzaei-Aghsaghali, A.[1], Maheri-Sis, N.[2], Mansouri, H.[3], Razeghi, M.E.[4] and Alipoor, K.[5], [1]Shabestar University, Department of Animal Science, Islamic Azad University, Shabestar Branch, Shabestar, Iran, 15, 7 allay, Moalem St., Apadana Ave., Urmia city, West Azerbaijan, 5716814758, Iran, [2]Shabestar University, Department of Animal Science, Islamic Azad University, Shabestar Branch, Shabestar, 5716814758, Iran, [3]Animal Science Research Institute, Karaj, 5716814758, Iran, [4]Agricultural and natural resourccs Research Center, Urmia, 5716814758, Iran, [5]Shabestar University, Department of Animal Science, Islamic Azad University, Shabestar Branch, Shabestar, 5716814758, Iran; afshar.mirzaii@gmail.com

The present study was conducted to determine the nutritive value and estimation metabolizable energy and organic matter digestibility of some food industrial by-products with gas production method. In this experiment three canulated steers (Holestein) were used. The amount of gas production for feedstuffs at 0, 2, 4, 6, 8, 12, 24, 48, 72 and 96 hours were measured. The food industrial by-products that used in this study include: grape marc, pomagranate seed pulp and pomagranate peel. The results showed that the Crude Fiber (CF) contents of white grape marc, pomagranate seed pulp and pomagranate peel were 22.8, 32.3 and 15.1% and gas production for feedstuffs at 24 hours were 30.92, 22.92 and 47.42%; respectively. The organic matter digestibility (OMD) of white grape marc, pomagranate seed pulp and pomagranate peel were 50.50, 42.34 and 59%; respectively.The Metabolizable Energy (ME) contents of white grape marc, pomagranate seed pulp and pomagranate peel were 7.4, 6.2 and 8.85 MJ kg-1 DM, respectively.

Alfalfa hay replacement with Kochia scoparia and its effects on early lactation Brown Swiss dairy cows
Shahdadi, A.R.[1], Zaherfarimani, H.[1], Saremi, B.[2] and Naserian, A.A.[3], [1]Education center of Jihad Agriculture, Animal Sci Department, Between Jihad and Jomhouri Sq., 91875, Iran, [2]National ellite foundation, Research and planning, Zafaranieh-shirkouh-11th block, 91875, Iran, [3]Ferdowsi Uni, Animal Sci Department, Azadi sq., 91875, Iran; behsa2001@yahoo.com

In order to investigating dairy cows performance fed different levels of Koshia scoparia hay (Ks) instead of Alfalfa hay, nine early-lactation Brown Swiss cows with 28±0.65 kg milk production and 71.4±6.3 days in milk were allocated randomly to one of three treatments 1) 0% Ks 2) 15% Ks 3) 30% Ks replacement (as fed based) in a change over (3×3 Latin square) design. This study was done at Higher Education Center of Jihad-e Agriculture in Mashhad city. This experiment had three periods consist of 14 days adaptation and 7 days sampling. DMI and milk yield was recorded at sampling days. Milk samples were taken in last two sampling days to determine its ingredients (Fat, Protein, Lactose, Solid non fat, and total solids). At the end of sampling phase, rumen liquid was taken using stomach tube and pH was determined individually. Data were analyzed using SAS 9.1. Means were compared using Duncan test ($P<0.05$). Results showed that Koshia scoparia had no effect on milk yield, fat corrected milk 4% and milk ingredients except Lactose% which was significantly reduced when 30% of Alfalfa was replaced ($P<0.05$). Generally, there was a numerical increase in milk ingredients in 0% in respect to 15 and 30% substitution of alfalfa with Koshia. In addition, Rumen pH was significantly increased by the amount of Koshia scoparia elevated in cows' diet ($P<0.05$) which was significantly higher in 15 and 30% in compare with 0%. DMI and feed conversion rate were not affected by treatments. To sum up, it seems that Koshia scoparia can replace Alfalfa hay in Dairy cows' diet up to 15% in early lactation without adverse effects on milk yield and composition, DMI, and feed conversion rate. In addition, the rumen environment improved as can be seen by pH increase.

Suckling lambs meat quality influenced by oil-supplemented ewe diet

Vieira, C.[1], Manso, T.[2], Bodas, R.[3], Díaz, M.T.[1], Castro, T.[4] and Mantecón, A.R.[3], [1]Consejería de Agricultura y Ganadería de Castilla y León. Instituto Tecnológico Agrario, Estación Tecnológica de la Carne, Filiberto Villalobos s/n Guijuelo, 37770 Salamanca, Spain, [2]ETS Ingenierías Agrarias, Universidad de Valladolid, Avd. Madrid s/n, 34004, Palencia, Spain, [3]Instituto de Ganadería de Montaña (CSIC-ULE), Grulleros, 24346 León, Spain, [4]Dpto. Producción Animal. UCM., Avda Puerta de Hierrro s/n, 28040 Madrid, Spain; vieallce@itacyl.es

Forty eight lactating Churra ewes were used to investigate the influence of feeding four dietary vegetable oils (hydrogenated palm oil (Control), olive oil (OL), soybean oil (SO) and linseed oil (LI)) on carcass and meat quality of suckling lambs. After lambing, all lambs stayed with their dams and were raised exclusively on maternal milk until slaughter at 11 kg live weight. Animal performance and carcass quality were evaluated. Muscle colour (L*, a*, b*) and lipid oxidation (TBARS) were measured in M. longissimus at 24 h, 5 days and 8 days after slaughter. Animal performance and carcass characteristics of suckling lambs were not affected by ewe diet composition. However, muscle colour parameters were affected by treatment. Lambs suckling ewes in OL group showed higher a* values at 24 h, 5 and 8 days ($P<0.05$) than those in groups Control and SO. The effect of ewe feed composition only affected lipid muscle oxidation 8 days after slaughter, the greatest TBARS values being observed for lambs from LI group ($P<0.05$), and the lowest for OL lambs. Therefore, the type of oil feed to lactating ewes affect colour and lipid stability of suckling lambs.

Breeding for improved hunting performance in Norwegian Elkhound and Swedish Jämthund populations in Finland

Liinamo, A.-E., MTT Agrifood Research Finland, Biotechnology and Food Research, Biometrical Genetics Group, FIN-31600 Jokioinen, Finland; anna-elisa.liinamo@mtt.fi

The aim was to develop a routine breeding value estimation for hunting traits of the two most popular elkhound breeds in Finland based on official field trial records. In the first stage, genetic parameters were estimated using REML methodology for the most important traits evaluated at the field trials, as well as several other measures describing dogs' overall trial career derived from the actual field trial records. Data in parameter estimation included field trial records from 1991 to 2006, with 41,321 trials from 6,694 Norwegian Elkhounds and 13,731 trials from 2,395 Jämthund. Heritability estimates for most traits evaluated at the field trials were very low (<0.05), but they were somewhat higher for the derived career measures. The highest heritability estimate was obtained for the age of the dog at the first successfully completed trial round (0.48), which can be thought to describe the early maturity of the dog. In the second stage, seven measures were selected for routine breeding value estimation, based on the heritability estimates and so that they cover the most important aspects of elkhound performance. EBVs have been estimated annually since 2007 using BLUP methodology for these seven measures for both breeds.

Basic behaviour trait characterization: creating a tool to select for temperament in dogs

Arvelius, P., Fikse, W.F. and Strandberg, E., Swedish Univ. of Agricultural Sciences, Dept. of Animal Breeding and Genetics, P.O. Box 7023, S-750 07 Uppsala, Sweden; per.arvelius@hgen.slu.se

Since 1989, the Swedish Working Dog Association (SWDA) has carried out a standardized behavioural test called Dog Mentality Assessment (DMA). The test was originally devised to help breeders in controlling their genetic material, mainly with a focus on traits important for 'working ability'. In previous studies, broader behavioural traits, so-called personality traits, have been defined using factor analysis of the DMA variables, and additive genetic variation has been found for these personality traits (h^2 0.10-0.25). The personality traits have been shown to correlate well with everyday life behaviour. Two thirds of the Swedish dog population is registered by the Swedish Kennel Club (SKC). Each club associated to the SKC is obliged to establish a breed-specific breeding strategy for its breed. Many of the clubs have expressed that they will increase the emphasis put on systematic selection for behavioural traits. For most breeds, the only tool available to accomplish this is the DMA. Today, almost 8,000 dogs representing more than 180 breeds are being tested annually. Given the resources required per tested dog, and the non-profit nature of the DMA, it is burdensome for the SWDA to meet the demand. Also, for many breeds everyday life behaviour is of primary interest rather than 'working ability'. For these reasons, a reduced version of the DMA - comprising measurements showing high heritabilities and high correlations with everyday life behaviours of importance - is being developed. Using a BLUP animal model, heritabilities for, and genetic correlations between, all 33 behavioural variables measured in the DMA have been estimated for the 13 most commonly tested breeds. Further analyses, e.g. correlations between predicted breeding values for the DMA variables and questionnaire answers on everyday life behaviour, will be made. Our intention is that the new Basic Behaviour Trait Characterization will become a common and cost-effective test of value for all dog breeds.

Genetic and environmental factors affecting behaviour test results in Rottweilers

Liimatainen, R.[1], Liinamo, A.-E.[2] and Ojala, M.[3], [1]Finnish Guide Dog School, Siltaniitynkuja 1, FIN-01260 Vantaa, Finland, [2]MTT Agrifood Research Finland, Biotechnology and Food Research, Biometrical Genetics Group, FIN-31600 Jokioinen, Finland, [3]Helsinki University, Department of Animal Science, P.O. Box 28, FIN-00014 Helsinki, Finland; anna-elisa.liinamo@mtt.fi

The aim of this study was to evaluate the genetic variability present in various temperament traits in Rottweiler dogs based on data collected at official canine behaviour tests in Finland in 1980-2003. During this time, altogether 2,327 Rottweilers had been officially behavior tested, which corresponds to 15% of all registered Rottweilers in the time period. The studied traits included courage, sharpness, defense drive, play drive, nerve stability, temperament, hardness, affability and reaction to gunfire. Dogs' temperament traits were influenced by their sex, age at testing, testing place, testing year-season, breeder and the evaluating judge. All traits showed genetic differences between the dogs, but the heritability estimates were in general low, ranging from 10-13% (sharpness, play drive, hardness, reaction to gunfire) to 5-7% (courage, defense drive, nerve stability, temperament and affability). Genetic correlations between the traits were strong and positive (>0.75) between courage, nerve stability and hardness, and strong and negative (<-0.6) between affability and sharpness/temperament.

Genetic parameters and predictive capacity of behavioural puppy test in guide dogs
Tenho, L.[1], Liinamo, A.-E.[2] and Juga, J.[1], [1]Helsinki University, Department of Animal Science, P. O. Box 28, FIN-00014 Helsinki, Finland, [2]MTT Agrifood Research Finland, Biotechnology and Food Research, Biometrical Genetics Group, FIN-31600 Jokioinen, Finland; laura.la.hautala@helsinki.fi

The aim of the study was to estimate heritabilities for eleven behavioural measures that are evaluated in a puppy test used by the Finnish Guide Dog School. In addition, the relationship between the dogs' puppy test results and later success in guide dog training was also studied. The data included 686 Labrador Retrievers born in years 1991-2008 at the guide dog school. Preliminary t-test analysis did not show any statistically significant differences in puppy test results between later selected and culled dogs. Results of the study will be used to further develop the puppy test to be more informative and accurate in selecting guide dog puppies for training and for breeding a working guide dog.

Genetic parameters and predictive capacity of a behavioural aptitude test in guide dogs
Nikkonen, T.[1], Liinamo, A.-E.[2], Juga, J.[1] and Tenho, L.[1], [1]Helsinki University, Department of Animal Science, P.O. Box 28, FIN-00014 Helsinki, Finland, [2]MTT Agrifood Research Finland, Biotechnology and Food Research, Biometrical Genetics Group, FIN-31600 Jokioinen, Finland; laura.la.hautala@helsinki.fi

The aim of the study was to estimate heritabilities for twelve behavioural measures that are evaluated in an aptitude test used for selecting dogs for training at the Finnish Guide Dog School. The environmental factors affecting the tested traits and the relationships between the different traits were also analysed. In addition, the relationship between the dogs' individual test results and success in guide dog training was also studied. The data included 462 Labrador Retrievers tested in years 1997-2008 at the guide dog school. Results of the study will be used to further develop the aptitude test to be more informative and accurate in selecting dogs for guide dog training and breeding.

High prevalence of canine autoimmune lymphocytic thyroiditis in giant schnauzer and hovawart dogs

Ferm, K.[1], Björnerfeldt, S.[2], Karlsson, Å.[3], Andersson, G.[2], Nachreiner, R.[4] and Hedhammar, Å.[5], [1]Swedish University of Agricultural Sciences, Department of Animal Breeding and Genetics, P.O Box 7023, 75007 Uppsala, Sweden, [2]Swedish University of Agricultural Sciences, Department of Animal Breeding and Genetics, BMC, P.O Box 597, 75124 Uppsala, Sweden, [3]Swedish University of Agricultural Sciences, University Animal Hospital, Section of Clinical Pathology, P.O Box 7040, 75007 Uppsala, Sweden, [4]Diagnostic Centre for Population and Animal Health, P.O. Box 30076, Lansing, MI 48909-7576, USA, [5]Swedish University of Agricultural Sciences, Department of Clinical Sciences, P.O Box 7054, 75007 Uppsala, Sweden; Katarina.Ferm@hgen.slu.se

Canine Lymphocytic thyroiditis (CLT) is one of the most common endocrinopathies in dogs, and affects multiple breeds in high frequency. Clinical signs of CLT usually appear in dogs from their early middle age (4-6 years), when most breeding dogs have already made their breeding debut. As a result, conventional selection of breeding stock has not been effective in reducing CLT prevalence. Our intention was to estimate prevalence of CLT and hypothyroidism within two previously indicated high-risk dog breeds, and to investigate the occurrence of earliest evidence of CLT within these breeds. By screening two birth cohorts of giant schnauzer and hovawart dogs (3-4 and 6-7 years old, respectively) for elevated serum levels of autoantibodies to thyroglobulin (TgAA) and thyroid stimulating hormone (TSH), we estimated a very high prevalence of CLT; in total about 16% in the giant schnauzer and 13% in the hovawart. In the young cohorts, only 3% of the giant schnauzers and 0% of the hovawarts had shown clinical signs of hypothyroidism. However, elevated TgAA- and/or TSH levels were identified in more than 10% of the dogs in both breeds. Many breeding animals make their breeding debut in the age of 2-3 years, i.e. before any clinical signs of CLT usually appear. Therefore, screening of TgAA and TSH in potential breeding animals would be advisable in high-risk breeds, in order to reduce CLT prevalence.

BLUP animal model for prediction of breeding values for hip dysplasia in Norwegian dog breeds

Madsen, P.[1], Indrebø, A.[2] and Lingaas, F.[3], [1]University of Aarhus, Faculty of Agricultural Sciences, Dept. of Genetics and Biotechnology, Genetics and Biotechnology, P.O Box 50, DK-8830 Tjele, Denmark, [2]Norwegian School of Veterinary Medicine, Small Animal Sciences, P.O. Box. 8146, Oslo, Norway, [3]Norwegian School of Veterinary Medicine, 3Section of Genetics, P.O. Box 8146, Oslo, Norway; per.madsen@agrsci.dk

Hip dysplasia (HD) is the most common cause of rear leg lameness in dogs. Diagnostic technique is with x-Ray and hip scoring tests. Many studies have shown that HD is heritable with a polygenic background, shows heritability (h^2) from ~0.1 to 0.8. In Norway as in other countries selection for improved HD status has been used for several years. The selection has been on own performance or on a selection index including information from close relatives. Inspired by the success of using BLUP Animal Models (AM) in commercial livestock breeding programs, BLUP AM has now been introduced for dog breeding in several countries. The Norwegian Kennel Club has introduced BLUP AM in 2008. The implementation included data validation, model development and estimation of genetic parameters. HD data from 1986 to 2008 for 30 breeds, with number of registration from 1,180 to 22,345 were available. A linear model including year of birth (YOB), sex and sex * YOB interaction as fixed effects, and litter and animal as random effects was chosen. Heritability estimates range from 0.08 (Newfoundland) to 0.48 (Saint Bernard). Model validation was based on EBV's predicted on the complete data (EBV_F) for each breed and EBV's predicted on a data set, where the last 2 years HD-registrations were discarded (EBV_R). For the group of dogs with discarded HD records, EBV_R will be based on pedigree information only. A significant deviation from 1.0 for the regression of EBV_F on EBV_R indicates bias. For most breeds the regression coefficient did not differ significantly from 1.0, but for 5 breeds with more than 100 HD registrations in the last 2 years, the regression was significantly below 1.0 indicating that EBV_R is biased upwards. Analysis to disclose the reason for this bias is ongoing.

Genetics of radiographic signs related to degenerative lumbosacral stenosis and its correlations to canine hip dysplasia in the German shepherd dog

Stock, K.F.[1], Ondreka, N.[2], Tellhelm, B.[2] and Distl, O.[1], [1]University of Veterinary Medicine Hannover, Institute for Animal Breeding and Genetics, Buenteweg 17p, 30559 Hannover, Germany, [2]Justus Liebig University Giessen, Department of Veterinary Medicine (Small Animal Clinic), Frankfurter Strasse 94, 35392 Giessen, Germany; Kathrin-Friederike.Stock@tiho-hannover.de

Degenerative lumbosacral stenosis (DLSS) includes alterations in the lumbosacral (LS) region of the spine which lead to narrowing of the vertebral canal and compression of nerval structures. DLSS may cause severe clinical signs, but radiographic signs related to DLSS can also be found in clinically sound dogs. Results of radiographic examinations of the spine were available for 572 German shepherd dogs (GSD), 95% of which were born in 2000-2007. Radiographic screening for canine hip dysplasia (CHD) is routinely done in the GSD. CHD information on all dogs from birth years 2000-2007 (n=50,273) was considered for multivariate genetic analyses. CHD was analyzed as categorical trait with 5 levels (1 = unaffected to 5 = severely affected), together with 2 continuous and 7 binary (0 = unaffected, 1 = affected) traits related to DLSS. Genetic parameters were estimated in linear and mixed linear-threshold animal models using Gibbs sampling. Heritability estimates were moderate for CHD and binary DLSS traits (0.14-0.50) and high for continuous DLSS traits (0.86-0.97). CHD was genetically uncorrelated with the continuously recorded DLSS traits LS malalignment and relative width of LS intervertebral disc. However, negative additive genetic correlations of -0.30 to -0.52 were estimated between CHD and 5 of the binary DLSS traits, including LS transitional vertebra. There were indications of positive genetic correlations around 0.2 of CHD with diffuse dorsal contour of the sacrum and sacral osteochondrosis dissecans (OCD). Relevant additive genetic correlations between CHD and radiographic signs related to DLSS may indicate involvement of same genes in causing disposition for important skeletal diseases in the GSD.

Dog breeding and breeding practices: example of French breeds

Leroy, G., Danchin-Burge, C., Verrier, E. and Rognon, X., INRA/AgroParisTech, UMR1313 Génétique Animale et Biologie Intégrative, 16 rue Claude Bernard, 75231 Paris Cedex05, France; gregoire.leroy@agroparistech.fr

Among features that characterise French dog breeding, the main one is the fact that most dog breeders are not professional ones: according to French Kennel Club, in 2007, 95% of purebred breeders produced less than 10 litters, yet they still represented 65% of the whole litters production. Dog breeders differ according to their breeding goals and practices, which are more or less related to the breed raised. Using French Kennel Club data base and results from a survey, relations between some breeding practices and the demographical characteristics of 55 breeds and their breeders were investigated. Even if the main traits selected in dog breeds are linked with standard conformation or behaviour, working skills remain an important breeding goal in a large number of breeds, and particularly among most hunting breeds. Generation interval T, which was on average equal to 4.2 years, was clearly lower in working breeds than in non-working breeds, whatever the working discipline: the 10 breeds with the largest proportion of working dogs had T values of about one year more than non-working breed. T was also found to be lower in breeds whose births increased during the most recent years, which is linked to a larger demand of puppies in fashionable breeds. On average, 6% of the litters were produced by mating close relatives, showing that inbreeding is a common breeding practice in dog species. However this rate ranged between 1 and 16% according to the breeds, and was generally lower in breeds where there are a larger number of new breeders. Finally, some other breeding practices may be specifically found in some breeds, like registration of dogs without pedigree, which is particularly frequent in scenthounds breeds, where a large number of breeders do not register their dogs. To conclude, some of these results may have an interest regarding current categorization of breeds to an international nomenclature.

A survey on morphological traits of Basset Hound dogs raised in Italy
Carlini, G.[1], Ciani, E.[2], Ciampolini, R.[3], Bramante, A.[3] and Cecchi, F.[3], [1]ENCI-FCI registered breeder, via dello Olivuzzo 30, 50143 Firenze, Italy; [2]University of Bari, General and Environmental Physiology Department, Via Amendola 165/a, 70126 Bari, Italy, [3]University of Pisa, Animal Production Department, Viale delle Piagge 2, 56124 Pisa, Italy; elenaciani@biologia.uniba.it

The aim of the present work was to report the first results of a survey on morphological traits of Basset Hound dogs reared in Italy, as a part of a more comprehensive study whose objective is to identify morpho-functional attributes having the greatest potential of being genetically improved. Body measures were taken from 48 adult (mean age 2.78±1.71 years) Basset Hound dogs (24 males and 24 females). The animals belonged to five different farms. For each animal, the following biometrical measurements were considered: withers height, chest height, chest depth, trunk length, rump length, ischium width of the rump, ear and nose length, chest and cannon circumference. ANOVA was used to test the differences between males and females for morphological measurements (sex and farm as fixed factors and age at measurements as covariate). A simple correlation method was thus applied to the least square means. No significant differences were observed among farms and between sexes. The average values of withers height were 35.9±1.35 cm and 36.0±1.80 cm, for females and males, respectively; this value was consistent with those reported in the breed standard. In this study withers height was positively related to chest height ($r=0.40$; $P<0.01$) and to cannon circumference ($r=0.29$; $P<0.05$) while it was inversely related to trunk and nose length ($r=-0.33$ and -0.32 respectively, $P<0.05$). Significant and positive correlations were also observed with trunk length and nose length ($r=0.384$; $P<0.01$) and with chest circumference and chest depth ($r =0.299$ and $r=0.354$ respectively; $P<0.05$).

Morphological analysis of Dachshund in Czech Republic during half a century
Šebková, N., Andrejsová, L., Vejl, P., Čílová, D., Vrabec, V., Klokočníková, M. and Lebedová, L., Czech University of Life Sciences Prague, Department of Animal Science and Ethology, Kamýcká 129, 16521 Prague 6 - Suchdol, Czech Republic; sebkova@af.czu.cz

The aim of this work was the morphological analysis of Dachshunds in Czech Republic, which were measured in 1960 and in 2004. In 1960, 144 dogs were measured (46 smooth-haired, 35 long-haired, 63 wire-haired), in 2004 188 dogs were measured (29 smooth-haired, 45 long-haired, 89 wire-haired, 25 miniature). Basic statistical parameters (overall means, standard deviations and variation coefficient) were estimated for both populations. Highly significant differences were found for weight - average weight in 1960 was 8.00 kg for males and 7.23 kg for females. Today the average weight is 8.42 kg for males and 7.42 kg for females. The differences between thorax circumference in 1960 (45.84 cm males, 43.85 cm females) and 2004 (42.26 cm males, 40.44 cm females) were significant. It is interesting, that the relation between length of head and length of mouth was the same in 1960 and 2004. During that time the popularity of individual type of hair was changing fundamentally. Today the wire-haired Dachshund is the most popular.

Evaluation of temperament characteristics of Czechoslovakian Wolfdogs in Czech Republic 1982 - 2004

Šebková, N., Andrejsová, L., Čílová, D., Vejl, P., Vrabec, V., Čapková, Z., Hartl, K., Jedlička, J. and Nováková, K., Czech University of Life Sciences Prague, Department of Animal Science and Ethology, Kamýcká 129, 16521 Prague 6 - Suchdol, Czech Republic; sebkova@af.czu.cz

The Czechoslovakian Wolfdogs is a breed that was bred from crosses between a German Shepherd Dog and wolf. The breed has been listed in the pedigree register since 1982. The aim of this work was to analyze temperament traits. The basis of analysis was data of 1,031 dogs, collected within the period 1982-2004. Genetic progress in the observed trait can be attributed especially to positive selection, which is more intensive in the male part of the population, i.e. to preference for breeding males with desired temperament. The selection differential was determined by the difference between the frequency of fathers and mothers desired characters in comparison to the total assessed population. With male selection differential $d_m=12=$ and female $d_f=1=$, a genetic progress of 16% was achieved during 22 years of breeding, i.e. on average 2.6-3.2% per generation. This corresponds to an effective heritability of the observed trait of 40 - 49% - higher than in previously published cases.

Conformation selection in the German shepherd dog

Stock, K.F., Dammann, M. and Distl, O., University of Veterinary Medicine Hannover, Institute for Animal Breeding and Genetics, Bunteweg 17p, 30559 Hannover, Germany; Kathrin-Friederike.Stock@tiho-hannover.de

Conformation traits represent important selection criteria in dog breeding. Working dogs like the German shepherd dog (GSD) must have a certain stature to be able to fulfill their duties. Current means of withers height (WH) and body weight (BW) in GSD are close to the upper limits defined in the breeding standard. Therefore, strategies to avoid further increase of size and to maximize the proportion of dogs fitting the breeding standard with respect to WH and BW were to be compared. Results of conformation evaluations from breeding approvals, arranged by the German GSD breeding association in 1994-2005, were used. Body measurements were available for 14,416 male and 21,612 female GSD from 26,155 litters. WH and body mass index (BMI), i.e. the quotient of BW and squared WH, were considered as traits to directly select for certain size and stature. Using information on 17,154 GSD from litters with at least 2 dogs with conformation data, within litter variances of WH (vWH) and BMI (vBMI) were defined as traits to select for conformational homogeneity of litters. Officially recorded scores of canine hip dysplasia (CHD) of all dogs were used to monitor possible side effects of conformation selection on CHD. Genetic parameters were estimated multivariately in linear animal models using Gibbs sampling. Heritabilities ranged between 0.19 and 0.34 for all traits with additive genetic correlation of -0.11 between WH and vWH and 0.11 between BMI and vBMI. Breeding values (BV) for WH and BMI were negatively correlated with -0.20 in sires and -0.17 in dams. Expected selection response was studied using relative breeding values (RBV) of parents, assuming exclusion of sires and/or dams with RBV larger than 120, and comparing means of WH, BMI and CHD score between offspring of all and selected parents. Concurrent selection for small WH and vWH was found to most efficiently reduce mean WH in males and females. Little impact on CHD distributions implies compatibility with selection against CHD in the GSD.

Gene flows in dog breeds and their impact on heterozygosity

Leroy, G., Danchin-Burge, C., Verrier, E. and Rognon, X., INRA/AgroParisTech, UMR1313 Génétique Animale et Biologie Intégrative, 16 rue Claude Bernard, 75231 Paris Cedex05, France; gregoire.leroy@ agroparistech.fr

Despite the fact that most dog breeders are hobby breeders, there can be a large number of exchanges of reproducers within dog breeds and in particular between countries. Here we investigated reasons why such exchanges may vary from one breed to another, by using the French kennel club registrations data for 182 breeds. International exchanges were negatively correlated with registrations of dog of unknown origins, while several parameters might explain variations in gene flows, such as population size, breed origin, and specific breeding practices in some breeds. By analysing the effects of those gene flows on the genetic diversity of 60 breeds, it was shown that the percentage of stud service made in foreign countries was actually negatively linked to heterozygosity. It might be explained by the fact that breeders tend to use frequently similar foreign origins.

Utilization of pedigree breeding values for selection against hip and elbow dysplasia in different dog breeds

Stock, K.F., Dammann, M., Heine, A., Engler, J. and Distl, O., University of Veterinary Medicine Hannover, Institute for Animal Breeding and Genetics, Buenteweg 17p, 30559 Hannover, Germany; Kathrin-Friederike. Stock@tiho-hannover.de

Canine hip dysplasia (CHD) and elbow dysplasia (ED) represent important skeletal diseases of the dog, for which relevant genetic determination has been shown. Routine radiographic screening of breeding animals for CHD and/or ED is practiced in several dog breeds to allow phenotypic selection. In addition, genetic evaluation for CHD and/or ED has been introduced in some dog populations to use breeding values (BV) not only for selection to improve the population, but also for planning of matings, i.e. choice of mates to limit the risk of future offspring developing CHD or ED. However, selection responses as measured by proportions of affected dogs were mostly smaller than expected. Therefore, the predictive value of pedigree breeding values (pBV) for CHD and ED phenotypes was investigated in different dog populations. Considered dog breeds were German shepherd dog (GSD), Rottweiler (RO), Labrador Retriever (LR) and German Drahthaar (GD). CHD scores were available for 184,489 GSD from birth years 1985-2007, 2,867 LR from birth years 2000-2004, and 7,303 GD from birth years 1995-2005. ED scores were available for 2,386 RO from birth years 1997-2005. Proportions of dogs free of CHD were 60.7-70.0%, proportion of RO free of ED was 68.5%. BLUP BV were predicted for CHD in GSD and LR and for ED in RO. Gibbs sampling was used for prediction of BV for CHD in GD. In each case, CHD or ED information on dogs from previous birth years was used to predict pBV for next year's progeny. Across breeds, diseases and methods of BV prediction, the proportion of explained phenotypic variance of CHD or ED (r^2) was only 3% for pBV, implying very limited predictive value of pBV. As opposed to other species in which pBV allow reliable predictions of phenotypes of future progeny, pBV are of little value for planning of matings in the dog.

The colour inheritance of Pumi and Mudi Hungarian dog breeds

Rózsa Várszegi, Z.S., Posta, J. and Mihók, S., University of Debrecen, Institute of Animal Husbandry, Böszörményi str. 138., H-4032, Debrecen, Hungary; postaj@agr.unideb.hu

The colour affects the dog's appearance. The Pumi and the Mudi is as multicolour as the Puli. All colours can be crossed with the others, so the level of inbreeding can be decreased. If inheritance of the colour is known, the colour of the offspring can be predicted, and incorrect litter announcement can be detected. In our study we were analyze: 1) What kind of colours exists in the Hungarian population? 2) What are the combinations and proportion of the colours in the population? 3) Is the colour specifying correct? In case of the Pumi the colour of 1023 dogs from 193 litters was examined. Data of 599 puppies from 114 litters were analyzed in Mudi breed. The colours of the Pumi are: grey in various shades, black, fawn, white, brown (not allowed by the official standard). The colours of the Mudi: black, blue-merle, ash colour, brown, white, fawn. In case of the Pumi breed more than the half of the examined population was grey (56.2%). The percentage of the white dogs (13%) was quite high. This can be the result of wrong colour denomination, so a part of them must be fawn. A part of the fawn dogs with mask was denominated grey, or turning into grey from black. Probably the epistatic row in colour inheritance by Pumi is the next: black – brown – grey – fawn with mask - white – fawn. In case of Mudi, the 2/3 of the population was black. The rarest Hungarian herding dog is luckily really multicoloured. The epistatic row of the colour inheritance is the same as for the Pumi. The brown colour must be accepted as a primal colour in this breed. The brown dogs, and the fawn dogs with mask probably were not written in the litter announce, or these were named incorrectly. The blue-merle gene does not occur in other Hungarian breeds, the appearance of it could permit of disperse. Accurate colour denomination should be used by the pedigrees, because the examination opened up several failures, for example only those dogs are white, which was born white.

Research on potentials for improving the quality of rural livelihood in Norway through diversification of the small ruminant sector

Eik, L.O., University of Life Sciences (UMB), Department of Animal and Aquacultural Sciences, P.O.Box 5003, 5003 Aas, Norway; lars.eik@umb.no

Sheep and goat farming systems in Norway are based on extensive use of non-fertilised natural pastures and a long barn-feeding period. One million ewes are kept for the production of meat and wool while 70,000 goats are kept traditionally for milk production, although the number of goats kept for other purposes is increasing rapidly. Nearly all lambs and surplus ewes are normally sent for slaughter directly from pastures in fall, resulting in shortage of lamb meat in off-seasons. The paper discusses strategies for more even annual deliveries of small ruminants to abattoirs. Meat from lambs raised on mountain pastures without any supplementary feeding or treatments is often considered to be of superior quality. Due to natural variation in weight and body condition, however, autumn grazing of some of the lambs on cultivated pastures, supplemented with concentrates, has become a common practice in most parts of the country. Significant differences between the groups were found in sensory traits, grading, fat content and fatty acid composition, meat color, and meat flavor. The results suggest that meat from lambs raised in extensive systems on mountain range have certain qualities that might be used in promotion of local and regional products. Recent studies indicate quality difference in meat from ewe and ram lambs, when slaughtered beyond September. Therefore it is recommended that ram lambs be slaughtered directly from pastures in fall. In recent years, new multipurpose systems (milk + meat + landscaping or meat + landscaping + cashmere fibre) have been introduced alongside the normal intensive milk production system. The new systems which also include payment to farmers for up-keeping of an open and diverse landscape are promising, and may lead to a more sustainable and profitable small-ruminant sector in Norway.

The challenge of sustainability for local breeds and traditional systems: dairy sheep in the Basque Country

Ruiz, R., Díez-Unquera, B., Beltrán De Heredia, I., Mandaluniz, N., Arranz, J. and Ugarte, E., Neiker, P.O. Box 46, Vitoria, Spain; rruiz@neiker.net

Livestock farming, particularly sheep and goats, has played for centuries a key role for the livelihoods of people in the Mediterranean basin. In particular, dairy sheep can be envisaged as a paradigmatic activity in this area and the implications and importance of these systems are unquestionable in relation to landscape configuration, diversity of products, cultural heritage, etc. A broad diversity of breeds, management practices and production systems were shaped to fit to the local specificity, conditions and resources available. However, the sustainability of many of these production systems is seriously endangered nowadays, and the remaining ones will have to face adaptation strategies to get adapted to changing policies, consumer demands, social, economic and environmental conditions. The Latxa sheep system traditionally carried out in the mountainous areas of the Basque Country is presented as a case study of this situation: initially a local breed reared upon the base of natural resources, but nowadays enjoying a successful breeding programme, and production being protected under quality and denomination of origin labels. However, some of the flocks end up disappearing despite being economically profitable, showing that sustainability does not definitely depend only upon productivity parameters. Several pressures have been identified in this system: access to land, standards of living expectations, growing dependency from inputs, low meat and milk price, wolf predation. But there are also interesting tools, actors involved and opportunities to carry out in order to improve sustainability: cheese making, close access to consumers, environmental services, training and education initiatives (shepherds school), agro-ecology approaches for farming practices, growing sensibilization to quality food consumption. Therefore, social and environmental indicators have to be considered as they can be crucial for adaptability, resilience, self-sufficiency and equity attributes.

Socio-economic benefits from Bedouin sheep farming in the Negev

Valle Zárate, A.[1], Albaqain, A.[1], Gootwine, E.[2], Herold, P.[1] and Albaqain, R.[1], [1]Institute of Animal Production in the Tropics and Subtropics, Universtität Hohenheim, Garbenstrasse 17, 70593 Stuttgart, Germany, [2]The Volcani Center, Agricultural Research Organization, Ministry of Agriculture, P.O. Box 6, Bet Dagan, 50250, Israel; valle@uni-hohenheim.de

About 250,000 heads of sheep are kept by Bedouin families in about 1,200 flocks in the semi-desert Negev region in the south of Israel. Traditional extensive sheep farming practices still predominate, but get increasingly under pressure for intensification of production in view of increasing shortages of grazing areas, draughts and feed costs. The study describes current Bedouin sheep farming systems and analyses the effects of socio-economic factors and sheep breeding, feeding and management on animal performance, flock output and economic success. Semi-structured interviews were conducted in 30 Bedouin families from 11 tribes throughout the Negev, covering flocks with 51 to 1,106 sheep. Information on prices was crosschecked by key person interviews in institutions and on markets. Flock outputs ranged from 13 to 58 kg of marketable live weight per ewe and year and differed significantly by the factor combination of tribe-distance to the market and by the interaction between genotype of ewes and age of lambs at sale. Gross margins ranged from -147 to 296 NIS per ewe and year and net benefits (NB) were negative for 43% of the Bedouin flocks. Farms with negative NB had lower herd sizes, lower prolificacy, higher mortality but similar cost structures as economically successful farms. Traditional extensive sheep farming in low input systems based on the local Awassi breed was not positively contributing to family income and nutrition, however still regarded as a cultural benefit.

Introducing the prolific Afec-Awassi strain to semi extensive Bedouin sheep flocks in Israel as a mean to secure their economic success

Gootwine, E.[1], Abu Siam, M.[2], Reicher, S.[1], Al Baqain, A.[3] and Valle Zarate, A.[3], [1]ARO, The Volcani Center, Institute of Animal Science, POB 6, Bet Dagan, 50250, Israel, [2]Siam Veterinay Clinic, POB 519, Rahat, 85357, Israel, [3]Universität Hohenheim, Institute of Animal Production in the Tropics and Subtropics, Garbenstr. 17, Stuttgart, 70593, Germany; gootwine@agri.gov.il

Bedouin farmers in the southern dry region of Israel raise some 250,000 low-prolific local Awassi sheep under semi-extensive conditions. Shortage in grazing land has forced the farmers to invest more on feedstuff, making traditional sheep production largely unprofitable. Economic assessment revealed that increasing lamb production in the Bedouin flocks is the key factor for economic success. Afec-Awassi is a prolific Awassi strain developed by introducing the Booroola mutation to the Improved Awassi dairy sheep maintained under intensive management. Disseminating of Afec-Awassi BB rams in Bedouin's flocks (n=4) resulted in the production of 549 B+ females kept as replacements. Prolificacy of B+ ewes (n=167) lambed so far was 1.7 and 1.9 lambs born/ewe lambing (LB/L) in the 1st and the 2nd parities, respectively. Overall lambs' post-natal survival rate at birth was 0.94. Prolificacy of multiparous local Awassi ewes (n=2,260) in nine control local Awassi flocks was 1.2 LB/L, with post-natal survival rate of 0.96. Average birth weight of singles, twins and triplets was 5.3, 4.2 and 3.4 kg respectively, with no difference between lambs born to B+ or to local Awassi ewes. The breeding work was assisted by genotyping for carrying the B allele of the FecB locus and by applying advanced reproductive techniques like hormonal synchronization and ultrasonic pregnancy diagnosis. Our results show that prolific Afec-Awassi ewes managed under semi extensive conditions manifest high reproductive performances, similar to Afec-Awassi ewes kept under intensive conditions. Economic assessment based on actual prices and performance records showed that introducing the Afec-Awassi to Bedouins flocks can successfully and profitably be accomplished.

Foraging selectivity of three goat breeds in a Mediterranean shrubland

Glasser, T.A.[1], Landau, S.[1], Ungar, E.D.[1], Perevolotsky, A.[1], Dvash, L.[1], Muklada, H.[1], Kababya, D.[1] and Walker, J.W.[2], [1]Volcani center - ARO, Natural Resources, P.O.B. 6, 50250, Bet-Dagan, Israel, [2]Texas A&M University, Texas AgriLife Research, 7887 U.S. Hwy. 87 N. San Angelo, 76901, Texas, USA; Tglasser@gmail.com

Foraging behaviours of the Damascus, Mamber and Boer goat breeds were compared on the South Carmel mountain ridge of Israel. Dietary choice was determined for a group of 10 or 11 yearling animals of each breed, housed and grazed separately to prevent social facilitation, during a total of 4 (Mamber) or 5 (Damascus and Boer) 4-day periods in fall 2004 and spring 2005. The proportions of the three main dietary components – P. *lentiscus* L. (20% tannins), *Phillyrea latifolia* L. (3% tannins) and herbaceous species (as one category) in the diet (which also included concentrate) were determined by application of near-infrared reflectance spectroscopy to faecal samples ('faecal NIRS') (n=124). On average, P. *lentiscus* formed 13.0% of the DM ingested by Damascus goats, but only 5.0 and 4.9%, respectively, of that ingested by Mambers and Boers (Damascus > Mambers = Boers, $P<0.0001$). Damascus goats ingested diets richer in tannins (5.4 vs. 4.2%, respectively, $P<0.0001$). The contribution of herbaceous species to ingested DM in the spring was higher in Mambers than in the other breeds (33 vs. 27%, respectively). Boer goats selected the most nutritious diets in terms of Crude Protein (CP) content and *in vitro* Dry Matter Digestibility (IVDMD). In spite of their differences in foraging selectivity, the local Damascus and Mamber goats had similar dietary percentages of CP and similar IVDMD. Our data suggest that the Damascus is a preferred candidate to control P. *lentiscus* encroachment and is the least likely to compete with cattle for green grass in the spring. These findings may contribute in the attempt to find ecologically sound ways of controlling the spread of the tannin-rich shrub *Pistacia lentiscus* L., which threatens rangeland biodiversity and amenity values in the Mediterranean climatic region of Israel.

Link local breeds with territories: some contrasted relationships

Lauvie, A., Lambert-Derkimba, A. and Casabianca, F., INRA UR LRDE, Quartier Grossetti, F-20250 Corte, France; anne.lauvie@corte.inra.fr

Local breeds are often presented as genetic resources well adapted to a specific territory. This focus on adaptation is emphasizing the question of linkage between a breed and a territory. In this meaning, such link is often seen in a positive way. Our work shows how the link between local breeds and territory could have either positive or negative aspects. We compare two contrasted relationships of local breeds to a territory: The Nustrale pig breed in Corsica and the Blanc de l'Ouest pig breed in the western part of France. Through those cases we underline potentialities and constraints in the link between breed and territory. In order to express those potentialities and constraints, our analysis distinguishes several dimensions: -Biogeographical: use of the local feeding resources and geographical scattering of the breeders. -Breeding system: livestock practices and production management. -Product: specificity and image of the products stemming from the breed. -Food chain: organizational aspects of the food chain on the territory (including cohabitation with pig industry for the Blanc de l'Ouest). -Symbolic and regional identity: crosswise all the dimensions mentioned above and the involvement of several actors within the society. As a major issue, we show that for a same dimension, a constraint for one breed may be considered as a potentiality for the other. So, when talking about links between a breed and a territory, one must detail this link and distinguish restricting dimensions from valorising ones.

Seasonal organic pig production with a local breed

Kongsted, A.G.[1], Claudi-Magnussen, C.[2], Horsted, K.[1], Hermansen, J.E.[1] and Andersen, B.H.[1], [1]Aarhus University, Faculty of Agricultural Sciences, P.O. Box 50, 8830 Tjele, Denmark, [2]DMRI Consult, Danish Meat Research Institute, Maglegårdsvej 2, 4000 Roskilde, Denmark; anneg.kongsted@agrsci.dk

It is important that organic pork differs markedly from conventional pork regarding taste, appearance and production methods in order to overcome the heavy price competition. That is the hypothesis behind the current project. A seasonal outdoor rearing system based on a traditional and local breed is believed to be a feasible strategy for producing organic pork with high credibility and superior eating quality. The study included a comparison between a modern genotype and the Danish Black-Spotted pig which does almost not exist in the modern pig production of today. 17 gilts farrowed outdoors in April 2007 and 12 gilts in May 2008. The offspring were slaughtered at 40 kg (male pigs) and 100 kg to 140 kg (female pigs and castrates). The two breeds were compared regarding daily gain, feed conversion rate, animal behaviour, the thickness of the back fat and the meat, meat colour and sensory profile. Preliminary results from the first year of study showed e.g. 37% lower daily gain from birth to slaughter, 30% higher back fat thickness and 11% lower meat percentage in the Black-Spotted female pigs compared to the modern genotype. In contrary, the Black-Spotted pig produced redder and darker meat compared to the modern genotype and the fat of the Black-Spotted pig was characterised as having a special nutty taste. In conclusion, preliminary results indicate that the local breed differs markedly with respect to several meat quality aspects compared to the modern breed but also shows clear disadvantages regarding growth performance and lean content.

Effect of slaughter weight on carcass quality and fatty acid composition of subcutaneous fat of Porc Negre Mallorquí (Majorcan Black Pig)

Gonzalez, J.[1], Jaume, J.[2], Gispert, M.[1], Tibau, J.[1] and Oliver, M.A.[1], [1]IRTA, Camps i Armet, 17121 Monells, Spain, [2]IBABSA, Esperanto 8, 07198 Son Ferriol, Spain; joel.gonzalez@irta.es

Majorcan Black Pig is an autochthonous breed from Mallorca reared in extensive conditions, with free access to outdoor pasture. Carcass of pure breed is used to produce Sobrassada de Mallorca de Porc Negre, a high quality fermented sausage. The aim of this study was to characterize carcass quality and fatty acid profile of this breed, and determine if animals slaughtered at lighter weight could be used to produce Sobrassada maintaining its technological and nutritional quality. Twenty-seven females were split in two groups according to their live weight: light animals (L) from 85 to 120kg, and heavy animals (H) from 125 to 185 kg. The latter is the most common slaughter weight. The minimum backfat thickness was measured at Gluteus medius level, being thicker in heavy animals (L42.3 vs. H61.1mm; $P<0.0001$). Regarding fatty acid profile, saturated fatty acids were significantly higher in heavy pigs (L37.7 vs. H39.3%; $P<0.001$), whereas monounsaturated fatty acids content was higher in light pigs (L51.8 vs. H50.0%; $P<0.001$). No differences in polyunsaturated fatty acids content were observed between groups (L10.5 vs. H11.0%). These results indicate that fat from light animals could modify the technological quality of the final product, and it is needed to know to which extent. Assessment of nutritional quality was based on n6/n3 and PUFA/SFA ratios, and also on atherogenicity (AT) and thrombogenicity (TH) indexes, which estimate the probability of athero and thrombogenesis. Since part of consumers has shown a growing interest in healthy food, meat industry demands for fat with proportions of fatty acids more adequate from the human health point of view. Heavy pigs had a significant more adequate n6/n3 ratio (L10.3 vs. H8.8; $P<.0001$), but there were no significant differences regarding PUFA/SFA (L,H0.28). Otherwise, TH (L1.09 vs. H1.16; $P<0.05$) and AT (L0.48 vs. H0.51; $P<0.05$) indexes were significantly more convenient in lighter pigs.

Sustainability assessment of sheep farming systems in the north of Spain: a methodologic approach

Diez, B.[1], Ripoll, R.[2], Beltrán De Heredia, I.[1], Molina, E.[3], Olaizola, A.[4], Bernués, A.[2], Villalba, D.[3] and Ruiz, R.[1], [1]Neiker, Apdo. 46, 01080 Vitoria, Spain, [2]CITA, Av.Montañana 930, 50059 Zaragoza, Spain, [3]Univ. Lleida, Avda. Rovira Roure 191, 25198 Lleida, Spain, [4]Univ. Zaragoza, Miguel Servet 177, 50013 Zaragoza, Spain; bdiez@neiker.net

In Spain there are 24 million sheep (26% of EU-25), who supply 12% of the national gross product coming from livestock. The huge diversity of environmental and climatic features have moulded the characteristics of the local breeds, condition the natural resources available and the chances to adopt feeding and reproduction strategies. There are other factors that have played a crucial role in the evolution of management practices, and in the intensification level accomplished: support programmes, social trends, economic changes, etc. In this context, the assessment of sustainability has an increasing interest. A sample of the main farming systems existing in the Basque Country, Aragon and Catalonia were taken as case studies to test the MESMIS framework; to assess sustainability of sheep farming in these areas; and to identify critical points, constraints and opportunities. First, SWOT analysis were carried out for each system, and then agreed by participatory approaches. Then relevant indicators were identified and assessed. Obviously not every indicator has the same weight in the assessment of sustainability, so significant efforts were required to agree through a multidisciplary approach. The Latxa dairy system yielded the highest productive results in terms of net margin/sheep, basically due to milk production and cheese-making activities. These farmers rated better their level of satisfaction for the activity. Meat production systems either based upon local or foreign breeds, have an uncertain continuity due to economic challenges (evolution of prices of inputs and meat) and social factors. Taking into account that the consumption of natural resources supplied from 45% to 68% of the average annual energy requirements os sheep, the environmental implications of sheep in some areas is relevant.

Assessment of the threats to the Heritage Sheep breeds in Europe using a geographical information system

Ligda, C.[1], Mizeli, C.[2], Tsiokos, D.[3] and Georgoudis, A.[3], [1]National Agricultural Research Foundation, P.O. Box 60 458, 57 001 Thessaloniki, Greece, [2]Holstein Association of Hellas, Derveni, Thessaloniki, Greece, [3]Aristotle Universtity of Thessaloniki, University Campus, 54 124 Thessaloniki, Greece; chligda@otenet.gr

The present work has been developed in the frame of the Heritage sheep project funded under the EU Regulation 870/04. The project aims at the identification of the potential threats and values of the genetic resources of heritage sheep breeds and the development of a scoring system. Furthermore, case studies for the *in situ* conservation and strategies for the *ex situ* conservation of the heritage sheep breeds were elaborated. A specific WP is devoted on the establishment of a website and a database for the dynamic presentation of the information for the Heritage Sheep breeds. The data collected referred to the assessment of the threats to the breeds, the definition of regions of high risk and economical, social and environmental pressures. Geographical information linked with the breeding regions of the breeds was collected in order to present the results of the analysis using the ArcGIS 9.3. The Geographical Information System was used as a tool in order to increase the awareness on the local sheep genetic resources of the relevant stakeholders, but also of the general public.

An integrated view of technical, economic and social factors influencing sustainability of sixteen local European cattle breeds

Martín-Collado, D.[1], Díaz, C.[1], Choroszy, Z.[2], Duclos, D.[3], Bay, E.[4], Hiemstra, S.J.[5], Kearney, F.[6], Mäki-Tanila, A.[7], Viinalass, H.[8] and Gandini, G.[9], [1]INIA, Mejora Genética Animal, Apdo 8111, 28080 Madrid, Spain, [2]NRIAP, Sarego 2, Krakow, Poland, [3]IE, Rue de Berci, 75595 Paris, France, [4]FUSAGx, Passage des Deportes 2, 5030 Gembloux, Belgium, [5]CGN, Edelhertweg 15, 8200AB Lelystad, Netherlands, [6]ICBF, Co Cork, Bandon, Ireland, [7]MTT, Agrifood Research, 31600 Jokioinen, Finland, [8]EMÜ, Kreutzwaldi 64, Tartu, Estonia, [9]UNIMI, Via Festa del Perdono 7, 20122, Italy; martin.daniel@inia.es

This study aims to identify common patterns influencing dynamics and sustainability of 16 local European cattle breeds from Belgium, Estonia, Finland, France, Holland, Ireland, Italy and Spain. A broad range of factors related to technical (e.g. farm size, cows/ha, land ownership), economic (e.g. cattle importance on farm and family income, type of market) and social (e.g. farmers and stakeholders attitude) aspects were surveyed on a total of 401 farms. Discriminant Analysis has been used to study the implications that heterogeneity within and across countries and breeds may have in making inferences in across country or breed analyses. Thus, farms were classified according to country and breeds. Then the objective was to evaluate how all factors available allow us to distinguish such classes. The analyses provided the percentage of observations (farms) that should not be included in the pre-defined groups (country and breed) according to the variables considered. When only economic variables were considered in the analyses, 36% and 59% of farms were incorrectly assigned to their country and breed, respectively. However, considering technical variables the proportions went down to 33% and 45%. Finally, when both groups of variables were considered 20% and 35% of farms were incorrectly assigned. To what extent these mismatches may allow us to identify general patterns has to be evaluated.

Effect of mountain pasture versus indoor breeding system on somatic cell count in cow milk
Frelich, J. and Šlachta, M., University of South Bohemia, České Budějovice, 37005, Czech Republic; frelich@zf.jcu.cz

In mountain areas of the Czech Republic the breeding of dairy breeds of Holstein (H) and Czech Fleckvieh (C) relies on two feeding strategies: (1) the seasonal pasture and fresh-cut herbage supplementation during May – October period, followed by a grass-silage feeding indoor in the rest of a year, and (2) all-year-through grass and maize silage feeding without any access to pasture. The somatic cell counts (SCC) in monthly collected individual milk samples were measured and the GLM analysis of STATISTICA was applied in order to evaluate the significance of Breed and Breeding system on SCC. Next variables with fixed effects were also included in the model: Parity, Year (1–3), Season (October–April; May–September) and Days in milk. In total 103,503 records of 4,965 Holstein and 3,315 Czech Fleckvieh cows were analysed. Except the variable of Year all the other variables revealed significant effects on SCC, higher SCC beeing found in H-breed than in C-breed (F=1,190;$P<0.001$) and in Indoor than in Pasture system (F=147;$P<0.001$). The interaction between Breed and Breeding system was significant (F=52;$P<0.001$), which indicated a more pronounced effect of Breeding system on SCC by H than by C breed.

Added value of milk for cheese yield in local and cosmopolitan dairy cattle breeds
Pretto, D., Penasa, M. and Cassandro, M., University of Padova, Department of Animal Science, Viale dell'Università 16, 35020, Legnaro, Padova, Italy; denis.pretto@unipd.it

The aim of this study was to compare milk added value from two local Italian cattle breeds (Alpine Grey and Rendena) and two cosmopolitan dairy breeds (Holstein Friesian and Brown Swiss) reared in Trentino mountain area. Several studies evidenced that milk from local dairy breeds is more suitable to be processed into cheese, so the development of payment systems that take into account the added value of milk for cheese production could play an important role in the conservation and valorization of local animal genetic resources (AnGR). The value of raw milk (€/kg) was estimated for each breed from chemical contents (protein, fat and whey) and somatic cell count, simulating the current milk pricing system. Cheese yield, using as reference Trentingrana cheese-chain, was estimated accounting for milk coagulation properties (rennet clotting time and curd firmness) and chemical contents. The value of cheese was calculated multiplying cheese yield by market price (€/kg) of Trentingrana cheese. Added value was obtained as difference between value of cheese and value of milk. Results showed that added values of milk from the two local AnGR (Alpine Grey and Rendena) were 20% and 63% greater than Holstein Friesian, respectively; while Brown Swiss showed an intermediate added value (51%). Cheese yield of local AnGR might partially compensate the lower level of milk production, therefore a cheese market oriented strategy with payment systems including milk added value could enhance profitability and interest in rearing and safeguarding local AnGR.

Performances of Bruna dels Pirineus beef cattle breed in the mountainous areas of Catalonia

Piedrafita, J.[1], Fina, M.[1], Casellas, J.[2], Quintanilla, R.[2], Orriols, M.[3] and Tarrés, J.[1], [1]Universitat Autònoma de Barcelona, G2R, Ciència Animal i dels Aliments, Campus de Bellaterra, 08193-Cerdanyola del Vallès (Barcelona), Spain, [2]IRTA, Genètica i Millora Animal, Rovira Roure 191, 25198 - Lleida, Spain, [3]FEBRUPI, Carretera de Solsona 7, 08600-Berga (Barcelona), Spain; jesus.piedrafita@uab.es

The Bruna dels Pirineus is a medium-sized meat type cattle breed located in the Eastern Pyrenees, having an excellent maternal aptitude. It comes from an early crossing between a local Pyrenean breed and the old dual purpose Brown Swiss cattle breed. A total of 46,807 calving records belonging to 25,915 cows and collected during the last 20 years were processed to study the evolution of breed performance. Age at first calving was 1023.7 ± 108.1 d, the length of productive life, from first calving until culling, being 9 years on average, with a corresponding replacement rate of 11%. On average, annual calving rate was 0.92 ± 0.16 (397 d calving interval). Birth weight was 45.4 ± 6.5 kg for males and 42.1 ± 6.0 kg for females. Calving difficulty varied according to sex of calf. Calvings without assistance were the most frequent (74.0% for males and 82.6% for females, respectively). A lower proportion of calvings needed slight assistance (17.7% and 13.7% for males and females) or a strong assistance of the farmer (5.8% and 2.8%), and only 2.4% of male calvings and 0.8% of female calvings needed veterinary assistance or caesarean. Calf mortality until weaning was 3.1% and was more pronounced during the first month (2.5%). Weaning weight was 227.1 ± 46.0 kg for males and 216.7 ± 44.9 kg for females, with weaning ages of 177.0 ± 38.4 d and 186.9 ± 44.0 d, respectively. Yearling weight of males was 540 kg. A program to monitor muscular hypertrophy has started recently.

Heat stress response of some local and European breeds of feedlot beef cattle grown under the Mediterranean climate conditions

Bozkurt, Y. and Ozkaya, S., Suleyman Demirel University, Faculty of Agriculture, Department of Animal Science, Isparta, 32260, Turkey; ybozkurt@ziraat.sdu.edu.tr

In this study, data from Holstein (11), Brown Swiss (27), Simmental (8) cattle as European breeds (EB) and Boz (12) and Gak (48) as local breeds (LB) grown under feedlot conditions were used to evaluate and compare performance differences in the Mediterranean type of climate in response to heat stress. Initial mean weights of cattle were 200, 196, 213, 203 and 223 kg for Holstein, Brown Swiss, Simmental, Boz and Gak repsectively. There were statistically significant ($P<0.05$) differences in daily live weight gains (DLWG) of both type of cattle. Altough there was no significant ($P>0.05$) interaction between temperature and breed types, liveweights of both type of cattle were affected by the heat and the weight gains were decreased as the temperature and humidity increased. There were no statistically significant ($P<0.05$) differences in performance between EB cattle and between LB cattle themselves. However, Simmentals tended to perform better than the rest.The results showed that under the Mediterranean conditions EB cattle were better suited to the feedlot beef systems than LB cattle.

Spanish specially-protected autochthonous breed Black Morucha
Alvarez, S.[1] and Sánchez Recio, J.M.[2], [1]Universidad de Salamanca, Área de Producción Animal, Filiberto Villalobos 119, 37007 Salamanca, Spain, [2]Asociación Nacional de Criadores de Ganado Vacuno de Raza Morucha Selecta, Santa Clara 20, 37002 Salamanca, Spain; salvarez@usal.es

Black variety of Morucha breed is a specially-protected autochthonous breed in Spain. Morucha is an autochthonous Spanish breed with two varieties distinguished by their colour: one of them is grey (cárdena) and it is a promoted breed; the other one is black. It is very well adapted to its environment, and it has very good maternal characteristics. An extensive production system is used in farming, very linked with its natural basis, which are dehesas, pastures with Quercus trees. It is probably a good example of environmentally friendly systems in livestock. This work aims to present Black variety, its census in controlled genealogical registers and not controlled animals which are actually in production. Its 2,921 heads in 2008 were 17% of total Morucha animals in controlled genealogical register, and there were Black animals in 16% of controlled farms. Farm size is slightly higher in Black variety that considering the whole Morucha (83 vs. 78 cows). The evolution in the last years is presented, and whereas the Morucha census considered as a whole has decreased, Black Morucha census has increased, probably due to recently developed conservation work. Number of farms and geographical distribution are analysed. The activities developed in order to preserve this breed are exposed. Black Morucha census is compared with other Spanish specially protected breeds, and we conclude that this breed is in an intermediate position: there some other breeds with less animals (Frieiresa, Vianesa, Albera…), with a similar census (Cachena, Alistana Sanabresa) and breeds with more animals (Asturiana de la Montaña, Bruna dels Pirineus).

Comparison of fattening performance and slaughter value of local Hungarian cattle breeds to international breeds
Holló, G., Somogyi, T. and Holló, I., Kaposvár university, Guba S. street 40., 7400, Hungary; hollo. gabriella@sic.hu

For the elaboration of a breeding strategy aiming at the enhancement of Hungarian slaughter cattle production it is essential to know the performance of Hungarian breeds. The objective of the trial was thus to compare fattening, slaughter and meat quality parameters of fattening bulls from different genotypes fattened under same conditions. Altogether 62 growing bulls from different breeds - Angus (A), Charolais (CH), Holstein (H), Hungarian Grey (HG), Hungarian Simmental (HS), Charolais x Hungarian Grey (CH x HG) - were fattened in small groups. The diets consisted of maize silage ad libitum, grass hay 2 kg /day, and 2-4 kg concentrate. After 430 kg live weight the concentrate contained 25% linseed supplementation. The target final live weight was 600 kg. The daily gain during the fattening period varied from 897 (HG) to 1240 g/day (A). Based on EUROP classification data, the conformation scores of CH (9.50) and HS (9.33), and the fatness scores of A (8.33) and CH x HG (7.56) were significantly higher than those of the other genotypes. CH had the highest carcass weight (362.7 kg), whereas the lowest one was the HG (322.9 kg). The proportion of kidney fat changed together with fatness score (A: 1.47%, HG: 0.75%). The dressing percentages of CH (59.50%) and HS (57.84%) were significantly higher than those of HG (55.13%) and H (55.03%) bulls. Analysing the tissue composition of carcass it was found that the lean meat yields in all genotypes (HS: 74%, CH: 72%, HG, CH x HG, H: 71%) were significantly higher than that of A animals (67%). The highest intramuscular fat level of three examined muscles was detected in beef of A and HG genotypes. The ratio n-6 to n-3 was lowest in beef of A and HG. The content of conjugated linoleic acid was the highest in purebred HG group.

Effect of different diets on carcass composition and fat quality of Mangalica pigs
Holló, G.[1], Nuernberg, K.[2], Somogyi, T.[1] and Holló, I.[1], [1]Kaposvár University, Guba Sándor street 40., 7400 Kaposvár, Hungary, [2]Research Institute for the Biology of Farm Animals, Wilhelm-Stahl-Alle 2, 18196 Dummerstorf, Germany; hollo.gabriella@sic.hu

The effect of free range pasture fattening with barley and tritical supplementation (E) versus conventional intensive indoor feeding (I) on carcass composition and fat quality of 51 Mangalica pigs (initial live weight 40 kg) was investigated. Pigs were slaughtered at 105 kg live weight. Both manual cutting and Computer tomography (CT) data showed that the carcasses of the extensive group were leaner, had less fat and bone than those of the intensive group however the differences were not significant. The pork cut belly contained a high amount of muscle tissue in the extensively kept pigs compared to the intensively kept animals (33% vs 27%, respectively). Different measurements (manual cutting and CT) for the estimation of the back fat proportion are closely related (r=0.91), whilst for the subcutaneous fat of the belly, the correlation coefficient was lower (r=0.58). The intramuscular fat content in longissimus (9.9% vs 10.9%) and semimembranosus (4.2% vs 2.6%) muscle is considerably high in both groups compared to conventional breeds. The sum of SFA in back fat, belly fat, as well as in longissimus muscle was significantly decreased by the extensive keeping system compared to that of the indoor keeping system. The n-3 fatty acid proportions were significantly increased in subcutaneous fat of the belly and in intramuscular fat of both muscles by free range pasture fattening. The Mangalica breed kept free range and supplemented with concentrate can supply high quality carcasses for the production of original Hungarian salami sausage.

Which place for territorial projects in the decision-making about local breeds' future?
Lambert-Derkimba, A.[1], Casabianca, F.[1] and Verrier, E.[2], [1]INRA SAD - LRDE, Quartier Grosseti, 20250 Corte, France, [2]AgroParisTech - INRA UMR GABI, 16 rue Claude Bernard, 75005 Paris, France; derkimba@corte.inra.fr

Our work is focussing on the way local breeds are mobilized by some territorial projects. We try to understand the effects of such projects on the development and the orientation of these breeds. Among several projects, we decided an in-depth analysis of the impact of PDO products, showing anchorage in the territories by the resources they mobilize (breeds, livestock systems, local know-how), on the decision-making processes for the management of concerned local breeds. We addressed two major questions: (1) to what extent association between breeds and PDO can impact the management of genetic resources from a quantitative (development and diffusion of the breed) and qualitative point of view (objectives of selection, genetic structure of the herds), and (2) what configurations give some room to the PDO rules of production within the definition of the objectives of breed selection? The analysis is based upon the cases of the Northern Alps (PDO cheeses Abondance, Beaufort, Reblochon and Tome des Bauges/cattle breeds Abondance, Tarentaise and Montbéliarde, the latter not being local), and of Corsica (project of PDO for pork-butchery/pig breed Nustrale). We highlight an impact of these PDO projects on the quantitative development of the breeds by the mobilization of the local authorities of extension bodies and the inter-professions. In the particular case of the Tarentaise breed, the syndicate of the PDO Beaufort takes an effective part in the development of the objectives of selection (construction of synthetic index) of the local breed. As a main issue, two types of criteria are proposed: 1) interactions between the collectives (breed vs products) and 2) room for leeway of the stockbreeders. This set of various criteria, not being a prediction tool, enable us to propose keys of interpretation of the consequences of the coupling breed/PDO.

Morphogenetic characterization and germplasm cryopreservation of Churra Tensina, an endangered local ovine breed

Sanz, A., Alvarez-Rodríguez, J., Folch, J., Martí, J.I., Joy, M., Alabart, J.L. and Calvo, J.H., CITA de Aragón, Avda Montanana 930, 50059 Zaragoza, Spain; asanz@aragon.es

Churra Tensina (CT) is a coarse-wooled hardy breed raised traditionally in Aragon Pyrenees due to their great coping ability to harsh environments. In the framework of a breed conservation project, population structure and farming system were characterized through surveys to farmers. In 2007, the census was 7,600 individuals. Flocks have medium size (average 182 animals), with incomes complemented with other activities. Feeding is based on grazing high mountain areas in summer and forest pastures and meadows the rest of the year. The main lambing season is autumn, followed by winter, and the commercial product is a light lamb (20-25 kg live-weight). The zoometric variables of the animals registered in ATURA (CT farmers association) were studied to morphologically characterise the breed and to provide basis for a detailed *in situ* conservation genetic programme of the breed. These variables showed homogeneity and an evident sexual dimorphism. According to these variables, CT may be classified as a brevilineous, dolicocephalous and eumetric breed. Thirty microsatellites proposed by FAO and ISAG for biodiversity studies have been used to characterise CT breed. All microsatellite loci analysed were found to be polymorphic, showing this breed good diversity values despite its low effective population size. Finally, a programme for the cryogenic conservation of embryos and semen has been established. The germplasm banks contain 366 doses of semen from 17 males and 153 embryos, obtained from 11 males y 35 females. Both banks include material from animals of different scrapie PRNP genotypes, to assure their long term availability. Concurrently, carcass and meat quality of CT light lambs has been studied and certain diversification alternatives for labelled lamb market have been prospected (lambs slaughtered at 9-12 kg and 28-35 kg) in order to promote this breed, showing high quality products and no commercial constraints in the lamb market.

Development of a sustainable conservation programme for the mountainous sheep breed of Katsika in Greece

Ligda, C.[1], Koutsotolis, I.[2], Kotsaftiki, M.[3] and Zotos, I.[4], [1]National Agricultural Research Foundation, P.O. Box 60458, 57001 Thessaloniki, Greece, [2]Assosiation of Pastoral Farmers, Dossiou 2, Ioannina, Greece, [3]Aristotle University of Thessaloniki, University Campus, 54 124 Thessaloniki, Greece, [4]Animal Genetic Improvement Centre, 28th October, 32, Ioannina, Greece; chligda@otenet.gr

The paper describes the *in situ* conservation program for the Katsika sheep breed that is raised in the mountains of Epirus in the West part of Greece. The total population that reach the 1,865 heads is raised by 6 farmers, who are supported by the agro environmental measures, as the breed is included in the 'Program of the breeds that are endangered'. The farming system of the breed is based on grazing. In the majority of the flocks transhumance is practiced, with traditional installations in the mountains. The breed is very well adapted in the particular environment and has sufficient production, as the average milk yield is higher than the performance of other local mountain breeds. In addition, the establishment of the Association of the Pastoral Farmers that aims towards the improvement of the status of living of their members by promoting the breed and its products contributes to the development of the breed. One of their main priorities is the development of a system of valorization of the products connected with the breed. The objective of the Association is to include in the network all the involved bodies, including the local industry in order to have a separated production. The Katsika breed, due to its adaptation in the specific environment and its production system, could be a successful model for the development of a sustainable breeding program aiming to the recognition of the breed contribution to the environment, the diversification of the production and the products typicality.

The Olkuska sheep: a case study of a successful conservation programme
Martyniuk, E., Warsaw University of Life Sciences, Department of Genetics and Animal Breeding, ul Ciszewskiego 8, 02-786 Warszawa, Poland; elzbieta_martyniuk@sggw.pl

The Olkuska sheep is the native Polish long-wool breed, traditionally kept in the southern region of the country, near Olkusz and Cracow. The breed was created on the basis of the local primitive long-wool stock, upgraded with Pomeranian sheep since the early 1930s; after the World War II, the Romney Marsh rams were introduced to improve wool yield and wool quality. The main feature of the breed is its exceptional prolificacy, well over 200%, with the wide range of litter size, from 1 to 7, resulting from the segregation of a putative major gene increasing ovulation rate. The high litter size of Olkuska sheep is combined with early pubescence, and a very good maternal ability based on the high milk yield. Traditionally Olkuska sheep were kept on small private farms, in flocks of a few heads only. In the 1950s Olkuska sheep population accounted for 10,000 heads; since 1960s, the number kept decreasing, reaching only about 200 heads in the early 1980s. Over the next 10 years the population size dropped to about 100 ewes; in this critical time flocks belonging to Cracow and Warsaw Agricultural Universities were the ones that ensured the survival of this breed. The introduction of conservation measures for endangered local livestock breeds within agri-environmental programme in 2005 resulted in a renewed interest in native breeds and establishment of new flocks of Olkuska sheep. The population has been increasing rapidly, from 174 ewes (11 flocks) in 2004 through 259 ewes (15) in 2007, reaching 456 ewes (29) in 2008. As the population of Olkuska sheep went through the prolonged bottle-neck period, there is a need to control inbreeding level and provide breeders with sound and timely advice on exchange of the breeding stock. The current task is also to ensure that Olkuska sheep will find its place in the production system and that the breeders will be able to maintain profitability of its utilisation if support system is not in available any longer.

Production of organic ewe milk with an authoctonous sheep breed (Lacha) of Spain
Eguinoa, P., Saez, J.L. and Maeztu, F., Instituto Técnico y Gestión Ganadero S.A., Avda. Serapio Huici, 22, 31610 Villava, Spain; peguinoa@itgganadero.com

The diversity of animal production systems is currently an added value and not a development constraint like in previous decades. In this sense, the present study tries to advance in the knowledge of a sustainable organic production system with dairy sheep on the basis of the utilization of autochthonous breeds, such as Lacha, in grazing conditions. During two campaigns (March to July) raw milk quality during four management periods has been analysed from a 300-sheep flock from an organic farm in the Navarre Pyrenees (900 m of altitude). Four different feeding management periods have been identified: (1) start of lactation, with ewes fed hay and concentrate respecting 60/40 ratio of the regulation on organic production; (2) turnout to pastures in April, with concentrate supplement; (3) only grazing; (4) end of lactation in July. The results obtained allow to conclude that the physicochemical quality of the milk changes along lactation and with the diet type, but it is never lower than that of conventionally managed Lacha flocks. Microbiological quality has been ideal in all the periods. In conclusion, a good management of the natural resources (either to be preserved or grazed) is very neccesary in an organic production system, and it allows to obtain high quality products.

Investigation of the physicochemical characteristics and identification of the fatty acids and volatile flavour compounds in ewes' milk of the indigenous sheep breeds Boutsiko and Karamaniko (Epirus, Greece)

Tzora, A.[1], Fotou, K.[1], Anastasiou, I.[1], Giannouli, A.[1], Metsios, A.[2], Kantas, D.[3], D'Alessandro, A.G.[4], Giannenas, I.[5], Goulas, P.[3] and Tsinas, A.[1], [1]Tei Of Epirus, Animal Production, Kostakioi, 47100, Arta, Greece, [2]University Of Ioannina, Faculty Of Medicine, 45110, Ioannina, Greece, [3]Tei Of Larisa, Animal Production, 41110, Larisa, Greece, [4]University Of Bari, PRO.GE.S.A., 70126 Bari, Italy, [5]University Of Thessaly, Faculty Of Veterinary Medicine, Trikalon 224, 43100 Karditsa, Greece; tzora@teiep.gr

This study aimed to determine and evaluate the physicochemical characteristics, the fatty acids and the volatile flavour compounds of the milk yield from ewes belonging to the indigenous, low performance yet well locally adapted breeds Boutsiko and Karamaniko. A total of 270 milk samples were analyzed, 190 samples originated from ewes of Boutsiko breed and 80 from ewes of Karamaniko, representing the 11% and 5.3% of the total breed population respectively. Physicochemical analysis included proteins, fat, lactose, total solids and total solids without fat, using the Milkoscan 120, and pH. Fatty acids and volatile flavour compounds were identified by GC/MS and SPME-GC/MS respectively. The results showed that: a) the values of the physicochemical parameters of the milk of the two indigenous breeds were significantly higher than those of other high performance domestic breeds. b) Protein, fat and total solids content determined in Karamaniko breed's milk, were higher than in the Boutsiko breed, while lactose and total solids without fat content were lower. c) The pH values in both breeds milk were similar and within the normal standards. d) The matrix of some free fatty acids and volatile flavour compounds proved to be different among the two breeds raised locally. Conclusively, the results indicate that milk originating from animals belonging to these indigenous breeds acquire special quality characteristics and may contribute to the improvement of local dairy products.

***Ex vivo* effects of goat milk from indigenous breeds and its components linoleic and linolenic acids on human platelet aggregation**

Metsios, A.[1], Skoufos, I.[2], D'Alessandro, A.G.[3], Verginadis, I.[1], Simos, I.[1], Voidarou, C.[2], Bezirtzoglou, E.[4], Arsenos, G.[5], Galamatis, D.[5] and Karkabounas, S.[1], [1]University Of Ioannina, Faculty Of Medicine, 45110 Ioannina, Greece, [2]TEI Of Epirus, Animal Production, Kostakioi Artas, 47100 Arta, Greece, [3]University Of Bari, PRO.GE.S.A., Via Amendola 165, 70126, Bari, Italy, [4]Democritus University Of Thrace, Food Science And Technology, 68200 Orestiada, Greece, [5]Aristotelian University Of Thessaloniki, Veterinary Medicine, 54124 Thessaloniki, Greece; jskoufos@teiep.gr

The objective was to evaluate the ex-vivo properties of goat's and cow's milk by comparing the effects of their component linoleic and linolenic acids on human platelet aggregation induced by two aggregation agonists (ADP, PAF) and LMS cells. Milk samples from different goat breeds (Capra Prisca, Capra Ionica, and Saanen) were used. Those goats were reared under semi-extensive husbandry conditions. Platelets were prepared from blood samples that were obtained from healthy humans. Plasma Rich in Platelets (PRP), obtained by centrifugation, was kept in water bath until the termination of the aggregation test. Assays for platelet aggregation were conducted by an aggricometer calibrated on Plasma Poor in Platelets (PPP), obtained after a second centrifugation. Samples of diluted milk were added to PRP and left for incubation for few minutes. Thereafter, the percentage of platelet aggregation was observed by adding each of the three platelet agonists. Overall, both types of milk exhibited ex-vivo inhibitory effects on platelet aggregation. Milk from indigenous Greek goats showed the strongest response, causing a complete inhibition of platelet aggregation at the dose of 10 µl/0.5 ml PRP. Linolenic acid showed a strong dose-dependent inhibitory effect on platelet aggregation, while linoleic acid had no effect at all. The results indicate that the milk of indigenous Greek and Italian goats, used in the present study, has nutraceutical properties. The later is interesting concerning human health issues.

Evaluation of Florina (Pelagonia) sheep breed for milk yield, partitioning and composition

Skapetas, B.[1], Bampidis, V.A.[1], Christodoulou, V.[2], Katanos, I.[1], Laga, V.[1] and Nitas, D.[1], [1]Alexander Technological Educational Institute, Department of Animal Production, P.O. Box 141, 57400 Thessaloniki, Greece, [2]National Agricultural Research Foundation, Animal Research Institute, Giannitsa, 58100 Giannitsa, Greece; bampidis@ap.teithe.gr

In an experiment with 48 Florina (Pelagonia) breed ewes (16 of the first, 16 of the second and 16 of the third and subsequent lactations), effects of lactation stage, parity and milking hour on milk yield, partitioning and composition were determined. The experiment started on 60±5 days postpartum and lasted 24 weeks. Twelve ewes had twin type births, while the others had single type births. Ewes were milked twice daily, at 8:00 and 16:00 h, in a 1×24 side by side milking parlour of Casse type with 12 milking units and a low milk line and air pipeline. The functional characteristics of milking machine were: vacuum level 38 kPa, pulsation rate 120 pulsations/min and pulsation ratio 50:50. Milk yield and milk fractions (total machine milk and hand stripping milk) were recorded twice daily every 4 weeks. Milk composition was examined for the morning and afternoon pooled milk samples. Results showed that total machine milk was 684±18 ml (82.69% of milk yield), whilst hand stripping milk was 143±5 ml (17.31% of milk yield). The effect of lactation stage and parity were significant ($P<0.05$) for both milk fractions. Milking hour did not significantly influence the total machine milk. In the whole experimental period, total machine milk contained 7.12±0.07% fat, 6.09±0.03% protein, and percentage 4.79±0.03% lactose, while hand stripping milk contained 8.36±0.08% fat, 5.98±0.03% protein, and 4.74±0.02% lactose. In both fractions, stage of lactation and parity significantly influenced ($P<0.05$) fat, protein and lactose contents. It's concluded that Florina (Pelagonia) breed, as a local breed, has a satisfactory milk yield and composition, but a relatively unsatisfactory machine milking ability.

Evaluation of Florina (Pelagonia) sheep breed for meat quality

Bampidis, V.A.[1], Christodoulou, V.[2], Ambrosiadis, J.[3], Arkoudelos, J.[4], Labrinea, E.[4] and Sossidou, E.[2], [1]Alexander Technological Educational Institute, Department of Animal Production, P.O. Box 141, 57400 Thessaloniki, Greece, [2]National Agricultural Research Foundation, Animal Research Institute, Giannitsa, 58100 Giannitsa, Greece, [3]Aristotle University, School of Veterinary Medicine, Thessaloniki, 54006 Thessaloniki, Greece, [4]National Agricultural Research Foundation, Institute of Technology of Agricultural Products, Athens, 14123 Athens, Greece; vchristodoulou.arig@nagref.gr

In an experiment with 40 male and 40 female growing Florina (Pelagonia) lambs, effects of slaughter weight on meat quality were determined. The lambs were allocated at random to four equal treatment-groups (42d, 30BW, 45BW and 60BW) of 10 males and 10 females each, and assigned for slaughter at 42 day (weaning), 30, 45 and 60% of the mature body weight (BW), respectively. The 60 lambs in treatments 30BW, 45BW and 60BW were grouped in 6 pens, 10 each, separating males from females, and fed a high concentrate diet. After slaughter, samples of the longissimus lumborum et thoracis muscle from the right half of the carcass of all lambs were removed for chemical composition analysis, color evaluation, sensory evaluation, and fatty acid (FA) and cholesterol analysis. Results showed that degree of maturity affected ($P<0.05$) slaughter weight, as well as all meat quality characteristics, except protein and ash percentage, yellowness and hue values, and C15:0, n3 fatty acids and the ratio n6/n3 which did not differ. Florina lambs can be slaughtered in heavier carcass weights, heavier than the 15 kg as the present practice. The heavier carcasses can appeal to wider range of consumers due to the excellent muscle development and the reduced visible fat cover. Florina sheep raised for heavy carcass meat production may be more economically beneficial for the breeders and the increased meat produced locally can help reduce need for importing meat from abroad.

Characterization of carcass and meat traits in 'Serpentina' goat kids

Bressan, M.C.[1], Jerónimo, E.[1], Cachatra, A.[2], Babo, H.[2], Saraiva, V.[2] and Santos-Silva, J.[1], [1]INRB, Fonte Boa-Santarém, 2005-048, Portugal, [2]Assoc. Portuguesa de Caprinicultores da Raça Serpentina, R. Diana Liz ÉVORA, 7005-413, Portugal; mcbressan1@hotmail.com

The development of native livestock breeds depends on the acceptance of their products, and the recognition and certification of meats under a PDO label can represent an important contribution to the maintenance of autochthonous breeds. The Serpentina is a native breed of goats raised under extensive conditions in the Alentejo (southern Portugal) and its products are traditionally consumed, either as carcasses or in different cuts. In order to characterize carcass and meat traits, 12 kids (Serpentina breed) were slaughtered and carcasses were evaluated at 48h post mortem. The half-carcasses had a mean weight of 2648.2g, with yields (%) of lean, bone, subcutaneous fat and intermuscular fat of 64.48, 25.62, 3.89 and 6.00, respectively, with a lean:bone ratio of 2.53. The mean weight for cuts was 33.1% for the leg, 21.2% for the shoulder, 15.9% for the ribs, 11.4% for the flank, 9.7% for the neck and 8.81% for the loin. Meat characteristics were assessed in the L. dorsi (LD) and semimembranous (SM) muscles. The shear force were similar in LD (6.01 kg) and SM (6.54 kg). The same pattern was observed for water holding capacity (29.6 vs. 29.0%, respectively). The SM had a lighter (L*=49.0) and redder (a*=12.1) colour than the LD (L*=46.0, a*=10.9), but b* was similar in the two muscles. The external surface of the carcass had a lighter colour, with higher redness and lower yellowness than the inner surface. A positive association was found between the degree of L* (r=0.76) and a* (r=0.75) in the two muscles, but no significant associations were found between b*, water holding capacity and shear force in the LD and SM. Serpentina kids offer a high quality product, with high lean meat yield, light pink meat and shear force values similar to products obtained from other Portuguese native breeds, such as Serrana (5.5 kg carcass weight) or Bravia (4 kg carcass weight).

Performance of the Ripollesa sheep breed under current production systems in Catalonia

Caja, G.[1], Bach, R.[2], Piedrafita, J.[1], Rufí, J.[2] and Casellas, J.[3], [1]Universitat Autònoma de Barcelona, G2R, Campus de la UAB, 08193 Bellaterra, Spain, [2]ANCRI, Finca Camps i Armet, 17121 Monells, Spain, [3]IRTA, Genetica i Millora Animal, Av. Alcalde Rovira i Roure 191, 25198 Lleida, Spain; rabanet@grn.es

Ripollesa is a local sheep breed, characterized by spotted face and legs (black or brown), and intended for lamb ('xai lletó', 11 kg BW; and 'xai de ramat', 25 kg BW) and wool (medium quality) production. It is exploited in the NE and central Catalonia (Spain). Current production systems are semi-extensive (grazing during spring and autumn) and semi-intensive (fed with hay and concentrate). Average flock size for 31 farms totalling 14,492 sheep was 468 sheep, 70.5% of which were registered as pure breed in the ANCRI (Associació Nacional de Criadors d'Ovins de Raça Ripollesa) flock book. Flock book contained data of 9,556 ewes and rams which were genotyped for scrapie resistance (Programa Aries) and tagged with 2 ear tags and 1 electronic bolus. Lambing data was recorded using either a notebook or an electronic reader by the farmer itself and uploaded to a data base program (AncriData). A total of 35,034 lambing records, collected during the last 6 yr from 25,524 ewes, were processed to evaluate and study the evolution trends of breed performance. On average, annual lambing frequency was 1.32±0.01 (277±3 d lambing interval) and prolificacy was 1.26±0.01 lambs/ewe. Prolificacy trend was +0.063 lambs in 10 yr, needing an effective increase in most flocks. Lamb mortality averaged 13.6±0.5% (perinatal, 5.9%) and birth weigh 3.8±0.1 kg BW. Culling rate was 15.1±1.0%. Annual productivity (lambs sold per ewe present) was 1.31±0.04 lambs/ewe, being the harvesting weight 24.1±0.10 kg BW, on average. Scrapie genotyping for ARR homozygote and heterozygote showed frequencies of 13.4 and 43.1%, respectively. Current genetic selection for breed improvement focus on prolificacy, average daily gain during the first 90 d and ARR genotype, being supported by research (UAB and IRTA) and autonomic government (DAR) institutions.

Meat quality traits of Apennine suckling-lambs traditionally reared in two different seasons in Abruzzo region, Italy

Mazzone, G., Lambertini, L., Giammarco, M., Angelozzi, G. and Vignola, G., University of Teramo, Food Science Dept., Viale Crispi 212, 64100 Teramo, Italy; gmazzone@unite.it

In Abruzzo region, as in other Regions of Central Italy, a semi-extensive farming system, mainly of Apennine breed or its crossbreeds, is still widely practiced to produce a traditional lamb slaughtered at about 60 days of age. The rearing system is quite well characterized, planning slaughtering around Easter and Christmas and envisaging the use of pasture from midmorning to evening till December for the sheep while lambs, reared permanently into the sheepfold, are fed at evening and night on their mothers milk, receiving a hay and concentrates supplement from 25-30 days to slaughter. Only during winter sheep receive meadow hay and some supplementary foods. To contribute to the characterization of lambs produced by this typical rearing system and to evaluate the influence of the breeding season on meat characteristics, 80 carcasses of Apennine male lambs, from single birth, 60 ± 3 days old, were purchased, half at Christmas and half at Easter, from the same slaughterhouse. On L. lumborum of each lamb pH1 and pHu and meat colour were measured, water losses were determined, chemical composition and muscle fatty acid profile were analysed. Lambs slaughtered at Easter had a similar carcass weight to those reared in autumn (9.4 kg vs 10.0 kg, $P=0.15$), but showed a pHu significantly higher (5.81 vs 5.70, $P<0.01$). Their meat lightness (L*) and cooking loss were significantly higher ($P<0.05$) while their fat content tended to be lower ($P=0.10$). The L. lumborum fatty acids profile evidenced a similar composition in total SFA while PUFA concentration was higher in the carcasses of lambs slaughtered at Christmas (20 32% vs 16.65%, $P<0.01$) with a more favourable PUFA/SFA ratio and a significant reduction of their n-6/n-3 ratio (2.60 vs. 3.98, $P<0.01$). These results are to relate to the typical farming system and to the larger amount of milk the lambs bred in Autumn received when the sheep milk production is stimulated by the availability of good pasture.

Effects of pre- and post-partum feeding system on the performance of autumn-lambing ewes raising suckling lambs

Joy, M., Alvarez-Rodriguez, J., Blasco, I., Ripoll, R. and Sanz, A., CITA de Aragón, Avda. Montañana, 930, 50.059, Zaragoza, Spain; mjoy@aragon.es

The aim of this study was to analyse the effects of the pre- and post-partum feeding system on milk production and composition of Churra Tensina ewes. A 2x2 factorial design was carried out. Two feeding systems were evaluated: permanent pasture grazing vs indoors grass hay-fed; and two periods: pre- partum (last third of pregnancy) vs post-partum period (5 weeks). Forty-nine autumn-lambing ewes were used (5.7 years old, 52.3 kg LW, 3.23 BCS). Ewes and lambs were weighed weekly. Milk production was estimated weekly by the oxitocin and machine milking technique (4-h interval). Milk composition was determined with infrared. Lambs were slaughtered as suckling lambs commercial category (11.2 kg LW). Diet did not affect either LW or BCS of ewes ($P>0.05$) in any period. No interactions were observed ($P>0.05$) except for lactose content ($P<0.05$). Pre-partum feeding system only affected the milk protein content (5.41 vs 5.10% for grass hay vs pasture grazing; $P<0.05$), and post-partum did not affect any studied parameter. In contrast, week of lactation influenced the standard milk production, fat content, fat yield and protein yield ($P<0.001$). Greater milk production was obtained on week 1 to 3 (1.20 kg/d), being standard milk yield peak on week 1 (1.39 l/d). Fat and protein content and their yields were greater on week 1 (9.64 and 5.92%, 115 and 70 g/d; $P<0.05$) than the rest of lactation. Ewes grazing during the pre-partum period showed greater lactose content on week 1 than those hay-fed indoors (4.95 vs 4.75%; $P<0.05$), but it was similar thereafter. Lamb LW at birth was similar across feeding systems (3.6 kg). ADG was lower and the slaughter age was slightly greater in lambs whose dams were hay-fed indoors during the post-partum period than in those that grazed green forage outdoors (220 vs 259 g/d and 36 vs 32 days for hay and pasture grazing; $P<0.05$). In conclusion, feeding system around parturition did not affect ewe performance but the offspring from hay-fed ewes showed a lower performance.

Quality products from the local Spanish Breeds under extensive conditions
Fernández, J.A. and Luque-Cuesta, M., Federación Española de Asociaciones de Ganado Selecto, Technical Services, c/Castelló, 45, 28001, Spain; feagas@feagas.es

The gene pool of Spanish local breeds is very rich in terms of quality and diversity, and is composed of 149 different livestock breeds: 38 bovine breeds, 43 ovine breeds, 22 caprine breeds, 11 porcine breeds, 20 equine breeds and 15 avian breeds. Livestock farming using local breeds is characterized by its link with a natural environment, which is specific to the breed, to the farmer, to a craft production system, etc. Such livestock farming systems are based on the exploitation of natural and local resources under extensive conditions: among others, meadows, mountains and valleys. The products obtained from the Spanish local breeds give extra quality to other related goods produced in Spain (meat, milk, cheese, leather). The future of the Spanish local breeds is based on different aspects that are connected with the very effective relationship between breeders and animals. Indeed, the variety of the livestock species and breeds are essential to achieve very important goals, such as avoiding the desertification of our territory. In order to prevent the extinction of these breeds it is necessary to increase their profitability through Protected Geographical Indication (PGI), Protected Designation of Origin (PDO) and Traditional Speciality Guaranteed (TSG) which add extra value to the product.

Yield of carcass and meat cuts in Paraíso Pedrês and Label Rouge chicken raised in semi-extensive system
Faria, P.B.[1], Bressan, M.C.[2], Souza, X.R.[1], Rossato, L.V.[3], Rodrigues, E.C.[3], Botega, L.M.G.[3], Cardoso, G.P.[3], Pereira, A.A.[3], Ramos, E.M.[3] and Gama, L.T.[2], [1]Instituto Federal de Educação, Ciência e Tecnologia, Santo Antonio do Leverger, MT, 78106-000, Brazil, [2]INRB, Fonte Boa-Santarém, 2005-048, Portugal, [3]UFLA, C.p.37, Lavras, MG, 37200-000, Brazil; peterbfvet@yahoo.com.br

An experiment was carried out to evaluate carcass and meat cut yields in male and female chicken of the Paraíso Pedrês and Label Rouge lines, raised in a semi-extensive system, and slaughtered at 65, 75, 85 and 95 days of age. Each combination of treatments (sex*line*age) had 3 replicates, with 4 birds per replicate. Traits evaluated were live weight at slaughter (LW), carcass weight (CW), carcass yield (CY) and yield of different cuts (breast, thigh, drumstick, back, neck, thigh and wings). Males had higher LW and CW than females ($P<0.05$) and the Paraíso Pedrês strain had higher ($P<0.05$) mean values than the Label Rouge strain for LW (2511.3 vs. 2146.3 g) and CW (1866.7 vs.1607.1 g). For CY, an interaction between strain, sex and slaughter age was observed ($P<0.05$), but CY was always higher in male than in female birds. Breast yield was higher ($P<0.05$) in the Paraíso Pedrês strain and in females, and it increased with slaughter age. Thigh yields were higher ($P<0.05$) in the Label Rouge (13.71%) than in the Paraíso Pedrês strain (13.37%), and in male (14.06%) than in female birds (13.01%). The drumstick yield was affected by the interaction between strain, sex and slaughter age ($P<0.05$), but the yield was higher in males. Wing yield was affected ($P<0.05$) by strain, with a higher mean in Label Rouge(11.66%) than in Paraíso Pedrês (11.30%), and reduction was observed in wing yield as slaughter age increased. Overall, male chickens of the Paraíso Pedrês strain reach higher LW and CW at 85 and 95 days and, in this strain, breast yield is higher than in Label Rouge. On the other hand, male chickens of the Label Rougestrain show higher yield of thigh and drumstick at all ages.

Physico-chemical characteristics of meat in chicken of the Paraíso Pedrês and Label Rouge lines
Faria, P.B.[1], Bressan, M.C.[2], Souza, X.R.[1], Rodrigues, E.C.[3], Rossato, L.V.[3], Botega, L.M.G.[3], Cardoso, G.P.[3], Pereira, A.A.[3], Ramos, E.M.[3] and Gama, L.T.[2], [1]Instituto Federal de Educação, Ciência e Tecnologia, Santo Antonio do Leverger MT, 78106-000, Brazil, [2]INRB, Fonte Boa-Santarém, 2005-048, Portugal, [3]UFLA, C.p.37 Lavras, MG, 37200-000, Brazil; peterbfvet@yahoo.com.br

An experiment was carried out to evaluate physico-chemical characteristics of meat from male and female chicken of the Paraíso Pedrês and Label Rouge lines, raised in a semi-extensive system, and slaughtered at 65, 75, 85 and 95 days of age. Each combination of treatments (sex*line*age) had 3 replicates, with 4 birds per replicate. Breast and leg cuts were deboned at 24 h post mortem and frozen at -18 °C. The determinations were made after thawing of the cuts at 4 °C for 24 h. Variables assessed in the breast and leg of experimental animals were humidity, protein, fat, ashes, colour (CIE L*a*b*), pH and cooking loss. The readings were taken at three different points on the inner surface of the cranial position of the pectoralis major muscle and three different points in the leg inside the fibular longus muscle. Fat content of the breast was higher ($P<0.05$) in Paraíso Pedrês females (0.86%), than in males of the Paraíso Pedrês (0.63%) or Label Rouge (0.57%) lines. In the leg, Label Rouge chicken (85 d), have higher ($P<0.05$) humidity and lower ($P<0.05$) fat, than Paraíso Pedrês, while Paraíso Pedrês females at 95 days have more fat ($P<0.05$) than males of the same age. The mean pH and cooking loss of the breast were similar between the strains. In the breast, the yellowness (b*) was higher ($P<0.05$) in Label Rouge (6.24), than in Paraíso Pedrês (5.41). In the leg, Label Rouge had higher yellowness at 95 (6.87), than at 85 days (5.90). Overall, leaner cuts are obtained from the breast, in males and in Label Rouge chicken. Meat with higher yellowness can be obtained from the breast, in females and in Label Rouge chicken.

The effect of genotypes and housing conditions on tenderness of chicken meat
Konrád, S.Z., Szücs, E. and Kovácsné Gaál, K., University of West Hungary, Vár 2, 9200 Mosonmagyaróvár, Hungary; Szucs.Endre@mkk.szie.hu

The aim of this study was to establish tenderness of chicken meat in relation to genotype and housing. Seven genotypes such as purebred local Hungarian Yellow chickens, Hungarian Yellow pullets crossed by S 77, Foxy Chick, Redbro, Hubbard Flex and Shaver Farm meat-type cocks and Ross 308 commercial broiler chickens were used in the study. Cross-bred hybrids and Hungarian Yellow chickens were reared under free-range conditions for 84 days. The rearing for the intensively fattened Ross 308 broilers lasted for 42 days. Textural characteristics of samples were taken from three locations of breast and analysed for palatability by Stevens QTS 25 penetrometer using the non-destructive TPA method for evaluation of hardness, gumminess and chewiness values. The hardness, gumminess and chewiness values of breast meat were higher for free-range chicken than the broilers reared intensively. For Ross 308 chicken reared under free-range conditions vs. intensively reared the hardness values were 865.5 grams and 209.3 grams, respectively. Values for gumminess and chewiness were in the same order: 353.4 and 151.8 grams, vs. 1154.7 and 476.7 units, respectively. For all three parameters the influence of genotypes and housing conditions was statistically significant.

Comparison of egg weight between two quail strains
Vali, N., Animal science, Islamic Azad University Shahrekord Branch, 8813733395 Shahrekord, Iran;
nasrollah.vali@gmail.com

In order to investigate egg weight of two quail strains 2,550 eggs of Japanese quail (Coturnix japanese) and 1975 eggs of Range quail (Coturnix ypisilophorus) were weighed individually at three age groups (first group: 60-145, second group: 145-230, and last group: 300-385 days of age). Body weights of two strains were not significantly different ($P>0.05$). Body weights at 60 days age were significantly different ($P<0.01$), but there were not any significant difference between the ages of 145 and 300 days ($P>0.05$). Egg weight of Japanese quails and Range quails were 11.23 ± 0.03 and 11.17 ± 0.05 respectively which were not significantly different ($P>0.05$). Effects of the interaction of strain, age and sex for egg weight and body weight were significantly different ($P<0.01$). Minimum and maximum egg weight for Japanese quail were 7.08 and 13.84gr respectively, however these records were 7.01 and 13.84gr for Range quail. Individual variation in egg weight of two strains was significant ($P<0.01$).

Amino acids composition and their variations during lactation in donkey milk of Martina Franca breed
D'alessandro, A.G., Martemucci, G. and Trani, A., University of Bari, PRO.GE.S.A., Via G. Amendola, 165/A, 70126 BARI, Italy; dalex@agr.uniba.it

In recent years, local breeds have been the subject of numerous studies directed towards the characterization of their products. In Italy, Martina Franca donkey is a local breed at risk of extinction, but donkey milk have attracted considerable interest in its use for human consumption, especially for children intolerant to cow's milk, for most complicated cases of food multiple intolerance, and for elderly people. The knowledge of nutritional, functional and nutraceutical characteristics of its milk can contribute to productive and economic exploitation of this breed and thus to its preservation. In this study amino acids composition and their variations during lactation were quantified in donkey's milk of Martina Franca breed. Milk samples were collected once a month throughout the lactation period of 7 months from fourteen healthy jennies, reared in South Italy under semi-extensive conditions and subjected to a dally machine milking. Milk samples were analyzed by a CG/FID methodology for determination of amino acids content. The results indicate that amino acids composition of donkey milk changes during lactation ($P<0.01$). The average of total amino acids ranged from 92.5 mg/l (3rd month) to 164.9 mg/l (5th month; $0.05>P<0.01$). Within the nutritionally essential amino acids, threonine was the most abundant one (60.5 to 119.8 mg/l), showing the peak on 5th-6th month of lactation, while triptofhan, methionine and phenylalanine were the least abundant. Considering the non-essential amino acids, the most abundant were serine (172.0 to 237.8 mg/l) and glutamic acid (151.6 to 196.3 mg/l), with the highest value ($0.05>P<0.01$) at 3rd and 5th month of lactation, respectively. In conclusion, the study contributes to the knowledge of the nutritional value of donkey's milk in Martina Franca breed.

Fatty acid profiles in alligator (Caiman yacare) meat from animals raised in the wild or in captivity

Vicente-Neto, J.[1], Santana, M.T.[1], Bressan, M.C.[2], Faria, P.B.[3], Rodrigues, E.C.[3], Kloster, M.[1], Ferrão, S.B.P.[3], Andrade, P.L.[3], Vieira, J.O.[3] and Gama, L.T.[2], [1]Escola Agrotécnica Federal, C.p.244 Cáceres-MT, 78200-000, Brazil, [2]INRB, Fonte Boa-Santarém, 2005-048, Portugal, [3]UFLA, C.p.37-Lavras-MG, 37200-000, Brazil; mcbressan1@hotmail.com

An experiment was conducted to evaluate fatty acid (FA) in Jacaré do Pantanal meat originating from animals raised in the wild (WA, n=6) or in captivity (CA, n=6). Alligators had a live weight of 5.93-6.78kg at slaughter, and all experimental methods were approved by the IBAMA. Samples were collected from the tail (ílio-ischio-caudalis muscle) and neck (occipito-cervicalis medialis muscle) cuts. Fat content was influenced by raising system ($P<0.05$), with a higher amount of fat in WA (2.98%) than in CA (0.66%) animals. In both systems, the tail cut had a higher amount of fat (3.13%) when compared to the neck (0.51%). The CA animals had higher ($P<0.05$) levels than WA of C18:0 (13.03 vs. 10.20%) and C18:1cis-9 (33.21 vs. 27.61%), but lower ($P<0.05$) levels of C16:1cis-9 (3.61 vs. 4.81%), C18:3n-3 (0.96 vs. 2.63%), C22:4n-6 (0.78 vs. 2.80%) and C22:6n-3 (0.49 vs. 0.98%). The neck and tail cuts differed ($P<0.05$) in their mean content of C14:0 (2.12 vs. 1.15%), C16:1cis-9 (4.91 vs. 3.5%), C18:1cis-9 (35.06 vs. 25.75%), C18:2n-6 (10.24 vs. 15.24%), C18:3n-3 (2.06 vs. 1.53%), C20:4n-6 (6.71 vs. 10.80%), C20:5n-6 (0.28 vs. 1.27%), C22:4n-6 (1.06 vs. 2.52%) and C22:6n-3 (1.18 vs. 0.25%). The total amount of saturated and monounsaturated FA was similar among systems and cuts. However, samples from WA had a higher amount ($P<0.05$) than CA for n-3 FA (3.61 vs. 1.43%) and total polyunsaturated FA (31.04 vs. 23.62%). Furthermore, the neck had more total n-3 FA (3.25%) and polyunsaturated FA (32.59%), than the tail (1.79 and 22.07%, respectively). Meat from alligators raised in the wild, even though it has a higher amount of fat, is richer in n-3 FA and total polyunsaturated FA. Regardless of the raising system, the neck cut is leaner and healthier for consumers than the tail cut.

Comparison of feedlot performance between dromedary and Bactrian×dromedary camels: 1- After a 5-month fattening duration

Asadzadeh, N., Sadeghipanah, H., Sarhadi, F., Babaei, M., Banabazi, M.H. and Khaki, M., Animal Science Research Institute of Iran, Animal Production and Management, Shahid Beheshti Street, 3146618361 Karaj, Iran; hassansadeghipanah@yahoo.com

The objective of this study was comparing feedlot performance of dromedary and Bactrian×dromedary hybrid camels after a 5-month fattening duration. 14 Iranian Kalkoohi dromedary and 14 Bactrian×dromedary camels (12-month-old; 7 male and 7 female in each group) were used in a 2×2 factorial experiment. Main effects were level of genotype (dromedary or hybrid) and sex (male or female). Camels in individual pens were fed a diet containing 25% alfalfa, 25% wheat straw, 50% concentrate as a total mixed ration (TMR) and ad libitum. To determine feed intake, daily feed offered and refusals were weighed. Camels were individually weighed, after water and feed 12-h-restriction, at 30-d intervals. Statistical analysis carried out using general linear models of the SPSS. Interaction effect genotype×sex on any assessed characteristics were not significant ($P>0.05$). Final live weight in hybrids (381.31±12.85 kg) were significantly ($P=0.01$) higher than in dromedaries (332.71±8.25 kg) and in males (379.50±11.07 kg) were significantly ($P=0.01$) higher than in females (330.92±10.09 kg). Average daily gain had a tendency ($P=0.07$) to be higher in hybrids (684.6±47.48 g) in comparison with dromedaries (532.1±59.17 g), but there was no significant difference between males (654.8±59.28 g) and females (552.6±52.83 g). Feed conversion ratio in hybrids (7.80±0.62) were better than in dromedaries (9.62±0.96) and in males (8.45±0.84) were better than in females (9.06±0.86), but nonsignificantly ($P>0.05$). These results suggest that during a 5-month fattening period from 12 until 17-month-old, hybrid camels with Bactrian sires and Kalkoohi dromedary dams in comparison with Kalkoohi dromedary camels, and also males in comparison with females have a better feedlot performance.

Comparison of feedlot performance between dromedary and Bactrian×dromedary camels: 2- After a 8-month fattening duration

Sadeghipanah, H., Asadzadeh, N., Sarhadi, F., Banabazi, M.H., Babaei, M. and Khaki, M., Animal Science Research Institute of Iran, Animal Production and Management, Shahid Beheshti Street, 3146618361 Karaj, Iran; hassansadeghipanah@yahoo.com

This study was conducted to compare feedlot performance of dromedary and Bactrian×dromedary hybrid camels after a 8-month fattening duration. 8 Iranian Kalkoohi dromedary and 8 Bactrian×dromedary camels (12-month-old; 4 male and 4 female in each group) were used in a 2×2 factorial experiment. Main effects were level of genotype (dromedary or hybrid) and sex (male or female). Camels in individual pens were fed a diet containing 25% alfalfa, 25% wheat straw, 50% concentrate as a TMR and ad libitum. To determine feed intake, daily feed offered and refusals were weighed. Camels were individually weighed, after water and feed 12-h-restriction, at 30-d intervals. Statistical analysis carried out using general linear models of the SPSS. Interaction effect genotype×sex on any assessed characteristics were not significant ($P>0.05$). Final live weight in hybrids (415.57±16.61 kg) and males (423.86±11.10 kg) were significantly ($P<0.05$) higher than in dromedaries (368.00±12.36 kg) and females (359.71±12.32 kg), respectively. There was no significant ($P>0.05$) difference in average daily gain (ADG) and feed conversion ratio (FCR) between hybrids (345.2±37.16 g and 16.50±2.38) and dromedaries (412.4±96.53 g and 14.33±2.43) and between males (431.0±62.30 g and 13.29±1.84) and females (326.7±79.28 g and 17.54±2.67). These results show that during a 8-month fattening period from 12 until 20-month-old, feedlot performances of hybrid camels with Bactrian sires and Kalkoohi dromedary dams in comparison with Kalkoohi dromedary camels, and also feedlot performances of male in comparison with female camels were not significantly different. Finally, ADG and FCR in last months of a 8-month-fattening period become undesirable; therefore, it is seems that a 8-month-fattening period is too long and uneconomical for both genotype and both sexes.

Sensory meat quality of some bovinae species (beef, buffalo and yak) aged for 2 or 8 days

Sañudo, C.[1], Campo, M.M.[1], Muela, E.[1], Ficco, A.[2] and Failla, S.[2], [1]University of Zaragoza, Animal Production and Food Science Department, Miguel Servet 177, 50013-Zaragoza, Spain, [2]CRA, PCM, Monterrotondo, Italy; csanudo@unizar.es

Meat market requires alternatives to reach some niches of consumers. Several species are not well known although their breeding and production could be a good choice in some areas. In several European countries yak and buffalo have been introduced, representing a valuable income to local producers, especially when the quality and specificity of their products, such as mozzarella from buffalo milk, has been enhanced. The aim of the present study was to compare the sensory quality of bovine meat from some domestic species that are reared across Europe. Fifteen animals composed of five of the following species: yak (Bos grunniens), buffalo (Bubalus bubalus) and Maremmana breed (Bos taurus) were reared in the same farm (Gran Sasso Mountain, Italy), using alfalfa hay ad libitum and 1 kg for 100 kg of live weight of concentrate and slaughtered at different ages, when they reached the commercial maturity (18, 14 and 16 months respectively for yak, buffalo and beef). Some Longissimus dorsi slices were obtained 24 hours after slaughtering, vacuum packaged and kept at -18 °C until analysis. A nine-member trained panel assessed odour, textural and flavour descriptors, as well as overall acceptability in meat aged for 2 or 8 days. None significant interactions Specie x Ageing were found. Specie effect was only significant on beef odour intensity, having buffalo meat lower score than yak or Maremmana breed. Ageing effect was significant on most textural parameters and on beef and liver flavor intensities, as well as on overall acceptability, being better considered the most aged meat. We could conclude that ageing is much more important than species effect. However a distinctive quality should be investigated in Yak and Buffalo meat (comparing with other breeds of Bos taurus, rearing end points, productive systems or technological procedures) to be able to enhance meat peculiarities besides the processing effect.

Water buffaloe farming systems in west azarbaijan

Mirza Aghazadeh, A., Urmia University, Animal Science, Urmia, 57153-165, Iran; a.aghazadeh@urmia. ac.ir

The water buffalo is a native animal of Iran, with over 80 percent of its population concentrated in the North West (Azerbaijan) and 18 % in the south of the country. Official neglect and pro-Holstein propaganda have caused a severe reduction in buffalo numbers in Iran in recent years. The design for this study was descriptive research. Qualitative and quantitative data were collected from 96 farms in order to gain a deeper understanding of the characteristics of buffalo production system and the issues and constraints faced by the buffalo farmers. Priority buffalo production identified in Azerbaijan was buffalo production integrated with fodder and cash crop production (61% irrigated land and 39% dry land). Sixty one and half percent of farmers cultivating less than 5.0 hectares. Small farms devote 32.9% of area for fodder production. Small holding farms manage their animals according to the opportunities offered by the environment, pasture, stubble, shrubs and grass. 91.7% of the farmers owned adult buffalo, ranging from 1 to 9 head (an average of 2.61 buffalo per house-hold). 5.2% of farmers hold buffalo, crossbred and native cattle together. The average number of cows per farm was 0.82. Data regarding cropping pattern and animal ownership showed that the number of large animals per household and per hectare of wheat were 9.0 and 2.63. Big farmers own more animals, but given the fact that most fodder is provided by specialty fodder crops and crop residue, a more relevant measure is the number of animal equivalents per hectare of cultivated land, which on small farms is more than on large farms. It can be concluded that buffalo is an integral part of sustainable agricultural production system in W-Azerbaijan. This system is based on the low cost agricultural by-products as nutritional input to animals for producing quality food of high biological value. The concerted efforts are needed to conserve and improve this system by undertaking proper research programmes at farm level.

Hormonal and neurotransmitter mechanisms regulating feed intake

Sartin, J.L.[1], Wilborn, R.[1], Whitlock, B.[2] and Daniel, J.[3], [1]Auburn Univ, Anatomy, Physiology, Pharmacol & Clin Sci, Coll Vet Med, Auburn, AL 36849, USA, [2]Univ Tennessee, Large Anim Clin Sci, CVM, Knoxville, TN, USA, [3]Berry Coll, Anim Sci, Coll Ag, Mt Berry, GA, USA; sartijl@auburn.edu

Appetite control is a major issue in normal growth and in suboptimal growth performance settings. A number of hormones selectively activate or inhibit key neurotransmitters within the arcuate nucleus of the hypothalamus, where feed intake is regulated. Examples of appetite regulatory neurotransmitters are the stimulatory neurotransmitters neuropeptide Y, agouti-related protein (AgRP), orexin and melanin concentrating hormone, and the inhibitory neurotransmitter, melanocyte stimulating hormone (MSH). Examination of messenger RNA (using *in situ* hybridization and real time PCR) and proteins (using immunohistochemistry) for these neurotransmitters in ruminants has indicated that physiological regulation occurs for several of these critical genes and proteins, especially AgRP. Moreover, intracerebroventricular injection of each of the four stimulatory neurotransmitters can increase feed intake in sheep and may also regulate either growth hormone, luteinizing hormone, cortisol or other hormones. Development of melanocortin-4 receptor antagonists (MC4R; mediates feed intake inhibition by MSH) provides an opportunity to increase feed intake by reversing the appetite inhibitory effects of MSH. Utilizing endotoxemia as a tool to inhibit appetite provides a mechanism to test the functionality of these antagonists. The natural MC4R antagonist, AgRP, can prevent the inhibition of feed intake after injection of endotoxin. Synthetic antagonists of the MC4R increase feed intake in normal animals and may prevent the effects of disease to reduce feed intake. Thus, knowledge of the mechanisms regulating feed intake in the hypothalamus may lead to mechanisms to increase feed intake in normal growing animals and prevent the wasting effects of severe disease in animals.

Body temperature and heart rate increase with oxygen consumption in sheep although the response differs with level of feed intake
Robertson, M.W., Dunshea, F.R. and Leury, B.J., The University of Melbourne, School of Land and Environment, Royal Parade, Parkville, Victoria 3010, Australia; fdunshea@unimelb.edu.au

A human exercise physiology system (ADinstruments Pty Ltd, Sydney) with modifications was used to obtain acute, real time measurements of gas exchange in sheep fed at different levels of feed intake. Three Merino lambs (28.8±0.8 kg) and 3 crossbred lambs (30.2±1.0kg) were allocated to a 72 h fast, maintenance or *ad libitum* diet, in a 2x3x3 (breed x treatment x time) Latin square design. Gas exchange was measured at 07:00, 08:00, (animals fed at 09:00, no measurement) 10:00, 11:00, 12:00, 14:00, 16:00, 20:00, 01:00 and 06:00 h, for a period of 15 min. Heart rate and core temperature were also measured at 3 times. Prior to feeding average O_2 consumption did not differ between the planes of nutrition (8.1 vs 8.7 and 11.9 mL/min/kg$^{0.75}$ for fasted, maintenance and *ad libitum* fed lambs, respectively, s.e.d=3.03). Oxygen consumption in the fasted animals did not change throughout the 24 h period. In maintenance fed lambs O_2 consumption increased ($P<0.05$) to 192% and peaked at 197% of pre feeding levels, 3 h and 5h respectively, post feeding. Ad libitum post-feeding O_2 consumption increased ($P<0.05$) 73% from pre-feed levels within 1 h and O_2 remained elevated ($P<0.05$), peaking 7 h after feeding at 207% above pre-feeding levels. Core temperature increased with increasing O_2 consumption in a similar manner (0.0124/ ml/min/kg$^{0.75}$) although the intercepts were different (38.9 vs 39.3 and 39.7, $P<0.001$, R^2=0.627). On the other hand, heart rate increased with increasing O_2 consumption in fed sheep but not in fasted sheep. In conclusion the PowerLab exercise physiology system is an excellent tool for measuring differences in O_2 in response to various planes of nutrition or physiological interventions in small ruminants. Also, body temperature and heart rate both increase with oxygen consumption in sheep although the response differs with level of feed intake.

Amino acids and insulin as regulators of muscle protein synthesis in neonatal pigs
Davis, T.A., Suryawan, A., Orellana, R.A., Fiorotto, M.L. and Burrin, D.G., USDA/ARS Baylor College of Medicine, Childrens Nutrition Research Center, Pediatrics, Houston, Texas 77030-2600, USA; tdavis@bcm.edu

During the neonatal period, the rate of growth is higher than at any other stage of postnatal life and a majority of the mass increase is skeletal muscle. The rapid growth of skeletal muscle in the neonate is driven by an elevated rate of protein synthesis. Feeding profoundly stimulates muscle protein synthesis in neonates and the response decreases with age. The feeding-induced stimulation of muscle protein synthesis is modulated by enhanced sensitivity to the postprandial rise in insulin and amino acids. The high rate of protein synthesis in neonatal muscle is in part due to enhanced activation of the insulin signaling pathway. Thus, the feeding-induced activation of the insulin receptor, insulin receptor substrate 1/2, phosphatidylinositol 3-kinase, phosphoinositide-dependent kinase 1, protein kinase B, mammalian target of rapamycin (mTOR), ribosomal protein S6 kinase-1 (S6K1), eukaryotic initiation factor (eIF) 4E-binding protein 1, and eIF4E associated with eIF4G decrease with development. The reduced activation of negative regulators of insulin signaling also contributes to the high rate of neonatal muscle protein synthesis. These include protein tyrosine phosphatase 1B, phosphatase and tensin homologue deleted on chromosome 10, protein phosphatase 2A, and tuberous sclerosis 1/2. The amino acid signaling pathway converges with the insulin signaling pathway at mTOR to regulate the activation of translation initiation factors, including S6K1 and 4EBP1, that regulate the binding of mRNA to the 40S ribosomal complex. The enhanced activation of components in the insulin and amino acid signaling pathways in response to the postprandial rise in insulin and amino acids contributes to the high rate of protein synthesis and rapid gain in skeletal muscle mass in neonatal pigs.

Early postnatal skeletal myofibre formation in piglets of low birth weight is stimulated by L-carnitine supplementation during suckling
Loesel, D., Kalbe, C., Nuernberg, G. and Rehfeldt, C., Research Institute for the Biology of Farm Animals, Research Units Muscle Biology and Growth and Genetics and Biometry, Wilhelm-Stahl-Allee 2, 18196 Dummerstorf, Germany; loesel@fbn-dummerstorf.de

To study the effect of L-carnitine supplementation to suckling piglets on early postnatal myofibre formation, muscle growth, and body composition, 48 piglets of low (LW) and middle (MW) birth weight from 9 German Landrace gilts received 400 mg L-carnitine (n=25) or a placebo (n=23) once daily from d 7 to 27 of age and were slaughtered on d 28 of age (weaning). The proportions of perirenal ($P=0.10$) and intramuscular fat ($P=0.05$) were lowest in carnitine-treated piglets. Circulating glucose concentration were highest in treated LW piglets ($P=0.13$). The proportion of esterified carnitine in semitendinosus (ST) muscle was highest ($P<0.001$) in treated female piglets. The ratio of lactate dehydrogenase to isocitrate dehydrogenase was lowest ($P=0.12$) in ST muscle of treated piglets indicating a more oxidative muscle metabolism. Total number of ST myofibres was 13% higher ($P=0.02$) in treated compared with placebo LW piglets thereby reaching the unchanged level of MW littermates. Treated LW piglets had a 2.4-fold higher mRNA expression of the gene encoding the embryonic isoform of myosin heavy chain in ST muscle compared with placebo piglets ($P=0.05$). Treated piglets exhibited the highest DNA:protein ratio ($P=0.02$) in ST muscle, which resulted from a higher DNA concentration ($P=0.04$). However, because myofibre size, creatine kinase activity, and protein concentration remained unchanged, the ST muscle of treated piglets was not less mature than placebo. It seems that intensified fatty acid oxidation improved energy balance and stimulated myogenic proliferation, which in treated LW piglets may have contributed to a compensatory increase in myofibre number. Thus, particularly in LW piglets an early postnatal L-carnitine supplementation may attenuate the negative consequences of low birth weight on body composition.

Molecular basis for impaired insulin action during myogenesis
Pijet, B.[1], Pijet, M.[1], Pogorzelska, A.[1], Pajak, B.[2] and Orzechowski, A.[1,2], [1]Warsaw University of Life Sciences, Physiological Sciences, Nowoursynowska 159, 02-776 Warsaw, Poland, [2]Mossakowski Medical Research Center Polish Academy of Sciences, Cell Ultrastructure, Pawiński ego 5, 02-106 Warsaw, Poland; arkadiusz_orzechowski@sggw.pl

Insulin is a primary anabolic hormone that responds to meal and determines the animal muscle mass. Some cytokines reported as catabolic antagonize insulin action and impair insulin-dependent myogenesis. Previously, we have shown that mitochondrial activity limits insulin action. In this study we decided to verify the hypothesis that leptin, TNF-alpha and IFNs affect myogenesis through the specific targeting of certain kinases and transcription factors. Acute and chronic experiments were performed on C2C12 clonal line of satellite cells. The cells were encouraged to differentiate upon insulin addition (10 nM). Myogenesis was monitored at the myogenin level. Viability (mitochondrial activity) was evaluated with methyl tiazolyl blue (MTT) method. Real-time RT PCR and Western Blots were used to appraise gene and protein expression, respectively. Genes controling mitochondrial functions (mfn2, mtSSB, mtTFA, cox-1) as well as proteins under insulin control were investigated. Insulin stimulated by twofold ($P<0.01$) muscle cell viablility similarly to GSK-3beta inhibitors SB216763 and LiCl. No additive affect was observed when insulin was used with SB216763 or LiCl ($P>0.05$) suggesting that with respect to mitochondria GSK-3beta is the ultimate insulin target. In turn, TNF-alpha and IFN-gamma acted in accordance to stimulate NF-kappaB activity (increased by 50%, $P<0.05$). Consequently, myogenesis was retarded. Similar antimyogenic effect was observed when leptin (50 ng/ml) was added ($P<0.05$). Moreover, leptin potentiated proapoptotic effect of staurosporin ($P<0.05$). Phosphatidylinositol 3-kinase (PI-3K) inhibitor LY294002 but not protein kinase kinase/extracellular-signal-regulated kinase (MAPKK/ERK - MEK) inhibitor PD98059 retarded mitochondrial respiration and abrogated insulin-mediated myogenesis ($P<0.05$).

Endocrine and metabolic regulation of growth and body composition in cattle

Hocquette, J.F., INRA, UR1213, Herbivore Research Unit, Theix, 63122 Saint-Genes Champanelle, France; hocquet@clermont.inra.fr

Muscle metabolism (in interaction with other organs and tissues, including adipose tissue) plays an important role in the control of growth and body composition. Muscle ontogenesis has been well described in different genotypes of cattle for myofibers, connective tissue, as well as intramuscular adipocytes. The ontogenesis or the action of putatively important factors controlling muscle development (IGF-II expression, IGF receptors, GH receptor, myostatin, bFGF, TGF-beta 1, insulin, thyroid hormones) has also been studied on bovine foetal muscle samples and satellite cells. The glucose/insulin axis has been specifically studied in both bovine adipose tissue and heart. Clearly, cattle, as sheep, are mature species at birth based on their muscle characteristics compared to other mammalian or farm species. The different myoblast generations have been well characterised in cattle, including the second generation which is liable to be affected by foetal undernutrition at least in sheep. Interesting genotypes e.g. double-muscled genotype, have been characterised by an altered metabolic and endocrine status associated with a reduced fat mass, specific muscle traits and different foetal characteristics. Finally, the recent development of genomics in cattle has allowed the identification of novel genes controlling muscle development during foetal and postnatal life. Generally, a high muscle growth potential is associated with a reduced fat mass and a switch of muscle fibres towards the glycolytic type.

Relation between calf birth weight and dam weight in Belgian Blue double-muscled cattle

Fiems, L.O. and De Brabander, D.L., ILVO, Animal Sciences, Scheldeweg 68, 9090 Melle, Belgium; leo.fiems@ilvo.vlaanderen.be

Belgian Blue double-muscled (BBDM) cows frequently calve with cesarean. There is an increasing aversion against this practice, so called for reasons of animal welfare. One of the main causes of dystocia is the relatively high calf birth weight (CBW). Therefore, the relation between CBW and dam body weight (DBW) was studied using 374 dams and their offspring involving 916 parturitions. Only full-term gestation periods and single births were investigated. Dam birth weight and CBW were correlated, going from 0.148 for female calves ($P<0.003$) to 0.173 for all calves ($P<0.001$) and 0.207 for males ($P<0.001$). However, large variations from 0.439 to -0.454 were observed within parities and calf gender. A higher correlation was found between CBW and dam weight before (0.325) than after calving (0.301). Postpartum DBW increased significantly up to 5th parturition while CBW increased up to 2nd and 3rd parturition for females and males, respectively. Consequently, postpartum DBW:CBW ratio increased from 11.3 at 1st parturition to 13.7 at 5th parturition and then leveled off. Dam birth weight was moderately correlated with daily gain during the first 4 months of life: 0.178 ($P=0.001$) and BW after 1st calving 0.279 ($P<0.001$). DBW after 1st calving was moderately correlated with BW gain during the first 4 months of life: 0.143 ($P=0.008$). However, especially a daily BW gain below 0.6 kg resulted in a lower postpartum DBW. Only 4.4% of the calvings occurred without cesarean. These calvings were characterized by a higher cow age and parity ($P<0.001$), a lower CBW, a lower ratio of CBW to postpartum DBW, and a lower frequency of male calves ($P<0.01$) than calvings with cesarean. Five cows had 59% calvings without cesarean. From the low correlations between weight of dams and birth weight of offspring and between half-sibs, and the occurrence of dams with several calvings without cesarean, we suggest that there may be a possibility to select for BBDM cows that calve with a lower frequency of cesarean.

Postnatal growth requirements of dairy calves and long term productivity measured by a test day model and from field observations

Raffrenato, E.[1,2], Soberon, F.[2], Everett, R.W.[2] and Van Amburgh, M.E.[2], [1]CoRFiLaC, Regione Siciliana, 97100 Ragusa, Italy, [2]Cornell University, Ithaca, NY 14853, USA; er53@cornell.edu

For many years, early life management of the calf has focused on survival rates and rumen development. Recent studies suggest that nutrient intake during the pre-weaning phase may have long term carry-over effects on milk yield potential. The objective of this study was to investigate this relationship in the Cornell Dairy Herd and a commercial dairy farm using a Test Day Model (TDM). The management objectives of the calf program have been to double the birth weight by weaning through increased milk replacer supply. Lactation residuals from the TDM were generated from 790 and 450 heifers from the Cornell Herd and the commercial farm, respectively. Linear regressions were run on measures of pre-weaning growth performance, management factors and TDM milk yield solutions. Significant correlations were found for pre-weaned average daily gain (ADG), weaning weight and breeding weight. Pre-weaning ADG, ranging from 0.11 kg to 1.34 kg, had the greatest correlation with first lactation yield for the Cornell Herd. Using the TDM solutions, for every 1 kg of pre-weaning ADG, heifers produced 856 kg more milk during their first lactation ($P<0.01$) on average among the two farms. Using the TDM residuals, pre-weaning ADG accounted for 22 % of the variation in first lactation milk yield. Further, regression of pre-weaning and pre-pubertal growth on milk yield over three lactations demonstrated a strong positive effect of ADG on milk yield with regression coefficients of 3,868 kg and 13,712 kg, respectively (R^2 =0.15 and 0.25, respectively) for the commercial farm. These results demonstrate that increased growth rate prior to weaning has positive effects on lactation milk yield and longevity and results in some form of imprinting that is yet to be understood. These data differ greatly from the observations made over the last 20 years concerning the effect of growth rate prior to puberty on milk yield.

Correlation between growth curve parameters and plasma metabolites in replacement dairy heifers

Abeni, F.[1], Calamari, L.[2], Stefanini, L.[3] and Pirlo, G.[1], [1]CRA-FLC Centro di Ricerca per le Produzioni Foraggere e Lattiero-casearie, Sede distaccata di Cremona, Via Porcellasco 7, 26100 Cremona, Italy, [2]Università Cattolica del Sacro Cuore, Istituto di Zootecnica, Via Emilia Parmense 84, 29100 Piacenza, Italy, [3]Azienda Sperimentale Vittorio Tadini, Loc. Gariga, 29027 Podenzano (PC), Italy; fabiopalmiro. abeni@entecra.it

This paper considers the relationships among growth curve parameters and plasma metabolites in dairy heifers. Sixty Italian Friesian heifers were randomly assigned on two experimental feeding groups to obtain a moderate (0.70 kg/d; M) or a high (0.90 kg/d; H) average daily gain from 5 to 15 mo of age. Every 28 d, heifers were weighed, scored for BCS, and measured for: wither height, hip height, body length, and heart girth. Blood samples were collected at 9 and 15 month of age, then at 14 and 7 days before calving, to be analyzed for plasma metabolites and enzymes. Laird's growth function was estimated for body size measurements, and Pearson's correlation coefficients were calculated between b_1 (which reflects the initial specific growth rate), b_2 (which reflects the maturation rate of the animal), and plasma metabolite concentrations during growth and before first calving. Among the explored blood metabolites and enzymatic activities at 9 and 15 mo of age, those with high correlation with growth curve parameters were glucose, urea, and ceruloplasmin. Both at 9 and 15 mo of age plasma cholesterol and ceruloplasmin concentrations were positively correlated with their respective values before calving ($P<0.001$). Plasma glucose at 15 mo was negatively correlated with plasma albumin before calving ($P<0.001$). The curve parameters from skeletal height measurement were correlated with prepartum plasma creatinine concentration, suggesting that this item was affected by the development of skeletal muscles. Before calving, plasma NEFA and BHBA were positively correlated with enzymatic activities and positive acute phase proteins, mainly related to liver function.

Effect of birth weight on contractile types and numbers of fibers in semitendinosus muscles of sheep
Sirin, E.[1], Sen, U.[1], Aksoy, Y.[1], Ulutas, Z.[1] and Kuran, M.[2], [1]Gaziosmanpasa University, Department of Animal Science, 60250 Tokat, Turkey, [2]Ondokuz Mayis University, Department of Animal Science, 55139 Samsun, Turkey; mkuran@omu.edu.tr

The aim of this study was to investigate the effects of birth weight on contractile types and numbers of muscle fibers in semitendinosus muscles of sheep. A total of 21 male lambs of the Karayaka breed was allocated to three groups based on their birth weights (L; 3.4±0.6, M; 4.2±0.7 and H; 5.1±1.1 kg). Lambs were subjected to a standard fattening period for 60 days following weaning on day 90 and slaughtering on day 150. Live weight of lambs at slaughter, carcass yield and weight of semitendinosus muscle of the L group were lower than that of M and H groups ($P<0.05$). There were no significant differences between birth weight groups in terms of proportion of type I, type IIA and type IIB muscle fibers stained by ATPase. A reduced muscle fiber area and fiber numbers (per μm^2) of type IIA were observed in lambs from the L group compared to lambs from M and H groups ($P<0.05$). There was no difference between birth weight groups in terms of the amount of total DNA in the muscle samples. These results indicate that lambs with low birth weight have larger muscle fiber area and reduced numbers of fibers per μm^2 in type IIA fibers of semitendinosus muscle compared to lambs with high birth weights.

Do serum IGF-I and leptin concentrations monitor growth rate and muscle composition in yearling bulls?
Blanco, M.[1], Casasús, I.[1], Sauerwein, H.[2] and Joy, M.[1], [1]CITA - Gobierno de Aragón, Unidad de Tecnología en Producción Animal, Avda. Montañana 930, 50059 Zaragoza, Spain, [2]ITW - Universität Bonn, Physiology & Hygiene, Katzenburgweg 7-9, D-53115 Bonn, Germany; mblanco@aragon.es

Three fattening strategies were tested in 21 autumn-born Parda de Montaña male calves from weaning (224 kg) to slaughter at 450 kg. One group of calves were concentrate-fed until slaughter (CON), another group of calves rotationally grazed in lucerne paddocks supplemented with 1.8 kg DM/d of barley until slaughter (LUC), and the third group of calves had the same management as LUC calves for 3 months (Period 1) and were then finished on concentrates (Period 2) until they reached the slaughter weight (LUC+CON). During Period 1, CON calves presented slightly greater weight gains than their grazing counterparts ($P<0.10$). During Period 2, LUC calves had the lowest, CON calves an intermediate, and LUC+CON calves the highest ADG. Overall weight gains were greater for LUC+CON, intermediate for CON, and lowest for LUC calves. During Period 1, IGF-I and leptin concentrations were greater in CON calves than in their grazing counterparts ($P<0.05$). During Period 2, IGF-I and leptin concentrations increased in all fattening strategies, and LUC+CON calves presented the greatest increases in the concentration of both hormones. Furthermore, serum IGF-I and leptin concentrations were related to live live weight but not to ADG. Finally, only serum leptin at slaughter was related to intramuscular fat content.

Maternal low and high protein diets during gestation in gilts have long-lasting effects on the endocrine and immune response in their offspring
Otten, W., Kanitz, E., Tuchscherer, M., Graebner, M., Stabenow, B., Metges, C.C. and Rehfeldt, C., Research Institute for the Biology of Farm Animals (FBN), Wilhelm-Stahl-Allee 2, 18196 Dummerstorf, Germany; ekanitz@fbn-dummerstorf.de

An inadequate nutrient supply during gestation can cause fetal growth retardation, metabolic changes and alterations of immune and stress systems in the offspring. In this study, we determined the effects of maternal low and high protein diets during gestation of gilts on endocrine and immune function in their offspring. Forty-two German Landrace sows (first parity) were fed isocaloric diets with high (HP, 30%), low (LP, 6%) or control (CP, 12%) protein levels throughout gestation. The offspring was reared by foster sows until weaning on postnatal day (PND) 28 and fed according to recommendations. Salivary cortisol and serum immunoglobulins were measured in gilts over the course of gestation. Plasma cortisol and catecholamines, serum immunoglobulins, CD4+ and CD8+ cells as well as TNFalpha and IL-6 were determined in the offspring at basal and challenging conditions (weaning, LPS-, ACTH- and insulin-challenge). In gilts, the LP diet reduced growth performance, increased salivary cortisol, and decreased total serum protein ($P<0.001$), without affecting immunoglobulin levels. In the offspring, the LP diet reduced the mean birth weight of piglets ($P<0.01$). Serum IgG was decreased in LP offspring on PND 27 ($P<0.05$) and 180 ($P=0.08$), whereas serum IgA was decreased in HP offspring on PND 27 ($P=0.09$) and 180 ($P<0.01$). In addition, a decrease of CD4+ cells was found in HP piglets at PND 27 ($P<0.01$). In stress and challenging situations, LP piglets showed an increased cortisol response to weaning ($P<0.05$), an increased IL-6 response to a LPS challenge at PND 47 ($P<0.05$) and an increased adrenaline response to an insulin challenge at PND 70 ($P<0.05$). The present results indicate that a dietary protein deficiency during pregnancy of gilts has long-lasting effects in the offspring with excessive immune and stress responses to challenges.

A reference proteome map of the gastrocnemius muscle in the rabbit: a model research tool for larger species with special reference to physiology and meat science
Almeida, A.M.[1], Campos, A.[2], Van Harten, S.[1], Cardoso, L.A.[1] and Coelho, A.V.[2,3], [1]IICT & CIISA, CVZ, FMV Av. Univ. Técnica, 1300-477 Lisboa, Portugal, [2]ITQB/UNL, MS Laboratory, ITQB Av. Republica (EAN), 2780-157 Oeiras, Portugal, [3]Universidade de Évora, Dep. Química, Largo dos Colegiais 2, 7004-516 Évora, Portugal; aalmeida@fmv.utl.pt

In several species, the establishment of a proteome reference map of a specific tissue has been accomplished in order to allow a better knowledge of the distribution of the proteins in a specific species and tissue. Such studies are of particular significance in laboratory animal species such as rat or mice but in recent years studies have also focused on farm animals particularly swine and cattle. The rabbit (*Oryctolagus cuniculus*) is a widely used species as both a production and a model animal. Although muscle physiology studies (especially using the gastrocnemius) exist for this species, no reference proteome map seems to be available. In this work we describe the first reference map of the rabbit's gastrocnemius muscle using both 2D electrophoresis and the identification of proteins through Peptide Mass Fingerprinting (PMF). A total of 45 proteins were localized and identified: 1) Cell structure and contractile apparatus proteins (actin, myosin and troponin); 2) proteins associated with metabolic pathways (creatine kinase and enolase) and 3) Cell Defense proteins like α-crystallin. A reference map for most of the major proteins expressed in this muscle is described enabling possible comparisons with studies under diverse physiological situations with relevance to studies in meat science.

Protein metabolism of Nellore steers (Bos indicus) with low and high residual feed intake

Gomes, R.C.[1], Sainz, R.D.[2] and Leme, P.R.[1], [1]Faculdade de Zootecnia e Engenharia de Alimentos, Universidade de São Paulo, Departamento de Zootecnia, Duque de Caxias Norte 225, 13635-900 Pirassununga, SP, Brazil, [2]University of California, Department of Animal Science, One Shields Avenue, 95616-8521 Davis, CA, USA; prleme@usp.br

Residual feed intake (RFI) is a feed efficiency trait independent of growth and mature weight. Genetic improvement in RFI may reduce the costs of feeding cattle, however a better understanding of biological processes underlying variation in RFI is necessary in zebu cattle. In this sense, it was aimed to evaluate myofibrillar protein metabolism in high- and low-RFI zebu (Bos indicus) cattle. Seventy-two Nellore steers (16 to 21 month-old, 334±19 kg initial body weight [BW]) were fed a finishing ration (74.5% TDN, 14.3%CP) on an ad libitum basis, for 70 days. Daily dry matter intake (DMI) and average body weight gain (ADG) were measured individually. RFI was calculated as the difference between actual DMI and the predicted DMI determined by linear regression of DMI on mid-test $BW^{0.75}$ and ADG. The lowest and highest 12 RFI steers were classed as low- (most efficient) and high-RFI (least efficient) groups, respectively. Total urine was collected for determination of daily 3-methylhistidine (3MH) excretion and myofibrillar protein breakdown rates. Initial and final skeletal muscle masses were estimated based on the BW. There were differences ($P<0.01$) between low- and high-RFI groups for DMI (9.3 vs. 11.1 kg/d), feed:gain (6.4 vs. 7.6) and RFI (-0.80 vs 0.89 kg/d), but not for ADG (1.48 vs. 1.48 kg/d) and final BW (441 vs. 448 kg). High- and low-RFI cattle showed similar ($P>0.05$) skeletal muscle protein gain (57.2 vs. 55.2 g/d), total 3MH in muscle (113 vs. 110 mmol) and total 3MH excretion in urine (1.96 vs. 2.06 mmol/d). There were no differences between high- and low-RFI cattle for fractional rates of myofibrillar protein degradation (1.76 vs. 1.85%/d), synthesis (2.01 vs. 2.09%/d) and accretion (0.24 vs. 0.24%/d). Myofibrillar protein metabolism did not differ between low- and high-RFI steers.

Phenotypic correlations between residual feed intake, growth, carcass traits and reactivity in Nellore bulls (*Bos indicus*)

Stella, T.R., Leme, P.R., Silva, S.L., Ferraz, J.B.S., Gomes, R.C., Nogueira Filho, J.C.M. and Silva Neto, P.Z., Faculdade de Zootecnia e Engenharia de Alimentos, Universidade de São Paulo, Departamento de Zootecnia, Pirassununga, 13635-900, Brazil; prleme@usp.br

Residual feed intake (RFI) is a feed efficiency trait phenotypically independent of growth rate and mature weight. Breeding beef cattle for improved RFI may reduce feeding costs and increase profitability of beef business. However, associations between RFI with production traits and animal behaviour were not studied in Nellore cattle. A total of 96 Nellore young bulls, born in 2005 and 2006, that had measures on weaning weight (WW), birth weight (BW), yearling weight (YW), post weaning weight gain (WG) and reactivity score (animal agitation when restrained in the squeeze chute), were evaluated for RFI in 2007 and 2008, respectively. Bulls were fed individually for 68 days with a diet containing 68% of TDN and 14.3% CP. Body weight (BW) during the test and daily feed intake (DMI) were measured individually and RFI was calculated as the difference between actual DMI and predicted DMI for a common mid-test $BW^{0.75}$ and daily BW gain (ADG). All animals were weighed and ultrasonic measurements of Longissimus muscle area (REA) and backfat thickness (BF) on the 12th and 13th ribs were obtained every 21 days. Pearson correlation analysis was performed and probabilities lower than 5% were considered statiscally significant. There were no correlations between RFI and WW (r=-0.04), YW (r=-0.05), and weight gain from 205 to 550 days of age (r=-0.02). Also, RFI was not associated with ADG (r=-0.70), backfat thickness (r=0.13) and Longissimus area (r=0.15). RFI was also not correlated to reactivity of the bulls (-0.001). RFI was positively correlated with feed conversion ratio (DMI:ADG) (r=0.39), feed intake (r=0.56) and scrotal circumference (r=0.22) and negatively correlated with feed efficiency (r=-0.42) and birth weight (r=-0.22). Therefore, RFI was not associated with growth, body weight and carcass traits in Nellore bulls.

Effect of a match or mismatch of maternal and offspring environment on the development and behavior of the offspring

Van Der Waaij, E.H.[1,2], Van Den Brand, H.[1], Van Arendonk, J.A.M.[2] and Kemp, B.[1], [1]Wageningen University, Adaptation Physiology Group, P.O. Box 338, 6700 AH Wageningen, Netherlands; [2]Wageningen University, Animal Breeding and Genomics Centre, P.O. Box 338, 6700 AH Wageningen, Netherlands; liesbeth.vanderwaaij@wur.nl

It is generally accepted that who we are is determined by the combination of our genes and the environment we have experienced. However, research has shown that we not only inherit genes, but also part of the environmental experience from our parents, especially our mother. The maternal environmental experience is passed on prior to birth or hatch and affects gene expression in the offspring, thereby adjusting their developmental tract. This inheritance of environmental experience may exist to prepare the offspring for the environment in which they will live. Likewise, a mismatch between the environment of the mother and her offspring would result in suboptimal development to that environment. This may affect behavior and increase health problems, such as obesity and cardio-vascular diseases, in the offspring. So would any mismatch in maternal and offspring environment have negative consequences for the offspring? To test this, an experiment was set up where broiler breeders were kept on one of two diets for 6 weeks: 30 g/d less or 30 g/d more than the standard diet. After 4 weeks the hens were inseminated and egg collection was started. Fresh eggs were tested for some egg components and for embryo development (morphology and cell number). The rest of the eggs were incubated and the chicks of each group of hens were raised on either ad lib or 25% less than ad lib diet (i.e. in a 2x2 design). Body weight and behavior were recorded once weekly for 6 weeks, after which the chicks were slaughtered. At slaughter, blood samples were collected and a number of traits were recorded related to physiological development and potential cardio-vascular problems. The experiment is currently running and no results were available at the time of abstract submission, but will be presented at the conference.

The effects of docking on growth and blood biochemical parameters of Sanjabi fat-tailed lambs

Moeini, M. and Nooriyan Sarvar, E., Razi University, Animal Science, Kermanshah, 67155, Iran; mmoeini2008@yahoo.com

The effects of docking on growth traits and blood biochemical parameters were investigated using 24 fat-tailed Sanjabi single-born male lambs raised from a large commercial sheep herd. The lambs were randomly divided into two groups. One group (n=12) were docked at two days of age with rubber-rings using elastrator. The second group (n=12) were left intact. After weaning (90 days), all lambs were moved to rustic rangelands for 45 days. Then all the lambs were fed concentrates ad libitum during a 60 day fattening period. The blood biochemical parameters including urea, total protein, glucose, triglycerides, cholesterol, low-density lipoproteins (LDL) and high-density lipoproteins (HDL) were measured during the suckling and fattening periods. The Fat-tail docking had no effect on lamb growth from birth to weaning. However, BW and ADG of docked lambs were significantly higher ($P<0.05$) compared with intact lambs at the end of fattening period. In the fattening period, blood cholesterol and LDL of docked lambs were lower than in intact lambs ($P<0.05$).

Streptozotocin-induced hypoinsulinemia caused hyperphagia and alterations in blood constituents and leptin concentrations, and performance of Zel lambs

Moslemipur, F.[1], Torbatinejad, N.M.[1], Khazali, H.[2], Hassani, S.[1] and Goorchi, T.[1], [1]Gorgan University of Agricultural Sciences and Natural Resources, Animal Science, Shahid Beheshti ave., 49138-15739, Gorgan, Iran, [2]Shahid Beheshti University, Biology, Kaj Sq., 19615-1178, Tehran, Iran; farid_6543@yahoo.com

The importance of Insulin in ruminants is different from that of monogastrics. In this study, effects of permanent hypoinsulinemia on appetite, performance as well as serum constituents and leptin concentrations were investigated in Zel lambs. Twenty male feedlot lambs were divided into four treatment groups and maintained individually. Treatments were control and single, intravenous injection of three doses of streptozotocin; 25, 50 and 75 mg/kg BW, named low, middle and high dose, respectively. The duration of the experiment was eight consecutive weeks, and injection was performed at the end of third week. Blood samples were collected weekly via jugular vein at fasting and post-prandial times. Feed and water intakes and also animals' weight changes were measured weekly. The high-dose group could not continue the experiment because of abnormalities. Results showed the occurrence of hypoinsulinemia with injection of middle-dose with significant decrease in fasted and post-prandial insulin and also fasted leptin concentrations vs control ($P<0.05$). The middle-dose also caused marked and significant increases in blood glucose, triglycerides, total protein and ketone bodies levels vs control ($P<0.05$). Animals receiving middle-dose showed hyperphagia and enhanced water intake as compared to controls ($P<0.05$) but idespite increased feed intake, they did not gain more weight than controls. Urine sugar and protein levels in middle-dose group were significantly greater than others. Results suggest a regulatory role of insulin in energy metabolism of ruminants by exerting two opposing effects; central catabolic and peripheral anabolic. Furthermore, insulin was linked to leptin secretion.

Effect of terbutaline on carcass traits and blood metabolites in boiler chickens

Golzar-Adabi, S.[1], Kamali, M.A.[2] and Moslemipur, F.[3], [1]University of Ankara, Turkey and Organization of Agricultural Jahad, Animal science, East Azarbaijan, Tabriz, Iran, [2]Agricultural Research and Education Organization, Animal science, Tehran, Iran, [3]Gorgan University of Agricultural sciences and natural resources, Animal science, Shahid Beheshti Ave., 49138-15739, Gorgan, Iran; farid_6543@yahoo.com

An experiment was conducted in order to study the physiological effects of a beta-adrenergic agonist (Terbutaline) on carcass traits and blood parameters in broiler chickens. A total of 120 one-day-old male Cobb broiler chicks were assigned to 3 treatment groups with 4 replicates and 10 birds/replicate. All treatments were fed the same base ration but with 3 doses of terbutaline; 0 (control), 7.5 (LD) and 15 (HD) mg terbutaline/kg of feed for 4 weeks. Performance, carcass weight, carcass efficiency of chicks, skeletal muscles composition and blood parameters were determined at 49 days of age. Terbutaline had significant effects on feed intake, weight gain and feed conversion ratio ($P<0.05$). Specifically, feed conversion ratio was lowest in response to LD ($P<0.05$). The highest weight of the whole body was seen in LD group ($P<0.05$). Terbutaline also increased carcass weight, carcass efficiency, and breast and thigh muscles weight ($P<0.01$). On the other hand, LD caused a decrease in abdominal fat ($P<0.05$). LD increased breast and thigh muscles protein content ($P<0.01$). The blood glucose, T_4, triglycerides and cholesterol levels were increased, and insulin ($P<0.05$), BUN, and uric acid were decreased as a result of adding terbutaline to the ration ($P<0.05$). Different levels of terbutaline had no effect on blood concentrations of alanine and glutamine ($P>0.05$). Results suggest that the beta-adrenergic agonist terbutaline can affect performance of chicks. It can also change blood constituents' levels resulting in altered energy and protein metabolism.

Precision dairy farming: the next dairy marvel?

Bewley, J.M., University of Kentucky, 407 W.P. Garrigus Building, Lexington, KY 40546-0215, USA; jbewley@uky.edu

Precision Dairy Farming (PDF) is the use of technologies to measure physiological, behavioral, and production indicators on individual animals to improve management strategies and farm performance. Many PDF technologies, including daily milk yield recording, milk component monitoring, pedometers, automatic temperature recording devices, milk conductivity indicators, automatic estrus detection monitors, and daily body weight measurements, are already being utilized by dairy producers. Other theoretical PDF technologies have been proposed to measure jaw movements, ruminal pH, reticular contractions, heart rate, animal positioning and activity, vaginal mucus electrical resistance, feeding behavior, lying behavior, odor, glucose, acoustics, progesterone, individual milk components, color (as an indicator of cleanliness), infrared udder surface temperatures, and respiration rates. The main objectives of PDF are maximizing individual animal potential, early detection of disease, and minimizing the use of medication through preventive health measures. Perceived benefits of PDF technologies include increased efficiency, reduced costs, improved product quality, minimized adverse environmental impacts, and improved animal health and well-being. Real time data used for monitoring animals may be incorporated into decision support systems designed to facilitate decision making for issues that require compilation of multiple sources of data. Technologies for physiological monitoring of dairy cows have great potential to supplement the observational activities of skilled herdspersons, which is especially critical as more cows are managed by fewer skilled workers. Moreover, data provided by these technologies may be incorporated into genetic evaluations for non-production traits aimed at improving animal health, well-being, and longevity. The economic implications of technology adoption must be explored further to increase adoption rates of PDF technologies. Precision Dairy Farming may prove to be the next important technological breakthrough for the dairy industry.

Are Jerseys better keepers: body condition and ultrasonographically evaluated fat depth in Jerseys and Holsteins

Szendrei, Z., Vígh, Z., Harangi, S. and Béri, B., University of Debrecen, Institute of Animal Science, Böszörményi út 138., 4032 Debrecen, Hungary; harangis@agr.unideb.hu

We condition scored and ultrasound examined second lactation Jerseys (n=64) and Holsteins (n=58) in two occasions. Body condition (BCS) was scored on a five-point scale with 0.25 divisions and ultrasonographic measurements were taken on four parts of the body (abdominal region, ribeye, rump: P8 and rump fat). Based on the results it can be concluded that BCS and the fat depth measured on the four body parts are in close correlation (r=0.79-0.88), therefore scoring is a reliable method to estimate energy reserves of dairy cows. Whereas fat depth does not differ on the four measurement sites, to ultrasound only one is enough to BCS. After grouping the animals into three condition categories (poor-, good- and over) it was confirmed that fat depth is significantly different (P=0.05) among them. When the two breeds were compared by categories, they differed only in one case: P8 in the poor category. This suggests that the smaller Jerseys deposit relatively more fat than a big framed Holsteins. Regression analysis showed that fat depth explains 75% of the BCS in Jerseys and 82% of BCS in Holsteins. This value could be higher if fat depth of the pins can be measured with ultrasound. From the regression analysis the following BCS estimating equations were calculated: Jersey: BCS=1.463+(2.586×fat depth (cm)), Holstein: BCS=1.206+(3.065×fat depth (cm)). Weak but positive correlation (Jersey r=0.369, Holstein r=0.481) was found between the BCS and the days in milk (DIM). Consequently, with increase in DIM, BCS increases, too. Milk production and BCS is in strong negative correlation in Jerseys (r=-0.608), on the contrary we found no correlation in Holsteins.

The effect of breed on heat detection and fertility of dairy cows maintained on 2 feeding levels

Cutullic, E.[1], Delaby, L.[1], Michel, G.[2] and Disenhaus, C.[1], [1]INRA, UMR 1080 Dairy Production, 65 rue de Saint Brieuc, 35000 Rennes, France, [2]INRA, UE326, Le Pin-au-Haras, 61310 Exmes, France; catherine. disenhaus@agrocampus-ouest.fr

In seasonal calving systems, dairy cows have to be pregnant within a short period. High ovulation detection and fertility are both needed. The objective of this study was to evaluate the adaptation to such systems of 2 breeds within 2 feeding levels. 105 Normande (NO, dual-purpose) and 98 Holstein (H) cows were assigned to a low or high feeding level (L-group: 50% grass silage and 50% hay lage in winter, no concentrate at grazing; H-group: 55% maize silage, 15% alfalfa hay and 30% concentrate, 4kg concentrate at grazing). Thrice-weekly milk progesterone assays and ultrasonography were used to determine ovulation detection rate (ODR) and late embryo mortality (LEM). In both breeds, H-group cows produced more 100-day average daily milk yield (MY) than L-group ones but lost less body condition (+9.4 kg/d and -0.44 unit; $P<0.001$). In both breeds, ODR was higher in the L-group than in the H-Group (77% vs. 60%, N=254 and 282; $P<0.001$). Feeding level effect on ODR was greater for NO than for H cows (-21% vs. -14%, N=303 and 233; $P<0.01$). Following 1st and 2nd services, conception rate was improved in the H-group but only for H cows (73% vs. 48%; $P<0.01$). Conversely, LEM was higher in the H-Group only for H cows (30% vs. 9%, N=64 and 64; $P<0.01$). Finally NO cows had higher calving rate than H cows (53% vs. 38%, N=135 and 128; $P<0.05$). Difference in calving rate was not significant between feeding groups ($P>0.25$). In a 3-month service period, pregnancy rates were 72% vs. 57% for H and NO respectively ($P<0.05$) and were not different between feeding groups ($P>0.80$). However, in NO cows, as high feeding level decreased ODR without affecting fertility, the 6-week pregnancy rate was lower in the H-Group (29% vs. 52%; $P<0.05$). In conclusion, NO breed appeared well adapted to seasonal calving systems even though feeding level leading to high MY did not allowed very short breeding period.

Bivariate analyses of body condition score and health traits in Canadian Holstein cattle using a random regression model

Neuenschwander, T.F.-O.[1], Miglior, F.[2,3], Jamrozik, J.[1] and Schaeffer, L.R.[1], [1]University of Guelph, CGIL, Dept. of Animal and Poultry Science, 50 Stone Rd E, Guelph, ON, N1G 2W1, Canada, [2]Agriculture and Agri-Food Canada, Dairy and Swine Research and Development Centre, 2000 College St, Sherbrooke, QC, J1M 1Z3, Canada, [3]Canadian Dairy Network, 660 Speedvale Ave W, Guelph, ON, N1K 1E5, Canada; miglior@cdn.ca

Body Condition Score (BCS) has been recorded by VALACTA DHI field staff since 2000 in the province of Quebec. Recording is made several times throughout lactations. Eight diseases (mastitis, lameness, cystic ovarian disease, left displaced abomasum, ketosis, metritis, milk fever and retained placenta) are recorded by dairy producers in Canada since April 2007. The objective of this study was to estimate genetic correlations between BCS and 8 diseases in Quebec Holsteins using random regression analyses. Each disease (coded as a binary trait: 0=sick, 1=healthy), was analyzed with BCS using bivariate linear animal models, including genetic and environmental random regressions (Legendre polynomials of order 3 in the model for BCS. Minimum number of disease recordings per herd was applied to ensure a sufficient quality of disease recording in the herds included in the analysis. Data subsets were created to estimate variance components. Dataset size varied from 2,384 cows for metritis to 11,558 cows for mastitis. Genetic correlations between BCS and health traits were generally low and slightly positive at the beginning of the lactation (high BCS, low susceptibility to diseases). This result shows that cows with good body reserves were less susceptible to diseases. Later in the lactation, the correlations generally became negative (-0.02 to -0.18). The exceptions were left displaced abomasum and ketosis, for which the genetic correlations were always negative (-0.02 to -0.49). A high BCS was generally related to cows with a lower feed intake at the onset of lactation and, therefore with a higher risk of contracting any of these two diseases.

Single nucleotide polymorphisms for German Holstein cows with exceptional long productive life

Distl, O.[1], Rohde, H.[1], Stock, K.F.[1], Reinhardt, F.[2], Reents, R.[2] and Brade, W.[3], [1]University of Veterinary Medicine Hannover, Institute for Animal Breeding and Genetics, Buenteweg 17p, 30559 Hannover, Germany, [2]Vereinigte Informationssysteme Tierhaltung w.V., Heideweg 1, 27283 Verden/Aller, Germany, [3]Landwirtschaftskammer Niedersachsen, Versuchswesen Tier, Johannsenstraße 10, 30159 Hannover, Germany; ottmar.distl@tiho-hannover.de

A serious decrease in longevity since more than 40 years is obvious in dairy cattle breeds in Germany. A survey of milk recording data has shown that approximately 3-5% of the dairy cows have a three-fold longer productive life than an average cow. The objective of our study is to collect detailed information on German Holstein (GH) cows with an exceptional long productive life at moderate to high lifetime performance and then to assess the predictive ability of single-nucleotide polymorphisms (SNPs) genotyped on the Illumina bovine 50K SNP chip for longevity. Models employed include linkage disequilibrium (LD) methods and Bayesian least absolute shrinkage and selection operator. We sampled two groups of cows. The first group consisted of 972 cows from 33 paternal half-sib groups with at least eight lactations. The second group contained 497 cows with at least twelve lactations. These cows needed less veterinary treatments and were at lower risk to production diseases than their herd mates. Genotypes of 41,518 SNPs were available for more than 450 cows. Training and testing sets were generated by drawing random samples from these cows. An optimized SNP genotype subset was developed using the coefficient of determination of the SNP genotypes for the phenotypes of the cows. The phenotypic variance explained by the best subset of SNPs in strong LD with longevity and lifetime performance was >0.65. Further verification will be done in samples of >1,000 cows and bulls. These results demonstrate the potential for detecting loci in strong LD with longevity at good lifetime performance and genome-guided selection for increasing length of productive life in dairy cattle.

Assessment of breeding strategies in genomic dairy cattle breeding programs

König, S.[1] and Swalve, H.H.[2], [1]University of Göttingen, Institute of Animal Breeding and Genetics, Albrecht-Thaer-Weg 3, 37075 Göttingen, Germany, [2]University of Halle, Institute of Agricultural and Nutritional Sciences, Adam-Kuckhoff-Str. 35, 06108 Halle, Germany; skoenig2@gwdg.de

The availability of genomic breeding values (GEBV) provides the possibility of a modification of dairy cattle breeding programs. Selection index calculations including genomic and phenotypic observations as index sources were used to determine the optimal no. of offspring per genotyped sire with a focus on functional traits and the design of co-operator herds, and to evaluate the importance of a central station test for genotyped bull dams. The no. of required daughter records per sire to achieve a predefined correlation between index and aggregate genotype (rTI) strongly depends on the accuracy of GEBV (rMG) and the heritability of the trait. For a desired rTI of 0.8, heritability=0.10, and rMG=0.5, at least 57 additional daughters have to be included in genetic evaluation. Daughter records of genotyped sires are not necessary for optimistic scenarios where rMG>rTI. Especially for lowly heritable (functional) traits and rMG<0.7, there still is a substantial need of phenotypic daughter records. Phenotypic records from genotyped potential bull dams have no relevance for increasing rTI, even for low rMG of 0.5. Hence, genomic breeding programs should focus on recording functional traits within progeny groups, preferably in co-operator herds. As shown in further scenarios, the availability of highly accurate GEBV for production traits and lowly accurate GEBV for functional traits increase the risk to widen the gap between selection response in production and functionality. Counteractions are possible, e.g. via higher economic weights for the lower heritable traits. Finally, an alternative selection strategy only considering two pathways of selection for genotyped male calves and for cow dams was evaluated. This strategy is competitive with a four-pathway genomic breeding program if the fraction of selected calves is below 1%, and if selection is focussed on functionality.

Estimates of genetic parameters among body condition score and calving traits in first parity canadian ayrshire cows

Bastin, C.[1], Loker, S.[2], Gengler, N.[1,3] and Miglior, F.[4,5], [1]Gembloux Agricultural University, Animal Science Unit, Gembloux, B-5030, Belgium, [2]University of Guelph, CGIL, Dept. of Animal and Poultry Science, Guelph (ON), N1G2W1, Canada, [3]National Fund for Scientific Research, Brussels, B-1000, Belgium, [4]Dairy and Swine Research and Development Centre, Agriculture and Agri-Food Canada, Sherbrooke, J1M1Z3 (QC), Canada, [5]Canadian Dairy Network, Guelph, N1K1E5 (ON), Canada; miglior@cdn.ca

The objective of this study was to estimate genetic correlations between body condition score (BCS) and calving traits using random regression animal models. Calving traits were a) calving ease (CE) scored from 1=unassisted to 4=surgery and b) calf survival (CS) scored from 0=dead to 1=alive. The data analyzed included first parity Ayrshire BCS records collected between 2001 and 2008 by field staff in herds from Québec. BCS observations were available from 100 days before the calving to 335 after the calving. Calving records were extracted for herds with at least one BCS record. Data included 9,944 BCS observations; 12,011 CE records and 11,600 CS records. (Co)variances were estimated by REML using 2 two-traits models. For BCS, regression curve of genetic and permanent environmental effect were modelled using Legendre polynomials of order 3. For calving traits, no covariance between maternal and direct effects was assumed. The genetic correlation between the maternal effect of CE and the BCS during the 100 days before and after calving ranged between -0.40 and -0.25; a low BCS seemed to increase the chance of the cow to calf with difficulty. For direct CE and maternal and direct CS, the highest correlations with BCS occurred in mid and late lactation. The genetic correlations between BCS and direct and maternal CS ranged from 0.2 to 0.4 and the genetic correlation between BCS and direct CE was around 0.6 at 200 days in milk. It indicated that the ability of the cow to recover its body reserves after the postpartum period would increase the chance of the calf to born easily and to survive.

Associations among BCS, milk production and days open in Walloon primiparous dairy cows

Laloux, L.[1], Bastin, C.[2], Gillon, A.[2], Bertozzi, C.[1] and Gengler, N.[2], [1]Association Wallonne de l Elevage asbl, R&D, Rue des Champs Elysées 4, 5590 Ciney, Belgium, [2]Faculté des Sciences Agronimiques de Gembloux, Unité de Zootechnie, Passage des Déportés 2, 5030 Gembloux, Belgium; llaloux@awenet.be

Many studies showed that the reproductive performance of dairy cows may be affected by negative energy balance (NEB). The evolution of the body condition score (BCS) in early lactation may reflect the NEB. For this study, BCS of 1,296 primiparous dairy cows were recorded in 76 Walloon dairy herds between April 2006 and December 2008. Associations among milk production, BCS and days open (DO) were studied using logistic regression models in order to find NEB indicators. DO were transformed in binary data (0.1) based on days in milk (DIM) intervals. Two dependant variables were created: DO3579 for DO between 35 and 79 DIM and DO80119 for DO between 80 and 119. Two sets of independent variables were built. The milk production set contained variables such as milk production/composition and breeding values associated, peak yield, estimated 305-d yields, nadir milk protein and fat content, DIM at nadir/peak, etc. The second set concerned the BCS information. It contained BCS and BCS changes during the 120 firsts d of lactation, nadir BCS and DIM at which it occurred, BCS at calving, etc. Two logistic regression models were developed, one for each dependant variables by using SAS stepwise logistic regression. Significant associations were identified. Except for the herd effect, the only measure significantly associated with the likelihood of DO3579 was the BCS loss between calving and 75 d of lactation. Regarding the second model, two genetic milk yield measures and the mean BCS between 60 and 105 d of lactation were significantly associated with the likelihood of DO80119. In both models, the association between DO and the severity of NEB was negative. These preliminary results show that novel functional trait such as BCS could help Walloon breeders to identify cows that present a risk of poor fertility.

Selection of bull dams for production and functional traits in an open nucleus herd

Hansen Axelsson, H.[1], Johansson, K.[2], Petersson, K.-J.[1], Eriksson, S.[1], Rydhmer, L.[1] and Philipsson, J.[1], [1]Swedish University of Agricultural Sciences, Department of Animal Science and Genetics, P.O. Box 7023, 75007 Uppsala, Sweden, [2]Swedish Dairy Association, Box 7023, 75007 Uppsala, Sweden; Helen.Hansen@ hgen.slu.se

A breeding scheme with an open nucleus herd for dairy cattle is used by the breeding company Viking Genetics. The aim of this study was to compare different scenarios for bull dam selection in a nucleus herd. A deterministic simulation study using selection index methodology was made. In the scenarios studied, different amounts of information on functional traits were available when selecting the bull dams, and the resulting genetic response in these traits was compared. Field-recorded fertility traits used in the scenarios included conception rate at first insemination, interval between calving and first insemination and number of fertility treatments. In addition, field recorded udder health traits and protein yield were included. In a nucleus herd it is possible to record more indicator traits that may improve the selection for functional traits. In the scenarios heat intensity and progesterone levels were considered as new indicator traits recorded in the nucleus herd and the interval between calving and first insemination and somatic cell score were assumed to be better recorded with higher heritability in the nucleus herd than in ordinary herds. Economic weights as practiced in the Nordic Cattle Genetic Evaluation were adapted and used in the scenarios. The results showed that these economic weights currently used lead to undesirable genetic changes in functional traits for bull dams selected in a nucleus environment. Restriction index technique was used to find the economic weights that gave desired response in functional traits. To accomplish this, the economic weights for functional traits had to be raised considerably.

Phenotypic and genetic variation in milk flow for dairy cattle in automatic milking systems

Carlström, C.[1], Pettersson, G.[2], Johansson, K.[3], Stålhammar, H.[4] and Philipsson, J.[1], [1]Swedish University of Agricultural Sciences, Department of Animal Breeding and Genetics, Box 7023, S-750 07 Uppsala, Sweden, [2]Swedish University of Agricultural Sciences, Department of Animal Nutrition and Management, Kungsängen Research Centre, S-753 23 Uppsala, Sweden, [3]Swedish Dairy Association, Box 210, S-101 24 Stockholm, Sweden, [4]Viking Genetics, Box 64, S-532 21 Skara, Sweden; caroline.carlstrom@hgen.slu.se

Increased dairy herd sizes have led to installment of automatic milking systems in many herds. Such systems put new demands on the cows as regards ease of milking. They also enable on-line recording of a number of traits. The aim of this study was to estimate the phenotypic and genetic variation in milk flow rate, udder conformation and teat placements recorded for dairy cattle in automatic milking systems. The first step was to investigate how data directly extracted from automatic milking systems should be edited in order to be usable for genetic analyses. Data on average milk flow (kg/min) and peak flow (kg/min) per udder quarter from 25 Swedish dairy herds with automatic milking systems were used. In total data from three million milkings of 6,200 cows of the two breeds Swedish Holstein and Swedish Red from the years 2004-2008 were used. After an extensive editing process 67% of the milkings were considered suitable and used for a first analysis. Both milk flow traits were normally distributed and showed large variation among cows. The total means were 3.44 kg/min (SD=1.05) for the average milk flow and 5.03 kg/min (SD=1.31) for the peak flow. Genetic parameters were estimated with a bivariate animal model. Preliminary estimates of the heritabilities were 0.52 and 0.51 respectively for the two milk flow traits. The genetic correlation between them was 0.98. As a following step genetic parameters for milk flow traits across lactations and breeds will be estimated as well as correlations with udder and teat conformation.

Hoof diseases in dairy cows have different genetic components

Buch, L.H.[1,2], Sørensen, A.C.[1], Lassen, J.[1], Berg, P.[1], Jakobsen, J.H.[3], Eriksson, J.Å.[4] and Sørensen, M.K.[1,2], [1]Aarhus University, Dept. of Genetics and Biotechnology, Box 50, 8830 Tjele, Denmark, [2]Danish Cattle Federation, Udkærsvej 15, 8200 Aarhus N, Denmark, [3]SLU, Dept. of Animal Breeding and Genetics, Box 7023, 75007 Uppsala, Sweden, [4]Swedish Dairy Association, Box 210, 10124 Stockholm, Sweden; Line. HjortoBuch@agrsci.dk

Hoof diseases (HD) are an increasing problem in many Nordic dairy herds. It is uncertain to what extent previous breeding strategies have contributed to this development. To study the problem genetic correlations (r_g) between four HD, protein yield, mastitis, days from calving to first insemination (CFI) and number of inseminations (NI) were estimated in first-parity Swedish Red cows using tri-variate linear animal models. Occurrence of dermatitis (DE), heel horn erosion (HH), sole hemorrhage (SH), and sole ulcer (SU) were reported by hoof trimmers during their routine visits. After editing the dataset contained about 64 000 hoof records. Heritabilities were low for all HD (0.03 to 0.05). The HD fall into two groups: (1) DE and HH are often related to hygiene (r_g=0.87) and (2) SH and SU are often related to feeding (r_g=0.73). The r_g between the two groups were low ($r_g \leq 0.23$). These results indicate that the hygiene-related HD (HRHD) and the feed-related HD (FRHD) are not influenced by the exact same genes. The r_g between all four HD and protein yield were low to moderate and unfavourable. In other words, uncritical selection for production may result in more cases of HD. The r_g between FRHD and mastitis were moderate and favourable (0.35 and 0.32) whereas the r_g between HRHD and mastitis were low and not significantly different from zero. Low to moderate r_g were found between FRHD and CFI (0.10 and 0.33), and low r_g were found between HRHD and CFI. The r_g between HRHD and NI were moderate and favourable (0.32 and 0.22) whereas the r_g between FRHD and NI were low and not significantly different from zero. In general, the two groups of HD show different patterns of correlations to the other functional traits, but both are unfavourably correlated to production.

Genetic analysis of reproductive disorders and their relationship to fertility in Fleckvieh dual purpose cattle

Koeck, A.[1], Egger-Danner, C.[2], Fuerst, C.[2] and Fuerst-Waltl, B.[1], [1]University of Natural Resources and Applied Life Sciences, Department of Sustainable Agricultural Systems, Division Livestock Sciences, Gregor-Mendel-Strasse 33, 1180 Vienna, Austria, [2]ZuchtData EDV-Dienstleistungen GmbH, Dresdner Straße 89/19, 1200 Vienna, Austria; astrid.koeck@boku.ac.at

The objective of this study was to estimate genetic parameters for various reproductive disorders based on veterinary treatments for Austrian Fleckvieh (Simmental) dual purpose cattle. The analysed health traits included metritis (MET) within 150 d after calving, cystic ovaries (CYST) within 150 d after calving, retained placenta/puerperal diseases (RP/PUERP) till 14 d after calving and overall reproductive disorders (REPR) till 150 d after calving. Disease frequencies were 3.7, 8.3, 4.4 and 13.4% for MET, CYST, RP/PUERP and REPR, respectively. Heritabilities were estimated with threshold sire models (TM) and linear animal models (LM). Both the TM and LM analyses were based on the generalized linear mixed model framework. The models included parity as fixed effect and herd-year-season of calving, genetic, permanent environmental and residual effects as random effects. The TM estimates for heritability ranged from 0.064 to 0.088, the LM estimates were lower (0.015 to 0.041). Furthermore measures of fertility were calculated from calving and insemination dates and included non-return-rate 56 (NR56), interval from calving to first insemination (CFI) and interval between first and last insemination (FLI). Genetic correlations between health and fertility traits were estimated with bivariate linear animal models. The estimates of genetic correlations were low to moderately high (-0.109 to 0.756). Thus, the incorporation of reproductive disorders into the Austrian fertility index may offer the potential to improve the accuracy of breeding value prediction for fertility in Fleckvieh dual purpose cattle.

Profile of direct and maternal genetic variance components of body measurement traits in Japanese Black calves
Oikawa, T., Koyama, K., Munim, T. and Masamitsu, T., Okayama University, Graduate School of Natural Science and Technology, 1-1-1 Tsushima-naka, Okayama-city, 700-8530, Japan; toikawa@cc.okayama-u. ac.jp

Variance components were estimated for 9 body measurements in Japanese Black calves. The measurement traits on 932 calves at early stage of growth (0-8 mo.) recorded monthly (Total number =14,681) for 32 years at an experimental station of Okayama University. Additive direct and maternal genetic variance component were estimated by REML using VCE5 for the measurement traits. Body weight also recorded every month. The direct genetic variance of two traits of body height, wither height and hip height showed high heritability at the early period (0.1 mo.) and low at middle of the period (4 mo.) and again high at the later period(8 mo.). As a result it showed concave curve. During early stage of growth due to the fluctuation of heritability maternal genetic variance component showed convex curve. The direct genetic variance components of 3 traits measuring body width, hip width, thurl width, and pin bone width showed increasing trend where the maternal variance components showed decreasing trend. The direct genetic variance component of canon circumference of right fore leg was consistently high where the maternal component showed hill like curve. During weaning stage the genetic correlation between direct and maternal genetic effect were weekly negative for body weight, hip length, hip width, pin bone width and body length. There was no genetic correlation for hip width, thurl width, and canon circumference and was slightly positive for wither height, hip height, chest depth and chest girth.

Distribution of the polled allele in Austrian Fleckvieh
Baumung, R.[1], Stingler, N.[1], Fuerst, C.[2] and Willam, A.[1], [1]University of Natural Resources and Applied Life Sciences Vienna, Dep. Sust. Agric. Syst., Div. Livestock Sciences, Gregor-Mendel-Strasse 33, A1180 Vienna, Austria, [2]ZuchtData EDV-Dienstleistungen GmbH, Dresdner Strasse 89/19, A1200 Vienna, Austria; roswitha.baumung@boku.ac.at

Modern cattle housing systems result in an increased interest for polled cattle. However, in Austria the use of polled Fleckvieh (Simmental) is still quite low. Therefore the study aimed at finding how many polled animals are already available in the Austrian Fleckvieh population. For this purpose a list consisting of 498 polled sires from Austria and Germany was generated where the bulls' name, ear tag number, birth year and exact horn status were recorded. This list was compared to the current top list of dual purpose Fleckvieh sires ranking the 100 best bulls according to their total merit index. The first and only polled sire appeared on rank 56. Further a list consisting of the 500 genetically most important ancestors (i.e. showing the highest marginal genetic contribution to a certain reference population) was generated and compared with the list including all polled bulls. In a reference population consisting of all dual purpose animals born between 2002 and 2006 two polled sires appeared with a marginal genetic contribution of less than 0.1% each. However, in a list of the 500 most important ancestors for a reference population consisting of animals bred mainly for beef, 15 polled sires were found. The first nine explained 1.2% of the genes in this reference population. The other six contributed less than 0.1% each. Further, allele frequencies of the polled allele 'P' were estimated using a modification of the gene dropping approach. Allele frequencies in two reference groups consisting of animals bred for beef and dual-purpose differed significantly and were 0.84% and 0.05%, respectively.

Genetic relationships between metabolic diseases and fertility in Fleckvieh dual purpose cattle
Koeck, A.[1], Fuerst-Waltl, B.[1], Fuerst, C.[2] and Egger-Danner, C.[2], [1]University of Natural Resources and Applied Life Sciences, Department of Sustainable Agricultural Systems, Division Livestock Sciences, Gregor-Mendel-Strasse 33, 1180 Vienna, Austria, [2]ZuchtData EDV-Dienstleistungen GmbH, Dresdner Straße 89/19, 1200 Vienna, Austria; astrid.koeck@boku.ac.at

The objectives of this study were to investigate the genetic variability of some metabolic diseases in Austrian Fleckvieh (Simmental) dual purpose cattle and to estimate the genetic relationship to fertility traits. The recording of metabolic diseases was based on veterinary treatments and included milk fever (MF) within 10 d before to 10 d after calving, ketosis (KET) within 10 d before to 100 d after calving and all metabolic diseases (META) from 10 d before to 100 d after calving. The overall incidences of MF, KET and META were 2.2, 0.9 and 3.3%, respectively. Heritabilities for metabolic diseases were estimated with threshold sire models (TM) and linear animal models (LM). Both the TM and LM analyses were based on the generalized linear mixed model framework. The models included parity as fixed effect and herd-year-season of calving, genetic, permanent environmental and residual effects as random effects. The TM estimates for heritability were 0.161, 0.175 and 0.123 for MF, KET and META, respectively; the LM estimates were lower (0.030, 0.023 and 0.03). Furthermore, genetic correlations between metabolic diseases and non-return-rate 56 (NR56), interval from calving to first insemination (CFI) and interval between first and last insemination (FLI) were estimated with bivariate linear animal models. Genetic correlations between metabolic diseases and fertility were low to moderate.

On-farm welfare assessment in cattle and risk factors for selected welfare problems: the Welfare Quality® approach
Winckler, C.[1] and Knierim, U.[2], [1]University of Natural Resources and Applied Life Sciences (BOKU), Department of Sustainable Agricultural Systems, Gregor-Mendel-Strasse 33, 1180 Vienna, Austria, [2]University of Kassel, Faculty of Organic Agricultural Science, Department of Farm Animal Behaviour and Husbandry, Nordbahnhofstrasse 1a, 37213 Witzenhausen, Germany; christoph.winckler@boku.ac.at

One of the aims of Welfare Quality®, a European research project on integration of animal welfare in the food quality chain, is the development of on-farm welfare assessment systems that are scientifically sound and feasible at the same time. The full on-farm monitoring systems emphasise the animal's point of view by placing much importance on animal-based measures. However, environmental and/or management-related measures have also been included where applicable, i.e. if no feasible animal-based measure was available. Four animal welfare principles (good housing, good feeding, good health, appropriate behaviour) have been defined and within these principles twelve distinct but complementary animal welfare criteria such as 'absence of injuries' or 'expression of social behaviours' were highlighted. The Welfare Quality® assessment shall allow welfare specific product information. At the same time it may facilitate advice on possible improvements. This paper focuses on the assessment systems developed for dairy and beef cattle. They have been tested on commercial farms in several EU countries. In terms of feasibility, it is possible for a single observer to carry out the farm assessment within a one-day visit. Because effective improvement strategies require sound knowledge on the underlying causes and risk factors, we will further present results of epidemiological studies with regard to lameness and hock lesions in dairy cattle.

Risk based indicators used for assessing animal welfare in livestock herds

Nielsen, T.R., Bonde, M.K., Thomsen, P.T. and Sørensen, J.T., Aarhus University, Faculty of Agricultural Sciences, Department of Animal Health, Welfare and Nutrition, P.O. Box 50, DK-8830 Tjele, Denmark; tine.rousingnielsen@agrsci.dk

Animal based and environmental indicators reflect two different but equally important aspects of on-farm welfare assessments. Animal based indicators can at their best be seen as direct indicators of animal welfare. However, they are expensive to record and will often only reflect animal welfare at a certain point in time or only a fraction of the animals and the measurement may be of low validity or precision. Environmetal indicators more reflect risk factors for potential welfare problems and will often be precise and problem oriented. However, animals with similar environmental conditions may have very different animal welfare. Expert opinion studies used for selecting the most important indicators for on farm welfare assessment systems often led to rather complex protocols consisting of animal based indicators including health and behaviour indicators and environmental indicators including systems based and management based indicators. This complexity is a major challenge in the development of operational assessment systems. A way forward for on-farm welfare assessment systems for decision support is to develop systems with updating welfare assessment. Such systems will exploit previous knowledge and optimize the time used on new recordings. For systems relevant for authorities and maybe also for certification systems it may be an idea to develop risk based systems where expensive recordings are directed by cheaper external indicators used to point out risk farms. A concept for development of risk based indicators will be described using examples from recent research in the area.

Potential of routinely collected farm data to monitor dairy cattle welfare

De Vries, M.[1], Bokkers, E.A.M.[1], Dijkstra, T.[2], Van Schaik, G.[2] and De Boer, I.J.M.[1], [1]Wageningen University, Animal Production Systems group, Marijkeweg 40, 6709 PG Wageningen, Netherlands, [2]Animal Health Service, Arnsbergstraat 7, 7400 AA Deventer, Netherlands; marion.devries@wur.nl

In the EU Welfare Quality® (WQ) project, validated measures have been developed to assess the welfare status of dairy farms. The on-farm welfare assessment, however, is a time consuming and therefore expensive activity. Routinely collected data of various farm variables relating to health, fertility, productivity and product quality, may have potential to signal impaired animal welfare on dairy farms. To determine this potential, an analysis of the associations between such farm variables and the thirty-seven variables that are part of the WQ dairy cattle welfare assessment is required. Three WQ variables, i.e. somatic cell count, mortality and culling rate, are already routinely collected as part of regular farm records in the Netherlands. The other WQ variables are not collected as such, but it was expected that they could be associated with farm variables. The objective of this study was to explore associations between farm variables and WQ variables in a literature review. Literature showed that nearly all WQ variables are associated with farm variables. Often, single WQ variables are related to multiple farm variables. For example, the WQ variable prevalence of lameness can be associated with the farm variables milk yield, fat and protein contents of the milk, calving to first service interval, non-return rate, number of services, average age of the herd, and culling rate. Inversely, single farm variables are associated with multiple welfare variables. Similar association patterns were found for other variables and will be presented. It is concluded that routinely collected farm data show potential to signal impaired dairy cattle welfare. Due to the multifactorial nature of dairy cattle welfare, intelligent combinations of farm variables are most promising to detect farms at risk. In a next step, associations will be verified in a field study.

Development of a welfare assessment system on farm and at the slaughterhouse in pigs (welfare quality® project)

Temple, D.[1], Velarde, A.[2], Manteca, X.[1], Llonch, P.[2], Rodriguez, P.[2] and Dalmau, A.[2], [1]Veterinary School (UAB), Ethology and Animal Welfare, Cerdanyola del Valles, 08193, Spain, [2]IRTA, Animal Welfare, Camps i Armet, 17121 Monells (Girona), Spain; Deborah.Temple@irta.es

A protocol to assess animal welfare was developed for weaners and growing pigs on farm and for growing pigs at the slaughterhouse according to the 4 animal welfare principles of the Welfare Quality® project. Twelve criteria related to good feeding, good resting, good health and appropriate behaviour were assessed using mainly animal-based measures. A total of 30 intensive, 10 extensive farms and 10 slaughterhouses were assessed. Differences were found among different intensive farms, among different extensive farms and among different slaughterhouses for most of the measures. On intensive farms the highest prevalence of a measure was found for moderate bursitis with a mean of 45.5% animals affected (range 7.2-83.1%). In extensive conditions the highest prevalence of a measure was also found for moderate bursitis with a mean of 7.5% of animals affected (range 3.2-13.8%). For most of the measures, differences between slaughterhouses were found to be caused by the installations and management of the slaughterhouse, such as general fear, slipping and falling or stunning effectiveness, but not for the measures taken to assess transport conditions or farm origin, such as lameness or sick and dead animals In the slaughterhouse, the highest prevalence of a measure was found for slip with a mean of 28.0% of the animals unloaded affected (range 8-59%).

The welfare of sheep under extensive management systems

Goddard, P.[1], Stott, A.[2], Waterhouse, A.[2], Milne, C.[2] and Phillips, K.[3], [1]Macaulay Institute, Craigiebuckler, Aberdeen, AB15 8QH, United Kingdom, [2]Scottish Agricultural College, Kings Buildings, Edinburgh, EH9 3JG, United Kingdom, [3]ADAS, Woodthorne, Wolverhampton, WV6 8TQ, United Kingdom; p.goddard@ macaulay.ac.uk

Across Europe there is growing concern for the welfare of farmed animals. Welfare is important for system sustainability: Eurobarometer surveys indicate increasing levels of public interest. For more intensive systems, welfare assessment has historically centred on provision of resources related to animal needs such as space and bedding. A noticeable shift in approach to consider 'output' indicators of welfare state now emphasises a more animal-centred view. These indicators are relatively poorly developed for sheep but in extensive systems there is good opportunity for sheep to express normal behaviour and utilize natural resources, features rated highly by consumers. To be viable financially, stockpersons must be able to deliver high levels of welfare in a cost-effective way. Much of the 'provision' of welfare requires a high labour input yet the trend in extensive sheep systems is often to reduce this. There is thus a tension between the capacity of animals within the system to respond to 'natural' challenges of their extensive environment and the ability of the system and its people, to support sheep at key times when their genotypic survival traits are at risk of failure. This can be through planned intervention at key handling events and tactical management of nutrition based around natural herbage. Variation in the popular view of what constitutes 'extensive' and 'natural' reflects this dichotomy between animals left to their own devices or supported by the stockperson when necessary; the public view that welfare could be higher under extensive management may much depend on the external conditions and farm systems. Welfare assessment in extensive systems needs to combine elements related both to intensive activities (handling during management operations) and also the long-term experiences of the sheep when they range largely unsupervised.

Risk factors related with health and behaviour traits in welfare of horses

Distl, O., University of Veterinary Medicine Hannover, Institute for Animal Breeding and Genetics, Buenteweg 17p, 30559 Hannover, Germany; ottmar.distl@tiho-hannover.de

Risk factors for welfare in horses are strongly dependent of the use of the horse. Sport horses are subjected to fatalities and injuries while racing. Epidemiological approaches were of great value to identify risk factors for fatalities and injuries in this competitive sport and could initiate efforts to safeguard equine welfare. Improved schooling and training of horses for the type of race and providing good-to-soft ground decreased the number of horses with fatalities. Duration of the racing career of the horse, the number of starts per time period accumulated in a short time before the race with the fatality and over the racing career are further important risk factors. Orthopaedic and traumatic disorders have the largest proportion of diseases across different horse breeds, followed by gastrointestinal (colic), respiratory and heart diseases. Orthopaedic and traumatic disorders are often related to degenerative conditions, feeding practices and diet composition, possibilities for movements and the intense use of the horse. Diseases of the locomotory system are the most important causes of premature retirement and culling of older horses as well as young horses at the beginning of their careers. For colics many risk factors have been identified including feeding practices, parasite control and horse related factors such as sex, age and breed. There are many different diseases and specific risk factors. Crib-biting, wood-chewing, weaving and box-walking are commonly observed stereotypic behaviours in horses. Changes in feeding, housing and weaning practices substantially lower the incidence of abnormal behaviour in young horses. Housing and management conditions allowing tactile contact with other horses, daily free movement on paddocks or pasture and sufficient amounts of roughage but less or no concentrates should prevent most of the stereotypies. A multidisciplinary approach seems most useful to identify the main contributing factors for equine welfare benchmarks.

Herd- and sow level risk factors for shoulder ulcers in lactating sows

Bonde, M., Faculty of Agricultural Sciences, University of Aarhus, Animal Health, Welfare and Nutrition, Blichers Alle 20, P.O.Box 50, 8830 Tjele, Denmark; marianne.bonde@agrsci.dk

Shoulder ulcers in lactating sows are commonly observed in intensive pig production systems. The ulcers are characterised as pressure injuries, and they are presumed to be affected by multiple factors related to sow condition as well as environmental and management factors in the herd. Therefore, shoulder ulcers can be a key indicator of sow welfare in the herd, reflecting suitability of the housing system, management of the herd as well as the general condition of the sows. The relationship between housing system, herd management, clinical condition of the sows and the prevalence of shoulder ulcers has been investigated in this survey, including 3,831 lactating sows from 98 herds. The sows in all herds were housed in conventional farrowing crates during lactation, while pregnant sows were housed in various group housing systems. Each herd was visited twice in the spring and summer 2008, respectively. Data were collected regarding housing and management of the sows, and clinical examinations of sows with respect to presence of shoulder ulcers and general health were carried out. Overall 17,2% of the sows were diagnosed with shoulder ulcers. The herd prevalence ranged from 2,5% to 42,5%, inter-quartile range: 10-24,7%. The variation in prevalence of shoulder ulcers between herds suggests that herd level risk factors may be present. Potential risk factors could relate to flooring, bedding, space allowance, pen hygiene, and feeding management, and results from the analysis of these factors will be presented at the meeting. Further, sows housed in similar environments, for example in the same herd obviously differ in their susceptibility to shoulder ulcers. The association between shoulder ulcers and other clinical characteristics of the sow will also be presented and discussed at the meeting.

The identification of farm characteristics, and welfare associated management practices, as potential risk factors for the presence of pathologies and *Salmonella* in slaughtered finishing pigs

Sanchez-Vazquez, M.[1], Smith, R.[2], Gunn, G.[1], Lewis, F.[1] and Edwards, S.A.[3], [1]Scottish Agricultural College, Inverness, IV2 4JZ, United Kingdom, [2]Veterinary Laboratory Agency, Weybridge, KT15 3NB, United Kingdom, [3]Newcastle University, Newcastle, NE1 7RU, United Kingdom; Sandra.Edwards@ncl.ac.uk

In recent years, animal welfare (AW) has become an essential topic in animal production, especially in Great Britain, with an increasing demand from the consumers for raising welfare standard. Practices such as the use of bedding material and outdoors production have been implemented aiming to promote AW in the pig production. Our study combined existing information from abattoir monitoring health schemes and quality assurance programmes (QAPs), to help identify welfare associated management practises and other farm characteristics that influence pig health. Information from farms on: pig stocking levels; feeding practices; housing systems and geographical location were collected from the QAPs. Health assessment results on eight different conditions (on the skin, in the thorax and in the abdomen) and Salmonella serology taken from slaughtered pigs were also collected for those farms. GLMM were used for multivariable analysis, allowing for clustering at the batch or herd level. Partly-slatted floors appeared as a potential risk factor for the presence of two respiratory conditions (Enzootic pneumonia-like lesions and pleurisy), whereas the use of solid flooring with bedding appeared to be protective. In contrast, solid flooring was associated with a higher odds of a sample being Salmonella positive. Farms that carried out all stages of production indoors had a lower risk of pleurisy in comparison with those reporting some stages of production outdoors (e.g. farrowing). By exploring these data, this project has led the British pig industry to a better understanding of how farm characteristics and management practices, especially those associated with animal welfare, influence the health of farmed pigs.

Optimising use of labour at lambing time for more profit and higher welfare in extensive sheep farming systems

Kirwan, S.[1] and Stott, A.W.[2], [1]Scottish Agricultural College, Land Economy and Research Group, Ferguson Building, Craibstone Estate, Aberdeen AB21 9YA, United Kingdom, [2]Scottish Agricultural College, Land Economy and Research Group, Kings Buildings, West Mains Road, Edinburgh EH9 3JG, United Kingdom; susanne.grund@sac.ac.uk

The economic situation of extensive sheep farming is at risk, with reduced Government subsidy and low returns for hill sheep. Labour has been shown to be a key input factor for both productivity and welfare in extensive production systems. Despite that little research has been done to try and optimise labour input for productivity and welfare. This study is a first attempt to link labour input, productivity and welfare using field studies on both research and commercial farms as well as linear programming (LP) modelling. Preliminary data analysis highlighted that Three high labour demand tasks (e.g driving – 196 min/day) were not closely associated with productivity and welfare, while critical productivity and welfare tasks (e.g. lambing) had a very low labour demand (< 5min/day). Modelling showed that small increases in the risk to productivity lead to high labour savings per ewe/day 7.07min, 6.97min, and 4.80min respectively, however the welfare risk for individual sheep is much higher. The model results indicate that hill sheep live to a high welfare standard with low risks to productivity, there is scope for beneficial marginal reductions in labour but larger reductions within legal guidelines could seriously endanger the welfare of individual sheep without great risk to productivity.

Farmers talking about animal welfare

Karkinen, K.T. and Saarinen, K., University of Joensuu, Sociology and Public Policy, Yliopistokatu, P.O. Box 111, Joensuu, Finland; katri.karkinen@joensuu.fi

The article offers a discursive approach to the study of animal welfare. The data comes from interviews of farmers rearing dairy cattle and of other specialists in animal husbandry from Southeast Finland. All interviewees shared a concern for animal welfare. The farmers regarded the conditions in which the animals are kept as a key element in animal welfare. The agricultural advisor regarded animal welfare as something that already exists but could be improved. The representative of dairy plant took up the importance of hygiene. The veterinary surgeon's concern for animal welfare emphasised complying with the law. Finally, the vendor of stock feed saw it as the responsibility of the farmer to find solutions to any problems with cattle breeding. The European Union offers farmers subsidies for improving cattle welfare. The subsidy is an incentive to build new cowsheds. Dairy farmers do not find the enlargement of their facilities as the only way to improve animal welfare. Instead, they talk of other upgrading measures on the farm, such as improved techniques of growing silage, as possible contributions to animal welfare. At the beginning of the paper, we will briefly introduce the social scientific and ethological approaches to animal welfare.

Welfare evaluation system in goats bred under extensive conditions

Rodero, E.[1], Perez Almero, J.L.[2], Cruz, V.[3], Sanchez, M.D.[4] and Alcalde, M.J.[5], [1]Univ. Cordoba, Campus de Rabanales, 14071, Spain, [2]IFAPA, Ctra Sevilla - Cazalla Km. 12,2, 41200, Spain, [3]Asociación caprina Negra Serrana y Blanca Andaluza, Juan Carlos I, 9, 23200, Spain, [4]Asociación caprina Blanca Celtibérica, Camino Angosto, s/n, 02530, Spain, [5]Universidad de Sevilla, Ctra. Utrera, Km1, 41013, Spain; aldea@us.es

The Spanish local meat goat breeds are bred under extensive conditions. It is believed that these breeds do not need lots of handling and their welfare levels are invariable. Nevertheless, they need the right manage to ensure their animal welfare and their adaptation to the environment. An evaluation system is proposed through this paper, which has into account - in 13 sections and 120 items -, for instance, vigilance, buildings, environment and resources, food, healthiness and behaviour, handling of adult animals and kids, records and freedom of movement. The sections are adjusted to the global marking (maximum: 135 points for the worse), the sanitary and behavioural issues are more appreciated (x2) and the patio exterior characteristics, - which is scarcely used - are less appreciated (x0.75). The method was put into practice on a total of 53 farms belonging to the Blanca Serrana Andaluza Goat Breed (BSA, 18), the Blanca Celtibérica Goat Breed (BC, 20) and the Negra Serrana Goat Breed (NS, 15). The average results were enough good (table). No significant differences were noted among breeds (0.05). Although large coefficients of variation were observed, only to farms exceed the 40 points (<34% of the maximum). The Blanca Celtibérica Goat Breed deserves different results in certain aspects, for instance in those related to the norms of grazing (0.1) and the behavioural aspects (0.05). However, compared to Negra Serrana Goat Breed, their results were worse for records (0.1) and sanitary state and resources availability (0.05).

Pre-slaughter conditions, genetic-stress susceptibility, meat quality, and animal welfare in pigs: results from a Spanish survey

Guàrdia, M.D.[1], Estany, J.[2], Balasch, S.[3], Tor, M.[2], Oliver, M.A.[1], Gispert, M.[1] and Diestre, A.[4], [1]IRTA, Finca Camps i Armet, 17121, Spain, [2]Universitat de LLeida, Rovira Roure 191, 25198, Spain, [3]Universidad Politécnica de Valencia, Camino de Vera, 46022, Spain, [4]PIC, Avinguda Argüí 80, 08190, Spain; jestany@prodan.udl.es

Meat consumers increasingly demand that animals are reared, transported and slaughtered in a humane way. A total of 116 deliveries, comprising 15,695 commercial pigs delivered from commercial farms to five Spanish pig commercial abattoirs were surveyed. Polychotomous logistic regression models were used to identify and assess the risk factors for pork becoming PSE or DFD meat, as well as for skin damage occurrence. The effect of pre-slaughter conditions on blood stress indicators (cortisol, lactate and CPK) were analysed using a mixed model. The relationship of blood stress indicators with pork quality and skin damage was also assessed. Abattoirs should be especially careful in summer, when the risk of PSE was double than in winter. The effect of transportation time decreased the risk of PSE but not irrespective of stocking density. For transits longer than 3 h, the PSE risk increased with stocking density during transport, while the opposite occurred for shorter transits. The risk of DFD increased with stocking density, lairage time, and with on-farm fasting times longer than 22 h. Carcass skin damage was associated with increased risk of obtaining PSE and DFD pork and with higher levels of blood stress indices. The use of a skin damage score as a pork quality index may help to monitor both animal welfare and pork quality. The results confirm that pre-slaughter conditions and the RYR1 genotype influence the risk of producing PSE pork, skin damage incidence, and blood stress indicators, while the risk of DFD pork is only affected by the ante mortem handling conditions. However, only 10% of the remaining variability associated to these traits was explained by the delivery. Therefore, there is still considerable scope for improving meat quality and welfare conditions.

Risk factors for sow and piglet welfare indicators in an on-farm monitoring system

Scott, K.[1], Binnendijk, G.P.[2], Edwards, S.A.[1], Guy, J.H.[1], Kiezebrink, M.C.[2] and Vermeer, H.M.[2], [1]., Newcastle University, NE1 7RU, United Kingdom, [2]., Wageningen University and Research Centre, Netherlands; Sandra.Edwards@ncl.ac.uk

Many existing welfare monitoring systems are based predominantly on environmental and production-based descriptors, rather than directly assessing the welfare of the animals themselves. The aim of the EU Welfare Quality project is to use predominately animal-based indicators of welfare, including amongst others body injuries, body condition and stereotyped behaviour, in an on-farm monitoring system. A prototype monitoring system to evaluate sow and piglet welfare has been devised and evaluated on a total of 82 farms in the UK and The Netherlands. These farms encompassed a wide variety of production systems, enabling risk factors for poor welfare outcomes to be explored. Welfare standards on the farms visited were generally good, with 1.2, 0.8 and 1.1% respectively of pregnant and lactating sows observed as having substantial body injuries, bursitis and vulval lesions. Risk factors associated with housing and feeding systems were identified for certain indicators. Significant effects of feeding method were observed on body injury scores ($P<0.001$), with farms where sows were fed outdoors in groups tending to have lower median body injury scores than farms using indoor electronic sow feeding (ESF) systems. Vulval lesions were significantly affected by feeding method ($P<0.001$); farms using ESF had significantly higher median vulval lesion scores compared with farms where sows were fed individually ($P<0.05$), group-fed indoors ($P<0.001$) or group-fed outdoors ($P<0.01$). Median bursitis scores were higher on farms with fully-slatted flooring than other floor types ($P<0.001$). Stereotyped behaviours tended to be more prevalent on farms where sows had no straw bedding. Levels of withdrawal response to human approach were generally low (little withdrawal behaviour) on all farms visited; no significant effect of housing system (stalls, indoor loose housing or outdoors) was observed on median response scores.

Impact of heat stress on production, energy efficiency and fertility of Holstein cows in pasture based systems in North West Germany

König, S., Schierenbeck, S., Schmitz, C., Brügemann, K., V. Borstel, U.U. and Gauly, M., University of Göttingen, Institute of Animal Breeding and Genetics, Albrecht-Thaer-Weg 3, 37075 Göttingen, Germany; skoenig2@gwdg.de

Heat stress in dairy cattle is caused by combinations of temperature, relative humidity, air movement and solar radiation. Daily meteorological data, recorded for the years 2000 to 2003, were merged with respective test-day records and insemination dates of 41,275 Holstein cows located on 989 farms in North West Germany. The aim was to assess the impact of temperature (T in °C), relative humidity (RH in %) and temperature-humidity indices (THI) on production and fertility. The THI was calculated as $THI = 0.8T + RH(T - 14.4) + 46.4$. Linear models (Gaussian traits) and GLMMs with probit link functions (categorical traits) were used for the analysis. An increase of T, RH, and THI was only associated with a significant decrease ($P<0.001$) in test-day yields for milk (kg), protein (kg), and protein (%) when analyzing a subset of 140 farms keeping their 5,371 cows on pasture from May to September. A substantial reduction of observations in favourable classes for combinations of milk urea nitrogen and protein (%), and an increase of milk urea nitrogen (ppm), both indicating problems in energy efficiency, was noted with higher T, RH, and THI. Analyzed fertility traits were non-return rates after 56 d (NR56) and 90 d (NR90), calving interval (CI), and calving ease (CE). No universal trends were observed for fertility traits in relation to T and THI. Lower conception rates as measured by NR56 and NR90 were significantly ($P<0.05$) associated with an increase of RH on days within the interval of 3 weeks before the insemination date, but unrelated to later RH measurements. Longer CI with increasing RH was mainly due to the detrimental impact of RH on NR, because the association between RH observed directly after calving and the interval from calving to first service was negligible. In conclusion, the RH component of heat stress was more important than T or THI for all analyzed traits.

Microclimatic indexes to compare different dairy cows housing systems in different climatic conditions

Menesatti, P.[1], D'andrea, S.[1], Cavalieri, A.[1], Baldi, M.[2], Lanini, M.[3], Vitali, A.[4], Lacetera, N.[4], Bernabucci, U.[4] and Nardone, A.[4], [1]CRA-ING, Monterotondo (Roma), Italy, [2]CNR-IBIMET, Roma, Italy, [3]CNR-IBIMET, Firenze, Italy, [4]UNITUS-DIPA, Viterbo, Italy; paolo.menesatti@entecra.it

In the last few years, global warming increased frequency and magnitude of heat stress in dairy cattle. The aim of this work was to investigate how different housing systems in different climatic conditions, affect both temperature (T) and Temperature-Humidity Index (THI) inside the cow buildings. The THI is recognized as an accurate and useful indicator to evaluate heat stress in dairy cattle, since it takes into consideration both air temperature and relative humidity. Three dairy cow farms were considered: A, cooling-ventilation system with concrete roofing; B, no cooling-ventilation system with concrete roofing; C, no cooling-ventilation system with isolated roofing. All the barns were open structures. To compare the housing system performances, it was necessary to setup indices to measure T and THI differences between inside and outside the buildings. Climatic data were recorded from May to September 2007 and 2008 outside and inside the buildings by using weather stations and thermohygrometric data-loggers, respectively. The sum of the values exceeding T>24 °C and THI>72, inside (SDI, SHDI) and outside (SDO, SHDO) the buildings were calculated together with their ratios. In practice, higher ratios indicated a lower microclimatic efficiency of the structures. The farms A, B and C had SDI/SDO values equal to 1.09, 1.16, and 1.01, and SHDI/SHDO values equal to 1.25, 1.45, and 1.32, respectively. In synthesis, results show that isolated roofing highlights the lowest increase of the internal temperature, while the cooling-ventilation system is the most effective for the limitation of THI increment. Hence, this study suggests that the presence of a cooling-ventilation system is more efficient than thermal isolation to realize microclimatic conditions which may permit to reduce welfare decline during hot seasons.

Variation of milk yield in relation to temperature humidity index in different dairy cow housing systems: a preliminary time-lagged approach

Aguzzi, J.[1], Vitali, A.[2], D'andrea, S.[3], Cavalieri, A.[3] and Menesatti, P.[3], [1]ICM-CSIC, Barcelona, Spain, [2]UNITUS-DIPA, Viterbo, Italy, [3]CRA-ING, Monterotondo (Roma), Italy; paolo.menesatti@entecra.it

In this study we surveyed variation in milk yield of Holstein Friesian cows according to a climatic temperature-humidity index (THI). We considered four farms located in the area of centre Italy: two equipped with sprinklers and fans, and two without any cooling system. We recorded continuously over 24 months (2007-2008), at a frequency of 30, temperature and humidity by thermohygrometric data-loggers placed inside the barns. In the same farms, we also recorded monthly average of milk yield. THI index was computed by Kelly and Bond (1971) formula. As synthetic global microclimatic index, we used THI-hours per month (THI-hM, as the sum of hours per month where THI >72). We observed that THI-hM increases more than the 600% and 400% from June to August when compared to May and September, respectively, for all farms typologies. Milk production was compared to THI-hM June-August (2007+2008 years) by means of time lagged correlation models. Respect to THI-hM June-August (2007+2008 years), monthly milk yield was lagged in a first model of zero-one month (0-1 lag), and in a second model of zero-one-two months (August-October, 0-2 lag). Significant negative correlations were found only for farms without sprinklers and fans both for 0-1 lag and 0-2 lag. These results indicate that decline of milk yield during the hot season, hold over until fall season and that prolongation of the negative effects of hot could be linked to the microclimatic control inside the barns. The presence of cooling systems seems to smooth this effect, but more detailed analysis based on daily milk production, management strategies, feeding and genetic aspects are necessary in order to better understand the short and long term consequence of heat stress on milk production in dairy cows.

How ruminants adapt their feeding behaviour to diets with high-risk for acidosis

Mialon, M.M.[1], Commun, L.[1,2], Martin, C.[1], Micol, D.[1] and Veissier, I.[1], [1]INRA, UR1213, Theix, 63122 St Genès-Champanelle, France, [2]ENVL, Av Bourgelat, 69280 Marcy L'Étoile, France; mrichard@clermont.inra.fr

Subacute ruminal acidosis (SARA) is a common digestive problem with a large economic impact in cattle fed diets rich in readily fermentable carbohydrates like finishing systems. This study focused on the ways ruminants receiving high-concentrate diets can adapt their feeding behaviour to reduce the risk of acidosis. In a first trial, 24 weaned calves from Blonde d'Aquitaine breed were fed individually ad libitum and allocated to 3 experimental diets (92% concentrate - 8% straw, 56% concentrate - 44% hay, 43% concentrate - 57% maize silage) during 160 days. In a second trial, 11 Texel wethers with ruminal cannulae were fed 28-day a diet based on alfalfa hay and wheat offered ad libitum in a free choice situation. Individual feed intake and feeding behaviour were recorded in the 2 trials. Ruminal pH was measured from a ruminal fluid sample collected by rumenocentesis in the first trial and continuously in the second one. The bulls fed the high-concentrate diet spent less time eating than the others and spread their meals over the day. Their ruminal pH after feeding was lower during the first 2 months of finishing. Their feed intake became very irregular at mid-finishing period, probably resulting from digestive disorders for some animals. In trial 2, the sheep fed ad libitum ate forage and concentrate in a ratio of 50/50. They spent quite 8 hours per day with a ruminal pH under 5.6 although they spread their intake during the day. After an occurrence of SARA, they decreased significantly their wheat intake during 2 or 3 days but did not modify other components of feeding behaviour. Large variations in feeding behaviour were observed between individuals in the 2 trials. In conclusion, ruminants seem to adapt their feeding behaviour to high-concentrate diets either by spreading their intake during the day or limiting their intake of concentrate. Such adaptations are likely to limit acidosis.

Management strategies for inbreeding control in unselected and selected populations

Fernandez, J. and Consortium, E.U.R.E.C.A., INIA, Ctra. Coruña Km 7,5, 28040 Madrid, Spain; jmj@inia.es

The avoidance of high levels of inbreeding is crucial in the success of a breeding or conservation program due to the harmful effects it produces on productive and fitness related traits (inbreeding depression). Managers can act at two different points to reduce the increase of inbreeding: they can select which individuals are providing offspring to the next generation (and how many) and also decide the mating pairs. Based on the idea of equalising the parents' contributions, regular hierarchical methods were developed to reduce the rate of inbreeding. A different approach recalls the importance of coancestry in the measurement and control of the genetic diversity and the magnitude of genetic drift. Consequently, the idea is arranging parents' contributions as to minimise the weighted global coancestry (the so called Optimum Contribution strategy). This methodology has very good properties including its dynamic and flexible nature which makes it able to adapt to practical constraints. In selection schemes, two opposite targets have to be balanced: obtaining the highest response on the selected trait and reducing the increase of inbreeding levels. Suboptimal strategies are based on reducing the importance of relatives' information by using inflated heritabilites in the estimation of breeding values or the implementation of non optimal indices, so less related individuals are to be selected. The Optimum Contribution strategy can also be applied to manage breeding programs by maximising the expected genetic gain while imposing a restriction on the global coancestry (and thus on the expected inbreeding in the next generation). Regarding the mating design, several strategies have been proposed: using factorial schemes, avoiding the mating between relatives (through minimum coancestry mating), circular and compensatory mating. The effectiveness of each of them depends on the particular characteristics of the population and the time horizon we are interested in.

Development of a genetic indicator of biodiversity for farm animals

Villanueva, B.[1], Sawalha, R.M.[1], Roughsedge, T.[1], Rius-Vilarrasa, E.[1] and Woolliams, J.A.[2], [1]Scottish Agricultural College, West Mains Road, Edinburgh EH9 3JG, United Kingdom, [2]Roslin Institute and Royal (Dick) School of Veterinary Studies, Roslin Biocentre, Midlothian, EH25 9PS, United Kingdom; villanueva.beatriz@inia.es

In 2002, Parties to the UN Convention on Biological Diversity made a commitment to 'achieve by 2010 a significant reduction of the current rate of biodiversity loss at the global, regional and national level'. In order to assess progress towards the 2010 target a limited number of indicators of biodiversity need to be developed using existing data. The objectives of this study were 1) to produce an indicator of genetic diversity for livestock species; and 2) to evaluate the proposed indicator in UK sheep and cattle. The indicator proposed is the species average effective population size (Ne) for the lower tail of the distribution of Ne across breeds. It is sensitive to 1) genetic variation within breeds, as it is based on the Ne of individual breeds, and 2) what is happening to breeds most at risk of disappearing. A total of 31 sheep and 20 cattle UK breed societies provided the information required for estimating Ne. This represents 53% of sheep and 58% of cattle breeds native to the UK. For breeds with pedigrees available, Ne was estimated from rates of change in inbreeding. For breeds with no pedigree information, Ne was estimated from predictive equations assuming mass selection. For each species, the indicator was calculated by 1) estimating Ne for each breed; 2) calculating the distribution of Ne; iii) finding the average Ne for the lower 20% tail of the distribution. In step 3), 20% was chosen because, given the number of breeds with available information, it provides a good compromise between giving high weight to the breeds most at risk, without being too sensitive to events surrounding a single breed. For sheep, indicator values were 36 and 41 animals for years 2001 and 2007, respectively. Equivalent values for cattle were 22 and 23. The small increases in the indicator of diversity from 2001 to 2007 were not significant.

Software tools for management of genetic variation

Maki-Tanila, A.[1], Fernandez, J.[2], Meuwissen, T.[3], Toro, M.A.[4] and Hiemstra, S.J.[5], [1]MTT, Biotechnology and Food Research, Jokioinen, Finland, [2]INIA, Animal Breeding, Madrid, Spain, [3]UMB, Animal Science, As, Norway, [4]UPM, Animal Science, Madrid, Spain, [5]WUR, CGN, Lelystad, Netherlands; Asko.Maki-Tanila@mtt.fi

Risks for loosing genetic variation (drift) or increasing homozygosity (inbreeding depression) in small populations are due to increased coancestry and inbreeding. Changes in these parameters depend on the effective population size N_e. The effects of breeding practices can be assessed in pedigree recorded populations. Estimates of inbreeding coefficient are influenced by the traceable depth of genealogical information and are not therefore comparable between populations. A useful picture is given by the dynamics of variation or rate of inbreeding. Even better would be the change in coancestry, as it is more stable and gives early signals about possible undesirable development. For computing the changes over time, first the individual inbreeding coefficients and pairwise coancestries are computed. The available software is relying on tabular, contribution, gene dropping or Wright's path method. Using the output, the parameter changes over years or generations can then be calculated to deduce an estimate for N_e (e.g. software ENDOG). When the pedigree information is missing, census numbers (preferably augmented by the variances of family sizes) of male and female parents give some indication about N_e. The best strategy for the management of genetic variation is to maximise N_e. This is achieved by equalising the contributions from the parents or, in selected populations, optimising the contributions with respect to genetic gain and rate of inbreeding (e.g. software GENCONT). There are many simplified mating strategies. In long-term choosing parents and their contributions is more relevant than mating design. When pedigree information is missing, e.g. circular mating could be practiced over herds or coancestries could be approximated by resorting to genomics.

Demonstration of the opportunities offered by a software for inbreeding control

Meuwissen, T.H.E., Norwegian University of Life Sciences, Inst. of Animal & Aquacultural Sciences, Box 5003, 1432 As, Norway; theo.meuwissen@umb.no

Given a desired rate of inbreeding, the (multi-trait) EBV of selection candidates, their pedigree and possible additional constraints of the breeding plan, optimal contribution selection optimises the contribution of each candidate such that genetic gain is maximised while attaining the desired inbreeding rate. In conservation schemes inbreeding is minimised. The optimised contribution of a candidate determines whether the candidate is selected, and how many offspring the candidate should obtain. The usage of a software package, GENCONT, for this task will be demonstrated. It will be shown how it controls or minimises the rate of inbreeding in a number of cases.

Genetic conservation of small populations from theory to practice in the Netherlands
Windig, J.J., Wageningen UR, Animal Breeding and Genomic Centre, P.O. Box 65, Lelystad, Netherlands; jack.windig@wur.nl

Effective population size can be increased by setting up proper breeding programs. Tools are available to assist the breeder and gene bank manager to select parents for breeding or select animals for storing material from in the gene bank. Optimal contributions are theoretically the best solution possible. In practice, however, it is often hard to realize the required contributions. Examples will be given for a commercial nucleus in cattle breeding, a rare sheep breed with a reliable pedigree and a rare goat breed with a less reliable pedigree, selection of rams for storage in the gene bank, and a free roaming cattle population with marker estimated kinships. When no pedigree is available and a breed is split into herds various rotational mating schemes are a good alternative for optimal contributions. Two example in sheep breeding will be given, one successful, the other not. Less effective measures, such as limiting the number of offspring per sire, are also available and several examples will be given. From all the examples given here guidelines for when to use what approach can be derived. No matter what method is used a strong breeding organization with committed members is needed.

Management of genetic variability in French small ruminants with and without pedigree information: review and practical lessons
Danchin-Burge, C.[1,2], Palhiere, I.[3] and Raoul, J.[1], [1]Institut de l Elevage, 149 rue de Bercy, 75595 Paris Cedex 12, France, [2]INRA/AgroParisTech, UMR 1313 GABI, 16 rue Claude Bernard, 75231 Paris Cedex 05, France, [3]INRA, UR631 SAGA, 31326 Castanet-Tolosan, France; palhiere@toulouse.inra.fr

There are numerous ways to manage the genetic variability of a breed but a main difficulty is to find a suitable method for a given situation. The purpose of this paper is to review different methods used in French small ruminants. Then, from these results, some practical advices will be given to help breeds managers to choose the right method, depending on the type of constraints they are facing: demography trends, pedigree knowledge, selection or conservation, breeders' willingness to apply strict mating rules … For rare breeds, the main goal is to limit the increase of inbreeding. A pedigree analysis performed on 16 small sheep populations showed that the main factors that impacted their genetic variability were the number of active males and their replacement rate. From these results, different management ways are suggested: the GENCONT software is well adapted for a breed raised in a single flock with good pedigree knowledge; rotational schemes should be preferred for breeds a maximum of 5 herds; finally the SAUVAGE software is well suited for breeds with more flocks and no or low pedigree information. In selected populations, the goal is twofold and conflicting: genetic progress and preservation of genetic variability. Two different methods are used in the French breeding programs. The first method aimed at equalizing family size of active sires at each step of selection. The second one weighs up genetic progress and variability by minimizing the increase of inbreeding and optimizing the contribution of each selection candidate. Whatever the method and the breed, it should be emphasized that only a collective organization such as gathering young sires in breeding centres, enables effective management of genetic variability.

Design of community-based sheep breeding programs for smallholders in Ethiopia

Duguma, G., Mirkena, T., Sölkner, J., Haile, A., Tibbo, M., Iñiguez, L. and Wurzinger, M.,; gdjaallataa@ yahoo.com

There are numerous failures of genetic improvement programs in developing countries. One of the reasons is that farmers as final beneficiaries are not included in the design of such programs from the beginning. This paper describes a new approach where farmers are involved in all steps of design and implementation of a breeding program. Four locations representing different agro-ecologies that are habitat to four indigenous sheep breeds were included: Afar (pastoral/agro-pastoral), Bonga, Horro and Menz (mixed crop-livestock system). After a detailed description of the production systems, breeding objectives were defined with communities. Different methodologies such as workshops, surveys, hypothetical choice experiments, individual own flock and group ranking of live animals were employed to explore the preferences of sheep owners. The application of various methods allows a validity cross-check of results and ensures that all traits are captured. Six traits for ewes (body size, coat colour, mothering ability, twinning, lambing interval, tail type) and five traits for rams (body size, coat colour, tail type, libido, presence/absence of horn) were identified as breeding objectives of smallholders and one additional trait (milk yield) for the Afar pastoral/ agro-pastoral farming communities. Based on this information alternative breeding programs involving different levels of recording were simulated. In discussion with the researchers the communities then decided which alternative they will implement.

Genetic structure of the French Blonde d'Aquitaine, Charolais and Limousin beef cattle populations

Bouquet, A.[1], Fouilloux, M.N.[2], Renand, G.[1] and Phocas, F.[1], [1]INRA, UMR 1313 GABI, 78352 Jouy-en-Josas, France, [2]Institut de l Elevage, Genetics department, 149, rue de Bercy, 75595 Paris, France; alban. bouquet@jouy.inra.fr

The genetic diversity of the French Blonde d'Aquitaine (BA), Charolais (CH) and Limousin (LI) populations was assessed by a pedigree analysis using the PEDIG software. Data included 807,221 BA, 2,992,246 CH and 2,142,788 LI animals born between 1972 and 2006. The complete generation equivalent numbers reached 5.5, 6.4 and 8.6, respectively for BA, LI and CH calves born in 2006, versus 1.5, 1.5 and 4.9 for BA, LI and CH calves born in 1972. The average inbreeding level (F) in the population and its rate of increase (dF) were calculated with the algorithm by Meuwissen and Luo (1992). In 2006, F was found to be low for BA (0.8%), CH (0.4%) and LI (0.5%) populations. On the last decade, dF slightly rose by +0.024% per year in the BA population whereas it remained null in CH and LI populations (+0.002% per year). To have a deeper insight in the genetic structure of populations, numbers of efficient founders (nfe) and of efficient ancestors (nfa) were calculated. In the BA population, the ratio nfa / nfe pointed out marked bottlenecks in the 1970s due to an artificial insemination (AI) rate above 60%. However, bottlenecks did not worsen, owing to AI rate declining to 35% in 2006. In the CH and LI populations, the declining ratio nfa/nfe indicated occurrence of bottlenecks from the 1990s. The heaviest use of some sire lines explained this phenomenon. It is due, 1) to the use of a BLUP animal model for beef genetic evaluation since 1993 and, 2) to a strong increase in AI rate between 1991 and 2006, from 8% and 15% to 19% and 33%, respectively for LI and CH populations. Due to very low dF, unlikely effective sizes (N_e) were found for the LI and CH populations by formula $N_e=1/2dF$. Ne was then derived from the loss of heterozygosity over the last decade. Compared to French dairy cattle breeds where Ne usually varies between 50 and 80, high Ne values were obtained for the French BA (313), CH (585) and LI (417) populations.

Mating animals by minimising the covariance between ancestral contributions generates more genetic gain without increasing rate of inbreeding in breeding schemes with optimum-contribution selection

Henryon, M.[1], Sørensen, A.C.[2] and Berg, P.[2], [1]Danish Pig production, Genetics R&D, Axeltorv 3, 1609 Copenhagen, Denmark, [2]Aarhus University, Department of Genetics and Biotechnology, Faculty of Agricultural Sciences, P.O. Box 50, 8830 Tjele, Denmark; mhe@dansksvineproduktion.dk

We reason that mating animals by minimising the covariance between ancestral contributions (MCAC mating) will generate more genetic gain than minimum-coancestry mating without increasing the rate of inbreeding in breeding schemes where the animals are selected by optimum-contribution selection. We will test this hypothesis by stochastic simulation and compare the mating criteria in hierarchical and factorial breeding schemes, where the animals are selected based on breeding values predicted by animal-model BLUP. Random mating will be included as a reference mating criterion. We expect that MCAC mating will generate 5 to 10% more genetic gain than minimum-coancestry mating in the hierarchical and factorial breeding schemes without increasing the rate of inbreeding. Moreover, we expect that it will generate up to 25% more genetic gain than random mating. Should this be the case, it will highlight the benefits of MCAC mating over minimum-coancestry mating, particularly because these benefits can be achieved without extra costs or practical constraints. MCAC mating merely uses pedigree information to pair the animals appropriately and should, therefore, be a worthy alternative to minimum-coancestry mating and probably any other mating criterion.

Addressing longissimus dorsi ultrasonic scans in large animals by image collaging

Baro, J.A.[1], Ramos, R.M.[2] and Carleos, C.E.[3], [1]Universidad de Valladolid, CC. Agroforestales, Campus de la Yutera, avda. de Madrid s/n, 34004 Palencia, Spain, [2]ASEAVA, Abarrio 33, Llanera, 33202 Asturias, Spain, [3]Universidad de Oviedo, Estadística e Investigación Operativa, Facultad de Ciencias, Avda. Calvo Sotelo s/n, 33007 Oviedo, Spain; baro@agro.uva.es

Linear ultrasonic scanners have been used by pork breeders, and lately by beef breeders, to predict carcass quality *in vivo* at low costs. The largest available probes are 17 cm long, while longissimus dorsi muscle sections are much larger in mature cattle. As a consequence, current practice allows only young animals to be measured, or else estimates areas from muscle depth, resulting in incomplete or inaccurate estimates. We present a system for *in vivo* estimation of muscle section area in large animals with ultrasonic image collages. This note describes the collaging protocol and discusses some issues regarding the problems encountered.

Evaluating breeding strategies in the Sorraia Horse endangered breed by pedigree data analysis
Oom, M.M.[1,2], Kjöllerström, H.J.[1] and Pinheiro, M.[2], [1]Universidade de Lisboa, Faculdade de Ciências, Departamento de Biologia Animal/Centro de Biologia Ambiental, C2-3 Piso, Campo Grande, 1749-016 Lisboa, Portugal, [2]Assoc. Intern. Criadores do Cavalo Sorraia, Barbacena, 7350-431 Elvas, Portugal; mmoom@fc.ul.pt

The Sorraia Horse represents a primitive equine breed with a continuous presence in the Iberian Peninsula since early Pleistocene. Recovered in 1937 from only 12 founders (7mares/5 stallions), the breed has been managed as a closed population, with only ~200 living animals worldwide. The small number of founders, the reduced effective population size and the complete genetic isolation led inbreeding to steadily increase, reaching extremely high levels. The breed is considered by FAO as in 'critical maintained risk status', and as 'rare/particularly endangered' by Portuguese authorities. A good deal of genetic information has been obtained to characterize the breed, either from distinct genetic markers analysis, or from pedigree ones. All molecular markers showed low levels of genetic variation and heterozygosity, indicating that the genome may be widely affected by founder effect, genetic drift and inbreeding. Only two mtDNA haplotypes were found, according to pedigree data. Pedigree analyses revealed extremely high levels of inbreeding (F),mean kinship (mK) and average relatedness (AR) in the living population. Two of the founders are no longer represented and genetic contribution of underrepresented ones is at a high risk of loss. It is crucial to define breeding plans for the breed in order to retain the genetic diversity still available, so breeders are yearly advised for the most suitable stallion to each herd. Minimizing inbreeding and using a larger number of stallions have been breeding priorities in the last years. We analyzed these breeding strategies on the genetic and demographic evolution of the breed in the last decade, in order to define breeding plans in the future. Genealogical data were recorded in SPARKS software for studbook management and genetic and demographic analysis was complemented by ENDOG 4.5 and PM2000 computer programs.

A primer extension assay to use for cattle genotype assisted selection: laboratory preliminary results
Sevane, N., Crespo, I., Cañón, J. and Dunner, S., Facultad de Veterinaria (Universidad Complutense de Madrid), Producción Animal, Av. Puerta de Hierro, s/n, 28040 Madrid, Spain; nata_sf@hotmail.com

Genotype Assisted Selection can be a cheap and effective way to improve the genetic progress in cattle breeds, specially in the case of high cost phenotypic recording which is the case for many economic traits in beef cattle. Here we develop a Capillary Primer-Extension Assay to identify differences between individuals in genes previously described to influence meat, milk, coat and sex performance traits. From a wider study, 15 loci were selected among those associated directly or potentially to meat tenderness, marbling and muscle growth, milk yield, protein and fat content or coat colour. Different breeds, most of them beef cattle, one dairy and one semi-feral (non selected) have been genotyped for these mutations and frequencies are discussed.

National genebank information repositories in FABISnet

Duchev, Z.I.[1], Gandini, G.[2] and Groeneveld, E.[1], [1]Institute of Farm Animal Genetics, FLI, Animal breeding and genetic resources, Höltystraße 10, 31535 Neustadt, Germany, [2]University of Milan, Department of Veterinary Sciences and Technologies for Food Safety, Via Celoria 10, 20133 Milano, Italy; Zhivko. Duchev@fli.bund.de

With the increasing demand for food and the usage of highly productive breeds the conservation of domestic animal diversity is recognized as priority task on international level. The monitoring, sustainable use and the conservation of livestock diversity will play important role in the future development of new breeds, adapted and performing well in various conditions. One of the relatively cheap ways of preserving the autochthonous breeds is the *in vitro ex situ* conservation of germplasm in genebanks. There are already available collections for many breeds in Europe, however these are disconnected and in the most countries there is no central national genebank information system. In the FABISnet project, in cooperation with EU, EAAP, FAO and partners from 14 countries, an European network of national biodiversity information systems for monitoring the national farm animal genetic resources was created and linked to the EAAP database (EFABIS) and world database (DAD-IS) at FAO. In the second part of the project the network is extended with 10 national genebanks management information systems based on the CryoWEB software. These systems are intended as a central documentation of the samples stored in various repositories. Each system collects data about the animal donors and their pedigree, the production of the samples, the distribution of the samples in the storage facilities, all movements and status changes,along with their history. A protocol for data exchange was developed for the upload of the cumulated yearly statistics for the conserved germplasm in EFABIS database.

Breeding for high welfare in outdoor pig production: a simulation study

Gourdine, J.L.[1], De Greef, K.[2] and Rydhmer, L.[3], [1]INRA, UR143, Animal Genetics Department, Domaine Duclos, 97170 Petit Bourg, Guadeloupe, France, [2]WUR, Animal Sciences Group, Box 65, 8200 AB Lelystad, Netherlands, [3]SLU, Department of Animal Breeding and Genetics, P.O. Box 7023, 75007 Uppsala, Sweden; Lotta.Rydhmer@hgen.slu.se

The present simulation study was conducted within the EU-supported project Q-Porkchains. Concern for animal welfare has increased. Consumers often associate outdoor production with high welfare, and good health is an important aspect of welfare. To our knowledge, there is no special breeding program aiming for high welfare in outdoor pig production. The aim of this study was to compare the genetic progress in simulated breeding programs for a dam-line. Selection for high welfare was compared to a conventional breeding program. The deterministic simulation program SelAction was used to calculate the genetic progress. Three schemes were defined: 1) a conventional scheme that aimed at improving sow's production and reproduction traits (litter size, piglet mortality (PM), mean piglet weight at weaning, weaning-to-mating interval (WMI), ADG from 0 to 20 kg, ADG from 20 to 100 kg and lean content); 2) extension of the first scheme with welfare considerations (leg condition of sows after first lactation (LEGw) as an extra trait and double weight on PM and WMI); 3) a breeding program for high welfare in which the weights of LEGw, PM and WMI were increased until genetic progress was achieved. Results show that those traits (LEGw, PM and WMI) should contribute to more than 40 % weight in the index in order to improve all considered welfare traits. This is at the expense of economically important traits. The implementation of breeding for high welfare requires other prerequisites, such as a high willingness to pay for welfare among consumers.

Management of local genetic resources: the case of the Creole breeds of Guadeloupe

Naves, M.[1], Quenais, F.[1], Farant, A.[2], Arquet, R.[2], Gourdine, J.L.[1] and Mandonnet, N.[1], [1]INRA, UR143 URZ, Domaine Duclos, 97170 Petit Bourg, Guadeloupe, [2]INRA, UE1294 PTEA, Domaine de Gardel, 97160 Moule, Guadeloupe; nathalie.mandonnet@antilles.inra.fr

Few studies deal with the management of local breeds in the tropics, although they are in an urgent need for preservation and improvement. The Creole breeds of cattle, goat and pig of Guadeloupe offer a great interest, because of there productivity and natural resistance to diseases related to ticks, in cattle, genetic variability for resistance to strongyles, in goat, and meat quality and heat tolerance in pig. Our experiments on the characterisation and preservation of these breeds aim to understand their adaptation traits, and to implement sound breeding programs for both production and adaptation. Experimental nucleus herds gather about 250 goats, 90 cows, and 25 sows, managed to maintain the consanguinity as low as possible. The number of known generations and the inbreeding coefficient are of 12 and 2.3% in goat, 3 and less than 1% in cattle, and 16 and 13% in pig. The goat nucleus is closed, while the cattle and pig stocks are open to sires from private herds, where the population is managed by professional organisations. Cryopreservation is also implemented. In goat, 256 embryos from 16 donors are stored by the French National Cryobank, and 2,500 doses of semen from 32 bucks are stored locally in a Biological Resource Centre. In cattle, 8000 doses of semen from 21 bulls are stored. Collections of DNA and other biological samples are also maintained for experimental studies. More than 7,500 samples are stored, with an increment of about 900 samples per year. Subpopulations representative of the diversity of each breed are characterised for a panel of markers. Researches of QTL for production and adaptation traits are also undertaken. Through these activities, combining *in situ* and *ex situ* methods for research and breeding purpose, INRA is highly committed in the characterization, preservation and improvement of local genetic resources valuable for the humid tropics.

Relatedness among cryo-bank bulls of the Yakutian Cattle breed as estimated with microsatellite data

Tapio, I.[1], Tapio, M.[1], Li, M.-H.[1], Popov, R.[2] and Kantanen, J.[1,3], [1]MTT Agrifood Research Finland, Biotechnology and Food Research, ET-Building, 31600 Jokioinen, Finland, [2]Yakutian Research Institute of Agriculture, Yakutian Research Institute of Agriculture, 677002 Yakutsk, Russian Federation, [3]NordGen - Nordic Genetic Resource Centre, Animal Sector, P. O. Box 115, 1431 Aas, Norway; juha.kantanen@mtt.fi

The Yakutian Cattle are the unique last remnants of the Siberian Turano-Mongolian cattle, with 1,200 purebred animals left, and are well adapted to the extreme sub-arctic conditions. Semen of 6 Yakutian bulls is stored in a cryo-bank. However, due to the traditional free herding style of Yakutian Cattle in summer pastures, with several randomly mating bulls within a herd, pedigree records of these 6 bulls are not available. We analysed 30 autosomal microsatellites in order to clarify genetic relatedness between these bulls and provide recommendations for the use of their semen in conservation and breed management. Pairwise relatedness among the bulls was computed using MER v3.0 program. In addition, we studied the value of the cryo-bank bulls for the preservation of genetic variation of the contemporary Yakutian Cattle by calculating allelic and gene diversity estimates and mean molecular coancestries. Although our simulation results indicated that 30 loci are insufficient for an unequivocal determination of relatedness among individuals, the data suggested four cryo-bank bull-pairs as potential half-sibs. We propose a breeding scheme based on the rotation of breeding females and that the cryo-bank bulls are divided into three groups. Based on the mean molecular coancestries, the cryo-bank bulls were less related to the cow population compared with the contemporary breeding bulls. They added to the allelic variation in the contemporary population by 3% and in the male subpopulation by 13%. No significant increase in gene diversity was recorded. Our results clearly demonstrate the importance of *ex situ* cryo-banking of genetic material in the conservation of rare domestic animal breeds.

Optimising smallholder pig breeding organisation in villages in Northwest Vietnam
Roessler, R., Herold, P., Momm, H. and Valle Zárate, A., Universität Hohenheim, Institute of Animal Production in the Tropics and Subtropics, Garbenstr. 17, 70593 Stuttgart, Germany; roessler@uni-hohenheim.de

Village breeding programmes using appropriate pig breeds are currently developed to help improving smallholder pig production in the Son La province in Northwest Vietnam. The success and long-term sustainability of such programmes depend not only on technical appropriateness, but also on the organisational feasibility under given environmental conditions. This study evaluates the organisational feasibility of village breeding programmes and possibilities for their integration into provincial, regional and national structures. An institutional analysis is used to provide a descriptive assessment of breeding institutions in Vietnam. Information was collected from group discussions with small-scale pig producers in Son La and interviews in various public and private institutions across northern Vietnam, complemented by information from legal documents. Findings suggest that breeder cooperatives at village level could be a promising option to improve smallholder pig breeding and to ensure the sustainability of village breeding programmes, tightening the relatively weak links to other breeding institutions and counterbalancing commercial farms and public breeding institutions.

An in-depth pedigree analysis of Canadian Holstein Cattle
Stachowicz, K.[1], Sargolzaei, M.[1], Miglior, F.[2,3] and Schenkel, F.S.[1], [1]University of Guelph, CGIL, 50 Stone Rd E, N1G 2W1, Guelph, ON, Canada, [2]Agriculture and Agri-Food Canada, DSRDC, 2000 College St, J1M 1Z3, Sherbrooke, QC, Canada, [3]Canadian Dairy Network, 660 Speedvale Ave W, N1K 1E5, Guelph, ON, Canada; kstachow@uoguelph.ca

The pedigree of 8,764,141 purebred Canadian Holstein cattle born between 1883 and 2008 was analyzed to assess past and current levels of inbreeding and genetic diversity. The dataset included 10,328 proven bulls born between 1975 and 2006. Analyses were performed using CFC, EVA and Pedig software. The completeness and depth of pedigree were very good. The percentage of animals with both parents known was above 90% in last two decades (100% for all proven bulls). The average discrete generation equivalent was greater than 14 in 2004 and later, and was 1 generation longer for proven bulls. Pedigree completeness index for animals born after 2005 was 0.90, 0.87, 0.64, and 0.43, considering 5, 10, 20, and 30 generations back in time, respectively. The population reached an average level of inbreeding of 5.9% in 2008. Proven bulls were on average more inbred, 7.1% in 2006. The average coancestry between cows born between 2000 and 2008 and proven bulls born between 1998 and 2006 was 6.8%, which indicates future expected level of inbreeding. When missing parents were accounted for, inbreeding was 10.4% for animals born in 2008, and 10.8% for proven bulls born in 2006. Effective population size for animals born between 1998 and 2008 was around 60. For animals born in 2008 the effective number of founder genomes and the effective number of ancestors were 7.7 and 16.0, respectively. The 6 ancestors with highest genetic contributions to the current population explained 50% of the gene pool in 2008, while 50 ancestors explained 90%. When looking at causes of genetic diversity loss, genetic drift contributed to more than 90% of the total loss since 1970. As genetic drift is a function of effective population size, management of this parameter will be crucial for the future genetic diversity in Canadian Holstein population.

Estimation of genetic parameters using fixed regression model in Holstein cattle of Iran
Razmkabir, M., Moradi-Shahrbabak, M., Pakdel, A. and Nejati-Javaremi, A., Department of Animal Science, University of Tehran, Karaj, Iran; m.razmkabir@uok.ac.ir

Test-day data of milk yield provide an example of longitudinal data or repeated measurements that several statistical models have been proposed for the genetic evaluation of these records in dairy cattle. Also, some countries have already implemented test-day models for genetic evaluation of dairy cattle. In this study in order to compare genetic parameters from test-day and lactation records, data were obtained from the Animal Breeding Center of Iran spanning 1997 to 2007. Data were restricted to first lactation milk yield records with days in milk between 5 and 305 d, age at first calving between 18 and 42 month and daily milk yield between 3 and 70 kg. Estimation of variance components and genetic evaluation were performed using a fixed regression model and the Matvec program (Kachman, 2001). Under the fixed regression model, test-day samples from the same lactation are considered as repeated observations on the same traits, and a permanent environmental effect accounts for environmental similarities between different test days within the same lactation. In this study the Wilmink Function was used to describing the shape of the lactation curve. Temporary environmental effects were considered independently distributed. Estimates of heritability for day and 305-day records were 0.17 and 0.27, respectively. Also the relative proportion of permanent environment variance to total variance was 0.37 for test-day records. A spearman rank correlation between breeding values for TD and 305-day milk yields indicated that more extreme changes occurred in ranking of cows than for sires. These changes seemed to be associated with cows whose lactation curves deviated from the standard lactation curve.

Practical aspects of utilising gene bank stocks in commercial dairy cattle population
Juga, J.[1] and Mäki-Tanila, A.[2], [1]University of Helsinki, Department of Animal Science, P.O.Box 28, 00014 University of Helsinki, Finland, [2]MTT Agrifood Research Finland, Biotechnology and Food Research, ET-talo, 31600 Jokioinen, Finland; jarmo.juga@helsinki.fi

Gene banks have been proposed as a source of genetic variation to commercial breeding populations when running into problems or in responding to changes in production environment or as a source of single genes improving disease resistance or product quality. Marker assisted introgression of a gene via backcrossing scheme has been shown to be the most efficient strategy for this. Backcrossing for several generations is, however, very time consuming and expensive, when dairy cattle is considered and even more so, if the difference in economically important traits between the conserved and commercial populations is large and the commercial population has a steady genetic increase in these traits. In Finland a gene bank strategy for the native Finncattle with its three different types Western, Northern and Eastern, has been running for many years. The Western type is the most common with some 2000 cows in milk production herds while the two other types are less common with less than 500 live cows in each. The progeny testing scheme for Finncattle AI bulls has stopped being effective quite a few years ago, while the strategy is to use young bulls and to avoid inbreeding. The multi-breed genetic evaluation gives a difference of 48.4 kg, 113.8 kg and 79.1 kg in the 1st lactation protein yield between Ayrshire and Western, Eastern and Northern type Finncattle cows born in 2004-2006, respectively. According to economic values used in the Nordic Total Merit index these differences are worth 180-300 €. The genetic change in Ayrshire population is about 2 kg protein per year. This means that the crosses never really catch up the difference. The aim of this paper is to illustrate the challenges we face and options we have in introgressing a single gene from a gene bank population to genetically improving commercial dairy cattle population.

The influence of genetic selection and feed system on dry matter intake of spring-calving dairy cows
Coleman, J.[1,2], Pierce, K.M.[2], Berry, D.P.[1], Brennan, A.[1] and Horan, B.[1], [1]Teagasc, Moorepark dairy production research centre, Fermoy, Co. Cork, Ireland., N / A, Ireland, [2]University College Dublin, School of agriculture, food science and veterinary medicine, Belfield, Dublin 4, Co. Dublin, Ireland., N / A, Ireland; john.coleman@teagasc.ie

The objective of pasture-based milk production systems post EU milk quotas must be to maximise profitability per hectare through excellence in grassland management practices. Such systems will facilitate increased overall farm stocking rates without increasing exposure to high cost external feed sources. The objective of this study was to investigate the influence of genetic selection using the Irish total merit index (Economic Breeding Index; EBI) and feeding system on milk production and dry matter intake (DMI) throughout lactation. In 2006, three genetic groups were established from 126 animals: 1) LowNA, national average herd genetic potential; 2) HighNA, high genetic potential North American Holstein-Friesian (HF) and 3) HighNZ, high genetic potential New Zealand HF. Animals were randomly allocated to one of two feeding systems: Moorepark (MP) blueprint pasture system (2.6 LU/ha and 500 kg concentrate/cow) and a high output per hectare (HC) pasture system (2.9 LU/ha and 1,200 kg concentrate/cow). HighNA had the highest milk and solids corrected milk (SCM) yields, HighNZ were lowest for milk yield with similar SCM yield to LowNA group. Genetic group had no significant effect ($P>0.05$) on grass DMI or total DMI (TDMI), however, grass DMI per 100 kg bodyweight was greatest for the HighNZ group. The HC system realised greater milk yield, SCM yield, lower grass DMI, reduced grass intake per 100 kg bodyweight and a higher TDMI. Results suggest that genetic selection using EBI has no significant effect on grass DMI. Increasing stocking rate combined with concentrate supplementation increases reduces grass DMI, increases TDMI and increases overall milk production per hectare while maintaining high levels of pasture utilisation.

Real time report for managing breeding populations
Groeneveld, E.[1], V.d.westhuizen, B.[2], Maiwashe, A.[2] and Voordewind, F.[2], [1]Institute of Farm Animal Genetics (FLI), Animal Breeding and Genetic Resources, Höltystr. 10, 31535 Neustadt, Germany, [2]ARC - Animal Production Institute Agricultural Research Council - Animal Production Institute, Quantitative Animal Breeding, Private Bag X2, Irene 0062, South Africa; eildert.groeneveld@fli.bund.de

Management of populations requires detailed knowledge about the population structure, usage of sires, generation intervals and rates of inbreeding. All of this should be available at regular intervals, preferably at least once a year which necessitates an automated system. PopReport as a subsystem based on the APIIS database framework provides such a packages which allows loading, computing and generating type set reports. On the basis of pedigree data and possibly reproductive information two reports are generated on a per breed level broken down by years to assess the development of the parameters over time. The Population structure report gives information on the use and representation of parents in the population together with their age structure.Exact generation intervals are computed for the four paths. The inbreeding report computes statistics on pedigree completeness and the development of inbreeding and additive genetic relationship. On this basis a number of estimates of the effective population size is computed and presented in tabular and graphical for for each breed in the APIIS database. Such a documentation of parameters relevant to biodiversity issues could then be included in the yearly reporting on the breeding progress for a given population, along with genetic trends as are already reported now for many populations. As such it can then serve as an early warning system, if trends go in an undesirable direction. The Population Report is available as a subset in the RapidAPIIS package underthe GPL license free of charge.

Breeding goals for sustainable animal breeding: views and expectations of stakeholders from the production chains and the general public

Dockes, A.C.[1], Magdelaine, P.[2], Daridan, D.[3], Tocqueville, A.[2], Remondet, M.[4], Selmi, A.[4] and Phocas, F.[5], [1]French Livestock Institute, 149 rue de Bercy, 75595 Paris, France, [2]ITAVI, 75009 Paris, France, [3]IFIP, 75595 Paris, France, [4]INRA, UR1216, 94205 Ivry, France, [5]INRA, UR1313, 78352 Jouy en Josas, France; anne-charlotte.dockes@inst-elevage.asso.fr

To contribute to the management of European farm animal genetic resources in Europe, the project 'COSADD' identified and ranked relevant sustainable breeding goals, according to the preferences of food chain and general public stakeholders. Four species - cattle, fish, pig and poultry - were studied, as representatives of the diversity of meat production systems and breeding schemes. In 2008 30 semi-structured interviews were carried out with representatives of the breeding schemes and agro-industrial actors. Stakeholders' priorities were analyzed in both current economic and sustainable development contexts. The interviewees shared some views: the priority of the economic and marketing aspects, the necessity to produce animals easy to raise and, resilient to sanitary risks. Environment and animal welfare were also important traits even if considered as external demands. However, many differences in priorities were associated to the specificities of each species production system (intensive or not, link to the soil) and breeding schemes (private or public). In 2009 interviews were carried out with retailers and representatives of general public associations (consumers, animal or environment protection). They focused on public views and expectations toward animal production, definition of sustainable development and opinion on some controversial questions about the relations between animal production and environment or animal welfare. Public opinion about different selection methods were also tested and are presented in the communication.

Test of methods for measuring efficiency of selection in dairy cattle breeding schemes, with examples from Danish Holstein Friesian

Berg, P.[1] and Sørensen, M.K.[1,2], [1]Aarhus University, Genetics and Biotechnology, P.O.Box 50, 8830 Tjele, Denmark, [2]Danish Cattle Federation, Udkærsvej 15, 8200 Aarhus N, Denmark; Peer.Berg@agrsci.dk

Dairy cattle breeding schemes are characterised by long generation intervals and thus long time intervals between selection and mating decisions and the resulting response to selection. HACCP (Hazard Analysis and Critical Control Points) principles could be used to monitor the hazards and thus efficiency of the breeding scheme. HACCP is based on the definition of Critical Control Points (CPPs) to monitor and reduce the hazards. The current project aims at proposing CPPs, in the terminology of HACCP, for monitoring and controlling the efficiency (hazards) of dairy cattle breeding schemes. Critical control points are proposed for evaluating biases and efficiency in selection of bull dams for production of young bulls, efficiency of optimising genetic contributions of bull sires and average relationships of bull dams and bull sires. These critical control points are illustrated on Danish Holstein Friesian using historical data on bull dam selection. Data from annual genetic evaluations from 1999 to 2008 on bull dam candidates and bull dams are used. Based on these data evaluations of each cohort (year) were computed. There are large variation in the change of predicted breeding values (EBV) and mendelian sampling terms (MS) of bull dams, indicating that this can be used to detect biases in cow evaluation. More generally trends in EBV and MS differ among age groups of bull dams. Across years it is shown that only 40% of the selection differential of mendelian sampling terms of bull dams that could have been obtained given updated information from the following years are realised. Implications of genomic selection on these efficiency measures are discussed, in addition to its impact on the definition of CPPs.

Strategic management of pig genetics

Kremer, V.D.[1], Newman, S.[2], Kinghorn, B.P.[3], Knap, P.W.[2] and Wilson, E.R.[2], [1]Genus/PIC, The Roslin Institute, Roslin, EH25 9PS, United Kingdom, [2]Genus/PIC, Hendersonville, TN 37075, USA, [3]University of New England, Armidale, NSW 2350, Australia; valentin.kremer@pic.com

Genetic improvement has a wide range of actions but only two really critical control points – animal selection and mate allocation. The best animals to select depend on the pattern of mate allocation, and vice versa. Mate Selection allows this to be done simultaneously. The number of factors involved (e.g., breeding objectives, selection pressure, crossbreeding, inbreeding avoidance, migration) and the complexity of their interrelations require a strategic management approach. Mate Selection based on Optimal Contribution Theory (OCT) emerged in the late 1990s and established itself around 2000 as software became available for implementation. The information required to implement Mate Selection is collected in most animal breeding programs, including pedigree data; sex of the animal; trait EBVs and/or index value; candidate status; and, if available, marker/breed genotypes. Weekly analyses for each population include all current and future candidates (active animals, juveniles and unborns). The optimization system is customized for farm routines and requirements. The outcome is a proposed set of future virtual matings, based on OCT, farm management requirements and population-specific desired ΔF and ΔG as strategic target parameters. This paper shares the experience accumulated during this period within PIC and shows examples of achieved results. Factors contributing to the decision process and their interrelations are shown graphically. A balanced approach of managing short term system reactivity and long term objective targeting insures achievement of genetic gains over one genetic standard deviation per year as well as maintaining genetic diversity.

Allele effects of genetic markers on meat quality traits in Swiss beef breeds

Hasler, H.[1,2], Dufey, P.-A.[3], Kreuzer, M.[1] and Schneeberger, M.[1], [1]ETH Zurich, Institute of Animal Sciences, 8092 Zurich, Switzerland, [2]Swiss College of Agriculture, Länggasse 85, 3052, Switzerland, [3]Agroscope Liebefeld-Posieux Research Station ALP, P.O. Box 64, rte de la Tioleyre 4, 1725 Posieux, Switzerland; heidi.hasler@bfh.ch

At present, meat quality is not incorporated in the Swiss breeding programme for beef cattle because its recording is labour intensive and expensive. The aim of this study was to elaborate basic information for considering meat quality through marker assisted selection (MAS) in Swiss beef cattle breeding. Therefore, effects of commercially available genetic markers (marker T1 in the Calpastatin gene and markers T2, T3 and T4 in the Calpain gene) on meat quality traits were analysed using samples and quality data of 270 animals, collected in various trials. A linear model including fixed effects of treatments and number of favourable alleles was applied. Effects on tenderness traits were detected for markers T1 and T3. The homozygous unfavourable genotype was associated with the lowest sensory tenderness. A significant effect of marker T1 on Warner Bratzler Shear Force was detected. The lowest shear force was associated with the heterozygous genotype. The number of favourable alleles added over all four investigated markers was also positively associated with sensory tenderness. Depending on the trial, significant effects were found between tenderness markers and intramuscular fat content, flavour and juiciness. It was concluded that analyses using data from a pedigreed population and the investigation of other markers are necessary before these markers can be definitely implemented in the Swiss breeding programme for beef cattle.

Detection of quantitative trait loci affecting carcass and meat quality in two French beef breeds

Allais, S.[1], Levéziel, H.[2], Lepetit, J.[3], Hocquette, J.F.[4], Rousset, S.[5], Denoyelle, C.[6], Bernard-Capel, C.[6], Journaux, L.[1] and Renand, G.[7], [1]UNCEIA, 149, rue de Bercy, 75595 Paris cedex, France, [2]INRA, UMR 1061, 123, avenue Albert Thomas, 87060 Limoges, France, [3]INRA, UR 370, Centre de Theix, 63122 Saint Genes Champanelle, France, [4]INRA, UR 1213, Centre de Theix, 63122 Saint Genes Champanelle, France, [5]INRA, UMR 1019, 58, rue Montalembert, 63122 Saint Genes Champanelle, France, [6]Institut de l Elevage, 149, rue de Bercy, 75595 PARIS cedex, France, [7]INRA, UMR 1313, Domaine de Vilvert, 78350 Jouy en Josas, France; sophie.allais@jouy.inra.fr

Breeding organizations in charge of selecting the AI sires in France implemented a research program 'Qualvigène' where sensory attributes and related muscle characteristics would be systematically measured in addition to slaughter beef traits. For the QTL detection, we used 248 Limousine and 243 Blonde d'Aquitaine purebred young bulls in 6 sire families (3 in each breed). These animals were genotyped for 186 markers (160 microsatellites and 26 SNPs in candidate genes). The QTL detection was realized with the QTLMAP software developed at INRA to map QTL with linkage analysis approaches. The QTL detection technique is an interval mapping applied to half-sib families. On the thirteen chromosomes studied at the moment, we found 19 QTL of carcass quality (carcass conformation, carcass yield), 5 QTL of fat traits (intramuscular fat content and carcass fat) and 4 QTL of meat quality (tenderness, flavour and shear force). After achieving the one-QTL and one-trait analysis, we will perform a multi-QTL and multi-traits analysis to get a fine description of the QTL.

Sow productivity and piglet survival in Norsvin Landrace

Zumbach, B.[1], Madsen, P.[2] and Holm, B.[3], [1]Norsvin, P.B. 504, 2304 Hamar, Norway, [2]Aarhus University, P.O. Box 50, 8830 Tjele, Denmark, [3]Norsvin USA, 2768 Superior Drive NW, Rochester, MN 55901, USA; birgit.zumbach@norsvin.no

The study examined the genetic background of piglet mortality as a litter trait and its relationship with sow productivity traits in the Norsvin Landrace. Data included 20,531 and 7,332 1st and 2nd parity litters, respectively, collected from 2001 to 2008. Litter traits analyzed were no. piglets born alive (LB), no. stillborn piglets (SB), no. piglets dying during the suckling period (DS), total no. piglets dying until weaning (TD), no. weaned (W), and litter weight at 3 wk corrected by no. piglets weighed at 3 wk (LW3). Heritability (h2) estimates for litter traits in 1st and 2nd parities were 0.11 ± 0.01 and 0.10 ± 0.02 for LB, 0.08 ± 0.01 and 0.09 ± 0.02 for SB, 0.07 ± 0.01 and 0.05 ± 0.01 for DS, 0.11 ± 0.01 and 0.11 ± 0.01 for TD. Based on h2 estimates, TD is as well apt as a selection trait as LB. Genetic correlations (rg) between LB and SB in 1st and 2nd parity were -0.04 ± 0.10 and 0.29 ± 0.17, between LB and DS 0.66 ± 0.06 and 0.47 ± 0.12; rg between TD and SB in 1st and 2nd parity were 0.75 ± 0.05 and 0.79 ± 0.08, between TD and DS 0.90 ± 0.02 and 0.92 ± 0.03. While selection for LB increases DS, selection for low TD should reduce both, SB and DS. In 1st and 2nd parity rg between LB and W were 0.79 ± 0.04 and 0.80 ± 0.06; rg among TD, W and LW3 were not significantly different from 0 in both parities; Selection for LB increases efficiently W at the cost of TD. LW3 as a mother ability does not counteract this relationship. Thus, piglet mortality should be included into the breeding goal.

Heritability of longevity in Landrace and Large White populations using Weibull and grouped data models

Poigner, J.[1], Mészáros, G.[1], Ducrocq, V.[2] and Sölkner, J.[1], [1]University of Natural Resources and Applied Life Sciences, Gregor Mendel Straße 33, A-1180 Vienna, Austria, [2]Institut National de la Recherche Agronomique, Domaine de Vilvert, 78352 Jouy en Josas, France; gabor.meszaros@boku.ac.at

Length of productive life in sows was analyzed with animal, sire, sire-maternal grandsire, sire-dam and sire-maternal grandsire-dam within maternal grandsire Weibull and grouped data models, as the first step towards breeding value evaluation in Austria. Records from 7,204 Landrace and 8,138 Large White purebred sows were used. Length of productive life was defined as the interval between first farrowing and culling or censoring, where culling was assumed at last weaning or farrowing. Animals alive at the end of data collection or at tenth farrowing were considered as right censored. Grouped data models are adapted to longevity measures expressed on a discrete scale with few possible time values. Here time changes occurred at each farrowing and each weaning. Time dependent interactions of herd and year, as well as parity and number of piglets born alive relative to herd mean, along with time independent effect of age at first farrowing and the genetic component were included to all models. Increased risk of culling was observed for animals with first farrowing in before 300 or after 420 day of life. There was a strong interaction between the effects of parity and relative number of piglets: the very limited culling for low litter size in first parity can be related to apparent exclusion from the data set of low productive sows in first parity. Genetic variance differed depending on the model used. Effective heritabilities for Weibull model ranged from 0.05-0.1 for Landrace and from 0.08-0.17 for Large White sows; for the grouped data models from 0.08-0.12 for Landrace and from 0.04-0.09 for Large White. Sire and sire-dam models gave slightly higher heritabilities compared to sire-maternal grandsire and sire-maternal grandsire-dam models.

***In vitro* variability of the pig innate immune response**

Botti, S.[1], Rapetti, V.[1], D'andrea, M.[2], Davoli, R.[3], Mari, F.[4], Lunney, J.K.[5] and Giuffra, E.[1], [1]Parco Tecnologico Padano-CERSA, V. Einstein, 26900 Lodi, Italy, [2]Univ. of Molise, V. De Sanctis, 86100 Campobasso, Italy, [3]Univ. of Bologna, V. Zamboni 33, 40126 Bologna, Italy, [4]Univ. of Siena, V. Bianchi 55, 53100 Siena, Italy, [5]ANRI, ARS, USDA, Animal Parasitic Diseases Laboratory, Beltsville, MD, 20705, USA; sara.botti@tecnoparco.org

Collection and refinement of disease phenotypes are essential for the effective outcome of genome wide analysis studies aimed at unravelling the genetic basis of disease susceptibility. Livestock research requires efficient protocols for biobanking similar to those employed in human medicine. The goal of this project is to phenotype the major pig commercial breeds versus South-European traditional breeds and pure wild boar populations in terms of time of onset and intensity of the innate immune response to Salmonella infection. As our biological model, we used porcine PBMCs (peripheral blood mononuclear cells) cultured *in vitro* and stimulated with LPS (bacterial lipopolysaccharide) derived from Salmonella enterica serotype typhimurium. We developed a system to isolate PMBCs on Ficoll-Hystopaque gradients and freeze them with 95% preservation of cell viability. Blood samples collected so far include Landrace (n: 32), Large White (n: 41) and Duroc (n: 25); sampling of traditional breeds (Cinta Senese and Casertana) and pure wild boar populations from Central Italy is in progress. Cells are stimulated with LPS or mock-stimulated; Trizol extracts and culture supernatants are collected after 0, 2, 4, 6, 8, 24h. A microarray approach for the detection of TNF-α, INFγ, IL-1β, IL2, IL4, IL-6, IL-8,IL-10, IL-12 is being used to first select which cytokines will be profiled on a large number of animals by classic ELISA and real time PCR of related genes. These results will add to the current knowledge of pig biodiversity. It will serve as a reference for full genome scan investigations on pig resistance for multiple disease traits that are currently in progress. These results are obtained through the EC-funded FP6 Project 'SABRE'.

The contribution of social effects to heritable variation in average daily gain of piglets from birth till weaning

Bouwman, A.C.[1], Bergsma, R.[1,2], Duijvesteijn, N.[2] and Bijma, P.[1], [1]Animal Breeding and Genomics Centre, Wageningen University, P.O. Box 338, 6700 AH Wageningen, Netherlands, [2]IPG, Institute for Pig Genetics, P.O. Box 43, 6640 AA Beuningen, Netherlands; aniek.bouwman@wur.nl

The aim of this study was to investigate whether there is heritable social variation in average daily gain from birth till weaning in piglets. Milk and the establishment of a teat fidelity are sources of social interaction between suckling piglets nursed by the same sow. If a heritable social effect is present, but ignored, the selected animals might be the most competitive ones with a negative effect on the growth of their group members, resulting in a lower response to selection than expected. The social interaction model was extended with a maternal component for the foster dam to analyze this. Four different animal models were compared: a basic model with a direct heritable effect; a social model accounting for direct and social heritable effects; a maternal model with the foster dam as heritable maternal effect added to the basic model; and a social-maternal model accounting for direct, social and maternal heritable effects. Both maternal models were significantly better than their equivalent non-maternal models ($P<0.01$). The social model was not significantly better than the basic model ($P=0.07$), and the social-maternal model was also not significantly better than the maternal model ($P=0.39$). Results show that direct, maternal and social heritability were 0.07, 0.06 and around 0.0009, respectively. The total heritable variance, including direct, social and maternal heritable variance and their covariances, expressed relative to the phenotypic variance (T^2) ranged from 0.07 through 0.15. There was no evidence for a heritable social effect between piglets in a foster group and the generally used maternal model performed best. For breeding purposes this means that there is no evidence for competition effects of selection candidates on the growth of their group members.

Divergent selection for residual feed intake in the growing pig: correlated effects on feeding behaviour and growth and feed intake profiles in Large White pigs

Gilbert, H.[1], Al Aïn, S.[1], Sellier, P.[1], Lagant, H.[1], Billon, Y.[2], Bidanel, J.P.[1], Guillouet, P.[3], Noblet, J.[4], Van Milgen, J.[4] and Brossard, L.[4], [1]INRA, UMR1313, GABI, 78350 Jouy-en-Josas, France, [2]INRA, UE967, GEPA, 17700 Surgères, France, [3]INRA, UE88, UEICP, 86480 Rouillé, France, [4]INRA, UMR1079, SENAH, 35000 Rennes, France; helene.gilbert@jouy.inra.fr

Residual feed intake (RFI) is defined as the difference between the observed daily feed intake (DFI) and the 'theoretical' DFI predicted from maintenance and production requirements. Divergent selection for RFI has been conducted for five generations in Large White male pigs recorded from 35 to 95 kg body weight (BW) in collective pens equipped with single-place electronic feeders. The individual visits to electronic feeders were grouped as meals (meal criterion: 2.5 min) and feeding behaviour traits such as eating time per day (ETD), rate of feed consumption (RFC) and meal characteristics were computed. For each of 920 boars weighed weekly, DFI and BW were jointly analysed to estimate parameters for each individual using models describing growth (Gompertz function of age) and feed intake (power function of BW). Heritability estimates, from 859 boars and 819 sibs (females and castrated males), were around 0.15 ± 0.02 for RFI, 0.30 ± 0.04 for DFI, and 0.50 ± 0.03 for feeding behaviour traits. Heritability estimates for the model parameters describing growth and feed intake were 0.18-0.19 ± 0.03 for the shape parameters and were higher (0.27 ± 0.05 to 0.43 ± 0.03) for the parameters directly related to the magnitude of feed intake and growth. Correlations of these parameters with RFI, DFI and ETD showed that a high early feed intake and a high early growth rate were genetically related to higher RFI, i.e. lower feed efficiency. Correlations implying eating rate and meal characteristic were close to zero. The expected feed intake at 50 kg BW displayed a heritability of 0.41 ± 0.03 and a fairly high genetic correlation with RFI (0.61 ± 0.05), so that it could be a potentially interesting predictor for RFI.

Differentially expressed genes for prolificacy in a F2 intercross between Iberian and Meishan pigs

Martínez-Giner, M.[1], Pena, R.N.[1], Fernández-Rodríguez, A.[2], Tomàs, A.[3] and Noguera, J.L.[1], [1]IRTA, Genètica i Millora Animal, Lleida, Spain, [2]INIA, Mejora Genética Animal, Madrid, Spain, [3]UAB, Ciència Animal i dels Aliments. Facultat de Veterinària., Bellaterra., Spain; maria.martinez@irta.cat

We have generated an F2 cross between Iberian and Meishan pigs, in order to study the genetic base of reproductive characters of economic interest such as prolificacy, maternal capacity and piglet survival. These two breeds are characterized for their high differences in their phenotypic performance for reproduction traits. Meishan sows present high prolificacy, while Iberian sows show a significantly minor litter size. RNA from ovary, uterus and pituitary gland from F2 sows were classified by high (11.48 piglets born alive) and low (5.78 piglets born alive) prolificacy at three different reproductive stages (heat, 15 day and 45 days of gestation). RNA was hybridized in porcine microchips (GeneChip® Porcine Genome Array, Affymetrix, 24,123 probes) to characterize gene expression in sows differing in prolificacy. Cluster analysis shows that the major variation is present amongst tissues, followed by reproductive stage in ovary and uterus. Analyzing separately each tissue, the ovary presents differences between heat and pregnant sows, while in uterus the differences are between sows at 45 days of pregnancy and the rest. In pituitary gland there is no a clear distinction among reproduction stages. Comparing high and low prolificacy, we have obtained 293 differentially expressed genes in pituitary gland, 471 in ovary and 61 in uterus, taking the posterior probability corresponding to the cut point given by the FDR<0.05. According to the classification in biological function, nearly all genes from each tissue belong to Cellular process group (210 in pituitary gland, 307 in ovary and 40 in uterus). Nevertheless, genes involved in reproductive function were mainly categorized within two groups: Reproduction (with 14 genes in pituitary gland, 17 in ovary and 5 in uterus) and Developmental process (65 in pituitary gland, 117 in ovary and 17 in uterus).

Genome-wide association analyses for loci controlling boar taint

Karacaören, B.[1], De Koning, D.J.[1], Velander, I.[2], Haley, C.S.[1] and Archibald, A.L.[1], [1]The Roslin Institute, University of Edinburgh, Roslin, Midlothian EH25 9PS, United Kingdom, [2]Danske Slagterier, Axeltorv 3, 1609 København V, Denmark; burak.karacaoren@roslin.ed.ac.uk

Boar taint, which is an offensive urine-like odour detected in the cooked meat of some mature boar carcasses, is caused by the accumulation of high levels of skatole and/or androstenone in backfat. Levels of these compounds are subject to genetic influence and quantitative trait loci (QTL) with effects on androstenone and skatole levels have been detected by linkage analysis in Large White x Meishan crosses. The aim of this study was to exploit a SNP (single nucleotide polymorphism) chip with up to 6,500 SNPs to map taint QTL at higher resolution in a commercial population. One thousand pigs comprising 500 pigs with high skatole levels (>0.3 µg/g) each matched by a litter mate with low skatole levels (a discordant sib-pair, DSP, design) were selected from ~6,000 Danish Landrace pigs recorded for skatole levels at slaughter and genotyped. After quality control to remove uninformative markers and animals with anomalous genotype patterns the data comprised 817 individuals and 2,715 SNPs. We used the GRAMMAR approach to detect association using a linear model, while taking the pedigree structure into account. In addition, we used a case-control design to detect association between sib pairs with 10,000 permutations to deal with multiple hypothesis testing. We detected possible associations for several six SNPs within and close to the CYP2E1 gene (P=9.999e-05). This locus explains about 5% of phenotypic variance in skatole levels. We further used discordant sip pair test to examine associations based on different allele counting scheme. Whilst these results should be confirmed in other populations they are consistent with the known role of CYP2E1 in the breakdown of skatole. We are currently undertaking a second scan with the recently developed pig 60K SNP chip which should provided greater genome coverage. These results were obtained through the EC-funded FP6 project 'SABRE'.

Nutritional effects on epigenetic modifications and their inheritance in pigs

Braunschweig, M.H. and Stahlberger-Saitbekova, N., Institute of Genetics, Vetsuisse Faculty, Bremgartenstrasse 109A, CH-3001 Berne, Switzerland; martin.braunschweig@itz.unibe.ch

In a three generation pig feeding experiment we are investigating whether differential nutrition of F0 boars affects gene expression and carcass traits in the next but one F2 generation. It is hypothesized that the nutrition related transgenerational effects are epigenetic in nature and the transmission is through the paternal line. This implies that nutrition induced epimutations are not reprogrammed in gametogenesis or in the early embryo resulting in transgenerational epigenetic inheritance. Two groups of eight F0 boars received either a control diet or a methyl-supplemented diet consisting of cofactors and methyl donors required for the one-carbon metabolism. F0 boars were mated to sows to produce the F1 male generation which was fed the control diet. These F1 boars, progeny of the F0 boars which received either a methyl-supplemented diet or a control diet, produced then 36 and 24 F2 pigs, respectively. We measured IGF2 and IGF2R gene expression in muscle, kidney and liver tissues of F0 boars. The imprinting status of the IGF2 gene in these tissues was analyzed by means of the microsatellite SWC9. DNA methylation patterns were determined in differentially methylated regions (DMR), in a CpG island and in the QTN region of the IGF2 gene. The expression levels of IGF2 and IGF2R genes do not differ between the two groups of F0 boars. No difference was found in the DNA methylation patterns around the IGF2 locus in tissues of F0 boars receiving either the methyl-supplemented diet or the control diet. We further analyzed gene expression of IGF2 and IGF2R in the F1 boars and in the F2 generation as well as the global DNA methylation level in muscle and liver tissues. We compared the carcass traits of the F2 progeny between the two groups. Results of the three generation pig feeding experiment to study transgenerational inheritance of epigenetic modifications will be presented. These results are obtained through the EC-funded FP6 Project 'SABRE'.

Analysis of RNA stability in pig post mortem skeletal muscle with Affymetrix GeneChip® Genome Arrays: implications for gene expression studies

Colombo, M.[1], Galimberti, G.[2], Calò, D.G.[2], Astolfi, A.[3], Russo, V.[1] and Fontanesi, L.[1], [1]University of Bologna, DIPROVAL - Sezione Allevamenti Zootecnici, Via F.lli Rosselli, 107, 42100 Reggio Emilia, Italy, [2]University of Bologna, Department of Statistics, Via Belle Arti, 41, 40126 Bologna, Italy, [3]University of Bologna, Interdepartmental Centre of Cancer Research 'G. Prodi', Via Massarenti, 9, 40138 Bologna, Italy; michela.colombo@studio.unibo.it

In general the belief that post mortem delay critically influences RNA and mRNA stability is still a widely held notion despite contrasting evidences that mRNAs might be stable for variable period after sampling of death of the animals. Here, with the aim to evaluate the RNA stability of post mortem porcine skeletal muscle, his degradation level and his possible use in gene expression analysis experiments, we analysed RNA extracted from Semimembranus muscles at different post mortem times using microfluidic capillary electrophoresis and by microarray analysis with Affymetrix GeneChip® Genome Arrays. RNA integrity number (RIN) and 28S:18S ratio showed no degradation for RNA obtained from muscle samples collected at 20 min, 2 h, 6 h, and 24 h after slaughtering, but degradation occurred in samples collected at 48 h *post mortem*. Microarray analysis of the first four post mortem time points confirmed these results indicating that the gene expression level is not affected by any degradation process. According to these results, studies that involve gene expression analysis in porcine skeletal muscle did not need to collect specimens just after slaughtering opening new perspectives for forensic evaluation and authentication of meat products based on gene expression profiles.

Session 14

Theatre 12

Comparison of the porcine gene expression profiles between the major sites for lipid metabolism, liver and fat

Fernández, A.I.[1], Óvilo, C.[1], López-Bote, C.[2], Barragán, C.[1], Rodríguez, M.C.[1] and Silió, L.[1], [1]INIA, Mejora Genética Animal, Madrid, 28040, Spain, [2]Facultad Veterinaria, UCM, Producción Animal, Madrid, 28040, Spain; avila@inia.es

Liver and adipose tissue are the major sites for lipid metabolism and differences among species have been reported for both tissues. The aim of the present study was to investigate the porcine lipid metabolism profile of these tissues by comparing their gene expression patterns. Hepatic and backfat samples were collected at slaughter from eight Iberian pigs, four females and four males. RNA was isolated using RiboPure kit and hybridized with Affymetrix porcine GeneChip™. Expression data analysis was carried out with BRB software, using an ANOVA test including sex and tissue as fixed effects. A FDR <0.01 was assumed. A total number of 7,847 probes showed expression differences between both tissues: 4,584 probes appeared overexpressed in fat, while 3,263 probes showed overexpression in liver. We identified, using David database, 334 differentially expressed probes, which corresponded to 243 genes related to lipid metabolism. Adipose tissue showed 113 upregulated lipid metabolism genes, while 130 genes appeared upregulated in liver. These results indicate that both tissues are of potential interest for studying the genetic basis of porcine traits related to lipid metabolism. However, it was also observed differences between both tissues in expression patterns of genes implicated in specific processes of lipid metabolism: genes codifying for key enzymes of de novo and unsaturated fatty acids synthesis such as ACACA, FASN, LPL or SCD appeared overexpressed in adipose tissue (3.5x-830x). Besides, genes codifying proteins in charge of fatty acid transport (FABP3, FABP4, FABP5) also appeared overexpressed in fat (more than 40-fold). In liver it was observed overexpression of a number of genes belonging to the CYP family (1.2x–314x), which participate in lipid synthesis, as well as genes codifying for apolipoproteins (1.8x-1740x) and key fatty acid oxidation enzymes: GPAT (7.7x) and CPT1 (4.7x).

Session 14

Theatre 13

High-density genome association study for androstenone in a terminal sire line

Harlizius, B., Duijvesteijn, N. and Knol, E.F., IPG, Institute for Pig Genetics, P.O. Box 43, 6640 AA Beuningen, Netherlands; naomi.duijvesteijn@ipg.nl

To avoid boar taint in fresh pork meat, male piglets are castrated in many EU countries. However, this poses an animal welfare issue and is not accepted by consumers. Androstenone and skatole are known as the major components causing unpleasant boar taint in entire males of pig. A commercial terminal sire line has been phenotyped for androstenone, skatole, and indole levels in fat to identify QTL related to boar taint. By selective genotyping 500 high and 500 low androstenone animals within full sib groups were selected. The average androstenone level (µg/g) of the low group was 0.9 and 2.9 for the high group. These animals were selected for genotyping of 64,232 SNPs on the Illumina Infinium assay. The mean call rate was 98.6% for individuals and 94.9% for the SNPs. Among those, 43,687 SNPs showed a MAF >0.10 and a mean heterozygosity of 0.32. Results on androstenone will be presented using genome-wide association methods. Though the design of the study is not ideal (less power compared with androstenone) for skatole, indole and some slaughter traits, a genome-wide association study will be conducted for these traits. Also 45 fathers and 11 dams were genotyped for further linkage studies to confirm possible regions found by genome-wide association for androstenone. These results are obtained through the EC-funded FP6 Project 'SABRE'.

Meat quality evaluation of different rabbit genotypes

Ribikauskiene, D. and Macijauskiene, V., Institute of Animal Science of Lithuanian Veterinary Academy, Animal Breeding and Genetics, R. Zebenkos 12, 82317 Baisogala, Radviliskis distr., Lithuania; daiva@lgi.lt

The study was carried out in 2006. Two experimental groups of rabbits were formed: group 1 – purebred New Zealand (NZ), group 2 – Hyplus hybrid and French Lop and New Zealand crossbreds ((HHxFL)xNZ). The rabbits were raised under the same feeding and housing condition. The fattening of rabbits started at an average weight of 1,242 g and slaughtering was at 3,147 g weight. The rabbits have been fattened for on average 64 days. The purpose of the study was to analyze the meat quality of purebred and commercial crossbred rabbits. The physicochemical indicators of meat and content of fatty acids were determined at the Analytical Laboratory of the Institute of Animal Science of LVA. The analysis was carried out on samples of the M. Longissimus dorsi. Lean meat analysis indicated that the content of myristoleic (C14:1) and heptadecenoic (C17:1) acids were significantly lower and those of linoleic (C18:2) and α-linolenic (C18:3) higher in the meat of (HHxFL)xNZ crossbred rabbits in comparison with purebred rabbits. Meat quality of crossbred rabbits was of higher value with lower fat content and higher quality of protein and intramuscular fat.

Genetical genomics for technological meat quality in chicken: targeted approach on a QTL affecting breast meat pH

Nadaf, J.[1], Le Bihan-Duval, E.[1], Dunn, I.C.[2], Berri, C.M.[1], Beaumont, C.[1], Haley, C.S.[2] and De Koning, D.J.[2], [1]INRA, UR83, Recherches Avicoles, Nouzilly, 37380, France, [2]Roslin Institute and R(D)SVS, University of Edinburgh, Genetics and Genomics, Midlothian, EH25 9PS, United Kingdom; javad.nadaf@roslin.ed.ac.uk

Technological quality of meat is usually defined as its ability to be further processed and stored, with muscle pH as an important indicator of quality. We previously reported the first QTL for breast meat initial pH (pH15) in an F2 population from a cross between divergent lines for body weight. One of the QTL on chromosome 1 affecting pH15, was recently considered for further fine mapping and 11 new microsatellite markers (a total of 28 markers on the chromosome) were developed and all animals (698) were genotyped for the new markers. In this study, QTL mapping analyses with all markers were done to better characterize the region for a genetical genomic approach focused on this QTL. The significance level and confidence interval of the QTL were improved and it was found that the effect of QTL was more important in females than in males (about a two-fold difference). Breast muscle samples of 32 female birds (16 from each genotype with a contrast of about 1.4 phenotypic SD) were utilised in a dye-balanced design to compare the gene expression levels for the two genotypes. Selecting samples based on information from both genotype and phenotype should improve the power of study compared to simple microarray study, and focusing on only one QTL is economically more efficient (working on 32 rather than 698 F2 animals). This study is expected to provide better knowledge about the genes and mechanism underlying or contributing to the variation of pH in chicken meat. This work was conducted as part of the SABRETRAIN Project, funded by the Marie Curie Host Fellowships for Early Stage Research Training, as part of the 6th Framework Programme of the European Commission. This Publication represents the views of the authors, not the European Commission.

Lifetime litter production after 130 generations of selection for first parity litter size in mice
Vangen, O.[1], Zerabruk Feseh, M.[1] and Wetten, M.[2], [1]Norwegian University of Life Sciences, Department of Aquaculture and Animal Science, POBOX 5003, N-1432 Aas, Norway, [2]AQUA GEN A/S, Norsvinsenteret, N-2300 Hamar, Norway; odd.vangen@umb.no

In a long time selection experiment for litter size in mice it has been selected for first parity litter size for 130 generations, the longest lasting selection experiment in mice ever reported. In the high line the litter size was 23.5 live born pups, while in the corresponding litter size in the control line was 10.5. Inbreeding ranged from 22 to 66 % in the different lines. To measure lifetime performance in the last generation of the experiment (generation 131), 50 females in each of the two lines were allowed to produce several litters. After four parities it turned out that the litter size was so low and so many females were not producing litters (were empty) that the measure of lifetime performance was defined in summarizing four parities. Over these four parities the litter size in the high line dropped to 8.5 pups, while it dropped to 3.0 pups in the control line. Only 6 females produced litters this last parity, the corresponding figure in the control line was 5. Selection for first parity ltter size has changed the production curve. Genetic parameters and effect of inbreeding on lifetime performance will be presented.

Practical application of DNA parentage test for scrotal hernia problems
Higuera, M.A.[1] and Carrión, D.[2], [1]ANPS, Goya, 115, Madrid, Spain, [2]PIC España, S. Cugat, Barcelona, Spain; mahiguera@anps.es

Developmental abnormalities such as inguinal and scrotal hernia and cryptorchidism are observed at low frequency in commercial pig production units. Under particular environmental conditions high frequency of such defects are observed. This work presents the usage of novel DNA technology by the Spanish Pig Breeders Association (ANPS) in finding solutions to scrotal hernia cases. Scrotal hernia cases are multi-factorial problems: developmental abnormalities, multi-gen genetic patterns, infectious and chemical factors. In Nucleus units, with excellent management conditions and healthy, developmental abnormalities are not normally observed. Once those genes reach commercial farms environment, in some particular situations there are farms experiencing significant problems with high frequency of scrotal hernias, while other farms using identical genetic are not showing this developmental abnormalities problem. Single sire mating and individual identification of the animals from birth are not ordinarily implemented at commercial farm level. So, it is not possible to trace back pedigree information under those scenarios. It is here where DNA paternity test is proven effective. The ANPS implemented a DNA based identification program for boars inscribed into the national heard books. This program has two goals: to determine if piglets that show some kind of developmental abnormalities are the offspring of boars registered in the heard books and if under particular environmental situations those defects may be associated to specific sires. DNA samples were collected form crossbreed animals that show scrotal hernia defects in two commercial units. We can conclude that paternity test it is a useful analytical tool to determine if genes from specific boar are present in the problematic populations. This tool it is also effective identifying boars with higher incidence of the problem in his progeny by facing the frequency of usage of specific boars to the scrotal hernias incidence in his progeny.

Genetic relationship between ascites-related trait and body weight at different ages in broilers

Closter, A.M.[1], Elferink, M.[1], Van As, P.[1], Vereijken, A.L.J.[2], Van Arendonk, J.A.M.[1], Crooijmans, R.P.[1], Groenen, M.A.M.[1] and Bovenhuis, H.[1], [1]Wageningen University, Animal Breeding and Genomics Centre, P.O. Box 338, 6700 AH Wageningen, Netherlands, [2]Hendrix Genetics B.V., Research & Technology Centre, P.O. Box 114, 5830 AC Boxmeer, Netherlands; Ane-Marie.Closter@wur.nl

The objective of this study was to estimate genetic parameters for the ascites-related trait, heart ratio and body weight at three different ages. Traits were measured on 7,800 broilers, which were kept under cold temperatures and increased levels of CO_2 in order to challenge their resistance to ascites. Weight of the right ventricle as a percentage of the total ventricle weight (RATIO) was determined on all chickens, including early mortality. Chickens were weighed at 2 weeks (BW_2), 5 weeks (BW_5), and 7 weeks of age (BW_7). Heritabilities and genetic correlations were estimated using model that adjusted for maternal environmental effects. The heritability for RATIO was 0.35 and maternal environmental effect was 0.02; heritability for BW_2 was 0.33 and maternal effect was 0.06; heritability for BW_5 was 0.22 and maternal effect was 0.05, and heritability for BW_7 was 0.18 and maternal effect was 0.04. Genetic correlation between body weights at consecutive measurements were high and positive; 0.85 (0.04) between BW_2 and BW_5, and 0.82 (0.05) between BW_5 and BW_7. Genetic correlation between BW_2 and RATIO was 0.16 (0.12), indicating that chickens with higher genetic potential for BW_2 tend to be slightly more susceptible to ascites. Genetic correlation, however, between BW_7 and RATIO was -0.37 (0.12). This indicates that birds with high genetic potential for BW_7 are the ones which tend to be less susceptible to ascites when challenged for ascites. Consequently, the fraction of the chickens that suffer from ascites effects the genetic correlation between RATIO and BW.

Multivariate analysis of founder specific inbreeding depression effects in Pirenaica Beef Cattle

Varona, L., Moreno, C. and Altarriba, J., Unidad de Genética Cuantitativa y Mejora Animal, Universidad de Zaragoza, 50013. ZARAGOZA, Spain; lvarona@unizar.es

Inbreeding has been traditionally associated with reduction of fitness. Recently, some studies have detected founder-specific variability in terms of inbreeding depression, which can be modelled via parametric distributions. The aim of this study is to expand this parametric modelling of founder specific inbreeding depression to a multivariate framework and to apply it to birth weight (BW), weight at 120 days (W120), weight at 210 days (W210), cold carcass weight (CCW) and conformation (CONF) and fat cover (FAT) scores in the Pirenaica beef cattle breed. The model of analysis included systematic effects for herd, season, year, sex and a covariate with age plus random founder specific inbreeding depression and additive genetic effects. The posterior mean estimate (and posterior standard deviation) of the average inbreeding depression for F=0.10 was -0.31 (0.15), -2.67 (1.27), -4.91 (2.43) and -6.24 (1.53) kg for BW, W120, W210 and CCW and 0.008 (0.018) and -0.028 (0.016) units for CONF and FAT, respectively. The posterior mean estimate (and standard deviation) of the variance between founders for F=0.10 was 1.69 (0.64), 111.78 (44.64), 414.04 (200.03) and 125.57 (52.43) kg^2 for BW, W120, W210 and CCW and 0.016 (0.010) and 0.010 (0.007) score $units^2$ for CONF and FAT, representing from 6 to 17% of the phenotypic variation. The higher posterior densities at 95% for correlation between founder specific inbreeding depressions did not included zero for BW and W120, BW and W210 and W120 and W210. The sense of these correlations were consistent with the additive genetic correlations although in some case their posterior distributions were not overlapped, such us BW and W210, illustrating an alternative genetic regulation for inbreeding depression and additive genetic action.

Genetic parameters for beef carcass cuts in Ireland

Pabiou, T.[1], Fikse, W.F.[2], Nasholm, A.[2], Cromie, A.R.[1], Drennan, M.J.[3], Keane, M.G.[3] and Berry, D.P.[4], [1]Irish Cattle Breeding Federation, Highfield House Bandon, Co Cork, Ireland, [2]Swedish University of Agricultural Sciences, Animal Breeding and Genetics, Ultuna, Uppsalla, Sweden, [3]Teagasc, Grange Research Center, Dunsany, Co Meath, Ireland, [4]Teagasc, Moorepark Research center, Fermoy, Co Cork, Ireland; tpabiou@ icbf.com

The objective of this study was to estimate genetic parameters for the weights of different primal cuts, using an experimental and a commercial dataset. The experimental and commercial dataset included 413 and 635 cross bred Belgian Blue, Charolais, Limousin, Angus, Holstein, and Simmental animals, respectively. Univariate analyses using a mixed linear animal model with relationships were undertaken to estimate the heritability of cold carcass weight, carcass conformation and fat, and the cut weights while a series of bivariate analyses were used to estimate the phenotypic and genetic correlations between carcass weight, carcass conformation, carcass fat, and the major primal cuts. Heritability estimates for cold carcass weight in both datasets were moderate to high (>0.48) while heritability estimates for carcass conformation and fat grading were higher in the commercial dataset (>0.63) than in the experimental study (>0.33). Across both datasets heritability estimates for wholesale cut weight in the forequarter varied from 0.03 to 0.79 while heritability estimates of carcass cut weight in the hindquarter varied from 0.14 to 0.86. Genetic correlations were strong amongst the different carcass cut weights within the two different datasets. Genetic correlations between selected carcass cut weights and carcass weight were moderate to high (minimum 0.45; maximum 0.88). Positive genetic correlations were observed in the commercial dataset between the different wholesale cut weights and carcass conformation, while these were both positive and negative in the experimental dataset. Selection for increased carcass weight will, on average, increase the weight of each cut. However, the genetic correlations were less than unity suggesting a benefit of more direct selection on high value cuts.

Using QTL studies to screen candidate genes involved in meat quality differences in Avileña Negra Ibérica within a microarray context

Moreno-Sánchez, N.[1], Rueda, J.[2], Carabaño, M.J.[1] and Díaz, C.[1], [1]INIA, Mejora Genética Animal, Crta. de la Coruña km 7.2, 28040 Madrid, Spain, [2]Facultad de Biología, Universidad Complutense de Madrid, Genética, José Antonio Novais 2, 28040 Madrid, Spain; natalia@inia.es

Gene expression and immunohistochemical studies, along with biochemical, instrumental and organoleptic measurements have been accomplished in two muscles (Psoas major, PM, and Flexor digitorum, FD) of Avileña Negra Ibérica calves to investigate the mechanisms underlying meat quality differences among commercial cuts. Meat quality traits, fiber type composition and histological properties showed significant differences between muscles. The microarray experiment detected a total of 66 differentially expressed (DE) genes. In many of them the pattern of differential expression is in accordance with the differences in the fibre type composition, metabolic activity and meat quality traits. Positioning the DE genes within previously identified QTL regions of interest would indicate their potential involvement in the biological processes underlying meat quality differences. The aim of this study was to identify which of the DE genes are contained within QTL for meat quality traits and to evaluate their potential role in meat quality differences. QTL positions were downloaded from available databases and literature or searched by using STS maps. Twenty three out of the 66 DE genes were located inside significant QTL regions for a variety of traits related to tenderness, intramuscular fat content and composition. The role of some of these genes in processes involved in tenderisation or adipogenesis is already known. However, their involvement in the meat quality differences needs to be understood.

Genetic analisys and growth curve of certain meat cuts in cattle

Baro, J.A.[1], Carleos, C.E.[2], Menéndez-Buxadera, A.[3] and Cañón, J.[3], [1]Universidad de Valladolid, CC. Agroforestales, La Yutera, avda. de Madrid s/n, 34004 Palencia, Spain, [2]Universidad de Oviedo, Estadística e Investigación Operativa, Facultad de Ciencias, Avda. Calvo Sotelo s/n, 33007 Oviedo, Spain, [3]Universidad Complutense de Madrid, Genética, ctra. de la Coruña s/n, 27040, Spain; baro@agro.uva.es

Beef cattle is sacrificed after achieving the fastest growth rates, and morphology and composition undergo great changes at this stage. Genetic analysis of such data requires age adjustment by means of regression terms and, recently, with random coefficients, in order to account for the growth curve. Preliminary results are presented for the analysis of data form an slaughter house cutting room that consisted of weight of cuts from 1131 Asturiana de los Valles calves sacrified between 2004 and 2007. Genetic components of variance, heritabilities, and genetic and phenotypic correlations were estimated for cuts from different body regions. Data was analysed under multivariate animal models that included different choices of age and weight covariates. It was found that models without covariate terms resulted in higher genetic variance component (Vg) estimates. Introduction of a age term reduced Vg, and the inclusion of carcass weight as covariate reduced Vg but not for cuts form the dorsal area. Estimates were similar when both age and carcass weight were included.

Preliminary study on lipogenic genes expression in diaphragm tissues of Japanese black heifers in association with GH gene polymorphism

Ardiyanti, A.[1], Abe, T.[2], Shoji, N.[3], Nakajima, H.[3], Kobayashi, E.[2], Suzuki, K.[4] and Katoh, K.[1], [1]Tohoku University, Animal Physiology, Miyagi, 981-0855, Japan, [2]National Livestock Breeding Center, Molecular Genetics, Fukushima, 961-8511, Japan, [3]Yamagata General Agricultural Research Center, Animal Husbandry, Yamagata, 996-0041, Japan, [4]Tohoku University, Animal Breeding and Genetics, Miyagi, 981-0855, Japan; astrid@bios.tohoku.ac.jp

Bovine growth hormone (GH) gene polymorphism (SNP) at exon 5 has been shown to alter intramuscular (i.m.) fatty acid composition in Japanese Black heifers. GH alters insulin sensitivity in adipocytes, in which insulin stimulates the lipogenesis by regulating levels of FASN, SCD, and SREBF-1mRNA expression. In the present study, therefore, we investigated plasma insulin concentration and tissue levels of FASN, SCD, and SREBF-1 mRNA in Japanese Black heifers (31.2 mo+0.9) with GH genotype AA (n=6), BB (n=6) and CC (n=4). Diaphragm tissues were ground into powder before analyzed for genes expression using real time PCR. Gene expression levels were normalized using fat percentages for statistical analysis. CC-typed heifers had a lower insulin level than AA and BB heifers. Also, CC-typed heifers had the lowest mRNA expression levels for FASN, SCD, and SREBF-1 in diaphragm tissues among three genotypes, although the differences did not reach significant levels. Although any lipogenic gene expression was not correlated with percentages of C18:0, C18:1, C18:2, and the ratio of USFA/SFA, FASNmRNA level was correlated with SREBF-1mRNA level, whereas insulin level was correlated with SCDmRNA level ($P<0.05$ and $P<0.10$, respectively). In conclusion, SNP in GH gene may alter plasma insulin levels and affect the lipogenic gene expression levels in diaphragm tissues. However, a larger number of measurements may give a clearer result for statistical analysis.

Maternal protective behaviour of German Angus and Simmental beef cattle after parturition and its relation to production traits

Hoppe, S.[1], Brand, H.[2], Erhardt, G.[2] and Gauly, M.[1], [1]Department of Animal Science, Albrecht Thaer Weg 3, 37075 Göttingen, Germany, [2]Department of Animal Science, Ludwigstrasse 21b, 35390 Giessen, Germany; mgauly@gwdg.de

A total of 390 German Angus (Aberdeen Angus x German dual-purpose breeds) and Simmental cows were tested in seven consecutive years (2000-2006) for maternal protective behaviour which was assessed by categorising behavioural response of the dams during earmarking their calves. The test was conducted within 24 h after parturition by the same person. Analysis of variance of maternal protective behaviour scores (MBS) was performed using a model including breed, lactation-number and calving month as fixed effects as well as the interaction between breed and lactation-number. The cow was included as a random effect. Breed, lactation-number and the interaction breed x lactation-number highly affected MBS. German Angus was scored higher than Simmental as well as cows with higher lactation-numbers in comparison to younger cows. Heritability was estimated under consideration of the whole relationship matrix and differed between 0.14 (S.E. 0.08) for German Angus and 0.42 (S.E. 0.05) for Simmental. Repeatabilities for MBS were 0.24 (S.E. 0.04) for German Angus and 0.42 (S.E. 0.05) for Simmental, respectively. Weaning weights and average daily weight gains of the calves were not correlated with maternal protective behaviour scores.

Genetic parameters for fatty acid composition and feed efficiency traits in Japanese Black Cattle

Inoue, K.[1], Kobayashi, M.[2] and Kato, K.[3], [1]National Livestock Breeding Center, Fukushima, 961-8511, Japan, [2]Shonai Animal Hygiene Service Center, Yamagata, 999-7781, Japan, [3]Livestock Improvement Association of Japan, Tokyo, 104-0031, Japan; k1inoue@nlbc.go.jp

Genetic parameters were estimated for feed efficiency traits (feed intakes (FI), feed conversion ratios (FCR) and residual feed intakes (RFI) of DCP and TDN), melting point of fat (MP) and fatty acid composition, which have a relation to the taste and flavor of beef. Fat and meat (M. trapezius) samples were taken from the carcasses of 863 Japanese Black steers derived from 65 sires for determination of MP and fatty acid composition of total lipid of intra-muscular adipose tissue. Genetic parameters were estimated using uni- and bivariate animal models with AIREMLF90 program. In addition, the pedigree information for 4,841 animals was used. Heritability estimates for MP, each fatty acid, monounsaturated fatty acids (MUFA), the ratio of saturated fatty acids to MUFA (MUS)and the ratio of elongation (ELONG) were high (0.63 to 0.86) except C18:2(0.34). Those for FIs were also high (0.70) but FCRs and RFIs were low (0.09 to 0.23) in this study. Genetic correlations of MP with C18:1, MUFA, MUS and ELONG were negative (favorable) and high (-0.85, -0.98, -1.00 and -0.66, respectively). The correlation estimates for feed efficiency traits of DCP were quite similar to those of TDN. The estimates for FIs and RFIs also showed a similar trend.Genetic correlations of C18:1, MUFA, MUS and ELONG with FIs and RFIs were positive (unfavorable) but all low (0.06 to 0.17), while those with FCRs were all negative (favorable) (-0.38 to -0.10). These results suggest that it is possible to improve the quality of beef fat (fatty acid composition) directly or indirectly with MP; furthermore, selecting MP or fatty acid traits does not affect feed efficiency significantly.

Selection method of a performance test considering carcass data in Japanese Black cattle

Sugimoto, T.[1], Sato, M.[2], Hosono, M.[3] and Suzuki, K.[1], [1]Graduate School of Agricultural Science, Tohoku University, 1-1 Amamiya-machi, Tsutsumidori, Aoba-ku, Sendai, Miyagi, 981-8555, Japan, [2]Miyagi Prefecture Animal Industry Experiment Station, 1 Iwadeyama Minamisaza Aza Hiwatashi Ohsaki Miyagi, 989-6445, Japan, [3]National Livestock Breeding Center Hyogo Station, 954-1 Issaicho Haji Tatsuno Hyogo, 679-4017, Japan; tsugi@bios.tohoku.ac.jp

The use of a new method to estimate breeding value (BV) effects on the performance test traits of young male Japanese Black cattle was investigated. Records of 514 Japanese Black bulls from a sire population born during 1978-2004 and records of 30,746 of their field progeny (during 1988-2004) were used to estimate genetic parameters for bulls' performance test traits and their respective relations with field progeny carcass traits. The bulls' performance test traits were daily gain (DG) and final grade (FG). The field progeny carcass traits were carcass weight (CW), beef marbling score (BMS), rib eye area (REA) and rib thickness. To estimate genetic parameters on carcass and performance traits, carcass data during 1995-2004 and all performance trait data were used. The breeding values (BV1) of the performance-tested bulls were estimated by combining the performance test traits and the carcass trait of the fattening cattle slaughtered previously at the end of performance tests, according to the pedigree relation. Second breeding values (BV2) of the performance tested bulls were estimated by combining the performance test traits and the carcass trait of the fattening cattle of half-sibs and full sibs of the bulls during 1988-2004. Genetic correlation of CW with performance traits was low to moderate (0.16-0.53). That of BMS with DG (0.12) was low; that of REA with FG (0.33) was moderate. Correlation coefficients between BV1 and BV2 for carcass traits were very high (0.74-0.89), except for REA, which was slightly high (0.65). Results of this study indicate the effectiveness of the selection concerning the carcass traits of young bulls using performance test traits and carcass trait information from pedigree.

New Alleles Detection in BoLA-DRB3 Locus of Iranian Sistani Cattle by molecular Based Testing

Ghovvati, S., Nassiry, M., Soltani, M., Sadeghi, B. and Shafagh Motlagh, A., Ferdowsi University of Mashhad, Department of Animal Science, Azadi Sq., Mashhad, P.O. Box: 91775-1163., Iran; Ghovvati@stu-mail.um.ac.ir

This study describes the identification new alleles of BoLA-DRB3 gene in Iranian Sistani cattle. Cows (n=200) were genotyped for bovine lymphocyte antigen BoLA-DRB3.2 alleles by polymerase chain reaction and restriction fragment length polymorphism method. Bovine DNA was isolated from aliquots of whole blood and a two-step PCR followed by digestion with restriction endonucleases RsaI, HaeIII and BstX2I was conducted on the DNA from Iranian Sistani Cattle. In the herd studied, we identified 24 alleles which 21 alleles were similar to those reported in earlier studies. The remaining 3 alleles (DRB3.2 *obc, * ibc, and *eac) with frequency of 0.33%, 1.33% and 8.1% had not been reported in studies carried out previously. For confirmation of new alleles, sequencing was conducted (sequence base typing or SBT). The obtained sequence of new patterns were submitted to the NCBI with accession numbers of DQ486519, EU259858, EU259857 and compared with published sequences in the GenBank database using BLAST. Results from sequencing and BLAST of these new patterns demonstrated a high degree of homology with reported sequences of other DRB3 alleles in NCBI GenBank. Based on the results we recommend this pattern to be considered as a new allele for BoLA-DRB3.2 system. Our results indicate that exon 2 of the BoLA-DRB3 Locus is highly polymorphic in Iranian Sistani Cattle and can be used as selective index and breed marker in whole of Sistani population.

Effect of Calpain1, Calpastatin and Cathepsin genes and polygene on beef shear force in Piemontese young bulls

Ribeca, C.[1], Maretto, F.[1], Gallo, L.[1], Albera, A.[2], Bittante, G.[1] and Carnier, P.[1], [1]University of Padova, Animal Science, viale Università 16, 35020 Legnaro (PD), Italy, [2]ANABoRaPi, Via Trinità 32A, 12061 Carrù (CN), Italy; cinzia.ribeca@unipd.it

Interest of researchers and meat industry for beef quality has grown as a consequence of specific requirements of consumers, and tenderness is a well-known major attribute of meat quality. Double muscle cattle breeds, such as Piemontese, can probably fit well with some consumers requests. Current evidence suggests that proteolysis of key myofibrillar and associated proteins is the cause of meat tenderization. Although several proteolytic systems related to the tenderization process have been described in the literature, interest of meat scientists has been mainly focussed on Calpain1, Calpastatin and Cathepsin genes The aim of this study was to evaluate the effect of Calpain1(CAPN1), Calpastatin (CAST) and Cathepsin (CAT) genes on beef shear force of Longissimus Thoracis muscle collected from 990 Piemontese young bulls, progeny of 109 AI sires. Shear force was measured on 5 cylindrical cores by a TA.HDi Texure Analyzer. Single nucleotide polymorphisms (SNPs) were investigated by RFLP-PCR technique. Animals were genotyped for the CAPN1 SNP g.4558 G>A and SNP g.6545C>T, CAST SNP g.2870A>G, SNP g.2959A>G and SNP g.282C>G and CATD SNP g.77G>A. Animal model procedures will be used to estimate haplotypes effects, by regressing phenotypes on haplotype probabilities, while accounting for polygenic effects on the trait.

Prediction of beef meat cuts using video image analysis in Ireland

Pabiou, T.[1] and Berry, D.P.[2], [1]Irish Cattle Breeding Federation, Highfield House Bandon, Co Cork, Ireland, [2]Teagasc, Moorepark Research Center, Fermoy, Co Cork, Ireland; tpabiou@icbf.com

The objective of this study was to assess the potential of video image analysis (VIA) for the prediction of various beef carcass primal using an experimental and a commercial dataset of carcass cuts. Regressions of carcass cut weights were also tested using cold carcass weight and the current EUROP carcass classification scheme. The experimental and commercial dataset included 436 and 281 cross bred Belgian Blue, Charolais, Limousin, Angus, Holstein, and Simmental steers and heifers, respectively. The different beef joints were clustered into four groups based on the retail value of the cuts in each of the datasets: Low Value Cuts (LVC), Medium Value Cuts (MVC), High Value Cuts (HVC), and Very high Value Cuts (VHVC). Multiple regression techniques were used and the accuracy of the regressions were established using R^2 and Root Mean Square Error (RMSE). In the experimental dataset, the R^2 ranged from 0.79 (VHVC) to 0.90 (LVC) using carcass weight as the sole predictor of the primal groups, and increased to 0.85 (VHVC) and 0.93 (LVC) when VIA variables were used as predictors. Conjointly, the RMSE of the prediction across the meat groups decreased from 2.025- 5.892 using the carcass weight predictor to 1.690 - 4.868 when using the VIA variables as predictors. Higher R^2 were observed when predicting overall weights such as hind- forequarter weights, and total meat, fat, and bone weights (min R^2=0.82; max R^2=0.99) using VIA variables. VIA technology offered in overall better predictions than carcass weight and EUROP grading predictors in all the traits considered in this study.

Relation between Neuropeptide Y and Melanocortin 4 receptor genes and milk yield and milk quality of Sarda breed sheep

Luridiana, S., Mura, M.C., Daga, C., Vacca, G.M., Pazzola, M., Dettori, M.L. and Carcangiu, V., Università degli Studi di Sassari, Biologia Animale, Via Vienna 2, 07100, Sassari, Italy; endvet@uniss.it

Appetite regulation is a very important aspect of animal production but the process is still not well known in sheep. Briefly, it could be described as controlled by an appetite stimulatory or inhibitory pathway. Neuropeptide Y (NPY) is the strongest feed intake stimulant in sheep. Its gene is composed of three exons separated by two introns. Among the receptors involved in the appetite regulation, Melanocortin-4 receptor (Mc4r) plays an important role. This receptor binds some endogenous neuropeptides that cause tonic inhibition of appetite or those that increase feed intake. Mc4r gene is composed of a single exon of 1,794 bp, whose sequence is partially unknown in sheep. The aim of this work was to examine the codificant region of NPY and Mc4r genes. To this purpose, 200 Sarda breed sheep were chosen and for each animal milk yield and milk quality were estimated. From each ewe 10 ml of blood were collected and then used for DNA extraction. All the three NPY exons and the Mc4r exon were amplified. Amplicons were visualized by electrophoresis on 2% agarose gels. A 2 micro l aliquot of each amplicon was denatured and, then, submitted to SSCP (Single-Strand Conformational Polymorphism Analysis). After SSCP analysis, no NPY exon exhibit polymorphism, while Mc4r exon showed two different pattern. The Mc4r exon showed in 15 animals a single nucleotide polymorphism (T192C) resulting in a aminoacid substitution (Val65Ala). These animals were heterozygous for this SNP (T/C). Statistical analysis showed no association between genotype T/C and milk yield and milk fat. Instead, a positive association was found between genotype heterozygous and milk protein yield ($P>0.001$). However, it would need to expand the research examining a greater number of animals, so that evaluate if mutations in these genes, involved in appetite regulation, could influence milk production in dairy sheep.

The association of stearoyl-CoA desaturase (SCD1) and sterol regulatory element binding protein-1 (SREBP-1) genotypes with the adipose tissue fatty acid composition in Fleckvieh cattle

Barton, L., Kott, T., Bures, D., Rehak, D., Zahradkova, R. and Kottova, B., Institute of Animal Science, Pratelstvi 815, 104 00 Prague - Uhrineves, Czech Republic; barton.ludek@vuzv.cz

Stearoyl-CoA desaturase (SCD) is an enzyme responsible for the conversion of saturated fatty acids (SFA) into monounsaturated fatty acids (MUFA). It is also involved in the endogenous production of the cis-9 trans-11 isomer of conjugated linoleic acid (CLA). The SCD1 gene is regulated at the transcriptional level by sterol regulatory element binding proteins (SREBPs). We investigated the previously reported genetic polymorphisms of the SCD1 and SREBP-1 genes in Fleckvieh cattle using the PCR-RFLP and AS-PCR methods, respectively. The genomic DNA was obtained from a total of 370 bulls. The frequencies of alleles A and V of the single nucleotide polymorphism in exon 5 of the SCD1 gene (SNP 878C>T) were 0.555 and 0.445, respectively. In the 84-bp Ins/Del polymorphism in intron 5 of the SREBP-1 gene, the frequency of the L allele (insertion) was markedly higher (0.920) than that of the S allele (deletion; 0.080). Fatty acid profile was determined in a total of 367 samples of intramuscular fat (IMF) and 150 samples of subcutaneous fat (SCF). The AA genotype of SCD1 polymorphism showed a lower content of C18:0 ($P<0.01$) and higher contents of C14:1 cis-9 ($P<0.001$) and C18:1 cis-9 ($P<0.05$) in IMF compared to the VV genotype. As a result, the bulls with genotypes AA or AV had lower SFA ($P<0.01$), higher MUFA ($P<0.05$) and higher MUFA/SFA ($P<0.01$) than VV animals. The SREBP-1 polymorphism was associated with a higher content of C14:1 cis-9 ($P<0.01$) in the LS genotype in SCF. The results of this study demonstrated the existence of the polymorphisms in the SCD1 and SREBP-1 genes in the population of Fleckvieh cattle and their associations with the concentrations of some of bovine adipose tissue fatty acids. The study was supported by the project NAZV QH 81228.

Effect of TG5 gene polymorphism on a basic chemical composition of beef

Riha, J.A.N.[1], Vrtkova, I.R.E.N.A.[2] and Dvorak, J.O.S.E.F.[2], [1]Research Institute for Cattle Breeding, Ltd., Vyzkumniku 267, 788 13 Vikyrovice, Czech Republic, [2]MZLU in Brno, UMFG, Zemedelska 1, 613 00 Brno, Czech Republic; jan.riha@vuchs.cz

The goal of this study was to assess its effect on a basic chemical composition parameters (protein, dry matter, nitrogen matters and fat content (%)) of beef. The TG5 gene codes thyroglobulin, which influences fat metabolism and growth of fat cells. A lot of autors remained its effect on beef marbling. Our dataset consisted of 66 unrelated crossbreeds (young bulls; Aberdeen Angus, Charolais, Galloway breeds). Amplification of the PCR product and PCR-RFLP test was performed according to standard protocol. The effect of the genotypes was calculated using the GLM procedure. The estimated genotype frequencies consisted of CC (40.92%), CT (54.54%) and TT (4.54%). Unfortunatelly, using the post-hoc Tukey HSD test, no significant differences between TG5 polymorphisms for watched parameters were revealed, probably according to small dataset and high variability in TT genotype group of animals. The protein content decreased according to genotypes in the order: TT (x=21.453, sx=0.306)>TC (x=21.336, sx=0.088)>CC (x=21.200, sx=0.102). The dry matter content decreased in the order: TC (x=25.242, sx=0.278)>CC (x=24.827, sx=0.321)>TT (x=23.787, sx=0.964). The nitrogen matters content decreased in the order: TT (x=3.433, sx=0.049)>TC (x=3.413, sx=0.014)>CC (x=3.393, sx=0.016). The dry matter content decreased in the order: TC (x=25.242, sx=0.278)>CC (x=24.827, sx=0.321)>TT (x=23.787, sx=0.964). The fat content decreased in the order: TC (x=2.462, sx=0.291)>CC (x=2.042, sx=0.337)>TT (x=1.087, sx=1.009). We formulated the hypothesis about the positive effect of allele T on protein and however on nitrogen matters content. Second hypothesis about positive effect of allele T on fat and dry matter content seemed to be false (TT<TC), but it could be caused by non-equal number of individuals in present dataset (and CC<TC vice versa). This work was created with support of projects of MŠMT ČR – 2B08037 and INGO LA330.

Meat quality traits throughout ageing in Nelore and Red Norte cattle

Andrade, P.L.[1], Bressan, M.C.[2], Gonçalves, T.M.[1], Ladeira, M.M.[1], Ramos, E.M.[1], Lopes, L.S.[1], Oliveira, L.M.F.S.[1], Rossi, R.O.D.S.[1], Oliveira, T.T.O.[1], Ricardo, C.F.[1], Vital, N.[1] and Gama, L.T.[2], [1]UFLA, C.p.37, 37200-000, Brazil, [2]INRB, Fonte Boa, Santarém, 2005-048, Portugal; ltgama1@yahoo.com

Changes in meat quality characteristics throughout the ageing process were assessed in Nelore (NE, n=22) and Red Norte (RN, n=22) intact bulls slaughtered at 24 months of age. The animals were feedlot-finished (120 days) with corn silage (50.0%) and concentrate (50%) fed ad libitum, and then transported for 60 km to a commercial abattoir. The pre-slaughter and slaughter procedures were carried out according to Brazilian official standards, and included electrical stimulation and chilling of the carcasses at 0-5 °C. Samples of the longissimus thoracis muscle were taken at 24 hours post mortem, kept at 2 °C and analyzed at 24 hours, and after 7, 14 and 21 days of ageing. The mean pH at 24 h, as well as cooking loss, moisture, protein, fat and ash were similar in meat samples from RN and NE (P>0.05). In general, meat color was similar (P>0.05) among breeds throughout ageing. Both in RN and NE samples, the major change in color during the ageing process took place from 7 to 14 days for L*, a* and b*. The Warner Bratzler shear force was lower (P<0.05) in RN than in NE at 24 hours (4.44 vs. 5.36 kg), 7 days (3.28 vs. 4.35 kg) and 14 days (3.12 vs. 4.10 kg), such that the reduction in shear force was about 26% in RN at 7 days of ageing, and 24% in NE at 14 days of ageing. The myofibrillar fragmentation index was higher (P<0.05) in RN than in NE at 24 hours and at 21 days of ageing. A negative association existed between shear force and redness, with a correlation of -0.56 (P<0.01) at 24 hours and -0.41 at 7 days (P<0.01). Overall, meat from RN and NE animals was generally similar in pH, moisture, protein, fat, ash and color parameters in different phases of ageing. However, meat from RN had lower shear force throughout ageing than meat from NE.

Genetic correlations between beef quality and breeding goal traits of Piemontese double muscled cattle

Cecchinato, A.[1], De Marchi, M.[1], Albera, A.[2], Gallo, L.[1], Bittante, G.[1] and Carnier, P.[1], [1]University of Padova, Department of Animal Science, Viale dell'Università 16, 35020 Legnaro, Italy, [2]Associazione Nazionale Allevatori Bovini di Razza Piemontese, Strada Trinita 32a, 12061 Carru, Italy; alessio.cecchinato@unipd.it

Beef quality traits (BQ) are important aspects of beef production both for consumer appreciation and for the related effects on long term profitability of the breed. The Piemontese is the most important Italian beef cattle population, exhibiting double muscling. Current breeding goal traits are daily gain (DG), live fleshiness (LF), bone thinness (BT), and direct and maternal calving performance. Assessment of BQ is difficult to be carried out routinely on a large scale because of costs and time-consuming lab methods. The genetic relationship between BQ and breeding goal traits is of particular interest to predict correlated response of BQ to selection for other traits. This study aimed to estimate genetic correlations between DG, LF or BT, measured on breeding candidates in a central testing station and BQ measured on progeny of AI sires. Measures of BQ were available for 1,208 young bulls. Animals were fattened in 124 farms and slaughtered at the same commercial slaughterhouse from March 2005 to July 2006. Share force, cooking loss, drip loss, and color traits were BQ traits. Performance testing information were from 2,432 animals which were evaluated for DG, LF, and BT. The genetic correlations between BQ and performance traits were estimated trough Bayesian analyses using Gibbs sampling. Results on genetic parameters of BQ and performance traits are discussed.

Genetic parameters for feed efficiency in Irish performance tested beef bulls

Crowley, J.J.[1,2,3], Mcgee, M.[3], Kenny, D.A.[1], Crews Jr., D.H.[4], Evans, R.D.[5] and Berry, D.P.[2], [1]University College Dublin, School of Agriculture, Food Science and Veterinary Medicine, Belfield, Dublin 4, Ireland, [2]Teagasc, Dairy Production Research Center, Moorepark, Fermoy, Co. Cork, Ireland, [3]Teagasc, Beef Research Center, Grange, Dunsany, Co. Meath, Ireland, [4]Colorado State University, Dept. Animal Sciences, Fort Collins, Colorado 80523, USA, [5]Irish Cattle Breeding Federation, Bandon, Co. Cork, Ireland; john.j.crowley@teagasc.ie

The objective of this study was to determine the phenotypic and genetic variation for feed intake, bodyweight (BW), average daily gain (ADG), and measures of feed efficiency including feed conversion ratio (FCR), relative growth rate (RGR), Kleiber ratio (KR), residual gain (RG) and residual feed intake (RFI) in Irish performance tested beef bulls. Observations on all traits were available on up to 2,605 bulls from one test station across 21 years; breeds included in the analyses were Aberdeen Angus (AA), Charolais, Hereford, Limousin (LM) and Simmental. The test period was at least 70 days. Bulls were individually offered concentrates ad libitum, with a restricted forage allowance. Large differences in performance and feed efficiency existed among breeds. For example, AA, on average, ate 1.6 kg DM/day more than LM breed and gained 0.04 kg/day more over the 70 day test period. Results showed LM was the more efficient breed when efficiency was defined as FCR, RG or RFI. Genetic parameters were estimated across breeds using linear animal mixed models. Heritability estimates for feed efficiency traits ranged from 0.24 ± 0.05 (RG) to 0.46 ± 0.06 (RFI); maternal heritability estimates ranged from 0.02 ± 0.04 (RG) to 0.12 ± 0.05 (RGR). Genetic correlations between feed efficiency measures were strong. Results from this study indicate significant genetic differences in performance and some measures of feed efficiency among performance tested beef bulls. Results from this study clearly identified significant genetic variation in feed efficiency traits as evidenced by significant breed effects and heritability estimates.

Detection of QTL for beef fatty acid composition in bovine chromosome 22

Gutiérrez-Gil, B.[1], Williams, J.L.[2], Richardson, R.I.[3], Wood, J.D.[3] and Wiener, P.[4], [1]University of León, Department of Animal Production, Campus de Vegazana, 24071, León, Spain, [2]Parco Tecnologico Padano, Polo Universitário, 26900, Lodi, Italy, [3]University of Bristol, Langford, BS40 5DU, Bristol, United Kingdom, [4]The Roslin Institute and Royal (Dick) School of Veterinary Studies, Roslin, EH25 9PS, United Kingdom; beatriz.gutierrez@unileon.es

The fatty acid (FA) composition of meat is important because of the impact of diet on human health. Marker-assisted selection could help to improve FA profiles of beef, which show low heritabilities and are difficult to routinely measure. A search for QTL affecting FA composition in beef has been conducted in a Charolais x Holstein cross-population. Measurements of FA composition were collected for the 235 bull calves from the second generation. We present here the results obtained for bovine chromosome 22. A panel of 8 microsatellite markers evenly distributed across this chromosome was genotyped across the whole population, and a QTL analysis was performed with the QTL Express software. For the 24 analysed traits, a total of 9 overlapping QTL were identified in the first half of chromosome 22, close to marker BM3406 (chromosome-wise p-values <0.01-0.05). All the significant associations identified showed an additive mode of inheritance, with the Charolais allele being associated with increased levels of the individual fatty acids showing significant effects (myristic, palmitic, stearic, palmitoleic, oleic and conjugated linoleic fatty acids) when compared to the Holstein allele. Significant effects were also found to influence two FA indexes (total intramuscular FA content and total saturated FA) and the Polyunsaturated:Saturated fat (P:S) ratio. The Holstein allele was associated to decreased levels of the two FA indexes and an increased P:S ratio. The position of these QTL overlap with a QTL for intramuscular fat content previously reported in this population, which showed the same mode of inheritance. These results suggest that a single gene localised on bovine chromosome 22 is underlying the genetic effects reported here.

Signature of selection around the myostatin locus in Piedmontese cattle typed by a 54,000 SNP panel

Valentini, A.[1], Pariset, L.[1], Bongiorni, S.[1], Williams, J.L.[2], Ajmone Marsan, P.[3], D'andrea, M.[4], Pilla, F.[4], Quaglino, A.[5], Albera, A.[5] and Nardone, A.[1], [1]UNITUS, via de Lellis, 01100 Viterbo, Italy, [2]PTP, via Einstein, 26900 Lodi, Italy, [3]UNICATT, via E. Parmense, 29100 Piacenza, Italy, [4]UNIMOL, via De Sanctis, 86100 Campobasso, Italy, [5]ANABORAPI, Strada provinciale per Trinità 32/A, 12061 Carrù (CN), Italy; alessio@unitus.it

Besides applications in breeding, high density markers can be useful for investigating the genetic structure of populations and revealing chromosomal regions that might be under selection. We used the Illumina panel of 54,000 SNP to investigate population parameters within the Piemontese and Marchigiana beef breeds. Genomic DNA was extracted from meat samples of 228 Marchigiana male individuals and from semen of 379 Piemontese bulls. The latter represent most of the bulls approved for AI of this breed currently available. An F statistics was calculated for each SNP to investigate the genetic diversity across the genome within and between these breeds. Analysis of SNPs on chromosome 2 showed that the Fis value close to the myostatin locus reaches almost 0.2 for Piemontese, for the Marchigiana the value is about -0.07 in the same region. The Piemontese is a 'double muscled' breed and has been selected to be almost fixed for a mutation in the myostatin locus on Chromsome 2. The difference in Fis in this region between these breeds is possibly a result of selection carried out on myostatin mutation in the Piemontese breed. The peak covers about 6 Mbp, a quite large span that indicates a recent selection for this trait, therefore confirming historical records.

Positional cloning of the causative mutation for bovine dilated cardiomyopathy (BDCMP)
Owczarek-Lipska, M.[1], Schelling, C.[2], Eggen, A.[3], Dolf, G.[1] and Braunschweig, M.H.[1], [1]Institute of Genetics, Vetsuisse Faculty, University of Berne, Bremgartenstrasse 109 a, 3001 Berne, Switzerland, [2]Veterinary Genetics, Vetsuisse Faculty, University of Zurich, Tannenstrasse 1, 8092 Zurich, Switzerland, [3]Laboratoire de Génétique biochimique et de Cytogénétique, UR339, INRA-CRJ, 78350 Jouy-en-Josas, France; marta. owczarek@itz.unibe.ch

Bovine dilated cardiomyopathy (BDCMP) is a heart muscle disorder which was observed during the last 30 years in cattle of Holstein-Friesian origin. In Switzerland BDCMP affects Swiss Fleckvieh and Red Holstein breeds. BDCMP is characterized by a cardiac enlargement with ventricular remodeling and chamber dilatation. The common symptoms in affected animals are subacute subcutaneous oedema, congestion of the jugular veins and tachycardia with gallop rhythm. A cardiomegaly with dilatation and hypertrophy of all heart chambers, myocardial degeneration and fibrosis are typical post-mortem findings. It was shown that all BDCMP cases reported worldwide traced back to a red factor carrying Holstein-Friesian bull, ABC Reflection Sovereign. An autosomal recessive mode of inheritance was proposed for BDCMP. The disease locus was mapped to a 6.7 Mb interval MSBDCMP06-BMS2785 on bovine chromosome 18 (BTA18). Currently, using a combined strategy of homozygosity mapping and association study we were able to narrow down the interval of interest to approximately 1.0 Mb flanked by microsatellite markers DIK3006 and MSBDCMP51. A BAC contig of 2.9 Mb encompassing the crucial interval was constructed to establish the correct marker order on BTA18. Genotyping of additional BDCMP affected animals with SNP makers revealed further recombinant chromosomes and narrowed down this 1.0 Mb interval to 184 kb distance. Ongoing studies are focused on the fine mapping and re-sequencing of this interval using a shotgun-based strategy to sequence long PCR products. The aim of present study is to identify a mutation responsible for BDCMP.

Genetic parameters for cow weight at calving and at calf weaning in South African Simmental cattle
Crook, B.J.[1], Neser, F.W.C.[2], Bradfield, M.J.[2] and Van Wyk, J.B.[2], [1]Agriculture Business Research Institute, UNE, Armidale, 2351 NSW, Australia, [2]University of the Free State, Animal Wildlife and Grassland Sciences, P.O. Box 339, 9300 Bloemfontein, South Africa; neserfw.sci@ufs.ac.za

A study was conducted to compare mature cow weight in the South African Simmental population when defined as the weight of the cow at calving or the weight of the cow at weaning of the calf. Data included in the analysis were 14,458 records for cow weight at calving (CWT-C) representing 6,534 cows and 18,871 records for cow weight at weaning (CWT-W) representing 8,395 cows. All cows were born between 1968 and 1996, while all calves were born between 1977 and 1998. The following effects had a significant influence on the data and were included in the genetic analysis: cow age in years fitted as a covariate term (linear and quadratic) and contemporary group fitted as a fixed effect. Contemporary group was defined as the unique combination of herd, birth year of calf, month of weighing, breeder-defined management group code for the calf and supplementary feeding code for the cow (for CWT-W). All analyses were done using ASREML, first fitting uni-trait and then bi-variate animal models that made provision for up to 4 weights per cow. The estimated genetic correlation obtained between the two cow weight traits was 0.95 ± 0.03, with a residual correlation of 0.61 ± 0.02. The heritability estimates for CWT-C and CWT-W from this analysis were 0.29 ± 0.04 and 0.37 ± 0.04 respectively. From a breeding perspective, these results confirm that little benefit is to be gained from weighing cows at calving if cows are to be weighed at weaning. If cow weights are to be recorded, then the weight at weaning is the more reliable and practical measure to record.

Genetic structure of the Iranian buffalo population

Aminafshar, M.[1], Amirinia, C.[2], Vaez Torshizi, R.[3] and Hosseinpour Mashhadi, M.[4], [1]Islamic Azad University, Science & Research Branch, Faculty of Agriculture & Natural Resource, Department of Animal Science, Tehran, Iran, [2]Animal Science research Institute, Department of Biotechnology, Karaj, Iran, [3]Tarbiat Modares University, Faculty of Agriculture, Tehran, Iran, [4]Islamic Azad University, Mashhad Branch, Mashhad, Iran; aminafshar@srbiau.ac.ir

In order to study genetic structure of the Iranian buffalo population, 360 blood samples were collected from North (Mazandaran & Guilan provinces), Southwest (Khuzestan province) and Northwest (Azerbaijan province) of Iran. DNA was extracted using salting out method. Fifteen microsatellite markers, CSSM019, CSSM029, CSSM033, CSSM038, CSSM041, CSSM043, CSSM047, CSSM057, CSSM060, CSRM061, CSSM062, CSSM070, BRN, BMC1013 and ETH003 were selected according to their polymorphism and amplified by PCR. PCR products were run on Acrylamid gels, and genotype of animals was detected. The mean of observed and effective number of alleles has been higher in Azerbaijan populations (4.36 and 3.52) than Khuzestan (4.28 and 3.32) and North Iranian buffalo (4.21 and 3.31) populations. They were equal to 4.36 and 3.74 respectively in all Iranian buffalo population. The mean of expected heterozygosity in North, Azerbaijan, Khuzestan and all Iranian buffalo population were equal to 0.68, 0.69, 0.68 and 0.71 respectively. The Shannon's information index of the Iranian buffalo population was equal to 1.34 ± 0.31. It has been higher in Azerbaijan (1.31 ± 0.29) than North (1.25 ± 0.25) and Khuzestan (1.25 ± 0.28) buffalo populations. The mean of polymorphism information content of the Azerbaijan (0.64 ± 0.03) has also been higher than North (0.62 ± 0.03) and Khuzestan (0.62 ± 0.03) buffalo populations. The mean of F_{is}(related to inbreeding coefficient) was equal to -30.7 in Azerbaijan, Khuzestan and North subpopulation and F_{it} was equal to -26.7 in all Iranian buffalo population. Results showed the enough genetic variation in all above populations. The value of F_{st}(0.034) showed a little difference between all subpopulation of the Iranian buffalo.

Identification of Short Interspersed Elements (SINEs) in Iranian River Buffalo

Shokrollahi, B.[1,2,3], Amirinia, S.[3], Dinparast Djadid, N.[1], Mozafari, N.[4] and Kamali, M.A.[3], [1]Pasteur Institute of Iran, Malaria and Vector Research Group, Biotechnology Research Center, Pasteur sq., Tehran, Iran, [2]Islamic Azad University, Sanandaj Branch, Department of Animal Science, Pasdaran Street, Sanandaj, Iran, [3]Animal Sciences Research Institute, Biotechnology Group, Heidar Abad, Karaj, Iran, [4]Medical Sciences University of Iran, Department of Microbiology, Tehran, Iran; Borhansh@yahoo.com

The aim of this research was to identify repeat sequences in Iranian river buffalo, which may be a basis for future genetic studies. Short Interspersed Element (SINE) is one of repeat sequences that mainly have used for evolutionary and phylogenetic studies. SINEs are non-viral retro transposable repetitive sequences with a length of 70–500bp that are widespread among eukaryotic genomes. While some SINEs are derived from 7SL RNA or 5S rRNA, most SINEs are derived from tRNA. Hence, the tRNA-like secondary structure as well as the conserved RNA polymerase III–specific internal promoter sequences (designated A and B boxes) allows new SINE elements to be distinguished from other repetitive elements in the genome. In this study 3 Short Interspersed Elements (SINEs) -like sequences referred to as BOVTA, CHR-I and CHR-II were identified. The isolation of SINEs began with RAPD-PCR enrichment. Several RAPD primers were used to amplify fragments in separate reactions. Many obtained intense bands were gel-extracted and ligated into PTZ57R/T vector and the plasmids transformed into DH5α cells. Plasmid DNA from each cloned fragment was purified and then DNA inserts were sequenced. Sequences masked with repeat masker program, and result showed the presence of 3 short interspersed elements including BOVTA, CHR-I and CHR-II. RNA polymerase III internal promoter and flanking short direct repeats commonly found in short interspersed elements. BLAST search reveals significant homology with other previously described SINEs and tRNAs.

SNP analysis of the buffalo (*Bubalus bubalis*) CFTR gene

Castellana, E.[1], Ciani, E.[1], Guerra, L.[1], Barile, V.[2] and Casavola, V.[1], [1]University of Bari, 1Department of General and Environmental Physiology, Via Amendola 165/a, 70126 Bari, Italy, [2]CRA – PCM, Via Salaria 31, 00015 Monterotondo, Italy; elenaciani@biologia.uniba.it

CFTR (Cystic Fibrosis Transmembrane Conductance Regulator) is a glycosylated protein that functions as a cAMP-regulated anion channel, known to conduct both Cl- and HCO3- and to regulate several transporters and proteins. Its possible role in mammalian sperm physiology has been recently pointed out by studies realized on heterozygous mutant mice carrying a CFTR disrupting mutation which resulted to have a significantly reduced decrease in sperm fertilizing ability compared to homozygous wild-type mice. In addition, *in vitro* CFTR inhibition has been shown to significantly reduce mouse sperm capacitation and capacitation-associated events, such as the increase in intracellular pH and the membrane hyperpolarization, by blocking the influx of chloride and bicarbonate ions into the cell. The buffalo species represents a crucial livestock resource in many countries, occupying a critical niche in many marginal agricultural systems where it provides milk, meat and draught power. Several constraints still limit the application of modern-day reproduction techniques in the buffalo species; among them, a severe reduction in frozen/thawed sperm quality has been observed, mainly in an individual-dependent fashion. The genomic characterization of the buffalo CFTR gene may provide useful markers to search for association with sperm fertility traits. We formerly characterized part of the buffalo CFTR gene (corresponding to a region of about 3 kb), highlighting the presence of 12 novel polymorphic sites both in intron (7) and exon (5) regions. In particular, two SNPs (C143T and C453T) in exon 13 have been predicted to change the amino acid sequence (R143W and P453L, respectively). Here we present the results of a SNP survey conducted on a larger population sample in order to confirm the previously observed polymorphic sites and to estimate the population allele frequencies.

Identification of Bov-A2/HpaII polymorphism of the CYP21 gene in Brazilian Zebu breeds

Martins Da Silva, A.[1], Rios, Á.F.L.[1], Ramos, E.S.[1,2], Lôbo, R.B.[1,3], Galerani, M.A.V.[1], Vila, R.A.[1] and De Freitas, M.A.R.[1], [1]University of São Paulo, Genetics, Bandeirantes Avenue 3900, 14049-900, Brazil, [2]University of São Paulo, Obstetrics and Gyneacology, Bandeirantes Avenue 3900, 14049-900, Brazil, [3]ANCP, João Godoy Street 463, 14020-230, Brazil; amsilva@genbov.fmrp.usp.br

The CYP21 (Steroid 21-hydroxylase gene) is involved in the synthesis of steroid hormones and their disability is associated with several metabolic disorders such as congenital adrenal hyperplasia in humans. The promoter region of the bovine CYP21 presents a Bov-A2 (a SINE, Short Interspersed Nucleotide Element) element in its sequence. The region studied in this work encompasses the sequence of this retroelement. The main aim of this study was to verify the occurrence of a polymorphism in this specific Bov-A2 element in Gir, Guzerá and Nelore cattle, Zebu breeds. Samples of DNA from 133 animals were genotyped through PCR reaction with specific primers, amplifying a fragment of 351pb that was digested with the enzyme HpaII. The results identified the polymorphism Bov-A2/HpaII (T/C changed) with allelic frequencies: 0.18, 0.30 and 0.34 (C); 0.82, 0.69 and 0.65 (T) respectively for Gir, Guzerá and Nelore. The averages of genotypic frequencies were 0.06 for CC, 0.51 for TT and 0.44 for CT, where a high frequency of heterozygotes and TT homozygotes were observed. In mice, there are some examples where the presence of transposable elements inserted into regulatory regions could affect gene expression and phenotype development. It is not clear if this Bov-A2 element and the polymorphism in it can affect the expression of CYP21 gene. However, this SNP is overlapped with a putative Sp1 transcription factor binding site and, the T>C mutation could also create a CpG motif, which is possible to be a DNA methylation site. Future functional studies on this region should clarify if this T>C polymorphism has some functional effect and, if there is some correlation with the higher incidence of the T allele. This is the first study on occurrence of this polymorphism in these zebu breeds.

Novel polymorphisms to detect Alectoris introgression: a multiplex-primer extension approach

Sevane, N., Cortés, O., García, D., Cañón, J. and Dunner, S., Facultad de Veterinaria (Universidad Complutense de Madrid), Producción Animal, Av. Puerta de Hierro, s/n, 28040 Madrid, Spain; nata_sf@hotmail.com

The red-legged (*Alectoris rufa*) and rock (*Alectoris graeca*) partridge populations have decreased dramatically in Europe in the last few decades, due partially to the high cynegetic pressure which leads to the release of captive-bred individuals to reinforce hunting areas. Also, the fact that breeders use species selected for productivity traits, such as growth rate, and well adapted to captivity (e.g. *Alectoris chukar*), has promoted the release of differing degrees of hybridized individuals for these purposes. However, the introduction of hybrids in wild populations is not legal and must be avoided. For this purpose, we developed an allele-specific primer extension assay to identify hybrid partridges before restocking. As a first step and starting with *Gallus gallus* genome information, 73 SNPs and 5 INDELs were identified between *A. rufa* and *A. chukar* by PCR-SSCP, 27 of which (as well as 1 mitochondrial SNP) were optimized in 3 multiplex PCRs (a 13-plex, a 10-plex, and a 5-plex). The selected polymorphisms were genotyped and their allelic frequencies estimated on red-legged partridges sampled from non-restocking Spanish areas, and chukar partridges from Greek and Spanish farms. Power calculations to determine an optimum subset of markers for a given significance level were performed. The simple and efficient SNP typing assay developed here (compared with other methods based on STR or RAPD markers) should find an application in the control of the genetic integrity of partridges in hunting areas and on farms.

Global gene expression changes associated with meat quality traits in the gluteus medius of Duroc pigs

Cánovas, A., Quintanilla, R., Díaz, I. and Pena, R.N., IRTA, Genetica i Millora Animal, 191 Rovira Roure, 25198, Spain; angela.canovas@irta.cat

In the last decades, reducing back fat thickness and increasing carcass lean content has been part of the pig breeder's selection objectives. However, this selection has also resulted in a decrease in intramuscular fat content, with adeverse effects in meat juiciness, tenderness and flavour. In order to detect and identify genes involved in muscle lipid deposition and metabolism we have used a microarray approach over muscle samples from an experimental Duroc population of 385 castrated males distributed in five half-sib families. A total of 70 samples of gluteus medius were selected from animals with extreme values for an index composed by cholesterol and lipid metabolism parameters, such as plasma lipoprotein and triglycerides levels, percentage of intramuscular fat and muscle fatty acid composition (HIGH and LOW groups; 35 animals per group). Each sample was individually hybridized using GeneChip Porcine Genome® arrays (Affymetrix). After normalizing data with the gcRMA algorithm, class comparison between groups was performed with a global t-test obtaining 1060 probes (879 single genes) whose expression levels differed significantly (p-value<0.01) between the HIGH and LOW groups. The list of differentially expressed genes was used to investigate over-representation of functions and processes through Gene Ontology (GO) analysis using a variety of GO exploration Web Tools. There were 19 pathways significantly affected by the list of resulting genes ($P<0.01$). Amongst these, there were the biosynthesis of unsaturated fatty acids, the adipocytokines and PPAR gene transactivation pathways. Thus, many of the significant pathways were very relevant to the muscular and/or adipose physiology. With all this information in hand, we selected a total of 31 genes to validate by qPCR. Validation of microarray data by qPCR revealed a high correspondence between both assays and confirmed differential expression for 19 out of 25 genes tested.

Structural and functional study of the coding region and the proximal promoter of the pig HMGCR gene

Cánovas, A.[1], Quintanilla, R.[1], Reecy, J.M.[2], Marqués, M.[3] and Pena, R.N.[1], [1]IRTA, Genetica i Millora Animal, 191 Rovira Roure, 25198, Spain, [2]Iowa State University, Animal Science, 2255 Kildee Hall, 50011, USA, [3]INDEGA, Universidad de León, Campus de Vegazana, 24071, Spain; angela.canovas@irta.cat

The 3-hydroxy-3-methylglutaryl-CoA reductase (HMGCR), which converts HMG-CoA into mevalonate, catalyzes the rate limiting step of de novo cholesterol biosynthesis. In order to contribute towards a better knowledge of genes that control cholesterol and lipid deposition in pigs, we have evaluated the potential role of HMGCR gene with respect to genetic variation in serum lipid levels and fat deposition in pigs. We characterized the mRNA expression profile of the porcine HMGCR in six tissues involved in lipid metabolism. HMGCR is expressed at high levels in the duodenum and fat. In contrast, heart and liver had lower HMGCR mRNA expression levels. Subsequently, we quantified the muscle and liver expression levels in 70 individuals corresponding to the most extreme animals for several lipid deposition and cholesterol related traits from a population of commercial Duroc pigs. In addition, we amplified the proximal promoter (600 bp) and coding region (2,159 bp) of this gene and identified four SNP polymorphisms: two in the promoter at positions -239 bp and -16 bp, and two synonymous SNP in the coding region at positions +807 and +2247. The results indicate that HMGCR expression differ between the two groups of pigs, and is affected by HMGCR genotypes. In silico analysis of the promoter SNPs revealed that the distal polymorphism could potentially affect the binding myogenic bHLH family members (e.g. myoD, myogenin, myf5 and myf6), while the more proximal mutation at position -16 a potential Ets-2 binding-site. Futhermore, we evaluated the interaction between -239 promoter SNP and promoter activators and inhibitors in hepatic (HepG2) and muscle (C2C12) cell lines. Our findings indicate that HMGCR may play an interesting role in the genetic variability of lipid and cholesterol metabolism in pigs.

Efficient production of transgenic piglets using ICSI-SMGT in combination with reca protein

Garcia-Vazquez, F.A.[1], Ruiz, S.[1], Grullon, L.[1], De Ondiz, A.[1], Aviles-Lopez, K.[1], Carvajal, J.A.[1], Matas, C.[1], Gutierrez-Adan, A.[2] and Gadea, J.[1], [1]Facultad de Veterinaria, Fisiologia, Universidad de Murcia, 30100, MURCIA, Spain, [2]INIA, Reproducción Animal y Conservación de Recursos Zoogenéticos, Carretera A Coruña, km 5.9, 28040, MADRID, Spain; fagarcia@um.es

The sperm mediated gene transfer (SMGT) is a method for transgenic animal production based on the intrinsic ability of the spermatozoa to bind exogenous DNA and to allow its transference to the oocytes. Numerous attempts have been made to improve the SMGT method such as the use of intracytoplasmic sperm injection (ICSI) to deliver transgene containing sperm cells directly into the egg. Recombinases or transposases have been used to increase the transgene integration into the genome. RecA from *E. coli* is one of the best characterized recombinases and plays an important role in homologous recombination and DNA repair. The objective of this study was to investigate whether integration and expression of exogenous DNA into the pig genome is improved by RecA in a SMGT-ICSI system. A total of 386 oocytes were injected with spermatozoa incubated with RecA-EGFP complexes and transferred into the oviducts of 3 prepuberal crossbred sows. Two sows were diagnosed pregnant (66% of success) and farrowed 7 piglets, two of them were born dead. The DNA integration by PCR in samples from different tissues was analyzed. In 6 animals (4 lives and 2 stillborn) were positive for this test. All analysed tissues (liver, spleen, epiplon, kidney, blood, fat, tail and ear) were positive in live piglets and only skin sample was positive in stillborn animals. In samples from positive animals, Western Blot analysis was used to detect the expression of EGFP protein, and was detected in all of tissues analyzed (100%). This method of 'active transgenesis' using recombinases improve transgenic rate animals (86%) and reduce the mosaicism (66.7%) in comparison to passive transgenesis methods. Supported by 10BIO2005/01-6463 and AGL2006-03495.

The effects of single and epistatic QTL for fatty-acid composition in a Meishan × Duroc crossbred population

Uemoto, Y., Sato, S., Ohnishi, C., Terai, S., Komatsuda, A. and Kobayashi, E., National Livestock Breeding Center, Technology, 1 Odakurahara, nishigo-mura, nishi-shirakawa-gun, Fukushima, 961-8511, Japan; y0uemoto@nlbc.go.jp

We performed a whole genome quantitative trait locus (QTL) analysis to confirm the existence of QTL affecting fatty acid composition, and to investigate the effects of additive, dominance, imprinting, and epistatic interaction between QTLs in an F2 resource population. The F2 population, comprising 166 pigs, was obtained by crossing a Duroc boar and Meishan sow. It was measured for fatty acid composition, and was used for whole genome QTL analysis, using to total of 180 microsatellite markers. For single QTL analysis, a total of two suggestive QTLs and one significant QTL were detected. Suggestive QTLs for C14:0 and C16:1 were identified on chromosomes 12 and 7, respectively, and a significant QTL for C18:2 was detected on chromosome 5 with the greatest LRT of 22.9. For C14:0, a significant QTL with paternal imprinting effect was also detected on chromosome 12, which locus was in the same region as an additive QTL effect, with a high LRT of 24.2. The suggestive QTL on chromosome 7 was not significant when correction for back fat thickness was included. For epistatic QTL analysis, a total of five epistatic pairs were located on chromosomes 4, 5, 9, and 16. The same epistatic pairs were significant when correction for back fat thickness was included. The individual QTL in single QTL analysis and the QTLs with epistatic interactions were not the same loci except for C18:2. For C14:0, an epistatic QTL pair was detected on chromosome 16, with the lowest p-value of 4.9×10^{-12}. Epistatic interactions affecting fatty acid composition did not show clear preeminence in this study. The present study also constitutes one of the first reports on the mapping of imprinting QTLs and epistatic pairs between QTLs affecting fatty acid composition in a swine population.

Quantitative trait loci for chronic respiratory diseases and immune traits in Landrace purebred swine

Okamura, T.[1], Kadowaki, H.[2], Shibata, C.[2], Suzuki, E.[2], Awata, T.[3], Uenishi, H.[3], Hayashi, T.[3], Mikawa, S.[3] and Suzuki, K.[1], [1]Graduate School of Tohoku University, Agricultural Science, 1-1 Tsutsumidori-Amamiyamachi,Sendai, 981-8555, Japan, [2]Miyagi Prefecture Animal Industry Experiment Station, 1 Iwadeyama Minamisawa aza Hiwatasi, Osaki, 989-6445, Japan, [3]National Institute of Agrobiological Science, Genome Reserch Department, 2-1-2 Kannondai, Tsukuba, Ibaraki, 305-8602, Japan; okamura@bios.tohoku.ac.jp

This experiment investigated whether QTL for chronic respiratory disease and immune traits were segregated in a Landrace purebred population that had been selected for meat production and the extent of Mycoplasma hyopneumoniae (MPS) lesions in the lung. Nasal bone lesion by atrophic rhinitis (AR) and lung lesion by MPS were assessed for about 600 pigs after slaughter. Various immune traits – phagocyte activity (PA), complement alternative pathway activity (CAPA), white blood cell count (WBC), granulocyte-to-lymphocyte ratio (GLR), cortisol concentration (CORT), and antibody production (AP) – were measured in peripheral blood of about 1,300 pigs at 7 weeks of age and at 105 kg body weight. Polymorphism of 99 microsatellite markers covering all autosomes was investigated in these animals. Using the the population's multi-generation pedigree structure, QTL analyses were done for a full-sib family population. We analyzed data from the population using IBD scores, as implemented in LOKI software. Subsequently, QTL analyses were performed using SOLAR based on a variance component method. Genome scan results showed evidence for significant QTL (<1% chromosome wide error rate) affecting the MPS on Sus scrofa chromosome (SSC) 2, PA at 7 wk on SSC14, PA at 105-weight on SSC5, WBC at 7 wk on SSC2, WBC at 105-weight on SSC2 and 3, and CORT at 105-weight on SSC7. Additionally, we identified 11 putative QTL (<5% chromosome wide error rate) on affecting AR, MPS, CAPA, WBC, and AP. The QTL analyses via a variance component method within a purebred population showed that QTLs were segregated in a population of purebred Landrace.

Search of candidate genes for porcine prolificacy traits based on gene expression differences in ovary tissue

Fernández-Rodríguez, A.[1], Rodríguez, M.C.[1], Fernández, A.[1], Pena, R.[2], Balcells, I.[3], Óvilo, C.[1] and Fernández, A.I.[1], [1]INIA, Mejora Genética Animal, Madrid, 28040, Spain, [2]IRTA, Lleida, 25198, Spain, [3]UAB, Barcelona, 08193, Spain; amanda@inia.es

Several genome scans have reported some porcine chromosome regions related with number of piglets born alive and total number of piglets born, but few genes have shown significant associations. In the present study we have performed a gene expression study using Affymetrix microarrays on ovary RNA of Iberian x Meishan F2 sows to identify potential candidate genes for prolificacy traits. F2 sows were ordered according to their estimated breeding values (EBV) for prolificacy traits. Six animals with higher and six with lower EBV were selected for microarray expression analyses. Expression data were analysed between high and low prolificacy using a mixed model (325 probes differentially expressed) and a t-Student test (only two). Three genes (Ssc.19416.1.A1_at (unknown), ERBBIP2 and VTN) were selected to be validated by quantitative real time PCR in a higher number of animals. We could not confirm the microarray results, probably because there are very low abundance levels of these transcripts in ovary tissue. However, VTN semiquantitative PCR showed the differential expression detected with the microarrays methodology, being the expression level higher in low than in high prolificacy class.In a previous study we detected two epistatic QTL on SSC12 (QTL1 at 15cM and QTL2 at 97cM) for piglets born alive and total number of piglets born in the same F2 cross and so far, we did not detect the causal mutations. Therefore, using these expression results and the likely VTN gene position on SSC12, we propose it as a potent candidate gene to underlay QTL2. Moreover, it has been reported an interaction between VTN and the protein encoded by $\alpha_v\beta_3$ integrin gene, which is located within the QTL1 confidence interval, this interaction could be the responsible of the detected epistatic effect. Besides, future validations will be carried out for other interesting genes related with prolificacy.

Assessing porcine structural variation with array-CGH and massively parallel sequencing

Santos Fadista, J.P., Holm, L.E., Thomsen, B. and Bendixen, C., University of Aarhus, Faculty of Agricultural Sciences, Genetics and Biotechnology, Blichers Allé 20, P.O. Box 50, DK-8830 Tjele, Denmark; joao.fadista@agrsci.dk

Genomic structural variation, including copy number variation (CNV) and indels (insertion/deletion), is known to be genome-wide present in primates and rodents but much less is known in farm animals. CNVs and indels can be used for phenotypic mapping in genome-wide association studies and have been linked to various diseases in humans. In this study we set ourselves to compare two platforms for assessing CNVs at 1kb resolution: Array comparative hybridization (array-CGH) and massively parallel sequencing (MPS). Despite MPS being the most expensive platform, it provides a better view of structural variation since it can detect not only CNVs (in a comparable way to the arrays), but it can also detect indels and SNPs, which the array-CGH platform cannot. We envisage that with the drop of prices for the MPS reagents, this will be the platform of choice for assessing genomic structural variation and corresponding putative associated traits.

Bi-polar imprinting of IGF2 affects litter size in Meishan-F2 crossbred sows

Heuven, H.C.M.[1,2], Coster, A.[2], Madsen, O.[2] and Bovenhuis, H.[2], [1]Utrecht University, Faculty of veterinary medicine, Yalelaan 104, 3584 CM Utrecht, Netherlands, [2]Wageningen University, Animal Breeding and Genomics Centre, PO-Box 338, 6700 AH Wageningen, Netherlands; h.c.m.heuven@uu.nl

Chromosome 2 in pigs harbors several paternally and maternally imprinted genes, e.g. IGF2. In mice, an effect of IGF2 on litter size has been shown. Our goals were to study the effect of the IGF2 on litter size in pigs and to elucidate the imprinting mode of this gene. Litter size records of first and second parity obtained on 210 F1 sows and 256 F2 sows from a Meishan x 'White' population were analyzed. IGF2 was sequenced and 13 additional SNP markers surrounding the IGF2 locus were identified and genotyped. Three SSRs (Sw2443, SwC9 and Sw256) which had been genotyped previously were included in the analysis. Haplotypes, i.e. ordered genotypes, were determined using the Cluster Variation Method. Genotype contrasts were redefined in terms of a mean, additive, dominance and imprinting effect. These orthogonal contrasts were fitted in a linear model together with fixed effects, a random polygenic effect and a random maternal effect. There was a significant effect of IGF2-genotype on litter size. The contrasts showed no additive or dominance effects but only an imprinting effect. This (bi-polar) imprinting effect indicated that the paternally transmitted IGF2 'A' allele as well as the maternally transmitted IGF2 'G' allele increased litter size by approximately one piglet per litter. Bi-polar imprinting occurs when the two heterozygotes are significantly different from each other but the two homozygotes have a similar effect and are not different from each other. It has been hypothesized that bi-polar imprinting can be caused by two genes in complete LD where one gene is paternally imprinted and the other gene is maternally imprinted. The paternally imprinted gene in our study is most likely IGF2. Identification of the maternally imprinted gene is the objective of an ongoing experiment.

Genetic and transcriptomic analysis of intramuscular fat content and composition in two pig muscles

Solé, M.[1], Pena, R.N.[1], Amills, M.[2], Cánovas, A.[1], Gallardo, D.[1,2], Reixach, J.[3], Díaz, I.[4], Noguera, J.L.[1] and Quintanilla, R.[1], [1]IRTA, Genètica i Millora Animal, Rovira Roure 191, 25198 Lleida, Spain, [2]UAB, Dept. Ciència Animal i dels Aliments, 08193 Bellaterra, Spain, [3]Selección Batallé, 17421, Riudarenes, Spain, [4]IRTA, Tecnologia dels Aliments, 17121 Monells, Spain; marina.sole@irta.cat

Intramuscular fat (IMF) content and fatty acid composition are key traits influencing the sensory, technological and nutritional properties of meat. This feature, together with the strong impact of dietary fat composition on human health, has generated an increasing interest in elucidating the genetic architecture of meat composition traits. We have used an approach combining QTL and microarray expression analysis to identify genes influencing IMF content and composition in two pig muscles. The animal material consisted of 350 barrows belonging to a commercial Duroc line and distributed in 5 half-sib families. All individuals were genotyped for 110 microsatellites covering the 18 porcine autosomes. The percentage of IMF, cholesterol content and fatty acids profile were determined in the Longissimus dorsi (LD) and Gluteus medius (GM) muscles. Several genome-wide significant QTL with effects on these traits were detected at SSC 3, 6, 7, 12, 14 and 18. Moreover, a gene expression analysis of 70 individuals with extreme values (35 individuals per group) for several parameters related with lipid metabolism and fat meat traits was carried out by using the GeneChip Porcine Genome Array (Affymetrix). We have detected 350 genes differentially expressed in the LD and GM muscles, along with 579 genes differentially expressed in the two groups of animals with high and low levels of lipid metabolism. The performance of an ontological analysis showed that genes differentially expressed between muscles as well as between groups could be classified in different functional categories. Moreover, a list of candidate genes has been elaborated taking into account expression data and the positional information provided by the QTL confidence intervals.

Genetic parameters for individual piglet survival in Norsvin Landrace in 1st and 2nd parity
Zumbach, B.[1], Madsen, P.[2] and Holm, B.[3], [1]Norsvin, P.B. 504, 2304 Hamar, Norway, [2]Aarhus University, P.O. Box 50, 8830 Tjele, Denmark, [3]Norsvin USA, 2768 Superior Drive NW, Rochester, MN 55901, USA; birgit.zumbach@norsvin.no

The study examined the genetic background of individual piglet survival in the Norsvin Landrace which was measured at birth (SVB), after birth until the age of 3wk (SV0-3), and at 3 wk (SV3). Data included 229,651 and 84,272 records on piglets in 1st and 2nd parity, respectively, collected from 2001 to 2008. The binary data were analyzed with DMU using Generalized Linear Mixed Models with a probit link function. The sire-dam model included herd-year, month of birth and no. of piglets born (linear covariate) as fixed effects and sire genetic, dam genetic, litter and residual as random effects. Direct heritability estimates (h^2_d) for SVB were 1.9% and 2.4%, maternal heritability estimates (h^2_m) 2.1% and 0.5% in 1st and 2nd parity, respectively. For SV0-3, h^2_d was similar in 1st and 2nd parity (2.4% and 2.3%, respectively), while h^2_m decreased slightly (1st parity: 1.3%; 2nd parity: 0.9%). Estimates of h^2_d for SV3 were 2.0% and 2.1%, those of h^2_m were 1.4% and 1.0% in 1st and 2nd parity, respectively. Total h^2 for all the traits ranged from about 3% to 4%, and tended to be higher in 2nd parity. Genetic correlations between direct and maternal effects were around 0 in 1st parity and slightly positive in 2nd parity. Both, sow and piglet genotype contribute to piglet survival. While the contribution of the piglet genotype was similar across survival traits and parities, sow genotype seemed to have a higher impact in 1st parity.

Molecular analysis of apolipoprotein D (APOD) and low density lipoprotein receptor-related protein 12 (LRP12) genes in pigs
Melo, C.[1], Zidi, A.[1], Quintanilla, R.[2] and Amills, M.[1], [1]Universitat Autonoma de Barcelona, Ciencia Animal i dels Aliments, Campus UAB, 08193 Bellaterra, Spain, [2]IRTA, Genetica i Millora, Av. Alcalde Rovira Roure 191, 25198 Lleida, Spain; Carola.Melo@uab.cat

In the current work, we have performed the molecular characterization of pig apolipoprotein D (APOD) and low density lipoprotein receptor-related protein 12 (LRP12) genes which map to quantitative trait loci affecting serum triglyceride (SSC4) and high density lipoprotein (SSC13) concentrations, respectively. Apolipoprotein D is a member of the lipocalin superfamily and participates in the reverse transport of cholesterol from body tissues to liver. Sequencing of 0.8 kb of the APOD gene in eight Duroc pigs yielded a nucleotide sequence with a high identity with other mammalian orthologous sequences, although we did not find any polymorphism. Analysis of the amino acid sequence with Scan Prosite allowed us to to identify a lipocalin domain typical of proteins binding small hydrophobic molecules like steroids, retinoids and lipids. The LRP12 gene belongs to the low density lipoprotein (LDL) receptor family and it is mainly involved in the uptake of circulating LDL. We have amplified 2.9 kb of the pig LRP12 coding region in 8 Duroc pigs with the aid of four sets of primers. In silico analysis of the amino acid sequence revealed five cysteine-rich LDL-receptor class A domains that are typical of molecules binding LDL. Moreover, we have found four synonymous mutations at C516T, C771T, A780G and A1110G. A next goal would be to perform an association analysis between these polymorphisms and serum lipid concentrations.

Estimates of genetic parameters for within-litter variation in piglet birth weight and for survival rates during suckling in Duroc

Ishii, K.[1], Takahashi, Y.[2], Arata, S.[3], Ohnishi, C.[3], Sasaki, O.[1] and Satoh, M.[1], [1]National Institute of Livestock and Grassland Science, Tuskuba, 305-0901, Japan, [2]National Livestock Breeding Center Miyazaki Station, Kobayashi, 886-0004, Japan, [3]National Livestock Breeding Center, Nishigo, 961-8511, Japan; kazishi@affrc.go.jp

Data were obtained from National Livestock Breeding Center Miyazaki Station, and included 7,992 piglets in 918 litters by 682 Duroc sows selected by average dairy gain, back fat thickness, and eye muscle area during 9 generations in a closed herd. Piglets were weaned at about 28 days of age. Records of individual piglet birth weights and number of piglets in a litter during suckling were used to estimate the genetic parameters. The following ten traits were analyzed in a multi-trait animal model: average, minimum, and maximum piglet weights in a litter at birth, within-litter variance, within-litter standard deviation (SD), and within-litter coefficient of variation (CV) in birth weight, and survival rates in a litter at days 0, 10, 20, and 30. The heritability estimates for average, minimum, and maximum piglet weights, within-litter variance, SD, and CV in birth weight were 0.31, 0.16, 0.42, 0.15, 0.23, and 0.15, respectively. The estimates of heritability for survival rates during suckling were low. Minimum piglet weight in a litter at birth was estimated to have moderately negative genetic correlations with survival rates during suckling. The estimated genetic correlations between maximum piglet weight in a litter at birth and survival rates at days 0, 10, 20, and 30 were -0.34, -0.56, -0.70, and -0.68, respectively. Within-litter CV in birth weight was estimated to have moderately positive genetic correlation with survival rate at day 0 of 0.28. However, the estimated genetic correlation decreased consistently with age. Because the genetic correlations between survival rate at day 30 and maximum piglet weight and within-litter CV at birth were high, number of piglets at weaning could be improved using within-litter variation in piglet birth weight.

Microsatellite mapping of QTL affecting meat quality and production traits in Duroc x Large White F2 pigs

Sanchez, M.P.[1], Iannuccelli, N.[2], Bidanel, J.P.[1], Billon, Y.[3], Gandemer, G.[4], Gilbert, H.[1], Larzul, C.[1], Riquet, J.[2], Milan, D.[2] and Le Roy, P.[5], [1]INRA, UMR1313 GABI, 78350 Jouy-en-Josas, France, [2]INRA, UMR444 LGC, 31326 Castanet-Tolosan, France, [3]INRA, UE967 GEPA, 17700 Surgères, France, [4]INRA, UAR2 SDAR, 17700 Surgères, France, [5]INRA-Agrocampus, UMR598 GA, 35042 Rennes, France; marie-pierre.sanchez@jouy.inra.fr

An F2 cross between Duroc and Large White pigs was performed in order to detect QTL for 15 meat quality traits (including intramuscular fat content - IMF - and fatty acid composition - FAC - of longissimus dorsi muscle), 13 production traits and 3 stress hormones levels traits. Animals from the 3 generations of the experimental design (including 456 F2 pigs) were genotyped for 91 microsatellite markers covering all the porcine chromosomes. We have detected 67 QTL located on all chromosomes. Three suggestive QTL were detected on chromosomes 1, 13 and 15 for IMF and 9 QTL were detected for FAC traits. Two of these QTL, located on chromosomes 10 and 14, acting on, respectively, the percentages of mono-unsaturated and saturated fatty acids, were significant at the genome-wide (GW) level. We have also detected 20 suggestive QTL for other meat quality traits. Production traits were influenced by 33 QTL, including 7 GW significant QTL. Among these 7 QTL, a QTL located on chromosome 3, with an effect of about 0.5 phenotypic standard deviation on post-weaning growth, has not been previously described. Finally, 3 suggestive QTL were found for cortisol, noradrenaline and adrenaline levels, on, respectively, chromosomes 7, 10 and 13. For each QTL, only 1 to 5 of the 6 F1 sires were identified as heterozygous. It means that all QTL are segregating in at least one of the founder populations used in this study. These results suggest that both meat quality traits and production traits can be improved in purebred Duroc and Large White pigs through marker-assisted selection. It is of particular interest for meat quality traits, which are difficult to include in classical selection programmes.

Comparison of different genetic models in genetic parameter estimation for litter size at birth in pigs

Satoh, M., Sasaki, S. and Ishii, K., National Institute of Livestock and Grassland Science, Animal Breeding and Reproduction Research Team, Ikenodai 2, Tsukuba-shi, 305-0901, Japan; hereford@affrc.go.jp

The first and later parities of litter size at birth are treated as either separate traits or repeated records in genetic parameter estimation. The objective of this study was to compare two different genetic models for litter size by using a stochastic computer simulation. The base animals were mated at random (10 females per male) to produce 60 males and 600 females of generation 1. Selection was either at random, on phenotypic performance of the sow or on BLUP of breeding value for the first litter size. From each litter, maximum two gilts and one boar were reared to weaning for breeding stock. Six generations were generated, including the base population. All sows have records of the first and second litter size. Then, the total number of records was 6,000. Two datasets were generated using as following models: i) a repeatability model with heritability (h^2) of 0.10 and repeatability (r^2) of 0.15, and ii) a multibariate model with h^2 of 0.10 in each parity, genetic correlation (r_G) of 0.90 and environmental correlation (r_E) of 0.05 between two parities. In the latter, either the value of h^2, r_G or r_E was varied from 0.05 to 0.15, from 0.75 to 0.95 and from -0.20 to 0.20, respectively. The genetic parameters were estimated using the different genetic model. Two hundred replicates were simulated for each selection scheme, model and various parameters. i) Estimates of h^2, r_G, and r_E were approximately 0.11, 0.91 and 0.06, respectively. ii) Estimate of r^2 decreased slightly with increasing true h^2 and estimate of h^2 was slightly lower than true h^2. The estimates of h^2 and r^2 increased with true r_G. When true r_E was negative, the estimate of h^2 was lower than 0.1 and the estimate of r^2 was 0, respectively. However, when true r_E was positive, estimate of h^2 was consistently 0.09 and estimate of r^2 increase with r_E. Effect of selection scheme on genetic parameter estimates was small.

Novel epistatic genetic interactions of QTL associated with lean and fat tissue characteristics in pigs

Duthie, C.[1], Simm, G.[1], Doeschl-Wilson, A.[1], Kalm, E.[2], Knap, P.W.[3] and Roehe, R.[1], [1]Scottish Agricultural College, Edinburgh, EH9 3JG, United Kingdom, [2]Christian-Albrechts-University of Kiel, Kiel, D-24118, Germany, [3]PIC International Group, Schleswig, D-24837, Germany; carol-anne.duthie@sac.ac.uk

Epistatic QTL analysis was carried out to investigate the contribution of epistasis to the genomic regulation of body composition traits in pigs. Data were available from a three generation full-sib design created from crossing Pietrain sires with a crossbred dam line. Phenotypic data were available for 315 F_2 animals for carcass cuts, lean tissue and fat tissue weights measured at slaughter weight (140 kg body weight). In total, 386 animals were genotyped for 88 molecular markers covering 10 chromosomes. From the genomic analysis, 24 significant epistatic QTL pairs were identified, each accounting for 5.8% to 10.2% of the phenotypic variance. All types of epistatic effect were identified; however, the additive-by-additive effect was the most prevalent. A large proportion of the identified QTL did not show significant additive or dominance effects and therefore only express their effects through interactions with other QTL. The epistatic QTL pairs with the highest effect were identified between two locations of SSC1 for carcass length and between two locations of SSC7 for entire loin weight, accounting for 9.5% and 10.2% of the phenotypic variance respectively. Additive-by-dominance and dominance-by-additive effects were identified for entire neck weight between two QTL on SSC6, one of which is in the vicinity of the RYR1 gene. A dominance-by-dominance effect was identified for entire belly weight between QTL on SSC7 and on SSC1 around the location of the MC4R gene. An additive-by-dominance effect was identified for entire ham weight between QTL on SSC9 and on SSC2 around the location of the IGF2 gene. This study shows that epistasis plays an important role in the genomic regulation of body composition of pigs. Information about epistasis can add to our understanding of the genomic networks which form the fundamental basis of biological systems.

Detection of epistatic QTL for meat quality and carcass composition in a porcine Duroc x Pietrain population

Große-Brinkhaus, C.[1], Phatsara, C.[1], Jonas, E.[1,2], Tesfaye, D.[1], Jüngst, H.[1], Tholen, E.[1] and Schellander, K.[1], [1]Institute of Animal Science, University of Bonn, Animal Breeding and Genetics, Endenicher Allee 15, 53115 Bonn, Germany, [2]Faculty of Veterinary Science University of Sydney, ReproGen- Centre for Advanced Technologies in Animal Genetics and Reproduction, 425 Werombi Road, Camden NSW 2570, Australia; cgro@itw.uni-bonn.de

Meat quality and carcass composition play an important role in the economy of pig production. The heritability for most of these traits is low to moderate and moreover the evaluation of the parameters occurs at the end of each production cycle. Therefore the development of marker assisted selection strategies are of considerable importance for these traits. Previous studies in a Duroc x Pietrain population revealed single Quantitative Trait Loci (QTL) for meat quality and carcass composition traits. The aim of the present study was to identify and characterize epistatic effects between already known (detected) and not known (not yet detected) QTL. For this purpose 300 F_2 animals of the Duroc x Pietrain resource population were analyzed using 142 markers (microsatellites and SNP) over all 18 autosomes. The linkage analysis was performed in three steps with a maximum likelihood approach using the program QxPak. Single QTL detected in the first step were fixed in the second step, and additional regions were screened. At least not associated areas were evaluated with a simultaneous interval mapping. Different networks of interacting QTL for different traits were observed between chromosomal regions. A number of epistatic QTL were found for traits associated with meat quality, meat area/content and fat-related traits. Among them previously detected QTL could be verified and additional regions were identified.

QTL detection on SSC12 for fatty acid composition of intramuscular fat and evaluation of porcine ATP Citrate Lyase (ACLY) as candidate gene

Muñoz, M.[1], Alves, E.[1], Sánchez, A.[2], Varona, L.[3], Díaz, I.[4], Barragán, C.[1], Rodríguez, M.C.[1] and Silió, L.[1], [1]INIA, Mejora Genética Animal, Madrid, 28040, Spain, [2]UAB, Ciència Animal i dels Aliments, Barcelona, 08193, Spain, [3]IRTA, Genética y Mejora Animal, Lleida, 25198, Spain, [4]IRTA, Tecnología de los Alimentos, Monells, 17121, Spain; mariamm@inia.es

Our objective was to realize a QTL scan on SSC12 for fatty acid profile on intramuscular fat (IMF) of M. longissimus in an experimental cross between Iberian and Landrace parental pig lines. Fatty acid (FA) composition influences fat and meat quality and, usually, the analysis are performed on samples of backfat (BF). However the analysis of IMF is justified since differences in FA profile and its genetic basis have been reported between BF and IMF. The content of 23 different FA was measured in IMF samples from 56 F3 and 79 backcross (BC) animals. Twelve SSC12 markers were genotyped in order to perform the QTL scan, including six microsallites and six polymorphisms of functional candidate genes for FA composition, namely: FASN, ACOX1, ACLY, GIP, ACACA and SREBF1. One significant QTL for C18:1 (n-9) (3.31±1.10) and C18:2 (n-6) (-3.04±0.96) was detected in positions 49 - 50 cM, between markers ACLY (48.6 cM) and GIP (62.6 cM). The ACLY:c.3757T>C SNP was located in 3'UTR region and presented intermediate frequencies in parental lines. The effect of ACLY:c.3757T>C on C18:1 (n-9) and C18:2 (n-6) acids contents was significant in both standard (-1.14±0.45 & 0.94±0.39 respectively) and marker-assisted association tests (-1.27±0.44 & 1.04±0.39 respectively). However, when the effects of QTL and polymorphism were analyzed together in MAAT, none of them disappeared, suggesting that the tested polymorphism is not causal but could probably be in linkage disequilibrium with the causal mutation. This QTL effect is not coincident with none of the QTL for FA backfat composition previously mapped on SSC12, in a F2 cross of the same parental lines. These results support the hypothesis of a different genetic control of FA metabolism in IMF and BF.

Within litter variation in birth weight in pigs: sources of variation

Knol, E.F., IPG, Institute for Pig Genetics, P.O. Box 43, 6640 AA Beuningen, Netherlands; egbert.knol@ ipg.nl

Variation in birth weight in pigs (variation), expressed as within litter standard deviation, is relevant because of: (1) survival, (2) increased efficiency in later production. Genetic trend in litter size reduces birth weight and, with birth weight, reduces survival chances of the average piglet. Feeding strategies during gestation can help to maintain birth weight for a number of generations of selection. Variation is, relatively, unexplored and unexploited. In a descriptive study we analyzed 2500 litters of (terminal sire lines * commercial dam crosses) to estimate the effect of different sources of variation. All relevant farrowing traits were analyzed with ASReml and all independent effects were treated as random. Estimates for variation in birth weight (between brackets the estimates for average birth weight and total mortality from litter) as percentage of total phenotypic variance were: parity 8.0% (3.5/3.2), dam cross 3.6% (0.0/0.1), line of terminal sire 0.0% (1.9/6.5), individual sire 2.8% (3.7/1.5), permanent environment 2.2% (7.3/7.9), genetic animal effect 8.3% (28.8/5.3) and unexplained 74.3% (51.8/73.9). Repeatability of variation is 10.5% with a, relatively, large influence of the service sire. Average birth weight is largely influenced by the dam (repeatability of 36.1%) and little by the sire. Variation in birth weight in pigs increases with parity, selection against variation is possible and choice of dam crosses or sire lines is of minor importance in reducing variation.

Genetic correlations between lactation performance and growing-finishing traits in pigs

Bergsma, R.[1], Kanis, E.[2], Verstegen, M.W.A.[2] and Knol, E.F.[1], [1]IPG, P.O. Box 43, 6640 AA Beuningen, Netherlands, [2]Wageningen University, P.O. Box 338, 6700 AH Wageningen, Netherlands; rob.bergsma@ ipg.nl

Through genetic selection and improved environment, productivity of sows has increased over the past decades. Various authors suggested that genetically increasing the feed intake in lactating sows will facilitate sows to wean larger litters. In integrated pork production systems, the cost prize of a slaughter pig is predominantly determined by the costs during the grower-finishing phase. Because the mother accounts for 50% of the genetics of her offspring, it is economically worthwhile to include growing-finishing characteristics in a breeding goal for dam lines. There is a hesitation in doing so because selection for lean and efficient grower-finishers will decrease their feed intake capacity and might decrease feed intake capacity as a lactating sow. Goal of this study was to contradict or confirm this by estimating the genetic correlations between lactation performance and growing-finishing traits. Data on ad libitum and restricted fed lactating sows and ad libitum fed grower-finishing pigs from various experiments were analyzed. The genetic correlation between feed intake as a lactating sow and as a grower-finisher was mainly based on a mother-offspring comparison. Animals with a higher genetic merit for feed intake as a grower-finisher are heavier at start of parturition (irrespective of parity) and have a higher fat mass. No genetic correlations were found between feed intake as a grower-finisher and feed intake as a lactating sow (ad lib and restricted) neither with weight loss during lactation nor with litter weight gain during lactation. These results suggest that from the onset of parturition different genes are involved as during growing-finishing and that selection for lean and efficient grower-finishers will not lead to a decreased feed intake capacity of lactating sows.

Inbreeding and effective population size of Piétrain pigs in Schleswig-Holstein

Gonzalez Lopez, V.[1], Habier, D.[1], Bielfeldt, J.C.[2], Borchers, N.[3] and Thaller, G.[1], [1]Animal Breeding and Husbandry, Christian-Albrechts-University, Olshausenstr. 40, D-24098 Kiel, Germany, [2]Schweineherdbuchzucht Schleswig-Holstein e.V., Rendsburger Str. 178, D-24537 Neumünster, Germany, [3]Landwirtschaftskammer Schleswig-Holstein, Futterkamp, D-24237 Blekendorf, Germany; vgonzalez@tierzucht.uni-kiel.de

Since 1980 the number of Piétrain sows in Schleswig-Holstein decreased from 1,160 to 546 in 2007. Unless unrelated breeding pigs are imported from other populations, inbreeding is expected to increase with detrimental effects, such as inbreeding depression, genetic defects, and loss of usable genetic variability for genetic improvement. The objective of this study was to analyze trends in inbreeding and effective population size (N_e) of Piétrain pigs in Schleswig-Holstein. The pedigree data available for analyses contained Piétrain pigs of the herd book population in Schleswig-Holstein born between 1980 and 2006. Only ancestors of pigs born between 2004 and 2006 were used (4,884 animals) to estimate N_e as they contributed successfully to the breeding population today. Two methods were used to estimate effective population size: 1) the formula given by Hill (1979) to estimate drift N_e for each year and 2) inbreeding coefficients obtained from the numerator relationship matrix to estimate inbreeding N_e. Drift N_e fluctuated between 100 and 333 pigs between 1990 and 1998, but declined drastically to 8-65 pigs in the following years due to an increased variance of family size in the male paths. The harmonic mean across years resulted in a drift N_e of 43 pigs. Inbreeding N_e, however, was much higher with 206 pigs, which might be due to low average inbreeding in years after boars were imported. Pedigree analyses showed that those boars were mated extensively, resulting in a high variance of family size and low drift N_e. The underlying assumptions of both methods have to be analysed in subsequent studies using the pedigree data in order to explain the different estimates obtained with the two methods.

Strategies to improve growth in lactation without inflating within-litter heterogeneity in piglet weight

Canario, L.[1], Lundgren, H.[2] and Rydhmer, L.[2], [1]INRA, Animal Genetics, UMR1313 Animal Genetics and Integrative Biology, Jouy-en-Josas, 78352, France, [2]Swedish University of Agricultural Sciences, Animal Breeding and Genetics, Uppsala, 75007, Sweden; laurianne.canario@jouy.inra.fr

Genetic parameters of piglet weight heterogeneity at birth and 3 weeks were estimated on data from Norwegian Landrace herds (Norsvin). Individual weight at three weeks (W3) was collected on 146,572 piglets from 14,045 litters in 58 herds. Both birth and 3 week weights were registered on 20,008 piglets from 5 nucleus herds. Litter data were studied with multivariate trait models including litter size. The heritability values for the standard deviation of weight at birth (SDWB) and 3 weeks (SDW3) were 0.10 and 0.08, respectively. The genetic correlation between SDWB and SDW3 was moderate (0.53±0.31). The genetic correlation of number of piglets born alive with mean weight at 3 weeks (MW3) was negative (-0.40±0.07), and zero with SDW3 (-0.05±0.11). MWB was genetically correlated with MW3 (0.59±0.16) but independent from SDW3 (0.08±0.27). The estimates of direct and maternal heritability for W3 were 0.03 and 0.07, and the genetic correlation between these components was negative (-0.43±0.10). The genetic correlation between SDW3 and W3-direct was low (-0.18±0.14) and unfavourable between SDW3 and W3-maternal (0.66±0.08). These results suggest that selecting for MWB without increasing SDW3 is possible. Another strategy would be to consider both direct and maternal effects of W3 together with SDW3 in the genetic evaluation. Unfavourable correlations could then be taken into account, thus limiting the increase in within-litter heterogeneity that would occur if selecting on the maternal component of piglet growth.

Genetic parameters and genetic trends for litter size at birth and at weaning and teat number in French Landrace and Large White pigs

Guéry, L.[1], Tribout, T.[2] and Bidanel, J.P.[2], [1]IFIP, Pôle génétique, 35651 Le Rheu, France, [2]INRA, UMR1313 GABI, 78352 Jouy-en-Josas, France; jean-pierre.bidanel@jouy.inra.fr

Genetic parameters of the number of piglets born alive (NBA), nursed by a sow until weaning (NN), weaned from a sow (NW) and of the number of functional teats (NFT) were estimated in Large White dam line (LWd) and French Landrace (FL) pig breeds using REML methodology applied to a multiple trait animal model. Genetic trends from 1988 to 2007 were then estimated by computing average estimated breeding values on a yearly basis. The data consisted of 198,267 LWd and 110,034 FL litters, and of 374,121 LWd 200,036 FL records for NFT. Heritability estimates ranged from 0.07 to 0.12 for litter size and were 0.29 (LWd) and 0.30 (FL) for NFT. Though positively correlated, NN and NW appeared to be genetically different traits (rG=0.68±0.03 and 0.85±0.02 in LWd and FL, respectively). NFT had slightly negative genetic correlations with NBA in both breeds. Genetic correlations became null or positive in LWd (0.00±0.0x for NW and 0.24±0.07 for NN), but were less favourable in FL breed (-0.08±0.05 for NW and 0.06±0.05 for NN). Genetic trends for NBA exceeded 3.5 and 2.5 piglets/litter, respectively, in LWd and FL. Trends were slightly lower at weaning (2.7 and 2.0 piglets/litter, respectively). Using teat number to improve NN may be of interest in LWd, but is more questionable in FL breed.

Effects of the CTSD g.70G>A and IGF2 intron3-g.3072G>A polymorphisms on meat production and carcass traits in pigs: evidences of the presence of additional QTN close to the imprinted IGF2 region of chromosome 2

Fontanesi, L.[1], Speroni, C.[1], Scotti, E.[1], Buttazzoni, L.[2], Dall'olio, S.[1], Nanni Costa, L.[1], Davoli, R.[1] and Russo, V.[1], [1]University of Bologna, DIPROVAL, Sezione di Allevamenti Zootecnici, Via F.lli Rosselli 107, 42100 Reggio Emilia, Italy, [2]Associazione Nazionale Allevatori Suini, Via L. Spallanzani 4/6, 00161 Roma, Italy; vincenzo.russo@unibo.it

An imprinted QTL affecting muscle mass and fat deposition was reported on the telomeric end of the p arm of porcine chromosome 2 (SSC2p) and the insulin-like growth factor 2 (IGF2) intron3-g.3072G>A substitution was identified as the causative mutation. In the same chromosome region, we assigned by linkage mapping, the cathepsin D (CTSD) gene, a lysosomal proteinase, for which we previously identified a single nucleotide polymorphism in the 3'-untranslated region (g.70G>A). The strong effects of the CTSD mutation on several production traits suggested the hypothesis of a possible independent effect of this marker in affecting fatness and meat deposition in pigs. We, therefore, refined the mapping position of the CTSD gene by radiation hybrid mapping. Then, we analysed the IGF2 and the CTSD polymorphisms in Italian Large White and Italian Duroc pig populations, for which estimated breeding values (EBVs) for several traits were calculated. For IGF2, highly significant results were obtained for all the EBVs in Italian Large White breed, with the most relevant one for lean content ($P=2.2e^{-18}$). In the Italian Duroc pigs, after excluding the possible confounding effects of the IGF2 intron3-g.3072G>A polymorphism, significant association was evidenced for the CTSD marker ($P<0.0001$) against all carcass and production traits considered. The effects of the CTSD g.70G>A mutation was also confirmed in Italian Large White animals having the IGF2 intron3-g.3072GG genotype. Overall these results indicated that the IGF2 intron3-g.3072G>A mutation is not the only quantitative trait nucleotide (QTN) affecting fatness and muscle deposition on SSC2p.

Effect of IGF2 gene on sow productivity traits

Jafarikia, M.[1], Maignel, L.[1], Wyss, S.[1], Vanberkel, W.[2] and Sullivan, B.[1], [1]Canadian Centre for Swine Improvement, Central Experimental Farm, Building 54, Ottawa, Ontario, K1A0C6, Canada, [2]Western Swine Testing Association, R.R. 1, Box 1, Site 5, Lacombe, Alberta T0C 1S0, Canada; laurence@ccsi.ca

The IGF2 gene is located on porcine chromosome 2 and a single nucleotide polymorphism (SNP) in intron 3 of this gene has been reported to have large effects on certain carcass quality traits such as lean meat content. Several studies have also described the use of this paternally expressed gene to to produce leaner hogs from fatter sows. Moreover, it has been reported in certain studies that the gene could have an effect on sow productivity traits and longevity. The aim of this study was to investigate the effect of IGF2 gene on several sow productivity traits such as litter size at birth, litter weight at weaning, number of piglets weaned and farrowing interval in Canadian swine populations. Daughters of heterozygous (AG) boars that inherited either the paternal A or G allele were studied. Results show that a difference between Landrace and Yorkshire breeds in the effects of the IGF2 gene on sow productivity traits exists. Yet, for both breeds, a similar tendency was shown: sows that inherited the paternal G allele weaned more piglets and heavier litters in comparison to daughters of AG sires that inherited the paternal A allele. Further research is required to investigate other potential candidate genes located in close proximity to the IGF2 gene which may affect litter size.

Detection of epistatic QTL for prolificacy in commercial pig populations

Noguera, J.L.[1], Varona, L.[1], Ibañez, N.[1], Cánovas, A.[1], Quintanilla, R.[1], Buys, N.[2] and Pena, R.[1], [1]IRTA, Genètica i Millora Animal, Avda. Alcalde Rovira Roure 191, 25198 Lleida, Spain, [2]University of Leuven, Department of Biosystems, Kasteelpark Arenberg 30, B3001 Leuven, Belgium; joseluis.noguera@irta.es

The aim of this study was to identify QTL that contribute to the female prolificacy in two outbred commercial pig populations. A three generation F_2 intercross was created between Landrace (Line A) and Large-White (Line B) breeds for quantitative trait locus (QTL) mapping. In total, 1,804 parities from 656 F_2 sows were recorded for total number of piglets born (TNB) and number of piglets born alive (NBA). Purebred grandparents, F_1, and F_2 sows were genotyped for 197 markers covering the 18 porcine autosomes (SSC). Two statistical models were used to analyze the experimental data. The first model was one-dimensional QTL mapping performed using a regression model. One genome-wide significant QTL ($P<0.05$) with effects on TNB was located on SSC5 between the markers SW1319 and S0005. The second model allowed for epistatic interactions following Cockerham's model. Six (three at $P<0.01$ and three at $P<0.05$) and three (one at $P<0.01$ and two at $P<0.05$) bi-dimensional genome wide significant epistatic QTL were found for TNB and NBA, respectively. The following pig chromosomes were involved in these interactions: SSC1, SSC3, SSC4, SSC6, SSC8, SSC9, and SSC16. These results indicate that epistasis could play an important role on pig prolificacy and that complex interactions could be more important that the independent main effects.

SNPs of MYPN and TTN genes in Italian Large White and Italian Duroc pigs are associated to meat production traits

Braglia, S.[1], Comella, M.[1], Zappavigna, A.[1], Buttazzoni, L.[2], Russo, V.[1] and Davoli, R.[1], [1]University of Bologna, DIPROVAL, Via F.lli Rosselli 107, 42100 Reggi Emilia, Italy, [2]Associazione Nazionale Allevatori Suini, Via L. Spallanzani 4, 00161 Roma, Italy; silvia.braglia@unibo.it

Myopalladin (MYPN) and titin (TTN) play key roles in skeletal muscle physiology and they may have a relationship with meat production traits in pigs. We analysed by PCR-RFLP 272 Italian Large White (ILW) and 207 Italian Duroc (ID) breeds for two polymorphisms previously described and identified in the 3'UTR of the genes. Moreover an additional group of animals with extreme divergent breeding values (EBVs) for some productive traits (average daily gain, lean cuts, backfat thickness and visible intermuscular fat) was genotyped for both breeds. The results showed significant effects of both genes on all considered traits in ID population. An association between MYPN and backfat thickness (P=0.0392) was reported in the Italian Large White breed. In the group of divergent animals significant allelic differences were observed in the ILW for lean cuts at the MYPN locus and in the ID breed for visible intermuscular fat at MYPN and TTN. These results indicate that both genes can be considered as useful candidates for meat and carcass quality traits in pig.

***In vivo* predictability of egg yolk ratio in hen's eggs by means of computer tomography**

Milisits, G.[1], Donkó, T.[2], Sütő, Z.[3], Bogner, P.[2] and Repa, I.[2], [1]Kaposvár University, Faculty of Animal Science, Dean Office, Guba Sándor str. 40., 7400 Kaposvár, Hungary, [2]Kaposvár University, Faculty of Animal Science, Institute of Diagnostic Imaging and Radiation Oncology, Guba Sándor str. 40., 7400 Kaposvár, Hungary, [3]Kaposvár University, Faculty of Animal Science, Department of Poultry and Companion Animal Breeding, Guba Sándor str. 40., 7400 Kaposvár, Hungary; milisits.gabor@ke.hu

The aim of this study was to examine the predictability of egg yolk ratio in intact hen's eggs using computer tomography. The experiment was carried out with 60 hen's eggs which originated from a 36 week old ROSS-308 hybrid parent stock and collected on the same day. All of the eggs were weighed before the CT measurements and positioned thereafter for the scanning in standing/upright position. Eggs were scanned with a SIEMENS Somatom Plus 4 Expert spiral CT scanner at the Institute of Diagnostic Imaging and Radiation Oncology of Kaposvár University. The images were analysed using the Medical Image Processing V1.0 software developed by the above mentioned institution. Following the CT measurements, all of the eggs were broken and their yolk and albumen were separated. After weighing the yolk, its ratio to the whole eggs was calculated. For predicting the egg yolk ratio *in vivo*, prediction equations were created by means of the linear regression method using the CT data as independent variable in the model. It was confirmed that due to the overlapping of the X-ray density values of the yolk and albumen, evaluation based on the X-ray absorption does not seem to be useful for the *in vivo* prediction of the egg yolk ratio. Depending on the number of scans involved in the evaluation, the determination of the surface of the egg yolk on the cross-sectional images resulted in a 69.3-74.1% accuracy of prediction. Based on the results it was concluded, that computer tomography seems to be precise enough for using it in further investigations in order to examine the effect of egg composition on the egg's hatchability and hatched chick's development.

Effect of hen's eggs composition on the slaughter characteristics of hatched chicks

Milisits, G.[1], Pőcze, O.[1], Ujvári, J.[1], Kovács, E.[1], Jekkel, G.[2] and Sütő, Z.[1], [1]Kaposvár University, Faculty of Animal Science, Guba Sándor str. 40., 7400 Kaposvár, Hungary, [2]Research Institute for Animal Breeding and Nutrition, Gesztenyés út 1., 2053 Herceghalom, Hungary; milisits.gabor@ke.hu

The experiment was carried out with 1,500 hen's eggs originated from a 36 weeks old ROSS-308 hybrid parent stock and collected on the same day. Eggs' electrical conductivity was measured by means of the TOBEC (Total Body Electrical Conductivity) method. Based on the measured values eggs with extreme high, extreme low and average electrical conductivity values (10-10%) were chosen for further examinations. After the TOBEC measurements the dry matter, crude protein and crude fat content of 15-15 eggs in each experimental group was analysed chemically. Remaining eggs were incubated thereafter. Hatched chicks were reared till 42 days of age and slaughtered thereafter. During the slaughter procedure, the following traits were recorded: liveweight at slaughter, grillfertig weight, the weight of breast with skin and bones, the weight of thighs with skin and bones, the weight of breast muscle and the weight of abdominal fat. It was established that eggs with different electrical conductivity values differ from each other also in their chemical composition. It was observed that the slaughter weight of chicks hatched from eggs with low electrical conductivity was significantly higher than that of the chicks hatched from eggs with high electrical conductivity (3,264 g vs 3,125 g). It was found that the weight of the examined slaughter traits showed highest values in the case of chicks hatched from eggs with low electrical conductivity, while the lowest values could be observed in the case of chicks hatched from eggs with high electrical conductivity. The differences between the two extreme groups were statistically proven ($P<0.05$) almost in all cases. Similar results were obtained also in the case of the ratio of the examined traits to the slaughter weight, but in this case the differences were not statistically proven ($P>0.05$).

Mapping of loci affecting shell quality in egg layers

Tuiskula-Haavisto, M.[1], Honkatukia, M.[1], Wei, W.[2], Dunn, I.[3], Preisinger, R.[4] and Vilkki, J.[1], [1]MTT Agrifood Research Finland, Biotechnology and Food Research, Myllytie 1, 31600 Jokioinen, Finland, [2]Medical Research Council, Human Genetics Unit, Western General Hospital, Grewe Road, Edinburg EH4 2XU Scotland, United Kingdom, [3]The Roslin Institute and Royal (Dick) School of Veterinary Studies, Roslin Midlothian, EH25 9PS Scotland, United Kingdom, [4]Lohmann Tierzucht GmbH, P.O.Box 460, 27454 Cuxhaven, Germany; maria.tuiskula-haavisto@mtt.fi

One of the challenges in poultry breeding is to develop better tools to select for enhanced resistance to structural failure and bacterial penetration. We have searched the chicken genome for quantitative trait loci affecting egg shell quality to understand the genetic background of these traits and to develop markers to be used in breeding for improved egg shell quality. The mapping population consists of 1,800 individuals in an F2 line cross between commercial egg layer lines. In the genome scan, 27 autosomes and the Z chromosome have been analyzed using 162 microsatellite markers. Autosomal chromosomes were analyzed with QTLExpress. In addition, we searched for epistatic effects between loci using a new module of GridQTL. In all, 23 QTL affecting eggshell quality were found on the autosomes. Each QTL explains 2-5% of the phenotypic variance of the trait. Significant QTL effects were found on chromosomes 2, 6 and 14 and suggestive QTL on chromosome 3. On the Z chromosome, a cluster of 5 QTL affecting both eggshell breaking strength and deformation was found within one marker interval. Only one locus pair with epistatic effects was detected for egg quality traits. The QTL regions on chromosomes 2, 3, 6, 14 and Z are being fine-mapped using additional 768 SNP markers. These results are obtained through the EC-funded FP6 Project 'SABRE'.

Detection and confirmation of QTL for an adaptive immune response to KLH
Siwek, M.[1], Slawinska, A.[1], Knol, E.F.[2], Witkowski, A.[3] and Bednarczyk, M.[1], [1]University of Technology and Life Sciences, Animal Biotechnology, Mazowiecka 28, 84 - 085 Bydgoszcz, Poland, [2]Institife for Pige Genetics, P.O. Box 43, 6640 AA Beuningen, Netherlands, [3]University of Agriculture, Animal Biology and Breeding, Akademicka 13, 20-950 Lublin, Poland; siwek@utp.edu.pl

Improvement of health of chickens by selection for enhanced general resistance to pathogens is an attractive alternative for veterinary health treatments. Therefore quite often non pathogenic antigens are applied as a selection criterion to increase overall immune response. QTL to non pathogenic antigens which represent different types of immune responses have already been detected in various experiments. To verify the existence of a QTL observed in an initial genome scan, confirmation is necessary, preferably in an independent population. Keyhole Lymphet Heamocyanin (KLH) represents a novel antigen for birds, which they never encounter during lifetime, and which results in a TH-2 dependent (antibody) immune response. QTL for an adaptive response to KLH antigen on GGA14 has been originally detected in a cross of two lines selected for high and low response to SRBC. After first detection of the QTL region, this QTL has been validated in a cross of two lines selected against feather pecking. Hereby we present a second validation of this QTL for an adaptive response to KLH in a F2 cross of a commercial layer (White Leghorn) and native polish breed (Green-Legged Partridgelike). The experimental population, which consisted of 559 individuals, was typed with 7 microsatellite markers evenly spaced on GGA14. Titers of antibodies binding KLH were measured for all individuals by ELISA. Three genetic models were applied: a half-sib model (sire common parent /dam common parent), a line cross model using the regression interval method and a combined LDLA analysis. This study confirms the QTL for an adaptive response to KLH to GGA14 and makes a case for localization of genes related to immune response to this antigen.

PCR-restriction endonuclease analysis for strain typing of *Mycobacterium avium* subsp. *paratuberculosis* based on polymorphisms in IS 1311
Soltani, M., Nassiry, M., Sadeghi, B. and Ghovvati, S., Department of Animal Science, Faculty of Agriculture, Ferdowsi University of Mashhad, Mashhad, P.O. Box: 91775-1163., Iran; Ghovvati@stu-mail.um.ac.ir

Point mutations in the IS 1311 sequences from sheep and cattle strains of *Mycobacterium avium* subsp. *paratuberculosis* (Map) were targeted to develop a polymerase chain reaction (PCR) that would be useful in the diagnosis and control of Johne's disease. PCR/REA strategy based on amplifying a 268 bp fragment of IS1311 and digestion by HinfI was developed. Results showed that all of positive results were assigned to cattle strain (C). This simple and rapid test can be used on a range of diagnostic samples for the confirmation of Johne's disease and will be of benefit in control and eradication programs for this disease.

QTL detection and estimation of the epistasis extent for chicken production and resistance to disease traits

Demeure, O.[1], Bacciu, N.[1], Carlborg, O.[2], Bovenhuis, H.[3], De Koning, D.J.[4], Le Bihan, E.[5], Bed'hom, B.[6], Pitel, F.[7], D'abbadie, F.[8], Le Roy, P.[1] and Pinard, M.H.[6], [1]INRA, UMR598, 65 rue de Saint Brieuc, 35042 Rennes, France, [2]Uppsala University, Linnaeus Center for Bioinformatics, SE-751 24 Uppsala Universit, Sweden, [3]WUR, Animal Sciences, 6700AH Wageningen, Netherlands, [4]Roslin Institute, Roslin, Midlothian, United Kingdom, [5]INRA, UR83, Centre de Tours, 37380 Nouzilly, France, [6]INRA, UMR1313, Domaine de Vilvert, 78352 Jouy-en-Josas, France, [7]INRA, UMR444, Chemin de Borde Rouge BP 27, 31326 Castanet Tolosan, France, [8]SASSO, Route de Solférino, 40630 Sabres, France; olivier.demeure@rennes.inra.fr

Recently, approaches targeting the whole genome (genome wide selection, GWS) have expanded. However, using an additive model without possible interaction (epistasis) between loci, in particular between the few QTLs which explain a large part of the genetic variance of the traits, would be likely to reduce the efficiency of this strategy. In chicken, coccidiosis susceptibility is an economically important trait which could make the most of GWS. Therefore, we proposed to search for QTLs affecting coccidiosis susceptibility, to test if the selection of this trait could have an impact on production traits and to estimate the extent of the epistasis for all these traits. QTL detection and epistasis estimation are performed on two INRA experimental crosses (900 and 1,200 F2 animals phenotyped for coccidiosis susceptbility and composition traits, respectively) genotyped by 1,536 SNPs covering most part of the genome. Our aim is to implement these results for GWS in commercial lines. These lines represent 10 families totalizing 1000 animals which will be genotyped for 384 SNPs targeting the regions identified in the experimental designs. Detecing interacting QTLs is challenging for two reasons: the computation time necessary for testing all the possible interactions and the genetic model used for the detection itself. Especially, one of the most challanging part of the project is to develop a model for outbreed crosses.

The potential of egg shell crystal size and cuticle coverage for genetic selection in laying hens

Bain, M.M.[1], Rodriguez-Navarro, A.[2], Mcdade, K.[1], Schmutz, M.[3], Preisinger, R.[3], Waddington, D.[4] and Dunn, I.C.[4], [1]University of Glasgow, Faculty of Veterinary Medicine, Glasgow, G61 1QH, United Kingdom, [2]Universidad de Granada, Departamento de Mineralogia y Petrologia, Granada, 18002, Spain, [3]Lohmann Tierzucht, P.O. Box 4602, 7454 Cuxhaven, Germany, [4]Roslin Institute, Roslin, EH25 9PS, United Kingdom; ian.dunn@roslin.ed.ac.uk

In an attempt to improve selection for egg shell quality traits linked to prevention of bacterial ingress, we wished to better define aspects of egg quality. Two new measurements were evaluated, cuticle coverage and crystal size. The cuticle is a protein layer deposited on the surface of the eggshell which prevents bacterial ingress. Although stains are available to detect the cuticle it has never been quantified before. In the case of crystal size it is generally believed that smaller crystal size is associated with stronger material properties. Methods to determine the coverage of cuticle and crystal size were developed and their potential for genetic selection evaluated. A Rhode Island Red pedigree line was used in this study. The study population comprised between 32 and 38 sire families and around 880 female offspring. The % reflectance at 650 nm of cuticle stained eggs was measured using a spectrophotometer. For crystal size the total average intensity (TA) of diffracted X-rays which is related to calcite crystal size was measured. Heritability and genetic correlations were estimated for both traits The estimates of heritability for the cuticle were moderate (0.27 ± 0.13) and for TA high (0.61 ± 0.18). There was some evidence that there was genetic correlation between cuticle coverage and shell colour (0.31 ± 0.29) but with high error. TA was genetically correlated with thickness of the palisade layer (0.51 ± 0.20) and breaking strength (0.45 ± 0.25). We conclude that measurement of cuticle quality and crystal size show considerable promise for use in genetic selection programs aimed at improving egg safety and quality. These results are obtained through the EC-funded FP6 Project 'SABRE'.

Genetic parameters of survival to avian pathogenic Escherichia coli in laying hens (APEC)

Sharifi, A.R.[1], Preisinger, R.[2], Ewers, C.[3] and Simianer, H.[1], [1]Institute of animal breeding and genetics, Albrecht-Thaer-Weg 3, 37075 Göttingen, Germany, [2]Lohmann Tierzucht GmbH, Am Seedeich 9 -11, 27472 Cuxhaven, Germany, [3]Institute of Microbiology and Epizootics, Philippstrasse 13, 10115 Berlin, Germany; rsharif@gwdg.de

Avian pathogenic *Escherichia coli* (APEC) infection results in a significant reduction of performance and liveability of chicken, leading to a distinct economical loss in poultry production. One of the strategies to combat and control this disease is to breed more resistant chickens to APEC. The main objective of this research work was to estimate genetic parameter of APEC diseases for development of breeding programmes. 353 70-weeks old White leghorn chickens (Lohmann) were infected intra-tracheally with a characterized APEC field strain IMT5155 (O2-K1-H5) using 10^8 doses corresponding to the established models of infection. The time of mortality occurrence and the number of animals which survived at the end of 10 days of post infection period was recorded. Statistical and genetical analysis of recorded mortality data was carried out by application of a linear logistic model including the effects of tier and hatch. There was no significant effect of hatch on survivability. However, there was a highly significant effect of tier on survivability suggesting that the environmental condition have a distinct effect on susceptibility to APEC diseases. The estimated heritability for resistant to APEC was 0.13 +- 0.07, meaning that 13 per cent of the total variability of the susceptibility to the disease are based on additive genetic reasons. The estimated heritability is a low value in general, but in the usual range for fitness traits. Taking in consideration the very high reproduction ability and very short generation interval in chicken, it is feasible to select animal successfully for resistance to APEC disease.

Evaluation of genetic diversity within and between Native and Khaki Campbell Duck breeds using RAPD markers

Ghobadi, A., Sayyah Zadeh, H. and Rahimi, G., Mazandaran University, Faculty of Agriculture, Department of Animal Science, No 5.Nima Youshij Alley. Taleghani Blv. Mazandaran, 48189-39439 Sari, Iran; ghobadi_abbas@yahoo.com

A total of 100 genomic DNAs were isolated from two breeds of duck: Native and Khaki Campbell, through a modified salting out procedure. The samples were used in a Polymerase Chain Reaction (PCR) with 27 RAPD Markers. Amplified PCR-products with the markers were separated on a 2% agarose gel and stained with ethidium bromide. To evaluate the bands, polymorphic and monomorphic bands were described. The RAPD analysis data from 5 primers were utilized in estimating genetic diversity and genetic distance. The genetic distance between two population was 36/06. The genetic diversity within Native and Khaki Campbell breeds was 51/08 and 66/01, respectively.

Pre-weaning growth in rabbits: the maternal effects

Garcia, M.L., Muelas, R. and Argente, M.J., Universidad Miguel Hernández de Elche, Tecnología Agroalimentaria, Ctra. Beniel km 3.2, 03312 Orihuela, Spain; mariluz.garcia@umh.es

Variance and covariance components for growth traits during the lactation were estimated from individual records of 3,807 young rabbits. The analyzed traits were the individual weight (g.) at birth, at 7, 10, 14, 21 and 28 days of age and the daily gain (g/day) during lactation (28 days). The model included the litter size at birth as covariate; the parity order (1st, 2nd or more), the intake of milk before being weighed at birth (whether the kit sukled or not), and the season (4 levels) as fixed effects, and the common litter, the individual additive genetic, and the maternal additive genetic as random effects. Analyses were performed by Bayesian methods using Gibbs sampling. All Monte Carlo standard errors were small and the Geweke test did not detect lack of convergence in any case. The posterior means of direct heritability were 0.16, 0.22, 0.14, 0.13, 0.18, 0.16 and 0.17 for weight at birth, at 7, 10, 14, 21 and 28 days of age, and daily gain. The estimates of maternal heritability for these traits were 0.11, 0.13, 0.17, 0.10, 0.10, 0.14 and 0.13, respectively. The proportion of phenotypic variance attributable to the common litter effect was high and ranged from 0.33 for weight at 10 days to 0.54 for weight at 21 days. The posterior mean of the genetic correlations between direct and maternal effects were always negative and there was a tendency to increase with kit age (-0.09 for weight at birth and -0.51 for weight at 28 days), but the null correlation included in the highest posterior interval at 95% for all the traits. The negative correlation between direct and maternal additive effects led to a low total heritability for all the traits. These results suggest that effective genetic improvement in rabbit weight before weaning by selection should be based on both direct and maternal additive genetic effects. This study was funded by projects by the Comisión Interministerial de Ciencia y Tecnología AGL2008-05514-C02-02 and the Generalitat Valenciana GVPRE2008/145.

A microsatellite markerset for parental assignment in turbot

Arfsten, M., Tetens, J. and Thaller, G., Institute of Animal Breeding and Husbandry, Christian-Albrechts University, Olshausenstrasse 40, 24098 Kiel, Germany; marfsten@tierzucht.uni-kiel.de

The turbot (*Scophthalmus maximus*) is a flatfish with great commercial value for the marine aquaculture, but information about genetic structure of commercial breeding populations is lacking. Aiming at the development of advanced breeding programs, it is necessary to have pedigree information in order to be able to apply performance testing schemes. This is not trivial in aquaculture livestock species, because animals cannot be individually marked until they have reached a certain size and offspring from several matings is normally raised together. The aim of the study was therefore to develop a set of microsatellite markers applicable for parental assignment in turbot. A total of 43 microsatellites were analysed for this purpose in three commercial turbot populations from Scotland, Norway and France comprising 312 animals in total. The software package CERVUS 3.0 was applied to calculate the number of alleles, heterozygosity and polymorphism information content (PIC), as well as the average exclusion probability at each locus. Furthermore, we performed a simulation of parentage assignment based on these results using different marker sets for the entire material or for each population. We applied different confidence levels of 80% and 95%, respectively, and the genotyping error rate was adjusted to 1%. Ten polymorphic markers were identified as being suitable for inclusion in the parentage marker set based on their combined ability to assign parentage for each commercial turbot population tested. An assignment rate of 100% was obtained for all three populations with only five markers for the relaxed confidence level. When using the more stringent level of 95%, six markers were needed to assign all progeny to their true parents in each commercial population. These results demonstrate the suitability of the marker set for reliable parentage assignment in commercial turbot populations. This can be regarded as a crucial step towards the development of advanced breeding programs in turbot.

Estimation of the genotype x environment interaction for the broiler rabbits
Vostrý, L., Mach, K., Dokoupilová, A. and Majzlík, I., Czech Univ. of Life Sci., Prague, 16000, Czech Republic; vostry@af.czu.cz

The influence of the genotype x environment interaction of two HYPLUS broiler rabbit (n=184) genotypes F_1 ($\male$PS59 × $\female$PS19) and F_2 [($\male$PS59 × $\female$PS19) × ($\male$PS59 × $\female$PS19)] was analyzed for the following traits: body weight, average daily gains, average daily consumption of feed and feed conversion. The rabbits from multiple litters bought in the commercial farms were weaned at the age of 34-35 days. These animals were located in a wire cage in the Experimental and demonstration stable of the Czech University of Life Science or in fattening farm. The broiler rabbits were fattened during the interval from 42 to 84 days of age. The genotype x environment interaction was analyzed by the least-squares analysis using the linear model with fixed effect: genotype, replication, environment and interaction genotype × environment. The results of this study suggested, that the average daily consumption and feed conversion of rabbits were influenced by genotype x environment interaction. The effect of interaction was expressed as a covariance; the environment changes influenced differences between the expected genotype values. The genotype x environment interaction had no significant effect on growth performance of broiler rabbits. Furthermore, the results of our trials we can conclude; in the first part of fattening period the animals of genotype F_2 were much better adapted to environmental condition than the animals of genotype F_1.

Genetic analysis of fertility and average daily gain in rabbit
Tusell, L.L.[1], Rekaya, R.[2], Rafel, O.[1], Ramon, J.[1] and Piles, M.[1], [1]Institut de Recerca i Tecnologia Agroalimentàries, Unitat de Cunicultura, Torre marimón s/n, Caldes de Montbui, Barcelona, 08140, Spain, [2]University of Georgia, Dpt. of Animal and Dairy Science, Athens, GA, 30602-2771, USA; llibertat. tusell@irta.cat

A Bayesian bivariate Linear-Threshold Animal Model was implemented to determine the genetic correlation between doe fertility (F), defined as success or failure to conception, and average daily gain (ADG) in a rabbit line selected for growth rate during the fattening period. A total of 27,234 data of fertility from 7,895 females and 114,135 data of ADG, which included all the information of the selection process, were used for the analysis. The pedigree included 114,485 animals. The model used for ADG included the systematic effects of year-season, parity order and number of kids born alive, the animal additive effect, the maternal genetic and the permanent environmental effects, the environmental permanent effect of litter, and the random residual effect. The model for the liability for the binary trait (F) included the systematic effects of year-season and physiological status of the female, the female additive genetic effects, the female non additive genetic plus permanent environmental effects and the residual, which was divided in an environmental permanent effect related with the environmental permanent effect of litter for ADG, and a random residual term. The obtained heritabilities were 0.04 and 0.14 for F and ADG, respectively. The genetic correlation was low and negative (-0.12) with a probability of 88% of being lower than 0. Thus, it is not expected that female reproductive performance is affected by selection for growth traits in rabbit lines.

Effect of genetic line on lipid characteristics of rabbit muscle

Zomeño, C., Blasco, A. and Hernández, P., Universidad Politécnica de Valencia, Institute for Animal Science and Technology, Camino de vera s/n P.O.Box 22020, 46022 Valencia, Spain; crizose@posgrado.upv.es

The effect of genetic line on intramuscular fat, perirenal fat and activity of some enzymes related to lipid metabolism was studied. Longissimus (LD) and Semimembranosus proprius (SP) muscles were used in this experiment. A total of 60 animals from three lines (A, V and R) selected for different criteria were slaughtered at 9 and 13 weeks of age. Line A showed higher lipid content in LD (0.87 g/100g) than V and R (0.61 and 0.67 g/100g) at 9 weeks of age. At 13 weeks of age, line A had higher muscle lipid content (1.27 g/100g) than R (0.65 g/100g) and line V showed an intermediate value (0.83 g/100g). Perirenal fat content was also influenced by genetic line, showing line A higher values. Intramuscular and perirenal fat content were positively correlated in line A (r=0.56) and V (r=0.70), but no relationship was found in line R (r=0.06). Perirenal fat increased between 9 and 13 weeks in three lines, whereas lipid content of LD increased in line A and V but in line R remained stable. The SP muscle showed higher activities of glucose-6-phosphate dehydrogenase (G6PDH) and fatty acid synthase (FAS) than LD, while malic enzyme (ME) activity was higher in LD. Some line effect was observed for lipogenic activities. In LD, line A and V had higher G6PDH activity than line R. In SP, line R and V had lower G6PDH and ME than line A. An increase of G6PDH and ME activities with age was observed in SP. In SP muscle, a higher 3-hydroxyacyl-CoA (HAD) activity was found in line R, while citrate synthase (CS) activity was higher in line R and A than in line V. Glycolitic activity (LDH) was higher in LD, whereas SP showed higher oxidative activity (HAD and CS). The LDH activity increased with age in both muscles, while the HAD activity decreased in LD. Results from this study indicate that genetic line has an effect on intramuscular fat deposition and related traits that could lead to differences in meat quality.

Genetic correlations between weights and ovulation rate in rabbits

Quirino, C.R.[1,2], Laborda, P.[1], Santacreu, M.A.[1] and Blasco, A.[1], [1]Universidad Politécnica de Valencia, Institute for Animal Science and Technology, P.O. Box 22020 - Valencia, 46022, Spain, [2]CAPES, Coordenação de Aperfeiçoamento de Pessoal de Nível Superior, Brazil; cquirino@dca.upv.es

The aim of this study was to estimate the genetics correlations between weaning weight (WW) and slaughter weight (SW) with ovulation rate (OR) in rabbits. Data come from a synthetic line of rabbits selected during eight generations by ovulation rate. Selection was based on the phenotypic value of ovulation rate with a selection pressure of 30%. OR was estimated by laparoscopy in the second gestation, 12 days after mating. A total of 679 records were used to analyze OR while a total of 14,805 records were used to analyze WW and SW. The genetic parameters were estimated using Bayesian methods and marginal posterior distributions for all unknowns were estimated by Gibbs sampling. The heritability of OR, WW and SW was 0.26, 0.18 and 0.22, respectively. Genetic correlations between OR and WW was -0.02 and between OR and SW was 0.03. Phenotypic correlations between OR and WW was 0.21 and between OR and SW was 0.20.

Selection of important traits for breeding values estimation of the linear described type traits of the Old Kladruber horses
Vostrý, L.[1], Přibyl, J.[2], Šimeček, P.[2], Majzlík, I.[1] and Jakubec, V.[1], [1]Czech Univ. of Life Sci., Prague, 16000, Czech Republic, [2]Inst. Anim. Sci., Uhříněves, 10400, Czech Republic; vostry@af.czu.cz

Estimation of the genetic parameters and breeding values for 36 conformation linear described traits was evaluated in 977 Old Kladruber horses with repeated description within the period of 16 years (1990-2006). Genetic parameters were estimated using REML animal model. Out of all models tested using Akaike's information criteria, residual variance and heritability the model with fixed effects was selected: variety, stud, variety x stud interaction, sex, age at classification, year of classification and random effects: animal and permanent environment of animal. The direct heritabilities of traits studied ranged from 0.04 to 0.65 and values of genetic correlation from -0.68 to 0.96. The reduction of traits (from primary 36 to eventual 24) was carried out using combinatorics n-tuple of traits according to: the criterion of genetic similarity (cluster analysis), measure of uncertainty of multidimensional values, values of the variance of aggregate genotype, values of the selection index variance and correlation of trait to the first principal component of the genetic matrix. The value of the selection index variance was the most appropriate parameter for the selection of important traits out of 36 described. This method rejects the traits with low heritability and low genetic correlation with other traits. The breeding values estimated using BLUP multi-trait model showed a normal distribution. The standard deviations of breeding values were estimated within a range of 0.13 to 0.99.

Monitoring the genetic variability in different horse populations from the Czech Republic and the Netherlands
Kourkova, L.[1], Vrtkova, I.[1], Van Bon, B.J.[2] and Dvorak, J.[1], [1]Mendel University Brno, Animal Genetics, Zemedelska 1, 613 Brno, Czech Republic, [2]Het Nederlandse Fjordenpaarden Stamboek, Hoenderloo, 7351 TL, Netherlands; irenav@mendelu.cz

The study is monitoring the extent of genetic variability in different horse from the CZ and the Netherlands. Genetic research was aimed at the Czech gene reserve of Old Kladruby (n=159), Silesian Noric–SN (n=105), Hucul–H (n=294), Czech-Moravian Belgik–CMB (n=66) and Dutch Fjord Horse (n=133). Samples were from 757 horses. There were genotyped 17 microsatellites (AHT4, AHT5, HMS1, HMS2, HMS3, HMS6, HMS7, HTG4, HTG6, HTG7, HTG10, VHL20, ASB2, ASB17, ASB23, CA425 and LEX3) recommended by ISAG. The number of allele per each locus ranged from 4 (HTG7, HTG6, HMS1) to 12 (ASB17, LEX3) with a mean of 7.26 alleles. The allele frequencies, observed and expected heterozygosity, polymorphism information content, exclusion probabilities and combined exclusion probabilities were calculated. The alleles found with the highest frequency across all tested equine breeds were as follows: HMS3 – allele P, HTG4 – allele M, HTG6 – allele O and LEX3 – allele L. The highest heterozygosity was observed for locus VHL20, AHT4, HTG10, ASB17 and ASB23 – over 0.70 in all five breeds. The lowest value was determined for locus HTG6 in Fjord (0.20), CMB (0.29) and SN (0.50); for locus HTG7 in Old Kladruby (0.24) and for locus HMS1 in CMB (0.46). The probabilities of paternity exclusion/one parental genotype unavailable/ and parentage exclusion were in Old Kladruby = 99.88%/99.67%/99.99%, SN = 99.99%/99.98%/99.99%, H = 99.99%/99.99%/99.99%, CMB = 99.99%/99.97%/99.99% and Fjord = 99.99%/99.93%/99.99%, respectively. The main objective of this study was to show the extent of genetic variability in different horse populations from the Czech Republic and the Netherlands. The research concerns the variability of microsatellite DNA in genotypes of horses. The results have revealed that the Hucul has quite a high genetic variability, as shown by the allele number and heterozygosity level. Supported by MZe ČR-1G58073, QH92277.

Genetic diversity in rare versus common breeds of the horse

Cothran, E.G., Texas A&M University, Veterinary Integrative Biosciences, 107D VMA Bldg., 77843 College Station, TX, USA; gcothran@cvm.tamu.edu

Genetic diversity based upon 15 STR loci was estimated for 40 horse breeds. About half of these breeds are considered as rare based upon FAO guidelines while the rest are common breeds. Diversity was analyzed using the methods of Weitzman (1992) and of Caballero and Toro (2002) as well as by standard population genetic statistics. Levels of diversity were nearly equal for rare breeds as compared to common breeds with 52% of diversity in the rare breeds based upon Weitzman statistics. Contributions to total diversity of single breeds were generally higher for rare breeds than for common breeds with the exception of the Thoroughbred which had the highest individual contribution to total diversity. Even lumping common breeds into groups of similar types of breeds did not bring contributions of diversity up to the percentages for most single rare breeds. This is likely due the shared alleles and shared ancestry of the common breeds which may not be as great for the rare breeds. The higher percentage contribution to overall diversity of the rare breeds may be due to higher frequencies of uncommon allelic variants. Variability levels within rare breeds was on average nearly as high as that for breeds with larger population sizes although there was more variation in values among the rare breeds. Heterozygosity values for rare breeds were frequently high but allelic diversity was usually lower that seen in common breeds. The lowest values of both heterozygosity and allelic diversity were seen in rare breeds. The pattern of diversity overall suggest relatively recent bottlenecks in rare breeds that have not yet impacted heterozygosity. Most tests for recent bottlenecks did not show significant evidence for a botleneck. In some cases allelic diversity was high but this was seen mainly in breeds that have had know outcrosses in recent times. Additional breeds and analyses are currently underway which should shed more light upon the distribution of genetic diversity within breeds of the domestic horse.

Loss of genetic variability due to selection for black coat colour in the rare Mallorquí horse

Royo, L.J.[1], Alvarez, I.[1], Fernandez, I.[1], Perez-Pardal, L.[1], Payeras, L.[2] and Goyache, F.[1], [1]SERIDA, Area de Genetica y Reproduccion, Camino de los Claveles 604, 33203-Gijon-Asturias, Spain, [2]AECABMA, Inca, 07300 Mallorca, Spain; ljroyo@serida.org

The Mallorquí horse is a rare black-coated breed kept in the Mallorca Island. A recovery program started in early 70's with the foundation of the studbook. Population size is very small, with less than 300 individuals registered at December 2006. The breeders association do not register chestnut individuals in its studbook and, therefore, carriers of the chestnut allele tend to be not used for reproduction. The aim of this work is to assess the possible loss of genetic variability caused by a selection regime consisting in only using for reproduction homozygous black individuals. A total of 68 samples of Mallorquí horse reproductive individuals were obtained and genotyped for a set of 15 microsatellites. Presence of the chestnut allele was asses following Royo *et al.* A total of 14 individuals were carriers. Losses of genetic variability were assessed using the program MolKin 3.0. freely available at http://www.ucm.es/info/prodanim/html/JP_Web.htm. Genetic variability was assessed in terms of expected heterozygosity or gene diversity; and rarefacted average number of alleles per locus. Rejecting carriers for reproduction would give losses of 1.1% in gene diversity and 0.4% in allelic richness. These values can be interpreted by comparing them with those obtained for the homozygous black individuals: losses of 2.6% in gene diversity and increase in allelic richness of 2.1%. Within the Mallorquí horse, those individuals carrying the chestnut allele gather a significant part of the genetic variability of the breed. Particularly, these individuals seems to own a significant part of the rare alleles of the breed. This scenario suggests the need to implement a mating policy aiming at maintaining the genetic background represented by the group of individuals carrying the chestnut allele. Partially funded by a grant MICIN-INIA RZ2008-00010.

A genetic study on Turkish horse breeds based on microsatellite and mtDNA markers and inferences for conservation

Denizci, M., Aslan, O., Koban, E., Aktoprakligil, D., Aksu, S., Balcioglu, K., Turgut, G., Erdag, B., Bagis, H. and Arat, S., TUBITAK MAM, Genetic Engineering and Biotechnology Institute, PK21, 41470 Gebze, Kocaeli, Turkey; sezen.arat@mam.gov.tr

The increasing loss of animal genetic resources has become an important issue in the last decade and many countries has their own action plans following FAO's global management programme. Conservation studies and successful implementation of management plans first requires understanding the genetic composition of the populations. The studies on sheep, cattle and goat have shown that Anatolia is an important place in animal domestication process. Horses have served man in battle, at work, on the hunt, and in sports. They have been an important companion for the populations lived in Anatolia as they did in other parts of the world. There are studies on native Anatolian livestock animals based on DNA markers, but not on horses. This study analyses the present diversity within and between four native Anatolian horse breeds using 14 microsatellite loci and mtDNA diversity based on D-loop sequence. The preliminary results revealed relatively low allelic diversity with a mean of 6.68 alleles/locus/population and no detectable population structing. The average observed heterozygosity is 0.75/locus/population and the average unbiased expected heterozygosity is 0.774/locus/population. mtDNA analysis of 271 individuals revealed 111 haplotypes partitioned in 12 groups not associated with the breeds' phenotypic characteristics as also observed in livestock species. The high motility of horses, horse trading habits, keeping all the horses of a village in one place over winter without any control on their breeding and lack of proper breeding strategy for horse in the country are reflected in the results. Acknowledgement: This project, namely TÜRKHAYGEN-I, is supported by Turkish Scientific and Technical Research Council (grant no: KAMAG 106G005).

Analysis of inbreeding and generation interval of Silesian Noriker and Czech-Moravian Belgian Horse

Čapková, Z., Majzlik, I., Vostrý, L. and Andrejsová, L., CULS, Prague6, 165 21, Czech Republic; majzlik@ af.czu.cz

Silesian Noriker (SN) and Czech-Moravian Belgian Horse (CMB) are draught breeds bred in the area of Czech Republic. We analysed data of 917 SN foals (322 colts and 595 fillies) and 2323 CMB foals (801 colts and 1,522 fillies) born in the period 1990-2007. The inbreeding coefficient in stallions of SN is 3.55%, in mares reaches 3.38%. In CMB males, there is a value of 3.21% and in females 2.82%. In group of SN the coefficient of inbreeding is 3.39% and 2.88%. The average value of inbreeding coefficient of SN is rising, if the animal were born between 1990-1992, the average value reached 2.23% and for those born between 2005-2007 it is 4.06%. Animals CMB born in time period 1990-1992 proved an average value of inbreeding coefficient 1.93% and animals born in years 2005-2007 as high as 4.16%. For both parents, generation interval was 7.8 years for SN and 8 years for CMB.

Antibody production against Aleutian mink disease virus without viral replication

Farid, A.[1], Ferns, L.E.[2] and Daftarian, P.M.[3], [1]Nova Scotia Agricultural College, Plant and Animal Sciences, Truro, NS, B2N 5E3, Canada, [2]Veterinary Services, Nova Scotia Department of Agriculture, Truro, NS, B2N 5E3, Canada, [3]University of Miami, Wallace H. Coulter Center for Translational Research and Department of MIC & IMM, Miami, FL, 33136, USA; hfarid@nsac.ca

Immune response of adult mink to infection by the Aleutian mink disease virus (AMDV), development of the disease and severity of the diseases symptoms depend on the mink genetics, the strain of AMDV and environmental factors. The level of antibody in some persistently infected mink may be below the threshold level of detection by the commonly used counter-immunoelectrophoresis (CIEP) test (false negatives). Here we report three cases where viral-specific antibody production persisted long after apparent viral clearance or cessation of the viral replication. A group of 2005-born black female mink that were seronegative in Dec. 2005 was bred in single-sire mating in 2006 and 2007. Animals were CIEP-tested in nine other occasions between Feb. 2006 and Feb. 2008 at the ages of 10, 15, 18, 19, 22, 27, 30, 31 and 34 months. Kits were CIEP tested twice in each year at 5 and 7 months of age. Two females became seropositive by 15 months of age and one by 22 months of age. These mink remained seropositive until 34 months of age when they were killed. None of the carcasses showed any gross or microscopic lesions characteristic of Aleutian disease in kidneys, lungs, heart, brain or liver. Viral DNA was not detected by PCR in the spleen or lung tissues of these mink. All the 16 progeny of two of the females that whelped in 2007 remained seronegative by 7 months of age (Nov. 2007), indicating that transplacental infection by AMDV did not occur. It was concluded that AMDV was not replicating in these animals since 2007 breeding season and that the continued production of detectable levels of viral-specific antibody for a year could be the result of the presence of a cross reactive antigen, dormant ADMV, or involvement of a dominant cell mediated immune response.

Selection experiment for immune response to enhance disease resistance in mice

Sakai, E., Otomo, Y., Narahara, H. and Suzuki, K., Graduate School of Tohoku University, Agricultural Science, 1-1 Amamiyamachi Tsutsumidori, Aoba-ku, Sendai, 981-8555, Japan; erisa@bios.tohoku.ac.jp

Selection for peripheral immune traits was done in mice to investigate the possibility of breeding for disease resistance. Selections for high phagocyte activity (PA), high antibody production (ABP), and high PA, and high ABP (aggregate breeding value weighted by standard deviation of each breeding value) were produced. They were respectively named N line, A line, and NA line. Furthermore, a control line without selection was named C line. Selective breeding has continued: data for 14 generations are now available. Immune traits of about 5,500 mice from each line were measured. Moreover, to ensure the disease resistance of these mice, they were infected by i.v. route with Shiga-like toxin at 11 generations of selection. We compared weight loss and some immune traits such as leukocyte number, phagocyte activity, and antibody production between each selection line and the control line. The heritability estimate for PA in N line was 0.35 ± 0.03. That for ABP in A line was 0.34 ± 0.07; PA and ABP in NA line were 0.57 ± 0.05 and 0.33 ± 0.06. The phenotypic values of immune traits varied greatly, but the breeding values showed a substantial increase with selection. In an exposure experiment with Shiga-like toxin, immune responses of the selection line were significantly higher than that of the control line. Line differences were also found among three selection lines. These results suggest that establishment of a selection line related to the immune traits is possible: selection for peripheral immune traits is effective to enhance disease resistance.

Melanocortin 1 Receptor: polymorphisms analysed in alpaca (*Lama pacos*)

Guridi, M., Soret, B., Alfonso, L. and Arana, A., Universidad Pública de Navarra, Producción Agraria, Campus Arrosadía s/n Edificio de los Olivos, 31006 Pamplona, Spain; maitea.guridi@unavarra.es

Coat colour is a trait of important value for alpaca fibre quality and the characterization of the genes controlling colour patterns will contribute to develop a more efficient conservation breeding programme of these animals. Melanocortin 1 Receptor (MC1R), encoded by the Extension locus, is responsible for colour pigment synthesis in melanocytes: black eumelanine and red phaeomelanine. MC1R gene structure is well known in other mammals but not in alpacas, and the purpose of this work is to characterize polymorphisms associated to phenotype present in MC1R gene in alpaca. We chose 8 samples of alpaca hair from Huancavelica (Peru) herds, 2 of each main defined colour: black, brown, fawn and white. Samples were analysed with a spectrophotometer to obtain objective parameters of colour variation. DNA was extracted from hair bulbs, and primers were designed to amplify 1,172 pb of the MC1R gene, which included the whole unique exon. The sequences obtained revealed 2 new mutations besides the 11 described before, all of them in the non-coding region (C142G, T1142C). Even though no relation was found between mutations and colours, we added new information for future studies aimed to better understand the genetic background of coat colour in the alpacas.

A comparative approach to the identification of growth related QTLs in European sea bass (*Dicentrarchus labrax*)

Louro, B.[1,2], Hellemans, B.[3], Massault, C.[1], Volckaert, F.A.M.J.[3], Haley, C.[1], De Koning, D.J.[1], Canario, A.V.M.[2] and Power, D.M.[2], [1]Roslin Institute and R(D)SVS, Univ. of Edinburgh, Roslin, EH25 9PS, United Kingdom, [2]CCMAR, Univ. do Algarve, Faro, 8005-139, Portugal, [3]Katholieke Univ. Leuven, Leuven, B-3000, Belgium; bruno.louro@roslin.ed.ac.uk

Two projects funded by the European Commission, BassMap and Aquafirst (contracts Q5RS-CT-2001-01701 and 513691 respectively), recently identified QTLs affecting six morphometric traits and body weight in European sea bass, a commercial aquaculture species. A comparative approach was used to link the sea bass genetic maps and available sea bass genomic sequences. This facilitated the identification and characterization of the genomic regions in which the QTLs were located and permitted the development of new markers to increase resolution of the QTLs. The fishes Gasterosteus aculeatus and Tetraodon nigroviridis, which have well annotated genomes, were used as a scaffold to anchor sea bass genetic markers in linkage with the QTLs and sea bass BAC end sequences. Genomic regions of interest that have conserved synteny between sea bass and other fishes were assessed in silico by BAC scaffold alignments, supported by the conserved synteny observed between G. aculeatus and T. nigroviridis and Oryzias latipes. PCR was used to confirm whether BAC clones were overlapping. The comparative physical map based on BAC end sequences allowed the development of an 11 loci multiplex, which covered a QTL affecting several morphometric traits in a breeding population of sea bass from the BassMap project. The 11 loci multiplex was used to genotype the Aquafirst sea bass mapping population, but failed to confirm the QTL. The integration of physical and linkage maps using a comparative approach accelerates the fine mapping and functional characterization of previously identified QTLs of interest. Acknowledgments: This work was funded by the SABRETRAIN, Aquafirst and Aquagenome (resource and mobility grants) EC-funded FP6 Projects. BL benefited from a Portuguese FCT PhD scholarship (SFRH/BD/29171/2006).

Genetic analysis of reproductive diseases and disorders in Norwegian Red cows

Heringstad, B.[1,2], [1]Department of Animal and Aquacultural Sciences, Norwegian University of Life Sciences, P. O. Box 5003, N-1432 Aas, Norway, [2]Geno Breeding and A.I. Association, P. O. Box 5003, N-1432 Aas, Norway; bjorg.heringstad@umb.no

Fertility related diseases and reproductive disorders have been recorded routinely in the Norwegian health recording system since 1978. Heritability of and genetic correlations among silent heat (SH), cystic ovaries (CO), metritis (MET), and retained placenta (RP) were inferred. Records of 503,683 first lactation daughters of 1,059 Norwegian Red sires, with first calving from 2000 through 2006, were analyzed with a 4-variate threshold sire model. Presence or absence of each of the four diseases was scored as 1 or 0 based on whether or not the cow had at least one veterinary treatment. The mean frequency was 3.1% SH, 0.9% MET, 0.5% CO, and 1.5% RP. The model for liability had effects of age at calving and of month-year of calving, herd, sire of the cow, and a residual. Posterior mean (SD) of heritability of liability was 0.06 (0.01) for SH, 0.04 (0.01) for MET, 0.07 (0.02) for CO, and 0.07 (0.01) for RP. The genetic correlation between MET and RP was strong, with posterior mean (SD) 0.61 (0.09). The posterior distributions of the other genetic correlations included zero with high density, and could not be considered to be different from 0. The frequency of fertility related diseases and disorders is very low in the Norwegian Red population at present, so there is not much scope for genetic improvement. However, this study indicates that reasonably precise genetic evaluation of sires is feasible for these traits.

Relationship between the estimated breeding values of test day milk fat-protein ratio and the fertility evaluations of Nordic Red dairy sires

Liinamo, A.-E., Negussie, E. and Mäntysaari, E., MTT Agrifood Research Finland, Biotechnology and Food Research, Biometrical Genetics Group, FIN-31600 Jokioinen, Finland; anna-elisa.liinamo@mtt.fi

The motivation of this study was to quantify if the estimated test-day breeding values for milk production traits in early lactation (TD EBVs), especially the ratio of fat and protein production EBVs (TD FPR) that is related to the energy balance status of the lactating cow, could be used to improve accuracy of the evaluations for fertility traits. For this purpose, the EBVs for milk, fat and protein production in test days 15 to 300 were obtained from August 2008 joint Nordic breeding value estimation (NAV) for Nordic Red Cattle bulls born in 1999-2003, altogether 516 animals. Similarly the fertility EBVs were obtained for the bulls from the corresponding August 2008 NAV evaluation for interval from calving to first insemination (CFI), interval from first to last insemination (FLI), number of inseminations (NI), non-return rate (NRR) and treatments for fertility disorders (FT). The correlations of the TD FPR and the fertility EBVs across different test days were all negative and differed significantly from zero for CFI, and FLI and NI in cows. The highest correlations between TD FPR (with corresponding days in milk) and EBVs of CFI, FLIc, and NIc were -0.17 (dim 15), -0.15 (dim 180), and -0.16 (dim 180), respectively. The absolute values of correlations with fat and protein TD EBVs were somewhat higher than for TD FPR especially in late lactation, but they are probably related to general production stress and still too low for reliable prediction of fertility EBVs.

Genetic variability for days to first insemination in dairy cattle accounting for voluntary waiting period

Cue, R.I.[1], Carabaño, M.J.[2], Diaz, C.[2], Gonzalez-Recio, O.[2] and Ugarte, E.[3], [1]McGill University, Department of Animal Science, Macdonald Campus of McGill University, Ste Anne de Bellevue, Quebec, H9X 3V9, Canada, [2]INIA, Mejora Genética Animal, Crta de la Coruña, km 7,5, 28040, Madrid, Spain, [3]Unida Innovación Agraria, Arkuate Granja-Eredua 46 Post, 01080, Vitoria, Spain; Roger.Cue@McGill.ca

Breeding and test-day milk production records were obtained from the milk recording program of the Basque Country, Gerona and Navarra regional Holstein Associations. The date of first breeding (within each lactation) was matched with the closest test-day record (within 16 days). Edits for both breeding records and test-day records, were imposed, for age at calving, interval from calving to first breeding (25 to 160 days) and for test-day production. Sires had to have at least 50 progeny, with daughters in at least 5 herd-years, herd-years had to have daughters of at least 3 sires. Herd-years had to have at least 40 animals, and herds had to be represented for at least 3 consecutive years. There were 301 sires with 31,403 daughters, and 63,567 observations (1 record per lactation per cow). Interval from calving to first breeding was analysed using both a linear model and a Weibull fraility model (using the Survival Kit); separate analyses were done for records from each parity (parities 1 to 3). The model included effects of region-year-season of calving, month of calving, age at calving, voluntary waiting period (VWP) and test-day milk yield (standardized for fat% and protein%) and the random effects of herd-year of calving and sire of the cow. VWP was computed separately for each herd-year and separately for primiparous cows and multiparous cows. There was a positive relationship between length of VWP and days to first insemination. Heritability of days to first insemination was between 4% and 6% depending on the model and the parity, but was not affected by VWP. The estimated genetic correlations, computed from sire proofs, in different parities ranged from 0.5 to 0.6.

Mapping of fertility traits in Finnish Ayrshire by genome-wide association

Schulman, N.F.[1], Sahana, G.[2], Iso-Touru, T.[1], Schnabel, R.D.[3], Lund, M.S.[2], Taylor, J.F.[3] and Vilkki, J.H.[1], [1]MTT Agrifood Research, Biotechnology and Food Research, ET-talo, 31600 Jokioinen, Finland, [2]Aarhus University, Department of Genetics and Biotechnology, Research Centre Foulum, 8830 Tjele, Denmark, [3]University of Missouri, Division of Animal Sciences, 65211 Columbia MO, USA; nina.schulman@mtt.fi

We performed genome-wide association (GWA) mapping in order to detect markers associated with seven different fertility traits in Finnish Ayrshire cattle. The phenotypic data consisted of de-regressed EBVs calculated without correlations between traits from 340 progeny-tested bulls. The traits analyzed were time from calving to first insemination, non-return rate for cows and heifers, number of inseminations for cows and heifers and time from first to last insemination for cows and heifers. Genotypes were obtained with the Illumina BovineSNP50 panel and a total of 35,630 informative, high-quality SNP markers were used in the analysis. We performed the association analysis using a mixed model approach which fitted a fixed effect of the SNP and a random polygenic effect. We detected 10 regions on different chromosomes associated with fertility traits with a point wise p-value threshold of 0.00005. Based on Fisher's product test the GWA results confirmed QTL on four chromosomes for fertility traits reported earlier from the same breed. There was marked improvement in narrowing the QTL region in GWA compared with earlier linkage analysis. With GWA the QTL region ranged from a few hundred Kb to a few Mb.

Genomic instability following Somatic Cell Nuclear Transfer in cattle

De Montera, B., Oudin, J.-F., Jouneau, L., Noé, G., Chavatte-Palmer, P., Vignon, X., Eggen, A., Amigues, Y., Duranthon, V., Pailhoux, E., Boulanger, L., Jammes, H. and Renard, J.-P., INRA, Domaine de Vilvert, 78 352, France; beatrice.de-montera@jouy.inra.fr

SCNT cloned animals often display several developmental abnormalities and post-natal phenotypical variations which evidence that cloned animals may not share exactly the same genome. The poor survival rate of clones could be due either to genetic or epigenetic deregulation or defects. Yet, the genetic identity issue remains very poorly documented. To date, somatic cell nuclear transfer (SCNT) animals derived from the same donor genome were not demonstrated to be genetically identical. We applied a modified RDA (Representational Difference Analysis) subtractive technique in order to reveal possible genetic differences between 10 SCNT Holstein cows of the same genotype including 5 living adult cows and 5 perinatal aborted fetuses. We chose as a reference of non cloned animals 6 experimentally generated monozygotic twin pairs. The RDA subtractions displayed highly variable differential patterns between SCNT cows but identical patterns between twin couples. 155 differential SCNT unique sequences were localized within the centromeric regions of all bovine chromosomes showing that cloning might establish rapid sequence changes. Caryotyping and genotyping were performed on the same SCNT genomes. In parallel, DNAs were hybridize on a 54K SNP bovine array. The preliminary results show that despite chromosome number stability, one clone displays instability in microsatellite sequences known to induce genomic diversity in centromeres through sequence or length modification. These observations imply that cloning is not genetically neutral while reproducing a given genome. Such exploration can help defining SCNT genomes characteristics. Studying structural variation occurrence in animals derived from the same genome is a good strategy to determine genetic variables and to evaluate their impact on quantitative trait loci in cattle. These results are obtained through the EC-funded FP6 Project 'SABRE'

Gene expression patterns in anterior pituitary associated with quantitative measure of oestrous behaviour in dairy cows

Kommadath, A.[1], Mulder, H.A.[1], De Wit, A.A.C.[1], Woelders, H.[1], Smits, M.A.[1], Beerda, B.[2], Veerkamp, R.F.[1], Frijters, A.C.J.[3] and Te Pas, M.F.W.[1], [1]Wageningen UR, Animal Breeding and Genomics Centre, Animal Sciences Group, P.O. Box 65, 8200AB Lelystad, Netherlands, [2]Wageningen University, Department of Animal Sciences, Marijkeweg 40, 6709PG Wageningen, Netherlands, [3]CRV, Research & Development, P.O. Box 454, 6800AL Arnhem, Netherlands; arun.kommadath@wur.nl

Intensive selection for high milk yield in dairy cows has been at the cost of reduced fertility. Genomic regulation of oestrous behaviour, the expression of which is also reduced, is largely unknown. We aimed to identify and study those genes that were associated with oestrous behaviour among genes expressed in the bovine anterior pituitary either at the start of oestrous cycle or at mid cycle, or regardless of the phase of cycle. Oestrous behaviour was recorded in 28 primiparous cows from 30 days in milk onwards till the day of their sacrifice (between 77 and 139 days in milk) and an average heat score was calculated for each cow. A microarray experiment was designed to measure gene expression in the anterior pituitary of these cows, 14 of which were sacrificed at the start of oestrous cycle (day 0) and 14 around day 12 of cycle (day 12). Gene expression was modelled as a function of the orthogonally transformed average heat score values using a Bayesian hierarchical mixed model on data from day 0 cows alone (analysis 1), day 12 cows alone (analysis 2) and the combined data from day 0 and day 12 cows (analysis 3). Genes whose expression patterns showed significant linear or non-linear relationships with average heat scores were identified in all 3 analyses. Gene ontology terms enriched among genes identified in analysis 1 revealed processes associated with expression of oestrous behaviour whereas the terms enriched among genes identified in analysis 2 and 3 were general processes which may facilitate proper expression of oestrous behaviour at the subsequent oestrus. These results are obtained through the EC-funded FP6 Project 'SABRE'.

Transcriptional profiling of bovine endometrial and embryo biopsies in relation to pregnancy success after embryo transfer

Salilew-Wondim. D.[1], Hoelker, M.[1], Peippo, J.[2], Grosse-Brinkhaus, C.[1], Ghanem, N.[1], Rings, F.[1], Phatsara, C.[1], Schellander, K.[1] and Tesfaye, D.[1], [1]Institute of Animal Science, Endenicher Allee 15, 53115 Bonn, Germany, [2]2MTT Agrifood Research Finland, Biotechnology and Food Research, ET building, 31600 Jokioinen, Finland; dsal@itw.uni-bonn.de

In the present study a large scale gene profiling approachwas applied in bovineendometrial and embryo biopsies based on pregnancy outcomes following embryo biopsy transfer. For this the endometrial biopsies were recovered from 56 Simmental heifers at day 7 and 14 of the oestrus cycle. In the next oestrous cycle, 70% of the embryo biopsies were transferred to heifers. The remaining 30% embryo biopsies were retained for analysis. The endometrial and embryo biopsies were classified as follows. Those endometrial biopsies taken at days 7 and 14 of the oestrous cycle from those resulted in calf delivery (CD) were classified as CDd7 and CDd14, respectively, and those endometrial samples collected at days 7 and 14 of the oestrous cycle from those resulted in no pregnancy groups were classified as NPd7 and NPd14, respectively. The result showeda total of 1115 transcripts were found to be differentially expressed between CDd7 and NPd7, but only 14 transcripts were found differentially expressed between CDd14 and NPd14. Among differentially expressed transcripts between CDd7 and NPd7, some were found to be involved in regulation of transcription, in extracellular matrix structural constituent, G-protein coupled signalling pathway. The differentially expressed genes were found to be involved in different pathways. While embryo biopsies from CD group were enriched with gene like AMD1, RYBP, MIR16, RL6IP and FABP5, those from NP group were enriched with EEF1A1, STX8, SGK1, KRT8, and KPNA4. The major differences in gene expression profiles between non pregnant and calf delivery group animals were observed at day 7 than at day 14 of the pre transfer cycle. Acknowledgements: These results are obtained through the EC-funded FP6 Project 'SABRE'.

Homozygosity mapping of a Weaver like recessive disorder in Tyrol grey cattle

Sölkner, J.[1], Gredler, B.[1], Drögemüller, C.[2] and Leeb, T.[2], [1]BOKU - University of Natural Resources and Applied Life Sciences, Division of Livestock Sciences, Gregor Mendel Str. 33, A-1180 Viennna, Austria, [2]University of Berne, Insititue of Genetics, Postfach 8466, CH-3001 Bern, Switzerland; johann.soelkner@boku.ac.at

High throughput single nucleotide polymorphism (SNP) genotyping has made it feasible to search for a region of the genome where cases of a recessive disorder have overlapping homozygous strings of SNPs. Here we augment homozygosity mapping with cases by information from carriers which are not parents of the cases. This substantially narrows the candidate region where the mutation causing the recessive disorder resides. In 2003, a recessive disorder was found in Tyrol grey cattle, a local breed with a population size of ~5000 registered cows. The clinical signs of the disorder are similar to those of Weaver in Brown Swiss cattle but occur much earlier in life. Weaver marker tests indicated that the disease is caused by a different mutation. All cases were traced to Gusti, a cow born 1973 and being the mother of two important bull sires in this population. SNP genotyping with the Illumina 54k SNP chip was performed for 15 cases, 15 carriers that were not parents of these cases and 8 animals not related to Gusti. An ovelapping region of 47 homozygous SNP was found on chromosome 16, located at Mb 36.876 - 39.747. To narrow the region, we checked carrier animals for compatibility with the SNP haplotype derived from cases. Two carriers were not compatible with the case haplotype. They were homozygous for the alternative allele for 2 and 3 SNPs at one end of the region. This narrows the candidate region to Mb 38.334 - 39.747. The region hosts 3 candidate genes that have been related to function of the nervous system in human. Sequencing of the coding regions of these genes for cases and controls did not reveal a mutation. More sequencing is currently under way to locate the mutation responsible for the disorder.

Genetics of tuberculosis in Irish dairy cattle

Bermingham, M.L.[1], More, S.J.[2], Good, M.[3], Cromie, A.R.[4], Higgins, I.M.[2], Brotherstone, S.[5] and Berry, D.P.[1], [1]Teagasc, Moorepark Production Research Centre, Fermoy, Co. Cork, na, Ireland, [2]Centre for Veterinary Epidemiology and Risk Analysis, University College Dublin, Belfield, Dublin 4, na, Ireland, [3]Department of Agriculture and Food, Kildare St, Dublin 2, na, Ireland, [4]The Irish Cattle Breeding Federation, Bandon, Co. Cork, na, Ireland, [5]Institute of Evolutionary Biology, University of Edinburgh, West Mains Road, Edinburgh, EH9 3JT, United Kingdom; mairead.bermingham@teagasc.ie

There is a lack of information on genetic parameters for tuberculosis (TB) susceptibility in dairy cattle. Mycobacterium bovis is the primary agent of tuberculosis in cattle.The objective of this study was to quantify the genetic variation present among Irish Holstein - Friesian dairy cattle in their susceptibility to M. bovis infection. A total of 15,182 cow and 8,104 heifer single intradermal comparative tuberculin test (a test for M. bovis-purified protein derivative [PPD] responsiveness) records from November 1, 2002, to October 31, 2005 were available for inclusion in the analysis. Data on abattoir TB carcass lesions (confirmed M. bovis infection) were also available for inclusion in the analysis. Linear animal models, and sire and animal threshold models were used to estimate the variance components for susceptibility to M. bovis-PPD responsiveness and confirmed M. bovis infection. Heritability estimates from the threshold sire models were biased upwards. The threshold animal model produced heritability estimates of 0.14 in cows and 0.12 in heifers for susceptibility to M. bovis-PPD responsiveness, and 0.18 in cows for confirmed M. bovis infection susceptibility. Therefore, exploitable genetic variation exists among Irish dairy cows for susceptibility to M. bovis infection. A favorable genetic correlation close to unity was observed between susceptibility to confirmed M. bovis infection and M. bovis-PPD responsiveness, indicating that direct selection for resistance to M. bovis-PPD responsiveness will indirectly reduce susceptibility to confirmed M. bovis infection.

Quantitative trait loci detection for milk protein composition in Dutch Holstein-Friesian cows

Schopen, G.C.B., Koks, P.D., Van Arendonk, J.A.M., Bovenhuis, H. and Visker, M.H.P.W., Wageningen University, Animal Breeding and Genomics Centre, P.O. Box 338, 6700 AH Wageningen, Netherlands; ghyslaine.schopen@wur.nl

The Dutch Milk Genomics Initiative aims at identification of genes that contribute to natural genetic variation in quality traits of bovine milk. The objective of this study was to perform a whole genome scan to detect quantitative trait loci (QTL) for milk protein composition in 849 Holstein-Friesian cows. One morning milk sample was analyzed for the major milk proteins using capillary zone electrophoresis. A genetic map was constructed with 1,341 single nucleotide polymorphisms, covering 2,829 centimorgans (cM) and 95% of the cattle genome.The chromosomal regions most significantly related to the six major milk proteins ($P_{genome}<0.05$) were found on BTA6, and 11. The QTL on BTA6 was found at about 80 cM, and affected α_{S1}-casein, α_{S2}-casein, β-casein and κ-casein. The QTL on BTA11 was found at 124 cM, and affected β-lactoglobulin. The proportion of phenotypic variance explained by the QTL was 3.6% for β-casein and 7.9% for κ-casein on BTA6, and was 28.3% for β-lactoglobulin on BTA11. The QTL affecting α_{S2}-casein on BTA6 and 17 showed a significant interaction. We investigated the extent to which the detected QTL affecting milk protein composition could be explained by known polymorphisms in β-casein, κ-casein and β-lactoglobulin genes. Correction for these polymorphisms decreased the proportion of phenotypic variance explained by the QTL previously found on BTA6, and 11. Thus, several significant QTL affecting milk protein composition were found, of which some could partially be explained by known polymorphisms in milk protein genes. Fine mapping of the QTL regions to reduce confidence intervals of the detected QTL for milk protein composition, to facilitate new candidate genes that affect milk protein composition, is in progress by using the 60K bovine SNP chip.

Development of a genome-wide SNP marker set for selection against left-sided displacement of the abomasum in German Holstein cattle

Mömke, S.[1], Sickinger, M.[2], Brade, W.[3], Rehage, J.[4], Doll, K.[2] and Distl, O.[1], [1]University of Veterinary Medicine Hannover, Institute for Animal Breeding and Genetics, Bünteweg 17p, 30559 Hannover, Germany, [2]Justus-Liebig-University Giessen, Clinic for Ruminants and Swine, Frankfurter Str. 110, 35392 Giessen, Germany, [3]Chamber of Agriculture (LWK), Lower Saxony, 30159 Hannover, Germany, [4]University for Veterinary Medicine, Clinic for Cattle, Bischofsholer Damm 15, 30173 Hannover, Germany; stefanie. moemke@tiho-hannover.de

Left-sided displacement of the abomasum (LDA) is a common disease in various dairy cattle breeds. The prevalence of LDA in German Holstein dairy cattle is at 3.6% and the heritability was estimated at up to 0.53 in this breed. Milk performance and productive life are significantly reduced in affected cows. Five quantitative trait loci (QTL) for LDA were identified on bovine chromosomes (BTA) 1, 3, 21, 23, and 24 in German Holstein cattle. A genome-wide association analysis for LDA was performed using the Illumina 50K bovine SNP chip and a random sample of 96 LDA-affected cows and 120 controls. Cows in the control sample were older than nine years and never showed signs of LDA. The results of the SNP chip analysis confirmed the previously identified QTL. Further genome-wide significant associations with LDA were found on BTA 6, 8, 10, 13, 14, 20, and 26. Based on these results, a SNP marker set was developed to be used for identification of cows and bulls which hold a high genetic disposition for LDA. Verification of the power of this marker set was done in a representative sample of more than 1000 German Holstein cattle individuals. This marker set should support the breeders to reduce LDA efficiently.

DGAT1 K232A polymorphism greatly affects mammary tissue activity of milk component synthesis

Faucon, F.[1,2], Rebours, E.[1], Robert-Granie, C.[3], Bernard, L.[4], Menard, O.[5], Miranda, G.[1], Dhorne-Pollet, S.[1], Bevilacqua, C.[1], Hurtaud, C.[6], Larroque, H.[1], Gallard, Y.[7], Leroux, C.[4] and Martin, P.[1], [1]INRA UMR1313, Jouy-en-Josas, 78352, France, [2]Institut Elevage, Paris, 75012, France, [3]INRA UR631 SAGA, Auzeville, 31326, France, [4]INRA UR1213 Herbivores, Saint-Genès-Champanelle, 63122, France, [5]INRA AgroCampus Ouest UMR1253 STLO, Rennes, 35042, France, [6]INRA AgroCampus Ouest UMR1080 Production du lait, St-Gilles, 35590, France, [7]INRA UE326 Pin-au-Haras, Exmes, 61310, France; felicie.faucon@jouy.inra.fr

Animal feeding, genetics and milking frequencies are the main factors affecting significantly milk biosynthesis and secretion in the healthy udder, with a direct impact on the composition, and the techno-functional and nutritional properties of milk. The non conservative K232A mutation occurring in the bovine DGAT1 gene, which encodes the enzyme catalyzing the last step of triacylglycerol biosynthesis, is responsible for a fat content reduction (AA) and for a modification of its composition. To determine cellular mechanisms underlying these differences, we have compared, using microarray, gene expression profiles from mammary tissue biopsies of cows reared in well defined and controlled experimental conditions. To each homozygous cow KK corresponded its full-sib (n=4) or half-sib (n=2) homozygous AA, at the same lactating stage and day in lactation. Our study confirmed a significant reduction of milk fat content (41.6 vs. 51.6 g/kg) and unsaturated-middle chain fatty acid content (C14:1, C16:1), and a significant increase in saturated middle chain (C14:0) and unsaturated-long chain (C18:1 cis11, C18:1 cis12, C18:2n-6 and C18:3n-3) fatty acid content, for the AA compared to KK genotype. In addition, milk fat globules were significantly smaller. Gene expression profiling revealed up-regulation of genes involved in lactose, lipid and protein biosynthesis for AA individuals, suggesting that the mammary tissue of AA cows compared to KK is subjected to increase milk component synthesis.

Breed effects and heritability for bovine milk coagulation time determined by Fourier transform infrared spectroscopy

Lopez-Villalobos, N.[1], Davis, S.R.[2], Beattie, E.M.[2], Melis, J.[3], Berry, S.[2], Holroyd, S.[4], Spelman, R.J.[3] and Snell, R.G.[2], [1]Massey University, Private Bag 11222, Palmerston North, New Zealand, [2]ViaLactia Biosciences, P O Box 109-185, Auckland, New Zealand, [3]Livestock Improvement Corporation, Private Bag 3016, Hamilton, New Zealand, [4]Fonterra Research Centre, Private Bag 11 029, Palmerston North, New Zealand; N.Lopez-Villalobos@massey.ac.nz

Reduction in milk coagulation time can improve cheese processing. The objectives of this study were, first, to derivate calibration equations to predict time to start coagulation (ST) and time to cutting (CT) based on partial least squares using Fourier transform infrared spectroscopy (FTIR) data; second, to use the calibration equation to predict ST and CT in a representative sample of milk from 776 Holstein-Friesian (HF), 922 Jersey (JE) and 838 HFxJE crossbred cows; and third, to estimate breed effects and heritabilities for ST and CT. Data used to derivate the calibration equation comprised of 154 milk samples from second lactation crossbred cows collected during mid lactation in the season 2003-04. The concordance correlation coefficients of the calibration equations were estimated at 0.78 and 0.77 for ST and CT, respectively. Based on these values the calibration equations were used to predict the ST (pST) and CT (pCT) in 3,835 milk samples collected during the season 2007-08 from 251 herds used for the progeny test of young bulls. Heritability, repeatability and least squares means of pST and pCT for each breed group were obtained using a mixed model. Average and standard deviation of pST and pCT were 22.7±4.4 and 30.6±11.5. Estimates of heritability and repeatability were 0.23±0.05 and 0.45±0.02 for pST and 0.22±0.05 and 0.25±0.03 for pCT. There were no significant differences between breed groups for pST and pCT. Results from this study confirm previous studies showing that milk coagulation characteristics can be predicted using FTIR spectroscopy and that there is significant animal variation that can be exploited in a breeding program.

Does the fatty acid profile in Churra milk sheep show additive genetic variation?

Sánchez, J.P., Barbosa, E., San Primitivo, F. and De La Fuente, L.F., Univerity of León, Producción Animal, Campus de Vegazana s/n, 24071 León, Spain; jpsans@unileon.es

A total of 4,100 test-day milk samples, from 976 Churra ewes, sired by 15 IA rams, were recorded in 14 herds. From these samples FA contents were quantified by gaseous chromatographic. 12 FAs were considered to test for additive genetic determination. A repeatability animal model, with fixed effects of HTD, Age and DIM, was used for fitting data and parameters were inferred using a Bayesian MCMC procedure, assuming those priors standardly considered for these models. DIC and Bayes Factor (BF) were the criteria used for testing the hypothesis of additive genetic determination; BF was computed using the repeatability animal model parametrized in terms of the additive genetic standard deviation, considering a uniform distribution bounded between [0, crude standard deviation of the data] as prior for this parameter. For all saturated FAs both DIC and BF clearly supported the hypothesis of additive genetic determination, with heritability estimates ranging from 0.04(0.02) to 0.08(0.02), C16:0 and C18:0, respectively. Regarding mono-unsaturated FA only trans-vaccenic acid can be said to show additive genetic variation, with an estimated heritability of 0.02 (0.01). Among poli-unsaturated FA only linoleic acid showed significant additive genetic variation, having an estimated heritability of 0.11(0.04). This essential FA can only be modified by rumen bacteria; thus, finding significant additive genetic variation for it, but not for others FAs which are synthesized de-novo in mammary gland (i.e. CLA) was totally unexpected. This fact could indicate additive genetic determination in the composition and activity of rumen bacteria, and a lack of allelic variation in genes involved in desaturation of FAs in mammary gland (i.e. Delta9-desaturase gen). Our low or null estimated heritabilities, could discourage usage of breeding as a tool for modifying FA profile in Churra sheep milk; particularly for those FAs with major dietetic interest: CLA, Omega3 and Omega6 (except linoleic acid).

The use of genetic parameters for the standardization of linear type classification in Italy
Canavesi, F., Finocchiaro, R. and Biffani, S., ANAFI, Research & Development, Via Bergamo 292, 26100 CREMONA, Italy; fabiolacanavesi@anafi.it

In Italy around 260,000 first parity Holstein cows are yearly each year by a group of about 25 classifiers. The linear classification data are subsequently used in genetic evaluation for type traits. It is very important that all classifiers rank animals consistently and use a consistent data definition at all times. It is therefore important to develop tools that allow to evaluate the repeatability of scores for all traits within and between classifiers. The use of genetic parameters to assess the quality of work of classifiers within a country was first documented by Veerkamp *et al.* in 2002. One year of routine classification data were used to determine genetic parameters within and between classifiers in a series of bivariate analysis for all the conformation traits that are currently used in the Italian selection index. The objective was to verify if the methodology could be applied in Italy where classifiers work within an assigned geographical area. In turn, observation from one classifiers were compared with observations from the whole group of classifiers and heritability and genetic correlation estimated were used to assess repeatability and consistency of classifiers respectively. Results for genetic correlations varied from 0.73 to 0.99. More subjective and complex traits lile angularity did show a higher number of cases of genetic correlations lower than 0.80 compared to traits like stature that are easy to standardize. The procedure will soon become part of a routine that will be used to assess the quality of the work of classifiers. It will be used once a year by the responsible of the classification service in Italy to discuss with each classifiers specific issues related to specific traits. The final objective is to improve the overall quality of data provided for genetic evaluation of Holstein cows and bulls in Italy.

Association between mtDNA gene variants and fertility in progeny tested bulls
Ajmone Marsan, P.[1], Malusà, A.[2], Colli, L.[1], Nicoloso, L.[3], Milanesi, E.[3], Crepaldi, P.[3], Valentini, A.[4], Santus, E.[5], Ferretti, L.[2] and Negrini, R.[1], [1]UNICATT, via E. Parmense, 89, 29100 Piacenza, Italy, [2]UNIPV, via Ferrata, 1, 27100 Pavia, Italy, [3]DSA, via Celoria, 2, 20133 Milano, Italy, [4]UNITUS, Via C. de Lellis, 01100 Viterbo, Italy, [5]ANARB, Loc. Ferlina, 37012 Bussolengo (VR), Italy; paolo.ajmone@unicatt.it

In dairy cattle reproductive efficiency is decreasing worldwide while milk production and quality are steadily increasing due to a combination of improved management, better nutrition, and strong genetic selection. Here, we focus on cattle male fertility investigating the role of mitochondrial DNA (mtDNA) variation on sperm quality traits depending from energy metabolism, as motility and viability, that potentially influence semen fecundation ability. In collaboration with the Italian Brown Cattle Breeders' Association, we investigated a set of phenotypic parameters related to semen mobility and viability evaluated on 48 young bulls by CASA (Computer Assisted Semen Analysis) along with their complete mtDNA obtained from frozen semen. Genomic DNA was extracted using a commercial kit modifying the standard protocol to optimize lysis and increase yield. Complete mitochondrial DNA of all bulls were amplified using a customized set of primers designed on the bovine mtDNA Genbank reference sequence (BRS) to cover the entire mitochondrial genome. The amplicons were sequenced, assembled and aligned to RBS to built the complete mtDNA and to detect polymorphisms. A total of 750 Kbp were sequenced and 274 mutations were scored. Among these, 104 were within the coding regions, 97 in ribosomal RNA genes and 12 within tRNAs sequences. Association between individual SNPs, haplotypes and the male fertility traits were tested using uncorrected Single-point analysis, Haplotype analysis and computation methods that optimize the procedures for controlling false discoveries through an acceptable ratio of true- and false-positives. Acknowledgment: The research was supported by PRIN 2006 Prot. 2006077053.

Genetic parameters of milk yield and reproductive traits of Holstein-Friesian cows in Tunisia

Aloulou, R.[1], M'hamdi, N.[1], Ben Hamouda, M.[2], Kraïem, K.[1] and Garrouri, M.[3], [1]ISA Chott Mariem, BP 47, 4042, Sousse, Tunisia, [2]IRESA, 30, Rue Alain Savary, 1002, Tunis Belvédère, Tunisia, [3]OEP, 30, Rue Alain Savary, 1002, Tunis Belvédère, Tunisia; aloulou.rafik@iresa.agrinet.tn

To study the relationship between female fertility, one of the most important functional traits in recent years regarding to its economic importance, and milk production, a total of 39,747 lactations representing 20,671 Holstein-Friesian cows, the main exotic breed used for milk production in Tunisia, having calved between 1994 and 2002 in 93 dairy herds, were used. Data were available for official milk recording collected by the Livestock and Pastures Agency (Office de l'Elevage et des Pâturages). Studied variables included the standard 305 days milk yield (M_{305}), the number of insemination by conception (NAI), the interval from calving to first insemination (ICI_1), the days open (DO) and the 56-day non-return rate (NR_{56}). Calculated means were 6,390 kg (SD=2191), 2.15 (SD=1.33), 79 (SD=36), 128 (SD=63) and 0.27 (SD=0.44), for M_{305}, NAI, ICI_1, DO and NR_{56} respectively. Analysis of the non genetic sources of variation revealed highly significant effects ($P<0.001$) of herd, year and month of calving and age*parities on all the studied traits. Genetic parameters were estimated according to a repeatability two traits animal model from variance components using derivative-free Restricted Maximum Likelihood Method (DF-REML). Heritability coefficients estimated were low, 0.11, 0.031 and 0.012 for M_{305}, DO and NR_{56}, respectively. Corresponding repeatabilities were 0.4, 0.047 and 0.031. Genetic correlation between M_{305} and DO was medium and positive (0.314) while it was high and negative (-0.41) between M_{305} and NR_{56}. Genetic antagonism exists between fertility and milk production. An optimal balance between these two types of traits should be pursued and must be considered when designing selection strategies.

A systems biology approach of oestrous behaviour in dairy cows

Boer, H.M.T.[1,2], Veerkamp, R.F.[1] and Woelders, H.[1], [1]Animal Sciences Group of Wageningen UR, Animal Breeding and Genomics Centre, P.O. Box 65, 8200 AB Lelystad, Netherlands, [2]Department of Animal Science, Wageningen University, Adaptation Physiology, P.O. Box 338, 6700 AH Wageningen, Netherlands; Marike.Boer@wur.nl

Expression of oestrous behaviour in dairy cows has declined in the last decades, but the underlying mechanism is unclear. The aim of our project is to describe and understand the genomic regulation of oestrous behaviour in dairy cows, using systems biology approaches. We have prepared a review paper on genes and mechanisms involved in the regulation of oestrous behaviour. Oestrous behaviour is controlled by interactions between hypothalamus, pituitary and ovaries. Genomic studies in a number of animal species have shown that oestradiol induced gene expression in hypothalamus and other brain areas plays a pivotal role in the regulation of oestrous behaviour. Analysis of our own data of gene expression profiles in brain samples from heifers in oestrous or in the luteal phase, have indicated a number of genes associated with oestrous behaviour. However, it is difficult to derive understanding of the underlying mechanisms from lists of differentially expressed genes. Systems biology combines statistics, bioinformatics and mathematical modelling to integrate 'omics' and physiological data in order to improve the biological interpretation of the vast amount of available 'omics' data and to predict the behaviour of biological systems. Currently, we try to develop a mechanistic mathematical model of the regulation of oestrus and oestrous behaviour in dairy cows, involving the key factors involved and their intricate web of interactions. The model will be based on existing models, like the recent mechanistic model of the human menstrual cycle, and will be combined with our bovine gene expression data and mechanisms described in literature. These results are obtained through the EC-funded FP6 Project 'SABRE'

Polymorphism of the kappa-casein gene in Bosnian autochthones cattle races

Brka, M.[1], Hodžić, A.[2], Reinsch, N.[3], Đedović, R.[4], Zečević, E.[1] and Dokso, A.[1], [1]Institute of Animal Breeding, Faculty of Agriculture and Food Sciences, Zmaja od Bosne 8, 71000 Sarajevo, Bosnia and Herzegowina, [2]Institute of physiology, Veterinary Faculty, Zmaja od Bosne 90, 71000 Sarajevo, Bosnia and Herzegowina, [3]Research Institute for the Biology of Farm Animals (FBN), Wilhelm-Stahl-Allee 2, 18196 Dummerstorf, Germany, Germany, [4]Institute Zoo Techniques, Faculty of Agriculture, Nemanjina 6, 11080 Belgrad – Zemun, Serbia; mbrka@web.de

Fifteen purebred Busa cattle and thirteen Gatacko cattle were genotyped for the kappa-casein gene by Polymerase Chain Reaction-Restriction Fragment Length Polymorphism (PCR-RFLP). The allele A,B and C were found and the allelic frequencies were A=0.58 and B=0.42 in the purebred cattle and A=0.46, B=0.46 and C=0.08 in the crossbreed cattle. The genotypes composed of AA, AB, BB and BC was found in this study. The Kappa-casein genotype EE was not found in this study. The differentiation for allelic and genotypic frequencies between the purebred and the crossbred sires was not significant. Various alleles of the kappa-casein gene and their ratio were revealed. The allele B found in this small population will be useful for a sire selection program in the future.

Estimation of genetic parameters for calving ease for Simmental cattle in Croatia

Bulić, V.[1], Ivkić, Z.[1], Špehar, M.[1] and Cadavez, V.A.P.[2], [1]Croatian Livestock Center, Cattle breeding, selection and development, Ilica 101, 10000 Zagreb, Croatia, [2]CIMO - Escola Superior Agrária de Bragança, Departamento de Zootecnia, Campus de Santa Apolónia, Apartado 1172, 5301-854 Bragança, Portugal; vbulic@hssc.hr

The objective of this study was to estimate the genetic parameters for calving ease of Simmental cattle in Croatia. Data consisted of 135,892 calves, and pedigree included 639,999 animals. Calving ease was scored from 1 to 5 (1 = unassisted, 2 to 5 = various levels of assistance). Calving ease at first calving, and second or later parities where treated as different traits. Variance components were estimated by REML as implemented in VCE-5 package. Statistical model included age at calving, sex, region-year interaction as fixed effects and owner and direct additive genetic effect as random effects. The estimated heritabilities were 0.105±0.005 for calving ease at first calving and 0.078±0.005 for calving ease at second and later parities. These results suggest that model tested can be used for cattle evaluation for calving ease of Simmental cattle in Croatia.

SREBP-1 polymorphism in Brown cattle and its effect on milk fatty acid composition

Conte, G.[1], Chessa, S.[2], Mele, M.[1], Castiglioni, B.[3], Serra, A.[1], Pagnacco, G.[2] and Secchiari, P.[1], [1]University of Pisa, DAGA, via San Michele degli Scalzi 2, 56100, Pisa, Italy, [2]University of Milan, VSA, via Celoria 10, 20133, Milan, Italy, [3]CNR, IBBA, via Bassini 15, 20133, Milan, Italy; gconte@agr.unipi.it

Sterol regulatory element binding proteins (SREBPs) are transcription factors that play a central role in energy homeostasis by promoting glycolysis, lipogenesis, and adipogenesis. SREBPs belong to the original basic helix loop-helix-leucine zipper family of transcription factors. Differences in expression level and/or mutations of the SREBP gene may affect the expression level of Stearoyl CoADesaturase (SCD), leading to differences in fatty acid composition in the fat tissue of cattle. Two form of SREBP-1 gene were revealed in bovine specie. The L type differ from S type for a 84 bp insertion into the 5th intron. Work conducted on Japanese Black cattle revealed that S type was associated with a higher MUFA percentage than L type in Longissimus dorsi muscle, suggesting an influence of SREBP on desaturase activity. Studies on Holstein cattle report that SREBP gene was monomorphic (only the L allele was revealed). The aim of this work was to investigate the presence of L/S polymorphism in Italian Brown cattle (IBC) and to check an effect on milk fatty acid composition and SCD activity. Results revealed the L/S polymorphism in IBC with prevalence of the L allele. On 351 cows analysed, the genotype frequencies were 71.79%, 24.50%, 3.70% for LL, LS and SS respectively. Milk from SS cows showed higher values of product/substrate ratios for SCD enzyme and higher content of C11 and C15 fatty acids Since the absence of effects on monounsatured fatty acids content could be due to the unbalanced structure of data (very low SS cows), further studies on larger population are needed in order to better arise the role of SREBP on milk fatty acids composition and mammary lipid metabolism.

Genetic parameters for growth traits in Braunvieh cattle reared in Brazil

Cucco, D.C., Ferraz, J.B.S., Eler, J.P., Balieiro, J.C.C., Mourão, G.B. and Mattos, E.C.,; diegocucco@yahoo.com.br

The object of this research was to estimate direct and maternal heritability coefficients and genetic correlations for growth traits from birth to 550 days of age in Braunvieh cattle reared in Brazil. The weights analyzed were birth weight (BW, kg, N=9,955), weight at 120 days of age (W120, kg, N=5,901), weaning weight at 205 days (WW, kg, N=6,970), yearling weight at 365 days of age (W365, kg, N=4,055), weight at 450 days (W450, kg, N=3,453), and at 550 days (W550, kg, N=1,946). A complete animal model was used for estimate the variance components in single-trait analyses, with the software MTDFREML. Relationship matrix had 35,188 animals, with 18,688 animals with phenotypic measures. Direct and maternal heritability increased from birth to weaning, with estimates of 0.23±0.037, 0.25±0.050, 0.41±0.059 for direct heritability for BW, W120 and WW, 0.08±0.012, 0.15±0.032, 0.22±0.036 for maternal heritability, and 0.18, 0.14 and 0.16 for total heritability estimates. Heritability coefficients estimated for post-weaning weights decreased with age. For W365, W450 and W550 estimations for direct heritability were 0.29±0.061, 0.25±0.057, 0.16±0.060, while maternal heritability were 0.20±0.035, 0.18±0.035, 0.13±0.052, and total heritability were 0.30, 0.35, 0.26. Direct and maternal heritability estimates reached the maximum values at weaning, increasing before and decreasing after that age. However maternal influence is important in this breed until the 550 days of age, maybe due to high milk production of cows. Higher genetic correlations between weights were observed for close ages. Maternal effects should be considerate in genetic evaluation of growth traits until 550 days of age in Braunvieh cattle in the population studied.

Milk protein genetic variation and casein haplotype structure in the Pinzgauer cattle
Erhardt, G.[1], Mitterwallner, I.[2], Zierer, E.[3], Jäger, S.[1] and Caroli, A.[4], [1]Institut für Tierzucht und Haustiergenetik, Justus-Liebig-Universität Giessen, Ludwigstr. 21b, 35390 Giessen, Germany, [2]Rinderzuchtverband Salzburg, Mayerhoferstr. 12, 5751 Maishofen, Austria, [3]Landeskuratorium der Erzeugerringe für tierische Veredelung in Bayern e.V., Haydnstr. 11, 80336 München, Germany, [4]Dipartimento di Scienze Biomediche e Biotecnologie, Università degli Studi di Brescia, Viale Europa 11, 25123 Brescia, Italy; Georg.Erhardt@ agrar.uni-giessen.de

A total of 485 Original Pinzgauer cows sampled in Austria (n=275) and in Germany (n=210) were typed at milk protein genes to analyze the genetic variation affecting the protein amino acid charge. The milk protein genes α_{S1}-casein (CSN1S1), β-casein (CSN2), κ-casein (CSN3), α-lactalbumin (LAA), and β-lactoglobulin (LGB) were typed by isoelectrofocusing. From the 20 alleles analysed, a total of 15 alleles were found. The most polymorphic gene was CSN2, with 4 alleles detected. The prevalent alleles at each gene were CSN1S1*B, CSN2*A^2, CSN1S2*A, CSN3*A, LGB*A, and LAA*B. The breed is mainly characterized by a high frequency of CSN1S2*B (0.202 in the whole data-set). The wide distribution of this allele in European Bos taurus breeds has been assessed recently. Pinzgauer cattle can provide useful biological material for developing knowledge on this milk protein variant. The casein haplotype frequencies expected under the independence hypothesis were strongly different from the haplotype frequencies estimated taking association among genes into account. Of the 48 haplotypes expected, only 8 were found at a frequency equal or higher than 0.04. CSN1S2*B occurred mainly in the casein haplotype C-A^2-B-A (in the order: CSN1S1-CSN2-CSN1S2-CSN3), found with the following frequencies: 0.16 (whole data-set), 0.20 (Austria data-set), and 0.12 (Germany data-set). The prevalence of CSN1S2*B in C-A^2-B-A could indicate this combination as specific for the Pinzgauer, possibly as a consequence of Bos indicus introgression or a survival of an ancestral haplotype of the ancient population.

Effect of the Thyroglobulin (TG) and Leptin gene polymorphisms on the milk production traits in Hungarian Simmental cows
Farkas, V.[1], Kovács, K.[2], Zsolnai, A.[2], Fésüs, L.[2], Polgár, J.P.[1], Szabó, F.[1] and Anton, I.[2], [1]University of Pannonia, Georgikon Faculty of Agriculture, Deák F. u. 16., 8360 Keszthely, Hungary, [2]Research Institute for Animal Breeding and Nutrition, Genetics, Gesztenyés u. 1., 2053 Herceghalom, Hungary; istvan. anton@atk.hu

Leptin is the hormone product of the obese gene synthesized and secreted predominantly by white adipocytes. Polymorphisms in the leptin gene have been associated with serum leptin concentration, feed intake, milk yield and body fatness. In addition to this the effect of polymorphism in the 5´-untranslated region of TG gene –which product is the precursor of hormones that influence lipid metabolism- has been concluded to affect intramuscular fat and milk fat content in cattle. The objective of this study was to estimate the effect of Leptin and TG loci on milk production traits and to determine the distribution of the different genotypes and allele frequencies in the Hungarian Simmental population. 300 blood samples have been collected from different Hungarian Simmental herds; Leptin and TG genotypes were determined by PCR-RFLP (polymerase chain reaction-restriction fragment length polymorphism) assay. Milk production data have been registered throughout three consecutive lactations and statistical analyses have been carried out to find association between individual genotypes and milk production traits. The project was supported by the Hungarian Scientific Research Fund and National Research and Development Programme.

Estimation of allele frequency for Arachnomelia in Austrian Fleckvieh cattle

Fuerst, C.[1], Heidinger, B.[2], Fuerst-Waltl, B.[2] and Baumung, R.[2], [1]ZuchtData EDV-Dienstleistungen GmbH, Dresdner Strasse 89/19, A1200 Vienna, Austria, [2]University of Natural Resources and Applied Life Sciences Vienna, Dep. Sust. Agric. Syst., Div. Livestock Sciences, Gregor-Mendel-Strasse 33, A1180 Vienna, Austria; roswitha.baumung@boku.ac.at

Until the end of the year 2005, when for the first time since the seventies cases of Arachnomelia were reported in the Austrian and German dual purpose Fleckvieh (Simmental) population, this breed was considered to be free from the hereditary disease Arachnomelia. The objective of this study was to give an overview of the current situation of the allele frequency for Arachnomelia in the Austrian Fleckvieh population and its development over time. Furthermore it was clarified whether identified male carriers appear in a top list (i.e. ranking bulls according to their total merit index). It was shown whether and to which extent carrier animals contributed genetically to the current Fleckvieh population. A comparison between different programme versions for estimating allele frequencies was carried out. Nine identified male carriers could be detected in the top list of total merit indices. A pedigree analysis to identify the 500 genetically most important ancestors for a certain reference population was carried out. For a reference population including all animals born between 2002 and 2006 two carriers were found in the pedigrees (genetic contribution 0.1% and 0.6%, respectively). The mean allele frequencies for two sub populations including the birth cohorts 1992-1996 and 2002-2006 were 0.08% and 0.49%, respectively. An increasing trend for allele frequencies between 1970 an 2005 could be observed. However, since 2005 allele frequencies are decreasing and are generally on a very low level. The comparison of different programme versions revealed the importance of taking all available information on carriers but also on identified non-carriers into account.

Genetic analysis of veal traits in Austrian dual purpose cattle

Fuerst, C.[1], Gredler, B.[2], Kinberger, M.[2] and Fuerst-Waltl, B.[2], [1]ZuchtData EDV-Dienstleistungen GmbH, Dresdner Strasse 89/19, A-1200 Vienna, Austria, [2]University of Natural Resources and Applied Life Sciences Vienna, Department of Sustainable Agricultural Systems, Gregor Mendel Strasse 33, A-1180 Vienna, Austria; fuerst@zuchtdata.at

Veal production plays an important role in Austrian cattle industry. Currently only data from slaughtered bulls are used in the joint Austrian-German genetic evaluation. The aims of this study were to estimate genetic parameters and to implement a genetic evaluation for veal traits in Austrian dual purpose cattle. Data from slaughtered calves were carcass weight (CW), EUROP carcass conformation score (CCS) and fat score (FS) collected from Austrian slaughterhouses. Applying multivariate linear animal models, genetic parameters for the traits net daily gain (NDG) and CCS for Fleckvieh (Simmental), Braunvieh (Brown Swiss) and Pinzgauer were estimated. The fixed effects herd*year, slaughterhouse*year*season, sex and birth type, age as linear and quadratic covariate and the random genetic effect of calf were taken into account. Two different data sets were analyzed for each breed, comprising purebreds (<25% foreign genes) and crossings with beef breeds, respectively. Datasets ranged from 2,803 to 11,466 records. Heritabilities from the purebred datasets were 0.13, 0.19 and 0.24 for NDG and 0.20, 0.29 and 0.24 for CCS, for Fleckvieh, Braunvieh and Pinzgauer, respectively. Genetic correlations between NDG and CCS ranged from 0.06 (Pinzgauer) to 0.67 (Fleckvieh). Heritability estimates for the crossbred datasets were markedly higher and ranged from 0.26 (Fleckvieh) to 0.30 (Pinzgauer) for NDG and from 0.52 (Pinzgauer) to 0.68 (Fleckvieh) for CCS. Additional analyses including FS as a correlated trait did not influence results significantly. Breeding values based on purebred veal data were estimated using the same models. Correlations to the official breeding values based on bull data were low for NDG (0.1 to 0.3) and moderate for CCS (0.3 to 0.5). Routine genetic evaluation for veal traits will start for Pinzgauer in 2009, other breeds will follow.

Unravelling the mechanisms of mammalian ovarian follicular development and atresia: a cattle vs pig transcriptome and proteome study

Gerard, N.[1], Tosser-Klopp, G.[2], Benne, F.[2], Bonnet, A.[2], Delpuech, T.[1], Douet, C.[1], Drouilhet, L.[2], Fabre, S.[1,2], Izart, A.[2], Monget, P.[1], Monniaux, D.[1], Mulsant, P.[2], Robert-Granie, C.[3], San Cristobal, M.[2], Vignoles, F.[2] and Bodin, L.[3], [1]INRA, UMR85, 37380 Nouzilly, France, [2]INRA, UMR444, 31326 Castanet-Tolosan, France, [3]INRA, UR631, 31326 Castanet-Tolosan, France; gwenola.tosser@toulouse.inra.fr

In mammals, ovarian folliculogenesis leading to the ovulation of completely mature oocytes is a long and complex process that is regulated at different levels. However, it is already known, that between cattle and pig, the pattern of expression of some well-known genes either are very similar or strongly differ during follicular growth or atresia. Our strategy to discover genes or gene network involved in follicular antral development or atresia was to compare antral small, large, healthy or atretic follicles from cattle and pig both at the transcriptomic and the proteomic levels. Follicles from ten cows and ten sows were individually dissected. The granulosa cells and the follicular fluid were then pooled within the same animal to generate 54 RNA samples and 40 proteic samples, respectively. The transcriptomic analysis evidenced 254 genes that allowed a perfect classification of the 54 samples according to the three parameters we tested: species, size and status of the follicles. Specific species expression of some genes was highlighted, whether other genes were expressed in both species in a similar way or with a different pattern. Proteins from follicular fluids were analyzed by 2D-PAGE as follows: proteins were loaded onto ReadyStrip IPG strips pH3-10 that were then subjected to 10% SDS-PAGE. Using Progenesis Samespot software, 54 differential spots were selected and protein spot identification is underway. Do these results allow us to establish a link between the gene expression pattern and the mono vs polyovulation character of cattle and pigs? These results are obtained through the EC-funded FP6 Project 'SABRE' and the ANR-funded 'Genovul' project.

Analysis of quantitative trait loci affecting female fertility and twinning rate in Israeli Holsteins

Glick, G.[1], Golik, M.[1], Shirak, A.[1], Ezra, E.[2], Zeron, Y.[3], Seroussi, E.[1], Ron, M.[1] and Weller, J.I.[1], [1]ARO, The Volcani Center, Cattle and Genetics, P. O. Box 6, 50250 Bet Dagan, Israel, [2]Israel Cattle Breeders Association, Caesaria Industrial Park, 38900 Caesaria, Israel, [3]Sion, Shikmim, 79800 Shikmim, Israel; shiraka@agri.gov.il

Female fertility and twinning rate were analyzed by the multitrait animal model with parities 1 through 5 considered correlated traits. Fertility was scored as the inverse of the number of inseminations to conception at each parity. Negative genetic correlations between all combinations of parities 1 through 3 between twinning rate and fertility were found by multitrait REML analysis. We have previously reported the existence of QTL affecting these traits segregating on BTA7. The objective of this study was to test if the overall genetic relationship between these two traits was maintained at the level of individual genes affecting the traits. In the preliminary analysis, 384 Israeli Holstein bulls were genotyped by the BovineSNP50 BeadChip (Illumina, Inc.), and three chromosomes: BTA 6, 7, and 14 were included in the analysis. After edits there were 5,160 valid SNPs on these three chromosomes. Three hundred to 700 bulls were genotyped for an additional 225 SNPs on the first half of BTA7. Significance of SNP effects on both traits was tested by a linear model that included the effects of allele and the bulls' birth year on the bulls' genetic evaluations. A total of 49 SNPs had significant effects for both traits ($P<0.05$). Maximum effects were 2.7% conception and 2.0% twinning. Assuming independent association among the 49 effects, 13.5 significant effects were expected by chance, for a false discovery rate of 0.27. Of these 49 SNPs the effects associated with the two traits were in opposite directions in all but 7. Thus the null hypothesis of random association can be rejected by a binomial test with significance of $P<0.001$. These results indicate an antagonistic relationship between multiple ovulation and probability of conception. These results are obtained through the EC-funded FP6 Project 'SABRE'.

Effect of length of productive life on genetic trend of milk production and profitability

Honarvar, M.[1], Nejati Javaremi, A.[2], Miraei Ashtiani, S.R.[2] and Dehghan Banadaki, M.[2], [1]Islamic Azad University Shahriar- Shahr-e-qods Branch, Animal Science, 0098262-Shahr-e-qods, Iran, [2]The University of Tehran, Animal Science, Faculty of Agriculture-shahabbasi square-postal code:31587-11167 P.O.box:4111, 0098261-Karaj, Iran; honarvar.mahmood@gmail.com

The aim of this study was to evaluate relationship between length of productive life (LPL) and genetic trend of milk production and profitability of herds. LPL was defined as time from first calving to culling. Dynamic stochastic model was used to simulate dairy herd system. This model was consisted of biological characteristics such as reproduction, genetic and economic components. Both discrete (time-oriented) events such as freshening and breeding as well as continuous processes such as milk production and feed consumption were simulated individually for each animal. The basic characteristics of the animal component include pedigree, genetics, age at calving, number of service per conception, number of lactations and herd life. Other characteristics include time-oriented characteristics such as weight, age, physiological status, lactation stage, open days, days of pregnancy, estrus cycle, service date, and feed requirements. The herd was described as several animal groups: young stock <1 yr old, heifers' >1 yr old, and several groups of lactating and dry cows. Increasing LPL mean of herd from 35 to 65 months over 20 yr was resulted to decrease herd genetic merit of milk from 2,025 to 1,751 kg and mean of herd genetic trend per year was decreased from 101.24 to 87.56 kg, because of prolongation in generation interval. Increasing LPL was resulted to increase profit because: herd structure was changed and the proportion of mature cows in the herd was increased. This factor was led to increased milk production of the simulated herds. Increasing LPL was associated with decreased costs for raising replacement heifers and surplus heifer selling was increased.

Detection of B k-casein variant in Romanian dairy cattle breeds

Ilie, D.E.[1], Magdin, A.[1], Salajeanu, A.[1] and Vintila, I.[2], [1]Research and Development Station for Bovine Raising - Arad, Animal Genetics, Calea Bodrogului 32, 310059, Arad, Romania, [2]USAMVB, Faculty of Animal Science and Biotechnologies, Timisoara, Animal Genetics, Calea Aradului 119, 300645, Timisoara, Romania; danailie@yahoo.com

The aim of present study was to evaluate the allele and genotype frequencies of the k-casein gene in a dairy herd from the west part of Romania in order to have breeding programs that target an increase in the frequency of the favorable allele (B) of this gene in the dairy cattle population. The effects of the different k-casein alleles on the quality and the quantity of cow's milk have been widely reported and the availability of accurate and reliable protocols for the identification of the most common alleles is of great interest in breeding projects. Typing alleles A and B of k-caseinis of practical importance, because allele B is correlated with commercially valuable parameters of milk productivity (protein content and milk yield) and improves the cheese yielding capacity. The frequencies of the k-casein gene alleles and genotypes have been determined in three Romanian cattle breeds (Romanian Holstein-Friesian, Romanian Brown and Romanian Simmental) by means of PCR– RFLP analysis. At the k-casein locus, in Romanian Holstein-Friesian and Romanian Simmental breeds, we observed a relatively low frequency of B allele in comparison with A allele. The frequencies of the B allele in those two breeds vary from 0.14 to 0.27. In the Romanian Brown breed we observed the highest frequencies of k-caseinallele B (pB=0.62) and genotype BB (BB=0.33) that indicating that intensive selection for dairy production have been done and indirectly influenced k-casein frequencies. Results presented in this study are applicable in the selection of Romanian cattle breeds analyzed. This study is supported by project no. 299392/MADR and World Bank.

Genetic parameters for intermediate optimum type traits
Jagusiak, W.[1], Morek-Kopec, M.[1] and Zarnecki, A.S.[2], [1]University of Agriculture, Dept. of Genetics and Animal Breeding, al. Mickiewicza 24/28, 30-059 Krakow, Poland, [2]National Research Institute of Animal Production, ul. Krakowska 1, 32-083 Balice n. Krakow, Poland; rzjagusi@cyf-kr.edu.pl

The purpose of this paper was to study the scoring system of type traits with intermediate optimum. Six type traits with intermediate optimum were evaluated on the linear and modified scale in which the most desirable form of trait received the highest score. Data were type evaluations of 8,041 primiparous cows in Western Poland, daughters of 359 sires. The cows calved for the first time in 2008 and were classified in 1,110 herd-year-season-classifier subclasses. Six type traits with intermediate optimum (body depth, rump angle, rear leg-side view, foot angle, fore udder attachment and udder depth) were evaluated by classifiers using linear and modified scales. The descriptive type traits were also included in the analysis. A multi-trait animal model was used and a Bayesian approach with Gibbs sampling was applied to estimate (co)variance components for all traits. Heritabilities for linear scores ranged from 0.16 for foot angle to 0.44 for udder depth. The highest heritabilities were obtained for height at rump (0.62) and two descriptive traits: size (0.53) and overall conformation score (0.42). Modified evaluations of traits with intermediate optimum showed lower heritabilities than for linear scores (from 0.09 for rump angle to 0.31 for udder depth). Genetic correlations between modified scores and the respective descriptive trait were large for udder and leg traits (0.71 for fore udder attachment, 0.66 for udder depth, 0.67 for rear leg-side view and 0.58 for foot angle). Modified score of body depth showed low correlations with all descriptive traits (from 0.22 with feet and legs to 0.29 with final score) and correlations between modified score of rump angle and descriptive traits were close to zero.

Identification of the new polymorphisms in the promoter region of the MEF2C gene in cattle
Juszczuk-Kubiak, E.[1], Flisikowski, K.[2], Starzyński, R.[1], Wicińska, K.[1] and Połoszynowicz, J.[1], [1]Institute of Genetics and Animal Breeding, Polish Academy of Science, Jastrzebiec, Postępu 1, 05-552 Wólka Kosowska, Poland, [2]Lehrstuhl fuer Tierzucht Technische Universitat in Muenchen, 85354 Hochfeldweg 1, Fresing, Germany; e.kubiak@ighz.pl

Myocyte Enhancer Factor 2 (MEF2) proteins are a small family of transcription factors that play pivotal role in morphogenesis and myogenesis of skeletal, cardiac, and smooth muscle cells. In this study, on the basis comparing of the contig sequences of the human, bovine MEF2C gene, available in the GenBank database, sets of PCR primers were designed and to amplify the bovine MEF2C gene 5' region. Four overlapping fragments of the 5' region of the bovine MEF2C gene were amplified and then sequenced. The sequence analysis these fragments in individual animals representing different Bos taurus breeds revealed three polymorphic sites: C/T substitution (pos.-1604) is recognized by the BsrI, and InDelG (pos. -1434/-1433), InDelA (pos.-613/-612) are detected by any enzymes. These polymorphisms were identified for first time using this sequence.

Relationships among semen quality traits in Holstein bulls

Karoui, S.[1], Diaz, C.[1], Serrano, M.[1], Cue, R.[2], Celorrio, I.[3] and Carabaño, M.J.[1], [1]INIA, Dpto. Mejora Genetica Animal, Apdo 8111, 28080 Madrid, Spain, [2]McGill University, Department Animal Science, Faculty of Agricultural and Environmental Sciences, McGill University, Ste Anne de Bellevue, Quebec, H9X 3V9, Canada, [3]ABEREKIN, Centro de inseminación, Parque Tecnológico, Edificio 600, 48160 Derio, Spain; mjc@inia.es

Semen traits routinely collected in Holstein bulls in an AI centre were studied jointly in order to establish the phenotypic and genetic relationships among them. Records from 44,344 ejaculates from 511 bulls of 12 to 136 months of age collected between 1990 and 2007 were used. Records included information on volume per ejaculate, concentration, global motility, individual motility and post-thawing motility. Traits were jointly analysed using a multi-trait model that included the effects of number of ejaculate on collection date, week of collection, linear and quadratic regressions on age and days to previous collection, additive genetic and permanent environmental effects. REML estimates of the (co)variance components were obtained using VCE 6.0. Heritability estimates were moderate (from 6 to 18%). Genetic correlations were low between volume and concentration and post thawing motility (under 0.2), moderate between volume and fresh semen motility (around 0.3) and from moderately high (near 0.6) to high (over 0.8) among the rest of the traits. Post thawing motility, which might be considered more closely related to the bull's fecundity, showed a large correlation with fresh semen motility and moderate correlation with fresh semen concentration. Principal components analyses on the genetic correlation matrix revealed that the first two components explained 90% of the total variation in semen quality. The first component included all traits, with a very small weight on volume, and the second component heavily relied on volume. The phenotypic correlation estimates were lower than the genetic correlations. Four principal components were required to explain 92% of the total variation. Weights for traits in each component were similar in the phenotypic and genetic analyses.

Frequencies and patterns of outliers in dairy cow test day records when using different definition of outliers

König Von Borstel, U.U.[1] and Swalve, H.H.[2], [1]University of Göttingen, Animal Breeding and Genetics, Albrecht-Thaer-Weg 3, 37075 Göttingen, Germany, [2]University of Halle-Wittenberg, Institute of Agricultural and Nutritional Sciences, Group Animal Breeding, Adam-Kuckhoff Str. 35, 06108 Halle, Germany; uta.vonborstel@agr.uni-goettingen.de

Outliers in test day (TD) records occur frequently and may bias breeding value estimation and management decisions based on TD records. Therefore, reliable methods to detect and to treat outliers in TD records are required. The present study was undertaken to quantify outliers within cows while examining different definitions of outliers. Subsequently, the outliers found were analysed with regard to frequencies and their patterns. In a set of 36,068 TD records from Holstein cows in Northern Germany, a total of 23 thresholds were used to define outlying values in milk yield (MY), and 25 thresholds were used for both protein content (PC) and fat content (FC). Definitions of outliers were based either on departure from previous TD's observation (xT) or on deviation from the TD's value predicted by one of two random regression models having three (3GS) and five (5AS) regression coefficients. Reasonable outlier definitions, identifying outliers in 1.4% to 1.9% of the records, were found to be, e.g. thresholds based on mean plus three standard deviations of residuals (5AS/3GS-predicted minus observed value) calculated separately for high and low deviations. Obvious associations of outliers, e.g. in PC with high somatic cell count and in MY with farms were observed, suggesting an impact of management on TD records. The above information, along with associated variables such as diseases and magnitude and direction of the outlying value's deviation from the expected value should be used to determine the nature of an outlier (deterministic or probabilistic) and subsequently the most suitable treatment of the outlier.

Correlation vs Covariance matrices for rank reduction in test-day models?
Leclerc, H.[1] and Ducrocq, V.[2], [1]Institut de l Elevage, Département Génétique, Bat 211, 78352 Jouy en Josas, France, [2]INRA, UMR 1313 GABI, Bat 211, 78352 Jouy en Josas, France; helene.leclerc@jouy.inra.fr

The development of genetic evaluations on dairy traits based on individual test-day records represents a major computing challenge. Indeed, test-day models need a large number of parameters to describe the lactation curve accurately and moreover, the number of records to analyse can reach several hundred millions of test-day in some countries. To reduce computer requirements, one proposed approach is to use reduced rank test-day models where the smallest eigenvalues of the random effects dispersion parameters matrices are set to zero. In France, three types of random effects are included in the model: genetic (G), permanent environment (P) and herd-year (HY) effects, with initially 6, 6 and 9 elements respectively. To limit computations, it was initially decided to study a reduction of the number of elements for each type of random effects by one third. The aim of this study was to compare two strategies commonly used for rank reduction: a first approach based on the canonical decomposition of the correlation matrices (COR) and a second one based on the canonical decomposition of the covariance matrices (COV). To evaluate the impact of reduced rank approaches on the estimated effects on milk, fat and protein yields and contents, a regional dataset from the Montbéliarde breed was used, including 2.4 millions of test-day records from 135 743 cows. The estimated random effects obtained with each reduced-rank approach COR and COV were compared with the effects estimated with the full rank approach, used as reference (REF). With the COR approach, the correlations with the REF estimated effects were between 0.974 and 0.999 for G, 0.948 and 0.993 for P and 0.901 and 0.994 for HY, according to the trait evaluated. With the COV approach, they were between 0.999 and 1.000 for G, 0.992 and 0.999 P and 0.986 and 0.996 for HY. These results clearly show the better relevance of the COV approach and validate the use of a reduced rank approach in a test-day model genetic evaluation.

Factors associated with copper concentration and spontaneous oxidized flavour in cow's milk
Näslund, J.[1], Fikse, F.[1], Örde-Öström, I.-L.[2], Barrefors, P.[3] and Lundén, A.[1], [1]SLU, Department of Animal Breeding and Genetics, Box 7023, 750 07 Uppsala, Sweden, [2]SVA, Department of Virology, Immunology and Parasitology, Ulls väg 2B, 751 89 Uppsala, Sweden, [3]SLU, Department of Food Science, Box 234, 532 23 Skara, Sweden; jessica.naslund@hgen.slu.se

Spontaneous oxidized flavour is a quality defect in cow's milk caused by auto-oxidation of the unsaturated milk fatty acids by activated free radicals. Copper is one of the major pro-oxidants in milk, facilitating the oxidative process. To analyse the factors contributing to the variation in copper concentration and incidence of oxidative flavour in milk, a data set was used comprising 1,406 milk samples from 120 Swedish Red cows and 74 Swedish Holsteins belonging to the SLU experimental herd at Jälla, Uppsala. Individual morning milk samples were collected monthly during 25 months, and were analysed for flavour (three grades), copper concentration, and content of protein, fat, lactose and somatic cells. Copper concentration was analysed with a mixed animal model using PROC MIXED (SAS Institute Inc, USA). The heritability estimate for copper concentration in milk was 0.35. Copper concentration was lowest during spring and summer seasons, and it was highest during early lactation whereafter it showed a continuous decrease until week 14-39, followed by a small increase at the end of lactation. Occurrence of oxidized flavour was analysed with a generalized linear mixed animal model using the PROC GLIMMIX of SAS, with a multinomial ordinal response variable. The odds ratio for developing oxidized flavor was markedly higher in samples with a copper concentration above 0.082 mg/kg, a concentration that characterized more than 13% of the samples. However, almost 65% of these samples showed no oxidative flavour. Oxidative flavour was most common during the first three weeks of lactation. Cows in earlier parities produced milk that was more prone to develop oxidative flavour, with a linear decrease in risk with increasing lactation number.

Database of cattle candidate loci for milk production and mastitis

Ogorevc, J., Kunej, T. and Dovc, P., University of Ljubljana, Biotechnical Faculty, Department of Animal Science, Groblje 3, SI-1230, Domzale, Slovenia; jernej.ogorevc@bfro.uni-lj.si

In order to facilitate development of new genetic markers for milk production and mastitis resistance we used genome-wide comparative approach to review all known loci associated with milk traits. The database of cattle candidate loci for milk production traits and mastitis contains genes involved in mammary gland development and function and consists of 897 loci. These data represent a new research tool integrating different types of data and offer genetic background for subsequent functional studies. Candidate loci were identified considering different research approaches. Currently, our data base includes 142 genes that, when mutated or expressed as transgenes in mouse, result in specific phenotypes associated with mammary gland, 415 quantitative trait loci (QTL) for milk traits and mastitis in cattle, 24 single nucleotide polymorphisms (SNPs) showing specific allele-phenotype interactions affecting mammary gland traits and 10 causing mastitis resistance/susceptibility, 27 AFLP markers associated with clinical mastitis, 207 mammary gland development/production and 107 mastitis related genes with expression patterns which are associated with mammary gland function in cattle and mouse, 9 milk protein genes with a number of genetic variants, and one gene under epigenetic regulation during mastitis. The data was presented in a form of genetic map revealing regions with high density of candidate loci and positional overlaps between QTL and candidate genes. Cross-comparison of data obtained by different approaches was performed and the most promising candidates were analyzed in silico for expression in lactating mammary gland and potential genetic variation.

Milk protein k-casein genotypes and milk productivity of Latvian Native breeds

Paura, L., Jonkus, D. and Jemeljanova, V., Latvia University of Agriculture, Liela str. 2, LV-3001, Jelgava, Latvia; liga.paura@llu.lv

Latvian Brown (LB) and Latvian Blue (LZ) are Latvian native breeds. According to LatvianAgriculture Data Processing centre data under recording in 2008 were 63,163 the LB and 643 the LZ cows. Selection of LB dairy cattle has been based on quantitative traits such as milk yield, fat yield and in the last time selection for protein yield too. Average milk production in 2008 were 5,103 kg milk yield per lactation with fat content 4.44% and protein content 3.36%. LB cows were improved by Europe breeds Danish Red, Angler, Holstein Red, and Swedish Red-and-White. Now we should work not only in direction to improve LB breed production traits, but us well to preserve Latvian brown cattle breeds. LZ cows were popular in private farm due to they modesty and hair colour (grey-blue). The breed is characterised by longevity, resistance to the local conditions, and low food consumption. Average milk production in 2008 were 4,292 kg milk yield per lactation with fat content 4.40% and protein content 3.37%. Genotypes of 30 LB and 65 LZ cows were estimated for k-casein locus by restriction fragment length polymorphism analysis (PCR-RFLP) of amplified DNA. In LZ and LB cattle were found A, B alleles. The k-casein A and B allele frequencies were 0.816 and 0.184 in LB and 0.97 and 0.03 in LZ breed, respectively. The k-casein E allele was not found in population. The k-casein A allele were associated with highest milk yield and low protein content.

Genotype-environment interaction for milk yield between grass-based and conventional dairy cattle production in Portugal

Pavão, A.L.[1], Carolino, M.I.[2], Carolino, N.[2] and Gama, L.T.[2], [1]Univ. Açores, Campus A. Heroísmo, 9701-851 Angra Heroísmo, Açores, Portugal, [2]URGRMA-INRB, I.P., Fonte Boa, 2005-048 Vale de Santarém, Portugal; ana.lm.pavao@azores.gov.pt

Nearly one-third of the Portuguese dairy cows are in the Azores (AZ) region, where they are kept almost exclusively on grass-based systems, while in mainland Portugal (MP) more intensive production systems are generally used. Until now, the genetic evaluation of dairy cattle in Portugal has only considered records collected in MP, but the importance of dairy cattle in AZ encourages the inclusion of information from this region in national genetic evaluations. The very diverse production systems prevailing in MP and AZ may reflect genotype-environment interactions (GEI), and this work was conducted to assess the importance of these GEI for milk yield (MY). Data consisted of records on 17,608 lactations by 7,975 cows in 833 herd-year combinations (HY) in AZ, and 450,687 lactations by 191,935 cows in 22,039 HY in MP. First, a univariate analysis of MY was carried out for each region, with an animal model including the fixed effects of HY, month and age of calving, and the random effects of breeding value and permanent environment. The estimated heritability (h2) and repeatability (re) of MY were 0.29 and 0.49 in AZ, and 0.21 and 0.40 in MP. Predicted breeding values of 70 bulls with publishable results in AZ and MP had Pearson and Spearman correlations of 0.59 and 0.56, respectively. In a second approach, a bivariate analysis was used with the same effects, but considering the results in AZ and MP as correlated traits. The estimated h2 and re of MY were similar to those obtained in univariate analyses, while the estimated genetic correlation between MY in AZ and MP was 0.85. This suggests that a slight GEI may exist between the grass-based system typical of AZ and the more conventional production system used in MP, and a joint evaluation considering the two regions as separate traits might be advisable. Further research including more information, especially from AZ, should be conducted to confirm this.

Inbreeding impact on milk production in Spanish Holstein cows

Pérez-Cabal, M.A., Cervantes, I. and Gutiérrez, J.P., Facultad de Veterinaria - UCM, Producción Animal, Avda. Puerta de Hierro s/n, 28040 Madrid, Spain; mapcabal@vet.ucm.es

Inbreeding effect on milk production has been studied in Spanish Holstein cows using 520,945 test-day from 54,218 first-lactation cows. Monthly records of animals that calved from September 1993 to July 2008 were used. Individual inbreeding rate (ΔF) obtained from a pedigree involving 79,766 animals using ENDOG software was included in the animal model as a linear covariate. Significance of inbreeding coefficient was tested using the likelihood ratio test. Maximum inbreeding in the studied population was 30% and the average inbreeding 1.50%. The estimated regression coefficient for individual inbreeding rate was -0.15 ($P<0.001$). A 25% inbred primiparous cow (i.e., from a sire-daughter mating) would lead to 264 kg milk reduction per 305d-lactation and losses for a mate between animals with a common grandsire, i.e., 6.125% inbred animals, would be 62 kg. Results suggest that an effort should be made when mating animals avoiding common ancestors up to grand-grandsire level, because if annual genetic trend for milk yield in Spanish Holstein cows was 79 kg in the last 15 years, inbreeding depression due to mating sires and dams with a common grandsire would reduce a 78% the genetic progress.

Milk production efficiency of lactating Holstein-Friesian, Jersey and Jersey × Holstein-Friesian cows under grazing conditions

Prendiville, R.[1,2], Pierce, K.[2], Byrne, N.[1] and Buckley, F.[1], [1]Teagasc, Moorepark Dairy Production Research Centre, Fermoy, Co. Cork, N\A, Ireland, [2]University College Dublin, School of Agriculture, Food and Veterinary Medicine, Belfield, Dublin 4, N\A, Ireland; robert.prendiville@teagasc.ie

Data from 110 animals were available: 37 Holstein-Friesians (HF), 36 Jerseys (J) and 37 Jersey × Holstein-Friesian (F[1]). Sixteen HF, 10 J and 9 F[1] were in parity one while the remainder were in parity two. Milk yield was recorded daily, milk composition and live weight (LW) weekly and body condition score (BCS) every 3 weeks. Estimates of intake were measured at days 51, 110, 149, 198 and 233 in milk; using the n-alkane technique. The lactation was subsequently divided into 6 stages: less than 61 DIM, 60 to 121 DIM, 120 to 161 DIM, 160 to 191 DIM, 190 to 231 DIM and greater than 230 DIM. Large differences in milk yield were observed ($P<0.001$). Milk yield ranged from 18.3 kg per day for the HF to 13.8 kg/day for the J. Milk fat and protein content were highest for J at 5.33% and 4.06%, 4.75% and 3.84% for the F[1] and 3.96% and 3.49% for the HF. Yield of milk solids (MS) was lowest with the J, intermediate with HF and highest with F[1], 1.28, 1.33 and 1.41 kg respectively ($P<0.05$). Average LW during these periods was 498 kg for HF, 369 kg for J and 448 kg for F[1] ($P<0.001$). Total dry matter intake (TDMI) was 16.9 kg, 14.7 kg and 16.2 kg for the HF, J and F[1], respectively ($P<0.05$). Gross efficiency variables, MS per 100 kg LW, MS per kg TDMI and TDMI per 100kg LW were 0.35 kg, 0.088 g and 3.99 kg for the J, 0.32 kg, 0.087 g and 3.63 kg for the F[1] and 0.27 kg, 0.079 g and 3.39 kg for the HF ($P<0.001$). Greatest efficiency in terms of production was observed with J.

Heritabilities and genetic correlations of lactational and daily somatic cell score with conformation traits in Polish Holstein cattle

Ptak, E.[1], Jagusiak, W.[1] and Zarnecki, A.S.[2], [1]University of Agriculture, Dept. of Genetics and Animal Breeding, al. Mickiewicza 24/28, 30-059 Krakow, Poland, [2]National Research Institute of Animal Production, ul. Krakowska 1, 32-083 Balice n. Krakow, Poland; rzptak@cyf-kr.edu.pl

The objective of this study was to estimate heritabilities and genetic correlations for lactational (LSCS) and daily somatic cell scores (DSCS) with descriptive and linear type traits of Polish HF cows. Data were test day SCS and conformation evaluations of 24,599 primiparous cows, daughters of 802 sires. Cows calved from 2005 through 2006. LSCS was calculated as the average of at least four test day SCS. The DSCS was the test day SCS closest to the date of type evaluation. A multi-trait animal model was used to estimate genetic parameters. (Co)variance components were estimated with a Bayesian algorithm via Gibbs sampling. Heritability for LSCS was 0.20 and was much higher than for DSCS (0.12). Among type traits, heritabilities were high to moderate for height at rump (0.45), size (0.39), overall conformation score (0.30), two linear rump traits (0.28-0.29) and three linear teat traits (0.26-0.29). The genetic correlation between LSCS and DSCS was 0.85. In most cases DSCS showed higher genetic correlations with type traits than LSCS. Descriptive udder and feet and legs scores were negatively genetically correlated with both LSCS (-0.22 and -0.20) and DSCS (-0.29 and -0.33). SCS traits were positively genetically correlated with rump angle (0.20 for LSCS, 0.23 for DSCS) and negatively correlated with fore udder attachment (-0.25 for LSCS, -0.28 for DSCS), udder depth (-0.24 for LSCS, -0.16 for DSCS) and central ligament (-0.15 for LSCS, -0.17 for DSCS). Due to higher heritability, direct selection to lower LSCS would be more effective than to lower DSCS. The magnitude of obtained heritabilities and the favourable genetic correlations indicate that selection utilizing some type traits could improve resistance to mastitis.

Effect of GHR allele variants on milk production traits in a German Holstein dairy cow population
Rahmatalla, S.A., Mueller, U., Reissmann, M. and Brockmann, G.A., Humboldt-University of Berlin, Insitute of Animal Science, Breeding Biology and Molecular Genetics, Invalidenstr. 42, 10115 Berlin, Germany; rahmatas@agrar.hu-berlin.de

Recently, the growth hormone receptor (GHR) gene located on BTA20 has been identified as another strong positional and functional candidate gene influencing bovine milk production. A non-synonymous single nucleotide polymorphism (SNP) in exon VIII of the bovine GHR gene leads to a phenylalanine to tyrosine amino acid substitution in the transmembrane domain of GHR protein (F279Y). These gene variants showed highly significant effects on milk composition and milk yield. The objectives of our study were (1) the determination of the genotype and allele frequencies of the F279Y mutation of the growth hormone receptor gene in an active Holstein dairy cattle population and (2) the estimation of the allele effects on milk, fat, and protein yield, as well as fat and protein content. For genotyping the SNP, we developed a gene test for pyrosequencing. We genotyped 1370 dairy cows of three herds, which were kept under similar management conditions, and performed an association test between genotypes and individual phenotypic data. The gene frequencies in the analysed population were 83% and 17% for the F and Y alleles, respectively. This study confirms that the F279Y polymorphism is associated with a strong effect on milk yield and composition in our population. The allele substitution effects for the phenylalanine encoding variant were $0.127\% \pm 0.025$ for fat content and $0.092\% \pm 0.010$ for protein content respectively. A negative effect of -322 kg of the phenylalanine variant was found for milk yield. The data show the high potential of this polymorphism for use in selection.

Review of methods for fertility evaluation
Šafus, P., Institute of Animal Science, Genetics and breeding, Přátelství 815, 104 00 Prague, Czech Republic; safus.petr@vuzv.cz

The purpose of this paper is to give the review of methods used for fertility evaluation and shortly describe theirs principles. Not for all methods formula will be given and only the principle is explained. The evaluation of fertility can be made by means of different measures; hence the choice of mathematical-statistical model must follow specialties and nature of every different measure of particular trait. On the other hand, because fertility is complex phenomenon the separate evaluation of one measure can't satisfactorily describe whole reality. On this account the construction of evaluating models sometimes consider more then one trait, hence it seems to be difficult to provide compact information about evaluation and properties of single trait as well as about results of its evaluation. These traits can be evaluated in principle in the same way as every continues response in mixed linear regression model – animal model. The main difficulty is that they sometimes are not normally distributed and ignoring this fact can influence the results in a bad way. Big problem of every evaluation is bias due to not-considering of data incompleteness. In this respect an important problem in genetic analysis of fertility is how to handle cows that do not become pregnant or that are culled with unknown pregnancy status (i.e., censored records, term censoring). The Best linear unbiased prediction is extensively used for estimation of genetic merits for normally distributed traits, because it yields the maximum likelihood estimator of the best linear predictor. In this paper are described some suitable models – threshold model, longitudal data analysis for normal distributed traits, and multiple trait model.

Selection indexes in cattle breeding in the czech republic

Šafus, P. and Přibyl, J., Institute of Animal Science, Genetics and breeding, Přátelství 815, 104 00 Prague, Czech Republic; safus.petr@vuzv.cz

Selection indexes are used for sires of dairy and dual-purpose cattle in the Czech Republic. Total selection index (SIH) and sub-indexes were constructed for bulls of the Holstein breed according to groups of production traits – production sub-index for milk (IPH), sub-indexes for legs (IKC) and udder (IUC). Every index for selection for a group of traits applies all available information – breeding values for traits of milk performance, fertility and linear type trait classification. Total selection index (SIC) and sub-indexes for the selection of Czech Fleckvieh sires for A.I. – for groups of production traits – production index for milk (IPC), meat (IMC), reproduction (IRC), longevity (IDC) and resistance (IOC) were constructed. The sub-indexes were compared with the total index. Sub-indexes can be used for potential intensive selection aimed at a group of desired traits. Total index can be expressed as a combination of sub-indexes. A change in the weights of sub-indexes in the total combination makes it possible to make up customized index according to economic conditions of the own herd. The construction of indexes is based on available sources of information – breeding values of all traits in animal recording. Variants according to the traits included in breeding objective and in performance recording were tested – breeding objective comprised milk, meat, health, reproduction, longevity; production traits and linear classification of body conformation or production traits and general characteristics of body conformation and/or production traits, and general characteristics of body conformation and body measurements were used as source of information (in performance recording).

Estimation of genetic parameters for length of productive life using survival analysis in Holstein population of Japan

Sasaki, O.[1], Aihara, M.[2], Hagiya, K.[3], Ishii, K.[1] and Satoh, M.[1], [1]National Institute of Livestock and Grassland Science, Tsukuba, 305-0901, Japan, [2]Livestock Improvement Association of Japan, INC., Tokyo, 104-0031, Japan, [3]National Livestock Breeding Center, Nishigo, 961-8511, Japan; sasa1@affrc.go.jp

Length of productive life is a considerable trait for dairy cattle. This character is included in the national selection index of various countries. The length of productive life is usually evaluated by using a survival model or a best linear unbiased prediction (BLUP) model. The objective of this study was to estimate genetic parameters for length of productive life using a survival analysis in Holstein population of Japan. The data on the first 10 lactation records from January of 1991 to December of 2005 was obtained from Livestock Improvement Association of Japan. The data included records from 1,163,466 cows in 8,305 herds and 5,817 relative sires of the cows. The analytical model included (1) a fixed effect of the year and season of first parturition, (2) a time-dependent fixed effect of the annual change of herd size within a herd-size class, (3) a time-dependent fixed effect of group classified by milk yield of each herd within a prefecture-parity-stage class, (4) a fixed effect of age at first calving, (5) a random effect of herd, and (6) a random effect of sire. The effect of herd was assumed to follow a log-gamma distribution (γ=9.0). The Weibull model was used for estimating genetic parameters. The analysis was carried out by using the computer software Survival Kit V3.12. The estimates of the scale and shape parameters of the Weibull distribution were 1.92 and 0.000916, respectively. The heritability estimates on the log scale and the original scale were 0.0984 and 0.191, respectively. These heritability estimates were higher than those in some previous reports. These parameter estimates indicate the possibility of the genetic improvement in length of production life using survival analysis in Holstein population of Japan.

CSN1S1 gene: allele frequency, and the relationship with milk production traits in three Iranian indigenous cattle and Holstein breeds of Iran

Zakizadeh, S.[1], Reissmann, M.[2], Reinecke, P.[2] and Miraei-Ashtiani, S.R.[3], [1]Hasheminejad High Education Center, Animal Science, Kalantari Highway, 91769-94767, Iran, [2]Humboldt University of Berlin, Invalidenstrasse 42, 10115, Germany, [3]University of Tehran, Animal Science, Karaj- Agriculture complex, 11111, Iran; sonia_zaki@yahoo.com

CSN1S1 is one of the major proteins in milk of mammals. Milk protein genes are supposed to influence on milk yield and its compositions. In this study we aimed to determine allele frequencies of CSN1S1- 5` and B variants and their effect on milk traits in three indigenous (two Bos indicus and one Bos taurus) and Holstein breeds in Iran. DNA samples were gathered from 400 animals. ALF, SSCP and PCR-SSCP were used for genotyping of promoter and coding region of exon 17. CSN1S1*B variant was nearly fixed in Holstein but it was intermediate in indigenous breeds. All four alleles of promoter were found in breeds but in different frequencies. Allele B was found in combination with all four promoter alleles. Allele '4' promoter was not found in any cow having the allele 'C' in all breeds. Statistical analysis performed for Holstein and one of indigenous breed (Golpaygani). BC/23 genotype yielded the highest fat percentage ($P<0.05$) in Holstein but it had no significant effect on Golpaygani. There was not any homozygous CC for CSN1S1 in Holstein to show whether C variant would be advantageous for fat percentage. BC/22 genotype had no significant effect on milk production of Golpaygani but they tended to produce higher milk than the other genotypes. Differences of allelic frequencies and milk production traits between these breeds might be due to differences in origin of breeds or selection breeding programs.

Detection and correction of outliers for fatty acid contents measured by mid-infrared spectrometry using random regression test-day models

Soyeurt, H.[1], Dardenne, P.[2] and Gengler, N.[1,3], [1]Gembloux Agricultural University, Animal Science Unit, Passage des Deportes 2, 5030 Gembloux, Belgium, [2]Walloon Agricultural Research Centre, Quality Department, Chaussee de Namur 24, 5030 Gembloux, Belgium, [3]National Fund for Scientific Research, Rue Egmont 5, 1000 Brussels, Belgium; gengler.n@fsagx.ac.be

Calibration equations were developed to quantify fatty acids (FA) in milk using mid-infrared spectrometry. Belgian equations used showed a good ability to predict the contents of saturated (SAT) monounsaturated FA (MONO). Cross-validation coefficient of determination were 99.54% and 98.23% for SAT and MONO, respectively. Some outliers can appear and have impact on the estimation of breeding values. The aim of this study was to develop methods of outlier detection. The first method determined limitations based on individual examinations of extreme values. The second model was based on the residue estimated from random regression test-day models. The dataset contained 58,443 test-day records collected from 16,470 first parity Luxembourg Holstein cows in 699 herds. Models included as fixed effects: herd*test date, days in milk, and age. Random effects were regressed using Legendre polynomials of order 2 and were: herd*year of calving, permanent environment, and animal effect. The residual effect was assumed to be constant through the lactation. Detection by a random regression approach gave better results. The replacement of outlier values by the corresponded expected values was the better approach to calculate reliable breeding values.

Model evaluation and genetic parameters for milk urea nitrogen in Holstein dairy heifers

Stamer, E.[1], Junge, W.[2], Brade, W.[3] and Thaller, G.[2], [1]TiDa Tier und Daten GmbH, Bosseer Str. 4c, 24259 Westensee, Germany, [2]Institute of Animal Breeding and Husbandry, Christian-Albrechts-University, Olshausenstr. 40, 24098 Kiel, Germany, [3]Chamber of Agriculture Lower Saxony, Johannssenstr. 10, 30159 Hannover, Germany; gthaller@tierzucht.uni-kiel.de

Milk urea nitrogen (MUN) routinely recorded by infrared might be used as a selection trait to prevent higher nitrogen excretion. Therefore, model evaluation and estimation of both heritability and genetic relationships to the important milk production traits are needed. Data of 589 heifers tested between the 8th and the 180th day of lactation from the bull dam performance test at the dairy research farm Karkendamm were analysed. Cows were fed a total mixed ration ad libitum twice daily. Additionally, a fixed amount of concentrates was given via feeding stations. Milk ingredients were analysed weekly, and milk yield was recorded automatically for each milking. In model evaluation five well established parametric functions of days in milk were chosen for modelling both fixed and random regression coefficients. Model fit was judged by two information criteria and by illustration of the residuals against days in milk. Subsequently, on the basis of the best fitted random regression model variance components were estimated by REML. In addition to the fixed and random regressions the model includes the fixed effects test day and age of first calving. Estimations were based on univariate (heritabilities) and bivariate (genetic correlations) runs. The function of Ali and Schaeffer seems to be most suitable for modelling both the fixed and the random regression part of the mixed model. A serious alternative is the Legendre polynomial of fourth degree. Heritability for MUN ranges between 0.28 and 0.48. The yield traits milk, fat and protein are unfavourably genetically correlated with MUN. In contrast, fat and protein content are favourably correlated with MUN. So, a consequent selection on fat and protein content could reduce nitrogen excretion.

Variability of major fatty acid contents in Luxembourg dairy cattle

Soyeurt, H.[1], Arnould, V.M.-R.[1], Dardenne, P.[2], Stoll, J.[3], Braun, A.[3], Zinnen, Q.[3] and Gengler, N.[1,4], [1]Gembloux Agricultural University, Animal Science Unit, Passage des Deportes 2, 5030 Gembloux, Belgium, [2]Walloon Agricultural Research Centre, Quality Department, Chaussee de Namur 24, 5030 Gembloux, Belgium, [3]CONVIS herdbuch, Zone Artisanale et Commerciale 4, 9085 Ettelbruck, Luxembourg, [4]National Fund for Scientific Research, Rue Egmont 5, 1000 Brussels, Belgium; gengler.n@fsagx.ac.be

Common human health concerns and imminent needs for more sustainable nutrition patterns require from dairy industry and farmers a. o. a closer look at milk fatty acid (FA) profile. Therefore up to date calibration equations using mid-infrared (MIR) spectrometry were developed permitting the estimation of FA contents in bovine milk. The aim of this study was to estimate the variability of the major FA from data collected during the Luxembourg routine milk recording. A total of 148,296 milk samples with MIR-spectra were collected from October 2007 to January 2009 on 36,522 cows belonging to 5 breeds in 718 herds and scanned by Foss MilkoScan FT6000. The contents of saturated FA, monounsaturated FA, omega-9, short chain FA, medium chain FA, and long chain FA were obtained using Belgian MIR calibration equations. Analyzes were done by a multi-trait multi-lactation animal mixed models. Fixed effects were herd*test date, lactation stage*lactation number, age*lactation number, and breed effect. Random effects were herd*year of calving, permanent environment within and across lactation, animal effect, and residual effect. Breed differences as well as lactation effects were observed. Our results showed moderate heritability values suggesting the existence of a FA genetic variability. The variability of the first Luxembourg breeding values was large enough to develop selection tools for improving the nutritional quality of bovine milk fat.

Modelling effects of selected candidate genes on milk production traits as variable during a lactation
Szyda, J.[1,2], Komisarek, J.[3] and Antkowiak, I.[3], [1]Wroclaw University of Life Sciences, Department of Animal Genetics, Kozuchowska 7, 51-631 Wroclaw, Poland, [2]Wroclaw University of Life Sciences, Institute of Natural Sciences, Grunwaldzki 24, 50-365 Wroclaw, Poland, [3]Poznan University of Life Sciences, Department of Cattle Breeding and Milk Production, Wojska Polskiego 71A, 60-625 Poznan, Poland; joanna.szyda@up.wroc.pl

The major goal of the study is to test whether selected candidate genes have a constant or a variable effect throughout lactation. The analysed data set consists of 190 Jersey cows genotyped for seven functional SNP polymorphisms located within four unlinked candidate genes: the leptin receptor, the leptin gene, the acyl-CoA:diacylglycerol acyltransferase1 gene (DGAT1), and the butyrophilin gene. The production data comprise test day records for milk yield as well as fat and protein contents from the first three lactations. For the 305-day record of the first lactation the average milk yield is 4,186±469 kg, the average protein content is 3.84±0.33% and the average fat content amounts to 5.52±0.52%. The estimated additive effects on milk yield, constant for the whole first lactation, were estimated to 40.18 kg for the leptin receptor, 101.43 kg for the butyrophilin gene, 339.67 kg for the DGAT1 gene, and 140.80 kg for the leptin gene. In order to test whether the SNP effect is constant or variable during a lactation two mixed models are fitted to the data: 1) a model with a random additive polygenic cow effect and fixed effects of candidate genes constant across a lactation, 2) a model with a random additive polygenic cow effect constant across a lactation and fixed effects of candidate genes variable across a lactation. The differences in model fit are then compared via the likelihood ratio test.

Genetic parameters for milk coagulation properties in the first lactation Estonian Holstein cows
Vallas, M.[1,2], Pärna, E.[1,2], Kaart, T.[1,2] and Kiiman, H.[1,2], [1]Eesti Maaülikool, the Estonian University of Life Sciences, Kreutzwaldi 1, 51014 Tartu, Estonia, [2]Bio-Competence Centre of Healthy Dairy Products, Kreutzwaldi 1, 51014 Tartu, Estonia; mirjam.vallas@emu.ee

Milk coagulation properties (MCP) are an important aspect in assessing cheese-making ability. In milk-to-cheese steps, coagulation of milk is important and sensitive because it is the first phase and affects the following phases in the process. Genetic factors play an important role in defining milk quality for cheese-making. MCP were found to be heritable, but little scientific literature is available about their genetic aspects. The aim of this study was to estimate heritability, repeatability and herd effect on MCP and milk production traits in first lactation Estonian Holstein dairy cattle. A total 10,722 coagulated measurements of 2,608 Estonian Holstein cows (progeny of 196 sires) reared in 92 herds in Estonia were sampled from April 2005 to August 2007 at least 3 times during the lactation (7-305 day in milk) and the database COAGEN™ of Bio-Competence Centre of Healthy Dairy Products was formed. Individual milk samples were analyzed for milk coagulation time (RCT), curd firmness (E_{30}), milk yield, fat percentage, protein percentage and somatic cell count. Only 0.3% of individual milk samples did not coagulate in 31 min and were excluded from analyses. Estimates of heritability for RCT and E_{30} were 0.34±0.06 and 0.43±0.01, respectively.

Endometrial receptivity in lactating dairy cows

Peippo, J.[1], Räty, M.[1], Ahola, V.[1], Grosse-Brinkhaus, C.[2], Salilew-Wondim, D.[2], Sorensen, P.[3], Taponen, J.[4], Aro, J.[5], Myllymäki, H.[5], Tesfaye, D.[2] and Vilkki, J.[1], [1]MTT Agrifood Research Finland, Biotechnology and Food Research, ET Building, FI-31600 Jokioinen, Finland, [2]University of Bonn, Institute of Animal Science, Endenicher Allee 15, 53115 Bonn, Germany, [3]University of Aarhus, Department of Genetics and Biotechnology, P.O. Box 50, DK-8830 Tjele, Denmark, [4]University of Helsinki, Department of Production Animal Medicine, Paroninkuja, FI-04920 Saarentaus, Finland, [5]Embryocenter Ltd, P.O. Box 40, FI-01370 Vantaa, Finland; jaana.peippo@mtt.fi

The objective of this study was to evaluate factors affecting endometrial receptivity in lactating dairy cows. After calving cows were subjected to endometrial biopsy on days 0, 7 and 14 (day 0 = standing heat) of oestrous cycle (the monitoring cycle) followed by transfer of a biopsied blastocyst on day 7 of the subsequent cycle (the ET cycle). The both cycles were monitored by milk progesterone measurements to ensure normal cyclicity. The collected endometrial and embryo biopsy specimens were analysed in pools of three with the Affymetrix oligo arrays and Blue Chip cDNA arrays, respectively. According to the production data, recipient cows that conceived and delivered a calf (the calf delivery group) had higher overall breeding values and Fatkg indices compared to cows that did not get pregnant (the non-pregnant group). The overall progesterone profiles did not differ between the groups. Negative correlations were observed between progesterone concentrations and energy corrected milk yields (ECM) on the days 7, 13 and 14 of the monitoring cycle, as well as, on the day 7 of the ET cycle. Comparison of gene expression in the endometrial biopsy specimens indicated significant differences between the calf delivery and non-pregnant groups on the day 0 of the monitoring cycle. Integration of the phenotypic data in micro array data analysis will be discussed. Acknowledgements: These results are obtained through the EC-funded FP6 Project 'SABRE'.

Heritability of body condition score and relationships with milk production traits in Canadian Holsteins and Ayrshires

Loker, S.[1], Bastin, C.[2], Miglior, F.[3,4], Sewalem, A.[3,4], Fatehi, J.[1], Schaeffer, L.R.[1] and Jamrozik, J.[1], [1]CGIL, University of Guelph, Animal and Poultry Science, 50 Stone Rd E, Guelph, ON, N1G 2W1, Canada, [2]Gembloux Agricultural University, Passage des Déportés, 2, Gembloux, 5030, Belgium, [3]Agriculture and Agri-Food Canada, 2000 College Street, Sherbrooke, QC, J1M 1Z3, Canada, [4]Canadian Dairy Network, 660 Speedvale Ave W, Guelph, ON, N1K 1E5, Canada; sloker@uoguelph.ca

This study is a first step in the eventual development of a genetic evaluation for body condition score (BCS) in Canadian dairy cattle breeds. Specific objectives of the current study were to estimate genetic parameters of BCS using DHI data collected in Quebec and to estimate relationships between BCS and production traits. Breeds analyzed were Canadian Holstein (HO) and Ayrshire (AY). Phenotypic correlations were estimated for each breed based on data from all parities (1 to 3) after preliminary edits. Overall, BCS was significantly correlated with all traits for both breeds except for MUN and somatic cell count for Holsteins. For Holsteins, BCS was significantly negatively correlated with milk, fat and protein yield, and fat:protein and fat:lactose ratios across nearly all lactation stages (correlations between -0.03 and -0.16). Similar relationships occurred in AY data for milk, fat and protein yield, and fat:protein ratio, though only in the final stages of lactation. Significant positive relationships between BCS and protein % occurred across all stages of lactation for both breeds. Data used for variance component estimation was edited further. This left 4,641 first lactation BCS records on 1,338 AY cows of 251 sires and 39,125 BCS records on 10,000 HO cows of 1,632 sires. The overall heritability for BCS was 0.30±0.06 for AY and 0.20±0.03 for HO. Heritability across lactation ranged between 0.17 at 5 days in milk (DIM) and 0.36 at 275 DIM for AY and between 0.13 at 5 DIM and 0.32 at 290 DIM for HO.

Associations of growth hormone gene polymorphisms with milk production traits in South Anatolian and East Anatolian red cattle

Türkay, G.[1], Yardibi, H.[1], Paya, İ.[1], Kaygısız, F.[2], Çiftçioğlu, G.[3], Mengi, A.[1] and Öztabak, K.[1], [1]University of Istanbul, Faculty of Veterinary Medicine, Department of Biochemistry, 1University of Istanbul, Faculty of Veterinary Medicine, Department of Biochemistry, 34320-Avcilar-Istanbul, Turkey, [2]University of Istanbul, Faculty of Veterinary Medicine, Department of Animal Breeding and Husbandry, University of Istanbul, Faculty of Veterinary Medicine, Department of Animal Breeding and Husbandry, 34320-Avcilar-Istanbul, Turkey, [3]University of Istanbul, Faculty of Veterinary Medicine, Department of Food Hygene and Technology, University of Istanbul, Faculty of Veterinary Medicine, Department of Food Hygene and Technology, 34320-Avcilar-Istanbul, Turkey; gturkmen@istanbul.edu.tr

The current study was undertaken to determine the relationship between milk production traits of Eastern Anatolian Red (EAR) and South Anatolian Red (SAR) breed cows and polymorphisms of growth hormone gene (GH) which is a potentially effective Quantitative Trait Loci (QTL) on milk production traits. 50 cows that were newly delivered calves from each of EAR and SAR breeds were used. Triplicate milk samples were obtained between 0-30, 50-180 and 270-300 days of lactation period. Milk samples were analyzed for milk fat, protein, dry substance, refraction indices, and somatic cell count. In addition, DNA samples were obtained from blood samples of each cow and AluI and MspI polymorphisms in GH were determined using PCR-RFLP method. In both breeds, AluI polymorphism with VV genotype cows had higher milk fat percentage compared to other genotypes. Similarly, in SAR cows, those with MspI polymorphism and -/- genotype had higher milk fat percentage compared to other genotypes. The relationship between GH gene polymorphisms and other milk quality parameters could not be established. As a result, it can be concluded that GH gene polymorphisms can be of a valuable parameter to be used for selection of EAR and SAR cows for improving milk fat percentage.

Breed and heterosis estimates for milk production, udder health and fertility traits among Holstein and Norwegian Red Dairy Cattle

Begley, N.[1], Evans, R.D.[2], Pierce, K.M.[3] and Buckley, F.[1], [1]Moorepark Teagasc, Dairy Production Research Centre, Fermoy, Co. Cork, Ireland, [2]Irish Cattle Breeding Federation, Bandon, Co. Cork, Ireland, [3]School of Agriculture, Food and Veterinary Medicine, UCD, Belfield, Ireland; nora.begley@teagasc.ie

The objective of this paper is to present initial estimates of breed and heterosis effects for the Norwegian Red (NR) and the Norwegian Red×Holstein (NR×HO), relative to the Holstein (HO) dairy cows for milk production, udder health (defined as somatic cell score (SCS)) and fertility based on first and second lactation records (n=3711). Breed and heterosis effects were estimated by regressing the breed fractions and proportion of heterozygosity, respectively, on the phenotypic data using the statistical package DMU. Though not significant, the NR had numerically lower 305 d yields of milk (-64 kg), fat (-6.2 kg) and protein (-0.6 kg) compared to the HO. Significant heterosis was observed for 305 d yields of milk (+178 kg), fat (+8.5 kg) and protein (+7.7 kg). A genetic superiority for lower SCS (-0.15) was observed for the Norwegian Red compared to the Holstein. No significant heterosis was observed for SCS (-0.01). Superior fertility was evident for the NR compared to the HO with breed effect in favour of the NR for pregnancy rate to first service (PRFS; +6.4%), proportion of cows in calf after 6 weeks breeding (INCALF6; +9.5%), proportion of cows in calf after 13 weeks breeding (+7.7%), number of services per cow (-0.13), calving interval (CI; -11 days) and survival from 1st to 3rd lactation (+8.7%). Heterosis estimates observed for the NR×HO tended to be favourable and realistic in magnitude (PRFS +5.7%, INCALF6 +4.0%, CI -6.4 days). However, these were not significantly different from zero. In conclusion, crossbreeding with the NR appears to be a viable option for dairy farmers, resulting in similar milk production to that of the HO but improved herd reproductive efficiency/survival. The latter driven primarily by the additive genetic superiority of the NR breed.

Session 18

Theatre 1

Marker-assisted selection in PIC

Knap, P.W., PIC, Ratsteich 31, 24837 Schleswig, Germany; pieter.knap@pic.com

The typical initial structure of MAS in BLUP-based animal breeding systems matches single-trait phenotypic records to the genotypes of a few DNA markers through single-marker analysis. Such markers are mostly developed through a candidate gene approach. With a growing number of markers in the system (typically developed through genome scans), integrative actions will logically take place: 1: Realization that fixation of a few marker alleles should not override quantitative index selection. 2: Integration of MAS and index selection into a single step. 3: Reduction of false positives in marker development through Bayesian multiple marker analysis (BMMA). 4: Implementation of random-effect statistics for analysis of many effects with variable information content. 5: Inclusion of the above into the BLUP analysis. 6: Integration of all this into a single analysis that encompasses genotypic evaluation, estimation of any leftover polygenic components, and marker filtering. Since the introduction of HAL-1843® (Innovations Foundation, Toronto, CND), the number of DNA markers used to support breeding value estimation in the PIC system has increased truly exponentially to a 2008 number of 140. Most of the more recent among these markers stem from genome scans. Marker genotypes were initially included into BLUP models as fixed class effects, and the resulting BLUEs were added back to the associated polygenic estimates to form the final EBV. To deal with missing data due to incomplete population genotyping, marker genotypes were replaced by marker genotype probabilities, estimated from pedigree-based segregation analysis. BMMA removes indeed a large proportion of false positives, and the resulting filtering process leads to a strong reduction of the marker volume that needs to be handled in the EBV system. Random analysis weights the impact of an effect according to its variance component (VC) and its underlying number of observations, and VC estimation is a bottleneck that can be overcome by shrinkage of fixed effects through integrating marker heritabilities into the MME.

Session 18

Theatre 2

The use of a parentally imprinted QTN in differential selection of sire and dam lines

Buys, N.[1,2], Stinckens, A.[1], Mathur, P.[3], Janssens, S.[1], Spincemaille, G.[2], Decuypere, E.[1] and Georges, M.[4], [1]K.U.Leuven, Department Biosystems, Kasteelpark Arenberg 30, 3001 Leuven, Belgium, [2]RATTLEROW SEGHERS, gentec, Oeverstraat 21, 9160 Lokeren, Belgium, [3]Canadian Farm Animal Genetic Resources Foundation, 60 Fardon way, Ottawa, Ontario, K1G 4N4, Canada, [4]University of Liège, Unit of Animal genomics, GIGA tower, Avenue de l'hôpital 1, 4000 Liege, Belgium; nadine.buys@biw.kuleuven.be

Longevity of modern sows has been compromised as a result of the genetic selection for improved lean and faster growth. Modern genotypes are not always capable of sustaining the body reserves needed to support lactation and subsequent high reproductive performance.Since body fat deposition is necessary to sustain sow reproduction performance, for example to support adequate milk production and to limit body weight loss, the selection for leaner carcasses, demanded by the packing industry and consumers, may conflict with the prolificacy and longevity of the sow and lead to increased replacement costs of sows in pig production. The QTN in the IGF2 gene provides a possibility to overcome this conflict. The imprinting character of the gene might be used to produce lean slaughter pigs from fatter dams, that inherited the wild type allele from their father (genotype of grand parent boar = GG), crossed with terminal sires being homozygous for the lean allele (AA). This selection scheme was tested in different populations. The results show an influence of the IGF2-intron3 G3072A mutation on prolificacy and longevity in sows. The imprinted QTN is used in opposite selection in sire and dam lines. Because of the imprinted character of the gene, selection for the fatter allele in sow lines will not influence the carcass quality of the offspring.

Increasing genetic *E. coli* F18 resistance in Swiss pigs

Luther, H.[1], Vögeli, P.[2] and Hofer, A.[1], [1]SUISAG, Allmend, 6204 Sempach, Switzerland, [2]Institute of Animal Science, ETH, 8092 Zurich, Switzerland; hlu@suisag.ch

Oedema disease and post-weaning diarrhoea are mainly caused by enterotoxigenic and enterotoxaemic E. coli expressing F18 fimbriae. A mutation in the FUT1 gene was shown to be causative for these diseases. A/A pigs are resistant and A/G and G/G pigs are susceptible to infection with *E. coli* F18. So, it is possible to increase the number of resistant pigs by genotyping and selection of breeding candidates. SUISAG, the Swiss herd book and AI organisation, genotypes about 800 Large White dam line sows and young boars for E. coli F18 resistance annually. SUISAG administrates all genotypes in a combined database with the traditional pedigree and performance data. A majority of pigs in the herd book is still not genotyped, but the Swiss database system allocates them one allele in case of a known homozygous parent or progeny and thereby uses the genotyping information in an optimum way. Since 2006, SUISAG runs an elite[1] mating program and a subsequent station test for dam line boars to improve the selection of the AI boars in general. Known homozygous susceptible (G/G) or heterozygous (A/G) Large White dam line sows are not considered for elite matings. Boars out of the elite matings enter the test station at 25 kg. They are genotyped promptly afterwards if their genotype is not known from their homozygous parents. Less than 10% of the boars are G/G. They are slaughtered at the end of the test. Homozygous resistant boars (A/A) are preferred for AI. E. coli F18 resistance of the Large White dam line AI boars improved considerably. On 1.1.2009, there were 25 (15) homozygous resistant, 10 (12) heterozygous and 0 (12) homozygous susceptible boars at the AI station (in brackets: AI boars on 1.1.2006). So, the frequency of the resistance allele increased from 54% to 86% within the Large White dam line AI boars. SUISAG will continue the selection to eliminate the E. coli F18 susceptible allele from the Large White AI boars and thus increase the number of resistant piglets in the herd book and piglet producer farms.

Effect of selection for *E. coli* F4ab/ac resistance in pigs

Nielsen, B.[1], Jørgensen, C.B.[2], Vernersen, A.[1] and Fredholm, M.[2], [1]Danish Pig Production, Genetic Research & Development, Axeltorv 3, 1609 Copenhagen V, Denmark, [2]University of Copenhagen, Faculty of Life Sciences, Department of Basic Animal and Veterinary Sciences/Genetics & Bioinformatics, Grønnegårdsvej 3, 1870 Frederiksberg C, Denmark; BNi@danishmeat.dk

Diarrhoea caused by enterotoxigenic *Escherichia coli* (ETEC F4ab/ac) is a problem in pig production and ETEC F4ab/ac is responsible for more than 30% of the E. coli diarrhoea cases in piglets. An effort to reduce the prevalence of ETEC F4ab/ac infection will thus have a significant impact on pig welfare and greatly diminish the need for antibiotic treatment. In pig production, infection by enterotoxigenic *Escherichia coli* (ETEC F4ab/ac) is linked to a single recessive allele and homozygote recessive animals are resistant. The desired genotype is homozygous for the resistant allele and a selection programme among males for this genotype was started in 2003 in the three Danish nucleus breeds. Before selection, the prevalence of the recessive F4ab/ac genotype was 0.01, 0.19, and 0.88 in the nucleus of Landrace, Large White and Duroc. After three years none of the homozygous dominant animals were permitted in the performance tested young animals in Landrace. Investigations of faeces samples from production herds show that the prevalence of samples with *E. coli* O149 diarrhoea decreased from 20% in 2003 to 5% in 2008. The associations between the F4ab/ac recessive allele and other traits in the breeding program were obtained by BLUP estimation of single and multitrait animal models that include among others the fixed effect levels of the F4ab/ac genotypes. The analysis is complicated by the genetic trends in the population during the observed period. Generally, the effect of the resistant genotype was low and unfavourably associated to the production traits, eg. growth rate and meat content in carcass. However, the genetic loss is calculated to be regained within one year. The results indicate that the recessive F4ab/ac genotype has a favourable effect on piglet survival.

Marker- and gene-assisted selection and causes for the rapid proliferation of the immotile short tail sperm defect within the Finnish Yorkshire pig population

Sironen, A.[1], Uimari, P.[1], Serenius, T.[2] and Vilkki, J.[1], [1]MTT Agrifood Research Finland, Biotechnology and Food Research, 31600 Jokioinen, Finland, [2]Faba Breeding, P.O. Box 40, 01301 Vantaa, Finland; anu. sironen@mtt.fi

Previously, we have identified and characterized a defect causing male infertility within the Finnish Yorkshire pig population. Due to the specific phenotype of affected spermatozoa, the condition was termed the immotile short tail sperm (ISTS) defect. The causal mutation for ISTS involves insertion of long interspersed nuclear element (Line-1, L1) within the intron 30 of SPEF2 (also known as the KPL2) gene. This L1 insertion alters the splicing pattern of SPEF2 in two ways: in most of the transcripts of ISTS affected animals, exon 30 is skipped, while in a small number, exon 30 is included in the transcript along with part of the insertion. Importantly, both changes in the splicing pattern cause premature translation stop codons, and result in truncated protein products. The ISTS defect became an economic problem at the end of the 1990's. Nine affected boars were identified in 1998 and the carrier frequency reached 36% in 2001. Based on the initial whole genome scan we developed a DNA-test for marker assisted selection, which was used in pig breeding programmes between 2001 and 2005. Application of this test reduced the carrier frequency to 18% by the end of 2005. Since 2006, a test based on the L1 and SPEF2 sequences providing 100% certainty of the disease status has been available for Finnish pig breeders. The rapid proliferation of the defect indicates selection pressure favoring ISTS affected animals. To elucidate the possible factors contributing to the extensive use of ISTS carrier boars, we have analyzed the association between the ISTS locus and reproduction and production traits. Some statistically significant results indicate that the cause for high frequency of the ISTS defect may be a specific association with reproduction traits. Preliminary studies of the molecular mechanisms underlying these effects have also been elucidated.

Implementation of a marker-assisted selection program in the Chinese-European Duochan pig population

Schwob, S.[1,2], Riquet, J.[3], Bellec, T.[4], Kernaleguen, L.[4], Tribout, T.[2] and Bidanel, J.P.[2], [1]IFIP - Institut du Porc, La Motte au Vicomte - BP 35104, 35651 Le Rheu Cedex, France, [2]INRA, UMR1313 Génétique Animale et Biologie Intégrative, Domaine de Vilvert, 78352 Jouy-en-Josas Cedex, France, [3]INRA, UMR444 Laboratoire de Génétique Cellulaire, Chemin de Borde Rouge, 31326 Castanet-Tolosan Cedex, France, [4]ADN, Rue Maurice de Trésiguidy, 29190 Pleyben, France; sandrine.schwob@ifip.asso.fr

A marker assisted selection (MAS) program has been set up since 2001 by the French pig breeding organisation ADN to select 25% Chinese Meishan (MS)/75% European (EU) crossbred boars used to produce parental sows. In a first step, the crossbred boars were produced using a discontinuous crossbreeding scheme. Then, they originated from a 25% Meishan composite line. Boars specifically selected to produce parental sows using the MAS program differed from those used within the composite line. Four quantitative trait loci (QTL) affecting growth and carcass composition traits and located on SSC1, 2, 4 and 7 were considered. Boars were first selected for performance traits and then on marker-based information. The phenotype and genotype data obtained on candidate boars were analysed to a posteriori estimate the effects of the four regions considered. Results confirmed a significant effect of the SSC7 QTL on backfat thickness. The MAS program also allowed more homogeneous parental females to be produced, resulting in an increased proportion of gilts retained for breeding. MAS had no impact on the efficiency of within-line selection, as specific boars were used for crossbreeding. The interest of MAS within-line selection remains to be investigated.

Opportunities and limitations in the use of MAS for a commercial pig genetics company
Walling, G.A., JSR Genetics Limited, Research & Genetics, Southburn, Driffield, East Yorkshire, YO25 9ED, United Kingdom; Grant.Walling@jsrgenetics.com

Considerable resources have been invested in the identification of major genes in pigs. Despite such efforts only a limited number of markers or genes have proven to be commercially viable to include in breeding programmes through Marker Assisted Selection or Marker Assisted BLUP. Of greatest benefit have been genes with a major effect on meat quality such as the Halothane and RN loci where specific variants can be directly correlated with superior or inferior meat eating quality. Favourable alleles for major genes associated with traits that have historically undergone conventional selection have typically been fixed or very close to fixation. An example of which is the single nucleotide substitution (G-A) located at position 3072 in intron 3 of the IGF2 gene which increases muscle growth whilst restricting back fat. Upon investigation the more muscular A allele was found to be fixed in JSR Pietrain and Large White sire lines. MAS for traits not under direct selection is therefore a more desirable area for commercial pig populations. Meat eating quality traits are desirable however the power of detecting putative QTLs is reliant on the size of the QTL effect, number of progeny and allele frequency (power=66% when effect is 0.5 standard deviations and 50 progeny per sire with an allele frequency of 0.5). Sub-optimal allele frequencies have a major effect on power e.g. 47% at an allele frequency of 0.25 and hence such exploratory analyses can look unfavourable for commercial investment appraisals. Perhaps the most promising area is that relating to segregation analyses of disease traits. Recent studies of JSR data have highlighted strong evidence for major genes segregating for postweaning multisystemic wasting syndrome and for a leg defect problem at birth. In both instances the amassed phenotypic data allowed the calculation of breeding values for resistance or susceptibility to the specific condition based on the probabilities of carrying the major gene.

The potential of genomic selection to improve litter size in pig breeding programmes
Simianer, H., Georg-August-University Goettingen, Department of Animal Sciences, Albrecht-Thaer-Weg 3, 37075 Goettingen, Germany; hsimian@gwdg.de

High density SNP arrays are becoming available for pigs, allowing implementation of genomic selection programs for economically important traits. A typical breeding program based on a two breed cross to produce crossbred sows providing replacement for a total of 250,000 sows in the piglet production is used as a reference scenario. The breeding objective is litter size (weaned piglets per litter). Due to the fast turnover of breeding animals in the nucleus herd, the main potential to increase genetic progress are 1) increased accuracy of genomic breeding values at the time of selection and 2) increased selection intensity based on a large number of genotyped selection candidates. A genomic selection scheme is modeled deterministically to assess the impact of different factors. The achievable genetic progress is critically depending on the size of the training set and the number and distribution of genotyped selection candidates. With a training set of 1,000 progeny tested boars with 40 daughters each, and genotyping of 1,000 young boars as selection candidates per year (to select 80 boars in total), the genetic progress per year is expected to increase by 37 per cent compared to the conventional reference scenario. The return on investment is a gain of 6.63 Euro per Euro genotyping cost in that case. The optimum number of genotypings in the selection step decreases with an increase in the size of the training set. The accuracy of genomic breeding values is modeled as a function of the size of the training set and the amount of information per animal in the training set which needs to be verified empirically. The limiting factor for a practical implementation in many cases will be the availability of DNA of a sufficient number of progeny tested boars for the calibration step. Despite the fact that breeding goals are more complex in real life than assumed here, genomic selection appears to be a powerful tool to increase genetic progress even for traits with low heritability.

Genomic prediction for backfat in pigs

Janss, L.L.G.[1], Nielsen, B.[2], Christensen, O.F.[1], Bendixen, C.[1], Sørensen, K.K.[1] and Lund, M.S.[1], [1]Aarhus University, DJF Department of Genetics and Biotechnology, P.O. Box 50, DK-8830 Tjele, Denmark, [2]Dansk Svineproduktion, Axeltorv 3, DK-1609 Copenhagen V, Denmark; Mogens.Lund@agrsci.dk

To show the applicability of genomic selection in pig breeding a pilot study was carried out to predict phenotypic measurements on 978 Landrace boars using a medium-dense SNP panel. The phenotypic trait used was backfat thickness. In a nucleus breeding animal model the heritability of backfat was estimated to be 0.53, with additional environmental factors included in the model. For modelling and prediction within the genotyped boars raw phenotypes were used. Genotypes of the boars were obtained using a 6K Illumina bead chip, from which 2373 SNP's had good quality and minor allele frequency >1% and were used for building prediction models. Prediction models were constructed using a Bayesian variable selection method, in which a Normal mixture prior was used on SNP scaling factors (standard deviations) to divide the SNPs in one group with small (negligible) effects and one group with large (important) effects. The mixture proportions were kept fixed, varying from 90/10 to 70/30; a variance that determines the size of the small-effect SNPs was fixed at a small value so that all small-effect SNPs explain <1% of variance in the data; and a variance that determines the size of the large-effect SNPs was estimated from the data. A cross validation study was carried out by randomly dividing the boars into 6 groups and 6 analyses were performed that used 5 of the 6 groups to fit the model and predict phenotypes of the boars that were excluded from the estimation. The correlation between genomic predictions and raw phenotypes was 0.43, which corresponds to covering about half of the genetic variance with the SNP panel used. The fitted and predicted variances were very similar for different mixture proportions, but fitted variances were always higher than predicted variances.

Implementation of genome wide marker assisted selection, size of reference population

Huisman, A.E.[1], Vereijken, A.[2], Van Haandel, B.[1] and Albers, G.[2], [1]Hendrix Genetics, Hypor, P.O. Box 30, 5830 AA Boxmeer, Netherlands, [2]Hendrix Genetics, Research and Technology Centre, P.O. Box 30, 5830 AA Boxmeer, Netherlands; abe.huisman@hendrix-genetics.com

The introduction of genomic selection is one of the most significant changes in animal breeding since the introduction of BLUP. Genomic selection uses dense (SNP) marker maps to accurately predict breeding values for animals with unmeasured phenotypes. In order to estimate these marker effects, a reference population has to be put together. The structure and size of reference population as well as the number and spacing of markers and LD influence the reliability of the estimated marker effects. One of the unsolved issues is the composition and size of the reference population, i.e. how many and which animals do we need to phenotype and genotype to get reliable estimates of SNP effects? In dairy cattle, typically proven bulls are used to estimate SNP effects, the number of animals in the reference population starts at approximately 1,000 animals. It was shown that including more animals in the reference population will increase accuracy of estimated effects, and the dairy industry is now setting up reference populations of over 10,000 animals. Analyses of internal chicken datasets, where we used 20K SNPs in a pure broiler line, suggest that 1,000 animals would be enough for one line. High correlations between true and estimated breeding values were achieved with 500 records, due to pedigree structure and high LD within this line. In a breeding program with multiple lines, such as ours, do we need different reference populations for each line/subpopulation, or can we use one population and estimate the marker effects across lines. Earlier analyses suggest that not all markers have the same effects within and across lines.

Genomics in Broiler Breeding: the cost of doing business?
Avendano, S., Watson, K. and Kranis, A., Aviagen Ltd, 11 Lochend Road, EH28 8SZ, United Kingdom; savendano@aviagen.com

After the first release of the chicken genome sequence, Aviagen started 2005 with a can-didate gene approach relating SNPs to immunity, production and fitness traits. Subse-quently, genome-wide association (GWA) explores fast development of public-domain SNP panels and reduced high-throughput genotyping costs. We have completed 3 rounds of GWA with 6k, 12k and 42k SNP panels, based on public and own R&D on genome regions associated to economically important traits, speeding up high-throughput GWA addressing statistical power issues. Further development of SNP panels was optimized based on the acquired insight in the structure of genome-wide LD within and between populations. A key component is parallel evaluation of approaches: from standard MAS (assuming population-wide LD between SNPs and causative polymorphisms) with regression of in-dividual and grouped SNPs and haplotypes, to approaches combining high and low den-sity panels with non-parametric and Bayesian statistics. We focus on broiler and breeder performance traits, liveability and fitness traits across all strains. Case-control studies with samples of individuals and pooled individuals allows the study of resistance to dis-eases that cannot be recorded in bio-secure conditions, and for exploitation at the nucleus of data recorded in commercial conditions. Experience with the statistics of GWA and the structure of LD reveals that the complex-ity of the genetics of selection traits is always underestimated. The balance between 1) risk of fitting spurious information and 2) opportunity of exploiting genuine genomic effects on relevant traits, requires ongoing re-assessment of R&D strategy and validation of results. We collaborate with Iowa State University, Wisconsin University, University of Edinburgh and Roslin Institute. Combining a wide R&D outlook with our global ex-pertise is essential for capitalization on the opportunities of genomics in routine breeding: accuracy of selection, generation intervals, genepool characterisation, breeding program management, and strain security.

Markers on pig chromosome 5 for arthrogryposis multiplex congenita (AMC), a recessively inherited disease
Haubitz, M.[1], Genini, S.[2], Bucher, B.[1], Wettstein, H.-R.[1], Neuenschwander, S.[1] and Vögeli, P.[1], [1]Institute of Animal Sciences, ETH Zurich, Tannenstrasse 1, 8092 Zürich, Switzerland, [2]School of Veterinary Medicine, University of Pennsylvania, 3900 Delancy St., 19104-6010 Philadelphia, PA, USA; monika.haubitz@inw. agrl.ethz.ch

In mammals, arthrogryposis multiplex congenita (AMC) is a common malformation and can be caused by extrinsic or genetic factors. A genetic variant for the porcine AMC was identified in the Swiss Large White population. AMC diseased piglets show symptoms of persistent flexion of the limbs, overbite, deformation of the spinal column and perinatal death. The disease is autosomal recessively inherited and the mutation was mapped to a 5 cM region between two microsatellite markers, SW152 and SW904, on porcine chromosome 5. The two functional and positional candidate genes contactin 1 (CNTN1) and PDZ domain containing RING finger 4 (PDZRN4) were partially sequenced in order to locate new markers and to identify the causative AMC mutation. In these sequences 80 SNPs, two deletions and five inserts were found. Some of these mutations were in linkage disequilibrium with AMC. The sequence data were used to refine the map in the AMC region. The two flanking microsatellite markers, bE77 and SW904, were used to test for AMC. Two additional markers USP18 and CE17 were tested if the results were ambiguous. In combination with these microsatellite data newly developed markers should enable to discover AMC carriers with higher certainty. By consequently excluding putative AMC carrier boars from artificial insemination, the number of cases decreased from five families in 2006, with 16 pigs tested positive for AMC, to one family in 2008, with two pigs tested positive for AMC. Therefore by improving the existing AMC test we expect to eliminate all AMC carriers from the Large White population in Switzerland.

Effects of porcine LEPR and MC4R genes on productive and meat and fat quality traits in a Duroc x Iberian commercial cross

Muñoz, G.[1], Alcazar, E.[2], Fernández, A.[1], Barragán, C.[1], Carrasco, A.[3], De Pedro, E.[4], Silió, L.[1], Sánchez, J.L.[2] and Rodriguez, M.C.[1], [1]INIA, Mejora Genética Animal, Crta de la Coruña Km 7.5, 28040 Madrid, Spain, [2]SAT Valleheromoso, Crta. Villanueva - La Solana, 13248 Ciudad Real, Spain, [3]CSIC, Ingeniería, José Antonio Novais, 10, 28040 Madrid, Spain, [4]ETSIAM, Producción Animal, Campus de Rabanales CN IV km 396, 14014 Córdoba, Spain; munoz.gloria@inia.es

Leptin (LEPR) and melanocortin 4 (MC4R) receptors play a critical role in the regulation in mammals of feed intake, body weight, and energy balance. Hence both genes are candidates for body composition and growth-related traits in pigs. Polymorphisms LEPR c.2002C>T and MC4R c.1426A>G have been associated with feed intake, growth and fatness. Our aim was to analyze how these non-synonymous polymorphisms are affecting different productive and quality traits in a commercial Duroc x Iberian cross (n=530 crossbred animals). Data for premium cuts yield, growth and fatness traits, as well as meat and fat quality were recorded and analyzed using an animal model. Results showed significant associations between LEPR nucleotide substitution and most of the analyzed traits. The c.2002T allele increased body weight at 225 d (6.3±0.9 kg), backfat thickness at 130 kg (1.5±0.4 mm), intramuscular fat (1.0±0.3%), and Minolta lightness (1.7±0.4) of m. longissimus; however it was decreasing premium cuts percentage on carcass weight (-0.8±0.2%). Likewise the fatty acid profile in subcuatenous fat was analyzed. The LEPR c.2002T allele increased SFA (0.58±0.12%), and decreased MUFA (-0.33±0.12%) and PUFA (-0.25±0.06%). These effects were mainly due to changes in the content of palmitic, oleic and linoleic acids. Results showed no significant effects of MC4R c.1426A>G SNP or interaction between LEPR and MC4R SNP. A genetic test based on the LEPR c.2002C>T SNP may be useful to the choice of Duroc boars in order to reduce the undesirable heterogeneity of carcasses and premium cuts proceeding of crossbred animals.

Genomic tools for horse health

Brooks, S.A., Cornell University, Animal Science, 129 Morrison Hall, Ithaca, NY 14853, USA; samantha.brooks@cornell.edu

The era of genome sequencing has incalculably advanced the field of animal breeding in the past decade. The horse, though not a livestock species traditionally under intensive agricultural selection, is no exception. With the recent completion of a 6.8x full genome sequence, equine genetics now benefits from a full range of state of the art technology. Freely available sequence speeds the discovery of novel alleles and varients. Mapping efforts, powered by the 56K element EquineSNP50 chip by Illumina, can now move beyond single gene traits and in to whole genome association studies of complex disease and ETLs. Next generation transcriptome sequencing has increased the number of published horse transcripts by nearly 30-fold. Finally, these transcripts are incorporated in to a number of microarray platforms for in-depth analysis of gene expression profiles. I will provide an overview of these new tools in equine genomics and illustrate their application in my own work as well as recent exciting publications impacting horse health and production.

Inherited disorders and their management in some European warmblood sport horse breeds
Nikolić, D.[1], Jönsson, L.[1], Lindberg, L.[1], Ducro, B.[2] and Philipsson, J.[1], [1]Swedish University of Agricultural Sciences, Department of Animal Breeding and Genetics, Box 7023, S-750 Uppsala, Sweden, [2]Wageningen University, Department of Animal Breeding and Genetics, Marijkeweg 40, 6709PG Wageningen, Netherlands; danica.skulic@gmail.com

The aim of this study was to determine what strategies are employed to manage inherited disorders in European warmblood sport horses. An online survey was sent to 37 breed organisations in 29 countries, from which 9 countries replied. Regarding management of inherited disorders, the breed association was named as the primarily responsible organisation for formulating restrictions in the selection and use of horses for breeding. All countries had examinations for skeletal and joint disorders in breeding stallions but muscular disorders were examined for to a lesser extent. Reproductive and respiratory disorders in breeding stallions mostly resulted in automatic exclusion from breeding. Many of the conformational deviations were considered in the selection of breeding stallions only when severe, or could be compensated for with good performance. Four countries recorded disorders on both breeding stallions and young horses; this usually being done at official events or prior to sale. If disorders of stallions and young horses were recorded during private veterinary visits, there was little obligation to report them. Four countries recorded disorders in foals and also monitored fertility and reproductive disorders in stallions. Most countries with small populations had restrictions on the breeding of mares and were more likely to summarise and evaluate disorders of stallions and young horses. Although most countries had similar types of management strategies for breeding stallions and their progeny, there was little consensus on how specific disorders were considered. If it is not possible to create unified considerations, perhaps the answer would be to make data available to breeders who can then make knowledgeable decisions on which stallions to use.

Heritability of insect bite hypersensitivity in Dutch Friesian breeding mares
Schurink, A., Ducro, B.J. and Van Arendonk, J.A.M., Animal Breeding and Genomics Centre, Wageningen University, P.O. Box 338, 6700 AH Wageningen, Netherlands; anouk3.schurink@wur.nl

Insect bite hypersensitivity (IBH) is the most common allergic skin disease in horses, caused by bites of certain Culicoides species. IBH causes an intense itch, which results in self-inflicted trauma. Welfare of affected horses is therefore seriously reduced and some affected horses are even unsuitable for riding and showing purposes. Horse owners encounter economic losses due to possible veterinary costs and a reduced commercial value of affected horses caused by disfiguration. Currently, there is no effective treatment for or prevention against IBH available. However, we expect that selection against IBH is possible and therefore estimated the heritability of IBH in Dutch Friesian breeding mares. Mares (n=3,530) were visually scored for clinical symptoms during organized foal inspections by 8 inspectors in 2004 and 4 other inspectors in 2008. About 9% of the mares (n=313) were scored in both years, resulting in a total of 3,843 IBH-scores. Mares descended from 145 sires and 2,609 dams and average IBH prevalence was 17.8% (2004: 18.1%; 2008: 17.0%). We analyzed IBH as a binary trait with a threshold model and estimated heritability and various variance components. Preliminary results on a subset of the data using a linear model revealed a heritability between 0.08-0.12 on the observed scale and between 0.20-0.35 on the underlying scale. Because IBH, based on clinical symptoms, is a heritable trait in the Friesian horse population it is possible to reduce the number of affected Friesian horses by selection.

Animal hospital data for studies of prevalence and heritability of osteochondrosis (OC) and palmar/plantar osseous fragments (POF) of Swedish Warmblood horses (SWB)

Jönsson, L.[1], Dalin, G.[2], Egenvall, A.[3], Näsholm, A.[1], Roepstorff, L.[2] and Philipsson, J.[1], [1]Swedish University of Agricultural Sciences, Dept. of Animal Breeding and Genetics, Box 7023, S-750 07 Uppsala, Sweden, [2]Swedish University of Agricultural Sciences, Dept. of Equine Sciences, Box 7043, S-750 07 Uppsala, Sweden, [3]Swedish University of Agricultural Sciences, Dept. of Clinical Sciences, Box 7054, S-750 07 Uppsala, Sweden; Lina.Jonsson@hgen.slu.se

The objective of this study was to evaluate animal hospital data as source of information for estimation of the prevalence and heritability of osteochondrosis (OC) in stifle, hock and fetlock and palmar/plantar osseous fragments (POF) in fetlock. Data were obtained from Helsingborg animal hospital, south Sweden, of horses examined in years 1992 to 1998 in all ages. Reports from radiographic examinations of 879 'healthy' horses screened before sale and 3639 horses with clinical symptoms resulting in radiographic examinations were included in the prevalence study. For the heritability study the data was pooled and 3672 examined horses with pedigree information were included in a linear animal model analysis. The prevalence of OC at the animal level was 13% (stifle 9%, hock 6% and POF 11%). The heritability of OC at the animal level was 0.05 on the visible all-or-none scale. The corresponding heritabilities for OC in stifle were 0.02, in hock 0.06 and for POF 0.13. These values corresponded to heritabilities of 0.06-0.37 on the underlying quantitative scale. Animal hospital data is a source of health information that is not extensively used today due to risk of selected individuals and non-harmonized systems for reporting among different clinics. This study showed that disease data obtained from animal hospitals may be a valuable asset in studies of inherited disorders and possibly even for genetic evaluations. To facilitate the use of animal hospital data, there is, however, a need for improvements in documentation of pedigrees and availability of the data.

Simulating experimental designs to compare and select methods based on linkage disequilibrium and linkage analysis for studying osteochondrosis in horses

Teyssèdre, S.[1], Ytournel, F.[2], Dupuis, M.C.[3], Denoix, J.M.[3], Guérin, G.[4], Ricard, A.[1,4] and Elsen, J.M.[1], [1]INRA, SAGA, BP52627, 31326 Castanet Tolosan, France, [2]University of Goettingen, Dept. of Animal Genetics, Animal Breeding and Genetics Group, Albrecht-Thaer-Weg 3, 37075 Goettingen, Germany, [3]ENVA, CIRALE, D675, 14430 Dozule, France, [4]INRA, GABI, Domaine de Vilvert, 78350 Jouy-en-Josas, France; simon.teyssedre@toulouse.inra.fr

Low performances in racing horses can be due to several pathologic entities. Among those, osteochondrosis (OC), which is a disorder of cartilage in growth affecting foals, has a major sportive and then economical impact because 30% of sport horses have cartilage ossification disorders. To identify the genes underlying this affection, and in collaboration with CIRALE and CEMESPO, the project Genequin was created to obtain data type case/control on 600 French trotters ($h^2 \sim 0.25$). Each father of this design has at least one case and one control and we got the pedigree for each horse on 6-7 generations. Phenotypes are recorded on offspring by the study of 25 radiographic sites, and genotypes are available for each father and offspring with a chip SNP of 60K. In order to detect genes of susceptibility to osteochondrosis, various methodologies based on linkage disequilibrium (LD) and simultaneously based on linkage analysis and linkage disequilibrium (LDLA) are available. Different questions have to be addressed, e.g.: Given a total population size, what is the influence of the number of sire families on the statistical power? What methods are most effective for this type of population structure and data? To answer these questions, we used the program LDSO which was developed to generate data for QTL mapping. LDSO simulates the history of a single or two populations for T generations taking into account various evolutionary forces (mutation, selection, random drift, and bottleneck). We extended this program to fit as close as possible our experimental design.

Use of competition results for genetic evaluation of longevity in Swedish warmblood horses

Braam, Å.[1], Näsholm, A.[1], Roepstorff, L.[2] and Philipsson, J.[1], [1]Swedish University of Agricultural Sciences, Department of Animal Breeding and Genetics, P.O. Box 7023, S-750 07 Uppsala, Sweden, [2]Swedish University of Agricultural Sciences, Department of Equine Studies, P.O. Box 7046, S-750 05 Uppsala, Sweden; Jan.Philipsson@hgen.slu.se

The aim of the present study was to investigate the possibilities to use 'number of years in competition' as a measure of longevity in the genetic evaluation of the Swedish warmblood horses. Male horses not used in breeding born between 1967 and 1991 were included in the study. Competition results recorded 1971-2006 in the disciplines dressage, show jumping and eventing were used to estimate genetic parameters for number of years in competition. The study revealed that horses with placings in more than one discipline at an early age had a longer competition career. This result suggests a positive effect of all-round training of young horses on their longevity. For estimation of genetic parameters for number of years in competition different linear mixed animal models were tested. Depending on the model the heritability for number of years in competition varied between 0.07 and 0.17. The lower value was obtained when the number of competition years was adjusted for breeding values of gaits and jumping ability. High genetic correlations were also estimated between number of years in competition and competition results as young horses (4 and 5 years old). These results indicate that the trait number of years in competition not only represents the longevity of the horses but also their talents for performance. If number of years in competition is used for genetic evaluation of longevity independent of the sports talent, the model must account for this, e.g. by adjusting for age at first successful competition and/or other measures of sports talent.

Ranking in competition: an efficient tool to measure aptitude?

Ricard, A.[1] and Legarra, A.[2], [1]INRA, GABI, 78352 Jouy en Josas, France, [2]INRA, SAGA, Auzeville B.P. 52627, 31326 Castanet-Tolosan, France; anne.ricard@toulouse.inra.fr

Ranks in competition are now widely used in horse genetic evaluation. But is it a good tool to measure true ability in the discipline (jumping, dressage, eventing, races…)? To answer this question a simulation study was conducted. The model postulates a latent structure, with an underlying performance realized by each horse involved in the event. The rank observed is assumed to reflect the order of the values of the unobserved variables. Genetic parameters were estimated using self-made software based on a Bayesian approach and on the Gibbs sampler. Simulation involved 1,000 horses, with a variable number of event per horse and horses per event. The underlying model included one fixed effect and one random effect (the horse, i.e. genetic and permanent environmental effect). Two structures of competition were analysed: one where the distribution of horses among events were random and another where, as it is in reality, different levels (3) of competition were simulated. In the second structure, the higher the simulated ability of the horse (accuracy 0.70) the higher the probability to participate to the highest level. For unstructured competition, the underlying model for ranks was perfectly capable to estimate variance components and horse effects whereas other criteria as normal score or raw ranking underestimated repeatability, especially with low number of horses per event and variable number of horses per event. For structured competition, the underlying model for ranks was again the best model compared to other criteria but underestimated repeatability (0.18 versus 0.25 simulated) as it underestimated the true level of each level of competition. The multiple trait normal score model (one trait by level of competition) did not solve the problem: for each trait the differences between level were also underestimated to a larger extend and repeatability underestimated too (0.12 to 0.16). Improvement of underlying model for ranks must be made to overcome this underestimation.

A Thurstonian model for the analysis of ranks in the genetic evaluation of horses

Gómez, M.D.[1], Varona, L.[2], Molina, A.[1] and Valera, M.[3], [1]University of Cordoba, campus Rabanales, 14071 Cordoba, Spain, [2]University of Zaragoza, C,Miguelservet,177, 50013Zaragoza, Spain, [3]University of Seville, CTRA.utrerakm1, 41013Seville, Spain; lvarona@unizar.es

A thurstonian model for genetic analysis of ranks was used to analyze the performance of horses, using the Spanish Trotter Horse (STH) as an example. Traditionally, equines are being selected for the racing performance using BLUP animal models. In trotting races, the ranking place in a race is used as a measure of performance, because it allows the direct comparison between races hold in different countries and between the horses participating in the same races related to other horses participating in other events. However, the heterogeneity in the quality of competitors may produce substantial biases in genetic evaluation. The importance of competitive information (level of the race and level of one rank in this race) in the performance results supports the use of thurstonian models. The thurstonian model assumes an underlying variable associated with the horse performance that is transformed into the ranking of the animals within the competition. This underlying variable was assigned to be zero for the winner of each race, a lower value than the first to the second classified, a lower than the second for the third, and so on until the last classified. This model allows fixing the race effect, including a correction of the predicted breeding values by the quality of the animals competing in a race (level of the horses that participate in the same race). The model of analysis includes sex, age and race as systematic effects and rider-trainer, horse permanent and genetic effects. In this work, the genetic parameters and variance values were estimated for the STH. The heritability value was 0.09 for the underlying variable associated with ranking place. Adequate results were obtained. The performance for the top and bottom individuals and races are presented. The procedure avoids the effects of the heterogeneity between competition levels in the genetic evaluation.

An all-or-none trait to account for pre-selection in Icelandic horse breeding

Albertsdóttir, E.[1], Eriksson, S.[2], Sigurðsson, Á.[1] and Árnason, T.[1], [1]The Agricultural University of Iceland, The Faculty of Land and Animal Resources, Hvanneyri, 311 Borgarnes, Iceland, [2]The Swedish University of Agricultural Sciences, Department of animal breeding and genetics, Box 7023, 75007 Uppsala, Sweden; elsa@lbhi.is

Breeding values for Icelandic horses are estimated from breeding field-test scores on body conformation and riding ability traits. Less than 20% of mares born in Iceland attend breeding field-test events. There is a growing concern that the tested horses are not a random sample from the population, which might cause biased genetic evaluations. The attendance of a horse at breeding-field tests can be regarded as an all-or-none trait 'test-status'. The aim of this study was to estimate the heritability of test status and its genetic correlations with the breeding field-test traits. Breeding field-test data were collected from the international Icelandic horse database, Worldfengur, and included 39,443 mares born in Iceland between 1990 and 2001 whereof 7,431 mares were tested in the period 1994 to 2007. Variance and covariance components for breeding field-test traits and test-status were estimated with linear and threshold models using Gibbs sampler procedures. The results indicate that test-status is highly heritable (h^2=0.70) and moderately strongly genetically correlated (r_A=0.04-0.79) with most of the field-test traits. There is a tendency that pre-selection has become stricter over studied period. This underlines the importance of incorporating test-status in the genetic evaluation in order to reduce selection bias.

Splitting breeding goals and breeding programs: the example of Oldenburg

Schöpke, K. and Swalve, H.H., Institute of Agricultural and Nutritional Sciences, Group Animal Breeding, Adam-Kuckhoff-Str. 35, 06108 Halle, Germany; kati.schoepke@landw.uni-halle.de

The Oldenburg Horse (*Das Oldenburger Pferd*) has a long history and an outstanding record as a breed of Warmblood sport horses. Since 2001, two separate breeding goals were defined for two distinct sub-populations and breeding organisations, the *Verband der Züchter des Oldenburger Pferdes e.V* [OLD] and the *Springpferdezuchtverband Oldenburg-International e.V.* (O-INT), respectively. The latter organization focuses on breeding for show jumping. The data from both organisations is included in the national estimation of breeding values for Warmblood sport horses which is carried out by VIT, Verden, Germany. Based on the breeding values for mares of both organisations (OLD: 5,455 mares; O-INT: 833 mares) the effects of this specialisation were studied. A comparison of the average total index for dressage showed a considerably advantage of the OLD mares (OLD : O-INT = 102.38 : 87.55). This superiority in dressage was also proven by the maxima. The average total index for show jumping, however showed contrary results: a higher value for the O-INT (OLD : O-INT = 94.73 : 107.44), although the distribution of estimated breeding values demonstrates that there exist superior jumping horses in both populations. An analysis of the kind of performance records included in both populations revealed distinct differences with an emphasis on show jumping records for OLD-INT whereas OLD focuses on mare's performance tests. Genetic trends demonstrated the emphasis of both breeding goals with strong trends in dressage for OLD and evenly strong trends in show jumping for OLD-INT. For OLD-INT, trends in dressage were considerably lower than those for OLD. However, some stallions still are in common on top rank lists of sires of mares from both organisations. Even in the highly specialised OLD-INT program, stallions can be found whose daughters do well in both disciplines.

Heritabilities of kinematic traits measured with a 3D computerised motion analysis system in Lusitano horses at hand led trot

Santos, R.[1], Molina, A.[2], Galisteo, A.M.[2] and Valera, M.[3], [1]Escola Superior Agraria de Elvas, Av. 14 de Janeiro s/n, 7350 Elvas, Portugal, [2]Universidad de Córdoba, Campus Universitario de Rabanales, 14071 Córdoba, Spain, [3]Universidad de Sevilla, Ctra de Utrera Km1, 41013 Sevilla, Spain; rutesantos@esaelvas.pt

Dressage performance is becoming more important among selection criteria in the Lusitano breed. However, performance data are still scarce in this breed, and in other breeds have presented low heritability estimates, due to strong weight of environmental factors. In this work we studied kinematic characteristics of hand led trot using a computer assisted three-dimensional videographic system, since gait quality can be an indicator of dressage performance. Data of 21 kinematic variables were collected in 88 male Lusitano horses. Heritability estimates were obtained with a multivariate animal model, using a REML procedure. Speed had a significant effect, mainly on linear and temporal variables. Age and ability were also significant factors, indicating that training has an important effect on the locomotion pattern. The Lusitano horses in this study presented kinematic similarities to Andalusian horses, when compared to other sport horse breeds, showing a higher degree of flexion of the joints and shorter stride lengths. These results were expected, since both breeds have the same origin. In a principal component analysis, the first component, which accounted for 28.39% of the observed variance, was mainly influenced by linear variables, while the second component, which accounted for 17.02% of variance, was more influenced by angular variables. Heritability estimates were moderate to high, ranging from 0.12 ± 0.07 (for maximum height of the hindlimb hoof) to 0.88 ± 0.12 (for range of motion of the carpal joint). Even though more studies with larger samples are needed, these results seem to open good prospects for the introduction of kinematic traits in the selection of the Lusitano horse.

Effects of racing on reproductive performance in Standardbred trotters and Finnhorses

Sairanen, J.[1], Katila, T.[2] and Ojala, M.[1], [1]University of Helsinki, Dept. of Animal Science, P.O.Box 28, 00014 Helsinki, Finland, [2]University of Helsinki, Dept. of Production Animal Medicine, Paroninkuja 10, 04920 Saarentaus, Finland; Jenni.Sairanen@helsinki.fi

Racing stress may have temporary or permanent effects on equine fertility. Analysing these effects and finding the genetic correlation between fertility and racing performance were the main foci of this study. The data consisted of 33,679 Standardbred (SB) and 32,731 Finnhorse (FH) matings. The foaling outcome was 1 if a foal was born and 0 in other cases. A threshold model would suit this trait best, but the preliminary analyses were done with a linear mixed model. Fixed factors in the models were year and month of mating, mating type, stallion age group, age and type class of the mare, and inbreeding class of the expected foal. The fixed factor for racing stress of mares or stallions was based on either the number of races, best racing time, or earnings, during the mating year or the entire career. Random factors were the permanent environmental effects of the mare and stallion and additive genetic effects. Heritabilities for fertility were presented in an earlier part of the study. BLUE-estimates were calculated for estimable functions of different fixed effect classes. Genetic correlation between foaling outcome as a trait of the mare and the best time of the racing career was estimated using the REML procedure and a bivariate model: fixed and random factors for fertility were the same as above except that no fixed factor for racing was included; gender, year, age and country of birth were the fixed factors, and additive genetic effect was the random factor for racing. For mares in both breeds, it was beneficial to race during the mating year, as long as the number of races was low. SB mares with the best career times had the best foaling outcomes, but racing success of stallions had an opposite effect. Genetic correlation between foaling outcome and best racing time was favourable but low (r_g=-0.26±0.08) in the FH, and negligible (r_g=-0.06±0.12) in the SB.

Molecular characterization of Italian Heavy Draught Horse (IHDH) breed using mitochondrial DNA and microsatellite markers

Maretto, F. and Mantovani, R., University of Padua, Department of Animal Science, AGRIPOLIS, Viale Universita, 16, 35020 Legnaro (PD), Italy; fabio.maretto@unipd.it

The Italian Heavy Draught Horse (IHDH) breed counts more than 6,000 registered animals half of which are mares distributed in about 900 stud farms. The breed was established in the middle XIX century by the Italian government and originated mainly from crosses of Norfolk-Breton stallions with local derived Hackney, Percheron and Bretons mares. The breed was initially developed for agricultural and draught uses as well for artillery transport by the Italian army; nowadays it is mainly used for meat production and heavy draught works. To assess genetic diversity in the IHDH breed as a prerequisite for efficient management decisions, we analyzed 55 unrelated individuals using genotypic information from 23 microsatellite loci. Nineteen unrelated individuals of Italian Haflinger (IH) breed and 21 unrelated individuals of Quarter Horse (QH) breed were genotyped as reference populations for comparison purposes. A 410 bp mitochondrial (mt) DNA D-loop fragment was also analyzed for comparison with sequences from other European heavy draught horse breeds. For the IHDH breed the total number of alleles (N_A) was 161 ranging from 4 (locus Htg4) to 13 (locus Tky343) with a mean of 7.00±1.85. The average observed heterozygosity was 0.68±0.16 similar to the expected value of 0.72±0.12. Average gene diversity was 0.716 ranging from 0.362 (locus Htg3) to 0.855 (locus Vhl20). Wright's F-statistics in the entire sample revealed a moderate homozygote excess (F_{IT} = 10.8%) due only partially to a homozygote excess within breeds (F_{IS}=0.03). Phylogenetic analysis of mtDNA D-Loop fragments is in progress and will be illustrated. Microsatellite markers and mtDNA D-loop haplotypes data will provide a better understanding and characterization of the IHDH breed; population genetic data analysis will contribute to the conservation and implementation of selection programme of the breed and for comparison with other European draught horse breeds.

Opportunities and limits for the use of Equine SNP50 Bead chips

Langlois, B., INRA, Animal genetics, INRA-CRJ-SGQA, 78 350-Jouy-en-Josas, France; bertrand.langlois@ jouy.inra.fr

Sufficient SNP (# 750,000) were mapped to the Equ Cab 2.0 assembly to obtain a 60,000 SNP illumina array. Nearly 50,000 of these SNP were polymorphic (MAF>0). This resulted in an Equine SNP 50 Beadchip. The cost for genotyping is currently about 250 Euros per animal. The experimental design needed to answer different questions is therefore of great economical concern. In this paper we will check these problems for horse populations. Three levels will be considered: -1- The estimation of allele frequencies allowing the calculation of genetic distance between breeds or individuals (kinship with markers) -2- Detecting signatures of selection for those loci not significantly in Hardy Weinberg equilibrium. -3- The possibility to check for differences in linkage disequilibrium between chromosome regions and breeds. The conclusion is that for purpose1, N # 50-60 is sufficient. However N #100-130 is needed for purpose 2 and N # 2,000-3,000 for purpose 3. A low N would allow approaching all three levels but only for SNP with high MAF values.

Breeding values for longevity in jumping horse competition in France

Ricard, A.[1,2] and Blouin, C.[1], [1]Institut National de la Recherche Agronomique, Génétique Animale et Biologie Intégrative, Domaine de vilvert, 78352 Jouy en Josas, France, [2]Les Haras Nationaux, Direction des connaissances, 61310 Le Pin au Haras, France; anne.ricard@toulouse.inra.fr

Breeding values for longevity in jumping were calculated from results in competition from 1972 to 2008, i.e. 205,863 horses which competed 839,811 years and were issued from 11,673 sires. The trait was the difference between the last and first year in competition. The model included the fixed effects of year, month of birth, region of birth, age at first start, interaction between level of performance (preceding year), sex and year in competition and random effects of sire and maternal grand sire (all their relationships included). The analysis was performed using the survival kit with Cox proportional model adapted to discrete measurement. The major fixed effect was the influence of performance on longevity in competition with a complex pattern: the less succes in competition, the higher the chance for leaving the competition, especially for horses without earnings or with years without performances inside the career (relative risk 2.0 for females without earnings compared to geldings with average performance). But above a threshold, (approx. 0.5 standard deviation above the mean) performance has almost no effect on culling. Another interesting effect was that the younger the horse start competition, the longer it stays in competition which is against common belief: the half life was 5.5 years in competition for horse which began at 4 years old, versus 4.3 for horses which began at 6 (with the same average performances). Heritability was only 7% but the standard deviation of breeding values of stallions expressed in half life with accuracy higher than 0.75 was 0.44 year and difference of half life between the best and the worst stallion was more than 2 years. This implies possibility of selection, especially parental selection of stallions before approval.

Repeatability of free jumping parameters on young stallions performance tests of different duration
Lewczuk, D., Institute of Genetics and Animal Breeding PAS, Jastrzebiec, 05-552 Wolka Kosowska; ul.Postepu 1, Poland; d.lewczuk@ighz.pl

Repeatability of free jumping parameters in young stallions performance tests of different duration The material consisted of 1,771 jumps of 141 stallions filmed during 11-month tests, 301 jumps of 50 stallions filmed in 8-month test and 221 jumps of 43 stallions filmed in 100-day tests. The measurements of juming parameters were obtained by using a camera operating at 25 fr/sec and a manual program for video image analysis. The free jumping obstacles were constructed in the compared manner. The height of the filmed doublebarre was from 0.9 m to 1.2 m. Jumps were described by measurements of taking off and landing distances, heights of limbs lifting above obstacle, heights of elevation of the selected points of horse silhouette and position of head above obstacle. The repeatability was calculated using Proc MIXED of SAS program. The statistical model was similar for all tests using fixed effects of investigation/year/test, height of the obstacle, successive number of the jump and random effects of the horse and residual effect. The repeatabilities in the 11-month test were high for taking off, landing and bascule parameters and reached values from 0.50 to 0.59. The repeatabilities of the heights of limbs lifting in 11-month test were lower, but of the same value about 0.3 for front and hind limbs. In the test of the same duration but for horses selected as the best ones – the repeatabilities of the limbs lifting were higher for front limbs and almost the same for the hind limbs. The repeatabilities of the parameters that characterised the length of jump and the bascule of horse were higher in the test for best horses and reached 0.82 for landing and 0.69-0.77 for bascule points. The repeatabilities calculated for shorter 100-day test were lower for all parameters except landing distance.

Estimating genetic parameters for dressage performance in Belgian sport horses based on results from multiple competition levels
Peeters, K.[1], Ducro, B.J.[1] and Janssens, S.[2], [1]WUR, Animal Breeding adn Genomics Centre, P.O. Box 338, 6700 AH Wageningen, Netherlands, [2]K.U. Leuven, Livestock Genetics, Dept. Biosystems, Kasteelpark Arenberg 30, 3001 Leuven, Belgium; Katrijn.Peeters@wur.nl

The objective of this study was to estimate genetic parameters for dressage performance in Belgian sport horses. The two major associations that organize dressage competitions in Belgium, the Belgian Equestrian Federation (KBRSF-FRBSE) and the Rural Riders Association (LRV), provided 100,303 and 173,917 repeated measurements on 7,620 and 12,248 horses, respectively. Dressage performances were recorded on different levels, varying from starters' level until Grand Prix level. Level-adaptations were performed in order to rate the dressage performances by their true value; implicating that the original competition results were upgraded according to the level on which the animal competed. Heritabilities (h^2) and repeatabilities (r) for dressage performances were estimated, for original and level-adapted scores, using an animal model. In addition, influence of the inclusion of a rider-effect into the model was investigated. In general, applying level-adaptations onto scores resulted in an increase of h^2 and r estimates, indicating that level-adaptations uncover a stronger genetic predisposition for 'dressage talent'. Estimates of h^2 and r decreased when the effect of the rider was accounted for in the analysis. Confounding of the rider-effect with the genetics of the horse resulted in an overestimation of h^2 and r when the rider-effect was ignored and an underestimation of h^2 and r when the rider-effect was included. In conclusion, the suggested level-adaptations allow a proper estimation of the breeding values based on records from horses performing on different competition levels. Additionally, confounding of the rider-effect with the genetics of the horse is a problem for which no obvious solution could be found.

Genetic parameters of some linear conformation measurements assessed with a three-dimensional motion analysis system in Lusitano horses

Valera, M.[1], Santos, R.[2], Galisteo, A.M.[3] and Molina, A.[3], [1]Universidad de Sevilla, Ctra de Utrera km1, 41013 Sevilla, Spain, [2]Escola Superior Agrária de Elvas, Av 14 de Janeiro s/n, 7350 Elvas, Portugal, [3]Universidad de Córdoba, Campus Universitario de Rabanales, 14071 Córdoba, Spain; mvalera@us.es

Computer assisted methods of conformation analysis based on three-dimensional systems are both accurate and less time consuming than traditional methods of conformation assessment in horses, and should thus be considered when measuring large numbers of horses. In this study, we studied 12 segment lengths obtained using a three-dimensional motion analysis system in 88 Lusitano male horses, and then estimated the heritabilities and genetic correlations of these variables. Pearson's correlations were low to moderate, with the highest values being those of body length with croup length, and between the lengths of front and hind cannon bones, and front and hind fetlocks. As to the genetic parameters, with the exception of the length of the hind fetlock ($h^2=0.07\pm0.08$), all heritability estimates were moderate to high, ranging from 0.24 to 0.88. The genetic correlations which converged had generally moderate to high values. The combination of variables which presented the highest values, both for Pearson's correlation and genetic correlation, was body length with croup length. In spite of the reduced sample size, the results show that conformation measurements assessed by this method can be used as selection criteria in Lusitano horses.

Genetic analysis of morphological traits in two French draft horses: Ardennais and Cob Normand and in Haflinger breed

Danvy, S.[1] and Ricard, A.[1,2], [1]Les haras Nationaux, Direction des connaissances, 61310 Le Pin au haras, France, [2]INRA, GABI, 78352 Jouy-en-Josas, France; sophie.danvy@haras-nationaux.fr

Linear scoring is already used for breeding in many species. However, particular programs have only recently been developed in France for 2 French draft horses Ardennais and Cob Normand (since 5 years). Genetic analysis of such data required specific software able to analyse simultaneously large number of traits: 27 traits for Ardennais, 39 for Cob Normand. The genetic analysis was performed with REML using Wombat (K. Meyer, up to 32 traits simultaneously). Measured horses were 1,223 Ardennais horses and 607 Cob. With the pedigree, 4,410 and respectively 1,808 horses were included in the analysis. The model takes into account effects of sex, age, jury, year of the evaluation and region of birth. The optimal number of principal component was 8. The more heritable traits were skeletal development (0.50 for Ardennais and 0.48 for Cob) and straightness of trot gaits (0.30 and 0.34) and represented most of the variability of the first 2 principal components. Genetic correlations proved that standard conformation for Ardennais is a developed horse (skeletal and muscular), with large and deep chest, long thigh, short neck and active trot. For Cob Normand, the development is also important (with large joint also for legs) but the neck must be vertical rather than short and the back must be short. The head must be expressive. Genetic correlation showed some difficulties for the selection plan: the larger the horse the more it walks on 4 tracks (not suitable), the larger the joints, the worst was the consistency of the hock. Results will also be presented on Haflinger horses.

Early selection of Spanish Trotter horses: genetic correlations between race performance in young and adult horses

Gomez, M.D.[1], Menendez-Buxadera, A.[1], Valera, M.[2] and Molina, A.[1], [1]University Of Cordoba, Campus Rabanales, 14071 Cordoba, Spain, [2]University Of Seville, Ctra.Utrera km 1, 41013 Seville, Spain; pottokamdg@gmail.com

The Breeding Program of the Spanish Trotter Horse (STH) includes the genetic improvement of performance results on national and international races as main goal. Nowadays, the breeding evaluation is based on competition results, as in other European Trotters. Nevertheless, breeders require more information to maximise genetic progress of the population and early selection is very important to increase the efficiency of selection procedures. Genetic correlations for racing time between age-group 1 (2-4 years) and 2 (5-8 years) over the whole distances (1,600-2,700m) were estimated, to evaluate its predictive value for early selection. A database with 71,522 records from 4,380 horses, in races held between 1991 and 2007 in Spain, was used. The results from each age-group was considered as 2 different traits and a bi-character Random Regression Model were used to estimate the genetic (co)variance for all the trajectory of distance between age. The Hippodrome-date of racing (405 levels) and sex (3) were included as fixed effects. The animals (9,201), the jockey (1,009) and the permanent environmental effects due to repetitions of records from the same animals were included as random factors. Genetic correlations for racing time at different distances ranged between 0.17-0.99 for age-group 1 and between -0.68-0.99 for age-group 2. The higher correlations were obtained between the nearest distances and the lower ones between the more separated distances. Genetic correlations between age-group 1 and 2 at the same distance had medium-high magnitude, ranging between 0.47 (2,700m) and 0.78 (2,100). The higher values were obtained for medium distances and the lower values for the largest ones. These results are very important for early selection of the animals according to the distance they are going to participate.

Development of a scheme for the estimation of breeding values based on foal inspections, mare inspections, and performance tests of mares

Schöpke, K., Wensch-Dorendorf, M. and Swalve, H.H., Institute of Agricultural and Nutritional Sciences, Group Animal Breeding, Adam-Kuckhoff-Str. 35, 06108 Halle, Germany; kati.schoepke@landw.uni-halle.de

Since 2002 the two German horse breeding associations Pferdezuchtverband Brandenburg-Anhalt e.V. and Pferdezuchtverband Saxony-Thuringia e.V. have implemented a common breeding program for their Warmblood populations. Based on the performance data from foal and mare inspections as well as performance tests of mares, a system for the estimation of breeding values was developed. This system will be a supplement to the national estimation of breeding values which comprise data from performance tests and especially from dressage and show jumping tournaments. For our data, 26,490 animals with records, and 71,848 animals in the pedigree could be included. Variance components were estimated using REML animal models. Correlations between traits from foal and mare inspections were considerable, e.g. 0.48 to 0.85. This underlines that foal inspections indeed have their value in relation to later stages of selection although jumping abilities cannot be tested for foals. For the performance test of mares, the judges evaluate walk, trot, canter, riding ability and free jumping. For these traits we estimated heritabilities of 0.29, 0.46, 0.28, 0.18 and 0.41. In the national system as published by the Fédération Equestre Nationale (FN), breeding values are estimated for dressage and show jumping as well as for specific traits within these trait blocks. In the system presented here, also similar breeding values are estimated based on different records. Correlations between the estimated breeding values from both systems ranged from 0.69 to 0.79.

Pedigree analysis of the Lusitano horse breed

Vicente, A.[1], Carolino, N.[2,3] and Gama, L.[2,4], [1]Escola Superior Agrária Santarém, Qta Galinheiro, 2001-904 Santarém, Portugal, [2]INRB, IP, Un. Recursos Genéticos Reprodução Melhoramento Animal, Fonte Boa, 2000 Santarém, Portugal, [3]EUVG, Estrada Conraria, 3040-714 Castelo Viegas, Portugal, [4]FMV-UTL, Av Univ.Técnica, 1300-477 Lisboa, Portugal; apavicente@gmail.com

Lusitano is the most important native equine breed in Portugal, where other autochtonous breeds are also recognized, including the Garrano and Sorraia horse breeds and the Miranda Donkey. The Lusitanian registered population includes 4,000 breeding mares, of which about one-half are kept in Portugal and the remaining animals are spread throughout the world. Pedigree information on 50603 individuals born in the period of 1824 to 2007 was obtained from the national horse data base maintained by *Fundação Alter Real* and used to compute different demographic indicators. Up until 2007, the accumulated number of registered breeding animals was 15,496 (4,195 sires and 11,301 dams). The estimated generation interval was 10.4 years (11.2 in males and 9.6 in females) with an average number of known generations of 10.1. The mean inbreeding coefficient for animals born in 2006 was 9.91%, and 98.4% of them were inbred. The rate of inbreeding/year was 0.20% and effective population size was 24.5 for animals born in the period 1995-2007. Over time, the rate of inbreeding has increased, with a higher frequency in recent years of individuals in the categories of higher inbreeding levels. The genetic contribution of ancestors and founders to the reference population (represented by animals born between 2000 and 2007) was studied. Nearly 50% of the genetic pool in the reference population was contributed by 7 ancestors and 21 founders, with one sire-ancestor alone contributing with about 23%. Also, 75% and 90% of the current genetic pool was contributed by 23 and 73 ancestors. Overall, the estimated effective number of ancestor and founders was 13.8 and 37.5, respectively. These results confirm the strong influence of some families in the pedigree structure of the Lusitano breed, and suggest the need to carefully manage the genetic diversity currently existing.

Inbreeding trend in a closed nucleus of Lipizzan horses

Catillo, G., Carretta, A. and Moioli, B., CRA-PCM, Animal Genetics, via Salaria 31, 00015 Monterotondo, Italy; gennaro.catillo@entecra.it

The Italian Lipizzan stud consists of 160 horses, of which 43 mares and 8 stallions. All horses directly derive from ancestors of the Asburgic Imperial stud of Lipizza before the First World War. The nucleus possesses all the six male lines: Conversano, Neapolitano, Pluto, Favory, Maestoso and Siglavy; and eleven of the fifteen classical female families: Sardinia, Spadiglia, Argentina, Africa, Almerina, Fistula, Ivanka, Deflorata, Djebrin, Europa, Theodorosta; respecting in this way the genetic heritage of the Asburgic Empire. One of the major purposes of the stud is to maintain a pure breed, therefore respecting male and females families that were present at the Asburgic stud before 1915. In order to cope with the possible increase of inbreeding, breeding schemes are performed through mating groups, after simulation of the inbreeding coefficients of all potential newborn, then assigning to each mare the stallion based on the minor inbreeding that could be obtained, so to maximise genetic variability within the closed nucleus. Breeding animals are chosen among those that fully show the typical morphological and biometrical standards of the breed. Inbreeding coefficients were calculated for all individuals in the Herdbook archives, tracing back to 1,738, including about 965 males and 2,076 females. Inbreeding trend was estimated for the individuals born from 1945 to 2008, separately for males and females. Analysis of results indicated that percentage of homozygosity increased by 6% and, in detail, by 6.6% for stallions, and by 5.3% for the mares. These results, on one side, could be positively judged, because homozygosity might involve the fixation of desired traits; on the other side, they make evident that, in a closed nucleus, inbreeding trend should be regularly evaluated before the mating season, so to prevent an excess of inbreeding in the following generation.

Analysis of the degree of inbreeding in Holstein stallions

Pikuła, R.J. and Werkowska, W., West Pomeranian University of Technology, Department of Horse Breeding, ul. Doktora Judyma 24, 70-466 Szczecin, Poland; ryszard.pikula@biot.ar.szczecin.pl

Present-day Holstein horse is a horse with outstanding sports predispositions. It is a result of well-considered breeding work with respect to mating selection, breeding material selection and increased genetic consolidation. The present study aimed at analysing the degree of inbreeding in Holstein stallions. Research was carried out on 2.5-year old Holstein stallions which had been presented for qualification to breed in Neumunster in 1978-2006. Data were collected based on the Holsteiner Korung und Reitpferde – Auktion 1978 - 2006 Yearbooks. Based on the analysis of 927 pedigrees (for all stallions, down to the 5th generation inclusive), inbreeding coefficients (F) were calculated for each stallion (in %) using OptiMate computer software. Additionally, stallion numbers in respective years were presented in groups according to the value of inbreeding coefficient, with its every 3rd value being an interval. By means of Statistics 7.0 software package, statistical analysis of the obtained values of inbreeding coefficients (F) was carried out. The results indicate an increase in mean inbreeding coefficients in the examined time period. In 1978, F amounted to 0.55, whereas the highest value in 2003 and 2006 was 2.05 and 1.32, respectively. Statistically significant differences were found between F values from respective years. Accomplishment of the breeding assumptions resulted in a decrease in participation of non-inbred stallions from 75.9% in 1978 to 27.0% in 2004. In 2006, 17.9% of stallions was within a range F=3.1-6.0, while 5.3% of them was within a range F=6.1-9.0. Inbreeding was most frequently carried out on stallion Cor de la Bryere A.N. and Ladykiller xx. In order to keep control over a further increase of inbreeding and consequences resulting from it, breeders of the Holstein Horse Association have started to introduce into breeding the stallions of other breeds, e.g. Selle Francais, descending from well-known and valuable breeding and sport lines.

Influence of the foreign Trotter populations in the Spanish Trotter Horse assessed via pedigree analysis

Gómez, M.D.[1], Cervantes, I.[2], Molina, A.[1], Medina, C.[1] and Valera, M.[3], [1]UnivCórdoba, Campus Rabanales, 14071Córdoba, Spain, [2]UnivComplutense of Madrid, Avda Puerta del Hierro, 28040Madrid, Spain, [3]UnivSeville, CtraUtreraKm1, 41013Seville, Spain; pottokamdg@gmail.com

The Studbook of Spanish Trotter Horses (STH) was created in 1979, with 17,165 animals registered till mid year 2007. It was analysed to assess the available genetic variability and to ascertain the influence of different Trotter populations in the formation of the breed. The software ENDOG 4.5 and specific programmes were used. The analyses were made for a reference population including animals born in last generation interval (14.8). The number of complete equivalent generations was 4.9 and the average inbreeding was 1.1%. A 37.3% of the registered animals were horses imported from foreign populations. The number of founders was 2,381, 82.5% of them identified as imported horses. The influence of foreign founders was estimated. USA and France were the most represented countries of origin in the pedigree (23.3% and 18.7% of the animals have more than 75% of influence of USA and French founders, respectively), whereas in general the influence of Sweden, Italy and Spain populations was limited to values lower than 50%. The effective number of founders and ancestors was 191 and 68, respectively; the number of founder genomes equivalent was 40.1. Despite of STH' Studbook remains open to include animals from other Trotter populations, and the exchange between them is high, the available genetic variability (by the value of founder genome equivalents) has been reduced due to the use of determined reproducers (the 2 most important ancestors contributing about 14% of total genetic variability). These results will be useful to implement a selection program for STH, because of its condition of open population. Moreover, the connectedness with foreign populations is ensured and it could be measured to evaluate the usefulness of competition data across countries for an international genetic evaluation.

Genetic parameters and inbreeding effects on semen quality of Friesian horses
Ducro, B.J.[1], Boer, M.[2] and Stout, T.A.E.[3], [1]WUR, Animal Breeding and Genomics Centre, P.O. Box 338, 6700 AH Wageningen, Netherlands, [2]WUR, Animal Breeding and Genomics Centre, P.O. Box 65, 8200 AB Lelystad, Netherlands, [3]Utrecht University, Equine Sciences, Yalelaan 114, 3584 CM Utrecht, Netherlands; Bart.Ducro@wur.nl

The Friesian horse breed is a closed breed for many generations and, despite its current size, effectively it is a small population. Average semen quality is relative low, possibly due to inbreeding depression. The objective of this study was to evaluate the importance of genetic variance and inbreeding on semen quality of Friesian horses. Semen quality was investigated for 1146 stallions submitted for breeding soundness examination in September-November of the years 1987-2002. Inbreeding calculations were based on 6 generations of known pedigree. Heritabilities and genetic correlations of the semen quality traits were estimated using a multivariate animal model. Inbreeding coefficients were included as linear covariables into the genetic model. Heritabilities were moderate and varied between 0.16 (volume) and 0.27 (motility), except for percentages normal cells and abnormal acrosomes with h^2 of 0.52 and 0.60. Inbreeding had only a significant negative effect on (log)sperm cell concentration. It can be concluded that semen quality can be improved in Friesian stallions by selection, because of both moderate to high heritabilities and substantial variances of semen quality traits. Most of the genetic correlations among the traits were favorable. Inbreeding had only a significant effect on (log)sperm cell concentration.

Effect of strenous exercise on gene-expression in endurance horse PBMCs
Capomaccio, S.[1], Cappelli, K.[1], Barrey, E.[2], Felicetti, M.[1], Silvestrelli, M.[1] and Verini-Supplizi, A.[1], [1]Università degli Studi di Perugia, CSCS-DPDCV, Via S.Costanzo 4, 06126 Perugia, Italy, [2] INRA, Unité de Biologie Intégrative des Adaptations à l'Exercice, INSERM 902, Genopole Evry, France; katia.cappelli@unipg.it

Moderate physical activity may have beneficial effects on the overall health while strenuous physical effort induces an inflammation-like state; the molecular mechanisms underlying the cellular response to this phenomenon are still unclear but it is known that the immune system plays a key role leading to the hypothesis that physiopathological condition develops in athletes subjected to heavy training, are based on a derangement of cellular immune regulation. The purpose of the study is to have a snapshot of transcription in strenuous conditions, in order to identify candidate genes for such immune system derangement for planning an appropriate training schedule to obtain better performance and preserve athlete welfare. We choose peripheral blood mononuclear cells (PBMCs), to perform a wide gene expression scan using microarray technology on 10 endurance horses. Gene expression analysis was performed using time course strategy (basal, immediately after the race and 24 h after the race) and an equine microarray which included 384 equine transcripts of the mitochondrial and nuclear genome. After filtering, 110 genes were significant ($P<0.05$) for the comparison 'immediately after race VS basal' and 108 for the 24 h VS basal. Modulated genes data is presented and discussed. A biological interpretation of the whole data set was performed using two types of analysis to identify the main biological functions and the main metabolic pathways revealed by the expression profile. The categorization of the gene functions was generated through the use of Ingenuity® Pathway Analysis: the significant genes associated with biological functions were considered for the analysis. A GO (Gene Ontology) annotation was performed using Blast2GO software. qRT-PCR to validate changes in expression are ongoing.

Effect of adjustment of coancestries by pedigree depth on contributions to diversity assessed using genealogical information: a preliminary application to pure and derived Spanish horse breeds

Cervantes, I.[1], Valera, M.[2], Goyache, F.[3], Molina, A.[4] and Gutiérrez, J.P.[1], [1]University Complutense of Madrid, Avda. Puerta de Hierro s/n, 28040, Madrid, Spain, [2]University of Sevilla, Ctra Utrera km1, 41013, Sevilla, Spain, [3]SERIDA, Somió, 33203, Gijón, Spain, [4]University of Córdoba, CU Rabanales, 14071, Córdoba, Spain; icervantes@vet.ucm.es

At the genealogical level, methodologies quantifying genetic diversity in livestock populations are affected by differences in pedigree depth. This fact is due to overlapping generations and differences in genealogical recording. Therefore, genealogical information is (often) not used to quantify genetic diversity and set conservation priorities. Here we propose a simple method to adjust coancestries according to pedigree depths (number of generations occurred from the foundations of the pedigrees). This methodology allows to asses consistently within- and between-populations differences in gene diversity useful to set priorities for conservation in order to maximise overall gene diversity. The proposed method is tested on the whole pedigrees of six Spanish horse populations, three parental horse breeds (Arab-A-, Spanish Purebred -SPB, Andalusian- and Thoroughbred-TB) and three derived horse breeds: the Anglo-Arab (AA), Hispano-Arab (HA) and Spanish Sport horse (SSH).AA and HA are composite populations formed by continuous crossing between the SPB and TB and the SPB and A, respectively. The SSH is a recently formed population with large genetic contributions from the aforementioned breeds. The total available records were: 18,880 for A, 140,629 for SPB, 33,463 for TB, 8,289 for AA, 3,394 for HA and 7,099 for SSH. The complete equivalent generations ranged from 3.7 (SSH) to 9.9 (SPB). Contributions to genetic diversity of each of the horse breeds were computed.The highest losses of within-population diversity arose after removing SSH (-0.30%), the TB (-0.15%) and the other Arab-derived breeds (AA and HA) whilst gains in within-diversity arise if A (0.20%) and SPB (2.05%) were removed.

Genetic variability of Mezőhegyes horse breeds using genealogical and molecular information

Mihók, S., Posta, J., Komlósi, I. and Bodó, I., University of Debrecen, Institute of Animal Science, Böszörményi str. 138, H-4032, Debrecen, Hungary; postaj@agr.unideb.hu

The aims of the study were to analyze microsatellite variability and genetic diversity of Mezőhegyes horse breeds. Three horse breeds were founded in the Mezőhegyes National Stud, namely the Nonius, Gidran and Furioso-North Star. Blood samples were collected from Nonius, Gidran and Furioso-North Star breeds, 136, 138 and 58, respectively. Additionally, pedigree information of horses with blood samples were collected from the studbooks and were analyzed. Microsatellite data (twelve markers) were analyzed using Molkin, pedigree analysis was done using EnDog. Microsatellite markers were compared based on their PIC value. The most informative marker was 'ASB2' marker (15 alleles) with 0.818 PIC value, less informative was 'HTG4' marker (8 alleles) with 0.428 PIC value. Genetic distance based on molecular information was highest (Nei's minimum distance=0.0449; Reynold's distance=0.0308) between Gidran and Nonius, and was the lowest (Nei's minimum distance =0.0396; Reynold's distance=0.0225) between Gidran and Furioso-North Star. Each distance based on molecular data was significant. The distances among breeds based on pedigree information gave the same results as presented for molecular information. The close relationship between Gidran and Furioso might be the result of using Thoroughbred stallions during the history of the breed.

Genetic variability of Hutzul horse breed families

Mihók, S.[1], Posta, J.[1], Kusza, S.Z.[1] and Priskin, K.[2], [1]University of Debrecen, Institute of Animal Science, Böszörményi str. 138, H-4032, Debrecen, Hungary, [2]Biological Research Center of the Hungarian Academy of Sciences, Temesvári krt. 62., H-6701, Szeged, Hungary; postaj@agr.unideb.hu

The aims of the study were to analyze genetic diversity and microsatellite variability among six Hutzul horse breed families ('1 Panca', '12 Sarata', '4 Kitca', '5 Plosca', 'Árvácska' and 'Aspiráns'). Blood samples were collected from 63 mares. Additionally, pedigree information of these horses were collected from the studbook and were analyzed. Microsatellite data (twelve markers) were analyzed using Molkin and pedigree analysis was done using EnDog. Furthermore, ten of these mares (from five families) were compared based on mitochondrial DNA data. Microsatellite markers were compared based on their PIC value. The more informative marker was 'ASB2' marker (10 alleles) with 0.868 PIC value, less informative was 'HTG7' marker (4 alleles) with 0.357 PIC value.Genetic distance based on molecular information was highest between '5 Plosca' and '4 Kitca' (Nei's minimum distance=0.361; Reynold's distance=0.4) families, and was the lowest between 'Árvácska' and 'Aspiráns' (Nei's minimum distance=0.0248; Reynold's distance=0.019) families. The genetic distance based on pedigree information showed the following values: Nei's genetic distance and average distance were the highest (0.351 and 0.503) between '5 Plosca' and '4 Kitca'. Nei's genetic distance was the lowest (0.0404) between 'Árvácska' and 'Aspiráns' families, average distance was the lowest (0.421) between '1 Panca' and '5 Plosca' families, respectively. The DNA results sorted the ten mares into four separate groups which do not followed the family data.

Measure of association between assessed traits and final rank for a sport horses performance evaluation system within the UK

Whitaker, T.C., Clausen, A. and Mills, A.A., Writtle College, Centre for Equine and Animal Science, Chelmsford, Essex, CM1 3RR, United Kingdom; amy.mills@writtle.ac.uk

The evaluation of potential young sport horses has been routinely conducted via performance evaluations within many European breeding systems. Within the UK historically such an approach has had limited attendance and engagement. In 2005 the British Equestrian Federation established a national scheme for horses up to the age of four years old – The Young Horse Futurity. Evaluation of a variety of traits relating to potential suitability for sport horse competition is performed. This study undertook an evaluation between assessed traits and final rank within competition for all horse assessed at one year of age. Measures of association were established using spearman's ranks correlations. The population consisted of a total of 244 animals assessed over a four year period (2005 n=23, 2006 n=36, 2007 n=62, 2008 n=123). Changes and developments to assessed traits have been made through the four year period; this study only considered traits that were used consistently throughout the four year period. The study indicated moderate to high correlations between assessed traits and final rank in competition; conformation r_s=0.720 ($P<0.001$), correctness of passes r_s=0.584 ($P<0.001$), athleticism r_s=0.596 ($P<0.001$). In 2007 veterinary examination was incorporated within the assessment procedure, this was additionally considered. For 2007 and 2008 a veterinary examination relating to general soundness and health (n=185) returned r_s=0.422 ($P<0.001$). In 2008 an additional veterinary score was introduced relating to potential (n=123) r_s=0.537 ($P<0.001$). This study indicates some degree of differentiation between assessed traits and final rank at the assessment. Further work needs to be undertaken relating to differences and similarities between other age groups within the assessment process. Additionally the subsequent competition performance of the assessed animals needs to be captured.

Variation of linear described type traits in Czech-Moravian Belgian horse and Silesian Noriker

Majzlik, I., Vostrý, L., Čapková, Z. and Hofmanová, B., CULS, Prague 6, 165 21, Czech Republic; majzlik@ af.czu.cz

The analysis of 22 linear described traits of body conformation was carried out in two breeds of draught horses, namely Czech-Moravian Belgian horse (CMB – 580 heads) and Silesian Noriker (SN – 282 heads) using linear model with fixed effects (sex, year of birth and blood line). Both breeds are included in gene ressource.Statistical significant influence of sex was determined for 4 traits at CMB and 5 traits at SN only. Neither of the two effects (sex and age during description) reached statistical significance at prevailing number of traits of either group. Therefore neither sex nor age effects is significant for the breeding and CMB or SN selection on the linear description basis – no correction is necessery for these effects. Conversely, within 14 CMB traits and 16 SN traits high statistical significance has been reached at the year of birth effect. The year of the birth has proved high statistical significance for body conformation development of both breeds. A rather higher intrapopulation phenotypic variation occured at sires and lines – 46% of parameters reached the CV value higher than 15%.

Occurrence and origin of flea-bitten pattern in grey Purebred Arabian Horses

Stachurska, A.[1], Pieta, M.[2], Brodacki, A.[3] and Tomczyk, P.[1], [1]University of Life Sciences in Lublin, Department of Horse Breeding and Use, Akademicka 13, 20-950 Lublin, Poland, [2]University of Life Sciences in Lublin, Department of Sheep and Goat Breeding, Akademicka 13, 20-950 Lublin, Poland, [3]University of Life Sciences in Lublin, Department of Biological Foundations of Animal Production, Akademicka 13, 20-950 Lublin, Poland; anna.stachurska@up.lublin.pl

The data concerned 2,136 descriptions of 1,140 grey Purebred Arabian horses. The occurrence, intensity and colour of flea-bitten pattern were recorded. The intensity of the pattern was classified on a scale of 1-6. Gender, age and kind of main colour were taken into account. The impact of these factors and significance of differences between groups regarding the intensity of the pattern were determined with analysis of variance. The possibility of simple inheritance of the property was analyzed considering phenotypes in parents and mating results, i.e. phenotypes in the progeny, minimum 6 years old. In order to examine if a polygene mechanism can be responsible to some extent for the pattern, the heritability (h^2) was estimated (29 sire groups, 8.03 half-sib, on average). The results show the flea-bitten pattern occurs in 24.0% of grey Arabians: 9.7% of stallions and 31.9% of mares. It is strongly related to age: it appears for example in 0.5% of yearlings and 30.7% of 4-6-year-olds, whereas over 80% of horses more than 10 years old are flea-bitten. The colour of the spots is reddish brown in most horses (96.7%). The pattern is more intensive in mares (LSM=1.94±0.04) than in stallions (LSM=1.74±0.04). Usually, there are single spots solely on the head in young horses and the flea-bits spread over the whole body in successive age groups up to 13-17 years. The difference in the intensity between horses of various main colours is insignificant. Minimum two loci may control the appearance of the flea-bitten pattern. The heritability of the property amounts to 0.43.

The sustainability of organic dairy production in the U.S.
Rotz, C.A., USDA-Agricultural Research Service, 3702 Curtin Road, University Park, PA 16802, USA; al.rotz@ars.usda.gov

Both organic and conventional dairy farming practices in the U.S. have unique challenges for maintaining sustainable production. Economic sustainability may be most important because milk will not be produced in our economy unless there is profit for the producer. The demand for organic milk has created a more stable price that is normally substantially greater than the commodity price of conventional milk. As long as this price difference is maintained, organic dairy production offers an economically viable option, particularly for smaller farms. A readily available and relatively inexpensive supply of organic fertilizer is a challenge for organic producers. In our region, poultry manure provides an organic source of crop nutrients. When applied to meet nitrogen requirements, phosphorus requirements are normally exceeded. Over time, this can lead to high soil phosphorus levels and greater phosphorus runoff in surface water. When well managed perennial grassland is used as the primary feed in organic production, soil erosion and nutrient runoff are small. Organic farms that use row crops for feed production rely heavily on the use of tillage for weed control. With greater tillage, sediment and nutrient losses can be much greater than that found with conventional no-till practices and herbicide use. Organic practices may also increase the net greenhouse gas emission from farms. This again is linked to the use of poultry manure or other organic nutrient source for plant nutrition. The additional carbon brought onto the farm builds soil organic matter, which over years leads to greater soil respiration and loss of carbon. This additional emission, along with a lower milk production per cow, can increase the carbon footprint (net emission per unit of milk produced) of organic systems. The sustainability of both organic and conventional dairy production systems varies considerably dependent upon the management practices used. Organic systems have the greater challenge since available resources and management options are more limited.

Effects of increasing the farm produced content in organic feeds on pig performances
Royer, E.[1], Cazaux, J.G.[2], Maupertuis, F.[3], Crepon, K.[4] and Albar, J.[1], [1]Ifip-Institut du Porc, 34 bd de la Gare, 31500 Toulouse, France, [2]Adæso, Station Pau, 64121 Montardon, France, [3]Chambre d'Agriculture des Pays de Loire, BP 70510, 49105 Angers, France, [4]UNIP, 12 av George V, 75008 Paris, France; eric. royer@ifip.asso.fr

In three experimental facilities (Exp.1, 2 and 3), two organic diets, one complex including processed feedstuffs as wheat bran and heat-treated soya beans (control), the other simplified and containing over 80% of cereal and pulses (CP), were compared for growing-finishing pigs. The base components of the CP diets were moist maize grain, wheat and faba beans in Exp.1, triticale plus coloured-flowered peas in Exp.2, and triticale, oats, white-flowered peas and faba beans in Exp.3. The control and CP diets were formulated with 0.75 g ileal digestible lysine per MJ NE. Diets were given from 35 to 115 kg according to a restricted feeding plan in Exp.1 and Exp.3, more liberally in Exp.2. Respectively 96, 100 and 80 pigs were used in Exp.1, 2 and 3 and were blocked in straw bedded pens of 4, 25 and 40 pigs. In Exp.1, pigs receiving the control diet had a lower average feed intake than those offered the CP diet ($P<0.01$), whereas feed intake was similar between treatments for Exp.1 and Exp2. Daily weight gains (g/d) for pigs offered the control and the CP diets were respectively of 673 and 669 in Exp.1 ($P>0.05$), 747 and 719 in Exp.2 ($P=0.05$) and 684 and 677 in Exp.3 ($P>0.05$). The feed conversion rate (g/g) was high and reached respectively 3.28 and 3.41 in Exp.1 ($P=0.01$), 3.13 and 3.30 in Exp.2, 3.30 and 3.35 in Exp.3. The lean meat rate did not differ significantly in Exp.1, 2 and 3, for pigs given control and CP diets. The study underlines that with a moderate growth objective, an organic feed with a low energetic and protein concentration can yield a satisfying lean meat rate. In spite of a tendency for lower performances, especially concerning the feed conversion rate, a simplified diet based on cereal and pulses can be used, taking into account its economical interest for organic pig production.

Breeding replacement gilts for organic pig farms

Leenhouwers, J.I.[1], Ten Napel, J.[2], Hanenberg, E.H.A.T.[1] and Merks, J.W.M.[1], [1]Institute for Pig Genetics, P.O. Box 43, 6640AA Beuningen, Netherlands, [2]Animal Breeding and Genomics Centre, P.O. Box 65, 8200AB Lelystad, Netherlands; jascha.leenhouwers@ipg.nl

The objective of this study was to evaluate breeding structures and genetic lines for the organic pig industry. Reproduction results from over 2,000 crossbred sows and their slaughter pigs from 2006/2007 were collected on 15 Dutch organic pig farms and compared with results of conventional farms. Farrowing rate and preweaning survival were 6.4% and 9.3% lower, but litter size was 0.5 piglet/litter higher on organic farms. Organic farms with rotational crosses had higher preweaning survival (81.5% vs 76.8%) and number of piglets weaned/litter (10.3 vs 10.2) than organic farms with F1 sows. These results were used to estimate economic values for reproduction traits for several purebred Dutch sow lines and their crosses, as well as expected genetic progress within three potential breeding structures for the organic sector: a two-line rotation breeding system (RotBS), organic breeding farms producing F1 gilts (OrgBS), and a flower breeding system (FlowerBS). In FlowerBS, a dedicated organic sow line is developed based on data of a large number of organic farms with purebred sows for both replacement breeding and production of slaughter pigs. The OrgBS with Yorkshire line and Yorkshire/synthetic cross had the highest margin per sow place (€779), followed by RotBS with Yorkshire/Landrace cross (€706) and FlowerBS with Yorkshire line (€677). Considering a population of 5000 sows, FlowerBS gave the highest genetic progress in terms of cost price reduction (€3.72/slaughter pig/generation), followed by RotBS and OrgBS (€3.60/slaughter pig/generation). For FlowerBS, additional costs will be involved for maintaining a dedicated breeding program. In conclusion, OrgBS using conventional genetics is economically the most viable option for the organic pig sector. FlowerBS using a dedicated organic line may only be cost effective if sow population size is sufficiently large.

Self-sufficiency with vitamins and minerals on organic dairy farms

Mogensen, L.[1], Kristensen, T.[1], Søegaard, K.[1] and Jensen, S.K.[2], [1]Aarhus University, Department of Agroecology and Environment, P.O. Box 50, DK-8830 Tjele, Denmark, [2]Aarhus University, Department of Animal Welfare and Nutrition, P.O. Box 50, DK-8830 Tjele, Denmark; Lisbeth.Mogensen@agrsci.dk

Self-sufficiency of nutrients is a central element in the organic farming principles. In a project involving five private organic dairy farms, we aimed to achieve self-sufficiency in vitamins and minerals at farm level. All the herds are fed 100% organically grown feed, but so far supplements of minerals and vitamins based on inorganic and synthetic products are imported to all farms. The same level and type of supplement was used for the cows all year round, even though all cows were on grass for at least 150 days during the summer period. The average daily intake from the supplement for a lactating cow was 751 mg E vitamin, 111 mg Cu, and 558 mg Zn. The content of vitamin and minerals in the home-grown feeds was modelled taking into account the effect of choice of crops; conservation method; season, plant development and climate conditions at harvest; quality of the silage production, and duration of storage. The modelled contents of vitamins in the main ingredients in the feed ration were verified by measuring the actual vitamin content in the silage at harvest as well as losses during storage. As an example, at one of the farms, where the feed intake was based on 85% grass clover crops during the summer but only 68% during the winter, the home-grown feed could supply the cows with enough vitamin E according to the requirement (800 mg/day) during the summer feeding but not during the winter period. The Cu requirement (10 mg/kg DM) could not be met from home-grown feed during any season. However, supplements of vitamins and minerals secure that requirement was met. The final outcome of the project will result in strategies for achieving self-sufficiency in vitamins and minerals at individual farms through optimization of the choice of forage crops and management of feed production.

Energy consumption, greenhouse gas emissions and economic performances assessments in suckler cattle farms. Impacts of the conversion to Organic Farming

Veysset, P. and Bébin, D., INRA, UR1213, URH, Clermont-Theix, F-63122 St Genes Champanelle, France; veysset@clermont.inra.fr

The conversion to OF of four specialized suckler cattle farming systems were simulated by coupling an economic optimization model ('Opt'INRA') with a model assessing non-renewable energy (NRE) consumption ENR and greenhouse gas GHG emissions ('PLANETE'). According to the average prices observed in 2004-2007, and after adaptation of the production system, we will analyze the productive, environmental and economic impacts of this conversion. Farm structure (size and labour force) was considered a constant. The ban on chemical fertilisers entails a drop in the farm area productivity. For these specialized farms, meat production (total and per hectare) decreased by 15 to 35% depending on the initial level of intensification. The reduced use of inputs results in a 22 to 38% drop of NRE consumption per hectare, and a 7 to 16% drop per tonne of live weight produced. Because of its methane production, the cow is the biggest driver of GHG emissions at the suckler cattle farm scale, the shift to OF do not affect significantly GHG emissions per tonne of meat produced, but, the lower productivity per hectare (less animals reared per ha) allows a 11 to 26% reduction of GHG emissions per hectare of farm area. Economically, the drop in the productivity is not compensated by the gain on the meat selling price (+5 to +10%), the gross farm product falls 8 to 19% and the reduced use of inputs entails a strong drop in operational costs: -17 to -37%. The farm income falls more than 20% (-7 to -35%).

Milk performance, energy efficiency and greenhouse gases of dairy farms: case of Reunion Island

Vigne, M.[1], Bochu, J.L.[2] and Lecomte, P.[1], [1]CIRAD, Pôle KAPPA, 7 Chemin du IRAT, 97410 Saint Pierre, Reunion, [2]SOLAGRO, 75 Voie du TOEC, 31076 Toulouse, France, Metropolitan; mathieu.vigne@cirad.fr

The aim of this study was to assess non-renewable energy use and greenhouse gases emissions (GHG) of dairy farms from Reunion Island, a french territory (2,500 km²) situated in the Indian Ocean. The energy balances on 14 dairy farms for 2000 and 2007 have been established according to PLANETE methodology adapted to local context. Based on Life Cycle Analysis, PLANETE is a tool to quantify at the farm scale the level of energy consumptions and GHG emissions. Direct and indirect energy use are estimated according to standards all along the cycle and translated in fuel equivalent litre (EQF). GHG emissions (CO_2, CH_4 and N_2O) are estimated and summarized as CO2 equivalents (teqCO2) based on their global warming potential. From 2000 to 2007, the energy efficiency (input-output ratio) increased from 0.38 in 2000 to 0.44 in 2007 (+ 11%), energy consumption per 100 liters of milk decreased of 6.0 EQF (- 21.4%) and emissions of GHG decreased from 2.32 to 1.73 teqCO2 by 1000 liters of milk produced (- 25%). The proportion of energy consumption and GHG emission to tranport inputs from France to Reunion Island increased of 24% to reach respectively 21% and 26% of the total consumption and emission at the farm scale. But imported feeds represent still the most important post in the total energy consumption (57.3 and 60%). As a result, energy use and GHG emissions have decreased, improving the sustainability of dairy farms. However, to increase milk production with limited land ressources (from 4,900 to 7,100 liters per cow per year), farmers import more and more concentrates, principally from metropolitan France, which maintain the energy cost of the insularity and the impact of dairy activity on the environment at a high level. This underline the interest to improve quality local ressources and to shift to closer Indian Ocean regional markets for inputs, which would reduce impact of transport on the environment.

Effects of space allowance on gas emissions from group-housed gestating sows

Philippe, F.X.[1], Cabaraux, J.F.[1], Canart, B.[1], Laitat, M.[1], Vandenheede, M.[1], Wavreille, J.[2], Bartiaux-Thill, N.[2] and Nicks, B.[1], [1]University of Liège, Faculty of Veterinary Medicine, Boulevard de Colonster, 20, Bât. B43, B-4000 Liège, Belgium, [2]Walloon Centre of Agronomic Research, Department of Animal Productions and Nutrition, Rue de Liroux, 8, B-5030 Gembloux, Belgium; fxphilippe@ulg.ac.be

This study aims to compare ammonia (NH_3), nitrous oxide (N_2O), methane (CH_4) and carbon dioxide (CO_2) emissions from group-housed gestating sows allowed either 2.5 m^2 or 3 m^2 sow-1. Four successive batches of 10 gestating sows were divided into 2 homogeneous groups housed in 2 separated rooms equipped with a straw-bedded pens which surface was 12.6 m^2 (2.5 m^2 sow^{-1}) and 15 m^2 (3 m^2 sow^{-1}), respectively. The gas emissions were measured by infra red photoacoustic detection during 3 periods of 6 consecutive days. The ventilation rates were measured continuously and the hourly means were recorded with an Exavent apparatus (Fancom®). The differences between groups were tested by a mixed model for repeated measurements with 2 criteria (proc MIXED, SAS): space allowance (1 df), period (2 df) and interaction, with 144 data per period (24 h × 6 d). The daily gas emissions per sow were 6.5 and 7.5 g NH_3, 3.8 and 2.7 g N_2O, 15.2 and 10.1 g CH_4 and 2.4 and 2.1 kg CO_2, for sows allowed 2.5 and 3 m^2 respectively. All the differences were significant. Therefore, increased space allowance favours NH3-emissions probably by increasing emitting surface, but decreases emissions of greenhouse gases (N_2O, CH_4 and CO_2) probably by reducing litter anaerobic conditions required for their synthesis.

Estimation of methane emission from commercial dairy herds

Kristensen, T., University of Aarhus, Department of Agroecology and Environment, P.O. Box 50, DK-8830 Tjele, Denmark; Troels.Kristensen@agrsci.dk

Emission of methane (CH4) from the digestion process in the dairy cow is an important contribution to the total green house gas (GHG) emission from a dairy farm. Methane emission was estimated by five models from the literature based on information about dry matter intake (DMI) (total and divided in roughage and concentrates) or milk production (kg energy corrected milk (ECM)). The dataset consisted of 123 commercial farms with annual production of 7,906±742 kg ECM and DMI of 6,534±694 kg of which 44±11% was concentrate. The annual methane emission showed a large variation between the five models from 105 kg CH_4 to 165 kg CH_4 pr cow as average of the 123 farms, and the variation (SD) between farms ranged from 7-14 kg CH_4. In the following the methane emission was calculated as an average of the five model estimates within each farm, given an emission of 136±10 kg CH_4. The variation between farms was analyses by GLM. There was no effect of productions system (conventional vs. organic), but a significant effect ($P<0.001$) of breed (Holstein Frisian=141 Jersey=119 kg CH_4). If methane emission was calculated as g CH_4 pr kg ECM there was a significant effect ($P<0.001$) of both production system (conventional=16.4 organic=17.6 g CH4 pr kg ECM) and breed (Holstein Frisian=17.8 Jersey=16.2 g CH4 pr kg ECM). The variation in emission pr kg ECM was further analyses in a model (r^2=0.47) with average daily milk yield ($P<0.001$), variation in daily yield between cows within the herd ($P=0.05$), milk yield in 1. lactation relative to average of the herd ($P=0.004$). The estimates for higher daily yield and higher relative yield in 1. lactation were negative indicating a lower emission pr kg ECM. The estimates for variation were positive which indicate that a large variation between the cows will increase the emissions. These effects can be explained by the feed conversion rate, as the milk yield pr kg DMI explained 86% of the variation in the predicted methane emission pr kg ECM. Methane emission can be reduced 8% if the feed conversion is increased 10%.

Fatty acid composition of organic and conventional milk from UK farms

Stergiadis, S., Seal, C.J., Leifert, C., Eyre, M.D. and Butler, G., Newcastle University, School of Agriculture, Food and Rural Development, Nafferton Ecological Farming Group, Nafferton Farm, Stocksfield, Northumberland, NE43 7XD, United Kingdom; sokratis.stergiadis@ncl.ac.uk

This study investigated differences in fatty acid composition between organic and conventional milk from UK farms between March and October 2007. Milk samples from the bulk tank and management details were collected every eight weeks, from 10 organic and 10 conventional farms. Fatty acid (FA) methyl esters were separated and quantified by gas chromatography using a Varian CP-SIL 88 fused silica capillary column (100m x 0.25mmID x 0.2μm film thickness). Analysis of variance using a linear mixed effects model was used to compare results using R statistical software with 'location' (Wales or North East), 'management system' (conventional or organic) and 'month' (March, May, July, August, October) as fixed effects and 'farm' as a random effect. There was no significant effect ($P>0.05$) of management on oleic acid and linoleic acid concentration in milk, although α-linolenic acid and trans-11 vaccenic acid concentrations were significantly higher ($P<0.001$) in organic compared with conventional milk (+66.01% and +69.50%, respectively). Cis-9, trans-11 conjugated linoleic acid concentrations were also significantly higher in organic than in conventional milk (+43.58%; $P<0.01$). Concentrations of total ω-3 FA were significantly ($P<0.001$) higher by +60.47% in organic milk samples compared with conventional milk samples but ω-6 FA amounts were not significantly affected ($P>0.05$) by management. The ratio of ω-3:ω-6 was increased by +51.79% ($P<0.01$) in organic milk compared with conventional milk. Saturated FA and monounsaturated FA concentrations did not significantly differ ($P>0.05$) between organic and conventional milk although polyunsaturated FA concentrations were significantly higher (+21.45%; $P<0.01$) in organic milk. In conclusion, organic milk is likely to contain higher concentrations of certain beneficial FA than milk from herds under conventional management.

Farming factors affecting food safety and quality in NW Spain

Blanco-Penedo, I.[1], López-Alonso, M.[1], Shore, R.[2], Miranda, M.[3], Castillo, C.[1], Hernández, J.[1], Pereira, V.[1], García-Vaquero, M.[1] and Benedito, J.L.[1], [1]University of Santiago de Compostela, Animal Pathology, Estrada da Granxa s/n, 27002, Spain, [2]Lancaster Environment Centre, Centre for Ecology & Hydrology, Library Avenue, Bailrigg, Lancaster, LA1 4AP, United Kingdom, [3]University of Santiago de Compostela, Clinic Veterinary Science, Estrada da Granxa s/n, 27002, Spain; victor.pereira@usc.es

In beef farming, it is known that, to improve meat quality, it is necessary to examine the whole production chain from breeding to meat processing. Farm processes and resultant food product quality are linked through (amongst other things) the health of the animal and its disease status. Furthermore, improvement of husbandry conditions is expected to improve animal welfare and food product quality. The objective of this study was to analyse how beef-cattle farming in NW Spain on organic farms compares with intensive and conventional systems in terms of impacts on the safety and quality of cattle products. Data on the hygiene and quality of 244, 2,596 and 3,021 carcasses of calves from organic, intensive and conventional farms, respectively, were collected at the slaughterhouse. Organic calves generally had fewer condemnations for liver, kidney and heart pathologies. Liver parasitic infections were 2 fold higher in organic calves than those from other types of farm. Farm processes and resultant food product quality are linked through the health of the animal and its disease status. Overall better health status was not reflected by carcass performance as this was significantly lower for organic calves than for calves from conventional and intensive farms. Carcass performance seemed to be more determined by breed and dietary component than by health status of animals in our study.

The effects of sustainable dairy farming system on milk composition and quality

Jovanovic, S.J.[1], Popovic Vranjes, A.[2], Vegara, M.[3] and Savic, M.[1], [1]Faculty of veterinary medicine, Animal breeding and genetics, Bul. Oslobodjenja 18., 11000 Beograd, Serbia, [2]Faculty of Agriculture, Trg D. Obradovica 8., 21000 Novi Sad, Serbia, [3]Norwegian University of life Sciences (UMB), Department of International Environment and Development Studies, NORAGRIC, P.O. Box 5003, N-1432 Aas, Norway; msavic02@gmail.com

During the period of transition from conventional breeding to organic dairy production, the effects of sustainable farming practices, herd health management, nutrition and housing on milk production, milk composition and quality were studied. The research was done in five farms raising dual purpose Simmental breed, the most important well adapted cattle breed in Serbia. The basis of summer feeding of animals was grazing of pastures or meadows. In the winter, the animals were fed with hay, partly concentrates. The research was done in indoor season. These studies focus mainly on milk yield, milk constituents (fat and fatty acids, protein, casein, lactose), suitability for processing and on food safety (contents of pesticides, heavy metals and radionucleids). The average milk yield was 4,800 kg. The mean content (%) of total solids, fat, protein, casein and lactose in milk was 12.75, 4.08, 3.31, 2.83 and 4.64, in the sustainable farming system, respectively. The milk fat composition was examined. Somatic cell count was 287,000 cells/ml. The pH value was 6.52. A short coagulation time was established. Aflatoxins (AFM1) contamination, residues of antibiotics, organochlorine pesticides, heavy metals and radionuclides were not found. Results of the present study show some favourably milk traits of milk obtained in sustainable system compared with the results of conventional farming systems. Further research is necessary to explain opservation and to show its importance.

Diagnosis of the organic production system in Catalonia: from farm to fork

Panella, N.[1], Gispert, M.[1], Fernàndez, X.[1], López, F.[2], Bartolomé, J.[2] and Fàbrega, E.[1], [1]IRTA, Spain, [2]UAB, Spain; nuria.panella@irta.cat

The organic meat production is an emergent option in Catalonia but a reality in other European countries. An accurate knowledge of the organic meat production and distribution is of importance to establish the most adequate strategy to encourage this alternative production system. The main goal of this study was to perform a diagnosis of the organic meat industry in Catalonia, including all types of species (cattle, sheep, goat, pork and poultry), from the farm to the retailers. A sample of 69 farmers and 29 companies (slaughterhouses, cutting plants and meat factories) were surveyed. All pigs and poultry reared under organic conditions, from 38.8 to 54.9% of beef and 30% of lambs kept the organic label until the abattoir, whereas meat from organic horses is not sold as such in Catalonia. All the surveyed abattoirs slaughtered more than one species of ruminants (from 1.5% in beef to 2.9% in goats). More than a 95% of the distribution of the ruminant meat remains in Catalonia. Conversely, the organic pork providers purchase this meat outside of Catalonia (about 64.17% of the cases). In general, only 57.62% of the meat reaching the slaughterhouse as organic ends up in the consumer's hands. Basically, these losses occurred in the slaughterhouses because few of them have the certification and the demands are not enough. Only 35% of the surveyed companies value positively the subsidies to the sector. The study also illustrated that the main challenges are the improvement of the marketing and the organization of commercialization channels to reduce the prices. Additionally, 58% of the interviewed would support the tourist industry to sell more, and/or new products. From the present data, it was concluded that in order to better develop this sector, taking profit of the subsidies should be encouraged, the introduction of organic products to new markets should be more promoted, and a sound analysis of the reasons of the loss of the organic label from farm to fork should be carried out.

The performance and meat quality of Bonsmara steers raised in a feedlot or on organic pastures

Groenewald, I.B., Esterhuizen, J., Hugo, A. and Strydom, P.E., University of the Free State, Centre for Sustainable Agriculture and Rural Development, P O Box 29555, DANHOF, 9310, South Africa; groenei. sci@ufs.ac.za

The effects of production system on growth performance, yield and economics and the effects of feeding regime, pre-slaughter treatment and electrical stimulation on meat quality were evaluated. Sixty steers were divided into three treatment groups, viz. feedlot, organic pasture and conventional pasture feeding. The feedlot and conventional pastures treatments recieved the same finishing diet whilst the organic treatment received a diet with approved organic components. Initial weight, final live weight, warm carcass weight, cold carcass weight, warm and cold dressing percentage, average daily gain (ADG), pH at one and 24 h post mortem, intramuscular fat content of the lion and subcutaneous back fat thickness were measured. Feedlot cattle had significantly ($P<0.0001$) higher final weights, warm and cold carcass weights and derssing percentages, ADG, intramuscular fat content and back fat thickess measurements than organic and conventional pasture treatment groups. Pre-slaughter resting for a week prior to slaughter at the abattior had no effect on meat tenderness but electric stimulation showed a signicant ($P<0.0001$) positive response. Growth and carcass results were used to calculate price and feed margin for the different production systems. Feedlot cattle showed a higher profit than conventional and organic groups, mainly due to the faster and more efficient growth. The organic group showed higher profit than the conventional group as a result of the premium being paid for organic produced meat.

Comparison of mycotoxins residues in animal products produced organically and conventionally

Denli, M., Dicle University, Agricultural Faculty, Department of Animal Science, 21280, Diyarbakir, Turkey; muzaffer.denli@gmail.com

Mycotoxins are secondary toxic metabolites which are produced by several genera of *Aspergillus*, *Fusarium* and *Penicillium* occurring in foods and feeds. The contamination of mycotoxins in foodstuffs and feedstuffs is a significant problem in worldwide. Diseases caused by exposure to foods or feeds contaminated with mycotoxins are known as mycotoxicosis. In animals, toxicity is generally revealed as decrease performance. In addition, they have variety biological effects in animals such as liver, kidney toxicity and estrogenic properties. Mycotoxins and their derivates may detect in edible animal products such as meat, milk, eggs. The rate of their transfer from feeds to animal products is quite variable. The relationship between the amount of mycotoxins ingested and quantity of their metabolites in products is depending the metabolize pathways, the chemical structure of mycotoxins and the animal species exposed. Their presence in animal products has a potential health risk in human because of their synergetic properties and diversity toxic effects. Mycotoxins have been reported in variety of feedstuffs produced organically and conventionally. In addition, their residues in edible animal products produced organically are considerable. Therefore, there is a profound discussion for organic and conventional products with regarding to mycotoxin contents. This paper considers comparing mycotoxins residues in animal products produced organically and conventionally and also possible strategies promising with regard to reducing their presence in animal products for both production systems.

Viability of organic dairy farms located in northwest of Spain in different subsidy reduction scenarios
García, A., Perea, J., Acero, R. and Gómez, G., University of Cordoba, Animal Production, Edificio Produccion Animal, 14071, Spain; pa1gamaa@uco.es

Spanish organic dairy farms will be affected in the short term by lower subsidies. It is essential to ensure their future to analyse the viability of organic dairy farms in different subsidy reduction scenarios. The area of study includes three Spanish regions (Asturias, Cantabria and Galicia) that concentrate the 60% of the organic dairy farms registered in Spain. The sample of farms was obtained using a stratified sampling by regions and comprised 42% of the official census. The pay-back was determined for each farm, which establishes the level of milk production required to reach the fixed costs, with a unitary margin which is the difference between the weighted income per production unit and the average variable cost. A simulation analysis was developed to determine the economic results in three subsidy reduction scenarios. Scenario I represents the real situation in 2007. The level of production necessary to reach a balance between cost and income is 234,000 l, while the real production is 25% higher, with a mean unitary margin of 0.36 €/l. In all farms the pay-back is exceeded by the real production, so the economic profit is positive in all farms. Scenario II simulates the effect of the cessation of subsidies to compensate for loss of rent, which accounts 7.5% of total incomes. The cessation of subsidies reduces the unitary margin around 10%, reaching 0.321 €/l, and increases the pay-back to 205,418 l. In this scenario, the net result would be negative in 14% of farms. Scenario III simulates the cessation of all subsidies to farms (agri-environment and loss of rent), which accounts 13.6% of total incomes. The cessation of all subsidies reduces the unitary margin to 0.285 €/l, which represents a reduction of 20% compared to the initial situation. Consequently, the pay-back raised to 237,527 l. In these conditions, only 42% of farms continue to produce benefits. These results show the high dependence of the organic dairy sector to European financial model.

Characterization of the organic beekeeping in Spain: preliminary results
Perea, J.[1], García, A.[1], Acero, R.[1], Santos, M.V.F.[2] and Gómez, G.[1], [1]University of Cordoba, Animal Production, Edificio Produccion Animal, Campus Rabanales, 14071, Spain, [2]University Federal Rural of Pernambuco, Campus Pernambuco, 52171-900, Brazil; pa1gamaa@uco.es

Spain has a great potential for honey production, due to its rich flora and the climate allows the activity of bees most of the year, so the organic bee farm census has been increasing year after year. The aim of this study was to characterise the organic bee farms located in Spain from a survey including 45 farms (25% of official census). The results show that the organic honey is produced in family farms (60% of the workforce), which generate 2.25 annual work unit per farm on average. The farms have an average size of 358 hives, but with high variability. 25% of farms are small (less than 150 hives) while another 25% are large (over 594 beehives). The farms have an average productivity of 3,800 kg per year (10 kg per hive) and 56% of beekeepers have their main source of income in beekeeping. Beekeepers that have their main source of income in beekeeping follow a transhumant production system, while the remaining 44% develop a multifunctional system with cattle or ovine in extensive farms. The average unitary cost amounts to 9.28 €/kg honey and is formed mainly by labour cost (54.62%), amortizations (13.94%), supplies (13.33%) and taxes (7.76%). The average income per kg of honey is 10.54 € and is sufficient to generate profits in the activity. Incomes are comprised of: sale of honey (41.32%), sale of pollen (36.9%) and subsidies (21.78%). The honey is sold through two channels: the conventional channel, which is predominant and accounts for 74% of production at low price (2.5 €/kg) and the organic channel, which absorbs only 26.1% of production at high price (7.25/kg).

Competitiveness of organic dairy farms in northwest of Spain: subsidy reductions

Perea, J., García, A., Angón, E., Acero, R. and Gómez, G., Universitiy of Cordoba, Animal Production, Edificio Produccion Animal, Campus Rabanales, 14071, Spain; pa1gamaa@uco.es

It is essential to ensure the future of Spanish organic dairy farms to identify those aspects that differentiate competitive and non-competitive farms in a subsidy reduction scenario. The area of study includes three Spanish regions (Asturias, Cantabria and Galicia) that concentrate the 60% of the organic dairy farms registered in Spain. The sample of farms was obtained using a stratified sampling by regions and comprised 42% of the census. The farms were classified according to their economic profits (positive or negative) in a scenario without subsidies and an ANOVA test was used to identify the principal differences between both groups. The results show that competitive farms are larger than non-competitive farms ($P<0.05$), in both livestock (64.3 vs 39.1 animals) and surface (60.0 vs 39.3 ha), although land ownership is the main strategic difference due to determine the production system. 92.2% of the land in the competitive farms is owned by the farmer, and they apply improvements in pasture and cultivated 90% of the land in order to feed livestock, which reduces dependence on external food. While the non-competitive farms only the 39% of the land is owned by the farmer and it is atomised into many parcels of small size. This makes more difficult the development of crops and improved pastures, thus decreasing the stoking rate and increases the use of external food ($P<0.05$, 0.21 vs 0.48 kg/l milk). Consequently, the management of the herd in non-competitive farms is intensified and the feeding cost is duplicated ($P<0.05$, 491 vs 816 kg/cow), while the productivity is 22% lower than in competitive farms ($P<0.05$, 6,403 vs 4,990). The lowest productivity is also due to poor management of the replacement (replacement rate 24.4 vs 15.8%) and an increased incidence of intensive diseases. Furthermore, the non-competitive farms are equipped with a lower technological level. The lack of technology increases the workforce by 53% compared to competitive farm ($P<0.05$, 27.8 vs 18.1 cows/UWM).

Management of beef and dairy herd replacement

Agabriel, J.[1], Le Cozler, Y.[2] and Dozias, D.[3], [1]INRA, UR 1213 Herbivores, Theix, 63122 Saint Genes Champanelle, France, [2]AGROCAMPUS OUEST, UMR 1080, Dairy Production, Rennes, 35000 Rennes, France, [3]INRA, UE 326, Le Pin au Haras, Borculo, 61310 Exmes, France; Jacques.agabriel@clermont.inra.fr

Cattle herd replacement management should be revisited because of the combined evolutions of animal type, size, and level of production, and of the evolutions of the economic and environmental context. This presentation will deal with both dairy and beef productions. In these two cases, the strategy of herd replacement has to be considered not only at the animal scale (rearing strategies and optimal lifespan duration), but also at the herd and farm levels according to the various objectives assigned to sustainable livestock systems. During this presentation, an update of the main effects of heifers rearing strategies on subsequent performances will be given, with focuses on the effects of feeding strategies on puberty attainment, fertility of heifers, production and reproduction performances when adults, calving difficulties, longevity of the animals and carcass characteristics at slaughter. Interactions with breed maturity will be highlighted. In the second part, considerations on the sustainability of the different ways of production will be developed, taking into account the various dimensions of sustainability. A special grazer role is generally assigned to heifers, which are used to make the most of the least productive pastures. Consequently they have a positive environmental impacts which could vary with breed and age. Short- and/or long-time perspectives will be included. This presentation will mainly focus on some common production systems, such as intensive Holstein dairy production or Charolais beef herd production. However, attention will be paid on some alternative production systems (i.e. organic farming for example) or on some other systems that are greatly influenced by the characteristics of their land (mountains or Mediterranean herds for example).

Estimation of genetic parameters for heifer mortality in Austrian Fleckvieh

Fuerst-Waltl, B.[1] and Fuerst, C.[2], [1]University of Natural Resources and Applied Life Sciences Vienna, Department of Sustainable Agricultural Systems, Division of Livestock Sciences, Gregor Mendel-Str. 33, A-1180 Vienna, Austria, [2]ZuchtData EDV-Dienstleistungen GmbH, Dresdner Str. 89/19, A-1200 Vienna, Austria; waltl@boku.ac.at

Especially in dairy breeds, mortality rates of cows but also stillbirth and postnatal mortality rates have increased during the last years. This is not only relevant with regard to economical losses but also to animal health and welfare. Thus, the aim of this investigation was to explore the genetic background of postnatal mortality in calves and replacement heifers until first calving in Austrian dual purpose Fleckvieh (Simmental). The following periods were defined for analyses: P1 = 2-30d, P2 = 31-180d, P3 = 181-365d, P4 = 366d-age at first calving or a maximum age of 1200d if no calving was reported, P5 = 2d - age at first calving or a maximum age of 1200d if no calving was reported. Data of the federal state Lower Austria were utilized. Records of animals which were slaughtered or exported within a defined period were set to missing for this and consecutive periods while their records were kept for preceding periods. After further data restrictions (e.g. complete records on life history, discarding records of twins and multiples, foreign gene proportion <=12.5%, no transfer between herds), a total number of 28,044 heifers born in the years 2003-2006 were analysed. Their pedigree comprised of 120,592 animals. With 2.06, 1.67, 0.38, 0.95, and 6.11% in the defined periods P1-P5, respectively, mortality was low. For the estimation of genetic parameters, a threshold sire model with fixed year*month, number of lactation (lactations >5 were set to 5), calving ease, random herd*year and random genetic effect of sire was applied. The estimated heritabilities were 0.11, 0.06, 0.05, 0.03 and 0.06 for P1-P5, respectively. Even if postnatal mortality rate is lower than in other specialized dairy breeds, a genetic selection for this trait should be possible in Fleckvieh.

Genetic analysis of dystocia and stillbirth in Holsteins based on data including calf weights

Waurich, B.[1], Schafberg, R.[1], Rudolphi, B.[2] and Swalve, H.H.[1], [1]Institute for Agricultural and Nutritional Sciences, Group Animal Breeding, Adam-Kuckhoff-Str. 35, 06108 Halle, Germany, [2]State research Institute for Agriculture of Mecklenburg-Western Pomerania, Wilhelm-Stahl-Allee 2, 18196 Dummerstorf, Germany; benno.waurich@landw.uni-halle.de

The cattle breeding organization of Mecklenburg-Western Pomerania implemented a co-operator herd testing scheme with 22 large farms (ave. herd size = 750 cows) in 2005. With respect to recording of calvings, the system implemented comprises weighing of calves at birth and an accurate recording of the sex of the calf, be it alive or stillborn. The data analysed here included 39,617 calvings of Holstein cows from calving years 2005 to 2008. A calf was considered as stillborn if dead at birth or 48 h *post partum*. Dystocia was analysed as a binary 0/1 trait with 1's defined as more than one person assisting birth. A univariate Sire-MGS threshold model with random effects of sires and maternal grandsires and the fixed effects of herd, year-season, and parity was applied. This basic model was augmented alternatively by the fixed effects of sex, classes of birth weight, or both. Under the most complete model, direct heritabilities for stillbirth and dystocia were 0.08 and 0.06, respectively, and maternal heritabilities were 0.07 and 0.09. Genetic correlations between direct and maternal effects for stillbirth and dystocia were -0.10 and -0.25, respectively. Further examinations included comparisons of estimated breeding values from the above model with simple sire models in which the sire either was defined as the sire of the calf or the sire of the dam. Hardly any differences were found in ranking of sires comparing these models.

Comparison of F1 cows (North American Jerseys x Holsteins) with purebred Holsteins at a different milk production level

Brade, W.[1], Jaitner, J.[2] and Reinhardt, F.[2], [1]Chamber of Agriculture and University of Veterinary Medicine Hannover, Animal Research, Johannssenstr. 10, D- 30159 Hannover, Germany, [2]VIT, Heideweg 1, D- 27283 Verden, Germany; wilfried.brade@lwk-niedersachsen.de

A comprehensive crossbreeding experiment with North American Jersey bulls and German Holstein cows was initiated and conducted in Saxony and information of these animals is meanwhile available. Data of crossbred offspring of twelve North American Jersey bulls and their respective herd mates were used for analysing fertility, calving and production traits. First results of first lactation cows (N=848 F1 animals) showed, that the F1 cows were considerably better than the purebred Holsteins with respect to fertility and calving traits (1st lactation: interval from calving to first insemination: F1: 76.3 days, Holsteins: 84.5 days; days open: F1: 105.1 days, Holsteins: 120.3 days; stillbirth rate: F1: 6.9%, Holsteins: 10.5%). Based on the herd mean of the sum of fat-kg+protein-kg of the purebred Holsteins in the 1st lactation, three production classes were defined: 1st level (= high level): >570 kg fat+protein, 2nd level: 540 -570 kg fat+protein, 3rd level: <540 kg fat+protein. The standard deviations of yields as well as the average breeding values of the sires of purebred Holsteins are comparable in all three herd classes. The differences in the herd production levels are therefore mainly caused by environment and management. With increasing production level of the herds the inferiority of the F1 animals increased in milk yield. An interaction between herd milk production level and breed could be observed and will be further analysed.

Embryo survival: key to reproductive success in cattle

Diskin, M.G. and Morris, D.G., Teagasc, Mellows Campus, Athenry, Co. Galway, Ireland; michael.diskin@teagasc.ie

Embryo survival is a major factor affecting production and economic efficiency in both milk and beef production. For heifers, beef and moderate yielding dairy cows, it appears that fertilisation lies between 90 and 100%. In high producing dairy cows it would appear that it is somewhat lower and possibly more variable. The major component of embryo loss occurs before day 16 following breeding with some evidence of greater losses before day 8 in high-producing dairy cows. Late embryo-foetal loss, while numerically much smaller than early embryo losses, nevertheless, causes serious economic losses to producers because it is often too late to rebreed females when they repeat. Systemic concentrations of progesterone during both the cycle preceding and following insemination affect embryo survival rate with evidence that too-high or indeed too-low a concentration being negatively associated with survival rate. Uterine expression of mRNA for progesterone receptor, oestradiol receptor and retinol binding protein mRNA appears to be sensitive to changes in peripheral concentrations of progesterone during the first week after AI. Energy balance and dry matter intake during the 4 weeks after calving are critically important in determining conception rate when cows are inseminated at 70-100 days post calving. Concentrate supplementation of cows at pasture during the breeding period has minimal effects on conception rates though sudden reduction in dietary intake should be avoided. For all systems of milk and beef production more balanced breeding strategies, with greater emphasis on fertility and feed intake, must be developed. There is sufficient genetic variability within the Holstein and other breeds for fertility traits. Genomic technology offers the potential to identify genes responsible for improved embryo survival. Their incorporation into breeding objectives could increase the rate of genetic progress for embryo survival and thereby efficiency of milk and beef production.

Survey in the artificial insemination network on health status, production and reproduction performance in cattle herds as influenced by management

Knapp, E.[1], Chapaux, P.[2], Istasse, L.[1] and Touati, K.[1], [1]Liege University, Nutrition Unit - Ruminants and pigs clinics, Bd de Colonster 20, 4000 Liege, Belgium, [2]Association Wallonne de l'Elevage, Ch du Tersoit 32, 5590 Ciney, Belgium; eknapp@ulg.ac.be

A degradation of the reproduction performances is observed from many years in different countries. The aim of this work was to describe the impact of the health status, production and management on the reproduction performance in Walloon farms. An inquiry was carried on with the veterinarians of the artificial insemination network. There were 3,495 dairy and 5,598 beef herds involved. The age at first calving (AFC) was significantly different (27.0 vs 28.9 months, $P=0.002$) in dairy and beef herds. In beef herds, the AFC, the calving interval (CI) and the apparent fertility index were negatively correlated to herd size (r=- 0.183, r=-0.267, r=-0.324; $P<0.05$) indicating that in farms with large herds, high quality management seemed to be a factor of importance to achieve good reproduction. For the dairy farms, the CI only was correlated to the size (r=-0.152, $P<0.05$). In dairy herds, the AFC was reduced when the annual milk yield increased (31.2 months for production <5,000 kg milk vs 28 months for production >8,000 kg milk). For CI, an average production of about 6,000-7,000 kg seemed to be the best for cow fecundity (CI of 397 days). In the beef farms, the heifer and cow fecundities seemed to be improved with the increase of production expressed in total live weight gain/ha. Heat detection appeared to be the most important source of problems accountable for partial failure in the reproduction performance for both dairy and beef herds (28.7% and 35.8%). Nutrition induced also a large frequency of reproduction disorders with equal propositions in both herd types (30.6% and 28.4%). The dairy farmers seemed to invest more in reproduction than the beef farmers. The major acts requested by farmers to inseminators were the diagnosis of pregnancy made by hand (27.0 and 27.6%) or by sonography (19.1 and 17.2%).

Effect of body condition and suckling restriction with and without presence of the calf on cow reproductive performance on range conditions

Quintans, G., Banchero, G., Carriquiry, M., López, C. and Baldi, F., National Institute of Agricultural Research, Beef Production, Ruta 8, km 282, 33000, Uruguay; gquintans@inia.org.uy

Nutrition and suckling are the most important factors affecting the anoestrous period (AP). The presence of the calf and lactation are recognized as the most important factors involved in the suckling-induced suppression of LH secretion. The effects of body condition score (BCS) and suckling restriction with and without presence of the calf on reproductive efficiency were evaluated. Sixty three Angus x Hereford multiparous cows were managed to maintain different BCS at calving and thereafter (low v. moderate; L, n=31 and M, n=32). Within each group of BCS, cows were assigned to three suckling treatments at 66 d postpartum: 1) suckling ad libitum (S, n=20); 2) calves fitted with nose plates during 14 days remaining with their dams (NP, n=22); 3) calves were completed removed from their dams for 14 days, and thereafter returned with them (CR, n=21). Cows were bled monthly from 98 d prepartum until 66 d postpartum and weekly thereafter until 128 d postpartum (end of mating period). Plasma insulin concentrations and presence of corpus luteum (CL) were determined. At 94 d postpartum, presence of CL was greater ($P<0.001$) for NP and CR than for S cows (68, 57 and 21% for NP, CR and S, respectively). Also, more M BCS presented CL than L BCS cows (77 vs. 25; $P<0.0001$). The length of the AP was longer ($P<0.05$) in S than in NP or CR cows (108±3.4 vs. 95±3.4 and 91±3.5 days for S, NP and CR, respectively). Insulin concentrations were less for L BCS than for M BCS cows (1.46±0.06 mUI/ml vs. 1.82±0.06 mUI/ml; $P<0.0001$) and for S than for NP and CR treatments (1.48±0.08 mUI/ml vs. 1.74±0.07 mUI/ml and 1.71±0.08 mUI/ml for S, NP and CR cows, respectively). Suckling restriction with and without presence of the calf had an improved effect on reproductive performance compared to suckled cows and this could be mediated through increased insulin concentrations.

Effect of sexed-semen use on Holstein conception rate, calf sex, dystocia, and stillbirth in the United States

Norman, H.D. and Hutchison, J.L., Animal Improvement Programs Laboratory, Agricultural Research Service, US Department of Agriculture, Beltsville, MD, 20705-2350, USA; duane.norman@ars.usda.gov

Most artificial-insemination organizations in the United States now market sex-sorted semen. For 10.8 million US Holstein breedings with conventional semen since January 2006 and 122,705 sexed-semen breedings, data were available from all breedings for conception rate, 12 and 9% of breedings for calf sex and dystocia (births reported as requiring considerable force or extremely difficult), and 10 and 9% for stillbirth (born or died within 48 h). Statistical differences were determined by chi-square tests. Conception rate and calf sex ratios differed ($P<0.001$) with use of conventional and sexed semen. Conception rate was 57% for heifers and 30% for cows with conventional semen and 43 and 25%, respectively, with sexed semen. For heifers, 50% of calves were single females; 49%, single males; and 1%, twins with conventional semen; corresponding percentages for cows were 45, 49, and 5. With sexed semen, 90% of calves were single females; 9%, single males; and 1%, twins, for heifers and 85, 11, and 5%, respectively, for cows. Significance of differences for dystocia and stillbirth incidences with use of conventional and sexed semen varied. For births from conventional semen, incidence of dystocia was 4% for single female calves, 8% for single male calves, and 8% for twin calves for heifers ($P<0.001$) and 2, 3, and 5%, respectively, for cows ($P<0.001$); corresponding incidences of dystocia with sexed semen were 4, 9, and 4% for heifers ($P<0.001$) and 1, 1, and 2% for cows ($P>0.05$). Stillbirth incidence with conventional semen was 9% for single females, 11% for single males, and 15% for twins for heifers ($P<0.001$) and 4, 4, and 8% for cows ($P<0.001$); corresponding incidences of stillbirth with sexed semen were 10, 15, and 13% for heifers ($P<0.001$) and 3, 4, and 8% for cows ($P>0.05$). Differences between conventional and sexed-semen breedings for dystocia and stillbirth incidences may have been affected by herd recording practices.

Pelvic opening and calving ability of Charolais cattle

Renand, G.[1], Vinet, A.[1], Saintilan, R.[1] and Krauss, D.[2], [1]INRA, UMR 1313 GABI, Domaine de Vilvert, 78352 Jouy en Josas, France, [2]INRA, UE Bourges, Domaine de Galles, 18520 Avord, France; gilles.renand@jouy.inra.fr

The pelvic opening has been systematically recorded in the purebred Charolais herd of INRA at Bourges on heifers and adult cows as well as on the young males. The herd was founded in with a representative sample of 347 Charolais heifers. During the next 20 years a sample of 83 Charolais sires was used to randomly inseminate the females. All the progeny integrated the herd without any selection unless physiology disorders: 1,430 female and 1,508 male calves respectively. Each heifer was bred for calving at 3 years of age and each female was kept in the herd for 3 more breeding years, unless of accidents. Pelvic opening was measured on males just before slaughtering at 15 or 19 months and on growing heifers at 12, 18 and 24 months. Pelvic opening was also measured on cows the next day of each calving. The calving ease score and calf birth weight were also recorded. These two latter traits are analysed in a mixed linear model with direct and maternal genetic effects, while the pelvic opening at calving is analysed in a mixed linear model with repeated recording. Pelvic opening of male calves and of growing heifers are separate traits analysed in a mixed linear model with direct genetic effects only. In all the analysis the random effects are the animal genetic effects taking into the whole pedigree of the experimental animals. Variance components will be estimated jointly with the VCE software. Direct and maternal heritability and genetic correlation coefficients will be then estimated. The estimation of the correlation between the maternal effect of calving ease and pelvic opening at calving will be of primary interest for unravelling the contribution of the cow ability to the calving process. The correlation between the maternal effect of calving ease and pelvic opening of the growing heifers and particularly of the young males will be estimated for quantifying the accuracy of the precocious selection criteria for improving calving ability.

Investigations on the drinking water intake of growing heifers

Meyer, U., Grabow, M., Janssen, H., Lebzien, P. and Flachowsky, G., Friedrich-Loeffler-Institute, Federal Research Institute for Animal Health, Institute of Animal Nutrition, Bundesallee 50, 38116 Braunschweig, Germany; peter.lebzien@fli.bund.de

This study was carried out to investigate relationships between water intake of cows and factors such as ambient temperature, relative humidity, body weight, dry matter intake, dry matter content of the diet, roughage proportion of the diet, and Na and K intakes. Sixty-four German Holstein heifers with an initial mean body weight of 175 kg were used. The animals were kept in a slatted floor, thermally non insulated stable in groups of 8 per pen. The experimental diet consisted primarily of grass silage and restricted amounts of concentrates. The concentrate mixtures mainly comprised of wheat, barley, oats, soybean meal, dried sugar beet pulp, soybean oil and minerals. Water was freely available. Live weight and feed intake of each animal were electronically recorded continuously. The water intake of cows was logged electronically by weighing the water vat before and after drinking. The experiment ended when animal live weight was approximately 500 kg. Subjection of the data (n=19485) to multiple regression analysis (SAS stepwise procedure) yielded the following equation: drinking water intake (kg/day) = - 5.206 + 0.038 x body weight (kg) + 0.610 x average ambient temperature (°C) + 0.098 x roughage proportion of the diet (%) - 0.086 x relative humidity (%) + 0.530 x dry matter intake (kg/day), R^2=0.31. It is presumed that this equation considers the most significant factors for predicting the water consumption of growing heifers under feeding conditions existing in Central Europe.

Genetic analysis of calf diseases in Danish Holstein

Fuerst-Waltl, B.[1] and Sorensen, M.C.[2], [1]University of Natural Resources and Applied Life Sciences Vienna, Dep. Sust. Agric. Syst., Div. Livestock Sciences, Gregor Mendel-Str. 33, A-1180 Vienna, Austria, [2]Faculty of Agricultural Sciences, Dep. Of Genetics and Biotechnology, P.O. Box 50, DK-8830 Tjele, Denmark; waltl@boku.ac.at

In Denmark, routine registration of cattle diseases started in 1990 aiming at the reduction of disease frequencies by both, management and breeding. Yet in routine breeding value estimations, only data of lactating cows are utilized. Continued non-consideration of calfhood diseases eventually lead to increased disease incidences. The aim of this study thus was to estimate genetic parameters of different diseases in female calves and replacement stock. Data from the Danish health recording system were used, combining different diagnoses to the 6 main disease categories: Udder, Reproductive, Feet and Legs, Digestive, Other infectious and Total (any of the disease categories 1-5). Diagnoses of health traits were considered as all-or-non traits resulting in a binary data structure. In total, data from 87,757 Holstein heifers born in the years 1998 to 2007 were investigated. For the estimation of genetic parameters a linear and a threshold sire model with herd*year*season effect (random), and year*month, number of dam's parity, calf size as fixed effects and the random genetic effect of the sire were applied. Disease occurrences were relatively low with 1.74%, 2.89%, 2.01%, 1.59%, 2.55% and 10.2 for Udder, Reproductive, Feet and Legs, Digestive, Other infectious, and Total, respectively. Applying threshold models, corresponding heritabilities were 0.06, 0.03, 0.01, <0.01, 0.03 and 0.01, respectively, while by linear models all estimated heritabilities were below 0.01. The low heritability estimate for digestive diseases may partly be caused by group treatment of some younger calves. Correlations between breeding values estimated by linear and threshold model were >0.98 for all traits. Further monitoring and research towards including these traits in the breeding program are recommended.

Nutrigenomics: providing molecular tools that will revolutionize the animal production industry
Dawson, K., Alltech, Research, Catnip Hill, Nicholasville, KY, USA; kdawson@alltech.com

Functional genomic approaches using high-density microarrays and real-time PCR techniques are providing new tools for evaluating the effects of nutritional and management strategies on the growth and performance of livestock. These basic molecular techniques for monitoring gene expression patterns in various tissues are providing new ways for rapidly evaluating the effects of environment, disease processes, intoxication and dietary nutrients on the key processes that regulate animal metabolism. They will play a key role in developing a basic understanding of physiological regulation and in the development of new basic nutritional concepts. However, their value for improving animal production is also apparent in studies evaluating nutritional management strategies where their ability to minimize the need for long-term nutritional studies gives them a unique role in accelerating innovation in modern animal agriculture. These studies have given rise to increased interest in the use of Animal Nutrigenomics in both basic nutritional research and applied animal nutrition. Recent studies have used Nutrigenomics approaches to gain insight into many poorly understood or even undefined nutritional processes and have used gene expression patterns to define nutritional relationships that were previously unknown. These approaches have made it possible to better define the most appropriate forms for individual nutrients and more accurately determine their practical nutritional value. Such work has also allowed nutritionists to more clearly define production livestock responses to such minerals as selenium which has a complex global impact on numerous physiological functions. This has made it possible to develop novel nutritional strategies and formulation schemes that are more effective at increasing the efficiency of animal production. The net result will not only be a more definitive description of basic nutritional processes but also a clear definition of methods for decreasing production costs as novel feed ingredients and combination nutrients become available.

Differential gene expression in poultry, fed a deoxynivalenol-contaminated diet
Dietrich, B., Bucher, B. and Wenk, C., Institute of Animal Sciences, Department of Agricultural and Food Sciences, Swiss Federal Institute of Technology (ETH), Universitätsstrasse 2, 8092 Zürich, Switzerland; brunod@student.ethz.ch

The mycotoxin deoxynivalenol (DON) is a secondary metabolite from Fusarium species and is frequently present on wheat and other cereals. The effects of DON intake range from reduction of feed consumption to vomiting, diarrhea and gastroenteritis. DON binds to the 60S ribosomal subunit and inhibits subsequently protein synthesis at the translational level. It has been suggested that cells and tissues with high protein turnover rate, like the liver, immune system and small intestine, are most affected. The co-occurrence of several mycotoxins in cereals is frequently encountered and interactive effects are expected. Therefore, a control diet and diets with naturally DON-contaminated wheat were fed to five broilers per group at the moderate concentrations of 1.0, 2.5 and 5.0 mg DON/kg for 23 days. Three microarrays were used per group to determine the alterations of gene expression in the liver. Totally 368 genes were upregulated and 114 genes were downregulated in groups with a DON-contaminated diet compared to the control group. The genes were related to mRNA stabilization, signaling, apoptosis, DNA and protein repair, protein modification, complex I related genes and tight junctions. The genes AKR1B1, EIF2AK3 and MIA2 were downregulated and gene IFT57 was significantly upregulated due to DON contamination in the feed. This was shown in the microarray and the real-time PCR analysis. AKR1B1 seems to be a biomarker for cytotoxic substances and was also significantly reduced in HEPG2 cells challenged with DMNQ and heavy metals. The decrease of the expression of EIF2AK3 seems to be related to the translation inhibition. MIA2 is related to the immune system and might be a biomarker for immune suppression. IFT57 is a transcription factor of apoptosis related genes. The results indicate, that already at low DON levels, alterations of the gene expression occur in the liver of broiler chicken.

Nutrigenomics: methods and applications

Puskás, L.G.[1] and Kitajka, K.[2], [1]Hungarian Academy of Sciences, Szeged, Temesvari krt. 62., H-6726, Hungary, [2]Avidin Ltd., Szeged, Közép fasor 52., H-6726, Hungary; puskas.szbk@gmail.com

Traditional approaches cannot give us a detailed picture on how different diets and nutritional components exert their complex effects on biological systems, because they focus on only a few biomarkers. For example, dietary lipids not only influence the biophysical state of the cell membranes but, via direct and indirect routes, they also act on multiple pathways including signalling and gene and protein activities resulting in either positive effects with their disease-preventing and even therapeutic potential, and negative effects by causing severe diseases. Therefore, to understand the molecular basis of the effects and roles of different nutritional interventions global screening techniques such as DNA- or protein microarrays, high-throughput PCR are used. The technological backgound of these methods will be discussed to give us an overview how we can assess the changes, in a global way, at the genome, at the transcriptome and at the proteome levels. Besides methological details of the latest technologies, several nutrigenomics studies will be presented with major emphasis on lipids in health and disease. In the future systems biology and system networks in nutritional research will help us to understand the genomic background and the concerted changes in gene and protein activity level, as well as the metabolome level, caused by different diets. Defining novel 'system fingerprints' in nutrition will help us suggesting personalized diets and identification and improvement of novel bioactive ingredients, including dietary lipids.

Using nutrigenomic approaches to determine responses to mycotoxin in pigs

Pinton, P.[1], Gallois, M.[1], Grenier, B.[1], Gao, Y.[2], Rogel-Gaillard, C.[2], Martin, P.G.P.[1] and Oswald, I.P.[1], [1]INRA, UR66, Laboratory of Pharmacology-Toxicology, 180 chemin de Tournefeuille, BP93173, 31027 Toulouse cedex 3, France, [2]INRA, UMR Animal Genetic and Integrative Biology, Domaine de Vilvert, 78350 Jouy-en-Josas, France; ioswald@toulouse.inra.fr

Food safety is a major issue. In this respect much attention needs to be paid to the possible contamination of food and feed by mycotoxins. Mycotoxins are secondary metabolites produced by fungi, especially by those included in the genus Aspergillus, Fusarium and Penicillium. Mycotoxins are very common contaminants of cereals. Global surveys estimated that 25% of the world crop production is contaminated with such toxins. Moreover, most mycotoxins are heat-resistant and could remain after the disappearance of the producing fungi. Through a transcriptomic approach with DNA arrays, we analyzed the effect of deoxynivalenol on piglets. We set up a long oligonucleotide DNA array composed of the Qiagen-NRSP8 set (13,297 probes) a series of control element and 3773 unique probes that includes all genes and putative transcripts localized in the swine leukocyte antigen (SLA) complex region (826 probes) as well as immune response gene outside SLA (2,947 probes). Twelve piglets received for 5 weeks either mycotoxin-free diet or DON-contaminated diet (2.85 mg DON/kg feed). At necropsy, samples from jejunum were obtained. Total RNA was extracted reversed to cDNA and hybridized on the above mentioned DNA array using a dye-swap protocol. This parallel monitoring of the expression level of around 18,000 gene, indicate that 364 genes were differentially expressed in control and DON treated intestinal samples. This includes gene involved in immune response, immune development and tissue morphology. In conclusion, DNA array represent a powerful toll (i) to identify immune bio-markers in the course of a dietary exposure to mycotoxins and (ii) to determine the mechanism of action of mycotoxins through the analysis of the activated signaling pathways.

Modulation of adipose tissue gene expression and fatty acid composition by dietary fat in pigs

Mohan, N.H.[1], Sarmah, B.C.[2], Tamuli, M.K.[1], Das, A.[1], Bujarbaruah, K.M.[3], Naskar, S.[1] and Kalita, D.[2], [1]National Resarch Centre on Pig, ICAR, Rani, Guwahati, 781131 Assam, India, [2]College of Veterinary Science, AAU, Guwahati, 781022 Assam, India, [3]Indian Council of Agricultural Research, Krishi Bhavan, 110114 New Delhi, India; mohannh.icar@nic.in

One of the major areas of swine research is increasing lean to fat ratio of pig carcasses with an importance for body composition traits. In the present study, the experiments were designed to understand the effect of type of dietary fat on serum lipid profile, adipose tissue fatty acid composition and gene expression (Fatty Acid Synthase, FAS, Stearoyl CoA Desaturase, SCD,Sterol Regulatory Element Binding protein-1c, SREBP-1c, (Peroxisome Proliferator Activated Receptorγ, PPARγ), CCAAT Enhancer Binding Protein β, CEBP β). The pigs were fed with isocaloric ration for sixty days with 10% of the total energy contributed by either sunflower oil (SF) or coconut oil (CO). The FAS, SCD and SREBP-1c gene expressions in subcutaneous adipose tissue were significantly ($P>0.05$) reduced in SF fed animals by 0.84 ± 0.08, 0.39 ± 0.06 and 0.47 ± 0.03 times, respectively in comparison to CO fed pigs. Increased abundance of PPARγ mRNA was observed in SF fed group with no significant change in the CEBPβ gene expression in both CO and SF fed pigs. The palmitic and stearic acid content was 1.31 and 1.09 times respectively higher in CO fed compared to SF fed pigs. The SF feeding increased oleic, linoleic and linolenic acid contents in the subcutaneous adipose tissue. The type of dietary fat influenced subcutaneous adipose tissue fatty acid composition and the feeding of unsaturated fatty acid rich SF may decrease lipogenesis as well as adipogenesis through reduced expression of lipogenic genes. On the other hand, feeding of saturated fatty acid rich CO did not change lipogenic gene expression but decreased CEBPβ expression, suggesting that saturated fat may decrease adipogenesis. It is suggested that interventions based on nutrient genome interaction may be applied to develop designer pork.

Differential gene expression study in porcine adipose tissue conditional on oleic acid content of diet

Ovilo, C., Fernández, A.I., Rodrigañez, J., Martín-Palomino, P., Daza, A., Rodríguez, C. and Silió, L., INIA, Mejora Genética Animal, Ctra Coruña Km 7.5, 28040 Madrid, Spain; ovilo@inia.es

The study of the relation between diet composition and transcriptional profiling is a useful tool to understand the mechanisms involved in the gene expression and regulation of lipid metabolism. The objective of this work was the evaluation of the effect of high-oleic sunflower oil as dietary fat source on gene expression in porcine backfat. Twenty seven Iberian castrated males with 28 kg of body weight were distributed in two groups fed with isocaloric diets differing on high-oleic sunflower oil content (0% and 6%). Growth and fatness traits were similar in the two groups. Tissue samples were collected from pigs slaughtered with 110 kg of live weight. Fatty acid profile of subcutaneous fat showed significant differences between groups, with the high oleic diet group showing a greater percentage of MUFA (61.3 vs 52.7%) and PUFA (7.5 vs 6.8%), and lower percentage of SFA (31.2 vs 40.5%). Nevertheless the fatty acid profile of intramuscular fat was not affected by the treatment. Inner-layer subcutaneous backfat RNA samples from eight pigs of each diet were hybridized with the Affymetrix porcine GeneChip™. Expression data quality evaluation was carried out using affyPLM package in Bioconductor. Normalization was performed with the BRB software. Statistical analysis of differentially expressed (DE) genes was performed using GEAMM software, and a false discovery rate of FDR<0.01 was assumed. A scarce number of significant expression differences was found between diets: five transcripts appear overexpressed in the high oleic diet group (1.12 to 1.81 fold change) and only one in the low fat group (2.74 fold change). None of the six DE genes are known to be related with lipid metabolism. Differences in expression levels of some interesting genes related with fatty acid metabolism or transport (as RXRG or FABP5 genes) are close to the significance threshold assumed. Validation of some relevant results is being done by qPCR.

Development of a custom microarray platform for nutrigenomics studies in sheep

Stefanon, B.[1], Sgorlon, S.[1], Colitti, M.[1], Asquini, E.[2] and Ferrarini, A.[3], [1]University of Udine, Via delle Scienze, 208, 33100 Udine, Italy, [2]University of Trieste, P.le Valmaura, 9, 34148 Trieste, Italy, [3]University of Verona, Strada Le Grazie, 15, 37134, Italy; bruno.stefanon@uniud.it

To gain a better understanding of molecular mechanisms of plant bioactive compounds activities, a custom ovine microarray was developed. Thirty-six Sarda sheep were randomly assigned to six groups: CTR (without ACTH and supplementation), ACTH (with ACTH and without supplementation), LD (ACTH and 50 g/head/day of Larix decidua), ECHI (ACTH and 3 mg/kg live weight/day of Echinacea angustifolia), ANDRO (ACTH and 1 mg/kg live weight/day of Andrographis paniculata) and POLI (ACTH and 3 mg/kg live weight/day of Polinacea^{TM}). After 22 days of adaptation to the experimental diets, all the groups, except CTR, were injected twice a day with 0.5 mg of ACTH for 3 consecutive days. Sheep of CTR group received a dose of saline. Blood was sampled before (T0) and after 3 (T3) and 51 (T51) hours from the first injection. Experimental protocol was approved by local laws and regulations. A pool of the RNA extracted samples was prepared for each group and time of sampling. The custom oligoarray was synthesized on a CombiMatrix 90K platform, using 24384 35-40mer probes designed from 12194 UniGenes (NCBI - Build #13). Cy5 labelled pools were hybridized in duplicate on the microarray. Normalized data were analyzed with a two-way (treatment and time) ANOVA (MeV software v 4.1), using a statistical threshold of p-values <0.001. Two-way ANOVA identified 688, 1,063 and 165 genes differentially expressed respectively for treatment, time and the interaction. Hierarchical clustering (HCL) of differentially expressed genes was used to generate heatmaps for each of the main factors and iteraction (MeV software v 4.1). HCL showed a strong effect of ACTH treatment on expression patterns regulation. Bioactive compounds showed specific activities in counteracting the ACTH induced stress, indicating their involvement in cell communication, response to stress and immune system process.

Alteration of foetal hepatic gene expression due to gestation diet and breed

Wimmers, K., Nuchchanart, W., Murani, E. and Ponsuksili, S., Research Institute for the Biology of Farm Animals (FBN), Wilhelm-Stahl-Allee 2, 18196 Dummerstorf, Germany; wimmers@fbn-dummerstorf.de

Effects of maternal nutrition during pregnancy on gene expression, growth performance, and health of offspring were shown epidemiologically and experimentally in humans and model animals. We aim at identifying pathways and genes sensitive to foetal programming and evidencing and quantifying of the role of DNA-methylation in this phenomenon. Using two divergent breeds as model organisms enables to illuminate whether different genotypes react differentially on exogenous factors known to modulate gene expression via epigenetic mechanism. Knowledge of how nutrients target specific regulatory genes and thereby influence the overall energy homeostasis allows elucidating molecular routes of genotype-environment (diet) interaction. Therefore, 36 sows were fed methionine supplemented vs. control gestation diets and foetuses were sampled at 35, 63, and 91 dpc. Gestation diets had a significant effect on whole genome methylation and foetal weight. Also differences due to breed and interaction `breed×diet´ were significant. Affymetrix microarray expression profiles of foetal livers revealed large effects of diets at 35 and 91 dpc, whereas at 63 dpc differences were less pronounced - as were the effects of diets on DNA methylation. Global analysis of genes regulated at all prenatal stages in both breeds points to genes of metabolism of lipids, amino acids, carbohydrates, cell growth, proliferation, differentiation, and communication. In particular, significant differences in gene expression were associated with methionine metabolism (35 dpc), IGF-1 (63 dpc) and BMP signalling (91 dpc). Diet-dependent expression was confirmed exemplarily (BHMT, DNMT1, AMD1, IGF2R, IGFBP3, IGFBP5, SHC1, BMP6, SMAD5, MAP3K7) by real-time PCR. Quantification of DNA methylation at selected loci will shed light on the role of this epigenetic mechanism as a molecular basis of genotype-environment (diet) interaction. Results were obtained through the EC-funded FP6 Project 'SABRE'.

Underfeeding affects IGF-1 and gene expression in genital tract tissues in high producing dairy cows

Valour, D.[1], Degrelle, S.[1], Marot, G.[2], Dejean, S.[3], Dubois, O.[1], Hue, I.[1], Germain, G.[1], Humblot, P.[4], Ponter, A.A.[1], Charpigny, G.[1] and Grimard, B.[1], [1]INRA, UMR 1198, BDR, Domaine de Vilvert, 78350 Jouy en Josas, France, [2]INRA, UR337, SGQA, Domaine de Vilvert, 78350 Jouy en Josas, France, [3]Univ. Paul Sabatier, Institut de mathématiques, 118 rte Narbonne, 31062 Toulouse CEDEX 9, France, [4]UNCEIA, 13 rue Jouet, 94704 Cedex Maisons Alfort, France; damien.valour@jouy.inra.fr

In high producing dairy cows, the decline in fertility is a major cause of economic losses. This decrease is linked to genetic (milk production) and environmental factors (alimentation) whose interactions lead to the negative energy balance observed during postpartum. The purpose of this project was to evaluate the influence of negative energy balance during the postpartum period on the gene expression profiles in the genital tract (oviduct, endometrium and corpus luteum). Prim' Holstein cows were randomly assigned to control diet (100% requirements, n=4) or underfed (40% requirements, n=4) at calving. There was no difference between the groups at calving for: age, body condition score, previous milk yield, sire fertility index and milk index. Blood samples were taken regularly postpartum and plasma analysed for selected metabolites and hormones. Oestrus was synchronised 80d postpartum and tissue samples were taken 4, 8, 12 and 15d later. Metabolic and transcriptomic data were then generated. Principal canonical pathways altered in the different tissues were investigated and statistical correlations between gene expression and blood metabolites were calculated. The main modifications in gene expression were found in the oviduct and in the endometrium whereas the corpus luteum was not affected. Energy balance mainly altered the genes pathways related to lipid metabolism or potentially linked to fertility. Data were integrated using statistical correlation analyses that highlighted IGF-1 as a good predictor of energy balance in dairy cows. Acknowledgments: ANR GENANIMAL 03.P.406.

A review of scientific literature on transportation of the horses

Marlin, D., Hartpury College, Equine, Hartpury House Hartpury Gloucester UK, GL19 3BE, United Kingdom; dm@davidmarlin.co.uk

Almost every domesticated horse will experience some form of transportation at some time in its life. The most common reasons for transport are likely to include transport to competitions, for breeding purposes, when bought or sold, for veterinary treatment and for slaughter. The predominate means of transporting horses is by road, with trailers and trucks, vans or articulated vehicles. In contrast to sport and racehorses which are transported frequently, horses destined for slaughter for human consumption may or may not have had previous experience of transport. There is a growing trend for horses to be 'farmed' in Europe and these animals are unlikely to have had significant experience of transport before long distance transport for slaughter. The scientific literature on horse transport is not extensive, but has addressed issues such as stress, dehydration, behaviour and disease. The wealth of the literature concerns the pleasure, sport or racehorse with a smaller number of studies concerned with horses transported for slaughter. There is no doubt that the majority of horses that are moved distances from as little as a few miles to thousands of miles experience little if any discomfort, stress or ill effects. However, the effects of transport can have a noticeable negative effect on performance without obvious clinical signs of illness or disease or the effect being directly linked to preceding transport. More serious consequences for health can result from transport related effects on various body systems, most notably effects on the gastro-intestinal, musculoskeletal and respiratory tracts. Potential stressors or insults that horses may experience during transport can include separation, thermal stress, space restriction, exposure to noise, accelerations and decelerations of the transport vehicle, reduced air quality, reduced availability of feed and water, fatigue and infectious disease. This review will cover the published, peer-reviewed scientific literature on both non-slaughter and slaughter horses.

Heart rate variability as a quantitative index of stress during transport in horses

Jones, J.H.[1], Ohmura, H.[2], Takahashi, T.[2], Aida, H.[2] and Hiraga, A.[2], [1]University of California, Vet Med:Surgery & Radiology, 2112 Tupper Hall, Davis, CA 95616, USA, [2]Japan Racing Association, Sports Medicine Division, Equine Research Institute, 321-4 Tokami-cho, 320-08 Tochigi Prefecture, Japan; jhjones@ucdavis.edu

Transport of horses is frequently associated with the development of medical complications, one of the most common being 'shipping fever,' or transport-related respiratory disease. Immune suppression by the stress response is thought to be a primary factor in the development of this problem. Numerous studies have evaluated how environmental factors in the transport vehicle related to its design and the management of horses during transport affect the development of transport-related respiratory disease. The single most significant factor has been shown to be transport time, a variable over which a shipper may have little control. Horses shipped over longer periods of time are at higher risk for developing shipping fever. A method with potential for monitoring the degree to which horses are experiencing transport stress is the measurement of heart rate variability (HRV), or variations in the intervals between heart beats that occur over periods of several seconds to several minutes. These changes have been shown to be related to the degree of stimulation of a horse's parasympathetic and/or sympathetic nervous systems, subconscious controllers of body function that may be modified by the stress an animal experiences. Measurements of HRV responses of horses during transport suggest they are markedly different than those of the same horses at rest, offering possibilities for further refinement to develop a technique sensitive to changes in HRV that indicate when a horse is having a strong stress response to a travel episode. Because data for HRV analysis can be collected with commercially-available equine heart rate monitors, further evaluation of this technique offers the potential to monitor the stress animals being transported are experiencing in real time if appropriate software is developed.

Effect of transport time up to 8 hours on physiological and biochemical stress indicators and resulting carcass and meat quality in cattle: an integrated approach

Von Holleben, K.[1], Henke, S.[1] and Schmidt, T.[2], [1]bsi -Training and consultancy institute for careful handling of breeding and slaughter animals, postbox 1469, 21487 Schwarzenbek, Germany, [2]Consulting Company for Animal Husbandry and Biometrics, Fritz Sackewitz Str. 18, 31535 Neustadt, Germany; kvh@bsi-schwarzenbek.de

A field study was performed between 2000 and 2002 in northern Germany including 580 bulls cows and heifers on 63 commercial transports of less than 8 hours. The aim was to evaluate the impact of transport time within the context of stress factors during transport. Factors included in the mixed model statistical analyses were sex, breed, keeping system, clinical and behavioural findings (during loading, unloading and lairage), transport time, road quality, temperature and humidity, loading density, use of mounting prevention, mixing and lairage time. Response parameters were heart rate, biochemical stress indicators, carcass and meat quality. After 2 hours of transport time body temperature, cortisol and heart rate features indicate that animals have calmed down from loading and got used to some extent to the situation on the moving vehicle. Blood CK concentration goes up with increasing transport time, being higher for animals transported longer than 6 hours ($P<0.01$), showing the beginning of muscular fatigue. Heart rates for single animals rise again after 6 hours, however not reaching critical values. Energy metabolites (NEFA and ß-hydroxybutyrate) go up with increasing transport time but within the physiological range. Quality was significantly affected by transport time. Carcasses of cattle transported longer than 6 hours had a higher bruising score ($P<0.001$) and an increased pH24 in the longissimus dorsi ($P<0.05$) compared to cattle transported less than 6 hours. It was concluded that under German conditions (collecting of small groups, variable vehicle standard, moderate climatic conditions) a maximum transport time of 6 hours would be advantageous.

Long duration cattle transport: impact of 3 different stocking densities on physiological, behavioural and zootechnical indicators

Mounaix, B.[1], Brule, A.[1], Mirabito, L.[2], David, V.[2] and Lucbert, J.[2], [1]Institut de l Elevage, Bien-être animal Santé Hygiène et Traçabilité, Monvoisin, 35652 Le Rheu, France, [2]Institut de l Elevage, M.N.E. 149 rue de Bercy, 75595 Paris CEDEX 12, France; beatrice.mounaix@inst-elevage.asso.fr

From the 95/29/CE European directive on animal protection, stocking density of cattle during transportation is under regulation. Revision of current allowed densities is planned by EC 1/2005. This study aims to provide scientific information on the effects of changes in space allowances during commercial transportation of cattle on physiological, zootechnical and behavioural indicators. Three transportations of 29 h each (14 h – 1 h – 14 h) with feeding and watering of cattle inside the truck during the 1 h resting period are currently realized between France and Italy. Three different space allowances are compared: 1.16 m^2 (conforming to current regulation, group size: 6-7), 1.74 m^2 (+50% space allowance, group size: 4-5) and 1.05 m^2 (-10% space allowance, group size: 7-8), with 2 replicates of each density in each of the 3 transports. A total of 111 young cattle (Charolais, W approx 400 kg, 37 cattle per transport) are transported during the experiment. Blood samples are collected to measure physiological indicators of muscular tiredness (AST, CPK), of dehydration (total proteins, Na, K, Cl), of liver damages (urea, ALT), of glycogen reserves (glucosis, fructosamin) and haptoglobin as an indicator of inflammatory stress. These measurements are cross-analysed with behavioural indicators (lying versus standing time-budget, numbers of falls, feeding and drinking duration, agonistic behaviour) and zootechnical observations (weight loss, cleanliness grade and number of injuries). SAS multivariate analysis will be used to analyze data. Results are currently collected and analyzed. They will be available to be presented to EAAP. They will provide data on the effects on cattle welfare of the recommended decrease of density proposed by SCAHAW (2002) for long duration cattle transportation.

Study on temperatures during animal transport

Natale, F., Hofherr, J., Fiore, G., Mainetti, S. and Ruotolo, E., EU Commission-JRC, CI-Animal&Food Action, Via Marconi, 21020 Ispra (Varese), Italy; fabrizio.natale@jrc.it

The study compares the temperature standards in force and the standards proposed by EFSA scientists with the actual practices of commercial transport in the EU, providing a realistic picture about the temperatures experienced in livestock transports along the most important trade flows and climatic conditions in Europe. Over a 12 months period from February 2008 to February 2009 temperature and humidity records from 21 vehicles were collected and analysed. So far, data from February 2008 to Decenber 2009 are analyzed. Temperature and humidity were measured at 4 different positions in the vehicles, and differences were recorded between the highest and lowest temperature at any given moment. The position of the worst climatic conditions may, in fact, change within an animal compartment due to a number of factors. The number and position of temperature sensors are important for representing the most temperature conditions in the different parts of a vehicle. Beside the sensor location, the study indicates that in certain environmental conditions the presence of animals in a transport vehicle could be detected by the difference between the sensor in the front down part of the vehicle and the external sensor. Overall for all journeys and all animal categories a relatively low percentage of non-compliant journeys can be observed regarding the temperature thresholds in force. The EFSA thresholds would result in general in a higher percentage of non-compliant conditions. The inclusion of a tolerance rate would mitigate the mentioned effect.

Thermal stress in livestock during transportation: continuous recording of deep body temperature

Mitchell, M.A.[1], Kettlewell, P.J.[2], Villarroel, M.[3] and Harper, E.[4], [1]SAC, Bush Estate, Penicuik, Midlothian, EH26 0PH, United Kingdom, [2]ADAS, Alcester Road, Stratford Upon Avon, C37 9RQ, United Kingdom, [3]Escuela Técnica Superior de Ingenieros Agronomos, Avenida de Complutense, 28040 Madrid, Spain, [4]Livestock Transport Consultant, Bruton, Somerset, BA10 0JT, United Kingdom; malcolm.mitchell@sac.ac.uk

European regulations define thermal envelopes for the transport of livestock. As yet the validity of the legislation during commercial transport and in the pertinent thermal conditions has not been determined. In order to assess the physiological consequences of thermal challenges in transit it is necessary to monitor the animal deep body temperature (DBT) continuously with minimal human intervention. Both radio-telemetry and data logging can achieve these objectives. A study has employed both techniques to record deep body temperature of pigs and lambs on journeys under hot weather conditions typical of those encountered in southern Europe in summer. The journeys were of 8 hours duration and are typical of those associated with the transportation of animals to slaughter. Four journeys were undertaken with pigs and two with lambs. The patterns of DBT observed indicate that despite elevated ambient temperatures during the journeys and potential thermal challenge the DBT values for both pigs and lambs did not increase during the journeys and may decrease reflecting convective cooling in transit. Mean control and during-journey values for DBT indicate significant decrease ($P<0.05$ – ANOVA), typical values being 39.2 ± 0.4 °C and 38.9 ± 0.2 °C for pigs and 39.7 ± 0.3 °C and 39.2 ± 0.2 °C for lambs respectively. The results demonstrate that continuous monitoring of physiological variables in 'real world' transport conditions is an essential tool for assessing physiological stress and welfare. It is proposed that more detailed physiological information facilitates comprehensive assessment of the adequacy and pertinence of current and proposed animal welfare legislation.

Effects of the thermal micro-environment on breeder pigs on 72 hour export journeys under summer conditions

Mitchell, M.A.[1], Kettlewell, P.J.[2], Villarroel, M.[3] and Harper, E.[4], [1]SAC, Bush Estate, Penicuik, Midlothian, EH26 0PH, United Kingdom, [2]ADAS, Alcester Road, Stratford Upon Avon, Warwickshire, CV37 9RQ, United Kingdom, [3]Universidad Politecnica de Madrid, Escuela Tecnica Superior de Ingenieros Agronomos, Avenida Complutense, 28040 Madrid, Spain, [4]Transport Consultsant, 3 Fairview, North Brewham, Bruton, Somerset, BA10 0JT, Spain; malcolm.mitchell@sac.ac.uk

Long distance road transportation of breeding pigs may expose the animals to thermal challenges in transit. A study has been undertaken on female pigs (100 kg BWt) transporting them by road from the UK to Spain in summer conditions. On 7 replicate experimental journeys (72 hours), on a standard route and with commercial practices, the vehicle thermal micro-environment and ambient conditions were correlated with physiological and behavioural responses of the pigs. The average journey temperatures ranged from 14.8 °C to 22.2 °C and the maxima from 23.6 °C to 34.2 °C. The hotter conditions invariably occurred during the last stage (22 hours) of the journey in transit through Spain. The journeys covered a typical range of thermal conditions and transport micro-environments for Southern Europe. Significant differences ($P<0.01$) in deep body temperature ($P<0.01$) and in drinking and resting behaviours ($P<0.05$) were found between journeys. The journeys could be classified as mild, warm and hot reflecting both the imposed thermal loads and the physiological and bevavioural responses to the transportation stress. On none of the journeys was severe thermal stress identified. The data facilitated assessment of the severity of transport stress experienced by the pigs under a range of thermal conditions. The results suggest that if transportation is undertaken in a manner consistent with current legislation (temperature <35 °C), on appropriate vehicles and with high standards of personnel and practice there is little threat to the welfare of the pigs even in relatively hot conditions.

Health and welfare of horses transported for slaughter within the European Union

Marlin, D.[1], Meldrum, K.[2], Kettlewell, P.[3], Heard, C.[2], Parkin, T.[4], Kennedy, M.[5] and Wood, J.[6], [1]Hartpury College, Gloucester, GL19 3BE, United Kingdom, [2]World Horse Welfare, Snetterton, Norfolk NR16 2LR, United Kingdom, [3]ADAS, Ceres House, Lincoln LN2 4DW, United Kingdom, [4]Inst. of Comp. Medicine, University of Glasgow, Glasgow G61 1QH, United Kingdom, [5]Anglia Ruskin University, East Road, Cambridge CB1 1PT, United Kingdom, [6]Cambridge Infectious Diseases Consortium, Dept. of Veterinary Medicine, Cambridge CB3 0ES, United Kingdom; dm@davidmarlin.co.uk

Observational anecdotal evidence collected in the field by World Horse Welfare has shown evidence of poor welfare in horses transported for slaughter within the European Union. To date there has been limited scientific study of horses transported commercially to slaughter. An epidemiological investigation into the health and welfare of horses transported long distances to slaughter within the European Union was conducted between March and September 2008 by four veterinary surgeons. A total of 1519 horses in 64 separate shipments were observed prior to transport in Romania and 1,271 horses in 63 separate shipments were observed after transport in Italy. A high proportion of the horses observed in this study, either at origin, or at destination had evidence of poor health and welfare. Acute and chronic injuries were prevalent and lameness was common. On arrival in Italy, many horses exhibited clear signs of disease. There were frequent instances recorded of non-compliance with EU Regulation 1/2005 on the Welfare of Animals during Transport. The lack of provision of food and water to these animals in many instances is of extreme concern. Almost 2/3rds of horses leaving Romania had some form of external injury and 1 in 7 were considered unfit to be transported. Around 1/3rd of the horses arriving in Italy had cuts on their body and 1 in 3 were considered unfit to be transported. These results also support the contention that horses have distinctly different needs and responses to transport when compared with other farm/food animals, as has clearly been recognised in transport of pleasure, sport and the racehorses.

Animal transport navigation system an events recording automation

Fiore, G.[1], Natale, F.[1], Hofherr, J.[1], Bonavitacola, F.[1], Di Francesco, C.[2], Di Pasquale, A.[2] and Zippo, D.[2], [1]EU Commission-JRC, CI-Animal&Food Action, Via Marconi, 21020 Ispra (Varese), Italy, [2]Istituto Zooprofilattico Sperimentale, Via Campo Boario, 64100, Teramo, Italy; gianluca.fiore@jrc.it

The EU legislation for animal welfare during transports requires that vehicles should be equipped with a navigation system. Although arising from the animal welfare legislation such a navigation system could be used for improving and supporting the traceability of animal transports. In addition to the current systems of traceability, mainly based on documentary notifications and authorisation of transports, the navigation system would allow to track in real time animal movements not only by identifying the origin and destination of the movement but also its real geographical path and timing. Furthermore such system could be used to partially automate the movement notifications reducing in this way the administrative burden for the holdings of origin and destination. In order to archive such advantages the minimum technical requirements of a purely animal welfare navigation system already set out by JRC need to be expanded by including the identification of individual batches or animals transported. A new prototype of navigation system has been developed which includes such additional functionalities and the integration with animal identification trough RFID readers at the loading and unloading of the animals.

Effects of transportation on the physiophatological response of light lambs raised under pasture or concentrate feeding systems

Álvarez-Rodríguez, J.[1], Uriarte, J.[2], Sanz, A.[1] and Joy, M.[1], [1]Centro de Investigación y Tecnología Agroalimentaria, Gobierno de Aragón, Tecnología en Producción Animal, Av. Montañana, 930, 50059, Zaragoza, Spain, [2]Centro de Investigación y Tecnología Agroalimentaria, Gobierno de Aragón, Sanidad Animal, Av. Montañana, 930, 50059, Zaragoza, Spain; mjoy@aragon.es

The aim of this study was to analyse the effects of transportation on the immune and haematologicalparameters of light lambs raised under pasture or concentrate-feeding systems. Male Churra Tensina lambs (n=38) suckled their dams and grazed until slaughter (GR) or were weaned at 7 weeks and fattened on a concentrate-based diet (CO). They were transported to the abattoir when attaining 23 kg (86 vs. 74 days old in GR and CO, respectively). Blood samples were collected just before and after 2 h transport to determine leukocyte and erythrocyte populations as well as thrombocytes with a cell counter. Any leukocyte except eosinophils were affected by the feeding system (mean white blood cell number 9.7 x 10^3/μl, $P>0.10$). Eosinophils tended to be greater in GR than in CO (0.7 vs. 0.4%, $P=0.09$). Transportation increased neutrophils (20.1 vs. 28.5%, $P<0.05$) and monocytes (9.6 vs. 10.6%, $P<0.05$), but decreased lymphocytes (69.7 vs. 60.1%, $P<0.05$). Erythrocytes were not altered with the feeding system nor transport (mean red blood cells 10.8 x 10^6/μl, haematocrit 28.5%, haemoglobin 9.8 g/dl, $P>0.10$), but platelets tended to be lower in GR than in CO (599 vs. 1010 x 10^3/μl, $P=0.06$). There were slight stress-related effects of 2-h lamb transportation on the immune function, which were similar in both feeding systems, without important alterations in the haematological parameters.

Effects of loading methods on welfare of rabbits transported to the slaughterhouse

Giammarco, M., Vignola, G., Manetta, A.C., Lambertini, L. and Mazzone, G., University of Teramo, Food Science Dept., Viale Crispi 212, 64100 Teramo, Italy; mgiammarco@unite.it

The effect of different loading methods on welfare of rabbits transported to the slaughterhouse was investigated. A total of 384 rabbits, 82 days old, were transported from farm to the abattoir for a mean transport time of 100 min. At farm, 192 rabbits were loaded on the truck in a smooth way (S) (rabbits from the farm crates were placed in a wide trolley and carried gently into the transport cages, at a density of 12 animals per cage) and 192 rabbits were loaded, as is usually done, in a rough way (R) (rabbits from farm crates were carried all together in the same trolley and loaded hurriedly into the transport cages). Blood samples from 80 male rabbits were collected before and after transport and analysed for haematological and biochemical parameters. Independently from loading method, rabbits after transport showed a significant leukocytosis (11.20 vs 14.04 x1000/mcl, $P<0.001$) with neutrophilia (35.40 vs 50.96%, $P<0.001$) and lymphocytopenia (56.81 vs 38.56%, $P<0.001$); moreover a significant increase of AST (aspartate amino transferase) (25.39 vs 32.58 IU/l, $P=0.001$), ALT (alanine amino transferase) (31.84 vs 38.04 IU/l, $P<0.001$) and CK (creatine phosphokinase) activities (885.86 vs 2,905.76 IU/L, $P<0.001$) was recorded. Also glucose increased significantly after transport (139.98 vs 160.64 mg/dl, $P<0.001$) as well as total proteins (5.80 vs 6.01 g/dl, $P<0.01$) and osmolality (309.28 vs 334.79 mOsm/l, $P<0.001$). The stress effect exerted by transport was evidenced by the significant upsurge of corticosterone level (6.23 vs 14.88, $P<0.001$) and by the increasing of the neutrophils/lymphocytes ratio (0.67 vs 1.50, $P<0.001$). Haematological and biochemical data were not influenced by the different loading methods. Results obtained showed that stress parameters analyzed were more influenced by transport and handling itself rather than by the different loading methods on the truck.

Effect of different moving devices at loading on heart rate and blood lactate in pigs

Correa, J.A.[1], Gonyou, H.[2], Torrey, S.[3], Devillers, N.[3], Laforest, J.P.[1] and Faucitano, L.[3], [1]Université Laval, Quebec City, G1K7P4, Canada, [2]Prairie Swine Centre, Saskatoon, S7N5A9, Canada, [3]Agriculture & Agri-Food Canada, Sherbrooke, J1M1Z3, Canada; faucitanol@agr.gc.ca

The objective of this study was to evaluate the effects of alternative moving devices, such as the paddle and compressed air prod, compared to electrical prod at loading on heart rate and blood lactate concentration in pigs. A total of 360 pigs (120 ± 7 kg live weight) were randomly sorted out from the finishing pen and distributed according to three moving procedures: 1) moving with an electric prod and board from the finishing pen to the truck (E); 2) moving with a board and a paddle from the finishing pen to the truck (P); 3) moving with a board and a paddle from the finishing pen and using a compressed air prod in the ramp before going into the truck (A). A sub-population of 144 pigs (48 pigs/treatment) was equipped with heart rate monitors (Polar Electro Canada). Heart rate was recorded at 5 sec intervals and averaged over the following events: loading (L), transport (T), unloading (U) and lairage (LA). Blood samples were collected at exsanguination for the analysis of lactate. Data were analysed using an ANOVA for factorial design, with the animal as the experimental unit. Pigs handled with E showed a higher heart rate than those moved with P and A (187 vs 163 and 155 b/min, respectively; $P<0.0001$) at L, at U (168 vs 159 and 158 b/min, respectively; $P<0.05$) and in LA (120 vs 116 and 115 b/min, respectively; $P<0.05$). Heart rate did not differ between treatments during T. Pigs loaded with E had higher ($P<0.05$) lactate concentration in blood at exsanguination compared to pigs handled with A (18.3 vs 15.9 mmol/l), with pigs loaded with P being intermediate (17.4 mmol/l). Overall, these results suggest that from an animal welfare standpoint the board in combination with compressed air prod or paddle may be a valid alternative to the electric prod for pig handling at loading.

Animal welfare risk assessment during transport of pigs

Fuentes, M.C., Otero, J.L., Dalmau, A. and Velarde, A., IRTA, Finca Camps i Armet s/n, 17121 Girona, Spain; carmen.fuentes@irta.cat

Animal transport is considered to be one of the most suffering processes during the lifecycle of farm animals and it also influences meat quality and safety. In 2007, nearly sixty million of farm animals were transported, of which 33 million were pigs (source: TRACES). Since there has been a lot of discussion about transport of farm animals within Europe the European Food Safety Authority (EFSA) launched a call on the development of general guidelines and working methodology on risk analysis for animal welfare during transport, as non specific international guidelines are currently available. The risk assessment (RA) approach is based on 1) identification of hazards (environmental factors that may compromise animal welfare), 2) the characterization of the hazards identified (estimation of the impact of each hazard on the individual animal), 3) exposure assessment (estimation of the percentage of animals in the population exposed to each hazard), and 4) risk characterization, where the risk of each hazard is characterized in terms of both hazard characterization and exposure assessment. A separate risk analysis was performed for long (> 8 h) and short transports (< 8 h) and for different categories of animals (postweaning piglets, sows and boars, and slaughter pigs). The total number of hazards identified were 30 for postweaning piglets, 29 for sows and boars, and 27 for slaughter pigs. The hazards characterized as serious or very serious in these categories were: 40% for postweaning piglets, 35% for sows and boars, and 30% for slaughter pigs. Further results concerning animal welfare RA will be presented in the congress. The most significant difficulty in animal welfare risk assessment consists on the quantification of the severity and of the probability of exposure of the identified hazards to welfare. However, the use of a RA approach to evaluate issues related to animal welfare can be useful to better identify and rank welfare risk factors, and to prioritise possible management measures.

Selection in harsh environments: stakes and strategies for which results?

Mandonnet, N. and Alexandre, G., INRA-URZ, Animal genetics, Domaine Duclos, 97170 Petit-Bourg, Guadeloupe; Nathalie.Mandonnet@antilles.inra.fr

In developing regions livestock production is threatened by major climatic and biotic constraints. However the stake of animal production is to fulfill the future human food demand. The objectives of selection in harsh environments are then to increase animal productivity while assuring sustainability of breeding systems. Climate changes and go back to more extensive production systems make these tropical objectives acute for temperate areas as well. We notice that sustainable livestock improvement cannot be guaranteed for most harsh areas without utilization of the natural adaptation of the indigenous livestock breeds or at least without including adaptation and behavior traits as breeding objectives for indigenous or specialized breeds. A methodology is proposed to develop appropriate breeding strategies in these contexts. Successes and failures of breeding programs in the world are overviewed, highlighting key elements (breed choice, reliability of data recording, need of holistic approach, involvement of farmers, ...) for a sustainable development of small ruminant productions in the future.

Selection of beef cattle in harsh environments

Burrow, H.M., Cooperative Research Centre for Beef Genetic Technologies, CJ Hawkins Homestead, UNE, Armidale 2351, Australia; Heather.Burrow@une.edu.au

Cattle grazed at pasture in harsh environments are subjected to numerous stressors including ecto- and endo-parasites, seasonally poor nutrition, extreme (hot and cold) temperatures and/or humidity and endemic diseases often transmitted by parasites. The impact of each stressor on production and animal welfare is often multiplicative rather than additive, particularly if animals are already undergoing physiological stress (e.g. lactation). Under extensive production systems common in harsh environments, it is generally not feasible to control the stressors through management strategies alone. The best way to reduce the effects of these stressors to improve productivity and animal welfare is to breed cattle that are productive in their presence, without the need for managerial interventions. The availability of diverse breed resources with large differences in productive and adaptive attributes allows selection of the most appropriate breeds for use in harsh environments, thereby maximising the opportunity for high productivity. Research results from tropical beef cattle show that most adaptive traits are moderately to highly heritable, indicating they can be directly improved by genetic selection within breeds. Those results also suggest that adaptive traits are largely independent of productive attributes, meaning that selection to improve them genetically is unlikely to compromise growth, efficiency of feed utilisation, reproduction or product quality. However adaptive traits are difficult to include in genetic evaluation systems because of the difficulty of their measurement. Use of DNA markers associated with productive and adaptive traits therefore provides significant new opportunities in future to simultaneously improve all traits in the breeding objectives of livestock reared in harsh environments.

Comparison of local and crossbred cattle populations for resistance to high altitude disease in Ethiopia

Wuletaw, Z.[1], Wurzinger, M.[1], Holt, T.[2] and Sölkner, J.[1], [1]BOKU-University of Natural Resources and Applied Life Sciences, Gregor-Mendel-Str. 33, 1180 Vienna, Austria, [2]Colorado State University, Fort Collins, CO 80523-1678 Fort Collins, USA; johann.soelkner@boku.ac.at

High altidude disease (brisket disease) of cattle is a noninfectious, congestive heart failure caused by pulmonary hypertension. The incidence in the Rocky Mountains is 2 percent. Two studies were conducted to assess the prevalence rate of brisket disease and to compare adaption of indigenous cattle populations and their crosses with European types to high altitude in the Simien Mountains of Ethiopia, where cattle are kept up to 4,000 m. Pulmonary arterial pressure (PAP) is used as reliable predictor of susceptibility of an animal to the disease. In the first study a total of 218 animals, local breeds and crosses with Holstein Friesian and Jersey, found in the range of 1,700-3,500 m were tested for their PAP. Results show that no sign of brisket disease is observed. All PAP scores (21-47 mm Hg) fall within the range of low to moderate risks. Differences in means were not significant. Subsequently, a second study was conducted with 13 animals (< 1 year of age) composed of three local breed types and a crossbred of local x Holstein Friesian found in the range of 550-3,500 m altitude. The animals were transported to the high altitude area, 3,500 m, and kept for two months. PAP scores taken at the end of testing period ranged between low to high risk (22-7 mm Hg), but difference in means were not significant The lowland cattle group (550-750 m) shows the highest mean value making it a high-risk candidate for use in high elevation environment. The good adaptation of breeds developed in high altitude areas is most likely due to natural selection. Crosses with European cattle developed at low altitude do not seem to be affected. Yak and crosses of yak and cattle are tolerant to altitude. Action of a dominant major gene is hypothesised there. To get an insight on the mechanism of adaptation an histological study on respiratory systems is currently being undertaken.

Interactions between genotype and environment for cattle growth in the tropics: implications for selection

Naves, M.[1], Vallee, A.[1], Farant, A.[2], Mandonnet, N.[1], Menendez Buxadera, A.[1] and De La Chevrotière, C.[1], [1]INRA, UR143 URZ, Domaine Duclos, 97170 Petit Bourg, Guadeloupe, [2]INRA, UE1294 PTEA, Domaine de Gardel, 97160 Moule, Guadeloupe; claudia.chevrotiere@antilles.inra.fr

This study investigates the genotype by environment (GxE) interaction on post weaning growth, and variations in phenotypic plasticity in tropical conditions on the local Creole beef cattle. The experiment was carried out with 644 animals separated after weaning in two different systems. The first group received an intensive feeding regime (I), while the other was conducted at pasture (P). Animals were weighted every fifteen days. Samples of faeces have also been collected to evaluate the digestibility of the regime. Finally, the males were sacrificed and measurements have been conducted on carcasses. Different statistical analyses have been used according to the variable: GLM and mixed model multivariate analyses, and Random Regression Model to estimate the genetic (co)variance over the whole growth curve. As expected, the growth rate was 40% lower in P than in I, and the heritability of the weights was 60% smaller in P than in I. However, genetic correlations (rg) between weights in both systems tended to decrease with the age, supporting the evidence of the GxE interaction. Moreover, the relationship between the breeding values for the final weights at 15 months in I and 18 months in P shows a low rg. This GxE interaction affected the ranking of the animals. Only 1/3 of the 10% best animals in one fattening system were also the best in the other, and about 20% of the animals contributing to the genetic gain of one trait make the other trait decreased. The correlated response to selection in the alternate system is only 60% of the response expected in the same system. Therefore, these results should be taken into account in selection programs in the tropics, for their consequences on genetic gain expected in various systems. The existence of similar GxE interaction on traits linked to nutrition or tissue development is under investigation, but some preliminary results will be discussed in the paper.

Adaptive traits for sheep selection in harsh environment

Francois, D.[1], Boissy, A.[2], Foulquie, D.[3], Ligout, S.[2], Autran, P.[3], Allain, D.[1], Bibé, B.[1] and Bouix, J.[1], [1]INRA, UR 631 SAGA, chemin de Borde Rouge, 31326 Castanet-Tolosan, France, [2]INRA, UR 1213 Herbivores ACS, Theix, 63122 St Genès Champanelle, France, [3]INRA, UE 321 Domaine de la Fage, St Jean et St Paul, 12250 Roquefort sur Soulzon, France; Dominique.Francois@toulouse.inra.fr

An experiment has been conducted about selection of sheep in dry environment on the Larzac plateau (South of France). The Romane (INRA 401) flock has been raised outdoors all along the year and fed on rangelands. Adaptive abilities have been investigated following behavioural and fleece traits. Genetic effects have been estimated on three crops of 1,111 lambs issued from 15 sires. The survival rate has been related to fleece composition collected on lambs at birth. Hairy birth coat influenced favourably birth survival rate from 3 points to woolly one (93% vs 90%). Heritability of birth coat type was 0.56. Behavioural traits have been measured in weaned lambs (i.e. 75 day-old) during two standardised tests assessing attractiveness to social partners and to human (arena test and corridor test). The 30 original variables have been synthesised into 8 behavioural traits from PCA. Heritability estimates ranged from 0.04 (social attraction) to 0.41 (high bleats after isolation), avoidance of human estimate being medium with 0.29. A preliminary primo-detection of QTL with 72 microsatellites markers on 4 families of 361 lambs led to 7 QTL on 6 chromosomes for 5 traits. Favourable genetic correlations between the 8 behavioural traits and weak phenotypic correlations with growth and carcass traits have been found. Genotype by Environment Interaction has been estimated from outdoors and indoors lambs issued both from same sires, it seemed weak since sires ranged similarly for 6 of the 8 traits. Despite a need for additional investigations, hairy birth coat, reactivity to human and social attractiveness could be included in selection program in the purpose of improving the adaptation of farm animals to harsh environments. Low GxE to be confirmed allows selection in better environment.

The birthcoat type: an important component of lamb survival in the French Romane breed raised under permanent exposure outdoors

Allain, D.[1], Foulquie, D.[2], François, D.[1], Autran, P.[2], Bibe, B.[1] and Bouix, J.[1], [1]INRA, UR631, SAGA, Chemin de borde rouge, BP 52627, 31326 Castanet Tolosan, France, [2]INRA, UE321, Domaine de la Fage, Saint Jean et Saint Paul, 12250 Roquefort, France; daniel.allain@toulouse.inra.fr

Sheep meat industry in France is currently changing and leaving intensive agricultural areas for less favourable lands. Rejecting the classical choice of hardy low-performing animals in harsh environment, opposite to high-productive animals in favourable conditions, the French Romane breed (INRA401) was proposed as an interesting genetic strategy allowing good economical results in harsh areas such as permanent exposure outdoors in mountains areas. This breed is known to have a high productive potential and show a large variability in its fleece type leading to a high variable birthcoat type in lambs. The aim of the present study was to characterize the coat of the lamb at birth and to quantify lamb survival in relation to the birthcoat type and its protective aptitude concerning heat loss in the French Romane breed raised under permanent exposure outdoors from birth. A total of 5,702 lambs were used in a 10 years experiment. Birthcoat type, coat surface temperature, coat depth, lamb survival and growth were measured from birth to weaning. At birth two types of coat were observed: a hairy coat (64.5% of lambs) with a long coat depth (22.9 mm) or a woolly one (35.5% of lambs) with a short coat depth (8.4mm). Heritability estimate of birthcoat type (hairy or woolly) is high (0.56). It was shown that hairy-bearing coat lambs are more adapted to survive around lambing time (10.5% vs 13.8% and 14.5% vs 18.1% for total mortality rate at 10 and 50 days, respectively) due to a better coat protection with less heat losses at coat surface (21.1 °C vs 26.1 °C) and show better growth performances up to the age of 10 days than fine-bearing coat lambs. Birthcoat type has to be taken into account as an adaptation trait in any genetic strategies for sheep production in harsh conditions.

GxE interactions in Churra dairy sheep: heterogeneity across lactations and reaction norms

Sánchez, J.P., De La Fuente, L.F., Carriedo, J.A. and San Primitivo, F., Univerity of León, Producción Animal, Campus de Vegazana s/n, 24071 León, Spain; jpsans@unileon.es

Churra sheep has shown lower estimates of genetic parameters for production traits than other breeds, when the same repeatability animal models were employed to infer them. Our hypothesis is that GxE interactions and heterogeneity across lactations have enough magnitude in Churra to explain the lower estimates of genetic parameters for production traits in this breed. To investigate this hypothesis 4 models were considered to fit 77,251 test-day milk yield records from 7,998 ewes: Repeatability Animal Model (RAM); RAM with RR on HTD effect (RAM-RR), for fitting reaction norms on HTD effect; MultiTrait Repeatability Animal Model (MT-RAM); and MT-RAM-RR. In MT models 1st, 2nd and >=3rd lactation records were treated as different but correlated traits. Model parameters were inferred using an EM-REML algorithm. AIC and LRT clearly supported the hypothesis of heterogeneity across lactations and significant variation on reaction norms. From RAM estimated heritability was 0.13, and from MT-RAM variation on this parameters was observed across lactations, 0.16, 0.15 and 0.11, for 1st, 2nd and >=3rd lactations, respectively. RAM-RR revealed that heritability estimates for milk yield ranged from 0.14 in an average environment (0 effect of HTD), to 0.39 in the best situation (+2 s.d.) and 0.07 in the worst one (-1.5 s.d.). From model MT-RAM-RR similar results were observed for each lactation separately, being stronger the effect of GxE interaction during the 1st lactation. Performance under the average environment shown positive and high (>0.9) genetic correlations with production under positive or no-strongly negative environments (>=-0.4 s.d.), but it reached negative values in 1st and 2nd lactations with performances in harshest environments (<-0.5 s.d.). Considering GxE interactions and heterogeneity across lactations in evaluation models may improve the efficiency of selection schema.

Production systems of Creole goat in Guadeloupe and farmers' selection criteria

Jaquot, M.[1], Mandonnet, N.[1], Arquet, R.[2], Naves, M.[1], Mahieu, M.[1] and Alexandre, G.[1], [1]INRA, UR 143 URZ, Domaine Duclos, Prise d eau, 97170 Petit-Bourg, Guadeloupe, [2]INRA, UE 1294 PTEA, Domaine de Gardel, 97160 Le Moule, Guadeloupe; Nathalie.Mandonnet@antilles.inra.fr

Creole goat is a local population well adapted to its tropical environment and devoted to meat production. A selection scheme for this breed will be implemented in Guadeloupe, a French island in the Caribbean. A survey of 47 farmers was carried out in May 2008 to identify the main production systems and farmers' selection criteria. Farmers had on average 7.4 ha of land for 31 does and a total of 60 goats. Only 9% of them relieved on goat farming. The other had either off-farms activities or some other vegetal or animals productions. A small number of farmers (4%) kept only Creole goats. Most of them (58%) had mixed herd of Creole and crossbreds. A third of them (34%) reared only crossbreds goats. Crossbreds resulted mostly from uncontrolled mating between Creole, Boer and Anglo-Nubian breeds. Farmers appreciated Creole's rusticity and resistance but considered its growth as too slow. Goat selection criteria were conformation and growth for males (answer frequency 77%). These criteria were also important for females (answer frequency 30%). Nevertheless, maternal qualities were frequently cited (maternal behaviour 23%, reproduction 20%, milk production 17%). Health and diseases resistance were ranked poorly even though they had important economical consequences. A typology of farmers was created using a Multiple Factor Analysis followed by a Ward hierarchical classification. Five groups were found: 1) average farms (32%), 30 does, 5 ha; 2) big and traditional farms (11%), 60 does, 11 ha; 3) big extensive and highly diversified farms (32%), 20 does, 9 ha; 4) big and well-organised farms (15%), 40 does, 11 ha; 5) small and highly-specialised farms (11%), 30 does, 2 ha. Farmers of the first, third and fifth group were the most interested in a breeding program. Creole goat has a key-role to play for the sustainability of Guadeloupean and Caribbean rearing systems. A selection scheme will emphasize its importance.

Study on performance and estimation of genetic parameters for milk and fat yields in crossbred goats

Abbasi, M.A., Animal Science Research Institute, Dept of Animal Breeding and Genetic, Dehghan Villa 1, 3146618361, Iran; pmaz_abbasi@yahoo.com

In order to compare milk yield and fat yields of imported Sannan, Iranian Sannan, Najdi and Sannan*Najdi goats, 3,113 test day data collected from Garacharian station of Zanjan agricultural research center were used. The effects of genetic groups, year, season and litter size on milk and fat yields were tested by fixed model using SPSS software. Variance components and genetic parameters of traits were estimated by animal model (model 1 and 2 presented by Meyer) using DFREML package. The effect of genetic groups on milk and fat was significant but the effect of other fixed effects were not significant. The differences between milk and fat yields of Sannan and crossbred goats as compared with Najdi were considerable. The crossbreeding of native goats with high milk yield goat breeds such as Sannan caused high milk yield in crossbred goats. The phenotypic, additive genetic and residual variances of milk yield were estimated as 7,853.56, 3,470.96 and 4,112.60 with model 1 and 7,774.36, 3,200.86 and 4,093.36 with model 2. The permanent environmental variance due to animal was estimated as 460.14 with model 2. Including permanent environmental effects in model 2 decreased additive genetic variance. Estimates of the phenotypic, additive genetic and residual variances of fat yield were 12.25, 3.17 and 9.09 with model 1 and 12.25, 3.17 and 9.06 with model 2 resp. The heritability of milk yield was estimated as 0.48±0.12 and 0.41±0.31 with model 1 and 2 resp. The heritability for fat yield was 0.26±0.11 with both models. Generally, the results of this study showed that, crossbreeding of native goats with Sannan will increase milk and fat yields in crossbred goats. Also high hritabilities of these traits especially for milk yield suggested that the application of suitable breeding program based on intra herd selection will increase milk yield of native goats.

Determination of the suitable selection index for Baluchi sheep using computer simulation

Nemati, I.[1], Abbasi, M.A.[2], Eskandari Nasab, M.P.[1] and Jalali Zenos, M.J.[1], [1]Zanjan University, Animal Science, Zanjan city, 3146618361, Iran, [2]Animal Science Research Institute, Dept of Animal Breeding and Genetic, Dehghan Villa 1, Karaj, 3146618361, Iran; pmaz_Abbasi@yahoo.com

The objective of this study was the comparison of different selection indexes for Baluchi sheep with combination of 4 traits including litter size(LS), weaning weight, average daily gain and grease fleece weight using computer simulation. Relative economic value of these traits were 240, 8, 0.424 and 1. Breeding population simulated with the structure of sire referencing strategy included ten herds with 100 ewes per herd. Simulation program was written by visual basic 6. Selection indexes based on one trait were litter size (I1), weaning weight (I2), average daily gain (I3) and grease fleece weight (I4). The aggregate genotype of I1 was significantly higher than I2, I3 and I4 ($P<0.01$). Indexes with two traits were litter size and weaning weight (I5), litter size and average daily gain (I6), litter size and grease fleece weight (I7), weaning weight and average daily gain (I8), weaning weight and grease fleece weight (I9) and average daily gain and grease fleece weight (I10). The aggregate genotype of I5 (61.64) was significantly higher than the other two traits indexes ($P<0.01$). Selection index including litter size, weaning weight and grease fleece weight (I12) showed a high aggregate genotype (61.77), but the difference wasn't significant with aggregative genotype of I5. Application of selection index with four traits (I15) caused the aggregate genotype to increase to 66.64. This variant was significantly higher than other Indexes. In general, the results showed, I15 was a suitable selection index in view of the aggregate genotype among the fifteen indexes, but for the reasons of high inbreeding coefficient in I15 rather than I5, to involve expense recording for two traits such as average daily gain and grease fleece weight and the low economic value of these traits, I5 can be introduced as a suitable index for Baluchi sheep.

Genetic parameters for growth traits in Romane sheep

David, I.[1,2], François, D.[1,2], Bouvier, F.[3], Bodin, L.[1,2], Bibé, B.[1,2] and Bouix, J.[1,2], [1]INRA, UR631 SAGA, chemin de borde rouge, 31320 Castanet Tolosan, France, [2]INRA, UMT GENEpR, chemin de borde rouge, 31320 Castanet Tolosan, France, [3]INRA, UE332, Domaine de la Sapinière, 18390 Osmoy, France; ingrid. david@toulouse.inra.fr

Estimates of genetic correlation between direct and maternal effects for growth traits are often negative in field data. The biological existence of this genetic antagonism has been the point at issue. Some researchers perceived such negative estimate to be an artefact from poor modelling and poor data structure. Data from 19203 Romane (INRA401) lambs born between 2000 and 2009 in the experimental farm of la Sapinière (INRA-FRANCE) were used to estimate genetic parameters of different growth traits. In this particular dataset, animals were bred in the same indoor system, the pedigree information contains minimal sire misidentification, data were balanced between sires and measurements were perfectly controlled and standardized. Therefore data structure problems are limited. Traits analysed were weight at birth (BW), 45 days (W45), 64 days (W64) and 90 days (W90), average daily gains from birth to 45 days (ADG0-45), 45-64 days (ADG 45-64) and 64-90days (ADG64-90). REML estimates of variance and covariance components were obtained assuming animal-maternal, paternal-maternal grand sire, maternal-paternal models. Depending on the model, preliminary results for direct heritability range from 0.20 to 0.22 (BW), 0.12 to 0.17 (W45), 0.16 to 0.19 (W64), 0.19 to 0.24 (W90), 0.10 to 0.15 (ADG0-45), 0.25 to 0.28 (ADG45-64) and 0.16 to 0.19 (ADG64-90). Estimates of maternal heritability vary from 0.19 to 0.27 (BW), 0.09 to 0.10 (W45), 0.07 to 0.09 (W64), 0.06 to 0.08 (W90), 0.06 to 0.07 (ADG0-45) and 0.04 to 0.06 (ADG45-64). Direct-maternal genetic correlation estimates range from -0.19 to -0.12 (BW), 0 to 0.03 (W45), -0.02 to 0.07 (W64), -0.24 to -0.08 (W90), 0.08 to 0.11 (ADG0-45) and -0.54 to -0.31 (ADG45-64) which is less negative than expected. An experimental protocol is set up for better understanding these results.

Additional casein variants in sheep breeds with potential to be included in animal breeding

Giambra, I.J., Jäger, S. and Erhardt, G., Institut für Tierzucht und Haustiergenetik, Justus-Liebig-Universität Giessen, Ludwigstr. 21b, 35390 Giessen, Germany; Isabella.J.Giambra@agrar.uni-giessen.de

Sheep milk is in many countries the basis of regional dairy products but the knowledge about milk proteins and their genetic variation is limited. In a first step isoelectric focusing was applied for screening milk protein variants in Black Faced Mutton (n=57), East Friesian Milk Sheep (n=254), Gray Horned Heath (n=190), Merino Land Sheep (n=363), Merino Mutton Sheep (n=88) and Rhön Sheep (n=126). Beside the known genetic variants of α_{s1}-casein (CN) (A, C, D), α_{s2}-CN (A, B), and ß-lactoglobulin (A, B, C) additional variants could be demonstrated in α_{s1}-CN (H, I) and α_{s2}-CN (C, D). Their genetic basis was confirmed by examination of pedigree data with CFC Release 1.0. Alpha-lactalbumin and κ-CN were monomorphic at protein level in the breeds investigated. Further molecular genetic characterization of the ovine α_{s1}-casein gene (CSN1S1) revealed missing of complete exon 8 in mRNA of CSN1S1*H while at DNA level the sequence corresponding to exon 8 is present. Sequence differences concerning 5´-donor splice site of intron 8 were detected and determined as reason for alternative splicing in silico with NNSPLICE 0.9. Within the amino acid sequence of the protein this leads to an abbreviation of 8 amino acids. This will be associated with a reduced milk protein content as already shown in other ruminants. The first screening of different sheep breeds with the identification of further genetic variants shows the variability in milk proteins which will result in additional casein haplotypes. This offers a potential for association studies with milk yield and composition to improve milk production by breeding but also for phylogenetic and diversity studies.

Variability of beta-casein gene and relationships with milk traits in the Sarda goat

Dettori, M.L., Pazzola, M., Piras, G., Dhaouadi, A., Carcangiu, V. and Vacca, G.M., Università degli Studi di Sassari, Dipartimento di Biologia Animale, via Vienna 2, 07100, Italy; gmvacca@uniss.it

In goat milk the beta-casein is one of the most abundant proteins. Seven alleles have been identified in goat beta-casein gene (CSN2): A, A1, C, C1 and E, associated with a normal content (about 5g/l for allele) of this protein in milk, and 0 and 01, associated with a non-detectable amount. The CSN2 gene spans nine exons ranging from 24 bp (exon 5) and 492 bp (exon 7). In the exon 7, which codifies for about 82% of the mature protein, are localized the nucleotide changes characterizing the alleles A, C, E and 01. The aim of this research was to evaluate allele frequencies of beta-casein in Sarda goats and their influence on milk yield and composition. Blood samples were collected from 220 lactating goats and 80 bucks belonging to 20 flocks. Genomic DNA was analysed through a PCR single strand conformation polymorphism (SSCP) analysis, which allows the simultaneous detection of the CSN2 A, C, E and 01 variants. Milk samples were collected from each female goat in middle lactation, and milk yield was registered at the same time. Percentage of fat, protein and lactose were determined by IR spectroscopy. Allele and genotype frequencies and HW equilibrium were analysed using the GenePop software package. GLM was performed to show the links between genotype and milk traits. Female goats displayed the following genotype frequencies: AC = 45.45; CC = 35.91; AA = 15.46; C01 = 2.27; A01 = 0.91. The most frequent allele was C (0.598), followed by A (0.386) and 01 (0.016). Bucks gave the following results: AC = 46.25; CC = 37.50; AA = 7.50; C01 = 3.75; A01 = 5.00. The most frequent allele was C (0.625), followed by A (0.331) and 01 (0.044). No subject carried the E allele. Both examined populations were in HW equilibrium at this locus. Genotype CSN2 AA evidenced higher values ($P<0.05$) for milk protein percentage compared to the genotypes CC and AC. Data shows that also in this casein fraction strong alleles prevail indicating good cheese making properties of the Sarda goat milk.

SNP discovery in the ovine ABCG2 gene

García-Fernández, M., Gutiérrez-Gil, B., García-Gámez, E. and Arranz, J.J., University of León, Department of Animal Production, Campus de Vegazana s/n, 24071, Spain; beatriz.gutierrez@unileon.es

Breast Cancer Resistance Protein (ABCG2), a member of the ATP-binding cassette (ABC) transporters superfamily, is involved in the transport of xenobiotics from cells. This protein is strongly induced in the mammary gland during lactation. In dairy cows, a single nucleotide mutation in exon 14 of the ABCG2 gene has been proved as a QTN that results in decreased milk production but increased milk fat and protein concentration and yield. The underlying mechanism of that association is not yet understood. As an initial step to assess the influence of ABCG2 in milk production traits in dairy sheep, we have investigated the genetic variability of this gene. With this purpose, and based on the bovine genomic sequence of this gene, a primer pair was designed to amplify each of the complete 16 exons of the ovine ABCG2 gene, as well as part of the intronic flanking sequences corresponding to each exon. A total of 15 unrelated rams from the Selection Nucleus of Spanish Churra breed are being sequenced for the resulting amplicons. Up to now, we have sequenced 11 exons and have identified a total of 27 SNPs. Twelve polymorphisms were detected in the coding region of the gene (exons 6, 7, 9 and 16), one in the gene 3´UTR region and ten were found in the intronic sequences. Four additional allelic variants were detected at the 5' upstream and 3' downstream sequences of the gene. Similar number of transversions and transitions were detected. Across all the polymorphisms, the minor allele frequency (MAF) ranged from 0.09 (for the SNP located in the 5´-upstream region) to 0.5 (for one SNP found in exon 9). In general, MAF was higher for the SNPs localised in the coding sequence of the gene than for those identified in the non coding sequence analysed. Future research will address association analysis between some of the allelic variants reported herein and milk production traits recorded in a commercial population of Churra sheep.

Characterization of the ovine diacylglycerol acyltransferase1 (DGAT1), long chain fatty acid elongase (LCE) and sterol regulatory element binding protein-1 (SREBF1) genes

Dervishi, E.[1], Martinez-Royo, A.[1], Serrano, M.[2], Joy, M.[1] and Calvo, J.H.[1,3], [1]CITA-Aragón, Avda. Montañana 930, 50059, Spain, [2]INIA, Ctra. La Coruña km 7.5, 28040, Spain, [3]ARAID, Avda. Montañana 930, 50059, Spain; edervishi@aragon.es

Milk fat is an important factor influencing the nutritional and technological quality of dairy animal products. In this work, we have characterized three functional candidate genes related to milk fat content and fatty acids profile. Partial DNA fragments of sheep Long chain fatty acid elongase (LCE) and sterol regulatory element binding protein-1 (SREBF1) genes have been isolated from mRNA extracted from mammary gland. Primers were designed from the bovine contig and sequences deposited in GenBank. Genomic DNA from five domestic sheep breeds (Latxa, Manchega, Churra Tensina, Assaf, and Rasa Aragonesa) was used to search polymorphisms. Four partial fragments of these genes have been amplified and sequenced. Subsequently, we have used the ClustalW software to identify polymorphisms. The total coding region sequence of the diacylglycerol acyltransferase1 gene (DGAT1) was sequenced. Six partial fragments have been amplified, sequenced and aligned using ClustalW. Three SNPs were found: one in exon 13, one in exon 17 and one in intron 15. The SNP in exon 17 was a silent mutation while the SNP in exon 13 was an Alanine for Proline substitution. Finally, the specific expression of these genes in different tissues was investigated by RT-PCR.

Prediction of sperm production using hormonal factors in ram by artificial neural networks and linear regression

Hosseinnia, P.[1], Soltani, M.[2], Doosti, M.[2], Moladoost, K.[3], Karimi, D.[2], Abbasi, H.[2] and Taheri, A.[2], [1]Isfahan University of Technology, Department of Animal science, Isfehan, 86154, Iran, [2]Ferdowsi University of Mashhad, Department of Animal science, Azadi Sq., Mashahd, 91775-1163, Iran, [3]Agriculture engineering organization of Iran, Tehran, 1677836394, Iran; pouria_hossinniavaliseh@yahoo.com

Aim of this study was to investigate and compare the ability and accuracy of artificial neural networks (ANN) and linear regression models for prediction of sperm production in Kurdi rams using the information derived from hormonal factors. Artificial neural networks like biologic neural networks make the neurons and use parallel algorithm for analysis of data set, but the linear models use sequencely algorithm for it. In comparison with linear regression models, ANN can work with both linear and nonlinear variables simultaneously. Datasets used in this study included hormonal information and sperm concentration of Kurdi rams and were analyzed using SAS and MATLAB softwares. Results showed that the accuracy of prediction using ANN was significantly higher than linear regression models and it was because of simultaneous application of linear and nonlinear relationship between variables in prediction of output variable. Linear regression was not able to compute non linear relationship between input and out put variable and poor results for linear regression model may be due to the existence of independent variable that have nonlinear relationship with dependent variables.

Preliminary genetic (co)variances between wrinkle score and breech strike in Merino sheep, using a threshold-linear model

Scholtz, A.J.[1], Cloete, S.W.P.[2], Van Wyk, J.B.[3], Misztal, I.[4] and Van Der Linde, T.D.E.K.[5], [1]University of the Free State, Centre for Sustainable Agriculture and Rural Development, Faculty of Natural and Agricultural Scien, University of the Free State, Bloemfontein 9300, South Africa, [2]University of Stellenbosch, Department of Animal Sciences, University of Stellenbosch, Matieland 7599, South Africa, [3]University of the Free State, Department of Animal, Wildlife and Grassland Sciences, University of the Free State, Bloemfontein 9300, South Africa, [4]University of Georgia, Department of Animal and Dairy Science, University of Georgia, Athens GA 30605, USA, [5]University of the Free State, Department of Zoology and Entomology, University of the Free State, Bloemfontein 9300, South Africa; vanwykjb.sci@ufs.ac.za

Heritability estimates for and genetic correlations among wrinkle score and breech strike were determined on 2,327 16-month Merinos. Progeny present at hogget shearing were subjected to visual appraisal of wrinkles on the neck, body and breech. The occurrence of blowfly strike in the breech area (breech strike) was recorded in all animals at shearing. A three-trait animal model was fitted with breech wrinkle score, total wrinkle score and breech strike as traits. Estimates of heritability were consistently larger than double the corresponding standard errors, and amounted to 0.31 for breech wrinkle score and 0.33 for total wrinkle score. Breech strike on the underlying scale was also heritable (0.26) even at a fairly low incidence of 6.6%. The genetic correlations of both measures of wrinkleless with breech strike on the underlying scale were negative, but only approached significance at the 10% level for total wrinkle score. It was concluded that South African Merino sheep would respond to selection against breech strike under natural conditions. Selection for a reduced breech wrinkle score is also expected to result in a favourable correlated response in breech strike.

Freemartin detection and BMP15 genotype determination in replacement ewes of Rasa aragonesa sheep breed by PCR-duplex

Martínez-Royo, A.[1], Dervishi, E.[1], Alabart, J.L.[1], Folch, J.[1], Jurado, J.J.[2] and Calvo, J.H.[1], [1]CITA, Producción Animal, Av. de Montañana, 930, 50059 Zaragoza, Spain, [2]INIA, Mejora Genética Animal, Ctra. La Coruña Km. 7,5, 28040 Madrid, Spain; jhcalvo@aragon.es

A new naturally occurring mutation related to prolificacy in BMP15 gene has been recently described in Rasa aragonesa sheep breed. This mutation (FecXR allele) is a deletion of 17 base pairs that leads to an altered amino-acid sequence, which has being associated with both increased prolificacy and sterility in heterozygous and homozygous ewes, respectively. But increased prolificacy by co-twins or multiple births rises up sterility cases on born females because of freemartin phenomenon during pregnancy. In this sense, an accurate, sensitive, and quick method was developed by multiplex PCR for sex determination, freemartins detection and BMP15 mutation genotype in replacement ewe lambs. Freemartinism determination was performed using primers derived from an ovine-specific Y chromosome repetitive fragment (GenBank accession number U65982 and U30307). Rams carrying BMP15 deletion fragment provide an internal positive control for amplification reaction. FecXR and FecX+ alleles of BMP15 gene were determined by different size PCR amplification products (143 and 160 bp for FecXR and FecX+, respectively). Results showed 2.6% of freemartin in 150 replacements ewes. Results confirm usefulness of this multiplex PCR for detecting phenotypic sexed females, freemartins and heterozygous carriers for BMP15 deletion in order to detect high prolific ewes in commercial flocks or MOET and IVF programs and do replacement breeding management or market more efficient.

Genetic variability and phylogenetic relationships among ovine Algerian breeds using microsatellites

Gaouar, S.[1], Tabet Aoul, N.[2], Ouragh, L.[3], Khait Dit Naib, O.[4], Boushaba, N.[4], Brahami, N.[4], Hamouda, L.[4], Rognon, X.[5], Tixier-Boichard, M.[5] and Saïdi-Mehtar, N.[4], [1]University of Tlemcen, Department of Biology, Tlemcen, 13000, Algeria, [2]University of Oran Es-sénia, Department of Biotechnology, Oran, 31000, Algeria, [3]Institute of Agronomical and Veterinary, Hassan II, Department of Genetic and Veterinary Analyses, Rabat, 10 000, Morocco, [4]University of Science and Technology of Oran, Department of Applied Molecular Genetics, Oran, 31000, Algeria, [5]Institut National Agronomique Paris-Grignon, UMR Génétique et Diversité Animales, Paris Grignon, 75231, France; nmehtar2002@yahoo.fr

The genetic variability was studied in 158 unrelated individuals from six populations representing six sheep Algerian breeds: Hamra, Ouled-Djellal, Sidaoun, Taadmit, D'men, and Rumbi. These studies were carried out using 22 microsatellites. Four microsatellites were analyzed by PCR method followed by electrophoresis on polyacrylamid gel and silver staining DNA and the others with PCR followed by genotyping with automatic sequencer. Statistical methods were used to estimate the genetic variability within breeds. We also analyzed variability between breeds and constructed phylogenetic trees using two genetic distances. The preliminary results showed a great genetic variability for most of these breeds, Hamra breed seemed to be the most original one. The phenograms and multidimensional analysis (PCA) showed that 3 breeds: Ouled-Djellal, Taadmit and Rembi were grouped.

Additive mutational variance for litter size in the Ripollesa sheep breed

Casellas, J.[1], Caja, G.[2] and Piedrafita, J.[2], [1]IRTA-Lleida, Genètica i Millora Animal, Alcalde Rovira Roure, 191, 25198 Lleida, Spain, [2]Universitat Autònoma de Barcelona, Grup de Recerca en Remugants, Departament de Ciència Animal i dels Aliments, Facultat de Veterinària, 08193 Bellaterra, Spain; jesus.piedrafita@uab.cat

During the last decades, the importance of new mutations on polygenic variability has been revealed in several experimental species. Nevertheless, little is known about mutational variability in livestock where only a few mutations with large effects have been reported. In this manuscript, mutational variability has been analyzed on 1,765 litter size records from 404 Ripollesa ewes, in order to characterize the magnitude of this genetic source of variation and check suitability of including mutational effects in the genetic evaluations of this breed. The threshold animal model accounting for additive genetic mutations was preferred in front of the model without mutational contributions, with an average difference of 5.3 deviance information criterion units. Moreover, both models were compared through a Bayes factor showing that the model with mutational variability was 10.9 times more probable in average. The mutational variance component reached a modal estimate of 0.013 with the highest posterior density region ranging from 0.007 to 0.044. Mutational variance was almost 13 times smaller than the additive genetic variance (0.168) and 7 times smaller than the permanent environmental variance (0.092). Within this context, the mutational heritability was 0.9%, falling within the range of estimates reported in experimental species. The inclusion of mutational effects in the genetic model for evaluating litter size in the Ripollesa ewe originated moderate rearrangements in the genetic merit order of evaluated individuals, and suggested that the continuous uploading of new additive mutations must be accounted for in order to optimize the selection scheme. Moreover, this study is the first attempt to estimate mutational variances in a livestock species, contributing to a better characterization of the genetic background of productive traits of interest.

Study of genetic relationship in ten autochthonous Hungarian sheep flocks based on microsatellites

Kusza, S.[1], Nagy, I.[2], Bősze, Z.[2], Oláh, J.[1], Németh, T.[1], Jávor, A.[1] and Kukovics, S.[3], [1]University of Debrecen, Institute of Animal Science, Böszörményi str 138., 4032 Debrecen, Hungary, [2]Agricultural Biotechnology Centre, P.O. Box 411, 2100 Gödöllő, Hungary, [3]Research Institute for Animal Breeding and Nutrition, Gesztenyés út 1., 2053 Herceghalom, Hungary; timea.nemeth@atk.hu

Tsigai breed is existed and considered as autochthonous breed in many countries in Middle-, East- and South European countries, however, their phenotypic characters, production traits are widely differ. Two kinds of Tsigai sheep are bred in Hungary: autochthonous Tsigai and Milking Tsigai. In our study genetic relationships among ten Hungarian Tsigai populations were investigated using sixteen microsatellite markers (recommended by FAO and ISAG). The total number of alleles was 262 at the examined locuses. 15 population specific alleles were also detected. The mean number of alleles per locus ranged from 4,3 (OarAE119) to 11,9 (MAF70). The expected heterozygosities (H_{exp}) varied from 0.567 (BM6506) to 0.846 (BM1314). The observed heterozygosities (H_{obs}) varied from 0.323 (OarAE119) to 0.714 (INRA127). The F_{IS} values showed high inbreeding within populations. All examined population were less heterozygous than it was expected. The heterozigosity deficit was the highest in one of milking type populations and the least was in one of the indigenous populations. Nei's minimum genetic distance (D_A) value among examined populations were obtained,and used to construct a phylogenetic tree using UPGMA algorithm. The results indicated that the genetic difference was negligible between the following populations pairwise: two separate Hungarian autochthonous populations; another Hungarian autochthonous population and one of Milking Tsigai populations; another Hungarian autochthonous population and a transitional type population. From this study it could be concluded that microsatellite genotyping might be an efficient tool for studying the genetic relationships among Hungarian Tsigai populations.

Molecular-based estimates of effective population size in the rare Xalda sheep

Goyache, F.[1], Alvarez, I.[1], Fernandez, I.[1], Royo, L.J.[1], Perez-Pardal, L.[1] and Gutierrez, J.P.[2], [1]SERIDA-Somio, Camino de los Clavels 604, 33203 Gijon-Asturias, Spain, [2]UCM, Produccion Animal, 28040 Madrid, Spain; ljroyo@serida.org

Empirical evidence of the usefulness of different molecular-based methods to estimate effective population size (Ne) for conservation purposes in endangered livestock populations is reported. The rare Xalda sheep pedigree (1851 individuals) was available and the polymorphism of 21 microsatellites in 285 Xalda individuals was analyzed using two different approaches: a) individuals were assigned to a base population (BP) or 4 different cohorts (from C1 to C4) according to pedigree information; and b) individuals were assigned to groups G1 and G2 mimicking two random samplings separated by more than 1 generation interval. Molecular Ne was computed using: 1) linkage disequilibrium (Ne(D)); 2) a temporal method based on F-statistics (Ne(T)); 3) an unbiased temporal method (Ne(JR)); and iv) a Bayesian temporal method (Ne(B)). The estimates of Ne(D) decreased with pedigree depth from 68.4 for BP to 34.5 for C4 and from 87.9 for G1 to 70.4 for G2. The estimates of Ne(T), Ne(JR), and Ne(B) obtained using the first approach only presented consistent confidence intervals when temporal samplings involved BP, whilst those obtained for the sampling G1-G2 were similar for Ne(T) and Ne(B) (37.2 and 31.7) and lower for Ne(JR) (18), all of them showing narrow confidence intervals. Even though Ne(D) gathered the population changes due to pedigree accumulation, it was strongly affected by sampling size. Accordingly, repeated sampling would be beneficial. The temporal methods were strongly affected by a weak drift signal, particularly when samplings are not spaced sufficient generations apart. The use of molecular-based estimates of Ne is not straightforward and their employment in livestock conservation programs should be carried out with caution. Sampling strategies (including sampling sizes, sampling periods and age structure of the sampled individuals) must be carefully planned to ensure that robust estimates of Ne are obtained.

Pedigree analysis of Iranian Baluchi sheep breed
Sheikhloo, M., Tahmoorespur, M., Shodja, D., Pirany, N. and Rafat, A., Department of animal science, Ferdowsi university of Mashhad, 91775-1163, Iran; m_sheikhloo@yahoo.com

This work was aimed to analyze the pedigree information of the Iranian Baluchi sheep included in its herdbook from 1980 to 2006 in order to quantify the genetic structure and variability in the population. Data included 20,681 (9,681 males and 11,000 females) registered animals. Generation interval, the proportion of Known parents till the 5th parental generation, number of equivalent generations, effective number of founders, effective number of ancestors, Inbreeding (F) and average relatedness (AR) coefficient, were computed using the ENDOG v4.0 software. Generation intervals across the ram-son, ram-daughter, ewe-son and ewe-daughter pathways are 3.22, 3.15, 3.80 and 3.65, respectively. This parameter was lower in the ram-offspring pathway because rams are replaced earlier than ewes. For the first generation of ancestors, the pedigree knowledge was 89%. It drops to 67% in the second generation and to 10% after the 6th parental generation. Number of equivalent generation was 2.65 for whole pedigree. Number of founders was 1,819 while Effective number of founders was 106.96. This reflects the excessive use of some individuals as parents. The number of ancestors (founders or not) that explained 100% of genetic variability of the population was 1956, although only 36 individuals were necessary to explain 50%. The effective number of ancestors was 81. This value was lower than the effective number of founders and reflects the bottlenecks of pedigree. The average values of F and AR for the whole pedigree were respectively, 0.98% and 1.49% that were higher for males than for females. The mean F in the inbred animals was three times that found for the whole pedigree (3.21%), and their mean AR was 2.38%. The results of this study indicate that selection of animals increased the representation of some individuals in the whole pedigree and this must be considered in the breeding plane of the Baluchi sheep.

D-loop region of mtDNA sequence diversity in Iranian Domestic sheep and goats
Sadeghi, B., Shafagh Motlagh, A., Nassiry, M., Ghovvati, S. and Soltani, M., Department of Animal Science, Faculty of Agriculture, Ferdowsi University of Mashhad, Mashhad, P.O. Box: 91775-1163., Iran; Ghovvati@ stu-mail.um.ac.ir

Domestic animals such as sheep and goat have played a key role in food resource in Iran and These are important livestock species. Because they provide a good source of meat, milk, fiber and skin. Despite their importance, however, the origin and genetic study of this species remain poorly in our country. Molecular genetic data have greatly improved our ability to study the evolution of domestic mammals. Among the genetic markers mtDNA sequencing is one of the most useful and commonly employed methods for inferring phylogenetic relationships among closely related species and population. The mtDNA control region contain variable block that evolve four to five time faster than the remainder of the mtDNA molecule. In total of 43 unrelated sheep and goat blood and meat samples were collected (ovis aries [baluchi breed] N=14, ovis orientalis arkal N=8, capra hircus N=14 and capra aegagrus N=7). Nucleotide diversity and phylogenetic relationship calculated by sequencing the HVR1 segment of control region. An standard criteria for the definition of the different haplogroups was proposed based on the result of mismatch analysis and on the use of sequences of reference. Such a method could be also applied for clarifying the nomenclature of mitochondrial haplogroups in other domestic species.

Polymorphism of five lipid metabolism genes and their association with milk traits in Murciano-Granadina goats

Zidi, A.[1], Fernández-Cabanás, V.[2], Amills, M.[1], Polvillo, O.[2], Jordana, J.[1], González, P.[2], Carrizosa, J.[3], Gallardo, D.[1], Urrutia, B.[3] and Serradilla, J.M.[4], [1]Universitat Autònoma Barcelona, Departament Ciència Animal i dels Aliments, Campus UAB, 08193 Bellaterra, Spain, [2]Universidad de Sevilla, Departamento Ciencias Agroforestales, Ctra. de Utrera. Km. 1, 41013 Sevilla, Spain, [3]Instituto Murciano de Investigación y Desarrollo Agrario y Alimentario, Estación Sericícola, La Alberca, 30150 Murcia, Spain, [4]Universidad de Córdoba, Departamento de Producción Animal, Campus de Rabanales, 14014 Córdoba, Spain; ali.zidi@uab.es

Milk fat is an important factor influencing the nutritional and technological quality of dairy animal products. The moderate heritabilities (from 0.2 to 0.4) of milk fat content and fatty acids profile evidence the segregation of allelic variants with important additive effects on lipid metabolism traits. In this work we have characterized the polymorphism of five functional candidate genes in a Murciano-Granadina goat population. Genes under analysis were stearoyl-CoA desaturase (SCD), malic enzyme 1 (ME1), CD36 antigen or fatty acid translocase (CD36), prolactin receptor (PRLR) and hormone sensitive lipase (LIPE). Sequencing of the coding region and/or the 3'UTR of these genes allowed us to identify 20 single nucleotide polymorphisms (SNP) mapping to SCD (1 synonymous, 2 in the 3'UTR), ME1 (4 synonymous), PRLR (2 synonymous, 2 non-synonymous) and LIPE (2 synonymous, 1 non-synonymous). Moreover, we have detected a duplication of the CD36 gene with two copies displaying a nucleotide identity of 88%. Similarly, a duplication of the CD36 gene had been reported in cattle with two paralogous loci mapping to chromosomes 4 and 21. According to this, we have named the two caprine CD36 loci as CD36-4 and CD36-21. Both genes happened to be polymorphic, with 2 synonymous SNP at CD36-4 and 4 SNP at CD36-21 (1 non-synonymous, 3 in the 3'UTR). Currently, we are performing association analyses between these SNP and milk fat content and composition.

Paternal haplogroups in European and African domestic goats

Royo, L.J.[1], Perez-Pardal, L.[1], Traore, A.[2], Wisniewska, E.[3], Azor, P.J.[4], Alvarez, I.[1], Fernandez, I.[1], Plante, Y.[5], Kotze, A.[6], Ponce De Leon, F.A.[7] and Goyache, F.[1], [1]Serida, Genetica y Reproduccion, Camino de los Claveles 604, 33203-Gijon, Asturias, Spain, [2]INERA, 04 BP, 8645 Ouagadougou, Burkina Faso, [3]University of Technology and Agriculture, Bydgoszcz, 85-084, Poland, [4]Universidad de Cordoba, Campus Rabanales, 14071-Córdoba, Spain, [5]University of Saskatchewan, 6D62, S7N 5A8, Saskatoon, Canada, [6]National Zoological Gardens, Pretoria, Pretoria 0001, South Africa, [7]University of Minnesota, Animal Science, St Paul, MN 55108, USA; ljroyo@serida.org

Population genetic studies on goats are mainly based on mitochondrial DNA sequencing, thus only describing the female legacies. So far, using Y-chromosome specific SNPs, two common (C1 and C2) and one rare (C3, just found in one sample) Y-chromosome haplogroups have been described in domestic goats. Apparently, there is no geographic structure in the distribution of the two common haplogroups. In this preliminary work we have developed a dual fluorescent multiprobe assay for the genotyping of 3 SNPs located in two genes (AmelY and ZFY), which are able to differentiate among the 3 domestic haplogroups. A total of 257 bucks belonging to 18 different populations (12 in Europe and 6 in Africa) were genotyped and assigned to Y-haplogroups. No C3 samples were identified. Although the two major haplogroups (C1 and C2) were found in both continents, their distribution seems to show some geographic structure. C2 haplogroup is the only one found in Central Europe, with very low frequency in the Iberian Peninsula (4.7%) as well as in Sub-Saharan Africa (3.4%). C1 is the most frequent haplogroup in Sub-Saharan Africa (0.966) and Iberia (0.953), and is absent in central Europe. Mediterranean Africa goat could have an intermediate situation (C1: 0.27, C2: 0.73). The use of Y-specific microsatellites and a sampling over the world, will allow us to ascertain the paternal phylogeography of the species. Founded by CGL2008-03949/BOS and RZ07-00002.

Effects of CSN1S1 and CNS3 genotypes on technological properties and rheological parameters of milk in Murciano-Granadina goat

Caravaca, F.[1], Ares, J.L.[2], Carrizosa, J.[3], Urrutia, B.[3], Baena, F.[4], Jordana, J.[5], Amills, M.[5], Badaoui, B.[5], Sánchez, A.[5], Angiolillo, A.[6] and Serradilla, J.M.[4], [1]Universidad de Sevilla, Ciencias Agroforestales, Ctra Utrera Km 1, 41013 Sevilla, Spain, [2]Instituto Andaluz de Investigación y Formación Agraria y Pesquera, Alameda del Obispo s/n, 14004 Córdoba, Spain, [3]Instituto Murciano de Investigación y Desarrollo Agrario y Alimentario, Estación Sericícola, 30150 La Alberca, Spain, [4]Universidad de Córdoba, Producción Animal, Ctra N IV Km 396, 14014 Córdoba, Spain, [5]Universitat Autònoma de Barcelona, Ciència Animal i dels Aliments, Edifici V, 08193 Bellaterra, Spain, [6]Università del Molise, Scienze e Tecnologie per lAmbiente e il Territorio, C da Fonte Lappone, 86090 Pesche (IS), Italy; pa1semaj@uco.es

The effects of the caprine CSN1S1 polymorphisms on milk quality and cheese yield have been widely studied in French and Italian goat breeds. However, much less is known about the consequences of CSN3 genotype on technological and rheological properties of goat milk. An association analysis between polymorphisms at the CSN1S1 and CSN3 loci and milk technological properties (pH, Dornic acidity, coagulation time and cheese yield) and milk rheological parameters measured with a coagulometer Optigraph, was carried out. In this analysis we included 193 records from 74 Murciano Granadina goats distributed in three herds, that were collected bimonthly during a whole lactation. These goats were genotyped for the CSN1S1 and CSN3 loci by means of allele-specific PCR, PCR-RFLP and primer extension analysis. Data analysis, using a linear mixed model for repeated observations, revealed significant associations between CSN1S1 genotypes and the speed of the curdling process. Moreover, CSN3 had a significant effect on the time elapsed since rennet is added to the time point at clotting starts. No interaction between the CSN1S1 and CSN3 genotypes was observed.

Genetic parameters for milk yield, composition and rheological traits in Murciano-Granadina goats

Benradi, Z.[1], Ares, J.L.[2], Jordana, J.[3], Urrutia, B.[4], Carrizosa, J.[4], Baena, F.[1] and Serradilla, J.M.[1], [1]Universidad de Córdoba, Producción Animal, Crta N IV Km 396, 14014 Córdoba, Spain, [2]Instituto Andaluz de Investigación y Formación Agraria y Pesquera, Alameda del Obispo s/n, 14004 Córdoba, Spain, [3]Universitat Autònoma de Barcelona, Ciència Animal i dels Aliments, Edifici V, 08193 Bellaterra, Spain, [4]Instituto Murciano de Investigación y Desarrollo Agrario y Alimentario, Estación Sericícola, 30150 La Alberca, Spain; pa1semaj@uco.es

958 monthly milk yield records and samples were taken from 321 Murciano-Granadina goats. Milk components (protein, casein, CSN1S1, CSN1S2, and lactose contents) were analyzed using a near infrared spectrophotometer. Rheological (clotting time and curd firmness) values were determined with a coagulometer Optigraph©. 151 of these goats were genotyped for the CSN1S1 gene. The VCE5 software was used for the estimation of variance components, with a model including herd-year-season, parity ordinal and number of kids born as fixed factors. For the estimation of CSN1S1 heritability we employed a model including the CSN1S1 genotypes as a fixed factor. Heritabilities estimated for milk yield and composition (except lactose content) were within the range of values reported in other works. Those of total caseins, CSN1S1 and CSN1S2 contents were low (0.13, 0.10 and 0.08, respectively). CSN1S1 genotype explained 16.7% of the additive variance of CSN1S1 content. Heritabilities for curd firmness traits (from 0.12 to 0.20) were higher than those of clotting time traits (from 0.09 to 0.12). Genetic correlations between protein, casein, CSN1S1 and CSN1S2 contents were generally positive (from 0.57 to 0.91). Curd firmness and clotting time traits were negatively correlated (from -0.65 to -0.40). With a few exceptions milk protein and caseins contents showed positive (from 0.30 to 0.87) and negative (from 0.65 to -0.39) correlations with curd firmness and clotting time traits, respectively.

Genetic parameter for wool weight of lambs in Iranian Baluchi sheep

Hosseinpour Mashhadi, M.[1], Aminafshar, M.[2], Jenati, H.[3] and Tabasi, N.[4], [1]Islamic Azad University Mashhad Branch, Animal Science, 91735-413. Mashhad, Iran, [2]Islamic Azad University Science & Research Branch, Tehran, Iran, [3]Animal Breeding Center of Abas Abad, Mashhad, Iran, [4]Immonology Reseasrch Center, Mashhad, Iran; mojtaba_h_m@yahoo.com

Baluchi is the most common native breed of sheep in eastern part of Iran. The Baluchi is a double-purpose sheep for meat and wool. The colour of Baluchi wool is white so, the wool of Baluchi is suitable for carpet industry. The data were collected from a Baluchi flock in the Animal Breeding Station of Abbas Abad in northeast Iran. The data were for the first wool weight of lamb in spring on fifteen month of age. The number of records, mean of trait and coefficient of variation were 3,624, 1.5 ($\pm$0.33) kg and 22 percent, respectively. The fixed effects were investigated with JMP package under this model: $Y_{ijkln} = m + A_i + S_j + T_k + R_l + b (X_{ijkln} - X) + E_{ijkln}$. Where A is age of dam (10 level), S is sex (2 level), T is type of birth (3 level), R is year of birth (12 level) and b (Xijkln – X) is a covariate of age of lamb on the day of recording and E is residual effect. Co-variance component and genetic parameter were estimated by REML method under six different models of DFREML software. Direct heritability for wool weight based on model 1 was estimated 0.21($\pm$0.028). This value was estimated low for this trait compared with reports in literature. Estimated maternal heritability was low and is in agreement with other reports. Results of variance component of model seven and eight showed that the effect of maternal environment on wool trait was zero. The highest estimated heritability based on different models was for model one.Co-Variance component, genetic and phenotypic parameters of wool weight could be important parameters for genetic improvement. It is important to use wool records of different age, also collecting records of qualitative of wool trait for next study.

Polymorphism of calpastatin gene in makoei sheep using PCR- RFLP

Moradi Shahrbabak, H.[1], Moradi Shahrbabak, M.[1], Mehrabani Yeganeh, H.[1], Rahimi, G.H.[2] and Khaltabadi Farahani, A.H.[3], [1]University College of Agriculture & Natural Resources, University of Tehran, Animal Science, Karaj, 31587-11167, Iran, [2]Sari agricultural sciences and natural resources university, Animal Science, Sari, 3158711167, Iran, [3]Faculty of Agriculture, university of Arak, Department of animal Science, 3158711167, Iran; hmoradis@ut.ac.ir

Calpastatin has been known as candidate gene in muscle growth efficiency and meat quality. This gene has been located to chromosome 5 of sheep. In order to evaluate the calpastatin gene polymorphism, random blood sample were collected from 120 Makoei sheep from IRAN. The DNA extraction was based on DNA extraction Kit (2008 Cina Gene Kit). Exon and entron I from L domain of the ovine calpastatin gene was amplified to produce a 622 bp fragment. The PCR products were electrophoresed on 1% agarose gel and stained by etidium bromide. Then, they were digested with restriction enzyme MspI and then electrophoresed on 1.5% agarose gel with ethidium bromide and revealed two alleles, allele M and allele N. Data were analysed using PopGene32 package. In this population, MM, MN, NN genotype have been identified with the 46, 30, 24% frequencies. M and N allele's frequencies were 0.61, 0.39, respectively. The population was found to follow Hardy-Weinberg equilibrium.

Impact of a restrictive use of hormones on breeding and selection management in small ruminants

Fatet, A.[1], Pellicer-Rubio, M.T.[1], Palhière, I.[2], Bouix, J.[2], Lagriffoul, G.[3], Piacère, A.[3], Martin, P.[4], Boué, P.[4], Leboeuf, B.[5] and Bodin, L.[2], [1]INRA, UMR PRC, 37380 Nouzilly, France, [2]INRA, UR SAGA, 31320 Castanet-Tolosan, France, [3]Institut de l Elevage, 31320 Castanet-Tolosan, France, [4]Capgènes, 2135 route de Chauvigny, 86550 Mignaloux-Beauvoir, France, [5]INRA, UEICP, 86480 Rouillé, France; alice.fatet@ tours.inra.fr

The use of steroids in animal productions is suspected to be involved in the increase of hormone-dependant cancers. It is expected that European regulations might restrict or even ban their use in the future. In Europe, diverse breeding and selection strategies have been adopted in sheep and goat productions. In some countries, artificial insemination (AI) has been developed in sheep and, to a lower extent, in goats but this activity fully relies on hormonal treatments to synchronize oestrus during breeding season and induce oestrus out of the season. AI is a very efficient tool for several selection steps: planned mating, progeny testing, improving genetic connections between flocks, disseminating of genetic merit and rapid diffusion of specific genes of interest (i.e. scrapie resistance). AI is also used to spread production throughout the year and meet the constant needs of consumers. It is obvious that insemination activity will drastically decrease if it has to be realised exclusively on natural oestrus. Different scenarios will be explored regarding evolutions of breeding and selection schemes if restrictive use of hormones were imposed by European regulations. Technical and economical consequences will be evaluated depending on the strategies adopted for selection to remain efficient: impact on genetic progress, on seasonality of production. A wide range of situations could coexist from a completely AI-free selection to a strategy where AI or embryo transfer could be used in few selected animals and selection spread through male exchanges. Alternatives to hormonal synchronisation (i.e. male effect) are currently under study and may play an important role in the future for selection and production needs.

Foetal losses in early to mid-pregnancy in Icelandic ewe lambs

Dyrmundsson, O.R.[1], Eythorsdottir, E.[2], Jonmundsson, J.V.[1], Sigurdardottir, O.G.[3], Gunnarsson, E.[3] and Sigurdarson, S.[4], [1]Farmers Association of Iceland, Baendahoellinni v. Hagatorg, IS-107 Reykjavik, Iceland, [2]Agricultural University of Iceland, Keldnaholt, Is-112 Reykjavik, Iceland, [3]Institute for Experimental Pathology, University of Iceland, Keldum v. Vesturlandsveg, IS-112 Reykjavik, Iceland, [4]The Icelandic Food and Veterinary Authority., Austurvegur 64, IS-800 Selfoss, Iceland; ord@bondi.is

Ewe lambs of the Iceland breed are generally mated at seven months of age on Icelandic sheep farms and most farmers expect a lambing rate of 85-90%. The use of pregnancy diagnosis by ultrasonic techniques in recent years has uncovered the occurence of foetal death before or around day 70-90 of pregnancy in Icelandic ewe lambs. In some cases losses up to 70% of the ewe lambs that have conceived are observed with dead foetuses at scanning but the problem occurs irregularly between farms and years. Clinical signs of abortions are not visible in those animals and the foetuses are resorbed. A study of the the lambing rate of one year old ewes over a period of 14 years (1995-2008) in the Icelandic sheep recording scheme, showed a great variation in lambing rate between flocks and years during the whole period. A total of 793 farms had experienced one or more years with less than 75% lambing rate of ewe lambs. Analysis of 178 blood samples from ewe lambs with dead foetuses at scanning and from ewes and ewe lambs with normal foetuses in 2008 showed normal GPX activity (indirect measure of Selenium status) in all animals. Serum samples from 40 ewe lambs with dead foetuses at scanning were further tested for the presence of antibodies against Brucella ovis, Coxiella burnetti, Chlamydophila abortus, BDV virus, and Toxoplasma gondii, all of which are known causes of foetal death or abortions. All tests returned negative results and the problem remains unexplained.

Segregation of FecX and FecL genes influencing ovulation rate in different populations of Lacaune sheep

Drouilhet, L.[1], Lecerf, F.[2], Mulsant, P.H.[1] and Bodin, L.[3], [1]INRA, UMR444, Laboratoire de Génétique Cellulaire, Castanet Tolosan, France, [2]INRA, UMR598, Génétique Animale, Rennes, France, [3]INRA, UR631, Station d Amélioration Génétique des Animaux, Castanet Tolosan, France; Laurence.Drouilhet@toulouse.inra.fr

In the Lacaune sheep population, two major loci influencing ovulation rate and litter size are segregating: FecX and FecL. The $FecX^L$ mutation is a non conservative substitution (C53Y) in BMP15 that prevents the processing of the protein. The FecL locus is localized on sheep chromosome 11 within a region of about 1.1 megabases. In this interval, a unique haplotype is associated with the $FecL^L$ mutation. This particular haplotype could be detected by the DLX3:c.*803A>G SNP in the 3' UTR sequence of the DLX3 gene. This SNP provided accurate classification of animals (99.5%) as carriers or non-carriers of the mutation and therefore may be useful in marker assisted selection. The presence of the two mutations was studied in dairy and suckling Lacaune populations. Rams of the suckling population had an estimated breeding value (EBV) for litter size which was estimated through the records of related females. We found a correlation between the presence of $FecX^L$ and/or $FecL^L$ and the EBV value. This result is in accordance with the positive effect of $FecL^L$ and $FecX^L$ mutations on ovulation rate of females carrying different genotype combinations.

Premature luteal regression in superovulated goats induced to ovulate with GnRH or hCG

Saleh, M., Gauly, M. and Holtz, W., Department of Animal Science, Albrecht-Thaer Weg 3, 37075 Goettingen, Germany; mhdsaleh@maktoob.com

Premature luteal regression is a problem commonly encountered in superovulated goats. The purpose of the present investigation was to ascertain whether this can be alleviated by substituting the ovulation inducing agent gonadotropin releasing hormone (GnRH) with human chorion gonadotropin (hCG). Pluriparous Boer goat does, 2-6 years of age, were randomly assigned, 18h after the end of the superovulatory FSH-treatment, to one of three treatment groups. A group of 17 does were treated with GnRH (0.004 mg Buserelin, Receptal®), another group of 17 does with 500 I.U. hCG (Chorulon®), and a third group of 17 does with 1 ml physiological saline solution to serve as controls. To assess the time of ovulation, ovaries were monitored ultrasonographically at 12h intervals from 6h before to 36h after ovulation induction. To monitor corpus luteum function, plasma progesterone concentrations were determined at 2 day intervals until the onset of the superovulatory treatment and at daily intervals until embryo collection 7 days after ovulation induction. Embryos were flushed nonsurgically (Small Rumin Res 36: 195-200). The GnRH-treated does had well synchronized ovulations within 24.0 (±2.8) h vs. 34.7 (±6.4) and 43.4 (±9.9)h in the hCG- and NaCl-treated groups, respectively ($P<0.01$). Premature luteal regression became evident by day 4 after ovulation when serum progesterone concentration receded. Both GnRH and hCG treatment significantly increased the incidence of failure of normal corpus luteum formation: 100% and 88%, respectively, as compared to 56% in the NaCl group ($P<0.01$). No significant differences among GnRH-, hCG- and NaCl-treated groups were found with regard to number of transferable embryos (3.2, 1.9 and 4.6, respectively). Substitution of hCG for GnRH as ovulation inducing agent was of no avail. Follow-up studies addressing the discouragingly high incidence of premature luteal regression will be set about.

Ovarian functionning and its relationship with changes in live weight of maiden fat-tailed Barbarine ewes

Ben Salem, I. and Rekik, M., Ecole Nationale de Médecine Vétérinaire, Production Animale, ENMV, 2020 Sidi Thabet, 2020 Sidi Thabet, Tunisia; bensalemimen@yahoo.fr

This paper reports two experiments designed to investigate ovarian functioning of young female sheep reared under semi arid environment. The first experiment assessed the effect of the pattern of live weight change on the ovarian function of Barbarine ewe lambs at an age of approximately 1 year. A total of 171 weaned ewe lambs were grouped into three classes LWCI (n=46), LWCII (n=91) and LWCIII (n=34) with live weight loss being highest in LWCI and lowest in LWCIII. At 13 months of age, the proportion of ewe lambs found cycling in LWCIII (85.3%) was higher in comparison to animals in LWCI (43.4%; $P<0.001$) and tended to be superior to those in LWCII (61.5%; $P<0.05$). Following synchronisation with progestagen of the females found cycling, levels of plasma IGF-I and oestradiol concentrations were not different between the three classes. In the second experiment, 30 maiden ewes were retrospectively classified in three groups after synchronisation with a prostaglandin analogue. Group O+L+ was composed of maiden ewes showing oestrus and lambing thereafter (n=16); the group O+L- were maiden ewes showing oestrus but failing to lamb (n=7); the group O- were ewes not displaying oestrus (n=7). Diameter of the ovulatory follicles increased during the follicular phase, reaching a higher mean diameter in group O+L+ (6.4±0.23, $P=0.08$) than in groups O+L- and O- (5.7±0.36 and 5.9±0.55, respectively). In addition, the number of large follicles at oestrous tended to be higher in O+L+ animals (1.4±0.1, $P=0.073$) than in O+L- and O- ewes (1±0.2 and 0.9±0.2, respectively).

Effect of rbst treatmen on productive and reproductive performance of ewes

Abd El-Khalek, A.E. and Ashmawy, T.A., Animal Production Research Institute, El-Said Str. Dokki, Giza, 002, Egypt; tarek1101@yahoo.com

The current work aimed to study the effect of rbST treatment during pre- and postpartum period on live body weight, milk production, composition and lambing rates of crossbred ewes (1/2 Finnish Landrace x 1/2 Rahmani). A total number of 40 mature healthy crossbred ewes (½ Finnish Landrace x ½ Rahmani) having 3-4 years of age were divided into two equal groups (control and treatment), according to their live body weight (LBW). Ewes in the first group were served as a control group (C) without injection, while those in the second group were treated with a subcutaneous injection of 160 mg recombinant bST (rbST) at 14-day interval during one-month prepartum and 4 months postpartum. During May mating season, ewes were monitored for sign of oestrus and those observed in heat were natural mating using 4 fertile rams. Milk yield and composition were determined throughout 8 lactation weeks. Results show that LBW of ewes at prepartum and early postpartum suckling period was not affected by rbST treatment. However, treated ewes were heavier ($P<0.05$) by about 5% than controls only during late postpartum (breeding season from April to June). Treatment with rbST increased ($P<0.05$) average daily milk yield of ewes during the 1st eight weeks of lactation. The magnitude of increase ranged between 16 and 33% during different lactation weeks. Treatment with rbST decreased ($P<0.05$) contents of fat, protein and in turn total solids in milk of ewes during the 1st two months lactation. However, lactose and ash contents in milk of ewes were not affected significantly by rbST treatment. Treatment with rbST increased ($P<0.05$) oestrus/mating rate (40 vs. 70%) and reduced postpartum period of lambing ewes by about one month as compared to untreated ewes (65 vs. 93 days). Lambing rate based on number of mated ewes was significantly ($P<0.05$) lower in control than in treated ewes (75 vs. 85.7%). In conclusion, injection of 160 mg rbST at 14-day interval during one-month prepartum and 4 months postpartum improved milk yield and lambing rate of ewes.

Hemoglobin type and some productive and reproductive traits in Ossimi, Chios and their f1 cross ewes

El-Barody, M., Sallam, M., Marzouk, K., Haider, A. and Morsy, A.,; profelbarody@yahoo.com

Relationships among hemoglobin (Hb) types and some reproductive and productive traits were examined in 100 ewes from Chios (C), Ossimi (O) and F_1 crossbred ½ C ½ O ewes. In C and O ewes with Hb AB were the highest in fertility, While, Hb AA ewes were intermediate and Hb BB were the lowest. However, ½C ½O ewes with Hb AB had the highest fertility value, followed by ewes with Hb BB and Hb AA. In C and ½ C ½ O sheep, litter size at birth and at weaning were the highest among Hb BB ewes, followed by ewes with Hb AA and Hb AB. While in O sheep, ewes with Hb AA had the highest values for litter size at birth and at weaning followed by ewes with Hb AB and Hb BB. In general, ewes with Hb AB had heavier grease fleece weight and longer staple length, followed by ewes with Hb BB and Hb AA. Ewes with Hb AA produced finer wool, Hb AB ewes were intermediate and Hb BB ewes produced ticker wool. Also, ewes with Hb AA produced more milk (135.7±9.5 mg) than ewes with Hb AB (110.0±9.5 kg) or ewes with Hb BB (104.8±9.5 kg). Ewes with Hb BB had longer lactation length (96.9±2.3 days) followed by ewes with Hb AA (94.6±2.3 days) and ewes with Hb BB came in the last rank (90.7±2.3 days).

Sperm viability in frozen goat semen as affected by egg yolk level, dilution rate, type of freezing and month of the year

Sallam, A.A., Ashmawy, T.A. and El-Saidy, B.E., Animal Production Research Institute, Dokki, Giza, 002, Egypt; tarek1101@yahoo.com

Two hundred and fourteen ejaculates of good quality (>70% motility) were collected from 10 sexually mature Damascus x Baladi bucks having 3-5 years of age and with proven fertility. Two ejaculates from individual bucks were pooled before use. The buck semen was previously not tested for the phenomenon of coagulation with egg yolk. The concentration of spermatozoa in the ejaculates was $3.1-4.4 \times 10^9$ per ml and the proportion of progressively motile cells was 70-85%. Semen was diluted with Tris-citric acid-glucose-glycerol to evaluate different egg yolk levels (2.5, 5, 7.5, 10, 15 and 20%) on sperm motility in fresh and post thawed semen. Semen was diluted with 2.5% egg yolk (resulting the best level) in Tris-based extender at different dilution rates (0, 1:4, 1:5, 1:8, 1:10 and 1:16). Results revealed that sperm motility in diluted semen was the highest (54.75%, $P<0.05$) with egg yolk level (2.5%) and the lowest (39.0%, $P<0.05$) with the highest egg yolk level (20%) as compared with 83.13% in fresh semen. Sperm motility percentage in semen extended with 2.5% Tris-egg yolk extender at a rate of 1:5 semen: diluent showed the highest sperm motility percentage (54.5%, $P<0.05$), but did not differ significantly than those diluted at rates of 1:4 or 1:8 (53.68 and 51.97%, respectively). The current study indicated that adding egg yolk in Tris-based extender at a level of 2.5% and dilution rate of 1:5 for goat semen frozen in pellets form yielded the highest sperm motility in post thawed semen. Also, collection of semen for freezing is appropriate during autumn months (Sept. and October). Higher sperm concentration are required for AI of does to obtain higher fertility in goats.

The effect of hCG on the reproduction performance in Lory ewes outside the breeding season
Moeini, M.[1], Allipour, F.[1] and Sanjabi, M.R.[2], [1]Razi University, animal science, Kermanshah, 67155, Iran, [2]Irost, animal science, tehran- shhryar- asre enghlab, 15815, Iran; msanjabii@gmail.com

This study aims to determine the effects of supplementing human chorionic gonadotrophin (hCG) at insemination time or 12 days after artificial insemination (AI) on the reproductive performance in estrus-induced mature Lory ewes. A total of 390 Lory anestrous ewes (4-5 years of age, 35±4 kg) were randomly divided into three groups and after synchronization with progestagen sponge (Fluorogestone acetate, FGA) and 400 IU PMSG, the ewes in first group (T1) were injected 200IU hCG at AI time, second group (T2) were injected 200IU hCG, 12 days after AI and (T3) were kept as the control group. Estrous were determined using 35 teaser rams to calculate estrous rate. Prolificacy (No. of lambs born alive per ewe in lambing) and fertility (% ewes lambing per ewes inseminated) were assessed. Serum progesterone (P4) concentrations were measured in days 12, 14 and 16 after AI. The hCG injection increased the prolificacy in T1 group compared with control group ($P<0.05$) and fertility tended to be greater in hCG treatments groups but it was not significantly different. The hCG injection in days 0 and 12 increased the weight of single lambs at birth day ($P<0.05$). The P4 concentrations increased in days 14 and 16 in T1 and T2 compared with control ($P<0.05$). It can be concluded that treated with 200IU hCG given at the time of AI or 12 days after AI, increased serum progesterone concentrations and could improve reproductive performance in Lory ewes.

The anatomy of the Sanjabi ewe's cervix at different stage of cycle and age
Souri, M., Darabi, S., Moghadam, A.A. and Moeini, M., Razi University, Animal Science, Kermanshah, 67155, Iran; msouri@yahoo.com

The aim of the present study was to examine the gross anatomy of the cervix in slaughtered adult and non-adult ewes of the Sanjabi breed and its influence on the transcervical passage of an inseminating pipette into the uterine lumen. 200 reproductive tracts were excised immediately from carcasses of Sanjabi ewes which were identified as luteal or non-luteal based on the presence of a corpus luteum. The morphology of the cervical external os was classified as duckbill, flap, rosette or papilla. The cervical canal of each tract was filled with a silicone sealant for casting the mould. The cervix was opened longitudinally, its length recorded, the number of cervical rings counted. The mean lengths of the cervical mould of non-adult and adult ewes were 3.8±0.12 and 5.3±0.15 cm, respectively. The average number of funnel shaped folds in the cervical mould of non-adult and adult ewes was 3.2±0.19 and 3.4±0.22. However, the second and third-folds from the os were observed to be eccentric in both non-adults and adult ewes. The mean number of cervical rings (±Sem) was 4.0±0.06 with a range of 2-7 rings per cervix. Postmortem cervical penetration was deeper when the cervices were longer and wider, and with fewer folds. The information generated in this study would be useful for increasing the success rate of penetration in ewes exhibiting estrus in order to improve the lambing rate of tropical ewes following transcervical AI.

Exogenous melatonin improves embryo viability of undernourished ewes during seasonal anoestrus

Vazquez, M.I., Forcada, F., Abecia, J.A. and Casao, A., Faculty of Veterinary, University of Zaragoza, Animal Production and Food Sciences, Miguel Servet 177, 50013, Zaragoza, Spain; isavazq@unizar.es

This study investigated the effect of exogenous melatonin and undernutrition on embryo viability in postpartum ewes during seasonal anoestrus. At parturition, 36 adult Rasa Aragonesa ewes were assigned into two groups: treated (+MEL) or not (-MEL) with a subcutaneous implant of melatonin (Melovine®, CEVA) the day of lambing. After 45 days of suckling, lambs were weaned, and ewes were synchronized with intravaginal pessaries and fed to provide 1.5 (Control, C) or 0.5 (Low, L) times daily maintenance requirements. Therefore, ewes were divided into four groups: C-MEL, C+MEL, L-MEL and L+MEL. At oestrus (Day=0), ewes were mated and embryos were recovered by mid-ventral laparotomy on Day 5 and classified according to their developmental stage and morphology. No effect of diet or melatonin treatment was observed either on ovulation rate or number of recovered ova per ewe. Melatonin treatment improved significantly the number of fertilized embryos/corpus luteum (CL) (-MEL: 0.35 ± 0.1, +MEL: 0.62 ± 0.1; $P=0.08$), number of viable embryos/CL (-MEL: 0.23 ± 0.1, +MEL: 0.62 ± 0.1; $P<0.01$), viability rate (-MEL: 46.6%, +MEL: 83.9%; $P<0.05$) and pregnancy rate (-MEL: 26.3%, +MEL: 76.5%; $P<0.05$). Particularly, exogenous melatonin improved embryo viability in undernourished ewes (L-MEL: 40%, L+MEL: 100%, $P<0.01$). In conclusion, this study shows that melatonin treatment, improves ovine embryo viability during anoestrus particularly in undernourished postpartum ewes. Supported by grants AGL2007-63822 from CICYT and A-26 from DGA.

Ten years of embryo transfer applied to a selection program for prolificacy in sheep

Folch, J.[1], Cocero, M.J.[2], Marti, J.I.[1], Lahoz, B.[1], Olivera, J.[3], Ramon, J.[4], Roche, A.[5], Dervishi, E.[1], Calvo, J.H.[1], Echegoyen, E.[1], Sánchez, P.[1] and Alabart, J.L.[1], [1]CITA, Montañana 930, Zaragoza, Spain, [2]INIA, Ctra. Coruña K.5,9, Madrid, Spain, [3]Fac. Veterinaria, Ruta 3, k.363, Paysandú, Uruguay, [4]CIGA, Ctra. Mérida-Motul, Mérida, Mexico, [5]Oviaragón, C °Cogullada sn, Zaragoza, Spain; jfolch@aragon.es

Rasa Aragonesa breed is semiextensively exploited in flocks from 100 to 4,000 heads, to produce lambs slaughtered at 3 months of age. As prolificacy determines the farms economic viability, since 1998 a selection program for prolificacy is being developed in which males to be tested are produced by MOET. Embryo donors are the ewes having higher genetic values (GV) in the recorded flocks (115,000 ewes), which are purchased after checking for health status, scrapie genotype, and are in accordance to breed standard. Donors are inseminated with semen from select rams. Due to the number of lambings necessary to calculate the GV, donors are older than 6 years at the first embryo collection. A treatment of 8.8 mg of oFSH (Ovagen) was used in 134 from a total of 308 superovulated donors. Reproductive problems, such as metritis, regressed CL or no response to superovulation, occurred in 21.6% of them. The number of lambs obtained per donor (LB/Donor) was very low (0.1 ± 0.1; n=29). In the remainder 105 healthy ones, results increased with time due to the improvement of the technique. In this way, while embryo recovery rate and LB/Donor in the 1998-2001 period were 72.2 ± 3.2% and 3.6 ± 0.5 (n=70), in 2002-2007 they were 82.8 ± 3.9% and 4.9 ± 0.8 (n=35). When embryo recovery was repeated in suitable donors, LB/Donor decreased after the 2nd time (4.0 ± 0.4, n=105; 4.3 ± 0.7, n=36; 1.2 ± 0.5, n=11; and 2.5 ± 1.9, n=5, in the 1st, 2nd, 3rd and 4th recovery, respectively). Since 2007, embryos are sexed and checked for scrapie genotype before being transferred, resulting on 34.0% (n=97) LB/transferred embryo compared to 55.8% (n=52) in non-sexed group. In our conditions, embryo transfer is a useful tool to get males to be tested in the selection programme.

Onset of puberty in D'man and Timahdite breeds of sheep and their crosses

Derqaoui, L.[1], El Fadili, M.[2], Francois, D.[3] and Bodin, L.[3], [1]Institut Agronomique et Veterinaire, Hassan II, Rabat Instituts, Morocco, [2]INRA, CRRA, BP 415, RP Rabat, Morocco, [3]INRA, UR 631, SAGA, 31326 Castanet Tolosan, France; Dominique.Francois@toulouse.inra.fr

This study was carried at El Koudia station of INRA of Morocco with the main objective to determine age at puberty in D'man (D) and Timahdite (T) ewe and ram lambs and their crosses (DxT) raised under natural photoperiod. Eighty six ewe lambs (7 D; 15 T, 14 F1; 14 F2; 18 F3; 18 F4) and 23 ram lambs (4 D; 7 T; 4 F2, 1 F3; 7 F4) born in late December to early January were measured. Ewe lambs were checked for estrus by overnight exposure to vasectomized DxT rams with marked briskets. Laparoscopic examination of the ovaries was performed 4 to 10 days after the onset of estrus to confirm the occurrence of ovulation. Ram lambs were subjected to semen collection and testicular size measurements every other week. Semen was collected using electroejaculation and examined for volume and the occurrence of the first spermatozoa. The onset of puberty was considered as the age at the occurrence of the first ovulatory estrous in females and spermatozoa in males. All sheep involved in this study displayed signs of puberty during the course of the study. The overall mean age at puberty was 252.19 ± 29.84 days in ewe lambs. Genotype had a significant effect ($P<0.01$) on the age at puberty in T and DxT. Thus, D'man females were the first to exhibit signs of estrus (220 days) whereas Timahdite were the last (267 days) and DxT crosses being intermediate, i.e.: 236, 256, 260 and 253 days for F1, F2, F3 and F4, respectively. On the other hand, ram lambs reached puberty earlier than females, i.e. at a mean age of 132 days, younger for D'man (124 d), later for Timahdite (144 d) and close to D'man for crosses (127 d). In addition, measured testicular parameters were 6.09 cm; 4.13 cm; 2.25 cm, 21.37 cm, 0.43l for testicular length and diameter, diameter of the epididymal tail, scrotal circumference and volume of ejaculate, respectively.

Influence of selenium, vitamin E and Zn on semen quality of Blochi rams

Hedayati, M.[1], Tahmasbi, A.M.[2], Falah Rad, A.[2] and Vakili, R.[1], [1]Islamic Azad University of Kashmar, Dept. of Animal Sci., Kashmar, P.O.Box 9177 94897, Iran, [2]Depatrment of Animal Sci. Excellence Center for Animal Sci., College of Agr., Ferdowsi University of Mashhad, P.O. Box 917794897, Iran; madira_info@yahoo.com

Selenium has an important role in spermatogenesis and is a constitute tail of the sperm as a seleno-flagellate. Vitamin E also is an antioxidant reagent which plays a critical role in preventing reactive oxygen spoils (ROS). Over 200 protein and enzyme contain Zn and some of these enzymes may be of particular importance in function of reproductive tissues, but little information is known about that. An experiment conducted to evaluate the effect of selenium, vitamin E and Zn on reproductive performance on 20 Blochi ram (2 year old) which matched for weight and condition and were given A: basal diet, B: basal diet plus selenium vitamin E, C: basal diet plus Zn and D: basal diet containing selenium vitamin E and Zn for 63 days. Semen was collected 47 and 63 days after start of experiment and was assessed *in vitro* for volume of ejaculate, mortality, number of live spermatozoa and number of abnormal spermatozoa. No significant effects in growth rates between treated ram were found. Abnormal sperm were found in much greater number in the semen from the untreated rams. However, injection of selenium vitamin E significantly improved viable and normal sperm and differences was statistically significant ($P<0.05$). Although supplementation of Zn in the ram improved semen quality compared to control group but when Zn and selenium vitamin E were applied together more significant results were obtained ($P<0.05$). Although in this experiment admission of selenium vitamin E plus Zn had a good effects on semen quality when applied at least 47 days semen collection, however, itis recommended that further trails be conducted on this important subject using large number of animals.

Vitrification of caprine morulae, blastocysts and hatched blastocysts by the open pulled straw (OPS) method

Al Yacoub, A.N., Gauly, M. and Holtz, W., Department of Animal Science, George August University, Albrecht Thaer Weg 3, 37075, Goettingen, Germany; mgauly@gwdg.de

The objective of this study was to investigate the applicability of the open pulled straw (OPS) vitrification method found to be suitable for caprine blastocysts to morulae and hatched blastocysts. Blastocysts, as a control, were either vitrified by the OPS method or frozen by the conventional method. Of 11 recipients receiving OPS-vitrified blastocysts, 9 (82%) became pregnant and all of them kidded. The corresponding values for conventional freezing were 50% pregnant and 40% (4/10) kidding. Overall embryo survival was 70% (16/23) for vitrified and 42% (8/19) for conventionally frozen blastocysts. Only kidding rate was significantly different ($P<0.05$). Of 9 recipients receiving OPS-vitrified hatched blastocysts, 3 (33%) got pregnant and 2 (22%) kidded, resulting in 13% (2/15) embryo survival. With conventional freezing 33% (3/9) got pregnant and all of these went to term, resulting in 19% (3/16) embryo survival. The superiority of blastocysts over hatched blastocysts was significant ($P<0.05$) in terms of pregnancy, kidding and embryo survival rates. Morulae were cryopreserved by the OPS method only because they are known to rarely ever survive conventional freezing. Of the 9 recipients that received OPS-vitrified morulae not a single one got pregnant. In conclusion, OPS vitrification, being an efficient and low-cost alternative to embryo freezing can be recommended for caprine blastocysts but will not bring about improvement in hatched blastocysts or morulae.

Fixed-time insemination in estrus synchronized goats induced to ovulate with GnRH or hCG

Al Yacoub, A.N., Gauly, M. and Holtz, W., Department of Animal Science, George August University, Albrecht Thaer Weg 3, 37075, Goettingen, Germany; mgauly@gwdg.de

The objective of this study was to investigate whether it is possible to replace Gonadotropin Releasing Hormone (GnRH) as ovulation inducing agent in prostaglandin F2α ($PGF_{2\alpha}$)-synchronized does by human Chorionic Gonadotropin (hCG). It was hoped that, due to the long half-life of hCG, a more extended LH surge will bring about the formation of corpora lutea that are less prone to premature regression. Sixty does that had kidded at least once were randomly assigned to three treatment groups. All does were treated with 5 mg Dinoprost (Dinolytic®, Pfizer) during the luteal phase of the estrous cycle. The 20 does constituting the control group were inseminated 12-14 h after the onset of estrus. Another 20 does were treated, 48 h after $PGF_{2\alpha}$, with GnRH (0.004 mg Buserelin, Receptal®, Intervet), the remaining 20 does with 500 I.U. hCG (Chorulon®, Intervet). The ovulation inducing treatments were followed, 16 h later, by fixed time artificial insemination. There were no group differences with regards to the time passing between $PGF_{2\alpha}$ treatment and the onset of estrus (44.5, 46.6 and 41.6 h, respectively). The average duration of estrus in the GnRH group was approximately 10 hours shorter than in the other two groups (37.1 versus 46.4 and 48.4 h, respectively, $P<0.05$). The incidence of premature luteal regression was significantly higher in both the GnRH and hCG groups compared to the control group (40 and 35 versus 5%). In the control group, 60% of the inseminated does got pregnant as compared with 50% in the GnRH group and 35% in the hCG group. Corresponding kidding rates were 60, 40 and 35% and prolificacy 1.83, 1.88 and 1.71 kids respectively. In conclusion, substituting hCG for GnRH as a means of terminating ovulation in $PGF_{2\alpha}$-synchronized goats is possible, however, the hoped-for reduction in the incidence of premature luteal regression was not accomplished.

Fertility of Rasa Aragonesa rams carrying or not the FecXR allele of BMP15 gene when used in artificial insemination

Lahoz, B.[1], Blasco, M.E.[2], Sevilla, E.[3], Folch, J.[1], Roche, A.[2], Quintin, F.J.[3], Martínez-Royo, A.[1], Galeote, A.I.[2], Calvo, J.H.[1], Fantova, E.[2], Jurado, J.J.[4] and Alabart, J.L.[1], [1]CITA, Av. de Montañana 930, 50080-Zaragoza, Spain, [2]Carnes Oviaragón S.C.L., C° Cogullada s/n (Mercazaragoza), 50014-Zaragoza, Spain, [3]ATPSYRA, Av. de Movera 580, 50194-Zaragoza, Spain, [4]INIA, Ctra. de La Coruña Km 7.5, 28040-Madrid, Spain; blahozc@aragon.es

A new mutation in BMP15 gene (FecX^R allele) responsible for increased ovulation rate and prolificacy has been recently found in the Rasa Aragonesa sheep breed. Carrier and non-carrier sires are used in cervical artificial insemination within a selection programme for prolificacy carried out in farms of the cooperative UPRA-Carnes Oviaragón SCL. The objective of the present study was to compare the fertility between FecX^R carrier and non-carrier select sires. A total of 17516 ewes from 116 farms were inseminated from 2003 to 2007 with 5 unrelated FecX^R carrier and 42 non-carrier sires. A group of 10-24 ewes inseminated the same day, with the same sire and in the same farm, was considered as a lot. The number of lots inseminated with carrier and non-carrier sires was 440 and 963, respectively. Only were considered: lots with at least one lambing ewe, sires used in at least 3 lots, and farms where both types of sires were used. Ewes were synchronized with FGA sponges for 12-14 days and 480 IU eCG. Insemination was carried out with refrigerated semen (15 °C; 400x10^6 spermatozoa/ewe). Fertility of each lot was analyzed by a mixed-model ANOVA, using SAS. Factors included in the model and their levels of significance were: year ($P<0.05$), six-month period (NS) and their interaction (NS), type of sire (carrier or non-carrier) (NS), inseminator ($P<0.0001$) and age of sire ($P<0.05$), as fixed; sire (nested to type of sire) ($P<0.01$) and farm ($P<0.0001$), as random. These results show that fertility of FecX^R carrier sires (53.2±2.1%) was not significantly different to that of non-carrier ones (54.3±2.0%; lsmeans±standard error) when used in cervical insemination.

Determination of Pregnancy-Associated Glycoproteins (PAG) in goat by ELISA with two different antisera

Shahin, M., Friedrich, M., Gauly, M. and Holtz, W., Department of Animal Science, Albrecht-Thaer Weg 3, 37075 Goettinegn, Germany; mazhar_shahin@yahoo.com

Pregnancy-associated glycoprotein (PAG) is a macromolecule produced by the placenta and released to the maternal circulation. In cows, PAG serves as a useful means of pregnancy detection. The present study addresses the question to what extent plasma PAG determination may serve a similar purpose in goats and whether the bovine PAG-test can be utilized to this end. Blood samples were drawn by jugular venipuncture in 8 synchronized and artificially inseminated multiparous Boer goat does twice weekly during the first 7 weeks and the last 4 weeks of pregnancy, and once a week during the remaining part of pregnancy and 4 weeks postpartum. Plasma PAG concentrations were determined by applying a competitive enzyme-linked immunosorbent assay. Assays were conducted with a) caprine (AS 706) and b) bovine antisera (AS 726). In both assay systems purified bovine PAG served as standard and tracer and sheep anti-rabbit immunoglobulin g served as coating antibody. The PAG profiles generated with the two antisera were strikingly different: With the caprine antibody (AS 706) a drastic increase was recorded reaching a maximum of 72 ng/ml between day 56 and 63 of pregnancy. Thereafter, the PAG level decreased gradually to about 23 ng/ml at parturition and basal level of 0.3 ng/mL about 4 weeks postpartum. With the bovine antiserum (AS 726), the PAG level increased only to about 3.0 ng/ml and dropped to 1.8 ng/ml within a week after parturition. Goats carrying more than one kid had higher PAG concentrations than those carrying singletons, though the differences were not statistically significant except, in the case of caprine antiserum (AS706), between days 49 and 91 of pregnancy. In conclusion, plasma PAG determination can serve as a suitable means of pregnancy detection in goats from day 21 of conception onward. The assay for bovines may be applied, yet more dependable results are arrived at when employing an assay based on caprine antiserum.

Molecular genetics of neuroendocrine stress responses and robustness in pigs

Mormede, P. and Foury, A., Université Bordeaux 2, PsyNuGen, INRA UMR1286, 146 rue Léo-Saignat, 33076 Bordeaux, France; pierre.mormede@bordeaux.inra.fr

The hypothalamic-pituitary adrenocortical (HPA) axis is the the most important stress-responsive neuroendocrine system. Cortisol released by the adrenal cortices exerts a large range of effects on metabolisms, the immune system, inflammatory processes, and brain function, for example. Large individual variations have been described in HPA axis activity, with important physiopathological consequences. In terms of animal production, higher cortisol levels have negative effects on growth rate and feed efficiency and increases the fat/lean ratio of carcasses. On the contrary, cortisol has positive effects on traits related to robustness and adaptation. For instance, newborn survival was shown to be directly related to plasma cortisol levels at birth, resistance to bacteria and parasites are increased in animals selected for a higher HPA axis response to stress, and tolerance to heat stress is better in those animals which are able to mount a strong stress response. Intense selection for lean tissue growth during the last decades has reduced concommitantly cortisol production, which may be responsible for negative effects of selection on piglet survival. One strategy to improve robustness is to select animals with higher HPA axis activity. Several sources of genetic polymorphism have been described in HPA axis. Hormone production by the adrenal cortices under stimulation by ACTH is a major source of individual differences. Several candidate genes have been identified by genomic studies and are currently under investigation. Bioavailability of hormones as well as receptor and post-receptor mechanisms are also subject to individual variation. Integration of these different sources of genetic variability will allow developing a model for marker-assisted selection to improve animal robustness without negative side effects on production traits. Part of these results was obtained through the EC-funded FP6 Project 'SABRE'.

Biomarker development for recovery from stress in pig muscles

Te Pas, M.F.W.[1], Keuning, E.[1], Kruijt, L.[1], Van De Wiel, D.J.M.[1], Young, J.F.[2] and Oksbjerg, N.[2], [1]ASG-WUR, ABGC, P.O. Box 65, 8200AB Lelystad, Netherlands, [2]Univ Aarhus, Agric Sci, P.O. Box 50, 8830 Tjele, Denmark; marinus.tepas@wur.nl

Pork quality is negatively influenced by stress due to pre-slaughter exhaustion of energy metabolism components in muscle fibres. Changes in the proteome are expected due to related protein breakdown. Thus proteome profiles may indicate the status of stress and recovery from stress of muscle tissue. Therefore, pigs were exercised on a treadmill for 30 minutes and slaughtered immediately or after a recovery period of 1 or 3 hours. Control pigs were treated as regular slaughter pigs. Each group consisted of 10 pigs. The longissimus and biceps femoris muscles were sampled. Proteomics profiles were determined for each individual sample using the SELDI-TOF-MS equipment and 3 different array types analyzing positively or negatively charged proteins, or phosphorylated proteins, respectively. Analyses were performed with the software provided with the equipment. The analyses compared the expression levels of individual peaks for each of the animals per muscle and per treatment group. All proteomics experiments were done in duplicate. Each proteomics experiment was individually analyzed and only peaks that appeared in both analyses were considered significant. The first results indicate that the changes of the expression of a number of peaks may be related to stress and recovery of stress. Compared to non-stressed control pigs the expression of these proteins is either decreased or elevated after stress and the changes are reversed during the recovery period. However, peaks differ in the timing of this process. Therefore, at present we cannot conclude on the association between peak expression profile and stress recovery. Further analyses on the repeatability of the analyses are ongoing. Finally, it may be necessary to associate the proteomics expression profiles with physiological indicators of muscle response to stress before conclusions about possible biomarkers can be done.

Differentially expressed genes for aggressive pecking behaviour in laying hens
Buitenhuis, A.J., Hedegaard, J., Janss, L. and Sørensen, P., Aarhus University, Faculty of Agricultural Sciences, Department of Genetics and Biotechnology, Blichers allée 20, P.O. Box 50, DK-8830 Tjele, Denmark; bart.buitenhuis@agrsci.dk

Aggressive behaviour in group living animals is an important aspect of their daily life. Animals which are aggressive have advantages such as better access to food or territories and produce more offspring than the low ranked animals. In chickens the social hierarchy is measured using the number of aggressive pecks given and received, however little is known about the underlying genetics. In this experiment 60 34-week-old White Leghorn hens of the 8th generation of a high feather pecking selection line were kept in groups of 20. The pens were 2 m x 4 m with wood-shavings on the floor. Feed and water were provided ad libitum. All aggressive pecks to other chickens were recorded by video from 14:00 to 17:00. The hens were grouped into three groups based on their pecking behaviour: peckers (P), peckers and receivers (P&R) and receivers (R), respectively. The next day the birds were decapitated. The brain was frozen in liquid nitrogen and stored at -80C. The 20K chicken oligo array (ARK genomics) was used for gene-expression profiling on whole brain tissue of the hens. Differential expression of individual genes was assessed using linear modeling and empirical Bayes methods (Limma version 2.10.0). Results showed that there was a clear contrast between the R-P group comparison with 337 differentially expressed (DE) genes, i.e. clustering of the animals divided the individuals in their assigned group, whereas the contrast between the P&R-R group (180 DE genes) and P&R-P group (342 DE genes) were less clear. Furthermore it was shown that 1) the gene-expression was independent of either the number of aggressive pecks performed or the number of aggressive pecks received, and 2) the DE genes were among other things involved in muscle development, lipid metabolism, and memory formation. If confirmed in further studies these genes may contribute to a better understanding of aggressive pecking behaviour in laying hens.

An attempt at alleviating heat stress infertility in male rabbits with some antioxidants
El Tohamy, M., National Research Centre, Department of animal Reproduction, NRC, Dokki, Cairo, 126022, Egypt; eltohamymagda@yahoo.com

The objective of the present study was to study the effect of summer heat stress on spermogram, serum and seminal plasma biochemical and endocrinal parameters, and an attempt at alleviating heat stress using selected antioxidants of NZW rabbits buck. The study was performed in winter and summer each period lasting 12 weeks. In winter, one winter control group. Other five groups reared in summer and serve as heat-stressed groups, one summer control group and the other four groups received the following four antioxidants, Ascorbic acid, Zinc, Co enzyme Q_{10} and L-carnitine. The temperature–humidity index (THI) was calculated. Summer heat stress reversely affected both qualitative and quantitative traits of spermogram TBARS showed a significant increase while TAC and catalase showed significant decreases as affected by heat stress. Serum and seminal plasma component showed an identical pattern in response to heat stress. Results indicate that blood components were more sensitive to the effect of heat stress than the seminal biochemical components. Summer heat stress significantly increased glucose, AST, ALT, AcP, TL, total cholesterol, HDL and LDL-cholesterol and significantly decreased TP, AlP and TG. Antioxidants administration alleviated some of heat stress effects on biochemical parameters, zinc and l-carnitine were found to be the most beneficial antioxidants in the relief of spermogram, and they cause significant improvement in rabbit sperm characteristics. In such concern, ascorbic acid and co enzyme Q_{10} administrations have no considerable enhancement In conclusion; an adequate reproductive performance can be achieved in summer as well as in winter.

Behavioural genetics important for pig welfare

Rydhmer, L.[1] and Canario, L.[2], [1]Swedish University of Agricultural Sciences, Dept. of Animal Breeding and Genetics, Box 7023, S-750 07 Uppsala, Sweden, [2]INRA UMR1313, Animal Genetics and Integrative Biology, Pig Genetics, F-78350 Jouy-en-Josas, France; Lotta.Rydhmer@hgen.slu.se

Freedom from pain, fear and stress are key elements of animal welfare. Selection for behavioural traits could, together with changes in management and environment, improve the welfare of pigs. This review covers genetic studies of aggressive and social behaviour, fear and maternal behaviour. Aggressive behaviour after mixing causes welfare problems in many pig production systems. Although group housing during gestation improves average sow welfare, it can decrease the welfare of small, young and low-ranking sows. To perform aggressive behaviour (attack) has a much higher heritability than to receive aggressions, and selection for less aggressiveness seems possible. The behaviour of one animal depends, however, not only on its own genotype but also on the genotypes of all group members. Genetic effects can be separated into a direct effect and a group effect describing an animal's ability to influence the result of group members. Thus, production results can be used for an indirect genetic evaluation of behaviour. An alternative way is to record skin lesions. Number of lesions is a heritable trait correlated to aggressive behaviour. Fear of humans is a heritable trait in many species, including both growing pigs and sows. In sows, there is a genetic correlation between fear and piglet survival; less fear is associated with higher survival. Many other sow behaviours are important for piglet survival and consequently welfare: nest building, activity during farrowing, carefulness and reaction to screaming piglets, piglet savaging and nursing behaviour. Genetics of these behaviours will be reviewed and the possibilities of selection for pig behaviour will be discussed.

Pigs' aggressive temperament affects pre-slaughter mixing aggression, stress and meat quality

Turner, S.P.[1], D'eath, R.B.[1], Kurt, E.[2], Ison, S.H.[1], Lawrence, A.B.[1], Evans, G.[3], Thölking, L.[2], Looft, H.[2], Wimmers, K.[4], Murani, E.[4], Klont, R.[2], Foury, A.[5] and Mormède, P.[5], [1]SAC, Edinburgh, EH9 3JG, United Kingdom, [2]PIC Germany, Schleswig, D24837, Germany, [3]PIC, Abingdon, OX13 5FE, United Kingdom, [4]FBN, Dummerstorf, 18196, Germany, [5]INRA, Bordeaux, F33076, France; simon.turner@sac.ac.uk

Pre-slaughter stress negatively affects animal welfare and meat quality. Aggression when pigs are mixed for transport to, or on arrival at the abattoir is a major factor in pre-slaughter stress. This study examined the impact of aggressive temperament in young pigs on aggression, stress and meat quality at slaughter. Ten week old pigs were mixed to identify individuals of high (H) or low (L) aggressiveness. For transport and slaughter, single-sex groups of 8 pigs were mixed according to their temperament into HH, HL or LL combinations or left unmixed (U) in 4 slaughter batches (n=271). At slaughter, HH pigs showed more evidence of aggression (carcass skin lesions) and stress (higher plasma cortisol) and had loin muscle with higher pH at 24 hours (particularly in males), and lower redness (a*) and yellowness (b*) compared to other treatments. There were interactions between sex and treatment, particularly for stress physiology (lactate and creatine kinase) and meat pH. Genetic factors (dam and sire line and halothane locus) also affected production and meat quality parameters as expected. 'Commercially normal' levels of social stress were studied in 4 further slaughter batches where mixing was not controlled (n=313). The amount of mixing was a better predictor of skin lesions at slaughter than temperament. Pigs with more lesions had higher plasma cortisol and lactate and lower glucose, but meat quality did not differ. Aggressiveness, in the social context studied, was stable over 14 weeks. Skin lesions from aggression were associated with activation of the HPA axis and metabolic stress and, when aggressive pigs were mixed together, with greater risk of poor meat quality. These results were obtained through EC-funded FP6 Project SABRE.

Identification of genes involved in the genetic control of aggressiveness, stress responsiveness, pork quality and their interactions

Murani, E.[1], D' Eath, R.B.[2], Turner, S.P.[2], Evans, G.[3], Foury, A.[4], Kurt, E.[5], Thölking, L.[5], Klont, R.[5], Ponsuksili, S.[1], Morméde, P.[4] and Wimmers, K.[1], [1]Research Institute for the Biology of Farm Animals, (FBN), Dummerstorf, Germany, [2]Sustainable Livestock Systems, SAC, Edinburgh, United Kingdom, [3]PIC UK, Kingston Bagpuize, Oxfordshire, United Kingdom, [4]Université de Bordeaux 2, PsyNuGen, INRA, UMR1286, Bordeaux, France, [5]PIC Germany, Ratsteich 31, Schleswig, Germany; murani@fbn-dummerstorf.de

Pre-slaughter stress, e.g. due to aggressive interaction among pigs when mixed during transport or at the abattoir, has a profound impact on meat quality via physiological effects of stress hormones. Knowledge of genetic factors controlling stress hormone synthesis/secretion or mediating and modulating their effects on target cells/tissues will help to improve pork quality and animal welfare. In order to identify and evaluate genes related to aggressiveness and stress responsiveness as well as pork quality, animals of a commercial herd with different aggressive temperament were mixed and parameters of these traits were determined. Genes of the hypothalamus-pituitary-adrenal axis (CRH, AVP, CRHR1, CRHR2, AVPR1B, CRHBP, POMC, MC2R, NR3C1), that are involved in the control of the neuroendocrine stress response and behaviour, were analysed for polymorphisms and association with pork quality and stress response (cortisol, creatine kinase, lactate, glucose). The most consistent effects were found for AVP and AVPR1B, which are known to be involved in the expression of aggressive behaviour. This result indicates that aggressive interactions have a significant effect on pre-slaughter stress and thus on meat quality. These analyses are complemented by expression profiling of adrenal gland and longissimus dorsi muscle of individuals which experienced aggressive interactions of different intensity (according to lesion scores) and which showed differential stress response (creatine kinase, cortisol) and meat quality (pH, color). Results were obtained through the EC-funded FP6 Project 'SABRE'.

Temperament, adaptation and maternal abilities of Meishan and Large White sows kept in a loose-housing system during lactation

Canario, L.[1], Billon, Y.[2], Morméde, P.[3], Poirel, D.[1] and Moigneau, C.[1], [1]INRA, Animal Genetics, UMR1313 Animal Genetics and Integrative Biology, 78352 Jouy-en-Josas, France, [2]INRA, Animal Genetics, UE967 GEPA, 17700 Surgères, France, [3]INRA, Animal Genetics, UMR1286 PsyNuGen, 33076 Bordeaux, France; laurianne.canario@jouy.inra.fr

Changing to loose-housing could negatively affect piglet survival and well-being. Also, breed differences in sow behaviour can influence performance during lactation. Temperament, adaptation and maternal abilities of 16 Large White (LW) and 16 Meishan (MS) gilts were compared. Temperament was assessed by behavioural tests performed during post-weaning (reactivity to novel environment, voluntary human approach) and gestation (within-group aggressivity) periods. The neuroendocrine reactivity (plasma level cortisol) was measured during the first test. Females produced crossbred LWxMS piglets. Video recording started from the day gilts entered the individual farrowing pens (~111d of gestation). LW gilts produced larger litters (15.2 vs 12.9 piglets born) and heavier piglets (1.33 vs 1.14 kg) than MS gilts. In both breeds, stillbirth was remarkably low (0.6 stillborn per litter) as well as mortality during lactation. LW gilts produced more milk over lactation (38.2 vs 48.6 kg piglets). The day after entering the farrowing unit, video recording showed that MS and LW gilts did not differ in activity during the hour following a human approach test. However, within MS and LW breeds, gilts which spent more time standing, showed higher (r=0.64) and lower (r=-0.77) early growth of their litter, respectively. At farrowing, MS gilts compared to LW gilts, spent more time standing, nesting and having nose contacts with piglets. LW gilts may had a better nursing ability than MS gilts, but within LW breed, standing more during the adaptation period would be the sign of higher anxiety and predicted poorer maternal abilities. Relationships with temperament are currently investigated.

Genetic relationships across four criteria of Limousin calf temperament in restrained or unrestrained conditions

Benhajali, H.[1,2], Boivin, X.[2], Sapa, J.[1], Pellegrini, P.[1], Lajudie, P.[3], Boulesteix, P.[3] and Phocas, F.[1], [1]INRA, UMR1313 GABI, domaine de vilvert, 78352 Jouy-en-Josas, France, [2]INRA, UR1213 URH, Theix, 63122 Saint-Genès-Champanelle, France, [3]Institut de l elevage, Genetics, Boulevard des Arcades, 87069 Limoges, France; benhajali@hotmail.com

Genetic relationships between four on-farm criteria related to beef calf temperament in restrained or unrestrained conditions were estimated on 24 French Limousine farms. In order to ensure good family structure and connectedness across farms, 12 bulls selected for artificial insemination were used to produce an average progeny size of 56±23 weaned calves. Behavioural records were registered at an average age of 8 months for all the calves born in the farms between August 2007 and March 2008. Data editing consisted of selecting records from sires with at least 5 progeny records and from contemporary groups with at least 3 records per herd-management group. In total, data of 1,271 calves, bred by 65 sires including the 12 experimental bulls were analysed. Calves were submitted to a restrain test during weighing when three criteria were studied: an Australian subjective crush test score (AS) on a 5-point scale ranging from 1 (docile) to 5 (aggressive), the number of rush movements (RM) and the total number of movements (TM) were also scored using a 6-point score ranging from 1(<2 movements) to 6 (continuous movement). The reaction to human approach of the calves free to move was also assessed during a routine morphological scoring of animals using a 6-point score (HA) ranging from 1 (come near) to 6 (charge). Heritabilities were estimated at 0.18, 0.23, 0.29 and 0.17 respectively for AS, RM, TM and HA scores with standard errors ranging between 0.07 and 0.09. The 3 scores given during weighing were highly correlated all together (about 0.90). On the contrary, HA was insignificantly genetically correlated to the 3 previous scores suggesting that free animals' reaction to human approach could be independent of animal responses in restrained situations.

High milk production changes lying behaviour of dairy cows

Duchwaider, V.[1,2], Munksgaard, L.[2] and Løvendahl, P.[2], [1]University of Copenhagen, Faculty of Life Sciences, DK 1870 Frederiksberg, Denmark, [2]Aarhus University, Faculty of Agricultural Sciences, Research Centre Foulum, P.O. Box 50, DK 8830 Tjele, Denmark; Vibeke.Duchwaider@agrsci.dk

Increasing milk yield, along with competition in modern loose house systems, may limit time available for performing necessary needs, such as lying, and cows prevented from lying down show enhanced symptoms of stress. Information is therefore needed about how continued selection for higher milk yield (MY) may alter expression of important behaviours such as lying time (LT) and the number of lying periods (LP). The aim of this study was to estimate genetic parameters for lying time, number of lying periods and milk yield, and their correlations. A cohort study including 7 dairy herds with a total of 357 Holstein cows was used. LT and LP were obtained using electronic sensors (IceTag®) attached to the hind limb of the cows for four consecutive days. MY records were obtained as test day energy corrected milk (ECM). Two trait linear mixed models were used to obtain estimates of co-variance components. Mean LT, LP and ECM were 687 min/day, 11.6 periods/day and 28.5 kg ECM/day respectively. Repeatability for LT and LP were 0.66 (±0.08) and 0.69 (±0.11), and estimated heritabilities were 0.08 (±0.11) and 0.29 (±0.12) for LT and LP respectively. A strong genetic correlation was found between MY and LP. These results show that both LT and LP are partly under genetic control, and that selection for milk yield will give a correlated response in both traits, leading to more lying periods. The consequences of continued selection for higher yield may therefore be shorter lying time which in turn may decrease the welfare of dairy cows.

Cellular and genetic background of the adrenal sensitivity to adrenocorticotropic hormone in Large White and Meishan pigs

Rucinski, M.[1], Tyczewska, M.[1], Ziolkowska, A.[1], Mormede, P.[2] and Malendowicz, L.K.[1], [1]Poznan University of Medical Sciences, Histology & Embryology, 6 Swiecicki Street, 60-781 Poznan, Poland, [2]University of Bordeaux, Neurogenetics and Stress INRA UMR1286, CNRS UMR5226, 146 Léo-Saignat Street, F-33076 Bordeaux, France; marcinruc@amp.edu.pl

Available literature data indicate higher basal cortisol levels and augmented responses to ACTH in Meishan (MS) than Large White (LW) pigs. The aim of the present study was to investigate cellular and genetic backgrounds of these differences at the adrenal level. As evidenced by stereology, volumes of adrenal cortex and its zones and of adrenal medulla (M) of immature piglets is notably higher in LW than MS strain while the average cell volumes are similar in both strains. Adrenal cortex of LW pigs contains approximately 25% more parenchymal cells when compared with MS. Having a list of earlier identified genes, expression of which is affected by genotype and/or ACTH (CHCHD2, EIF1B, GADD45B, A2M and ACOX1), we performed *in situ* hybridisation (ISH) studies. Furthermore, by means of QPCR we also studied expression of that genes in individual compartments of the adrenal gland (ZG – zona glomerulosa; ZF/R – fasciculata/reticularis and M). Expression of studied genes is notably higher in the adrenal cortex than in M. In the cortex, expression levels of CHCHD2, EIF1B and GADD45B genes are higher in ZF/R than in ZG, while opposite is true for A2M and ACOX1. In MS pigs, expression levels of CHCHD2 were higher, and A2M and ACOX1 lower than in LW, while no differences were seen in expression of EIF1B, and GADD45B genes. ACTH injection resulted in an increase in expression levels of CHCHD2 and EIF1B of LW pigs, and decrease in expression levels of GADD45B, A2M and ACOX1 of LW pigs. Thus, the present study provides new data on cellular and genetic aspects of the adrenal sensitivity to adrenocorticotropic hormone in LW and MS pigs.

Changes in physiological and blood parameters in water stressed Awassi ewes supplemented with different levels of Vitamin C

Hamadeh, S.K., Hanna, N., Barbour, E.K., Abi Said, M., Rawda, N., Chedid, M. and Jaber, L.S., American University of Beirut, Riad El Solh 1107-2020, Beirut, Lebanon; shamadeh@aub.edu.lb

Vitamin C administration is not common in ruminant production. The current study was designed to assess the effect of oral vitamin C supplementation on alleviating water stress in Awassi ewes. Experiment 1 included four groups, four animals in each, with one group as control receiving daily water, the remaining three groups were watered once every four days, with the third and fourth groups supplemented with a daily oral dose of 3 g and 5 g vitamin C, respectively. This experiment extended for 24 days. Experiment 2 was similarly designed except that the third group was given 3g Vitamin C daily, while the last group was administered one 10g vitamin C oral dose at the beginning and at the middle of the experiment, and it extended for 21 days. The vitamin C treatments seemed to slightly reduce weight loss induced by water stress. Serum protein, albumin, globulin urea, creatinine, Na^+ and Cl^- concentrations were increased in all water stressed animals. Inconclusive results were observed for the effect of vitamin C on urea, creatinine, and electrolytes concentrations, under water stress. On the other hand, albumin concentration seemed to be higher in vitamin C supplemented animals, while protein and globulin concentrations were not significantly different in all water stressed groups. Vitamin C supplementation did not seem to induce major physiological changes under water stress, except for a reduction in weight loss. Further research is warranted to elucidate the different aspects of the role of vitamin C in water stress alleviation.

Estimation of genetic trends from 1977 to 2000 for stress-responsive systems in French Large White and Landrace pig populations using frozen semen

Foury, A.[1], Tribout, T.[2], Bazin, C.[3], Billon, Y.[4], Bouffaud, M.[5], Gogué, J.M.[6], Bidanel, J.P.[2] and Mormède, P.[1], [1]Université de Bordeaux, PsyNuGen, INRA UMR1286, CNRS UMR5226, 33076 BORDEAUX, France, [2]INRA, UMR1313 GABI, 78350 JOUY-EN-JOSAS, France, [3]IFIP, Pôle Génétique, 35650 LE RHEU, France, [4]INRA, UE967 GEPA, 17700 SURGÈRES, France, [5]INRA, UE450 Testage de Porcs, 35650 LE RHEU, France, [6]INRA, UE332 Domaine Expérimental de Bourges, 18390 OSMOY, France; aline.foury@bordeaux.inra.fr

An experimental design aiming at estimating realized genetic trends from 1977 to 1998-2000 in the French Large White (LW) and Landrace (LR) pig populations was conducted by INRA and IFIP-Institut du Porc. Large White sows were inseminated with semen from LW boars born in 1977 (frozen semen) or in 1998 and their second generation offspring were station tested. Landrace sows were inseminated with semen from LR boars born in 1977 (frozen semen) or in 1999-2000, and their progeny was station tested. Urinary concentration of stress hormones (cortisol and catecholamines) and traits related to carcass composition (estimated carcass lean content (ECLC) and global adiposity) and meat quality (pH 24 h) were measured. For the two populations, selection carried out since 1977 led to an increase in ECLC and a decrease in carcass adiposity. Between 1977 and 1998-2000, urinary concentrations of stress hormones were unchanged in the LR breed, but were decreased in the LW breed. Moreover, for the animals generated from LW boars born in 1977 and in 1998, urinary cortisol levels were negatively correlated with ECLC. Therefore, in the LW breed, selection carried out for higher ECLC resulted in a decrease in cortisol production, as well as a reduction of catecholamine production that may be responsible for the lower ultimate pH of meat. Therefore, selection carried out for increased carcass lean content led, in this breed, to large modifications in the functioning of the stress-responsive systems, thereby influencing a large range of physiological regulations and technical properties such as meat quality.

Genetic analysis of cross- and intersucking in Austrian dairy heifers

Fuerst-Waltl, B.[1], Rinnhofer, B.[1], Fuerst, C.[2] and Winckler, C.[1], [1]University of Natural Resources and Applied Life Sciences Vienna, Dep. Sust. Agric. Syst., Div. Livestock Sciences, Gregor Mendel-Str. 33, A-1180 Vienna, Austria, [2]ZuchtData EDV-Dienstleistungen GmbH, Dresdner Str. 89/19, A-1200 Vienna, Austria; waltl@boku.ac.at

Cross-sucking and intersucking are considered abnormal behaviours in cattle and constitute a common problem in dairy farming. Cross-sucking in calves is defined as sucking any body parts of another calf whereas intersucking in heifers and cows is defined as sucking the udder or udder area in all age groups. The aim of this study was to determine the genetic variability for abnormal sucking behaviour by estimating genetic parameters and examining individual differences between sires with large progeny groups. By means of a questionnaire, cattle breeders in the federal state Lower Austria were requested to identify all currently kept animals which are known of either inter- or cross-sucking being defined as the binary trait 'sucking' with 0 and 1 referring to the absence and presence of this abnormal behaviour. After restriction to the Austrian main breed Fleckvieh (dual purpose Simmental) and further data editing, records of 1,222 farms and 13,332 heifers aged between 21 and 700 days were investigated. In total, 8.6% of all calves/heifers in the data set were observed sucking. A heritability of 0.116 ± 0.041 was estimated applying a threshold sire model. Effects being taken into account were the random herd*year*season effect and the random genetic effect of sire. Breeding values were estimated applying the same model. When ranked by EBV, 2.1 to 3.0% and 13.4 to 16.6% of best and worst three bulls' offspring, respectively, were recorded as sucking. The results suggest that a reasonable genetic variability of sucking exists so that this trait could theoretically be used as selection criterion. Even if the trait sucking will not be included in future breeding goals, identification of extreme bulls and monitoring aiming at optimal designs of management and environment will be valuable.

Factors affecting the peripartal stress response in beef cows

Álvarez-Rodríguez, J.[1], Palacio, J.[2], Casasús, I.[1] and Sanz, A.[1], [1]Centro de Investigación y Tecnología Agroalimentaria, Gobierno de Aragón, Tecnología en Producción Animal, Av. Montañana, 930, 50059 Zaragoza, Spain; [2]Universidad de Zaragoza, Departamento de Patología Animal, c/ Miguel Servet, 177, 50013 Zaragoza, Spain; jalvarezr@aragon.es

This experiment tested the hypothesis that calf management and breed may be factors affecting the adrenal response around parturition. Fourteen winter-calving beef cows (live-weight 590 kg and body condition score 2.52), 7 Parda de Montaña and 7 Pirenaica, were assigned within breed to restricted suckling once-daily (RS1) or ad libitum (AS) from the day after calving. Individual faecal samples were collected on approximately 12, 48 and 72 hours (h) after delivery and freeze-dried. Glucocorticoid metabolites (GM) were analysed in duplicate by RIA. Results were assessed through ANOVA. Faecal GM concentration were similar in both breeds throughout the 3 days following parturition (mean 14.3 ng/g, $P>0.10$). Faecal GM concentration tended to be affected by the interaction between suckling system and time post-partum ($P=0.07$). Both groups had similar GM around 12 h post-partum (mean 19.9 ng/g), but on 48 h (the day after calf separation in RS1), the values were greater in RS1 than in AS (16.3 vs. 9.5 ng/g, $P<0.05$). Such differences were overridden on 72 h after calving (mean 10.1 ng/g, $P>0.10$). The percentage of decrease of GM from 12 h to 48 h was similar to that from 12 h to 72 h (-24.3 vs. -43.4%, $P>0.10$). There was a trend towards an overall lower decrease of GM in RS1 than in AS (-15.7 vs. -52.1%, $P=0.09$). Faecal glucocorticoid metabolites did not vary according to genotype and allow detecting the post-partum adaptative challenge and a potential acute stress of restricted suckling.

Age-dependent changes in behavioural and neuroendocrine responses of neonatal pigs exposed to a psychosocial stressor

Hameister, T., Puppe, B., Tuchscherer, M., Tuchscherer, A. and Kanitz, E., Research Institute for the Biology of Farm Animals (FBN), Wilhelm-Stahl-Allee 2, 18196 Dummerstorf, Germany; hameister@fbn-dummerstorf.de

Stressful early life events may have lasting effects on behavioural and neuroendocrine mechanisms of adaptation with consequences for health and welfare. In this study, we used a single social isolation as a model of psychosocial stress in suckling piglets to investigate the impact of these stressor on behavioural alterations in open-field tests, stress hormones and modifications in the expression of genes regulating glucocorticoid (GC) response in stress-related brain regions at 7, 21 or 35 days of age. The mRNAs of glucocorticoid receptor (GR), mineralocorticoid receptor (MR), 11ß-hydroxysteroid dehydrogenase 1 and 2 (11ß-HSD1 and 11ß-HSD2) and c-fos were analyzed by real-time RT-PCR in the hypothalamus, hippocampus and amygdala. The social isolation caused a significant increase in ACTH and cortisol concentrations and an enhanced behavioural arousal. The increased behavioural and neuroendocrine activity was associated with distinct changes in gene expression in the limbic system. Whereas the hypothalamic GR, MR and 11ß-HSD1 mRNA levels and the hippocampal 11ß-HSD1 were significantly higher in piglets exposed to isolation stress, in the amygdala isolation caused a significantly decrease in the MR mRNA expression. A marker of neuronal activity, c-fos mRNA, was significantly increased in both hypothalamus and amygdala of isolated piglets. The mRNA alterations of GC regulating genes, in conjunction with the behavioural and hormonal pattern, show a stronger impact of social isolation on days 7 and 21, indicating that younger piglets are more vulnerable to isolation stress. In conclusions, the present results emphasise that the GC regulating genes are involved in mediating emotional experience in pigs. With respect to welfare and health, the data also suggest that psychosocial stress effects should be considered for the assessment of livestock handling practices such as weaning of piglets.

Effect of carbon dioxide stunning (CO2) on animal welfare in lambs

Rodriguez, P.[1], Dalmau, A.[1], Llonch, P.[1], Manteca, X.[2] and Velarde, A.[1], [1]IRTA, Finca Camps i Armet S/N, 17121 Monells (Girona), Spain, [2]UAB, Ciencia animal i dels aliments, Facultat de Veterinaria, 08193 Bellaterra (Barcelona), Spain; pedro.rodriguez@irta.cat

The effect of the inhalation an atmosphere of 90% carbon dioxide on animal welfare was assessed in 32 male lambs of 22.38±0.525 live weight. Lambs were loaded in pairs into a crate and lowered into a pit (260 cm) prefilled with 90% of CO_2 measured 50 cm above the bottom of the well. The CO_2 exposure cycle lasted 106s, and consisted of the fist 23 s (time the cage was lowered), the following 60s (while the cage remained at the bottom of the well at the highest concentration), and the final 23 s (time the cage was raised). All lambs showed headshake and sneezing at 8.3±1.91 s after the start of descend of the crate. Thereafter, 100% of the lambs lost posture at 19.4±2.30 s and exhibited hyperventilation from 20.1±0.69 s until 36.0±0.53 s after the end of the exposure to the gas. On the other hand, during CO_2 exposure two animals showed retreat attempts at 5.5±0.50 s, one vocalization at 14s and gasping at 17 s. After loss of posture, 75% of the lambs showed gagging at 43.3±2.17 s. The study concludes that when lambs are stunned with high concentrations of CO_2 animals perform behavioural reactions that could be signs of aversion to the gas before the loss of consciousness.

Altered gene co-expression profile in extreme feather-pecking behaviour

Abreu, G.C.G., Labouriau, R. and Buitenhuis, A.J., Aarhus University, Department of Genetics and Biotechnology, Blichers Allé 20, P.O. Box 50, 8830 Tjele, Denmark; Gabriel.Abreu@agrsci.dk

Feather pecking (FP) in laying hens is characterized by pecking at and pulling out of feathers of the receiver animal, and poses a significant problem to industry. Although FP has been intensively studied in the last decades, the genetic mechanisms related to this behaviour are still under discussion. In a previous study, we identified a group of animals presenting extremely high FP. Studying the evolution of the proportion of this extreme behaviour along a sequence of eight generations selected for high FP; we found evidences of a strong genetic component associated with this very high FP behaviour (in fact a dominant major allele). This hypothesis was further supported by the fact that the gene transcription profile of the animals performing very high FP differ from the profile of the other animals performing FP (456 genes differentially expressed from a total of the 14,077 investigated genes). Here we present further details about this gene co-expression network and show that the transcription factors of these 456 differentially expressed genes are highly correlated. Using suitable models for complex networks (covariance selection models), we demonstrate that these differentially expressed genes occupy a relative central position in the network. Moreover, we identify some of those genes that occupy indeed key positions (hubs) in the gene co-expression network. This suggests that animals presenting extremely high FP might have basic processes altered.

Abnormal behaviours in therapeutic riding horses

Li Destri Nicosia, D.[1], Sabioni, S.[1], Facchini, E.[1], Ridolfo, E.[1], Cerino, S.[2], Giovagnoli, G.[2] and Bacci, M.L.[1], [1]University of Bologna, DIMORFIPA, Via Tolara di Sopra 50, 40064 OZZANO EMILIA BO, Italy, [2]FISE, V.le Tiziano 74 ROMA, 00196, Italy; dora.lidestrinicosi3@unibo.it

Therapeutic Riding (TR) is a global therapeutic method. Human-horse relationship is central in TR intervention. The horse becomes an active component provided that his features are recognized and respected. Nevertheless TR is recognized as potential stressor. The aim was identification/characterization of abnormal behaviours in TR horses. Firstly we conducted an investigation sending questionnaires to TR centres in North Italy. Then we made direct ethological observations on 4 TR horses. Those approaches highlighted: a late identification of stress-related behaviours; frequent onset of physical-behavioural problems (due to features of TR as repetitiveness, constrictiveness and interference with horse motor dynamics and as precariousness of human/animal relationship). Such conditions, however, were often under-estimated or not declared, until these compromised the use of TR horse. Therefore further ethological studies were performed on 3 experimental groups (n=4): TR (A), destined to sport activity (B) and semi-wild breeding (C) horses. They were observed 25 consecutive days, at fixed time (20 min, twice a day) and at random time (once every 3rd day) period, and a data sheet of behavioural assessment were utilised. Compared to referring ethogram, group A showed a prevalence of reactive type anomalies (tendency toward hypo-reactivity, with sporadic demonstrations of hyper-reactivity), in group B motor/oral-ingestive anomalies resulted more frequent. Group C didn't show abnormal behaviours. Our results confirm previous observations and seems to reflect that specific and various stressors, conflicting with horse's motivations and adaptive skills, may determine the arise of disturbs related to a state of constraint or anxiety. This research was supported by grants from Bologna University (RFO 60%).

Gene expression and ontological analyses reveal a key role of LINEs in endurance horse race-induced stress conditions

Cappelli, K.[1], Capomaccio, S.[1], Galla, G.[2], Barcaccia, G.[2], Felicetti, M.[1], Silvestrelli, M.[1] and Verini-Supplizi, A.[1], [1]Università degli Studi di Perugia, CSCS-DPDCV, Via S. Costanzo 4, 06126, Perugia, Italy, [2]Università di Padova, DAAPV, Viale della Università 16, 35020, Legnaro, Padova, Italy; vete5@unipg.it

Long Interspersed Nuclear Element L1 (LINE-1 or L1) are replicating repetitive elements that form over one-third of mammalian genomes. Compelling evidence has implicated L1 sequences in the regulation of genome-wide gene expression by acting as a molecular fine-tuner of the transcriptome. As a matter of fact, about 79% of mammalian genes contain at least one L1 segment in their transcription unit, mainly within intronic regions and in poorly expressed genes, influencing the transcription of surrounding sequences. A DD-cDNA-AFLP analysis was performed to isolate transcripts of differentially expressed genes related to exercise-induced stress in endurance horses. This approach led us to the identification of 75 TDFs with a contrasting expression pattern at the analyzed time points. An extensive use of bioinformatic tools, including a specific analysis with RepeatMasker software, revealed that more than the 20% of the identified fragments are L1 derived sequences. The modulation of expression of the selected TDFs was confirmed in 6 endurance horses by qRT-PCR using SyberGreen technology, normalizing the assay with appropriate housekeeping genes. Since most of the L1 containing TDFs are positioned within non-coding regions, all recovered LINE-1 fragments were used as query to detect L1-containing and L1-flanking genes. In addition to TDFs, and the whole set of L1-associated sequences were annotated according to the three structured GO vocabularies. Moreover, querying the KEGG database with the horse datasets enabled the identification of modulated metabolic pathways during exercise-induced stress. Overall results obtained by integration of the horse sequence datasets within overview metabolism maps and regulatory networks of the GenMapP2 platform will be presented and critically discussed.

Navigating the dynamics: resilient farms through adaptive management

Darnhofer, I., BOKU - Univ. of Natural Resources and Applied Life Sciences, Vienna, Dept. of Economic and Social Sciences, Feistmantelstr. 4, 1180 Vienna, Austria; ika.darnhofer@boku.ac.at

The complex interplay between consumer preferences, market prices, and national as well as international policies create an ever-changing framework for livestock farming systems. Farmers need to navigate both this uncertain and dynamic environment as well as accommodate the on-going changes on the farm itself (family life-cycle, droughts, diseases, etc.). To better understand these dynamics - and thus the challenges faced by farmers - it might be helpful to build on the insights derived from the study of complex adaptive systems as well as the resilience of social-ecological systems. These indicate that there are qualitatively different types of changes, i.e. changes that are different in speed and predictability. A farm that has gone through a prolonged period of relative stability and gradual changes is likely to suddenly face a shock, followed by a phase of chaotic and unpredictable change. Under such circumstances, the assumptions of conventional farm management, i.e. a focus on predictability and stability for a system-near-equilibrium, no longer hold. Unpredictable and dynamic environments require strategies that enable farmers to keep their farm resilient yet adaptive. Indeed, a farm needs to persist, i.e. maintain its identity in the face of internal change as well as external shocks and disturbances. At the same time the farm needs to be able to transform itself when an attractive opportunity arises or when it is needed, e.g., when the current system is no longer viable. The presentation will explore strategies that allow farmers to resist disruptions while at the same time organizing and influencing transitions, i.e. simultaneously fostering the attributes necessary for stability, adaptability and transformability. The focus will be on livestock farming systems, that tend to have long-term investments (e.g., animal housing, milking equipment) and may thus be less able to adapt quickly compared to farms with arable crops.

Farm, family and work: new forms, new adjustments? Strategies of agricultural households to changing socio-economic environment

Madelrieux, S., Nettier, B. and Dobremez, L., Cemagref, DTM, Domaine Universitaire, 2 rue de la Papeterie, BP 76, 38402 St Martin Hères, France; sophie.madelrieux@cemagref.fr

Livestock farming systems (LFS) meet great transformations due to the evolution of the agricultural context but also to profound changes of family and work. To maintain a farm activity does not depend only on technical and economic performances but also on the 'liveability' of the system for the farmer as for his/her family. It appears indeed more and more difficult to articulate farm, family and work (increase of divorces, celibacy, distress, suicides…). To better understand the LFS dynamics, we propose to analyse the way changes in agricultural activities, non agricultural activities and family events are articulated in households' trajectories. We surveyed 20 livestock farms in a same agricultural area of the Northern French Alps, so as to piece together the histories of farm, family and activities of each member of the household. We identified 7 logics of 'farm-family-work articulation' (several of them can be linked together in a same trajectory), the reasons of logics' changes and the links with changes in the agricultural activity. We assess the adjustments concerning the LFS. In most cases, these are linked to a change in the activities (agricultural or non agricultural) of the man, the woman or both of them (due to family events, problems of work…). Less frequently in our cases, adjustments are carried out to face a market chain's problem. To end, we discuss the use of this kind of knowledge to understand LFS dynamics, better taking into account the importance of family and work, and their relations with the farm.

Modelisation of work organisation of livestock farms to simulate changes

Hostiou, N.[1], Poix, C.[2] and Cournut, S.[2], [1]INRA, UMR 1273 Metafort, centre de Theix, F-63122 Saint Genès Champanelle, France, [2]ENITA, UMR 1273 Metafort, Marmilhat, F-63370 Lempdes, France; nhostiou@clermont.inra.fr

The work is crucial for ensuring the sustainability of livestock farms, and for that it is necessary to design tools and methods to assist farmers in the process of changing their system. The goal of the study was to formalize a conceptual framework of work organisation for livestock farms, considering the farmer as a work organiser as well as a worker. The purpose of the framework was to characterize the durations and rhythms of work according to various time scales. We aimed at describing the work organisation at farm level, using expert knowledge and literature to build the ontology defining the concepts and their relationships. The workgroup was characterized according to their rhythm of involvement, their status and their role in the decisions. We defined concepts to represent work organisation at different scales such as day, week, period and year, enabling us achieve a conceptual data model which linked concepts such as task, activity and workgroup taking rhythms into account (daily and seasonal work). For example, hay making involved specific workgroups and weeks and might modify the work organisation of the day (sequence of daily activities). We implemented a database (under MySQL and Ms Access) in order to represent actual farms. The conceptual data model was refined and the processing was defined: quantitative data could be produced on working times by worker, workgroup, activity, period and qualitative data were also available such as division of labour between workers. A schedule of the work organisation could also be obtained (forms of daily organisation, sequences...). This model may be used to evaluate scenarios of adaptation (development of grazing, employment of a paid worker...), and it will be improved by taking another element of work organisation into account: the buildings and equipment. In order to simulate the process of change, we also have to integrate rules of work organisation.

Analyzing trade-offs between production, economics, land use and labour in mountain farming systems through long-term stochastic simulation

Villalba, D.[1], Ripoll, G.[2], Ruiz, R.[3] and Bernués, A.[2], [1]University of Lleida, Rovira Roure 191, 25198 Lleida, Spain, [2]CITA de Aragón, Avda Montaña 930, 50059 Zaragoza, Spain, [3]NEIKER Tecnalia, Arkaute - Apdo 46, 01080 Vitoria, Spain; dvillalba@prodan.udl.es

Mountain cattle farms play diverse roles: food production, biodiversity and landscape conservation, rural development. Extensification promoted by CAP means greater utilization of natural resources, but farms can undergo uncertain reproductive and economical consequences in the long term. Farmer working conditions become also crucial in areas with high labour opportunity cost. A range of herd management strategies are possible within a particular location, and simulation models allow considering the interactions between diverse production factors and analyzing ex-ante consequences of these strategies. A stochastic simulation model was used to evaluate the calving periods in the Central Spanish Pyrenees: winter calving (WC); autumn calving (AC), 8 months lenght calving (8MC), and with a herd with two calvings in three years (2C3Y, hypothetical scenario of maximum extensification); each strategy was tested for two types of production systems: cow-calf farms and cow-calf/finishing farms. A herd of 100 cows was simulated during 15 years, but only data after reaching the steady state (year 6) was used in the analysis. The percentage of pregnant cows at the end of the mating season was higher for 8MC and 2C3Y (between 92% and 94%), intermediate for AC (88%) and lower for WC (78%), which also showed higher variability between years. Although AC and 8MC rendered higher productive performance, their higher labour requirements and winter feeding costs resulted in lower economic margin; they also implied lower utilization of natural resources. In economic terms, WC was the best strategy for cow-calf/finishing farms, whereas 2C3Y was the worse for the two types of production systems, although implied the highest use of natural resources. The extensification strategies were less sensitive to changes of feed prices.

Evaluation of Ankole pastoral production system in Uganda: systems analysis approach

Mulindwa, H.[1,2], Galukande, E.[1,3], Wurzinger, M.[1,4], Mwai, A.O.[4] and Sölkner, J.[1], [1]BOKU-University of Natural Resources and Applied Life Sciences, Gregor-Mendel-Strasse 33, 1180 Vienna, Austria, [2]National Livestock Resources Research Institute, P.O.Box 96, Tororo, Uganda, [3]National Animal Resources Center and Data Bank, P.O.Box 186, Entebbe, Uganda, [4]International Livestock Research Institute, P.O. Box 30709, 00100 Nairobi, Kenya; esaugalu@yahoo.com

The Ankole pastoral production system was formerly characterized by a nomadic life style. However, there is a current shift from nomadic to sedentary system and pastoralists are now keeping two separate herds of both pure Ankole and Friesian-Ankole cattle. The ecological and economic sustainability of the emerging system is currently being assessed through system analysis approach using a modelling approach (STELLA software). Eighteen farmers were selected and on-farm production records, management and feeding strategies are obtained from these farms on a regular basis. The model is further parameterized using existing literature as well as historical meterological data. The dynamic model is used to identify constraints to the current dairy production system and conditions under which either one or both herds can be kept on a sustainable basis. The model is aimed at producing long-term simulations of the dynamic interaction between ecnomic and climatic variations, and farm management at monthly, seasonal and annual scales in order to evaluate the performance of the emerging production system. In this paper, we present the structure of the model and results of the current productivity of the system as well as its constraints.

Integrating feeding and reproductive management: a modelling approach to assess dairy goat herd performance

Puillet, L.[1,2], Martin, O.[2], Sauvant, D.[2] and Tichit, M.[1], [1]INRA UMR 1048 SAD-APT, AgroParisTech 16 rue C. Bernard, 75231 Paris cedex 5, France, [2]INRA UMR 791 PNA, AgroParisTech 16 rue C. Bernard, 75231 Paris cedex 5, France; laurence.puillet@agroparistech.fr

In a context of feed cost fluctuation, a critical issue for the sustainability of dairy systems is to achieve an efficient diet valorisation. In dairy goat herd, feeding is mainly managed through groups of animals. Diet valorisation into milk is strongly dependant on the diversity of production potential and physiological states within each group; this latter is a direct consequence of reproductive management. Our objective was to develop a dairy goat herd model to assess the efficiency of feeding and reproductive management in terms of milk valorisation. The model is individual-based and represents technical operations of replacement, reproduction and feeding. It simulates herd productivity and demography over 20 years. Farmer's production project (i.e. a targeted milk production pattern) defines herd structure into female groups. Technical operations are represented by a set of discrete events formalising decision rules at the level of each female group. The animal level integrates the variability of biological responses driven by genetic parameters (milk production and body weight) and energy partitioning regulations. Simulations were run with three strategies of reproductive management, leading to contrasted distributions of physiological states. Genetic level and feeding management were defined by the same parameters in each experiment. Results were analysed at herd scale regarding to mean goat productivity and dry matter per kg of milk. These indicators show the effect of reproductive management on individual production and the efficiency of the global conversion of dry matter into milk. This study emphasizes the interest of herd modelling to analyse complex interactions between decisional and biological components. It opens promising perspectives to assess their relative contribution for an adaptive herd management.

Development of a tool to diagnose economic, social and environmental sustainability of animal husbandry systems: application to dairy farming

Arandia, A.[1], Intxaurrandieta, J.M.[2], Santamaria, P.[3], Mangado, J.M.[2], Icaran, C.[3], Nafarrate, L.[4], Del Hierro, O.[5], Lopez, E.[3] and Pinto, M.[5], [1]Universidad Publica de Navarra, Gestión de Empresas, campus arrosadia, 31006 pamplona, Spain, [2]ITG Ganadero., Avda. Serapio Huici, 22 Edif. Peritos., 31610 Villava (Navarra), Spain, [3]IKT, Granja Modelo s/n, 01192 Arkaute, Vitoria-Gasteiz, Spain, [4]SERGAL, c/ La Paloma, 4 bajo, 01002, Vitoria-Gasteiz, Spain, [5]NEIKER, Berreaga 1, 8160 Derio (Bizkaia), Spain; amaia.arandia@unavarra.es

European society sees agriculture not only as food production, but also as provider of other goods and services, difficult to quantify, but key to the future viability of farm holdings. With the intent of quantifying these externalities, a range of social, economic and environmental indicators are proposed. These indicators are used to characterise both farms and agricultural systems and to assess their long term sustainability. A software tool has been developed for data collection and processing. This tool also produces a graphical output that facilitates understanding and analysis of the results. In this communication the proposed indicators are described. After some preliminary tests carried out with specialised dairy holdings, the following aspects may be commented: 1) Initial data collection is relatively simple for economic indicators, since chosen farms are attached to technical-economic management programs. 2) Environmental information has also been easily collected, due to the previous completion of the 'Environmental Diagnosis' and the energy and greenhouse gas balance, with the DIALECTE® and PLANETE® tools, respectively. 3) Social indicators are the most original section, since aspects of a fundamental relevance, as well as complexity and difficulty of measurement are valued. The final objective of this tool is its application to the 800 farms attached to management programs in the CAV and Navarra (Spain).

Using modeling to assess the sensitivity of sheep farming systems to hazards

Benoit, M., Sirben, E. and Tournadre, H., INRA, Unité Recherches Herbivores (1213), Centre de Theix, F-63122, France; marc.benoit@clermont.inra.fr

The breeding activity is more and more subject to hazards, climatic (with impacts on technical performances), or concerning the animal health or economics (price volatility of raw materials). We used a model for setting up an experimental design based on systemic approach to compare the sensitivity of 2 sheep-for-meat systems faced to hazards. The first system ('2P') based on two lambing periods per year (spring and autumn), is compared to a '3P' system with an additional lambing period either in the early spring or late autumn. The '2P' aims at maximizing fodder self-sufficiency knowing that periods of high feeding needs of the flock correspond to periods of greater forage availability. However, this system is very dependent on fodder resources and could be very sensitive to climatic conditions and to seasonality variations of lamb prices. The '3P' would be less sensitive to certain hazards thanks to spreading animals feeding needs and lamb sales. A first simulation dealt with the impact of a 30% fall in fertility for spring mating. The '3P' returned more quickly to equilibrium in terms of lambing seasonal distribution (2 years needed) compared to the '2P' (3 to 6 years) and the loss in number of lambings is globally lower. Then, the effects of variations of some factors on the gross margin per ewe were simulated: price of the lambs (±15%), prolificacy (±15%), mortality of lambs (±50%), and price of purchased cereals (±20%). The '3P' is less sensitive than the '2P' to hazards studied individually (for April lambings for example), except for the price of cereals because of their larger use (for March or November lambings). Considering the range variation of the factors retained, it is the factor 'selling price of the Lambs' that had the strongest impact on gross margin per ewe.

Modelling of dairy cow feeding system to assess farms' adaptability to technical changes

Thénard, V., Martin, G. and Duru, M., INRA, UMR AGIR, BP 52627, 31326 Castanet Tolosan, France; vincent.thenard@toulouse.inra.fr

Grasslands are now being acknowledged for their multifunctional role. But the intensification of livestock production has led to a decrease in grassland use. This trend is less pregnant when cheese production occurs within Geographical Indication labels. Indeed, farmers' practices are subject to stricter specifications, with particular attention to grass use in the dairy cow diets. Accession to cheese production under GI specifications questions the adaptability of the farms and farmers' practices. Animal feeding is strongly interlinked with land use, resulting in a coordinated set of grassland and livestock practices. This set leads to a particular consistency of the feeding system which we use to assess farms' adaptability to technical changes. Two projects were conducted in French mountainous regions. Grassland and herd management's practices were recorded in 37 dairy farms and computed with Multivariate Correspondence Analysis to define feeding system's patterns. They were a sound base to assess farms' adaptability to 4 levels of specifications' requirement: maize or grass silage abandonment, and milk yield or concentrate limitation. Adaptability was assessed by technical changes necessary to adapt the feeding system to be in accordance with specifications' requirements. The main factors explaining the feeding system's patterns were the earliness of grass use for grazing and cutting (early vs. late), the period of milk production (spring vs. winter). These factors defined the room for manoeuvre of the technical changes. Also we discriminated feeding system's patterns through a production trade off between grass and milk intensification. Each pattern's position in this four-level trade off defines the room for manoeuvre. The traditional feeding system is adapted to stricter specifications; other patterns can require changing of grassland and/or livestock intensifications to be in accordance with levels of specifications' requirement.

Genome-wide evaluation: genomic BLUP or BayesB?

Daetwyler, H.D.[1,2], Pong-Wong, R.[1], Villanueva, B.[3] and Woolliams, J.A.[1], [1]The Roslin Institute and R(D)SVS, University of Edinburgh, Roslin Biocentre, EH25 9PS, Roslin, United Kingdom, [2]Wageningen University, P.O. Box 338, 6700 AH Wageningen, Netherlands, [3]Scottish Agriculture College, West Mains Rd, EH9 3JG, Edinburgh, United Kingdom; hans.daetwyler@roslin.ed.ac.uk

Genome-wide evaluation combines traditional breeding methods with genomic data to predict breeding values. In this simulation study we compared the accuracy of best linear unbiased prediction (GBLUP) and Bayesian (BayesB) genome-wide evaluation at varying numbers of loci with effect (Nqtl), ranging from 50 to 2,800. Populations of effective population size (Ne) 500 and 1,000 were randomly mated with recombination and mutation for 5,000 and 10,000 generations, respectively, to attain mutation drift balance. The genome was 10 Morgans long and approximately 1,800 and 3,000 bi-allelic loci were segregating in either Ne. In the last generation a training population of size 500 (1,000 for Ne 1,000) was generated to estimate loci effects and breeding values. Random loci subsets corresponding to the Nqtl desired were chosen to compute true breeding values and phenotypes. The heritabilities (h2) considered were 0.05, 0.2 and 0.5. Fifty replicates were run for each scenario. In GBLUP a realised relationship matrix was calculated based on all loci and in BayesB exact priors were chosen to equal Nqtl. GBLUP had a constant accuracy (correlation of true and predicted breeding values) at each h2 regardless of Nqtl. BayesB had higher accuracy than GBLUP when Nqtl was low. However, this advantage diminished as Nqtl increased and when Nqtl became large GBLUP may outperform BayesB. For the parameters studied the relative difference between the methods was most pronounced at h2 of 0.5 and became less as h2 decreased. These trends were similar for Ne 1000 and suggest that a high degree of linearity exists between the factors that affect accuracy. Our results illustrate that the performance of genome-wide evaluation methods depends significantly on population genetic architecture.

Comparison of Bayesian methods for genomic selection using real dairy data

Verbyla, K.[1,2,3], Bowman, P.[2], Hayes, B.[2], Raadsma, H.[4], Khatkar, M.[4] and Goddard, M.E.[1,2,3], [1]University of Melbourne, Parkville, VIC, Australia, [2]Department of Primary Industries, Bundoora, VIC, Australia, [3]Cooperative Research Centre for Beef Genetic Technologies, Armidale, NSW, Australia, [4]Cooperative Research Centre for Innovative Dairy Products, University of Sydney, NSW, Australia; klara.verbyla@wur.nl

Genomic selection refers to selection based on genomic breeding values (GEBV), where the genomic breeding values are calculated from marker effects located across the whole genome. This study evaluated seven Bayesian approaches for predicting SNP effects for genomic selection by assessing their ability to accurately predict GEBV in real dairy data set containing 1,498 Australian Holstein-Friesian bulls genotyped for 39,048 SNPs. The methods included BLUP and a Bayesian approach in which each SNP had a SNP specific variance (Bayes A). Each had three variants using a selected subset of SNPs (weighted and unweighted) and all SNPs. The other approach, Bayes C used stochastic search variable selection for all SNPs. The different approaches were applied to estimate GEBV for 9 core traits. Validation populations containing bulls proven in each of 2005, 2006 and 2007 were used to assess the accuracy of the GEBV by comparing the estimated GEBV with the published Australian Breeding Values. The approaches produced accuracies that were not significantly different for most traits. This suggests that a reduced set of SNP could be used to produce equally accurate GEBV. However, there was a significant difference between BLUP and the other methods for the trait fat percentage. This difference can be explained by the true distribution of QTL effects compared to the prior distributions for the SNP effects and variances for each method. BLUP does not seem to perform well for those traits that have a known gene or genes responsible for explaining a large percentage of genetic variation for the trait. This suggests that any prior information about a trait's QTL effect distribution should be used to decide which model will produce the most accurate GEBV.

Use of the elastic-net algorithm for genomic selection in dairy cattle

Croiseau, P. and Ducrocq, V., INRA, Génétique Animale et Biologie Intégrative, Domaine de Vilvert, 78352 Jouy-en-Josas, France; pascal.croiseau@jouy.inra.fr

The availability of the 54K SNP array in dairy cattle makes possible to envision the use of genomic prediction instead of classical genetic evaluations in selection programmes. However, this requires suitable statistical approaches as the number of observations is often much lower than the number of predictor variables. In animal breeding, it is clear that for a given trait, not all SNP across the whole genome are close to QTLs for the trait, whatever their size. In this context, we applied the Elastic-Net method to simultaneously select among all SNP the ones which are related with the studied trait and estimate their effect. The Elastic-Net approach consists of a combination of Ridge Regression and of a LASSO algorithm. A parameter λ varying between 0 and 1 indicates the relative weight given to Ridge Regression in the Elastic-Net procedure (pure Ridge Regression algorithm if $\lambda=1$, pure LASSO algorithm if $\lambda=0$). We present here an application of the Elastic-Net approach on a set of 694 bulls in Montbéliarde breed and 1827 bulls in Holstein breed using data available in 2004. The phenotypes of these animals are Daughter Yield Deviation (DYD) calculated for the French Marker Assisted Selection program for milk, fat and protein yields and contents. To evaluate the performance of the Elastic-Net algorithm, we use a validation set of 227 bulls for the Montbéliarde breed and 549 bulls for the Holstein breed with phenotypes (DYD) available in 2008. The Elastic-Net approach directly applied to the whole data set was very time consuming (more than 30 days for the Holstein breed). An alternative two-step approach where relevant SNPs were first selected by looking at moving intervals of 500 SNP and then combined into a single analysis was found very promising (1 day for the Holstein breed). Our results suggest that the optimum Elastic-Net procedure is very close to a full LASSO algorithm. A comparison with other methods is under way.

Performance of genomic selection for traits in mice using Bayesian multi-marker models

Kapell, D.N.R.G.[1], Sorensen, D.[2], Su, G.[2], Janss, L.L.G.[2], Ashworth, C.J.[3] and Roehe, R.[1], [1]Scottish Agricultural College, Sustainable Livestock Systems Group, West Mains Road, Edinburgh EH9 3JG, United Kingdom, [2]University of Aarhus, Faculty of Agricultural Sciences, Department of Genetics & Biotechnology, Research Centre Foulum, Tjele, Denmark, [3]The Roslin Institute, The University of Edinburgh, Roslin, Midlothian EH25 9PS, United Kingdom; dagmar.kapell@sac.ac.uk

Genomic selection uses genome wide dense markers to predict breeding values, as compared to conventional evaluations which estimate polygenic effects based on phenotypic records. Different models have been proposed for genomic prediction of breeding values. The aim of this study was to compare the predictive ability (PA) of models with different mixtures of marker effects, with and without inclusion of a polygenic effect, across a range of traits. Phenotypes and genotypes of 2,281 mice (10,946 SNPs) were analyzed for various traits with low to high heritability ($h^2=0.21-0.67$). The models covered a range of mixture percentages, from 5% to 100% of the markers being switched on. PA was determined by a cross validation and defined as correlation between predicted and observed phenotype. For each scenario a group of animals, selected randomly within or across families, was left out to be predicted by the remaining animals. Results showed a high PA for models with a 100% mixture (i.e. a common prior) for selection across families (0.89/0.54 for high/low h^2, respectively). Lower mixture percentages allowed fewer markers to have an effect, thus reducing the PA (0.84/0.36). Within family different mixtures showed little difference in PA (0.83/0.27 decreased to 0.82/0.25 for high/low h^2). Adding a polygenic effect improved the PA slightly. Genomic prediction showed a high PA in comparison to traditional polygenic prediction, especially benefiting traits with a low heritability when selection was across families. The results are obtained through the EC-funded FP6 Project SABRETRAIN.

Reducing redundancy using information theory in genomic predictions from high-density SNP genotypes

González-Recio, O.[1], Weigel, K.A.[2], Gianola, D.[2], Naya, H.[3] and Rosa, G.J.M.[2], [1]INIA, Mejora Genetica Animal, Ctra La Coruña km 7.5, 28040 Madrid, Spain, [2]University of Wisconsin-Madison, Dairy Science, 1675 Observatory Dr., 53076, USA, [3]Institut Pasteur de Montevideo, Unidad de Bioinformática, Mataojo 2020, 11400 Montevideo, Uruguay; gonzalez.oscar@inia.es

Information from high density SNP genotyping typically contains potential explanatory variables that are either unimportant or redundant, generating noisy information. Two-stage genome assisted prediction has been suggested for genomic selection. In the first-stage, information theory and machine learning may be used to detect informative SNPs and to reduce redundancy. Genotypes for 32,611 SNPs from 3,304 Holstein bulls born before 1998 were used. Information gain (IG) was calculated for every SNP based on productive lifetime PTA (PL). Four subsets were formed with the 500, 1000, 1,500 or 2,000 SNPs with highest IG. Redundant SNPs or in strong LD may have been selected within subsets. Thus, mutual information (MI) between pairs of SNPs was calculated to reduce redundancy. If a pair of SNPs had >33% of MI (corresponding to the 95th percentile of the empirical distribution of MI statistics), the SNP with lower IG within the pair was dropped from its subset. Thus, four new subsets were formed with 500, 1,000, 1,500, and 2,000 SNPs with highest IG but reduced redundancy. In a second stage, the 8 subsets were used to predict a testing set consisting of genotyped bulls born after 1998 with known progeny test PTA for PL, using the Bayesian Lasso and nonparametric (reproducing kernel Hilbert spaces regression: RKHS) approaches. The correlation between genomic predictions and PL was higher when SNP selection accounted for redundancy using MI, regardless of the method of prediction. Differences between including or excluding redundant SNPs were less marked for RKHS, which was more stable over subsets. The Bayesian Lasso outperformed RKHS only when redundancy was reduced or when the number of SNPs was 2,000.

Accuracy of breeding values estimated with marker genotypes as affected by number of QTL and distribution of QTL variance

Coster, A., Bastiaansen, J. and Bovenhuis, H., Wageningen University, Animal Breeding van Genomics Centre, P.O Box 338, 6700 AH Wageningen, Netherlands; albart.coster@wur.nl

Estimation of breeding values with the use of marker genotypes is expected to result in increased response to selection. In this simulation study, we evaluated the effect of number of QTL and distribution of QTL variance on accuracy of estimated breeding values. Breeding values were estimated with three distinct methods: a) Bayesian model (BayesB; b) Partial Least Square Regression (PLSR); c) combination of stepwise regression and shrinkage of regression coefficients (LASSO). Accuracies of breeding values estimated with LASSO were close to 0 in all circumstances. Accuracy obtained with method BayesB was higher than that obtained with PLSR when the total genetic variance was concentrated on a small number of QTL. Accuracies obtained with both methods were similar when total genetic variance was distributed over a larger number of QTL. We concluded that methods BayesB and PLSR were suitable for estimation of breeding values with marker genotypes and that number of QTL and distribution of QTL variance were important factors determining accuracy of these breeding values.

Accuracy of genomic evaluations depends on distance to the reference data

Su, G.[1], Guo, G.[1,2] and Lund, M.[1], [1]University of Aarhus, Faculty of Agricultural Sciences, Department of Genetics and Biotechnology, Blichers Alle 20, DK-8830, Tjele, Denmark, [2]China Agricultural University, College of Animal Science and Technology, Department of Genetics and Breeding, Yuanmingyuan West Road 2, 100094 Beijing, China; guosheng.su@agrsci.dk

A simulation study was carried out to investigate the efficiency of genomic evaluations based on different types of reference data. Two populations with common fonder animals were simulated. The simulated genotypes comprised 25,000 equally spaced SNP markers and 500 randomly assigned QTLs in a genome of 1,500 cM. A Bayesian approach with a common prior distribution of scaling factor (BayesA) was used to predict marker effects. Genomic estimated breeding values (GEBV) was calculated as the sum of all marker effects. Breeding value of candidates was predicted using reference data from the same population ($GEBV_{R-own}$), from the other population ($GEBV_{R-foreign}$), and combining the two populations ($GEBV_{R-combine}$). The accuracy of GEBV was measured as the correlation between GEBV and true (simulated) breeding value. Prediction error variance was measured as the variance of posterior distribution of GEBV. Results show that $GEBV_{R-foreign}$ had a low accuracy, compared with $GEBV_{R-own}$. In addition, the posterior variance of $GEBV_{R-foreign}$ underestimated prediction error variance. These undesired properties depended on the genetic distance between the reference population and the predicted population. However, $GEBV_{R-combine}$ had a higher accuracy than $GEBV_{R-own}$, due to increasing the size of reference data. In addition, the posterior variance was an appropriate measure of the prediction error variance of $GEBV_{R-combine}$. These results suggest that combining marker data and phenotypic data from different populations with common origin would increase the accuracy of GEBV. On the other hand, it does not seem feasible to use reference data of a population to predict breeding values of the individual in other populations.

Accuracy of genome-assisted breeding values for German Holstein Friesian bulls

Habier, D. and Thaller, G., Institute of Animal Breeding and Husbandry, Olshausenstraße 40, 24098 Kiel, Germany; gthaller@tierzucht.uni-kiel.de

Linkage disequilibrium (LD) due to historic mutation events affects accuracy of genome-assisted breeding values (GEBVs), persistence of accuracy over generations without further phenotyping, the potential to reduce breeding costs, effective selection intensity and inbreeding. The objective of this study was to analyze how additive-genetic relationships affect accuracy of GEBVs and to estimate accuracy due to LD from historic mutations using real cattle data from 773 German Holstein bulls. These bulls were genotyped for 54K SNPs and daughter yield deviations were available for milk, fat and protein yield of the first and second lactation of their daughters. GEBVs were estimated with Ridge-Regression BLUP (RR-BLUP) which models each SNP with identical prior variance and BayesB which fits only a small number of SNPs in each iteration of the MCMC-algorithm and allows for different prior variances of SNP effects. To analyze the effect of genetic relationships on accuracy and to estimate accuracy due to LD from historic mutations, bulls were assorted to training and validation data sets such that the maximum additive-genetic relationship between bulls in both sets was 0.25, 0.125, and 0.0625. For most traits, accuracy of GEBVs ranged between 0.5 and 0.6 for a relationship of 0.25. Accuracy declined substantially with decreasing additive-genetic relationship between training and validation set, where the amount of decline varied between traits. RR-BLUP tended to have lower accuracies and higher decline than BayesB, where the accuracy of some traits obtained by RR-BLUP was close to zero for a genetic relationship of 0.0625. In conclusion, although standard errors were high, a large part of the accuracy of GEBVs resulted from information other than LD due to historic mutations which is unfavorable for effective selection intensity, inbreeding and persistence of accuracy over generations, meaning that a reduction of performance testing can not be recommended with the statistical methods used in this study.

Multi-trait genomic selection: comparison of methods

Calus, M.P.L. and Veerkamp, R.F., Animal Sciences Group, Wageningen University and Research Centre, Animal Breeding and Genomics Centre, P.O. Box 65, 8200 AB Lelystad, Netherlands; mario.calus@wur.nl

Genomic selection is becoming common practice in animal breeding. It uses genome-wide dense marker maps, to accurately predict the genetic ability of animals, without the need to record phenotypic performance from the animal itself or from close relatives. Therefore, it is particularly beneficial for selection of traits that are difficult or expensive to measure. Without genomic selection, selection depends on predictor traits in multi-trait breeding value estimation. To get the benefit of recording of predictor traits in genomic selection, our objective is to develop a multi-trait genomic selection method. Four different multi-trait models are considered: 1) a model with a traditional pedigree based relationship matrix, 2) a model where the traditional pedigree based relationship matrix is replaced by a genomic relationship matrix based on markers, 3) a model that both includes polygenic effects related through pedigree information and SNP effects without considering the covariance between traits when estimating the SNP effects, and 4) the same as model 3, with considering the covariance between traits when estimating the SNP effects. The second model assumes equal contribution of each SNP to the total additive genetic (co)variance, while model 3 and 4 allow unequal contributions per SNP. Those 4 models will be compared for 2 simulated traits, having a genetic correlation of 0.2, 0.5 or 0.8 between them. Scenarios are included where some animals have phenotypes for one trait, but not for the other.

Computing procedures for genetic evaluation including phenotypic, full pedigree and genomic information

Misztal, I.[1], Legarra, A.[2] and Aguilar, I.[1,3], [1]University of Georgia, Animal and Dairy Science, Athens 30605, USA, [2]INRA, UR631 SAGA, BP 52627, 32326 Castanet-Tolosan, France, [3]Instituto Nacional de Investigación Agropecuaria, Las, Brujas, Uruguay; ignacy@uga.edu

Currently the genomic evaluations use multiple step procedures, which are prone to biases and errors. A single step procedure may be applicable when genomic predictions can be obtained by modifying the numerator relationship matrix A to H=A+Δ, where Δ includes deviations from original relationships. To avoid computing inv(H), mixed model equations due to the additive effect, say [Z'X Z'Z+ k inv(H)] can be expressed in an alternate Henderson form as [H Z'X H Z'Z+ Ik]. The modified equations have a nonsymmetric left-hand side where H may be poorly conditioned numerically. When those equations are solved by the conjugate gradient techniques, the only computations involving H are in the form of Aq or Δq, where q is a vector; the product Aq can be calculated efficiently in linear time using Colleau's indirect algorithm. Several alternative H are possible. One is to substitute the relationships of genotyped animals with the genomic relationship matrix. Another one is derived by conditioning the genetic value of ungenotyped animals on the genetic value of genotyped animals. Comparisons involved the regular equations and the modified equations with simulated Δ. Solutions were obtained by the Preconditioned Conjugate Gradient, which only works with symmetric matrices, and by Bi-Conjugate Gradient Stabilized, which works with nonsymmetric matrices. The convergence rate associated with the nonsymmetric solver was slightly better than that with the symmetric solver for the original equations, although the time per round was twice as high. The convergence rate associated with the modified equations was similar to that with the regular one for the nonsymmetric solver. When computation of terms with Δ can be done efficiently, it may be possible to modify the existing evaluation to incorporate the genomic information at approximately double the cost of the original evaluation.

How to combine pedigree and marker information into a single estimator for the calculation of relationships?

Bömcke, E.[1,2], Soyeurt, H.[1], Szydlowski, M.[1,3] and Gengler, N.[1,4], [1]Gembloux Agricultural University, Animal Science Unit, Passage des Déportés 2, 5030 Gembloux, Belgium, [2]FRIA, Rue d Egmont 5, 1000 Bruxelles, Belgium, [3]Poznan University of Life Sciences, Wojska Polskiego 28, 60-637 Poznan, Poland, [4]FNRS, Rue d Egmont 5, 1000 Bruxelles, Belgium; bomcke.e@fsagx.ac.be

Relationship coefficients are particularly useful to improve genetic management of endangered populations. These coefficients are traditionally based on pedigree data. In case of incomplete or missing pedigree, they are replaced by coefficients calculated from molecular data when this information is available. However, genotyping a complete population for a sufficient number of markers can be impossible. The main objective of this study was therefore to develop a new method to estimate relationship by combining molecular with pedigree data. It will be useful for specific situations, where neither pedigree nor molecular data are complete. In a companion paper, the compatible coefficients were determined. In this study, based on simulations of pedigree and marker data, the method to combine the selected coefficients was determined. Various parameters were taken into account in the model: number and quality of the marker (e.g. marker informativeness), mutation rate, quality of the pedigree (e.g. generation-equivalents), … The combined estimator has several advantages. Especially, negative relationship values obtained in literature with molecular-based estimators in case of small inbred populations can be avoided. In conclusion, if this combined estimator was originally developed for the management of an endangered horse population, it should also be a promising alternative to traditionally used estimators, e.g. for the management of small and/or rare breeds; especially in case of inbred populations, with both incomplete pedigree and partial molecular information.

Genomic relationship matrix when some animals are not genotyped

Christensen, O.F. and Lund, M.S., Aarhus University, Faculty of Agricultural Sciences, Dept. of Genetics and Biotechnology, Blichers Alle 20, P.O. Box, DK 8830 Tjele, Denmark; OleF.Christensen@agrsci.dk

The use of genomic predictions in breeding programs may increase the rate of genetic improvement, reduce the generation time, and provide higher accuracy of estimated breeding values (EBVs). The computational method used may be a linear BLUP or a non-linear MCMC method. Using genomic markers, a given SNP panel only captures part of the genetic variation. To capture remaining genetic variation, the model should also contain a polygenic genetic effect with the usual pedigree derived additive relationship matrix A. A proper decomposition of the variance explained by markers and residual polygenic variance requires that genomic genetic effects for non-genotyped animals are included in the model. In this paper we provide a natural extension of a linear BLUP method to the situation with non-genotyped animals. The method is based on a linear mixed model where the genomic genetic random effects are correlated with a genomic relationship matrix constructed from markers and the polygenic genetic random effects are correlated with relationship matrix A. The total genetic effect would then be the sum of the genomic and the polygenic genetic effects, and the combined estimated breeding value would be a blending of the traditional EBV and a genomic EBV. Variance component parameters in the model are estimated using REML and estimated breeding values are best linear unbiased prediction (BLUPs). Large SNP panels provide information about genomic relationships, but the situation where all animals of interest are genotyped may not be a likely scenario. The paper presents a natural extension of the genomic relationship matrix to the situation where not all animals have been genotyped. The method therefore integrates the information from animals without genotypes, and provides combined EBVs for all animals. The method is tested on simulated and real data.

EBV and DYD as response variable in genomic predictions

Villumsen, T.M., Janss, L., Madsen, P. and Lund, M.S., University of Aarhus, Faculty of Agricultural Sciences, Department of Genetics and Biotechnology, Blichers Alle 20, 8830 Tjele, Denmark; trinem. villumsen@agrsci.dk

In a dairy cattle dataset it was investigated if breeding values (EBV) or daughter yield deviations (DYD) was the most efficient response variable (RV) for genomic evaluation (GE). EBVs are often available for all traits, whereas DYDs aren't. DYD are independent and accurate measures of the performance of a bull's daughters since they are not regressed as the EBV. However, there is more information in EBV than DYD since there is also pedigree information. It was also investigated whether estimation of RV in an animal model (AM) or sire model (SM) gave the highest accuracies, and if a weighting of the EBV and DYD according to their accuracies improved the results. The approach was tested on protein yield and interval from first to last insemination (IFL). Accuracy of the genomic estimated breeding value (GEBV) was measured as the correlation between the official EBV and the GEBV. Official EBV for each trait were estimated in a Bayesian AM from full data consisting of 3,138,743 305-day protein yield records and 2,396,337 IFL records for 1st lactation cows calving from Jan. 1982 to Sep. 2007. RV and their accuracies for developing the prediction model for GE were estimated from a reduced data set using the same Bayesian approach, but omitting data for the last 2.5 years. 1,775 sires were genotyped with the Illumina 50k chip, 36,387 markers were used in the GE. Marker effects were estimated in a Bayesian approach. 1,500 genotyped reference sires had offspring in the reduced data set. 275 test bulls only had daughters after Jan. 2005 and were not included in the reference data. Their GEBV were based on marker information only. For both traits the correlation between official EBV and GEBV with EBV as RV was 0.5-0.6 for the test group. When the DYD was used as RV the correlations were a little lower. EBV and DYD estimated in an AM gave higher accuracies than SM. The inclusion of a weight factor increased the correlations, however mainly where DYD was RV.

Devaluation of estimated marker effects in populations under selection

Bastiaansen, J., Coster, A. and Bovenhuis, H., Wageningen UR, Animal Breeding and Genomics Centre, Marijkeweg 40, 6709 PG Wageningen, Netherlands; john.bastiaansen@wur.nl

Breeding values estimated using marker information (MEBV) rely on associations between markers and QTL. Associations between markers and QTL change over time as a result of recombination and changes of allele frequencies. Consequently, effects estimated for specific markers at a specific moment devaluate over time. In this simulation study, we studied the devaluation of marker effects and compared the impact of selection for MEBV or MASS selection. The number of QTL contributing to the trait was varied to study the sensitivity of conclusions to this characteristic. Marker effects for calculating MEBV were estimated using bayesian method BayesB described by Meuwissen and co-autors (2001) and using data of individuals in one or more generations before starting selection. Mean breeding value increased over generations of selection. Initially, rate of increase was higher with selection for MEBV than with MASS selection. Rate of increase declined over later generations for both methods and declined faster with selection for MEBV than with MASS selection. Accuracy of MEBV declined with number of generations after estimating marker effects and rate of decline was higher with selection for MEBV than with MASS selection. Genetic variance declined over generations of selection and rate of decline was higher with selection for MEBV and than with MASS selection.

Accurate prediction of genomic breeding values in Norwegian Red Cattle using dense SNP genotyping

Luan, T.[1], Woolliams, J.A.[1,2], Lien, S.[3], Kent, M.[1,3], Svendsen, M.[4] and Meuwissen, T.H.E.[1], [1]Norwegian University of Life Sciences, Department of Animal and Aquacultural Sciences, Arboretveien 6, N-1432, Ås, Norway, [2]University of Edinburgh, The Roslin Institute (Edinburgh), Roslin, Midlothian EH25 9PS, United Kingdom, [3]Centre for Integrative Genetics, Arboretveien 6, N-1432, Ås, Norway, [4]Geno Breeding and A.I. Association, Arboretveien 6, N-1432, Ås, Norway; tu.luan@umb.no

Genomic Selection is a newly developed tool for the estimation of breeding values by using genome wide dense markers rather than phenotypes of the selection candidates and their relatives. Studies with simulated data have already shown that a good accuracy could be achieved in estimating genome wide breeding values (GW-EBV) by Genomic Selection. Here, we present a study that is among the first to apply genomic selection to real genotypic data for Norwegian Red Cattle. The study was performed on milk yield, fat yield, protein yield, first lactation mastitis traits and calving ease. By examining three methods for genomic selection, Best Linear Unbiased Prediction (G-BLUP), Bayesian statistics (BayesB) and a mixture model approach (MIXTURE), we found that G-BLUP achieved generally highest accuracy. For the traits under study, the accuracies of GW-EBV prediction were found to vary widely between 0.12 and 0.62. We observed a strong relationship between the accuracy of the prediction and the heritability of the trait. GW-EBV prediction for the trait with low heritability achieves lower accuracy than the trait with high heritability. Thus, low heritability trait needed more data to achieve similar accuracy than high heritability trait. Our findings are especially helpful to researchers in choosing methods and data sizes required for the practical implementation of genome-wide dense marker genotyping in selection programs. Note: the presented results are obtained through the EC-funded FP6 Project 'SABRE'.

Whole-population relationship matrix including pedigree and markers for genomic selection
Legarra, A.[1], Aguilar, I.[2,3] and Misztal, I.[2], [1]INRA, UR631 SAGA, 31326 Castanet Tolosan, France, [2]UGA, Dept of Animal and Dairy Science, Athens 30602, USA, [3]INIA, Las Brujas, Uruguay; andres.legarra@toulouse.inra.fr

Current genomic selection requires use of pseudo-data (e.g., daughter yield deviations) computed from full records and pedigree data. This results in bias and loss of information. Further, it is not usable for breeding schemes involving small families. One way to include genomic information into a full genetic evaluation is by modifying the relationship matrix. The relationships of genotyped animals can be described by a genomic relationship matrix, G. However, if the relationships among ungenotyped animals are not accordingly modified, this results in incoherencies and ill-conditioning because G includes information on relationships among ancestors and descendants. We condition the genetic value of ungenotyped animals on the genetic value of genotyped animals via the selection index (e.g. pedigree information), and then use G for the latter. This results in a joint distribution of all genetic values, with a pedigree-genomic relationship matrix H. In this matrix genomic information is transmitted to the covariances among all ungenotyped individuals. The covariance matrix of ungenotyped individuals is: $Var(u_1)=A_{11}+A_{12}A_{22}^{-1}(G-A_{22})A_{22}^{-1}A_{21}$, with covariances with genotyped individuals $Cov(u_1,u_2)=A_{12}A_{22}^{-1}G$. Matrix H is (semi)positive definite by construction, and suitable for big-scale genetic evaluations via PCG. A simulation was done with 200 sires and 1000 dams producing 1000 offspring; 2000 SNPs in 2M were considered, 10 being true QTLs. Offspring had a recorded trait (h^2=0.10), whereas only founders were genotyped. So, DYD's are inaccurate, in particular for dams. Average accuracies with an infinitesimal model were 0.37 for sires and 0.18 for dams, whereas with H these were 0.42 for sires and 0.29 for dams. Evaluations were unbiased. The use of the joint relationship matrix will result in higher accuracies, in particular for animals with little information like prospective bull dams.

R package for simulation of high density marker data
Coster, A. and Bastiaansen, J., Wageningen University, Animal Breeding van Genomics Centre, P.O Box 338, 6700 AH Wageningen, Netherlands; albart.coster@wur.nl

Simulation of high density marker data is commonly used for the evaluation of methods. We developed package HaploSim for R that provides users with functions for simulation of custom scenarios. HaploSim uses a sparse representation of marker data to allow for large simulations. Gametes are the basic simulation unit of HaploSim and can either consist of one or multiple chromosomes. To avoid unnecessary computational load, simulations can be performed without QTL and QTL can be assigned to specific marker loci after finishing the simulations. Currently, only additive QTL action is implemented in HaploSim but users can program specific non-additive QTL action if this is wished. HaploSim provides a function for simulation of marker data through existing pedigrees. Integration of HaploSim into R facilitates programming of additional simulation functions and evaluation of simulations using statistical and graphical functions available in R. HaploSim is available from the R repository CRAN (http://cran.r-project.org/).

Calculate relationships using pedigree and marker information: what to combine into a single estimator?

Bömcke, E.[1,2], Szydlowski, M.[1,3] and Gengler, N.[1,4], [1]Gembloux Agricultural University, Animal Science Unit, Passage des Déportés 2, 5030 Gembloux, Belgium, [2]FRIA, Rue Egmont 5, 1000 Bruxelles, Belgium, [3]Poznan University of Life Sciences, Wojska Polskiego 28, 60-637 Poznan, Poland, [4]FNRS, Rue Egmont 5, 1000 Bruxelles, Belgium; bomcke.e@fsagx.ac.be

The knowledge of relationships among individuals is an important topic in all fields of modern genetics, from selection to conservation. Estimation of relationships was traditionally based on pedigree data. Today, molecular data can replace pedigrees. Numerous methods already exist. However, genotyping a complete population for a sufficient number of markers can be impossible, e.g., in case of local and/or rare breeds. Therefore, the idea was to combine these 2 sources of information into a new single relationship estimator. Based on simulations, the aim of this study was to determine what could be combined. Following the principle that things have to be similar to be combined, we compared first two coefficients that have the advantage to be similarly defined: the additive relationship coefficient (a_{xy}) calculated from pedigrees and the total allelic relationship (ta_{xy}) obtained from markers. The similarity between these two coefficients was measured by means of linear regression and correlations. In order to highlight the influence of inbreeding, Wright relationship coefficient ($r_{ped,xy}$) were also compared to ta_{xy} transformed into $r_{mol,xy}$. The results showed that the correlation increases when the values are made independent from inbreeding. If the pedigree coefficients are considered as the true value, both molecular coefficients tended to overestimate the relationship among individuals. It can be expected because they do not distinguish identical-by-state and identical-by-descendent alleles. The influence of marker quality was also highlighted. In all cases, the correlation coefficient increased when only the more informative loci were used. Uninformative genetic markers potentially making animals appear more related than they were.

Genomic footprints of influential sires in the Holstein population

Pimentel, E.C.G.[1], König, S.[1], Qanbari, S.[1], Reinhardt, F.[2], Tetens, J.[3], Thaller, G.[3] and Simianer, H.[1], [1]Georg-August University Göttingen, Department of Animal Sciences, Albrecht-Thaer-Weg 3, 37075 Göttingen, Germany, [2]Vereinigte Informationssysteme Tierhaltung w.V., Heideweg 1, 27283 Verden / Aller, Germany, [3]Christian-Albrechts University Kiel, Institute of Animal Breeding and Animal Husbandry, Olshausenstraße 40, D-24098 Kiel, Germany; epiment@gwdg.de

High density SNP data were used to investigate genome regions of some influential sires that may have undergone selection in the Holstein population. A set of 620 animals genotyped for 52,252 SNP was used. Comparisons of allele frequencies between groups of highly and lowly related individuals to such sires were made and Chi-square tests performed. Genome regions showing significant differences in allele frequencies were identified. Regions with significant differences for influential sires that were highly and lowly related with each other were compared. Agreements were observed for highly related sires and disagreements for lowly related ones. Disagreements could also be observed for comparisons between sires with outstanding genetic merit for different classes of traits. Occurrences of successive significant tests and of 1 Mbp-segments with different counts of significant results were investigated. Observed frequencies differed substantially from expected frequencies under the assumption of a random distribution. In conclusion, selected offspring of influential sires have not inherited a random sample of the founder genome, but certain chromosome regions are significantly overrepresented. The length of shared segments exceeds what would be expected under random sampling. Bulls with different breeding value profile seem to inherit different chromosome segments.

First results on genome-wide genetic evaluation in Swiss Dairy cattle

Stricker, C.[1], Moll, J.[2], Joerg, H.[3], Garrick, D.J.[4] and Fernando, R.L.[4], [1]applied genetics network, Boertjistrasse 8b, 7260 Davos, Switzerland, [2]Swiss Cattle Breeders Association, Schuetzenstrasse 10, 3052 Zollikofen, Switzerland, [3]Qualitas AG, Chamerstrasse 56, 6300 Zug, Switzerland, [4]Iowa State University, Department of Animal Science, 225 Kildee Hall, Ames, IA 50011-3250, USA; stricker@genetics-network.ch

Around 50,000 SNP genotypes of 1,'050 Brown Swiss sires became available in February 2009. Genotyping was based on Illumina's Bovine50K SNP chip. The distance between adjacent SNPs was between 20,000 and 50,000 bases for 36,261 intervals, 11,320 intervals were between 50,000 and 100,000 bases, 2,524 between 100,000 and 150,000 bases and 1,666 intervals were larger than 150,000 bases. 54 intervals were larger than 500,000 bases. Minor allele frequency was below 0.03 for 11,993 SNPs. 35,033 SNPs had a minor allele frequency between 0.1 and 0.5. Linkage disequilibrium was captured by the squared correlation between SNP-genotypes. Correlations between all SNPs were calculated. Then, correlations for distances between SNPs within 1000 bases were averaged. The maximum of the average correlation per chromosome ranged from 0.26 on chromosome 17 for an interval length 27,000 bases up to 0.48 on chromsome 29 for an interval length of 26,000 bases. Correlations leveled off to around 0.1 at an interval length of 200,000 bases for all chromosomes. An additional 1,150 genotypes of Swiss Red & White and 430 genotypes of Swiss Holstein sires are expected to be delivered by the end of April 2009. Results presented from genom-wide genetic evaluation will concentrate on milk related traits such as milk and protein yield. Proofs and de-regressed proofs will be used as the dependent variable. SNP-effects will be estimated based on methods BAYES A and B. Analyses will be within and across breeds. Accuracy of prediction will be assessed using cross-validation by identifying a set of animals for validation that are as unrelated as possible to the training animals. These results will represent the current state of genome-wide genetic evaluation in Switzerland.

A brief history of equine breeding technologies

Allen, W.R., The Paul Mellon Laboratory of Equine Reproduction, 3 Tower Stables, Cheveley Park, Newmarket, Suffolk, CB8 9DE, United Kingdom; paulmellonlab@btconnect.com

Artificial insemination (AI) originated in the horse when an Arab chieftain 'stole' some semen from the vagina of his enemy's mated mare and inseminated his own mare with the contraband. Equine AI was developed further in Russia and China during the early 20th Century and it took a leap forward in 1938 when Walton and Hammond in Cambridge, UK cleverly used AI to cross Shire horses with Shetland ponies to study the effects of maternal size upon fetal and postnatal development. With the exception of the Thoroughbred breed, AI using fresh, cooled or deep frozen semen is nowadays used successfully throughout the equine world, but with the continuing problem of the variation in freezability of semen between stallions of otherwise normal fertility. Embryo technologies in the mare commenced slowly in the early 1970's with the birth of live foals from embryo transfer (ET) in Japan and Cambridge. They progressed rapidly thereafter, however, with the first international transport of horse embryos in the oviducts of rabbits in 1975, the deep freezing of horse embryos in Cambridge and Colorado in the early 1980's, the splitting of embryos to produce monozygotic (identical) twins in the same two centres a few years later, and successful extraspecies embryo transfer in Cambridge and Cornell, USA using horses, donkeys, mules and zebra. Further breakthroughs came with the freezing of horse embryos in Cambridge and Colorado in the mid-1980's, the birth of the first foal created by *in vitro* fertilisation (IVF) in Nouzilly, France in 1989 and vitrification of embryos in Colorado in the 1990's. These advances were crowned momentously with the birth of cloned horses and mules in Cremona, Italy and Moscow, Idaho in 2003, followed 3 years later by the live birth of 5 foals all cloned from a single karyoplast line in College Station, Texas. Thus, although ET and related technologies were originally slow to get off the ground in the horse, they have caught up with a flourish in recent years.

The past, present, and future of the breeding stallion: the forces of darkness and the gathering of light

Varner, D., Texas A&M University, Large Animal Clinical Sciences, College of Veterinary Medicine and Biomedical Sciences, College Station, Texas 77843-4475, USA; dvarner@cvm.tamu.edu

Like that of all mammals, ejaculated equine spermatozoa must complete a lengthy list of functions in order to fertilize an oocyte. Given this inventory of spermatozoal requirements and the arduous journey required of spermatozoa, it becomes understandable why billions of spermatozoa are present in a given ejaculate to accomplish the seemingly simple feat of fertilization of a single oocyte. The biochemical and biophysical features are so sophisticated that many of the cellular and molecular mechanisms remain unresolved to this day. The scientific community is charged with developing accurate methods with which to evaluate the fertilizing capacity of spermatozoal populations, formulating techniques for preserving the integrity of spermatozoal function following cooled or frozen storage, and devising ways to bypass certain functional requirements of spermatozoa in order to enhance the fertility of subfertile stallions. While traditional methods of evaluating semen quality and breeding mares will be a mainstay in horse breeding operations, it is likely that molecular techniques will play an increasingly important role in diagnostic assays and therapeutic strategies. Futuristically, evaluation of candidate genes may become an important tool for evaluating certain spermatozoal traits and functions, and targeted gene mutations or gene silencing agents may prove to be important ways of regulating gene function in stallions that possess genetic imperfections that would otherwise negatively impact their fertility, or lead to transmission of undesirable genetic traits to offspring. More immediate goals might include devising superior methods for long-term cooled preservation of semen, enhancing the fertility of cryopreserved semen, and improving methodologies of *in vitro* fertilization (both conventional *in vitro* fertilization and intracytoplasmic sperm injection techniques) such that these techniques can be reliably applied in commercial breeding programs.

A new freezing extender for stallion semen to get high fertility rates

Pillet, E.[1,2], Duchamp, G.[1], Batellier, F.[1], Yvon, J.M.[1], Delhomme, G.[2], Desherces, S.[2], Schmitt, E.[2] and Magistrini, M.[1], [1]INRA, UMR85 Physiologie de la Reproduction et des Comportements, 37380 Nouzilly, France, [2]IMV-Technologies, 10 rue Clemenceau, 61302 L Aigle, France; elodie.pillet@tours.inra.fr

Since the first insemination with frozen semen, the low or fluctuating fertility results have limited the use of this technology. Our objective was to develop a new freezing extender, easy to use and able to improve the success of artificial insemination with equine frozen semen. In a first fertility trial, we compared INRA82 extender (as a control) versus INRA96® extender, both supplemented with egg yolk and glycerol. INRA82 contains milk whereas INRA96® contains only the purified fraction of milk caseins. INRA96® was custom-formulated for long term semen storage at 4 °C or 15 °C. Semen from 3 stallions was frozen and a total of 72 mares (n=84 cycles) were inseminated. Pregnancy rate per-cycle was significantly improved with INRA96® supplemented with egg yolk and glycerol: 71% (17/42) versus 40% (30/42) ($P<0.01$). In a second fertility trial, we compared INRA96® supplemented with egg yolk and glycerol (as a control) versus INRA96® supplemented with sterilized egg yolk plasma and glycerol. Plasma is the fraction of egg yolk obtained after elimination of egg yolk granules by high speed centrifugations. Semen from 2 stallions was frozen and a total of 68 mares (n=70 cycles) were inseminated. Pregnancy rates per-cycle showed no significant difference between fresh egg yolk and sterilized egg yolk plasma: 60% (21/35) versus 69% (24/35) respectively ($P>0.05$). Both fertility trials demonstrated that fertility after insemination with frozen semen can be greatly improved using INRA96® extender supplemented with egg yolk, or sterilized egg yolk plasma, and glycerol. These results will lead to the commercialization of an extender available ready to use and called Inra-Freeze. Our next objective is to identify the cryoprotective molecule(s) in egg yolk plasma.

Functional characterization of equine sperm: a molecular evaluation of stallion fertility

Gamboa, S. and Ramalho-Santos, J., Centro de Neurociências e Biologia Celular, Departamento de Zoologia- Universidade de Coimbra, Coimbra, 3004-517 COIMBRA, Portugal; scgamboa@esac.pt

Since 1993, our laboratory has been dedicated to the study of the physiology and biochemistry of stallion semen in an attempt to contribute to the understanding of fertility determinants, namely in the Lusitano Horse breed, the major national breed of horse in Portugal. Levels of fructose, glucose, citric acid and reactive oxygen species -ROS, SNARE proteins, 'Raft organizers' and differential ubiquitination of sperm cells were measured. Oxygen consuptiom by sperm mitochondria was monitored with a Clark type electrode. Fluorescent dyes were used to assess vitality (IP/SyBr14), acrossomal integrity (PSA/FITC), membranes stability (MC540), mitochondrial membrane potential [($\Delta\psi_m$) JC-1] and apoptosis (TUNEL). Pregnancy results, as well as embryo recovery following insemination, allow us to evaluate the relationship between semen traits and fertility potential. Members of different SNARE families the (t-SNARE syntaxin, the v-SNARE synaptobrevin/VAMP, the calcium sensor synaptotagmin, and the ATPase NSF) as well as components of lipid rafts (Caveolin-1) can be detected in the acrosomal cap and equatorial segment. NSF and ubiquitin surface staining seemed to inversely correlate with stallion fertility[1]. Discreet levels of ROS and membrane instability seemed to be advantageous to sperm cells. Correlations between some fluorescent markers and classical traits were found according with stallion's fertility. No evidence of apoptotic mature spermatozoa was found. The diagnosis of male (in)fertility must depend on an integrated analysis. Our results suggest that structural and functional analysis of equine sperm may constitute useful tools in predicting stallion fertility because these techniques could distinguish seasonal effects better than traditional semen parameters. This is important because breeding soundness examinations and semen cryopreservation occur during short-day light period.

The multiple challenge of horse cloning: production, health, social acceptation, law, genetic application

Palmer, E., Chavatte-Palmer, P. and Reis, A.P., CRYOZOOTECH, 16 rue Andre Thome, F 78120 SONCHAMP, France; ericpalmer@cryozootech.com

Horse cloning is not only a scientific challenge, but it has to jump over multiple fences until becoming a tool for animal production. Since the birth of the first horse clone in 2003, the efficiency of the technology has been improved in different directions. However the overall efficiency is difficult to know, as academic papers deal with one step or another one, but all the attempts are not fully reported. A limiting factor for production is the availability of slaughter ovaries and the ban of horse slaughter in USA is a retarding factor. Following the production of a cloned embryo, the chances of getting a healthy foal are still low because of losses along pregnancy but the large offspring syndrome seems absent in the equine model, avoiding the animal welfare ethic concerns. Only a few horses have reached adulthood and larger numbers will be necessary to assess health status and phenotype identity of clones compared to their model. The clone may show some differences from the model for reasons related or not to cloning, but the transmission of genes make it equal to the model, so that making stallions is the only 'sure' application. The social acceptation of horse cloning is progressively improving and our first cloned stallion, Pieraz-Cryozootech-Stallion has successfully been registered as a foal in Zangersheide stud-book, as a stallion in Anglo European Stud-book, as a normal stallion allowed to perform A.I. by the French Haras Nationaux, and his first offspring has received a passport. Whether the clones should be allowed or even obliged to show before being approved as stallion or they should be registered in consideration of the performances of the model is not yet answered by horse authorities. From a genetic point of view, using stallions, clones of gelded champions would have a highest interest in disciplines where few stallions are tested and using clones in competition would delay their use as stallions and increase the generation interval.

Optimal reproductive policies for Italian Heavy Draught Horse

Mantovani, R.[1], Contiero, B.[1] and Pigozzi, G.[2], [1]Department of Animal Science - Padova, Viale Università, 16, 35020 Legnaro (PD), Italy, [2]Italian Heavy Draught Horse Breeders Association, Via Francia, 3, 37135 Verona, Italy; roberto.mantovani@unipd.it

The rearing systems (RS) of Italian Heavy Draught Horse (IHDH) differ in Italy moving from the north (animals in stables, S) to the south of the country (wild system, W, or mixed system, SW, depending on the season). In most stud farms, reproduction is mainly based on stallions' natural services. This study has aimed to identify the better age at 1st foaling for mares (AFF at 3 or 4 yrs) and the genetic lifespan (GL) of young stallions, in order to optimize the IHDH stud farms' reproductive policies. Reproductive performance at 1st and 2nd foaling of 1,513 mares were used. Mares had normally 1st foaled at 3 (n=745) or 4 yrs of age (n=768) and were from stud farms based on S (n=488), W (n=345) or SW (n=680) RS. Logistic analysis were carried out modelling the risk of unsuccessful reproduction in the subsequent season, considering both AFF and RS. As regard the GL of young stallions (YS), this was estimated by regression of the mean differences in EBV, at 1st and up to 6th evaluations, between young (mean no.=48) and proven stallions (mean n=478). Six generations of YS were used, and the differences in EBV were as global selection index. Results obtained were an higher risk of unsuccessful reproduction on the 2nd foaling season for 3 yrs 1st foaling mares (+40%, $P<0.01$). In comparison with the best reproductive success at 2nd foaling (S-4 yrs), the highest risk of unsuccessful reproduction was in W-3 yrs (+167%), followed by SW-3 yrs (+91%) and S-3 yrs (+62%). As regard the GL of YS, that showed a 0.52 standard deviation (sd) higher EBV at 1st evaluation than proven stallions, the estimated annual decrease in EBV was -0.07 sd/yr. Optimal reproduction policies could be obtained in IHDH stud farms by limiting foaling at 3 yrs, particularly in W-RS, and using a YS for 3-4 yrs to maintain a sufficiently high selection differential with proven stallions.

Transgene-induced reprogramming of equine somatic cells

Diaz, C., Parham, G. and Donadeu, F.X., Roslin Institute, Royal (Dick) School of Veterinary Studies, The University of Edinburgh, Roslin BioCentre, Midlothian EH25 9PS, United Kingdom; catalina.diaz@roslin. ed.ac.uk

Pluripotent stem cells are capable of differentiating into all three embryonic germ layers and because of this they have enormous potential for biomedical research and regenerative therapy. Embryonic stem (ES)-like cells have been generated from human and rodent somatic cells by forcing the ectopic expression of 4 transcription factors, Oct4, Sox2, Klf4 and cMyc. These induced pluripotent stem (iPS) cells have the greatest potential in medicine because they can be produced in a patient-specific manner. A species, other than humans, that is likely to benefit from this potential is the horse, particularly in regard to the treatment of musculoskeletal injuries. The objective of this study was to assess the reprogramming of equine somatic cells transduced with virus-encoded transcription factors. Cultures of fibroblasts derived from the skin of a foal were transduced with retroviral constructs coding for the mouse sequences of Oct4, Sox2, Klf4 and cMyc, and the transduced cells were cultured in either standard mouse or human ES cell conditions. Distinct colonies appeared with the two culture conditions beginning five days after transduction. Colonies that continued to grow were picked for expansion and some were stained with the early reprogramming marker, Alkaline Phosphatase (ALP). Most of those colonies were positive for ALP. The mouse ES cell line, Bruce-4, was also positive for ALP whereas non-transduced equine fibroblasts were negative. Colonies that emerged in human ES cell conditions after the second week post-transduction had the greatest resemblance with ES cells with well-defined edges and a large nucleus to cytoplasm ratio. Those colonies were successfully expanded and they continued to grow and to maintain their ES-like morphology by day 70 after transduction. More tests are being carried out to further characterise the reprogramming status of these colonies and the results will be presented.

Assessment of the post-thaw measurable parameters of equine spermatozoa taken for a comparison between amides and glycerol as cryoprotectants

Mills, A.A.[1], Whitaker, T.C.[1] and Matson, T.[2], [1]Centre for Equine and Animal Science, Writtle College, Essex, United Kingdom, [2]Stallion AI Services, Twemlows Hall, Shropshire, United Kingdom; Amy.Mills2@ writtle.ac.uk

Variability in the post-thaw quality of cryopreserved semen has been observed between stallions. The cryoprotectant currently used commercially during cryopreservation is Glycerol which has toxic effects on spermatozoa in high concentrations. Different kinds of amides (Methyl formamide (MF), Dimethyl formamide (DF)) have been used with success for freezing stallion spermatozoa and have been identified to be superior cryoprotective agents than Glycerol (GLY). The study aimed to assess amides as alternative cryoprotectants to GLY. One ejaculate from 10 commercial breeding stallions was split and frozen in an egg base extender containing either GLY (5%), MF (5%), DF (5%), MF + DF (2.5% and 2.5%, respectively) or GLY + DF (1% and 4% respectively) in 0.5ml straws. Post-thaw sperm motility parameters were evaluated immediately at 0hr and 1hr post-thaw. Cell morphology was evaluated using nigrosin-eosin staining. Data was analysed via a 2-way ANOVA. No significant difference ($P<0.05$) between treatments was observed for motility at 0hr or 1hr post-thaw. The percentage of live acceptable spermatozoa was significantly higher ($P<0.05$) when GLY (46.4±3.5) was used compared to MF (31.4±7.5) and GLY + DF (32.7±7.5). There was a significant difference ($P<0.05$) of the percentage of live normal spermatozoa between GLY (71.2±5.5) and MF (44.7±10.2). There was a significant main effect of treatments on the percentage of other live abnormal spermatozoa observed. The percentage of abnormal cells was significantly lower ($P<0.05$) when GLY + DF (1.9±0.7) was used compared with GLY (9.7±4.7). There was no significant main effect ($P<0.05$) on the outcomes of the remaining morphological parameters between treatments. It was concluded DF consistently performed as well as GLY and could replace GLY without a significant ($P<0.05$) decline in the measurable post-thaw parameters.

Seasonal semen freezeability from Lusitano stallions using flow-citometry

Agrícola, R.[1,2], Chaveiro, A.[2], Robalo Silva, J.[3], Lopes Da Costa, L.F.[3], Horta, A.E.M.[1] and Moreira Da Silva, F.[2], [1]INRB, Reproduction, Quinta da Fonte Boa, 2005-048 Vale de Santarém, Portugal, [2]UAC, Reproduction, Terra Chã, 9700 Angra do Heroismo, Portugal, [3]FMV, Reproduction, Lisbon, 1300, Portugal; ricardoagricola@hotmail.com

In the present study, seasonal changes in Lusitano stallion's semen quality were evaluated to establish the best time for semen cryopreservation. In this work we evaluated the seasonal semen freezability by flow cytometry. For such purpose five Lusitano stallions, in four periods of semen collection defined as spring, summer, autumn and winter. Ejaculates were collected once a week for 4 weeks, and cryopreserved. Viability, motility and HOST were also evaluated by microscopy. A flow cytometer was used to evaluate sperm viability by using the combination of fluorescent probes propidium iodide (PI) and SYBR-14 for sperm viability. In addition, acrosome integrity was evaluated using FITC-PNA and PI probe. Results were evaluated for statistical differences by a one-way ANOVA. Results from viability evaluated by flow citometry, in frozen-thawed semen from ejaculates collected in autumn (37.73±2.20%) and summer (40.99±4.08%) differ significantly from winter (54.26±2.10%) and spring (46.71±2.32%) ($P<0.05$). This viability was also significantly different among stallions ($P<0.0001$) The high number of dead spermatozoa after the freezing-thawing process in autumn/summer most likely indicates that seasonally related changes in seminal plasma compounds may affect semen preservation. Nevertheless, no significant differences were observed for acrosome integrity among seasons, but it was significantly different among stallions ($P<0.0001$). A high significant correlation was observed between viability and acrosome integrity (r=0.91, $P<0.03$). Viability, motility and HOST evaluated by microscopy were different among horses ($P<0.0001$) but not among seasons. The present study demonstrated that, in our climatic/latitude conditions, seasonal differences occur, being preferable to freeze in winter/spring period.

Proteomic analysis of mare follicular fluid

Nadaf, S.[1], Roche, S.[2,3], Tiers, L.[2,3], Lehmann, S.[2,3] and Gerard, N.[1], [1]INRA, PRC, Nouzilly, 37380, France, [2]Institut de Génétique humaine du CNRS, Montpellier, 34396, France, [3]Plateforme de Protéomique Clinique Hopital Saint Eloie, Montpellier, 34295, France; Somayyeh.nadaf@tours.inra.fr

Follicular fluid accumulates into the antrum of ovarian follicles during their growth. It is in part an exudate of serum, as surrounding cell layers permit the free diffusion of proteins of up to 50 kDa. This fluid also contains locally produced factors related to the metabolic activity of ovarian cells and reflects the physiological status of the follicle. It has been demonstrated that it contains essential substances implicated in oocyte maturation and fertilization, granulosa cells proliferation and differentiation, ovulation and luteinization of the follicle. Studies on its components may contribute to understand the mechanisms underlying these processes. The aim of our study was to investigate for the first time the protein profile of mare follicular fluid (MFF) and to remove several high-aboundance serum proteins (HAP) which can prevent the detection of lower-abundance proteins (LAP) that are of special interest for discovery of proteins that play important roles in ovarian physiology. The analyses were done in two parts. In the first part, we attempted to identify some of the proteins visualized in the pH range 3-10 and in the 10–200 kDa. About 1,000 protein spots were detected by the SameSpot program, of which 17 were identified by LC-Q-TOF Mass spectrometer and all of them were HAP. The second part was to demonstrate that human serum depletion kits were able to remove some HAP from MFF. We observed that unbounded and bounded protein profiles were different compared to the crude fraction. These preliminary results are encouraging and tend to demonstrate that an approach combining immunocapture of HAP can be applied to deplete follicular fluid. Such approach would present a major interest for proteomic analysis and profiling of this fluid, and may allow the discovery of some factors that would play essential roles during follicular development or maturation.

Is subfertility or infertility related to chromosome abnormalities in the Sorraia horse? Preliminary results

Kjöllerström, H.J., Collares-Pereira, M.J. and Oom, M.M., Universidade de Lisboa, Faculdade de Ciências, Departamento de Biologia Animal/Centro de Biologia Ambiental, C2-3 Piso, Campo Grande, 1749-016 Lisboa, Portugal; mmoom@fc.ul.pt

Extensive cytogenetic investigations in horses have shown that chromosomal abnormalities, especially those of sex chromosomes, are associated with infertility or subfertility, early embryonic death and abortion. The nuclear genome of the horse comprises 64 chromosomes (31 pairs of autosomes and the sex chromosomes) which are presently well characterized. The most commonly reported sex chromosome abnormality is the X monosomy (63, XO). Other frequently found abnormal equine karyotypes include XY male-to-female sex reversal, XXX trisomy, and different types of mosaicism and chromosomal rearrangements, all affecting reproductive success, though an occasional pregnancy may occur. Although it has been shown that such chromosome abnormalities are a common cause of infertility in horses, no cytogenetic studies have ever been performed in the Sorraia horse breed to determine its influence in the observed reduced fertility. We now present preliminary results of the first cytogeneticevaluation of the extant population, with special focus on individuals with fertility problems and ambiguous sexual phenotypes. We used standard methods for obtaining chromosome preparations from peripheral blood lymphocytes, combined with conventional chromosome banding techniques, especially G- and R-banding, to allow a more detailed diagnosis of specific chromosome structural changes and reliable comparisons of karyotypes. At least twenty-five metaphase spreads per specimen were screened to detect possible mosaicisms. This approach is of paramount importance, as karyotyping can save breeder's time and money by early detection of the animals that, due to chromosomal abnormalities, will later show none or poor breeding performance.

Broodmare demographics: a study of a United Kingdom sport horses stud during a stud season
Whitaker, T.C., Tyler, C.M. and Mills, A.A., Writtle College, Centre for Equine and Animal Science, Chelmsford, Essex, CM1 3RR, United Kingdom; amy.mills@writtle.ac.uk

Previous studies have highlighted the demographics of the thoroughbred broodmare population visiting stud farms. Limited study has been undertaken within the sport horse population. The study aimed to evaluate the demographic profile of broodmares visiting a large sport horse stud within the UK over a single stud year. Data was collated and discriminated by various measures of status and outcome. Frequencies were returned for discriminating variables, further cross tabulation comparison and chi square analysis was undertaken. The sample population consisted of 238 sport horse broodmares. The largest proportion of mares 51.6% (n=123) (χ=36.210, $P<0.001$) were ages between nine and fifteen years, 23.4% (n=55) were aged over fifteen years. 41.6% of mares were recorded as maiden, 38.2% as barren and 20.2% with foal at foot. 62.6% (n=147) of mares (χ=92.630, $P<0.001$) were inseminated with fresh semen; 25.6% (n=61) were inseminated with frozen semen. At the end of the season 66% (n=155) positive pregnancy rate was reported (χ=21.782, $P<0.001$). No significant effects were found between age and method of service, although only 19.7% of mares over the age of fifteen were inseminated with frozen semen. Age of mare and outcome of pregnancy diagnosis showed no significant effect; 70% of mares eight years or under returned positive diagnosis compared with 61% of mares aged over fifteen years. The relationship between positive pregnancy diagnosis and status of the mare showed: 70% of maiden mares in foal, 69% barren mares and 48% of mares with foals at foot (χ=7.845, $P<0.05$). This study indicted different demographics in the mare population when compared to previous studies in thoroughbred. The rate overall of positive pregnancy was consistent with others studies on sport horse populations. It is interesting to note the significant lower rate of positive pregnancy diagnosis observed in mares that have foals at foot.

Viability of epididymal fresh and freezing sperm stored at 0, 24, 48, 72 and 96 h after stallion death
Matás, C., Avilés-López, K., Vieira, L., Garriga, A., García Vázquez, F.A. and Gadea, J., University of Murcia, Dpto Physiology, Veterinary School, 30100, Spain; cmatas@um.es

Sudden death, catastrophic injury, castration or any other event that makes semen collection or mating impossible may prematurely terminate a stallion's reproductive life. Epididymal sperm may be the last chance to ensure preservation of genetic material after injury or death of a valuable male. Viable epididymal sperm can be harvested post-mortem for possible use in assisted reproductive technologies. The objective of this study was to identify a 'window of opportunity' to collect equine epididymal sperm for subsequent freezing and determine the influence of prolonged cold storage (4 °C) of stallions epididymides on post-thaw sperm viability. Fifty-eight testes with attached epididymides were collected during one year at slaughterhouse. Spermatozoa from 10 epididymis were immediately recovered by retrograde flushing with air, evaluated and frozen (control). The remaining epididymides were cooled to 4 °C and stored for 24, 48, 72 and 96 hours, after which spermatozoa were collected and frozen. Viability was evaluated by IP/carboxifluorescein stain in sperm samples before freezing, after dilution in criopreservation medium and after thawing (0, 30 and 120 min of incubation at 37 °C). The present study demonstrates that equine sperm can be stored within the testicle at 4 °C for up to 72 h post-mortem without loss of viability of fresh (82.5, 81.2, 78.2, 80.6 vs. 68.9%, $P<0.01$) and diluted epydidimal spermatozoa (76.8, 73.9, 71.8, 75.7 vs. 59.3%, $P<0.01$). However, the freezing procedures damaged significantly the sperm membrane and reduced significantly the viability with no differences between the experimental groups (mean values 0 min: 28.9, 30 min: 12.4 and 120 min: 7.2%). In conclusion sperm harvested from stallion tissues after storage for 72 hours at 4 °C remain viable.

Effect of stocking rate on sward characteristics and milk production in grazing dairy cows

Roca Fernández, A.I., González Rodríguez, A. and Vázquez Yáñez, O.P., Agrarian Research Centre of Mabegondo, Animal Production, Ctra. AC-542 Betanzos-Santiago km 7,5, 15080, Spain; anairf@ciam.es

The stocking rate (SR) is the most important parameter affecting per unit area of land due to its influence on herbage intake, sward quality and milk production. The objective of this study was to investigate the effect of SR on sward characteristics, milk yield and pasture dry matter intake (PDMI) of spring and autumn calving dairy cows in spring/early summer. Cows grazing in rotationally grazed pastures of perennial ryegrass and white clover were stocked at medium (M, 3.9 cows ha^{-1}) and high (H, 5.2 cows ha^{-1}). Forty-four spring calving (S, mean calving date 15th February) and twenty-eight autumn calving (A, mean calving date 30th October) primiparous and multiparous Holstein-Friesian dairy cows were randomly assigned to four treatments. Milk yield and composition were analyzed from March to August in 2008 and pasture production, quality and sward utilization were measured for each treatment. The HA daily herbage allowance (DHA) and pasture dry matter intake (PDMI) were significantly lower (14.9 and 10.3 kg DM d^{-1}, respectively) than the MA (18.4 and 14.1 kg DM d^{-1}, respectively). The lower HS herbage intake was compensated with a significantly higher sward quality (ADF, 291 g kg^{-1};NDF, 518 g kg^{-1}; IVOMD, 781 g kg^{-1}) and an increase on sward utilization (83%) enough to achieve a significantly higher milk production (25.3 kg day^{-1}). According to other studies the HA showed a detrimental effect on milk production per cow but increased the amount of pasture harvested per hectare. No significant differences were found for BW and BCS with an average value of 575.8±13.6 kg and 3, respectively. Milk urea content was satisfactory, 203.2±18.1 mg kg^{-1}. There were significant differences between treatments for milk protein (HA, 31.9 and MA, 30.6 g kg^{-1}) and milk fat (HA, 39.7 and MA, 36.9 g kg^{-1}). Further studies are necessary to establish the effects of SR on sward characteristics and animal performance in Galicia.

Effects of tissue partitioning on perennial ryegrass swards grazed at different herbage masses and pasture allowances

Roca Fernández, A.I.[1], O'donovan, M.[2], Curran, J.[2] and González Rodríguez, A.[1], [1]Agrarian Research Centre of Mabegondo, Animal Production, Ctra. AC-542 Santiago-Betanzos, 15080, Spain, [2]Moorepark Dairy Production Research Centre, Dairy Production Department, Fermoy, Co. Cork, Spain; anairf@ciam.es

The objective of this study was to investigate the effect of pre-grazing herbage mass (HM) and pasture allowance (PA) on the sward composition in the upper and lower sward horizon (> and <4cm). Dry matter (DM) yield and leaf, stem and dead (LSD) content were measured across the grazing season. Sixty-four spring-calving dairy cows were balanced and randomly assigned to four grazing treatments. The treatments were based on two HM, low (L- 1600 kg DM ha^{-1}) or high (H- 2400 kg DM ha^{-1}) and two PA, low (L- 15 kg DM cow^{-1} day^{-1}) or high (H- 20 kg DM cow^{-1} day^{-1}). Sward characteristics and herbage composition were determined in two periods. Data analysis was performed using the statistical program SAS. Herbage mass was higher in period II for the two low HM treatments and lower for the two high HM treatments. Sward density was significantly higher in both periods for the two high HM treatments. The two high PA treatments showed a higher sward density in period I. There was an interaction in both periods for post-grazing height between HM and PA, the HH treatment had a higher residual than the LH treatment and the two low PA treatments. Pre- and post-grazing heights were significantly lower in both periods for the low HM treatments. Post-grazing height was significantly lower when animals were allocated 15 kg DM cow^{-1} day^{-1}. Leaf proportion was significantly higher in period II for the two low HM treatments in the upper and lower sward horizons (> and <4cm). Herbage removed was significantly higher in both periods for the two high PA treatments. There was an interaction in both periods for herbage utilization between HM and PA, the HH treatment had significantly lower sward utilization than the LH treatment. The two low PA treatments had significantly higher sward utilization than the two high PA treatments.

Do organic farming practices lead to specific nutritive value of green fodder on upland dairy farms?

Capitaine, M.[1], Agabriel, C.[2], Boisdon, I.[1], Andanson, L.[1] and Dulphy, J.-P.[3], [1]Enita Clermont, Clermont Université, UR AFOS 2008.03.100, site de Marmilhat, BP 35, F-63370 Lempdes, France, [2]Enita Clermont, Clermont Université, UR EPR 2008.03.102, USC INRA 2005, site de Marmilhat, BP 35, F-63370 Lempdes, France, [3]INRA, URH, Theix, F-63122 Saint-Genès-Champanelle, France; capitaine@enitac.fr

The management practices applied on grasslands, like fertilization or harvesting time, have an impact on the fodder nutritive value. The organic farming specifications could lead to specific management practices and then to a fodder nutritive value different from conventional. The aim of our study was to test this hypothesis. The study was carried out during 5 years on 8 upland dairy farms of the Massif Central, between which 4 farms were run in organic farming and 4 in conventional farming. We surveyed 24 grasslands used for early cutting, late cutting or dairy cow pasture. Herbage samples were taken on the first and second growing cycles, and analysed to determine mineral content and the organic matter digestibility. Then we calculated the UFL, PDIN and PDIE content. We worked on the results of 112 samples: 60 from organic farming and 52 from conventional farming. In our conditions the fodders from organic and conventional farms did not differ in terms of nutritive value. The analysis was made on 4 groups of samples according to the growing cycle and the plants development stage at harvesting date. The plant development stage was defined with the sums of the day degrees based on Dactylis growth model. To confirm our first results, the same analysis will be made with the botanical composition of each grassland in order to consider the development stage of the main grass species and their proportion in the grassland cover. Then we are also running a multifactorial analysis to identify the factors affecting the nutritive value of our samples. The aim of this analysis is to understand why the organic farming specific management practices do not lead to a different nutritive value of fodder than in conventional farming.

Infrared thermography as a suitable technique to evaluate the quality of corn silage after the fermentation process

Miotello, S.[1], Stelletta, C.[2], Simonetto, A.[1], Cecchinato, R.[3], Tagliapietra, F.[1] and Bailoni, L.[1], [1]Animal Science Department, Viale Università,16, 35020 Legnaro (PD), Italy, [2]Veterinary Clinical Science, Viale Università,16, 35020 Legnaro (PD), Italy, [3]KWS Italia Spa, Via S.Casadei,8, 47100 Forlì, Italy; silvia.miotello@unipd.it

Infrared thermography (IRT) is an innovative technique based on the detection of infrared radiation from the surface of an object. IRT is largely used in different fields regarding animal health (i.e. prevention of the inflammatory conditions: mastitis and laminitis), but few researches report its utilization in feed evaluation. The objective of this study was to verify the suitability of the IRT to assess the quality of the corn silage after the fermentation process. Six different hybrids of corn were ensiled into mini-silos (width x height x depth: 3 x 2.5 x 6.2 m) and the IRT measurements of each silage were taken at three different periods corresponding to the end of the first, second, and third part of each bunker silo. All images were scanned using a hand-held portable infrared camera (ThermaCam P25, Flir System) which was calibrated to environmental temperature, humidity and absorptive conditions on each sampling day. The distance of the camera from the cutting face of the silos was 4 m and the images were taken before and after the silage removing. The images were elaborated by a specific software (ThermaCam Researcher basic 2.08) and all data were subjected to statistical analysis. The changes in the average temperature (DT) before and after the removing silage tended to decrease with the progressive emptying of silos (8.7, 6.7 and 4.9 °C; $P<0.001$) indicating an increasing stability of the corn silage. Significant differences of DT were detected among different hybrids from 5.4 to 8.4 °C, respectively for F and E hybrids ($P<0.001$). High correlations were detected between DT values and some chemical and nutritional characteristics of the silage. These preliminary results show that IRT can be useful to evaluate the quality of silage after the fermentation process.

Temperature and pH of silages processed in micro-silos and obtained from different substrata

Santos, M.V.F.[1], Perea, J.[2], Martínez, G.[2], Garcia, A.[2], Gómez, G.[2] and Ferreira, R.L.C.[1], [1]University Federal Rural of Pernambuco, Campus Pernambuco, 52171-900, Brazil, [2]University of Cordoba, Animal Production, Edificio Produccion Animal, Campus Rabanales, 14071, Spain; pa2pemuj@uco.es

The research was carried out at Rabanales farm from the University of Cordoba and it aimed to characterize different products ensiled in micro-silos covered with plastic. Evaluated products and their respective proportions were described as follows: MSI – maize without inoculum; MCI – maize (99.9%) + inoculum (0.1%); PSI – wheat straw (40%) + brewery yeast (60%); and PCI – wheat straw (39.9%) + yeast (59.3%) + inoculum (0.1%) + urea (0.7%). The inoculum was manufactured by a private company and the growing media was the rumen liquor. The micro-silos were kept outside, without soil contact. Evaluation was performed in January 2009, 90 days after ensiling. The following averages and standard deviations were obtained: 540.5 ± 107.93 kg of weight, 0.73 ± 0.02 m height, 0.82 ± 0.01 m length, 1.37 ± 0.05 m width, and 649.75 ± 151.10 kg/m^3 bulk density, respectively. Six temperatures were taken in each experimental unit using a digital sound. The pH reading were performed using a ph meter in the aqueous extract. A 4 x 2 factorial arrangement was used in a complete randomized design with four products and two layers in the micro-silo (superior and inferior), with three replications. Averages were compared using the Tukey test ($P<0.05$). Significant effects were observed for temperature and pH ($P<0.05$) regarding the product, but no interaction was observed between product and layer. The product PSI showed the lowest temperature (6.68 °C), statistically different from the other products ($P<0.05$). No differences were observed between MSI (9.56 °C) and MCI (10.35 °C), however, MSI was different from PCI (11.88 °C). The highest pH was observed for the micro-silo with PCI (6.23) and the lowest for MSI (3.82), being these two different from each other. No significant differences were observed ($P>0.05$) between MCI (4.28) and PSI (4.43).

Stability of fatty acids in grass and maize silages after exposure to air during the feed out period

Khan, N.A., Cone, J.W., and Hendriks, W.H., Wageningen Institute of Animal Sciences, Animal Nutrition, Zodiac (building 531), Marijkeweg 40, 6709 PG Wageningen, Netherlands; nazir.khan@wur.nl

Lipids in the forages are extensively hydrolyzed in the silo with a concomitant increase in the level of free FA (FFA). After opening of the silo, exposure of the FFA to air and light with a simultaneous increase in pH and microbial growth could induce oxidization. The present study investigated the stability of FA in grass and maize silages exposed to air for 0, 12 and 24 h. Eight maize silages were selected with varying dry matter (DM) contents, being very wet, wet, normal and dry. In addition, eight grass silages were chosen on the basis of ammonia (NH_3) concentration and pH level. Grass and maize silages were sampled 8-10 weeks after ensiling and transported anaerobically to the lab in cooled plastic bags. After mixing, each sample was divided into three subsamples and exposed to air for 0, 12 or 24 h. Thereafter concentrations of individual FA were quantified by gas chromatography (GC). Exposure to air up to 24 h significantly lowered ($P<0.01$) the contents of linolenic acid (C18:3), linoleic acid (C18:2), oleic acid (C18:1) and total FA in maize silages. In grass silages, a 24 h exposure to air decreased ($P<0.05$) the mean concentrations of C18:3, C18:2 and total FA ($P<0.01$). In both grass and maize silages a decline in the concentrations of major unsaturated FA (UFA) was associated with a concomitant increase ($P<0.01$) in the proportion (g/g total FA) of palmitic acid (C16:0). The relative decrease in total FA after 24 h exposure to air was higher in maize silages with a high moisture content and decreased progressively with increasing DM contents. In contrast, pH and NH_3 levels of grass silages had no effect ($P>0.05$) on the stability of FA during the feed out period. The present study demonstrated that extended exposure of silages to air during feeding increased the proportion (g/g total FA) of C16:0 and lowered the concentration of poly unsaturated FA (PUFA).

Chemical composition and *in situ* dry matter degradation of various organic acid treated whole crop barley silage

Vatandoost, M., Danesh Mesgaran, M., Heravi Moussavi, A. and Vakili, A.R., Ferdowsi University of Mashhad, Dept. Animal Science (Excellence Centre for Animal Science), P O Box 91775-1163, Iran; vatandoost_58@yahoo.com

The aim of this study was to evaluate the effect of formic acid or acetic acid on chemical composition and *in situ* dry matter (DM) degradation of whole crop barley silage. The forage (35% DM) was ensiled as untreated (UT) or treated with formic acid (3.4 or 6.8 ml/kg DM; F3 or F6, respectively) or acetic acid (3 or 4 ml/kg DM; A3 or A4, respectively) for 30 days (n=4). Silage extract pH was determined using pH meter (Metrohm 691, Swiss). NH3-N concentration was determined in acidified silage extract (5 ml of the extract + 5 ml of 0.2 N HCl) using distillation method. Crud protein was determined using Kjeldahl method (Kjeltec 2300, Foss Tecator. Sweden). Neutral detergent fiber (NDF) was expressed as the ash free residue after extraction with boiling neutral solutions of sodium lauryl sulfate and EDTA. Four sheep (44±5 kg body weight) fitted with rumen fistulae were used. Bags (10 × 12 cm) were made of polyester cloth with a pore size of 52 μm. About 5 g DM of each sample was placed in each bag, then incubated (n=4) for each time (2, 4, 8, 16, 24, 48, 72 and 96 h). For zero time, bags were washed using cold tap water. The equation of $P=a+b(1-e^{-ct})$ was applied to determine degradation coefficients (a= quickly degradable fraction, b= slowly degradable fraction, c = fractional degradation rate constant). Both additives did not have any significant effect on crude protein content of the silages. These additives caused a significant ($P<0.05$) decrease in pH (UT= 4.07, F3= 3.95, F6= 3.57, A3= 3.96 and A4= 3.83; SEM= 0.037) and increase in NDF content (UT= 553, F3= 585, F6= 640, A3= 650 and A4= 607 g/kg DM; SEM= 9.174). NH3-N concentration (mg/dl) was significantly decreased when formic acid was applied (UT= 9.10, F6= 8.29; SEM= 0.29). Slowly degradation fraction of DM of F6 (0.49±0.03) and A4 (0.51±0.02) was significantly increased compared with UT (0.46±0.03).

Variation in polyphenolic compounds in forages: amount and composition

Reynaud, A.[1], Cornu, A.[1], Fraisse, D.[2], Besle, J.M.[1], Farruggia, A.[1], Doreau, M.[1] and Graulet, B.[1], [1]INRA, UR1213 Herbivores, Centre de Clermont-Ferrand/Theix, F-63122 St-Genès-Champanelle, France, [2]Laboratoire de Pharmacognosie, Faculté de Pharmacie, F-63000 Clermont-Ferrand, France; areynaud@clermont.inra.fr

Polyphenolic compounds are present in significant amounts in forages and have a potential beneficial effect on human health after their transfer in milk. Little is known on the sources of variability of these compounds in cow's milk but the nature of the forage in the diet is highly probable. To assess this variability, the polyphenolic composition was determined in 8 permanent pastures (Arrhenatheretea and Festuco-Brometea classes), in 4 temporary pastures poorly diversified, both harvested in June 2007 in the same area, and in 4 maize silages. The comparison of chromatographic profiles obtained using liquid chromatography paired with photodiode-array (l= 275nm) did not underline any statistical difference in the number of peaks having a typical phenolic compound spectrum (43±3.6) or in the sum of peak areas. However, peak pattern was very different between forages ($P<0.001$; Wilks' test) and can be related to their botanical composition. All forages taken together, a total of 107 peaks were separated among which some of them were identified. Sixteen peaks were common (such as p-coumaric acid, homorientin, apigenin) to the 3 different types of forage, 30 (such as ferulic acid) were specific of maize silage and 61 (including 3 isomers of chlorogenic acid, chicoric and rosmarinic acids, verbascoside, daidzein, genistein, biochanin A, formononetin, hesperidin, luteolin-7-o glucoside, schaftoside, rutoside and quercetin-3-glucuronide) were specific of permanent or temporary pastures, respectively. Among the 16 common peaks, 5 had a significantly ($P<0.05$) higher content in maize silage than in permanent pastures. In conclusion, even though the peak number and sum of areas were not different, the polyphenolic composition strongly and significantly varied between forages. These differences have to be confirmed now in the corresponding milks.

Genetic resources and landscape management for animal fibre production under European conditions

Allain, D., INRA, UR631, SAGA, Chemin de borde rouge, BP 52627, 31326 Castanet Tolosan, France; daniel.allain@toulouse.inra.fr

The major exotic fibre-producing animal groups are sheep, goat, rabbit, camel, South American camelids and the bovine species. The profitability of animal fibre production will depend on prices in world markets, the existence of niche markets, high levels of fibre production and low production cost. Under present cost/price structures, European fibre-producing enterprises are no longer profitable except when producers have created niche markets to obtain a premium price of the farm output. These niche markets concern only the more valued fine fibre: mohair and cashmere by goats, Angora by Angora rabbits and fine or coloured wool by sheep. Mohair industry was developed by importation of Angora goat stock and improvement of fibre quality and production was then achieved by selective breeding in different countries. Due to the lack of source of quality stock, the development of cashmere enterprises was based on crossbreeding programmes between feral goats and various imported strains, followed with genetic selection from the resultant gene pool. For fine wool and angora production, stocks were available in Europe from centuries. Niche markets for fine wool are very small and wool production is generally not profitable and may indeed be undesirable compared to meat or milk production. Angora production is now confidential but there still remain a high quality stock. Small ruminant species devoted to fibre will easily contribute to landscape management. Fine wool and cashmere are best suited to utilise the physically most difficult land resources of the mountainous and hill areas in Europe. Mohair production will be better adapted to the drier and warmer areas. Angora production does not require for land but it contributes to rural development in less-favoured areas. The establishment of fine fibre enterprises from other species is technically possible, most of them have apparent abilities to adapt to a wide variety of land types and climatic conditions, but it will depend on the availability of suitable breeding stock.

Regulatory biology of hair follicle development, behaviour and expression of phenotype in animal fibre production

Galbraith, H.[1,2], [1]University of Aberdeen, Institute of Biological and Environmental Sciences, 23 St Machar Drive, AB24 3RY, United Kingdom, [2]University of Camerino, Department of Environmental Science, Via Circonvallazione 95, 62024 Matelica, Italy; h.galbraith@abdn.ac.uk

Hair fibre is an important commercial product principally of sheep, goats, camelids and rabbits. Knowledge of its biology has been derived from these animals and from genetically modified mice. The yield and physical characteristics of such fibre from hair follicles, embedded in skin, is dependent on developmental and other mechanisms mediated by genetic expression of morphogenic and mitogenic signalling molecules, their receptors and transcription factors in both epidermal (e.g. keratinocytes which constitute hair shaft and supporting structures and pigment-producing melanocytes) and dermal (fibroblasts) components. A range of autocrine (same cell), paracrine (neighbouring cell) or systemic/endocrine-like (distant) signalling molecules with stimulatory or inhibitory activity has been identified. These regulate numbers, patterning and anatomical structures of follicles (primary/secondary type: differentiation into specialist keratinocyte cell lineages of hair shaft, medulla, cuticle and root sheaths) prenatally, and postnatally, dynamics of their activity (e.g. follicle cycle and length of anagen/telogen in response to extrinsic environmental factors and intrinsic regulation). Secreted molecules from fibroblasts are known to be essential for normal proliferation and differentiation of keratinocytes, processes involving synthesis of enzymes and molecules of cytoskeleton. Expression of commercially important phenotypic properties (e.g. diameter/medullation/cuticle/lustre/crimp/ waviness (assymetric growth)/pigmentation) is increasingly being related to regulatory signalling by products, including polymorphic forms, of individual genes and gene families. This paper will identify candidate regulatory genes and their potential for use to improve selection of animals and production of fibre of superior yield and quality.

Multiple splice variants of the ovine Mitf, c-Kit genes and evaluation of its' role in the dominant white phenotype in Merino sheep

Saravanaperumal, S.A.[1], Pediconi, D.[1], Renieri, C.[2] and La Terza, A.[1], [1]University of Camerino, Dept. of Molecular, Cellular and Animal Biology, Via Gentile III da Varano, 62032 Camerino (MC), Italy, [2]University of Camerino, Dept. of Environmental Science, Via Gentile III da Varano, 62032 Camerino (MC) CR & ALT are joint last authors, Italy; siva.saravana@unicam.it

In this study, we investigated the role of the 2 candidate genes: Mitf and c-Kit for the genetic background of 'dominant white' phenotype in Merino sheep. Reverse transcription (RT)-PCR analysis of Mitf gene, revealed two splice variants hereinafter referred as SP1 and SP2, from skin biopsies. In particular, the variant SP2, commonly known as 'isoform-M', differs from SP1 with the insertion of a stretch of 18bp (CGTGTATTTTCCCCACAG, pos.560-578) in the coding region for the amino acids: ACIFPT, resulting in a Mitf isoform of 419 aa (+form) vs the 413 aa of SP1 (-form). At present, to characterize other skin-expressed Mitf isoforms, we are carrying out 5'and 3'RACE experiments. Preliminary results shows 3 PCR amplicons ranging in size from 1.1 to 0.45 kbp for the 5'UTR and 3 amplicons for 3'UTR ranging from 3.3 to 0.8 kbp. The cDNA encoding ovine c-Kit was also amplified (3.770 bp) from the same skin biopsies, of which the complete coding sequence comprises of 2828 bp, 2840 bp, respectively. In fact, here also we report two splice variants, characterized by the presence or absence of four-amino acid sequence 'VTAK' (i.e., GTAACAGCAAAG pos.1534-1545). By means of RACE strategy, we obtained PCR amplicons of 0.75 kbp and 0.7 kbp for the 5'and 3'UTR regions of the c-Kit gene, respectively. Our future study will focus on the physiological switching of these isoforms in context with the promoter and the cell type which might have a combined influence on the gene expression programs and thus, shed light on the molecular mechanism behind dominant white phenotype in sheep.

Genetic analysis of colorimetric parameters for the differentiation of coat colors in a Spanish Alpacas population

Bartolomé, E.[1], Peña, F.[2], Sánchez, M.J.[2], Daza, J.[2], Molina, A.[2] and Gutiérrez, J.P.[3], [1]University of Seville, Ctra. Utrera,1, 41013Seville, Spain, [2]University of Cordoba, Campus Rabanales, 14071Cordoba, Spain, [3]UCM, PuertadelHierro,s/n, 28040Madrid, Spain; v92bamee@gmail.com

The aim of this study was to estimate genetic parameters for coat color from the alpacas population established in Spain. Coat color classes were defined as white, brown, black and mixed (combinations of two colors). Phenotypes were also measured quantitatively according to standardized international procedures (Commission Internationales de l'Eclairage L*, a*, b*), where L* describes lightness, a* describes color saturation from red to green, and b* describes color saturation from yellow to blue. The total color saturation was derived from a* and b* and referred to as Chroma. 164 alpacas from the spanish population were measured, 44 from the Suri breed and 120 from the Huacaya. Each animal was measured in three different body regions: shoulder, ribs and croup, taking the colorimeter measure from the fiber base. As the pedigree depth was small (kinship matrix of 250 animals), due to the scarce time that the alpaca is being bred in Spain, heritabilities were estimated using a multivariate animal model with 3 genetics groups, regarding to geographic origin (Chile, Peru and the United Kingdom). Heritabilities estimated for L*, a*, b* and fiber diameter, were 0.41, 0.91, 0.79 and 0.51, respectively. Genetic and phenotypic correlations between measured parameters were calculated. Results were between -0.55 (L* and a*) and 0.68 (b* and a*) for genetic correlations and between -0.38 (L* and a*) and 0.89 (a* and b*) for phenotypic correlations. The results of this study show the existence of possible major genes involved in coat color inheritance.

The effects of different feeding patterns on the production performance
Lou, Y.J.[1,2] and Jiang, H.Z.[2], [1]China Agri.Univ., Colle. of Resources and Environmental Sci., Bejing, 100094, China, [2]Jilin Agri. Univ., Colle. of Ani. Sci. & Technology, Changchun, 130118, China; jianghuaizhi6806@126.com

With decreasing of grassland resources, the grazing has become a limiting factor of Cashmere goat husbandry developing. To change this feeding pattern in to barn feeding is a good way to develop down-bearing goat. In this paper adult Liaoning Cashmere goats were used in order to study the effects of grazing and barn feeding on production performance. The experiment was performed in Gaizhou of Liaoning province. In grazing group (A)14 male and 30 female adult goats were used, in barn feeding group (B)13 male and 29 female adult goats were used. Grazing land is regions of hills and mountains with luxuriant vegetation. The vegetation coverage rate was over 80% with Gramineae, Leguminosae, and Compositae. The goats in barn feeding group were feed in closed barn with slatted floor and a playing ground in front. The goats in grazing group mainly live on forage except supplementary feeding in spring and winter. The adult male goats in barn feeding group were feed 600 g concentrate, 300 g alfalfa hay, 500 g guinea grass, and maize straw adlibitum, While the adult female goats were feed 200 g concentrate, 300 g alfalfa hay, 300 g guinea grass, and maize straw adlibitum. The results showed that there was significant difference on the feedstuff species intake by goats between A and B groups. Feedstuff intake by goats in thee A group has a great variety and rich of nutrition. The species of feedstuff intake by B group goats was not very diverse. The numerical value of down length, down fineness, down curvature, down yield, and the net down rate of goats in the B group showed an increasing tendency compared with goats in the A group. The down fineness of male goats in the B group was about 0.49µm thicker than that in the A group (19.88±1.42 and 19.39±1.46µm), and the difference is significant ($P<0.05$). The down length of female goat in the B group was 4.49 mm longer than that of A group (76.88±12.84 and 72.39±13.49mm), and the difference was significant ($P<0.05$).

Phenotypic and genetic variation of fleece weight, fineness of fibre and its coefficient of variability in Peruvian alpaca
Valbonesi, A.[1], Pacheco, C.[2], Lebboroni, G.[1], Antonini, M.[1] and Renieri, C.[1], [1]University of Camerino, Environmental Sciences, Via Circonvallazione 93, 62024 Matelica, Italy, [2]DESCO, Centro de estudio y promocion del desarrollo, Calle Malaga Grenet 678, 1675 Arequipa, Peru; carlo.renieri@unicam.it

Four hundred and fifty alpaca, from an experimental herd in the Peruvian plateau ('Alpaquero Developing Centre' of Toccra, located in the Arequipa Plateau, Caylloma Province), were chosen for an investigation of phenotypic and genetic variation of fleece weight, fibre fineness and coefficient of fineness variability. The alpaca specimens comprised 388 huacaya (217 males and 171 females) and 68 suri (34 males and 28 females), ranging from 119 to 371 days old. Data were analysed by means of analysis of covariance, using type of fleece and sex as fixed factors and age at first shearing as covariate. The estimated mean values (at covariate value of 288.3) for huacaya and suri were, respectively: 1) fleece weight (kg) 1.47 (sd=0.28) and 1.46 (sd=0.30); 2) m) 20.10 (sd=2.03) and 21.08 (sd=2.04); 3) coefficient ofmfibre diameter (variation 20.32 (sd=4.30) and 22.34 (sd=4.29). Sex, as well as its interaction with the type of fleece, had no effect on the three traits. The type of fleece had no effect on the fleece weight but affected both the fibre diameter and the fineness coefficient of variability, at a significant level ≤0.006. The estimated heritability values were: 0.10 for the fibre diameter, 0.19 for the coefficient of variation, and 0.37 for the fleece weight. A significant genetic correlation was found between all the pair wise combinations of these traits. The environmental correlations were significant at $P<0.001$, with the exception of the pair fleece weight - coefficient of variability. This investigation is the first step towards the estimation of a genetic index for each of the 450 alpaca and, therefore, the establishment of a selected nucleus to be used in a breeding program aimed at improving quality and quantity of fibre within the experimental herd in Toccra.

Inheritance of white, black and brown coat colour in alpaca by segregation analysis

Valbonesi, A.[1], Apaza Castillo, N.[2], La Manna, V.[1], Gonzales Castillo, M.L.[2], Huanca Mamani, T.[2] and Renieri, C.[1], [1]University of Camerino, Environmental Sciences, Via Circonvallazione 93, 62024 Matelica, Italy, [2]INIA, ILLPA Puno, Rinconada Salcedo, Puno, Peru; vincenzo.lamanna@unicam.it

Coat colour heredity patterns in alpaca were investigated through segregation analyses on the offspring of 17 paternal half sib families. Crosses involving white, black and brown specimens, were carried out in an experimental herd in the Peruvian plateau ('INIA ILLPA Centre', Puno province). The goodness of fit for the monofactorial (single gene) hypothesis was evaluated applying both the G-test (as the log-likelihood ratio test is also called), with Williams's correction (Gadj) for single segregation, and the heterogeneity G-test for replicate segregations. Segregation analysis was applied only to segregating families showing at least one proband, an individual among the offspring with a supposedly recessive phenotype (truncate selection), and the expected frequencies for each family were statistically corrected with the *a priori* method proposed for truncate selection by Andresen. Dominance with complete penetrance of white over black was observed in the four crosses of white parents (Gtotal = 3.44; $P=0.33$), as well as in three of the four white male x brown females crosses (Gtotal = 3.48; $P=0.32$). Dominance with complete penetrance of black over brown was observed in the four crosses of black parents (Gtotal = 5.64; $P=0.23$), and in one of the two black male x brown females crosses (Gadj <0.001; $P=0.99$). This latter hypothesis was further supported by the results of three crosses between brown parents, where all the offspring (25 crias) was brown. Statistical tests seem to validate the monofactorial hypothesis of dominance with complete penetrance of full white over pigmented, and black over brown. Since full white no albino fleece is particularly appreciated by the textile industry, an understanding of the phenotypic relationship among coat colours is a basic step for establishing correct reproductive practices in alpaca breeding.

Evaluation of MC1R gene polymorphism in Vicugna pacos

Crepaldi, P.[1], Milanesi, E.[1], Nicoloso, L.[1], La Manna, V.[2] and Renieri, C.[2], [1]Università degli Studi di Milano, Animal Science, Via Celoria 2, 20133 Milano, Italy, [2]Università degli Studi di Camerino, Environmental Sciences, Via Circonvallazione 93, 62024 Matelica, Italy; vincenzo.lamanna@unicam.it

In alpaca (*Vicugna pacos*) fibre production and quality represent the most important characteristics from a market point of view and, as a consequence, current breeding programs are mainly based on these phenotypic traits. The identification of genetic markers associated to coat colour could lead to the set up of more reliable breeding programs, particularly aimed at the production of naturally coloured fibre. The high conservation of coat colour genes across the species previously studied allows to select candidate genes for this trait also in alpaca. In particular, Melanocortin 1 Receptor (MC1R) is known to play an important role in the differentiation of red versus black phenotypes in many species. In this work polymorphism of alpaca MC1R gene was evaluated by sequencing 1,190 base pairs comprising the whole coding sequence and part of the UTR regions in 6 alpacas characterized by different coat colours (2 solid black, 2 solid red, 1 brown-pied and 1 red-pied). Sequence comparison revealed the presence of 8 SNPs, 6 of which in the coding sequence. Four of these were silent mutations, whereas 2 (A82G and A258G) resulted in amino acid changes (thr/ala and met/val) in position NH2 terminal and in the second trans-membrane domain of the receptor. Analysis of the genotypes revealed that the 3 animals characterized by a black/brown coat colour were homozygous for the same allele at all the polymorphic loci, while the red animals showed an heterozygote status at all the polymorphic loci. These findings, although preliminary, agree with GenBank sequence data (EU135880 and EU220010) relative to alpacas described as 'fawn' and 'black/brown', and confirm the high level of polymorphism of alpaca MC1R gene described in a recent publication.

Asip and MC1R in coat colour variation in Alpaca

Bathrachalam, C.[1], La Manna, V.[2], Renieri, C.[2] and La Terza, A.[1], [1]University of Camerino, Dept. of Molecular, Cellular and Animal Biology, Via Gentile III da Varano, 62032 MC, Italy, [2]University of Camerino, Dept. of Molecular, Cellular and Animal Biology, Via Gentile III da Varano, 62032 MC CR & ALT are last joint authors, Italy; antonietta.laterza@unicam.it

Coat colour is an important objective in the selection of alpaca, especially in animals reared for fine fibre production. At present, very few information is available about the genetic basis of coat colour in alpaca. Results from other domesticated mammals suggested that Asip and MC1R genes plays a key role in the determination of coat colour by regulating the type, amount and distribution pattern of the pigments eumelanin and pheomelanin. Moreover, up to now, no many experimental segregation trials were performed in order to define the inheritance of coat colour in alpaca. In this context, cDNA encoding the alpaca Asip (this is the first report on Asip in alpaca) and MC1R were amplified from mRNA derived from skin biopsies which were from experimental segregation trails, realised with the collaboration of the National Institute of Agronomic Research, Perù. Here, we present Asip and MC1R alleles from the white and brown animals. The full coding region of Asip comprises of 402 bp and it codes for a protein of 133 aa. The full coding region for MC1R comprises of 954 bp and it codes for a protein of 317 aa. Comparison between Asip sequences shows 5 mutations (SNPs), i.e. C11G, A18C, C290A, T291C and G352A, in particular 2 of them (at position 18 & 290) are synonymous mutations whereas, the remaining 3 resulted in non synonymous mutations leading to amino acids substitutions, i.e. T for S11, C for R291 and R for H352. In MC1R we observed 6 mutations (SNPs) i.e. G82A, C126T, G259A, G376A, A618G and T901C and among these, 2 synonymous mutations (at position 126 & 618) and 4 non synonymous mutations, i.e. A for T82, V for M259, G for S376 and C for R901. This study will aim to identify more Asip and MC1R alleles in our population and thus, pose the basis for the development of marker assisted breeding programme for coat colour in alpaca.

Nitrogen distribution in llama milk (Lama glama)

Fantuz, F.[1], Pacheco, C.[2], Soza Vargas, A.[2], Lebboroni, G.[1] and Renieri, C.[1], [1]Univ. di Camerino, Di Scienze Ambientali, via Circonvallazione, Matelica, Italy, [2]Centro de Estudios y Promocion del Desarrollo, (DESCO), Arequipa, Peru; francesco.fantuz@unicam.it

Llama is mainly bred in South America, for work, meat and wool but is increasing the interest in breeding llama for fiber production and as pet animal also in Europe, North America and Australia. However there are only few data on llama milk composition. Aiming to increase knowledge about nitrogen distribution in llama milk, 5 lactating female llamas were used to provide milk samples which were obtained by hand milking on day 60 and 120 from parturition. Milk samples were then freeze-dried until analysis. Freeze dried samples were resuspended in distilled water at 14% dry matter for before analysis. Total nitrogen (TN), non protein nitrogen (NPN) and non casein nitrogen (NCN) were determined by Kjeldahl method and casein nitrogen (CN) and whey nitrogen (WN) were then calculated. From another aliquot of milk samples, whole casein and whey protein fraction were obtained by isoelectric precipitation. Separation of individual milk proteins was obtained by SDS-PAGE. The relative amount of main individual whey proteins was determined by image analyser. Values (mean±SD), expressed as mg/g DM of freeze dried milk, were 46.35±2.81, 2.80±0.58, 12.18±2.21, 34.18±3.0, 9.37±1.99 respectively for TN, NPN, NCN, CN and WN. When expressed as mg/100ml of resuspended milk, milk total nitrogen (649±39.98) as well as NPN (39.33±8.16) and NCN (170.47±30.98) were in the range reported for cow's milk. Casein fraction appear to be predominant in llama milk as in milk from other ruminants, being CN 478.60±42.07 and WN 131.14±27.94. CN/TN and NPN/TN ratio were respectively 0.73±0.04 and 0.06±0.01. Semi-quantitative analysis of individual whey proteins indicated that α-lactalbumin accounts for an average of 70.45% (±8.48) on total whey protein. Lactoferrin, serum albumin and immunoglobulins percentages were respectively 2.86±1.87, 12.77±4.85, 2.64±1.13. β-lactoglobulin was not detected in the analysed samples.

Global perspectives on animal trait ontology

Reecy, J.M.[1], Park, C.A.[1], Hu, Z.-L.[1], Hulsegge, I.[2], Van Der Steen, H.[3] and Hocquette, J.-F.[4], [1]Iowa State University, Ames, IA, 50010, USA, [2]Wageningen University, AB Lelystad, 8200, Netherlands, [3]EEAB, Abbotswood, WR11 4NS, United Kingdom, [4]INRA, Theix, 63122, France; jreecy@iastate.edu

With the advent of high-throughput genotyping, gene expression, proteomics, metabolomics, etc., we have entered an era where integration of information across species, experiments and disciplines will be paramount. Thus, there is an urgent need to precisely define terms (e.g. phenotypes, traits, diseases, methods) so as to capture biologically relevant distinctions. This need extends beyond the livestock community to include the model organism and human research communities. Ontologies, which identify and define entities and the relationships among them in specific domains of interest, offer a powerful approach for annotating biological data in a form that allows users and software tools to retrieve, inter-relate, and extract biological knowledge. In order for such ontologies to be broadly useful to the livestock community, they need to capture the knowledge and expertise of multiple experts from industry and research groups. Hence, we have begun to create a consortium, representing the relevant livestock communities, to develop, maintain, and update pertinent ontologies. We envisage a close collaboration between livestock consortia members and consortia already working on the development of animal trait ontologies in model organisms. The Animal Trait Ontology (ATO) and associated software tools for collaborative creation, editing, curation, and management of ontologies provide the infrastructure necessary for engaging livestock communities in the process of creating comprehensive genomic resources for annotating, integrating, and analyzing phenotype and genomic data. Development of these resources will facilitate future data consolidation, analysis, and comparisons. The development of an ATO is an ongoing process with changing approaches and objectives requiring input from several groups. Integration of useful components into one global holistic system is a responsibility for all involved.

The number of genes underlying a trait under selection affects the rates of expected and true inbreeding more than their distribution over the genome

Pedersen, L.D., Sørensen, A.C. and Berg, P., Faculty of Agricultural Sciences, Dept. of Genetics and Biotechnology, Aarhus University, Blichers Allé, P.O. Box 50, 8830 Tjele, Denmark; Louise. DybdahlPedersen@agrsci.dk

We investigated to what extent the number of loci and the distribution of the positions of loci affecting a low heritability trait (h^2=0.04) affected the level and rate of pedigree estimated as well as true inbreeding, i.e. identity-by-descent, when applying gene assisted selection (GAS) relative to traditional BLUP selection. The investigation was carried out using stochastic simulation of a population intended to resemble the breeding nucleus of a dairy cattle population. A finite locus model was applied and the number of loci investigated was 30, 100, and 300, respectively. Two different distributions of gene locations were compared: A uniform and a multinomial distribution and the autosomes of the bovine genome were modelled based on the MARC bovine linkage map. Favourable QTL allele effects were sampled from a gamma distribution (q=5.4; a=0.42) summing to an overall genetic variance of 0.04 and the frequencies were sampled from a continuous uniform distribution. The results showed that pedigree estimated inbreeding, as well as true inbreeding at the QTL, decreased when the number of loci under selection increased and this more so when using GAS than BLUP. Unlike the rate and level of inbreeding, the inbreeding in the region surrounding the QTL, was higher in GAS relative to BLUP and this more so when the number of genes were high. The distribution of the positions did not have a significant effect on any measure of inbreeding.

LD pattern and signatures of recent selection in Holstein cattle

Qanbari, S.[1], Pimentel, E.C.G.[1], Tetens, J.[2], Thaller, G.[2] and Simianer, H.[1], [1]Georg-August University, Animal Breeding and Genetics Group, Department of Animal Sciences, Göttingen, 37075, Germany, [2]Christian-Albrechts University, Institute of Animal Breeding and Animal Husbandry, Kiel, 24098, Germany; sqanbar@gwdg.de

Genotypic data assessed by 60K SNPs beadchip (Illumina, Inc) from 810 German Holstein–Friesian were used to characterize LD properties. A total of 41,398 markers were included into final analysis which cover 2632.58 Mbp of the genome with average adjacent marker space estimated as 63.59 kbp. The average observed heterozygosity and mean MAF were estimated as 0.37 ± 0.12 and 0.28 ± 0.13, respectively. A total of 717 haplo-blocks spanning 120,972 kb (4.26%) of the Holstein genome were detected. Mean block length was estimated as 173 ± 130.6 kb. The mean value of r2=0.29 ± 0.31 was observed in pair-wise distances of <30 kb and it fell to 0.23 ± 0.27 at 30 to 60 kb, which is close to the average inter-marker space. At ranges 0 to 30, 30 to 60 and 60 to 100 kb, 34.0, 27.4, and 4.8% of SNP pairs exhibited r2 larger than 0.3, respectively. In our analysis, 10 genes were tested for signatures of selection by means of the extended haplotype homozygosity (EHH) analysis. This analysis showed significant P-values for a slower decay of LD in DGAT1, Casein cluster, LPR, SST genes and approached significance for the GHR gene. This indicates that many of the functional candidate genes are subject to recent selection pressure. The results of this study describe a second generation of LD map statistics for the Holstein genome which has four times higher resolution compared to the available maps. Also the level of LD obtained in this study indicates that a denser SNP map is needed to capture completely the LD information required for whole-genome fine mapping and genomic selection.

The distribution of additive and dominant QTL effects in porcine F2 crosses

Bennewitz, J.[1] and Meuwissen, T.H.E.[2], [1]Institute of Animal Husbandry and Breeding, University of Hohenheim, Garbenstarsse 17, 70599 Stuttgart, Germany, [2]Institute of Animal and Aquacultural Sciences, University of Aas, Dröbackveien, 1432 Aas, Norway; j.bennewitz@uni-hohenheim.de

Having knowledge of the distribution of QTL additive and dominant effects would contribute to the understanding of the genetics of quantitative traits and is of interest in several fields. The present study used published QTL mapping data from three F2 crosses in pigs for 34 meat quality and carcass traits to derive the distribution of additive QTL effects as well as QTL dominance coefficients. The crosses were derived from European Wild Boar, Meishan and Pietrain. The published QTL effects were additive and dominance effects together with their standard errors. The dominance coefficients were calculated by dividing the observed dominance effects by the corresponding absolute additive effects. The standard errors of the dominance coefficients were approximated using the delta method. The additive effects were truncated due to the fact that only significant QTL were reported. We fitted mixtures of normals using a modified EM algorithm and additionally exponential distributions to the additive effects. The heterogeneous error variances of the estimates and the truncation of the data were considered. The distributions were compared by repeated cross validation. The results showed that the exponential distribution fitted the data best. The distribution was leptokurtic with a high density for small effects and a low density for large effects. Mixtures of normals were fitted to the dominance coefficient, again taking the heterogeneous error variances into account. The EM algorithm clearly suggested fitting only one component, which implies a normal distribution for the dominance coefficients. The mean was slightly positive. The derived distributions can be used for Bayesian analysis of genomic data.

Using dense marker maps to determine genetic diversity over the neutral genome

Engelsma, K.A.[1,2,3], Calus, M.P.L.[1], Hiemstra, S.J.[1,3], Bijma, P.[2], Van Arendonk, J.A.M.[2] and Windig, J.J.[1,3], [1]ASG Lelystad, ABGC, P.O. Box 65, 8200 AB Lelystad, Netherlands, [2]Wageningen University, ABGC, P.O. Box 338, 6700 AH Wageningen, Netherlands, [3]Centre for Genetic Resources (CGN), P.O. Box 65, 8200 AB Lelystad, Netherlands; krista.engelsma@wur.nl

Determination of the genetic diversity present within livestock breeds is of crucial importance for an efficient use of resources available for conservation. The objective in this study was to develop a method to estimate genetic diversity across the genome in a livestock population using dense marker maps, and which is more closely related to estimates of genetic diversity used in quantitative genetics. This method can give a better insight in the genetic diversity, compared to current methods such as heterozygosity. Genetic diversity was determined in a simulated population at each locus on a neutral genome, using SNP-marker information and IBD-matrices containing relationships between alleles. Information from groups of markers lying closely together was used by formulating haplotypes, in order to estimate IBD-probabilities more precisely. The obtained genetic diversity was compared to a classical genetic diversity measure, marker heterozygosity. Both heterozygosity and IBD relatedness varied considerably over the genome. Heterozygosity at single SNPs was a poor predictor of heterozygosity at neighbouring markers (r=0.11) while flanking markers predicted heterozygosity slightly better (r=0.28). Genetic diversity estimated with haplotype derived IBD matrices of flanking markers was not related to heterozygosity (r=0.04). Average genetic diversity over stretches of 40 SNPs was correlated to average heterozygosity (r=0.47). Heterozygosity in polymorphic markers can be quite different from SNPs. Estimation of genetic diversity at specific points in the genome under the neutral model is a challenge.

Prediction of haplotypes with missing genotypes and its effect on marker-assisted breeding value estimation

Mulder, H.A., Calus, M.P.L. and Veerkamp, R.F., Animal Breeding and Genomics Centre, ASG Wageningen UR, P.O. Box 65, 8200 AB Lelystad, Netherlands; herman.mulder@wur.nl

In livestock populations, missing genotypes on a large proportion of animals is a major problem when implementing marker-assisted breeding value estimation for QTL with a known effect. The objective of this study was to develop a method to include missing marker genotypes in breeding value estimation by predicting the number of haplotype copies (nhc) for ungenotyped animals, using 1, 2 or 4 markers. For genotyped animals the nhc represents the number of copies an animal carries for a certain haplotype, i.e. 0, 1 or 2 copies. For both genotyped and ungenotyped animals, the nhc were treated as phenotypic records in a mixed model framework using the additive genetic relationship matrix and the observed nhc of genotyped animals. This yielded predicted nhc for all animals. The predicted nhc were subsequently used in marker-assisted breeding value estimation by applying a random regression on these covariables. To evaluate the method, a population was simulated with one additive QTL and an additive polygenic genetic effect. The QTL was located in the middle of a haplotype based on SNP-markers. The accuracy of the total EBV increased for genotyped animals, but, as expected, for ungenotyped animals the increase was marginal unless the heritability was smaller than 0.1. Haplotypes based on 1 marker gave lower accuracy than using 4 markers. The accuracy of the total EBV approached the accuracy of gene-assisted BLUP when using 4-marker haplotypes with a distance of 0.1 cM between the markers. The proposed method is computationally very efficient and suitable to apply for marker-assisted breeding value estimation in large livestock and plant populations including effects of a number of known QTL. These results were obtained through the EC-funded FP6 Project 'SABRE'.

Incorporating genomic information into dairy international genetic evaluations
Dürr, J.W., SLU, Animal Breeding and Genetics, Box 7023, 750 07 Uppsala, Sweden; joao.durr@hgen.slu.se

Dairy genetics is undergoing a major transformation through the use of genomic information in genetic evaluations and genomic EBVs are already in use for both young sire selection and semen marketing. Affordable genotyping by means of the single-nucleotide polymorphism (SNP) technology has made possible incorporating genomic information from a large number of animals from reference populations into national selection schemes, allowing estimation of genetic merit with reasonably high reliability of young animals without progeny. Interbull is now facing the challenge to create a framework and develop methodologies to use genomic information into international genetic evaluations. The organization has appointed a task force of renowned scientists to lead developments and a successful workshop was carried out in January 2009. Major issues in debate are the extension of genomic data exchange between countries, the need to account for pre-selection bias, the need of proper validation procedures for genomic EBVs, the combination of direct genomic and progeny based breeding values, the development of Genomic-MACE (GMACE) procedures that allow multiple-country comparisons between genomic EBVs and the feasibility of an international genomic database for estimation of SNP effects in different country scales at Interbull. A stepwise implementation has been planned, starting with the use of conversion formulas from one country scale to another, continuing with implementation of GMACE and finally applying the importing countries prediction equations on genotypes of traded animals. Having a common genomic database at Interbull is particularly appealing for small populations, which would greatly benefit from having a larger reference population for SNP effect estimation. This breakthrough which finally approximated molecular and quantitative genetics was possible only because of the extensive and comprehensible phenotypic data from conventional breeding programs is available to allow interpretation of the genomic information.

The repercussions of statistical properties of interval mapping methods on eQTL detection
Wang, X.[1], Elsen, J.M.[2], Gilbert, H.[3], Moreno, C.[2], Filangi, O.[1] and Le Roy, P.[1], [1]INRA, UMR598, 65, rue de Saint Brieuc, 35042 Rennes, France, [2]INRA, UR631, Auzeville B.P. 52627, 31326 Castanet Tolosan, France, [3]INRA, UMR1313, Domaine de Vilvert, 78352 Jouy-en-Josas, France; xiaoqiang.wang@rennes.inra.fr

QTL detection on a huge amount of phenotypes, like eQTL detection on transcriptomic data, highlights the statistical properties of interval mapping methods. One of the steadiest outcomes is the high number of eQTL detected on markers locations. The aim of this communication is to describe QTL detection in this particular context through the use of simulated data. Designs of sib families were simulated and analysed using the QTLMaP software. Different parameters, such as the QTL effect, the QTL location, the number of markers or the density of the genetic map, were taken into account. Simulations under the no QTL hypothesis showed that, whatever the location, ie on a marker or between two markers, the nominal test statistics follows a x2 distribution with a number of degrees of freedom depending on the number of parents. Simulations under the one QTL hypothesis confirmed that the estimated location of the QTL is biased. Indeed, it is closer to markers locations than it should be, which is even more noticeable towards the bounds of the linkage group. The lower the QTL effect, the higher this bias. The repercussions of the above on eQTL detection are discussed. These results are obtained through the EC-funded FP6 Project 'SABRE'.

Removing undesired introgression

Amador, C.[1], Toro, M.A.[2] and Fernandez, J.[1], [1]INIA, Ctra. Coruña Km 7,5, 28040 Madrid, Spain, [2]ETSIA, UPM, Ciudad Universitaria, 28040 Madrid, Spain; amador.carmen@inia.es

Sometimes, besides the general objective of any program of genetic resources conservation, management of certain subpopulations requires the maintenance of its genetic pool differentiated from the rest of subpopulations. Within the domestic animal field, typical cases are those of species where breeds exist associated to quality products (Tajima or Kobe ox in Japan, Iberian pigs), to a particular activity (Spanish pure breed horses, bullfight cattle) or just for aesthetical reasons (dog breeds). Factors leading a population to suffer introgression may be diverse: incorrect management, when no awareness existed of the importance of keeping that unit independently; a high extinction risk, making necessary to use outgroup individuals to keep higher population census sizes (genetic rescue); regeneration of an extinct (or almost extinct) population from a semen bank, requiring the use of females from other breed as mothers (half of the genetic information in the first generation will not come from the target population). In all cases it would be crucial to eliminate a posteriori the exogenous genetic information and keep the characteristic genetic background of the population we are dealing with. In the present study, we have considered a simple scenario where genealogical information of the external individuals and the time they entered the population are available. Several parameters from the classical genealogical analysis (founder contribution, allele retention, mean coancestry) were used to determine the influence of foreign founders and to determine the management strategy. The most powerful method to reach our aim (in terms of speed of depuration, maintenance of original genetic diversity and inbreeding levels generated) seems to be minimising the mean coancestry of the external founders with the target population. Most important factors for the success of the depuration process are the number of external animals entering the population and the number of generations elapsed until the management started.

Reform and design of Masters programs in Latin America

Serradilla, J.M.[1], Villarroel, M.[2], Toro, M.[2], Wurzinger, M.[3] and Alfa Iii - Alas Consortium, A.L.A.S.[3], [1]Universidad de Cordoba, Ctra.N IV Km.396, 14014 Cordoba, Spain, [2]Universidad Politécnica de Madrid, Ramiro de Maetzu 7, 28040 Madrid, Spain, [3]BOKU-University, Gregor Mendel Strasse 33, 1190 Vienna, Austria; miguel.toro@upm.es

The Europeam Union has launched the ALFA programme to contribute to the economic and social development of Latin America and the more balanced development of the society, through cooperation between higher education institutions of the EU and Latin America. The overal objective of the 36 month action is the improvement and harmonisation of the higher education in the field of Animal Science at Masters level in 4 LA countries (Argentina, Bolivia, Mexico, Peru) using internationally recognised standards. The action consits of four major components. Component 1 is the reform of 4 Master Programs and the development of 3 new curricula, 2 is related to equip academic and administrative staff with knowledge for the implementation of the programs, 3 is the implementation of new teaching methods and the devolpment of joint courses. Finally 4 is the purchase of new equipment (IT- and laboratory equipment etc.) to support all other activities. The target group will include about 130 academic and administrative staff members. Courses will be offered on language training, didactics, project management and IT.Short-term internships at EU Universities will offer the opportunity to learn practices in management and administration. Direct beneficiaries will be about 3,000 undergraduates and 300 graduate students from the Latin American partner universities. They will benefit from new teaching methods and improved course contents which will expect to increase attractiveness for MSc programs.The ultimate beneficiaries will be thousands of livestock producers and rural communities.The newly established quality management system supports the institutional sustainability of the action.This new system will be introduced into the routine work of the universities. They will continue working towards a double degree program.

Characteristics and dynamism of beef cattle farming systems under Boeuf du Maine protected geographical indication

Couvreur, S., Schmitt, T. and Lautrou, Y., Groupe ESA, Laboratoire de productions animales, 55 rue Rabelais, 49007 Angers, France; s.couvreur@groupe-esa.com

In the changing context of beef production, a higher technico-economical adaptability of beef cattle systems is required. The knowledge of beef cattle farming under geographical indications is not good enough to assess their adaptation ability. This paper aims to assess the technical characteristics and the dynamism of beef cattle systems under an old protected geographical indication: the Boeuf du Maine (BM) PGI (created in 1996). Surveys were carried out in 86 farms producing BM animals and located in the Loire Region (France). Farm management was studied in relation to the level of animals sold under the PGI trademark using a multivariate approach. A high diversity of farming systems was observed: 32 farms are specialised in beef cattle production, 27 and 26 farms have also a dairy cattle and a poultry/pig production. Thirty-one farmers buy calves and fatten them as beef heifers and steers. Among them, 23 produce mainly milk, poultry or pigs. These farms produce 12.5±8.3 BM animals (93% are heifers) which represent 80±6.7% of the animals that can be sold under BM PGI. The production of BM animals is an easy opportunity for them to enhance the value of permanent grasslands located far from the farm. It's also a good way to maintain grasslands in a region where cereal crops are dominant and could give in the future a greater environmental value to the BM PGI. All the other farms have a beef herd with or without another animal production. Among them, 17 don't fatten the males, 29 fatten young bulls and 9 fatten steers. The size of the beef herd is variable (31 to 84 calving/year). In farms with others animal productions, beef production is often not dominant in the system. These farms produce 12.8±10.0 BM animals (53% are cows, 36% heifers) which only represent 47±33% of the animals that can be sold under BM PGI. Thus, the production of BM animals is one opportunity to enhance the value of the beef cows.

Efficiency of Swiss and New Zealand cows under roughage-based feeding conditions

Kunz, P.L., Piccand, V. and Thomet, P., Swiss College of Agriculture, Department of Animal Sciences, Laenggasse 85, 3052 Zollikofen, Switzerland; Peter.Kunz.1@bfh.ch

The high yielding dairy cow which is widespread in Switzerland is not suited to a production system based on roughage and very little concentrates. Investigations in Ireland and New Zealand have shown that the New Zealand Holstein Friesian population is well adapted to a pasture-based milk production system with seasonal calving. For this reason 72 pregnant heifers with at least two generations of New Zealand ancestry were imported from Ireland in 2006 and placed on twelve dairy farms in Switzerland. The objective was to investigate over 3 years (2007- 2010) the attributes of cows adapted to a roughage-based seasonal production system under Swiss conditions. The amount of feed over one year was composed of 65-70% grazed pasture, 20-25% conserved roughage and at most 300 kg of concentrates. For comparison pairs of Swiss (CH) and New Zealand (NZ) cows were established (86 cows in total) with similar age (±6 months) and calving date (±35 days) on the twelve farms. Body weight (BW) in the first lactation was higher in CH (544 kg) than in NZ (477 kg, $P<0.001$) cows, and body weight changes during the first five months after calving were different between CH (-10 kg) and NZ (+11 kg, $P<0.002$) cows. Milk yield was similar in both groups (CH 4987 kg energy corrected milk (ECM) per lactation, NZ 5089 kg ECM, $P<0.47$), milk composition was higher in NZ cows (fat CH: 4.05%, NZ: 4.24%, $P<0.02$; protein CH: 3.26%, NZ: 3.43%, $P<0.001$). Efficiency (kg ECM/kg metabolic BW) was higher in NZ (44.3) than in CH cows (50.0, $P<0.001$). The calving interval from first to second lactation was 368 days for both groups. The results show advantages for NZ cows, but there are also individual Swiss cows with similar attributes. Additional variables are currently being analysed, which will hopefully help to achieve the aim of finding the key attributes of cows adapted to a pasture-based seasonal production system.

Session 34

Theatre 3

EFSA's scientific assessment on the effects of current farming and husbandry systems on dairy cow welfare

Ribo, O., Candiani, D. and Serratosa, J., European Food Safety Authority (EFSA), Animal Health and Welfare (AHAW) Unit, Largo N. Palli 5A, 43100 Parma, Italy; oriol.ribo@efsa.europa.eu

EFSA was requested by the European Commission to make a scientific assessment, considering whether current husbandry systems comply with the welfare requirements of dairy cows from the pathological, zootechnical, physiological and behavioural points of view. In particular, the impact of genetic selection for higher productivity on animal welfare should be evaluated, considering the incidence of lameness, mastitis, metabolic and fertility disorders. A working group of experts was set up, encompassing expertise related to the specific issue, including Risk Assessment (RA). Working group meetings were held to compile an exhaustive Scientific Report containing all available scientific evidence and data in relation to dairy cow welfare, such as farming systems, genetics, management and disease, nutrition and metabolic disorders, housing and management, social and maternal behaviour, lameness, mastitis and reproductive disorders. Four separate RAs, focused on metabolic and reproductive disorders, udder disorders, leg and locomotion problems and behavioural problems, were carried out. In the RAs, hazards which have important consequences for dairy cow welfare, such as genetics, housing (e.g. space and pen design), feeding (e.g. liquid feed) and management (e.g. grouping) have been considered within each of the husbandry systems considered (cubicle houses, tie stalls, straw yards and pasture). Based on these 4 RAs, four Scientific Opinions have been prepared, which include conclusions from the RA plus the conclusions and recommendations from the Scientific Report related with the specific RA subject. A fifth Scientific Opinion will integrate all conclusions from the 4 RAs with the conclusions and recommendations from the scientific report, as an overall assessment of dairy cow welfare. The adoption of the five scientific opinions by the AHAW Panel and subsequent publication on EFSA's website is foreseen for June 2009.

Session 34

Theatre 4

Vines and ovines: using sheep with a trained aversion to grape leaves for spring vineyard floor management

Doran, M.P.[1], George, M.R.[2], Harper, J.H.[3], Ingram, R.S.[4], Laca, E.A.[2], Larson, S.[5] and Mcgourty, G.T.[3], [1]University of California, Cooperative Extension, 501 Texas St., Fairfield, CA 94533, USA, [2]University of California, Plant Sciences, 1 Shields Ave., Davis, CA 95616, USA, [3]University of California, Cooperative Extension, 890 Bush St., Ukiah, CA 95482, USA, [4]University of California, Cooperative Extension, 11477 E Ave., Auburn, CA 95602, USA, [5]University of California, Cooperative Extension, 133 Aviation Blvd., Suite 109, Santa Rosa, CA 95403, USA; mpdoran@ucdavis.edu

Traditional vineyard floor management practices have limitations and potentially undesirable effects. Herbicide applications, mowing and tillage rely on petroleum and can impair soil, water and air quality. Excessive rains that prevent tractor access into the vineyard can delay floor management and affect vine development by allowing vegetation to compete with the vines for soil nutrients and by increasing the risk of frost damage. Sheep grazing is a cultural practice to manage vineyard floor vegetation that is growing in use. Sheep can eliminate the need for herbicides, and they can be used in vineyards rain or shine. The biggest impediment to their use is that sheep like to browse the spring growth of grapevines. Two trials were conducted in 2006 and 2007 at Hopland, California, USA to determine the efficacy of training a grape leaf aversion in sheep and using those sheep for grazing vineyard floor vegetation. The first trial tested the persistence of a grape leaf aversion induced by orally administering lithium chloride (LiCl) to sheep. In a second trial vineyard plots were grazed with trained and untrained sheep. Aversion testing indicated that strong aversions can persist over 9 months. The grazing trial showed that trained sheep had almost no impact on the vines while untrained sheep removed an average of 50% of all vine shoot material. Sheep with a dietary aversion to grape leaves will extend the time sheep can graze in vineyards through the spring months when floor vegetation grows most vigorously.

Once daily milking and feeding level combined effects on goat welfare
Komara, M.[1], Giger-Reverdin, S.[2], Marnet, P.G.[1], Roussel, S.[2] and Duvaux-Ponter, C.[2], [1]INRA/Agrocampus-Ouest, UMR 1080, Domaine de la prise, 35590 Saint Gilles, France, [2]INRA-AgroParisTech, UMR 791, 16, rue Claude Bernard, 75231 Paris, France; moussa.komara@rennes.inra.fr

40 goats (Alpine and Saanen) in late lactation were housed in individual pens, milked twice a day (TDM) and fed ad libitum for 3 weeks. They were then allocated to two balanced groups according to their milk production and their parity. One group was milked only once a day in the morning (ODM) for 1 week (period P1). Each group was then divided into 2 other balanced groups according to their milk production: one group was fed *ad libitum* and the other received a quantity of feed adapted to milk production during 3 weeks (period P2). Twice a week, cortisol plasma concentration was measured as an indicator of a stress response and individual behaviour of each goat was recorded at milking during the first 5 s after teat cup attachment as a measure of welfare. Sixteen goats were also used to determine ODM effects on time-budget by video recording for 23 h once a week. Behaviour was recorded by scan sampling of five seconds every four minutes. During P1 and P2, there were no effects of milking frequency on plasma cortisol level, on the number of goats ruminating, back hunching, foot moving and kicking at milking and on the total time spent either standing or lying. During P2, the time spent eating was reduced by ODM and the number of goats ruminating at milking was lower in the feed adjusted group during the first week but this effect disappeared at the end of P2. This experiment did not demonstrate a modification in plasma cortisol level, behavioural response at milking and time-budget under ODM management.

Management simulation tool for evaluating individual identification of beef cattle
Gaspar, P.[1], Oltjen, J.W.[2], Drake, D.J.[2], Ahmadi, A.B.[2], Romera, A.J.[3], Woodward, S.J.R.[4], Bennett, L.N.[5], Haque, F.[5] and Butler, L.J.[5], [1]Universidad de Extremadura, Escuela de Ingenierías Agrarias, Badajoz, Spain, [2]University of California, Animal Science, Davis, CA, USA, [3]DairyNZ, Newstead, Hamilton, New Zealand, [4]Lincoln Ventures, Ruakura Research Centre, Hamilton, New Zealand, [5]University of California, Agricultural and Resource Economics, Davis, CA, USA; Jwoltjen@ucdavis.edu

This paper presents a decision support tool that will permit cattle producers to see how they may benefit when individual animal identification and records are used on their ranch. UC Davis has employed an object-oriented approach based on modifying an existing cow-calf simulation model (CCFARM) to simulate and evaluate the real management changes/options a rancher could adopt if he had individual animal identification. The resulting model has been embedded into a decision support tool called PCRANCH. PCRANCH consists of three components. The first component is the input interface which allows the user to specify the range (physical characteristics of the farm), herd (animal numbers and types), block (land allocation), weather (climate data), and management parameters of the farm. The second component is the run interface which launches the simulation, and the third component is the output interface which allows the user to view a series of reports and graphs, generated from the output files of the CCFARM simulation engine. In this way, PCRANCH makes it easy for ranchers to assess the advantages and disadvantages of identification assisted management systems projected into the future. The approach of wrapping a detailed research model in a user-friendly interface has great potential to make other research-oriented simulation programs accessible to general users.

The role of livestock and precision grazing for controlling noxious weeds invading annual rangelands

Doran, M.P.[1], Becchetti, T.A.[2], Larson, S.R.[2], George, M.R.[2], Cherr, C.[2], Kyser, G.B.[2], Ditomaso, J.M.[2], Harper, J.M.[2], Davy, J.[2] and Laca, E.A.[2], [1]University of California, Cooperative Extension, 501 Texas St., Fairfield, CA 94533, USA, [2]University of California, Plant Sciences, 1 Shields Ave., Davis, CA 95616, USA; mpdoran@ucdavis.edu

California's Mediterranean climate predisposes its natural ecosystems to invasion by plants from other regions of the world with similar climatic conditions. California, and many other states, struggle to maintain healthy rangeland plant communities that support sustainable livestock production, wildlife habitat and recreation for a growing population. Medusahead, Taeniatherum caput-medusae, is an invasive annual grass that has invaded millions of acres of rangelands in the western United States. Medusahead degrades the whole ecosystem, reducing biodiversity, commercial grazing, wildlife habitat, and recreation value of rangelands. We investigated the use of high-intensity precision grazing by sheep to control meduashead in California annual grasslands. We also evaluated the loss of biodiversity and forage value of land as a function of level of infestation prior to and after the control of medusahead. Well timed high-intensity grazing successfully reduced medusahead infestations approximately 90% compared to ungrazed areas and increased forb cover, native forb species richness and overall plant diversity. Quality of medusahead declined significantly with advancing phenological stage. However, the reduction in grazing value appears to have a strong behavioral component, as livestock reject this species after the boot stage. Calculated losses in grazing value depended on behavioral response to scale of infestation. Rangeland managers need to be well informed of the ecological and economical impacts of no management changes, as well as the application of precision grazing and other management tools that could be applied to their specific rangeland situation.

Spanish Ministry of the Environment and Rural and Marine Affairs project for the IPPC Directive implementation in Spain: results of 2003-2008 and future work

Pineiro, C.[1], Montalvo, G.[2], García, M.A.[2], Herrero, M.[3], Sanz, M.J.[4] and Bigeriego, M.[5], [1]PigCHAMP Pro Europa, Gremio de los Segovianos, 40195 Segovia, Spain, [2]Tragsega, S.A., Julian Camarillo, 28037 Madrid, Spain, [3]Feaspor, Coches, 40002 Segovia, Spain, [4]CEAM, Charles Darwin, 46980 Valencia, Spain, [5]Spanish Ministry of the Environment and Rural and Marine Affairs, Alfonso XII, 28071 Madrid, Spain; carlos.pineiro@pigchamp-pro.com

During the last decade the approach to environmental issues related to animal production is changing, by means of including concepts such as emissions from soil, water and air pollution, and more efficient use of energy and water resources. Latest regulations have been developed under this concept, such as the Directive 96/61/EC concerning integrated pollution, prevention and control (IPPC) in intensive pig and poultry production. In the EU Reference Document (BREF) on Best Available Techniques (BAT) for Intensive Rearing of Poultry and Pigs, several techniques were proposed for emissions abatement. In 2003, a group of Spanish expert, under the financing of the Spanish Ministry of the Environment and Rural and Marine Affairs (MARM), implemented a plan to evaluate the BAT proposed by the BREF under Spanish management systems and climatic conditions. The BATs selected were assessed for the pig and poultry sectors in the different production phases: laying hens, broilers, gestating sows, lactating sows, nursery, growers-finishers, manure storage and manure spreading. Cost calculations were also carried out according to the methodology suggested in the BREF. All the results obtained from different experiments performed were updated in the Spanish Guide Document, both technical and economical; were included in a software, developed for the MARM, to calculate pollutant emissions, resources consumption and BAT effects on emissions and consumptions from Spanish farms; and have been transferred to technicians and farmers. On going activities are focused on extend studies at bovine sector (cows and calves) and to include this species in the software tool.

IPPC best available techniques assessment under Spanish conditions

Pineiro, C.[1], Montalvo, G.[2], Garcia, M.A.[2], Herrero, M.[3] and Bigeriego, M.[4], [1]PigCHAMP Pro Europa, Segovianos, 40195 Segovia, Spain, [2]Tragsega, Julian Camarillo, 28037 Madrid, Spain, [3]Feaspor, Coches, 40002 Segovia, Spain, [4]Spanish Ministry of the Environment and Rural and Marine Affairs, Alfonso XII, 28071 Madrid, Spain; carlos.pineiro@pigchamp-pro.com

Integrated Pollution Prevention and Control Directive (IPPC) are aimed to decrease emissions and to save resources (water and energy) through the promotion of the Best Available Techniques (BAT). Spanish Ministry of the Environment and Rural and Marine Affairs, implemented a plan to evaluate the BAT proposed under Spanish management systems and climatic conditions. The studies were performed in commercial farms. The concentrations of NH_3 and CH_4 were monitored with an infrared photo acoustic multi-gas monitor, and the ventilation rate in each room was registered. Measurements were performed in central Spain plateau to evaluate the effect of BAT (2004-06), and in Pyrenees climate (2007), and in Mediterranean climate (2008) to measure the emission values of the reference techniques. Results of the efficiency of BAT effectiveness were as follows: in gestating sows, reduced manure pit was able to reduce NH_3 and CH_4 emissions by 49 and 28% respectively. In lactating sows, manure pan underneath reduced NH_3 and CH_4 emissions by 32 and 65% respectively. In nursery, frequent manure removal, manure channel with sloped side walls, and low protein diets were highly effective in the reduction of emissions: 24, 51 and 63% for NH_3, and 10, 65 and 63% for CH_4. In growers-finishers, frequent manure removal, manure channel with sloped side walls, partially slatted floor and low protein diets were effective in decreasing NH_3 (10, 36, 42 and 60%), and CH_4 (65, 52, 34 and 33%) emissions. The information provided is being used by farmers and technicians in order to understand better the effect of BAT and to promote their use in the production sector.

Productivity of the laying hybrids reared in different husbandry systems

Usturoi, M.G., Boisteanu, P.C., Radu-Rusu, R.M., Pop, I.M., Dolis, M.G. and Usturoi, A., University of Agricultural Sciences and Veterinary Medicine, Animal Science Faculty, 8 Mihail Sadoveanu Alley, 700489 IASI, Romania; umg@univagro-iasi.ro

The pressure exerted by the animals welfare organizations led to the establishment of certain new rearing systems for laying hybrids. However, these new systems do not always provide the optimal conditions for expressing the best yielding potential of the hens. The biological material comprised 4,698 Lohmann Brown hybrids, randomly allocated in 5 groups: a control group (Lc), which comprised hens reared within classical cages battery (500 cm^2/hen) and 4 experimental groups: L1exp (rearing in modified battery=1,000cm^2/hen); L2exp (rearing in opened panels batteries=500 cm^2 in the nesting+resting cage and 500cm^2 in the cage with feeding and water devices); L3exp (rearing on floor, permanent layer=0.17 m^2/hen) and L4exp (rearing on floor, permanent layer=0.13 m^2/hen and access to an external paddock =2.0 m^2/hen). During the 60 weeks of laying, the fowl in the classical battery (Lc) achieved a production of 325.05 eggs/hen which was 2.68-15.89% higher than those of the experimental groups. The yield level generated the feed conversion ratio values, which were 6.89-38.32% lower in Lc, comparing with the experimental groups. Casualty incidence was influenced by the amount of hens per surface unit, reaching just 7.46-11.61% in the experimental groups, comparing to 11.66% in the Lc group. The superintensive system (classical cages batteries) provides to the hybrids the better technological conditions, materialized in higher yield responses. Although the other rearing alternatives provide better welfare conditions, they also decrease the technical performances that could be achieved on the surface built unit.

How useful is the cast antler for determination of deer population status? Inferences from farm herds
Gallego, L., García, A.J., Landete-Castillejos, T., Gaspar-López, E., Olguín, C.A., Ceacero, F., López-Parra, J.E. and Estevez, J.A., Universidad de Castilla-La Mancha, Campus UCLM s/n Albacete, 02071 Albacete, Spain; jose.estevez@uclm.es

Desirable traits of domestic or captive-bred animals may be modulated by selective breeding through selection and management, culling undesirable attributes. A key point is to define the breeding objectives formulated on target market requirements for which the animals are being produced. On deer, these traits may include antler size, velvet production, growth rates or reproductive success. A higher weight and leanness are desirable traits for venison demands. However, the Spanish market is mainly interested on trophy values, and their improvement is a must. Despite c. 70,000 Iberian red deer (Cervus elaphus hispanicus) are shot every year in Spain, only few of them (<2%) are classified by the Official Trophy Board as quality trophies. Consequently, managers seek for useful, easily applicable diagnostic indices of health and condition state of deer herds. For technical baseline purposes, factors that affect the yearly renewable antlers must be properly assessed. The classical approach has been to get scientific data for studies supporting game management on observations (labour costly), and obtaining samples from hunted individuals (biased dataset). We have developed a technique consisting of using the mineral profile, bone structure variables and mechanical properties of cast antlers to assess physiological effort made to grow the antler, mineral deficiencies, and even indirect climatic influences. Differences observed among populations pointed out mineral deficiencies in the wild even under optimal management (e.g. Cu, Co, Mo, Mg, Na, Se and Sr) but also effects caused by improper management (e.g. excessively high content of S), which in turn affected the trophy quality. We have also observed that presence of some minerals (e.g. Se) in antlers is influenced by age (sub-adults/adults), because some minerals may be shifted to physiological functions which prevail over antler growth.

Characterisation and typology of Awassi and Assaf dairy sheep farms in the NW of Spain
Milán, M.J.[1], Caja, G.[1], González, R.[2] and Fernández, A.M.[2], [1]Universitat Autònoma de Barcelona, G2R, Campus de la UAB, 08193 Bellaterra, Spain, [2]Cargill SLU, Ctra. Pobladura Coomonte, 49780 Benavente, Spain; MariaJose.Milan@uab.cat

Awassi and Assaf sheep are important dairy sheep (more than 900,000 ewes) in Castilla-León (NW of Spain) where compete with dairy sheep local breeds. Aiming to study the structure and performance of the Awassi and Assaf farms a typology analysis was done. Data of 69 farms (70% Assaf, 30% Awassi) included in a technical advising network were used. Results showed that farms were fully oriented to dairy, although 25% sold other products, mainly cereals. Moreover, 25% farms did not have crop land and 75% had 55.4 ha, on average (pastures, 19%; cereals, 26%; forages, 48%). Farmers were tenant (84%) and young (<45 yr, 70%), had new houses and used 2.1 annual work units per farm (familiar, 90%). Flocks had 493 ewes, yielded 309 L/ewe-year (fat, 6.5%; protein, 5.3%), which was sent to dairy industries for cheese, and 1.35 lambs/ewe (10.4 kg/lamb). All farms had permanent shelters (only occasional grazing), milking machine, and 91% planned ewes matting. A 68% of the farms bought >50% forages, and 87% of them bought >50% concentrates. Artificial rearing was done in 38% of farms. Total mixed rations were used in 33% of farms, and the rest used forage and separated concentrate. Annual income was 276.7 €/ewe (milk, 83%; lambs, 17%). Estimated annual gross margin, was 179 €/ewe. According to results, 3 types of farms were differentiated. The first group were farms owned by young farmers with higher education level, the size of the flocks was medium and had the smallest land, being dependent of external resources but had high ewe productivity. The second group were farms with large flocks, producing and consuming their own forage. The third group was constituted by farms with small flocks and large crop land, producing forage and cereals used for consumption and selling, and with a low dependency of external resources; in this group a 33% of farms decreased flock size in the last years.

Productive results of primiparous rabbit does with different live weight at first insemination

Pinheiro, V., Outor-Monteiro, D., Lourenço, A. and Mourão, J., Universidade de Trás-os-Montes e Alto Douro, Animal Production, P.O. Box 1013, 5001 801 Vila Real, Portugal; analou@utad.pt

This study was performed to evaluate the relationship between body weight at first insemination and growth, feed intake, reproductive performance, and fertility rate of rabbit does. The ninety does used were fed ad libitum. At 17 weeks of age the does were inseminated and split, according to their body weight, among three groups of 30 animals each: Heavy (H; 3,900±97g), Medium (M; 3,661±54g), and Light (L; 3,554±98g). At kindling litters were equalized in number. Kits were weaned at 35 d of age. Does, kits and feed were weighted at the insemination, kindling, 18 d after kindling and weaning. The data from non-pregnant does were only considered for fertility rate. Data were analysed using the ANOVA procedures and Tukey test for means comparison. Does fertility and kits mortality rate were performed using c^2 test. The does body weight differed ($P<0.05$) at insemination, kindling and weaning, but the daily weight gain between insemination and weaning was similar ($P>0.05$; 10g/day) among groups. The feed intake of does was lower ($P<0.05$) in L group than in M and H groups (421, 466 and 463 g/d respectively) during the entire lactation period. The L group had higher fertility rate than H group (96.7 vs. 80.7%) but these results were not statistically different ($P>0.05$). Kits mortality during the first 18 days of life differ ($P<0.001$), was higher in L group and lower in H group (28.1, 19.1 and 8.3% in L, M and H group, respectively). A low body weight at first insemination did not affect the litter size (9.8 kits), nor the weight of kits (48g). The litter weight did not differ at kindling, but after 18 d of life the litter weight was lower in L group than M and H group (2,008, 2,394 and 2,370 for L, M and H does, respectively, $P<0.05$). The results suggest that, under an intensive rearing system, the body weight of rabbit does at first insemination can have a positive effect on productivity, measured by growth and mortality of kits, but their fertility decreases.

Effect of two management systems of piggery waste slurry on biogas yield

Moset, V.[1], Cambra-López, M.[2], Moya, J.[2], Lainez, M.[1], Ferrer, J.[3] and Torres, A.G.[2], [1]Instituto Valenciano de Investigaciones Agrarias, Centro de Tecnologia Animal, Poligono de la Esperanza s/n, 12400 Segorbe Castellon, Spain, [2]Instituto de Ciencia y Tecnologia Animal, Camino de Vera s/n, 46022 Valencia, Spain, [3]Instituto de Ingeniería del Agua y Medio Ambiente, Camino de Vera s/n, 46022 Valencia, Spain; moset_ver@gva.es

Methane producing capacity of by-products depends on characteristics of the substrate. The characteristics of livestock manure substrates are very variable. Furthermore, livestock manure management is an important factor influencing these characteristics. The objective of this study was to determine the composition and biogas yield of two pig slurries: one from slatted floor storage; and the other from a separation of solid and liquid fraction. To determine slurry composition, organic matter (DQO and DBOlim), nutrients (nitrogen and phosphorus) and solids (total and volatile) content of the soluble and suspended fraction was analyzed. To determine biogas yield, anaerobic biodegradability test in hermetically closed vessels was done. Biogas was measured by manometric methods and methane by gas chromatography. Organic matter, nutrients and solids content was lower in slurries from the separation treatment than from slatted floor storage. Biogas yield, expressed per amount of DQO added and the percentage of methane of the total produced biogas was also lower in these slurries. Others studies have demonstrated that a separation treatment can improve biogas production in the liquid fraction. However, anaerobic digestion was done immediately after the separation treatment in most of these studies. In farm conditions, it is difficult to synchronize anaerobic digestion and separation treatment in time, thus microflora can consume easily biodegradable organic matter in the slurries. Separation of solid and liquid fraction can improve anaerobic digestion in slurries whenever it is done immediately after the separation treatment; otherwise a fast degradation of organic matter can occur.

The competitive and sustainable stockman is labor efficient

Bostad, E. and Swensson, C., Swedish University of Agricultural Sciences, Rural Buildings and Animal Husbandry, P.O. Box 59, 230 53 Alnarp, Sweden; elise.bostad@ltj.slu.se

Facing a changing climate and fluctuating prices on the world market of agricultural products, a high competitiveness of the farmer is urgent to sustain in the business. With increased costs for labor and other accessible resources, planning and organization of labor use is an important tool for development within the sector. The aim of the study was therefore to enhance the sustainability of beef production through systems for optimal farm logistics and efficient use of labor. By higher work efficiency we also expected an effect of higher utilization of resources such as feed, litter, buildings, land and fuels. Through in-depth field studies during spring 2008 and spring 2009 on 30 farms producing barley beef calves and/or young bulls the labor use during predefined work task were investigated. Questionnaires sent out to all Swedish farmers specialized in young beef cattle production were complementing the field-studies. The predefined work tasks focused on the handling of animals, feed, litter and manure including cleaning procedures. High focus was put on efficient management routines for optimal calf health in the quarantine area for newly purchased calves. Preliminary results show that systems used for beef cattle production in Sweden are very diverse in terms of herd size, buildings, level of mechanization and fragmentation of farm units. The total work time per calf was affected by herd size, housing system and level of mechanization, as well as the frequency of the work task. Feeding, strewing of litter and weighing of calves were work tasks with major effects on the total time. Improvements within these areas were found to be highly achievable. It is so far concluded that a reformation of labor use in the Swedish beef cattle production is of large importance to meet the increasing national and global demand for meat. The complete results including the last field studies are expected to be presented during 2009.

Sustainable ostrich production system in Romania

Ciocîrlie, N.[1] and Kremer, V.D.[2], [1]Spiru Haret University, Faculty of Veterinary Medicine, 47 Masina de Paine Street, 021127 Bucharest, Romania, [2]Genus/PIC, The Roslin Institute, Roslin, EH25 9PS, United Kingdom; valentin.kremer@pic.com

Ostrich production is a relatively new industry in Romania. However, at about 50 km from Bucharest one may find probably the largest integrated ostrich production system in Eastern Europe. Part of the costs required by this large investment were covered by European funds allocated to Romania during the EU-joining process (SAPARD Program).The system includes: a modern incubator (3,000 eggs capacity), a breeding /grower /finisher farm (20,000 heads per year), and ostrich slaughter house (300 birds per day), a meat processing plant, a skin processing plant, sewage treatment station and incinerator. A farming area of about 300 ha provides all the needed feed for the birds, creating a fully integrated system. This paper is a case study on the specifics of ostrich production in Romania, the costs involved by the investment in an ostrich farm, the technological characteristics of an ostrich slaughter house and the social impact of the industry on the area, as well as presenting elements of marketing of ostrich meat locally. The study highlights the benefits of breeding ostrich: the successful use of degraded agricultural land, the reduced poluting impact on the environment, the high nutritional value of ostrige meat for human consumption, and the added value of the wide range of by-products that complement the meat production for increased efficiency. This is a clear example of a sustainable livestock production system.

Influence of the pluviometry in the pasture production in 'Dehesa' extensive system

Espejo Díaz, M., Espejo Gutiérrez De Tena, A.M., González López, F., Prieto Macías, P.M. and Paredes Galán, J., Centro de Investigación Agraria La Orden - Valdesequera, Producción Forestal y Pastos, Apartado 22, 06080 Badajoz, Spain; manuel.espejo@juntaextremadura.net

The 'Dehesa' is an ecosystem constituted by pasture and trees that extends by the Southwest of the Iberian Peninsula. The Mediterranean climate that conforms it has a very variable pluviometry between stations and between years, being rains the factor that more influences the production of pasture. The relations between both parameters have studied in order to quantify they influence. For it we have considered the production of dry matter data (kg DM/ha) collected in the project 'Montado/Dehesa II SP4.E127/03 (INTERREG IIIA)' in 48 pilot farms during 4 years measured by the method of grazing exclusion cages. The rain data of near stations came from the National Agency of Meteorology. The correlations have been studied using the SPSS program. The results with 181 data display an average pasture production of 2.347 kg DM/ha, although with great variations between years, types of pasture and zones. The average digestibility coeficient of pasture in 'Dehesa' is 0.55, then the annual energetics production is 3,505 Mcal ME/ha (2.026 maintenance ewe rations/ha). The relations study shows that exists a high significant correlation (R^2=0.551, **=P<0.01) between the total pluviometry and the total production of DM. The autumn pluviometry shows a significant correlation with the autumn and winter production (R^2=0.469**), and March, April and May rain influence on spring production (R^2=0.234**). The winter pluviometry is the best indication of the, expected total pasture production (R^2=0.618**). The September pluviometry is the best indication of autumn and winter pasture production (R^2=0.581**) and those from March is the best indication of spring pasture production (R^2=0.479**).

ALCASDE project: study on the improved methods for animal-friendly production in particular on alternatives to the castration of pigs and on to the dehorning of cattle

Oliver, M.A.[1], Mirabito, L.[2], Veissier, I.[3], Thomas, C.[4], Bonneau, M.[5], Doran, O.[6], Tacken, G.[7], Backus, G.[7], Cozzi, G.[8], Knierim, U.[9] and Pentelescu, O.N.[10], [1]IRTA, Monells, 17121, Spain, [2]IE, Paris, 75595, France, [3]INRA, St.Genès Champanelle, F63122, France, [4]EAAP, Rome, 00161, Italy, [5]INRA, St. Gilles, 35590, France, [6]UWE, Bristol, BS 161QY, United Kingdom, [7]LEI, The Hague, 2585 DH, Netherlands, [8]UNIPD, Legnaro, 35020, Italy, [9]UKA, Kassel, 34109, Germany, [10]UASVM, Cluj-Napoca, 400372, Romania; mariaangels.oliver@irta.es

ALCASDE is an European project with the aim of developing and promoting alternatives to the surgical castration of pigs and the dehorning of cattle. Subproject 1 will examine Alternatives to the surgical castration of pigs; Methods to detect boar taint at the slaughter line; Demand and Acceptance of consumers; and Integration of knowledge by economic modelling. The research will produce up to date evidence of consumer attitudes and acceptability of the meat from entire males as well as develop a rapid, simple and cost-effective method(s) to be applied on line to detect the major boar taint compounds, androstenone and skatole. A meta-analysis of available information regarding welfare and meat quality of entire and immunocastrated pigs will also be carried out. Subproject 2 will consider the state of the art of dehorning in the EU member states; Assessment of benefits and drawbacks of dehorning and alternatives to dehorning in dairy and beef cattle; and short- and long- term strategies for future development. Evidence will be produced of dehorning practices and farmer's attitudes across the EU member states and alternatives to cattle dehorning will be considered. The final aim of both subprojects is to provide the DG SANCO with recommendations that will support EU policy, as well as the competitiveness of animal friendly-high welfare production systems in Europe. This will be achieved on solid science and by the involvement of key stakeholders across the food chain to ensure that the results can be rapidly adopted in practice.

Dimensional features of the skeletal myocytes at the laying hens reared within different accommodation systems

Radu-Rusu, R.M., Usturoi, M.G., Usturoi, A., Boisteanu, P.C., Pop, I.M. and Dolis, M.G., University of Agricultural Sciences and Veterinary Medicine, Animal Science Faculty, 8 M. Sadoveanu Alley, 700489, Iasi, Romania; rprobios@gmail.com

The EU regulations in poultry welfare impose modification of the classical housing systems applied in laying hens exploitation. The study goal was to assess the dynamics of the somatic myocytes thickness at the hens reared in certain versions of accomodation systems. 1731 'Lohmann Brown' hens were used, divided in four groups: control Lc-conventional cages, L1exp.-enlarged cages (600 cm^2/hen), L2exp.-enlarged cages (1000 cm^2/hen) and L3exp.-conventional cages with opened front panels (summarized 1000 cm^2/hen in nesting+resting cage and feeding+watering cage)+free movement on the floor. 15 individuals/group were selected to sample tissue from four muscles: Pectoralis superficialis, Biceps brachialis, Quadriceps femoris and Gastrocnemius lateralis. The histological smears were examined by photonic digital microscopy, measuring the small and large diameters of the myocytes. The average thickness of these cells was calculated. Statistical analysis has been run by the ANOVA single factor method. Histometry revealed differences between the hens reared within classical and alternative housing systems. The myocytes average diameter in P. superficialis muscles varied between 39.24μ-Lc and 44.81μ-L3exp. The L2exp. had the thickest fibers in wings muscles-31.47μ, while the thinnest ones were noticed in Lc group-29.73μ. In rear limbs muscles the L3exp. hens had the thickest myocytes (40.62μ-Q. femoris, respectively 36.21μ-G. lateralis), while the thinnest myocytes in thighs and shanks have been measured at Lc group. Very significant differences occurred for P. superficialis myocytes diameter and distinguished significant ones for rear limbs muscles cells, between the fowl accommodated in classical cages and the L3exp group. The results suggest the existence of a relation between the freedom of movement and the hypertrophy of the skeletal musculature.

Education can improve broiler performance traits

Zarqi, H., Zakizadeh, S. and Ziaee, A., Hasheminejad High Education Center, Animal Science, Kalantari Highway, 9176994767, Iran; sonia_zaki@yahoo.com

Significant improvements in the performance of commercially reared broilers have been made during the last half of the twentieth century. The changes are believed to have resulted from genetic selection for faster growth, improved nutrition and flock health, and better management practices. In this survey we aimed to investigate the roll of human management on production performance of broiler flocks in northeast of Iran. Ninety four flocks were chosen randomly and their performance during summer of 2007 filled out in systematic questionnaires. Management information included farmer age, related education, and practical background of farmer in poultry science, short time education, having technical manager or not. Production performances such as livability, slaughter age, production unite, feed conversion, average weight, production per square meter, and daily gain (g) were measured. Data were analyzed in a standard least squares model with slaughter age as covariate variable. Results showed that related education had significant effect on livability ($P<0.05$) and slaughter age ($P<0.01$). Only twenty six percent of farmers had education over high school. Farmers who had master degree could maximize the livability percent and minimize the slaughter age. Also, farmers who took part in short time education training could increase the livability range ($P<0.05$). Having a technical manager could influence on slaughter age ($P<0.05$), so that slaughter age was increased in the case of not having a technical manager. Slaughter age could affect on livability, production unite, feed conversion, average of weight, and daily gain (for all traits; $P<0.01$). Feed conversion, average of weight had positive relationship with slaughter age, but livability, production unite, and daily gain were decreased by increasing of slaughter age. It seems that knowledge of related education, took part in short time training, and having a technical manager could have an important roll in improvement of production performances in broiler flocks.

Assessing the effectiveness of revegetation methods for mountain grazing resources preservation in the Montseny Biosphere Reserve (NE Spain)

Madruga-Andreu, C., Bartolomé, J. and Plaixats, J., Universitat Autònoma de Barcelona, Ciència Animal i dels Aliments, Edifici-V. Campus UAB, 08193-Bellaterra (Cerdanyola del Vallès), Spain; josefina.plaixats@uab.cat

High mountain grazing resources are declining in Europe as a result of climate and environmental changes and the impacts of tourism activities such as hiking, skiing and others recreational uses. Revegetation process in degraded systems is important in severe environments such as high-altitudes zones, where unassisted recovery of the plant cover can be a slow process due to short growing seasons and harsh climatic conditions. This study aims to evaluate the effectiveness of three revegetation treatments in high mountain pastures of the Montseny Biosphere Reserve in relation to feed supply, nutritional value and species diversity of the communities established after restoration. 2.5 ha of degraded pastures were revegetated during the spring of 1999, using a commercial seed mixture (30% *Lotus corniculatus*, 30% *Achillea millefolium*, 20% *Lolium perenne* and 20% *Festuca arundinacea*) and three different techniques: hand sowing, regular hydro-seeding, hydro-seeding and post-transplanting single native species of Festuca gautieri. Biomass production, chemical composition, nutritional quality and diversity indices in the revegetated samples and adjacent natural pastures were measured and compared seven years after restoration. Native pasture and the two hydro-seeding treatments presented the highest productivity (ranging from 1,800 to 2,500 kg DM/ha) and nutritional quality (ranging from 0.75 to 0.78 UFL/ kg DM). In terms of biodiversity, native pastures showed greater species richness (23) than treatments ($P<0.001$). However, the Shannon-Weaver index was significantly lower only in hand sowing treatment (1.72) compared to natural pastures (2.26). Taking into account productivity, nutritional quality and biodiversity results, the revegetation method with native fescue transplants was the most effective to restore high mountain pastures.

Cattle preferences for two forage tropical trees, *Calliandra calothyrsus* and *Erythrina berteroana*

López, K.[1], Pol, A.[1] and Bartolomé, J.[2], [1]Facultad Regional Multidisciplinaria de Estelí Leonel Rugama Rugama, Universidad Nacional Autónoma de Nicaragua en Estelí, Estelí, Nicaragua, [2]Universitat Autonoma de Barcelona, Ciencia Animal i dels Aliments, Edifici V, 08193 Bellaterra, Spain; jordi.bartolome@uab.es

Multiple food options with a simultaneously available test (cafeteria test) were carried out in order to discover cattle preferences for two forage trees (*Calliandra calothyrsus* and *Erythrina berteroana*) which were unknown by the animals in relation to a well know grass (*Paspalum notatum*). The study was performed with four female criollo cattle breed in a state placed at 1,200 m a.s.l. in the dry tropical region of North West Nicaragua. During a period of 10 days animals were placed in individual pens of 2 x 4 m and daily offered 0.5 kg of fresh forage of the three species simultaneously. Individual animals were tested separately and in turn, each one was allowed to consume the material undisturbed for 15 minutes. All groups of forages were weighted before and after being exposed to cattle in order to calculate the plant proportion consumed. After the feeding trial, the animals were reincorporated into the farm herd and they spent 8 hours grazing in Paspalum fields. A preference index (IP) was calculated using the formula: IP = Ci/SCn (Ci is the i species consume and SCn is the total forage consumed). Results showed a higher IP values for Calliandra and Erythrina (0.47 and 0.32 respectively) in relation to Paspalum (0.22). The mean forage browsed was higher ($P<0.05$) in the case of Calliandra (274±25 g) respect to Erythrina (128±22 g) and Paspalum (187±25 g). It is concluded that Calliandra could be an interesting option (better than Erythrina) in order to complement cattle diets during the dry season in tropical areas.

Effect of different feeding strategies on animal welfare and meat quality in Uruguayan steers

Del Campo, M.[1], Manteca, X.[2], Soares De Lima, J.M.[1], Hernández, P.[3], Brito, G.[1], Sañudo, C.[4] and Montossi, F.[1], [1]Instituto Nacional de Investigación Agropecuaria, Ruta 5 Km 386, 45000 Tacuarembó, Uruguay, [2]Universidad Autónoma de Barcelona, UAB, 08193 Bellaterra, Spain, [3]Universidad Politécnica de Valencia, Camino de Vera s/n, 46022 Valencia, Spain, [4]Universidad de Zaragoza, Miguel Servet 177, 50013 Zaragoza, Spain; mdelcampo@tb.inia.org.uy

Animal welfare determinations should integrate productive, physiological and behavioural indicators. Eighty Hereford steers backgrounded on pasture were finished on one of the following diets: T1: pasture (4% of animal live weight: LW), T2: pasture (3% LW) plus concentrate (0.6% LW), T3: pasture (3% LW) plus concentrate (1.2% LW), and to an ad libitum concentrate treatment, T4, to study the effects on animal welfare and meat quality. Productivity increased with the level of energy in the diet. Differences in average daily gain (T4>T3>T2>T1) were due to the energetic composition of the diet. Animals from T4 had the best performance but their health was compromised, a longer habituation process was evident through physiological indicators (acute phase proteins, APP, and fecal glucocorticoids metabolites) and a higher mortality rate was registered in that treatment. Animals from T4 also showed the highest values of preslaughter APP and behaviour was restricted throughout the experiment. It seems that T1, T2 and T3 would not compromise animal welfare, but special considerations must be taken into account with intensive feeding systems, especially preventing dietary diseases and animal deaths. Temperament had a significant effect on live weight. Regardless of the feeding strategy, temperament appears to be an important factor considering its influence on productivity and also on final pH values and meat tenderness. Stressors that lower muscle glycogen pre-slaughter had a significant effect on meat tenderness and animals with a more excitable temperament were clearly more susceptible, showing higher shear force values. Meat tenderness was enhanced in the pasture treatment.

Comparative reproductive performance of two captive collared peccary populations in the amazon region

Mayor, P. and Lopez-Bejar, M., Universitat Autonoma de Barcelona, Faculty of Veterinary, Edifici V, Campus Bellaterra, 08193 Bellaterra, Spain; manel.lopez.bejar@uab.es

Collared peccary (*Tayassu tajacu*) is a candidate for captive breeding programs because benefits for the local economy of meat and pelts trade in Latin-America. Reproductive performance of captive collared peccary populations in the Amazon region was analyzed from two experimental farms: under semi-extensive conditions (Peru) for 56 months (28 females and 77 parturitions), and under semi-intensive conditions (Brazil) for 65 months (29 females and 74 parturitions). Under semi-extensive conditions, parameters were: 28.5 months at first parturition; mean parturition–conception of 143 days, 0.96 births per year and female, mean litter size of 1.58 piglets, and yearly reproductive production of 1.37 piglets per female. Under semi-intensive conditions, first parturition was at 21.3 months, mean parturition–conception 58 days, mean production 1.03 births per female with a litter size of 1.85 piglets per birth, and yearly reproductive production was 1.86 newborns per female. Although both captive populations presented similar inefficient factors, the collared peccary shows interesting reproductive parameters for its introduction in captive breeding programmes in the Amazon region.

Hierarchical structure affects reproductive function of the captive collared peccary

Lopez-Bejar, M., Mayor, P., Coueron, E., Jori, F. and Manteca, X., Universitat Autonoma de Barcelona, Faculty of Veterinary, Edifici V, Campus Bellaterra, 08193 Bellaterra, Spain; manel.lopez.bejar@uab.es

The aim of this study was to study the relationship between dominance index status and reproductive function of collared peccary (Tayassu tajacu) females. Twelve collared peccary females were kept in captivity on an experimental farm (French Guyana). Collared peccaries were enclosed in groups of 3 individuals with an average occupied space of 1.52 m^2 per animal. During an experimental period of 90 days, both behavioral and reproductive data were simultaneously collected. Agonistic encounters were recorded and a social dominance index was assigned to each female per group. Sexual cyclicity of females was determined through the study of fecal progesterone. All experimental groups presented at least a cycling and a non-cycling female. All dominant females were cycling, and all but one subordinate female were non-cycling. This study suggests that there exist a strong hierarchical structure in groups of collared peccary females that largely affects the reproductive function of the females.

Organic bedding materials and their thermo-technical properties in different climate conditions

Lendelova, J., Mihina, S. and Pogran, S., Slovak University of Agriculture, Department of Building, Tr. A. Hlinku 2, 949 76 Nitra, Slovakia (Slovak Republic); Stefan.Mihina@uniag.sk

Thermo-technical properties of organic materials (straw, sawdust, separated slurry with thickness of 200 mm on concrete base) in comparison to rubber mats and mattresses used for bedding of cubicles for dairy cows were evaluated. Thermal resistance and thermal effussivity were calculated according to technical standards. Coefficient of thermal conductivity needed for these calculations were obtained in real conditions of experimental farms. Thermal resistance of straw varied from 0.91 to 2.91 m^2.K.W^{-1}, that of wooden sawdust from 0.63 to 1.724 m^2.K.W^{-1}, of separated slurry from 1.01 to 2.28 m^2.K.W^{-1}, and of rubber mattresses and mats from 0.76 to 1.31 m^2.K.W^{-1}. Data from thermal effussivity of straw ranged from 162.34 to 423.63 Ws$^{1/2}$m^{-2}K^{-1}, of wooden sawdust from 245.43 to 419.16 Ws$^{1/2}$m^{-2}K^{-1}, of separated slurry from 308.97 to 469.36 Ws$^{1/2}$m^{-2}K^{-1}, and of rubber mattresses and mats from 144 to 552 Ws$^{1/2}$m^{-2}K^{-1}. Data were collected both in summer and winter periods and using both dry and wet organic materials.

Incidence of winter feeding, parasitism, and housing on the growth performances of dairy heifers in Wallonia

Picron, P., Turlot, A., Froidmont, E. and Bartiaux-Thill, N., CRA-W, Walloon Agricultural Research Centre, Animal Productions and Nutrition, 8 rue de Liroux, 5030 Gembloux, Belgium; p.picron@cra.wallonie.be

With a replacement rate of 30% in dairy herds, it is essential to ensure the best rearing conditions of young cattle for the development of future dairy cows. The aim of this study was to quantify the incidence of key management aspects such as housing, parasitism and winter feeding on growth performances of dairy heifers. Thirty farmers (774 heifers), located in the main dairy regions in Wallonia, participated to the survey. Development and growth were determined using Heart Girth (HG) measurements and compared with a French reference. Feed intake was estimated from the French Fill Unit System. A prediction equation of BW from HG, related to the genetic features of Walloon dairy herds ($n=256$, $BW=0.0005$ $HG^{2.6161}$, $r^2=0.96$), was developed. The results suggested that 80% of the breeders could pretend to an early calving, at 24-26 month of age. For half of the others, a lack of net energy supply in the diet could explain a growth delay. On the other hand, the protein supply was too high in 90% of the diets compared to the animal's requirement. This was due to the large utilization of grass silage. Ventilation efficiency, estimated by the difference of hygrometry inside and outside the barns, was not optimal in 40% of the farms. This suggested that heifers were often accommodated in less adapted barns. The parasitism management did not differ fundamentally from one farm to another. Globally, heifers were treated systematically (1 to 4 treatments/year) and did not acquire a sufficient immunity at the end of the first grazing period (estimated by blood pepsinogen measurement). They should be treated again for the next grazing season. Knowing that the treatment cost varied between 5 and 21 € in drugs per heifer and per year, a better parasitism management in the future should improve the economical and ecological performance.

Impact of general anaesthesia and analgesia on post–castration behaviour and performance of piglets

Schmidt, T., Koenig, A. and Von Borell, E., Martin-Luther University Halle-Wittenberg, Animal Husbandry & Ecology, Adam-Kuckhoff-Str. 35, 06108 Halle, Germany; tatjana.schmidt@landw.uni-halle.de

Injection anaesthesia with ketamine/azaperone (K/A) is a presumably painless alternative to commonly used non- anaesthetised castration. To protect anaesthetised piglets from being crushed, they have to be separated from the sow for 3h following castration. The aim of this study was to test if this separation would affect behaviour and weight gain (WG). A combination of K/A anaesthesia (Ursotamin®, 25mg/kg; Stresnil®, 2mg/kg) and analgesia (Metacam®, 0.4mg/kg) was used on 5-7 d old piglets (anaesthesia+analgesia: Comb[n=29], analgesia: Met[n=24], control: Cont[n=29]). Behaviour was compared for a 3h period before castration and after reunion. A change in the preferred teat position (TP) occurred in 27.5% of the Comb animals (Met: 16%; Cont: 17.2%) with a smaller number changing to a lower ranked TP (Comb: 10.3%, Cont: 13.8%). None of the piglets receiving Met changed to a lower TP. The significantly higher number of teats used by Comb piglets (Wilcoxon, $P=0.004$) suggests a decrease in suckling order stability (SOS). Suckling duration differed among treatments (ANOVA, $F=13.3$, $df=2$, $P<0.001$) with an increase in Met ($+68.9\pm16.5\%$), but a decrease in Comb piglets ($-27.6\pm11\%$; Cont: $+5.9\pm8.3\%$). Despite these differences, the WG one day after castration showed no treatment effect. Within treatments, the WG was higher after, relative to before, castration in Cont ($P=0.003$) and Met piglets (paired t-test, $P=0.008$), but slightly lower in Comb piglets. The results suggest that Met affects behaviour, perhaps due to less post-castration (p.c.) pain. This advantage is not apparent for animals also receiving anaesthesia, probably because of impaired coordination. Although the behavioural changes did not affect weight gain significantly, a decrease in SOS indicates a certain degree of p.c.-stress due to fighting over teats. Thus, p.c.-behaviour must be taken into account when evaluating alternative castration methods.

Cattle farmers and BVDV: some key points that determine farmers's strategy

Frappat, B.[1], Fourichon, C.[2] and Pecaud, D.[3], [1]French cattle institute, 149 rue de Bercy, 75 595 Paris cedex 12, France, [2]Veterinary School - INRA, Department of Farm Animal Health and Public Health, BP 40706, 44307 Nantes Cedex 03, France, [3]University of Nantes, Risks and vunerabities department, rue Christian Pauc, BP 20606, 44306 Nantes cedx 3, France; brigitte.frappat@inst-elevage.asso.fr

Animal health is a major issue for farmers to get high yields and quality products with low production costs. This is also a professional obligation which engages farmers' responsibility towards pears, consumers and even animals. Yet recent surveys carried out in France showed that preventive attitudes and practices are not so common especially when we look at contagious diseases that are not under regulation such as BVDV. Why? 4 types of factors affect farmers decisions: knowledge, representations (the way they think, the images they have in mind), social and professional networks they belong to, economic and practical constraints. This work, based on more than 80 semi structured interviews explores the 3 first items. The biological knowledge of farmers varies much and may sometimes limit the adoption of safe behaviors. Practical knowledge and above all a previous experience of the disease in the herd turn to be decisive factors in favor of good practices. Representations play a major role through 4 themes that are: 1) the image of a healthy herd (how do farmers define it, how do they care? 2) the sanitary risks (how big seems the danger, what consequences may occur, is it possible and worth to prevent risks? 3) the actors involved in health management (who are the key actors, can they be trusted? 4) sanitary measures (how do farmers understand and consider them?). We also found that to a certain extent, the more farmers take part to technical and social networks the more they are susceptible to adopt good practices. All these elements interfere in complex ways to make final individual decisions. Nevertheless they have to be taken into account in order to better convince farmers to involve in voluntary sanitary schemes.

Integrating parameters to assess animal welfare using multicriteria decision analysis

Krieter, J., Institute of Animal Breeding and Husbandry, Olshausenstraße 40, 24098 Kiel, Germany; jkrieter@tierzucht.uni-kiel.de

Welfare is a multidimensional concept that recommends an integration of information produced by many parameters. Parameters obtained may vary in precision, relevance and also in their contribution to an overall welfare assessment. Often the size of scale of the parameters is different and the intervals between ordinal scales are not equivalent. This can lead to confusing results if common tools (e.g. weighted arithmetic mean, weighted sum) with all their well-known drawbacks are used for aggregation. In the present paper fuzzy integrals are applied for aggregation of the parameters to an overall assessment of animal welfare. Fuzzy measures and integrals are related to the framework of multicriteria decision making. The distinguishing feature of fuzzy integral is that it is able to model the interaction between the parameter (ranging from redundancy to synergy) and to identify the weights of the parameter in the overall score. The method is illustrated with a small example including good feeding, housing, health and appropriate behaviour as parameter (criteria).

Incidence and antibiotic susceptibility of bovine respiratory disease pathogens isolated from the lungs of veal calves with pneumonia in Switzerland

Rérat, M.[1], Albini, S.[2], Jaquier, V.[2] and Hüssy, D.[2], [1]Agroscope Liebefeld-Posieux Research Station ALP, Tioleyre 4, P.O. Box 64, 1725 Posieux, Switzerland, [2]Institute of Veterinary Bacteriology, ZOBA, Vetsuisse Faculty, University of Bern, Länggassstrasse 122, P.O. Box 8466, 3001 Bern, Switzerland; michel.rerat@alp.admin.ch

The present study was carried out to investigate the incidence and antibiotic susceptibility of the major bovine respiratory bacterial pathogens in veal calves with pneumonia. At the entry in the fattening unit the calves were prophylactically treated either by a single administration of tulathromycin (group A, n=20) or by a peroral administration of chlortetracycline, sulfamidine, and tylosine (group B, n=20) or not prophylactically treated (group C, n=19). All calves had the same housing and feeding conditions. Mucus samples from the lower respiratory tract were obtained by transtracheal lavage prior to first therapeutic treatment on the day of diagnosis of pneumonia. Samples were cultured to identify bacterial pathogens. During the study, 15 calves in group A, 12 in group B and 18 in group C suffered from at least one pneumonia episode. From the 91 isolated pathogens, the most prevalent were Pasteurella multocida (22% of isolated pathogens), Mycoplasma bovis (18%), and Mannheimia varigena (15%). Pasteurellaceae isolates were tested for their antibiotic susceptibility. All Pasteurella multocida (n=20), Mannheimia varigena (n=13), and Mannheimia haemolytica (n=4) isolates were susceptible to gentamicin and florfenicol. Resistances were found to trimethoprim/sulfonamid (35%, 8% and 0% resistant isolates, respectively), to penicillin (10%, 8% and 50%) and to tetracycline (10%, 15% and 50%). In conclusion, the majority of the bovine pathogens isolated from the lower respiratoy tract of calves with pneumonia in this study belonged to the commensal bacterial flora of the upper respiratory tract. Trimethoprim/sulfonamid, penicillin, and tetracycline should not be used as first choice for the therapy of pneumonia in veal calves in the area under study.

Preference behaviour of cows choosing a robotic milking stall

Gelauf, J.S.[1], Van Der Veen, G.J.[1] and Rodenburg, J.[2], [1]Vetvice, Moerstraatsebaan 115, 4614 PC Bergen op Zoom, Netherlands, [2]DairyLogix, RR#4, Woodstock, N4S 7V8, Canada; jack@dairylogix.com

When cows access 2 robotic milking stalls, their preferences provide an indication of how barn layout affects attendance. If both stalls are equally attractive, there is potential to reduce stress when a stall is out of service. With more than one group, lack of preference may result in easier adjustment to a new group. Cross use (CU) defined as % of cows that visit each of 2 milking stalls 40 to 60% of the time, and selective use (SU) or % of cows that visit 1 stall >90% of the time, was measured on 12 farms with 1,165 cows. Average Cu and SU were 38.7 and 19.7%. One herd with 2 robots parallel to the stalls, in opposite directions, 10 m apart had a CU of 10.4% and SU of 66.9%. After being housed in seperate groups for 2 years, a single group was made 1 year prior to data collection. Apparently, previously learned behaviour lead to lower CU and higher SU. Among the other 11 herds, CU was 41.6% and SU 14.8%. A herd with a right and left entry robot placed head to head had the lowest CU, 23.3% and the highest SU, 26.0%. Three herds with left and right entry stalls, side by side, sharing 1 robot room, had below average CU of 34.0% and higher SU of 23.4%. A herd with two right entry robots on the exterior of a right angle, with one stall more visible from the barn also had low CU of 33.0% but SU was 10.4%. These data suggest poor visibility from the housing area reduces cross use. Among 7 Farms with good visibility, 4 with opposite entry robots had both higher CU (54.0 vs 48.6%) likely due to entry points close together, and higher SU (11.1 vs 8.1%), likely due to a learned preference for right or left entry. 4 Herds with a single commitment pen for both robots, had higher CU (55.1 vs 44.6%) than 2 with free traffic. While the small number of herds prevents statistical analysis, the data suggests that prior training and barn layour both influence CU and SU. Robot visibility, close entry points and commitment pens increase CU, and matched entry direction increases SU.

Investigation on relative frequency and alteration of *Clostridium* spp. in poultry intestine

Seidavi, A.R.[1], Mirhosseini, S.Z.[2,3], Shivazad, M.[4,5], Chamani, M.[4], Sadeghi, A.A.[4] and Pourseify, R.[3], [1]Islamic Azad University, Rasht Branch, Animal Science Department, Rasht, 4185743999, Iran, [2]Guilan University, Animal Science Department, Rasht, 4185743999, Iran, [3]Agricultural Biotechnology Institute, North Regrion, Genomics Department, Rasht, 4185743999, Iran, [4]Islamic Azad University, Science and Research Branch, Animal Science Department, Tehran, 4185743999, Iran, [5]Tehran University, Animal Science Department, Tehran, 4185743999, Iran; alirezaseidavi@yahoo.com

The objective of this study was to develop a PCR based method for rapid quantification of *Clostridium* spp. and investigation on its relative frequency in duodenum, jejunum, ileum and cecum of broilers. The specific detection of the *Clostridium* spp. species was based on PCR amplification of the 16S rRNA gene using oligonucleotide primers. Relative frequency alteration of *Clostridium* spp. investigated by means of Gel-Proc Analyzer software based on linear regression with extrapolation method. In total rearing period, the highest relative frequency of *Clostridium* spp. belonged to cecum (66.96%) and the lowest relative frequency of *Clostridium* spp. belonged to jejunum (3.47%). At 4 day of ages, the highest relative frequency of *Clostridium* spp. belonged to cecum (75.69%) and the lowest relative frequency of *Clostridium* spp. belonged to duodenum and jejunum (approximately 0.0%). At 14 day of ages, the highest relative frequency of Clostridium spp. belonged to cecum (42.20%) and the lowest relative frequency of *Clostridium* spp. belonged to jejunum (15.84%). At 30 day of ages, the highest relative frequency of *Clostridium* spp. belonged to cecum (72.24%) and the lowest relative frequency of *Clostridium* spp. belonged to jejunum (approximately 0.0%). Relative frequency of Clostridium spp. consists 37.62, 21.89 and 40.48% at 4, 14 and 30 day of ages. The strategy of bacteria quantification using PCR and densitometry is fast, sensitive, specific and can be used as a reliable choice for routine quantification of these bacteria groups in chick gastrointestinal contents.

Improving information management in organic pork production chains

Hoffmann, C. and Doluschitz, R., Universität Hohenheim, Department of Farm Management (410C), Schloss Hohenheim, 70593 Stuttgart, Germany; c.hoffmann@uni-hohenheim.de

EU regulation 834/2007 harmonises the terms for the organic pig and pork production within the EU. Nevertheless, due to national distinctions, there are quite large variations in structure of their supply chains (heterogeneity of production and processing) at EU member state level. As a result of increasing requirements to quality assurance of products (such as EU regulation 178/2002), traceability is gaining more importance. Hence, the relevant information has to be selected and data have to be evaluated. A well structured information management, not only internal, but also inter-organisational is essential, also for organic pig producers. The main objective of the study is to improve the competitiveness of organic pork production and to contribute to higher standards in food safety by evaluating the status quo of information management in several European countries and to create a concept for improving efficiency of internal and inter-organisational information management, according to the special terms of organic pig production. The study takes place in several main organic pork production countries in Europe. It is divided into two parts: In part one about 30 expert interviews will provide an overview of the status quo of organic pork production chains and information management in Europe. In a further step outstanding results will be observed in detailed case studies. Using cost and benefit approaches the inter-organisational information systems in organic pork production chains will be analysed extensively. Finally, a concept for improving internal information management as well as inter-organisational information management in organic pork production chains will be designed. At the 60th Annual Meeting of the EAAP 2009 part one (status quo analysis) of the research will be completed. Not only deficits, but also potentials for optimisation of information management in organic pork production systems in Europe will be presented.

Electronic identification and management of eradication plan for Brucellosis, Leucosis and Tuberculosis in buffalo breeding: innovative system for recording of data collected by Veterinary Services

Marchi, E., Ferri, N. and Grosso, G., Istituto Zooprofilattico Sperimentale of Abruzzo and Molise, Campo Boario street, 64100, Italy; e.marchi@izs.it

Buffalo farming in Italy found favorable breeding condition in the southern part of Italy,where are located most of the buffaloes. A total amount of 258,000 buffaloes are recorded in the Italian cattle and buffaloes data base and almost 50% of the animals are present in the Caserta province. Identification of buffaloes, considering the particular environment conditions in which they live, may result difficult (dirty ear tags,consumption due to adverse atmospheric conditions,losses) and some breeders have decided to use an electronic device. The presence of brucellosis in buffaloes requires specific tools to ensure a unique,inalterable and permanent identification in order to apply measures for the eradication plans. Recently (2006), the Campania region has developed a specific regional plan to identify electronically buffaloes using a ceramic bolus provided with a transponder. The National Register and Identification Centre for cattle and buffaloes located in Teramo, at the Istituto Zooprofilattico Sperimentale of Abruzzo and Molise, developed an innovative system to record both identification and sanitary data of buffaloes collected from 'in field activities', in order to improve the sanitary management of the farms. A total amount of six thousand buffaloes have been tested for brucellosis and samples collected from each animal have been identified using an hand-held computer equipped with windows mobile software and provided with an application able to manage the data downloaded (identification number, sex, race, birth date) from the National Database. This application can be used to manage all the activities (identification, sampling, vaccination), including the printing of sticky labels for the identification of blood samples collected in field,thus ensuring a strong identification tool for Veterinary Services, able to improve the activities of the eradication plans.

Effect of zeolite A on the periparturient feed intake and mineral metabolism of dairy cows

Spolders, M.[1], Grabherr, H.[1], Lebzien, P.[1], Fürll, M.[2] and Flachowsky, G.[1], [1]Friedrich-Loeffler-Institute, Federal Research Institute for Animal Health, Institute of Animal Nutrition, Bundesallee 50, 38116 Braunschweig, Germany, [2]University of Leipzig, Faculty of Veterinary Medicine, Large Animal Clinic for Internal Medicine, An den Tierkliniken 11, 04103 Leipzig, Germany; peter.lebzien@fli.bund.de

Subclinical periparturient hypocalcaemia is a frequent disease of high yielding dairy cows and there exist different strategies to prevent this metabolic disorder. The addition of zeolite A as a calcium binder to the preparturient ration is one of these strategies. However each of these strategies could have negative side effects, also the feeding of zeolite A. The objectives of the present two experiments were to study the influence of different doses of zeolite A on feed intake and mineral metabolism, especially the incidence of hypocalcaemia. In the first experiment, a supplementation of 90 g zeolite A per kg dry matter (DM), and in the second experiment 12.5, 25 and 50 g zeolite A per kg DM of the total mixed ration (TMR), consisting of 48% maize silage, 32% grass silage and 20% concentrate (on DM basis), were tested. High zeolite A doses (50 and 90 g/kg DM) reduced significantly total DM-intake (- 35 and - 48%), which resulted in a reduction of energy and total calcium intake. The desirable preventing effect on the calcium metabolism was detected for zeolite A doses of 25, 50 and 90 g/kg DM. The majority of analysed calcium concentrations in serum around calving, which is the characteristical parameter in diagnosing a subclinical hypocalcaemia (< 2 mmol/l), were higher than 2 mmol/l. However, only the low zeolite A concentration of 12.5 g/kg DM had no stabilising effect on the serum calcium concentration, the hypocalcaemia incidence was as high as in the control group without zeolite supplementation (75%). A zeolite A addition of 25 g/kg DM ($\approx$200-300 g per cow and day) seems to be the optimal dose for an effective prevention of subclinical hypocalcaemia in combination with only marginal negative side effects.

Effects of Lysine on Ascaridia galli infection in grower layers
Daß, G., Kaufmann, F., Abel, H.J. and Gauly, M., Department of Animal Science, Albrecht Thaer Weg 3, 37075 Göttingen, Germany; mgauly@gwdg.de

A 2x2 factorial experiment with lysine level of the diet and Ascaridia galli infection as the main factors was performed in grower layers. Groups 1 and 3 were fed a high lysine diet (10.5 g Lys/ kg DM), whereas groups 2 and 4 received a lower lysine diet (8.5 g Lys/ kg DM). Birds in groups 3 and 4 were additionally infected with 250 embryonated eggs of A. galli at an age of 4 weeks and slaughtered 7 weeks post infection to determine worm. The diet with lower lysine level resulted in a reduced body weight development ($P=0.006$). Infected animals had lower body weight development when compared with un-infected animals ($P=0.002$). Infected birds on the lower lysine diet had the lowest feed and protein utilization ($P<0.0001$) and showed an obvious diverted nutrient utilization. The lower lysine diet resulted in higher infection rate ($P=0.05$) than the high lysine diet. However worm counts of the infected groups did not differ significantly ($P>0.05$). It can be concluded that dietary lysine level does not alter response of the host to the infection to a great extent; however animals fed on lower lysine diet seem to require higher amounts of feed and crude protein to perform on the same level.

Cost- and benefit aspects of IT-based traceability systems: results of a Delphi-survey
Roth, M. and Doluschitz, R., Universitaet Hohenheim, Farm Management, Institut 410c, 70599 Stuttgart, Germany; m-roth@uni-hohenheim.de

Food scandals have led to several approaches to guarantee traceability of food of animal origin. But most of these approaches are not compatible among each other and often don't include relevant data from livestock farms or animal transportation. Facing these challenges, the IT FoodTrace project was established to create an IT solution that ensures traceability along the supply chain of meat products. Although an integrated traceability solution will generate benefits for the stakeholders, it might be possible that the acceptance of such a solution is low, due to costs and labour input. The aim of this study is to identify and to quantify relevant cost- and benefit-aspects of stakeholders. Another aim is to detect the acceptance of IT-based traceability systems among potential stakeholders, such as livestock farmers, slaughtering facilities or food retailers. The study has been carried out by means of a Delphi-survey. In two successive surveys about 50 experts were asked about their opinion concerning cost and benefit aspects as well as acceptance issues of IT-based traceability and quality assurance systems. The results are Wilcoxon tested. First results show, that the willingness to pay an extra for meat products with guaranteed traceability is estimated low or even not existent in the eyes of experts. This result differs from the findings of a case study, where the interviewed persons stated a positive willingness to pay. More than 86% of the experts estimate license costs as well as general running cost as the most relevant cost factors of an IT-based traceability solution, whereas 56% don't think, that IT-based traceability solutions will lead to an increase in internal controls costs. Overall, 92% of the experts state acceptance of an IT-based traceability solution, if economical benefits will be certain. Experts see clear advantages in IT-based traceability solutions, but costs and labour inputs as well as data security issues are limiting factors.

Agonistic behaviour of weaned piglets

Stukenborg, A.[1], Traulsen, I.[1], Puppe, B.[2] and Krieter, J.[1], [1]Christian-Albrechts-University, Institute of Animal Breeding and Husbandry, Olshausenstr. 40, 24089 Kiel, Germany, [2]Research Institute for the Biology of Farm Animals Dummerstorf, Division Behavioural Physiology, Wilhelm-Stahl-Allee 2, 18196 Dummerstorf, Germany; astukenborg@tierzucht.uni-kiel.de

In pig production agonistic behaviour often results in injuries which can cause health problems and reduce pig performance. The aim of the study was to investigate the agonistic behaviour of piglets after weaning. The data were collected on a closed herd sow farm of the breeding company 'Hülsenberger Zuchtschweine' from October 2007 until August 2008. Immediately after weaning, the 297 female piglets were recorded for 48 hours. 10 piglets per pen were individually marked (on average 29.4 pigs per pen). Their agonistic behaviour traits were described by the time, the aggressor/receiver and the winner/looser (or undecided) of a fight. This information was used to calculate a dominance index (DI) (DI = (wins-defeats)/(wins+defeats)). The DI ranks from -1 (absolutely submissive) to +1 (absolutely dominant). In addition the skin lesions of the piglets were evaluated at weaning and one week later. The lesion score (LS) was recorded for the anterior, central and the caudal third of the body. The LS ranked from 0 (no wounds) to 4 (many, deep wounds). In total 10,531 fights were observed. The average number of fights per piglet (NF) was 42.2 (σ=22.6). The overall time being involved in agonistic behaviour (TA) varied between 50 and 8,921 seconds per piglet (μ=2,338 s, σ=1,663 s). Piglets with a high DI had a significant longer TA ($P<0.001$) and a higher NF ($P<0.05$). The number of being aggressor had no impact on TA, however it was positively related to NF ($P<0.001$). The analysis of the LS showed a significant increase of wounds for all three body areas ($P<0.001$). Further analysis of the agonistic behaviour from these piglets as growing pig and finally as gilt could indicate whether more aggressive piglets become more aggressive sows. Also genetic relationships will be considered.

Assessment of the activity of growing rabbits in a metabolic cage

Rodríguez Latorre, A.[1], Olivas Cáceres, I.[2], Estellés Barber, F.[1], Calvet Sanz, S.[1], Villagrá García, A.[2] and Torres Salvador, A.G.[1], [1]Universidad Politécnica de Valencia, Institute of Animal Science and Technology, Camino de Vera s/n, 46022 Valencia, Spain, [2]Instituto Valenciano de Investigaciones Agrarias, Centro de Tecnología Animal (CITA), Polígono La Esperanza nº 100, 12400 Segorbe (Castellón), Spain; villagra_ara@gva.es

General activity patterns of growing rabbits are supposed but they have not been deeply studied. The aim of this work was to assess these general patterns of activity at different ages during the growing period in individually caged rabbits and the tendency to stand up inside a cage (as they have difficulties to stand when they are housed under commercial conditions). In order to achieve this objective, ten growing rabbits from the experimental farm in the Universidad Politécnica de Valencia were used. The animals were one, two, three, four and five weeks old, being two animals in each of these age groups. Each day of the experiment, one rabbit was introduced into a metabolic cage in order to assess the general activity with continuous recording, and it remained inside during 24 hours. After this period, the animal was changed and the recording started again. The observed activities were lying, sleeping, sitting, eating, drinking, walking, standing and others, and the beginning and the finishing of each activity was registered in order to be able to define the duration of each activity during the 24-h period. Results were subjected to logistic regression analyses throughout SAS System using the glimmix procedure. These results demonstrated the nocturnal activity of rabbits when they grow up. A big difference was observed according to their age, as in young rabbits this pattern is not as clear as it is in older ones. Regarding the tendency to stand up, the average time that the animals were in this position was 8 minutes and 30 seconds, during a 24 hours period, although these results must be studied deeply.

Short-term changes of milk somatic cell count on individual udder quarters of dairy cows in different cell count ranges

Müller, A.B., Rose-Meierhöfer, S., Berg, W., Ammon, C., Ströbel, U. and Brunsch, R., Leibniz-Institute for Agricultural Engineering, Engineering for Livestock Management, Max-Eyth-Allee 100, 14469 Potsdam, Germany; amueller@atb-potsdam.de

Somatic cell count (SCC) is a main indicator of udder health and milk quality, it also reflects the quality of dairy management. Complete milk samples above 100,000 SCC/ml milk are a sign of subclinical or clinical mastitis. In Germany the limit for delivered bulk milk is 400,000 SCC/ml milk. The aim of the study was to evaluate short-term changes in SCC of individual quarters and possible consequences for milking technique and management. 30 dairy cows were arranged in three groups, each with ten cows, according to their SCC in the last monthly milk recording: group 1 (healthy) ≤100,000, group 2 (conspicuous) >100,000 - ≤400,000 and group 3 (diseased) >400,000 SCC/ml milk. Quarter foremilk samples were collected manually without pre-stripping over a period of twelve days. SCC of the unpreserved samples was measured with the Fossomatic 5000 (FOSS). Maximum and minimum SCC values of all quarters in each class, measured during the trial period, were averaged. The mean minimum was subtracted from the mean maximum to get information about the fluctuation of SCC per group. Data show that fluctuations of SCC in healthy quarters (group 1) are about 47,000 cells/ml milk. With 172,000 cells/ml milk conspicuous quarters show higher average fluctuations, while diseased quarters alternate even about 1.7 million cells/ml milk. Seven cows were classified as diseased considering averaged quarters, but looking at the SCC of the single quarters, 45 quarters of totally 24 cows had to be classified as diseased in the trial period. Also three of nine cows, which were presumed to be healthy, had at least one diseased quarter. Not all detected SCC peaks higher than 100,000 cells/ml milk indicate diseased quarters, but also management deficits. The study points out the need for measuring SCC on individual quarters in short-time intervals for better monitoring of udder health.

Using structural equation models to illustrate the relationship between herd characteristics and somatic cell counts

Detilleux, J., Theron, L. and Hanzen, C., University of Liege, Faculty of veterinary medicine, Bld de Colonster, 4000 Liege, Belgium; jdetilleux@ulg.ac.be

Several factors have been reported as risk factors for the increase in somatic cell counts in dairy cattle herds. In this study, we examined the relationship between a large number of herd characteristics and high bulk tank somatic cell counts using structural equation modeling (SEM) technique. Characteristics were collected between January 2006 and October 2007 on 348 farms in the Southern part of Belgium. Briefly, they included herd demographics, productive and reproductive indicators, regimen composition, feeding procedures, shearing, types of housing for heifers, milking and dry cows, strategies of mastitis prevention and treatment, milking practices and characteristics of milking machine. A key feature in SEM is that observed variables (characteristics) are understood to represent a small number of latent constructs that cannot be directly measured but are inferred from the observed variables. The SEMs have also the ability to provide separate estimates of the relations among latent constructs and observed variables (the measurement model) and the relations between constructs and bulk tank somatic cell counts (the structural model). In our analysis by the SEM (proc Calis on SAS), several models were contrasted using various goodness of fit indexes to obtain the best final model. Results are compared with a more conventional epidemiologic approach (general linear model). Both approaches give insight in the current techniques to keep bulk tank somatic cell counts below penalty threshold in Belgium.

Mastitis agents: frequencies and antibiotic resistances: an example from Saxony-Anhalt, Germany

Schwanke, I.[1], Schafberg, R.[1], Rösler, H.J.[2], Döring, L.[2] and Swalve, H.H.[1], [1]Institute of Agricultural and Nutritional Sciences, Group Animal Breeding, Adam-Kuckhoff-Str. 35, 06108 Halle, Germany, [2]State Agency for Milk Recording, Angerstr. 1, 06118 Halle, Germany; renate.schafberg@landw.uni-halle.de

The mastitis laboratory of the milk recording agency in Saxony-Anhalt offers bacteriological tests of milk samples and also resistance tests for the agents to their farmers as supplementary services. This study is based on the results of the past two years of data from the mastitis lab. The data set included 64,014 foremilk samples for bacteriological findings (2007: 26.13%/2008: 25.62% microbiological positive) and 1,544 isolates (591/953) for antibiotic resistance tests. Coagulase-Negative Staphylococci (CNS) are the most common mastitis agents (40.5% / 45.5%), followed by Esculin-Positive Streptococci (EPS) (28.4% / 32.4%) and S. aureus (14.5% / 15.1%). The relationship with somatic cell scores can be demonstrated very clearly, the average of infected samples was 4.2 (±2.7) vs. 2.8 (±2.2) for none infected in 2007 and 4.4 (±2.7) vs. 2.8 (±2.2) in 2008. The growth of the microbe colonies is rarely 'isolated', in 2007 often 'enhanced' and in 2008 mostly 'accumulated'. This underlines that milk from infected animals is highly contaminated with mastitis agents and thus poses a high risk for other cows. Contrary to expectations, the resistances of isolated agents against specific antibiotics (i.e. Amoxycillin, Ampicillin, Cefazolin, Cefoperazon, Cefoperazon, Cefquinom, Cloxacillin, Oxacillin, or Penicillin) did not change significantly over time. Clear differences between the effects of different antibiotics on identical agents were found. However, for the most frequent agents the susceptibility of the agents to common antibiotics is favourable.

Comparative study between sheep and goats on rumenic acid and vaccenic acid in milk fat under the same dietary treatments

Tsiplakou, E. and Zervas, G., Agricultural University of Athens, Nutritional Physiology and Feeding, Iera Odos 75, GR-11855, Greece; eltsiplakou@aua.gr

A number of studies have shown that the rumenic acid (RA= cis-9 trans-11 $C_{18:2}$ CLA) content of milk fat is usually higher in sheep than in goats, due partly to different dietary regimens. An experiment was conducted with 12 lactating dairy ewes and 12 goats with the objective to compare the two animal species (sheep/goats) fed diet with the same forage/concentrate (F/C) ratio, on their milk RA and vaccenic acid (VA) production. The experiment was carried out in three consecutive phases, lasted 3 weeks each, immediately after weaning of lambs and kids. In phase I, the ewes and the goats were fed according to their maintenance and lactation requirements, with 14 kg alfalfa hay, 4 kg wheat straw and 12 kg concentrate the 12 ewes (F/C ratio = 60/40), and with 14 kg alfalfa hay, 4 kg straw and 24 kg concentrate the 12 goats (F/C ratio = 43/57). In phase II, 14 kg alfalfa hay, 4 kg straw and 14 kg concentrate were offered daily to each group of sheep and goats, with a F/C ratio = 56/44. In phase III, all ewes and goats were fed individually with 0.8 kg alfalfa hay, 0.2 kg wheat straw and 0.8 kg concentrate daily with a F/C ratio = 56/44. The results showed that in phase I, the concentrations of RA and VA milk fat content did not differ significantly between sheep and goats. In phases II (group feeding) and III (individual feeding), where sheep and goats fed with the same amount of food of the same F/C ratio, the sheep milk fat had higher RA and VA content compared to goats. In conclusion, these findings support the hypothesis that there are species differences, as RA and VA production concerns, which needs further investigation.

Effects of alfalfa hay particle size and cottonseed hulls as nonforage fiber source on nutrient digestibility, serum glucose and BUN of holstein dairy cows

Abdi-Benemar, H., Rezayazdi, K. and Nikkhah, A., University of Tehran, Animal science departement, Faculty of agriculture, Karaj, Tehran, 31587-77871, Iran; abdi_1360@yahoo.com

Ration particle size has been observed to affect DMI, milk fat and nutrient digestibility. Nonforage fiber sources (NFFS) such as Cottonseed hulls (CSH) possess a large amount of NDF that can be used as a forage substitute. It was suggested that particle size of dietary forage can interact with NFFS. The objectives of this study were to investigate effects of CSH as NFFS and interaction between CSH and alfalfa hay (AH) particle size. Twelve Holstein dairy cows in midlactation were assigned to a change-over design with three periods. Diets differed in alfalfa hay particle length (short and long) and cottonseed hulls substituted for AH (0 and 9.63% DM). Dietary treatments were: 1) long AH no CSH, 2) long AH with CSH, 3) short AH no CSH and 4) short AH with CSH. Samples of feces were collected from all cows for 5 d and composited by cow within a period. On d 25 of each period, blood samples were taken from the tail vein and analyzed for Glucose and Blood Urea Nitrogen (BUN). Data were analyzed using the mixed model procedure of SAS with model effects for alfalfa hay particle size (APS), fiber source (FS) and two-way interaction of APS × FS. APS and FS did not affect DMI. DM, OM and NFC digestibility were not affected by APS or FS but significantly affected by interaction between APS and FS. EE and CP digestibility were not affected by APS but significantly decreased by FS. NDF digestibility was not affected by APS but significantly affected by FS and interaction between APS and FS. Blood glucose was not affected by APS or FS but BUN decreased by FS. Results of this study showed that inclusion CSH decreased BUN and did not decrease blood glucose. Therefore, CSH can be substituted partly for AH when forage sources are expensive or its stores are limited. Also, particle size of dietary forage can interact with NFFS, as earlier hypothesized.

Effect of dietary digestible fibre and fat level on the fatty acid content and composition of caecotrophes in rabbits

Papadomichelakis, G., Anastasopoulos, V. and Karagiannidou, A., Agricultural University of Athens, Nutritional Physiology and Feeding, 75 Iera Odos Street, 118 55, Athens, Greece; gpapad@aua.gr

Caecotrophes play an important role in the utilization of the caecal microbial activity products in rabbits and their chemical composition might be altered by nutritional treatments, thus affecting the recycling of nutrients. The aim of the present study was to examine the effect of dietary digestible fibre and fat level on excretion and fatty acid composition of soft faeces. Forty-four weaned rabbits, aged 35 days, were allocated in 4 treatments following a 2×2 factorial design. The animals were fed 4 diets with 2 levels of digestible fibre (DgF, 165 vs. 240 g/kg) supplemented with soybean oil (SO, 20 g/kg) or not. At 68 days of age a light plastic flat collar was put on animals for 24 h to prevent caecotrophy. The collected caecotrophes were lyophilized and their fatty acid content was determined by gas chromatography. Data were analysed by the GLM procedures (two-way ANOVA) of SPSS (v. 10.0). Soft faeces excretion was not affected by the treatments. High DgF level increased daily ether extract recycling ($P<0.05$) and fat content ($P<0.01$) of caecotrophes. Total contents of fatty acids, microbial membrane fatty acids (odd-numbered, methyl- and hydroxy- branched) and conjugated linoleic acid (CLA) were increased ($P<0.001$) with high DgF level. Addition of SO resulted in higher ($P<0.001$) total content of fatty acids, but decreased ($P<0.01$) the proportion of microbial fatty acids. In conclusion, alterations in the nutrient supply entering the caecum affect the fatty acid content and composition of the caecotrophes, thus changing significantly the recycling of microbial fatty acids and CLA in rabbits.

Selenium content in different parts of the gastrointestinal tract of fattening bulls offered a diet supplemented with selenite or a diet with feedstuffs high in selenium

Robaye, V., Dotreppe, O., Hornick, J.L., Istasse, L. and Dufrasne, I., Liege University, Nutrition Unit, Bd de Colonster 20, 4000 Liege, Belgium; listasse@ulg.ac.be

Dietary selenium (Se) could be offered either on a mineral (selenite) or an organic form provided from yeasts or from feedstuffs. A fattening diet based on maize silage and supplemented with spelt, barley, soja bean meal, sugar beet pulp, molasses and mineral mixture was used as control. Sodium selenite was added to the control diet for the mineral Se group. Spelt and barley grown with Se enriched fertilizer spread at a rate of 4g Se as selenate / ha at the 2nd and 3rd nitrogen applications and linseed meal high in Se owing to the Canadian origin of the seeds were used in the organic Se group. The 3 diets were offered to bulls fitted with rumen and duodenal cannulas. The measurements were carried on one month after the animals were on their specific diets. The dietary Se concentration was 51.0, 205.6 and 192.6 μg/kg DM in the control, mineral Se and organic Se groups respectively. The corresponding plasma Se concentrations were 40.3, 49.9 and 67.5 μg/l ($P<0.001$). The Se contents were different according to the gastrointestinal tract location owing to the type of Se supplementation. In the rumen content, it was with the organic Se that the Se concentration was the highest (405.3 vs 376.1 vs 223.9 μg/kg DM, ($P<0.001$) for the organic, mineral and control groups respectively). There were no differences between the two types of supplementation in the Se content at the duodenum (291.3 and 292.1 vs 141.9 μg/kg DM). Feces were characterized by Se contents of 428.5, 538.3 and 653.2 μg/DM ($P<0.002$) in the control, organic and mineral groups respectively. There were no differences at each location in the Se contents when the samples were taken before the morning meal or 4 hours later. The higher concentration in the feces with the mineral form indicated a reduced absorption which could be associated with the lower plasma Se concentration.

Changing diets to modify methane emission from sheep

Chaudhry, A.S., Newcastle University, Agriculture, Food & Rural Development, Agriculture Building, Newcastle upon Tyne, NE1 7RU, United Kingdom; a.s.chaudhry@ncl.ac.uk

Ruminant animals are partly blamed for causing environmental pollution primarily through the rumen fermentation of their diets. This fermentation process is vital to supply essential nutrients for the integrity and growth of ruminant tissues. It is recognised that the rumen fermentation process can generate up to 80% of the total energy requirement of these animals depending upon the quality and quantity of their dietary intakes. This dietary energy is converted into rumen microbial protein which accounts for over 70% of the amino acids required for post-rumen utilisation. The extent of energy production and microbial protein synthesis in the rumen are however influenced by the dietary components. Although the cell wall is the predominant energy source in a ruminant diet, its association with other nutrients such as soluble carbohydrates to influence rumen fermentation is also realised. Therefore, simple alteration of dietary components may help modify methane emission, minimise nutrient wastage, improve the environment and increase the farm incomes. This study was conducted according to a 4x4 Latin Square Design by involving 4 sheep and 4 diets representing 2 soluble carbohydrates to fibre ratios and 2 levels of intake. The results showed changes in the patterns of oxygen consumption and methane production by sheep when consuming diets containing different ratios of soluble carbohydrates and fibre. However, the rate and extent of methane production depended upon the level of intake of these sheep and the amount of soluble carbohydrate in their diets. It appears that simple dietary changes may help manipulate rumen fermentation, nutrient utilisation and methane emission in ruminants.

Measurement of palatability of 14 common ingredients used in feed mixes for lambs and ewes

Mereu, A.[1], Giovanetti, V.[2], Acciaro, M.[2], Decandia, M.[2] and Cannas, A.[1], [1]University of Sassari, Dipartimento Scienze Zootecniche, via de Nicola 9, 07100 Sassari, Italy, [2]AGRIS, DiRPA, Strada Statale Sassari-Fertilia, 07040 Olmedo, Italy; cannas@uniss.it

The palatability of 14 of the most common ingredients used in concentrate feed mixes was measured in 6-min tests, carried out supplying the ground feeds to 14 female lambs or 14 multiparous dry ewes, in two 14 (days) x 14 (feeds) Latin square experiments. The lambs had the following rank of palatability (expressed as DM intake during exposure time; DMI): soybean meal 49 (24.5 g), wheat grains (22.8 g), pea grains (17.4 g), corn grains (14.9 g), soybean hulls (13.1 g), beet pulps (11.9 g), wheat brans (11.4 g), soybean meal 44 (10.9 g), corn middlings (7.5 g), canola meal (5.0 g), sunflower meal (2.8 g), corn gluten meal (1.7 g), dehydrated alfalfa (0.4 g), and oat grains (0.0 g). The DMI of the 2 preferred feeds was significantly higher ($P<0.05$) than that of the 6 least preferred feeds. The ewes had a clear preference for four feeds (beet pulps: 62.8 g; wheat grains: 56.4 g; pea grains: 56.3 g; corn grains: 52.7 g), whose intake was significantly higher ($P<0.05$) in respect to the other feeds. Comparisons between lambs and ewes showed that lambs ate significantly ($P<0.05$) more soybean meal 49 (2829 vs. 188 mg/kg$^{0.75}$ BW), soybean hulls (1543 vs. 339 mg/kg$^{0.75}$ BW) and soybean meal 44 (1255 vs. 377 mg/kg$^{0.75}$ BW), while their DMI was lower ($P<0.05$) for beet pulps (1400 vs. 3360 mg/kg$^{0.75}$ BW), corn gluten meal (191 vs. 1346 mg/kg$^{0.75}$ BW) and oat grains (0.0 vs. 533 mg/kg$^{0.75}$ BW). The DMI of lambs varied from high to low values in a continuum, suggesting a marked influence of sensorial perceptions and, to a lesser extent, of previous feeding experience. In contrast, the ewes had a marked preference for 4 feeds often supplied as single ingredients (beet pulps and wheat grains, pea grains and corn grains) and low intake or complete rejection of the remaining feeds, including several commonly used in sheep feed mixes but rarely supplied alone.

Persistency of the effect of a whey protein emulsion gel on the proportion of poly-unsaturated fatty acids in milk

Van Vuuren, A.M.[1], Van Wikselaar, P.G.[1], Van Riel, J.W.[1], Klop, A.[1] and Bastiaans, J.A.H.P.[2], [1]Wageningen UR, ASG Animal Production, P.O. Box 65, 8200 AB Lelystad, Netherlands, [2]FrieslandCampina, P.O. Box 449, 8000 AK Zwolle, Netherlands; ad.vanvuuren@wur.nl

To study the persistency of the effect of supplementing a whey protein emulsion gel (WPEG) of soybean and linseed oil on milk fatty acids (FA), a 10-wk experiment was carried out with 32 lactating HF dairy cows. During the first 5 weeks, all cows were fed on a restricted grazing regime with pasturing during the daytime and stalling during night-time feeding a mixture of maize silage, grass hay, soybean expeller and minerals. During the last 5 weeks, all cows were kept indoors and fed a total mixed ration containing grass silage, maize silage, grass hay and concentrates. Cows were equally and randomly allotted to one of two treatments: a control treatment (no supplementation) and a treatment supplementing WPEG at a rate of approximately 1.5 kg/d per cow (=0.45 kg crude fat). Feed intake and milk production were monitored daily and milk was sampled weekly for fat, protein, lactose and FA analyses. Supplementing WPEG had no significant effect on monitored dry matter (DM) intake (14.1 kg/d during restricted grazing – grass intake not determined; 22.4 kg/d during the last 5 wk). WPEG had no effect on milk yield (35 kg/d) and on concentrations and yields of milk fat and milk protein. Supplementing WPEG increased lactose concentration (from 44.9 to 46.6 g/ kg milk; $P<0.01$)). WPEG decreased proportions of medium-chain FA and increased proportions of most C18 FA, except cis9 C18:1 and trans10, cis12 C18:2. Supplementing WPEG led to a twofold increase in cis9, cis12 C18:2 (from 1.8 to 4.0 g/100 g FA; $P<0.001$) and a fivefold increase in cis9, cis12, cis15 C18:3 (from 0.5 to 2.2 g/100 g FA; $P<0.001$). The calculated recovery of supplemented C18 FA in milk was 29%. From these results it is concluded that feeding vegetable oils as WPEG to dairy cattle is an adequate and persistent method to increase poly-unsaturated FA in milk.

Effect of concentrate allowance in an automatic milking system

Weisbjerg, M.R. and Munksgaard, L., Faculty of Agricultural Sciences, Aarhus University, Animal Health, Welfare and Nutrition, P.O. Box 50, DK 8830 Tjele, Denmark; martin.weisbjerg@djf.au.dk

Automatic milking is based on cows' voluntary visits to the automatic milking unit (AMU) for milking. To obtain an optimal visit frequency of 2.5-3.5, concentrate feeding is often used as a reward in the AMU. The aim of the present experiment was to examine the effect of concentrate allowance and the energy concentration in the basal mixed ration on milk production and visits to the AMU. 87 cows, involving Danish Holstein, Danish Red and Danish Jersey, entered the experiment for the first 70 days of lactation. The experiment was carried out in a research herd with automatic registration of intake of ad libitum fed mixed ration, and of concentrate offer and orts in the AMU. Treatments were combinations of basal mixed rations (MR) composed of maize silage, grass/clover silage, barley, rapeseed cake and sugar beet pulp, with low (6.63 MJ NEL/kg DM), medium (6.94 MJ NEL/kg DM) or high (7.42 MJ NEL/kg DM) energy concentration, and 3 or 6 kg daily concentrate offer in the AMU. After calving, cows were randomly divided on 4 treatments, L6 (low MR+ 6 kg conc.), M6 (medium MR+ 6 kg conc.), M3 (medium MR+ 3 kg conc.), and H3 (high MR + 3 kg concentrate). Responses to treatments were analysed for the period 20-70 days in milk. For L6, M6, M3 and H3, respectively, per cow per day, intake of basal ration was 13.0, 15.3, 15.4 and 18.1 kg dry matter ($P<0.0001$), concentrate orts 0.3, 0.5, 0.2 and 0.2 kg ($P<0.06$), intake of concentrate 4.4, 3.4, 2.3 and 2.3 kg ($P<0.0001$), milking frequency 3.13, 2.64, 2.52 and 2.66 ($P<0.004$), visits without milking allowance 2.44, 1.30, 0.57 and 0.91 ($P<0.0001$), resulting in energy corrected milk production of 31.4, 31.8, 30.2 and 34.9 kg ($P<0.01$). In conclusion, increased concentrate offer in the AMU increased AMU visit frequency, especially in combination with a low energy concentration in the basal ration.

Effects of long-term feeding of genetically modified maize (Bt-maize, MON 810) on dairy cows

Steinke, K.[1], Paul, V.[2], Gürtler, P.[2], Preißinger, W.[3], Wiedemann, S.[2], Albrecht, C.[4], Spiekers, H.[3], Meyer, H.H.D.[2] and Schwarz, F.J.[1], [1]Technische Universität München, Department Animal Science - Section Animal Nutrition, Hochfeldweg 4, 85350 Freising-Weihenstephan, Germany, [2]Technische Universität München, Department Animal Science - Section Physiology, Weihenstephaner Berg 3, 85350 Freising-Weihenstephan, Germany, [3]Bavarian State Research Centre for Agriculture, Prof. Dürrwächter Platz, 85586 Poing, Germany, [4]University Bern, Bühlstr. 28, 3012 Bern, Switzerland; schwarzf@wzw.tum.de

Over a period of 25 months (two consecutive lactations), 36 dairy cows (Simmental) divided into two groups (18 cows per group) were fed with either transgenic Bt-maize (MON 810) as silage, kernels and cobs or a control diet consisting of the corresponding isogenic components. Cry1Ab DNA was positively detected in transgenic kernels and cobs and Cry1Ab protein was positiveley detected in transgenic silage, kernels and cobs. Early lactating cows consumed a diet consisting of nearly 33% maize silage, 17% maize cobs and 15% maize kernels (in dry matter) completed with grass silage, straw, rapeseed meal, vitamin-mineral-mix, urea and molasses. During continuous lactation and during dry periods the diet was diluted with straw to adapt to the lower performance. The nutrient composition and the energy content of the maize components and the diets of the transgenic and isogenic groups revealed no major differences. Daily feed intake and milk yield did not differ between the groups within the 1st and 2nd lactation of the trial. Milk fat and protein were significantly higher for cows fed Bt-maize when comparing the lactation curves in the 1st lactation. However, the differences between the overall means were small and there were no significant effects in the 2nd lactation period between the groups. Regarding metabolic parameters (glucose, NEFA, BHBA), liver enzymes (AST, GLDH, GGT) and bilirubin in plasma, no differences between transgenic and isogenic fed cows were found, exept for glucose content, which was higher ($P<0.05$) for transgenic fed cows in the 1st lactation.

Effects of low protein diets and rumen protected CLA on growth performance of double-muscled Piemontese bulls

Dal Maso, M., Schiavon, S., Tagliapietra, F. and Bittante, G., University of Padova, Department of Animal Science, Viale dell'università 16, 35020 Legnaro Padova, Italy; matteo.dalmaso@unipd.it

Growth performance of double muscled Piemontese bulls kept on 2 TMR differing for crude protein density (diet HP: CP = 14.7% DM; diet LP: CP = 11.0% DM) and top dressed with 80 g/d of rumen protected CLA mixture ($_{CLA}$) or with 65 g/d of hydrogenated soybean oil ($_{HSO}$) were studied. Forty-eight bulls (279+24 kg initial LW), divided in the 4 experimental groups, were housed in 12 fully slatted floor pens. DM intake was measured weekly by weighting the amount of feed distributed and the orts of each pen. Individual LW, SEUROP and fatness scores were assessed monthly. Trial lasted 340 days. Dressing percentage and conformation indexes were performed at slaughter, too. Data were analysed by ANOVA. Over the whole trial, no significant differences due to the treatment on average daily gain (ADG = 1.186±0.14 kg/d) and DM intake (DMI = 8.5±0.27 kg/d) were found. With respect to HP, LP showed a significantly lower ADG ($P<0.001$) from 0 to 65 days on trial (279 to 370 kg LW), partially compensated by a significantly higher ADG from 121 to 148 days on trial ($P<0.05$; 450 to 480 kg LW), while no differences were found on later stages (480 to 670 kg LW). Over the whole trial feed conversion ratio (FCR) was significantly affected by the interaction CP x Additive (FCR: 6.90, 7.22. 7.49 and 7.11 kg DM/kg growth for HP_{HSO}, HP_{CLA}, LP_{HSO} and LP_{CLA}, respectively; $P<0.05$). Total N excretion was 59.0 kg/head with HP and it was reduced to 42.5 kg/head with LP ($P<0.001$). No significant differences due to the treatments were observed for the slaughter traits. Low protein ration can be used on Piemontese double muscle bulls from 370 to 670 kg LW without negative consequence on growth performance, FCR and carcass quality, whereas a reduction of 28% of N excretion can be achieved.

Characterisation of soiled water/dilute slurry on Irish Dairy Farms

Minogue, D.[1], Murphy, P.[1], French, P.[1], Coughlan, F.[1] and Bolger, T.[2], [1]Teagasc, Moorepark, Dairy Production Research Centre, Fermoy, Co. Cork., Ireland, [2]University College Dublin, School of Agriculture, Food Science & Veterinary, Belfield, Dublin 4, Ireland; denis.minogue@teagasc.ie

Soiled water/dilute slurry is produced through the washing-down of milking parlours and holding areas. These effluents contain nutrients that are potentially available to plants, but also pose a potential threat to water quality if not managed correctly. Current management in Ireland is regulated primarily by the Nitrate Regulations which do not distinguish between dilute slurry and more concentrated slurries and also provide no closed periods for spreading of soiled water and no guidance for nutrient management. Very little is known of the quantity and composition of soiled water/dilute slurry produced on Irish dairy farms. To assess this, a national farm survey was carried out, involving 60 farms, sampled every 28 days for 1 year. Total volumes, relevant farm characteristics and rainfall were recorded. Samples were analysed for: Total N, Total Oxidisable N, K, MRP, Total P, Ammonia, Biochemical Oxygen Demand (BOD) and percent dry matter (DM). Descriptive statistics were used to characterise soiled water quantities and composition. Correlation analysis was used to investigate relationships between chemical parameters. Mean concentration of N, P and K concentrations were 480, 70 and 959 mg/l, respectively. BOD levels ranged from 138 to 19,085 mg/l with a mean of 2494mg/l. Mean DM was 0.4% with a range of 0.015 to 3.9%. Variability in concentrations throughout the year was large. About 70% of the farms produced soiled water (B.O.D. <2,500 mg/l and DM% <1%) rather than dilute slurry. Although nutrient concentrations are low, in comparison to slurry, the large volumes produced offer a potentially significant nutrient source, particularly given evidence for increased N availability at lower DM content.

Performance, metabolic parameters and fatty acid composition of milk fat due to dietary CLA and rumen protected fat of dairy cows

Schwarz, F.J.[1], Liermann, T.[1], Möckel, P.[2], Pfeiffer, A.-M.[3] and Jahreis, G.[2], [1]Technische Universität München, Section Animal Nutrition, Hochfeldweg 4, 85350 Freising, Germany, [2]University Jena, Institute of Nutrition, Dornburgerstr. 24, 07743 Jena, Germany, [3]BASF SE, Charlottenstr. 59, 10117 Berlin, Germany; schwarzf@wzw.tum.de

Feeding rumen-protected CLA alone and in combination with rumen protected fat (RF) with regard to energy partitioning on performance, metabolic responses and fatty acid composition of milk fat was investigated. Special attention was given to the extent of post-effects on performance after finishing feeding the supplements. The trial lasted from first week p.p. until 14th week (period 1) and from 15th to 26th week (post-period). Dairy cows (Red Holstein x Simmental) were allocated as follows: I-control (n=17), II-CLA (n=18) and III-CLA+RF (n=18). CLA supplement (Lutrell®, BASF, 40g/cow/d, lipid-encapsulated, containing 10.7% CLA, cis-9,trans-11, 10.7% CLA, trans-10,cis-12) and RF (Dunafat 100, 700g/cow/d) were added with concentrate. Daily feed intake was unaffected by treatment. Milk yield tended to be higher in group II and III (35.9kg and 35.8kg) in contrast to control (34.8kg). Milk fat was significantly depressed by CLA (group II and III, 3.00% and 3.06% vs. group I 3.67%). After finishing the feeding of the supplements milk fat again increased (3.57% (II, III) vs. 3.83% (I)) whereas the milk yield stayed on a higher level (27.6kg (II, III) vs. 25.5kg (I)). Blood parameters (glucose, NEFA, BHBA), liver enzymes (AST, GLDH, γ-GT) and bilirubin showed no significant treatment differences. In week 14th the sum of trans C-18:1 and CLA fatty acids of milk fat had significantly increased only in group III (Σ C-18:1 trans (% of FAME): 1.88 (I), 2.04 (II), 3.64 (III), Σ CLA (% of FAME): 0.46 (I), 0.54 (II), 0.61 (III)). The proportion of CLA cis-9,trans-11 was in average about 80% whereas CLA trans-10,cis-12 only amounted to 3-4% of total CLA after CLA supplementation.

Nutritive value of Foxtail millet grown in different sowing date and density as a source of forage for animal nutrition

Torbatinejad, N.[1], Galeshi, S.[2], Ghoorchi, T.[1] and Moslemipur, F.[1], [1]Gorgan University of Agricultural Sciences and Natural Resources, Animal science, Shahid Beheshti Ave., 49138-15739, Gorgan, Iran, [2]Gorgan University of Agricultural Sciences and Natural Resources, Agronomy, Shahid Beheshti Ave., 49138-15739, Gorgan, Iran; N_torbatinejad@yahoo.com

An experiment was conducted in the Agricultural Research Station of Eraghi Mahalleh, Gorgan-Iran. The aim of this experiment was to measure the nutritive value of forage of foxtail millet grown at the north of Iran in different sowing date ant density. Foxtail millet (Setaria italica L.), was cultivated using 3×3 factorial arrangement of RCBD with three replications. The sowing date levels, were July 1, 16, 31 and August 15; and plant density levels were 30, 45 and 60 plant /m². millet that planted at August 15 were not germinated. Plants were harvested when the seeds were at about late maturity. The result showed that the effects of sowing date and plant density had a significant effect on the NDF and ADF content ($P<0.05$). Among the groups, millet planted at July 16 with density of 30 plant/m2 contained the greatest protein content (%11.93). The effect of date and interaction of date×density on the percentage of NFE was significant ($P<0.05$). The highest percentage of ash (10.0) was obtained from plant grown on date of July 1 and density of 45 plant / m². Main effects and their interaction had not any significant effect on the Ca and P content of millet forage ($P<0.05$). DMD of millet forage in different treatments were varied between 41.8% to 58.9%. Similarly DE was varied between 7.5 and 10.5, ME between 6.0 and 8.7 MJ/kg and TDN between 82.5% and 86.0%. The highest rate of rumen gas production related to the 12 hours incubation time (46.8-58.8 Mm). In conclusion, of the forage of millet planted at July 1 with density of 60 plant /m² is suggested for animal use because of their best rate of DM production and highest percentage of chemical composition (especially protein) in the climatic condition of north of Iran.

Optimizing of rapeseed meal use for broiler chicks

Golzar-Adabi, S.[1], Kamali, M.A.[2] and Moslemipur, F.[3], [1]Ankara University, Turkey and Organization of Agricultural Jahad, Animal science, East Azarbaijan, 49617-45176, Tabriz, Iran, [2]Agricultural Research and Education Organization, Animal science, Ministry of Agriculture, Tehran, Iran, [3]Gorgan University of Agricultural sciences and natural resources, Animal science, Shahid Beheshti Ave., 49138-15739, Gorgan, Iran; shahramadabi@yahoo.com

To evaluate effects of dietary replacement of soybean meal (SBM) with rapeseed meal (RSM), three hundreds female broiler chicks (61days old, of Cobb breed) divided into 6 treatment groups, with 4 replicates and 15 birds/replicate. The amount of metabolizable energy (AME_n), glucosinolates and erucic acid contents of RSM were determined with Sibald method and HPLC, respectively. Crude protein (CP) content of RSM was measured, then RSM was replaced at levels 0 (control), 20, 40, 60, 80 and 100% of SBM. Progoitrin, total glucosinolates, erucic acid, AME_n and CP contents of RSM were 39.9 µ mole/g, 69 µmole/g, 3.47 µmole/g, 1990.4 Kcal/kg and 32%, respectively. The feed conversion ratio in groups 60, 80 and 100% were significantly different vs control and also groups 20 and 40% ($P<0.01$). After slaughter, no significant differences were observed in relative weight of pancreas among treatment groups, but groups 80 and 100% caused significant increases in relative weights of gizzard, ingluvies full jejunum, empty cecum, heart and liver vs control ($P<0.05$). Serum concentrations of triiodothyronine and thyroxine in groups 80 and 100% were lower than control ($P<0.01$). Viscosity of jejunum and ileum raised significantly by increasing the level of RSM replacement, therefore groups 80 and 100% had the lowest apparent total tract digestibility of non-starch polysaccharides and dry matter ($P<0.05$). Economic efficiency improved by replacing, due to the lower cost of SBM. In conclusion, SBM can be satisfactorily replaced with RSM at 40% level in broiler chicks' diets, but for higher replacing levels, it is necessary to provide diets with better quality. Furthermore, high levels of RSM would impair the performance of broilers.

Effects of Calf Biogenic Anti-Scour (CBAS) supplementation of milk on serum concentrations of IgG, insulin, IGF-1, glucose and albumin, and on weaning date and TSFDM in newborn Holstein calves

Ghorbani, R.[1], Torbatinejad, N.M.[1], Rahmani, H.R.[2], Hassani, S.[1] and Moslemipur, F.[1], [1]Gorgan University of Agricultural sciences and natural resources, Animal science, Shahid Beheshti ave., 49138-15739, Gorgan, Iran, [2]Isfahan University of technology, Animal science, P.O.B. 84156, Isfahan, Iran; N_torbatinejad@yahoo.com

Synthetic milk supplements improve health of newborn calf. Eighteen newborn Holstein calves divided into three groups receiving one or two doses of Calf Biogenic Anti-Scour (CBAS) containing bovine growth hormone, IGF-1, prolactin, insulin and so on, or phosphate-buffered saline (as control) added to colostrum, and then to whole milk. Blood samples were collected at 0, 7, 14, 28, 42 days after birth and also at weaning date. Sera were analyzed for IgG, insulin, IGF-1, glucose and albumin concentrations. The time of cessation of feeding milk (as weaning date), and the time start to feed dry matter (TSFDM) were recorded, individually. All calves were fed equally throughout the experiment. Results showed that CBAS supplementation has no significant effects on serum insulin, IGF-1, glucose and albumin ($P>0.05$), while it resulted in a marked increase in IgG concentrations ($P<0.0001$). There was not significant difference in IgG concentrations between CBAS groups ($P>0.05$). Weaning date was shorter in CBAS groups vs control, especially in group receiving two doses ($P=0.035$). In other hand, TSFDM preceded in CBAS groups as compared to control ($P=0.008$). It was not different between two CBAS groups ($P>0.05$). In conclusion, we suggest that CBAS supplementation is effective to improve health of newborn calves, especially by enhancing IgG biosynthesis. Furthermore, CBAS can potentially improve gut development by shortening weaning date and also TSFDM.

Effects of feed form and nutrient requirements on performance, mortality and meat yield cost of broiler chicks

Samiei, R.[1], Yaghobfar, A.[2], Dastar, B.[3] and Moslemipur, F.[3], [1]Agricultural organization of Golestan, Gorgan, Iran, [2]Agricultural Research and Education Organization, Animal science, Ministry of Agriculture, Tehran, Iran, [3]Gorgan University of Agricultural sciences and natural resources, Animal science, Shahid Beheshti Ave., 49138-15739, Gorgan, Iran; samiei_25@yahoo.com

An experiment was conducted to compare the effects of feed form (pellet vs. mash) and dietary nutrients' density (min vs max of recommended requirements) on performance of broiler chicks, designed in a 2×2 factorial arrangement. Eight hundreds of Ross 308 male broiler chicks were assigned for 4 treatment groups with 8 replicates and 25 birds/replicate, and raised on litter floor pens for 42 days. Results showed that feed processing and dietary nutrients' density had a significant effect on feed intake for the first and third weeks of raising, respectively ($P<0.05$). Furthermore, max nutrients' density and pelleting of feed resulted in a significant increase of weight gain ($P<0.05$). No significant effects were observed for two dietary nutrients' density in feed conversion ratio, but pelleting significantly improved feed conversion ratio as compared to mash form throughout the experiment ($P<0.05$). Birds were fed pelleting feed had higher mortality than those were fed mash feed. Dietary nutrients' density had no significant effect on meat yield cost, but pelleting feed significantly decreased it. We introduce the feed form and dietary nutrients' density as effective factors on performance, health and economic aspects of broilers production.

Degradability characteristics of dry matter and crud protein of triticale silage treated with microbial additive, molasses and urea in fattening male calves

Kocheh Loghmani, M.[1], Forughi, A.R.[1] and Tahmasebi, A.M.[2], [1]Hasheminejhad high ejucational center, animal science, shahid kalantary street mashhad, 9178695398, Iran, [2]Ferdosi university, animal science, shahid kalantary street mashhad, 9178695398, Iran; afroghi@yahoo.com

This research was conducted to study, degradability characteristics of dry matter and crud protein of triticale silage treated with microbial additive, molasses and urea in fattening male calves, In this experiment triticale whole crop was harvested, chopped, and ensiled with urea, molasses, form aldehyde and microbial additive for 42 days in 12 experimental silage were included. Without additive (1), 1% urea (2), 0/5% urea (3), 3% molasses (4), 6% molasses (5), 3% molasses and 0/5% urea (6), 6% molasses and 1% urea (7), 65 molasses and 0/5% urea (8), 3% molasses and 1% urea (9), microbial additive (10), 1% formaldehyde (11), 2% formaldehyde (12), 2% (dm) formaldehyde. pH, air stability, dry matter and apparent quality were significant between experimental silage ($P<0/05$). Results showed there was significant difference between experimental silage for crud protein (cp), ADF. NDF, Ca, P and EE amount ($P<0/05$). The results of the incubation diets with using nylon bag technique in time 24 (h) showed that dry matter and protein rumen degradation in silage 8 in compare with other silage were increased significantly ($P<0/05$). The results of this experiment indicated that triticale silage treated with molasses and urea increased dry matter and protein rumen degradation and stability, dry matter and apparent quality.

Compensatory growth features after two periods of feed restriction in Dalagh rams
Safarzadeh-Torghabeh, H., Ghoorchi, T., Hassani, S. and Moslemipur, F., Gorgan University of Agricultural sciences and natural resources, Animal science, Shahid Beheshti Ave., 49138-15739, Gorgan, Iran; Ghoorchit@yahoo.com

To investigate effects of feed restriction on growth parameters and carcass composition, 21 Dalagh rams (5 month-old) divided into 3 treatment groups including; no restriction (control), 30 days under feed restriction (30R) and 60 days under feed restriction (60R). Control animals were fed at recommendations while 30R and 60R animals were fed under maintenance recommendations. Then in refeeding period, all animals were fed adlibitum with the same feedlot ration supplying beyond of their requirements for 8 weeks. Feed intake and weight gain were weekly measured, and meat samples were collected individually and then analyzed. Animals in groups 30R and 60R consumed more feed than controls, and also, effect of the length of restriction was significant ($P<0.05$). Furthermore, the ratio feed intake/body weight was greater in feed-restricted groups, especially in 60R ($P<0.05$). Weekly weight gain has a marked increase in feed-restricted groups vs control. This ratio was significantly greater in group 60R vs 30R ($P<0.05$). Feed restriction improved feed conversion ratio as compared with control, and it was lower in group 60R than others ($P<0.05$). Fat and protein contents of meat were affected by feed restriction where fat content decreased in feed-restricted groups while protein content increased instead ($P<0.05$). Fat/protein ratio of meat in group 60R showed a significant decline vs control and 30R groups. In conclusion, feed restriction can alter growth features and carcass composition of sheep and this depends upon length and intensity of restriction.

Protected protein sources may reduce nitrogen excretion of dairy cattle and soybean meal import
De Campeneere, S., De Boever, J.L. and De Brabander, D.L., Institute for Agricultural and Fisheries Research, Animal Sciences, Scheldeweg 68, 9090 Melle, Belgium; sam.decampeneere@ilvo.vlaanderen.be

Most dairy diets induce high rumen NH3 concentration peaks 1-2 h after feeding. With 2 dairy cow trials (T), we tried to flatten the rumen NH3 curve post-feeding, by reducing RDPB (rumen degradable protein balance), but also to provide sufficient NH3 throughout the rest of the day (by using formolated (F) rapeseed meal (R) and/or soybean meal (S)). In T1, treatment T1HS (high RDPB, S) was compared with T1LFS (low RDPB, FS) and T1LFSR (low RDPB, FSR) to evaluate if lowering the RDPB could be compensated by using FS or FSR. When RDPB decreased from 161 (T1HS) to -40 (T1LFS) and -54 g/day (T1LFSR), milk production (MP) decreased ($P<0.001$) with 2.3 and 2.5 kg. N excretion was lower ($P<0.01$) for T1LFS and T1LFSR: 9.9 and 9.6 g N per kg FPCM than for T1HS (10.3 g) and N-efficiency of T1LFSR (0.35) was significantly higher than that of T1HS (0.33), with T1LFS in between (0.34). T2 (with treatments T2HS, T2LS and T2LFS) also showed a decrease of the MP ($P=0.023$) when RDPB was lowered from high (150 g/d) to low (-60 and -105 g/d) with an average decrease of 1.2 kg. FPCM production of T2HS was 1.0 kg and 0.7 kg higher than for T2LS and for T2LFS, proving FS to be as effective as S at a low RDPB. N excretion was lower for T2LS and T2LFS (10.6 and 10.2 g/kg FPCM) vs. 11.2 for T2HS, with N-efficiency for T2LS and T2LFS, being 0.34 and for T2HS 0.33. Protection of protein sources, reduces the need for S import. Treatments of T1 received 68 (T1HS), 30 (T1LS) and 23 g (T1LFS) and those of T2 86 (T2HS), 82 (T2LS) and 32 (T2LFS) g of imported S per kg of FPCM, respectively. In conclusion, FS or FSR could not compensate for a reduction in RDPB. With lower RDPB, MP was reduced, but also N excretion was reduced and N efficiency was increased. Protection of S can importantly reduce the import of S without indications of adverse effects on cow performances. The use of protected local R can further reduce the need for S. This work was financially supported by IWT-Vlaanderen.

Effect of clinoptilolite zeolite on daily gain, carcass characteristics, blood characteristics, physiological reactions and feeding behaviors in Holstein beef steers

Yazdani, A.R. and Hajilari, D., Gorgan University of Agricultural sciences and natural resources, Animal science, Shahid Beheshti Ave., 49138-15739, Gorgan, Iran; aryazdani@yahoo.com

In this experiment, 27 crossbred steers (250 kg body weight) were fed a diet with clinoptilolite (CLN), a natural zeolite, substituted at 0, 2.5 and 5% of the diet dry matter. The objective of this study was to evaluate effects of natural zeolite on daily gain, carcass characteristics, blood characteristics, physiological reactions and feeding behaviors in Holstein beef steers. The animals were assigned randomly to 1 of the 3 groups having 7 animals in each with different ration. The experiment lasted for 210 days. Average Daily Gain of steers in T3 diets were highest ($P<0.05$) compare to control groups and T2 diets. Daily dry-matter intake for all the treatments diets was not significantly different. Hot carcass weight and dressing percentage in T3 was significant among others but lean cuts percentage and quality grade for all the treatments were not significant among each other. Overall physiological reactions like body temperature and respiration rate in all the treatments were not significant. Likewise on the observation of feeding behaviors lesser time was taken in rumination time in control groups than other treatments. In feeding time none of the experimental steers showed significant effect. In case of blood profile which was consist of Hemoglobin, Glucose, PCV and total protein also measured. Among all the treatments, were not significantly different.

Changes on milk fatty acid composition in cow´s milk after supplementation

Roca Fernández, A.I., González Rodríguez, A., Vázquez Yáñez, O.P. and Fernández Casado, J.A., Agrarian Research Centre of Mabegondo, Animal Production, Ctra. AC-542 Betanzos-Santiago km 7,5, 15080, Spain; antonio.gonzalez.rodriguez@xunta.es

Many dietary factors, which affect conjugated linoleic acid (CLA) concentrations in milk fat have been researched over the last years, being possible to modify the milk fatty acid composition through the ration, increasing the insaturated fatty acids. The CLA has several associated health promoting attributes, including anti-carcinogenic, growth promotion and anti-obesity activities. The effect of supplementation with oilseeds (cotton) compared to concentrates with cereal grains (barley) was studied in order to establish differences between the fatty acid composition and the CLA content in milk fat of dairy cows. Milk yield and composition were analyzed across 70 days in autumn with three herds of Holstein-Friesian dairy cows (n=36) at end of lactation (mean calving date, 19th February and milking, 200 days), two under cotton (C): at two levels (5 and 7 kg DM cow^{-1}day^{-1}) and one under barley (B) and 7 kg DM cow^{-1}day^{-1}. The fatty acid composition and the CLA content were determined in milk samples collected across ten weeks by gas chromatography. A data analysis was performed using the statistical program SPSS 15.0. There were significant differences for milk production between treatments (C5: 14.6, C7: 16.9 and B7: 17.4 kg day^{-1}, respectively). Milk urea content, protein and fat were significantly higher in the B7 than in the cotton treatments. There were not significant differences for CLA content, but at the high rate more CLA appeared in concentrate with cotton (C7: 3.9 and B7: 3.7 g kg^{-1}). However, linoleic acid was significantly higher in the C7 than in the B7 (23.7 and 21.1 g kg^{-1}, respectively). Saturated fatty acids, lauric and palmitic acid, decreased in both treatments with cotton, while stearic acid increased. The results confirm that milk fatty acid composition changes after supplementation and the CLA content and acid linoleic can be increased by offering supplements containing high rate of cotton.

Utilization of treated vegetable and fruit market waste silage in feeding lactating goats
Khattab, M.S.A.[1], Abo El-Nor, S.A.H.[1], Kholif, S.M.[1], El-Sayed, H.M.[2] and Khorsheed, M.M.[2], [1]national research centre, dairy science, al-behos street, dokki, cairo, egypt, 12622, Egypt, [2]faculty of agriculture, ain shams university, animal production, shoubra el-kheima, 11241, Egypt; msakhattab@yahoo.com

Effect of vegetable and fruit market wastes silage on productive performance of lactating goats was studied. Six lactating Zaraibi goats after 7 days of parturition were divided into three groups using 3x3 Latin square designs to study the effect of treatments on nutrients digestibilities, milk yield and composition. Animals were fed basal diets consisted of 50% concentrate: 50% roughage. The first group was fed concentrate feed mixture (CFM) and the roughage source was Darawa (T_1). The second group was fed 50% CFM + 50% vegetable and fruit market wastes (VFMW) silage treated with lactic acid bacteria (T_2). The third group was fed 50% CFM + 50% VFMW silage treated with formic acid (T_3). Digestibilities of DM, OM, CP, CF and EE were significantly ($P<0.05$) increased with treated silages compared to T_1. Yield of milk was slightly higher in T_2 followed by T_3 and then T_1. However, milk total protein, fat, and solids not fat contents did not differed significantly ($P>0.05$) among treatments. While, results of milk lactose recorded a significant increase ($P<0.05$) in T_2 followed by T_3 and then T_1. It could be concluded that supplementing LAB or formic acid to VFMW silage for lactating goats slight improved nutrients digestibility and slightly increase in milk production and composition with no deleterious effect on general health of the treated animals as compared to animals fed the control diet.

Juveniles of Pseudoplatystoma fasciatum fed with lyophilized bovine colostrum: IGF-I expression in muscle and intestine
Pauletti, P., Rodrigues, A.P.O., Cyrino, J.E.P. and Machado-Neto, R., University of São Paulo - ESALQ/ USP, Animal Science - Zootecnia, Av. Pádua Dias, 11, 13418900, Brazil; raul.machado@esalq.usp.br

The production of IGF-I in various tissues suggest paracrine and autocrine actions, that are involved in organ growth in fish. In relation to speckled catfish (Pseudoplatystoma fasciatum) no information exist regarding the cellular sites of IGF-I synthesis. The objective of this study was to evaluate effects of diets with partial replacement of protein source by lyophilized bovine colostrum (BC), a rich source of IGF-I, on muscle and intestine IGF-I expression of juvenile speckled catfish (35.14±2.23 g and 14.38±0.44 cm, n=3) fed ad libitum for 30 or 60 days with diets containing 45% crude protein and 4,000 kcal/kg with increasing levels of BC (0, 5, 10, 15 and 20%). IGF-I gene expression in tissues was detected by semi quantitative real-time PCR. No specific primers (Ictalurus punctatus) were used to amplify the IGF-I mature region, having 18S ribossomal as an internal standard. The identity of the PCR products was confirmed by direct sequencing. The substitution of the usual protein source by colostrum influenced the weight gain, specific growth rate and food conversion of the juveniles only at 30 days, probably due to an adaptation period to colostrum protein. The muscle and intestine IGF-I expression differed among periods and diets ($P<0.05$). The highest expression of muscle IGF-I was observed at 60 days. The lowest expression was observed in control and 5% BC diets. Differently of the muscle, the lowest expression of intestine IGF-I was observed at 60 days and no differences was observed in response to the highest level of BC. In the intestine, at 30 and 60 days it was observed lower expression compared to muscle. The results demonstrate that the protein source may influence the expression of the IGF-I in extrahepatic tissues. Supported by FAPESP - The State of São Paulo Research Foundation.

Influence of full-fat sunflower seed on performance and blood parameters of broiler chickens
Nassiri Moghaddam, H., Salari, S., Arshami, J. and Golian, A.G., Ferdowsi university of Mashhad, Excellent center of Animal science, Azadi square, 0098 Mashhad, Iran; hnassirim@gmail.com

Influence of full-fat sunflower seed on performance and blood parameters of broiler chickens As an alternative to fats and oils, full-fat oilseeds such as soybean seed are used to replace the supplemented fats and oils in broiler diets. However, soybean seed has anti-nutritional factors such as trypsin inhibitors, which need further processing, thus increasing the cost of soybean seed. Among the various oilseeds available on the market, full-fat sunflower seed (FFSS) contains more ether extract and is available at a relatively low price. This experiment was conducted to study the effect of FFSS on performance of broiler chickens. 176 day-old male broiler chickens were allocated to four treatments with four replicate in a completely randomized design for 7 weeks. Treatments were 0, 7, 14 and 21% of FFSS. The diets were isocaloric and isonitrogenous. At 28 d, blood samples were taken and during the experiment, performance parameters were recorded. Data for all parameters were subjected to an analysis of variance, using the general linear model procedure of SAS. Feed intake and weight gain increased significantly when increasing levels of FFSS was incorporated in the diet during the experiment. Except for 1 to 21 and 1 to 49 days of age, FCR improved significantly. The triglyceride concentrations tended to be lower in the birds fed increasing levels of FFSS, but this effect was not significant. Other factors including glucose, total cholesterol, HDL, LDL, alkaline phosphatase, protein, calcium and phosphorus were not significantly affected. Although a small reduction in LDL and an increase in HDL observed. FFSS was proven as a good source of CP and ME in broiler diets. The results from the current experiment indicated that substitution of FFSS for corn, soybean meal up to 210 g/kg of diet had positive effect on performance parameters.

Is mastitis occurence related to feeding management in dairy herds?
Froidmont, E.[1], Delfosse, C.[1], Planchon, V.[2], Bartiaux-Thill, N.[1], Hanzen, C.[3], Humblet, M.F.[3], Théron, L.[3], Beduin, J.M.[3], Bertozzi, C.[4], Piraux, E.[5] and Jadoul, T.[5], [1]Walloon Agricultural Research Centre, Animal Production and Nutrition Department, rue de Liroux 8, 5030 Gembloux, Belgium, [2]Walloon Agricultural Research Centre, Unit of Biometry, Data Processing and Agrometeorology, rue de Liroux 9, 5030 Gembloux, Belgium, [3]Liège University, Theriogenology of Animal Production, Sart Tilman B42, 4000 Liège, Belgium, [4]Walloon Association of the Breeding, Research and Development, Rue des Champs Elysées 4, 5590 ciney, Belgium, [5]Milkcomite, Route de Herve 104, 4651 Herve, Belgium; froidmont@cra.wallonie.be

The aim of our study was to correlate the feeding characteristics and the fulfilment of the animal requirements to the risk of mastitis by auditing 33 farms. These farms had maintained the same feeding scheme the last 3 months before auditing. All the feedstuffs were sampled for an infra-red spectrometry analysis. In 43% of the farms, the feed was provided under a total mixed diet form, while in the others, the breeders distributed concentrate in addition in amounts depending on the individual production level. The N efficiency of the diets fluctuated between 20.4 and 31.4%. The milk production estimated on the basis of net energy and digestible proteins contents of the diets (allowed milk production) was sometimes lower (until -6.6 l/d/cow), and sometimes higher (until +6.9l/d/cow) to the observed milk production; suggesting respectively a body reserve mobilization or an excessive nutrient supply. Main results suggested that herds receiving an excess of nutrient equivalent to more than 4 l/d of milk had a higher SCC (+158,000 cells/ml, $P<0.05$). They also showed that cows receiving a total mixed diet had a higher SCC compared to those receiving concentrate in function of their production (331,000 vs 248,000 cells/ml, $P<0.05$). These results could be explained by the fact that a misbalanced feeding is a source of stress that could weaken the immune system and induce mastitis.

Chemical composition and degradability parameters of raw and toasted Iranian *Lathyrus sativus* seed
Golizadeh, M., Riasi, A., Fathi, M.H., Naeemipoor, H., Allahrasani, A. and Farhangfar, H., University of Birjand, Department of Animal Science, Faculty of Agriculture, 97175/331, Birjand, Iran; Riasi2008@ gmail.com

The grass pea (*Lathyrus sativus*) is adapted to harsh and low rainfall environments and the seeds have considerable potential as a good quality and cheap protein source in animal nutrition. An experiment was conducted to test some nutritive value of raw and toasted Iranian *Lathyrus sativus* seed (120 °C for 1, 2, and 3 hours). Raw and processed dried samples were ground topass through a 2-mm screen and weighed (1.5 g DM) into the polyester bags (6*10 cm, 50 μm pore size).The bags were incubated in the rumen of fistulated cows (450 + 11 kg) for 0, 2, 4, 8, 16, 24, 48 hours. After removal from the rumen, the bags were washed using cold tab water and then dried in an oven (70 °C for 48 h). The equation of $p = a + b(1 - e^{-ct})$ was used for determination of degradability parameters. The CP, NDF, ADF, EE, and Ash content of Iranian *Lathyrus sativus* seed was 36.1, 18.6, 7.45, 1.32 and 4.24%, respectively. Results showed that raw *Lathyrus sativus* seed had higher rapidly degradable fraction (a) of DM and CP (0.545 ± 0.028 and 0.695 ± 0.025, respectively) than those the toasted seeds. But the seed toasted for 3 hour had the highest slowly degradable fraction (b) of DM and CP (0.678 ± 0.064 and 0.640 ± 0.15, respectively) and its degradation constant rate (c) was lower than the other treatments. It was concluded that the toasting for 3h at 120 °C may have beneficial effects on DM and CP degradability of *Lathyrus sativus* seeds.

The effect of vanilla flavoured calf starter on performance of Holstein calves
Fathi Nasri, M.H., Riasi, A., Arab, A., Kamalalavi, M., Vosoughi, V. and Farhangfar, H., The university of Birjand, Department of Animal Science, Birjand, Iran; mhfathi@gmail.com

Twenty one male Holstein calves were used to evaluate the effects of vanilla flavour added to starter on preweaning and postweaning calf performance. Following 2 d of colostrum and transition milk feeding, calves were assigned in a completely randomized design to 2 treatments including 1) unflavoured starter and 2) flavoured starter. Calves were fed whole milk at 10% of the initial body weight daily and had free access to starter and water. The weaning criterion was defined as the calf age at a daily intake of 0.80 kg of starter for 2 days, consecutively. Increased starter DMI ($P<0.05$) was observed for calves fed flavoured starter during the preweaning period (440 vs. 400 g/d). This effect did not carry over into the postweaning phase, but this increase was observed (($P<0.05$) over the entire experiment for calves fed flavoured starter from 3 d age to weaning (1250 vs. 1050 g/d). Weaning age was attained 2 to 3 d earlier (($P<0.03$) when flavoured starter was fed (57.7 vs. 60.3 d). Average daily gain (ADG) over the preweaning phase was significantly ($P<0.01$) higher for calves fed flavoured starter (400 vs. 330 g/d). No differences in ADG were observed postweaning or over the entire experiment due to treatment. Calves fed flavoured starter had better feed efficiency (ratio of ADG to DMI (starter and milk)) during the preweaning phase (430 vs. 380). These findings demonstrate that supplementing starter with vanilla as a flavour agent is advantageous to calf performance.

The effects of polyethylen glycol supplement on *in vitro* gas production of canola hybrids and canola meals

Kilic, U., Ondokuz Mayis University, Department of Animal Science, Ondokuz Mayis University, Faculty of Agricultural, Department of Animal Science, 55139 SAMSUN, Turkey; unalk@omu.edu.tr

The aim of the study was to determine the effects of polyethylen glycol on *in vitro* gas production and *in vitro* gas production kinetics of some canola hybrids and canola meals. In this study, four canola hybrids (Bristol, Eurol, Capitol and Licrown) and canola meals purchased from market were used. Two rams (SakızxKarayaka) aged 4 with ruminal cannulas were used in gas production technique. All of the feedstuffs were incubated for 3, 6, 9, 12, 24, 48, 72 and 96 hours. The highest total phenolic matter contents (2.03%) were determined for Licrown ($P<0.01$). The effects of PEG supplement on vitro gas production, potential gas production (b) and gas production rate (c) of all canola hybrids and canola meals for 3, 6, 9, 12, 24 and 48 hour incubations were not significant ($P>0.05$). However, PEG supplementation increased *in vitro* gas production and total gas production (a+b) for 72 and 96 hour incubations in Bristol hybrid ($P<0.01$). *In vitro* gas production and total gas production (a+b) were increased by PEG supplementation for 72 hour incubation in Capitol hybrid and for 96 hour incubation in Canola meal ($P<0.01$). Canola meal had higher gas production levels than all the canola hybrids for all incubation periods. ($P<0.01$). In conclusion, PEG supplementation affected *in vitro* gas productions for 72 and 96 hour incubations for all the feeds used in this study ($P<0.01$).

Evaluating rations for high producing dairy cows using three metabolic models

Swanepoel, N.[1], Robinson, P.H.[2] and Erasmus, L.J.[1], [1]University of Pretoria, Pretoria, 0002, South Africa, [2]University of California, Davis, 95616, USA; nanswanepoel@gmail.com

Improving efficiency of dietary nitrogen (N) capture in milk (often considered low at 25-35% in lactating cows) by increasing the utilization of intestinally absorbed AA through supplementation of amino acids (AA) in a ruminally protected (RP) form, is of worldwide interest since excretion of excess dietary N as urea, which is rapidly converted to ammonia in fecal/urine slurries, is volatilized and negatively impacts air quality. Total mixed rations (TMR) and commodity feeds from 16 California dairy farms were sampled and chemically analyzed to evaluate their nutrient profiles using three metabolic models in order to predict AA profiles of intestinally delivered protein, identify potentially limiting AA and to determine if the nutrient profiles of these rations were consistent enough to produce an RPAA package to provide cows with their 'ideal' intestinally delivered AA profile. Feed delivery records were used to calculate dry matter (DM) and N intake of the 16 early lactation groups of cows, while milk production and composition data were used to determine milk N output. Of all models, Shield estimated DM intake closest to the measured value (102%) and was the only model to suggest a correlation between intestinal AA delivery and production. A predicted decrease in Lys:Met ratio from 2.8-2.4, due to increased Lys and decreased Met % in MP, was correlated to increased milk yield (33-51 kg/d; $r^2=0.31$), milk N content (166-229 g/d; $r^2=0.40$) and N intake (572-882 g/d; $r^2=0.24$), suggesting that providing limiting AA (i.e., Lys), has a positive effect on production. There was a high consistency within model among cow groups in the predicted limiting AA sequence among TMR's, suggesting that there is sufficient consistency in the nutrient profiles among these TMR's to support production of a common RPAA complex. However the calculated RPAA packages varied sharply by model, with Amino Cow focused on Met and Lys, CPM Dairy on Ile and Leu and Shield on Lys and Ile.

Vitamin C supplementation did not affect performance and carcass characters of broiler chicks under normal conditions

Dashab, G.H.[1], Mehri, M.[2], Keikha-Saber, M.[1], Sadeghi, G.H.[1,3] and Alipanah, M.[1], [1]Faculty of Agriculture, Zabol University, Animal Science Department, Ferdowsi University of Mashhad, Mashhad, 91775-1163, Iran, [2]Faculty of Agriculture, Payam-e Noor University, Qeshm, Animal Science Department, Ferdowsi University of Mashhad, Mashhad, Iran, 91775-1163, Iran, [3]Faculty of Agriculture,University of Kurdistan, Animal Science Department, Kurdistan, 98617, Iran; mehran.mehri@gmail.com

An experiment was conducted to determine whether supplementation of Vit-C would affect performance of broiler chicks during a 42 day rearing period. Five levels of Vit-C (0, 100, 200, 300, and 400 ppm) were used for 1000 day-old male chicks in a completely randomized design with four replicates (50 birds per each). Vit-C supplementation had no ($P>0.05$) significant effect on body weight gain, feed intake, feed conversion, carcass percentage, thigh percentage, breast percentage and abdominal fat during the starter (0 to 21days) and grower (21 to 42 days) periods. However, body weight gain was numerically higher in chicks fed with a diet containing 300 ppm VC. Supplementation of Vit-C also improved FCR and decreased abdominal fat numerically. In conclusion, addition of vitamin C up to 400 ppm had no beneficial effect on broiler chicken performance and carcass characters and it could be related to normal rearing conditions and absence of stressors in this study.

Effects of vitamin C supplementation on blood variables in broiler chicks

Mehri, M.[1], Dashab, G.H.[2], Keikha-Saber, M.[2], Sadeghi, G.H.[3] and Ebadi-Tabrizi, A.R.[1], [1]Faculty of Agriculture, Animal Science Department, Payam-e Noor University, Qeshm, 91775-1163, Iran, [2]Faculty of Agriculture, Zabol University, Animal Science Department, Ferdowsi University of Mashhad, Mashhad, 91775-1163, Iran, [3]Faculty of Agriculture,University of Kurdistan, Animal Science Department, Kurdistan, 98617, Iran; mehri.mehran@gmail.com

This experiment was carried out to determine effects of vitamin C on blood variables of broiler chicks under normal condition in a 42 days period. One thousand day-old chicks were allocated to five treatments (0, 100, 200, 300, and 400ppm) and 4 replicates (50 birds per each), in a complete randomized design. Blood samples were collected into syringes from wing vein at 20 d and 40 d for measuring white blood cell (WBC), red blood cell (RBC), hemoglobin (HB), hematocrit (HCT), platelet (PLT), neutrophil (Nut), lymphocyte (Lym), monocyte (Mono), eosinophil (Eos), alkaline phosphatase (ALP), calcium (Ca), phosphorus (P), and albumin (Alb). In 20 day old chicks, RBC, HB, HCT, and Ca were affected significantly by supplementation of Vitamin C ($P<0.05$). The highest values of RBC, HB, and Ca were obtained with addition 300ppm of Vitamin C, and the lowest values were belonged to control group. However, the highest value of HCT was appeared in chicks who received 200ppm of Vitamin C in the diet. None of blood variables were affected by vitamin supplementation during the second three weeks of the experiment. Overall, the results of this experiment was not able to show anti-stress specifications of Vitamin C through the blood variables which may be due to absence of stress conditions such as high environmental temperature throughout the study.

Nitrogenous compounds balance and microbial protein production in crossbred heifers fed forage cactus, sugar cane bagasse and urea associated to different supplements

Pessoa, R.A.S.[1], Ferreira, M.A.[1], Leão, M.I.[2], Valadares Filho, S.C.[2], Silva, F.M.[1], Bispo, S.V.[1] and Farias, I.[3], [1]Universidade Federal Rural de Pernambuco/UAG, Recife, Pernambuco, Brazil, [2]Universidade Federal de Viçosa, Viçosa, Minas Gerais, Brazil, [3]Instituto Agronômico de Pernambuco, Recife, Pernambuco, Brazil; pessoa@uag.ufrpe.br

The objective of this work was to evaluate the effect of association of forage cactus (FC) to sugar cane bagasse (SCB) and urea on nitrogenous compounds balance (NCB) and microbial protein synthesis in milk heifers supplemented or not, in Pernambuco State, Northeast of Brazil. Twentyfive Holstein-Gir crossbred heifers with average of 227 kg of live weight (LW) were used, kept in feedlot system and assigned to a randomized block design. The control ration was composed of 64.0% of FC, 30.0% of SCB, 4.0% of urea:ammonium sulphate mix (U:A) (9:1) and 2.0% of mineral mix (MM), in dry matter basis. The heifers were supplemented based on the LW (0.5% of LW). The supplements had characterized the treatments, together with the control treatment. The tested supplements were: wheat meal, soybean meal, cottonseed meal or whole cottonseed. The proportion of ingredients in experimental rations for supplemented animals was: 57.0% of FC, 26.0% of SCB, 3.5% of U:A, 1.8% of MM and 11.7% of supplement. The NCB was not influenced, presenting average value of 49.3 g/day. The cottonseed meal or soybean meal supplementation increased the nitrogen urinary excretion, the urea concentration and urea nitrogen in the serum and the urea urinary excretion and urea nitrogen. The association of FC to SCB and urea, without the use of supplements allowed microbial synthesis efficiency (MSE) of 105.0 g crude protein microbian (CPM)/kg of consumed total digestible nutrients (TDN). The whole cottonseed supplementation provided greater alantoin and purine derivatives urinary excretion and better MSE (127,8 mmol/day; 149,0 mmol/day and 156,3 g CPM/kg of consumed TDN, respectively), being, therefore, the most indicated in such conditions.

Forage cactus, sugar cane bagasse and urea associated to different supplements in diets for crossbred heifers performance

Pessoa, R.A.S.[1], Ferreira, M.A.[1], Leão, M.I.[2], Valadares Filho, S.C.[2], Silva, F.M.[1], Bispo, S.V.[1] and Farias, I.[3], [1]Universidade Federal Rural de Pernambuco/UAG, Recife, Pernambuco, Brazil, [2]Universidade Federal de Viçosa, Viçosa, Minas Gerais, Brazil, [3]Instituto Agronômico de Pernambuco, Recife, Pernambuco, Brazil; pessoa@uag.ufrpe.br

The objective of this work was to evaluate the effect of association of forage cactus to sugar cane bagasse and urea on performance of milk heifers supplemented or not, in Pernambuco State, Northeast of Brazil. Twentyfive Holstein-Gir crossbred heifers with average of 227 kg of live weight (LW) were used, kept in feedlot system and assigned to a randomized block design. The control ration was composed of 64.0% of forage cactus, 30.0% of sugar cane bagasse, 4.0% of urea:ammonium sulphate mix (9:1) and 2.0% of mineral mix, in dry matter (DM) basis. The heifers were supplemented based on the LW (0.5% of LW). The supplements had characterized the treatments, together with the control treatment. The tested supplements were: wheat meal, soybean meal, cottonseed meal or whole cottonseed. The proportion of ingredients in experimental rations for supplemented animals was: 57.0% of forage cactus, 26.0% of sugar cane bagasse, 3.5% of urea:ammonium sulphate mix, 1.8% of mineral mix and 11.7% of supplement. The cottonseed meal or soybean meal supplementation increased the intakes of DM, organic matter (OM), crude protein (CP) and non fiber carbohidrate (NFC). The intake of total digestible nutrients and the digestibility of DM, OM, CP and NFC were not influenced by treatments, presenting average values of 4.2 kg/day and 60.9; 63.1; 77.9; 82.9%, respectively. The association of forage cactus to sugar cane bagasse and urea, without the use of supplements allowed weight gain (WG) of 430.0 g/day. The supplementation with soybean meal, cottonseed meal or whole cottonseed improved the WG (720.0; 840,0; 750.0 g/day, respectively) and feed:gain ratio (10.8; 9.8; 9.1 kg of DM intake/kg of WG, respectively). Therefore, the supplementation must be based on cust of supplement.

Effects of monensin and increasing crude protein in early lactation on performance of dairy cows
Ghorbani, B., Ghoorchi, T., Amanloo, H. and Zerehdaran, S., University of Gorgan, Animal science, Gorgan, 4763778686, Iran; ghorbani.behnam@yahoo.com

Twenty-four Holstein dairy cows were used to evaluate the singular and combined effects of different level of crude protein and monensin treatments during the early lactation on digestion and milk yield of dairy cows. The experiment was designed as completely randomized with a 3×2 factorial arrangement of treatments. The factors were three concentrations of CP supplement (19.5, 21.4, 23.4% of dry matter) and two levels of monensin (0 and 350 mg/cow/d).This experiment consist of three periods and each period were 3 wk in length. Monensin did not affect DMI, milk yield, lactose and SNF but it reduced milk fat and protein percentage. Monensin premix significantly decreased rumen ammonia but rumen pH and microbial protein synthesis was not affected by monensin treatment. Although, Monensin treatment inceased apparent digestibility of DM, NDF, ADF,CP, but they were not significantly. Increasing dietary CP, improved milk and protein production, but did not alter the other components of milk. DMI were reduced with increasing dietary CP, but digestibility of DMI was not affected. Digestibility of NDF, ADF, CP were improved by increasing dietary CP. Feeding 23.4% CP to early lactation dairy cows, decreased body weight loses in first 63 days postcalving. Increasing diet CP from 19.5 to 21.4% did not significantly increase ruminal ammonia, but increasing to 23.4% have significant effect on it. Monensin did not affect urine volume excretion, but there was a linear relationship between level of crud protein in the diet and urine volume excretion. Microbial protein synthesis was affected by increasing CP level, on this way maximum protein synthesis was achieved in 21.4% CP.

Evaluation of polyethylenglycol (PEG6000) as an indigestible marker for dairy ewes fed indoors or grazing
Caja, G.[1], Ralha, V.M.[2] and Albanell, E.[1], [1]Universitat Autònoma de Barcelona, G2R, Campus de la UAB, 08193 Bellaterra, Spain, [2]Universidade Católica Portuguesa, Escola Superior de Biotecnologia, Rua Dr. António Bernardino de Almeida, 4200-072 Porto, Portugal; elena.albanell@uab.cat

With the aim of estimating the individual feed intake of dairy sheep penned in small experimental groups, a total of 12 Manchega dairy ewes at late lactation (71.6 ± 2.8 kg BW; 0.5 ± 0.1 l/d) were used. Ewes were allocated in 2 groups of 6, milked twice daily and fed: 1) indoors with fescue hay ad libitum (90.9% DM; 10.6% CP, 61.3% NDF, DM basis) and concentrate (89.5% DM; 21.3% CP, 13.0% FND, DM basis), or 2) grazing Italian ryegrass (16.9% DM; 30.7% CP, 50% FND, DM basis) in early spring (6 h/d) and supplemented indoors with the same fescue hay ad libitum and concentrate. Concentrate (0.6 kg/d) was fed individually in the milking parlor at milking times (08.00 and 17.00 h). Polyethylene glycol (PEG6000) was diluted in water and dosed orally to each ewe (48 g/d), once daily (after the a.m. milking) during 10 d, and used as an indigestible marker for estimating the individual faecal excretion of DM. Faeces were collected twice daily during the last 5 d (by using total collection bags), sampled (5%), dried (60 °C) and analysed for DM and PEG. A previous calibration (44 faeces samples; 1-12% PEG; $R^2=0.997$) was done by near infrared analysis (NIR) between 1,100 and 2,500 nm. No faeces or intake alteration were reported during the experiment. For indoor feeding conditions PEG recovery was high (101%) allowing an accurate estimation of fecal DM ($R^2=0.94$; $P<0.01$), DM digestibility (62.5%) and total feed intake (1.587 ± 0.113 kg DM/d). PEG recovery was lower on grazing conditions (82%) being its use less accurate for faecal DM estimation ($R^2=0.41$; $P<0.01$). In conclusion, PEG6000 was propossed as indigestible marker for estimating DM faecal excretion in dry diets and for individual intake partitioning in group fed dairy sheep.

Importance of mountain forage for evaluation of feeds and biodiversity protection of protected landscape area of the Czech Republic

Koukolová, V., Homolka, P. and Jančík, F., Institute of Animal Science, Department of Nutrition and Feeding of Farm Animals, Přátelství 815, 104 00 Praha Uhříněves, Czech Republic; vkoukolova@seznam.cz

The objective of this study was to investigate relationship among *in vivo* digestibilities of dry matter (DM), crude protein (CP), neutral-detergent fibre (NDF) and acid-detergent fibre (ADF) of two forages (pasture forage and Deschampsia flexuosa) originated from Krkonoše Mts. National Park of the Czech Republic. The *in vivo* digestibilities of DM, CP, NDF and ADF were determined in metabolic trials using six wethers of the Romanovské breed (weighing 83+9kg). The chemical composition values (g/kg of absolute dry matter) of pasture forage and Deschampsia flexuosa were 238 and 220 g/kg for CP, 19 and 25 g/kg for ether extract, 60 and 52 g/kg for ash, 724 and 641 g/kg for NDF, and 309 and 282 g/kg for ADF, respectively. The *in vivo* digestibilities of DM, CP, NDF and ADF averaged 73, 79, 78 and 71% for pasture forage and 74, 85, 78 and 62% for Deschampsia flexuosa, respectively. Significant differences ($P<0.05$) between the pasture forage and Deschampsia flexuosa in the *in vivo* digestibilities of CP and ADF were observed. The study was supported by the Ministry of Agriculture of the Czech Republic (MZE0002701403 and MZE0002701404).

The effect of Akomed R and lipase on the performance and carcass characteristics in early weaned rabbits

Zita, L., Tůmová, E., Bízková, Z., Ledvinka, Z. and Stádník, L., Czech University of Life Sciences Prague, Dept of Animal Husbandry, Kamýcká 129, 16521, Czech Republic; stadnik@af.czu.cz

The objective of the present study was to investigate the effect of commercially available oil Akomed R[®], lipase addition and weaning age on growth, feed consumption and carcass value in broiler rabbits. In the experiment rabbits were weaned at the age of 21 and 34 days (20 ones per age) and were placed in individual cages. At each weaning age rabbits were split into 2 groups which received a commercial pelleted type feed mixture (9.5 MJ ME, 17.7% crude protein) and experimental type feed mixture included 1% of Akomed R[®] (60.8% caprylic acid, 38.7% capric acid and 0.5% lauric acid) and 0.5% lipase. Rabbits were so assigned into four groups. Temperature of 16 °C and relative humidity about 65% were maintained during the whole fattening period. Water and feed were available ad libitum. Results of the experiment did not show significant effect of supplement of Akomed R® and lipase on live weight of rabbits. There was a positive effect of Akomed R[®] and lipase on feed conversion where rabbits had a significantly ($P\leq0.001$) lower feed conversion (3.00 and 3.33 kg, respectively) in comparison with rabbits without Akomed R® (3.14 and 3.57 kg, respectively). There was no significant effect of weaning age on growth and weight gain, but the feed consumption was higher in rabbits weaned at 34 days of age. Final live weight at 77 days of age was non-significantly higher in both groups of rabbits weaned at 34 days of age (2,964 and 3,063 g, respectively) in comparison with rabbits weaned at 21 days of age (2,947 and 2,803 g, respectively). Dressing percentage was significantly ($P\leq0.001$) higher in both groups of rabbits weaned at 34 days of age. The proportion of renal fat was negatively affected by weaning age and positively by the supplement. Rabbits weaned at 21 days of age had a higher renal fat content. The study was supported by Ministry of Education, Youth and Sports of the Czech Republic (Project No. MSM 6046070901).

High dietary crude protein clearly affects dry matter intake, milk production and nitrogen metabolism in fresh cows
Ghelich Khan, M., Amanlou, H. and Mahjoubi, E., Zanjan University, Animal Science, Zanjan-Tabriz Road, 313, Iran; ghelich.khan@znu.ac.ir

The objective of this study was to investigate the effect of high dietary crude protein (CP) levels on dry matter intake (DMI), milk production and nitrogen metabolism in fresh cows. Solvent-extracted soybean meal (SSBM) containing high energy and CP was selected for this purpose and the top-dressing method was used for worker affair facilitation. Twenty-one Holstein fresh cows free of clinically diagnosed transition disorders were used in this experiment. The cows were randomly assigned to 1) basal diet (CP=20.3%), 2) basal diet + 1 kg of top-dressing SSBM (CP=21.8%) and 3) basal diet + 2 kg of top-dressing SSBM (CP=24.4%). The cows were individually fed immediately after parturition until wk 4 and were milked 6 times a day. DMI increased noticeably from treatment 1 to 3 (17.53, 18.02 and 20.58 kg/d, respectively; $P<0.05$). Average raw milk yields were 35.98, 36.87 and 42.27 kg/d, respectively and tended to be significant ($P=0.11$). Milk fat percentage decreased significantly (4.65, 4.51 and 3.86%, respectively; $P<0.01$). 3.5% fat corrected milk yield increased while the treatments did not have noticeable differences (42.35, 42.52 and 45.24 kg/d, respectively). Milk protein and fat yields were not affected by treatments. Ruminal concentration of NH_3-N and total VFA increased as dietary CP level increased but the differences in treatments were not significant. The treatments did not influence uric acid concentration, though a tendency was detected ($P<0.15$). Urinary urea excretion increased by adding to the dietary CP and treatments had noticeable differences (11.01, 16.01 and 19.72 mg/dl, respectively; $P<0.01$). These results demonstrated that by adding top-dressing SSBM and using high levels of CP, DMI and milk production increase and at the same time ruminal condition and score feces stay desirable.

Serum enzyme status of Chios ewes fed increasing amounts of copper from copper sulfate
Bampidis, V.A.[1], Christodoulou, V.[2], Chatzipanagiotou, A.[3], Sossidou, E.[2] and Salangoudis, A.[4], [1]Alexander Technological Educational Institute, Department of Animal Production, P.O. Box 141, 57400 Thessaloniki, Greece, [2]National Agricultural Research Foundation, Animal Research Institute, Giannitsa, 58100 Giannitsa, Greece, [3]Aristotle University, School of Agriculture, Thessaloniki, 54006 Thessaloniki, Greece, [4]Papanikolaou Hospital, Laboratory of Biochemistry, Thessaloniki, 57010 Thessaloniki, Greece; bampidis@ ap.teithe.gr

Eighteen lactating 3-year-old Chios ewes were used in an experiment to determine effects of orally administered copper on serum aspartate aminotransferase (AST), L-alanine aminotransferase (ALT), lactate deydrogenase (LDH) and alkaline phosphatase (ALP) levels. The experiment started after weaning on day 42 postpartum, and lasted 6 weeks. Ewes were allocated, after equal distribution relative to milk yield and body weight, into three treatments of 6 ewes each, and were accommodated in one floor pen/treatment. All ewes were offered daily a diet that contained a basal level of 16.4 mg copper. Ewes in treatment Cu0 received no additional copper (control), while those in treatments Cu60 and Cu95 received 60 and 95 mg additional copper, respectively, as an oral solution of copper sulfate. Therefore, ewes in treatment Cu0, Cu60 and Cu95 consumed daily 16.4, 76.4 and 111.4 mg Cu, respectively. The serum AST (132±8.9 U/l), ALT (22±1.6 U/l), LDH (557±15.5 U/l) and ALP (137±5.0 U/l) levels of ewes were similar ($P>0.05$) among treatments and all ewes remained clinically healthy until the end of the experiment. Results suggest that Chios ewes exhibit tolerance to copper supplementation for a period of 6 weeks.

Critical impacts of high dietary crude protein on dry matter intake, blood metabolites and energy balance in fresh cows

Ghelich Khan, M., Amanlou, H. and Mahjoubi, E., Zanjan University, Animal Science, Zanjan-Tabriz Road, 313, Iran; ghelich.khan@znu.ac.ir

This study was conducted to investigate the effects of high levels of crude protein (CP) on blood metabolites, liver Enzymes, BCS and BW in fresh cows. Solvent-extracted soybean meal (SSBM) consisting high energy and CP was selected for this purpose and the top-dressing method was used for worker affair facilitation. Twenty-one Holstein fresh cows free of clinically diagnosed transition disorders were assigned randomly to 1) basal diet (CP=20.3%), 2) basal diet+1 kg of top-dressing SSBM (CP=21.8%) and 3) basal diet+2 kg of top-dressing SSBM (CP=24.4%). The cows were individually fed immediately after parturition throughout 4 weeks and were milked 6 times a day. As dietary CP level increased, plasma glucose concentration increased (41.03, 42.46 and 46.75 mg/dl, respectively) but treatment differences were not noticeable. Increased dietary CP resulted in decreased plasma Non-Esterified Fatty Acid (NEFA) concentration (0.96, 0.92 and 0.49 mmol/l, respectively) which tended to be significant ($P<0.15$) and also causes the concentration of plasma β–hydroxybutryate (BHBA) to decline significantly among treatments (2.97, 2.65 and 1.49 mmol/l, respectively; $P<0.05$). Plasma concentrations of insulin, cholesterol, BUN, albumen, P and Ca were not affected by treatments. BW changes (BW losses) in this study respectively were -59.76, -72.61 and -42.61 kg, and noticeable ($P=0.04$). BCS changes (-0.48, -0.40 and -0.37, respectively) were not influenced by treatments. To make sure of liver integrity and function, the enzymes AST (Aspartate Aminotransferase), AP (Alkaline Phosphatase) and GGT (Gamma Glutamyl transferase) were measured at the beginning and end of the study which were not affected and remained in normal reference range. These results confirm that in a period with low appetite and DMI phase lag, by consuming high levels of CP, DMI increases significantly ($P<0.05$) and undesirable plasma metabolites, BW losses and BCS changes also decrease.

Use of textile industry by-products of boehmeria nivea as livestock feed

Pace, V., Carfì, F. and Contò, G., CRA-PCM, Via Salaria 31, 00016 Monterotondo (RM), Italy; vilma. pace@entecra.it

The aim of this study was to carry out a chemical and nutritional evaluation on different parts of ramie plant to verify a possible use of the textile industry by-product as feedstuff for ruminants. Four stocks of samples from different crops and different areas were analyzed. The standard chemical composition was determined on the whole plant and on its parts. The study was particularly focused on leaves and tops, not used in fibre production and available to feed livestock. Organic matter enzymatic digestibility was determined by a double step method: the first based on cellulosolitic enzymes (cellulase and hemicellulase) and the second on proteolytic enzymes (pepsin). Chemical analysis showed a good crude proteins content on leaves, tops and whole plant: 17.00±1.52%, 15.25±0.77% and 11.79±3.32% on DM basis respectively. A normal fibre fraction amount was also observed (leaves: ADF=30.82±6.48%, NDF=33.19±4.27% and ADL=11.27±4.58%; tops: ADF=36.18±3.98%, NDF=39.07±5.62% and ADL=13.07±0.58%; whole plant: ADF=43.59±9.50%, NDF=48.17±9.55 and ADL=11.45±2.20%). Ash content resulted high in all plant parts, with values of 23.08±2.62% in tops, 22.50±2.19% in leaves and 17.46±3.21% in the whole plant. Ash analysis showed high calcium concentration, especially in leaves, with values exceeding 4% of DM. The other minerals were in a normal range, except the lead, present at higher levels (from 4 to 7 mg/kg). Organic matter enzymatic digestibility resulted low: 34.23±7.24% in leaves, 29.96±6.76% in tops and 25.09±8.66% in the whole plant, and the energy values were 0.33, 0.28 and 0.21 MilkFU/kg DM respectively. The observed low digestibility levels might be due not only to the high content of minerals and lignin, but also to the presence of other unknown factors able to reduce the enzymatic activity. Palatability tests were carried out on sheep, offering dried leaves or the whole fresh plant, minced and fed alone or mixed to alfalfa hay: in the first trial the leaves were refused, in the second one the feedstuff was accepted without problems.

Effects of propionic acid, formic acid and molasses plus urea on corn silage microorganisms

Ghanbari, F.[1], Ghoorchi, T.[1], Khomeiri, M.[2], Ebrahimi, T.[2] and Hosseindoost, A.[1], [1]Gorgan University of Agricultural Science and Natural Resources, Animal Science, Beheshti Street, Gorgan 4913815739, Iran, [2]Gorgan University of Agricultural Science and Natural Resources, Food Science and Tchnology, Beheshti Street, 4913815739, Iran; farzadghanbari@yahoo.com

This study was conducted to determine the effects of some silage additives on corn silage microorganisms. Prior to ensiling, whole crop corn forage was choppd and treated with water(control), propionic acid (1%), formic acid (0.8%) and molasses plus urea (13%). Ensiling was carried out in 10 liter bucket. Silos were opened after 60 dayes for evaluation of total, acid lactic bacteria, yeast and mold counts. The data were analysed as a Completely Randomized Design with 5 replicates for each treatment. Total count of microorganisms in silages treated with water (control) and molasses plus urea was significantly higher than other treatments ($P<0.05$). Lactic acid bacteria count was significantly higher in silage treated with molasses plus urea compared to other treatments ($P<0.05$). Yeast pollution was not observed in any of the treatments. By using propionic acid, formic acid and molasses plus urea, the yeast count in the control group showed a drastic decrease, up to zero.

Effects of canola meal diets on growth performance, carcass characteristics and thyroid hormones in finishing lambs

Ghoorchi, T., Rezaeipour, V., Hasani, S. and Ghorbani, G.R., Gorgan university of agricultural sceinces and natural resources, Animal sceince, Gorgan, Golestan, 49138-15739, Iran; ghoorchit@yahoo.com

A completely randomize design was used with 4 treatments and 6 replications (lamb) in each treatment was conducted to investigate the effects of canola meal on growth performance, carcass quality and thyroid hormones in lambs. The experimental groups were different levels of canola meal (0, 33, 66 and 100 percent) instead of cottonseed meal. Lambs were weighed, and feed consumption was recorded for feed efficiency ratio computation in every month and whole of the experimental period. Finally carcass analysis was done for every treatment. The canola meal used in this experiment contained 14.75 micromole per gram DM aromatic glucosinolates.The results showed that the effects of different levels of canola meal on daily gain during the whole experimental period was not significant ($P>0.01$), but in 2nd month of gain recording was significant ($P<0.01$).feed consumption show a significant difference between treatments ($P<0.01$), however, feed efficiency does not affected by increasing of canola meal inclusion rate in the diets ($P>0.01$). Carcass characteristics, except for liver weight, was not statistically significant ($P>0.01$). Effect of different levels of canola meal on thyroid hormones secretion was not statistically significant ($P>0.01$). According to the results in this experiment, we can replace include canola meal in the diets of finishing lamb without any problem.

Strategies for decreasing environmental emissions and pollution through animal nutrition

Denev, S.A., Petkov, G.S. and Mihailova, G.S., Trakia University, Student Campus, 6000 Stara Zagora, Bulgaria; stefandenev@hotmail.com

The ability to manage modern animal production systems for minimal environment pollution has improved dramatically during the past decades, but continues to be one of the most significant challenges facing the animal industry around the world. Nutrition plays a significant role in environmental sustainability of non-ruminant production, but the development and implementation of 'eco-nutrition' swine and poultry feedings programs is in its infancy. The diet has a significant impact on the types and amounts of nutrients excreted and consequently the precursors of gas and odor emissions. The amount of nutrients and associated odors emitted from non-ruminant animals into the environment can be modulated by several different nutritional strategies. In general, nutrient excretion may be reduced by using nutritional manipulations to enhance nutrient utilization in the animals. Other strategies include: the development of environment friendly feeding programs; increasing the number of feed phases to better meet the animal's age-related requirements; formulating diets to include the minimal amounts of nutrients required to satisfy production goals; meeting the animal's amino acid requirements; using high quality protein sources and feed ingredients with high digestibility; formulating diets based on nutrient availability instead of total nutrient content and etc. Feed additives, such as acidifiers, probiotics, prebiotics, symbiotics, phytobiotics, enzymes, have received considerable attention about their ability to modulate the gastrointestinal microflora, to increase the digestibility and absorption of nutrients, to reduce the nutrient excretion, environmental emissions and pollution. This paper reviews the nutritional (dietary) strategies to reduce environmental emissions and pollution from non-ruminants.

Parcial replacement of soy bean meal by urea on Giralanda lactation cows ration

Saran Netto, A.S.[1], Fernandes, R.H.R.[2], Barcelos, B.[3], Conti, R.M.C.[1] and Lima, Y.V.R.[2], [1]FZEA, ZAZ, Duque de Caxias Norte 225, 13635045, Brazil, [2]FMVZ, VNP, Duque de Caxias Norte 225, 13635045, Brazil, [3]Anhanguera Educacional, Veterinária, Waldemar Silenci 340, 13614 370, Brazil; saranetto@yahoo.com

Rations with non protein nitrogen has its importance on milk production cost, although it has to be considered milk function on human health. Twelve Girolanda dairy cows were used to compare partially replacement of soybean meal by urea on lactation performance. Two treatments were used: Control (C) and Urea (U), the nitrogen sources of the then were soybean meal for control treatment and soybean meal partially replaced by urea (45%), 1/3 of total protein, on Urea treatment. Milk production, composition and physico-chemical characteristic were measured. Protein nutrition was observed by blood urea nitrogen. Somatic cell count was used to monitor udder health. There were differences ($P>0.05$) among treatments only for milk production 9.41 for Urea treatment vs. 7.98 kg/cow/day for Control treatment. Milk composition and physico-quimical characteristics were similar among treatments ($P>0.05$). This study indicated that using urea for lactating Girolanda dairy cows had no detrimental effect on lactation performance or on milk nutritional value. Keywords: milk composition, milk production.

Body condition scoring method for the blue fox (*Alopex lagopus*)

Kempe, R.[1], Koskinen, N.[2], Peura, J.[1], Koivula, M.[1] and Strandén, I.[1], [1]MTT Agrifood Research Finland, Biotechnology and Food Research, H-building, FI-31600 Jokioinen, Finland, [2]MTT Agrifood Research Finland, Animal Production Research, Tervamäentie 179, FI-05840 Hyvinkää, Finland; riitta.kempe@mtt.fi

During the last decade, blue fox has been bred to be large and fat, because price of the pelt is mostly determined by pelt size. In order to attain large pelt size, blue fox is fed unrestricted during growth period. This excessive dietary intake and inadequate utilisation of energy gives both very fat and obese foxes at pelting time. However, extreme obesity should be avoided because of its negative impact on animal health. Obesity is detrimental especially to young vixens used for breeding. Extreme fattening followed by fast and outstanding weight loss influences negatively fertility traits. New breeding strategies need to take into account obesity, and an easy way to record it is needed. Body condition scoring (BCS) is a subjective method for estimating subcutaneous fat cover. In this study BCS measure was assessed for 868 blue foxes on a scale of 1 to 5, where 1 is very thin and 5 obese. BCS measure is an indicator of body fat stores and its categories reflected the amount of subcutaneous fat, fat content of carcass, body weight and grading size. The BCS measure was compared with five other measurements: fat thickness measured with ultrasonic equipment, fat content of the whole carcass, body weight, grading size and animal length. BCS had the highest correlation with body weight. The BCS method developed in this study proved to be a useful and practical tool for assessing the degree of obesity in the blue fox. The method allowed distinguishing fatness in an animal from its large size. The BCS method can be applied to live animals during growth period. However, presence of a long heavy hair coat can complicate visual appraisal and then palpation of the animal determines the body condition more accurately. Assessing the nutritional status and degree of obesity of foxes via BCS is a convenient and time saving way in comparison to weighing a fox.

Composition and effect of feeding date by products on ewes and lamb performances

Najar, T.[1], Ayadi, M.[2], Casals, R.[3], Ben M'rad, M.[1], Bouabidi, M.A.[4], Such, X.[3] and Caja, G.[3], [1]INAT, P A, 43, av C. Nicolle, 1082, Tunisia, [2]ISBA Médenine, SCQ, km 22 Route Djorf Médenine, 4119, Tunisia, [3]UAB Barcelona, G2R, Facultat de Veterinària, 08193, Spain, [4]CSFPA - Ph., Dégache - Tozeur, 2260, Tunisia; najar.taha@inat.agrinet.tn

With the aim to evaluate the nutritive value of Tunisian date by products (DB), 2 experiments were conducted on D'Man ewes. Date by products are whole dates rejected for human consumption. A total of 11 DB varieties were sampled and analyzed. In Exp. 1, suckling ewes (n=20) were used during 90 d to compare a control diet (C; 1.0 kg oats hay and 0.7 kg concentrate, as fed) to a DB diet (1 kg oats hay, 0.35 kg concentrate and 0.35 kg DB mixture). Whole or seed less DB nutritive composition, feed intake and lamb BW were measured. In Exp. 2, concentrate was progressively substituted by the DB mixture during 21 d in 12 ewes. Initial daily offer of DB was 0.4 kg and final was 0.7 kg with an increase rate of 50 g every 3 d. Composition of DB changed according to variety, being the ranges: CP (1.99 to 4.50%), NDF (10.52 to 26.05%) and ADF (7.15 to 21.96%). Whole DB have higher contents than seedless DB: CP (3.24 vs. 2.88%), NDF (18.11 vs. 13.93%), ADF (13.01 vs. 9.77%) and ADL (5.99 vs. 5.63%), with a 5% net energy difference (1.92 vs. 2.02 Mcal NEL/kg DM, $P=0.071$). In Exp. 1, DB mixture composition was intermediate (3.04% CP, 15.24% NDF, 11.05% ADF) and lambs ADG between 10 and 45 d, did not vary according to feeding treatment (C, 113 + 22; DB, 128 + 13 g/d; $P>0.05$). In Exp. 2, ewes feed intake decreased as rate of concentrate substitution by DB increased. Total daily intake decreased from 1.64 to 1.31 kg (as fed) from the beginning to the end of the experiment, as a consequence of the decrease of hay intake (0.24 kg/d) and refuse of DB seeds (0.093 kg/d). In conclusion, DB can substitute concentrate at a minimum rate of 25% in sheep diets, although fill value increases with DB incorporation. New experiments are needed to determine the optimum and maximum levels of incorporation.

Effect of nitrogen supply on inter-organ urea flux in dairy cows

Røjen, B.A. and Kristensen, N.B., University of Aarhus, Faculty of Agricultural Sciences, Blichers Allé, P.O. Box 50, 8830 Tjele, Denmark; Betina.AmdisenRojen@agrsci.dk

The objective of the study was to investigate urea metabolism in lactating dairy cows supplied with decreasing amounts of N (15.0 to 12.6% CP). Eight Holstein cows (19±1 kg DMI/d, 34±3 kg milk/d) fitted with ruminal cannulas and permanent indwelling catheters in major splanchnic blood vessels were assigned to three treatments in an incomplete triplicate 3x3 Latin square design. Treatments were: Control (C; basal ration + water infusion), INF-L (basal ration + continuous ruminal infusion of 4.1 g urea/kg DMI), and INF-H (basal ration + 8.5 g urea/kg DMI). Cows were fed equally sized portions at 8 h intervals. Eight hourly sets of urine and arterial, portal, hepatic, and ruminal blood samples were obtained, starting 30 min before morning feeding on last day of each period. Data was analyzed using the mixed model procedure of SAS. Arterial blood urea N conc. decreased ($P<0.01$; 7.8 to 3.3±0.2 mmol/l) with decreasing N supply. The net hepatic flux and urinary excretion of urea N decreased linearly ($P<0.01$) from 587 to 334±47 mmol/h and 276 to 81 mmol/h, respectively, with decreasing urea infusion. The gut entry rate of urea N (net portal uptake + salivary urea N flux) was not affected by treatment ($P=0.34$; 317±29 mmol/h). Urinary urea N excretion increased from 25% of net hepatic flux with C to 52% with INF-H ($P<0.01$), meaning that the cows transferred as much as 75% of the urea N produced to the GI tract with C. Urinary urea N excretion in percent of N supply (dietary + infused) decreased with decreasing urea infusion from 23 to 8±1%. The GI tract clearance of blood urea N increased linearly ($P<0.01$; 650 to 1255±102 mL/min) with decreasing N supply. In conclusion, the cows are able to increase the GI tract clearance of blood urea N on a diet increasingly deficient in N. However, the total gut entry rate of urea N was not affected by treatment and cows were not able to sustain rumen microbes with increasing amounts of recycled urea N when N supply was decreased.

Nutrient composition of the Iberian sow's milk: effect of environmental temperature

Aguinaga, M.A., Haro, A., Castellano, R., Seiquer, I., Nieto, R. and Aguilera, J.F., Instituto de Nutrición Animal (IFNA), Estación Experimental del Zaidín (CSIC), Cno del Jueves s/n, 18100 Armilla, Granada, Spain; jose.aguilera@eez.csic.es

Breed differences are widely recognised to affect sow milk production and composition, but there is a lack of information concerning this subject in the Iberian sow. Although a negative influence of high ambient temperatures on milk synthesis at the mammary gland has been described, the magnitude of this effect on the Iberian sow has not still been investigated. Eight Iberian sows (136±2 kg BW) in their third pregnancy were involved in the study. Four of them farrowed in summer and the other four in autumn. One week before farrowing the sows were individually housed in pens placed in a farrowing room. The environmental temperature in the farrowing room ranged between 25-29 °C in summer and 20-24 °C in autumn. Once in the pens the sows were fed a commercial lactation feed (per kg, as-fed basis: 12.76 MJ ME, 144 g CP, 6.8 g Lys) at 1% BW. On the day of parturition they were offered 1.5 kg of the feed and thereafter, daily food allowance was increased by 0.6 kg to reach 4.5 kg per day the fifth day of lactation, which was maintained onwards. Litter size was adjusted to 6±1 by cross fostering. Milk samples from all functional teats of all sows were collected weekly from day 7 to day 35 of lactation. Samples were frozen and lyophilized until analysis. Fat and energy contents of milk (59.5 g/kg and 24.73 MJ/kg) were not affected by either season or day of lactation. The protein content, however, increased gradually from 53.0 on day 7th to 59.6 g/kg at the end of the lactation. No effect of the environmental temperature was noticed. On the contrary, significant increases ($\approx$20%) in calcium and phosphorus contents were observed at the highest temperature. Milk appeared progressively enriched in both minerals over lactation. Ca and P contents of milk at days 7 and 35 raised from 2.15 to 3.25 and from 1.79 to 2.35 g/kg, respectively. These values are notably higher than those reported for milk from sows of conventional breeds.

The non-volatile organic acids content in Italian green crops

Pezzi, P.[1], Fusaro, I.[1], Manetta, A.C.[1], Angelozzi, G.[1] and Formigoni, A.[2], [1]University of Teramo, Department of Food Science, 212, Viale Crispi. Teramo, 64100, Italy, [2]University of Bologna, DIMORFIPA, 50, Via Tolara di Sopra, Ozzano Emilia (BO), 40064, Italy; ifusaro@unite.it

The objective of this study was to assess the concentration of main organic acids content in several green crops that are administered to cattle after different preservative processes. These acids, except trans-aconitic, are intermediate of citric acid cycle. Malic acid is involved in preventing ruminal acidosis in cattle and it represents, with soluble sugars, a rapidly fermentable source of energy. 72 samples of green crops of different species, at different vegetative stadium, were collected. All samples were analysed by HPLC with an ion exchange stationary phase. Simultaneous determination of citric, malic, trans-aconitic and fumaric acid were performed. Also chemical composition was determined. The samples were divided in: 20 samples of Italian Ryegrass (*Lolium italicum*), 8 samples of fescue grass (*Festuca pratensis*), 8 samples of triticale (x Triticosecale), 15 samples of corn (Zea mays, 13 samples of alfalfa (*Medicago sativa*) and 8 samples of white clover (*Trifolium repens*). Mean total organic acids content were respectively 29.7, 22.1, 23.3, 11.5, 43.0 and 31.6 g/kg DM. Malic acid was the main non-volatile organic acid in all crops, except corn, and it represent respectively 83.8, 87.1, 65.8, 22.9, 72.1 and 89.9% of the total amount. Citric acid, except corn and triticale, was the second one and its concentration was respectively 15.0, 12.2, 13.4, 7.4, 22.4 and 8.5 g/kg DM. Trans-aconitic is the most important non-volatile organic acid in corn (7.3 g/kg DM) and the second one in triticale (6.2 g/kg DM) while it is very low in all other forages. Malic acid was the most correlated with total organic acids content (r=+0.949; $P<0.001$). Excluding corn and triticale, citric, trans-aconitic and fumaric acids were all positively correlated ($P<0.001$) with ADL, DM, ADF and soluble protein and negatively correlate ($P<0.001$) with crude fat.

Blood biochemical changes and sperm quality in bulls fed diet with dry extract from *Tribulus terrestris*

Petkova, M., Grigorova, S. and Abadjieva, D., Institute of Animal Science, Pochivka 1, 2232 Kostinbrod, Bulgaria; m_petkova2002@abv.bg

It is developed a Bulgarian product Vemoherb T on the basis of dry extract from annual herb *Tribulus terrestris* (TT) L (Zygophylaceae). The main active components of this herb are saponins of the furostanol type protodioscin and protogracilin. Many clinical investigations on the effect of TT as ecological product on humans and labor animals were carried out in Bulgaria and in other countries. There are also our investigations on the effects on rams, rabbits, laying hens, cocks, mini cocks. There exist no data concerning the effect of TT extract in bulls of service. This study was conducted to asses the effect of Vemoherb T on the blood biochemical changes and sperm quality in bulls of service. The experimental protocol shows that the study was carried out with 9 bulls of service from three different breeds. Bulls were fed daily rations, composed by meadow hay (65%) and compound feed (35%) and supplemented or not with Vemoherb T (3 mg/kg body weight daily) dissolved in the compound feed. The treatment continued 2x40 days. Blood samples were collected once day for the treatment and at end of the experiment, morning before feeding. The total cholesterol, also HDL, LDL, VLDL, glucose, total protein, urea, GOT, GPT, testosterone in blood serum were determined. The sperm was collected by artificial vagina two times per week and quality parameters (volume, concentration, motility, survivability) were estimated. The comparison of experimental results from feeding Vemoherb –T shows that the investigated parameters after the treatment are within the reference range for the corresponding animal species and categories. This experiment demonstrates suitability of Vemoherb T to improve bulls reproductive capacity and give opportunity to use greater amount from this product with experimental aim.

The effect of concentrate to forage ratios on *Ruminococcus albus* population in rumen fluid of Holstein steers determined by real-time PCR

Vakili, A.R.[1], Danesh Mesgaran, M.[1], Heravi Musavi, A.[1], Hosseinkhani, A.[2], Yáñez Ruiz, D.R.[3] and Newbold, C.J.[4], [1]Ferdowsi University of Mashhad, Dept. Animal Science, Exellence center for Animal Sicence, P.O. Box: 91775-1163, Iran, [2]University of Tabriz, Dept. Animal Science, 51757, Iran, [3]Unidad de Nutrición Animal Estación Experimental del Zaidín (CSIC), Granada, E-18008, Spain, [4]Institute of Biological, Environmental and Rural Sciences, Aberystwyth University, SY23 3AL, United Kingdom; Vakili_ar@yahoo.com

The objective of the present experiment was to investigate the effect of concentrate to forage ratios on the population of *Ruminococcus albus* in the rumen fluid of Holstein steers (300±15 kg, body weight) fitted with rumen canola. Animal were fed experimental diets (7 kg of DM/d) differing in their concentrate (155 g CP/kg DM; 30% maize, 34% barley, 8% soybean meal, 5% sugar beet pulp, 10% wheat bran, 12% cottonseed meal, 0.3% $CaCo_3$, 0.5% mineral and vitamin premix) to alfalfa hay (155 g CP/kg DM) ratios [60:40 (T1), 70:30 (T2), 80:20 (T3), and 90:10 (T4)] in a 4×4 Latin square design. Samples of rumen fluid were taken before the morning feeding and 4 h post feeding, then, DNA was extracted (Qiagen Ltd, Crawley, West Sussex, UK). *Ruminococcus albus* rDNA concentrations were measured by real time PCR relative to total bacteria amplification ($\Delta\Delta Ct$). The 16s rRNA gene-targeted primer sets used in the present study were forward: CCCTAAAAGCAGTCTTAGTTCG and reverse: CCTCCTTGCGGTTAGAACA. Cycling conditions were 95 °C for 10 min, 45 cycles of 94 °C for 10s, 55 °C for 20s and 72 °C for 15s; fluorescence readings were taken after each extension step. Data were analyzed using the GLM procedure of SAS and the means compared by the Tukey test ($P<0.05$). The results of the present study demonstrated that increasing the inclusion of concentrate in diets caused a change in the population of Ruminococcus albus in the free rumen fluid taken before and 4 h after the morning feeding [T1=35 and 33, T2=6 and 8, T3=17 and 23, T4=45 and 39, SEM=17 and 13 ($\times 10^{-5}$), respectively].

Effects of pellet size and sodium bentonite on growth and performance of milk fed Brown Swiss calves

Rastpoor, H.[1], Foroughi, A.R.[1], Shahdadi, A.R.[1], Saremi, B.[1], Naserian, A.A.[2] and Rahimi, A.[1], [1]Education Center of Jihad Agriculture, Animal Science, Asian High Way, between Jihad Sq. and Jomhuri Sq. Mashhad., 9176994767, Iran, [2]Ferdowsi University, Agriculture college, Animal Science, Asian High Way, Azadi Sq., 1111, Iran; afroghi@yahoo.com

In order to investigating the effects of pellet size and Sodium Bentonite (SB) on growth and performance of dairy calves, 24 milk fed Brown Swiss calves were offered different calf starters and allocated randomly to treatments immediately after birth. Experimental treatments were including: 1) Pelleted size (PS) (4mm diameter) contain 2% SB, 2) PS (4mm diameter) without SB, 3) PS (6mm diameter) plus 2% SB and 4) PS (6mm diameter) without SB. Calf starter (NRC 2001) prepared from birth to 75 days after birth adlibitum. Daily Dry Matter Intake (DMI), body weight and body characteristics (Body length, wither height, hip height, pin width, hip width, pin to hip, stomach size, heart girth, metacarpus and metatarsus size) were measured every 15 days. Data were analyzed in a completely randomized design with repeated measurement analysis using mixed procedure of SAS 9.1 and means were compared via lsmean ($P<0.05$). Data without time effect were analyzed in a completely randomized design and means were compared using Duncan test ($P<0.05$). Results showed that there were no significant difference between average daily gain (ADG) of calves pre and post weaning. DMI pre weaning in treatments 2 and 4 was significantly higher than treatments 1 and 3 ($P<0.05$). Although there were not any significant difference in feed conversion ratio between treatments pre weaning, but feed conversion ratio post weaning showed significant difference ($P<0.05$). Treatments had no effects on age, body weight, and body characteristics of calves at weaning. It seems that using of 4mm pellet size could improve feed conversion ratio. Also Sodium Bentonite could reduce DMI without any effect on ADG, and feed conversion ratio.

Effects of pelleted starter and different levels of sodium bentonite on blood and ruminal metabolites and nutrient digestibilitiy of neonatal dairy calves

Shahdadi, A.R.[1], Foroughi, A.R.[1], Rastpoor, H.[1], Saremi, B.[1], Naserian, A.A.[2] and Rahimi, A.[1], [1]Education center of Jihad Agriculture, Animal Science, Asian High way, Between Jihad Sq. and Jomhouri Sq.mashhad, 9176994767, Iran, [2]Ferdowsi University, Agriculture college, Animal Science, Asian High way, Azadi Sq. Mashhad, 1111, Iran; afroghi@yahoo.com

24 Brown Swiss dairy calves weighting 39.5 ± 1.2 kg were randomly assigned into four treatments in order to investigating the effects of pelleted starter and different levels of Sodium Bentonite (SB) on blood and ruminal metabolites and nutrient digestibility of dairy calves. Treatments were including: 1) Pelleted size (PS) (4 mm diameter) contain 2% SB, 2) PS (4 mm diameter) without SB, 3) PS (6 mm diameter) plus 2% SB and 4) PS (6 mm diameter) without SB. Diets were formulated to be isonitrogenous (NRC 2001). Blood and rumen liquid samples were obtained every 15 days. Nutrient digestibilities were determined 5 days at weaning and at the end of experiment. Data were analyzed in a completely randomized design with repeated measurement analysis using mixed procedure of SAS 9.1 and means were compared via lsmean ($P<0.05$). Results showed that treatments had no significant effect on blood metabolites (Glucose, Blood Urea Nitrogen, Total Protein and Triglyserides). Rumen pH in treament 1 was significantly higher than other treatments ($P<0.05$). Treatments had no significant effect on Dry Matter, organic Matter and NDF digestibility pre and post weaning, but crude protein digestibility pre and post weaning were significantly affected by treatments ($P<0.05$). Although crude protein digestibility pre weaning in treatments 1 and 2 was significantly higher than treatments 3 and 4, but crude protein digestibility post weaning in treatments 2 and 4 was significantly higher than treatments 1 and 3 ($P<0.05$). It seems that using of 4 mm pellet size could improve digestibility of crude protein. Also diets containing Sodium Bentonite could increase crude protein digestibility.

Prevalence of propanol fermentation in corn silage

Raun, B.M.L. and Kristensen, N.B., University of Aarhus, Faculty of Agricultural Sciences, Blichers Allé 20, 8830 Tjele, Denmark; birgitte.raun@agrsci.dk

The objective of this study was to investigate the prevalence of propanol fermentation in corn silages at Danish dairy farms and monitor the seasonal pattern of propanol levels in corn silage. Twenty randomly selected dairy farms feeding corn silage were visited 5 times at 2 months intervals from January to September 2007. Samples were obtained by drilling vertical cores (5 cm in diameter) one meter behind the bunker face. Water extracts of silages were analyzed for alcohols, low-molecular weight esters, ammonia, VFA, and L-lactate. Effects of time and correlations among variables were analysed using the mixed and corr procedures of SAS. Defining propanol fermentation as propanol content of at least 5 g/kg DM at any time point, the observed prevalence of propanol fermentation was 20% (4 out of 20 farms). Corn silage from farms designated as having propanol fermentation increased ($P=0.03$) from 3.1 ± 1.0 g/kg DM in January to 6.5 ± 0.9 in September and showed both a numerically greater content and greater increase compared with silage from the other farms (1.1 ± 0.4 in January to 1.8 ± 0.5 in September). The observed propanol content ranged from not detectable to 9.1 g/kg DM. Propanol content was correlated with contents of acetic acid (r=0.86), propionic acid (r=0.46), propyl acetate (r=0.88), propanal (r=0.60), 2-butanol (r=0.81), ammonia (r=0.49), and L-lactic acid (r=-0.38). Propanol fermentation had a relatively high prevalence (20%) at Danish dairy farms and the propanol levels in silage were increasing during the season. The fermentation profile of propanol silage indicates that these silages have a high aerobic stability partly due to the high acetic acid content. However, numerous reports from the dairy industry also indicate great concerns about the impact of propanol fermentation products on dairy cows. The high prevalence of propanol fermentation emphasizes the need for a better understanding of the influence of this fermentation profile on silage palatability and dairy cow metabolism.

Growth performance and feeding behaviour of cattle supplemented with different levels of babasu palm (*Orbignya phalerata*) silage

Faria, P.B.[1], Babilônia, J.L.[1], Bressan, M.C.[2], Rodrigues, M.C.O.[1], Silva, D.C.[1], Anjos, M.A.[1], Morais, S.B.[1], Pereira, A.A.[1] and Gama, L.T.[2], [1]IFECTMG, Santo Antonio do Leverger-MT, 78106-000, Brazil, [2]INRB, Fonte Boa-Santarém, 2005-048, Portugal; peterbfvet@yahoo.com.br

Babasu is a palm tree very common in northern Brazil. An experiment was conducted to evaluate the impact of replacing corn silage by babasu silage (BS) in confined cattle. Castrated Nelore males (n=25) were used, with initial live weight of 256±2.0 kg. Animals were given commercial concentrate (1% of live weight/d), and assigned to five treatments, where corn silage was either provided ad libitum (treatment B0), or replaced at 25, 50, 75 and 100% by BS (treatments B25, B50, B75 and B100, respectively). After an adaptation period of 14 d, cattle were kept in individual pens, where feed consumption was measured, and time spent ingesting feed, ruminating, resting and ingesting water was assessed through visual observation for periods of 12 h, with data collected at intervals of 15 minutes. Feed consumption declined linearly as the proportion of BS in the diet increased, with a reduction of about 3.2 to 5.3 kg in the ingestion of feed/d per 25% increase in BS, such that the mean feed intake/d was 23.3 kg in B0 and 7.6 kg in B100. Average daily gain was similar in B0 and B25 (about 1.1 kg) but dropped afterwards as BS increased in the diet, to reach a mean value of 0.2 kg in B100. Animals in B100 had higher resting time (6.68 h) than the other treatments (ranging between 5.09 and 5.70 h), but lower rumination time (2.54 vs. 3.28 to 3.65 h). Time spent ingesting feed was higher in B0 and B100 (2.56 and 2.60 h, respectively) than in the other treatments (3.03 to 3.15 h). No differences were observed between treatments in time spent drinking water. Overall, the inclusion of BS as a substitute of corn silage at a level above 25% of the roughage intake caused a decline in feed intake and growth rate and changed the feeding behaviour of cattle, with an increase in resting time and a reduction in time spent in rumination.

The effects of supplemented diet with fish oil and canola oil during transition period to early lactation on milk yield, dry matter intake, and metabolic responses of early lactating dairy cows

Vafa, T.S., Naserian, A.A., Heravi Mousavi, A.R., Valizadeh, R., Danesh Mesgaran, M. and Khorashadizadeh, M.A., Excellent Center for Animal Science, Faculty of Agriculture, Ferdowsi University of Mashhad, P.O. Box 91775-1163 Mashhad Khorasan Rzavi, Iran; vafa_toktam@yahoo.com

The study was designed to test the effect of including fish oil and canola oil from transition period to early lactation on milk yield and metabolic responses in Holstein dairy cows. Cows were randomly assigned in treatments: 1) 0% oil (control, n=9) and 2) 2% oil (supplemented, 1% fish oil-1% canola oil, n=9) from -2 to 7 weeks relative to calving. Cows were blocked by parity, previous 305-2x milk production and expected calving time. Dry matter intake (DMI) was recorded daily and feed sample was collected weekly. Cows were milked 3 times per day and daily yields were recorded. Using vacutainer tubes, blood samples were collected weekly before the morning feeding, kept on ice and centrifuged within 30 min at 3000 x g for 20 min. Aliquots of serum were stored at −20 °C until analysis for glucose, triglyceride, cholesterol, and serum urea nitrogen (SUN). The data repeated in time were analyzed by using a mixed model (PROC MIXED, SAS Inst. Inc., Cary, NC) for a completely randomized design with repeated measures. Inclusion of fish oil and canola oil in diet increase milk yield significantly ($P=0.042$). Dry matter Intake ($P=0.72$; 21.16 and 20.19±0.50), blood glucose ($P=0.92$; 57.79±1.46 and 57.61±1.39 mg/dl, respectively), cholesterol ($P=0.37$; 113.86±3.63 and 118.28±3.42 mg/dl, respectively), SUN ($P=0.45$; 18.21±0.45 and 18.68±0.43 mg/dl, respectively), and triglyceride ($P=0.45$; 23.61±1.22 and 24.86±1.09 mg/dl, respectively) were similar between control and supplemented diets in seven weeks after calving. The results show that including fish oil and canola oil had no apparent effects on metabolic responses, but milk yield was significantly increase in supplemented diet.

Aerobic preservation of brewer's grains with a blend of chemical substances
Acosta Aragón, Y.[1], Kreici, A.[2], Roth, N.[1], Klimitsch, A.[2] and Pasteiner, S.[1], [1]Biomin Holding GmbH, Industriestrasse 21, 3130 Herzogenburg, Austria, [2]Biomin Research Center, Technopark 1, 3430 Tulln, Austria; yunior.acostaaragon@biomin.net

A trial was conducted to determine the needed dosage to stabilize fresh brewer's grains under aerobic conditions for a week using a blend of propionic acid and sodium benzoate (PNB). The fresh material was transported to the laboratory and divided into two parts. One half of the material was preserved with its natural dry matter (DM) content (20- 22%), the other one was dried up to 30% of DM. Treatments for preservation in each DM range were 0, 2, 3, 4, 5 and 6 kg PNB/ ton. Samples were stored at 25 °C for 1 week in opened containers. The parameters evaluated were DM, pH value, temperature and colony count of yeasts and moulds (only in the treatments with 0, 3 and 6 l PNB/ ton). Each treatment was done in 3 replicates. The DM content as well as temperature in the different treatments did not vary markedly during the experimental week. Growth of moulds was evident after 5 days in the treatment with 20% DM and no PNB added. After 6 days, moulds were visible in treatments with no PNB, independently of the DM content, as well as in the treatment with 2 kg PNB/ ton and 20% DM content. PH development was very different and depended on DM level. After 3 days pH levels in treatments with 20% of DM increased considerably to 5.5, indicating instability and favorable conditions for the growth of bacteria mainly. PH values in treatments with 30% DM content never exceeded 4.7. Mould infestation tended to increase rapidly in the treatment with no PNB added and 20%DM, from less than 10^2 to $10^{4.5}$ (after 2 days) and $10^{6.9}$ (after 7 days). When the brewer's grain was treated with PNB, independently of the DM content or the PNB dosage, no growth of moulds was identified. After 7 days of the experiment, growth of moulds in materials with higher DM content (30%) was less (log cfu = 3.8 vs. 2.3 for 20 and 30% DM).

Feeding value of Kangar (*Gandelia tournefortii*) hay and the growth performance of Bluchi lambs fed by diets containing this hay
Valizadeh, R., Madayni, M., Sobhanirad, S. and Salemi, M., Ferdowsi University of Mashhad, Animal Sciences, P.O. Box 91775-1163, Mashhad, Iran; rvalizadh@yahoo.com

This experiment investigated the feeding value of kangar (*Gandelia tournefortii*) hay and the effect of diets containing this hay on growth performance of Baluchi lambs. Kangar hay was collected from the Khorasan natural rangeland at the stage of late maturity and evaluated in terms of chemical composition, in sacco degradability and a growth study with lambs. Twenty-four male and 24 female lambs were allocated to 4 dietary treatments in a feedlot condition. All groups received a concentrate mixture of 60% and 40% roughage including alfalfa hay (25%), dry what straw (15%), soaked what straw (15%), dry kangar hay (15%) and soaked kangar hay as treatments 1, 2, 3 and 4 respectively. Chemical composition mainly CP of kangar hay was much higher than wheat straw and even comparable to alfalfa hay. The average DM degradability of kangar hay was 67.2% after 120 h incubation, but more than 76% of the incubated DM degraded during the first 24 h. These values for CP and NDF were 73.5 and 54.5% respectively. These parameters ranked kangar hay as a medium-quality forage. Male lambs fed by TMR containing the soaked Kangar hay gained better than other groups. Average daily gains of male and female lambs on dietary treatments of 1, 2, 3 and 4 were 218, 237, 241, 276 and 197, 215, 229 and 259 g respectively. The best feed conversion ratio was also recorded for the male lambs on diet containing 15% soaked kangar hay (6.5 kg feed consumed per kg of weight gain). The mean dressing percentage for the lambs allocated to treatments 1, 2, 3 and 4 were 54.7, 54, 9, 53.4 and 53.5 for males and 53.1, 52.7, 54.0 and 55.7 for females respectively. It was concluded that inclusion of Kangar hay can be beneficial mainly for smallholder farmers during periods of low rainfall and forage scarcity.

Effect of supplementation of NSP enzymes in association with phytase on egg production parameters in laying hens fed maize and soybean meal based diet

Mathlouthi, N.[1], Massaoudi, I.[2], Sassi, T.[3], Majdoub-Mathlouthi, L.[4], Uzu, G.[5] and Bergaoui, R.[2], [1]Ecole Superieure Agriculture du Kef, Production Animale, Route Dahmani, 7119 Le Kef, Tunisia, [2]Institut National Agronomique de Tunisie, Production Animale, 43 avenue Charles Nicolle, 1082 Tunis, Tunisia, [3]CRISTAL, 21, rue de l'Usine, Z.I. Charguia II, 2080 Ariana Aéroport, Tunisia, [4]Institut Supérieur Agronomique de Chott-Mariem, Production Animale, Chott-Mariem, 4042 Sousse, Tunisia, [5]ADESSEO, 42, avenue Aristide Briand, 92160 Antony, France; lmajdoub@lycos.com

The aim is to evaluate the response of laying hens to multi-enzymes preparation containing xylanase, b-glucanase and phytase. 768 laying hens aged 25 weeks, were fed basal diet of maize (59.9%) and soybean meal (22%) without (BD-), or supplemented (MEP) with 50 mg of multi-enzymes preparation / kg of diet during 20 weeks (5 periods of 28 days each). They were fed a third diet based on (BD+) 63% of maize and 24% of soybean meal. BD+ was formulated to cover the nutritional requirements for laying hens. BD- contained less nutriment than BD+ taking into account the nutritional supplementation of 50 mg of multi-enzymes preparation. Egg production (% hen-day), egg weight, egg mass, feed intake, feed conversion ratio and body weight were recorded for 20 weeks. Results showed that the addition of the multi-enzymes preparation in diet increased egg production (93.96 vs. 92.18%, $P<0.05$) compared to BD-. Moreover, the egg mass (58.98 vs. 57.22 g/day/hen) and the feed conversion ratio (1.914 vs. 1.994) of layers fed MEP based diet were better than those fed BD- based diet. Feed intake in BD- layer group was higher than in MEP group. Nevertheless, layers fed BD- and MEP based diets had similar egg weight and changes in body weight. Egg production parameters of laying hens fed MEP and BD+ based diets were similar. Finally, the results indicated that the use of NSP enzymes in association with phytase in hen diets can reduce the dicalcium phosphate use without affecting the laying hen performances.

The effect of compost residue obtained from edible fungi (*Pleurotus Florida*) production in fattening performance of crossbred Taleshi-Holstein calves

Mirza Aghazadeh, A., Pasha Zanousi, M. and Taha Yazdi, M., Urmia university, Animal science, Urmia, 57153-165, Iran; a.aghazadeh@urmia.ac.ir

During the last decades production of fast growing edible fungi has been useful and economical activity. Some of these fungi such as *Pleurotus Florida* were effective in degradating grain straws and improving their nutritive value as animal feed. In order to study the effects of residue compost obtained from *Pleurotus Florida* fungi on the fattening performance of male crossbred (Taleshi × Holstein) calves, the production of fungi on the straw bed were done for the period of three month. After the harvesting period, the remained compost was air dried and stored. Sixteen crossbred (Taleshi × Holstein) male calves were selected and randomly distributed in a completely randomized block design with four treatments (T1 control group, T2 with 10, T3 with 20 and T4 with 30% compost residue of *Pleurotus Florida*, respectively) each with four replicates, for the period of 120 days (22 days as a pre experiment adaptation period). Significant differences were not found in feed intake, body weight gain and feed efficiency between different treatments ($P<0.05$). The cost of feed consumption per kg of body weight gain were 10217, 8122, 6486 and 5787 RL, respectively. The cost of feed was decreased by increasing the level of compost in the ration, and the group four was most economical. It was concluded that, the compost remained after biological treatment with *Pleurotus Florida* could be used up to 30% level of the fattening rations of male crossbred calves without adverse effect.

Effect of electron-charging of feed and water on the blood characteristics in growing lamb
Iwakura, T.[1], Sato, K.[2], Kojima, K.[3], Niidome, K.[3] and Tobioka, H.[1], [1]Tokai University, School of Agriculture, Minamiaso-mura, Aso-gun, Kumamoto 869-1404, 869-1404, Japan, [2]Kumamoto University, Graduate School of Medicine and Pharmacology, 5-1 Ohe Honmachi, Kumamoto-shi, 862-0973, Japan, [3]JEM Co, 341-2 Kamiyoshida, Yamaga-shi, 861-0524, Japan; 8anfm001@mail.tokai-u.jp

In order to establish the environmentally friendly system of animal production without any chemical, many studies have been conducted with the live bacteria like probiotics and acidifiers in the past 20 years. The electron-charging system of feed and water which is an alternative for them was applied to the growing lambs. Blood TBARS and free-radical scavenging ability and generative capacity were investigated. Three heads of lamb of 4 months were fed on concentrate feed under high (27 °C) and normal (22 °C) temperature (temp) for 4 months with 9 terms, referring to the 2 terms of roughage-based feed under high and normal temp. The 2 terms of electron charging where electron-charged feed and water were given to animals were placed between no-charging terms. The blood samples were taken before feeding and 4 hr after feeding in each term. The blood TBARS and free radical scavenging ability and generative capacity using electron spin resonance (ESR) method with spin trap agent were determined. Basement of free radical spin adduct tended to be lower for treatment periods of electron-charging. The intensities of free radical scavenging ability and generative capacity showed the higher tendency under high-temp conditions. The generative capacity of free radical tended to increase from 1 to 2 term under high temp and from 5 to 6 term under normal temp where feed and water were electron-charged. TBARS also tended to be lower in electron-charging periods. Therefore electron-charging of feed and water have the effects on scavenging and generative ability of free radical and decrease of lipid peroxide formation.

Use of PCR method for *Enteroccocci* identification in gastrointestinal tract of poultry after probiotics application
Kačániová, M., Fikselová, M., Nováková, I., Pavličová, S. and Haščík, P., Slovak University of Agriculture in Nitra, Microbiology, Tr. A. Hlinku 2, 94976 Nitra, Slovakia (Slovak Republic); Miroslava.Kacaniova@uniag.sk

The aim of our work was monitoring of *Enterococcus faecium* and *Enterococcus faecalis* gene from chyme of poultry by polymerase chain reaction after probiotics application. We isolated DNA from individual chyme samples, which we measured spectrophotometrically. We used specific primers (E.FAECALIS REV, E.FAECALIS FOR and E.FAECIUM REV and E.FAECIUM FOR). We investigated thirty chyme samples of chickens for detection of *Enterococcus faecium* and *Enterococcus faecalis* gene. The presence of *E. faecium* and *E. faecalis* in ceca of chickens was detected by means of the PCR. The amplification gene of *E. faecalis* in PCR produces 941 bp and *E. faecium* 550 bp. The DNA content by spectrophotometrically of *E. faecalis* was detected 2.59 to 13.24 and *E. faecium* from 6.74 to 24.17 $\mu g.\mu l^{-1}$ in ceca of chicken samples.

Affect the presence of 1B/1R translocation the nutritional value of wheat (*Triticum aestivum*)

Kodes, A.[1], Hucko, B.[1], Dvoracek, V.[2], Stehno, Z.[2], Mudrik, Z.[1] and Plachy, V.L.[1], [1]Czech University o Live Science, Microbiology, Nutrition and Dietetic, Kamýcká 129, 165 21 Prague 6 - Suchdol, Czech Republic, [2]Crop Research Institute Prague, Genetic bank, Drnovská 507, 161 06 Prague 6 – Ruzyně, Czech Republic; kodes@af.czu.cz

In animal nutrition it is necessary to calculate with lower nutritional value of wheat proteins in context of lower content of easier digestible albumin and globulin fraction and totaly lower content of essential amino acids lysine, threonine, tryptophane and sulfuric amino acids. A set of selected 18 DH wheat lines with a higher agronomical potentional and according to presence or absence of allele Gli 1B3 characterizing 1B/1R translocation was subsequently divided into two numerically comparable sub-sets (with 1B/1R translocation and without the translocation). In growth and balance experiments with laboratory rats, we were measured: Protein Efficiency Ratio - PER, Biological Value of Protein - BV, and Net Protein Utilization - NPU. Finally, we can conclude, that presence of 1B/1R translocation significantly decreased the values of balance parameters (NPU, BV). Nevertheless, a similar effect of the translocation was not proven in growing test (PER). It is possible to assume that individual rations between albumins + globulins and gluten protein composition of grains influenced values of PER more significantly than the presence of 1B/1R translocation which is only one of many genetic factors participating in protein composition of grains.

Improving utilization of agricultural by-products using biological treatments

Mohamed, M.I.[1], Boraie, M.A.[2], Salama, R.[2], Abd-Allah, S.A.E.[1], Fadel, M.[3] and El-Kady, R.I.[1], [1]National Research Center, Animal production, EL-Bohoth Street, Dokki, Giza, Egypt., 12622, Egypt, [2]Faculty of Agriculture Al-Azhar Universit, Animal production, Nasr City, Cairo, Egypt., 12622, Egypt, [3]National Research Center, Microbiology, EL-Bohoth Street, Dokki, Giza, Egypt., 12622, Egypt; shalaby1956@yahoo.com

This study aimed to improve the six crop residues namely, Corn cobs, Corn stalks, Peanut hulls, Banana wastes, Wheat straw and Rice straw as a results of its biological treatments with *Trichoderma reesi*, (T1), *Saccharomyces cereviciae*, (T2) and a mixture of fungi and yeast (50% of each) (T3). The different biological treatments increased significantly ($P<0.05$) CP and ash contents of all crop residues treated compared with control. The different biological treatments recorded a slightly decrease ($P<0.05$) for DM, OM, EE. The content of CF and NFE of all crop residues decreased significantly ($P<0.05$) with all biological treatments compared with control. Biological treatments significantly ($P<0.05$) decreased NDF, ADF ADL, Cellulose and hemicellulose specially for treatments by mixture of fungus and yeast or fungus followed by yeast. DM disappearance were significantly ($P<0.05$) increased for biologically treated crop residues in all ruminal incubation periods than those of the control. The different biological treatments indicate that DM disappearance increased significantly ($P<0.05$) in all treated crop residues under study compared with the control. DM disappearance for up to 72 hrs were recorded the highest values of means in all biological treatments. The highest values of TDN, DCP were recorded with G1 (Corn cobs treated with mixture of fungi +yeast) while the lowest values of TDN, DCP were recorded with G4 (penut hulls treated with mixture of fungi +yeast) compared with the control.

The effect of different levels of beef tallow in diet on microbial protein supply and protozoal population fluctuation in rumen of sheep

Ashabi, S.M.[1], Jafari Khorshidi, K.[1], Rezaeian, M.[2] and Kioumarsi, H.[3], [1]Islamic Azad University, Qaemshahr branch, Animal Science Department, Qaemshahr City, Mazandaran Province, 1666976113, Iran, [2]University of Tehran, Department of Animal Health and Nutrition, Qareeb Street, Azadi Av.Tehran, 1419963111, Iran, [3]School of Biological Science, Universiti Sains Malaysia (USM), Pinang, Malaysia, 11700 Pinang, Malaysia; kaveh.khorshidi@gmail.com

Fats added to ruminant diets can greatly change fermentation in the rumen. This research was carried out to assess the effect of different levels of beef tallow in diet on microbial protein supply and protozoal population fluctuation in rumen of sheep. Four ruminally fistulated sheep (Zel rams) were used in a 4×4 Latin square design to evaluate the effects of different levels of tallow on microbial protein synthesis and protozoal population changes in the rumen and to determine whether changes in microbial protein synthesis were related to protozoal changes. The diets (55% concentrate, 30% wheat straw, and 15% alfalfa hay) contained no added fat (control), 2% tallow, 4% tallow or 6% tallow on a DM basis. Urinary excretion of purine derivatives was determined to estimate the microbial protein synthesis. Moreover, ruminal holotrich and total protozoa were counted for 0, 0.5, 1, 2, 5, 8, 9, 13, 17, and 21 hours after morning feeding. Housing and management conditions were the same for all animals. The results revealed that microbial protein synthesis and excretion of purine derivatives were not affected by treatments but tended to decrease as fat level increased in the diet ($P<0.05$). Ruminal holotrich protozoa concentration was not affected by added fat in all times after morning feeding but after 5 hours after feeding, holotrich population in all times tended to decrease by increasing fat level ($P<0.05$). Rumen total protozoa concentration showed no significant differences between treatments at 0, 0.5, 1 and 2 hours after feeding but after this times decreased by increasing fat level ($P<0.05$).

Effects of yeast *Saccharomyces cerevisiae* on ruminal ciliate population in Ghizel sheep

Taghizadeh, A.[1], Ansari, A.[1], Janmohamadi, H.[1] and Zarini, G.[2], [1]University of Tabriz, Animal Science, 51664, Iran, [2]University of Tabriz, Dep. of biology, 51664, Iran; ataghius2000@yahoo.com

The purpose of this study was to evaluate effects of Yeast culture on rumen protozoa biodiversity and generic distribution. Sixteen Ghizel rams (33.8±6.5) were used and treatments were included in the experiment being the control diet (C) that was not supplemented with YSC and contained only lucerne hay, and the other treatments contained 2.5, 5 and 7.5 g YSC/head/day. The experiment period was 21 day and on last day, rumen fluid samples were collected from all rams just before feeding and 2 h after feeding by stomach tube and filtered through four layers of cheese cloth, then 5 ml of rumen fluid mixed with 20 ml of formalin solution (8.5 g NaCl, 100 ml formalin 40% and 900 ml distilled water) and stored for protozoa identification and counting. Protozoa generic distribution were counted using a Neubauer Improved Bright-Line counting cell. Data were analyzed as a completely randomized design by using the GLM procedure of SAS software. The mean rumen protozoal concentrations and generic distribution of ciliate protozoa in the four treatment groups was affected ($P<0.05$) by addition of YSC in both sampling time. The ciliate protozoa that belonged to Entodinomorph family were identified in before feeding samples, but in the 2h after feeding samples both of entodinomorph and holoticha family were identified. Generic composition of ciliate protozoa were affected by YSC and there was significant difference ($P<0.05$) between treatments. Entodinium genus was the most popular ciliate protozoa which identified and counted in the ruminal fluid. Number of total protozoa was higher in the 7.5 g YSC treatment ($P<0.05$). It was concluded that YSC were altered rumen microorganism population, resulting positive effects on cellulose digestibility and on nitrogen incorporation in microbial proteins.

Effects of chemical and physical treatments on antinutritional factors in grass pea (*Lathyrus sativus*)
Vahdani, N., Dehgan-Banadaki, M. and Moravej, H., University of Tehran, Animal science, Karaj, Tehran, 3413966733, Iran; vahdani.narges@gmail.com

Grass pea is one of the leguminous forages cultivated in many countries. This forage contains a neurotoxin amino acid, β-N-oxalyl -L- α, β -diaminopropionic acid (β-ODAP), which can cause a paralysis of lower limbs (lathyrism) and condensed tannin (CT), which can affect on digestibility. So reduction of these Antinutritional factors (ANFs) by some treatments can improve the nutritive value of this forage. The forage was harvested at late flowering stage. The sun dried forage was chopped 3-5 cm length, and then treated with different chemical solutions. The ratio of forage to reagent volume was kept at 1:4 (W/V) for water soaking, 0.03 M solution of $KMnO_4$, 0.05 M solution of NaOH and 0.1 M solution of $NaHCO_3$. The amount of polyethylene glycol (PEG) (50g/1 kg of DM) was based on a kg of DM. The experimental plant contained the high amount of ODAP (1.18%). ODAP is hypothesized to function as a carrier molecule for zinc ions, soils depleted in micronutrients or high iron content may be responsible for the high level of neurotoxin. So, the low level of zinc in soils can cause to high level of ODAP in experimental grass pea in Iran. Results showed that water soaking (80.48%) and NaOH (63.81%) can greatly reduce the ODAP concentration. ODAP is soluble in water and have an acidic structure, so the alkali pH (12.28 for NaOH) can reduce ODAP concentration while acidic treatments like PEG (pH= 5.91) decrease ODAP, alone 19.05%. CT content of grass pea was (0.02%). Forage treated with NaHCO3 (98.8%) and water soaking (70.18%) show highest reduction in CT content. In spite of the alkali pH (12.28) of NaOH, this treatment can not reduce CT concentration (7.32%) in compare to another alkali treatments. In conclusion water soaking treatment can reduce both ANFs and is a practical treatment in farms for improvement the nutritional value of grass pea.

The evaluation of three ruminal degradability models using the plot of residuals against predicted values
Fathi Nasri, M.H.[1], Danesh Mesgaran, M.[2] and Farhangfar, H.[1], [1]The university of Birjand, Department of Animal Science, Birjand, Iran, [2]Ferdowsi university of Mashad, Department of Animal Science, Mashad, Iran, mhfathi@gmail.com

The evaluation of different models as candidates to describe ruminal DM and CP degradation kinetics of raw and roasted whole soybeans from data obtained using the *in situ* polyester bag technique were conducted using the plot of residuals (y-axis) against predicted values (x-axis) approach. The candidate models were included: a segmented model with three spline-lines delimited by two nodes or break points, constraining splines 1 and 3 to be horizontal asymptotes, and follows zero-order degradation kinetics (model I); a simple negative exponential curve with first order kinetics and assuming a constant fractional rate of degradation (model II); and a rational function or inverse polynomial which assumes a variable fractional rate of degradation that declines with time (model III). The models were fitted to the DM and CP ruminal disappearance data by nonlinear regression using the PROC NLIN of the SAS (SAS, 1999) to estimate ruminal degradation parameters. The results showed that the plot of residuals against predicted values was a useful visually and provided an easy to use approach for assessing models goodness-of-fit. Based on this approach, the model III was less suited than models I and II to describing the degradability patterns of the experimental feeds because the residual points generated horizontal bands in plots which is an indicator of model inadequacy and the need for adding extra terms in the fitted equation (e.g., square or cross-product terms). Also, linear and quadratic relationships between residuals and predicted values were assessed and results showed statistically significant linear and quadratic trends ($P<0.001$) for model III, but not for models I and II.

Can the beet pulp feeding reduce BCS in fat cows?

Mahjoubi, E., Amanlou, H., Zahmatkesh, D., Ghelich Khan, M., Aghaziarati, N. and Siari, S., Zanjan University, Animal Sciemce, Zanjan, Iran; e_mahjoubi133@yahoo.com

Eighteen lactating Holstein cows were used in a randomized complete block design for investigation of increasing a lipogenic nutrient (beet pulp) at the expense of a glycogenic nutrient (barley) on decreasing fat cow's BCS. The cows were 171 ± 16 d in pregnancy and 289 ± 35 d in milk at the beginning of the experiment. The cow's BCS were 4.12 ± 0.35 at the beginning of the trial. The cows were assigned randomly to three dietary treatments containing: 1) 23.47% barley (0% beet pulp), 2) 14.87% barley (8.6% beet pulp), or 3) 6.27% barley (17.2% beet pulp). Beet pulp replacing barley grain increased milk fat percent (4.37 vs. 4.91 and 5.18, $P<0.002$) and milk energy (0.76 vs. 0.82 and 0.84 Mcal/kg, $P<0.01$) without affecting milk yield (17.9, 17.4, 17.9) and milk percents of protein and lactose. Increasing beet pulp tended to increase milk energy output (13.6 vs. 14.26 and 15.4 Mcal/d, $P=0.10$). By adding beet pulp, BCS (0.13 vs. -0.09 and -0.12, $P=0.01$) and back fat thickness (2.5 vs. -0.4 and -1.6 mm, $P<0.01$) reduced linearly. Plasma glucose (65.83, 58, and 57.16 mg/dl, $P<0.01$) and cholesterol (157.33, 122.4, and 120.8 mg/dl, $P<0.03$) decreased as beet pulp were substituted for barley grain, respectively. No difference was found in the plasma content of insulin and NEFA between treatments. In summary, these results suggest that inclusion of a lipogenic feed, especially BP (a relative cheap feed), in diet of fat cows can slightly reduce BCS, without compromise milk production and milk energy output.

Utilization of artichoke (*cynara scdymus* l.) by-product in feeding growing rabbits

Abdel-Magid, S. and Awadalla, I., National Research Centre, Animal Production Department, EL-Behoth street, Dokki, Giza, Egypt, 12622, Egypt; ibrahimawadalla@hotmail.com

The present study was conducted to investigate the effects of partial replacement of clover hay by dried artichoke by-product (DAbp) in growing New Zealand white (NZW) rabbits. Twenty seven of NZW rabbits of 6 weeks of age were allotted at random on three different diets contained clover hay (CH) at 30% of the total diet (control) as a source of roughage that was replaced by 25 and 50% of DAbp.The three pelleted experimental diets were formulated to be approximately isocaloric and isonitrogenous. The feeding experiment lasted 8 weeks. Results indicated that body weight of rabbits decreased with increasing DAbp levels in the diets, but differences was not significant, feed consumption of rabbits increased with increasing levels of DAbp in their diets, feed conversion of rabbits fed diets containing DAbp was not significant different when compared to the control group. Digestibility coefficient of different nutrient were improved for rabbits fed control diet as compared to those fed diets containing 25 or 50% DAbp. Group 3 achieved the best economic efficiency followed by group 2 and the last was the control group. It could be recommended to substitute up to 50% of with DAbp in the pelleted of growing rabbits.

Visualizing energy balance in murciano-granadina dairy goats by artificial neural networks

Fernández, C.[1], Cerisuelo, A.[2], Piquer, O.[3], Martínez, M.[4] and Soria, E.[4], [1]Universidad Politécncia Valencia, Ciencia Animal, Camino Vera S/N, 46022, Spain, [2]IVIA. Segorbe. Castellón. Spain, CITA, Segorbe, Castellon, Spain, [3]Universidad Cardenal Herrera Ceu, Producción Animal, Edificio Seminario, Moncada-Valencia, Spain, [4]Universdiad Valencia, Ingeniería Electronica, Campus Burjassot, Burjassot-Valencia, Spain; cerisuelo_alb@gva.es

This study is focused in Valencia Comunity (Spain) and the objective was to use a tool to visualize data using no linear multi variante analysis. Thrity four Murciano-Granadina dairy goats were selected from a farm during lactation. A total mixed ration was elaborated (17 MJ/kg DM and 16% crude protein) and body weight, dry matter intake, milk producion and composition were recorded. Energy balance was estimated using a simulation model SIMLECABRA (AnimalSim®) and using the records from this trial as inputs for the simulation model. The aim of this study is to analyze the relationship among different variables related with milk production and energy balance. This analysis is carried out by using an artificial neural network. The artificial neural network we used was the Self Organizing Map (SOM). This tool allows mapping high-dimensional input spaces into much lower-dimensional spaces, thus making much more straightforward to understand any representation of data. These representations enable to visually extract qualitative relationships among variables (Visual Data Mining). SOM were used to analyze data with the system identification toolbox of MATLAB v7. SOM considered in this study is formed by 3´10 neurons (3 neurons); the chosen architecture is given by the range of the input variables and the dimensionality of the proposed problem surveys and management practices were evaluated. The average dry matter intake was 1.5 kg per goat and day, the map shown that 32% of the goats studied has positive energy balance. The SOM obtained indicate a group of neurons with negative energy balance (68%) related with higher mil yield and number of kids per partum. The use of SOM in the descriptive analysis of this kind of data sets has proven to be highly valuable in extracting qualitative and quantitative conclusions and guiding in improving the performance of farms.

Proteosyntetic activity in the rumen of dairy cows

Mudrik, Z., Hucko, B., Kodes, A., Plachy, V. and Bojanovsky, J., Czech University of Life Sciences, Microbiology, Nutrition and Dietetics, Kamycka 129, 165 21, Prague 6, Czech Republic; mudrik@af.czu.cz

In a series attempting to dairy cows of Holstein bred we watched the proteosyntetic activity of ruminal microorganisms in relation to their production. We compared the two groups of dairy cows had the same ration in the ad libitum system. The comparison group were selected so that they are in approximately the same time, lactation and the comparable figures last lactation. Intensity of proteosyntsis was calculated according to noninvas methods for determinative of purine derivatives in urine of dairy cows as an indicator of the creation of nucleic acids in the rumen of biomass going into the small intestine, according to the method reported of Chan *et al.* and Gonda *et al.*. In the same structure of feed ration and unlimited access to feed, biomass formation was limited to milk production. Group with low production 16.1 kg milk, days of lactation 335, production in the last lactation 6,528 kg had calculated amount of generated biomasses 787 g and group with high production 37.6 kg, days of lactation 318, production in the last lactation 7,717 kg have calculated amount of generated biomasses 1,218 g.

Session 36

Effect of condensed tannins in the gas production *in vitro*
Borba, A.[1], Vieira, S.[1], Silva, S.[2], Rego, O.[1] and Vouzela, C.[1], [1]University of the Azores, DCA, CITA-A, Terra Chã, 9700 Angra do Heroismo, Portugal, [2]SRAF, Serv. Des. Agra. Terceira, Vinha Brava, 9700 Angra do Heroismo, Portugal; borba@uac.pt

Tannins are natural constituents of the pastures, with effects on the ruminal fermentation. In this assay the total content in tannins was determined in 3 species by the method of radial diffusion assay: *Lotus corniculatus*, *Lollium perenne* and *Trifolium repens*. We verified, in comparison with standard samples tannic acid, that tannins exist in the flower of *T. repens* (0.81 equivalents of tannic acid) and in *L. corniculatus* (1.07 equivalents of tannic acid) but not in *L. perenne* (0 equivalents of tannic acid). The method of acid-butanol was used for quantification of CT on the 3 species. Quebracho was used as standard. The concentration of tannins in the 3 species was: 0.34 mg/ml for *T. repens*; 0.83 mg/ml for L. corniculatus and 0 mg/ml for *L.perenne*. To determine the effect of the TC on the ruminal fermentation, assays of gas production *in vitro* were carried through, with inclusion of CT proceeding from the Quebracho in doses of 0, 2.5 and 5% DM. A significant reduction of gas production was verified for doses of 5% of CT, this expresses a reduction of methane emission to the atmosphere and an increase of exploitation of the protein effect among other positive effects on ruminants. The addition of Quebracho to fibrous food, on *in vitro* assay for gas production, diminishes the fermentation tax, diminishing the amount of produced gas, diminishing the extension of degradation and the concentration of the resultant products of this degradation (ammonia and volatile fat acid).

Session 36

Palatability of grass pea (*Lathyrus sativus*) as alternative forage in Iran
Vahdani, N., Rezayazdi, K. and Dehgan-Banadaki, M., University of Tehran, Animal science, Karaj, Tehran, 3413966733, Iran; rezayazdi@ut.ac.ir

Grass pea (*Lathyrus sativus* L.) is an annual leguminous crop. The legume can provide an economic yield under adverse environmental conditions and offers great potential for use in marginal low-rainfall areas. Indeed, this has made it a popular crop in subsistence farming in certain developing countries that have extreme weather conditions. Grass pea contains several antinutritional factors (ANFs). It seems that existence of these ANFs can affect on palatability of this forage. So in present study we determined the palatability of this forage in comparison with alfalfa. Six Varamini ewes (45±5 kg) were used for evaluation of palatability of grass pea over control (afalfa). These ewes were fed twice a day (08:00 and 20:00) and fresh water and salt licking blocks were available all times. Each meal covers 55% of maintenance energy requirement, where test plant and control were provided exactly half of the ME. Before starting the experiment, the pre-period was carried out and the sufficient time to consume half of the total feed in each meal was chosen (15 min). The experiment was elongated for 10 day. The allocation of the two different forages was switched between troughs to avoid association of place forage type and day time by animals. At last the palatability index (PI) was calculated according to Bensalem *et al.* (1994). Palatability of grass pea was 87% in compare with alfalfa. Palatability is defined as the result of the physical and chemical characteristics that evoke appetite. Also; ruminants do select feeds on the basis of flavor and color. So it seems that the existence of the large number of plant secondary compounds in *L. sativus* had negative influence on palatability index.

Evaluation of nutritional value of barley distillers' grain supplementing with different silage additives

Tahmasbi, A.M., Kazemi, M., Valizadeh, R., Danesh Mesgaran, M. and Gholami Hossein Abad, F., Department of Animal Science, Excellent Center for Animal Science, Ferdowsi University of Mashhad, P.O.Box 917794897, Iran; a.tahmasbi@lycos.com

Barley distillers' gain is a one of the most readily protein and fiber sources for dairy cattle. Increasing in energy costs have led to large amount of this agro- industrial by- product being market in wet form. One of the problems encountered in using wet barley distillers' grain is limited storage window and problem in handling wet material. Fresh BDG contain about 70%-80% moisture and it can store in silage to extend shelf life However, there is limited research information available on improving BDG silage with different silage additives. Triplicate samples of 3.5 kg of wet BDG were treated with molasses (2 and 4% DM), sulfuric acid (2 and 4% of DM) and urea (2 and 4% DM) and compacted with vacuum in double-lined plastic tube. Plastic tubes were sealed and stored in ambient temperature about 25 ℃ up to 60 days. After 6 days mini- silos were opened and samples were taken from each replication for analysis. Results indicated that ensiled BDG with 4% urea had a higher NH3-N, pH, crude protein and ADF than other treatments and differences were statistically significant ($P<0.05$). However, pH for silage containing sulfuric acid (4%) was lowest. The NDF percentage in urea treatment (2%) was higher than other treatments but organic mattered was higher in silage contained sulfuric acid.

Correlation of *in vitro* gas production and *in situ* technique for evaluation of tomato pomace degradability

Safari, R., Valizadeh, R., Tahmasbi, A. and Bayat, J., Ferdowsi University of Mashhad, Department of Animal science, 75818-94953, Iran; Rashid_safari@yahoo.com

This study was aimed to evaluate Correlation of *in vitro* gas production and *in situ* technique of tomato pomace degradability. Tomato pomace was obtained from five large factories and dried. Seeds and peels components were separated. *In situ* DM degradability was estimated for unground and ground (samples were ground through 2 mm screen) whole tomato pomace, seeds and peels components using the modified *in situ* polyester bag technique. Bags were incubated in the rumen of three fistulated steers fed ordinary diets. The bags were removed following 0, 2, 4, 8, 12, 24, 36, 48 and 72 h of incubation. *In vitro* gas production for the same components, with rumen liquor of same animals at same times, was measured. Regression model was made among the *in situ* and *in vitro* data by SAS 9.1. The degradability of whole pomace and its components showed that most of the DM was degraded after 24 h, although degradation increased up to 48 h but at a much lower rate. The grinding process was highly effective for improving the degradability measures for all samples especially seeds. The results of *in vitro* gas production after 72h incubation for grounds and unground pomace and compounds (229.74±7.8, 178.14.6±10.4, 225±9.5, 104.25±8.33, 81.37±8.8 and 199.3±10.7 ml/g of DM for whole pomace, seeds and peels as ground and unground respectively) were similar to the *in situ* findings (67.33±0.27, 79.6±0.8, 62.8±0.4, 36.07±1.53, 31.73±1.06 and 46.47±0.33% of DM for whole pomace, seeds and peels as ground and unground respectively). Correlations for DM degradability measured by two methods for whole tomato pomace and its components were 0.96, 0.98, 0.98, 0.98, 0.98, and 0.97 for whole pomace, seeds and peels as ground and unground respectively. According to the regression models, *in vitro* gas production for the DM contents of whole tomato pomace and its components could be used instead of the *in situ* technique at lower cost in shorter time.

Application of biophysical methods for soybean quality assessment
Caprita, R. and Caprita, A., University of Agricultural Sciences and Veterinary Medicine, Calea Aradului 119, 300645 Timisoara, Romania; rodi.caprita@gmail.com

The challenge in soybean meal (SBM) processing is to apply the optimum amount of heat to produce the most nutritious product. The proposed methods measure changes in protein solubility during heating of soybean meal by determination of the refractive index and the dynamic viscosity of dilute potassium hydroxide extracts of meals. SBM samples were ground and several particle sizes were obtained using a series of standard sieves: 65 µ, 125 µ, 200 µ, 315 µ, and 630 µ. SBM was heated in a forced air oven at 120 °C for varying periods of time: 5, 10, 15, 20, 25 and 30 minutes. The samples were analyzed for: the protein solubility in a dilute solution of KOH according to the procedure of Araba and Dale, the refractive index of the supernatant solution with an Abbe refractometer, the dynamic viscosity coefficient with a viscometer Brookfield LVDV, the urease index (UI) based on the pH increase from ammonia released from urea by residual urease enzyme. The experimental data show a negative correlation (r=-0.9218) between the protein solubility and the particle size. KOH protein solubility remains high, during initial heat treatment. Additional heat treatment decreased KOH protein solubility, and UI rapidly approached zero. A simple and rapid method for estimating protein solubility of soybean meals on the basis of changes in the refractive index and dynamic viscosity of dilute potassium hydroxide solution extracts was tested and we found to be highly correlated with the usual protein solubility in KOH test: r=0.9382 for refractive index and r=0.8943 for dynamic viscosity. UI is useful to determine if the soybean meal has been heated enough to reduce the anti-nutritional factors, but it is not very useful for determining if soybean meal has been over-processed. The refractive index and the dynamic viscosity of dilute potassium hydroxide solution extracts are reliable, rapid and non-polluting methods for evaluating the quality of soybean meal.

In vitro gas production technique to determining of nutritive value of various sources of citrus pulp
Bayat Koohsar, J., Tahmasebi, A.M., Nasserian, A.A., Valizadeh, R. and Safari, R., Ferdowsi University, Mashhad, 51667, Iran; javad_bayat@yahoo.com

his study was aimed to evaluate the pattern of fermentation of different citrus pulp sources including; orange (*Citrus sinensis*), tangerine (*Citrus reticulate*), lemon (*Citrus auranifolia*) and grapefruit (*Citrus paradisi*) by gas production method. Dried citrus pulp (DCP) is an agro-industrial by-product with high level of energy and low in CP and NDF. The DCP samples were collected from different citrus-processing industry in north part of Iran. The procedure of *in vitro* gas production method developed by Menke (1988) was used. Rumen fluid was collected from two ruminally fistulated, steers and mixed. The amounts of gas production were monitored at 0, 2, 4, 12, 24, 36, 72 and 96 h and cumulative gas production were calculated. Data were subjected to analysis of variance using SAS program. *C.reticulata* cul.younesi and *Citrus aurantifolia* had higher and lower Cumulative gas production compared to other DCP (487.9, 518.9, 522.9, 50.9, 492.2, 474, 530.5, 425.2, 458.1 and 466.3 ml/g DM for *C. Sinensis* cul. Tamson, *C. Sinensis* cul.siahvaras, *C. Sinensis* cul. Brohen, *Citrus unchiu*, *C.reticulata* cul. Page, *C.reticulata* cul. Celemantin, *C.reticulata* cul.younesi, *Citrus aurantifolia*, *Citrus aurantium* and *Citrus paradisi* respectively). Gas production rate of *C. Sinensis* cul.siahvaras was numerically higher than others (0.05, 0.08, 0.08, 0.05, 0.05, 0.07, 0.08, 0.04, 0.08 and 0.06 ml/g DM /h for *C. Sinensis* cul. Tamson, *C. Sinensis* cul.siahvaras, *C. Sinensis* cul. Brohen, *Citrus unchiu*, *C.reticulata* cul. Page, *C.reticulata* cul. Celemantin, *C.reticulata* cul.younesi, *Citrus aurantifolia*, *Citrus aurantium* and *Citrus paradisi* respectively). Total gas production was higher for *C.reticulata* cul and lower for *Citrus aurantifolia* (500.2 vs. 449.7 ml/g DM). Maximum gas production of different citrus pulp sources achieved with in 24 h and increased up to 48 h but at a much lower rate.Result indicated that various sources of dried citrus pulp have potential as feedstuff for ruminants.

Effects of protected fat supplements on total tract digestion, and plasma metabolites of early lactation Holstein Cows

Ganjkhanlou, M., Reza Yazdi, K., Ghorbani, G.R., Morravege, H., Dehghan Banadaki, M. and Zali, A., Tehran university, Animal science, Karaj, 45841, Iran; ganjkhanloum@yahoo.com

This study was conducted to evaluate the digestibilities of commercial fat supplements in early lactation cows. Twelve (nine multiparous and three primiparous) Holstein cows (26 ± 4 day in milk) were used in a replicated 3×3 Latin square design with 21-d experimental period and three treatments: control (no fat supplementation), and supplemented with 30 g/kg prilled protected fat (Energizer-10) or 35 g/kg Ca salt of protected fat (Magnapac). Cows were fed ad libitum a total mixed ration consisting of 200 g/kg corn silage, 200 g/kg alfalfa hay and 600 g/kg concentrate mix. Each period had 14 days of adaptation and 7 days for sampling. Ether extract digestibility was increased by 7% and 8% with supplementation of rumen protected fat in multiparous and primiparous cows respectively, but total tract digestibilities of DM, OM, CP, NFC, ADF, or NDF were not affected by fat supplements in all cows. Mean apparent digestibility of fat were not influenced by source of fat supplemented and the TMR containing Ca salt of protected fat had similar digestibility of fat than did TMR containing prilled fat. Plasma urea; glucose; triglyceride; LDL and plasma HDL were unaffected by supplemental fat ($P>0.05$). Total cholesterol and NEFA in plasma was greater in cows fed inert fat than in cows fed the control diet in multiparous and primiparous cows. In multiparous cow's plasma NEFA increased with fat feeding, about 1.28 times for prilled protected fat and 1.26 times for the Ca salt of protected fat. These results indicate that supplementation of early lactating diet with rumen protected fat increased ether extract digestibility but without altering digestibilities of DM; OM; CP; NFC; ADF; or NDF.

Examining the efficiency of the semi substitution of the maize with a by-products obtained by manufacturing vegetables and fruits in mixtures for growing and fattening pigs

Cilev, G.[1], Sinovec, Z.[2], Palashevski, B.[1] and Pacinovski, N.[1], [1]Institute of Animal Science, Ile Ilievski str. 92A, 1000 Skopje, Macedonia, [2]Faculty of Veterinary Medicine, Bulevar Oslobodjenja 18, 11 000 Belgrade, Serbia; goce_cilev@yahoo.com

The aim of this experiment was examine the possibilities of the maize's substitution as an energetic nutrient with by-products obtained by manufacturing tomatoes, peppers and grapes in the nutrition of swine on the production results and health condition. The experiment of the growing and fattening pigs is carried out on 48 pigs (in a period of about 60 days) divided into 3 groups with each group having 8 pigs of different sex with an average body weight of 27.00+0.64 kg (K); 27.69+0.71 kg (O-I) and 27.50+0.68 kg (O-II). For the experiment used mongrels of Swedish and Dutch races with equalized genetical potential. Each group in experiment consist of equal number of males and females. The experiment lasted for 100 days in 2 phases with 50 days each phase and used 2 mixtures, one for growing (25-60 kg) and one for fattening (60-100 kg) pigs formulated according (NRC, 2000). The pigs from the control group (K) from experiment are fed with mixture without a share from the examined by-products, whereas the experimental groups (O-I and O-II) were fed with food with substitution of the maize with different qualities of the above mentioned by-products. In the pig's feeding mixture – 6% (O-I) i.e. 9% (O-II) from the examined by-products. Average Body Weight (ABW) in the control group (K) and experimental groups (O-I and O-II) in the end of fattening period (100 day) were 96.20 kg, 98.10 kg and 99.50 kg, respectively. Average Daily Gain (ADG) and Average Feed Conversion (AFC) in the control group (K) and experimental groups (O-I and O-II) for all fattening period (1-100 days) were (712 g vs 3.05 kg-K), (724 g vs 2.87 kg-O-I) and (720 g vs 2.85 kg -O-II). No significant difference was observed in all three groups ($P<0.05$).

Viability of bovine embryos derived from serum-free oocyte maturation

Korhonen, K.[1], Pasternack, A.[2], Ketoja, E.[1], Räty, M.[1], Laitinen, M.[2], Ritvos, O.[2], Vilkki, J.[1] and Peippo, J.[1], [1]MTT Agrifood Research Finland, Biotechnology and Food Research, ET Building, 31600 Jokioinen, Finland, [2]University of Helsinki, Department of Bacteriology and Immunology, P.O. Box 21, 00014 Helsinki, Finland; kati.korhonen@mtt.fi

Serum is widely used in bovine *in vitro* embryo production (IVP) due to its ability to promote oocyte maturation and embryo development. On the other hand, it has also been connected to developmental perturbations such as large offspring syndrome, and it may act as possible source of animal transmitted pathogens (e.g. BVD). Studies on serum-free media have been going on for decades although often resulting decreased embryo development and viability. We, however, were recently able to develop a serum-free (PVA) oocyte maturation medium supplemented with growth factors resulting similar embryo yield (PVA 35% vs. serum 35%, $P=0.82$) and proportion of first quality embryos (PVA 69% vs. serum 67%, $P=0.66$) than serum containing medium. The growth factors (GDF9, BMP15, IGF-I, EGF, BDNF, FGF8 and LEP) used in the oocyte maturation were selected based on their known ability to enhance cell growth. Since the conventional morphological assessment of embryos is known to be inadequate, embryo viability was also evaluated on gene expression levels. The genes discovered in WP6 of the SABRE-project indicating good embryo quality were used in this study to compare the viability of embryos derived from PVA and serum groups. The gene expression results of these selected genes indicating the ability of an embryo to initiate pregnancy will be discussed. These results are obtained through the EC-funded FP6 project 'SABRE'

Effect of different diluents on Markhoz buck sperm quality: 1- In fresh semen

Mafakheri, S.[1], Sadeghipanah, H.[2], Salehi, S.[1], Khalili, B.[1] and Babaei, M.[2], [1]Kordestan Research Center of Agricultural and Natural Resources, Animal Science, Pasdaran Street, Sanandaj, Iran, 6616936311 Sanandaj, Iran, [2]Animal Science Research Institute of Iran, Animal Production and Management, Shahid Beheshti Street, Karaj, 3146618361 Karaj, Iran; mafakheri.shiva@gmail.com

This study was conducted to determine the effects of diluent type on fresh semen quality in Iranian Markhoz goats. Semen from eight young (1.5-year old) Markhoz bucks was collected once a week for eight weeks (as replicates) during breeding season (November and December) by artificial vagina. Ejaculates that had a acceptable quality (mass movement 4 or 5 and sperm concentration 1×10^9/ml) were pooled and diluted at ratio of 1:4 by the tris-yolk-fructose-citric acid-SDS (Tris) or skim milk (Milk) diluents. Immediately after dilution, samples were assessed for percent total motile, progressive motile, live and abnormal sperm. Statistical analysis carried out as a completely random design (2 treatments with 8 replicates) using SPSS software. Percent total motile, progressive motile and live sperm in Tris (67.06 ± 5.98, 45.00 ± 5.36 and 74.43 ± 3.93, respectively) were significantly ($P<0.01$) higher than in Milk (13.36 ± 4.32, 6.21 ± 2.54 and 35.00 ± 6.34, respectively). Percent abnormal sperm was not significantly different in two treatments ($P>0.05$). These results show that during breeding season, without seminal plasma removal (sperm washing), Tris diluent can be useful to dilution goat semen for freshly use, but skim milk can not be use as a diluent for Markhoz buck unwashed semen.

Effect of different diluents on Markhoz buck sperm quality: 2- in freeze-thawed semen

Sadeghipanah, H.[1], Mafakheri, S.[2], Salehi, S.[2], Mohammadian, B.[2] and Babaei, M.[1], [1]Animal Science Research Institute of Iran, Animal Production and Management, Shahid Beheshti Street, 3146618361 Karaj, Iran, [2]Kordestan Research Center of Agricultural and Natural Resources, Animal Science, Pasdaran Street, Sanandaj, 6616936311 Sanandaj, Iran; mafakheri.shiva@gmail.com

This study was conducted to determine the effects of diluent type on freeze-thawed semen quality in Iranian Markhoz goats. Semen from eight young (1.5-year old) Markhoz bucks was collected once a week for eight weeks (as replicates) during breeding season (November and December) by artificial vagina. Ejaculates that had a acceptable quality (mass movement 4 or 5 and sperm concentration 1×10^9/ml) were pooled and diluted at ratio of 1:4 by the tris-yolk-fructose-citric acid-SDS (Tris) or skim milk (Milk) diluents (both diluents including 5% glycerol). Diluted semen from each treatment were packed into 0.25 ml straws and frozen in nitrogen vapor. Thawing was performed by immersing in water bath at 37 °C for 30s. Thawed samples were diluted with the same diluent (but without glycerol) at a ratio of 1:1 and reposed again in water bath at 37 °C for 15min. Then samples were assessed for percent total motile, progressive motile, live and abnormal sperm. Statistical analysis carried out as a completely random design (2 treatments with 8 replicates) using SPSS software. In freeze-thawed semen samples, there was no significant ($P>0.05$) difference between two treatments in all assessed characteristics (8.75 and 8.14% total motile, 1.31 and 1.86% progressive motile, 14.69 and 10.93% live and 10.88 and 9.21% abnormal sperms, respectively for Tris and Milk). Ultimately, freeze-thawed semen qualities in both of diluents were very low and undesirable. These results show that, without seminal plasma removal (sperm washing), both of diluents that tested in current work were not acceptable for freezing Markhoz buck semen.

Sex ratio in swine and how it is affected by sire and conception time

Vatzias, G.[1], Asmini, E.[2] and Maglaras, G.[1], [1]Technological Educational Institute of Epirus, Animal Production, Kostakii, Arta, 47100, Arta, Greece, [2]Technological Educational Institute of Larissa, Project Management, Trikalon Ring Rd, 41110, Larissa, Greece; vatzias@teiep.gr

Sex ratio is expressed as the number of males per 100 females. Primary sex ratio is the ratio at conception and secondary sex ratio is the proportion of males at birth. The elucidation of the factors that influence sex ratio it is of great interest for the livestock industry. The objective of the present study was to identify if sire and time (month) at conception have an effect on sex ratio of the piglets. Data were obtained from 740 litters sired by 83 purebred and crossbred boars for a period of two years, from a commercial farm. The results indicated that sex ratio was not influenced by the sire ($P>0.05$). On the other hand, a clear effect ($P<0.05$) was shown for the time at conception. Sows bred during the coldest and warmest months of the year (December – March and June - August, with an average monthly atmospheric temperature ranging between 5.0° C - 8.5 °C and 22 °C - 27 °C, respectively), exhibited a higher number of male and female offspring respectively. The above results indicate that environmental factors, such as temperature may play an important role in determining the sex ratio in swine.

Pathogen specific response of the bovine mammary gland to lipopolysaccharide from *E. coli* and lipoteichoic acid from *S. aureus*

Arnold, E.T., Morel, C., Bruckmaier, R.M. and Wellnitz, O., Veterinary Physiology, Vetsuisse Faculty, University of Bern, 3001 Bern, Switzerland; emanuel.arnold@physio.unibe.ch

Lipoteichoic acid (LTA) and lipopolysaccharide (LPS) are cell wall components of *S. aureus* and *E. coli*, resp., which are common causes of clinical mastitis. This study was performed to investigate if intramammary challenge with LTA or LPS elicits a different immune response. In experiment 1, the intramammary challenge was performed with 10μg of LTA or LPS, diluted in 10ml saline solution (0.9%). At 0, 6, and 12 h after challenge biopsy samples of the mammary gland were taken for mRNA expression measurements by real-time RT-PCR. In experiment 2, the challenge was performed with 0.2μg LPS or 10μg LTA. The different dosage was chosen to reach a similar increase in somatic cell count (SCC) with both components. SCC was measured hourly. Milk samples were taken at 0, 6, and 12 h for determination of tumor necrosis factor alpha (TNFa) by radioimmunoassay and lactate dehydrogenase (LDH) by an enzymatic test. In experiment 1, TNFa mRNA abundance was increased ($P<0.05$) after 6 h (rel. expression $\Delta\Delta$CT 3.1±0.44) and 12 h (rel. expression $\Delta\Delta$CT 2.1±0.42) in LPS treated quarters but not in LTA treated quarters. The increment of Interleukin-8 (IL-8) mRNA was more pronounced in LPS treated quarters compared to LTA treated quarters. In experiment 2, TNFa in milk increased after 6 h and decreased to pre-challenge levels after 12 h in LPS treated quarters. In contrast, TNFa in milk did not change in quarters challenged with LTA. LDH was elevated 44 fold after 6 h, and was still 31 fold higher after 12 h in LPS treatment. In LTA quarters LDH tended to increase ($P=0.08$) after 6 h and recovered to baseline levels at 12 h. In conclusion, in both experiments LPS induced a stronger response of the measured factors than LTA despite similar SCC response. The present results are consistent with earlier investigations which showed a reduced and slower reaction of TNFa and IL-8 towards bacterial infection with *S. aureus* than with *E. coli*.

Physiological differences between metabolic stable and instable cows

Graber, M.[1], Kohler, S.[2], Müller, A.[2], Burgermeister, K.[2], Kaufmann, T.[1], Bruckmaier, R.M.[1] and Van Dorland, H.A.[1], [1]University of Bern, Vetsuisse Faculty, Veterinary Physiology & Clinic for Ruminants, Bern, Switzerland, [2]Swiss College of Agriculture, Animal Science, Zollikofen, Switzerland; marco.graber@physio.unibe.ch

Metabolic hepatic regulation during early lactation may vary between cows, and may be an underlying cause why some cows are more susceptible to suffer from metabolic and related disorders than others. A field study was carried out to understand the differences in plasma parameters and mRNA levels of hepatic parameters during the transition period up to mid lactation in dairy cows characterized as metabolically stable or instable. This characterization was based on a ranking system for the type and frequency of metabolic (ketosis, milk fever) and related disorders (e.g. reproductive disorders, mastitis) experienced by cows during their previous lactations. In total, 220 multiparous dairy cows were included in the study, from which 50 were characterized as metabolically instable, and 170 were characterized as metabolically stable cows. Blood and liver samples were obtained from each cow in week 3 ante partum (-wk3) and in week 4 (+4wk) and 13 post partum (+13wk). Blood plasma was assayed for concentrations of metabolites and hormones. Liver was analyzed for mRNA expression levels encoding for enzymes of the gluconeogenesis (PEPCKm, PC), citric acid cycle (CS), lipid metabolism (ACSL, CPT1A, CPT2, ACADVL), and ketogenesis (HMGCS2, BDH2). Metabolically stable cows had higher ($P<0.05$) plasma glucose concentrations than instable cows during the studied period. There was a tendency ($P<0.1$) for stable cows to have higher concentrations of albumin, IGF-I and T_3, but lower insulin concentrations than instable cows. Levels of HMGCS2 mRNA were higher ($P<0.05$), and mRNA levels of CPT1A were generally lower ($P<0.05$) in stable cows than in instable cows. The observed physiological differences between metabolically stable and instable cows based on their past health status, suggests the presence of a genetic component underlying metabolism that in part determines the incidence of disorders.

Effects of lipoxygenase metabolites of arachidonic acid on the placental matrix metalloproteinase activation

Kamada, H.[1], Nakamura, M.[2], Kadokawa, H.[3] and Murai, M.[2], [1]National institute of Livestock and Grassland Science, Ikenodai-2,Tsukuba,Ibaraki, 305-0901, Japan, [2]National Agriculture Research Center for Hokkaido region, Hitsujigaoka-1,Toyohira,Sapporo,Hokkaido, 062-8555, Japan, [3]The University of Yamaguchi, Yoshida1677-1,Yamaguchi, 753-8515, Japan; kama8@affrc.go.jp

The mechanism of fetus discharge at delivery is well understood, however, there is little information about the process of placenta discharge. Our previous presentation in this meeting has shown an evidence of activation of matrix metalloproteinase (MMP) of placental fibroblasts by arachidonic acid (Ara) and the inhibition of its activation by one of lipoxygenase (LOX) inhibitors. In this experiment, effects of LOX metabolites of Ara on the placental MMP activation were investigated using placental fibroblast cells. Hydroxy and hydroperoxy metabolites of Ara had a weak or no activity on MMP activation. However, 12-oxoeicosatetraenoic acid (12-oxoETE) had a strong activity of MMP activation. Other isomers of oxoETE have no activities of MMP activation at the same concentration of 12-oxo isomer. These results indicated that 12-oxoETE is one of candidates of signal for placenta separation after delivery. This work was supported by the Programme for Promotion of Basic and Applied Researches for Innovations in Bio-oriented Industry (2008), and Grant-Aid for Scientific Research of Japan Society for the Promotion of Science (2004-2007).

Timing of the LH peak and P4 blood concentration after the ovulations in buffalo cows during different superovulatory treatments

Terzano, G.M.[1], Malfatti, A.[2], Todini, L.[2], Barile, V.L.[1], Maschio, M.[1], Razzano, M.[1] and Mazzi, M.[1], [1]PCM, CRA, V.Salaria 31, 00016Monterotondo, Italy, [2]Unicam, DES, V.Circonv.93, 62024, Italy; giuseppinamaria. terzano@entecra.it

The aim of the trial was to assess the LH preovulatory peak and P4 blood concentration after ovulation in buffalo cows subjected to three different superovulatory treatments. On 12 adult buffaloes synchronized by a double PGF2α injection 12 days apart, a progesterone releasing intravaginal device (PRID) was inserted on the day of estrus (d0) and the animals were divided in three homogeneous groups: Group A received a 4-days decreasing dose (175-150-100-75 IU from d7 to d10) of an equal mixture of 1000 IU of FSH+LH and PGF2α on d9 at PRID withdrawal; Group B was treated by a single i.m. injection of 2000 IU of PMSG (d7) and PGF2α on d10 at PRID withdrawal; Group C received 2000 IU of PMSG (d7), PGF2α (d10 at PRID withdrawal) and a 2-days (d10=100; d11=75) decreasing dosage of 300 IU of FSH+LH; all the FSH+LH daily doses were administered 12 hours apart (half-half). Ultrasound examinations were performed daily starting on day 7 till the ovulations by a 7.5 MhZ linear rectal probe. Starting 12h after the PRID removal, every 4h took place 22 consecutive blood samplings to LH assays. Four days after the PRID removal blood samplings took place daily for 4 consecutive days to assess P4 concentration. All buffaloes had a preovulatory LH peak that occurred earlier ($P<0.05$) in Group C than in Groups A and B (19±3.8h vs 46±12h and 33±8.9h, respectively). After the preovulatory LH surge, concentration of P4 started increasing earlier ($P<0.05$) in Group A than in Groups B and C. There were no significant differences in themeanvalues of LH peak (6.33±3.64ng/ml) nor in ovulation rate (mean number of CL=2.67±1.97). These results suggest that the different superovulatory treatments affected the timing of LH peak and P4 concentration increase after ovulation but not the LH peak occurrence and the ovulation rate, and that treatment A seems could induce a earlier luteinization.

Physiophatological response of lactating ewes under different feeding systems
Álvarez-Rodríguez, J.[1], Sanz, A.[1], Uriarte, J.[2] and Joy, M.[1], [1]Centro de Investigación y Tecnología Agroalimentaria, Gobierno de Aragón, Tecnología en Producción Animal, Av. Montañana, 930, 50059 Zaragoza, Spain, [2]Centro de Investigación y Tecnología Agroalimentaria, Gobierno de Aragón, Sanidad Animal, Av. Montañana, 930, 50059 Zaragoza, Spain; mjoy@aragon.es

The aim of this study was to analyse the effects of the feeding system and lactation on the immune and haematologicalparameters of sheep. Churra Tensina ewes (n=20, body condition score 2.8) were allocated to two feeding systems from the week after lambing: Permanent grazing (GR) or rationed grazing (8 h/day) with 0.5 kg barley meal supplement (S). Blood samples were collected on weeks (wk) 2, 4 and 6 post-partum (pp) to determine leukocyte and erythrocyte populations as well as thrombocytes with a cell counter. Any leukocyte group was affected by the feeding system during lactation ($P>0.10$), being mean white blood cell (WBC) number 9.7 x 10^3/µl. WBC were lower on wk 2 than on 4 and 6 pp (8.7 vs. 10.1 and 10.4 x 10^3/µl, respectively, $P<0.05$), as well as did neutrophils (26.9 vs. 33.3 and 37.0%, $P<0.05$). However, lymphocytes decreased as pp advanced (61.8 vs. 56.2 and 54.8%, $P<0.05$), as did monocytes (9.8 and 9.0 vs. 6.6%, $P<0.05$). Although red blood cell (RBC) number were similar in both groups (mean 7.6 x 10^6/µl, $P>0.10$), haematocrit and haemoglobin tended to be lower in GR than in S (24.6 vs. 26.5% and 7.9 vs. 8.5 g/dl, $P<0.10$). RBC were greater on wk 2 than on 4 and 6 pp (8.0 vs. 7.4 and 7.5 x10^6/µl, $P<0.05$), as well as haematocrit (26.5 vs. 25.0 and 25.2%, $P<0.05$), whereas haemoglobin was lower on wk 2 and 4 than on 6 pp (8.1 and 7.8 vs. 8.6 g/dl, $P<0.05$). Platelets were lower in GR than in S (259 vs. 358 x 10^3/µl, $P<0.05$). The stage of lactation had major effects on ewe immune and haematological parameters, but not the feeding system.

Administration of recombinant bovine tumor necrosis factor-alpha affects hormone release in lactating cows
Kushibiki, S.[1], Shingu, H.[1], Moriya, N.[1], Komatsu, T.[2], Itoh, F.[3], Kasuya, E.[4] and Hodate, K.[5], [1]National Institute of Livestock and Grassland Science, Tsukuba, Ibaraki, 305-0901, Japan, [2]National Institute of Agricultural Research Center for Tohoku Region, Morioka, Iwate, 020-0198, Japan, [3]National Institute of Agricultural Research Center for Hokkaido Region, Sapporo, Hokkaido, 062-8555, Japan, [4]National Institute of Agrobioligic Sciences, Tsukuba, Ibaraki, 305-0901, Japan, [5]Kitasato University, Towada, Aomori, 034-8628, Japan; mendoza@affrc.go.jp

Tumor necrosis factor-alpha is a powerful macrophage cytokine releasing during infection. We examined the effect of recombinant bovine tumor necrosis factor-alpha (rbTNF) administration on hormone release in lactating cows. Twelve Holstein cows treated s.c. with rbTNF (2.5 ug/kg) or saline twice (1100 and 2300 h). At 1100 h the next day, the cows were given GHRH (0.25 ug/kg), TRH (1.0 ug/kg), TSH (10 ug/kg), or ACTH (500 ug/head) via the jugular vein. In the GHRH challenge, the plasma GH concentration was lower in the rbTNF group than in the control group. The GH and TSH responses to TRH were also smaller in the rbTNF group than in the control. The prolactin response to TRH was not affected by the rbTNF-treatment. Inthe TSH challenge, the rbTNF-treated cows had lower responses, as measured by plasma T_3 and T_4, than the control cows. The rbTNF treatment produced an increase in the basal plasma cortisol level, but the cortisol response to ACTH was the same level in both groups. The milk yield was reduced by the rbTNF administration during 4 days. These data demonstrate that TNF alters the secretion of pituitary and thyroid hormones in lactating cows. This effect may contribute to the suppression of the lactogenic function of the mammary gland observed in cases of coliform mastitis with high circulating TNF levels.

Peripartal metabolic adaptations in naturally scrapie-infected and healthy ewes

Álvarez-Rodríguez, J.[1], Sanz, A.[1], Monzón, M.[2], Garza, C.[2], Badiola, J.J.[2] and Monleón, E.[2], [1]Centro de Investigación y Tecnología Agroalimentaria, Gobierno de Aragón, Tecnología en Producción Animal, Av. Montañana, 930, 50059 Zaragoza, Spain, [2]Universidad de Zaragoza, Patología Animal, C/ Miguel Servet, 177, 50013 Zaragoza, Spain; jalvarezr@aragon.es

Scrapie is a neurodegenerative disease belonging to the transmissible spongiform encephalopaties. The hypothesis was that biochemical changes in blood metabolites due to peripartal adaptations might differ between naturally scrapie-infected (S) pre-clinical ewes and healthy (H) ewes. Rasa Aragonesa ewes (n=22, body condition score 3.2) were fed at maintenance and blood sampled on weeks -2,+1 and +2 relative to lambing. Plasma levels of triglycerides (TRIG), cholesterol (CHOL), NEFA, β-hydroxybutyrate (BHB) and urea were analysed with colorimetric kits. Data were assessed through ANOVA (previous log-transformation of TRIG, NEFA and BHB to meet normality). Plasma TRIG did not differ between S and H (0.36 mmol/l, $P>0.05$), but they were greater in ewes carrying twins than singles on week 2 of lactation (0.49 vs. 0.27 mmol/l, $P<0.05$). Plasma CHOL was similar in S and H (1.40 mmol/l, $P>0.05$), but it was higher on week -2 than subsequently (1.53 vs. 1.33 mmol/l, $P<0.05$). Plasma NEFA were lower in S-carrying twins than in their H counterparts (0.10 vs. 0.40 mmol/l, $P<0.05$), and these were greater on weeks -2 and +1 than on +2 (0.24 vs. 0.10 mmol/l, $P<0.05$). Plasma BHB was similar across S and H (0.54 mmol/l, $P>0.05$), but it was higher in ewes carrying twins than singles on week -2 (0.80 vs. 0.37 mmol/l, $P<0.05$). Plasma urea was similar across groups and remained steady (6.54 mmol/l, $P>0.05$). Pre-clinical prion disease hardly altered blood substrates around lambing.

Influence of porcine oviductal fluid on motion parameters in ejaculated and epididymal boar sperm

Avilés López, K., Carvajal, J.A., García Vázquez, F.A. and Matás, C., University of Murcia, Dept Physiology, Veterinary School, 30100, Spain; avileslopez@gmail.com

Some findings indicate that sperm motility was affected by oviductal fluid. Sperm binding to the epithelium cells plays an important role in preserve sperm fertility, stored and reduce the incidence of polyspermic fertilization. The use of epididymal and ejaculated spermatozoa led us the evaluation of the possible effect of seminal plasma interaction with the gametes. Computer-assisted sperm analysis (CASA) is a tool for the objective assessment of sperm motions. The aim of this study was to determine the influence of Porcine Oviductal Fluid (POF) on motions parameters of ejaculated (EJ) and epididymal (EP) boar spermatozoa under capacitating conditions. POF was collected from oviductal tube from slaughtered gilts in postovulatory stage of the oestral cycle and centrifuged at 7000g for 10 min at 4 °C. EJ and EP spermatozoa were processed by 3 different treatments: 1) spermatozoa without any treatment (control=C); 2) washed through Percoll® gradient (P); and 3) washed through Percoll® and incubated with POF (50μg/ml) for 30 min (P-POF). All the spermatozoa were incubated in TALP medium in capacitating conditions for 30 min before motility evaluation by CASA system. For EJ sperm the presence of FOP reduced the percentage of total motility (66.9 vs 58.2%) and modify the pattern of movement with reduced lateral head displacement (ALH, 3.6 vs 3.2 μm) and beat cross-frequency (BCF, 4.9±4.5) compared to P group ($P<0.05$). However, in EP spermatozoa the FOP did not affected the motion parameters. The POF modulates some motions parameters only in ejaculated sperm. Probably some interactions between proteins adhered to the ejaculated sperm membrane (from seminal plasma) and POF proteins could modulate sperm motility. Supported by Seneca 08752/PI/08 and AGL2006-03495.

Lactational pattern of milk yield and milk composition in dairy buffaloes
Ambord, S. and Bruckmaier, R.M., Veterinary Physiology, Vetsuisse Faculty University of Bern, Bremgartenstr. 109a, 3001 Bern, Switzerland; rupert.bruckmaier@physio.unibe.ch

The patterns of milk yield and milk composition including somatic cell count (SCC) during one entire lactation were investigated in the milk of 60 Mediterranean water buffaloes located at one farm in Switzerland. Milk yield was recorded during one morning and one evening milking and composite milk samples, proportionally mixed from morning and evening were analysed by using a Combifoss 6000 (Foss Electric, Denmark) for fat, protein, urea, and SCC. The duration of lactation varied considerably between buffaloes. Therefore, the herd was divided into 5 groups based on the individual duration of lactation, i.e. the number of samples taken (group 1 = 5 samples, group 2 = 6 samples, group 3 = 7 samples, group 4 = 8 samples, group 5 and herd = 9 samples). Despite different duration of lactation, daily milk yield, fat and protein followed a similar pattern in all groups. In contrast, no systematic pattern was observed for SCC and urea Their results varied between and within groups. The daily milk yield in month one was 11.2±0.3 kg (range: 9.9-11.6 kg) and reached a maximum in month two of lactation (11.4±0.3 kg; range: 9.5-12.0 kg). Thereafter, daily milk yield decreased continuously towards the end of lactation. Fat content was lowest in samples 1 and 2 (6.6±0.2%; range: 6.2-7.2%) and thereafter increased continuously to highest values in the last sample before the end of lactation (9.1±0.2, range: 8.0-10.1%). Milk protein content was 4.6±0.1% in sample 1 (range: 4.4-4.7%), 4.0±0.1% in sample 2 (range: 4.0-4.2%) and thereafter increased continuously in a range of 4.0-5.4%. SCC was highest in month one of lactation (96±30 cells/ ul; range: 84-156 cells/ ul) and shortly before drying off in all groups (186±23 cells/ul; range: 113-223 cells/ul). Inbetween SCC ranged between 20 and 70 cells/ul. Urea was always between 18 and 32 ppm with some outliers independent of lactational stage.

The effect of heat stress on oestrus intensity of heifers and multiparous Jersey cows
Soydan, E., Onder, H. and Kuran, M., Ondokuz Mayis University, Agricultural Faculty Department of Animal Science, Samsun, 55139, Turkey; esoydan@omu.edu.tr

Heat stress has an adverse effect on reproductive performance of dairy cows in addition to factors such as the number of lactation, physiological stage and the level of milk yield. In this study, we investigated whether environmental temperature during the year has any effect on oestrus intensity of heifers and multiparous Jersey cows. Records of average daily temperatures on a monthly basis and 510 Jersey dairy cows, classified as heifer and multiparous, over 14 years were used in the analyses. Oestrus intensity of heifers was higher at January and February ($P<0.05$), and also from July to September ($P<0.01$), and was lower from March to June ($P<0.05$) than those of multiparous cows. Spearman correlation coefficient was observed as 0.371 and -0.699 ($P<0.05$) between temperature and oestrus intensity for heifers and multiparous cows from January to December, respectively. Furthermore, during months with high temperature (from May to October) the relationship between temperature and oestrus intensity for multiparous cows was higher (-0.943; $P<0.01$) than that for heifers (0.657; $P>0.05$). These results indicate that multiparous Jersey cows are more sensitive to heat stress than heifers probably due to the physiological demand for milk and fat production.

Oral progestagen and colostrogenesis in primiparous sows

Foisnet, A. and Quesnel, H., INRA, UMR 1079, Domaine de la Prise, 35590 Saint Gilles, France; aurelie. foisnet@rennes.inra.fr

In ewes, a negative relationship between serum progesterone concentrations in late pregnancy and colostrum yield was reported. The aim of the present study was to determine the effect of an oral progestagen, altrenogest, administered in late gestation on colostrogenesis in sows. Sows were treated with altrenogest (20 mg/d) from d 109 to d 112 of gestation (ALT112, n=6) or from d 109 to d 113 (ALT113, n=8) or were not treated (Control, n=10). Colostrum production was estimated during 24 h starting at the onset of parturition using piglets' weight gains. Colostrum samples were collected at the onset of parturition then 6, 12, 24, 36 and 48 h later and were assayed for concentrations of Na+ and K+ (all samples) and immunoglobulin G (IgG, at t0, t24 and t48). Gestation length was longer in ALT113 than in Control group (116.3±0.2 vs 114.6±0.3 days; *P*<0.001) and intermediate in ALT112 group (115.8±0.5 days). Colostrum production (4.1±0.2 kg), litter size (13.9±0.7 piglets born alive) and litter birth weight (17.2±0.8 kg) were not different between groups (*P*>0.1). Concentrations of IgG in colostrum were greater (*P*<0.05) in Control than in treated sows (ALT112 + ALT113; 59.1±6.8 mg/ml vs 48.0±4.0; 8.0±2.2 vs 5.4±1.4; 2.4±0.4 vs 0.6±0.2 at t0, t24 and t48 respectively). In literature, concentrations of IgG in colostrum have been suggested to partly depend upon the permeability of tight junctions between epithelial mammary cells, which can reduce paracellular transfer of IgG. In the present experiment, Na+:K+ ratio in colostrum was lower (*P*<0.05) in ALT113 than in Control sows, suggesting a reduced permeability of the mammary epithelium. However, Na+:K+ ratio did not differ between ALT112 and Control sows (*P*>0.1). Further analyses, especially of prolactin concentrations, are needed to explain the reduction in IgG contents. To summarize, progestagen treatment did not influence colostrum yield but reduced IgG content.

The effects of antioxidant therapy on sheep pregnancies at high altitude are present before rapid fetal growing period

Parraguez, V.H., Carmona, K., Alegria, D., Urquieta, B., Galleguillos, M. and Raggi, L., Faculty of Veterinary Sciences, International Center for Andean Studies/University of Chile, Santa Rosa 11735, Santiago, 8820808, Chile; vparragu@uchile.cl

Sheep breeding is an essential activity for Andean highlands farmers. Newborn lambs at high altitude (HA, 3600 m) are smaller than at low altitude (LA, 500 m), due to hypoxia and oxidative stress. We demonstrated that the newborn weight in sheep natives from HA is higher than those of sheep natives from LA with pregnancy at HA. This difference appears to be associated to the higher placental weight and vascular area observed in HA natives sheep. We also demonstrated that antioxidant therapy during the entire pregnancy improve the fetal weight and normalize partially the placenta in pre-term pregnancies at HA. In this work we reported the effect of antioxidants on fetal and placental traits in sheep pregnancies, just before starting the rapid fetal growth period. Eight groups of pregnant sheep were used: HH, HA natives with gestation at HA; LH, LA natives with gestation at HA; LL, LA natives with gestation at LA; HL, HA natives with gestation at LA. The additional 4 groups were similar to the previous, but supplementation with antioxidants (Vitamin C 500 mg and E 350 IU per animal per day) was started 30 days before gestation begins (groups HHV, LHV, LLV and HLV, respectively). At about 100 days of pregnancy, fetuses and placentas were removed. Fetuses were weighted. Placentas were weighted and fixed for histological measurement of vascular area. Fetal weight in LL group was the highest (860.3±200.6 g), while in LH group the lowest (531.8±11.6 g). The other groups showed intermediate values without significant differences. Placental weight was higher in HH, LLV, LL and LH groups, being HH the highest (528.8±56.4 g). No differences in placental area surface occupied by vasculature were obtained. It is concluded that the effects of antioxidant therapy in pregnancies at HA is already present at about 100 days of pregnancy. Supported by Grant FONDECYT 1070405, Chile.

Effect of dietary energy level on follicle and oocyte quality during post-partum period in dairy cows
Räty, M.[1], Mikkola, M.[2], Dufort, I.[3], Gravel, C.[3], Tammiranta, N.[1], Taponen, J.[2], Mäntysaari, P.[1], Ketoja, E.[1], Sirard, M.-A.[3] and Peippo, J.[1], [1]MTT Agrifood Research Finland, ET Building, FI-31600 Jokioinen, Finland, [2]University of Helsinki, Paroninkuja, FI-04920 Saarentaus, Finland, [3]Laval University, Pavillon Paul-Comtois, Quebec G1V 0A6, Canada; mervi.raty@mtt.fi

Nutritional status is one of the major factors affecting reproductive performance in high producing dairy cows. Post partum negative energy balance affects morphological quality of oocytes, which is one of the key factors for subsequent embryo quality and viability. The aim of this study was to investigate the effects of energy-restricted diet on the quality of cumulus-oocyte complex (COC) during early postpartum period, and to find indicator genes specific for energy deprivation. At calving, the oocyte donors were assigned either to low energy diet or to control diet meeting the energy requirements. COC's were collected from ovaries of the donor animals (N=12) with ultrasound guided ovum pick-up technique (OPU) once a week for eight weeks time, starting 7-14 days after parturition. After collection, cumulus cells were separated from oocytes. Total RNA was extracted from the pool of oocytes and cumulus cells selected to be good representatives of feeding group according to serum β-hydroxybuturate analyses. Suppressive subtractive hybridization technique (SSH) was used to create two specific cDNA libraries, one from the oocytes and another from the cumulus cells, both enriched with genes associated with low energy diet. PCR products of over 500 clones from both of the libraries were spotted on glass slides to create a microarray specific for oocytes and cumulus cells. Oocyte and cumulus cell SSH samples were hybridized to microarray slides. According to the microarray results, 255 clones from both of the libraries were selected and sequenced. The sequences were submitted to a BLAST analysis for identification. From oocyte library 96 genes and from cumulus cell library 103 genes were identified. Altogether nine genes were selected as candidate genes for further analysis.

CFTR and NHE proteins: localization and functional significance in sheep (*Ovis aries*) sperm cells
Muzzachi, S., Guerra, L., Ciani, E., Castellana, E. and Casavola, V., University of Bari, Department of General and Environmental Physiology, Viale Amendola 165/a, 70126 Bari, Italy; elenaciani@biologia.uniba.it

Cystic Fibrosis Transmembrane Conductance Regulator (CFTR) is a glycosylated protein that functions as a cAMP-regulated anion channel, known to conduct both Cl- and HCO3-. Recent works suggest that CFTR plays a role during *in vivo* and *in vitro* sperm capacitation, preventing intracellular pH increase and membrane hyperpolarization by blocking the influx of chloride and bicarbonate ions into the cell. Recent investigations suggest that CFTR can also have a major role as regulator of several transporters, including Na/H exchangers (NHEs) that, in turn, have been suggested to regulate intracellular pH of spermatozoa in various species. In this work we have investigated CFTR and NHE1 expression and cellular localization by immunoblotting and immunofluorescence techniques, together with CFTR and NHE1 functionality by spectrofluorimetric analysis. For the CFTR protein, two bands have been recognized, one at around 180 kDa and a weaker band at around 130 kDa; two bands have been detected for NHE1 as well, one weaker band at about 95 kDa and one at about 80 kDa. Indirect immunofluorescence highlighted a positive staining in the mid-piece and a weak staining in the equatorial segment both for CFTR and NHE1 proteins. The functional activity of CFTR and NHE1 has been evaluated in sperm populations incubated, in different experimental conditions, with plasma membrane permeable fluorophores, MQAE and BCECF-AM, respectively. Inhibition of CFTR with CFTRinh172 resulted in a decrease (62%) in chloride influx, while the inhibition of NHE1 with dimethyl amiloride resulted in a reduction (60%) in the rate of intracellular pH recovery. The co-expression of CFTR and NHE1, together with the spectrofluorimetric results, suggests that a possible interaction between CFTR and NHE1 may have a functional role in sheep sperm cell physiology.

Effect of various levels of Vitamins E and C in milk and tris extenders on characteristics of Atabay ram semen in frozen condition

Jafari Ahangari, Y., Parizadian, B. and Zerehdaran, S., Gorgan University of Agricultural Sciences and Natural Resources, Dept. of Animal sciences, Gorgan, 49138-15739, Iran; yjahangari@yahoo.co.uk

The objective of present study was to investigate the effect of various levels of Vitamins E and C in milk and tris extenders on Atabay ram semen characteristics in frozen condition. Semen samples were collected from 6 Atabay rams using an artificial vagina. Suitable semen samples were mixed together and subjected to different treatments of vitamin E and C in milk and tris extenders. Semen characteristics including percentages of motile and viable spermatozoa were assessed. This experiment was carried out in 3×3 factorial arrangement with three replications on the basis of completely randomized design. First factor was various levels of Vitamin E (0, 30 and 60 μg/ml) and second factor was various levels of Vitamin C (0, 150 and 300 μg/ml). Results showed that in frozen condition, the effect of Vitamin E on motility and viability percentages of spermatozoa in milk and tris extenders were significant ($P<0.01$). The effect of Vitamin C on motility and viability percentages of spermatozoa in milk and tris extenders were not significant ($P>0.05$). Mean comparison on the basis of Duncan test showed that the highest motility of spermatozoa in tris (48.48±0.087) and milk extenders (46.45±0.085) were obtained at the level of 30 μg/ml Vitamin E. In conclusion, the use of Vitamin E at the level of 30 μg/ml in milk and tris extenders is recommended for long time storage of Atabay ram semen in frozen condition.

The maternal effect and different uterine body environments on reproductive traits of Cheviot and Suffolk male lambs

Jafari Ahangari, Y.[1], Smith, S.[2] and Blair, H.[2], [1]Gorgan University of Agricultural Sciences and Natural Resources, Dept. of Animal sciences, Gorgan, 49138-15739, Iran, [2]Massey University, Institute of Veterinary, Animal and Biomedical Sciences, Private Bag 11222, Palmerston North, New Zealand; yjahangari@yahoo. co.uk

The maternal effect and different uterine body environments on reproductive traits of 32 Cheviot and Suffolk male lambs were investigated. Experimental male lambs were born as a result of artificial insemination and embryo transfer techniques. Therefore, they were divided into four groups of Cheviot lambs bred by Cheviot ewes, Suffolk lambs bred by Cheviot ewes, Cheviot lambs bred by Suffolk ewes and Suffolk lambs bred by Suffolk ewes. At 120, 150, 180 and 210 days of age, body weight, scrotal circumferences, testis lengths and testicular diameters of lambs were assessed. From 150 days of age, semen samples were collected by the use of an electro-ejaculation and semen characteristics such as ejaculation volume wave motion, density, live rates and morphology assessments for major and minor spermatozoa defects were carried out. Results showed that lambs body weights in groups one to three was significantly different than group four ($P<0.01$). Testis measurements were significantly different between four groups of lambs ($P<0.01$). The increase on live rates of spermatozoa with increasing age of lambs from 150 to 210 days were not significantly different ($P>0.01$). But the major morphological defects of spermatozoa were decreased significantly ($P<0.01$). Therefore, further research works are needed to establish and develop an appropriate knowledge on the maternal and uterine effect on lambs body and reproductive traits.

Preimplantational embryo mortality in the rabbit: intrinsic or maternal factors
Lopez-Bejar, M., Mayor, P. and Lopez-Gatius, F., Universitat Autonoma de Barcelona, Faculty of Veterinary, Edifici V, Campus Bellaterra, 08193 Bellaterra, Spain; manel.lopez.bejar@uab.es

Superovulation was induced to provoke embryo overcrowding and factors affecting preimplantational embryo viability were studied. Ovulation rate, embryo number and quality, presence at oviduct or uterus and recovery rate were recorded. Embryos were cultured with or without the addition of maternal secretions. The mean number of recovered embryos per female was 23.3. Recovery rates were 104.3%, 92.6%, 90.1% and 65.7% at 26, 44, 64 and 85 hours postcoitum, respectively. Recovery rate and embryo number were significantly lower at 85 hpc. Ovulation rate and ratio of good quality vs bad quality embryos (up to 98%) were not different. At 64 hpc, 33% embryos and 97.9% at 85 hpc were obtained from the uterus. A reduction in the number of females with more than 8 embryos was recorded at 85 hpc (up to 50%). No significant difference in development after culture was observed (with or without maternal secretions). Only rate of development to blastocyst was lower from 2-cell embryos (26 hpc; 71% vs 94-100% for other intervals). Embryo mortality and adequacy of embryo number to normal prolificacy was detected from the interval of 64 to the 85 hpc in a relatively acute process, probably related to the stay at the uterus. The absence of intrinsic factors affecting embryo viability was demonstrated after *in vitro* culture where embryos developed successfully.

Effect of the addition of vitamins E and C on survival and motility characteristics of local fowls semen
Jafari Ahangari, Y.[1] and Karamzadeh Omrani, H.[2], [1]Gorgan University of Agricultural Sciences and Natural Resources, Dept. of Animal Sciences, Gorgan, 49138-15739, Iran, [2]Islamic Azad University, Dept. of Animal Sciences, College of Agriculture, Ghaemshahr, Iran; yjahangari@yahoo.co.uk

Twenty local fowls of seven months old were selected and trained for semen collection by the method of abdominal massage at a local farm of Behparvarane Amol, Mazandaran province, North of Iran. Semen samples were collected, pooled and extended with sterilized and homogenized cow skimmed milk using a dilution rate of one portion semen to four portions of extender at 30 °C . Experimental treatments were consisted of two levels of vitamin C (0 and 150 µg/ml) and three levels of vitamin E (0, 8 and 16 µg/ml). Extended semen samples were incubated for further assesment after 0, 6, 12 and 24 hours at 37 °C. Assesments of live and motility rates of spermatozoa and pH of semen samples were carried out with three replications. Results showed that addition of vitamin E had a significant effect on live and motility of spermatozoa ($P<0.05$), but had no significant effect on pH of semen ($P>0.05$). Concentrations of 8 and 16 µg/ml vitamin E increased live and motility rates of spermatozoa during 24 hours of incubation. The concentration of 150 µg/ml vitamin C caused an increase in live and motility rates of spermatozoa until 12 hours of incubation. Effect of interaction between vitamins E and C on spermatozoa motility and pH of semen samples was not significant ($P>0.05$). The addition of 16 µg/ml vitamin E and 150 µg/ml vitamin C in cow skimmed milk is recommended for liquid storage of Mazandaran local fowls semen.

Effect of transient hypo- and hyperthyroidism on reproductive parameters of Iranian Broiler Breeder hens

Akhlaghi, A.[1], Zare Shahne, A.[2], Zamiri, M.J.[1], Nejati Javaremi, A.[2], Rahimi Mianji, G.[3], Jafari Ahangari, Y.[4], Molla Salehi, M.R.[5] and Falahati, M.[5], [1]Shiraz University, Shiraz, 711, Iran, [2]Tehran University, Karaj, 261, Iran, [3]Tabarestan University, Sari, 151, Iran, [4]Gorgan University of Agriculture and natural Resources, Gorgan, 171, Iran, [5]Babol-Kenar Line Breeding Center, Babol-Kenar, 111, Iran; Amirakhlaghi837@yahoo.com

One hundred and thirty two 26-week-old broiler breeder hens (Arian) were randomly assigned into one of three treatments as control (CON), hypothyroid (HYPO; propylthiouracil (PTU)-treated) or hyperthyroid (HYPER; thyroxine (T_4)-treated) with 4 replicates of 11 hens for each. PTU and T_4 were administered at a level of 100 and 1 mg/lit into drinking water of HYPO & HYPER, respectively (starting at week 30 up to 33 of age). Blood sampling started at week 29, and repeated every week until the week 35 as well as weekly body weighing. Using ELISA, plasma levels of T_3, T_4 and estradiol were assayed. Egg number, fertility, hatchability, grading of day-old chicks and embryonic developmental stage of unhatched eggs were determined for individual artificially inseminated hen. Effects of PTU and T_4 treatment on plasma T_4 and T_3 was significant, but not on estradiol levels. Increased body weight following PTU treatment was not observed in other groups. Weekly egg number of HYPER was significantly lower than other two. Hatchability in HYPO was 0.000 while other groups did not significantly differ. 1st graded chick number in CON was greater than HYPER. Fertility in HYPER was lower than those of other two. In unhatched eggs, percent of pre-internal pipping stage of embryonic life in HYPO was more than other groups; while, that of internal pipping for HYPO was higher than others; but external piping was not different among groups. In conclusion, among the different reproductive parameters in this study, hatchability and weekly egg production are of highest sensitivity to thyroid hormone withdrawal and overdose, respectively.

Reproductive responses and Hepatic Gene Expression of GHR of Iranian Broiler Line hens to transient hyperthyroidism

Akhlaghi, A.[1], Zare Shahne, A.[2], Zamiri, M.J.[1], Nejati Javaremi, A.[2], Rahimi Mianji, G.[3], Jafari Ahangari, Y.[4], Daliri, M.[5] and Deldar, H.[3], [1]Shiraz University, Shiraz, 711, Iran, [2]Tehran University, Karaj, 261, Iran, [3]Tabarestan University, Sari, 151, Iran, [4]Gorgan University of Agriculture and Natural Resources, Gorgan, 171, Iran, [5]N.I.G.E.B., Tehran, 21, Iran; Amirakhlaghi837@yahoo.com

Seventy two 26-week-old of Iranian broiler line hens (Arian) were randomly assigned into one of four treatments as control (Con-B/Con-D) or hyperthyroid (Hyper-B/Hyper-D; thyroxine (T4)-treated) with 3 replicates of 6 hens in each line B or D. T4 was administered (1 mg/l) into drinking water in hyperthyroid group (week 30-33). Weekly blood sampling & body weighing conducted from week 29 to 35. Using ELISA, T3, T3 and estradiol were assayed. Egg number (EN), fertility (F), hatchability (H), day-old chick grading and embryonic developmental stage of unhatched eggs were determined for individual artificially inseminated hen. One hen/replicate was slaughtered to ovarian assay and hepatic gene expression of Growth Hormone Receptor (GHR) using real-time RT-PCR on week 33. EN recording continued to week 57. T4 treatment on plasma T4 levels was significant. Effects of treatment & age on BW & EN were significant as the same to their interaction on EN but not for BW. H percent was not significantly affected. 1st grade chicks for Con-D and Hyper-B was the highest and lowest, respectively; while T4 increased 2nd ones in line D, no difference was seen in line B. Fertility was not affected. In unhatched eggs, treatment effects on pre-internal and external pipping percentage was significant; but not on internal pipping. GHR gene expression differed too. Neither hierarchical and non-hierarchical follicles nor ovary weight was affected. In conclusion, although fertility or hatchability were not affected, weekly egg number decreased following postpubertal hyperthyroidism.

Effect of transient hypothyroidism on Hepatic Gene Expression of GHR and reproductive parameters of Iranian Broiler Line hens

Akhlaghi, A.[1], Zare Shahne, A.[2], Zamiri, M.J.[1], Nejati Javaremi, A.[2], Rahimi Mianji, G.[3], Jafari Ahangari, Y.[4], Daliri, M.[5] and Deldar, H.[3], [1]Shiraz University, Shiraz, 711, Iran, [2]Tehran University, Karaj, 261, Iran, [3]Tabarestan University, Sari, 151, Iran, [4]Gorgan University of Agriculture and Natural Resources, Gorgan, 171, Iran, [5]N.I.G.E.B., Tehran, 21, Iran; Amirakhlaghi837@yahoo.com

Seventy two 26-week-old of Iranian broiler line hens (Arian) were randomly assigned into one of two treatments as control (Con-B/Con-D) or hypothyroid (Hypo-B/Hypo-D; propylthiouracil(PTU)-treated) with 3 replicates of 6 hens in each line B or D. PTU was administered (100 mg/l) into drinking water in hypothyroid group (starting at 30 up to 33 weeks of age). Weekly blood sampling and body weighing (BW) were conducted from week 29 to 35. Using ELISA, T3, T4 and estradiol were assayed. Egg number (EN), fertility (F), hatchability (H), day-old chick grading and embryonic developmental stage of unhatched eggs were determined for individual artificially inseminated hen. One hen/replicate was slaughtered to ovarian assay and hepatic expression of Growth Hormone Receptor (GHR) using real-time RT-PCR on week 33. EN recording continued to week 57 of age. PTU treatment on plasma T4 and T3 levels was significant. Effects of treatment and age on BW & EN were significant, same to their interaction on EN but not for BW. H percent was affected by treatment and no eggs of HYPO group yielded chicks. GHR gene expression differed too; but F percent did not. In unhatched eggs, treatment effects on pre-internal and external piping percent was significant; but not in internal piping. Neither hierarchical & non-hierarchical follicles and ovary weight was affected. In conclusion, in contrast to fertility, hatchability and GHR gene expression can considerably be affected by postpubertal hypothyroidism.

Milk composition and fatty acid profile modelling along milking, lactation and lifetime

Baro, J.A.[1] and Martinez Villamor, V.[2], [1]Universidad de Valladolid, CC. Agroforestales, Campus de la Yutera, avda. de Madrid s/n, 34004 Palencia, Spain, [2]ASCOL, I+D, Polígono de ASIPO, Llanera, 33428 Asturias, Spain; baro@agro.uva.es

Changes in milk composition along physiological cycles are not well described, specially for fatty acids. In this study, a model has been developed to describe milk composition and fatty acid profile changes along the three main time scales affecting milk cattle productive life: lifetime trend, lactation curve and milking session. Response variables were milk yield, protein %, fat %, lactose %, and fatty acid profiles. Factors in the model were milking fraction (FRAC), day in milk (DIM) and calving age as an age-calving index (ACI). Two scenarios were studied: a single production unit with 90 cows, and the regional Milk Testing scheme of Castilla y Leon (Spain), with over one million cows. Both data sets produced coherent results, and the main conclusions from the modeling are: -Milk fractions are richer in protein and poorer in fat at the beginning, with maximum differences of 1% and 5%, respectively. -Effect of lactation: at peak yield, all response variables attained lowest values, but for lactose. -Effect of lifetime: highest values for response variables at intermediate ages. -Effect on fat %: main modeling factors included ACI, DIM, FRAC, and FRAC-ACI interaction. -Effect on protein %, lactose %, and SCC: best modeled with ACI, DIM, FRAC, and ACI-DIM interaction. -Fatty acids: C16:0 and C14:1 were the most abundant; C16:1 and C17:1 showed the most variable levels, as concluded from their coefficients of variation. Omega 3 and 6 fatty acids showed very small changes in every time scale.

Soybean lecithin is suitable cryoprotectant for cryopreservation of ram semen

Sharafi, M.[1], Nasr-Esfahani, M.H.[2], Nili, N.[1] and Nassiri Moghaddam, H.[1], [1]Isfahan University of Technology, Department of animal science, College of agriculture, 841568311, Iran, [2]Royan Institute, Embryology, Dept of Embryology, Reproductive Medicine Research Center, (Isfahan campus), ACER, 8158968433, Iran; hnassirim@gmail.com

The purpose of the present study was to evaluate ram semen *in vitro* fertility after the freezing–thawing process with extenders containing soybean lecithin. Soybean lecithin levels of 0.5, 1 and 2% (w/v) were assessed in combination with 7% glycerol in a basic Tris medium. Bioexcell was used as control treatment. Semen samples were diluted with extenders and then frozen. The sperm parameters were assessed after thawing for motility, viability and capacitation status. Fertility was recorded as cleavage rate at 3day and blastocyst rate at 8day after *in vitro* fertilization (IVF). Significant effects of various concentration of soybean lecithin were noted for the parameters investigated ($P \leq 0.05$). The percentage of motility was recorded to be 41.8%, 51.9% and 39.7% for 0.5, 1 and 2% lecithin respectively. Also the percentage of viable spermatozoa was estimated to be 36.08, 48.06 and 35.7 for 0.5, 1 and 2% lecithin respectively. Lecithin at 1% had more positive effect than other concentrations. Bioexcell produced 49.18% percentage of motile spermatozoa and 46.8% viable spermatozoa. No significant differences in the staining patterns of capacitation status were observed. In IVF experiment, the cleavage rate being significantly higher in oocytes fertilized with semen cryopreserved in 1% Lecithin. Different concentration of lecithin had no effect on embryo development. Results indicated that soybean lecithin is suitable cryoprotectant for cryopreservation of ram semen. Animal origin free extender based on soybean lecithin we have investigated here is a viable alternative to traditional egg yolk-based extenders.

The repeatability and accuracy of real-time ultrasound technique in measuring ribeye area of Charolais cattle

Harangi, S., Radácsi, A. and Béri, B., University of Debrecen, Institute of Animal Husbandry, Böszörményi str. 138, H-4032 Debrecen, Hungary; harangis@agr.unideb.hu

Nowadays, real-time ultrasound (RTU) can be used to gather live cattle data that subsequently can be used for the genetic prediction of carcass cutability and meat quality traits. Application of RTU in practice depends on the accuracy and repeatability of the technique. Thus, it's necessary to recognize these parameters, which have primary importance in selecting superior sires for carcass merit. The aims of this study were to evaluate the repeatability and accuracy of measuring ribeye area (REA) under Hungarian conditions. REA was measured for the first time with Falco 100 RTU equipment on live animals, before the day of slaughter in Charolais fattening bulls (ultrasound ribeye area – UREA). Ultrasound pictures were traced at two times by the same operator to assess repeatability of the RTU technique. After slaughter, REA was measured on the carcass by planimeter at the same anatomical point as the ultrasound measurements (carcass ultrasound ribeye area – CREA). Statistical analysis was carried out to examine the relationship between CREA and UREA to determine the accuracy of RTU. There were no significant differences between the records from repeated evaluations of the same UREA picture ($P > 0.05$). Strong positive correlation coefficients were found ($r = 0.95$-0.96) between the values for the UREA and CREA datasets. This means that these factors change in a similar way so it's sufficient to evaluate the ultrasound records only once. There were no significant differences between the CREA and UREA values ($P > 0.05$). Correlation between the magnitude of the differences between CREA and UREA, and CREA values was found to be low, and the error of the evaluation didn't depend on the real size of the ribeye area (CREA). Consequently, RTU can be an accurate technique to estimate CREA from UREA values.

Effect of body condition and suckling restriction with and without presence of the calf on milk production and calves performance on range conditions

Quintans, G., Banchero, G., Carriquiry, M., López, C. and Baldi, F., National Institute of Agricultural Research, Beef Production, Ruta 8 km 282, 33000, Uruguay; gquintans@inia.org.uy

Lactation and presence of the calf are important factors involved in the suckling-induced suppression of LH secretion. The effects of body condition score (BCS) and suckling restriction, with and without the presence of the calf, on milk production (MP) and calf performance were evaluated. Sixty three Angus x Hereford multiparous cows were managed to maintain different BCS at calving and thereafter (low (L) v. moderate (M); L, n=31, M, n=32). Within each BCS group, cows were assigned to three suckling treatments (ST) at 66 d postpartum (pp): 1) suckling ad libitum (S, n=20); 2) calves fitted with nose plates during 14 days while remaining with their dams (NP, n=22); 3) calves were completely removed from their dams for 14 days, and thereafter returned to them (CR, n=21). Cows were bled monthly from 98 d prepartum until 66 d pp and weekly thereafter until 128 d pp. Plasma non esterified fatty acid (NEFA) concentrations were measured. From 65 d pp until weaning, MP was assessed every 20-22 d. A BCS x time interaction on NEFA concentrations was found due to an increase 42 d prepartum in NEFA levels in L BCS compared to M BCS cows(1.59±0.06 and 1.08±0.06 ng/ml, $P<0.001$). From 66 to 122 d pp, cows in M BCS had greater ($P<0.05$) NEFA concentrations than cows in L BCS (0.37±0.008 and 0.31±0.008 ng/ml). Within M BCS cows, MP was similar among ST (4.24, 3.87 and 4.18 kg/d for S, NP and CR, respectively). In L BCS, cows in NP and CR produced less ($P<0.005$) milk than those in S (4.12, 3.4 and 3.29 kg/d for S, NP and CR, respectively). The correlation between calf daily live weight (LW) gain and MP was 0.49 ($P<0.0002$) and calf LW at weaning was greater ($P<0.001$) for S than for NP or CR (159.3±3.1, 150.1±2.9 and 147.0±3.1 kg for S, NP and CR, respectively). The reduction in MP and calf weaning LW was similar between suckling restriction treatments with and without presence of the calf.

Evaluation of beef genetic merit for growth rate in beef x dairy steers

Keane, M.G.[1], Campion, B.[1], Berry, D.P.[2] and Kenny, D.A.[3], [1]Teagasc, Beef Production, Grange Research Centre, Dunsany, Co. Meath, Ireland, [2]Teagasc, Dairy Production, Moorepark Research Centre, Fermoy, Co. Cork, Ireland, [3]University College, Dublin, Agriculture, Food Science and Veterinary Medicine, University College Dublin, Belfield, Dublin 4, Ireland; gerry.keane@teagasc.ie

In Ireland, beef genetic evaluations are carried out on an across-breed basis. For dairy beef production, growth rate comprises ~70% of the monetary value of the total merit index. The objective of this study was to evaluate genetic merit for growth rate expressed as expected progeny difference for carcass weight (EPD_{CWT}). Spring-born male progeny out of Holstein-Friesian cows and Aberdeen Angus (AA=10) and Belgian Blue (BB=13) sires of either high (H=13) or low (L=10) EPD_{CWT} were used. Pure-bred Friesians (FR=7) and Holsteins (HO=12) were also included. In total, 170 animals with sire verification, distributed across 6 genetic groups, namely AAH (n=32), AAL (n=24), BBH (n=31), BBL (n=27), FR (n=28) and HO (n=28) were reared to slaughter. Mean sire EPD_{CWT} values, weighted by number of progeny per sire, for these genetic groups were 3.4, -13.4, 26.7, 13.0, -8.1 and 0.9 kg, respectively. Slaughter weight and carcass weight were significantly greater for H than L, but because of significant genetic merit x beef breed interactions, the effects were evident for AA only. Slaughter weight, kill-out proportion and carcass weight were significantly higher for BB than AA. Compared to HO, m. longissimus depth, fat depth and kill-out proportion were significantly higher for FR. Overall, with increasing genetic merit for growth, live weight and carcass weight increased for AA but not for BB. Dry matter intake and feed efficiency were not affected by genetic merit and the extra weight due to H was not accompanied by any increase in fatness. BB were superior to AA for all important production traits. FR had greater m. longissimus and fat depths, and a higher kill-out proportion than HO, but slaughter weight and carcass weight were not significantly different.

Coarseness of grain or level of rumen by-pass starch had marginal effects on rumen environment and rumen wall conditions in concentrate-fed veal calves

Jarltoft, T.C., Kristensen, N.B. and Vestergaard, M., Aarhus University, Faculty of Agricultural Sciences, Foulum, DK-8830 Tjele, Denmark; Terese.Jarltoft@agrsci.dk

The effect of coarseness of grain and source of starch in concentrate on the rumen environment was studied in 9 rumen-fistulated bull calves. Three blocks of 3 bull calves were weaned at 2 months of age and fed one of 3 pelleted concentrate rations (N, R or S) and long barley straw *ad libitum* until slaughter at 11.5 months of age. Starch sources in N and R were identical and consisted of barley and wheat. Ingredients of N were finely ground whereas those of R were coarsely ground. In the S concentrate, half the barley and wheat was substituted with finely ground corn and sorghum to increase theoretical rumen escape starch from 30 in N and R to 100 g/kg dry matter (DM) in S. Crude protein (15%) and energy content was similar in all 3 concentrates. Calves were ruminally cannulated at 9 months of age. Two weeks post surgery 3 sampling periods were initiated (9 rumen samples per 24 h in the medial (M) and ventral (V) rumen sac) with a 3-week interval. Average daily concentrate intake (7.6 kg DM), straw intake (0.7 kg DM) and daily gain (1.35 kg) were not affected by treatments. Rumen pH in V was lower in S- compared with N- and R-fed calves ($P<0.05$). A similar tendency was observed in M. pH was 0.30 units lower in M compared with V. pH in V was below 5.8 for more hours in S- compared with N- and R-fed calves (16.9 vs. 12.0 h/d). Total concentration of volatile fatty acids in M was greater in S- compared with N- and R-fed calves ($P<0.01$) and S also had the lowest acetate ($P<0.10$) and highest propionate ($P<0.10$) proportion. Evaluated macroscopically at slaughter, the shape of rumen papilla was in favour of S ($P<0.10$) and papillae were longer in S and R compared with N ($P<0.05$). The results show that more slowly fermentable starch or more coarsely ground grain had only marginal effects on rumen environment and rumen wall conditions of concentrate-fed calves.

The use of visible and near infrared reflectance spectroscopy for prediction and improvement of meat quality characteristics in beef

Roehe, R.[1], Prieto, N.[1], Ross, D.W.[1], Navajas, E.A.[1], Nute, G.R.[2], Richardson, R.I.[2], Hyslop, J.[1] and Simm, G.[1], [1]Scottish Agricultural College, Edinburgh, EH9 3JG, United Kingdom, [2]University of Bristol, Division of Farm Animal Science, Bristol, BS40 5DU, United Kingdom; Rainer.Roehe@sac.ac.uk

A slaughter trial was carried out to identify improved measurement techniques for meat eating and carcass quality in beef. Data on 194 steer and heifer beef cattle from rotational crosses of Aberdeen Angus and Limousin were available. Meat eating quality characteristics measured were sensory traits (tenderness, flavour, juiciness and abnormal flavour, 14 days post mortem (pm)), instrumental tenderness (slice shear force, 3 days pm; Volodkevitch 10 days pm), fatty acid profiles (saturated and unsaturated fatty acids, 48 h pm), colour (48 h pm) and cooking loss (14 days pm). Visible and near infrared reflectance spectroscopy (Vis-NIR) measurements were taken in the abattoir on the M. longissimus thoracis between 12th/13th ribs at 48 h pm over the spectral range from 350 to 1,800 nm. Partial least square regression was used for prediction of meat eating quality traits from Vis-NIR spectra. Measurements of Vis-NIR showed correlations with sensory traits of 0.45 (juiciness) to 0.77 (flavour), physical measurements of 0.61 (Volodkevitch) and 0.74 (slice shear force), fatty acids in the range from 0.49 to 0.80, colour of 0.93 (red-green) to 0.95 (yellow-blue) and cooking loss of 0.60. The results indicate that based on the measurements of Vis-NIR spectra online in the abattoir, it was possible to successfully predict numerous (mostly lowly correlated) meat quality characteristics. This shows the high variation in absorption at different Vis-NIR wavelengths due to factors such as muscular fibre characteristics, chemical bonds and colour, which are associated with various meat quality characteristics. Generally, the results support the use of on-line Vis-NIR in the abattoir for early, fast and relatively inexpensive estimation of beef meat quality and its use in value based marketing systems and genetic improvement programmes.

Enteric methane emissions in extensive cattle in Salamanca (Spain)

Alvarez, S.[1], Sánchez Recio, J.M.[2], Alvarez, M.J.[3] and Jovellar, L.C.[4], [1]Universidad de Salamanca, Area de Producción Animal, Filiberto Villalobos 119, 37007 Salamanca, Spain, [2]Asociación Nacional de Criadores de Ganado Vacuno de Raza Morucha Selecta, Santa Clara 20, 37001 Salamanca, Spain, [3]Universidad de Navarra, TECNUN, Manuel de Lardizabal 13, 20018 San Sebastián, Spain, [4]Universidad de Salamanca, Area de Ingenieria Agroforestal, Filiberto Villalobos 119, 37007 Salamanca, Spain; salvarez@usal.es

Livestock is a source of greenhouse gases. In particularly, ruminants produce a high quantity of enteric methane. All these gases must be evaluated in national inventories, and there are specific guidelines set by the Intergovernmental Panel on Climate Change (IPCC). For enteric methane, the estimation is based on average feed gross energy intake and CH_4 conversion rates (Ym). This work aims to improve the estimation of annual emissions of CH_4 in extensive cattle farms in Salamanca (Spain).This Spanish province has the highest number of beef cows, and the production system is based on pasture combined with Quercus trees, with a low stocking rate (0.2-0.5 cow ha^{-1}) of an autochthonous breed, Morucha, and grazing all year around. Supplementary feeding is needed in summer (because of drought) and in winter (because of a cessation in growth due to low temperatures). This is a prototype of an animal production system integrated into the environment. Tier 2 method proposed in 2006 by IPCC is used for estimation of enteric CH_4 emission. Ym is estimated depending on digestibility of feed, as described in the bibliography, together with characteristics and intake of pasture and supplementary feeding. The ration for different stations in the study area is calculated and used in the estimation. The value of enteric CH_4 emission per cow obtained, 62 kg year^{-1}, is consistent with values obtained from other research work and the value recommended by IPCC in the Tier 1 method. Nevertheless, further research is needed in order to estimate properly the role of extensive livestock systems as a sink or a source of greenhouse gases.

Modelling CO2 footprints and trace gas emissions for milk protein produced under varying performance and feeding conditions

Dämmgen, U.[1], Brade, W.[1] and Döhler, H.[2], [1]University of Veterinary Medicine Hannover, Institute for Animal Breeding and Genetics, Buenteweg 17p, 30559 Hannover, Germany, [2]Association for Technology and Structures in Agriculture (KTBL), Bartningstr. 49, 64289 Darmstadt, Germany; ulrich.daemmgen@daemmgen.de

Trace gas emissions are likely to become a threat to milk production in Central Europe. Sustainability of milk production has to take emission reduction into account. Scenarios have to be constructed which allow for the evaluation of mitigation options for both greenhouse gases and air pollutants, in particular ammonia. Any change in production results in changes in the emissions of CO_2, CH_4, NMVOCs, NH_3, NO, N_2O and primary particles. Also, any measure aiming to reduce emissions of one single gas will have effects on almost all other emissions. The agricultural emission model GAS-EM is a combination of process and mass flow oriented modules. It is applied to describe in detail the entire production chain of milk protein from mineral fertilizer production to nitrous oxide emission from the continental shelves. Scenarios deal with the effects of varying milk yield and/or different milk protein and fat contents as well as varying feed composition and manure management options including the intensity of grazing, housing and storage types, spreading techniques and incorporation times. Emissions from tractors and other combustion engines are considered. Hence, detailed information can be obtained to evaluate emission reduction options by establishing both their CO_2 footprint and their contribution to atmospheric pollution. A combination of (low cost) measures can effectively reduce the adverse effects of milk production on ecosystems and air quality.

Assessment of the Australian system for the prediction of beef quality (MSA): which perspectives for the French beef sector?

Hocquette, J.F.[1], Moevi, I.[2], Jurie, C.[1], Pethick, D.W.[3] and Micol, D.[1], [1]INRA, UR1213, Herbivore Research Unit, Theix, 63122 Saint-Genes Champanelle, France, Metropolitan, [2]Institut Elevage, Route Epinay, 14310 Villers-Bocage, France, Metropolitan, [3]Murdoch University, School of Veterinary and Biomedical Sciences, Murdoch, Western Australia 6150, Australia; hocquet@clermont.inra.fr

Australia has developed the Meat Standards Australia (MSA) grading scheme to predict beef quality for consumers. This system is comprehensive, consistent and scientifically supported. It is based on the development and the exploitation of a huge database, including the use of a large-scale consumer testing system with cuts cooked in different ways as well as information on the corresponding animals, carcasses and meats. Statistical analyses were carried out to identify the critical control points of beef palatability which is indicated for individual muscles and for a specific cooking method from animal characteristics before and after slaughter and also from ageing time. The personalities questioned in France recognize quality numbers of the MSA system. It is relevant, finalized, serious and original in its conception, innovating towards the suggested segmentation of the beef market, without being prescriptive in regard to the factors that affect beef quality. It is also credible, flexible and open ended. But it possesses some weak points. Its development in Australia at the farmer and abattoir level has been strong, but the final delivery of precise quality grades to consumers is still lacking. Its adaptability to France would be difficult due to the complexity and the specificity of the French beef industry and market. But, the program is uniquely innovative and deserves consideration. It could facilitate awareness and could induce the indispensable changes for the preservation and the development of the French beef sector.

Remote sensing and geographical information systems applications to determine grassland types and grazing systems in highlands of the Eastern Turkey

Bozkurt, Y.[1], Basayigit, L.[2] and Kaya, I.[3], [1]Suleyman Demirel University, Faculty of Agriculture, Department of Animal Science, Isparta, 32260, Turkey, [2]Suleyman Demirel University, Faculty of Agriculture, Department of Soil Science, Isparta, 32260, Turkey, [3]Kafkas University,Faculty of Veterinary Medicine, Department of Animal Science, Kars, 36600, Turkey; ybozkurt@ziraat.sdu.edu.tr

In this study, the aim was to investigate the possibility of determining grassland boundaries, biomass quantitiy, stocking rates and pasture quality using Remote Sensing and Geographical Information Systems. The study was carried out between April 2005 and April 2008 in the East Anatolian Region (in Kars province) with an altitude above 2,000 m where the animal husbandry is dependent on grassland. For this purpose, 3 different test areas, having different altitudes (Low, Middle and High) were chosen as experimental locations and type of grassland and soil, and the pattern of grass species were determined. The amount of available biomass for grazing and stocking rates were then predicted, and according to the measured quality, the most efficient grazing systems were suggested. The possibility of monitoring of animal movements was also evaluated.

Accuracy of records collected on animals and carcasses at slaughterhouse
Lazzaroni, C. and Biagini, D., University of Torino, Dept. Animal Science, Via L. da Vinci 44, 10095 Grugliasco, Italy; carla.lazzaroni@unito.it

To verify the effectiveness of data collection in a cattle slaughterhouse, recorded data were collected before checking and official processing and analysed, to highlight possible mistakes. One year's records on animals (genetic type, age, sex, live weight) and carcasses (weight, dressing percentage, category, conformation, fatness) were analysed in detail (mean, minimum and maximum value, and/or frequency distribution) to verify their accuracy. Data were evaluated for 49,486 slaughtered animals, of 48 different genetic types (59.8% males and 40.2% females). The average slaughter age was 31.5 months, with a live weight of 539 kg and a carcass weight of 326 kg, corresponding to a dressing percentage of 61.4%. Almost half of slaughtered animals were uncastrated young males (A, <24 months of age; 46.5%), followed by cows (D, females that have calved; 24.7%), veal calves (V, 8-12 months; 15.2%), and other females (E; 13.0%). Concerning conformation, few carcasses were graded S (superior; 3.2%), more graded E (excellent; 28.9%), U (very good; 19.0%) and P (poor; 18.2%). For fatness, carcasses were evaluated as 1 (low; 8.4%), 2 (slight; 86.4%), or 3 (average; 5.2%). The results obtained confirm the need to improve data accuracy, especially regarding birth date (at least 1.20% wrong by default and 4.06% for excess), live weight (0.10% wrong by default and 0.35% for excess) and carcass weight (0.38% wrong by default and 3.47% for excess), the latter two influencing dressing percentage (15.84% wrong by default and 13.63% for excess). Several of those results could be attributed to mistakes in typing the real values (lack of one or more digits, repeat of digits or failure to enter the comma in the number), confirming the need to improve data collection both at the beginning of the slaughtering chain and at the end. If paying more attention would slow the processing line, then it could be better to check the data before official processing.

Use of milk feeders and group housing in veal calves production
Lazzaroni, C. and Biagini, D., University of Torino, Dept. Animal Science, Via L. da Vinci 44, 10095 Grugliasco, Italy; carla.lazzaroni@unito.it

To test the possibility of rearing veal calves in large groups using automatic calf feeders, a trial was conducted from June to November on 103 Friesian male calves, according the latest EU suggestion on cattle welfare. The calves were housed in 3 groups of about 30 each, in 3 pens of 102 m² in the same stable, and reared from 33 to 175 d of age (average period 142 days) at slaughter. In each pen, 2 automatic milk replacer dispensers, 8 m of linear manger for dried maize silage (as fiber feed) and 2 drinking troughs were available 24h per day, to ensure feed and water availability. At arrival, the calves had a large range in age (38 day) and live weight (39 kg). During the rearing period, the calves suffered gastro-enteric and respiratory diseases, causing a high mortality rate (27%), so that at slaughter only 75 animals remained. At the start of the trial, the animals weighted 4,900 kg, corresponding to an average of 52.6±7.7 kg/head, and increased to 9,671 kg at the end of the rearing period, corresponding to an average of 127.9±36.0 kg/head (20 animals weighed more than 150 kg). Thus, total gain was 4,771 kg, corresponding to an average of 75.1±31.3 kg/head and an average daily gain of 0.66±0.28 kg/head/d (9 animals gaining more than 1 kg/d). At slaughterhouse, carcass weight and dressing percentage were recorded; the former averaged 71.2±20.6 kg (17 carcasses weighted more than 90 kg), while the latter was on average 55.4±4.9% (20 animals dressed more than 57%). Total milk powder consumed in the whole period was 15,032 kg, or an average of 200.4 kg/head, while total dried maize silage consumption was 5,779 kg, corresponding to 77.1 kg/head. Thus, average feed consumption was 277.5 kg/head/period. While the overall results were not as good as expected, several calves achieved good performances, showing the feasibility to use automatic calf feeders for large groups of animals.

Effect of body condition changes and milk urea content after calving on reproduction in Czech Fleckvieh cows

Kubesova, M.[1,2], Frelich, J.[2], Stipkova, M.[1], Rehak, D.[1] and Marsalek, M.[2], [1]Institute of Animal Science, v.v.i., Cattle Breeding, Přátelství 815, 104 00 Praha-Uhříněves, Czech Republic, [2]University of South Bohemia, Special Livestock Breeding, Studentská 13, 370 05 České Budějovice, Czech Republic; kubesova. marta@seznam.cz

The Czech Fleckvieh is an original dual-purpose cattle breed in the Czech Republic and belongs to the Simmental cattle family. Dual-purpose cattle are believed to have different responses to negative energy balance from diary cows (Holstein) because of their lower genetic merit for milk production and for mobilization of body reserves. The objective of this study was to examine the relationship between body condition (BCS) change and milk urea content (MUC) post partum, and reproduction in Czech Fleckvieh cattle. The BSC was measured before calving and then at monthly intervals. Milk samples were taken monthly and milk urea content was determined. Next, data from the reproduction performance database of the Czech Fleckvieh herd book were used to record selected reproduction indexes (calving to first service interval, calving to conception interval, calving interval and number of services per conception). The dataset was analyzed by multifactorial analysis of variance using the procedures COOR and GLM. The significant effect of BCS change after calving on the length of the calving to conception interval was determined. In the cows with a BCS change in the 1st month of lactation of +0.25 to -0.25 points this interval was shortest; animals with BCS loss of more than 1.75 points had the longest interval. The milk urea content at conception was related to the calving to conception interval and calving interval. The shortest indexes applied to cows with the lowest milk urea content. In this study, it is obvious that both BCS change and milk urea content in Czech Fleckvieh cows didn't influence ovarian function and the onset of estrus after calving, but they did affect probability of conception or embryo development. This study was supported by the projects MZE 0002701404, MSM 6007665806.

Short co-incubation time in bovine IVF with OPU oocytes and sex-sorted/unsorted spermatozoa

Ruiz, S.[1], Zaraza, J.[2], De Ondiz, A.[1] and Rath, D.[2], [1]Murcia University, Physiology, Veterinary Faculty, 30100. Murcia, Spain, [2]Friedrich Loeffler Institut, Institut für Nutztiergenetik, Höltystrasse, 10, 31535. Mariensee, Germany; sruiz@um.es

The combination of ovum pick up (OPU) and *in vitro* fertilisation (IVF) techniques could be an alternative to traditional embryo production. Sex-sorted sperm have been successfully incorporated into IVF in cattle, but these sperm have altered patterns of motility and a reduced lifespan. The objectives of this study were to investigate the effects of reducing the duration of gamete co-incubation time on the performance of bovine IVF with OPU oocytes and sex-sorted/unsorted frozen-thawed semen. In total, 84 OPU sessions were carried out in 18 normal cyclic, dry and non stimulated cows (Holstein Friesian and Schwarzbunterind breeds). OPU and sex-sorting techniques have been previously described. Oocytes were fixed and stained at 4, 8 and 12 h post insemination (hpi) to evaluate penetration (PEN), monospermy (MON), male pronucleus formation (MPF) and performance (PERF, monospermic oocytes with 2 pronuclei from total matured oocytes). These parameters did not differ between sperm treatments. No interaction between sperm treatment and co-incubation time was observed ($P>0.05$). Co-incubation time affected fertilization. PEN increased progressively at 4 (9.09 and 25.0%), 8 (44.4 and 55.6%) and 12 (65.4 and 69.6%) hpi for sex-sorted and unsorted sperm, respectively ($P<0.01$). Differences in MPF and PERF were observed in co-incubation time ($P<0.01$), with the best rates obtained in 12 hpi with sex-sorted sperm (78.6 and 64.7%, respectively). Previous studies have reported a reduction of *in vitro* fertility of sex-sorted sperm compared with unsorted sperm in cattle, but we did not find differences in performance of bovine IVF with OPU oocytes between sperm treatments. It is concluded that for OPU oocytes, a reduction of gamete co-incubation time during IVF adversely affected PEN, MPF and PERF and, regardless sperm treatment used, best results were obtained for 12 hpi.

Influence of milking indicators on teat parameters
Ježková, A., Pařilová, M., Stádník, L. and Vacek, M., Czech University of Life Sciences Prague, Department of Animal Husbandry, Kamýcká 129, 16521, Prague 6 - Suchdol, Czech Republic; stadnik@af.czu.cz

The influence of milking vacuum and milk flow level when detaching the cluster, on milking performance and teat characteristics, were studied in four separate experiments using 51 cows (26 Danish Red and 25 Holsteins; 330 teats). In the 4 experiments, vacuum level and detachment level were adjusted to 39 kPa and 400 g.min^{-1}, 39 kPa and 100 g.min^{-1}, 45 kPa and 400 g.min^{-1}, and 45 kPa and 100 g.min^{-1} All the experiments were carried out in a free-stall barn with an automatic milking system using ultrasonograph Aloka with linear probe (7.5 MHz) for scanning teats, a Vernier calliper for taking measurements of teats, Lucia and SAS software. Both milking vacuum and over-milking influenced external and internal teat parameters. Change in teat length, measured before and after milking, was higher when a vacuum of 45 kPa was used (0.31 cm) and significantly ($P<0.05$) longer (by 72.2%) compared to teats milked with 39 kPa. Teat canal length, teat end width and teat cistern width were larger when using the higher milking vacuum. Milking time per cow was shorter when using vacuum of 45 kPa and the higher detachment level (212 s). Teats lengthened most when over-milking and vacuum of 45 kPa were applied compared to lower milking vacuum (4.62 vs 4.58 cm). Teat canal length was significantly ($P<0.05$) longer (12.28 mm) in over-milked teats compare to non over-milked teats (11.95 mm). Teat end of over-milked teats was significantly wider ($P<0.01$) than in non over-milked teats (21.44 vs. 21.13 mm). All observed internal teat parameters were affected by the interaction of vacuum and detachment level. The only change in internal teat parameters which was influenced by this interaction was the change in teat canal length measured before and after milking. The smallest change was found when milking vacuum of 39 kPa and no over-milking was applied to the teats.

Reproductive performance of dairy cows following estrus synchronization treatment with PGF2α and progesterone
Kaim, M., Lavon, Y. and Ertracht, S., Agricultural Research Organization, Institute of Animal Science, POB 6, 50250 Bet Dagan, Israel; kaim@agri.huji.ac.il

Studies have shown that high progesterone levels during the luteal phase preceding first artificial insemination (1st AI) were associated with higher conception rates in dairy cows. The objective of this study was to compare the reproductive performance of dairy cows following estrus synchronization treatment with PGF$_{2\alpha}$ and progesterone (CIDR), with that of untreated control cows. In total, 168 primiparous (PP) and 214 multiparous (MP) high-producing Holstein cows were studied over 6 months. Clusters of cows were formed at 2-week intervals. The cows in each cluster were categorized into one of two treatment groups according to parity and days in milk. Control cows (n=180) were inseminated following the first estrus detected after the end of the voluntary waiting period. The CIDR cows (n=202) received two PGF$_{2\alpha}$ injections given 14 d apart. An intravaginal device containing progesterone (CIDR), which was inserted 9 d after the first PGF$_{2\alpha}$ treatment for 5 d, was removed when the second PGF$_{2\alpha}$ treatment was administered, and cows that manifested estrus during the next 7 d were inseminated. Cows not observed in estrus were treated again. In both groups, AI was performed once daily a.m., following an observed estrus. There was 79% of PP and 85% of MP cows in estrus following first synchronization treatment. Conception rates at 1st AI for PP and MP cows were respectively, 35.4% and 30.3% in the control group vs. 54.7% ($P<0.02$) and 44.8% ($P<0.05$), in the CIDR group. Overall conception rates at 1st AI for control and CIDR groups were respectively, 32.6% and 49.0% ($P<0.01$). Pregnancy rates at 120 d after calving for PP and MP cows were respectively, 45.1% and 44.9% in the control group vs. 62.8% ($P<0.02$) and 52.6% (NS), in the CIDR group. Overall pregnancy rates for control and CIDR groups were respectively, 45.0% and 56.9% ($P<0.01$). In conclusion, synchronization with PGF$_{2\alpha}$ and progesterone improved the conception rate in high-producing dairy cows.

Utilization of bioelectrical impedance to predict *Longissimus thoracis et lumborum* muscle intramuscular fat in beef carcasses

Silva, S.R.[1], Morais, R.[2], Patricio, M.[1], Guedes, C.[1], Lourenço, A.[1], Silva, A.[1], Mena, E.[1] and Santos, V.[1], [1]CECAV-UTAD, POBox 1013, 5001-801 Vila Real, Portugal, [2]CITAB-UTAD, POBox 1013, 5001-801 Vila Real, Portugal; eligomes@utad.pt

Bioelectrical impedance analysis (BIA) has been used to assess body and carcass composition of several farm species. However, little information has been reported for meat quality traits such as intramuscular fat (IF), which is in accordance with consumer preferences. Our objective was to examine the usefulness of BIA to predict the *longissimus thoracis et lumborum* muscle (LM) IF of beef. Fifty two samples of LM muscle from 26 beef carcasses of 357 kg mean weight were used. A 3 cm thick slice sample of LM was removed 24 h after slaughter at the 4th and 5th lumbar vertebrae. After removal of all subcutaneous and intermuscular fat, the LM samples were placed over a flat surface for BIA measurements. A two-electrode bioelectrical impedance analyzer, built specifically for this purpose, based around a high precision impedance converter system (AD5933, Analog Devices), was used. After calibration, and at a frequency of 50 kHz, the magnitude of the impedance and relative phase of the impedance was calculated using a Discrete Fourier Transform (DFT) algorithm to obtain the resistance (Rc) and reactance (Xc) values. Two hypodermic needles were used as electrodes. The needles were inserted 2 cm into the LM samples with 10 cm distance between the needles. The IF content of LM samples was determined by chemical analysis using the Soxhlet method. Regression analysis was used to relate BIA measurements (Rc and Xc) and IF. The results showed that the IF content of LM was predicted accurately by Rc and Xc BIA measurements ($r^2=0.73$, and $r^2=0.70$, $P<0.001$, respectively). From the current data set it can be concluded that BIA was able to predict IF in LM samples. Further research is needed to develop BIA equipment for use in meat processing plants for objective measurement of meat characteristics.

Comparative study of milk quality from manual or mechanical milking

Saran Netto, A.[1], Fernandes, R.H.R.[2], Azzi, R.[3] and Lima, Y.V.R.[2], [1]FZEA, ZAZ, Duque de Caxias Norte - 225, 13635045, Brazil, [2]FMVZ, VNP, Duque de Caxias Norte 225, 13635045, Brazil, [3]UNIP, Veterinarian, cantareira, 05020000, Brazil; saranetto@yahoo.com

The aim of this study was to compare milk quality from different milking systems, manual and mechanical. The animals and equipament on a milk farm located at Cunha City, São Paulo were used. Fifteen cows were milked manually and another 15 were milked mechanically using vacuum equipment. Each milk sample was tested for milk composition, somatic cells count (SCC) and total bacterial count (TBC). Milk composition was similar for both treatments. However, the values for SCC were higher for animals milked manually (446, 366 and 342 x10^3/ml vs 49, 93, 125 x10^3/ml), while the TBC level were higher for animals milked mechanically (11.8, 119.4, 92.2 vs 4.6, 11.8, 29.6 x10^3CFU/ml). This indicates high bacterial contamination on the mechanical equipment and an absence of correlation between SCC and TBC.

Effect of different nonprotein nitrogen sources on performance and carcass characteristics of feedlot finished Nelore steers

Corte, R.R.P.S., Nogueira Filho, J.C.M., Brito, F.O., Leme, P.R., Pereira, A.S.C., Aferri, G. and Silva, S.L., USP, Duque de Caxias Norte, 13635900, Brazil; jocamano@usp.br

To assess the effects of soybean meal replacement by different nonprotein nitrogen sources on the performance and carcass characteristics of finishing cattle, 46 Nelore steers with a mean initial weight and age of 333 kg and 20 months, respectively were fed one of four diets: 1) CTL (control diet): 12% soybean meal and 1% Urea, 2) O: 6% soybean meal and 1.8% Optigen®II, 3) U: 6% soybean meal and 1.66% Urea, and 4)U+O: 6% soybean meal, 1.0% Urea and 0.72% Optigen®II. Diets had 78.5% concentrates and were isoproteic (15.5%) and had the same total digestible nutrients (77.4%) and rumen degradable protein (10.4%) values. Steers were allotted to four pens according to initial body weight (block) and were weighed at 25 day intervals after 18 hours fasting. After 75 days, animals were slaughtered in the experimental packing plant of USP, according to proper welfare guidelines. After slaughter, dressing percentage (live weight / hot carcass weight *100) were recorded. Twenty four hours later, Longissimus muscle area and backfat thickness at the 12th rib were determined. Dry matter intake, average daily gain and feed efficiency were not affected by treatments ($P>0.05$) with mean values of 2.6%, 1.56 kg and 0.15 kg gain/kg of dry matter, respectively. Dressing percentage, Longissimus muscle area and backfat thickness were not affected by treatment, with mean values of 58.7%, 71.8 cm^2 and 3.62 mm, respectively. These results show that replacement of SM by the nonprotein nitrogen sources used in this study did not affect performance or carcass traits of feedlot finished steers. Thus, in these conditions these nitrogen sources can be used to reduce the costs of feeding while still achieving similar results to traditional soybean meal-based diets.

Embryo transfer as a mean of increasing pregnancy rates in repeat breeder cows

Gacitua, H., Zeron, Y., Segal, A., Dekel, I. and Arav, A., Agricultural Research Organization, Animal Science, P.O.Box 6 Bet Dagan, 50250, Israel; Gacitua@Agri.Huji.Ac.Il

We transferred either *in vitro* fertilization (IVF) embryos or parthenogenic embryos and evaluated the cows 35 days after artificial insemination (AI). Only repeat breeder (RB) (between 3 and 6 AI) were chosen for this study. Primiparous (n=7) and multiparous (n=37) Holstein cows were selected for the experiment between April and June of 2008. Slaughter house ovaries were collected during the winter of the same year, taken to the laboratory, and oocytes were aspirated and *in vitro* matured for 24h. IVF was carried out with sperm from the same bull and chemical activation was achieved with ionomycin and 6 DMAP. Day 7 embryos were vitrified using ethylene glycol and trehalose and cooled using the VitMaster at ultra rapid cooling rate. The cryopreserved embryos were warmed prior to embryo transfer (ET) on location at the dairy farm one week after natural heat and AI was performed. After palpation, two parthenogenic embryos or one IVF embryo were transferred to the ipsilateral or contralateral horn, respectively. We checked with ultrasound the presence of pregnancy 35 days after AI. The overall pregnancy rate (PR) in this farm was 33% for multiparous (n=144) and 37% for primiparous (n=35) cows. For RB without ET (control), results were 19%, 30%, 15% and 35% for 3,4,5,and 6 AI, respectively, with an overall PR value of 24%. Results with ET showed increasing PR between 3, 4, 5 and 6 inseminations: 31%, 23%, 50% and 33%, respectively, with an overall PR value of 34% (15/44) which is 10% higher than the control. The PR for the parthenogenic and IVF embryos was 27% and 50% respectively. A paternity test (DNA) of the calves born from the ET showed that all were born (n=7) from the AI and not from the ET-IVF embryos. We conclude that ET at one week following AI can increase the PR of repeat breeder cows.

Analytical strategies for residue analysis of veterinary drugs in milk

Roca, M.I.[1], Althaus, R.L.[2], Borrás, M.[1], Beltrán, M.C.[1] and Molina, M.P.[1], [1]Universidad Politécnica de Valencia, Instituto de Ciencia y Tecnología Animal, Camino de Vera s/n, 46022 Valencia, Spain, [2]Universidad Nacional del Litoral, Facultad de Ciencias Veterinarias. Departamento de Ciencias Básicas, R.P.L. Kreder 2805, 3080 Esperanza, Argentina; pmolina@dca.upv.es

The presence of antibiotic residues in milk constitutes a potential human health risk and can interfere in technological processes in the dairy industry. For this reason, a program of milk quality control and traceability 'Letra Q' has been established in Spain (RD 1728/2007). This includes mechanisms for detecting the presence of antimicrobial residues in raw milk at levels above the Maximum Residues Limits (Regulation 2377/90/CEE). The first stage of the control system uses screening methods (microbiological, enzymatic, protein receptor-binding assays, etc.) to determine the presence or absence of veterinary drug residues. Due to the high number of screening methods available, this study aims to evaluate them and establish analytical strategies for antimicrobial residue analysis in raw cows' milk. The results showed, generally high values of specificity (81-100%) which indicates few 'false positive' results. The sensitivity was different according to the antimicrobial substances and methods analysed, which can be equivalent to non-reproducible results depending of the methods used in each laboratory and stage of control. From these results and the frequency with which the most commonly used antimicrobial substances are employed in cattle pathologies, different analytical strategies are proposed. These strategies combine methods to obtain the highest possible detection level in order to guarantee that the milk reaches the consumer free of antibiotic residues.

Effects of inbreeding on sperm quality in Fleckvieh bulls

Fuerst-Waltl, B., Gredler, B., Maximini, L. and Baumung, R., University of Natural Resources and Applied Life Sciences Vienna, Dep. Sust. Agric. Syst., Div. Livestock Sciences, Gregor-Mendel-Strasse 33, A1180 Vienna, Austria; roswitha.baumung@boku.ac.at

Currently, almost 96% of the cows in the Austrian Fleckvieh population are artificially inseminated (AI). The use of a low number of possibly related bulls might lead to increased inbreeding. Thus, inbreeding depression becomes an increasingly important issue in cattle breeding. The potential effect of individual inbreeding of Fleckvieh bulls on their sperm quality was therefore analysed. Using pedigree data, the inbreeding coefficients of 715 Fleckvieh bulls of two AI stations in Upper and Lower Austria were calculated and incorporated in a statistical model for analysis of sperm quality. The basis for all calculations was information on five sperm quality parameters (volume, concentration, motility, number of spermatozoa per ejaculate and percentage of viable spermatozoa) of approximately 30,000 ejaculates. Records of the two AI stations were analysed separately. Although most bulls were inbred to some extent, mean inbreeding coefficients of 1.2 and 1.5%, respectively, can be considered as very low. Despite the low inbreeding level, a negative effect of inbreeding on sperm quality was found. In both stations, all five sperm quality traits were affected by inbreeding depression, four significantly (exception concentration). In most cases a linear effect of inbreeding was found. Only for the trait motility was a non-linear relationship detected. Neither the actual inbreeding level nor the inbreeding depression seems to be alarming. However, monitoring of inbreeding depression effects on fertility traits is to be recommended since the low heritability of fertility traits only allows slow genetic improvement.

Effect of nutrition and time period of beef ageing on chosen meat quality parameters of bulls slaughtered in the same age

Filipcik, R., Hosek, M. and Kuchtik, J., Mendel University of Agriculture and Forestry, 2B06107, Department of Animal Breeding, Zemedelska 1, 61300 Brno, Czech Republic; xfilipci@mendelu.cz

The aim of this study was to evaluate the effects of different nutritional treatments after weaning, and period of beef ageing after slaughter, on meat quality parameters of crossbred bulls (n=144) of the Czech Fleckvieh and Charolais breeds. The feed ration for group A consisted of maize silage, clover-grass silage, meadow hay and concentrates, that for group B was maize silage, meadow hay, concentrates and urea, while group C grazed on permanent pasture. The periods of beef ageing were fresh, 2 weeks, 4 weeks and 6 weeks. At slaughter there were no significant differences in the age of bulls (mean 650±27 days). To evaluate the meat quality parameters the m. longissimus lumborum et thoracis was used. The nutritional treatments had a significant effect on contents of total protein (TP) and intramuscular fat (IMF). The highest contents of TP (21.51±0.67%) and IMF (1.88±0.65%) were in group A. On the contrary, the lowest contents of these were found in group C (TP=21.12±0.46%; IMF= 0.61±0.33%). The nutritional treatments did not have a significant effect on water retention (WR), but there was a significant effect of this factor on tenderness. The highest tenderness was found in meat of bulls from group A, whilst the lowest tenderness was found in group C bulls. Time of ageing had no significant effect on TP and IMF contents. On the contrary, it did significantly affect tenderness and WR. The most tender meat was fresh meat, while the most firm meat was obtained after 6 weeks of ageing. The highest WR was also found in meat after 6 weeks of ageing, but the lowest WR was found in fresh meat.

Improving repeat breeder cows' fertility by synchronizing ovulation and timed inseminations

Kaim, M.[1], Gal, Y.[2], Abramson, M.[3], Ben-Noon, I.[2] and Moalem, U.[1], [1]Agricultural Research Organizatio, POB 6, 50250 Bet Dagan, Israel, [2]Kibutz, Ashdot Yaakov Meuhad, 15150, Israel, [3]Hahaklait, POB 3039, 38900 Caesaria, Israel; kaim@agri.huji.ac.il

The prevalence of cows with 3 artificial inseminations (AI) out of the total in the Israeli dairy herd during 2007 was 36% for primiparous (PP) and 43% for multiparous (MP) cows. For cows with 4 or more AI, the corresponding values were 24% and 30%. On average, the conception rates of the repeat breeder cows were 20-30% lower than conception rates at first AI. The objective of the current study was to improve repeat breeder cows' fertility by synchronizing ovulation, followed by two fixed timed consecutive inseminations. Cows with 3 or more AI were defined as repeat breeder cows. The year-long study was conducted in a large commercial dairy farm. Sixty-four PP and 117 MP Holstein cows that were approaching 3 or more AI were randomly assigned to one of two treatment groups according to parity and AI number. The treatments were 1) Control–cows (n=93) that manifested estrus were re-inseminated, and 2) Treated–cows (n=88) that manifested estrus were not re-inseminated. Seven days after estrus, the cows were treated by $PGF_{2\alpha}$ injection, followed by a GnRH injection 2 d later. Cows were inseminated 24 h and 48 h after GnRH injection. Conception rates for PP and MP cows were respectively, 29.4% and 16.9% for Control vs. 43.3% (NS) and 34.5% ($P<0.05$) for Treated. Overall conception rates for the Control and Treated groups were respectively, 21.5% and 37.5% ($P<0.02$). In conclusion, synchronizing ovulation followed by two fixed time inseminations improved the conception rate of repeat breeder cows.

Impact of health data quality on breeding efficiency in Austrian Fleckvieh cows

Egger-Danner, C.[1], Koeck, A.[2], Obritzhauser, W.[3], Fuerst, C.[1] and Fuerst-Waltl, B.[2], [1]ZuchtData EDV-Dienstleistungen GmbH, Dresdner Strasse 89/19, A-1200, Austria, [2]University of Natural Resources and Applied Life Sciences, Division of Livestock Sciences, Gregor-Mendel-Strasse 33, A-1180, Austria, [3]Chamber of Veterinaries, Biberstrasse 22, A-1010, Austria; egger-danner@zuchtdata.at

A project to establish an Austrian wide health monitoring system for cattle commenced in 2006. At present, 11,800 farms with about 200,000 cows participate. The main aims of the project are the provision of support for herd management and estimation of breeding values for health traits. A precondition for efficient use of health data is strict data validation. Before recording the data into the database, routine plausibility checks concerning first treatments are carried out. The challenge is to distinguish between farms with low frequencies and incomplete documentation and/or recording. The impact of strictness of data validation was analysed for Fleckvieh (Simmental) dual purpose cows. A minimum number of 0.1 first diagnoses per cow, year and farm are general preconditions. The data sets originated from farms assisted by veterinarians providing a large amount of diagnoses and less stringent criteria concerning provision of diagnostic data. Heritabilities were estimated with threshold sire models. For mastitis (between 10 days before and 50 days after calving), strict and less strict validation resulted in heritabilities of 0.076 and 0.069,respectively, with corresponding average frequencies of mastitis of 5.0% and 4.6%. For fertility disorders (until 150 days after calving) less stringent validation decreased the heritability from 0.064 to 0.047 and frequencies from 13.3% to 10.6%. The correlations between breeding values ranged from 0.84 to 0.92. Differences in the frequencies especially for fertility disorders were also observed between direct electronic transmission of the data by the veterinarian and manual recording by the performance recording organisations. This might be due to differentiation between application of drugs with and without a withdrawal period.

Effect of grain processing on serum metabolites in finishing calves

Hernández, J.[1], Castillo, C.[1], Pereira, V.[1], Vázquez, P.[2], Suárez, A.[1], López Alonso, M.[1], García Vaquero, M.[1] and Benedito, J.L.[1], [1]Veterinary Faculty, USC, Animal Pathology, Campus Universitario, 27002 Lugo, Spain, [2]CESFAC, I+D+I Research Department, Diego de León, 28006, Madrid, Spain; joaquin.hernandez@usc.es

Grain processing methods are used to alter certain characteristics of the grain, which improves starch digestion by exerting effects on site and extent of digestion The predominant grain processing methods used in Spanish feedlots are grinding and pelleting The present study evaluated the effects of grain processing (pelleting and grinding) on different serum metabolic parameters (glucose, non-esterified fatty acids (NEFA), urea nitrogen (SUN), total proteins (TSP), albumin, L-lactate) of bull calves. Twenty Belgian Blue bull calves were utilized for a 77-day feedlot study. Animals were allotted randomly to one of two experimental groups: (1) calves fed a concentrate in a pelleted form (PF, n=10), (2) calves fed a ground concentrate (GF, n=10). During the study period, all the measured parameters fell within the physiological ranges for beef under intensive conditions. This finding, in addition to the lack of clinical symptoms of ruminal disturbances, suggests that neither of the processing methods caused detrimental effects to animal health. Excepting SUN and creatinine, none of the measured parameters was significantly affected by grain processing or time×grain processing interaction, and only serum creatinine could be considered as a valuable biomarker of pelleted high-grain consumption in bull calves.

Application of Lacto-Corder in the control of production and milkability properties of dairy cows in Croatia

Mijić, P.[1], Bobić, T.[1], Knežević, I.[1], Puškadija, Z.[1], Bogdanović, V.[2] and Ivanković, A.[3], [1]Faculty of Agriculture, University of J. J. Strossmayer in Osijek, Trg sv. Trojstva 3, 31000 Osijek, Croatia, [2]Faculty of Agriculture, University of Belgrad, Nemanjina 6, 11080 Zemun, Serbia, [3]Faculty of Agriculture, University of Zagreb, Svetošimunska c. 25, 10000 Zagreb, Croatia; aivankovic@agr.hr

Control of dairy cow production in Croatia consist of two steps, measurement of milk amount after milking and taking of milk samples for chemical analysis. This kind of control gives very little information about production. On the other hand, for several years other countries use sophisticated devices, such as the Lacto-Corder, for monitoring of many parameters important for production and selection of cows. The Lacto-Corder is a measuring device for production control and collecting of milk samples. It is acknowledged by the International Committee for Animal Recording. In Croatia, for research work and investigation of its use in production, the first Lacto-Corder was bought. Research work was conducted on Holstein cows (n=457). Cows were in the range from first to sixth lactation. Thirteen production parameters were measured. Duration of the main milking phase was 4.51 min. The average amount of milk in that time was 10.18 kg. Average milking rate was 2.27 kg/min. Research results have shown a justification for using a Lacto-Corder in everyday production control of dairy cows in Croatia. The amount of information and its precision for every examined animal was considerably higher. The only obstacle to the use of the Lacto-Corder in Croatia today is the very high purchase price.

Relative economic weight of some production and functional traits of dairy cattle

Szabó, F.[1], Fekete, Z.S.[1], Wolf, J.[2] and Wolfová, M.[2], [1]University of Pannonia, Animal Science and Production, Deák. F. 16., H-8360 Keszthely, Hungary, [2]Institute of Animal Science, Uhríneves, Prague, Czech Republic; szf@georgikon.hu

The economic value of 7 traits was calculated for the dairy cattle population in Hungary in 2008, using a bioeconomic model based on the program package ECOWEIGHT. The importance of the study derives from more than 70% of the cattle in the country belonging to the dairy industry. The study was based on the typical dairy farm size of 330 Holstein-Friesian cows, with a production level of 7,000 kg annual milk yield. Cows were managed in a loose-housing system with parlour milking, representing current commercial dairy enterprises. A total mixed ration based on maize silage and concentrates, with some alfalfa hay, was offered to 4 groups (first-, second-, third-phase of lactation and a dry group). Besides the dairy enterprise, calf and replacement rearing were also taken into consideration. Income came from milk, calves, culled cows and manure sale. About 50% of the total costs related to feed, with the remainder due to factors such as management, reproduction and health services, labor, interest and amortization. Annual revenues and costs were used for the economic calculations. Gross margin was taken as a difference between income and variable costs. Marginal economic value of a given trait was defined as the partial derivative of the profit function, which was standardized by multiplying by the genetic standard deviation of the trait. The relative economic values for traits were expressed as a percentage of the standardized economic value of 305-d milk yield. The relative economic importance of the evaluated traits were as follows: 305-d milk yield 100%, length of productive life 51%, conception rate of cows 36%, 305-d protein yield 35%, 305-d fat yield 20%, stillbirth 13%, pregnancy rate of replacements 3%.

Some economic aspects of milk production

Szabó, F., Buzás, G.Y. and Heinrich, I., University of Pannonia, Animal Science and Production, Deák F. u. 16., H-8360 Keszthely, Hungary; szf@georgikon.hu

The economics of milk production was calculated for Hungary using input-output analysis. A high (€0.30/kg) and a low (€0.20/kg) milk price were considered. Calculation was done for 6,000, 7,000 and 8,000 kg annual milk yield per cow. Price and cost data were collected in different farms in the country. Cows were managed in a loose-housing system with parlour milking, representing typical, current commercial dairy enterprises. A total mixed ration based on maize silage and concentrates, with some alfalfa hay, was offered to cows. Besides the dairy enterprise, calf and replacement rearing were also taken into consideration. Income came from milk, calves, culled cows and manure sales. Total variable costs including feed, replacement, and other variable costs (such as management, reproduction and health services, interest and amortization) were used for the economic calculations. Feed costs were calculated from the nutrient requirements of cows. Labor costs weren't taken into consideration. Annual revenues and costs were used for the profitability evaluation. Gross margin (GM), in which labor costs and profit of the farmer are included, was taken as the difference between income and total variable costs. Profitability was calculated as a percentage of total variable cost in the income. The ratio of the feed:cow replacement:other variable costs for 6, 7 and 8 thousand kg annual milk yield were 56:23:21%, 60:21:19%, and 62:20:18%, respectively. GM and profitability values, when milk price was high, were €513, 608, 667/cow, and 43.5%, 46.1%, and 45,1%, for 6, 7, and 8 thousand kg milk yield per cow, respectively. When milk price was low, the corresponding values were €322, 371 and 406/cow, and 26.9%, 28.1%, and 27.5%. Gross margin per cow increased with increasing milk production, but the profitability from 7 to 8 thousand kg milk yield decreased due to higher feed costs.

Comparative study between monensin or saccharomyces supplementation in finishing bull calves: effects on productive and metabolic parameters

Castillo, C.[1], Pereira, V.[1], Hernandez, J.[1], Mendez, J.[2], Vazquez, P.[3], Blanco, I.[1], Vecillas, R.[1] and Benedito, J.L.[1], [1]Veterinary Faculty, USC, Animal Pathology, University Campus, 27002 LUGO, Spain, [2]COREN, Research Department, Juan XXIII, 32003 Ourense, Spain, [3]CESFAC, 1+D+I Department, Diego de León, 28006 Madrid, Spain; cristina.castillo@usc.es

The aim of the present study was to evaluate the effects of two dietary supplements, the ionophore antibiotic monensin and a live S. cerevisiae culture, on metabolism in cattle during the finishing phase of the production cycle. A 77-day feedlot study was conducted using 42 Belgian Blue steers. The study period was from 23 to 34 weeks of age. Cattle were housed in a commercial feedlot farm (Coren SCL). Three experimental treatments were applied as follows: (1) control (no supplementation, n=10, C), (2) monensin (n=16, MON, Rumensin, Elanco Animal Health), at a concentration of 30 mg/kg concentrate on a dry matter (DM) basis, and (3) a live culture of S. cerevisiae strain NCYC Sc 47 (n=16, SACC, Biosaf Sc 47, Eurotec Nutrition) at a dose of 500 mg/kg concentrate DM. Blood samples were collected by jugular puncture on days 0 (just after the adaptation period prior to supplementation), 3, 7, 13, 51 and 77 (the last day of the finishing period prior to slaughter). The parameters measured were venous blood pH, bicarbonate, pCO_2 serum L-lactate, glucose, urea and non esterified acids. Production parameters were measured and can be considered as useful complementary information associated with supplementation. No significant differences in productivity were observed among groups. Mean serum L-lactate levels remained stable over time within physiological ranges, with no significant differences among groups, suggesting that supplementation did not influence lactate production. The lack of positive effects from S. cerevisiae supplementation on performance and energy metabolism, must be evaluated with caution, as the effect of yeast is intimately related to the yeast culture preparation, the dose and the components of the ration.

The freezing point of milk from individual cows in Latvia

Jonkus, D., Kairiša, D. and Paura, L., Latvia University of Agriculture, Liela str. 2, LV-3001, Jelgava, Latvia; daina.jonkus@llu.lv

In Latvia, freezing point is presently used as a quality indicator of cows' raw milk and its limit value is ≤-0.520 °C. Until 2008, there was no research in Latvia on cows' milk freezing point variations taking into account individual animal genetic and physiological factors together with environmental conditions. The objective of this study was to investigate the average freezing point value and the factors that influenced it. In the 6-month period from August 2008 to January 2009, 1,646 milk samples were analyzed from Latvian Brown and Holstein Black and White cows with an average milk yield of 21.4 kg per day (s.d.=6.54). The average freezing point of milk was -0.531±0.0106 °C, with a range from -0.402 °C to -0.659 °C. There were 16% of samples above the limit value. The effect of breed, parity, stage of lactation and month of observation were included in the model as fixed factors. Milk yield, fat, non-fat solids (NFS), urea and somatic cell count (SCC) were covariate factors. Influence of breed, parity and milk yield on freezing point variation were not significant. Statistically significant freezing point changes were found between months (lowest in August –0.543 °C, highest in January -0.523 °C), as well as between lactation phase - lowest in the third phase (-0.533 °C) and highest in the second phase (-0.529 °C) ($P<0.05$). Significant covariate factors were fat content, NFS, urea and SCC. Correlations between freezing point and milk yield and milk content were negative (r from – 0.345 to – 0.248).

Comparison of milk composition and quality between voluntary and conventional milking system

Kairiša, D., Jonkus, D., Muižniece, I. and Paura, L., Latvia University of Agriculture, Liela str. 2, LV-3001, Jelgava, Latvia; daina.kaisa@llu.lv

In the dairy industry, technological improvements are continuously being made to the mechanisation of the milking process in order to reduce manual labour. Voluntary milking systems are designed to ensure maximum welfare needs of animals - free choice of milking time and milking frequency, as well to facilitate human labour and influence on the cow. Besides the automatic recording of data, they also facilitate improvements in animal health control. The aim of this research was to analyse variation in milk yield and milk composition in voluntary and conventional milking systems. In the teaching and research farm 'Vecauce' of the Latvian University of Agriculture, dairy cows managed in a loose housing system were subjected to two different milking systems, voluntary and conventional. The research was carried out from July to October 2008 on 80 dairy cows, 40 per group. Milk samples from individual cows were collected. Average milking frequency was 3 in each group. Average cows parity in the voluntary group was 2.3, and in the conventional system was 2.9. Most of the cows from both groups were in the second phase of lactation. The highest daily milk yield in the voluntary group was in August at 26.1 kg, but in conventional system, it was in July at 24.8 kg. In the following months, daily milk yield decreased in both groups connected with changes in lactation days. The milk composition indices used were protein and fat contents. For the total research period, a statistically significant higher fat content was found in the conventional milking system which was up to 4.8%. For protein content, the tendency was opposite. The lowest protein content was in the voluntary milking system in August at 3.31%. In the total period, a higher protein content was found in the conventional milking group and it was highest in September and October at 3.9%. Somatic cell count (SCC) was used as an index of milk quality. In the voluntary milking system, SCC for the total period was at the limit value and did not increase above 400,000/ml.

Breeding for profit: an economic comparison of Jerseys and Holsteins

Szendrei, Z., Holcvart, M. and Béri, B., University of Debrecen, Institute of Animal Science, Böszörményi út 138., 4032 Debrecen, Hungary; szendreiz@agr.unideb.hu

Production and reproduction data of primiparous Jerseys imported to Hungary and domestic Holsteins were analyzed and compared. Jerseys calved significantly younger (25.9 months) than Holsteins (29.6 months), but there were no differences in insemination rate (1.62 vs. 1.78) and days open (109 vs. 94 days). There were significant differences in the 305 days milk production between the two breeds. Holsteins were superior in milk volume production (8,012 kg) to Jerseys (5,447 kg), but Jerseys had superior milk solids concentration (5.44% fat, 3.77% protein) compared to Holsteins (3.87% fat, 3.26% protein). Income per cow after the milk sold was corrected to the reference values was similar for both breeds (Jersey € (Euro) 1,240, Holstein €1,411), but due to the differences in milk component ratios, income for extra milk fat (€144 vs. €22) and protein (€78.9 vs. €10.3) was significantly higher for Jerseys. In conclusion, the two breeds differed not only in income (Jersey €1,499, Holstein €1,912) but in feed cost also (€425 vs. €908). Nevertheless the profit per cow, calculated from the difference of the former two did not differ (€1,074 vs. €1,003). Income per 1kg milk was similar for both breeds (Jersey: €0.27, Holstein: € 0.23), but since there was a difference in feed cost (€0.08 vs. €0.11), profit per 1kg milk (€0.19 vs. €0.12) was greater for Jerseys. Although it is difficult to draw detailed conclusion from this limited study, it can be seen that milk production of the two breeds is consistent with the literature, and even though we have not found statistically significant differences in profitability, the €70.7 profit difference per cow and €0.07 profit difference per kg milk is considerable.

Effect of semen batch on fertility of Austrian dual purpose Fleckvieh cattle

Gredler, B.[1], Fuerst-Waltl, B.[1], Fuerst, C.[2] and Sölkner, J.[1], [1]University of Natural Resources and Applied Life Sciences Vienna, Department of Sustainable Agricultural Systems, Gregor Mendel Str. 33, 1180 Vienna, Austria, [2]ZuchtData EDV-Dienstleistungen GmbH, Dresdner Str. 89/19, 1200 Vienna, Austria; fuerst@zuchtdata.at

The effect of semen batch on the fertility measures non-return-rate after 56 (NR56) and 90 (NR90) days of 31,476 artificial inseminations of Austrian dual purpose Fleckvieh (Simmental) cows was analysed. The semen batch (i.e. date of semen collection) was recorded at inseminations in the years 2004-2006 in Lower Austria. Thus, it was possible to combine information on semen characteristics with insemination data. The semen batch was specified by the following seven semen batch characteristics: number of viable sperm in the straw, the age of bull at time of semen collection, dilution factor, sperm motility, and the percentage of viable sperm before freezing and after thawing. All statistical analyses were carried out using SAS PROC MIXED. Seven different models for NR56 and NR90 were run including one of the seven semen batch characteristics at a time, and the fixed effects of farm within inseminator, days to first service, lactation number, and month and year at time of insemination. An eighth model was used to examine the effect of service sire on NR. For the latter model, the fixed effect of service sire and the random effect of semen batch within service sire were included in addition to the fixed effects already mentioned. The NR56 was significantly affected by the number of viable sperm in the straw ($P<0.05$), the percentage of viable sperm after thawing ($P<0.01$) and the service sire ($P<0.001$). For all other factors representing the semen batch, no significant effect could be observed. For NR90 similar results were obtained. Results indicate that additional information on fertility can be obtained by taking semen batch data into account.

Investigation of yearly changes for image analysis traits in M. *longissimus thoracis* in Japanese Black and Japanese Black × Holstein

Hamasaki, Y., Murasawa, N., Nakahashi, Y. and Kuchida, K., Obihiro University of A&VM, Obihiro, Hokkaido, 080-8555, Japan; hamasaki@obihiro.ac.jp

The degree of marbling in M. *longissimus thoracis* (ribeye) is economically very important in Japan. Previous studies of our group showed that not only the marbling percent of ribeye area (MP) but also the coarseness of marbling particles affected the grading of beef marbling standard number (BMS). They also showed that the correlation coefficient of BMS with MP was positive, and those of BMS with coarseness of marbling particles were negative. The aim of this study was investigate the yearly changes in image analysis traits that affect BMS. Digital images of the 6-7th rib cross section from 6,083 Japanese Black (JB) and 4,108 Holstein crossbreds (JB×H) were used. High quality digital images were taken from September 2005 to December 2008, using a special camera for beef carcass cross section. MP, coarseness index of marbling (CIM), and coarseness index of the largest marbling deposit (CLM) were calculated by image analysis. Analyses of variance were performed to calculate the least squares means of BMS and image analysis traits by breeds. Shipping year (4 levels), shipping month (12 levels) and sex (2 levels) were used in the statistical model as fixed effects. BMS of JB were significantly higher in 2007-2008 (5.26-5.42) than in 2005-2006 (4.91-5.06). BMS on JB×H were significantly higher in 2008 (3.51) than in 2005-2007 (3.14-3.20). Significantly higher MP were observed in JB in 2007-2008 (43.5%-44.8%) than 2005-2006 (41.9%-42.6%). MP in JB×H were significantly higher in 2008 (35.9%) than in 2005-2007 (33.6%-34.5%). CIM in JB were significantly coarser in 2008 (12.1%) than in 2005-2007 (10.6%-11.2%). The significant difference by the year for CIM was not seen in JB×H. In both breeds, CLM showed a similar tendency to CIM. MP showed desirable changes in both breeds by year. But CIM and CLM showed undesirable change in JB (i.e. marbling particles become coarser by year).

A multi criteria method for conforming homogeneous groups of cows for experimental purposes

Acosta Aragón, Y.[1], Jatkauskas, J.[2] and Vrotniakienė, V.[2], [1]Biomin Holding GmbH, Industriestrasse 21, 3130 Herzogenburg, Austria, [2]Institute of Animal Science of Lithuanian Veterinary Academy, Baisogala, Lithuania, Department of Animal Nutrition and Feeds, R Žebenkos St. 12, 82317 Baisogala, Radviliškis Distr., Lithuania; yunior.acostaaragon@biomin.net

A crucial aspect for comparison of different treatments in an experiment with animals is the formation of the animal groups. This aspect will decide the quality of the experiment and its results. The situation is further complicated if the researcher conducts experiments on milk cows, which differ in productive and reproductive aspects. For an experiment at the Institute of Animal Science of the Lithuanian Veterinary Academy, 24 Lithuanian Black-and-White dairy milking cows had to be selected from a larger group of 131. At the start of the experiment, the animals will be allocated to two groups of 12, each group having a similar mean age, live weight, lactation number, date of calving and therefore lactation time, current milk yield, last year's milk yield, milk fat, milk protein content and live weight. The cows were selected according to their proximity in the average values for each parameter. For this purpose, data were collected from the data base available in the farm, and a matrix was made with all the information. The first step was to calculate the average values of each parameter and the deviations from the average. A second step was to calculate the maximal values of the deviations and to divide each deviation by the maximal value for the parameter. This second intermediary step was carried out in order to avoid numerical differences in magnitude of, for example, milk production (from 2,000 to 7,000 liter) and number of lactations (from 1 to 8). In a final step, the values obtained after steps 1 and 2 were averaged for each cow and a multi criteria index was calculated. Finally, the cows were ordered upward and the first 24 animals with the lowest multi criteria index (lowest differences to the group average) were selected for the experiment.

Influence of pelvis shape detected by infrared thermography on automatic body condition scoring

Polák, P.[1], Halachmi, I.[2], Klopcic, M.[3], Peškovičová, D.[1] and Boyce, R.[4], [1]Animal Production Research Centre, Hlohovecká 2, 951 41 Lužianky, Slovakia (Slovak Republic), [2]ARO, Volcani Center, Bet Dagan 50250, Israel, [3]University of Ljubljana, Biotechnical Faculty, 1230 Domzale, Slovenia, [4]IceRobotics, Roslin BioCentre, EH25 9TT Roslin, United Kingdom; Marija.Klopcic@bfro.uni-lj.si

The hypothesis tested was that the shape of the cow's pelvis can influence the automatic body condition score when performed by means of infrared thermography. The pelvis length index (PLI = 100 * width of hips * length of pelvis-1) of 160 Holstein cows was detected on thermographic images of the rear half of the cow's body from a bird's eye perspective. Cows were divided into two groups by the PLI with a dividing limit of 0.9. In group 1, there were 95 cows with PLI less than 0.9. Average manual body condition score (MBCS) for group 1 was 2.25 and automatic body condition score by thermography (TBCS) was 2.27. In group 2, there were 55 cows with PLI higher than 0.9. Average MBCS in group 2 was 2.23 and average TBCS was 2.19. There was no statistically significant difference between average values of MBCS between groups. On the other hand, TBCS was significantly higher for the group with the smaller pelvis index. The difference between groups was also investigated for the average live weight of cows. Cows with a larger pelvis index (i.e. more square shape pelvis) had a statistically significant lower live weight (65.43 kg) than cows with a smaller pelvis index. It could be said that there is a tendency to under-predict the index in lighter cows. As this research was carried out on a limited number of cows, more research is needed. The PLI should be integrated into the automatic BCS algorithm.

Characterization of grass and extrusa of the heifers grazing Girolanda of pasture of *Brachiaria decumbens* stapf., under different stocking rates

Modesto, E.C.[1], Silva, A.M.[1], Lira, C.C.[1], Santos, M.V.F.[1], Lira, M.A.[2], Dubeux Jr., J.C.B.[1] and Mello, A.C.L.[1], [1]Universidade Federal Rural de Pernambuco, Zootecnia, Rua Dom Manoel de Medeiros S/N, Bairro: Dois Irmãos, 52171-900, Brazil, [2]Instituto Agronômico de Pernambuco - IPA, Av. Gal. San. Martin. 1371 Bonji. Caixa Postal 1022., 50761 - 000, Brazil; elisa@dz.ufrpe.br

The objective of this study was to evaluate the characteristics of the grass and the extrusa from heifers, under three stocking rates on grass pasture of *Brachiaria decumbens* Stapf., at different periods of evaluation, in the Woodland Zona of Dry Pernambuco. The experiment was conducted at Agronomic Institute of Pernambuco (IPA) in Itambé-PE, in the period March 2008 to July 2008. Six heifers of the breed Girolanda were used, of average body weight 400 kg. The treatments consisted of 3 stocking rates (2, 4 and 6 UA), and each block was formed by three paddocks. The design was a randomized block design in a split plot with repeated measures over time. Collection of extrusa from animals and manual simulation of grazing were used to assess chemical composition and fractions of the plant. Characteristics bromatological of forage obtained by extrusa of animals and the simulation of grazing ranged with the evaluation period, and days of grazing at different stocking rates. The fractions of the plant showed variation during the evaluation period, day of grazing and stocking rate. The fraction which represented most of the leaf extrusa, were higher at 1 day of grazing (90%) and at the stocking rate of 2UA (87%).

Comparison of steer breed types for muscle traits

Keane, M.G.[1] and Allen, P.[2], [1]Teagasc, Grange Beef Research Centre, Dunsany, Co. Meath, Ireland, [2]Teagasc, Ashtown Food Centre, Dunsinea, Castleknock, Dublin 15, Ireland; gerry.keane@teagasc.ie

The objective of this study was to compare Holstein-Friesian (HF), Piemontese x Holstein-Friesian (PM) and Romagnola x Holstein-Friesian (RO) steers for muscle chemical composition and colour traits. Spring-born calves, the progeny of sires from the breeds being evaluated, and out of Holstein-Friesian cows, were reared in a standard two year-old beef system. At the start of finishing, 40 animals per breed type were assigned to a 3 (breed types - HF, PM and RO) x 2 (feeding levels) x 2 (finishing periods) factorial experiment. The two feeding levels were 3 kg/day and 6 kg/day supplementary concentrates with grass silage ad libitum. The two finishing periods were 124 days and 207 days. Carcass weights were similar for HF and PM but significantly heavier for RO. Carcass fat class was similar for HF and RO but significantly lower for PM. Muscle chemical composition and drip loss did not differ between PM and RO but HF muscle had significantly lower moisture and protein concentrations and a higher lipid concentration. The higher feeding level increased ($P<0.001$) carcass weight by 18 kg but had no effect on muscle chemical composition or drip loss. Extending the finishing period increased ($P<0.001$) carcass weight by 39 kg and also increased ($P<0.001$) carcass fat class. Muscle moisture and protein concentrations were lower ($P<0.001$), and muscle lipid concentration was higher ($P<0.001$), for the extended finishing period. Blooming had little effect on L colour value, but a and b colour values were higher following blooming. It is concluded that HF had lower muscle moisture and protein concentrations, and a higher lipid concentration than PM and RO which were similar. There were no important muscle colour differences between the breeds. Feeding level had no effect on chemical composition and only minor effects on colour. Extending the finishing period increased carcass weight and all measures of fatness. It also reduced muscle moisture and protein concentrations, and increased muscle lipid concentration.

Efficacy of herbal medicine on on dermatic wounds in dairy cows

Ahadi, A.H.[1], Sanjabi, M.R.[1] and Moeini, M.M.[2], [1]Iranian Research Organization for science &Technology(IROST, Anim Science, No 71,forsat st, Enghelab Av, 15815-3538 Tehran, Iran, [2]Razi University, Anim science, Taghbostan, University St, 15815, Iran; amirahadi1964@yahoo.com

Large numbers of indigenous medicinal plants have been reported to possess wound healing properties. As Iran has been endowed with a rich variety of herbal medicine flora, this study was designed to determine the effect of different individual and combinations of herbal medicines on wounded skin in dairy cows. The chosen herbals were Maticaria Chamomila, Achilla, Rosa damascenamill and Lawsonia. The 32 cows were selected in 4 herds that had medium to serious skin injury. The cows were treated with different combinations and percentages of the above mentioned herbal medicine in the form of ointment for 5 months. The best treatments were Maticaria Chamomila, Achilla, Rosa damascenamill and Lawsonia with 40%, 40%, 10% and 10% healing rates, respectively. The combination had a significant effect ($P<0.5\%$) on healing wounded skin.

Sire x herd: interaction for milk performance traits and for culling by first-calf heifers in Saxonia
Hamann, H.[1,2] and Brade, W.[3], [1]Institute for Animal Breeding and Genetics, University of Veterinary Medicine Hannover, Bünteweg 17, D-30559 Hannover, Germany, [2]Center for Informatics Baden-Württemberg, Stuttgarter Straße 161, D-70806 Kornwestheim, Germany, [3]Chamber of Agriculture in Lower Saxony, Johannssenstraße 10, D-30159 Hannover, Germany; henning.hamann@iz.bwl.de

Genotype x environment (G x E) interactions in dairy cattle breeding are becoming more important because breeding organizations tend to sell semen to farmers operating in various environments. Since bulls must always be evaluated for the routine estimation of breeding values, a model was chosen which included this type of interaction. For culling and several milk performance traits (milk, fat and protein yield, fat and protein content), the data from first calf heifers in Saxonia were analyzed for the existence of G x E interactions. The data were restricted to sires that had daughters in at least 4 different herds. At least 4 different sires were used in each herd. In total, 150 sires and 207 herds were included. For the analyses of milk performance traits the data were restricted to heifers that had at least 60 days in milk. Therefore, the analyses were based on 20,404 heifers in the case of culling, and on 19,278 heifers in the case of milk traits. A total of 5,044 animals were culled within 500 days of first calving, 25.7% of them due to udder related problems. Culling was analyzed as a binary trait under a threshold model with calving season and age of first calving as fixed factors. Sires, herds and their interactions were considered as random effects. For the milk production traits similar models were used, but herd was considered as fixed and days in milk was used as a covariable. For most of the traits, the existence of G X E interactions was evident. Therefore, strategies which will reduce the number of test herds in a breeding program should be considered carefully, because the G X E interactions may lead to biased selection decisions.

Survey on antibiotic usage for mastitis in Spanish dairy cows
Zorraquino, M.A.[1], Berruga, M.I.[2] and Molina, M.P.[3], [1]Universidad Pública de Navarra, Departamento de Producción Agraria, Campus de Arrosadía, 31006 Pamplona, Spain, [2]IDR-ETSIA. Universidad de Castilla-La Mancha, Departamento de Ciencia y Tecnología Agroforestal y Genética, Avenida de España, 02071 Albacete, Spain, [3]Instituto de Ciencia y Tecnología Animal. Universidad Politécnica de Valencia, Camino de Vera, 14, 46022 Valencia, Spain; pmolina@dca.upv.es

Veterinary drugs, particularly the anti-microbials, are widely used in livestock. In the case of cattle, mastitis is one of the most significant infectious diseases and antibiotics are used for its treatment and prevention during lactation and the dry-off period. In order to know which antibiotics are most frequently used in the treatment of mastitis in Spanish dairy cows, two studies were conducted. In the first one, veterinarians were surveyed on the usage of antibiotics for mastitis employed as systemic or intramammary drugs (for lactating animals or for dry-off therapy), and furthermore on the management practices associated with the 'extra-label' use of antibiotics. A second study was carried out by compiling the sales data provided by Spanish pharmaceutical companies (Veterindustria) of the different systemic and intramammary antibiotics used to treat mastitis. Veterindustria data revealed that, in 2006, 10.3 millions euros of veterinary drugs were sold for mastitis treatments, of which 57% were intramammary antibiotics for lactating animals, 33% were for intramammary treatment for the dry-off period, and 10% were for parenteral injection. According to the veterinarians surveyed, the most commonly used antibiotics are the b-lactams (penicilins 73.6% and cephalosporins 26.4%). The aminoglycosides that are combined with b-lactams in a large number of drugs were also common (85.8%). Other significant substances were colistin (14.2%), macrolides (9.3%), novobiocin (7.9%), tetracyclines (5.1%) and sulfonamides (3.5%). These results agree with those obtained from the analysis of the sales data of the pharmaceutical companies.

Effect of animal welfare and feeding system on Charolais young bulls meat quality
Vincenti, F.[1], Iacurto, M.[1], Capitani, G.[2], Gaviraghi, A.[3] and Gigli, S.[1], [1]Council for Research in Agriculture, Research centre for meat production and genetic improvement, Via Salaria, 31, 00015 - Rome, Italy, [2]Agristudio Nutrition S.r.l., Via Gramsci, 56, 42100 - Reggio Emilia, Italy, [3]University of Milan, D.I.P.A.V., Via Celoria, 1022 -Milan, Italy; federico.vincenti@entecra.it

The aim of this study was to evaluate the effects of animal welfare and two different feeding systems on meat quality traits of conventionally processed Charolais young bulls (CH). Animal welfare was evaluated on 144 CH, using schedules based on evaluation of ethological and physiological indexes. Welfare evaluation was performed at two times, first at the beginning of the study (SBA) and secondly at one month before slaughter (SBR). Feeding systems were based on different maize grain treatments, and different starch and crude protein contents (I Group: maize crumb; crude protein (CP) %=13.76, starch %=42.75 on a dry matter (DM) basis; II Group: maize flour; CP %=15.56; starch %=30.83 on DM basis). Meat quality parameters (pH; colour; shear force and water loss) were measured on 19 samples of longissimus thoracis (10th-11th ribs) after 7days of ageing. Results of the animal welfare evaluation showed that I group animals had a declining trend (SBA=53% vs SBR=48%, animals without ethological and physiological problems). In the II group we observed that all animals had ethological and physiological problems (SBA and SBR). Furthermore at slaughter, liver and lung pathologies were evaluated and it was observed that I group had 9% and II group had 28%. Therefore, II group showed a lower welfare level than the I group. Meat quality data showed significant differences in pH (I=5.70 vs II=5.55); shear force (kg) both on raw meat (I=9.09 vs II=11.34) and on cooked meat (I=11.84 vs II=14.14); cooking loss (%) (I=20.84 vs II=31.29). In conclusion, this work shows that meat from II group had a lower quality than I group. Maybe, these differences were not due to the feeding system but to animal welfare.

Relation between bull sperm respiratory burst activity and the *in vitro* fertilization rate: a new approach to evaluate bull's fertility
Mota De Azevedo, S., Faheem, M., Carvalhais, I., Chaveiro, A., Habibi, A., Agricola, R. and Moreira Da Silva, F., University of the Azores, Agrarian Sciences - Animal Reproduction, Terra Cha, 9700 Angra do Heroismo, Portugal; jsilva@uac.pt

Sperm of 10 different bulls (3 ejaculates per bull) was used to evaluate sperm oxidative bust activity by flow cytometry, correlating this with the results of *in vitro* fertilization (IVF) as well as the further embryo development to the stage of blastocyst. After thawing, the straw content was split in two identical parts. One was employed for the *in vitro* fertilization, while the other was used for flow cytometry to evaluate sperm oxidative burst activity by an assay using 2',7'-dichlorofluorescin diacetate. Sperm viability was evaluated by contrast phase fluorescent microscopy using a solution of 4',6-Diamidino-2-<wbr></wbr>phenylindole dihydrochloride and by flow cytometry using a SYBR-14 and propidium iodide viability kit. For the IVF, cumulus oocytes complexes (n=3,250), were matured *in vitro* for 24 h and the embryos cultured for 9 days. Presumptive zygotes were evaluated 24 hours after fertilization and eight days after to evaluate the blastocyst production. Results of sperm viability demonstrated a strong correlation between data obtained by flow cytometry and by phase contrast microscopy (r=95%, $P<0.05$). As far as embryo production and the sperm metabolism oxidative burst activity are concerned, it was observed that bulls, in which the burst activity was higher, resulted in better results for *in vitro* fertilization and on further embryo production. The correlation between burst activity and fertilization rate and further embryo development to blastocysts was respectively 98% and 96% ($P<0.001$). This study allows the conclusion that there is a positive correlation between bull sperm H_2O_2 production and their ability to fertilize bovine oocytes as well as their development to the blastocyst stage. The sperm oxidative burst activity, measured by flow cytometry, can thus be an excellent method to predict the potential fertility of different bulls.

Session 39

Organisation and promotion plan of equine sector in Spain

Castellanos Moncho, M., S.G Conservation of resources and feed, Ministry of Environment, rural and marine affairs, Alfonso XII nº 62, 28071 Madrid, Spain; mcastell@mapa.es

Organisation and Promotion Plan of Equine Sector in Spain Castellanos, M. Ministry of the Environment and Rural and Marine Affairs Livestock based on equine production is one of the most traditional activities in Spain where the horse industry and legal framework has experimented several changes in the last years. In 2003, after a detailed analysis about the situation of the equine sector, the Ministry of the Environment and Rural and Marine Affair developed a specific plan for the Organisation and Promotion of Equine Sector in Spain. The aims of this plan were: to incorporate the equine production in the livestock policy, to promote the equine production within the Common Agricultural Policy and to expand the commercial uses of equines (tourism, therapeutic uses, leisure). These objectives were developed using four programs: 1. Zootechnical and sanitary program. An official Register for the Agricultural farms (REGA) was created to control the establishments related to horse breeding or management. The official registers and the establishment of sanitary rules favour the welfare and health of the animals. 2. Promotion of the equine sector program. The stud-farms have received economic support for the modernization and for the purchase of tested animals and semen. Equine tourism and international trade have also been promoted. 3. Conservation and breeding program. This topic is regulated by the Spanish zootechnical normative for all species and the legislation for horse performance recording. The support given to breeders and to associations has increased the participants in the horse performance recording. Nowadays, there are 24 breeds with breeding programme and there are Young Horse Competitions for different disciplines with genetic evaluations published. 4. Complementary actuations related to formation issues to attain specialised staff and promotion issues in order to increase the use of the products and services of the horse.

Session 39

Horse biodiversity and their contribution to rural development in Spain

Azor, P.J.[1,2], [1]University of Cordoba, Genetics, Edif. Mendel Pl. Baja. Ctra. N-IV km 396a. Campus de Rabanales, 14071. Córdoba, Spain, [2]Asociación Nacional de Criadores de Caballos de Pura Raza Española (ANCCE), Cortijo de Cuarto. Bellavista, 41014. Sevilla, Spain; ge2azorp@uco.es

According to the Official Catalogue of livestock breeds of Spain (Spanish law RD 2129/2008), there are fourteen autochthonous horse breeds in Spain. The Spanish Purebred Horse (PRE) is the main breed raised in Spain, with a census of 100.397 horses registered in the Studbook, and distributed in 8.948 stud farms. Around 50% of them are located in Andalusia. The rest thirteen native horse breeds are endangered: Asturcón, Burguete, Caballo de Monte del País Vasco, Pura Raza Gallega, Hispano-Arab, Hispano-Bretón, Jaca Navarra, Losina, Majorca, Marismeña, Menorca, Monchina and Pottoka. These breeds are mainly located in the north of Spain, except Majorca and Menorca Horses that are mainly located in the Balearic Island and Marismeña breed in Andalusia. The main aptitudes of these breeds are leisure activities, breeding, sport competitions, horse-shows, equestrian tourism, hippotherapy and meat production. There are other international breeds raised in Spain included in the official catalogue: Arab, Anglo-Arab, Thoroughbred and Spanish Trotter. Additionally, the development of the Spanish equine industry stimulated by the increasing use of horses for sports, has led to the creation of a new breed, the Spanish-Sport horse that is a composite breed created for sportive purposes. Nine of these breeds are developing an official Breeding Program, with different breeding objectives: Conformation, Show-jumping, Dressage, Eventing, Driving, Racing, (Thoroughbred and Trotter Horses) and Endurance races. And the endangered breeds are developing their conservation programs in order to maintain the genetic variability and to contribute to rural development. The last point is a main goal in the agrarian policies.

Morphological characterization of traditional maremmano horse

Tocci, R.[1], Sargentini, C.[1], Ciani, F.[2], Benedettini, A.[3], Lorenzini, G.[1], Martini, A.[1] and Giorgetti, A.[1], [1]Università di Firenze, Scienze zootecniche, via delle Cascine, 5, 50144 Firenze, Italy, [2]Consdabi, Loc. Piano Cappelle, 82100 Benevento, Italy, [3]Horse breeder, Loc. Poggio alle cavalle, 57028 Suvereto (Li), Italy; roberto.tocci@unifi.it

The traditional Maremmano horse, listed on the Registro Volontario Regionale delle risorse genetiche autoctone della Regione Lazio (L.R. n. 15 del 01/03/2000), is a Tuscan Latial equine population. At the origin of the breed contributed the Oriental horses, and larger north European equines. Within the end of XIX and the beginning of XX century the selection and the improvement of the breed began, through the introduction of British equines. The results of the improvement was an horse with good morphology united to the frugality of Maremmano. The modern and select sport Maremmano has a genealogical book run by ANAM. Together with the present improved breed survives a population of Traditional Maremmano, characterized by roman or moose nose, large and inclined croup with low tail junction; strong legs, often with feathering; wide and puissant hooves. The work aims the morphological characterization of the traditional Maremmano, considering a larger group in Grosseto province, and other individuals of Tuscan and Latial provinces. The biometrics are homogeneous among the farms; the measures and the main body indices didn't show significant differences. The biometrics of traditional Maremmano were: height at withers of 163 and 163 cm, chest circumference of 198 and 198 cm, frontshank circumference of 22.5 and 23.5, FI (Frontal Index) of 28.9 and 28.9, DTI (Dactyl Thoracic Index) of 11 and 12.4, for adult females and males respectively. The morphological characteristics were: bay or black coat color, Roman head (43%), Roman nose (20%), moose nose (16%), straight profile (10%). The legs were strong with wide hooves and often showed the feathering (26%). 10% of population showed the whiskers, a feature not found in Registro Volontario, but known among the breed experts.

Morphological body measurements, body condition score and ultrasound measurements of Portuguese Garrano horse breed

Santos, A.S.[1], Verhees, E.[2], Silva, S.R.[1], Quaresma, M.[1], Sousa, V.[3] and Pellikan, W.F.[2], [1]CECAV-UATD, Po-Box 1013, 5001-801, Vila Real, Portugal, [2]Animal Nutrition Group, Dep. of Animal Science, WUR, Marijkeweg 40, 6709 PG, Wageningen, Netherlands, [3]ANCRG, 4970-743, Arcos de Valdevez, Portugal; assantos@utad.pt

Garrano is an ancient pony breed from Portugal and exists mainly in free ranging systems. Such systems are characterized by the use of natural resources with seasonal shortage of pasture. The objective of this study was to evaluate body reserves of these animals. Data concerning morphological measurements (MM), body condition score (BCS), cresty neck score (CNS) and real time ultrasound (RTU) measurements were gathered at the end of spring on 21 Garrano horses 276±10 kg LW, living in semi wild environment. MM included height, girth, body length, side length, neck circumference, neck length and crest height. BCS was determined by visual appraisal and palpation on six particular parts of the body and scored from 1 to 9. CNS was scaled from 0 to 5. For RTU animals were scanned with an Aloka SSD 500V real time scanner using a linear probe of 5.0 MHz. Subcutaneous fat depth (SF) was determined by placing the probe at the back over the 3rd lumbar vertebrae and at rump. Tissue depth measurements (TD) were obtained by placing the probe at thoracic cage between the 12th and 13th rib and over the centre of the ribs on the right side of the horse. Descriptive statistics and correlations between RTU, BCS, CNS and MM were established. Values observed for BCS (5.5±1.4), CNS (2.8±1.4) indicate that the horses were in good body condition. The SF RTU measurements at the back (4.2±1.3mm) and at the rump (3.6±1.1mm) showed a variation (CV=29 and 32%, respectively) close to that observed for BCS (CV=26%). BCS was correlated with CNS ($P<0.01$) and with TD RTU measurements ($P<0.01$). Both BCS and CNS were not correlated with SF RTU measurements ($P>0.05$). For MM, it was found that the ratio neck circumference:crest height was the most suitable MM to assess the CNS (r=0.795; $P<0.01$) and BCS (r=0.739; $P<0.01$) and therefore the body reserves.

Using real-time ultrasound measurements to monitor small variations in body condition of horses

Santos, A.S.[1], Verhees, E.[2], Pellikaan, W.F.[2], Van Der Poel, A.F.B.[2] and Silva, R.S.[1], [1]CECAV-UTAD, Po-Box 1013, 5001-801, Vila Real, Portugal, [2]Animal Nutrition Group, Wageningen University, Dep. Animal Science, Marijkeweg 40, 6709 PG, Netherlands; assantos@utad.pt

19 kg LW) received changing dietary conditions through time in order to obtain small variations in body reserves. BCS and RTU measurements were made in 3 sessions: in the beginning of the experimental period (S1), on days 11(S2) and 34(S3). Two weeks prior to S1 animals had hay restricted (4.5 kg/d). Between S1 and S2 animals were fed at maintenance, and thereafter animals were fed to maintain BCS. RTU measurements were taken with an Aloka 500V equipment. Subcutaneous fat depth (SF) was determined by placing the probes over the 3rd lumbar vertebrae and at rump (SF- SFBack and SFRump respectively). Tissue depth measurements between (TDbR) and above (TDaR) the ribs were obtained by placing the probe at thoracic cage between the 12th and 13th rib and over the centre of the ribs. RTU data were analysed by repeated measures ANOVA using proc MIXED in SAS. The 7.5 MHz probe gave significant lower values for all RTU measurements ($P \pm$Although body condition score (BCS) is commonly used to estimate the nutritional status of horses, it doesn't identify small variations of the animal's fat reserves. Little information is available on the use of ultrasound for measuring fat thickness on horses. The aim of this study was to compare 2 ultrasonic high frequency probes (5 and 7.5 MHz) for monitoring small variations in fat thickness of horses. Five horses (347 < 0.05) except for SFBack ($P > 0.05$). RTU showed ability to identify, for the whole period, the increase in tissue depth for TDaR and TDbR (2.3 and 1.6 mm, $P < 0.05$, respectively) and also for SFBack and SFRump (1.9 and 0.8mm, $P < 0.05$, respectively). For SFBack between S2 and S3 it was possible to identify an increase close to 0.5mm($P < 0.01$). From these results it can be conclude that the RTU associated with image analysis show high ability in discriminate small variations in fat and tissue depths.

Kinematic differences between untrained, dressage and traditional bullfighting Lusitano horses

Santos, R.[1], Galisteo, A.M.[2], Molina, A.[2], Garrido Castro, J.L.[2], Bartolomé, E.[2] and Valera, M.[3], [1]Escola Superior Agrária de Elvas, Av. 14 de Janeiro s/n, 7350 Elvas, Portugal, [2]Universidad de Córdoba, Campus Universitario de Rabanales, 14071 Córdoba, Spain, [3]Universidad de Sevilla, Ctra de Utrera km1, 41013 Sevilla, Spain; rutesantos@esaelvas.pt

Dressage and bullfighting are probably the most popular uses of the Lusitano horse. These activities have different goals, and hence different training methods. In dressage, the amplitude, suppleness and rhythm of gait are the most valued characteristics, while in bullfighting, reunion, swift changes of pace and the ability to turn in small spaces are very important. The results of a canonical discriminant analysis for kinematic variables showed that the traits that have mostly influenced the first canonical variable (separating untrained from dressage horses) were variables concerning the hind limb, such as hind maximum and minimum protraction-retraction angles, maximum elevation of the hind hoof and overreach distance. As for the second canonical variable, that differentiates bullfighting horses from both the other groups, even though the separation is not as clear, it is mainly influenced by fore and hind maximum elevations of hooves and hind limb stance and swing phases durations. The analysis of variance, considering speed as a covariable, shows that dressage horses have a larger minimum protraction-retraction angle, a shorter stance phase and a longer swing phase of the hind limbs. On the other side, bullfighting horses lift their front and hind hooves significantly higher than the other 2 groups. Maximum hind protraction-retraction angle was significantly smaller in untrained horses. The results show that training should, therefore, be taken into account when evaluating kinematic characteristics of Lusitano horses.

Comparison of different weaning systems in horses

Münch, C. and Gauly, M., Department of Animal Science, Albrecht Thaer Weg 3, 37075 Göttingen, Germany; chmuench@gwdg.de

In this study the effects of traditional (one-step) weaning (TW, foals and mares were completely separated in one step) were compared with effects of a Trainer-Horse-Method (TH; one older mare was left with foals after weaning) and the stepwise removal of foals (SN). A total of 21 thoroughbred mares and foals were used in the study (TH=7; TW=6; SN=8). Behaviour of foals (staying, moving/agitation and vocalization) was recorded for five consecutive days after weaning by direct observation using a time-sampling-interval of ten minutes. Data were analyzed using the procedures MEANS, GLM and MIXED of SAS. TW foals showed significant behaviour modifications after weaning. Vocalization frequencies ($P<0.05$) and moving activities ($P<0.001$) in traditional weaned foals were highest when compared with other treatments. No significant differences were shown between TH and SN. In all groups an adaptation occurred after five days. In conclusion, the traditional weaning method elicited more behavioural changes. Therefore TH and SN seem to be more gentle for the foals than TW.

Foraging behaviour by horses facing a trade-off between intake rate and diet quality

Edouard, N.[1,2], Fleurance, G.[1,3], Dumont, B.[1], Baumont, R.[1] and Duncan, P.[2], [1]INRA, UR1213, 63122 Saint Genès Champanelle, France, [2]CEBC CNRS, UPR1934, 79360 Beauvoir sur Niort, France, [3]Les Haras Nationaux, Direction des Connaissances, 19230 Arnac-Pompadour, France; nadege_edouard@yahoo.fr

Grass represents a large part of the diet of horses, so a proper understanding of the tactics they use for extracting nutrients from swards is essential for their management. Since the number of horses is increasing in Europe, understanding the principles governing their feeding selectivity is also important for the management of grasslands. However, little is known of the way by which horses adjust their feeding choices and daily intake in response to variations in vegetation features. For grazers, patches of tall grasses that are ingested faster are generally poorly digested because of their high fibre contents; conversely, highly digestible patches are short and often impose low intake rates. The foraging behaviour of horses facing a trade-off between sward height and quality was thus explored to test whether horses would select their feeding sites to maximise their nutrient intake rate. Three groups of three 2-yr-old saddle horses were grazed on a pasture managed to produce three swards varying in both height and quality (vegetative to reproductive stage). They were offered binary choices in a Latin-square design to assess preferences; daily intake was measured to value the consequences of their choices. Instantaneous intake rate was determined from bite rate at pasture and bite mass estimated using trays indoors. The taller sward matured across time, so the differences of quality among swards increased. The horses selected the taller sward in the first period, and the shorter alternatives in the following ones. The rates of digestible dry matter and energy intake were higher on the tall swards; digestible protein was the best predictor of the horses' selection. The horses however expressed daily partial preferences: they may have been balancing protein and energy intake by feeding on both swards offered simultaneously and thus met their nutritional requirements.

Toxicity of Adonis Aestivalis (summer pheasant's eye) in horses

Gagliardi, D.[1], Guidi, L.[1], Pierni, E.[1], Esposito, V.[1] and Miraglia, N.[2], [1]Sammarco Consultant Associated, Contrada Coluonni, 1, 82010 Benevento, Italy, [2]Molise University, Dept. Animals, Vegetables, Environmental Sciences, Via De sanctis, 86100 Campobasso, Italy; docdan60@alice.it

In this study we reported the toxicity of Adonis Aestivalis (summer pheasant's eye, Ranunculaceae family) in twenty one horses from three studs between 2003 and 2008. Adonis spp. are considered unpalatable but in this cases no one stabled horse refused the hay offered. Adonis plants contain a series of cardenolides similar to oleander and foxglove. The toxic cardiac glycosides are adonitoxin, strophantin, vernadigin and cymarin and are highest in the leaves and flowers. Strophanthidin, the aglycone of several cardenolides in Adonis spp., is due to inhibition of the sodium potassium adenosine triphosphatase enzyme system pump resulting in hyperkalemia and degenerative changes in the heart. After 24-48 hours by administer of infested hay, the horses was presented with lethargy, collapse and colic signs. Physical examination revealed that the horse was dehydrated and had dark mucus membranes, tympany and generalized ileum with moderate gaseous distension of the small intestine and cecum. All horses in therapy were examined with blood analysis, comprensive of gas analysis and ultrasound examinations. At the beginning the medical therapy was symptomatic and with supportive care but, after individualized the toxic plant, started with repeated injections of atropine sulphate as a parasympathetic antagonist for management of bradycardia and arhythmias. No one horse had adverse effect by atropine use and only two horses died (one for each big stable); in both case there was a compromising of the general metabolic parameters and reserved prognosis was made before therapy started. Further studies need to follow this preliminary study to determine the toxicokinetics of Adonis cardiac glycosides in horses and the proper supportive therapy.

Fatty acids composition in plasma triglycerides, in plasma non esterified fatty acids and in red cells membranes as influenced by linseed or sunflower oils in horses compound feedstuffs

Robaye, V., Dufrasne, I., Dotreppe, O., Istasse, L. and Hornick, J.L., Liege University, Nutrition Unit, Bd de Colonster 20, 4000 Liege, Belgium; listasse@ulg.ac.be

Diets high in concentrate and therefore in starch are usually offered to sport horses. Oil has been suggested in horses diet to reduce starch and associated disturbances. Eight adult horses were used in 3 3x3 latin square designs, one being incomplete. There were 3 diets with a control, a linseed oil or a sunflower oil compound feedstuff. The diets were based on 50% grass hay and 50% compound feedstuff. The control compound feedstuff was made of 47.5% whole spelt and 47.5% rolled barley. Eight percents of barley was substituted by 8% linseed oil or 8% sunflower oil in the fat concentrate. The oil intake was 0.278 kg/d. On the whole, there were large differences in the individual fatty acids contents between triglycerides and non esterified fatty acids in the plasma in terms of number of acids (8 vs 5 detected acids) and of concentration (sum of total acids of 113.4 vs 2.6 mg/100ml). The number of detected acids was much larger (19 acids) in the red cells membranes. Linseed oil increased the content of C 18:3 n-3 in the triglycerides (6.32 vs 2.07 mg/100ml, $P<0.001$) and there were tendencies for higher contents in the non esterified fatty acids fractions and in the red cells membranes. There was also a tendency for a higher content in the C 18:2 n-6 in the triglycerides but not in the non esterified fatty acids and in the red cells membranes. When sunflower oil was offered, there was an increase in the C18:2 n-6 content in the non esterified fatty acids fractions (1.38 vs 0.64 mg/100ml, $P<0.001$). There was also a tendency for an increased content in the C 18:2 n-6 in the triglycerides but nearly no effects on the red cells. Both oils reduced the arachidonic acid contents in the triglycerides and in the red cells ($P<0.001$). Such reductions could be of interest owing to less proinflammatory reactions when these oils were included in the diet.

Altered welfare is linked with aggressiveness in horses

Fureix, C., Jego, P. and Hausberger, M., Université Rennes1, Ethologie Animale et Humaine, campus Beaulieu, 35042 Rennes, France; carole.fureix@univ-rennes1.fr

Behavioural problems in horses, in particular aggressions towards humans, are a common source of accidents in professionals (e.g. the third after dogs and bovid in veterinarians). Temperamental traits and experience-induced aspects are commonly mentioned when evaluating horses' reactions to humans. However, fewer studies focused on the relation between horses' welfare and their undesirable reactions towards humans. Yet there are some recent elements suggesting that an altered welfare (from a chronic discomfort to a painful experience) could affect animals' relational behaviour. In the present study, we hypothesized that part of the aggressiveness towards humans in horses may be linked with altered welfare. Thus, fifty nine horses from 3 riding centres were submitted to five standardized behavioural tests in order to evaluate their reactions to humans (a motionless person test, an approach-contact test, a sudden approach test with or without a saddle on the arm and halter fitting test). At the same time, a large set of potentially welfare-related indicators were recorded: health-related (e.g. vertebral problems…), postural (e.g. ears positions at rest…), physiological (e.g. stress hormones…) and behavioural (e.g. behavioural repertoire…) indicators. It appeared that about half (51%) of the horses showed at least once an aggressive reaction towards the experimenter. Interestingly, these aggressive horses displayed a characteristic ears laid back posture (Mann Whitney tests, $P<0.05$), were statistically more prone to suffer from vertebral problems (binomial tests, $P<0.05$) and had higher level of plasmatic cortisol rate (Mann Whitney tests, $P<0.05$). These promising results confirm a possible relation between altered welfare and part of the undesirable reactions towards humans in horses (here aggressiveness). Moreover, some of the indicators used are easy to record (e.g. ears positions at rest) and could be used to identify aggressive horses and consequently improve safety.

The importance of the setting in Therapeutic Riding with autistic patients

Cerino, S.[1], Bergero, D.[2], Miraglia, N.[3] and Gagliardi, D.[4], [1]FISE (Italian Equestrian Federation), Therapeutic Riding, Viale Tiziano 74, 00196 ROME, Italy, [2]University of Veterinary Medicine, Animal Production, Epidemiology and Ecology, Vial Leonardo da Vinci 44, 10095 Grugliasco (To), Italy, [3]University of Molise, Animal, Vegetables and Environmental Sciences, Via De sanctis, 86100 Campobasso, Italy, [4]Sammarco Veterinari Associati, C. da Coluonni 1, Benevento, Italy; s.cerino@alice.it

Following a therapeutic riding work with autistic patients,in the present paper we discuss the importance of a well determined setting as well. Autism is a pathology concerning the loss of contact with reality. Hippotherapy for these subjects, is the possibility to peceive the Other and an outside world, usually meaning nothing to to them. The horse is the mean by this is possible, allowing the establishing of an empathic communication between himself and the patient and then between the patient and therapeutist. So, in Hippotherapy riding sessions with autistic patients, the precise setting of a 'containing' structure (someway similar to the analytic setting) makes easier the Self structuring work. The setting area is filled up with words, silences, human bodies, animal bodies, horse movements and behaviours, representing the possibility of communication with the autistic world. The horse plays a basic role. It becomes the mean to establish an empathy with the rider-patients and to allow to open a window over closed and still autistic world. So it is essential to know properly and to understand correctly all the animal bahaviours, its way to express feeling and to 'answer' the patients' demands. The animal has therefore the dual and central function to stimulate the communication relationships and to get itself in relation with autistic patients. So, the therapeutist can look for some kind of communication enclosing in a completely Self centred horizon, like the autistic one, an object relationship with the external world.

Relationship between conformation and gait characteristics in Spanish Purebred Horses

Gómez, M.D.[1], Peña, F.[1], García-Monterde, J.[1], Rodero, A.[1], Galisteo, A.M.[1], Agüera, E.[1] and Valera, M.[2], [1]UnivCórdoba, Campus Rabanales, 14071Córdoba, Spain, [2]UnivSeville, CtraUtreraKm1, 41013Seville, Spain; pottokamdg@gmail.com

The value of the sport horses depends mainly on performance in competitions, a combination of conformational, physiological and behavioural traits, which are heritable. Body shape defines the limits for range of movement, the function of the horse and its ability to perform. Therefore, conformation traits (CT) can be used on indirect preselection for performance traits and to increase the repeatability in the selection on performance traits with low heritability. The efficiency of indirect preselection depends on the genetic correlations between CT and performance traits. So, the main aim of this work was to estimate the genetic correlations between 28 CT and 16 biokinematic traits (BT) (10 at walk and 6 at trot) to select most objective selection criteria for dressage ability. A total of 130 Spanish Purebred males (4.6±1.5 years-old) were evaluated under experimental conditions with treadmill. Genetic correlations were estimated by VCE 5 software, using a bivariate mixed animal model including age and stud of animal as fixed effects, and animal additive genetic effect and residual error as random effects. At walk, the 50% of CT were no correlated with BT, being temporal traits more correlated than linear ones. The 82.9% of the significant genetic correlations between CT and BT were higher than 0.5; and the 29.3% have negative sign, most of them with temporal traits. At trot, the 66% of the significant genetic correlations between CT and BT were higher than 0.5 and the 22% of them were negatives, most of them angular traits. In general, BT at trot were more correlated with CT than those at walk. Therefore, selection based on CT could ensure an indirect selection for some trot characteristics of interest for dressage ability in Spanish Purebred horses.

The effects of selenium source on distribution of selenium within the milk of lactating mares

Juniper, D.[1], Bassoul, C.[2] and Bertin, G.[3], [1]Univeristy of Reading, Agriculture, Earley Gate, RG6 6AR, United Kingdom, [2]EARL Chevalait, Ferme de la Moisière, 61500 Neuville près Sées, France, [3]Alltech, RU Regulatory Affairs, 14 Place Marie-Jeanne Bassot, 92300 Levallois-Perret, France; gbertin@alltech.com

Selenium (Se) is an important trace element in both animals and man and the augmentation of animal diets is one method of improving both animal Se status and the Se content of animal derived products. Recent studies in both large and small ruminants have shown significant effects of Se source on both total Se content and the distribution of Se species within milk. The aim of this study was to determine the effects of Se source on the total Se content of horse's milk and the distribution of Se species. Twenty lactating horses were randomly allocated to one of two dietary treatments augmented with either seleno-yeast (SY[Saccharomyces cerevisiae CNCM I-3060]) or sodium selenite (SS). Dietary treatments were identical except for Se source (Se contents of 0.63±0.11 and 0.64±0.11 mg/kg DM for SS and SY, respectively) and were offered for a period of 60 d. Milk was individually sampled at enrolment (T_0) and at 15 d intervals throughout the study (T_{15}, T_{30}, T_{45} and T_{60}) for determination of total Se. Se species were determined in pooled milk samples taken at T_{60}. There were no differences between treatments in total Se at T_0 with mean milk Se contents of 5.6 and 5.7±0.26 ng/g FW for SS and SY, respectively. By T_{15} total Se contents of milk were higher ($P<0.001$) in SY milk when compared to SS (8.5 and 5.5±0.45 ng/g FW, respectively), a trend that continued throughout the duration of the study. At T_{60} milk total Se contents were 10.7 and 6.7±1.01 ng/g FW for treatments SY and SS, respectively ($P<0.001$). Speciation data indicated that SeMet contents were greater in the milk of SY horses at T_{60} when compared to those receiving comparable doses of SS, indicating that differences in total Se and SeMet are most probably attributable to the preferential uptake and incorporation of dietary SeMet derived from SY.

Ass's milk yield and composition during lactation

Salimei, E.[1], Maglieri, C.[1], Varisco, G.[2], La Manna, V.[3] and Fantuz, F.[3], [1]Università del Molise, Di S.T.A.A.M., via De Sanctis, Campobasso, Italy, [2]Ist. Zooprofilattico Sperimentale, via Bianchi, Brescia, Italy, [3]Università di Camerino, Di. Scienze Ambientali, via Circonvallazione, Matelica, Italy; francesco. fantuz@unicam.it

Clinical evidences show that ass's milk is well tolerated by infants with cow's milk allergy and its use is reported to be useful for preventing or treating gastrointestinal diseases and degenerative pathologies. Aiming to study ass's milk production and composition during lactation, 16 Martina Franca jennies were used to provide milk samples obtained by machine milking. Asses were fed 8 kg of coarse hay and 2.5 kg of mixed feed daily. Asses were housed with the foals that were separated from the dam 3 hours before milking. Milk production (2 milkings/d) was recorded and milk samples were collected every 3 weeks. The study lasted 170 days and jennies were between 69 and 90 days in milk (DIM) at the 1st sampling time. Average milk yield ($\pm$SD) was 1384$\pm$518.04 g/d. Milk total solids, protein, fat, lactose and ash contents (g/100g) were respectively 9.11$\pm$0.44, 1.64$\pm$0.25, 0.28$\pm$0.23, 6.58$\pm$0.31, 0.39$\pm$0.08. Somatic cell count (SCC) was very low (3.97$\pm$0.53 logSCC/ml) when compared to classical dairy species. A significant decrease was observed for milk yield (-22%) and for total solids (-7%) and protein content (-16%) up to the 3rd (113-134 DIM) and 4th sampling time (135-156 DIM) respectively. Milk fat, lactose and SCC were not affected by the stage of lactation. Correlation coefficients were positive and significant between fat content and total solids (r=0.40), milk yield and fat (r=0.35) and milk yield and total solids (r=0.33). Protein content was positively correlated with total solids (r=0.50), fat (r=0.53) and ash (r=0.42). SCC was positively correlated with protein (r=0.39) and fat content (r=0.49) and negatively correlated with lactose (r=-0.57). Correlation coefficients were negative and significant between lactose and protein (r=-0.65) and lactose and fat (r=-0.61).

Nitrogen distribution in ass's milk during lactation

Fantuz, F.[1], Maglieri, C.[2], Varisco, G.[3], La Manna, V.[1] and Salimei, E.[2], [1]Università di Camerino, Di Scienze Ambientali, via Circonvallazione, Matelica, Italy, [2]Università del Molise, Di S.T.A.A.M., via De Sanctis, Campobasso, Italy, [3]Is Zooprofilattico Sprimentale, via Bianchi, Brescia, Italy; francesco.fantuz@unicam.it

Due to its composition ass's milk is considered a valid alternative to the available hypoallergenic formulas for infants affected by multiple food allergy. Aiming to investigate on ass's milk production and its nutritive value during lactation, 16 pluriparous Martina Franca asses were used to provide milk samples obtained by machine milking. Jennies were housed with the foals that were separated from the dam 3 hours before milking. Milk samples were collected every 21 days. The study lasted 170 days and jennies were between 69 and 90 days in milk (DIM) at the 1st sampling time. Milk samples were analysed for total nitrogen (TN), non protein nitrogen (NPN) and non casein nitrogen (NCN) by Kjeldahl method. True protein nitrogen (TPN), casein N (CN), whey N (WN) and casein index (CI) were then calculated. Milk urea was determined by IR. Milk TN, NPN, NCN, TPN, CN and WN contents (mean$\pm$SD, mg/100g) were respectively 258.75$\pm$37.90, 36.13$\pm$6.93, 143.89$\pm$26.24, 222.57$\pm$34.48, 118.35$\pm$29.09 and 107.46$\pm$23.55. CI (CN/TN) was 0.45$\pm$0.07. Milk urea was 32.48$\pm$5.58 mg/dL. During lactation, a significant decrease was observed for TN (-17%), NPN (-21%), TPN (-17%), CN (-20%) and milk casein percentage (-20%) between the 1st and the 4th sampling time (135-156 DIM) then differences were not statistically significant. Milk urea significantly increased in advanced lactation. NCN, WN, CI and NPN/TN ratio were not affected by the stage of lactation. The abovementioned parameters, except CI and NPN/ratio, were positively correlated with TN. CN was positively correlated with TPN (r=0.72) and NPN (r=0.27) but not with NCN. WN was positively correlated with NPN (r=0.23), NCN (r=0.96) and TPN (r=0.53). Correlation coefficient between milk urea and NPN was not significant.

Therapeutic and paralympic riding in Italian equestrian federation

Cerino, S., Italian Equestrian Federation, Therapeutic Riding, Viale Tiziano 70, 00196 Rome, Italy; s.cerino@alice.it

The Italian Equestrian Federation set up in 2003 the Therapeutic Riding Department, concerning the development of this activity all over the country area and particularly the training of qualified technicians able to work in the therapeutic section, one of the most important and delicate one of the whole riding activity. Nowadays there are more than 1000 members in the Therapeutic Riding area and about 120 Therapeutic Riding Centres acknowledged by FISE (Italian Riding Federation) and acting inside Riding Centres affiliated or associated to the Italian Fderation. In Italy the Therapeutic Riding went on especially in Hippotherapy and in Carriage Driving section, but there are also several athletesplaying International Dressage Competition. In Italy, however, the Carriage Driving competitions are not very popular and so they are rare. Since February 2009 FISE has been acknowledged by Italian Paralympic Commettee (CIP) as a Sport Paralympic Federation, and acquired all the expertise concerning the Paralympic riding competitions. So a new Department has been created inside FISE, the Paralympic Riding Department, consisting of a technical and organizational staff in order to follow up the athletes and their horses, especially the best ones. Both the Departments, Therapeutic Riding and Paralympic Riding, work together exchanging their expertise and professionalism:in fact starting fron therapeutic riding, it will be possible to increase the number of disabled riders and to achieve an international competitiveness by an ever greater number of atheltes.

Characterization of body condition changes in Lusitano broodmares fed pasture based diets

Fradinho, M.J.[1], Ferreira-Dias, G.[1], Correia, M.J.[2], Gracio, V.[2], Beja, F.[2], Perestrello, F.[3], Rosa, A.[4] and Caldeira, R.M.[1], [1]FMV-TULisbon, CIISA, Av. Univ. Tecnica, 1400-377, Portugal, [2]FAR, Coud. Alter, 7441-909 Alter Chão, Portugal, [3]Comp. Lezirias, Porto Alto, 2135-318 S. Correia, Portugal, [4]Qta Lagoalva Cima, Alpiarça, 2090-222 Alpiarça, Portugal; amjoaofradinho@fmv.utl.pt

Most of Lusitano (PSL) studs in Portugal are managed under extensive feeding systems, with pasture based diets. When grass production is scarce, supplementary feeds are generally used, but farm practices vary widely. Body condition (BC) scoring is a practical tool to access body reserves in livestock and its assessment could provide valuable information about the adequacy of feeding management. The main objective of this study was to characterize BC changes along the year in PSL broodmares under extensive conditions. BC was monthly assessed (three breeding seasons) in four groups of mares (A n=17, B n=19, C n=6; D n=17) on different stud farms, from the 9th month of gestation to the post-weaning period. All mares were kept on pasture: A and B groups were daily supplemented (compound feeds and preserved forages) according to pasture availability and farm practices, while C and D groups were rarely supplemented. Changes on BC throughout the breeding cycle were small (0.25 - 0.75 in 0-5 points scale) and seems to be affected by foaling month. Highest BC scores were recorded on the end of spring/beginning of summer, decreasing thereafter until weaning, in the fall. Early foaling mares (Feb) showed the lowest BC scores along the cycle. On the opposite, higher BC scores were observed on mares that foaled latter (May), but they generally were already loosing BC by then. Mares that were almost exclusively fed with pasture (C and D) showed higher BC changes along the year and pasture quality and availability extensively affected those changes. BC scoring data provide valuable information to identify critical phases when supplementation could be crucial and aid developing adequate feeding strategies for different systems.

Correlations between conformation of proximal limbs and biokinematics of trot in Lusitano horses

Santos, R.[1], Valera, M.[2], Galisteo, A.M.[3], Garrido Castro, J.L.[3] and Molina, A.[3], [1]Escola Superior Agrária de Elvas, Av 14 de Janeiro s/n, 7350 Elvas, Portugal, [2]Universidad de Sevilla, Ctra de Utrera km1, 41013 Sevilla, Spain, [3]Universidad de Córdoba, Campus Universitario de Rabanales, 14071 Córdoba, Spain; rutesantos@esaelvas.pt

Many authors have previously stated the role of conformation traits in the quality of the horse's gait, thus justifying the importance of morphological traits in most equine breeds' selection schemes. Particularly, the conformation of the proximal regions of limbs, such as the shoulder, the croup and the thigh, has often been associated with performance results. In this study, conformation and kinematic data were collected from 88 male Lusitano horses using a three-dimensional videographic system. The phenotypic and genetic correlations between 12 conformation variables of the proximal limbs and 13 kinematic variables were then estimated. Genetic correlations were estimated with a multivariate animal model, using a REML procedure. The results show that the angle of the scapula and the angle of the femur seem to be the most highly correlated with kinematic traits, in both phenotypic and genetic estimates. Radius length was also highly correlated, mainly with angular and linear variables of the trot, and pelvic length and angle also showed moderate to high correlation values. The angle of the tibia is directly proportional to tarsal angle, and this is one of the identified faults in Lusitano conformation. It has presented moderate to high genetic correlation values with most kinematic traits. The results show that these conformation traits should be taken into account in the definition of indirect criteria of gait quality in a future breeding scheme of the Lusitano breed.

Horse Kinský: current situation and future

Hofmanová, B. and Majzlik, I., CULS, Kamycka 129, 165 21 Prague 6- Suchdol, Czech Republic; hofmanova@af.czu.cz

Breed Horse Kinský is old czech warmblooder founded by family Kinský esp. by earl Octavian Kinský – the studbook was founded in 1834. This breed was primarily used for steplee-chase and crosscountry riding. A typical feature of this breed is high occurence of coat color palomino and buckskin. Between 1953 and 1989 the Horse Kinský came close to extinction.However, in 1991 a succesful process of gradual regeneration began and since 2007 the breed Horse Kinský is officially being recognised by Ministry of Agriculture. Also a breeders Society for Horse Kinsky has been established. Currently the studbook registers 173 mares and 13 stallions in 5 families. Despite the low number of animals in breed population the Horse Kinský is not included into genetic resources of the Czech Republic. Nowadays the breeding target is to establish and breed a sport horse of a high performance that can be used also as a horse for recreational riding. The qualities of Horse Kinský have been proved many times at sport events in Czech republic nad abroad.Many horses have been exported so far. The breed is very popular esp. in Germany.

Donkey emotional expression: preliminary study about empathy giving origin to emotional relationship with able and disabled children

Tralli, M.[1], Stanzani, F.[2], Giovagnoli, G.[3], Cerino, S.[4] and Bacci, M.L.[5], [1]University of Bologna, DIMORFIPA, Via Tolara di sopra 50, 40064 Bologna, Italy, [2]University of Bologna, Via Tolara di Sopra 50, 40064 Bologna, Italy, [3]Scuola Italiana di Fisioterapia Animale, Sr. dei Cappuccini 102, 53100 Siena, Italy, [4]FISE, Therapeutic Riding, Viale Tiziano 74, 00196 Roma, Italy, [5]University of Bologna, DIMORFIPA, Via Tolara di Sopra 50, 40064 Bologna, Italy; s.cerino@alice.it

The present paper aims to point out how children can evaluate the emotional expression of donkeys, used as well as horses in Therapeutic Riding, and that this understanding is the starting point of the empathic relationship establishing between human being and animals. The experiment took place in Verona Horse Fair 2006, involving 176 children, who were asked to indicate the mood of a donkey submitted to an appropriate ethological stimulus. The donkey behaviour had be evaluated with FAS (Facial Affective Scale), scheduled for men, and therefore it required a particularly specific Self-identification. The results showed that the children were disabled to evaluate the light pleasure stimuli, but they were able to identify and understand the light unpleasant ones.Moreover, thinking about the communication standars establishing between horses and autistic children, we see how often autistic patients express their mood to the animals by no-verbal communication, that is touching the animal (pulling the mane or strongly strolling its neck). The result of these actions (boring for the animal) is an avoiding answer, perceived by disabled children as 'positive' reaction. Finally we can point out how understanding tha animal correct behaviour is the starting point for both man and horse safety and well-being. Such understanding acts as the basis for the establishing of empathic relationships, allowing them a correct useful work even in Therapeutic Riding and in all The AAT. It is essential to deep ethological analysis to establish a proper interaction both with the donkey and the horse.

The difference between sexes in population of Old Kladruby horse

Andrejsová, L.[1], Čapková, Z.[1] and Vostrý, L.[2], [1]Czech University of Life Science, Animal Science and Ethology, Kamýcká 129, 165 21, Prague 6, Czech Republic, [2]Czech University of Life Science, Genetics and Breeding, Kamýcká 129, 165 21, Prague 6, Czech Republic; capkovaz@af.czu.cz

The aim of this work was determined the differences between sexes in population of Old Kladruby horse. The horses were measured within the period 1980-2004; all the 603 (66 stallion, 537 mares) measured horses were 3-10 years old. The data for performance test analysis were collected within the period of 1995-2004 of 372 (79 stallions, 293 mares) Old Kladruby horses. The stallions were higher, heavier and they have larger pastern perimeter then mares. The mares had larger thorax perimeter than stallion. The significant differences were found for height, pastern perimeter and thorax perimeter. The significant differences of performance traits scores were found for all traits scores beyond type and sex expression, general harmony and marathon. But in all traits stallions got higher score means then mares, hence stallions have more strictly pre-selection. The reason of better scores of stallions there is sex dimorphism, stallions are more temperament, robuster and they have better motion mechanic.

Onset of puberty in Lusitano fillies

Fernandes, R.[1,2], Fradinho, M.J.[2], Correia, M.J.[3], Mateus, L.[2] and Ferreira-Dias, G.[2], [1]Instituto Superior de Agronomia, Tapada da Ajuda, 1349-017 Lisboa, Portugal, [2]Faculdade de Medicina Veterinária, CIISA, Av. Univ. Técnica, 1300-477 Lisboa, Portugal, [3]Fundação Alter Real, Coudelaria Alter, 7441-909 Alter-do-Chão, Portugal; ritacfernandes@sapo.pt

In females, puberty is defined as the time of first ovulation and it can be determined by measuring plasma progesterone levels. Spring born fillies can reach puberty during the following spring, but the age of puberty depends on season of birth, photoperiod, immune status, breed and nutrition. One-year-old fillies (n=14) and two-year-old fillies (n=5) were kept under extensive management. Blood was collected by venopuncture every 11 days for progesterone analyses. Additionally, weight was assessed to determine the percentage of mature weight of the animals at the onset of puberty. Between April and June, 57.1% of one-year-old fillies, aged from 12.7 to 16.0 months (13.6±0.7), showed ovarian activity. The average concentration of progesterone in one-year-old fillies was 7.4±2.3 ng/ml. The average weight at the time of pubertal ovarian ciclicity was 308±6.3 kg, which corresponded to 61.6% of mature body weight for the Lusitano horse. The duration of the first period of ovarian ciclicity was approximately two months. All one-year old fillies entered anestrus by June. During the sampling period, all the two year old fillies ovulated twice and displayed an average progesterone concentration of 9.2±1.2 ng/ml. This value was not significantly different from the levels of progesterone registered for one-year-old fillies ($P>0.05$). In this study, data obtained for the Lusitano fillies was similar to those reported for other light breeds. However, given the multifactorial nature of the onset of puberty, these results should be looked upon as preliminary.

Management of therapeutic riding horses and animal welfare

Li Destri Nicosia, D.[1], Sabioni, S.[1], Cerino, S.[2], Giovagnoli, G.[2] and Bacci, M.L.[1], [1]Uiverity of Bologna, DIMORFIPA, Via Tolara di Sopra 50, 40064 Ozzano Emilia Bo, Italy, [2]FISE, V.le Tiziano 74, 00196 Roma, Italy; dora.lidestrinicosi3@unibo.it

The captivity contains many factors conflicting with horse's adaptive abilities. Some specific features of Therapeutic Riding added to such factors risk to compromise animal's welfare. Any defensive response toward stressors (organic or psychic) is expounded with the activation of NEI (neural-endocrine-immunological) response. We have studied horse's natural management (NM) in a TR centre in Emilia Romagna, finalized to NEI equilibrium of horses (n=8). A strategy for abnormal behaviour management must be directed to alternative behaviours expression, the more possible consistent with each interpretative levels (phylogeny, ontogeny, neural-endocrine, adaptive). NM is directed to create the possibility of expression of fully consistent behaviours. It interferes on different managerial levels: feeding/ nourishment; movement conforming to the physiology and biomechanics (correct equestrian techniques; functional trimming of un-shoed feet; harness not conflicting with the biomechanics of the horse); social behaviours (inter and intra specific). An ethological management has been provided, based on the use of communication and application of postural and expressive language techniques within defined protocols of behavioural therapy. Ethological concepts have been used for the specific training of horses too. Since 2002 till now animals have shown the possibility of expression of adaptive responses, suitable to the context and functional. Our behavioural observations suggest the homeostatic condition of the horses, supporting previous data (Placci, Thesis 2008) comparing different management typologies through endocrine and immunological parameters. More researches on endocrine and immunological parameters are requested. This research was supported by grants from Bologna University (RFO 60%)

Economical impact of chromosomical translocations on pig production

Higuera, M.A., Spanish Pig Breeders Association, Goya, 115, 28009 - Madrid, Spain; mahiguera@anps.es

Among different alteration in the structure of pig chromosome, the major significance has reciprocal translocations (RT). RT is not expressed in the phenotype of the animal, but has important impact on reproduction parameters. What's more, this defect is heritable and might be transmitted to the offspring. The problem affects sows and boars as well, what means that both of parents are carriers of this genetic defect. As the effect is observed a reduction of the litter size due to early embryonic death and a problem affects as an average 40% of embryos. Nowadays there is only one technique that enables detection of the gene defect and consists of cariotype screening. The test makes possible to define if any part of one chromosome has been translocated with the part of another chromosome. To evaluate the economical impact of RT the separate analysis of boars and sows is needed. In sows, the elimination of carrier animals is easy, due to strict requirements of breeding programs (elimination of sows with litters smaller than defined criteria). The boars affected by RT, do not present any alternation of sperm quality that during standard semen assessment might be detected. In case that carrier boar is not detected early, the economical impact on production might be dramatic. On the level of breeding farms, the economical impact of RT is expressed by the need of rejection of whole diminished litters, what produce looses in genetic potential. The risk appears in litters reduced only in 10%, when the genetic problem is not detected, and the offspring might be selected as future breeding gilt. On the level of production farms, is more complicated to detect the effect of the boar with RT problem, in case that situation continues, may produce huge reduction in the number of slaughter pigs. It is necessary to screen the cariotype of all boars that are used in a breeding program and terminal boars that we suspect that are producing litters of reduced number of piglets.

Genetic parameters for meat percentage, average daily gain and feed conversion rate in Finnish Landrace and Large White pigs

Peltonen, A. and Sevón-Aimonen, M.-L., MTT Agrifood Research Finland, Biotechnology and Food Research, H2, FI-31600 Jokioinen, Finland; marja-liisa.sevon-aimonen@mtt.fi

In 2006 a new Central Test Station for pigs was opened by Faba Breeding. At the same time feeding of tested pigs was changed from slightly restricted (RF) to ad libitum (AL). The objective of this study was to update the genetic parameters estimated from the central test data for meat percentage (M%), average net daily gain (ADG, g/d) and feed conversion rate (FC, FU /kg gain, 1 FU = 9.3 MJ NE) and to compare the results to previously used parameters. During Data included 2548 Finnish Landrace (FL) and 1,684 Large White (FW) pigs. Pigs were fed individually using electronic feeding system and slaughtered (except for the best boars) after the 13 week test period. (Co)variances were estimated using an animal model REML and DMU program package. Statistical model contained sex and rearing batch as fixed effects, start weight as a covariate and litter, pen, additive animal and error as random effects. Inbreeding was also accounted for. Phenotypic variances were larger in AL than previously estimated variances in RF, but heritabilities were at same level. Heritabilities for M%, ADG and FC in FL (FW) were 0.19, 0.18 and 0.31 (0.31, 0.30 and 0.34), respectively. Correlation between ADG and FC were zero in AL, whereas it was highly negative in RF (favourable). In AL, moderate negative correlation (-0.40 in FL and -0.22 in FW) was observed between ADG and M% (unfavourable), while in RF this correlation was moderately positive. In both AL and RF correlation between FC and M% varied from -0.34 to -0.79 (favourable). Changes in feeding practices between the old and new test procedure affected more the correlations between traits than their heritabilities.

Effect of terminal sire genotype, slaughter weight, and gender on growth performance and carcass traits in European-Chinese pigs

Viguera, J.[1], Peinado, J.[1], Flamarique, F.[2] and Alfonso, L.[3], [1]Imasde Agroalimentaria, S.L., C/ Nápoles 3, 28224 Pozuelo de Alarcón, Spain, [2]Grupo AN, Campo de Tajonar, s/n, 31192 Pamplona, Spain, [3]Universidad Pública de Navarra, Campus de Arrosadía s/n, 31006 Pamplona, Spain; jviguera@e-imasde.com

A total of 256 pigs of 30.9±4.9 kg of initial BW from crossbreds with Youna sows were used to evaluate the effects of terminal sire genotype (DUR, Duroc; PIE, Pietrain), slaughter weight (105 vs 115 kg BW), gender (B, barrows; G, gilts), and their interactions on performance and carcass traits. There were 8 experimental treatments in a 2 x 2 x 2 factorial design, and each treatment was replicated four times. Animals were allotted in pens (eight pigs/pen) according with terminal genotype sire, gender, and initial BW. No significant differences between PIE and DUR pigs were found for growth performance ($P>0.10$). Crossbreds from PIE boars had greater dressing percentage, and trimmed ham and loin yields than crossbreds from DUR boars (77.5 vs 77.2; 13.6 vs 13.1; 7.07 vs 6.59% respectively; $P<0.05$). However, DUR sired-pigs showed higher backfat depth at P2 and Gluteus medius muscle than PIE sired-pigs (19.5 vs 17.2; 29.1 vs 25.0 mm respectively; $P<0.001$). As we expected, pigs slaughtered with 115 kg BW had more dressing percentage than pigs slaughtered at 106 kg BW (78.1 vs 76.6%; $P<0.05$), but trimmed lean cuts proportion was similar for both slaughter weights. Also, gilts showed higher ham and loin yields (13.5 vs 13.2 and 7.06 vs 6.60% respectively; $P<0.001$) but lower backfat depth at P2 and Gluteus medius muscle than barrows (20.0 vs 16.7 and 29.4 vs 24.7 mm, respectively; $P<0.05$). The use of a crossbred between Youna sows and DUR boars and a slaughter weight of 115 kg BW improves the carcass traits of pigs destined for the dry-cured industry.

Meta-analysis of the halothane gene effect on seven parameters of pig meat quality

Salmi, B., Bidanel, J.P. and Larzul, C., French National Institute for Agricultural Research, Animal Genetics, INRA - GABI bâtiment 211 Domaine de Vilvert 78350 Jouy-en-Josas Cédex, 78350, France; catherine.larzul@jouy.inra.fr

Technological meat quality has a significant economic impact and many publications have shown that pig meat quality is strongly influenced by the effect of major genes and by rearing and slaughter conditions. The quality of meat is generally assessed by measuring meat pH at different times post mortem, colour or drip loss. A meta-analysis based on n=3,530 pigs from 23 publications was carried out in order to assess the possibility of predicting the effects of halothane gene, sex, breed and slaughter weight of animals on seven selected parameters: pH at 45 minutes post mortem (pH45), ultimate pH (pHu), reflectance (L*-value), redness (a*-value), yellowness (b*-value), drip loss (DL) and lean percentage. A comparison of two statistical methods was carried out: the method of effect-size and the more well-known random effects model. The meta-analytic method of effect-size was associated to Markov chain Monte Carlo (MCMC) techniques for implementing Bayesian hierarchical models, in order to avoid the problem of limited data and publication bias. The results of our meta-analysis showed that halothane genotype had significant effect on all the analysed parameters of pig meat quality. Meta-regression allows to explain this variability between studies by integrating co-variables such as breed, sex and slaughter weight in the different regression models. According to our results, the halothane gene effect was associated to the breed effect only for the following parameters: pH45, L*-value, b*-value and DL, while the slaughter weight had significant effect only to explain differences of pHu between the homozygous genotypes. In response to inconsistencies in the literature about the difference between the genotypes NN and Nn, the results of our meta-analysis showed that the difference between these two genotypes was significant for all the analysed parameters except for DL and a*-value.

Selection for intramuscular content and fatty acid composition in a Duroc pig line

Reixach, J.[1], Tor, M.[2], Díaz, M.[1] and Estany, J.[2], [1]Selección Batallé, Avda Segadors, s/n, 17421 Riudarenes, Spain, [2]University of Lleida, Animal Production, Rovira Roure, 191, 25198 Lleida, Spain; jestany@prodan.udl.es

Intramuscular fat (IMF) content and composition are two important traits influencing pork quality. It is known that IMF is unfavourably genetically correlated to carcass lean content but there is scarce information about the genetic correlation of IMF fatty acid composition with production traits. The aim of the present study is, first, to present a selection experiment aimed at decreasing backfat thickness (BT) at restrained intramuscular fat content and, second, to discuss the selection opportunities for selecting IMF fatty acid composition in a Duroc line. In the experiment, pigs in group S (n=172) were selected against BT while pigs in group C (n=188) were randomly chosen, while IMF content was kept similar between groups. Pigs were selected according to the mid-parent (litter) breeding values for BT and for IMF content in the gluteus medius adjusted for carcass weight. BT resulted to be lower in group S than in group C but not IMF content and fatty acid composition. However, body weight was lower in group S. Genetic parameters associated to fatty acid composition in the gluteus medius were estimated using data from 722 barrows. Estimates showed that there is enough genetic variation to change fatty acid composition through selection. Monounsaturated fatty acids resulted to be positively correlated with IMF content, uncorrelated with BT and negatively correlated with BW. It is concluded that, provided that there are enough records on IMF available, BT can be independently selected from IMF content, and that selection for oleic acid can be a good strategy for increasing both IMF and monounsaturated fatty acid content at no change in BT. However, caution should be taken for undesirable correlated responses in average daily gain.

Biological potential of fecundity of sows

Waehner, M.[1] and Bruessow, K.-P.[2], [1]Anhalt University of Applied Sciences, Agriculture, Strenzfelder Allee 28, 06406 Bernburg, Germany, [2]Research Institute of the Biology of Farm Animals, Reproduction Biology, Wilhelm-Stahl-Allee 2, 18196 Dummerstorf, Germany; m.waehner@loel.hs-anhalt.de

The profitability of pig production considerably depends on the number of born alive and fostered piglets. The development regarding the number of piglets born for German Landrace (DL), Large White (DE) and Pietrain (PI) in Germany during 1980-2006 is limited (DL: 10.4-11.0; DE: 11.0-11.0; PI: 10.2-10.0). Comparing international data, variations between 10.2 to 13.6 piglets born alive are observed. Reproductive performance in sows is mainly determined by (1) the number of ovulated follicles and fertilized oocytes, (2) the divvy of surviving embryos and fetuses, and (3) the morphological and functional performance of the uterus to support fetal development up to birth. The question whether the ovary and/or uterus are limiting factors can be answered as followed. The pool of ovarian follicles is not the limiting one, although only about 0.5% of oocytes present in the ovary are ovulated during the sows' lifetime. Selection for ovulation rate increases the number of ovulating follicles, but not of piglets born alive. Limiting is the uterine capacity. This physical, biochemical and morphological limitation of the uterus include space, nutrients, gas exchange and surface of the placenta. Although a relationship exists between uterine dimension and the number of fetuses/piglets, uterine length alone is not a prerequisite of higher uterine capacity. Placental efficiency (indicating how much gram fetus is supported by one gram placenta) and the degree of placental blood supply appear to be essentially for litter size. At present, the (really) assumed potential of fecundity is 15.0 piglets born alive, 2.4 litters/year, <10% losses and 32.5 piglets per sow/year (compared to current data of 11.1, 2.26, 13.8 and 21.5; respectively).

The relationship between birth weight of piglets, duration of farrowing and age of sows of German Landrace

Fischer, K. and Wähner, M., Anhalt University of Applied Sciences, Agriculture, Strenzfelder Allee 28, 06406 Bernburg, Germany; k.fischer@loel.hs-anhalt.de

A high number of live born and vital piglets is the basis for an efficient piglet production. Therefore the composition of the herd plays a crucial part. Investigations in a pig breeding farm with sows of German Landrace confirm that sows of the 4th and 5th litter do have the best results with 12.6 total born piglets and 11.9 live born piglets in the 4th litter. Gilts have the lowest results with 11.0 total born piglets and 10.3 live born piglets. The birth weight of piglets is dependent on several impact factors e.g. the litter size. At a litter size of 9 total born piglets the birth weight is 1.55±0.46 kg, at a litter size of 14 to 16 total born piglets the birth weight is lower with 1.36±0.37 kg. In addition the birth weight is dependent on the time interval between two piglets during farrowing, the heavier the piglet the longer the time interval. The part of piglets with a birth weight less than 1 kg gets higher with rising litter size. At a number of 14 to 16 total born piglets the proportion of light piglets is about 18.2%. The proportion of litters with dead piglets rises also according to litter size. If the litter size is higher than 12 total born piglets the proportion of litters with dead piglets rises up to more than 50%. In our investigation the proportion of dead piglets was the highest with 18.4% in the last quarter o farrowing. The duration of farrowing is dependent on the litter size as well as on the litter of the sow. Gilts need only little time for farrowing. They need 3h20min ± 1h28min. Sows of the 5th and 6th litter need a longer time with 4h 26min ± 1h 58min. As a result it is necessary to supervise the time of farrowing. The number of dead piglets needs to get reduced, especially the number of piglets of appropriate birth weight which died in the last part of farrowing.

Genetic analysis of functional teats in the sow

Chalkias, H. and Lundeheim, N., Swedish University of Agricultural Sciences, Department of Animal Breeding and Genetics, Box 7023, 75007 UPPSALA, Sweden; helena.chalkias@hgen.slu.se

Litter size is included in the breeding goal in most breeding programs. However, the selection for increased litter size is not beneficial if the number of functional teats, needed for high daily gain and uniform litters, is lower than litter size. Non functional teats are regarded to be of less value. The nursing and suckling behaviour of piglets follow a complex scheme and the milk ejection lasts for only 10-20 seconds. Thus the importance of having enough functional teats. Our purpose with this study was to analyse the genetic relationship between number of teats, age and backfat thickness at 100 kg. Our data included information on 24,600 purebred Yorkshire pigs from the Swedish breeding company Quality Genetics. We considered the records on males and females as different traits. The statistical model included the fixed effects of herd, year, birth parity number and random effects of animal, litter and pen during fattening period. Number of functional teats increased somewhat with parity number. Pigs born in small litters had somewhat less non functional teats than pigs born in larger litters. Significantly higher (+0.1) number of functional teats was found for pigs born in July-December compared with January-June. The heritability for number of functional teats was in the range 0.28-0.40 and for number of non functional teats in the range 0.04-0.22. The genetic correlation between number of functional teats for males and for females was high: +0.8. Number of functional teats had for male pigs, but not for females, a significant genetic correlation of 0.2 with age at 100 kg. There were no significant genetic correlations between number of functional teats and backfat thickness for any of the genders. Our result show that the number of functional teats can be improved by selection. Number of non functional teats is a complex trait and further studies including reproductive traits will be performed.

The behaviour of male fattening pigs following either surgical castration or vaccination with a GnRF vaccine

Baumgartner, J.[1], Grodzycki, M.[2], Andrews, S.[3] and Schmoll, F.[1], [1]University of Veterinary Medicine Vienna, Veterinaerplatz 1, 1210 Vienna, Austria, [2]Pfizer Animal Health, PSF 4949, 76032 Karlsruhe, Germany, [3]3Pfizer Animal Health, Ramsgate Road, CT13 9NJ, Sandwich, Kent, United Kingdom; johannes. baumgartner@vu-wien.ac.at

From an animal welfare point of view vaccination of male fattening pigs against gonadotropin-releasing factor (GnRF) is beneficial because it avoids a surgical procedure, which is associated with pain and stress even when performed under general or local anaesthesia. The objective of our study was to analyse the behaviour of male fattening pigs either surgically castrated without anaesthesia (T1) or vaccinated with a GnRF vaccine (Improvac^{TM}) at the beginning of the fattening period and 4-5 weeks prior to slaughter (T2). Each treatment comprised 8 groups of 12 male pigs, housed in fattening pens with partially slatted floor and liquid feed provided 3 times a day. Data on postures were scored from 24-hour videos recorded in every week of the fattening period (16 weeks) using scan sampling. Social behaviour was analysed in every second week by continuous behaviour recording of focus animals. Data were analysed using linear mixed models (postures) and the GEMOD procedure (social behaviour) with repeated measurement design. Overall, during the whole fattening period, vaccinates (T2) were more active than surgical castrates (T1), indicated by a higher proportion of pigs standing (T1: 10.4%; T2: 11.8%; $P=0.028$). No effects of treatment on the total number of agonistic interactions and on 'biting and fighting' were found. In T2 the prevalence of aggressive behaviours decreased after the second vaccination ($P<0.001$), which was not found in T1 during the same period. T2 animals showed a higher number of mounting behaviour compared with T1 animals ($P=0.005$), but on a very low level. It is concluded that housing of male pigs vaccinated against GnRF in single sex groups of 12 individuals does not increase behavioural problems in the fattening period compared to surgically castrated males.

Immunocastration decreases variation in growth of boars

Dunshea, F.R.[1], Cronin, G.M.[2], Barnett, J.L.[1], Hemsworth, P.H.[1], Hennessy, D.P.[3], Campbell, R.G.[4], Luxford, B.[5], Smits, R.J.[5], Tilbrook, A.J.[6], King, R.H.[1] and Mccauley, I.[7], [1]The University of Melbourne, School of Land and Environment, Parkville, Vic 3010, Australia, [2]The University of Sydney, Camden, NSW 2570, Australia, [3]Pfizer Animal Health, Parkville, Vic 3010, Australia, [4]Pork CRC, Roseworthy, SA 5371, Australia, [5]QAF Meat Industries, Corowa, NSW 2646, Australia, [6]Monash University, Clayton, Vic 3800, Australia, [7]Department of Primary Industries, Attwood, Vic 3049, Australia; fdunshea@unimelb.edu.au

A problem in producing boars to meat market specifications is variation in growth performance which may be exacerbated by negative behaviors which are reduced by castration. Sixty each of entire boars (E), immunocastrated boars (IM) and barrows (B) were housed in pens of 15 pigs of each sex with access to electronic feeders. IM boars were given the immunocastration vaccine (Improvac®, Pfizer Animal Health, Parkville) at 14 and 18 wks of age. From 18 to 23 wks daily gain was greater in IM boars than in E boars and B (1,090 v. 944 and 908 g/d, respectively, $P<0.001$). Feed intake was greater in IM than in E boars with B intermediate (3,068 v. 2,517 and 2,870 g/d, respectively, $P<0.001$). The SD of live weight increased with age but was lower (P from 0.032 to 0.09) in IM boars than in E boars and B between 18 and 22 and 17 and 21 wks of age. The CV of liveweight declined with age and was lower (P from 0.024 to 0.073) in IM boars than in the E boars and B between 18 and 22 and 14 and 21 wks of age. Carcass fighting damage was greater ($P=0.007$) for E entire boars than for IM boars or B. The frequency of pigs with fighting scores of 2 or greater (on a scale of 0-3 with 3 highest) were 35.6, 3.6 and 0% for E boars, IM boars and B ($\chi^2=45.0$, $P<0.001$). In conclusion, immunocastration increased growth rate and feed intake while decreasing variation in live weight and carcass damage.

Growth performance, meat quality and agonistic behaviour of immunocastrated pigs in comparison to boars and barrows

Albrecht, A.K.[1], Große Beilage, E.[2] and Krieter, J.[1], [1]CAU, Institute of Animal Breeding and Husbandry, Olshausenstr.40, 24098 Kiel, Germany, [2]University of Veterinary Medicine Hannover, Field Station, Büscheler Str.9, 49456 Bakum, Germany; aalbrecht@tierzucht.uni-kiel.de

The aim of the study was to evaluate the effects of vaccination against boar taint on average daily weight gain, feed conversion rate and carcass quality of Improvac®-treated boars (IC) to barrows (SC). Data recording was performed on the Institute´s research farm from April to September 2008. The study was carried out in two identical batches. Pigs within batches (n=224 each) were allocated to two different treatment groups: immunization versus surgical castration. Within treatment groups pigs were randomly assigned to two different diets (standard and high protein diet) in order to examine interactions of growth performance and castration technique. Pigs were housed in pens of two pigs each sorted by weight, castration technique and diet for a fattening period of approximately 110 days. Pigs were slaughtered with an average weight of 113 kg. All pigs were weighted on a weekly basis, feed intake recorded and the feed conversion rate calculated. At the slaughterhouse pH value, conductivity, colour, carcass length and weight, back fat thickness and lean meat percentage were recorded. Additionally drip loss, intramuscular fat content and shear force were measured. Compared to barrows (mean values from raw data) vaccinated pigs showed a better feed conversion rate (FCR_{IC}=2.3 and FCR_{SC}=2.6) and a higher lean meat percentage (LMP_{IC}=55.9% and LMP_{SC}=54.9%). Intramuscular fat content in IC (IMF_{IC}=1.34%) was lower than in SC (IMF_{SC}=1.55%). Shear force (SF_{IC}=8.2 and SF_{SC}=7.7) and drip loss (DL_{IC}=4,1% and DL_{SC}=4,5%) showed no differences among treatment groups. The agonistic behaviour is currently analysed in two batches including three treatment groups: IC (n=25), boars (n=25) and SC (n=25). All pigs will be videotaped and the percentage of time spend on agonistic behaviour will be examined. Results will be presented.

Wet scrubber: one way to reduce ammonia and odours emitted by pig units

Guingand, N., IFIP Institut du Porc, La Motte au Vicomte, 35651 Le RHEU, France; nadine.guingand@ifip.asso.fr

With the intensification of European regulation on atmospheric pollution, reduction of ammonia emitted by pig units become a priority for farmers. Conflicts between pig farmers and neighbourhood are generally based on odours emissions. The part of buildings in ammonia produced by pig production is estimated of 60% of the whole ammonia emitted. For odours, this part is around 70%. Sows and the rearing of piglets until 30 kg represent less than 30% of ammonia and around 40% of odours emitted by the building. Focus the means of reduction of ammonia and odours produced by rooms housing grower-finisher pigs appears to be a great way to decrease the whole quantity of ammonia and odours emitted by a pig farm. Nowadays, in pig production, the implementation of wet scrubber is probably the only means to reduce both. Wet scrubber equipment in pig production has a great development but several types of scrubber are proposed to farmers who not really know what are the main criteria illustrating the optimal efficiency of scrubbers on ammonia and odours. The aim of this article is to synthesise these main criteria based on studies achieved on experimental stations and on commercial units equipped with different kind of wet scrubbers. The incidence of some technical parameters of wet scrubber conception – volume, air and water flow rate - are illustrated on the efficiency on ammonia, odours and dust reduction. Global results of various types of wet scrubber are presented and their efficiency on ammonia, odours and dust emissions are showed in relation with season and physiological stages. In a last part, an economic comparison is realized between wet scrubbers and others ways of reduction proposed at present to pig farmers.

Ammonia concentration around pig farms in Spain

Sanz, F.[1], Sanz, M.J.[1], Montalvo, G.[2], Garcia, M.A.[2], Pineiro, C.[3] and Bigeriego, M.[4], [1]CEAM, Charles Darwin, 46980 Valencia, Spain, [2]Tragsega, Julian Camarillo, 28037 Madrid, Spain, [3]PigCHAMP Pro Europa, G Segovianos, 40195 Segovia, Spain, [4]Spanish Ministry of the Environmental and Rural and Marine Affairs, Alfonso XII, 28071 Madrid, Spain; pacosanz@ceam.es

A large proportion of NH3 emitted locally is deposited in the immediate neighborhood of the source. Quantitative information about the spatial location of emission sources, as well as estimation of the point source emissions is crucial for target oriented abatement. The aim of this study, financed and coordinated by the Spanish Ministry of the Environment and Rural and Marine Affairs, is to describe the variability of NH3 concentrations in the surroundings of several point sources (pig farms). Pig farms were selected as representative of most types of pig farms in Spain: grower finisher pigs, sows with piglets, one-site pig farm (in the central plateau), and one-site pig farm in Mediterranean area. Concentrations and depositions were estimated over an area of 1 km in each of the farms and ammonia concentrations were determined by Ferm passive samplers located at 3 meters above ground in four transects (N, S, E, and W) for the farms. Emission rates were calculated based on the Gaussian dispersion equation and the output of the model were compared with the experimental concentration measurements. The highest concentrations near the buildings for the sows with piglets and the one-site pig farm in the central plateau were similar (around 60 µg/m3), whereas around grower finisher pigs farm concentration reached 81 µg/m3. Concentration around one-site pig farm in Mediterranean area was 120 µg/m3, as could be expected due to the highest number of animals. The concentration predicted by the model show a good correlation (r>0.80) with the passive samplers measurements. Concentration fields decreased in all cases to levels of 2 and 5 µg/m3 within distances of less than 1 km (600 m), except in Mediterranean area, whose measured background was higher (30 µg/m3) probably due to the larger concentration of farms in the area.

Ammonia emission after application of pig slurry in Spain

Sanz, F.[1], Sanz, M.J.[1], Montalvo, G.[2], Garcia, M.A.[2], Pineiro, C.[3] and Bigeriego, M.[4], [1]CEAM, Charles Darwin, 46980 Valencia, Spain, [2]Tragsega, Julian Camarillo, 28037 Madrid, Spain, [3]PigCHAMP Pro Europa, Segovianos, 40195 Segovia, Spain, [4]Spanish Ministry of the Environment and Rural and Marine Affairs, Alfonso XII, 28071 Madrid, Spain; pacosanz@ceam.es

The land spreading of animal manure represents approximately one-third of the total NH_3 emissions from agriculture, so there has been much interest in the development of abatement measures in this area. The aim of this study, financed and coordinated by the Spanish Ministry of the Environment and Rural and Marine Affairs, was to evaluate the efficiency for reducing NH_3 loss of different pig slurry application techniques at field scale on grassland and arable soils in the Central Plateau of Spain. To test the reductions of NH_3 concentration in the air, several experiments were performed during 2004-2008 in Segovia (Spain). Ammonia concentration fields around and gradients above ground in the application plots were measured with passive samplers (Ferm type), that were collected at different times. To optimize the experimental setup and select a correct plot orientation in both experiments, a meteorological tower was installed one month before the experiments were started at each of the study sites. An approximation of the aerodynamic gradient method was employed to estimate the relative NH_3 emission rates for each plot at the different exposure times. In all experiments the NH_3 emissions for the mineral fertilizer application were near 0 µg/m2s. The broadcast spreading with splash plates resulted in the highest emissions. Percentage of reduction observed respect to splash plate was, for band spreader, 25-58%; for trailing shoe spreader 49%; and for injection 38%. We conclude that the slurry application systems evaluated were effective in the reduction of NH_3 loss from grassland and arable soils and that the results from different application techniques differences were of the order of 10%, although the variability was high for different experiments probably due to the different meteorological conditions.

Evaluation of Fe, Cu, Mn and Zn in manure of weaned pigs receiving high levels of organic or inorganic supplementation

Royer, E.[1], Granier, R.[2] and Taylor-Pickard, J.[3], [1]Ifip-Institut du porc, 34 bd de la Gare, 31500 Toulouse, France, [2]Ifip-Institut du porc, Les Cabrières, 12200 Villefranche de Rouergue, France, [3]Alltech, Summerhill Road, Dunboyne, Co. Meath, Ireland; eric.royer@ifip.asso.fr

A 40 days experiment has been undertaken to evaluate the Fe, Cu, Mn and Zn excretion of piglets given the upper concentrations allowed by European legislation in compound feeds. 40 male and female piglets blocked at an average weight of 8.0 kg (28 days of age) received ad libitum up to 28.6 kg control or experimental diets formulated without phytase. Inorganic (sulfates and oxide) or organic sources were respectively used in control and experimental diets as a supplementation of 110, 150, 50 and 110 mg of Fe, Cu, Mn and Zn elements per kg. The analyzed concentrations of Fe, Cu, Mn and Zn in the control and experimental diets were respectively of 228,161,72,120 and 239, 145, 76 and 145 mg/kg for phase 1 diets, and 279, 156, 70, 126 and 274, 143, 75, 150 mg/kg for phase 2 diets. A lower grade of the commercial zinc oxide used could explain that control diets had slightly lower contents in zinc than expected. There were no differences in average daily feed intake, daily gain and manure production by animals. Excretion of Fe, Cu and Mn did not differ for inorganic (respectively 5.6, 4.0 and 1.8 g/pig) and organic (respectively 5.6, 4.0 and 1.9 g/pig) sources, whereas the zinc excretion was slightly lower for pigs receiving the inorganic elements (3.3 g/pig) than for those given the organic form (3.9 g/pig), which was probably caused by the lower concentration in control diets. It can be concluded thatwhen high safety margins above the physiological requirements are applied, the excretion values per animal remain important. Thus, the environmental effects of trace elements could be reduced by a lower supplementation rate of these metals, while using dietary sources with better bioavailability values.

Influence of lean meat proportion on the chemical composition of pork

Okrouhla, M., Stupka, R., Citek, J., Sprysl, M., Kratochvilova, H. and Dvorakova, V., Czech University of Life Sciences, Animal Husbandry, Kamycka 129, 165 21, Prague 6, Czech Republic; stupka@af.czu.cz

The objective of this work was to verify the influence of lean meat proportion on the chemical composition of loin and ham of pork. A total of 116 finishing hybrid pigs commonlyused in the Czech Republic were fattened for this purpose. The pigs were divided according to the lean meat proportion criterion into 3 groups, resp. more than 60.0%, 55.0-59.9% and 50.0-54.9%. Representative muscle samples were taken from the right halves of these pigs. They were then homogenized and submitted to chemical analysis. The results of the measuring showed that the values of water content, total fat (TF), crude proteins and ash matter ranged at the loin at intervals of 72.50-72.80%, 1.56-1.96%, 23.20-23.40% and 1.37-1.40% and at the ham in intervals of 70.43-71.59%, 3.52-4.26%, 21.67-21.95% and 1.42-1.56%, respectively. In the carcass part of the musculus longissimus lumborum et thoracis (MLLT) it was demonstrated that the higher lean meat share, the lower the content of amino acids - threonine, isoleucine, lysine, aspartic acid, serine and proline. In the carcass portion of the musculus semimembranosus (MS) the values of valine, isoleucine, phenylalanine, lysine, serine, proline and glycine increased with an increasing lean meat proportion. Within the frameworkof statistical evaluation of differences between the groups, the values of IMF ($P<0.01$) in MLLT, water content, TF, ash matter, threonine, valine, phenylalanine, lysine, aspartic acid, serine, glycine and alanine in MS were highly significant ($P<0.05$; $P<0.01$ and $P<0.001$).

Suitability of post-slaughter parameters for determining carcass meatiness in Polish Landrace pigs

Żak, G. and Eckert, R., National Research Institute for Animal Production, Department of Genetic and Animal Breeding, Sarego 2, 31-047 Kraków, Poland; reckert@izoo.krakow.pl

The purpose of this research was to determine correlations between various post-mortem carcass parameters and carcass meat percentage determined based on detailed dissection of Polish Landrace (PL) pigs. A total of 16 traits were analysed, including 10 linear measurements of the carcass, 2 measurements of muscular and adipose tissue area, and weight of two primal cuts (ham and loin), dissected from the carcass by the EU reference method (Walstra and Merkus), which is used in Polish pig testing stations. Phenotypic correlations were then estimated between the analysed traits, slaughter parameters and meatiness calculated using two methods - the EU reference method and the Pig Testing Station (SKURTCh) method. It is concluded from the results obtained that out of linear traits, the measurements of backfat thickness at sacrum points II and III would be the most suitable for determining carcass meatiness using the Walstra and Merkus method, while the weight of ham and loin, dissected using the EU reference method, would be the most suitable out of ham and loin parameters to determine the weight of these cuts without backfat and skin. The weight of whole cut can also be used in the case of ham. The weight of whole loin is useless to estimate carcass meatiness. The most useful parameters to estimate carcass meatiness according to the SKURTCh method are loin eye height and width, and average backfat thickness from 5 measurements. The usefulness of the ham and loin parameters obtained using the SKURTCh method is similar to that obtained using the Walstra and Merkus method. When estimating carcass meatiness with the SKURTCh method, one could also use the weight of ham and loin without backfat and skin, obtained based on carcass dissection using the Walstra and Merkus method.

The influence of the gene MYF6 on selected indicators of the fattening capacity and the carcass values in pigs

Kratochvilova, H., Stupka, R., Citek, J., Sprysl, M. and Okrouhla, M., Czech University of Life Sciences, Animal Husbandry, Kamycka 129, 126 21, Prague 6, Czech Republic; stupka@af.czu.cz

The MYOG gene is MYOD family genes. The MYOD family consists of four related genes: MYOD 1 (MYF3), MYOG (MYF4, myogenin), MYF5 and MYF6 (MRF4, herculin). These genes control muscle cell determination and differentiation during embryonic development. The objective of this study was to determine the effect of MYF6 gene on selected traits of carcass value in pigs. To verify the associations between the polymorphisms and the selected carcass value traits, total of 102 pigs, including animals of Czech Large White breed, Czech Large White x Czech Landrace, Pietrain x (Czech Large White x Czech Landrace) and (Czech Large White x Duroc) x (Czech Large White x Czech Landrace) were used. The genotype BB was the most detected (52 pigs) in this group of pigs whiles the minimum pigs (22 pigs) were AA genotype so this genotype were excluded from this study. Within the genes MYF6 evidential statistical differences relating to the average daily weight gain were identified ($P \leq 0.05$) within the genotypes AA and AB. As the body weight concerned, evidential statistical differences were identified ($P \leq 0.05$) within the same genotypes. Statistically significant associations ($P \leq 0.01$) were observed for the fat content and the proportion of the main meat parts within the genotypes AA and BB. Contrary, statistically insignificant association were found for the weight and the main meat parts percentage. This research was supported by the Ministry of Agriculturae (QG 60045) and the Ministry of Education of Czech Republic (MSM 60460709).

The influence of the gene Myf-4 on the carcass value in pigs

Dvorakova, V., Stupka, R., Kratochvilova, H., Citek, J. and Sprysl, M., Czech University of Life Sciences, Animal Husbandry, Kamycka 129, 165 21, Prague 6, Czech Republic; stupka@af.czu.cz

The aim of this study was to determinate the influece of Myf-4 gene on selected production traits in pigs. This gene is allowed to be candidature for meat efficiency of pigs. For this purpose 67 head of pigs of three genotypes [LWD, LWD x L and PN(LWD x L)] were tested. We used the PCR-RFLP method. The fissure of the gene was made by help of restricted endonuklea MspI. Frequence of alela A and B were: A 0.66, B 0.34. By some traits of the carcass value statistically evidential differences depending on genotype were found out. Compared with the lean meat share in the carcass it was found, that alela A, which is for higher ratio of muscle, is overbearing, while the frequence of alela B has an contradictory tendency, so a lower ratio of muscle. There was noticed a statistically evidential difference ($P<0.01$) by the rate of muscle between the individuals AA and AB. This difference was confirmed partly by ZP method, it was measured 55.17% by AA and by AB it was 51.79%, the statistically evidential difference was found out also by an other method (FOM) so by AA 54.05%, AB 50.84% ($P<0.01$). The statistically evidential difference was also detected for an avarage daily gain, when the individuals who were in the polymorphic variant AB had 991.78 g in comparison with 949.35 g for AA ($P<0.05$). The statistically evidential differences were detected for the belly lean meat share namely for AA 51.15% resp. AB 47.76% ($P<0.05$). These results indicate the influence of Myf-4 genotype on selected production traits in pigs. This research was supported by the Ministry of Agriculturae (QG 60045) and the Ministry of Education of Czech Republic (MSM 60460709).

The impact of the plant protein source-substitution for meat-and-bone meal on performance and carcass value in growing-finishing pigs

Stupka, R., Sprysl, M., Citek, J., Trnka, M. and Okrouhla, M., Czech University of Life Sciences, Animal Husbandry, Kamycka 129, 165 21, Prague 6, Czech Republic; stupka@af.czu.cz

The aim of the study was to analyse the effect of substituting vegetable protein for meat-bone meal on the production potential in fattening pigs. The study included in total 72 hybrid pigs of the (LWsxPN) x (LWdxL) genotype of balanced sex at the age of 68 days and total average live weight of 24.15 kg. It has been found out that substituting vegetable protein for animal protein has almost no impact on the feed intake or on the growth intensity, i.e. the average daily weight gain (921 g and 914 g, respectively) was not proved, either. Animals fed with complete feeding mixture (CFM) without animal protein recorded lower values of meat formation (56,8% and 55.3%, respectively) and deposited more fat to the detriment of meat formation. In contrast, the pigs fed with animal protein achieved higher lean meat share, bigger loin eye area and higher height of the MLLT-meat as well as markedly lower back fat thickness (11.6 mm and 12.6 mm, respectively) throughout the fattening. During the test they also showed at individual monitored week intervals a higher protein-intake which was reflected in a higher meat-formation in the carcass. This research was supported by the Ministry of Agriculturae (QG 60045) and the Ministry of Education of Czech Republic (MSM 60460709).

The effect of carcass weight on carcass performance and proportion of tissues in main meaty cuts of pigs

Bahelka, I., Hanusová, E., Oravcová, M., Peškovičová, D. and Demo, P., Animal Production Research Centre, Animal Breeding and Quality Products, Hlohovska 2, 949 92 Nitra, Slovakia (Slovak Republic); peskovic@scpv.sk

Carcass traits of commercially produced hybrid pigs (n=241) in dependence on carcass weight were evaluated. Animals were divided into four weight categories: <80.0 kg, 81.0-90.0 kg, 91.0-100.0 kg and/or 101.0-110.0 kg. Day after slaughter, dissection of right half sides and detailed dissection into single body tissues were performed. Significant effect of carcass weight on all observed carcass traits (average backfat thickness - BF, weight of shoulder, of loin, of ham, of belly and of tenderloin) was found. The weights of these parts and also the weights of tissues in these parts increased with increasing carcass weight (BF: 24.05-31.90 mm, weight of shoulder: 4.90-6.85 kg, of loin: 5.97-8.39 kg, of ham: 9.19-12.74 kg, of belly: 4.30-5.80 kg and of tenderloin: 0.60-0.71 kg). Differences between categories in proportions of meaty parts from weight of half carcass were not significant with exception of tenderloin. Significant differences were determined in some proportion of single tissues in meaty parts between weight categories (e.g. percentage of meat and bones in ham, of fat with skin and bones in loin, of meat and fat in tenderloin). The results confirmed the changes in carcass comformation and proportions of single tissues in main meaty cuts of pig carcasses.

Breed-specific mechanisms of fat partitioning in pigs

Doran, O.[1], Bessa, R.J.B.[2,3], Hughes, R.A.[4], Jeronimo, E.[3], Moreira, O.C.[3], Prates, J.A.M.[2] and Marriott, D.[1], [1]University of the West of England, Institute of Bio-Sensing Technology, Frenchay Campus, Coldharbour Lane, Bristol, BS16 1QY, United Kingdom, [2]Technical University of Lisbon, Faculty of Veterinary Medicine, CIISA, 1300-477, Lisbon, Portugal, [3]INRB, Animal Production Unit, Fonte Boa, 2005-048 Vale de Santarem, Portugal, [4]University of Bristol, Clinical Veterinary Science, Langford, BS40 5DU, Bristol, United Kingdom; duncan.marriott@uwe.ac.uk

Pig breeds with similar subcutaneous fat (SF) content largely differ in intramuscular fat (IMF) level. This suggests that the mechanisms regulating fat partitioning are breed-specific. Our previous study demonstrated that the lipogenic enzyme, stearoyl-CoA desaturase (SCD) plays a key role in IMF formation in commercial breeds. The present study investigated whether the input of SCD in IMF and SF formation in pigs is breed-specific. The study was conducted on three genetically diverse breeds: Large White x Landrace (LWxL, with low IMF) and purebreds Bizaro (B) and Alentejano (A) (with high IMF). SCD protein expression was analysed in isolated microsomes by Western blotting. Fatty acid composition and fat content were determined by gas chromatography. It was demonstrated that SCD protein expression is positively related to monounsaturated fatty acid (MUFA) and total fatty acid content in muscles of LWxL but not in those from B or A breeds. In contrast to muscles, SF showed positive correlation between SCD expression, MUFA and total subcutaneous fat content in B and A breeds but not in LWxL cross. It was concluded that SCD expression and biosynthesis of MUFA may have a key input in the fat deposition in muscles of breeds with low IMF content but not in those with high IMF level. In the breeds with high IMF, SCD mainly contributes to the deposition of fat in subcutaneous adipose tissue.

Usefulness of live animal measurement of meat productivities for genetic improvement of carcass traits in Berkshire pig

Tomiyama, M.[1], Kanetani, T.[2], Tatsukawa, Y.[2], Mori, H.[2], Munim, T.[1], Suzuki, K.[3] and Oikawa, T.[1], [1]Okayama university, Graduate School of Natural Science and Technology, 1-1-1 Tsushima-Naka, Okayama, 700-8530, Japan, [2]Okayama Prefectural Center for Animal Husbandry and Research, Misaki-cho, Kume-gun, Okayama, 709-3494, Japan, [3]Tohoku University, Faculity of Agriculture, Sendai, 981-8555, Japan; dns17423@cc.okayama-u.ac.jp

Ultrasonic equipment is a useful tool to acquire subcutaneous information. The phenotypic relationships among measurements on live animals and on carcass traits were rather abundant, however, the results of genetic relationship were scarce. The objective of this study was to reveal accuracy of ultrasonic measurement of subcutaneous fat thickness (SCF) and loin eye area (LEA) in terms of efficiency for the genetic improvement of carcass traits. The data used in this study were collected on 4,773 purebred Berkshire (2,458 males and 2,315 females) pigs at the Okayama Prefectural Animal Husbandry and Research Center, Japan. Measurements on live animals were recorded at 105kg (at finish). Traits on live animal were age at finish, daily gain from birth to finish, back fat thickness (BFTF) and LEA (LEAF) at finish. Measurement of carcass traits were carcass weight, LEA (LEAC), SCF on midpoint (SCFB), shoulder (SCFS), loin (SCFL), half of carcass length (SCFH) and 10th rib (SCF10). Additive direct and maternal genetic (co)variance components for those traits were estimated by REML using VCE5. Estimates of genetic correlations for BFTF were highly positive with SCFB (0.91), SCFS (0.87), SCFH (0.97) and SCF10 (0.75), and were moderately positive with SCFL (0.51). These estimates indicated that measurement of fat thickness using ultrasonic equipment was quite accurate. Whereas the correlation of LEA was lower (0.37) between LEAF and LEAC. Therefore, technical improvement of measuring specialist is necessary to include LEAF into selection criteria. The phenotypic correlations among the traits were also compared with the previous reports.

Possibility of estimating belly and loin meat percentage in pigs based on data from dissection performer according to the Walstra and Merkus metod

Żak, G. and Tyra, M., National Research Institute of Animal Production, Department of Genetic and Animal Breeding, ul. Sarego 2, 31-047 Kraków, Poland; mtyra@izoo.krakow.pl

The aim of the study was to develop regression equations for estimating meat percentage in two valuable cuts: loin and belly. A total of 240 gilts of the Polish Large White and Polish Landrace maternal breeds were investigated. Data of calculations were obtained postmortem at the Pig Testing Station in Rossocha during a detailed dissection of left half-carcasses according to the EU method (Walstra and Merkus). Among several regression equations developed in the study, two are used to determine loin meatiness and two are designed to determine belly meatiness. The coefficient of determination (R^2) for two equations for estimating loin meatiness was 79.6 - 79.9, with RMSE = 2.02. Parameters of our third equation (R^2=57.6, RMSE=2.91) were slightly below the required minimum. The coefficients of determination for belly meatiness equations were high (R^2=99.3 for the first equation and R^2=98.9 for the second), but the RMSE coefficients of estimation accuracy were low (4.5 and 5.6, respectively). Further studies would be necessary to improve accuracy of estimating belly meat percentage using regression equations.

Possibility of improvement of lean meat content of ham and loin in pigs by selection for growth and feed conversion rate

Żak, G., Tyra, M. and Różycki, M., National Research Institute for Animal Production, Department of Genetic and Animal Breeding, Sarego 2, 31-047 Kraków, Poland; z1zak@cyf-kr.edu.pl

The aim of this investigation was to determine the effect of growth rate and feed conversion ratio in fatteners of two most common pig breeds in Poland on the weight and quality of loin and ham as defined by lean meat and by fat content. The research was carried out using records from the Pig Testing Station of 113 Polish Large White and 120 Polish Landrace gilts. Parameters of the tested cuts were assessed on the basis of weight of cuts, weight of lean meat (lean), weight of subcutaneous fat with skin, and percentage of lean and fat content. Based upon the level of indicators of their fattening performance the animals were divided into three groups. The Polish Landrace pigs occurred more susceptible to the effect of fattening traits on tissue content and at the same time on the quality of carcass cuts. The interrelation has become apparent between loin parameters of Polish Landrace pigs and their growth rate and between their ham parameters and feed conversion ratio (per 1 kg gain). In many cases the loin parameters examined differed significantly between breeds. Significantly wider variation as regards the composition of tested carcass cuts was found in Polish Large White than in Polish Landrace gilts. A trend was noted for the deposition of more fat in carcass cuts of pigs with highest and lowest live weight gain.

The use of the genetic potential of Polish Landrace boars for improvement of fattening and slaughter value

Żak, G. and Różycki, M., National Research Institute for Animal Production, Department of Genetic and Animal Breeding, Sarego 2, PL 31-047 Kraków, Poland; z1zak@cyf-kr.edu.pl

The analysis covered evaluation data for 6,414 Polish Landrace boars, which were performance tested between October 2004 and March 2005, and for 134 boars selected from the population, which were kept in nucleus breeding over the next years. The selection differential obtained by leaving 134 selected boars in breeding was compared to the differential that would be achieved by leaving best animals in terms of breeding value for the selection index of fathers of the next generation. It was shown that the individuals selected from the group of 6,414 boars were characterized by average breeding value exceeding the average breeding value for the population. However, the analysis of breeding value for individual boars chosen for fathers showed that individuals with negative evaluation results were also left in breeding. We found that it is possible to obtain selection differential that is 317% higher for daily gains, 396% higher for carcass meat content, and 258% higher for selection index, as compared to the selection differential obtained in the examined active population of Polish Landrace boars during the analysed period. For this to be achieved, the boars characterized by the highest breeding values should be left in nucleus breeding.

The model for profitability estimation in pig production

Citek, J., Sprysl, M., Stupka, R. and Kratochvilova, H., Czech University of Life Sciences, Animal Husbandry, Kamycka 129, 165 21, Prague 6, Czech Republic; sprysl@af.czu.cz

The aim of this work was to assess the model for estimation of economical parameters influencing the profitability in pig breeding based on the real reproduction and production efficiency. Firstly analyse of factors influencing the economy of fattening pig production were made. Following factors were included into economical model: fixed costs per 1 kg carcass body production (2 450 CZK/year), piglets number per sows/year (PIGLET), average daily gain (ADG), daily feed intake (DFI), cost of feed (CF), lean meat (LM), price per 1 kg of carcass (RP). The calculation results from production efficiency of 136 carcass pigs. The results of this work are following models enabling cost estimation, total income and profitability, pursuant to efficiency parameters and input resp. output prices. The models are for cost estimation in CZK per pig (R^2=0.73) y = 3633.006781 -0.543349*ADG -50.403655*PIGLEST +129.815364*CF +0.1063*ADG*CF, total income estimation in CZK per pig (R^2=0.88) y = -2250.8248 +2.622693*ADG +87.926498*RP, profitability estimation in % (R^2=0.83) y = -214.292164 +0.1333936*ADG +1.6983062*PIGLEST +0.3753065*CF +2.850505*RP -0.0094251*ADG*CF. The all models were established by procedure REG in SAS 9.1 Software. By help of this models a neutral profitability for individual efficiency parameters with average prices in the Czech Republic in a year 2008 was determined: CF=5000 CZK/t, RP=38.0 CZK/kg JUT, ADG=750 g/day, PIGLEST=24.1 was set on account of this models. This research was supported by the Ministry of Agriculturae (QG 60045) and the Ministry of Education of Czech Republic (MSM 60460709).

Economy in pig breeding with respect to genotype in the Czech Republic

Sprysl, M., Citek, J., Stupka, R. and Okrouhla, M., Czech University of Life Sciences, Animal husbandry, Kamycka 129, 165 21, Prague 6, Czech Republic; sprysl@af.czu.cz

The unfavourable economics of pork production in The CR is a result of many factors. The only way how to eliminate this situation is to use objective information from tests carried out in the station and field tests. The objective of this work is to determine firstly the yield rate of the selected genotypes of hybrid pigs by means of station tests, secondly the influence of the input and output prices for fattening profitability. For this 2 combinations of 72 heads of pigs of a uniform ratio of sex and PHx(LWDxL) resp. (LWSxPN)x(LWDxL) genotypes were tested. The course of indicators characterizing yield rate and production efficiency was regularly monitored in all pigs fed ad libitum. After sale, the basic economic indicators of the fattening were assessed. With a view to the achieved results, it may be stated that profitable fattening may be achieved after reaching 90kg live weight under the condition of either a reproduction yield rate of 25 reared piglets per sow a year or reaching the selling price per live pig at CZK 32 per kg. Profitable fattening in the weight range of 100-110 kg may also be achieved at the existing yield rate and grain price of CZK 3500 per ton. This research was supported by the Ministry of Agriculturae (QG 60045) and the Ministry of Education of Czech Republic (MSM 60460709).

A subset of candidate polymorphisms identified by 52 SNPs mini-array SNiPORK in two Duroc populations (differed by meat yield and quality) revealed significant differences in allele distributions

Olenski, K.[1], Sieczkowska, H.[2], Kocwin-Podsiadla, M.[2], Help, H.[3] and Kaminski, S.[1], [1]University of Warmia and Mazury, Animal Genetics Department, Oczapowskiego street 5, 10718 Olsztyn, Poland, [2]University of Podlasie, Department of Pig Breeding and Meat Science, Prusa street 14, 08110 Siedlce, Poland, [3]AsperBiotech Ltd, Oru, 58014 Tartu, Estonia; stachel@uwm.edu.pl

Although any breed is characterized by a set of established traits and qualities, in breeding strategies led by national programs or commercial companies animals belong to the same breed may differ significantly because of different origin. Comparing Duroc fatteners imported to Poland from Danish breeding companies versus Duroc pigs kept in Poland (based of import from different countries) we showed highly significant differences in many parameters of meat yield and quality. In this report we try to explain this difference by genotyping two unrelated Duroc populations (38 vs 40 fatteners) by mini-array SNiPORK capable to identify 52 candidate SNPs potentially associated with pork traits. Using chi square test we have found out that 23 SNPs have significantly ($P<0.01$) different allele distributions. Among them 11 showed adverse trend of allele frequency, namely CASP (calpastatin), CYP21 (cytochrome p450 steroid 21 hydroxylase), CYP2E1 (cytochrome p450 2E1), DECR1 (mitochondrial 2,4 dienoyl CoA reductase), GH (growth hormone), HAFABPdelT (heart fatty acid binding protein), PKM2 (muscle puruvate kinase 2), PPARG (peroxisome proliferator activated receptor gamma 1), SULT1A1 (phenol sulfating phenol sulfotransferase 1), TNNT3 (skeletal muscle troponin T3) and TYR (tyrosinase). The question is - whether these selected SNPs (particularly the above 11 SNPs) participate in phenotypic variation of meat traits? If the answer will be positive, functional analysis is needed to find out whether these SNPs change the level of mRNA, the quantity (or properties) of encoded protein and in effect the value of economically important traits.

The effect of creep feeding on genetic evaluation of Berkshire pig: possibility of earlier and simpler evaluation with favorable correlated response

Tomiyama, M.[1], Kanetani, T.[2], Tatsukawa, Y.[2], Mori, H.[2], Munim, T.[1], Suzuki, K.[3] and Oikawa, T.[1], [1]Okayama university, Graduate School of Natural Science and Technology, 1-1-1 Tsushima-Naka, Okayama, 700-8530, Japan, [2]Okayama Prefectural Center for Animal Husbandry and Research, Misaki-Cho, Kume-Gun, Okayama, 709-3494, Japan, [3]Tohoku University, Faculity of Agriculture, Sendai, 981-8555, Japan; dns17423@cc.okayama-u.ac.jp

The objective of this study was to choose selection traits by estimating genetic parameters for pre-weaning growth traits. The data used in this study were collected in Japan on 4,548 purebred Berkshire (2,344 males and 2,204 females) pigs at the experimental station in Okayama, Japan. The piglets were weaned on the nearest Thursday after reaching 25 days of age and raised with the same litter during birth to 60 days of age. The creep feeding for piglets was started after the birth. Additive direct and maternal genetic (co)variance components for those traits were estimated by REML using VCE5. Comparing variance components among heritabilities of direct and maternal genetic effects, heritabilities of direct genetic effect for weaning weight (WW) at 28.6 days of age dominated the maternal genetic effects. Whereas body weight at 14 days of age (W14) had similar heritabilities for these components. This result disagreed with those of previous reports. The estimated genetic correlations between direct and maternal genetic effect for W14 and WW were moderately positive (0.43 and 0.31, respectively). These results were different from negative estimates by the previous reports. This difference seems to be due to creep feeding program for piglets. Therefore the selection program based on direct genetic value of early growth traits on piglet can be carried out earlier with a simpler genetic model. Genetic correlation of direct genetic effects between WW and body weight at 60 days of age (W60) was highly positive. This result indicated that improvement of W60 is possible by including WW in selection criteria.

Impact of increased returns to estrus after mating sows on seasonal infertility

Maglaras, G.E.[1], Kousenidis, K.V.[2] and Kipriotis, E.A.[2], [1]T.E.I. of Epirus/School of Agricultural Technology, Animal Production, Kostakii, 47100, Arta, Greece, [2]National Agricultural Research Foundation, Komotini Agricultural Research Station, 18 Mer.Serron St., P.O. Box 63, 69100, Komotini, Greece; gmag@teiep.gr

The seasonal infertility syndrome was studied on the basis of determining whether the elevated numbers of Services induced by sows returning to estrus have an impact on the numbers of sows impregnated in the reduced fertility period of July-August-September. Numbers of sows mated (Services) and sows that farrowed (Pregnancies) were collected from 5 pig farms located in three different regions of mainland Greece. A total of 2,303 sows were involved in 6,809 matings and 5,606 farrowings grouped in monthly observations over a period of 12 months. The collected data was analyzed in respect of the Services-Pregnancies relation and the potential effect of farm and season on this relation. The thorough statistical analysis provided the proof for a close, non-linear relation between Services and Pregnancies ($R^2=0.96$, $P<0.01$). There was no farm effect established on the Services-Pregnancies relation. The effect of a season-specific factor was found to generate an inverse relation between pregnancies and services in the months of July, August and September ($R^2=0.97$, $P<0.01$) suggesting that the increased amount of services potentially consists of one of the factors that explain the decreased pregnancy levels. Excessive numbers of Services were then shown unable to relate to numbers of Pregnancies when the season-specific factor was ignored. It was therefore, concluded that elevated numbers of Services, which are, themselves, caused by unsuccessful matings of sows returning to estrus in the period of July-August-September, enhance the expression of reduced fertility and constitute an element of the seasonal infertility syndrome.

The effect of dietary herbs supplement in sow diets on the course of parturition, stress response and piglets rearing results

Paschma, J.M., National Research Institute of Animal Production, Dept. of Technology, Ecology and Economics Animal Production, Balice n. Krakow, Poland, 32-083, Poland; mtyra@izoo.krakow.pl

The aim of the present study was to determine the effect of using herbs mixture inidiets of periparturient sows on the course of parturition, urine cortisol levels and reproductive performance during two reproductive cycles. A total of 24 multiparous sows were assigned to 3 groups differing in the amount of dietary herbs: 0, 1.0 and 1.5% of the ration. The herb mix included: nettle leaf, chamomile leaf, caraway fruit and fennel fruit. This supplement was given to experimental sows from 100/day of pregnancy to 21/day of lactation. Sows behaviour was observed during parturition. On the three consecutive days before and after parturition, a total of 180 urine samples were collected to determine cortisol levels. The number and weight of piglets born and reared per litter was also evaluated. The use of 1.0% or 1.5% herb supplements in the diets reduced the stress response and shortened parturition by 280 and 360 minutes in groups II and III, respectively, compared to the control group. Cortisol levels in urine samples from the herb-supplemented sows were markedly lower than in the control group, both before and after parturition. The litters of sows of sows from experimental groups were larger and showed better survival.

Genetic relationships between meat productivity and reproductive performance in Berkshire pig

Tomiyama, M.[1], Kanetani, T.[2], Tatsukawa, Y.[2], Mori, H.[2], Munim, T.[1], Suzuki, K.[3] and Oikawa, T.[1], [1]Okayama university, Graduate School of Natural Science and Technology, 1-1-1 Tsushima-Naka, Okayama, 700-8530, Japan, [2]Okayama Prefectural Center for Animal Husbandry and Research, Misaki-cho, Kume-gun, Okayama, 709-3494, Japan, [3]Tohoku University, Faculty of Agriculture, Sendai, 981-8555, Japan; toikawa@cc.okayama-u.ac.jp

Many breeding programs were constructed including growth traits of piglet and reproductive traits of sows because those traits directly influences farmer's income. Genetic relationships between piglet growth and sows reproductive performance were often reported unfavorable. The objective of this study was to estimate the genetic parameters for meat productivity and reproductive performance in Berkshire pig. The data used in this study were collected in Japan on 4,773 purebred Berkshire (2,458 males and 2,315 females) pigs at the Okayama Prefectural Animal Husbandry and Research Center, Japan. Records of sows were collected on 564 litters from 114 dams. Additive direct and maternal genetic (co)variance components for those traits were estimated by REML using VCE5. Heritabilities for reproductive traits of sows were low. Within traits of sows, genetic correlations for the litter weight at birth (TWB) and weaning (WWL) with the litter size at birth (TNB) and weaning (TNW) were positively moderate to high. Genetic correlations of TWB and WWL with weaning weight, body weight at 60 days of age and age at 105 kg on growth traits showed favorable estimates. Genetic correlations of TWB and WWL with loin eye area and some subcutaneous fat thickness on carcass traits also showed favorable estimates. Whereas genetic correlations of TNB and TNW with growth and carcass traits were unfavorable. These results indicated that TWB and WWL should be chosen for improving reproductive performance of sows. The genetic correlations of the number of teat (TEAT) with TWB, TNB, WWL and TNW were slightly positive (0.17, 0.03, 0.06 and 0.01, respectively). Thus TEAT was independent with the litter traits of sows.

Effect of spray-dried plasma as a protein source on weanling pig performance

Polo, J.[1], Campbell, J.[2], Quigley, J.[2], Crenshaw, J.[2] and Russell, L.[2], [1]APC Europe, S.A., R&D department, Avda. Sant Julià 246-258. Pol. Ind. El Congost, E-08403. Granollers. Barcelona, Spain, [2]APC Inc., R&D department, 2425 SE Oak Tree Court, 50021. Ankeny. IA., USA; javier.polo@ampc-europe.com

An experiment with 216 weanling pigs (21 days of age at weaning; 7.5±0.7 kg BW) was conducted to evaluate the effects of two sources of spray-dried animal plasma (SDAP) on nursery pig performance compared with a control diet containing fish meal during the 14 d period post weaning. All pigs were fed a common diet from d 15 to d 35 after weaning. The three diets were balanced for energy, protein and lysine. There were 8 pigs per pen with 9 pens per dietary treatment. Comparing feed grade plasma respect to food grade plasma, the total plate count was 10^6 vs $<10^3$ cfu/g respectively and the ammonia level was 270 vs 117 ppm respectively. In the present experiment, pigs receiving diets containing either SDAP-Feed Grade or SDAP-Food Grade from d 0 to d 14 were heavier than the control (BW 10.1, 9.9 and 9.3 kg respectively. $P<0.001$) and had higher average daily food intake (ADFI 255, 254 and 215. $P<0.05$ g/d) average daily gain (ADG 196, 180 and 136 g/d. $P<0.001$) and better feed conversion ratio (FCR 1.46, 1.32 and 1.64. $P<0.001$) compared to those offered the control diet respectively. During d 15 to d 35 after weaning when pigs were fed a common diet; ADFI, ADG and FCR of pigs were not significantly different based on previous dietary treatment. There were no incidences of mortality due to diets throughout the duration of the experiment. There were no significant differences in performance between the two SDAP sources tested (Feed grade or Food grade), indicating that the observed plasma effect was due to the functionality of the product itself, and that the source of raw material had no consequence with regards to performance.

Influences of sows' activity in pre-lying behaviour patterns on the crushing of piglets

Wischner, D.[1], Kemper, N.[1], Stamer, E.[2], Hellbrügge, B.[1], Presuhn, U.[3] and Krieter, J.[1], [1]Institute of Animal Breeding and Husbandry, Christian-Albrechts-University, Olshausenstrasse 40, 24098 Kiel, Germany, [2]TiDa Tier and Daten GmbH, Bosser Strasse 4c, 24259 Westensee/Brux, Germany, [3]farm concepts GmbH & Co. KG, Heidmühlender Strasse/Moorhof, 23812 Wahlstedt, Germany; jkrieter@tierzucht.uni-kiel.de

In order to determine the influences of different posture patterns of sows on piglet losses, the present study aimed at the analysis of different traits of 'pre-lying' behaviour peri- and post partum. The behaviour of 386 German Landrace sows (with 438 pure-bred litters) in a nucleus herd was videotaped continuously starting with the beginning of farrowing until 48 hours post partum. From these animals, 40 sows were randomly sampled with a block data design. Twenty sows which crushed one or more than one piglet (C-sows) were compared to 20 sows which crushed no piglets (NC-sows). Posture patterns were analysed according to frequencies, duration and manner as well as the times of 'resting-activity' cycles of the offspring and their location to towards the sow. Each respective trait was calculated according to the difference between NC-sows and C-sows. Results showed that primiparous NC-sows performed 'sniffing' as an element of 'pre-lying' behaviour significantly more often and with longer durations (30 seconds) than C-sows. Furthermore, multiparous NC-sows 'looked around' more often, also towards the piglet nest site, before 'lying-down' by 'kneeling on front legs' ($P<0.05$). 'Nosing', often in combination with 'looking around', was significantly more frequent in primiparous NC-sows than in C-sows. The duration of the 'sleeping' and 'activity' behaviour at the mammary gland was significantly longer in piglets of C-sows than in those of NC-sows ($P<0.1$). Thus, the 'pre-lying' behaviour represents useful traits to characterise the maternal responsiveness of sows in relation to the crushing of piglets.

Effect of live weight at first breeding on reproductive and productive performances of gilts

Vouzela, C.F.M., Ferreira, G., Rosa, H.J.D., Rego, O.A. and Borba, A.E.S., University of the Azores, CITA-A, Angra do Heroísmo, 9701-851, Portugal; vouzela@uac.pt

The recent increase in costs of raw materials used in pig rations has dictated further research for high levels of reproductive efficiency. The main objective of this study was to determinate the ideal live weight of gilts at first breeding in order to obtain the best reproductive and productive performances. Three groups of LW x L x P gilts weighing 127.5±2.6 kg (A), 137.9±2.6 kg (B) and 147.2±2.6 kg (C) submitted to the same feeding, environmental and sanitary conditions were inseminated using the 'pressure method'. Gilts were inseminated three times with 24 h intervals following heat detection by the boar. Semen from one Pietrain boar was used at a dilution of 3x109 EPZ ml-1. Piglets were weighed immediately after born and weekly thereafter until weaning at 28 days of age. No significant difference was detected among groups in rates of fertility (100%, 96.6% and 93.3%, respectively for A, B and C), prolificacy (10.4, 11.4 and 11.1, respectively for A, B and C) and fecundity (95.4%, 110.4% and 104.0% respectively for A, B and C). The success rate (ratio between number of births and number of gilts detected pregnant at 21 days following insemination) was 100% in all groups. Mortality rates calculated either from born to 5 days (9.4%, 11.7% and 7.6%, respectively for A, B and C) or from 5 days to 28 days (6.9%, 10.4% and 4.1%, respectively for A, B and C) did not differ among groups ($P>0.05$). Weight of gilts at insemination had no effect ($P>0.05$) either on farrows weight or on piglets weight at born. At 28 days, farrows from group C (60.8±3.0 kg) had higher live weight than A (44.7±3.7 kg; $P<0.05$). Concerning live weight gain of farrows, differences were observed ($P<0.05$) between groups A (1.4±0.2 kg) and B (1.9±0.1 kg) and A and C (2.0±0.1 kg). It was concluded that the live weight of sows at first breeding only influenced ($P<0.05$) piglets weaning weight and mean daily weight gains of farrows, performing best the females with higher live weight.

Comparison between computerised liquid feeding and *ad libitum* dry feeding for sows during lactation
Ryan, T.P., Lynch, P.B. and Lawlor, P.G., Teagasc, Moorepark Research Centre, Pig Production Development Unit, Fermoy, Co. Cork, Ireland; tomas.ryan@teagasc.ie

The objective was to determine the effect of three feeding regimes during lactation on feed intake, sow weight change and piglet performance to weaning. At day 109 of gestation ninety sows were blocked on parity grouping (parity 1, parity 2 to 3 inclusive and parity 4 plus) and weight and allocated to treatment: A. Liquid feeding; curve 1 (25 MJ DE / day at farrowing to 98 MJ DE / day by day 21 of lactation), B. Liquid feeding; curve 2 (Curve 1 x 1.33), C. *ad libitum* dry feeding (Daltec A/S, Tybovej 1, Egtved, Denmark). Sows on curves 1 and 2 were fed twice daily a 4.1:1 mixture of feed (dry matter) to water by a computerised liquid feeding system (Big Dutchman, Vechta, Germany). The lactation diet contained 14.2 MJ DE / kg and 9.1g lysine / kg fresh-weight. The experimental curves were fed between farrowing and weaning (ca.28 days). Mean lactation feed intake was 71.2, 103.7 and 87.8 (s.e. 3.54 kg; $P<0.001$) for Treatments A, B and C, respectively. Sow back-fat loss during lactation was 3.8, 3.4 and 2.3 (s.e. 0.46 mm; $P<0.05$) for Treatments A, B and C, respectively. The number of pre-weaning deaths per litter was 0.55, 0.92, 1.66 (s.e. 0.21; $P<0.07$). Treatment had no effect on piglet weaning weight ($P>0.05$), piglet daily gain ($P>0.05$), within litter CV for weaning weight ($P>0.05$) and within litter CV for piglet daily gain ($P>0.05$).

Estimated possible and achieved the weaner production volume with the 28-day occupation period of the farrowing rooms
Sviben, M., The New York Academy of Sciences, Siget 22B, HR-10020 Zagreb, Croatia; marijan.sviben@zg.t-com.hr

When families of the pig producers became faced with problems during the globalisation crisis, in 2001 German experts showed possibilities of the development of the pig enterprise having 30 or 60 farrowing pens. It was written the suckling period could be shortened to 21 days so that 2.4 litters per sow a year would be achieved. The number of piglets weaned per sow a year would be bigger, the production costs lessened and the producer's income enlarged. To succeed the piglet producer would be obliged to increase the number of shifts per farrowing pen a year from 8.7 to 13 i.e. to shorten the occupation period of the farrowing room from 42 to 28 days. The 28-day occupation period of 8 farrowing rooms with a total of 500 pens in Croatian farm 021 was suggested in 1981, after establishing that drawn up plans for annual production of 5 millions kg of the hog live weight could not be attained. Taking into account 13.035 shifts per farrowing pen a year and 8 weaned piglets per litter, the weaner production volume of 52,140 piglets a year was estimated possible. The elaboration of the data published by Croatian Livestock Center for years 1986-1991, when in the farm 021 the swine reproduction method I was applied, resulted with the averages of 12.408 shifts per farrowing pen a year, 8.390 weaned piglets per litter and 52,055 weaned piglets annually. The method of prediction of the weaner production volume has been proven. The weaner production volume was determinated more than 60% by the number of shifts per farrowing pen a year.

To enlarge the weaner production volume increasing the number of sows or improving the exploitability of females

Sviben, M., The New York Academy of Sciences, Siget 22B, HR-10020 Zagreb, Croatia; marijan.sviben@ zg.t-com.hr

The quantity of commodities does the base of the income and wanted profit. Greater income will give greater profit, if input costs remain at the same starting level. To expand corporation managers buy the means for the work but not apply the procedures based on the achievements of technology. The statement that the design of the application of hyotechnology is an autonomous factor of production is still not accepted. Therefore the data, published by Croatian Livestock Center on the weaner production in the farm 004 from 1992 till 2006 and in the farm 016 from 1987 till 2001, were elaborated to be seen which way of enlargement of the production volume could give better prospects for the enlargement the profit but not only the income. In the farm 004 there were 264 pens in 12 farrowing rooms where 905 sows were used for 2,059 litters with 22,293 born and 17,720 weaned piglets in the starting year. During next 14 years following average numbers were achieved - sows 1,067 (+17.90%), litters 2,451 (+19.04%), litters per sow a year 2.296 (+0.92%), born piglets 25,927 (+16.30%), the survival rate 80.79% (+4.28%), weaned piglets 20,946 (+21.29%). In the farm 016 there were 360 pens in 6 farrowing rooms where 1,731 sows were used for 3,651 litters with 38,757 born and 31,341 weaned piglets in the starting year. During next 14 years following average numbers were achieved - 1,726 (-0.29%) sows, 4,006 (+9.73%) litters, 2.321 (+10.52%) litters per sow a year, 42,787 (+10.40%) born piglets, 82.32% (+1.81%) survived piglets, 35,221 (+12.38%) weaned piglets. In the farm 016 the weaner production volume was enlarged by improving the exploitability of females without costs for input females.

Multilevel approach to study boar fertility in commercial farm

Bacci, M.L.[1], Fantinati, P.[1], Alborali, G.L.[2], Zannoni, A.[1], Penazzi, P.[1], Bernardini, C.[1], Forni, M.[1] and Ostanello, F.[3], [1]University of Bologna, DIMORFIPA, Via Tolara di Sopra, 50, 40064 Ozzano Emilia, Italy, [2]IZS Lombardia and Emilia Romagna, Via Bianchi 9, 25124 Brescia, Italy, [3]University of Bologna, DSPVPA, Via Tolara di Sopra 50, 40064 Ozzano Emila, Italy; marialaura.bacci@unibo.it

Semen quality assessment represents a fundamental step for obtaining successful artificial insemination (AI) in pig industries, however the decline in boar fertility, non related to apparent causes, is a common and economically relevant problem. In commercial settings, the ejaculates were evaluated at collection, but traditional quality estimates are not able to foretell fertility outcome. New fertility parameters have been therefore studied *in vitro* (Popwell and Flowers, 2004; Turba *et al.*, 2007) and compared with traditional ones. The present research aimed to study the causes of fertility decline in boars not bound to clinical signs of disease, utilizing various approaches: study of *in vitro* fertility with traditional and new parameters, study of *in vivo* fertility and study of health status of subjects. Therefore nine boars of proven fertility have been monitored for 5 months from March and sperm and blood samples have been repeatedly collected for seminal and serological evaluations. At this level we researched ADV, PRRSV, PCV2, SIV (H1N1, H2N1, H3N2) antibodies. In order to evaluate boar fertility we utilized *in vitro* (motility, viability, acrosome condition, mitochondrial membrane potential, etc.), as well as *in vivo* parameters (Farrowing Rate and Litter Size outcome of 230 Artificial Insemination). The low percentage (<5%) of damaged acrosome in an ejaculate significantly correlates with high LS. On the contrary no correlations have been found among seroconversions for PRRSV (2 boars) and for ADV (2 boars) and *in vivo* fertility as well as positivity for SIV (H1N2 strain) (4 boars). This research was supported by grants from Bologna University (RFO 60%).

Effect of age and growth rate on intramuscular fat content of the longissimus dorsi muscle in Polish Landrace and Puławska pigs

Tyra, M., Żak, G. and Orzechowska, B., National Research Institute of Animal Production, ul. Sarego 2, 31-047 Kraków, Poland; mtyra@izoo.krakow.pl

Meat quality is not easy to define and depends on many factors. One of the characteristics of this parameter are sensory traits, including the related content of intramuscular fat. A total of 488 PL and 64 Puławska gilts, tested in a performance station were investigated. Animals were kept, fed, slaughtered and dissected in accordance with the station methods. Slaughter was conducted at 100 kg body weight. Analysis covered the content of intramuscular fat (IMF) determined using the method of Soxhlet. The IMF content was determined with regard to the rate of growth (daily intake) and age at slaughter. In terms of growth rate, animals were divided into the following groups: I – with gains exceeding 860 in the test, II – of 700 to 860 g; III – below 700 g in the test. In terms of slaughter age, group I included animals slaughtered at more than 170 days of age, II – between 140 and 170 days of age, III – animals slaughtered at less than 140 days of age. The results show that the increase in daily gains in the test is parallel to the decrease in the intramuscular fat content of the longissimus dorsi muscle of PL gilts. The differences observed between the group were significant ($P<0.01$). In Puławska gilts, no animals with mean daily gains exceeding 900 g were observed, and the difference between the animals of the other groups was statistically significant ($P<0.05$). In this breed, animals with higher daily gains also had a higher content of IMF. PL animals slaughtered at over 170 days of age were characterized by good quality of loin, as expressed by an over 2% content of IMF compared to the animals from the other groups. Similar relationships for this trait were found in Puławska animals. Results show the possibility of improving the taste of loin meat, which is the most valuable carcass cut, by controlling the age at slaughter, especially in the PL breed. However, this may produce some side effects such as increased carcass fatness.

Comparison of accuracy of intramuscular fat prediction in live pigs using five different ultrasound intensity levels

Tomka, J., Bahelka, I., Oravcová, M., Peškovičová, D., Hanusová, E., Lahučký, R. and Demo, P., Animal Production Research Centre Nitra, Hlohovecká 2, 95141 Lužianky, Slovakia (Slovak Republic); peskovic@ scpv.sk

The objective of this study was to evaluate the possibility of prediction of intramuscular fat (IMF) in live pigs using ultrasound method. The accuracy of prediction at five different ultrasound intensity levels was investigated. Cross-sectional images of longissimus dorsi muscle (LD) at right last rib area from hybrid pigs were taken. Each pig was scanned at the same frequency (3.5 MHz) and at the five different ultrasound intensity levels. The video image analysis was used to predict IMF content (UIMF70 to UIMF90). A sample of LD at the last rib was taken for laboratory analysis of IMF content (LAIMF). Correlations between LAIMF and UIMF were significantly different from zero (r=0.40-0.52), except for correlation between LAIMF and UIMF90 (r=0.14). Statistical model with LAIMF the dependent variable, UIMF and live weight the covariates, and sex the fixed effect was developed. Coefficients of determination (R2) were 0.33, 0.38, 0.34, 0.25 and 0.17 with UIMF at the intensity level 70, 75, 80, 85 and 90%. Root mean square errors (RMSE) ranged from 0.516 to 0.639%. Standard errors of individual prediction (SEP) ranged from 0.523 to 0.649%. Goodness-of-fit of the model was also justified by testing the residuals for normality. Although the results are not quite unequivocal in favour of the one intensity level, it seems that intensity levels 75 and 80% are the most suitable to predict IMF in live pigs. Further research is needed, mainly to increase accuracy of collecting, processing and evaluating the sonograms using video image analysis.

Relationships between muscle fibre characteristics and physico-chemical properties of pig *longissimus lumborum* muscle in different pig breeds

Orzechowska, B.[1], Wojtysiak, D.[2], Tyra, M.[1] and Migdał, W.[2], [1]National Research Institute for Animal Production, Department of Genetic and Animal Breeding, Sarego 2, 31-047 Kraków, Poland, [2]Uniwersytet Rolniczy, Al. Mickiewicza 24/28, 30-059 Kraków, Poland; orzech@agh.edu.pl

The aim of this study was to investigate the morphological and histochemical parameters of muscle fibres and meat quality traits of m. *longissimus lumborum* of different breed of fatteners, and to estimate the correlation of muscle fibre characteristics to daily gain, muscle pH and texture parameters. The study was carried out on 92 fatteners from three different breeds as follows: Polish Landrace (PL) (n=40), Polish Large White (PLW) (n=38) and Pietrain (n=14). Muscle samples were taken to categorize muscle fibre types (I, IIA and IIB) according to their dehydrogenase NADH activity and to determine muscle acidity (pH45, pH24). Texture properties were evaluated by Warner-Bratzler (WB) test and Texture profile analysis (TPA). Muscle fibre percentage, diameter, relative area (RA) and phenotypic correlation between muscle fibre traits and meat quality traits were estimated. The results obtained indicated that breed had no effect on meat quality traits, muscle fibre percentage and RA, but it affected the size of muscle fibres. Typically m. longissimus lumborum from Pietrain fatteners had larger diameter of all examined muscle fibre types. The phenotypic correlations between histological and physic-chemical traits were generally low. The percentage and RA of type IIB fibres, unlike that of type I fibres, positively correlated with daily gain. Moreover, increasing the daily gain is related to increasing size of type IIB fibres. Additionally, shear force, was negatively related to type IIA muscle fibre size. A similar tendency was found between pH24 and diameter of type IIA fibres.

Influence of slaughter weight on physical-chemical characteristics of dry-cured ham from heavy pigs

Rodríguez-Sánchez, J.A.[1], Ripoll, G.[1], Ariño, L.[2] and Latorre, M.A.[1], [1]CITA de Aragón, Avda. Montañana, 930, 50059 Zaragoza, Spain, [2]Integraciones Porcinas SL, C/ Portillo, 9, 44550 Alcorisa, Teruel, Spain; malatorreg@aragon.es

Teruel ham is a trademark of high quality dry-cured hams from heavy white pigs produced in a specific area of Spain. The Regulation of Denomination of Protected Origin of Teruel ham establishes some requirements to improve uniformity and quality of the end product. Among them, we can found the slaughter weight (SW) of pigs demanded (120 - 140 kg of body weight (BW)) and the minimum time of ripening process for hams ($\geq$ 14 months). A total of twelve hams were used to study the effect of slaughter weight (A: 120; B: 130; C: 140 kg BW) on ripening weight losses, color of subcutaneous fat and biceps femoris (BF) muscle and chemical composition of dry-cured product. All the hams were from Duroc x (Landrace x Large White) barrows that fed common diets. Each treatment was replicated four times. The average ripening period was 18 months. The ripening weight losses were lower for B and C hams than for A hams (38.7 vs 35.2 vs 34.6% for A, B and C, respectively; $P<0.05$) mainly due to the postsalting losses ($P=0.07$). No effect of SW was detected on subcutaneous fat color ($P>0.10$). In addition, the only color parameter of BF affected by SW was H° value being higher in C hams than in A or B hams (39.9 vs 42.7 vs 48.8 for A, B and C, respectively; $P<0.01$). Moisture content of dry-cured hams was similar for all the treatments ($P>0.10$) but the contents in sodium chloride (57.4 vs 53.0 vs 47.1 g/kg for A, B and C, respectively; $P=0.08$), potassium nitrate (2.23 vs 1.01 vs 1.23 g/kg; $P<0.01$) and sodium nitrite (153.6 vs 98.5 vs 67.8 g/kg; $P<0.01$) were higher for C hams than for A hams with B hams being intermediate. It is concluded that an increase in SW in pigs intended for Teruel ham manufacture affected slightly fat or muscle color of the ham. However, the ripening weight ham losses and the salt, nitrate and nitrite contents of the dry-cured product reduced as SW of pigs increased from 120 to 140 kg BW.

Differences between commercial barrows and gilts reared outdoor and intended for dry-cured products

Rodríguez-Sánchez, J.A.[1], Calvo, S.[1], Ripoll, G.[1], Iguácel, F.[2] and Latorre, M.A.[1], [1]CITA de Aragón, Avda. Montañana, 930, 50059 Zaragoza, Spain, [2]Centro de Técnicas Agrarias, Avda. Montañana, 176, 50080, Spain; malatorreg@aragon.es

Several studies have been conducted to characterise autochthonous pig breeds being raised in free-range production systems. However, the knowledge of the performance of modern pig genotypes in outdoors conditions is very limited and might be worth testing them in regions where no local breeds are raised. A total of seventy-four Duroc x (Large White x Landrace) pigs reared outdoor were used to study the effect of gender (B: barrows; G: gilts) on carcass characteristics. From 72.2 kg of body weight (BW) (in July) to the slaughter at 170.6 kg BW (in December), pigs were reared kept outdoor (290 m^2/animal) and had free availability of grass (Festuca glauca and Poa pratensis), shrubs (Ulex parviflorus and Lavandula angustifolia), and threes (Quercus ilex and Quercus faginea), and also a limited amount (although not quantified) of acorns. Also, animals had ad libitum access to water and a pelleted barley-corn-wheat-soybean meal concentrate (2,355 kcal NE/kg, 15.5% CP, and 0.75% total lys).After carcass measurements, a total of thirty-six carcasses, barrows and gilts at the same proportion, were randomly selected, processed and used to evaluate the yield of the main primal trimmed lean cuts (shoulders, loins, hams, ribs and bellies). No differences were found in hot carcass weight, fat depth or ham size between B and G ($P>0.10$). However, G tended to have longer carcasses than B ($P=0.09$). The weight of total primal trimmed lean cuts was higher for G than for B ($P=0.03$) mainly due to the heavier shoulders ($P<0.01$) and ribs ($P=0.07$). However, the yield of primal trimmed lean cuts was similar ($P>0.10$). It is concluded that both genders provided adequate carcasses quality after free-range management and slaughter at heavy weight. The differences in carcass characteristics between barrows and gilts were scarce but gilts had heavier shoulders and ribs than barrows at similar carcass weight.

The effect of cereal type and enzyme supplementation on carcass characteristics, volatile fatty acids, intestinal microflora and boar taint in entire male pigs

Pauly, C.[1], Spring, P.[1] and O'doherty, J.V.[2], [1]SCA, Länggasse 85, 3052 Zollikofen, Switzerland, [2]UCD, Belfield, Dublin 4, Ireland; peter.spring-staehli@bfh.ch

A 2×2 factorial experiment was conducted to investigate the effects of cereal type (barley v. oat) and exogenous enzyme supplementation (with or without) on intestinal fermentation and on indole and skatole levels in the intestinal content and the adipose tissue in Irish finisher boars. The experimental treatments were as follows: (1) barley-based diet, (2) barley-based diet with enzyme supplement, (3) oat-based diet, and (4) oat-based diet with enzyme supplement. The enzyme supplement (0.05 g/kg feed) contained endo-1,3(4)-β-glucanase and endo-1,4-β-xylanase. Animals were fed ad libitum from 76 to 114 kg. Feeding barley-based diets led to higher ($P<0.05$) volatile fatty acids concentrations in the large intestine. Proportions of propionic and butyric acids were higher and that of acetic acid lower in digesta from barley-based in comparison to oat-based diets ($P<0.001$). Consequently, pH in the large intestine was lower after feeding barley-based in comparison to oat-based diets. Animals fed unsupplemented oat-based diet had higher ($P<0.01$) indole concentrations in the digesta from the proximal colon than those fed barley-based diets. Feeding oat-based diets led to lower ($P<0.01$) skatole and higher ($P<0.001$) indole concentrations in the digesta from the terminal colon than barley-based diets. Skatole concentrations in the adipose tissue did not differ ($P>0.05$) between the experimental treatments. Pigs offered the barley-based diets had lower ($P<0.001$) indole concentrations compared with those fed the oat-based diet. In conclusion, feeding barley-based diets offered sufficient fermentable energy in the upper part of the colon to keep indole concentrations low; however, fermentable energy was not sufficient in the second part of the colon for efficiently controlling skatole. In oat-based diets, enzyme supplementation helps to limit indole synthesis in the upper part of the large intestine.

Effects of a perinatal zearalenone donation on mortality rate, weights during the suckling period and weights of reproductive organs from female piglets

Stephan, K.M.[1], Kauffold, J.[2], Bartol, F.F.[3] and Waehner, M.[4], [1]Anhalt University of Applied Science, Strenzfelder Allee 28, 06406 Bernburg (Saale), Germany, [2]University of Pennsylvania, 3800 Spruce St., Philadelphia, PA 19104, USA, [3]Auburn University, 236 Upchurch Hall, Auburn, AL 36849-5145, USA, [4]Anhalt University of Applied Science, Strenzfelder Allee 28, 06406 Bernburg (Saale), Germany; k.stephan@ loel.hs-anhalt.de

This study investigates the effect from an artificial Zearalenone (ZON) donation to gilts during the last time of gestation and the suckling period. In dependency on the mid feed intake 0.5 mg ZON per sow and day was given. Daily ZON intake rises up from 3.75 until 11.25 mg ZON per sow and day. Beside the experimental group of 8 sows a control group (CON) was arranged (n=7 sows). Diverse parameters from these 15 sows and their 161 life born piglets were collected to investigate an influence from ZON. To look for a transmission from ZON over the milk of sows the piglets were involved in a special cross-fostered system. So piglets from sows getting ZON were transferred to sows without a ZON exposition and vice versa. This generated 4 groups with piglets getting a different ZON influence. No influence from this ZON donation to the mortality rate from piglets during farrowing and the suckling period was located. Looking to the birth weights from all piglets there were no differences comparing piglets weight from ZON-sows with CON-sows found. On life day 19 there were detected higher values for piglets having ZON influence during the lactation time. After the 3 weeks suckling period all sows and selected female piglets were killed to collect the reproductive organs and the bile. There were no changes looking to the weights from uterus, cervix and ovaries from piglets with different ZON influence. In the bile from piglets, having no influence from ZON during lactation time, no ZON was found. Just α-Zearalenol (ZOL) was detected for 2 animals. Piglets, having influence from ZON-sows during lactation, accumulate ZON and also α-ZOL in the bile.

Comparison of bovine colostrum whey and defatted bovine colostrum supplementation on piglet post-weaning growth check

Boudry, C.[1], Gauthier, V.[1], Dehoux, J.-P.[2] and Buldgen, A.[1], [1]Gembloux Agricultural University, Animal Science Unit, Passage des Deportes 2, 5030 Gembloux, Belgium, [2]Faculty of Medicine, Catholic University of Louvain, Department of Experimental Surgery, Avenue Hippocrate 55/70, 1200 Brussels, Belgium; boudry.c@fsagx.ac.be

We showed previously that the incorporation of 2% of bovine colostrum (BC) whey in piglet diet reduces the post-weaning (PW) growth check. The objective of this new study is to reduce the costs of the treatment by replacing BC whey with defatted BC, a product which is 50% less expensive to produce (50 €/kg of defatted BC vs. 100 €/kg of BC whey). Ninety-six piglets weaned at 26±2 days of age (8.1 kg) were assigned to three treatments. Each group of piglets received a commercial diet supplemented with 10 g.kg^{-1} of: 1) milk ('Milk 1'), 2) defatted bovine colostrum ('Col 1') and 3) bovine colostrum whey ('Whey 1') for 10 days. Then, all the piglets received the commercial diet without any supplementation. Bodyweight and feed intake were measured two times a week for three weeks. Faecal Lactobacilli spp. and E. coli counts were determined the day before weaning and on days 2, 5 and 8 PW by real time PCR. A difference in the ADFI between the 'Whey 1' and 'Col 1' treatments was observed at the end of the first week PW (from days 4 to 7). During this period, piglets receiving the defatted BC showed a higher feed ingestion compared to piglets from the 'Whey 1' treatment (+ 26%, $P<0.05$). Concomitantly, the ADG tended also to be higher over this period (+ 23%, $P<0.1$) for the piglets receiving the defatted BC. The third week of the trial, the ADG of the 'Col 1' treatment was higher than the 'Milk 1' and 'Whey 1' treatments. No differences between the treatments were shown for the faecal flora. We may conclude from these results that defatted bovine colostrum is at least as good as BC whey to reduce post-weaning growth check, allowing to reduce the costs of the treatment by 50%.

Effect of bovine colostrums of 1st, 2nd and 3rd milking on growth performance and the immune system of newly-weaned piglets after an *E.coli* LPS challenge

Gauthier, V., Boudry, C. and Buldgen, A., Gembloux Agricultural University, Animal Science Unit, Passage des Deportes, 2, 5030 GEMBLOUX, Belgium; boudry.c@fsagx.ac.be

The aim of this work is to evaluate the influence of post-weaning (PW) diet supplementation with freeze dried bovine colostrums (BC) from 1st, 2nd or 3rd milking on growth performance and the response of the immune system of piglets. This is carried out in a context of an *E.coli lipopolysaccharide* (LPS) challenge. The experiment was performed on 100 newly-weaned piglets distributed between five treatments (Control-, Control+, Col1, Col2 and Col3). Control groups received a PW diet without any supplementation while colostrums groups received the same PW diet supplemented with 1% of defatted freeze-dried BC from the 1st (Col1), 2nd (Col2) or 3rd milking (Col3), until the 12th day PW. Then, all the piglets received the same PW diet without any supplementation. On day 5 PW, piglets of 'Control+', 'Col1', 'Col2' and 'Col3' treatments were injected IM whit 100 µg LPS/kg BW while 'Control -' received a solution of PBS. Average daily gain and feed ingestion measures were completed with blood analyses (IgA, IgM, IgG, INF-γ, TNF-α, IL-10). BC supplementation induced an increase of ADG in the 'Col1' group compared to both 'Col2' and 'Col3' groups before LPS injection. The LPS challenge induced severe skin inflammations, a decrease of growth performance for 20 days, and important changes of all measured blood parameters 3h post-injection. IgA and IgG concentrations where significantly higher on day 13, and IgM on day 9 and 13, compared to before injection in the 4 LPS treated groups. BW, ADG and ADFI of both 'Col1' and 'Col3' were higher than other two challenged groups at the end of the trial. In conclusion, the severe effects of LPS masked the potential benefits of BC on ADG and ADFI until day 20. However, higher BW, ADG, and ADFI of both 'Col1' and 'Col3' observed at the end of the trial suggest a restoration of gut damages promoted by the BC growth factors and a similar effect of 1st and 3rd milking BC on growth performance.

The effects of the probiotic containing *Bacillus licheniformis* and *Bacillus subtilis* spores on the health and productivity of growing pigs

Juskiene, V., Leikus, R., Juska, R. and Norviliene, J., Institute of Animal Science of LVA, R. Zebenkos 12, Baisogala, LT-82317, Lithuania; Violeta@lgi.lt

The purpose of this study was to investigate the influence of a probiotic on pig health, growth intensity, feed consumption and carcass quality. The study was conducted at the LVA Institute of Animal Science with two groups of German x Norwegian Landrace crossbreed pigs. The pigs in the probiotic groups were fed diets containing 0.04% probiotic (*Bacillus licheniformis* - $1,6 \times 10^9$ CFU/g, *Bacillus subtilis* - $1,6 \times 10^9$ CFU/g) additive from 42 to 120 days of age. From day 121 to the end of fattening, the probiotic was excluded from the diet of this group of pigs. The study indicated that 0.04%probiotic supplementation of the diet resulted in higher health indicators of pigs, i.e. the number of diarrhea cases was by 18%lower, and the average treatment length was 35.3 and even 44.6% ($P=0.053$) shorter at, respectively, older and younger ages. The study also indicated that there no statistically significant differences between the groups during the whole growing period. As regards pig growth rate younger age (42-91Days), there was a tendency towards higher growth intensity in the probiotic groups. The study also indicated that pig feeding with the probiotic had no significant influence on the carcass quality.

Dynamic readings of electronic identification devices in pigs

Santamarina, C.[1], Hernández-Jover, M.[2], Casaponsa, J.[1], Lopez, N.A.[1], Caja, G.[2] and Babot, D.[1], [1]Universitat de Lleida, Produccio Animal, Rovira Roure 191, 25198 Lleida, Spain, [2]Universitat Autonoma de Barcelona, Ciencia Animal i dels Aliments, Campus Bellaterra, 08193 Barcelona, Spain; clara@prodan.udl.cat

A total of 564 piglets from a commercial farm were electronically identified to compare the use of different stationary transceivers under on-farm conditions. Two types of identification devices from the two methodologies of information exchange (HDX, Half-Duplex; and, FDX, Full-duplex) were used: electronic ear tags (n=268) and injectable transponders (n=296). Ear tags were applied in the right ear of the animals and injectable transponders were applied in intraperitoneal position. Dynamic readings were conducted with four stationary transceivers from different commercial manufacturers. Pigs were moved through a corridor in both possible traffic directions in relation to the antenna's position. Results of each stationary transceiver varied according to the type of identification device, methodology of information exchange, body position of the transponder and traffic direction of the animals. Moreover, the behaviour and performance of the stationary transceivers with each of the identification devices differed between them. While three stationary transceivers showed the best results of reading frequency with only one specific identification device, there was one stationary transceiver that showed similar results between the different types of tested identification devices. Reading frequency, defined as [(number of read transponders / number of readable transponders) x 100], ranged between 0 and 100%. These results indicate that the selection of the most suitable combination between identification device and stationary transceiver is crucial to ensure efficient results of dynamic readings of electronically identified pigs under on-farm conditions.

Comparison of beef cattle and sheep production profitabilities in the Czech Republic

Milerski, M. and Kvapilík, J., Research Institute of Animal Science, Přátelství 815, 104 00, Prague 10 - Uhříněves, Czech Republic; m.milerski@seznam.cz

Permanent grassland covers nearly one million hectares, what is about 23% of the farmland in the Czech Republic and the process of transformation of some arable land to grassland is going on. The farmers newly involved in grassland management usually decide between using suckler cows or sheep for grazing. The aim of this study was to compare the economic efficiencies of beef cattle vs. sheep low input production systems. Profitability of both production systems is significantly influenced by subsidies from national or EU budgets. Before the Czech Republic joined EU (1.5.2004) sheep were in better position because of government subsidies per ewe in less favorable areas. In years 2005-2007 beef cattle production system benefited because of 30% increasing of prices of exported weaned beef calves caused by slaughter premiums and beef export refunds in some countries of EU-15. In 2008 different supplementary Top-Up payments per livestock unit (LU) of grazed ruminants were implemented, whereupon the payment per LU in beef cattle is by 53% higher than in sheep. This disproportion stems from EU regulations and negatively influences the perspectives of the sheep industry development in the Czech Republic.

New tools for the pastoral advising: GPS for grazing dairy Corsican goats

Bouche, R.[1], Aragni, C.[2], Duba, G.[1], Guéniot, F.[1] and Gambotti, J.[1], [1]INRA, SAD LRDE, Qtier Grossetti, 20250, France, [2]CRA Corse, Cl Ferracci, 20250, France; bouche@corte.inra.fr

In front of ecological intensification, the rangeland becomes again a strategic resource for the breeding of the dairy goats in Mediterranean mountains. But, these contributions remain difficult to estimate for breeders and their advisers. The areas covered by animals and the feeding value of numerous plants vary greatly depending on the season. In Corsica, with the aim of supplying new tools for the advice in traditional breeding, the researchers use the technologies of GPS to measure the movements and the behavior of animals on the territory. With a base of 25 GPS (for 200 goats/herd), they led a protocol in 4 stages: (1) Conception of Data collection. Choice of the equipment, fixation (horn, necklace), parameter of reception. (2) From Data to Goat. The following of Corsican goats all year round, on 25 herds situated in different regions (sea, mountain) enable to characterize their movements. How much km, at what speed, how occurs the sequences of different phases: walk, pasture or rest? (3) From Goat to Herd. A work linking videos observations, GPS measures and algorithms, allows to study and to formalize the occupation of the territory by the herd (forms, gregariousness, influence of guarding, topography,vegetation). (4) From Herd to Adviser's tools. A precise follow-up every 15 days of the productive practices and shepherd's know-how in 2 breedings allows to think the use of this technology between the adviser and the breeder to define the ration, establish zones of pastures. For all of these points, interesting results allow to envisage new manners to approach the advice in pastoral breeding of the small dairy ruminants (qualification and quantification of rangeland). The displaying to the breeders under a 3D format (like video game) allows approaching new questions compared to the traditional know-how.

Evaluation of two muscling QTL in sheep using ultrasound (US), computer tomography (CT) and video image analysis (VIA)

Macfarlane, J.M.[1], Lambe, N.R.[1], Rius-Vilarrasa, E.[1], Mclean, K.A.[1], Masri, A.[1], Haresign, W.[2], Matika, O.[3], Bishop, S.C.[3] and Bunger, L.[1], [1]SAC, West Mains Road, Edinburgh, EH9 3JG, United Kingdom, [2]Institute of Biological, Environmental and Rural Sciences, Aberystwyth University, Ceredigion, SY23 3AL, United Kingdom, [3]The Roslin Institute and R(D)SVS, University of Edinburgh, Roslin, EH25 9PS, United Kingdom; Lutz.Bunger@sac.ac.uk

The Texel muscling quantitative trait locus (TM-QTL), identified at the telomeric end of chromosome 18 in purebred UK Texel sheep, was originally reported to increase US measured muscle depth (M. *longissimus lumborum*, MLL) by around 4 to 7%. The LoinMax®-QTL (originally called Carwell) maps to a region near TM-QTL and callipyge. It was found to segregate in Australian Poll Dorset and was then introgressed into Poll Dorset in New Zealand, where it has been reported to increase MLL weight and area by 7 and 10%, respectively. The objective of this study was to comprehensively evaluate both QTL to determine whether they could benefit the UK sheep industry, through increased meat yield from crossbred slaughter lambs typically produced in this stratified system. Effects of these QTL on a range of carcass traits, including those measured *in vivo* and post-slaughter, were evaluated at 16w (LoinMAX®) or 20w (TM-QTL) of age in heterozygous carrier and non-carrier lambs, produced by crossing heterozygous carrier rams with non-carrier Mule (Bluefaced Leicester x Scottish Blackface) ewes from a lowland flock. Both TM-QTL and LoinMAX® significantly increased: US measured MLL depth, by 1.0 mm (4.53%)and 1.2 mm (4.98%), respectively; CT measured MLL width, by 2.2 mm (3.0%) and 1.1 mm (1.6%), respectively; and CT measured MLL area, by 0.9 cm^2 (5.14%) and 1 cm^2 (6.35%), respectively. QTL effects on a few other muscling, carcass dissection and VIA traits were also significant ($P<0.05$). These results indicate that both QTL have similar phenotypic effects and mainly affect muscling in the highly-priced loin region, suggesting that both QTL are of value and could be alleles of the same locus.

Identification of QTL associated with gastrointestinal nematode resistance in Creole goat

De La Chevrotière, C.[1], Bishop, S.[2], Moreno, C.[3], Arquet, R.[4], Bambou, J.C.[1], Schibler, L.[5], Amigues, Y.[6] and Mandonnet, N.[1], [1]INRA UR 143 URZ, Domaine Duclos, 97170, Petit-Bourg, France, [2]Roslin Institute, Roslin, EH25 9PS, Midlothian, United Kingdom, [3]INRA SAGA, Castanet-Tolosan, 31326, France, [4]INRA UE 1294 PTEA, Domaine de Gardel, 97160, Le Moule, France, [5]INRA Laboratoire de Génétique Bioch. Et de Cytogénétique, Jouy-en-Josas, 78252, France, [6]GIE LABOGENA, Jouy-en-Josas, 78352, France; claudia.chevrotiere@antilles.inra.fr

This study aimed to identify regions of the genome affecting resistance to gastrointestinal nematode in a Creole goat population continually exposed to a mixed nematode infection by grazing on irrigated pasture. A genome-wide QTL scan was performed on 383 offspring from 12 half-sib families. A total of 104 microsatellite markers were genotyped. Traits analysed using interval mapping were faecal egg counts (FEC), packed cell volume (PCV), eosinophils counts and bodyweight (BW) at 7 and 11 months of age. Levels of IgG, IgA and IgE anti-Haemonchus contortus L3 crude extracts and adult excretion/secretion products (ESP) were also analysed. This preliminary study identified 10 QTL associated with parasite resistance. QTL associated with FEC were found on chromosome 22 and 26. Three QTL were detected on chromosome 7, 8 and 14 for eosinophils counts. QTL associated with PCV were identified on chromosome 5, 9 and 21. QTL associated with BW at 7 months of age was found on chromosome 6. Lastly, QTL associated with level of IgA anti-ESP was found on chromosome 12. Analysis of remaining immunological traits is ongoing. Further studies with additional markers and animals will also be performed to confirm these first results in goat. Nonetheless, in comparison with sheep and taking into account chromosomal homology, most of the QTL detected in this study are located on chromosomes where QTL sheep were also found. The two QTL detected on chromosome 14 and 26 have no equivalent in sheep and could be specific to goat.

PRNP gene polymorphisms in Polish population of East: Friesian milk sheep

Lityński, R., Niżnikowski, R., Popielarczyk, D., Głowacz, K. and Strzelec, E., Warsaw University of Life Sciences, Division of Sheep and Goat Breeding, Nowoursynowska St 166, 02-787 Warszawa, Poland; roman_niznikowski@sggw.pl

The research was carried out on the flock of East-Friesian Milk sheep in Poland to genotype the prion protein gene (PRNP) polymorphisms. Blood samples were collected from 108 ewes and 7 rams and the DNA was isolated by the method of chromatography on silica mini-columns (A&A Biotechnology). The investigation of polymorphisms in the PRNP gene was done by the KASPar® technology. Frequencies of alleles ARR, ARQ and AHQ were 0.335, 0.391 and 0.274, respectively. Six genotypes of PRNP gene were observed. The ARR/ARR genotype was the most valuable at statistically significant level and the frequency of ARR allele was higher in rams in contrary to ewes what may be important in case of its possibly high frequency in offspring. It may be confirmed by the low frequencies of ARQ and AHQ alleles ($P \leq 0.05$) observed in rams in contrary to the ewes. No differences in the frequencies of other genotypes (exept ARR/ARR) due to the sex of animals were observed. The breeding work should be continued to increase the frequency of ARR allele. The selection of best rams with the ARR/ARR as the flock rams as well as the elimination of ARQ allele in flock ewes are prefered.

Inference of genotype probabilities and derived statistics for PrP locus in sheep

Gorjanc, G. and Kompan, D., University of Ljubljana, Biotechnical Faculty, Animal Science Department, Groblje 3, 1230 Domzale, Slovenia; gregor.gorjanc@bfro.uni-lj.si

The aim of our work was to infer PrP genotype probabilities in sheep to provide additional genotype identifications. The pedigree data consisted of 10,429 animals of Jezersko-Solcava sheep breed with 3,669 animals having PrP genotype data. There were 2,673 live non-genotyped animals. Five PrP alleles were present with the following frequencies: ARR 0.174, AHQ 0.074, ARH 0.083, ARQ 0.632, and VRQ 0.037. Iterative allelic peeling with incomplete penetrance model as implemented in the GenoProb program was used for inferring the genotype probabilities. There were only some additional identification of PrP genotype and national scrapie plan (NSP) type with high probability. We maintain that the main reasons for a low number of additional identifications can be attributed to the large number of alleles with moderate frequencies, incomplete penetrance model, uniform prior, and inherent pedigree and genotype data structure. In order to overcome the limits of additional genotype identifications we derived novel statistics (maximal NSP type, average NSP value and its variance and accuracy) to facilitate practical implementation of selection for scrapie resistance based on NSP types. Maximal NSP type can be used to infer maximal potential scrapie susceptibility of individual animals as well as for the entire flocks. The average NSP value encompasses all information contained in PrP genotype probabilities and is the most useful statistic for the selection on NSP type and therefore PrP genotype. These novel statistics can be used as a criterion for the selection against scrapie susceptibility for the whole population taking into the account the possible errors in the genotype and/or pedigree data.

Milk-fed kid (cabrito Transmontano) meat physicochemical quality

Rodrigues, S., Pereira, E. and Teixeira, A., Escola Superior Agrária, Instituto Politécnico de Bragança, Campus Sta Apolónia Apt 1172, 5301-855 Bragança, Portugal; teixeira@ipb.pt

This work aims to present some meat quality characteristics of a Protected Origin Designation product, cabrito Transmontano. Effects of sex and carcass weight were studied. Meat pH, colour characteristics and Warner-Bratzler shear force was evaluated in 60 milk-fed kids grouped by sex (males and females) and carcass weight (4, 6 and 8 kg). Carcass pH was measured one and 24 hours after slaughter in the 12th-13th ribs, using a portable pH meter. Meat colour was assessed by L*a*b* system using a colorimeter in the longissimus muscle at the 12th-13th thoracic vertebra. Hue and chroma parameters were also calculated. Texture was evaluated in cooked meat with an Instron press equipped with a Warner-Bratzler cell. Maximum load of shear force was measured in kgf. Results indicate that Transmontana milk-fed kids, at weight ranges determined by POD, did not show marked differences between sexes in meat physicochemical characteristics. Differences between carcass weights are more obvious. With carcass weight increase, the studied animals' meat became less luminous and of more vivid red colour and pH at 24 hours after slaughter decreased. At the same degree of maturity males and females are more similar than when compared at the same carcass weight. Generally, meat physicochemical characteristics were not correlated since only correlation coefficients between colour parameters were high and significant.

Lamb carcass and meat quality characteristics of improved and indigenous sheep breeds of north-western Turkey under an intensive production system

Yilmaz, A.[1], Ekiz, B.[1], Ozcan, M.[1], Kaptan, C.[2], Hanoglu, H.[2], Erdogan, I.[2], Kocak, O.[1] and Yalcintan, H.[1], [1]Istanbul University, Veterinary Faculty, Department of Animal Breeding and Husbandry, Avcilar, 34320 Istanbul, Turkey, [2]Marmara Livestock Research Institute, Bandirma, 10230 Balikesir, Turkey; yalper@istanbul.edu.tr

The aim was to determine the lamb carcass and meat quality characteristics under an intensive production system of sheep breeds of north-western Turkey and to produce knowledge for the breed choice in intensive production in this region. After weaning at approximately 85 days of age, 46 lambs from Turkish Merino, Ramlic, Kivircik, Chios and Imroz breeds were fattened for 56 days. Slaughter weights were 47.39 kg, 45.68 kg, 47.27 kg, 31.08 kg and 29.82 kg and chilled carcass weights were 23.35 kg, 22.33 kg, 23.51 kg, 14.33 kg and 13.75 kg, respectively. Improved Turkish Merino and Ramlic and indigenous Kivircik lambs had higher quality carcass production than indigenous Chios and Imroz lambs. Chios lamb carcasses had the highest tail root fat yellowness and tail weight. Turkish Merino lambs had high slaughter and carcass weights, shoulder and long leg percentages and lean and lean/total fat ratios in the leg and produced less fat than Ramlic and Kivircik lambs in certain carcass parts, showing that Turkish Merino can be used to obtain high quality lamb carcasses in intensive fattening in this region. Breed had no significant effect on pH at 45 min and 24 h post mortem, water holding capacity and cooking loss. Kivircik and Imroz lambs had lower Warner Bratzler shear force values than those of Ramlic and Turkish Merino lambs ($P<0.01$). Meat samples from Kivircik lambs had the highest redness value. Sensory tenderness scores given to meat samples of Kivircik lambs were significantly higher ($P<0.01$) than those of Turkish Merino, Ramlic and Imroz lambs. Among indigenous breeds Kivircik, which had high carcass quality close to those of improved breeds, can also be considered for production of better quality meat in north-western Turkey.

Evaluation of season effects on performance of Egyptian Zaraibi dairy goats

Salama, A.A.K.[1,2], Shehata, E.I.[1], El Shafei, M.H.[1], Hamed, A.[1], Caja, G.[2], Abdel Hakeem, A.[1], Abou-Fandoud, E.I.[1] and Such, X.[2], [1]Animal Production Research Institute, Sheep&Goat Dept, 4 Nadi El-Said, 12311 Dokki, Giza, Egypt, [2]Universitat Autònoma de Barcelona, G2R, Campus de la UAB, 08193, Spain; ahmed.salama@uab.cat

The Zaraibi (Egyptian Nubian) dairy goats, origin of the Anglo-Nubian breed, are common in the Nile Delta where are appreciated by their heat tolerance, prolificacy and milk yield (cheese production). The aim of this study was to evaluate the impact of milking season (summer vs. spring) on milk yield of Zaraibi goats under experimental farm conditions in Damietta governorate (Egypt). Summer group (53 goats; 36 primi- and 17 multiparous) kidded on March 11, lactated for 262 d (suckling, 95 d; milking, 167 d) and were fed with rice straw and concentrate. Spring group (68 goats; 36 primi- and 32 multiparous) kidded on November 9, lactated for 272 d (suckling, 95 d; milking, 177 d) and were fed with fresh Berseem clover, rice straw and concentrate. Machine milking was done twice daily and milk yield was recorded weekly. Despite the differences in environmental and feeding conditions, prolificacy (1.98±0.06 kids/goat) and litter weaning weight (18.8±0.7 kg BW) did not vary ($P>0.05$). Estimated milk yield during suckling was 1.04 l/d. Difference in daily milk yield between goat groups was moderate (13%; $P<0.07$); the spring goats tending to produce more milk than the summer goats (805±37 vs. 701±49 ml/d). Parity and prolificacy did not affect milk yield, and no interaction between parity and season was detected. On the contrary, interaction between lactation week and season was significant ($P<0.01$), and milk yield decreased as lactation advanced in the summer goats, whereas persistency was greater in the spring goats in which milk yield returned to rise at wk 20-23 of lactation (spring arrival). In conclusion, season effects on lactational performance of Zaraibi goats seem to be small and multifactorial (temperature, photoperiod and nutrition). Oriented research is needed to show the adaptability of Zaraibi goats to heat stress conditions.

An analysis of goat breeding on a selected dairy farm

Toušová, R., Stádník, L. and Krejčová, M., Czech University of Life Sciences Prague, Department of Animal Husbandry, Kamýcká 129, 16521, Prague 6 - Suchdol, Czech Republic; stadnik@af.czu.cz

The objective of this work was to evaluate zootechnical conditions in a selected goat dairy farm and subsequently focusing on the processing of goat milk. Czech White Shorthaired Goats (n=87) and Czech Brown Shorthaired Goats (n=6) are bred on this particular farm. Goats are kept on pasture during the summer and in the stable during the winter. The basis of the feeding ration consists of pasture, with the addition of oats, during the summer period. A hay, turnip, and production mixture is fed during the winter period. The fertility of the goats is about 180%, and the share of reared goatlings is 170%. The daily milk yield of the entire herd was 51 kg, ranging from 38 kg to 56 kg in individual months. The daily average milk yield per goat was 1.5 kg, while the highest daily milk yield was 4.3 kg. The average milk production per lactation was 406 kg of milk with a content of 3.09% fat, 2.73% protein, and 4.8% lactose in Czech White Shorthaired Goats and 471 kg of milk with content of 2.93% fat, 2.63% protein, and 4.7% lactose in Czech Brown Shorthaired Goats. From 70 to 100 kg of milk products – especially fresh cheese, goat milk, and kefír – are produced from the milk during one week.

Research regarding the situation of sheep size exploitations and sheep breed structure in Romania

Raducuta, I. and Ghita, E., University of Agronomical Sciences and Veterinary Medicine, Faculty of Animal Sciences, Street Marasti no. 59, district 1, 011464, Bucharest, Romania; elena.ghita@ibna.ro

The purpose of this work is to investigate the situation of sheep size farms and sheep breed structure in Romania after the integration in EU. At present, in Romania the total number of sheep exploitations is extra-large (479,972 units) compared to that of the old EU member states (France, Germany, UK, etc.), which is due to the fact that there are still a few large farms (6,377 units with over 100 head/unit) and many units where the number per herd is very small (377,811 units with 1-10 head/unit). In these last units, the sheep are kept only for family self-consumption. As regards the sheep breed structure there are five breed classes in Romania which detain in order the following percentage from the globally sheep livestock: Tsurcana (52.4%), Tzigaia (24.3%), Merinos (9.0), Crossbreeds (8.5%), Karakul (5.4%) and Other breeds (0.4%). In the last class are listed breeds which were imported in the recent years for improving the morpho-productive parameters of our local breeds such as Lacaune, Friesian, Texel, Suffolk, Bluefaced Leicester, Ile de France, Merinofleisch. From this situation it is pointed out that the Tsurcana breed has decreased from 65.0% (2003 year) to 52.4% (2008 year) of the total sheep livestock in Romania, the difference being taken mainly by the crossbreeds which have a superior rate of yield than the belated breed Tsurcana.

The goat welfare: consumer viewpoints

Alcalde, M.J.[1], Perez-Almero, J.L.[2], Ortega, D.[3], Diaz-Merino, A.[1] and Rodero, E.[3], [1]Univ Seville, Ctra Utrera Km1, 41013 Seville, Spain, [2]IFAPA, Las Torres, 41200 Alcala del Rio (Sevilla), Spain, [3]Univ Cordoba, C. Rabanales, 14014 Cordoba, Spain; aldea@us.es

A survey (n=125) on consumer's shopping habits regarding kid meat was conducted in South Spain. Its consumption is occasional in 40% of the them and another 40% do not consume this meat at all. It is considered a luxury and expensive meat (74%), and hard to find (71%), but healthy (89%), nutritive (90%), easy to cook (62%), with good sensory characteristics (75%) and animal friendly production (70%). When purchasing food, nutritional composition is not taken into account by 36% of the consumers and 63.6% do not care about the animal welfare-friendly meat. The incentive to buy is based on personal taste for the product (94%), the product appearance (72%) and on habit (46%), and it is not based on advertising (46%). However, consumers think that an advertising campaign could increase their consumption of kid meat (88%). Still, more information is needed on animal welfare conditions (97%), as a higher demand for animals farmed in welfare-friendly ways would improve their own welfare (93%) linked to better meat quality, and would give them more assurance and confidence (91%).

Regional sheep products

Kawęcka, A. and Sikora, J., National Research Institute of Animal Production, St. Krakowska 1, 32-083 Balice, Poland; akawecka@izoo.krakow.pl

Regional products are manufactured only in some regions of the European Union. They are bearing a protected trademark and the technology of their production is protected. Bryndza podhalańska (Podhale sheep cheese) is the first product from Poland to receive EU protection and join the group of products with a protected designation of origin (PDO), as a result of which it can be marked with an original product label. Bryndza podhalańska is a traditional product of Carpathian shepherds. Its production begins after milking, when warm milk is mixed with rennet, a milk-curdling enzyme. Curdled cheese mass is crumbled using a wooden mixer and then manually collected to mould a lump. It can be hung in a linen cloth to drain resulting in bundz – a delicate, sweet-flavoured cheese that can be used to make bryndza cheese after 1-2 weeks of maturation. At first, bundz is crumbled, moulded by hand into a uniform lump with the addition of 2-3% salt, and then pressed in wooden dishes to obtain uniform plastic mass. Oscypek cheese is the second Polish product registered as a regional product in the European Union. The initial stages of production are the same as for bryndza until obtaining the cheese lump, which is then repeatedly compressed and scalded in hot water. The cheese is formed when an internally carved ring is placed round the cheese, and the parts remaining outside are pressed by hand into conical shape, giving the whole a characteristic spindle-like appearance. It is then pierced with a skewer, soaked in brine for about 24 hours and after drying it is smoked on a chalet shelf over a fire. Smoked for 3-14 days, it acquires unique taste and shiny colour, ranging form yellow to light brown. Oscypek is a seasonal product. Milk used for its production comes exclusively from Polish Mountain Sheep. It is allowed to add cow's milk (40% at most) coming exclusively from Polish Red cows. The oscypek production season lasts from May to September because of the limited availability of sheep milk during the other months.

Importance of local sheep breeds
Sikora, J. and Kawęcka, A., National Research Institute of Animal Production, St. Krakowska 1, 32-083 Balice, Poland; jsikora@izoo.krakow.pl

Native breeds of sheep are very well adapted to local environmental conditions, undemanding in feed and highly resistant to adverse living conditions. The beneficial effect of native breeds on landscape architecture and conservation, especially in poor biotopes, enables their use as an alternative factor in environmental protection. Sheep also play an important role in rural tourism as a component of landscape and folk culture and supplier of many valuable products. To save local breeds from extinction and preserve valuable characteristics in the population, they were included in the sheep genetic resources conservation programme, while the national agri-environmental programme provides breeders with financial support. The multipurpose sheep known as Podhale Zackels accompanied man during the period when wild Carpathian areas were being settled becoming a permanent feature of highlander economy and culture. Coloured Mountain Sheep were kept by mountaineers due to their coloured wool used to produce regional dresses and decorative elements. Polish Corriedale sheep are characterized by good milk yield. Kamieniecka sheep produce uniform thick wool of good quality. Coloured Merino sheep show good slaughter performance and provide untypical, coloured thin wool. Olkuska sheep are characterized by high prolificacy, good milk yield and strong maternal instinct. Pomeranian sheep are docile and not skittish, making good use of pastures and producing exceptionally flavoursome meat. Świniarka sheep yield wool that is useful for carpet. Lowland sheep (Żelaźnieńska and Uhruska) give medium-thick wool of good quality and lambs with well-muscled carcasses. Wielkopolska sheep produce high-quality wool while showing good rate of growth and feed conversion when fattened. Wrzosówka sheep are fur-coat sheep giving skins of excellent quality. Wrzosówka meat has an exquisite taste resembling that of roe deer meat. Old-type Polish Merino is the precursor of the whole group of Merino sheep characterized by wool of excellent quality.

Restoration of the Carpathian goat in Poland
Sikora, J. and Kawęcka, A., National Research Institute of Animal Production, St. Krakowska 1, 32-083 Balice, Poland; jsikora@izoo.krakow.pl

The Carpathian goat is an old local breed found in the 19th century in the Polish Carpathian Mountains. The intensification of agriculture and breeding and the associated replacement of local breeds by more productive breeds has made the Carpathian breeds almost extinct. In 2005, small herds of Carpathian goats were found in Poland. The purchased goats, their offspring and herd bucks were moved to a farm belonging to the National Research Institute of Animal Production. Carpathian goats have harmonious body conformation with normal udders. The goats have shapely heads, long necks and thin horns directed upwards and backwards. They have beards and often characteristic wattle on the neck. Ears are long, narrow and lively. Many males and females have a characteristic fringe above the eyes. The trunk is well built, with an even back and sloping hindquarters. Bucks have wide horns twisted in a characteristic clockwise spiral as well as an abundant mane and beard. Carpathian goats are medium-sized animals. Females have a withers height of 60 cm and weigh 30-35 kg. The respective values for bucks are 70 cm and 50 kg. Carpathian goats are white with a semi-long hair coat and occasional down undercoat, which splits on mid-back, evenly falling on both sides of the trunk. Guard hair length averages 20.75 cm in goats and 31.0 cm in bucks. Daily milk yield at peak lactation exceeds 3 litres with lactation yield averaging 470 kg. The milk obtained contains 2.8% protein and 3.4% fat on average. Sexual maturation is early with prolificacy of 150-160%. This breed is resistant to adverse environment and very well adapted to living in submontane climate. It performs well in small backyard herds. Carpathian goats are very docile and friendly, which makes them suitable for agritourism farms. As regional and organic products, the cheese, milk, meat and skins obtained from Carpathian goats can the mark of the region and make its tourist offer more attractive.

Valorisation of the genetic diversity of the local sheep breeds in Romania, to produce quality carcasses according to market demands

Ghita, E., Pelmus, R., Lazar, C., Rebedea, M. and Voicu, I., INCDBNA Balotesti, Animal Biology, Calea Bucuresti no 1, 077015, Romania; elena.ghita@ibna.ro

In Romania, the suckling lamb meat is the preponderant type of production demanded on the market for sheep products. Carcass quality in suckling lambs depends on many factors such as breed, slaughtering weight, sex, feeding and weaning age. The purpose of the paper is to study the influence of the sheep breed on the carcass quality of the suckling lambs from three local breeds: Carabash, Tsigai and Tsurcan. The lambs (total number = 45) were weaned and then slaughtered close to the Easter. The Carabash lambs (n=15) had at 49 days the weight of 19.16±0.775 kg, the Tsigai lambs (n=15) had at 66 days the weight of 19.14±0.471 kg and the Tsurcan lambs (n=15) had at 81 days the weight of 17.72±0.519 kg. During the experiment we determined the slaughtering and commercial yield, proportion of body parts after slaughtering, proportion of the commercial parts of the carcass, meat to bone ratio for each region, specific carcass measurements, carcass grading, evaluation of the organoleptic characteristics and chemical composition of the meat. The experimental results have shown that the Carabash lambs have the carcass conformation closer to the conformation specific for meat breeds, the commercial yield was 54.38±0.624 and the highest fat percentage in the meat. Carcass grading in terms of fat was median to high, although the slaughtering age was much lower. The Tsurcan lambs were the latest, have the leanest carcasses, but the organoleptic traits are the best and the commercial yield was 47.42±0.429. Tsigai lambs are in between for most studied traits (the commercial yield was 51.68±0.708). Data analysis shows clearly that Carabash lambs are best suited for the production of suckling lambs because they have higher lambing weights, they gain more in weight during suckling and carcass characteristics are better than in other surveyed breeds.

Biodiversity characteristics of goats found in Poland based on microsatellite markers

Kawęcka, A. and Sikora, J., National Research Institute of Animal Production, St. Krakowska 1, 32-083 Balice, Poland; akawecka@izoo.krakow.pl

Goat breeding in Poland has a very long tradition but is still incidental to livestock production. The goat population is estimated to be 176,000. The greatest proportion of the national population is represented by Polish White Improved goats, derived from local and Saanen goats as well as from German White Noble goats. Less numerous are Fawn Improved goats derived from German Fawn and Alpine goats, and imported breeds such as Saanen, Alpine and Anglo-Nubian. There is also a large group of scrub goats, which are classified as the general-purpose type, vary widely in appearance and production traits, but are extremely resistant to adverse environmental conditions. The Polish goat population was shaped by old local breeds of the general-purpose type, such as Carpathian and the extant breeds Sandomierska and Kazimierzowska. Goats are most often kept in backyard systems, in large specialized herds or for hobby purposes. Dwarf goats, considered as representatives of feral goats, are kept in zoological gardens. The aim of the present study was to characterize the genetic structure of goats found in Poland based on microsatellite DNA polymorphism. Analysis was performed on samples of blood obtained from 215 goats of six breeds: Saanen, Anglo-Nubian, Alpine, Carpathian, White Improved and Dwarf. The polymorphism of microsatellite DNA sequences was analysed based on selected markers recommended by FAO for evaluation of goat biodiversity. Separation of the amplification products of the sequences analysed in the population studied helped to identify 47 polymorphic variants in 6 loci. The mean number of alleles per breed was the highest in Saanen (5.7) and the lowest in Dwarf goats (4.1). Heterozygosity was the highest in White Improved (0.7) and the lowest in Anglo-Nubian goats (0.5). Genetic differences between the breeds were calculated using the Unweighted Pair Group Method (UPGM). The greatest similarity was observed between Saanen and White Improved. The largest genetic distance was found between Carpathian and Dwarf breeds.

Use of veterinary antimicrobials for mastitis in small ruminants in Spain

Berruga, M.I.[1], Licón, C.[1], Rubio, R.[1], Lozoya, S.[1], Molina, A.[1] and Molina, M.P.[2], [1]IDR-ETSIA. Universidad de Castilla-La Mancha, Departamento de Ciencia y Tecnología Agroforestal y Genética, Campus universitario s/n, 02071, Spain, [2]Universidad Politécnica de Valencia, Instituto de Ciencia Animal, Camino de Vera 14, 46071, Spain; pmolina@dca.upv.es

The use of antimicrobial substances in small ruminants' milk can have serious effects on public health and dairy products quality; antibiotic residues in milk may cause antibiotic resistances, allergies in consumers or even defects in several fermented products. Knowledge of antibiotics usage in dairy small ruminants could avoid milk safety problems. A study was developed (March to November 2008) for collecting information on antibiotic usage in small dairy ruminants in Spain. A survey was conducted in different Spanish regions, with the participation of the veterinarians in charge of the sanitary control of dairy ewes and goats. A total of 102 questionnaires were received, corresponding to a 35.1% and 44.6% of the dairy sheep and goat Spanish census, respectively. The data were divided into six groups of questions, a general one with information regarding flock size and location, a second corresponding to the mean pathologies that are treated with antibiotics, concerning to the usage of antibiotics for mastitis therapy, related to antibiotic dry therapy, about antibiotic active ingredients, and the last one with reference to the 'extra-label' use of antibiotics. The results presented in this work only correspond to the usage of antibiotics for therapy of intramammary infections at lactation stage and at drying off. The survey revealed that almost the 75% of vets applied an antibiotic for mastitis during lactation. Data revealed that for this pathology the 80.3% of the veterinarians used betalactams and the 31.6% use macrolides, follow by quinolones. Antibiotic dry therapy was more frequently used in sheep (82%) than goats (73%). Antibiotics elected for dry therapy were betalactams (46.8%) and macrolides (36.7%), follow by aminoglycosides (17.7%) and quinolones (13.9%).

Colostrum as source of passive immunity in caprines: study of comparative antibody absorption in goat kids using bovine or caprine colostrum

Lima, A.L., Pauletti, P., Susin, I. and Machado-Neto, R., University of São Paulo, Animal Science - Zootecnia, Av. Pádua Dias, 11, 13418900, Brazil; raul.machado@esalq.usp.br

The objective of this study was to evaluate colostrum management alternatives, estimating the passive immunity acquisition efficiency in goat kids fed bovine or caprine colostrum. Thirty three animals, were randomly distributed in two groups that received at 0, 12, 24 and 36 hours after birth, caprine colostrum (group A) or bovine colostrum from Holstein cows (group B). Goat kids blood samples were collected at 0, 12, 24 and 48 hours, and at 5, 10, 15, 20, 25, 30, 40, 50 and 60 days of age. Serum variables analyzed were total protein (TP), using biuret method, and immunoglobulins (Ig), by zinc sulfate turbidity test (ZST). The bovine and caprine colostrum pools and serum references to adjust ZST test were analyzed by radial immunodiffusion method. In group B, the highest ($P<0.05$) concentration of TP was observed at 48.68±0.79 hours after birth, with mean of 7.16±0.28 g/dL, and the highest ($P<0.05$) concentration date for Ig at 48±0.73 hours, with mean of 37.56±2.38 ZST units. Group A showed maximum values of TP and Ig later than group B, at 20.05±1.36 and 20.11±1.72 days after birth, with means of 5.91±0.22 g/dl and 28.17±2.05 ZST units, respectively. The estimated correlation between units of ZST and mg/ml of Ig was r=0.83 ($P<0.05$). The results indicate that caprine colostrum can be used to replace bovine colostrum, and with advantage for the goat kid newborn initial immunoglobulin acquisition, consequence of the higher concentration of immunoglobulins present in cows colostrum compared to goats. Supported by FAPESP –The State of São Paulo Research Foundation.

Effect of an immunostimulant administration on goat kid immune system

Morales-Delanuez, A.[1], Juste, M.C.[1], Castro, N.[1], Sánchez-Macías, D.[1], Capote, J.[2] and Argüello, A.[1], [1]Las Palmas de Gran Canaria University, Animal Science Unit, Fac. Veterinaria, Transmontaña s/n, 35413, Arucas, Spain, [2]Instituto Canario de Investigaciones Agrarias, Carretera El Boquerón s/n, La Laguna, Tenerife, Spain; aarguello@dpat.ulpgc.es

The aim of the present study was to evaluate the effect of an immunostimulant administration on immune system (white blood cell, neutrophils, lymphocytes, eosinophils, monocytes counts and plasma IgG concentration). 20 Majorera goat newborn male kids were grouped into two lots. One group received two doses (at day 10 and 40 of life) of an immunostimulant (based on Corynebacterium parvum and Ocrobactrum intermedium) according to manufacturer recommendation (group IMM) and the control group received two doses (at day 10 and 40 of life) of saline serum (group SS). During the first two days of life, all goat kids received colostrum and after that, until day 70, they were fed with milk replacer. Blood samples were obtained at 10, 17, 24, 31, 38, 45, 52, 59 and 66 days of life and immediately white cell counts and differential counts were performed, after that blood was centrifuged and plasma was recovered. Goat IgG was measure using ELISA. No effects of immunostimulant were observed on any parameters. White blood cells ranged from 6,533 to 10,840 cells/ml, neutrophils percentage ranged from 38.2 to 62.2%, lymphocytes percentage ranged from 36.4 to 60.4%, eosinophils percentage ranged from 0 to 2.3%, monocytes percentage ranged from 0 to 3.4%, and plasma IgG ranged from 3 to 10 mg/ml. The use of immunostimulants must to be monitored because could be a waste by goat farmers.

Milk flow kinetic during two consecutive months of lactation in ewes

Macuhova, L., Uhrincat, M., Peskovicova, D. and Tancin, V., Animal Production Research Centre Nitra, Hlohovecka 2, 95141, Slovakia (Slovak Republic); peskovicova@scpv.sk

The aim of this study was to compare the milkability traits and the milk flow type stability in two consecutive months June and July. The trail was performed with 24 ewes of three breeds: Tsigai (TS, n=8), Improved Valachian (IV, n=8) and Lacaune (LC, n=8). Ewes were routinely milked twice a day in 1 x 24 milking parlour. Experimental milkings were performed during three successive days in the midlle of two month (June, July). During milkings an actual milk yield was recorded in one - second intervals using a graduated electronic milk collection jars. In total 285 milk flow curves were recorded. The curves were classified into three groups of types: 1 peak (1P, unimodal curves), 2 peaks (2P, bimodal curves), plateau (PL, peak flow over 0.4 l/min with having steady state phase longer than 10 s). PL type refers to ewes with larger emission curves and did not show clear differences between peaks. If ewe had all milk flow curves of the same type, the ewe was characterized as the ewe with the stabile milk flow type. Milk production varied according to milk flow curve type in both month (0.44±0.03, 0.50±0.03, 0.52±0.03 l in June and 0.39±0.03, 0.42±0.03, 0.41±0.03 l in July for 1P, 2P, PL; respectively). The frequency of different milk flow curve types (1P : 2P : PL) was 34 : 54 : 12% in June and 45 : 38 : 17% in July, respectively. It indicates more frequent the milk ejection occurrence in June than July. The same milk flow type was recorded in 50% ewes in both month. Within (single month) June and July 67% ewes had the same type of milk flow curve. In July, all milkability traits decreased except time of latency. In conclusion, we could demonstrate higher stability of milk flow types within month than between two consecutive months. The ewes with 1P had the most stabile milk flow curves.

Isolation of microbial pathogens from raw sheep's milk and their role in its hygiene

Fotou, K.[1], Tzora, A.[1], Voidarou, C.[1], Anastasiou, I.[1], Avgeris, I.[1], Giza, E.[1], Maglaras, G.[1] and Bezirtzoglou, E.[2], [1]TEI Of Epirus / School Of Agriculture, Animal Production, Kostakioi Artas, 47100, Arta, Greece, [2]Democritus University Of Thrace, Food Science And Technology, Orestiada, 68200, Orestiada, Greece; tzora@teiep.gr

The natural microflora of raw milk is a factor that determines its sensorial characteristics. The microbial load in the mammary gland of healthy animals is low, whereas the application of hygienic conditions in milking and cheese making procedures limits the possibility of contamination and helps maintaining the natural microflora and the particular characteristics of milk. The objective of the present study was to assess the Total Viable Count (T.V.C.) and the count of total psychrotropic bacteria of raw milk from sheep of two indigenous Greek breeds, Boutsiko and Karamaniko. Milk samples were collected from healthy animals and used for the isolation, identification and enumeration of pathogenic bacteria that are associated with the hygiene and quality of raw sheep milk (with a particular interest in bacteria that may cause human infection). During the study, a total of one hundred (100) samples of raw sheep milk were examined in order to isolate and identify the following bacteria: *Staphylococcus aureus*, *Salmonella* spp, *E. coli*, *Cl. perfringens* (vegetative cells and spores) and *Bacillus* spp. as well as to specify the Total Viable Count and the total number of psychrotropic bacteria. The methodology concerning milk sampling, preparation of samples and decimal dilutions was according to internationally accepted methods. The results showed that the Total Viable Count from milk samples of the two breeds used in the current study, were in accordance with the microbiological criteria of EU Legislation at approximately 97% accuracy. Despite the normal values for T.V.C. pathogenic bacteria were isolated in the milk, therefore, pasteurization of sheeps' milk is neccessary before human consumption and production of dairy products.

Omission of two weekend milkings in Manchega and Lacaune dairy ewes

Castillo, V., Such, X., Caja, G., Albanell, E. and Casals, R., Universitat Autònoma Barcelona, G2R, Campus Bellaterra, 08193 Bellaterra, Spain; vanesa.castillo@uab.cat

Forty two Manchega (MN) and 18 Lacaune (LC) dairy ewes were used to study the effects of omitting 2 milkings weekly in early- (wk 8 to 14) and mid-lactation (wk 15 to 22). Ewes submitted to milking omissions were milked twice daily from Monday to Friday (08:00 and 18:00 h), and once daily on Saturday and Sunday (16:00 and 14:00 h, respectively). Individual data were collected for milk yield, milk composition, and somatic cell count (SCC). Omitting 2 milkings weekly tended to decrease milk yield in MN ewes (-15%, $P=0.07$) in early-lactation, whereas no effects were observed in LC ewes. Milking omissions in mid-lactation did not affect milk yield in either breed. Milk composition and SCC were unaffected by milking omissions in both breeds and stages of lactation. Daily effects of milking omissions were evaluated in a sample of 22 MN and 11 LC ewes in early- (wk 12) and mid-lactation (wk 20). Milking omissions decreased milk yield, milk fat and milk lactose contents on the first omission day in both breeds, losses being more noticeable in early- than in mid-lactation. Milk protein content and SCC did not vary. After resuming the twice-daily milking routine on Monday, milk yield showed a compensatory increase that was greater in large- than in small-cisterned ewes and allowed milk yield to return to Friday values in LC and large-cisterned MN ewes. Milk fat content increased during Sunday and Monday, re-establishing Friday values in both breeds. Weekend milking omissions in early-lactation caused tight junction leakiness in both breeds, but mammary epithelium adapted to extended milking intervals when applied successively. In mid-lactation, mammary tight junction showed leakiness only in MN ewes. Omitting two milkings weekly could be an interesting management approach to reduce farm labor with no negative effects on milk yield and milk SCC values in dairy sheep. Losses in milk yield would be reduced if milking omissions were done during late lactation in small-cisterned ewes.

Lactational effects of once- versus twice-daily milkings throughout lactation in dairy ewes

Santibañez, A., Such, X., Caja, G., Castillo, V. and Albanell, E., Universitat Autònoma Barcelona, G2R, Campus Bellaterra, 08193 Bellaterra, Spain; joseantonio.santibanez@campus.uab.es

The effects of once- (1X) vs. twice-daily (2X) milkings throughout lactation on milk yield, milk composition, SCC, blood lactose, cisternal size and milk fractioning were studied in 2 breeds of dairy ewes differing in milk yield, udder compartments and milk ability (Manchega, MN, n=29; Lacaune LC, n=37). After the weaning of the lambs (wk 4), ewes were machine milked at 2X and blocked into 2 groups: 1X (MN, n=15; LC, n=18) and 2X (MN, n=14; LC, n=19). Individual milk recording was conducted weekly (wk 5 to 25) for milk yield, biweekly for milk composition, monthly for SCC, and 2 times (wk 5 and 14) for lactose, cisternal area and udder compartments. Reducing milk frequency throughout lactation impaired milk yield differently in each breed (MN, −46%; LC, −25%; $P<0.01$). A drop in milk yield was observed between wk 5 and 6 of lactation in the ewes passing from 2X to 1X. Milking frequency did not affect ($P>0.05$) percentages of major milk components (fat, protein, total solids and casein), the milk of MN ewes being richer in components than LC. During lactation, lactose in blood was not affected by milking frequency, but LC ewes had greater levels of lactose than MN (33.9 vs. 19.5 µml/l, respectively; $P<0.05$). The SCC was not affected ($P>0.05$) by milking frequency, but LC ewes had greater values than MN ($P<0.01$) and, for both breeds, a linear increase ($P<0.01$) in SCC from 1st to 3rd parity was observed (5.01, 5.02, 5.62 log10 cells/mL). Milking frequency had no effect on the fractioning and cisternal area. Cisternal and alveolar milk decreased ($P<0.05$) throughout lactation (503 to 270 ml; 450 to 164 ml, respectively). Area of the cistern decreased (44.5 to 34.6 cm^2) as lactation progressed, affecting ($P<0.01$) the fractioning of cisternal milk (small, 282 ml; large, 476 ml). Ewes LC showed greater cisternal area and cisternal milk ($P<0.05$) than MN (483 vs. 267 mL; 46.9 vs. 30.3 cm^2, respectively)

Mammary morphology of Sicilo-Sarde dairy sheep raised in Tunisia

Ayadi, M.[1], Ezzehizi, N.[1], Zouari, M.[1], Najar, T.[2], Ben M' Rad, M.[2], Such, X.[3], Casals, R.[3] and Caja, G.[3], [1]Institut Supérieur de Biologie Appliquée, SCQ, km 22.5 route Djorf, 4119 Médénine, Tunisia, [2]Institut National Agronomique de Tunisie, SPA, 43 Av. Charle Nicole, 2018 Cité Mahragène, Tunisia, [3]Universitat Autònoma de Barcelona, G2R, Campus de la UAB, 08193 Bellaterra, Spain; moez_ayadi2@yahoo.fr

Sicilo-Sarde sheep are a local breed used in North of Tunisia for cheese production. Ewes are characterized by coarse wool and black spotted body, face and legs. Production systems are semi-extensive (grazing during spring, summer and autumn) and intensive (fed with hay and concentrate) and lambing period is concentrated in October. A sample of 52 ewes from the flock of the OTD (Office des Terres Domaniales) at early lactation (68±10 d) and twice daily machine milked (down milk pipeline), were used to study the external and internal udder morphology and milk yield potential of the breed. External morphology (teat and udder) and milk yield potential (by using the double oxytocin injection method) were performed 4 h after milking and milk yield expressed on a day basis (×6). Cisternal area (by ultrasonography) and udder compartments (cisternal and alveolar milk, by using and using atosiban and oxytocin) were estimated at 8 h after milking. On average, udders were healthy (CMT, <1), small sized (volume, 509±184 ml), with medium sized teats (length, 18.5±5.5 mm; diameter, 10.6±2.4 mm) and inserted at 45±10. Cisternal area was 11.6±4.5 cm^2 and cisternal milk accounted for 57% of the total milk. Correlation between cisternal milk and cisternal area was medium (R^2=0.48; $P<0.05$) as a consequence of a multilocular structure. Lag time and total milking time were 1.9±0.1 s and 31±5 s, respectively, for 0.56±0.40 l/d. In conclusion, the Sicilo-Sarde breed showed an adequate udder morphology for machine milking (medium sized cisterns and teats), although milk yield needs to be improved. Project Cooperation Spain-Tunisia (AECI A/9275/07)

Microbiological profile of 'Conciato Romano', an artisanal cheese of Campania

Nuvoloni, R., Fratini, F., Ebani, V.V., Pedonese, F., Faedda, L., Forzale, F. and Cerri, D., University of Pisa, V.le Piagge 2, 56124 Pisa, Italy; rnuvola@vet.unipi.it

The aim of this study was to determine the microbiological profile of the Conciato Romano, an artisanal raw ewe's and/or goat's milk cheese produced in the province of Caserta (Campania, Italy). It's an ancient cheese, produced from sheep and goats living in a natural grazing land characterized by aromatic Mediterranean plants that confer typical organoleptic features to the milk. It's an irregular-cilindric-like shaped cheese of about 150 grams, with a smooth, thin and flavoured light-brownish rind; the inner part is white-yellow, with an agreeable odour and an intense and aromatic taste. The manufacturing process is characterized by typical phases as drying in a traditional mosquito-proof wooden structure in the open shade, washing with water used to cook a local home-made typical pasta, treating with a *concia* of olive oil, white vinegar, wild thyme and hot red pepper, aging inside clay vases from 6 months to 2 years. Eleven samples from two different batches of Conciato Romano were analyzed. Microbiological analyses targeted the presence and the evolution of: lactic acid bacteria (LAB), hygiene indicator and spoilage bacteria, main pathogen microrganisms. Sixty LAB isolates were submitted to phenotypic and genotypic identification. LAB appeared to be the dominant microflora throughout ripening. They were mainly represented by the species Lactobacillus plantarum, Lactococcus lactis subsp. lactis and Enterococcus faecium. Total coliforms, Escherichia coli and staphylococci showed the highest counts at 7th day of ripening then decreased, dropping to levels below 102 CFU/g at 4 months of aging. Low levels of yeasts and moulds were detected throughout the ripening period. Salmonella enterica, Campylobacter spp. and Listeria monocytogenes were absent in all samples. The results suggest that it could be advantageous to improve the level of hygiene during milk and cheese production in order to eliminate the undesirable microorganisms and to standardize quality of this unique cheese.

Effect of stage of lactation on composition and quality of organic milk of ewes crossbreeds of Lacaune, East Friesian and Improved Wallachian

Kuchtik, J., Pokorna, M., Sustova, K., Luzova, T. and Filipcik, R., Mendel University of Agriculture and Forestry in Brno, NPVII2B08069, Zemedelska 1, 61300 Brno, Czech Republic; kuchtik@mendelu.cz

The evaluation of the effect of stage of lactation (SL) on composition and quality of milk of crossbreeds of Lacaune (L), East Friesian (EF) and Improved Wallachian (IW), (L 50 EF 43.75 IW 6.25, n=10) was carried out on organic farm in Valašská Bystřice. The weaning of lambs was carried out in the end of April, whilst till the weaning the ewes were reared with their lambs indoors. After the weaning all ewes began to be machine-milked and since the weaning the ewes were reared, till the end of the study, on permanent pasture, supplemented with organic oat (0.1 kg/head/day) and organic mineral lick (ad libitum). The SL had a significant effect on milk yield (MY) and contents of total solids (TS), fat (F), protein (P), casein (C), lactose (L) and urea (U). The SL had also a significant effect on pH, titratable acidity (TA) and rennet clotting time (RCT). On the contrary the SL had not a significant effect on somatic cell counts (SCC) and rennet curd quality (RCQ). An average daily MY was 0.89 l and the average contents of TS, F, P, C, L and U per lactation were 20.32%, 8.79%, 6.20%, 4.97%, 4.51% and 58.51 mg/100 ml. The average SCC, pH, TA, RCT and RCQ per lactation were 107 670 in 1 ml, 6.66, 11.25 °SH, 144 seconds and 1.36 (very good quality). The highest contents of TS, F, P and C were found at the end of the lactation, on the contrary the lowest contents were found on the top of lactation. The L content grew in the beginning of pasture period, afterwards till the end of the lactation, there was found its gradually decreasing content. The contents of U and RCT were the most variable indicators within the frame of the study. On the other hand SCCs were very well-balanced during the lactation and on relatively very low levels. TA and RCQ gradually increased whereas RCT gradually prolonged with advanced lactation.

Milk characteristics of indigenous Hungarian Tsigai sheep populations

Kukovics, S.[1], Németh, T.[1], Molnár, A.[1] and Csapó, J.[2], [1]Research Institute for Animal Breeding and Nutrition, Gesztenyés u. 1., 2053 Herceghalom, Hungary, [2]Kaposvár University, Faculty of Animal Science, Guba S. u. 40., 7400 Kaposvár, Hungary; sandor.kukovics@atk.hu

Two kinds of indigenous Tsigai sheep are bred in Hungary: the so-called Autochthonous and the Milking Tsigai. Several differences were observed among the various Tsigai populations earlier. In the present study the milk production characteristics were examined in order to determine differences among them. Two Milking Tsigai and three Autochthonous Tsigai populations were selected for this study, which were kept under similar production system and nutritional conditions. Milk samples were collected from them individually once in a month (morning and evening), and group samples were also taken once in a month (morning and evening) over the lactation period. Samples were examined in official laboratories. Microsoft Excel 9.5 and SPSS for Windows 10.0 programs were used for processing the data received. Significant deviations were found in milk production among the Autochthonous and Milking Tsigai populations (45-68 litres versus 128-160 litres; $P<0.001$), and within the Autochthonous and also within the Milking Tsigai populations; as well as in the lactation length (74-113 versus 145-176 days; $P<0.01$). Differences were also found among the studied populations in dry matter (14.79-20.69%; $P<0.01$), in raw protein (4.24-6.56%; $P<0.01$), and in casein (3.08-5.04%; $P<0.05$) contents. Significant differences were also found in the fatty acid composition. The SFA (63.04-74.93%), MUFA (21.94-32.18%) and PUFA (3.16-5.18%) showed differences ($P<0.05$) among the Tsigai populations and within the Autochthonous and within the Milking Tsigai populations as well. Differences were also observed among the studied Tsigai populations in the amino-acid composition, especially in the case of glutamine (18.0-20.9%), proline (9.7-11.8%), asparagin (6.5-8.3%), while much smaller deviations were found in the case of other amino-acids. According to our results deviations among studied Tsigai sheep are also valid in milk composition.

Effects of omitting one daily milking on milk production in commercial flocks of Assaf dairy ewes

Such, X.[1], González, R.[2], Caja, G.[1], Aguado, D.[2] and Milán, M.J.[1], [1]Universitat Autònoma Barcelona, G2R, Campus Bellaterra, 08193 Bellaterra, Spain, [2]Cargill SLU, Ctra. Pobladura Coomonte, 49780 Benavente, Spain; xavier.such@uab.cat

Three commercial flocks of Assaf dairy ewes (n=426) in the Spanish region of Castilla-Leon were used to evaluate the effect of omitting one milking daily from early-mid lactation on lactational performance and udder health. Ewes were milked twice daily (2X) from lambing and lambs artificially reared. After 36, 41 or 98 d, according to flock, half of the ewes (n=220) were milked once daily (1X) until the end of lactation. The rest, were 2X milked and used as a control. Both groups were fed and managed under similar conditions. Throughout 4 mo, milk yield, milk composition (fat and protein) and SCC were monthly recorded. Lactational performance differed according to milk yield at the time of starting 1X. In 2 flocks, in which milk yield averaged 1.91±0.16 and 2.30±0.04 l/d, milk loss due to 1X was not significant (-7%; $P>0.05$) being milk yield average 173 and 205 l (in 120 DIM) respectively. Milk yield of ewes in the third flock before the treatment was higher (3.15±0.09 l/d) and 1X resulted in greater milk yield losses (-124 l, -39%; $P<0.01$) being total milk yield for 2X ewes 320 l in 120 DIM. In the first milk recording after starting the milking omission treatment, 1X ewes suffered 55% milk yield losses (1.36 vs.2.82 l/d for 1X and 2X, respectively; $P<0.01$). Once daily milking had no effect on average values of milk fat (6.23%), protein (5.13%) or SCC (319 x 10^3 cells/ml). However, these values increased ($P<0.05$) as lactation advanced (fat, 5.50 to 6.56%; protein, 4.91 to 5.29%; SCC, 233 to 413 x 10^3 cells/ml). In conclusion, once daily milking could be a recommendable practice in Assaf ewes when milk yield is <2.5 l/d. Under these conditions, milk yield losses are negligible and an improvement of farmer's quality of life is expected.

Effects of milking interval on secretion and composition of milk in three dairy goats breeds

Hammadi, M.[1], Ayadi, M.[2], Barmat, A.[1] and Khorchani, T.[1], [1]Arid Lands Institute, Livestock and Wildlife Lab., Km 22,5 Route el Djorf, Médénine, 4119, Tunisia, [2]Institut Supérieur de Biologie Appliquée de Médénine, SCQ, Km 22,5 Route el Djorf, Médénine, 4119, Turkmenistan; moez_ayadi2@yahoo.fr

Twenty-four lactating goats (Alpine, n=8; Damascus, n=8 and Murciano-Granadina, n=8) in mid lactation were used to study the short-term effects of different milking intervals (8, 16 and 24 h) on secretion and composition of milk. Milk was analyzed for physical (pH, density and acidity) and chemical (Total solid, fat, protein and ash) parameters. Murciano-Granadina produced less milk ($P<0.05$) than Alpine and Damascus breeds (1.30±0.36; 1.68±0.45; 1.64±0.42 l, respectively). Milk secretion rate reached the greatest values in the 8 to 16 h milking interval in Alpine (97 ml/h) and Murciano-Granadina (75 ml/h) goats, but it increased in Damascus goat after 16 h. The pH of milk does not change with milking interval and averages 6.6±0.05. Density of milk increased with milking interval in all studied breeds, and acidity increased only in Alpine and Damascus breeds. Milking interval affected milk fat content, which decreased markedly from 8- to 24-h, but no differences were observed in milk protein content which averaged 29.36±2.72 g/l. Milk ash content decreased from 8 to 24 h in Alpine (8.9±0.8 vs. 8.0±0.5) and Murciano-Granadina (8.6±0.9 vs. 7.8±0.6) goats. This parameter maintained constant in Damascus goat. No effect on the udder health was observed. In conclusion, this short-term study proved that Alpine and Murciano-Granadina dairy goats could maintain high secretion milk rates during extended milking intervals.

Goat milk cheeses

Sikora, J. and Kawęcka, A., National Research Institute of Animal Production, St. Krakowska 1, 32-083 Balice, Poland; jsikora@izoo.krakow.pl

In recent years, Europe has seen a growing interest in goat milk production and processing. The demand for milk goat products also increased in Poland once the health and taste properties of these products have become known. Goat milk is a source of many valuable nutrients. Until recently, farmers made most milk products for their own farm needs or in family-run processing plants. Today, goat milk is increasingly processed in dairies and the wide range of products includes fermented beverages, curd cheese, maturing cheese, blue cheese, UHT milk and even ice-cream. Many of these products were included in the List of Traditional Products, which was created to identify traditional products made in Poland and to increase the consumer knowledge of traditional foods and Polish culinary heritage. One of these products is Ser kozi podkarpacki (Subcarpathian goat cheese) produced in the Subcarpathian area. This product entered the list due to its unique properties, resulting from the use of traditional production methods. The aim of the study was to determine goat milk parameters at different stages of lactation from the cheese-making point of view. Milk and cheese produced from this milk originated from fawn goats kept on a farm in the Subcarpathian region. Control milkings were performed between May and October. Milk was collected using an automatic milking machine, stored in a cooler and used to make acid-rennet cheese according to an original formula. Milk yield per goat was calculated based on control milkings and ranged from 388 to 423 kg. The amount of cheese produced in particular months showed an upward tendency, because the same amount of milk enabled 8.9 kg cheese to be produced in May and 11.8 kg in October. Cheese production in the final stage of lactation was the highest and differed significantly from the mean amounts of cheese produced in previous lactation stages. This could be related to changes in the content of individual components in milk. The content of fat, protein and casein increased as lactation progressed, which had an effect on the amount of cheese made.

The effect of artificial rearing on kid growth and milk production of Damascus goats

Koumas, A.[1] and Papachristoforou, C.[2], [1]Agricultural Research Institute, Animal Production, P.O. Box 22016, 1516, Lefkosia, Cyprus, [2]Cyprus University of Technology, Agricultural Sciences, P.O. Box 50329, 3603, Lemesos, Cyprus; A.Koumas@arinet.ari.gov.cy

Yearling Damascus goats were allocated as they kidded on either natural suckling (NS, 22 goats) or were separated from their kids immediately after birth (24 goats); NS goats suckled up to two kids, while separated kids were artificially reared (AR) on milk replacer. Colostrum was given to AR kids by bottle feeding. All kids were weaned at 49±3 days of age. After weaning, 15 male kids from each NS and AR groups were fattened for 70 days. AR goats were milked twice daily. NS goats were milked once daily before, and twice daily after weaning. Birth weight of kids in both groups was similar (4.1 kg). NS kids had a faster ($P<0.01$) preweaning growth rate than AR kids (223 VS 191 g/day) and were heavier ($P<0.05$) at weaning (14.7 kg) than AR kids (13.8 kg). After weaning, males of both groups had similar growth rate (AR: 291, NS: 270 g/day). Final weight at 120 days of age of AR (36.0 kg) and NS kids (35.1 kg) was similar. Milk yield of AR goats (118 l) during the preweaning period was higher ($P<0.05$) than NS goats (31 L). Fat and protein content of milk was 3.69 and 3.71 for AR goats and 3.23 and 3.30% for NS goats, respectively. Post weaning milk yield (142 days) of both groups was similar (NS: 301 L, AR: 273 L). No differences were observed in milk fat and protein content (%) between suckling and non suckling goats (fat: 4.02 and 4.16, protein: 3.96 and 4.03, respectively). Total milk yield (190 days) was 392 L for AR and 332 L for NS goats. These results indicate, that artificially reared kids had satisfactory preweaning growth, slightly lower though than that of suckling kids; however, both groups reached the same final weight at 120 days of age. Goats on zero suckling, produced more commercial milk over the whole lactation than suckling goats. In dual purpose breeds as the Damascus, artificial rearing may increase farmer's income.

Goat milk technological parameters evolution through transition from colostrum to milk

Sánchez-Macías, D.[1], Castro, N.[1], Moreno-Indias, I.[1], Capote, J.[2] and Argüello, A.[1], [1]Las Palmas de Gran Canaria University, Animal Science Unit, Fac. Veterinaria, Transmontaña s/n, 35413, Arucas, Spain, [2]Instituto Canario de Investigaciones Agrarias, Carretera el Boquerón s/n, La Laguna, Tenerife, Spain; aarguello@dpat.ulpgc.es

The aim of present study was to evaluate the technological parameters changes during the transition from colostrum to milk. 30 Majorera goats were milked once a day during 90 days after partum and milk samples were obtained at partum, 1, 2, 3, 4, 5, 15, 30, 60, and 90 days. Density, pH, dry matter (DM), titratable acidity (TA), ethanol stability (ES) and rennet clotting time (RCT) were measured. Time after partum was used as main factor in the ANOVA model. Goat milk density, DM, TA and RCT significantly ($P<0.001$) dropped during the experimental time. pH and ES significantly ($P<0.001$) increased during the experimental time. Density ranged from 1026 to 1062 g/l, DM ranged from 125.3 to 288.8 g/l, TA ranged from 0.1 to 0.6% lactic acid, ES ranged from 30 to 92% water vs ethanol and RCT ranged from 2 to 35 minutes. In conclusion, technological parameters on goat milk spent more than 15 days to become stable.

Biological efficiency of goat milk production

Németh, T.[1], Szendrei, Z.[2] and Kukovics, S.[1], [1]Research Institute for Animal Breeding and Nutrition, Gesztenyés u. 1., 2053 Herceghalom, Hungary, [2]University of Debrecen, Centre of Agricultural Sciences and Engineering, Böszörményi út 138., 4032 Debrecen, Hungary; nemeth.timea@atk.hu

Concerning body size of various animal species several authors mentioned that the goat milk production is the most effective one. It could reach the ten times level of the body weight (50-60 kg body weight – 500-600 litres of milk), but limited number of exact data is available. Milk production efficiency of does from Hungarian Milking White (HMW, n=52), - Milking Brown (HMB, n=19), - Milking Multicolour (HMM, n=31), Alpine (n=68) and Saanen (n=127) breeds were analysed using body weight (BW, kg), wither height (WH, cm), lactation yield (litre) data. Body weight index (BWI) was derived from body weight and wither height, while lactation yield was corrected by two kind of methods (LY-1: 200-days production, LY-2: 200-days production corrected by number of lactation). Average BW were the highest in Saanen (54.63±10.96), followed by Alpine (54.02±7.71); HMB (50.95±7.22), HMM (48.63±9.39) and HMW (45.40±8.70); while the highest WH data was found in Alpine (67.34±4.96) and the lowest in HMW (64.35±4.28). The order in BWI was the same as in BW: Saanen (82.34±15.04), Alpine (80.44±11.81), HMB (77.76±9.23), HMM (73.84±11.57) and HMW (70.25±10.70). Alpine does gave the highest amount of milk by LY-1 (468.77±173.51), while 332.84±190.00 litre was produced by HMW. Using LY-2 method, Saanen's estimated production was 511.94±189.11 litre milk, followed by Alpine's, however, HMW gave the least amount (398.17±246.79) of milk. The efficiency received was more favourable using BW than using BWI. The HMB does produced milk the most efficiently according to the counting methods (EFF-1: 8.82; EFF-2: 9.81; and EFF-4: 6.45). The most efficient milk producing doe of HMB was 4 years old, had 53 kg BW, 60 cm WH and could produce 17 times more milk than her BW. The least efficient milk producing breed is the HMW, where an average doe produced 9 times more milk than her BW.

Muscle fiber morphology in purebred Pomeranian lambs and crossbreeds by meat-type rams

Brzostowski, H., Tański, Z., Milewski, S., Kosińska, K., Ząbek, K. and Zaleska, B., University of Warmia and Mazury, Department of Sheep and Goat Breeding, Olsztyn, Oczapowskiego 5, 10-719, Poland; kolihk@uwm.edu.pl

A morphological analysis was performed on samples of musculus longissimus lumborum collected from 50-day-old purebred Pomeranian (P) lambs and F1 crossbreeds by Suffolk (PS) and Texel (PT) rams, 12 animals per group. Muscle fiber area and diameter were measured, the percentage content of giant (hypercontracted) fibers with symptoms of degenerative changes was estimated, and the concentration of collagen – a fibrous scleroprotein - was determined. It was found that lamb genotypes had a significant effect on the examined characteristics. The smallest, i.e. the most desirable, area and diameter of muscle fibers was observed in meat from purebred lambs (346.29 μm^2 and 21.25 μm respectively), compared with meat from crossbreeds by Suffolk rams (404.82 μm^2 and 22.69 μm) and Texel rams (419.09 μm^2 and 22.98 μm). The highest percentage content of giant fibers was noted in meat from crossbred PT lambs (4.15%), followed by crossbred PS lambs (2.70%) and purebred P lambs (1.06%). Collagen content (mg/100 g meat) in particular groups of lambs was as follows: P - 223, PS – 236, PT - 247. The high area and diameter of muscle fibers, as well as the highest percentage content of giant muscles and the highest collagen concentration, determined in meat from crossbreeds by Texel rams, indicate that lambs of this group showed the highest susceptibility to stress, which resulted in the poorest quality of meat. The quality of meat from crossbreeds by Suffolk rams was slightly higher, while meat from purebred lambs was characterized by the best quality.

effect of the slaughter age of lambs on meat quality and the nutritional value of protein
Brzostowski, H., Tański, Z. and Milewski, S., University of Warmia and Mazury, Department of Sheep and Goat Breeding, Olsztyn, Oczapowskiego 5, 10-719, Poland; kolihk@uwm.edu.pl

Selected indicators of the nutritional value of protein and the quality of meat from 50- and 100-day-old Pomeranian lambs were studied. In order to evaluate meat quality, samples of musculus longissimus dorsi were taken to determine the proximate chemical composition, the water-to-protein ratio, the concentrations of cholesterol and collagen, and the energy value. The nutritional value of protein was estimated during a balance experiment on Wistar rats. The apparent digestibility (AD), true digestibility (TD) and biological value (BV) of protein, and net protein utilization (NPU) were determined. It was found that the slaughter age of lambs had a significant effect on some of the above indicators. Meat from 100-day-old lambs was characterized by a higher content of dry matter, fat, cholesterol and collagen, a higher degree of physiological maturity (estimated based on the water-to-protein ratio), and a higher net and gross energy value, compared with meat from 50-day-old lambs. Meat from younger lambs was marked by a higher apparent and true digestibility of protein and by a higher nutritional value, estimated based on BV and NPU. The low energy value, and the high values of AD, TD, BV and NPU indicate that lamb meat, in particular meat obtained from lambs slaughtered at the age of 50 days, may be an important component of a healthy low-fat diet.

Effect of the age of Ile de France lambs on the quality and properties of meat
Tański, Z., Brzostowski, H., Milewski, S., Ząbek, K. and Kosińska, K., University of Warmia and Mazury, Department of Sheep and Goat Breeding, Olsztyn, Oczapowskiego 5, 10-719, Poland

Selected quality indicators of meat from 50- and 100-day-old single ram lambs of the Ile de France breed were compared in the study. Samples of musculus longissimus dorsi were taken to determine the proximate chemical composition and physicochemical properties of meat, as well as the energy value, the content of cholesterol and collagen, muscle fiber size, amino acid concentrations in protein, the fatty acid profile of intramuscluar fat and the water-to-protein ratio. It was found that meat from older lambs (aged 100 days) was characterized by a higher content of dry matter, protein, fat and collagen, a larger diameter of muscle fibers, a higher calorific value, a darker color, a lower water-holding capacity and a more desirable water-to-protein ratio, which indicates that it was more physiologically mature, compared with meat from lambs slaughtered at the age of 50 days. Meat from younger lambs was marked by more desirable ratios of essential amino acids to non-essential amino acids (ESAA/NEAA), unsaturated fatty acid to saturated fatty acids (UFA/SFA), polyunsaturated fatty acids to monounsaturated fatty acids (PUFA/MUFA), and hypocholesterolemic fatty acids to hypercholesterolemic fatty acids (DFA/OFA). Due to lower concentrations of intramuscular fat, cholesterol and collagen, a lower calorific value, more desirable ratios between amino acids in protein and fatty acids in intramuscular fat, meat from younger lambs may be recommended as a valuable component of a low-calorie and low-fat diet.

Influence of a Moxidectin treatment in ewes on the growth rate of their lambs

Moors, E. and Gauly, M., Department of Animal Science, Albrecht Thaer Weg 3, 37075 Göttingen, Germany; mgauly@gwdg.de

Milk production of the ewe has significant effects on lambs' growth rates. Lower growth rates may be observed when ewes are affected by clinical or sub clinical infections, e.g. due to gastrointestinal nematodes. Even though the direct effect of an anthelmintic treatment is well known, little is known about the indirect effect. Therefore the aim of the present study was to evaluate the effect of a Moxidectin treatment of ewes on the average daily weight gains (ADW) of their lambs in two German sheep breeds (German Black Head Mutton and Leine sheep). In total 133 ewes were randomly divided into two groups: 1. no anthelmintic treatment (control group, n=63), and 2. Moxidectin treatment (n=70). In Leine lambs the anthelmintic treatment of the ewes did not influence growth rates, suggesting a higher nematode resistance in this breed. Average daily weight gains were significantly higher in German Black Head lambs compared to Leine lambs ($P<0.001$), and in single born lambs compared to multiples ($P<0.001$). Furthermore in single born German Black Head Mutton lambs average daily weight gains were significantly higher ($P<0.05$) in the Moxidectin treated group compared to single born lambs in the untreated group. However no differences could be observed between lambs from treated and untreated ewes that had more than one lamb. These findings might be explained as a result of differences in intakes of milk, concentrate and hay. In consequence of this, a selective anthelmintic treatment of ewes with single born lambs may be beneficial to increase productivity of lambs.

Mineral composition of meat from light lambs

Mioc, B.[1], Vnucec, I.[1], Prpic, Z.[1], Pavic, V.[1], Antunovic, Z.[2] and Barac, Z.[3], [1]Faculty of Agriculture University in Zagreb, Svetosimunska 25, 10000 Zagreb, Croatia, [2]Faculty of Agriculture University of J. Strossmayer, Trg Sv. Trojstva 3, 31000 Osijek, Croatia, [3]Croatian Livestock Centre, Ilica 101, 10000 Zagreb, Croatia; bmioc@agr.hr

The proximate composition and mineral content of light lambs muscle (derived from Istrian and Dalmatian Pramenka breeds) were studied. The musculus longissimus dorsi (MLD) muscle samples of 30 carcasses were analysed and the effects breed and sex were studied. Because no significant differences were found between sexes, data classified according to this factor were not included. A comparison of each variable (carcass weight (ranging 8.5-12.5 kg), moisture, protein, fat and mineral elements) between breeds was performed by Student's t-test and correlation coefficients between variables and carcass weight were determined by PROC CORR procedure of SAS statistical package (1999). Although lambs of investigated breeds were of similar slaughter age (2.5 months), mean value for carcass weight was higher ($P<0.05$) in Dalmatian Pramenka lambs than in Istrian lambs (11.5 kg vs. 10.4 kg, respectively). Breed had a significant influence on moisture and fat contents, whilst mineral composition (with the exception of selenium) was scarcely affected by breed and sex. Fat and moisture contents were significantly correlated with cold carcass weight (r=0.46 and -0.50, respectively). The magnesium, calcium, manganese and selenium contents were significantly ($P<0.05$) correlated (r=0.50, -0.46, 0.44 and 0.54, respectively) with carcass weight. This study contributes to characterization of lamb carcasses from Istrian Sheep and Dalmatian Pramenka breeds and provides new data on the composition of the MLD of light lambs.

Modulation of immunological status of piglets: influence of food contaminant such as mycotoxins

Oswald, I.P., INRA, UR66, Laboratory of Pharmacology-Toxicology, 180 chemin de Tournefeuille, BP93173, 31027 Toulouse cedex 3, France; ioswald@toulouse.inra.fr

The postnatal period is critical in terms of the development of an effective mucosal immune system that is capable of the adaptation to commensal bacteria, tolerance towards nutritional components and defence against pathogens in the gut lumen. Weaning is a period during which there is considerable change in the magnitude and diversity of exposure to environmental antigens derived from food and potentially pathogenic organisms. Under 'natural conditions' weaning is a gradual process, and in piglets it is not complete until 10-12 weeks of age. As a consequence, when weaned at 3-4 weeks of age neonatal pigs experience a transient immune hypersensitivity, often associated with clinical symptoms. At weaning, piglets are especially susceptible to feed contaminants such as mycototxins. Mycotoxins are secondary metabolites secreted by filamentous fungi produced on different raw material especially cereals. Very resistant to technological treatments, mycotoxins can be present in pig feed, where they present a wide spectrum of toxic effects including an alteration of immune function. The sensitivity of the immune system to mycotoxin-induced immunosuppression arises from the vulnerability of the continually proliferating and differentiating cells that participate in immune mediated activities and regulate the complex communication network between cellular and humoral components. Mycotoxin induced immunosuppression may be manifested as depressed T or B lymphocyte activity, suppressed antibody production and impaired macrophage/neutrophil-effector functions. Suppressed immune function by mycotoxins may eventually decrease resistance to infectious diseases, reactivate chronic infections and/or decrease vaccine and drug efficacy. At weaning piglets are especially sensitive to mycotoxin-induced immunosuppression which may have consequences in term of pig health.

Using mathematical models to unravel some mysteries of host-pathogen interaction in mammals: insights from a viral disease in pigs

Doeschl-Wilson, A.[1] and Galina-Pantoja, L.[2], [1]Scottish Agricultural College, Sustainable Livestock Systems, Sir Stephen Watson Building, Bush Estate, Penicuik EH260PH, United Kingdom, [2]Pig Improvement Company, 100 Bluegrass Commons Blvd., Hendersonville, TN 37075, USA; andrea.wilson@sac.ac.uk

Molecular techniques have provided valuable insight into the mechanisms underlying within host-virus dynamics and the host's immune response. Nevertheless many fundamental biological questions remain unanswered, as they concern less individual molecular mechanisms than the dynamics of the complex system as a whole. These questions usually require the help of mathematical models. The porcine reproductive and respiratory syndrome (PRRS), an endemic viral disease in pigs, causing large economic losses to the pig industry worldwide, is one such disease. Numerous *in vitro* and *in vivo* studies have elucidated key mechanisms for virus replication within the host and the host's immune defence. In particular, host genetics appears to play a vital role in the disease progression. Nevertheless important questions concerning the primary reasons for the typically observed long-term persistence of the infection and the high variation in host response have not been solved. Here we present mathematical models of host-virus interaction that were developed to explore the role of various components involved in the host response to PRRS virus infection identified by molecular studies, on the resulting infection characteristics. In particular, it is demonstrated how small differences between the hosts' immune systems can lead to drastic differences in the observed infection trends. The results of this work are relevant for the development of selection tools to enhance host disease resistance and for vaccine development.

Effect of spray-dried porcine plasma and plasma fractions on performance of Salmonella typhimurium-challenged weaned pigs

Bruggeman, G.[1], Rodríguez, C.[2] and Polo, J.[2], [1]Nutrition Sciences N.V., Booiebos 5, B-9031 Drongen, Belgium, [2]APC Europe, S.A., Avda. Sant Julià 246-258. PI El Congost, E-08403 Granollers, Spain; javier.polo@ampc-europe.com

Ninety-six weanling pigs (21 d of age; 4 pigs/pen; 4 pens/dietary treatment) were fed a control diet or separate diets containing 5% SDPP, an equivalent quantity of IgG from a concentrated IgG fraction (IC diet), or equivalent amount of albumin from a concentrated albumin fraction (AC diet). IC and AC fractions were derived from the same source of porcine blood used to produce the SDPP. The SDPP and plasma fractions replaced wheat gluten as the primary protein source in the control diet. Diets were balanced for protein, amino acids and energy content. At d 7 post weaning all pigs (except non-challenged control group) were challenged with Salmonella typhimurium (oral dose of 7.7 x 10^7 c.f.u). Health measurements (mortality and Salmonella shedding) and performance parameters (ADG, ADFI and FCR) were measured for 42 d. During the first week prior the challenge, only pigs fed the SDPP diet had improved ($P<0.05$) ADG and ADFI compared to the control group. During the following week after challenge, pigs fed diets with SDPP or plasma fractions (IC or AC) showed similar ADG and FCR values compared to the non-challenged control group and significantly higher ADG and better FCR than the challenged control group. From d 15 to 42, pigs fed SDPP or plasma fractions had higher ADG values than both control groups. Pigs fed the SDPP and plasma protein fraction diets showed lower Salmonella shedding than the challenged control group, especially in diets containing SDPP and IC. No occurrence of severe diarrhoea or mortality was observed for any treatments. These results suggest that the IC fraction had the greatest impact on shedding and growth performance, but some improvement was noted for the AC fraction as well, indicating that both of these protein fractions have a positive contribution to the SDPP effect noted under the conditions of this study. Supported by grant IWT010122 (Belgium)

Analysis of the pathogen spectrum and risk factors associated with coliform mastitis in sows

Gerjets, I.[1], Reiners, K.[2] and Kemper, N.[1], [1]Institute of Animal Breeding and Husbandry, CAU Kiel, Olshausenstraße 40, 24098 Kiel, Germany, [2]PIC Germany GmbH, Ratsteich 31, 24837 Schleswig, Germany; igerjets@tierzucht.uni-kiel.de

Coliform Mastitis (CM), an infection of the mammary glands, is an important disease in sows after farrowing. Due to the reduced productivity of the sows and high preweaning piglet mortality, the disease is associated with serious economic losses. CM is a multifactorial disease influenced i.e. by management, feeding and hygiene, and, moreover, by bacterial pathogens. This study is part of the FUGATO-plus-project 'geMMA' and examines the spectrum of pathogens involved in CM und analyses risk factors. Milk samples of 284 sows with CM and 284 non-infected sows of different age were obtained on six piglet rearing and fattening units. Sows were identified as CM-infected when the rectal temperature was above 39.5 °C. Bacteria involved in pathogenesis were identified by advanced bacteriological analysis of this milk including molecular techniques like PCR. The results were interpreted with regard to the lines, cycles, number of piglets born alive of the sows and the effect of the farm. A wide spectrum of pathogens was isolated, belonging mainly to coliform bacteria, with no explicit differences between healthy and diseased sows. The influence of the farm on the occurrence of specific bacteria was confirmed statistically. The cycles and numbers of piglets born alive had no effect on the incidence of CM, whereas the line presented a significant effect. Bacterial, environmental and animal factors may change the susceptibility for CM. These factors are interdependent and the relative influence of each factor depends on the type of pathogens. A holistic approach, considering husbandry, microbial influences and the genetic background, is needed to cope with future aspects of pig husbandry.

What matters to a sow in a farrowing system: the needs of lactating sows

Ursinus, W.W. and De Greef, K.H., Wageningen University and Research Centre, Animal Sciences Group, Edelhertweg 15, 8219 PH Lelystad, Netherlands; nanda.ursinus@wur.nl

Farrowing crates are commonly used in pig husbandry systems to ensure a high and efficient production level. Welfare of lactating sows is, however, negatively affected by this practice. A study based on existing knowledge is conducted to identify and describe the needs of lactating sows for determining what is important in a farrowing system to ensure optimal welfare. Needs are considered to be those requirements that a sow desires to experience optimal welfare. Concerning welfare it is important to consider if a sow's need gets satisfied or not. A distinction is made in needs of the sows before, during and after farrowing. Knowledge about sows in a natural environment, a facilitated environment, and a deficient environment is compared. To date, fifteen needs are listed and illustrated for the farrowing period and that are either basic needs for sows or specifically related to lactating sows: food intake and foraging, water intake, elimination, exploration including searching a nest site, nest building, behaviour related to parturition, maternal behaviour, resting behaviour, social contact, locomotion, body care, thermo comfort, safety, respiration and health. Special attention is paid to reproduction related behaviours and locomotion. Current systems comprising farrowing crates have insufficient 'degrees of freedom' to allow designs that meet the above mentioned requirements, implying a need for redesign. The identified requirements provide a solid science based basis to add the animal perspective to redesigns for economically, ecologically and societally acceptable housing systems.

Shoulder ulcers in sows: causes of variation

Mattsson, B.[1], Ivarsson, E.[2], Holmgren, N.[3] and Lundeheim, N.[2], [1]Praktiskt inriktade grisförsök (Pig), Filipdalsgatan 6, S-532 38 Skara, Sweden, [2]Swedish University of Agricultural Sciences (SLU), Box 7023, S-750 07 Uppsala, Sweden, [3]Swedish Animal Health Service, Filipdalsgatan 6, S-532 38 Skara, Sweden; barbro.mattsson@svenskapig.se

Shoulder ulcers (SU) among sows have increased in concern. Factors behind this animal welfare issue are reported to include the body condition of the sow, health problems of the sow and also floor condition. A study of the prevalence of SU among Swedish crossbred (L*Y) sows during lactation was performed. In total, 60 herds were visited, and 2668 lactating sows were scored for the presence and severity of SU on both right and left side of the body. At the herd visits, factors associated with the sow, farrowing pen and herd were recorded. The scoring of SU followed a 5 level scoring {0=no lesions; 4=deep wounds}. In the statistical analyses, the finding on each side of the sow was handled as a binary trait, where 0+1 (=0), and 2+3+4 (=1) were grouped together. The analyses were performed using the GLIMMIX procedure in the SAS software. The results showed significantly that: (1) sows in good body conditions have less SU; (2) prevalence of SU was higher on right side (19%) than on left side of the sow (14%); (3) agalactia increased the prevalence of SU; (4) prevalence of SU was higher when the slatted floor in dung area was made of plastic or cast iron, compared with concrete slatted floor; (5) prevalence of SU was higher when the solid concrete floor area, accessible for the sow, was small. Thus a healthy sow in good body condition in a friendly environmental has a lower risk to get SU.

Session 42

Shoulder sores are inherited

Lundgren, H.[1], Zumbach, B.[2] and Lundeheim, N.[1], [1]Swedish University of Agricultural Sciences, Dept. of Animal Breeding and Genetics, Box 7023, S-750 07 Uppsala, Sweden, [2]Norsvin, P.O. Box 504, N-2304 Hamar, Norway; Helena.Lundgren@hgen.slu.se

Shoulder sores in sows is a serious welfare issue in many herds. Main focus is often on environment and management factors, but our aim was to analyze the genetic background of shoulder sores. Data on 2,699 Norwegian Landrace sows were used. The sows were scored after weaning (~5 wks), from 0 (no sore) to 4 (severe open wounds). In total, 75% of the sows had no shoulder sores (score 0) and 14%; 8% and 3% had scores 1; 2 and 3+4 respectively. Variance components were estimated using a single trait animal model including the effects of parity, month of farrowing, herd and regressions on litter size and age at weaning. The heritability for shoulder sores was estimated at 0.14±0.04. This indicates that the problem of shoulder sores should be a matter of concern in breeding programs. Genetic correlations with other traits of interest (e.g. piglet growth and litter size) will be estimated.

Session 43

Electronic identification of animals: the EU's legal approach

Sprenger, K.-U., DG SANCO D1, Directorate General Health and Consumers, B232 03/43, B-1040 Brussels, Belgium; kai-uwe.sprenger@ec.europa.eu

For breeding purposes and also to ensure their livelihoods, livestock keepers from all over the world and not just from Europe have, for centuries, been identifying their animals in the most cost effective manner, with the tools available to them. However, with the establishment of the common market, harmonised rules for the identification of animals and the registration of their holdings were needed. The basic objectives for the EU rules on the identification of animals are the localisation, the tracing of animals and the linking animals with their veterinary documents, which are of crucial importance for the control of infectious diseases. The EU legal framework on identification of animals was mainly reinforced after the experience gained with several serious animal diseases, such as Bovine Spongiform Encephalopathy (BSE) and Foot-and-Mouth Disease (FMD). Depending on the individual needs of the various animal species, different legislation has been adopted for bovine animals (Regulation 1760/2000), sheep and goats (Regulation 21/2004), pigs (Directive 2008/71), equidae (Regulation 504/2008) and pet animals (Regulation 998/2003). The systems include several elements notably identifiers, databases, holding registers, passports or movement documents. The introduction of RFID (radio frequency identification) became a very important – and in some cases compulsory - element of the respective legal framework. The EU legislation allows a gradual introduction of RFID of different animal species. In addition to the aspect of disease control, the EU has always paid very particular attention to the multi-purpose use of livestock-RFID to allow additional benefits for keepers using the technology for their on-farm-management. Besides electronic eartags and boluses for ruminants (cattle, sheep and goats) and injectable transponders (mainly for horses, donkeys, pets and zoo animals), static and dynamic reading systems have been developed to fulfil different needs in practice.

Results of Spanish sheep and goat electronic identification in practice

Rinaldi Iii, A.R., Comité Español de Identificación Electrónica Animal - Tragsega, Julián Camarillo 6a-4b, 28037 Madrid, Spain; aernest@tragsa.es

Since the end of the IDEA Project in 2001, Spain has been specially committed to the implementation of electronic identification (EID) in all livestock species. In order to solve those questions remaining after the end of the EU project, and to obtain all the information required for an optimal implementation of EID in Spain, the Spanish Government has been carrying out several on-field studies in different species. After the publication of Council regulation (EC) 21/2004 most of the resources were focused on sheep and goat EID. During the last decade all available EID equipment has been tested in several field conditions and different species. Main data recorded, among many other relevant data, were: device retention, animal casualties, reading distance, dynamic reading efficiency of flocks wearing a unique EID device or combining different types of devices have been obtained and analyzed. Results obtained include studies on: 1) retention of heavy ruminal bolus (80 g) in Spanish autochthonous goat breeds (2002-2004); 2) retention rates and reading performance of 7 different combined EID devices (bolus and ear tags) by using different handheld and fixed readers in autochthonous sheep breeds (2004-2005); 3) retention rates and reading performance of injectable transponders applied in different positions in autochthonous goat breeds (2004-2006); 4) reading performance of different EID readers and 10 different combined devices under high mountain and extreme hot weather conditions in autochthonous sheep breeds (2004-2006); 5) retention of light bolus (20 g) in autochthonous sheep and goat breeds (2006-2007); 6) comparative retention rate analysis of 20 and 80g boluses in Canary Islands goat breeds (2006-2007); 7) manual recovery efficiency of 20g boluses in Spanish lambs in the abattoir (2006); and, 8) comparative retention rate of EID ear tags and light ruminal boluses (20 g) in autochthonous goat breeds (2007-2008).

Use of radio frequency ear tags in dairy herd management in Canada

Rodenburg, J.[1], Murray, B.B.[2] and Rumbles, I.[3], [1]DairyLogix Consulting, RR#4, Woodstock, Ontario, N4S 7V8, Canada, [2]Ontario Ministry of Agriculture, Food and Rural Affairs, P.O. Box 2004, Kemptville, Ontario, K0G 1J0, Canada, [3]Canwest Dairy Herd Improvement, 660 Speedvale Ave. W., Guelph, Ontario, N1K 1E5, Canada; jack@dairylogix.com

As part of the national livestock identification program, all Canadian dairy cattle have been tagged at birth with uniquely coded half duplex, 134.2 kHz radio frequency identification (RFID) ear tags since 2005. While designed to be a traceability tool, the tags provide a practical and inexpensive option for automated on-farm identification and for use in herd management systems. Field experiences in applications involving automatic calf feeding, parlor identification, and sorting systems are described. Because of lower cost and the regulatory requirement to use these tags, they have completely replaced commercial RFID neck tags in calf feeders. Applications involving older cattle have been less readily available and adoption is slower. Lessons learned from an informal field study of handling systems on one farm with a management rail and one farm with self locking headgates, using wand readers linked to handheld computers and headphones are presented. In one herd, time saved by the herdsman and veterinarian for creating work lists, applying multiple treatments following a single sort and describing and recording treatment events resulted in a 1 year payback. In the other herd, overcrowding, insufficient headlocks and inability to adapt to new technology and new work routines meant that the technology provided no measurable benefit. In addition to labour saving, other potential benefits include fewer treatment errors, ability to collect more detailed management information, and improved cow welfare when the need for restraint for handling is minimized. Practical challenges such as tag retention, tag positioning, reader design and location, and possible solutions and limitations to these issues were also studied.

Evaluation of identification devices in commercial pig farms

Santamarina, C.[1], Hernández-Jover, M.[2], Caja, G.[2] and Babot, D.[1], [1]Universitat de Lleida, Producció Animal, Rovira Roure 191, 25198 Lleida, Spain, [2]Universitat Autònoma de Barcelona, G2R, Campus de la UAB, 08193 Bellaterra, Spain; clara@prodan.udl.cat

With the aim of evaluating and compare traceability of identification devices under field conditions, over 1,800 pigs were tagged with different identification devices in 2 commercial farms in Spain. Identification devices used were: visual ear tags, electronic ear tags and injectable transponders injected intraperitoneally. Electronic devices were from the 2 methodologies of information exchange conforming ISO standards (HDX and FDX). Ear tags were applied, using specific tagger pliers, in the centre of the ear following manufacturer recommendations. Injectable transponders were applied using single-shot injectors with interchangeable needles. Application of devices was conducted at the rearing period and readability was checked at different times during the rearing and growing-fattening periods. Identification devices were recovered in the slaughterhouse previous to carcass cooling, without interfering the slaughter process and ensuring that all devices were recovered and none remained in the carcasses. Devices reported different results of traceability through the on-farm and slaughterhouse periods. During the on-farm period, all devices reported readability higher than 95%. However, during the slaughtering process, readability significantly differed between devices. Injectable transponders were not affected by the slaughtering process, maintaining the identification until evisceration. Contrarily, losses of visual ear tags (>5%) and losses and electronic failures of electronic ear tags (5 to 15%) were observed during the slaughter process. In conclusion, final traceability from farm to carcass release varied according to identification device. Visual ear tags reported traceability of approximately 95%, and electronic ear tags reported traceability up to 90%. Injectable transponders in intraperitoneal position reported the highest traceability value (>98%), achieving the recommendation of ICAR for official animal identification.

Performance evaluation of low radio frequency identification equipment for livestock

Bishop, J., Viaud, P. and Pinato, T., European JRC Ispra, G07-IPSC Tempest Lab (TP.361), Via E. Fermi 2749, 21027 Ispra (VA), Italy; james.bishop@jrc.it

Prediction of in-field performance of radio frequency identification (RFID) transponders and transceivers by laboratory measurements was investigated. Aims of the work were: 1) to define objective and reproducible tests demonstrating whether RFID products are fit for specific applications or not; and 2) to establish whether it is feasible to specify RFID systems for livestock in terms of simple, easily understood concepts such as minimum required reading distance (i.e. related to animal species) and/or throughput (i.e. animals per hour, per minute etc.). Some characteristics of RFID products, such as transponder resonance frequency and return signal spectra or transceiver activation field frequency, strength, and timing are easy to measure with appropriate equipment. These measurements are appropriate for 'black-box' testing (i.e. where a product is tested as received, and no connections to its internals are permitted), and are used in checking conformity to standards, e.g. ISO 11,785. The correlation between transponder minimum activation field strength, transceiver field strength, and observed reading distances under different laboratory conditions was calculated. A measure of a transceiver and antenna combination's ability to detect the signal generated by a transponder is required before predictions of in-field performance can be made. The effects on reading performance of a variety of electromagnetic disturbances were also measured. The disturbances themselves are defined by IEC (International electromagnetic immunity standards. All data were obtained with commercially available readers, but brands were anonymised.

Reading performance of animal radio frequency transponders
Hogewerf, P.H., Van Roest, H. and Van 't Klooster, C.E., Innovative Modern Agriculture - Wageningen, P.O. Box 397, 6700AJ Wageningen, Netherlands; ima-wageningen@hetnet.nl

Test procedures for low radio frequency identification (RFID) transponders and transceivers used for animal identification have been developed by the International Standardization Organization (ISO). This was done in close cooperation with the International Committee for Animal Recording (ICAR).The ISO 24631-3 (Radiofrequency identification of animals — Test procedure — Part 3: Evaluation of the performance of ISO 11784 and ISO 11785 RFID transponders) describes the procedure for measuring the following transponder performance characteristics: minimum activation field strength, dipole moment and bit length stability (for FDX-B transponders) or frequency stability (for HDX transponders). Conform the ISO 24631-3 protocol, the characteristics of 19 ICAR approved transponders were measured (15 ear tag and 4 bolus transponders), 5 of those transponders were of the HDX type and 14 transponders of the FDX-B type. The reading performance of these transponders was also estimated by measuring the reading distance with 3 different handheld transceivers capable of reading transponders complying ISO 11784 and ISO 11785. All measurements were done under laboratory conditions (ambient noise floor and ambient peak noise: <30 dBµV/m and bandwidth 2.7 kHz)). A good relationship between the electronic performances and the recorded reading distances was found (R^2=0.73 for FDX, and R^2=0.89 for HDX transponders). These results lead to the conclusion that the ISO 24631-3 measurements can be used as approval criteria for transponders for livestock applications were a certain reading desistance is required.

Performance tests of electronic identification devices and equipment under various beef cattle management scenarios
Fike, K.E., Rickard, B.A., Ryan, S.E. and Blasi, D.A., Kansas State University, Animal Sciences & Industry, 229 Weber Hall, 66506 Manhattan, Kansas, USA; dblasi@ksu.edu

Seven ISO 11785 transponder types (n=1,200) that were previously characterized for performance indices (read rate, read range, resonance frequency) under laboratory conditions were applied into ears of beef cattle and subsequently interrogated by transceivers under different cattle management scenarios (auction market, feedlot and abattoir). The objective of this study was to assess the effect of ear placement of transponders on read rate. The transponders were randomly assigned to be applied in one of two locations in the left ears of cattle (PP = primary position, between two main cartilage ribs of ears; and SP = secondary position, top, above the upper cartilage in the ear's curvature) nearest the base of the head. Logistic regression in the GENMOD procedure of SAS was used to develop a prediction model for the probability of a successful interrogation of a transponder for each production scenario. The model included the main effects of transponder type and ear location and their interaction. For the most part, the predicted read rate ranking of the different transponders at various beef cattle management scenarios was consistent with the laboratory findings. A transponder type x ear location interacted ($P<0.001$) to affected predicted read rates when interrogated at the feedlot receiving scenario only. The predicted read rate of only two transponder types was reduced when the transponder was in the SP position ($P<0.05$). In conclusion, location of the e-ID device in the bovine ear is not a critical factor for assuring maximum read rates.

Framework actions for implementing the use of electronic identification of sheep and goat in Spain
Maté Caballero, J., MARM, Alfonso XII 62, 28014 Madrid, Spain; fmate@mapya.es

Spain addressed many efforts for an early implementation of the electronic identification (EID) of sheep and goat according to the Council Regulation (EC) 21/2004. As a result of transposing the regulation into the Real Decreto 947/2005, Spain became the first EU state establishing a compulsory EID scheme for all sheep and goats older than 6 months. Consequently, a framework plan of actions was put in place during the following years. First step was the standardization of EID devices and equipment, which was done by the UNE 68402 standard. Next step was to interface the EID with the different data bases, for which information transfer systems needed to be created or adapted. Emerging problems were analyzed and the new needs of the sector covered, such as training of personnel. New structures were build up to support the developments. The most remarkable was RIIA (National database for animal EID), for registering all the EID animals. Other key database that were interconnected with RIIA were: REGA (National holding registration database), including GIS data (Geographic Information System), and REMO (National movement registration database). These 3 database systems were finally integrated in the SITRAN (national traceability of livestock system). Other databases created, not strictly needed for EID registration, but also based on the use of EID were: Unicity (national database which guarantees the uniqueness of each EID code) and ARIES (national ovine genotype database which gathers information on the genotypes for resistance to scrapie). Efforts made by the Spanish Government since 2002 to divulgate this new system have been remarkable: 2 international meetings (RIIE 2005 & 2007) with the participation of more than 20 countries and several publications. Moreover, in order to facilitate the implementation of the system, Central Administration subsidized up to 60% of the total cost of readers, EID devices and labour cost. Autonomous Communities also subsidized with different percentages, up to 40%. Farmers are supporting less than 10% of the total implementation costs.

Use of electronic identification for automated oestrus detection in livestock
Bocquier, F.[1], Viudes, G.[2], Maton, C.[2], Debus, N.[2], Gibault, L.[2] and Teyssier, J.[2], [1]Montpellier SupAgro, Place Viala, 34 060 Montpellier, France, [2]INRA, Place Viala, 34 060 Montpellier, France; bocquier@supagro.inra.fr

Electronic identification (e-ID) of livestock allowed the development of new equipments able to assist producers in farm management practices. Oestrus detection is a key for many breeding practices (e.g. artificial insemination) and strongly determines fertility results. With this aim, an autonomous radio frequency reading and recording equipment, automatically triggered by the male mounting activity, was developed (Pat. WO/2005/065574). The reader captures the e-ID codes of mated females which are stored jointly with the reading time. Captured data is further sent wireless to a remote computer and allows defining the oestrus characteristics of each female. In a first experiment, a group of 30 ewes were identified with 15 mm glass transponders fixed on the tail base and mated with 1 equipped ram. Few hours after ram introduction isolated readings indicated that the ram made mating attempts on non-receptive ewes (false positive). Thus, only ewes that had repeated readings were considered to be in oestrus (true positive). The number of readings per ewe were largely variable (3 to 205) as well as mean oestrus duration (mean = 20.4 h; SD=13.0 h). All detected ewes lambed. In a second experiment, 9 series of 4 ewes were oestrus synchronised (FGA sponges and eGH injection) from late summer to autumn. Sexual behaviour was more intense with the advancement of the reproductive season (i.e. the mean number of readings per ewe increased from 19.5 to 68.0). Oestrus activity started on average 18.5 h after sponge removal, duration evolved from 9.3±5.2 h in early September, to 31.6±23.3 h in mid-October. New equipment is under development for reading ewes identified with rumen boluses (grant INRA-DGER) and research is conduced to assess the best conditions for hormone-free insemination after oestrus detection. Similar devices for cows and goats are under development. The equipment may also be used for sorting pregnant and open females and parturition dates planning.

Electronic identification of Manchega dairy sheep: a balance after 10 yr of implementation (1999-2009)
Gallego, R.[1], Garcés, A.[1], García-García, O.[1], Domínguez, C.[1] and Caja, G.[2], [1]AGRAMA, Av. Gregorio Arcos s/n, 02006 Albacete, Spain, [2]Universitat Autònoma de Barcelona, G2R, Campus de la UAB, 08193 Bellaterra, Spain; rgallego@agrama.org

Manchega sheep is exploited in the centre of Spain (Castilla-La Mancha) for milk (DO Manchego cheese) and meat (IGP lamb) production. AGRAMA (Asociación de Ganaderos de Raza Manchega) drives the flock-book of the breed from 1969. Electronic identification (e-ID) was initiated in 1999 by enrolling in the IDEA project a total of 20,391 sheep. As a result, the AGRAMA board agreed in 2002 to use electronic boluses as official ID for all flock-book activities (registration, milk recording, artificial insemination, morphological evaluation, farm inventory, lambing recording, scrapie genotyping and health programs). A special information and training program on the use of e-ID was set up in 2005 for farmers and technicians. From the start, more than 250,000 sheep have been bolused and used for performance recording and management purposes. Cost of e-ID is estimated <2% milk yield. Specific software has been developed to interface with e-ID transceivers on the basis of simplicity, error control, access easiness and compatibility with other software. Software available for handheld readers are 'Idetec' (data up- and down-loading, data base updating by internet; PDA version is also available) and 'GenConsultas' (inventory). A software application for dynamic reading named 'Agramanga' (inventory, data and animal search) is also available. Moreover, a windows interface 'Agrama Smart Client' with a web structure and centralized data base, has been also implemented. This software is compatible with commercial software for milking parlour management and milk recording. Implementation results showed dramatic time reductions in milk recording and milk data processing, lambing data acquisition and management, and inventorying. Additionally, data errors are currently insignificant. In conclusion, electronic identification is today an irreplaceable tool for performance recording and data management in the Manchega dairy breed.

Extensive management of a beef breeding farm based on RFID technology in Argentina
Baldo, A.[1], Beretta, E.[2], Sorarrain, N.[1], Nava, S.[2] and Lazzari, M.[2], [1]Universidad Nacional de La Plata, Facultad de Ciencias Veterinarias, Zootecnia II, Casilla de correo 296, B1900AVW La Plata, Argentina, [2]University of Study Milan, Fac. Vet. Med., V.S.A. sez. Bioengineering, Via Celoria 10, 20133 Milano, Italy; ernesto.beretta@unimi.it

An integral identification and traceability system, based on the use of radio frequency identification (RFID) technology according to current European Union standards, was developed in an cattle farm in the area of El Salado (Province of Buenos Aires, Argentina). No electricity supply was available at the farm, which used car batteries as electric power source. The cow-calf breeding farm produced purebred Angus calves under a sustainable production system based on grazing improved pastures and maximizing the animal welfare concept. Experimental aims were to test different types of electronic identifiers (ceramic bolus and button ear tags) and transceivers (stationary and handheld readers) for the automated capture of breeding and management data of beef cattle under extensive conditions. Cows were identified with electronic ear tags and calves were identified with small boluses (52 g) at 1 to 25 d of age (26 kg BW on average). No bolus insertion problems or pain symptoms were observed at bolus administration. During the following year, both electronic identifiers were read by using a frame antenna fixed in the right side of a cattle crush equipped with an electronic scale. Retention and reading efficiency of the electronic identifiers were 100%. Specific breeding and management software was developed for on field data capture and key data querying. Data was transferred from the transceivers to a personal computer by Bluetooth or serial cable connection. Full system implementation costs per cow were estimated as equivalent to 5.93 kg BW of calf for the 210 cows herd (3.76 kg BW of calf for a 1,000 cows herd). The implemented system showed adequate functionality and economic sustainability on the Argentinean beef cattle scenario.

Using EID+DNA traceability system for tracing pigs under commercial farm conditions

Hernández-Jover, M.[1,2], Caja, G.[2], Ghirardi, J.J.[2], Reixach, J.[3] and Sánchez, A.[2], [1]University of Sydney, 425 Werombi Rd., 2570 Camden, Australia, [2]Universitat Autònoma de Barcelona, G2R, Campus de la UAB, 08193 Bellaterra, Spain, [3]Selección Batallé, Av. Segadors s/n, 17421 Riudarenes, Spain; mhernandez_jover@usyd.edu.au

With the aim of validating the EID+DNA traceability system a total of 2,108 Duroc male piglets were used. The system was based on the use of electronic identification (EID) by injectable transponders and DNA fingerprinting by analysis of a set of porcine microsatellites (n=12). Piglets were identified (9±3 d of age) using 32 mm half-duplex ISO transponders injected intraperitoneally (IP) in commercial farm conditions. Piglets were ear tagged and biopsied at the moment of IP injection, using 2 types of biopsying button ear tags (E1, n=979; and, E2, n=1,129). Grow-fattening up to 120 kg BW (7 mo of age) was conducted under intensive conditions. Harvesting was done in a high throughput abattoir (500 pigs/h) and pig EID was automatically transferred to the corresponding carcass, using a high frequency inlay label (45 × 76 mm; Tag-it, Tiris, Almelo, Holland). Samples were taken from carcasses using biopsying tubes and stored frozen until DNA analysis. On-farm losses of IP transponders were 0.6% (58.3% of them during wk 1 post-injection) and ear tag losses were 0.7 and 0.4% for E1 and E2, respectively. On-farm traceability results of alive pigs were: E1 (99.3%), E2 (99.6%) and IP (99.4%). Ear tag losses during harvesting (E1, 35.3%; E2, 37.6%) indicated their use for traceability was unsuitable in the slaughtering line. No losses of IP were reported during harvesting. Automatic transfer of EID to carcasses was 95.1% successful. Final pig traceability from birth to slaughter was 94.5% for IP transponders. A total of 100 pairs of samples (5%) were analyzed for DNA auditing. Results did not match in 4 pairs of samples, showing 96% traceability. In conclusion, the use of IP transponders improved pig traceability. Nevertheless, automatic transfer of EID to carcasses needs to be improved to provide a more reliable technique for the pig industry.

Economic analysis in relation with electronic identification of small ruminants

Hofherr, J., Natale, F., Solinas, I. and Fiore, G., EU Commission - JRC, CI-Animal&Food Action, Via Marconi, 21020 Ispra (Varese), Italy; johann.hofherr@jrc.it

Regulation (EC) 21/2004 requires individual identification of small ruminants in the EU Member States, and for animals born after 31 December 2009, electronic identification (EID) will become mandatory when sheep and goat population is >600,000 heads. This study calculated the costs of EID equipment, tagging and reading at national level for 4 case studies characterising different sheep production systems present in the EU. The costs and possible advantages of 4 different options for implementing EID were also compared. The results indicated that the major cost impact comes from reading and data processing equipment, and that the vast majority of the costs occurred on farm side. While holdings on which the animals were born supported the main EID costs, clear cost savings for reading were seen at market, assembly center and slaughterhouse levels. Even EID of traditional flocks should result in substantial cost savings when farm holdings delegated readings in connection with movements to control points. Although the slaughter derogation would reduce costs for farm holdings, it would also reduce traceability, in particular of frequent movements and batches of animals from different holdings of origin. This option would not unfold full advantage of automatic readings, confronting markets, assembly centers and slaughterhouses with 2 different identification systems (visual and EID) permanently. Additionally, EID is not only a cost, and its multi-purpose use may produce also benefits at farm and administrative level, as well as downstream in the production chain. Finally, for its primary aim regarding official identification and recording, EID will increase reading accuracy and, when fully implemented, will reduce time needed for movement and tracing quering and research, which benefits were not quantified in this study.

Cost-benefit analysis of the U.S. National Animal Identification System in California

Butler, L.J.[1], Oltjen, J.W.[2], Velez, V.J.[3], Evans, J.L.[3], Haque, F.[1], Bennett, L.H.[1] and Caja, G.[1,4], [1]University of California, Agricultural and Resource Economics, One Shields Av., Davis, California 95616, USA, [2]University of California, Animal Science, One Shields Av., Davis, California 95616, USA, [3]California Department of Food and Agriculture, Animal Health Branch, 1220 N Street, Sacramento, California 95814, USA, [4]Universitat Autònoma de Barcelona, G2R, Campus de la UAB, 08193 Bellaterra, Spain; butler@primal.ucdavis.edu

We studied benefits and costs of an animal identification (ID) system consistent with the US National Animal ID System (NAIS) in California. Both primary (animal health) and secondary (management and marketing) benefits depended on level of participation. We developed a large, detailed model for costs of animal ID technologies for beef (3 subsectors - cow-calf, feeder and stocker operations), dairy (3 subsectors - cow-heifer, cows only, and heifer raising operations), sheep and dairy goats. Technologies analyzed were electronic identification (RFID) ear tags and visual ID ear tags. The model consisted of 4 types of costs – labor, materials, equipment and services. Inputs were from survey results and data provided by State agencies or companies. Results showed that costs increase as one proceeds through the four steps in the system: premises registration, animal ID, event recording, and event reporting, with the largest increase for animal ID. Costs were directly proportional and inversely related to size of operation. Costs appeared higher for smaller operations mainly because of lumpy capital investments such as RFID readers, computers, and software. Visual ID was always less expensive, but was burdened by costs associated with recording information without benefit of RFID technology. RFID dynamic reading technology was more expensive and appropriate only for large operations. Labor costs were low and relatively insignificant. The largest costs were equipment costs. It is the lumpy nature of these costs that created the dramatic economies of size in the RFID technologies, and the shape of the cost function.

Cost and benefits analysis of electronic identification of sheep in England

Parsonage, F. and Warner, S., Department for the Environment Food and Rural Affairs (Defra), 1A Page Street, London SW1P 4PQ, United Kingdom; fred.parsonage@defra.gsi.gov.uk

The objective of this study was to estimate the cost and benefits of implementing new EU rules on the electronic identification (EID) and individual recording of sheep and goats (Council Regulation (EC) 21/2004) in England. The new rules have been introduced to improve traceability and the ability of Member States to effectively manage outbreaks of animal disease. The UK has the largest number of sheep in the EU (35 million head) and because of its stratified production system sheep generally move more than in other Member States (50-60 million movements per annum). The implementation of electronic identification and individual recording in the UK will therefore have a greater impact on the UK industry than in other Member States. Research commissioned by DEFRA on the implementation of EID indicates that the cost of implementing EID and individual recording in England outweighs the benefits. Depending on the level of derogations adopted, this ranges from: equipment start up costs, €11.60 million to €27.36 million with annual costs from €4.77 million to €12.78 million. Scaled up costs for the UK would be in the region of €26.98 million to €63.62 million for equipment and €11.10 million to €29.72 million for annual costs (Exchange rate calculated as at 27 March 2009 as £1 to €1.0689).

Implementation cost of sheep and goat electronic identification and registration systems in Spain according to European regulation

Milán, M.J., Caja, G., Ghirardi, J. and Sàa, C., Universitat Autònoma de Barcelona, G2R, Campus UAB, 08193 Bellaterra, Spain; MariaJose.Milan@uab.cat

A cost model was used to asses and compare different implementation strategies of the European Regulation for sheep and goat identification and registration in Spain (EC 21/2004 modified by EC 1560/07). Strategies were: 1) visual identification (VID) by 2 ear tags; 2) electronic identification (EID) by 1 bolus and 1 ear tag; and, 3) mixed VID and EID strategy (MID), consisting of VID for fattening stock and EID for breeding stock. Complete and simplified implementation of the regulation were considered as options. Total costs per animal identified for all strategies and options varied according to the implementation option, ranging between 2.14 and 3.26 €. The full EID was the most expensive strategy (3.19 to 3.26 €) for all implementation options. Cost of VID and MID strategies ranged from 2.30 to 2.59 € and from 2.14 to 2.56 €, respectively. The model was submitted to a sensitivity analysis without considering extra benefits of sheep and goat identification. Critical values for which the cost of MID equaled VID depended on strategy and option, and ranged from 3.5 to 5.0% for ear tag losses, and from 1.30 to 1.90 € for bolus price. In conclusion, price reduction of devices and reading equipment occurred in the last years make that electronic identification was a profitable alternative in cost terms to visual identification. The use of a mixed strategy combining visual ear tags (animals intended for slaughter) and electronic boluses (breeding stock) seems to be a currently profitable strategy to fulfill the EC Regulation requirements for the identification and registration of sheep and goats in Spain.

Comparison of leg bands and rumen boluses for the electronic identification of dairy goats

Carné, S., Caja, G., Rojas-Olivares, M.A. and Salama, A.A.K., Universitat Autònoma de Barcelona, G2R, Campus de la UAB, 08193 Bellaterra, Spain; sergi.carne@uab.cat

Murciano-Granadina dairy goats (n=220) intensively managed were used to evaluate the identification (ID) performance of: 1) leg bands (LB), consisting of plastic bands (181×39 mm, 21 g; n=220) closed with 2 types of button transponders (T1, 26.5 mm o.d. open female, 3.9 g, n=90; T2, 25 mm o.d. closed female, 5.5 g, n=130) on the metatarsus of the right leg; and 2) rumen boluses (RB, 68×21 mm, 75 g, n=220) that contained 32×3.8 mm transponders and were used as control. To decide on the inner perimeter of fastened LB, metatarsal perimeter of 47 replacement (6 mo of age) and 103 adult goats was previously measured. Time for LB tagging, reading of T1 or T2, and data recording with an ISO handheld reader was measured. Visual and electronic readability [(read/readable) × 100] was monitored during 1 yr in the milking parlour. Dead or culled animals were excluded (n=23). Metatarsus perimeter of replacement (70±1 mm) and adult goats (88±1 mm) were lower ($P<0.001$) than the inner perimeter of the fastened LB (110±1 mm). Replacement 6-mo goats were deemed inadequate for LB application. Time for LB tagging, and data recording was 53±3 s, similar to time previously reported for RB (49 s). At 1 yr, 2.5% RB (n=5) were lost, 3.6% T (n=7) were unreadable due to breakage, and 1.5% LB (n=3) were removed due to limping (1 inflamed leg was constricted by the LB, and 2 LB were too loose and got blocked on the pastern). Readability of T1 and T2, excluding the LB removed, was 93.9 and 98.3%, respectively ($P=0.11$). No readability difference between LB and RB was detected (98.5 vs. 97.5%, respectively; $P=0.48$). Only T1 and LB differed ($P<0.05$). In conclusion, leg bands were not adequate for the early ID of replacement goats although, suitably designed, may be a valid method for adult goats. In our conditions, electronic button tags placed on leg bands and standard sized rumen boluses did not meet the ICAR requirements for official goat identification (>98%).

Cost-benefit study of implementing electronic identification for performance recording in dairy and meat sheep farms

Ait-Saidi, A., Caja, G., Milán, M.J., Salama, A.A.K. and Carné, S., Universitat Autònoma de Barcelona, G2R, Campus de la UAB, 08193 bellaterra, Spain; ahmed.salama@uab.cat

Automated performance recording based on electronic identification (e-ID) proved to be a useful tool for increasing reliability of data collection and for saving labor time in sheep farms. Despite this, there is few data available on cost-benefit analysis of implementing e-ID in practice. With this aim, previous results comparing manual, semi-automatic and automatic performance recording in dairy and meat sheep farms were integrated in a cost-benefit analysis under Spanish conditions. The analysis scenarios considered: production aim (dairy, meat), intensification level (semi-intensive, extensive), flock size (400, 700), lambing rhythm (1, 1.5), milking or milk recording frequency (once-, twice-daily), and included weighing (2, 3 times/yr) and 1 annual flock inventory as stated in the current EU regulations. Annual extra costs necessary to buy the ID devices and the required reading equipment for e-ID ranged between 0.33 and 0.51 €/ewe according to scenarios. On the other hand, annual savings achieved by implementing e-ID ranged between 0.28 and 0.61 €/ewe according to scenarios. As a result benefits of using e-ID fully covered the annual extra costs in the case of dairy farms with 2X (or A4 milk recording) and in intensive meat farms having 1.5 lambing/yr. For dairy farms performing 1X (or AT method) or in extensive meat production, the savings only covered 93 and 86% of the extra costs for e-ID, respectively. Benefits of using automatic recording reached break-even points at: dairy sheep farms (477 ewes milked 1X, or 279 ewes milked 2X) and meat sheep farms (1,110 ewes under extensive, and 565 ewes under intensive management conditions). In conclusion, the implementation of electronic identification for performance recording showed to be cost-effective, especially for large flocks. Ongoing innovations and new software management will also make more profitable the use of e-ID in the future.

Study of preliminary conditions for a large scale use of cattle and small ruminants electronic identification

Duroy, S., Holtz, J., Rehben, E., Marguin, L., Balvay, B. and Gilain-Galliot, C., Institut de l Elevage, Génétique, 149 rue de Bercy, 75012 Paris, France; jacques.holtz@inst-elevage.asso.fr

As a result of the continuous improvement of livestock electronic identification, French farmers unions and the Ministry of Agriculture decided to analyze the possibility of applying this technology to official cattle, sheep and goats identification. With this aim, different pilot projects have been conducted since 2005. The projects involved different type of production, trade and slaughtering operators in cattle, sheep and goats and consisted on testing experiments on the use of EID for official identification under on-farm conditions. They dealt with the practical advantages for operators' daily work and the improvement of data processing. More than 250 farmers, 10 livestock traders, 3 sale yards, 12 slaughterhouses are currently involved, representing near 450,000 sheep, 50,000 cattle and 6,000 goats. Main on farm study topics focused on: (1) data exchange with herd management software; (2) links with automatic equipments; (3) automation of animal weighing; and, (4) their use for animal recording. In saleyards and slaughterhouses, the work was mainly addressed on the reliability of EID reading for moving animals. In conclusion, among the preliminary requirements for a large scale use of EID, it seemed to be necessary: (1) to be able of easy integration of stationary readers in existing management equipments (as race-ways and weighing cages); (2) to have a large range of ergonomic and interoperable identification and reading devices; (3) to get alternative solutions for non-read animals; and, (4) to guarantee the ISO standards recognition by automatic equipments on farm conditions. Moreover, development and implementation of new solutions during this experimental period, for instance in assistance during cow milking, have confirmed the expected potentialities of the electronic identification.

Electronic identification for individual water intake measurement in sheep

Ricard, E.[1], Weisbecker, J.L.[2], Aletru, M.[2], Duvaux-Ponter, C.[3] and Bodin, L.[1], [1]INRA, UR631 SAGA, Castanet Tolosan, France, [2]INRA, UE0065 Domaine Experimental de Langlade, Pompertuzat, France, [3]UMR INRA-AgroParisTech, PNA, Paris, France; Edmond.Ricard@toulouse.inra.fr

For more than 20 years, INRA SAGA has developed data collecting systems using radio frequency identification (RFID) devices for its own research needs in sheep and goat experimental farms. Besides the main goal, which was to provide a more reliable identification of the animals, RFID became a real technical tool for collecting new data (food intake, speed of milking, …) but also for simplifying data collection and decreasing the difficulty of animal management (tracking, automatic sorting, …). After several tests with different devices (ear tag, bolus, implant) it was decided to focus on an ear tag as the support of RFID. Reasons were the possibility of using this device from birth and its low loss rate(<1% per year) which makes it reliable for identification of sheep and goats. A system to measure individual water consumption has been developed. The objective was to succeed in the detection of animals which alter their water intake or stop drinking for a health problem or a change in their physiological status (oestrus...). The system uses electronic ear tag for identification and is based on a prototype previously developed for calves. Water trough provided with the measuring system was designed with dividing walls to prevent several animals drinking at the same time. A RFID reader (ISO standard) set up on a wall of this water trough recorded the animal identification code and time, while a water level sensor measured the quantity of water consumed. Data were recorded in a computer at the end of each visit during which sheep were able to drink ad libitum. The software recorded the details of each visit and allowed following the daily consumption of each animal on a given period of time. The device is also able of detecting individual changes in water intake.

RFID systems of cattle farm in Vojvodina

Trivunović, S.[1], Kučević, D.[1], Plavšić, M.[1], Stankovski, S.[2] and Ostojić, G.[2], [1]Faculty of Agriculture, Animal Science, Trg D. Obradovića 8, 21000 Novi Sad, Serbia, [2]Faculty of Technical Science, Industrial Engineering and Management, Trg D. Obradovića 6, 21000 Novi Sad, Serbia; tsnezana@polj.ns.ac.yu

This paper describes the use of basic elements of radio frequency identification (RFID) for cattle farms in Vojvodina. Practical considerations on the use of RFID elements and their reading and tracing ability for cattle farms is given. The RFID technology is an appropriate tool for animal identification in practice and the use of different transponder types, including the open questions for using UHF tags, is discussed. There are 4 main ways in which RFID can be used for animal identification: attaching a transponder to a collar, attaching a transponder in an ear tag (similar placement to current ear tagging, injecting small glass transponders in different body sites, or orally administering a bolus where the RFID transponder is encapsulated and which is permanently retained within the animals fore-stomachs. Obtained results of the conducted research showed that reading error averaged 0.51%. After 1 mo of usage, there was a tendency for increasing the errors to 1.1%. The most significant problem was identification device losses. The research showed that percentage of losses were higher by using the electronic ear tags when compared to conventional ear tags. It was assumed that the main reason for the higher losses was the greater weight of the electronic ear tags.

Use of electronic identification for milk production recording in dairy sheep farm

Štoković, I.[1], Sušić, V.[1], Karadjole, I.[1], Ekert Kabalin, A.[1], Mikulec, Ž.[1], Kostelić, A.[2] and Menčik, S.[1], [1]University of Zagreb, Faculty of Veterinary Medicine, Heinzelova ul. 55, 10000 Zagreb, Croatia, [2]University of Zagreb, Faculty of Agriculture, Svetošimunska c. 25, 10000 Zagreb, Croatia; igor.stokovic@vef.hr

Milk production recording is a very important activity for farm management in dairy sheep. Milk recording is done by collecting individual ewe identification (ID) and milk production data. Small ruminants are usually tagged with ear tags but this tagging system is not fully reliable in extensive production systems. As an alternative, the use of electronic ruminal boluses was chosen for tagging because they are easy to apply, easy to read and fraud resistant. Additionally to the improved ID for traceability, the use of automatic reading for electronically tagged animals gives new opportunities for recording animal performance data. Programmable hand held readers (Gesreader2, Rumitag) with stick antennas were programmed for milk recording with the following data being collected automatically (electronic ewe's ID and date) and manually by typing (milk yield data). Data were transferred from readers to computers in very simple format able to be used in any statistical software. In our trial milk recording was done in accordance with the A4 ICAR method (2 recordings in 24 h) every 30 d. Using this methodology, 4 milk recordings were made in 608 sheep from April to June 2008. Sheep were tagged with ruminal mini-boluses (Rumitag 20 g, half-duplex). Average milk yield in a.m. and p.m. milkings were 0.46 and 0.33 l, respectively. Reading failures were 5.88% on average. Most probable reason for failures was the use of hand held readers in crowded environment which caused double readings (reading same animal twice). In conclusion, the use of electronic ID was a useful tool for sheep milk recording in our conditions but the reading protocol should be improved.

Retinal image recognition for identifying and tracing live and harvested lambs

Rojas-Olivares, M.A., Caja, G., Carné, S. and Salama, A.A.K., Universitat Autònoma de Barcelona, G2R, Campus de la UAB, 08193 Bellaterra, Spain; mariaalejandra.rojas@campus.uab.es

A total of 152 lambs of 2 breeds (Lacaune, n=70; Manchega, n=82) were used to evaluate the use of retinal image (RI) recognition technology for identification and traceability. Retinal image and capturing time (CT) were recorded using an OptiReader device (Optibrand, Fort Collins, CO), at: 3 (live, n=152; slaughtered, n=50), 6 (n=58) and 12 mo of age (n=58). The first 2 wk (264 images) were used for operator training. Digitalized RI (1,272 images) were treated by using the Optibrand Data Management Software (v. 4.1.3), and the 3-mo enrolment RI were used as the reference for further analysis. Intra- and inter-age comparisons of pairs of RI were done using Optibrand's matching score (MS) as exclusion criterion (MS<80). Lambs wore ear tags and electronic boluses as a controls. Values of MS and CT during the training period averaged 92.1±0.8 and 144±15 s, and improved to 96.4±0.5 and 63±5 s when operator was already trained. Intra-age breed effects were detected in MS at 3-mo (P<0.10), but no effects were observed at 6 (96.2±0.6) and 12 mo (96.3±0.6); moreover, breed affected CT (P<0.001) being longer in Lacaune than Manchega lambs, and decreased by age to 34±4 s and 21±2 s, for 6 and 12 mo, respectively. Values in the slaughtered lambs were 67.4±2.0 and 45±7 s, both varying by breed (P<0.05). Live vs. slaughtered RI comparisons were unsatisfactory (MS = 58.4±1.8). Inter-age image analysis, used as a traceability indicator, showed no effects (P>0.05) by age (6 mo, 90.9±1.2; 12 mo, 90.6±1.1) and breed (Manchega, 91.7±1.7; Lacaune, 89.9±1.4). On average, MS and CT values were 90.8±1.1 and 27±2 s. Percentage of RI showing an inter-age MS higher than 70, 80 or 90 were 97.5, 83.6 and 68.1%, respectively, which indicates the convenience of using a MS = 80 as breaking value. In conclusion, retinal image recognition was a useful technology for auditing the identity of living lambs, but its use was not adequate in harvesting plants.

Electronic vs. visual identification for lambing data and body weight recording under farm conditions
Ait-Saidi, A., Caja, G., Carné, S. and Salama, A.A.K., Universitat Autònoma de barcelona, G2R, Campus de la UAB, 08193 Bellaterra, Spain; gerardo.caja@uab.cat

Manual (M), semi- (SA) and fully-automated (A) systems for sheep performance recording were compared in 2 experiments. System M used: visual ID (plastic ear tags), on-paper data recording and manual typing for data uploading. Systems SA and A used: electronic ID (20 g mini-boluses, 56 × 11 mm, containing 32 mm half-duplex transponders); performance data were recorded using a reader with keyboard in SA (i.e. lambing data) or automatically in A (i.e. BW), and automatic data uploading was done for SA and A. Each ewe wore an ear tag and a mini-bolus. Exp. 1 compared M and SA lambing recording in dairy (n=73) and meat ewes (n=80) processed in groups of 10. Time for lambing data recording was greater in dairy than meat ewes ($P<0.05$), due to the lower operator experience and because half of the dairy ewes needed ear tag cleaning, but M was greater than SA ($P<0.05$) for both dairy (1.11 vs. 0.80 min/ewe) and meat (0.78 vs. 0.68 min/ewe) ewes. Average time for data uploading was greater in M vs. SA (0.54 vs. 0.06 min/ewe; $P<0.001$). Consequently, overall time for lambing recording was greater in M than SA in dairy (1.67 vs. 0.87 min/ewe; $P<0.001$) and meat ewes (1.30 vs. 0.73 min/ewe; $P<0.001$). Data uploading errors were 4.9% in M, but no errors were detected in SA. In Exp. 2, ewes' BW was recorded by M and A systems in dairy (n=120) and meat ewes (n=120). Ewes were processed in groups of 20 using an electronic scale which was interfaced to a computer for A. Weighing time varied according to breed behavior and was greater in M than A (0.45 vs. 0.23 min/ewe; $P<0.05$). Average time for data uploading (0.18 vs. 0.02 min/ewe; $P<0.05$) and errors (8.8 vs. 0%) were greater in M than A. Overall time for BW recording in M and A was 0.63 and 0.25 min/ewe, respectively. In conclusion, the semiautomatic and automatic performance recording systems increased the throughput and reliability of sheep recording data, which resulted in significant savings in farm labor time.

Breaking resistance of lamb ears according to ear tag position and breed
Caja, G.[1], Xuriguera, H.[2], Rojas-Olivares, M.A.[1], González-Martín, S.[2], Salama, A.A.K.[1], Carné, S.[1] and Ghirardi, J.J.[1], [1]Universitat Autònoma de Barcelona, G2R, Campus de la UAB, 08193 Bellaterra, Spain, [2]Universitat de Barcelona, DIOPMA, Martí i Franqués 1, 08028 Barcelona, Spain; xuriguera@ub.edu

A total of 55 lamb ears tagged with official plastic ear tags were obtained after harvesting in a commercial slaughterhouse and used to study the resistance to breakage when submitted to tensile forces under laboratory conditions. Ears were cleaned with cold and warm tap water and classified according to breed (Manchega, n=23; Lacaune, n=11; other, n=21), side (left, right), ear tag insertion position (distal, central, proximal) and preservation method (<7 d in a refrigerator, 2 to 4 wk in a freezer). Breaking force was measured by submitting the ears to a tensile test using a computer-controlled universal testing machine (PCM Mecmesin). Ears were locked to a fixed clamp at the insertion base and ear tags fixed to a mobile clamp to be tested by pulling the ear tag at a constant displacement rate (500 mm/min) until the ears broke. On average, ears measured 115±2 mm long and 54±1 mm wide, have ear tags adequately inserted (36% distal, 24% central and 40% proximal) and broke longitudinally at 155±9 N (9.8 N = 1 kgf). No ear tags broke or opened during the test. Ear breaking force varied quadratically ($R^2=0.99$, $P<0.001$) according to ear tag insertion position (distal, 103±7; central, 136±8; proximal, 213±13 N) and by breed (Lacaune > Manchega > other), but did not vary according to side or preservation method ($P>0.05$). In conclusion, ear tag position was a key factor for ear breakage. New ear tag design taking into account sheep ear resistance may improve sheep welfare and ear tag retention for long-term identification according to the new EU regulations.

Session 43

Bolus features for the electronic identification of goats

Carné, S., Caja, G., Ghirardi, J.J. and Salama, A.A.K., Universitat Autònoma de Barcelona, G2R, Campus de la UAB, Bellaterra, 08193, Spain; sergi.carne@uab.cat

A total of 2,482 electronic identification boluses varying in dimensions were used to build-up a regression model for predicting their long-term retention in the forestomachs of goats. Goats were of dairy (Murciano-Granadina, n=1,536; Alpine, n=394) and meat (Blanca de Rasquera, n=552) breeds. Boluses consisted of 19 cylindrical capsules with a wide range of physical features: length (37 to 84 mm), outside diameter (9 to 22 mm), weight (W, 5 to 111 g), volume (V, 2.7 to 26.0 mL) and specific gravity (SG, 1.0 to 5.5). Each bolus contained an ISO half-duplex glass encapsulated transponder (32 × 3.8 mm). Boluses were administered by trained operators using adapted balling guns. Full-ISO handheld transceivers were used to perform static readings. Retention rate (RR = 100 × read/applied) was calculated from data recorded after 12 to 18 mo post-application. No administration problems were observed for any bolus type. Value of RR ranged from 0 to 100% depending on bolus features. Optimum RR (100%) was obtained with 4 bolus types that differed in W and V, but with a SG greater than 4.2; therefore, SG seems to be a key factor to reduce bolus losses. The rest of boluses used did not achieve the 98% RR required for official ID. Data recorded allowed to calculate a nonlinear regression model (R^2=0.98; $P<0.001$), although MSE (2.4) was greater than that previously reported in sheep (1.3). According to the obtained model, an increase in SG dramatically improved bolus RR, whereas slight increases were obtained in RR when bolus W increased. Therefore, the reduction of bolus size can be efficiently managed by increasing the SG, which needs to be greater than previously reported in sheep and cattle. In conclusion, suitable bolus retention can be achieved in goats by using specially designed boluses and different than those for sheep and cattle. In goats, the use of materials of high specific gravity is a key aspect to be taken into account in practice.

Session 43

Study and development of a RFID integrated automatic traceability system for the bovine meat chain

Beretta, E.[1], Nava, S.[1], Tangorra, F.[1], Baldo, A.[2], Dallan, E.M.[3] and Lazzari, M.[1], [1]University of Study Milan, Fac. Med. Vet., V.S.A. sez. Bioengineering, Via Celoria 10, 20133 Milano, Italy, [2]Universidad Nacional de La Plata, Facultad de Ciencias Veterinarias Zootecnia II, Casilla de correo 296, B1900AVW La Plata, Argentina, [3]Coldiretti MI-LO, Via Ripamonti 37/a, 20136 Milano, Italy; ernesto.beretta@unimi.it

Italian beef production chain is complex both under structural and organizational levels as a result of the great number of operators involved, large fragmentation of farm and meat industry phases, high imports of animals and meat, and complexity of the current market channels. Traceability is easily followed for each phase of the food chain (production, slaughtering, packaging and selling) but difficulties arise when trying to trace throughout all the productive subsystems. With this aim a research and demonstration project was conducted to implement an automated and integrated traceability system for the Italian bovine meat chain, based on the use of radio frequency (RFID). Electronic tags were used to identify: live animals, carcass hooks in the slaughtering line and meat cuts. RFID readers were installed in the farm (livestock gates), slaughterhouse (entrance, evisceration and carcass weighting areas), carcass cutting working stations, and meat shops (weighing place). The whole system was controlled by specific software integrated by 3 main components: livestock, slaughtering, storage/shop. Critical readings for tracing were: ear tag reading in the farm, ear tag reading in the slaughterhouse (before the stunning box), combined ear tag and hook tag reading at the evisceration area, carcass labelling and recording at the weighing area, registration of carcass entry in the cutting area, and using read-and-write electronic tags for adding meat cuts in the vacuum-packed batches. Selling actions in the store/shop area were also recorded. The system has been tested successfully in the slaughterhouse of Cooperativa Agricola San Rocco (Milan, Italy) were its different components proved to be functional and able to integrate all the subsystems.

Assessment of acute pain experienced by piglets from ear tagging, ear notching and intraperitoneal injectable transponders

Leslie, E.E.C., Hernández-Jover, M. and Holyoake, P.K., University of Sydney, 425 Werombi Rd, 2570 Camden, Australia; mhernandez_jover@usyd.edu.au

A total of 120 suckling piglets (4-12 d old) were used to evaluate the acute pain responses to different identification methods: 1) ear tag (Allflex, Laza Tag, Australia), 2) ear notch, and 3) intraperitoneal injection (IP) of 32 mm half-duplex transponders (Rumitag, Barcelona, Spain). Two groups of piglets according to the handling position for ear or IP treatments were used as control. Treatments were applied during 90 s and the pain response was determined by behavioral observations, vocalization recordings and physiological measurements. Saliva samples for cortisol analysis were collected 15 min pre- and post-treatment from all animals, and blood samples for lactate analysis were collected at treatment and after 10 min from the ear notched and IP groups. Only 43 pairs of saliva samples were suitable for analysis due to the low saliva production in young piglets. Piglet behavior was observed in the farrowing pen using instantaneous 5 min scan sampling across a 3 h period. Pain-related behavioral displays were higher ($P<0.01$) in ear notched and ear tagged animals than in other groups. Treated groups displayed more awake inactive behavior than control piglets. Ear tagged and IP piglets displayed more isolation than pigs in other groups. A tendency ($P=0.059$) was observed in vocalization differences across treatment groups, with the highest sound pressure among ear tagged pigs. Lactate levels increased ($P<0.05$) due to treatment in ear notched animals, with similar values post-treatment for ear notched and IP animals. Cortisol level increased ($P<0.05$) after treatment but no differences between groups were observed. Results suggest that identification practices of suckling piglets induce acute pain responses, these being more significant among ear tagged and ear notched animals. However, the non-specific behavioral displays of the IP piglets could be an indication of acute pain. The use of short-term analgesia may be warranted to reduce acute pain when identifying suckling piglets.

Electronic identification of game and ornamental species in practice

Encinas Escobar, A. and Rinaldi Iii, A.E., Tragsega S.A., Julián Camarillo, 6A 4 B, 28037 Madrid, Spain; aernest@tragsa.es

The main objective of the FaunIDE's research and development project was to study the use of radio frequency identification devices (RFID) in different game and ornamental species for population size, health and research purposes, among others. Different experiments were carried out on a long-term basis, consisting on: 1) Red partridge (Alectoris rufa); a total of 1,116 birds born in captivity were identified with RFID rings for a period of 14 months. Several prototypes of RFID rings were developed and tested until a suitable design, according to ISO 11784, was issued. Partridges identified were controlled through the whole productive cycle for hunting and most of them were also controlled after hunting. Moreover, birds used for breeding were also followed during the study. Retention and readability rates at the end of the experiment were 100%; no side effects were detected in the studied specimens. 2) Wild boar (Sus scrofa); a total of 94 ISO FDX-B 3 x 15 mm injectable transponders were applied in different body sites (ear, perianal region, and metacarpus region), and in 2 experiments done at different locations of the IREC (Instituto de Investigación en Recursos Cinegéticos; Zamudio and Ciudad Real, Spain) under laboratory and semi-captivity conditions. Evaluation of the application point, retention and readability rates was done during a period of 4 months. The necropsy of a sample of 23 dead wild boars was performed and the transponders recovered for study. The retention and readability rates varied markedly according the different injection body sites. 3) Ornamental birds; a total number of 477 ornamental birds of 30 different species belonging to the Town Hall of Madrid (Spain) and the Spanish Patrimonio Nacional were identified with 2 x 12 or 3 x 15 mm ISO FDX-B injectable transponders, depending on their size, and controlled for a term of 6 months. The great variability of the species identified provided a large amount of information on application procedures and injection body sites for birds.

EID and DNA traceability of animals and food

Cifcioglu, G.[1], Fiore, G.[1], Marchi, E.[2], Marcacci, M.[2], Camma', C.[2], Azzini, I.[3], Pagano, A.[3] and Ferri, N.[2], [1]EU Commission-JRC, CI-Animal&Food Action, Via Marconi, 21020 Ispra (Varese), Italy, [2]Istituto Zooprofilattico Sperimentale, Via Campo Boario, 64100 Teramo, Italy, [3]EU Commission-JRC, G9 - Econometrics and applied statistics, Via Marconi, 21020 Ispra (Varese), Italy; gurhan.cifcioglu@jrc.it

The past animal diseases crises, such as bovine encephalopathy and foot and mouth disease, have demonstrated that a reliable traceability system is needed in order to ensure food safety and proper control of animal health. Recently, electronic identification (EID) by radio frequency has become a binding standard in EU legislation (Regulation 21/2004) and DNA based techniques are becoming more promising tools in the traceability of the origin of animal products. International Society for Animal Genetics and the Food and Agriculture Organization have proposed sets of DNA microsatellites (STR) as individual identification markers for the study of animal genetic diversity and for conservation purposes in different species. The main objective of this study to develop an EID and DNA integrated system for traceability of cattle in both dairy and beef populations, as well as for their milk and meat, by using EID and DNA. The system integrates an EID code and the STR values of each animal, in order to ensure a tamper-proof trace-back methodology. According to the explored approach, integrated EID and DNA profile of cattle was used as tracing-back methodology for the identity of animals, carcasses, meat cuts and milk samples in a total of 200 animals and 200 animal product samples. Meat (n=40) and milk (n=40) mixtures from identified animals were also targeted by DNA analyses. Individual identification of the animals from which they come from, in beef and milk mixtures, prepared by mixing 3, 5, 7, 9, 11 and 15 samples of different individuals, were tested and discrimination of individuals in these mixtures were performed.

Social and cultural meanings of livestock farming in Western societies

Boogaard, B.K., Wageningen University, Rural Sociology, Hollandseweg 1, Wageningen, Netherlands; birgit.boogaard@wur.nl

In contemporary Western societies, animal farming has multiple - sometimes conflicting - images. For example, on the one side it is appreciated for the production of affordable animal products, but at the same time people are concerned about the treatment of animals and effect on nature and the environment. The images and meanings of animal farming systems are context-dependent and can differ between cultures. Consequently, the concept of sustainable animal farming systems can differ between cultures. The presented study focused on social and cultural meanings of animal farming systems in Western societies and the consequences for sustainable development. It explored the different collective meanings and underlying values of animal farming on the basis of farm visits with citizens in the Netherlands and Norway and two quantitative surveys in the Netherlands. The findings revealed three 'core values' of animal farming in the eyes of the public: Modernity, Traditions and Naturality. Citizens balanced and weighed these 'core values' against each other to form their opinion about different themes in animal farming, such as the animals and the rural landscape. In relation to sustainable development these findings showed that citizens did not think about sustainable development in terms of the economic, environmental and social pillar (the EES-concept) but instead their way of thinking reflects a so-called 'MTN-concept' (Modernity, Traditions, Naturality).

An inventory of pig production systems in Europe

Bonneau, M.[1], Dourmad, J.Y.[1], Phatsara, C.[2], Rydhmer, L.[3], Edge, H.[4], Fabrega, E.[5], De Greef, K.[6], Eidnes Sorensen, P.[7], Laugé, V.[8] and Ilari, E.[8], [1]INRA, France, [2]Univ. Bonn, Germany, [3]SLU, Sweden, [4]Univ. Newcastle, United Kingdom, [5]IRTA, Spain, [6]WUR, Netherlands, [7]DMRI, Denmark, [8]IFIP, France; michel.bonneau@rennes.inra.fr

The present study was conducted within the EU-supported project Q-Porkchains. A total of 84 production systems were identified in 23 European countries; Forty are conventional (CS) and 44 claim to be differentiated (DS) on Eating Quality (EQ 70%), Animal Welfare (AW 68%), Environment (EV 41%), Local (LO 30%), Nutritional Quality (NQ 25%), Organic (OR 25%). DS are much smaller (Min-Mean-Max million pigs/y: DS 0.01-1.0-12; CS 0.5-5.3-40). Compared to systems differentiated on other claims: (1) in EQ-DS, orientation towards high quality market segment, fatteners outdoors or on non-slatted floor, pure or local breeds are more frequent. Farmer-related QA systems and label/brands are less common. Technical performance is lower, carcasses are heavier and fatter. (2) in AW-DS, sows and piglets outdoors or on non-slatted floor, specific welfare rules and keeping sows loose are more frequent. Mortality rates of growing pigs and sows are higher. (3) in EV-DS, pigs outdoors or on non-slatted floor, solid manure are more frequent. Treating manure is less common. Lower technical performance and extra constraints result in higher production costs that are hardly compensated by higher selling prices. (4) in NQ-DS, the more common use of label/brands and QA systems is not reflected in selling prices that are little higher and do not compensate for higher production costs. (5) in OR-DS, pigs outdoors or on non-slatted floor, keeping sows loose, solid manure, specific welfare rules are more frequent. Treating manure is less frequent. Mortality rates of piglets and weaners are higher. Further work is currently ongoing within the Q-Porkchains project to investigate the sustainability of 15 contrasting production systems in Denmark, France, Netherlands, Spain and UK (one CS and two DS in each country).

Management and productivity in goat systems of less-favoured mountain areas in SW of Spain

Gaspar, P.[1], Mesías, F.J.[2], Escribano, M.[3] and Pulido, F.[2], [1]Universidad de Extremadura, Producción Animal y Ciencia de los Alimentos, Escuela de Ingenierías Agrarias. Ctra. Cáceres s/n, 06071 Badajoz, Spain, [2]Universidad de Extremadura, Departamento de Economía, Escuela de Ingenierías Agrarias. Ctra. Cáceres s/n, 06071 Badajoz, Spain, [3]Universidad de Extremadura, Producción Animal y Ciencia de los Alimentos, Facultad de Veterinaria. Avda. Universidad s/n, 10071 Cáceres, Spain; mescriba@unex.es

This paper analyses the main characteristics of traditional goat farming systems located in the Villuercas-Ibores, a mountainous region located in Extremadura (SW Spain). The territory shows low income indicators and also a very low density of population, both linked to the difficult orography, which allows modern agriculture and husbandry in a few specific areas. Historically, goat production has contributed remarkably to the economic and social development of this rural region, but a notorious recession of this sector has been observed in the last two decades, thus originating changes in the type and intensity of land utilization. There is one Protected Designation of Origin (P.D.O.) for the cheeses produced in this area, 'Ibores Cheese', which was developed as a tool to enhance the goat sector and improve the revenues of the farmers. The data were obtained through direct interviews with goat farmers (n=61) who could or not be producing under the P.D.O. 'Ibores Cheese'. Specific information on management was collected, together with data covering family characteristics, labour, livestock size, land use, facilities, continuity, recent changes in farming and farmers opinions. Principal Components Analysis was used to examine relationships between original continuous variables. Afterwards, a TwoStep Cluster analysis was applied using the factors obtained in the PCA and categorical variables. Three farm typologies were distinguished according to the use of land, the goat breed used and the P.D.O membership. The best management practices and productivity results were obtained by the farms further away from the traditional systems.

Characterizing of grassland use management for livestock farms using agricultural census databases
Thénard, V.[1], Jalabert, S.[2] and Thérond, O.[1], [1]INRA, UMR AGIR, BP 52607, 31326 castanet tolosan, France, [2]ENITA, UF Agrosystèmes et Forêts, CS 40201, 33175 Gradignan, France; vincent.thenard@ toulouse.inra.fr

Agricultural objectives have been changing and it involves a need to adapt agricultural systems to these new policy contexts. These evolutions are related to the new face of agriculture: the sustainability. Nowadays, we have to produce while protecting the environment. Biodiversity preservation has to be taken into account in agricultural system assessment. In the framework of grazing livestock systems, we have to consider the protection of grassland diversity. The characterization of grassland use management would permit to assess the impacts on the diversity. This characterization is used to be done for a small group of farms and needs interviews with the farmers. One of the objectives is to characterize grassland use management for agricultural areas. This study aims at setting up a methodology to create a farm typology with data from the Department of Agriculture. This methodology was applied to the farms' data which were extracted from the Agricultural Census 2000, so as to simply highlight their grassland management. The study area was set on four agricultural areas which cover 2,232 km² in central mountainous region of France. We used Multivariate Analysis to select descriptive criteria. The typology building has permitted to create and describe 5 groups of farms by several criteria related to the fodder system such as fodder areas, stocking density, etc. Some of them relate to productivity and to grassland use with variables such as silage or summer grazing practices. This dimension of the typology is close to the objective of grassland management characterization. The description of the groups in each of agricultural areas seems to present links between typology and natural environment. This land use management approach with Agricultural Census data could be use as preliminary step for sampling in a more detailed survey's work.

Intensification of integrated crop-livestock systems in the dry savannas of West Africa
Hansen, H.H.[1], Karbo, N.[2], Ouedraogo, T.[3], Bruce, J.[4] and Tarawali, S.[5], [1]Univ. of Copenhagen, Dept. of Large Animal Sciences, Groennegaardsvej 2, 1870 Frederiksberg C, Denmark, [2]Animal Research Institute (ARI), P.O.Box. AH20 Achimota, Accra, Ghana, [3]Institut de lenvironnement et des recherches agricoles (INERA), Production animales, 04 BP, 8645 Ouagadougou 04, Burkina Faso, [4]Animal Research Institute (ARI), P.O.Box. AH20 Achimota, Accra, Ghana, [5]International Livestock Research Institute (ILRI), P.O. Box 5689, Addis Ababa, Ethiopia, Ethiopia; han@life.ku.dk

Population changes and fluctuating climatic conditions in sub-Saharan Africa are predicted to result in intensification and expansion of agricultural production. Traditional production systems and the management of natural resources within them are breaking down. This appears to be occurring at a faster rate in the dry savannas in part because of the fragile, sandy soils, erratic rainfall and shortening growing periods. Unsustainable farming practices are emerging with potentially disastrous results for poor people, food security and the environment. This paper reports results of a Danish funded, ILRI led multi-institutional, inter-disciplinary project to promote intensification of sustainable agriculture and livestock production in Nigeria, Mali, Niger, Burkina Faso and Ghana. The project was based upon research that began in 1998 with funds from ILRI, International Institute of Tropical Agriculture (IITA) and International Crops Research Institute for Semi-Arid Tropics (ICRISAT), and the System-wide Livestock Programme (SLP). The use of improved and/or dual-purpose cowpea, millet, sorghum and groundnut varieties gave more fodder and more or equal amounts of grain in all countries. This increased plant production was only translated into increased animal gains in Nigeria where adjusting the level of cowpea and millet bran supplements increased ram daily gains compared to farmers' strategy. There is a need to find innovative ways of translating increased fodder production into livestock production.

Identifying farm typologies in an abandonment risk area in the Pyrenees (Spain)
López-I-Gelats, F., Milán, M.J. and Bartolomé, J., Universitat Autònoma de Barcelona, G2R, Campus UAB, 08193 Bellaterra, Spain; MariaJose.Milán@uab.cat

Recent trends in the agricultural market and technology prompt a rising interest on increasing productivity. Agricultural activity thus tends to be located on fertile and accessible land, what has lead to a decline in traditional labour intensive farming practices and abandonment of marginal agricultural land. Mountain regions are particularly vulnerable to this tendency. The aim of this study is to analyse the features of mountain livestock farms to disclose the different management and adjusting strategies. The research was conducted in the Eastern Pyrenees, in Catalonia (Spain), where the farming activity has been traditionally characterised by an extensive management of the herd. However, a drastic drop in the number of farms has been observed in the last decades, as well as changes in the forms of livestock farming and intensity of land management of the remaining farms. Structured interviews to farmers were conducted (n=59) in 2007. Principal Components Analysis is used to reduce the number of variables. Three typologies of farms have been uncovered by Cluster Analysis. Only intensified cattle farms (41%), most of them fattening calves, and large sheep operations (14%), with an average of 99.9 livestock units and 99.3 ha of meadows, show reasonable conditions for farm continuity. The rest of exploitations (45%) completely based on grazing are clearly not competitive in the present state of the market. These are horse or mixed farms with goats, which are often run by part-time or retired farmers and with a insignificant level of mechanisation. The results suggest that the current difficulties to guarantee the continuity of mountain farms trigger at the same time both intensification and extensification (or even abandonment) strategies. This process degrades the traditional extensive livestock raising systems of mountain regions and puts at risk their multifunctional nature.

Smallholder cattle production dynamics: a case study from South Africa
Stroebel, A.[1] and Swanepoel, F.J.C.[2], [1]University of the Free State (UFS), Centre for Sustainable Agriculture, P.O. Box 339, 9300, Bloemfontein, South Africa, [2]UFS and Fulbright Visiting Fellow, Cornell University, CIIFAD, Ithaca, NY, 14850, USA; stroebea.rd@ufs.ac.za

With the adoption of the Millennium Development Goals (MDGs), the international community has agreed to the eradication of extreme poverty and hunger as one of its primary targets. The objective of this limited investigation is to analyse the dynamics of cattle in smallholder livestock production systems at household and community level. A clear understanding of the role of livestock is essential to appreciate the choices made by the different actors, and to identify development pathways that are most likely to offer pro-poor benefits. One hundred smallholder communal cattle farmers in the Limpopo Province of South Africa were surveyed. The farmers owned between one and 72 cattle, with an average of 11.3 head of cattle per household. The average age at first calving was 33.9 months. The rate of calving, weaning, calf mortality, herd mortality and offtake were 49.2%, 33.3%, 27.1%, 15.7% and 8.7% respectively. Contrary to the situation in many other regions of Southern Africa, commercial enterprise, not social prestige, constituted the main reason for farming with cattle. It is concluded that a much more nuanced understanding of the role of livestock in the livelihoods of the poor is imperative. Without adequate understanding of the role of livestock in the livelihoods of the poor, and of the constraints to livestock production, it is unlikely that successful pro-poor impacts within the livestock sector could be implemented.

Local cattle breeds and performance potentials in rural areas of Iran
Safari, S.[1], Bokaian, J.[1], Qorbani, A.[1], Zakizadeh, S.[1] and Monazami, H.R.[2], [1]Hasheminejad Hig Education Center, Kalantati Highway, 9176994767, Iran, [2]Animal Science Organizatin, Khayam, 1, Iran; sonia_zaki@ yahoo.com

Local breeds play important roles in rural societies. Recently, National Breeding Center has started to evaluate performance traits of local breeds. The aim of this study was to identify potential rolls of local breeds in livelihood of rural villages in northeast of Iran. Data collected by interview of 65 farmers to record phenotypes, productions, feeding, reproduction, disease management, and housing systems. Most of farmers were not/or well educated with averagely 5 members who were breeding animals. Women were 58% of family workers and 90% were over 18 years. Dominant coat colors were black-white or black and they had horn. Average weight of matured animals was 289 kg with slaughtered weight of 147kg, and 6.9 years longevity. Dairy cow produced 7.85 kg/day milk with one calf annually. Most of families were consuming their own milk or milk products. Milking period was 7 months and calves were milked by the end of lactation. Except 10% of people, they were not consuming their own meat product. Frequent feedstuffs were straw and bran and few farmers could provide concentrate, hay, barley, or sugar beet pulp. Forage was supplied by farmer's agriculture or by-product. Animals were housed in permanent places, where floors were dusty without drainage. Light, ventilation, watering trough, and manger had poor quality, except density and water qualitys. The most frequent disease was tylerios and all of cattle were being vaccinated against foot-and-mouth disease, brucellosis and anthrax. Reproduction was natural with males/female ratio 0.33; only 15% of farmers were using artificial insemination. The first service of females was around 25th month and males could mate around 21.7 months old. Most farmers required facilities for providing feedstuffs, cow buying, or improving the housing conditions. Regarding to special geographical ecosystem surrounded by range of mountains, sustainable livestock system would be possible, if primarily needs could be fulfilled.

The Finnish fur farmer impaled in social change
Karkinen, K.T., University of Joensuu, Sociology and Public Policy, Yliopistokatu, P.O. Box 111, Joensuu, Finland; katri.karkinen@joensuu.fi

The 1960s in Finland saw the period of regional development policy. It was about creating new jobs, especially in Eastern and Northern Finland. One expanding source of livelihood was fur farming. At the beginning of the 1980s there were 500 fur farmers in Eastern Finland, who worked their farms part-time. By fishing, purchasing offal, and mixing in various grain products, they were able to keep the foxes, raccoon dogs, and minks alive and well. The subsidies for new building construction enabled the building of feed plants. The chief client was the fur auction arranged in the city of Vantaa.Fur farming benefited from the liberation of foreign trade, for most furs produced were exported. In the early 1990s, Finland was hit by an economic crisis, which drove small enterprises into difficulties.The crisis resulted in financial straits for the Eastern Finland fur farmers on account of lowered demand for furs and lowered public subsidies. About 90 per cent of the fur enterprises collapsed.New fur auction customers from the Far East and Russia encouraged farmers again to develop their trade. In 2008 there were 17 fur farms in Eastern Finland, operated by professionals. The rise of the animal rights movement brought fur farms to the fore in the media.The farmers were branded as tramplers of animal rights. One way of action among the activists was sabotage. Ten strikes were against farms in Eastern Finland.They are mostly unsolved cases of vandalism. In this study, the social practices and meanings attached to the fur farm are written out as an ethnography. It leans on a definition of the production system of the fur trade. The data used are statistics and archives as well as interviews and pictures. Interviewed people included farmers, family members, and public authorities. Interviewees were divided into two generations of fur farmers. The first one had practised their trade during the optimistic era of regional development policies, whereas the second one had suffered from financial straits and stigmatization.

Management of high ecological value grasslands: a way of agriculture diversification in Walloon Region

Turlot, A.[1], Rondia, P.[1], Stilmant, D.[2] and Bartiaux-Thill, N.[1], [1]Walloon Agricultural Research Center, Animal Productions and Nutrition, rue de liroux, 8, 5030 Gembloux, Belgium, [2]Walloon Agricultural Research Center, Farming Systems, Rue de Serpont 100, 6800 Libramont, Belgium; a.turlot@cra.wallonie. be

Through Natura 2000 network, European Union applies its environmental policy for the rehabilitation and the conservation of natural habitats. In this context, farmers can play a key role in maintaining open space of high ecological value. The aim of the project is to determine conditions allowing such pastoralism development and durability, especially with extensive grazing, in the Walloon area. To reach such a target, nineteen farmers, for whom this diversification represents a meaningful part of their agricultural activity, were followed. A survey has allowed to characterize these farms and to find out farmers motivations to start this activity as well as the brakes that could limit its development. The sustainability of this diversification was analysed using three approaches: (1) the economic dimension was characterised through a comparative analysis of the raw margin and the family's income, (2) the social dimension focused on the working time and on the arduous work and (3) a global farm sustainability approach performed with an adapted version of 'IDEA' (indicators system to characterise farms sustainability) system. This last part takes into account the three dimensions of farming activity sustainability in order to define targets and evolution plan specific to each system. At the moment, this method has been applied on 10 farms. This study has already showed that natural habitat management was mainly a sustainable activity from an agro-ecological point of view. It's less true on an economical point of view due to the strong dependency of this activity to subventions. We must keep in mind that the time necessary to perform this hard activity is very important.

Farming styles and local development: a European perspective

Garnier, A.[1], Nedelec, Y.[1] and Marie, M.[1,2], [1]Nancy-Université, ENSAIA, B.P. 172, 54505 Vandoeuvre, France, [2]INRA, SAD-ASTER, Av. L. Buffet, 88500 Mirecourt, France; Michel.Marie@Mirecourt.inra.fr

Fifty height farms from West (Ireland: 6, United Kingdom: 11, Netherlands: 6, Denmark:3), Center (Belgium: 5, Germany: 8, Swiss: 4), South (Italy:4, Slovenia: 2, Croatia: 1) or East (Czech Republic: 2, Slovakia: 3, Poland: 3) Europe have been surveyed regarding the attributes of sustainability on the basis of the IDEA method, and analyzed in a qualitative way with the multi-attribute decision making tool Dexi. These farms (35 from lowlands, 20 from hills, and 3 from highlands), conducted in a familial (48) or managerial (10) way, ran cultures or arboriculture (5), pork (2) or ruminants (dairy: 40, meat: 6, mixed: 5) in herb (9) or poly-culture and livestock (42). Multivariate analyzes have been performed on the basis of the characteristics of the production systems and of global sustainability, its agro-environmental, economical and socio-territorial components, and more precisely those related to the local development: quality of food, landscape, accessibility, social implication, short distribution circuits, multi-activity and services, employment, collective work, long-term sustainability, training, work load, quality of life, isolation, hygiene and security. A common assessment methodology applied in different contexts revealed the impact of public policies and of historical backgrounds on the local development. The territorially-linked signs of product quality are less developed in North (DK, NL) or East (PL) Europe. The results led to the identification of different of farming systems with interest in terms of sustainability, and rise a series of questions, such as the consequences of the development of zero-grazing systems, or of the scale-economy consecutive to the growing size of the farms on the territories, the agricultural heritage, the image of agriculture, or the employment. The transferability of farms is a major problem, due to different reasons (capital too high, low attractiveness of the profession, lack of training) among countries.

Sustainable livestock production in Serbia mountain regions

Savic, M.[1], Jovanovic, S.[1], Vegara, M.[2] and Popovic Vranjes, A.[3], [1]Faculty of veterinary medicine, Animal breeding and genetics, Bul. oslobodjenja 18, 11000 Beograd, Serbia, [2]Norwegian University of Life Sciences (UMB), Department of International Environment and Development Studies,NORAGRIC, P.O.Box 5003, N-1432 Aas, Norway, [3]Faculty of Agriculture, Trg D.Obradovica 8, 21000 Novi Sad, Serbia; mij@beotel.net

In order to enable protection and rational utilization of natural resources of the Central Serbia mountain ecoregions, sustainable livestock production program have been applied. Zackel breed, as a well adapted autochthonous breed, is a good basis for sustainable sheep production. Zackel breed is traditionally reared in mountain regions of Balkan Peninsula under very modest conditions. All types of Zackel breed are endangered.The most important type of Zackel breed is Sjenica sheep. Monitoring of animal genetic resources and habitats, the evaluation of productive and health characteristics of Sjenica sheep was performed. Sjenica sheep is a triple purpose, late maturing sheep. The average estimated body weight ranges from 50-70 kg. Lactation period lasts 5-6 months, average milk yield is 75 kg, with 6.8% milk fat. Milk is used for traditional manufacturing of cheese and kajmak, registered as autochthonous products. The lambing rate is 115%. Considering the FAO recommendations for Farm animal genetic resources characterization and conservation, the genetic characterization of Sjenica sheep has been performed, using microsatellite marker analysis. The program of conservation and rational utilization of Sjenica sheep in Central Serbia mountain regions is very important for the integration of livestock farming into the natural environment, as well as for the development of rural community. The general environmental conditions in the extensive farming system of Sjenica sheep should be improved for achieving better productive results. The selection should be focused on reproductive traits, along with preservation of the genetic resistance to diseases, the most prominent characteristic of Sjenica sheep which gives advantages in sustainable farming.

EADGENE, a European Animal Disease Genomics Network of excellence for animal health and food safety

Pinard-Van Der Laan, M.-H., INRA, Animal Genetics, UMR GABI, 78352 Jouy-en-Josas, France; pinard@ jouy.inra.fr

EADGENE is a EC funded NoE which coordinates a genomics approach to the unravelling of host-pathogen interactions, providing the basic knowledge necessary for the development of new or improved therapeutics and vaccines, improved diagnostics and the breeding of farm animals for disease resistance. Our innovative research projects complement existing projects between 15 leaders in this field. This research will impact upon human health and lifestyle choices by including on pathogens of importance in the food chain. Integrating activities are sustainable for sharing skills EADGENE has established a database of biological resources and provided microarrays for chicken, cattle and pigs, used as our common preferential tools. Our laboratories have adopted common QC standards and methods of analysis in genomics and bioinformatics. Four research areas Research activities are structured into multidisciplinary themes (1) structural genomics (genes, miRNA, regulating resistance mechanisms), (2) population genetics (models and relevant phenotypes, QTL), (3) functional genomics on mastitis (cattle, goats, sheep), salmonellosis (pig, poultry), E. coli EHEC (cattle) and viral models (IPNV and ISAV in salmon and trout), (4) operational genomics (meta-analysis). These collaborative studies compare and use the wealth of models (different hosts, pathogens, infections) using common genomic tools. Technology transfer and Dissemination to general public A special effort is dedicated to technology transfer bridging the gap between research and industry. Three concrete activities: the ontology of important traits; the development of databases from private companies; projects on poultry resistance. Ethical implications of animal health genomics are a fundamental component of EADGENE. Benefits for EADGENE researchers Benefits are increased access to resources, knowledge, staff recruitment and scientific cooperation. EADGENE has been a catalyst and unifier for the development of other national or European contracts.

The genetic basis of resistance to Infectious Pancreatic Necrosis in Atlantic salmon

Houston, R.D.[1], Haley, C.S.[1], Hamilton, A.[2], Guy, D.R.[2], Gheyas, A.[2], Mota-Velasco, J.[2], Tinch, A.E.[2], Taggart, J.B.[3], Bron, J.E.[3], Mcandrew, B.J.[3], Verner-Jeffreys, D.[4], Paley, R.[4], Tew, I.[4] and Bishop, S.C.[1], [1]The Roslin Institute and Royal (Dick) School of Veterinary Studies, University of Edinburgh, Roslin Biocentre, EH25 9PS, United Kingdom, [2]Landcatch Natural Selection Ltd., Alloa, FK10 3LP, United Kingdom, [3]Institute of Aquaculture, University of Stirling, Stirling, FK9 4LA, United Kingdom, [4]Centre for Environment, Fisheries and Aquaculture Science (Cefas), Barrack Road, Weymouth, DT4 8UB, United Kingdom; ross.houston@roslin.ed.ac.uk

The production of Atlantic salmon worldwide is currently facing the significant problem of outbreaks of the potentially fatal viral disease Infectious Pancreatic Necrosis (IPN). IPN can impact at two specific life-cycle stages; firstly on salmon fry in the freshwater environment, and secondly on post-smolts shortly after transfer to seawater. At both these stages, using natural challenge data in post-smolts and deliberate challenge data in fry, we have analysed the patterns of mortality in families of commercial salmon. We have demonstrated that resistance shows a significant genetic component, with moderate to strong estimated heritabilities. To search for individual loci underlying this resistance, we applied a QTL mapping strategy that accounts for the large differential in recombination rate between male and female salmon. A striking QTL was located on linkage group 21 that explains most of the variation in IPN mortality within segregating families. This QTL impacts at both the fry and post-smolt stages of the salmon lifecycle, and appears to be of paramount importance in the genetic regulation of IPN resistance. Therefore, the potential for the use of the QTL in marker-assisted selection is clear, and significant progress has already been made on its application. Furthermore, to gain insight into the physiological pathways mediating this QTL effect, a series of challenge experiments have been undertaken to allow us to compare the gene expression response to IPN challenge in fry of distinct QTL genotypes.

The chicken SNP Selector Database for genome-wide QTL Analysis

Fife, M., Watson, M., Howell, J. and Kaiser, P., Institute for Animal Health, Avian Genomics, Comton, Berkshire, RG20 7NN, United Kingdom; mark.fife@bbsrc.ac.uk

The accessibility of genome sequence and high density SNP maps for the chicken provide enormous opportunities to understand and exploit the genetic control of complex traits. We can now potentially correlate phenotypic variation with underlying genetic variation; however, successful mapping of these traits will only be possible with a substantial increase in the number of genetic markers. To maximize the precision of genetic selection, genotype data must be of sufficiently high resolution to identify the genes underlying quantitative trait loci (QTL). The 3 million publicly available SNPs for the chicken genome were generated from only a 'typical' layer, a 'typical' broiler and a Chinese Silkie. Most of the research lines, for which we have vast amounts of phenotypic data, are either of broiler or layer origin and therefore will only contain SNPs that are specific to these populations. Our objective is to identify genome-wide SNPs in each of the IAH lines allowing Linkage Disequilibrium patterns to be examined and relevant SNPs selected for mapping. We have now genotyped ~6000 genome-wide SNPs for up to 18 birds from each of the lines, including lines N, 6_1, 7_2, WI, C, P, 0, 15I, BrL and Sykes RIR. Taking advantage of divergent phenotypes between line $6_1^{(R)}$ and line N$^{(S)}$, which differ in disease resistance to Salmonella and Campylobacter, we used the 'SNP Selector' to identify informative SNPs for QTL analysis for a backcross mapping panel. Challenge experiments were repeated in triplicate with approximately 150 birds in each experiment. Genotyping was performed on 1536 fully informative genome-wide SNPs for birds with extreme resistance phenotypes. The use of the high density genome-wide markers has enabled identification of QTL in previously uncharacterised regions of the chicken genome, identifying loci that would have otherwise remained undetected. These results are obtained through the EC-funded FP6 Integrated Project SABRE–Cutting Edge Genomics for Farm Animal Breeding.

The application of modern and traditional research approaches to investigate salmonella susceptibility in chickens

Te Pas, M., Rebel, J. and Smits, M., Animal Sciences Group of Wageningen UR, P.O. Box 65, 8200 AB Lelystad, Netherlands; marinus.tepas@wur.nl

Contamination of poultry products with Salmonella enterica is an important problem in poultry meat producing chains. An effective way to reduce food poisoning due to salmonella would be to breed chickens with improved resistance to salmonella. Unfortunately, host responses to salmonella are complex with many different factors involved. To learn more about the susceptibility of chicken to salmonella we applied new and traditional research techniques, including microarray-, pathway-, and cellular influx analyses. Two chicken lines that differ in salmonella susceptible were used under control and salmonella infected conditions. A DNA microarray analysis was performed to compare gene expression profiles in the intestine of the two lines. Typically, microarray experiments produce long lists of differentially expressed genes. In order to better understand the biology behind these data, it is relevant to include the available biological information of the genes under study. With the pathways and biological network information that came from these integrated approaches, the intestines of the different chicken lines were re-investigated with traditional immunological techniques, in order to verify the findings. With this combination of techniques we were able to find differences between the chicken lines that lead to difference in host responses and possibly to differences in susceptibility to salmonella. These results may be of relevance for breeding and feed industry. These results are obtained through the EC-funded FP6 Project 'SABRE' and 'EADGENE'.

Across-line SNPs association study of innate and adaptive immune response in laying hens

Biscarini, F.[1], Bovenhuis, H.[1], Van Arendonk, J.A.M.[1], Parmentier, H.K.[1], Jungerius, A.[2] and Van Der Poel, J.[1], [1]Wageningen Universiteit, ABGC, Marijkeweg, 40, 6709 PG Wageningen, Netherlands, [2]Hendrix Genetics, Spoorstraat, 69, 5830 AC Boxmeer, Netherlands; filippo.biscarini@wur.nl

The present study was aimed at detecting QTLs for innate and adaptive immunity in laying hens. A set of 1534 SNP markers was used on about 600 hens from 9 different layers lines. A novel approach based on an across-line analysis and testing of the SNP-by-line interaction was adopted. The amount of LD conserved across lines has a shorter extent than in the individual lines and therefore SNPs significantly associated with immune traits are expected to be close to the functional mutations. The analysis was carried out in two consecutive steps. In the first step all SNPs were tested ignoring the polygenic effect. In the second step only SNPs with a significant and relevant effect on immunity and no significant SNP by line interaction were analysed taking into account the polygenic effect. Eventually, 59 significant associations between SNPs and immune traits were detected. Some of the results of this work confirmed QTLs in regions in which QTLs have been previously identified. Others may represent new QTLs for immunity. We found evidence for a role of the IL 17F gene on chromosome 3 on natural and acquired antibody titres and on the classical and alternative pathways of complement activation. The MHC genes on chromosome 16 showed significant effects on natural and acquired antibodies titres and classical complement activity. The IL 12 gene on chromosome 13 showed a possible effect on natural antibody titres.

Mapping of genes involved in susceptibility to *E. coli* and helminth infection in pigs

Jacobsen, M.J.[1], Nejsum, P.[1], Joller, D.[2], Bertschinger, H.U.[2], Python, P.[2], Edfors, I.[3], Cirera, S.[1], Archibald, A.L.[4], Churcher, C.[5], Esteso, G.[1], Burgi, E.[2], Karlskov-Mortensen, P.[1], Anderson, L.[6], Voegeli, P.[2], Roepstorff, A.[1], Goering, H.H.H.[7], Anderson, T.J.C.[7], Thamsborg, S.M.[1], Fredholm, M.[1] and Jorgensen, C.B.[1], [1]University of Copenhagen, Frederiksberg C, 1870, Denmark, [2]ETH, Zurich, 8092, Switzerland, [3]Univ of, Kalmar, 39182, Sweden, [4]Roslin, Midlothian, EH25 9PS, United Kingdom, [5]Sanger, Hinxton, Cambridge, United Kingdom, [6]Univ. Uppsala, Uppsala, 75124, Sweden, [7]SFBR, San Antonio, TX 78227-5301, USA; chj@life.ku.dk

E.coli and helminth infections cause significant health and welfare problems and compromise the sustainability of pig production systems. The enterotoxigenic *E. coli* O149, F4ac alone is responsible for more than 30% of the *E. coli* diarrhea cases in piglets. Studies have shown that susceptibility both to specific *E. coli* and helminth types are under genetic control. Resource families, where these susceptibility traits are segregating, have been constructed and our data demonstrate that genetic components are involved in resistance to *E. coli* F4ab/ac, Ascaris and Trichuris infections. Genome scans have been performed in order to locate genomic regions controlling susceptibility in the pig. For Ascaris and Thichuris infections a total of 195 pigs from 19 full-sib litters have been genotyped for 4890 SNPs and the genotypes are ready for linkage analysis. In relation to *E. coli* F4ac-susceptibility we have narrow down the candidate region to less than 3 Mb around the mucin 4 gene in the q41-region on pig chromosome 13. Haplotyping data of the mucin 4 region using more than 200 SNPs shows a large shared haplotype block on the chromosomes carrying the susceptible allele. These results were obtained, in part, through the EC-funded FP6 project 'SABRE'.

Global transcriptional response of porcine intestinal epithelial cell lines to *Salmonella enterica serovar Typhimurium* and *Choleraesuis*

Arce, C., Jiménez-Marín, A., Collado-Romero, M., Lucena, C. and Garrido, J.J., Universidad De Córdoba, Dpto.Genética, Cu Rabanales Edif C5, 14014 E, Spain; ge1arjic@uco.es

Salmonella typhimurium (ST) and *Salmonella choleraesuis* (SC) are two of the most frequent *Salmonella serovar* isolates from European pigs. SC causes in pigs a systemic disease while ST infection in the same species leads to a localized enterocolitis. Analysis of the different parts of the gut demonstrated that mRNA expression of inflammatory cytokines varied according to the site analyzed. Also, it has been demonstrated that ileum undergo a greater inflammatory cytokine response than jejunum after Salmonella infection. Microarrays offer the potential of revolutionise research in the management (diagnosis, treatment, assessment of prognosis and prevention) of many diseases. A first-generation Affymetric GeneChip Porcine genome array has been used to profile the gene expression in epithelial cell lines from different porcine gut sections (IPEC-J2 from jejunum and IPI-2I from ileum) over a time course of infection (2 and 4 h) with ST and SC. The results obtained showed more genes differentially expressed in IPI-2I cell line (66 and 405 genes after ST and SC infection) than in IPEC-J2 cell line (38 and 250 genes after ST and SC infection). With respect to the serovar used, in both cell lines infection with SC caused a higher variation in gene expression than ST. In addition, a temporally variation in the assay was observed, being the response opposite ST (4 h) later than the response to SC (2 h) infection. In all cases, genes related to immune response were found. In conclusion, the microarray experimental design used in this study provided direct comparison to identify differentially expressed genes due to Salmonella serovar or cell line variation. These results aim to contribute to explain the different response observed in the gut regions after Salmonella infection and the different inflammatory properties described to both Salmonella serovar. These results are obtained through the EC-funded FP6 Project 'SABRE'.

Identification of microRNAs in PK15 cells infected with Aujeszky disease virus
Tomas, A.[1], Rosell, R.[1], Segales, J.[1,2], Sanchez, A.[3] and Nunez, J.I.[1], [1]Centre de Recerca en Sanitat Animal, campus UAB, 08193, Barcelona, Spain, [2]Universitat Autonoma de Barcelona, Dept. Sanitat i Anatomia Animals, campus UAB, 08193, Barcelona, Spain, [3]Universitat Autonoma de Barcelona, Dept. Ciencia Animal i dels Aliments, campus UAB, 08193, Barcelona, Spain; ignacio.nunez@cresa.uab.cat

MicroRNAs (miRNAs) are a new class of small non-coding RNAs that negatively regulate gene expression. The study of miRNA-mediated host-pathogen interactions has emerged in the last decade due to the important role that miRNAs play in the antiviral defense. More than 100 miRNAs produced by viruses have been described, the majority of which are encoded by herpesviruses. The identification of miRNAs in swine viruses has not yet been explored. We have undertaken the characterization of miRNAs produced by Aujeszky disease virus (ADV), an alpha-herpesvirus causing pseudorabies in pigs. Cells from the PK15 cell line were infected with two different ADV strains (attenuated Begonia and pathogenic NIA3) at a moi of 0.05. A flask inoculated with cell culture medium was kept as negative control. Cells were lysed in Trizol at 24 h post-infection and total RNA was isolated. RNA integrity and quantity was assessed with the 2100 Bioanalyzer. Enrichment of small RNA fraction was done by excising the 15-25 nt size by PAGE. miRNA library was constructed in a two-step ligation procedure with the 3' and 5' adaptors from IDT technologies. Amplification by RT-PCR was carried out with fussion primers containing sequences complementary to the 3' and 5' adaptors and sequences complementary to the A and B adaptors used for high-throughput (HT) sequencing with the GS FLX 454 (Roche). Multiplex identifiers, including a five-nucleotide code, were used in fussion primers to allow distinguishing between the control, Begonia- and NIA3-infected samples. PCR products were cloned and sequenced as a quality control step of miRNA library construction. Validated amplicons were analyzed by HT sequencing. Several candidate miRNAs produced in response to infection with ADV have been identified.

Combining QTL and gene expression studies to identify important genes and genetic pathways underlying resistance to mastitis in Danish Holstein cattle
Sørensen, P.[1], Røntved, C.M.[2], Jiang, L.[1], Sahana, G.[1], Vilkki, J.[3], Thompsen, B.[1] and Lund, M.S.[1], [1]Århus University, Department of Genetics and Biotechnology, Blichers Alle 1, 8830 Tjele, Denmark, [2]Århus University, Department of Health, Welfare and Nutrition, Blichers Alle 1, 8830 Tjele, Denmark, [3]MTT Agrifood Research, Biotechnology and Food Research, ET building, FI-31600 Jokioinen, Finland; peter.sorensen2@agrsci.dk

The objective of this study is to identify important genes and genetic pathways underlying resistance to clinical mastitis in Danish Holstein cattle. In a previous study we identified a quantitative trait locus (QTL) affecting the risk of mastitis. Further investigations showed that the QTL exhibit specificity against one of the most common pathogens *E. coli* causing clinical mastitis. Haplotypes associated with high or low resistance to clinical mastitis were identified in the QTL region. Thirty two cows carrying high or low resistant haplotypes were tested for their disease susceptibility, degree and duration when exposed to *E. coli* intra-mammary. Results from the clinical data analyses indicate an effect of the QTL haplotypes. Genome-wide expression profiles of liver and udder samples were determined before and after pathogen challenge. These data gives us a unique opportunity to study DNA variations in the genome and the perturbations these variations give rise to at the expression level, which in turn lead to resistance to clinical mastitis. Results from expression analyses show that a large number of genes responded to the pathogen challenge in both liver and udder samples. Although high and low resistant cows had similar expression profiles we did identify a number of genes whose expression levels were different in the two groups. We will present a systems biology approach that through integration of diverse biological data rank the genes located in the QTL region with respect to being the most likely candidate gene. Acknowledgements: These results are obtained through the EC-funded FP6 Project 'SABRE'.

Response to mammary infection in two divergent lines of ewes selected on milk somatic cell counts: transcriptomic analyses of milk cells upon challenge

Bonnefont, C.[1], Rupp, R.[1], Caubet, C.[2], Toufeer, M.[2], Robert-Granie, C.[1] and Foucras, G.[2], [1]INRA, U631, Genetique animale, Chemin de Borde Rouge BP 52627, 31326 Castanet Tolosan cedex, France, Metropolitan, [2]INRA/ENVT, UMR1225, Sante animale, 23 Chemin des Capelles BP 87614, 31076 Toulouse Cedex 3, France, Metropolitan; cecile.bonnefont@toulouse.inra.fr

Since the 1990's, mastitis resistance has been included in European breeding objectives of dairy cattle. In France, selection programs for mastitis resistance have also been applied to dairy ewes. The selection criterion, defined as a log-transformation of milk somatic cell counts (SCS), is a well-known indicator of mastitis. Although it has a moderate heritability (0.15), its genetic correlation with bacterial infection is near the unit. This study compares two divergent lines of ewes selected on SCS breeding values of their parents. Previous research work already showed that half reduction of SCS in ewe milk (2.77±0.19 in the lower SCS line vs 4.46±0.25 in the higher SCS line) was associated with a reduced number of clinical mastitis by a factor of at least four, and with a shorter duration of mammary infections (OR=4.5). To provide enhanced insights into the genetic mechanisms involved in SCS-based selection, two groups of six ewes from the divergent lines were challenged twice with Staphylococcus bacteria. Two different species were inoculated, namely S. *epidermidis* on year 1 and S. *aureus* on year 2. Antibiotic treatment was used after the first experiment to clear the infection. Milk cells, essentially neutrophils, were collected 12 hours after mammary inoculation. RNA was extracted, and after an amplification-labelling step, hybridisation on an ovine 15K oligonucleotide microarray was performed. Although gene expression profiles differed between the two infections, statistical analyses enabled identification of a small group of genes differentially expressed between lines.

The French national breeding plan for scrapie resistance: an example of molecular information used in selection

Leymarie, C.[1], Astruc, J.M.[1], Barillet, F.[1], Bibe, B.[1], Bonnot, A.[1], Bouffartigue, B.[2], Bouix, J.[1], Dion, F.[2], Francois, D.[1], Jouhet, E.[2], Jullien, E.[1], Moreno, C.[1], Boscher, M.Y.[3], Palhiere, I.[1], Raoul, J.[1], Tiphine, L.[1], Bouchel, D.[4], Chibon, J.[4], Raynal, A.[4] and Tribon, P.[4], [1]INRA-IE (UMT genepR), BP27, 31326 CASTANET-TOLOSAN, France, [2]Races de France, 149 rue de Bercy, 75012 Paris, France, [3]LABOGENA, domaine de Vilvert, 78352 Jouy en Josas, France, [4]Ministère Agriculture, 78 rue de Varenne, 75007 Paris, France; Cyril.Leymarie@toulouse.inra.fr

The Ministry of Agriculture has launched in 2001 and funded (partly with EU support) a plan to eradicate scrapie in all French sheep breeds. The plan is based on the polymorphism of the PrP gene in order to 1) eradicate scrapie in affected flocks, 2) improve resistance to scrapie in the whole population in the framework of a large scale breeding programme based on the genotyping of reproducers in selection flocks. 650,000 reproducers have been yet genotyped and the ARR allele (resistant) frequency increased by about 45% in active rams. The key points of the plan success are: national collective organisation with rational choice of animals on farm, DNA analysis, centralized molecular database and diffusion of the molecular information linked with classical selection indices. About 15 laboratories have obtained the process agreement for PrP genotyping. Results are sent to the molecular database where each animal can be linked with the national genetic database. Thanks to data of genotypes and pedigree, a computation of genotype prediction method was developed. These predictions enable to have early information for lambs and thus to pre-select without DNA analysis. All genotypes and predictions are available for the selection organisations and breeders, in order to optimise the choice of the reproducers and help assortative matings. 1,000,000 predictions were yet stored. The French scrapie breeding program was a first example of a successful generalised gene assisted selection at population level and these tools could be easily adapted for others disease or production traits selection.

Gene prioritization using text mining and protein-protein interaction in livestock species

Jiang, L.[1], Sørensen, P.[1], Workman, C.[2] and Skarman, A.[1], [1]Faculty of Agricultural Sciences, Aarhus University, Department of Genetics and Biotechnology, Research Centre Foulum, Blichers Allé 20, 8830 Tjele, Denmark, [2]Technical University of Denmark, Department of Systems Biology, Center for Biological Sequence Analysis, Kemitorvet, Building 208, 2800 Lyngby, Denmark; li.jiang@agrsci.dk

The objective of this study is to rank important genes for diseases in livestock species (cow, pig, chicken) using text mining and protein-protein interaction data. Integration of these heterogeneous data has an increasingly important role in the gene prioritization of human diseases based on the similarity of phenotypes between diseases and the association of proteins with diseases. In livestock as compared to humans, the phenotypic descriptions of a limited number of diseases have been collected in databases although many more links exist between diseases and causative genes. Phenotypic descriptions of human diseases can be found in public databases such as Online Mendelian Inheritance in Man (OMIM). The phenotypic descriptions in OMIM records provide curated text reviewing relevant literature of which a majority of publications are found in PubMed. We hypothesized that PubMed abstracts could be used as substitute for the phenotypic description of the disease. We tested this hypothesis in human diseases for a set of causal genes that had been previously (correctly) identified using a Bayesian prediction model which integrates information on phenotypes from OMIM records and multiple sources of protein-protein interactions. We found that PubMed abstract worked equally well compared to the curated text in OMIM records in gene prioritization of human disease. We applied our approach to a number of livestock diseases and conclude that text mining of PubMed abstracts provides important phenotypic information that can be used for gene prioritization in livestock species. Acknowledgements: These results are obtained through the EC-funded FP6 Project 'SABRE'.

Analysis of the intestinal mucosal immune response to an experimental infection with *Salmonella typhimurium* in pigs

Collado-Romero, M.[1], Arce, C.[1], Carvajal, A.[2] and Garrido, J.J.[1], [1]Universidad de Córdoba, Genética, Campus de Rabanales, Edif. C5, 14071 Córdoba, Spain, [2]Universidad de León, Sanidad Animal, Campus de Vegazana, s/n., 24071, Spain; ge1gapaj@uco.es

Salmonellosis caused by *Salmonella typhimurium* is an important disease in animal safety and human health. The problem account with the economical loses in farms, but also with the public health risk of commercializing Salmonella infected pork products. The early immune response to Salmonella infection will be largely due to the mechanisms activated after bacteria interaction with intestinal mucosa, and will be crucial for successful host defence or pathogen colonization. In this work we have characterized the early immune response to S. *typhimurium* infection *in vivo* along the intestinal tract (ileum, jejunum and colon). For this, 16 piglets were orally infected with S. *typhimurium*. Disease progress and sampling of tissues were carried out during a time course comprising non inoculated, 1, 2, and 6 days after inoculation. Immune response was evaluated by quantifying the expression of mRNA of 28 immuno-related molecules using quantitative real-time PCR assays. The panel of molecules included cytokines, chemokines, pattern recognition receptors, intracellular signaling molecules, transcription factors and antimicrobial peptides. Changes in mRNA expression of these molecules were evaluated during the time course of the infection within an intestinal section, so as among the different sections at the same sampling time (including characterization of the basal immune state, non inoculated controls). Our results demonstrated that different intestinal sections respond differently to the infection. Thus, *ileum mucosa* was unable to mount such an appropriate response to S. *typhimurium* as did jejunum. On the other hand, colon, although able to mount the response, it did later. These results could help to understand the preference of S. typhimurium for ileum infection reported in previous works. These results are obtained through the EC-funded FP6 Project 'SABRE'.

Analysis of polymorphisms of Mx1 gene and influenza susceptibility: the case of Iberian pigs

Godino, R.F., Fernandez, A.I., Ovilo, C. and Real, G., INIA, Biotecnología y Mejora Genética Animal, Ctra La Coruña km 7.5, 28040 Madrid, Spain; fernandez.rosario@inia.es

Swine influenza is an important respiratory infectious disease in pig husbandry. In addition, pigs play an important role in the epidemiology of influenza infection, since they are considered as 'mixing vessels' where avian and human viruses can give rise to new variants of the virus that might cause pandemics. Iberian porcine is a Spanish autochthonous breed with elevated economic impact in many areas of Iberian Peninsula. They are mostly kept on free-range and semi-intensive husbandry systems which allow contact with wild birds and increase the risk of cross-transmission of avian Influenza viruses to pigs. Mx1 is an IFN-stimulated gene encoding a protein with GTPase activity that has been implicated in innate resistance to Influenza and other RNA viruses. It has been reported that its N-terminal region may be involved in intracellular protein sorting or motility while, the more divergent C-terminal region, may determine the specificity of antiviral activity. Up to now, two polymorphisms on the exons 13 and 14 (INDEL3 and INDEL11 respectively) of Mx1 have been reported in some domesticated porcine breeds. They produce amino acid changes in the C-terminal region of Mx protein that may affect to the susceptibility of pig to Influenza infection. Our objective is to study those and other potential polymorphisms in Iberian pigs with the aim of developing strategies of genetic selection directed to enhance innate resistance to influenza infection. The analysis of Mx1 polymorphisms has been done by sequencing exons 13 and 14 in ten Iberian, two Landrace and two wild boar samples. In order to determinate genotypic frequencies, we have genotyped the most interesting polymorphisms in a higher number of Iberian samples from different Iberian populations as well as other domestic pigs.

Campylobacter induces diverse kinetics and profiles of cytokines genes in human and swine intestinal epithelial cell lines

Jimenez-Marin, A. and Garrido Pavon, J.J., Unidad de Genomica y Mejora Animal, Universidad de Cordoba, Genetica, C.U. Rabanales, Edificio C5, 14071 Cordoba, Spain; gm2jimaa@uco.es

Infection with Campylobacter species is now considered to be the most common cause of acute bacterial gastroenteritis in humans worldwide. Inflammatory signals are initiated during interaction between these pathogens and human intestinal cells, but nothing is known about the stimulation signals of swine intestinal cells by Campylobacter. The objective of this study was examine the interaction between swine intestinal cells and Campylobacter sp analyzing the expression of various cytokines and chemokines in two porcine epithelial cell lines from different regions: IPEC-J2 (jejunum) and IPI-2I (ileum) and comparing their response to the human intestinal epithelial cell line INT-407 (jejunum and illeum). The cell types were stimulated with *C. jejuni* and *C.coli* employing a bacterium-cell-ratio of 1:1000. The samples of RNA and supernatants were harvested after total incubation time of 4, 6, 8 12 and 24 h. Expression of various cytokines and chemokines, between them IL-6, -8 and TNF-alpha was determined by quantitative real-time RTQ-PCR. The results obtained show up a similar response in the gene expression between the cell lines as response to Campylobacter sp. during the activation time; nevertheless IPI-2I and INT-407 present a different profile of IL-8 against both pathogens. The secretion of IL-8 protein during time course showed up that the expression of mRNA does not necessarily result in translation to protein. Therefore, the aim of this work was to investigate whether Campylobacter isolates are able to induce immune response in swine intestinal epithelial cells following *in vitro* incubation and to compare with responses of human intestinal epithelial cells after stimulation with these strains. Any differences in innate responses to this pathogen between the human and swine hosts should lead to a greater understanding of the disease process in human. These results are obtained through the EC-funded FP6 Project 'SABRE'.

Selection against congenital day blindness in Awassi sheep applying genotyping of the cone photoreceptor cGMP-gated channel α-subunit (CNGA3) gene

Reicher, S.[1], Seroussi, E.[1], Shamir, M.[2], Ofri, R.[2] and Gootwine, E.[1], [1]ARO, The Volcani Center, Institute of Animal Science, POB 6 Bet Dagan, 50250, Israel, [2]The Hebrew University of Jerusalem, Koret School of Veterinary Medicine, Rehovot, 76100, Israel; gootwine@agri.gov.il

Sporadic birth of lambs with impaired vision was reported in Improved Awassi flocks. Genealogical investigation suggested that the congenital malformation was inherited in an autosomal recessive mode. Clinical examination and behavioral assessments revealed that day vision but not night vision is impaired in the affected lambs. Electroretinography examination of these lambs showed diminished cone function and normal rod responses, which is characteristic of achromatopsia, a congenital, autosomal recessively inherited disorder described in human and dog. As mutations in CNGA3, CNGB3, and GNAT2 genes have been associated with achromatopsia, we sequenced their coding regions using RNA extracted from retinas of four affected and eight non-affected lambs. PCR primers design was based on the bovine orthologous genes. While the affected lambs were polymorphic for the CNGB3 and the GNAT2 genes, they were homozygous for a haplotype that carried nucleotide substitution at the CNGA3 gene, changing amino acid R236 to a stop codon. No such mutation was detected in the eight non-affected lambs. By PCR-RFLP based test, homozygosity for the stop codon mutation was detected in other twenty affected lambs. All non-affected individuals from the same flock (n=90) that were genotyped, were found to be non-carriers or heterozygous for the mutation. A selection program has been launched to eradicate the day blindness mutation from Improved Awassi flocks by genotyping breeding rams and ram lambs selected as replacements for the CNGA3 locus and culling all individuals that carry the day blindness mutation.

Angiopoietin-2 possibly affects SCS in HF cows

Tetens, J.[1], Baes, C.[2,3], Flisikowski, K.[4], Reinsch, N.[3], Fries, R.[4], Schwerin, M.[3] and Thaller, G.[1], [1]Christian-Albrechts-Univ., Olshausenstr. 40, 24098 Kiel, Germany, [2]Univ. of Hohenheim, Garbenstr. 17, 70599 Stuttgart, Germany, [3]Res. Inst. for the Biology of Farm Animals, Wilhelm-Stahl-Allee 2, 18196 Dummerstorf, Germany, [4]Technical Univ. Munich, Hochfeldweg 1, 85354 Freising-Weihenstephan, Germany; jtetens@tierzucht.uni-kiel.de

Mastitis in is one of the most important health problems in dairy cows causing economic losses and impairing animal welfare. Although management factors play an important role in mastitis control, susceptibility to udder infections has a genetic component. The selection for udder health in many countries is based on the somatic cell score (SCS), which is highly correlated with mastitis incidence. The aim of this study was the molecular dissection of a QTL for SCS on BTA27, which has previously been fine-mapped within the ADR granddaughter design (ADR-GDD) applying a combined LD/LE approach. The marker bracket containing the peak position spans the region from 3.3 to 4.1 Mb (Build 3.1) and includes the ANGPT2 gene, which is known to be directly involved in the extravasation and activation of neutrophils. We comparatively resequenced all exons and flanking intronic regions of this gene in 16 sires from the ADR-GDD showing extreme DYDs for SCS. We identified 9 polymorphisms within the ANGPT2 gene including a VNTR near the 5' splice site in intron 7. Subsequently, we genotyped 492 sires of the ADR-GDD descending from 6 grandsires for 3 SNPs and the VNTR within ANGPT2 and tested for association applying linear regression with the grandsire as fixed effect and DYDs (first to third lactation) as phenotypes. The VNTR was found to be significantly associated with SCS ($P=0.00007$, first lactation) indicating that ANGPT2 affects SCS in HF cows. We furthermore aimed to analyse the VNTR polymorphism in German Simmental and Brown Swiss, but in these breeds it was found to be monomorphic. Regarding the molecular function of ANGPT2, it could serve as a powerful tool to decrease SCS, but bearing the risk of impairing immune response in clinical mastitis.

An investigation into the perception of ethics and welfare of the elite competition horse

Van Dijk, S.[1], Gego, A.[2] and Wolframm, I.A.[1], [1]Van Hall Larenstein (Wageningen UR), Droevendaalsesteeg 2, 6700AK Wageningen, Netherlands, [2]Aachen School of Course Design, Laurensbergerstr. 130, 52072 Aachen, Germany; inga.wolframm@wur.nl

Recent years have seen an increase in levels of interest and scrutiny relating to all aspects of equine management and training. The welfare of the competition horse is consequently becoming one of the key priorities in equestrian sport. The current study investigates the perception of different aspect relating to the welfare of the elite competition horse. Seventy-four participants from four target groups (elite riders N=18, FEI veterinarians N=19, FEI show officials N=17, visitors N=20) were asked to complete a questionnaire designed to explore perceptions of welfare and related aspects of top competition horses. Chi-Square tests were used to test for significant differences between the four groups. Riders, more than other groups, consider regular feedings of concentrates to contribute most to equine welfare (χ^2=17.6; $P<0.05$). Officials, more than other groups, consider social contact to be the key contributing factor to equine welfare (χ^2=12.6; $P<0.05$). A significant difference was found for how groups rated the current welfare of the elite competition horse (χ^2=2.7; $P<0.05$) with riders, followed by officials, perceiving welfare as the highest and visitors rating it the lowest. Veterinarians more than other groups consider there to be the need for an annual limit on competitions (χ^2=12.4; $P<0.05$). Findings from the present study suggest that there exist some considerable discrepancies between the different target groups in the perception of equine welfare and contributing factors. Results also seem to reflect the existence of different sets of values and ethical norms for each of the four groups, which may originate from existing practices. Future implications include further increasing awareness of equine welfare requirements and the application of ethical standards derived from species-specific research rather than based on reasons of practicality.

P.R.E. horses, from leisure to the Olympic Games

Lucio, L., ANCCE, Sport & Competitions, Cortijo de Cuarto (Viejo), 41014 Sevilla, Spain; luislucio2000@yahoo.com

For centuries, Spanish horses have been part of the European history. Their power, arrogance, beauty and good character made them be present in the wars, the culture and the art of our continent for ages. Just 20 years ago, Spanish Equestrian Federation and the National Breeders Associations began an enthusiastic project in order to make their dreams become true; bring its beautiful PRE (Pure Spanish Breed) horses into the sport world. From 1992, when first Spanish stallion went to an international competition under F.E.I. rules, till now, the results could not be better. These impressive PRE stallions gave our country lots of success and happiness. Behind it, a strong breeding activity is creating every year many talented PRE horses ready to keep satisfying all the horse lovers' wishes. Nowadays, all over the world, thousands of leisure riders and sport competitors enjoy PRE horses as safe friends in hobby riding or strong partners in high performance activities as Dressage International Shows. The key of this success should be found in the team work done between the sport and breeding organisation and also in the value of the PRE horses in themselves. A quick tour around this process will give us a full picture of this transformation.

Prospects of equine genomics for performance horses

Distl, O., University of Veterinary Medicine Hannover, Institute for Animal Breeding and Genetics, Buenteweg 17p, 30559 Hannover, Germany; ottmar.distl@tiho-hannover.de

The information on the structure and organization of the horse genome increased very rapidly in the last few years. A whole genome shotgun (WGS) approach in conjunction with paired-end sequencing of BAC clones was employed to achieve a 6.8X coverage of the horse genome. The second version of the horse genome contains 2.68 Gb sequence and the sequence has been ordered and oriented on the horse chromosomes. Along with the horse genome project, several tools for whole genome analyses were developed. Using the sequenced female horse Twilight and about 100,000 WGS reads from seven horses of different breeds, an equine 50K SNP chip was developed. Several expression arrays for whole genome transcript analyses are now also available to study expression patterns of genes. The recent developments in equine genomics offer the tools to study complex genetic mechanisms and their interactions with environmental factors and stimuli. Thus, the performance horse is a valuable model organism to study the mechanisms influencing performance in different lifetimes and training phases. There are many research fields directly related to performance that need to be addressed. Interdisciplinary networks of researchers should be established to fully exploit these new technologies in an optimum way. All research fields such as nutrition, physiology, clinics, management and housing and horse clinics will greatly benefit from horse genomics and on the other hand genetics will understand the genetic mechanisms much better in the future. An open question may be if genomic selection procedures will be introduced like in cattle or pig industry. In this respect diseases or conditions limiting the performance capacities seem to be more important for selection purposes than the expected strengths and weaknesses of a horse in later life. Furthermore, a deeper understanding of the genetic determination of equine performance seems necessary before applying genomic selection in horses.

Economic impact of the horse industry: a special reference to Spain

Castejon, R., UNED, Economía Aplicada e Hª Económica, Senda del Rey 11, 28040 Madrid, Spain; rcastejon@cee.uned.es

The economy of the equestrian sector encompasses all the activities related to the equine world. All activities revolving around the use of a horse as entertainment, sport or business play a role in the 'horse industry'. These activities take place due to the existence of people demanding horses and a variety of goods and services associated with them on the one hand to offer those goods and services on the other. The former obtain the satisfaction they desire from practicing horseback ridding or from relating in some shape or form with horses and are willing to pay for it. The later obtain income from the sale of the goods and services they provide. Studies reveal that as income per capita becomes larger, the equestrian demand increases and, consequently, the global expenditure incurred by those individuals interested in horses. More than 500,000 horse, 9,000 breeders and 50,000 equestrian businesses, show the economic importance of the equestrian sector in Spain. That part of global expenditure that adds to the GDP of a country is what is generally defined as economic impact of the sector. Its relation will depend mostly on: the percentage of the expenditure that adds to national production and the participation in the country's global expenditure of those domestic goods and services associated with the equestrian demand. The different sub-sectors in the equestrian world have a different bearing as a function of the number of horses, the activity that is performed with them and the demand of each specific activity. The activities related to sports are those with most economic impact, about 8,000 € per horse, measured as total expenditure or with respect to the employment that they generate in the economy.

Nutrition of the elite horse: what practice is asking for, is a challenge for science

Coenen, M.[1], Vervuert, I.[1], Düe, M.[2] and Wehr, O.[3], [1]Institute of Animal Nutrition, Nutrition Diseases and Dietetics, Faculty of Veterinary Medicine, University Leipzig, Gustav Kühn Str. 8, 04159 Leipzig, Germany, [2]Fédération Equestre Nationale (FN), Freiherr von Langen-Straße 13, 48231 Warendorf, Germany, [3]Veterinary Clinic for Horses, Osterholz 2, 25524 Breitenburg, Germany; coenen@vetmed.uni-leipzig.de

Nutrition of elite horses is an underestimated part of a complex structure in the entire management of these animals driven by different people. The present communication is handled as a dialog from different positions. Veterinarians are in charge for the health management and recognize nutrition as an indispensible contribution to long lasting performance and to activate additional capabilities. Nutrition as a scientific discipline has to serve by the actual standard of knowledge. The need for roughage, the management of starch supply etc. are subjects of knowledge transfer into practice via veterinarians. Official institutions and trainers are evaluated referring to the success of the horse-rider unit. They need to adapt the knowledge in physiology, nutrition of themselves and of their clients. They additionally have to respond on actual challenges like climate conditions etc. The major abandonment of the scientific side is to drive an structure for continuous education offering science in an applicable format. The recommendations e.g. on roughage urgently need to be accepted on this level to become the status of an authoritative communication. The riders and their crew compose experience, estimates and knowledge. The success orientated structure in this section mostly takes high performance as a proof for health. The science is challenged to communicate confirmed scientific knowledge and to keep unproved measurements separate. The expanding utilisation of herbs as well as the amino acid supply to horses show a lack in scientific knowledge which is in part compensated by unproven practice. It becomes evident that all contributors to the management of the performance horse have to improve their input for the feeding strategy of the equine athletes.

Assessment of feed additives that improve the diet utilisation in the European Union

Anguita, M., Galobart, J. and Roncancio-Peña, C., European Food Safety Authority, FEEDAP Panel, Largo N. Palli 5/A, 43100 Parma, Italy; montserrat.anguita@efsa.europa.eu

In the European Union (EU), feed additives need to undergo an authorisation procedure as established in the Regulation (EC) No 1831/2003 in order to be placed on the market. The European Food Safety Authority (EFSA) is responsible of assessing the safety of feed additives for animals, humans and the environment as well as to evaluate its efficacy. Based on EFSA's opinion, the European Commission will grant or deny the authorisation of the product for its use in the EU market. One of the effects that an additive may have is to 'favourably affect animal production, performance or welfare, particularly by affecting the gastrointestinal flora or digestibility of feedingstuffs'. Most of the additives performing such effect fall under the category zootechnical additives, functional group digestibility enhancers. This functional group includes the present generation of feed enzymes. The assessment of an enzyme feed additive is based on the technical dossier prepared by applicants and focuses on: Identity and characterisation of the additive. The product should be identified and characterised. If produced by a GMO, the genetic modifications should be described to allow an assessment of the safety of the genetic modification. Possible risks associated to the production strains should be considered. Safety for: -the target animals, tolerance studies are required to demonstrate absence of adverse effects when fed to the target animals. -consumers of food derived from animals fed with the additive, is assessed by means of a set of genotoxicity studies and a subchronic repeated dose oral toxicity study. -users, it should be determined if the product is irritant to eyes/skin and whether it has a sensitisation potential. -environment, normally no risk for the environment is envisaged, but attention should be paid to the presence of recombinant DNA in the final product. Efficacy, three trials should show that the product is effective according to the claim made.

Flavours affect feed reward in lambs and ewes fed canola meal

Mereu, A.[1,2], Giovanetti, V.[3], Cannas, A.[1], Ipharraguerre, I.[2] and Molle, G.[3], [1]University of Sassari, Dipartimento Scienze Zootecniche, via de Nicola 9, 07100 Sassari, Italy, [2]LUCTA SA, Ctra de Masnou a Granollers, 08170 Montornés del Vallés, Spain, [3]AGRIS, S.S. 291 Sassari-Fertilia, 07040 Olmedo, Italy; alemerku@yahoo.it

The objective of this study was to enhance the acceptability of canola meal by altering its oronasal-sensorial profile through the addition of flavours. The palatability of canola meal fed alone (control) or in combination with 13 different flavours, formulated to elicit sweet (1 to 8), umami (9 to 12), or bitter (13) taste was assessed in 6-min palatability tests. Two experiments that followed a 14 (days) x 14 (treatments) Latin square design were carried out with 14 female lambs or 14 multiparous dry ewes. Mean dry matter intake of canola meal (DMI) did not differ among treatments in lambs. In the case of ewes, flavours affected ($P<0.001$) DMI resulting in higher consumption of canola meal treated with products 12 and 2 compared with products 6 and 9. For most treatments, DMI increased as the experiment progressed. There was a significant ($P<0.05$) relationship between DMI and experimental day for treatments 1 to 5 for lambs and for all treatments, except number 12 for ewes. The slopes of the regression of DMI on experimental days were always numerically higher for ewes than for lambs. The largest proportion of variation in DMI was accounted for by treatment 2 in lambs ($r^2=0.52$) and by treatment 5 in ewes ($r^2=0.85$). These changes represented respectively a 148% and 54% reduction in unexplained variation when compared with the unflavoured control ($r^2_{lambs}=0.21$ and $r^2_{ewes}=0.55$). Treatment 12 resulted in the highest DMI both in lambs (41.1 g) and ewes (65.1 g), a response that was non-adaptive given its consistency over time. As suggested by the reduced variability in DMI, some flavours (mostly sweet-based flavours) appeared to facilitate animal acceptance of canola meal. Therefore, flavours may improve the feeding management of sheep by reducing the variability in animal responses to new or unpalatable feeds.

Effects of medium chain fatty acids on rumen fermentation

De Smet, K.[1], Deschepper, K.[1], Van Meenen, E.[1] and De Boever, J.[2], [1]Vitamex N.V., Booiebos 5, 9031 Drongen, Belgium, [2]Institute for Agricultural and Fisheries Research, Animal Sciences, Scheldeweg 68, 9090 Melle, Belgium; ellen.van.meenen@vitamex.com

Major challenges of the cattle industry are to improve efficiency and reduce negative environmental impacts of animal production. In this context Medium Chain Fatty Acids (MCFA) have been suggested as effective dietary fats altering rumen fermentation through their antimicrobial activity. Former *in vitro* experiments with MCFA mixtures already showed a lower methanogenesis. Therefore two mixtures, containing different ratios of MCFA, were tested with regard to a control (C) for their effect on rumen fermentation (3 x 3 Latin square). Three lactating cows, provided with a rumen fistula, were used to evaluate the effect on rumen pH, rumen ammonia content and rumen degradability of organic matter (OM) and NDF. The results were processed with 2-factor-Anova and Fisher's LSD test ($P<0.05$). Preventing ruminal acidosis is a major challenge. Supplementing MCFA to the ruminant's diet resulted in a significant increase of the ruminal pH before feeding (on average 6.7 vs. 6.4 for C). One hour after feeding, when production of fatty acids reaches its maximum, the pH remains significantly higher when MCFA were added to the diet (on average 6.1 vs. 5.9 for C). Monitoring the ammonia concentration in the ruminal fluid, a persistent lower level is established adding MCFA to the feed. Since milk urea levels are a predictor of nitrogen utilisation efficiency, the lower milk urea levels (on average 254 mg/l vs. 293 mg/l for C) indicate higher nitrogen efficiency. With both products there was a tendency for a higher degradation of OM (on average 81.8% vs. 78.9% for C), and a lower degradation of NDF (on average 55.1 vs. 58.2% for C). The latter could be explained by a higher digestibility of the cell content. Lower nitrogen losses and a higher degradability of OM imply a better nutrient efficiency and hence a lower emission into the environment.

Performance, nutrient digestibility and blood composition in weaned pigs fed diets supplemented with antibiotics, zinc oxide or peptide binding complex minerals

Han, Y.K., Sungkyunkwan University, Food Science and Biotechnology, 300 Chunchun-dong, Jangan-gu, 440-746 Suwon, Korea, South; swisshan@paran.com

Two trials were conducted to evaluate high levels of Zn, antibiotics, or peptide binding complex minerals (PBCM) on performance, nutrient digestibility and blood composition responses of weaned pigs. Trial 1 was done to evaluate the performance responses of piglets fed basal diet without supplementation (negative control), supplemented with antibiotics (positive control), two levels of Zn (1,500 ppm and 2,500 ppm), or PBCM for a 28-d period. There were six pens per treatment and each pen housed five barrows. Growth performance improved for pigs fed feed additives supplemented diets compared with the control ($P<0.05$). The performance of pigs fed PBCM supplemented diets was essentially equal to that of the antibiotics or pharmacologic level of ZnO supplemented pigs indicating that PBCM can substitute for antibiotics and Zn in diets fed to nursery pigs. Trial 2 was done to evaluate the growth performance, nutrient digestibility and plasma composition between pigs fed feed additives: antibiotics (positive control), Zn 1,500 ppm, Zn 2,500 ppm, and PBCM supplemented diets fo a 14-d period. There were nine pens per treatment and each pen housed sixteen barrows. There was no difference in weight gain, feed intake and feed conversion between pigs fed diets with antibiotics, ZnO, or PBCM. The digestibility of energy, Ca, P, Cys, His, Lys, Thr, and Val was significantly ($P<0.05$) lower for pigs fed the ZnO supplemented diet than for pigs fed the PBCM diet. The level of plasma Zn, SGOT and SGPT was significantly ($P<0.05$) higher for pigs fed the ZnO supplemented diet than for pigs fed the other two diets. Somatomedin-C level was lower for pigs fed the 1,500 ppm Zn supplemented diet than for pigs fed the PBCM diet ($P<0.05$). Cortisol level was lower for pigs fed the 2,500 ppm Zn supplemented diet than for pigs fed the antibiotics diet. LDL-cholesterol concentrations in pigs fed antibiotics were substantially higher ($P<0.05$) than those in pigs fed antibiotics.

Efficacy of a probiotic (*Bacillus subtilis* C-3102) in weaned piglets

Medel, P.[1], Esteve-García, E.[2], Kritas, S.[3], Bontempo, V.[4], Marubashi, T.[5], Mc Cartney, E.[6] and Sánchez, J.[1], [1]Imasde Agroalimentaria, SL, Mad, Spain, [2]IRTA, Reus, T, Spain, [3]University of Thessaloniki, Th, Maced, Greece, [4]University of Milan, Mi, Milan, Italy, [5]Calpis, Co, Ltd, Japan, [6]Pen & Tec Consulting, SCP, Barna, Spain; pmedel@e-imasde.com

Four studies with 1,074 piglets in 132 replicates (single-sex pens) evaluated the efficacy of a probiotic feed additive, *Bacillus subtilis* C-3102 ($1x10^{10}$ viable spores/g, Calsporin®, Calpis Co. Ltd.). A completely randomized design was used in each study of 2 experimental treatments: 1) control basal diets, and 2) basal diets with $3x10^5$ CFU/g feed. In each study, male and female piglets (only males in Study 1) were fed prestarter feed for 14 days after weaning at ~26 days of age and starter feed from 14-42 days on trial. Two trials used pelleted feeds, 1 study mash feeds and 1 study pelleted prestarter then mash starter. Data were tested for homogeneity and pooled to enable statistical meta-analysis, where $P\leq0.05$ was judged significant, and $0.05<P\leq0.10$ a near-significant trend. Parameters chosen were body weight (g) at 14 and 42 days of trial, mortality (%), daily gain (g), feed intake (g/day) and feed efficiency (feed:gain) from 1-14, 15-42, and 1-42 days on trial. Probiotic addition and experiment were considered as main effects. Piglets fed supplemented diets were 2.6% heavier at 42 days of trial ($P=0.0351$). No significant differences between treatments were observed from 1-14 days of trial. Feed intake was decreased by 3.4% and feed efficiency was improved by 3.2% from 15-42 days of trial, but differences were not significant ($P=0.1226$ and $P=0.1190$, respectively). Data from the global period (1-42 days of trial) indicated a significant improvement in daily gain (3.8%, $P=0.0265$) and feed efficiency (5.0%, $P=0.0015$). Mortality (5.9%) was considered normal in the study models used and there were no significant differences in mortality between control and probiotic-fed piglets. These data provide evidence that Calsporin® improves weaned piglet performance at $3x10^5$ CFU Bacillus subtilis C-3102 per g feed.

The effectiveness of a recombinant cellulase used to supplement a barley-based feed for free-range broilers is limited by crop beta-glucanase activity and barley endoglucanases

Ponte, P.I.P.[1], Guerreiro, C.I.P.D.[1], Crespo, J.P.[2], Crespo, D.G.[2], Ferreira, L.M.A.[1] and Fontes, C.M.G.A.[1], [1]CIISA - Faculdade de Medicina Veterinária, Avenida da Universidade Técnica, 1300-477 Lisboa, Portugal, [2]Fertiprado, Herdade do Esquerdos, 7450-250 Vaiamonte, Portugal; pponte@fmv.utl.pt

The inclusion of exogenous cellulases and hemicellulases in wheat, barley and rye-based diets for monogastric animals decreases digesta viscosity, improves the efficiency of feed utilisation and enhances growth. In contrast, recent experiments suggest that polysaccharidases are inefficient for improving the nutritive value of pasture biomass used by free-range broilers. The feasibility of using recombinant cellulases to improve the utilisation of cereal-based feeds by pastured poultry remains to be established. A study was conducted to investigate the capacity of a recombinant cellulase from *Clostridium thermocellum*, to improve the nutritive value of a barley-based feed for free-range pastured broilers of the RedBro Cou Nu x RedBro M genotype (10 birds per unit, 4 replicates per treatment). Enzyme activities and stability were studied in the gastro intestinal (GI) tract of 8 birds per treatment. The data suggested that supplementation of a barley based diet with a beta-glucanase (BG) had no effect on the performance of the broilers, foraging in legume-based diets. In addition, zymogram analysis revealed that the lack of effect of the recombinant enzyme in improving the nutritive value of the barley-based feed does not result from enzyme proteolysis or inhibition in the GI tract, since the enzyme retains its full catalytic activity. Significantly, BG activity was identified in the crop of non-supplemented animals. The data suggest that endogenous cellulases are originated both from the barley-based feed and from the crop microflora. Together the data suggest that the moderate levels of cellulase activity observed in the crop of non-supplemented animals are sufficient to degrade, partial or totally, the anti-nutritive beta-glucans present in barley based diets.

Effect of rumen-protected methionine and choline on plasma metabolites of Holstein dairy cows

Ardalan, M., Rezayazdi, K. and Dehghan-Banadaky, M., University of Tehran, Department of Animal Science, University College of Agriculture and Natural Resources, 4111, Iran; rezayazdi@ut.ac.ir

Choline deficiency in transition dairy cattle may be associated with hepatic lipidosis. The only practical means of increasing choline to the dairy cows is to feed it in a rumen-protected form (RPC). Methionine is a sulfur-containing AA that is involved in many pathways including the synthesis of phospholipids, carnitine, creatine and the polyamines. One approach that has been used to supply additional methionine to the cow has been to protect it from ruminal degradation for subsequent absorption in the small intestine. Forty Holstein dairy cows in their first (n=24) and second (n=16) lactation were used in a lactation study from 4-week prepartum through 10-week postpartum to investigate the potential effect of feeding ruminally protected methionine (RPM) (16 g/d) and RPC (60 g/d) on plasma metabolites of Holstein dairy cows. Blood samples were obtained 3 h after feeding from the coccygeal vein at −15 d relative to expected calving, parturition day, and 7, 15, 30 and 60 d after parturition. The repeated measurements of plasma metabolites were analyzed using Proc Mixed of SAS. The statistical model included the effects of treatment, parity, time, and treatment by time interaction. The DM-based forage to concentrate ratio was 57:43 for the dry period diet and 44:56 for the lactation diet. Postpartum DMI was higher for RPM+RPC-fed cows ($P<0.05$; 23.3 vs. 16.1 kg/d). The treatments did not significantly affect plasma triglycerides, glucose, total protein, NEFA, BHBA, PUN, and AST across measurement times ($P>0.05$). The plasma total protein (6.4 vs. 5.8 gr/dl), glucose (59.5 vs. 56.9 mg/dl), triglycerides (10.1 vs. 8.8 mg/dl), and PUN (16.2 vs. 15.4 mg/dl) tended to be greater for RPM+RPC-fed cows than other groups. Parity effect was only significant for PUN ($P<0.05$). For plasma glucose concentrations, there was a tendency for an interaction of treatment and measurement time ($P<0.01$). NEFA concentrations were decreased after RPM+RPC feeding (0.26 vs. 0.33 mmol/l).

Impact of feeding fungal treated linseed hulls on nutrients digestibility, nutritive value, nitrogen balance and some rumen and blood parameters of sheep

Abedo, A.A.[1], Sobhy, H.M.[2], Mikhail, W.A.[3], Abo-Donia, F.M.[4] and El-Gamal, K.M.[5], [1]National Research Center, Animal Production Dep., Dokki, Giza, Egypt, [2]Institute of African Research, Natural Research Dep., Cairo Univ., Giza, Egypt, [3]Institute of African Research, Natural Research Dep., Cairo Univ., Giza, Egypt, [4]Animal Prod. Dep., By-Product Utilization Dep., Dokki, Giza, Egypt, [5]Ministry of Agric., Technical Office, Dokki, Giza, Egypt; abedoaa@hotmail.com

Six adult Ossimi rams were assigned into two groups. The first group was fed a concentrate feed mixture (CFM) to cover 50% of the maintenance requirements and untreated linseed hulls (LSH) were fed ad lib. The second group was fed CFM to cover 25% of the maintenance requirements and the fungal treated LSH (with T. harzianum for 9 days) was fed ad lib. Animals were fed for 21 days as a preliminary period and 7 days as a collection period for feces and urine. The experimental design was one way classification.The results indicated that the DM intake of fungal treated LSH was significantly ($P \leq 0.05$) higher than untreated LSH. Digestion coefficients of DM, OM, CF, and EE were numerically higher but not significantly affected for animals fed treated LSH compared with fed untreated LSH. While the digestibility of CP and NFE were significantly ($P \leq 0.05$) increased for treated LSH compared with untreated, no significant differences ($P > 0.05$) were recorded for TDN and DCP values for animals fed treated or untreated LSH. Nitrogen retention was numerically higher but not significantly affected for rams fed treated LSH than rams fed untreated LSH. Values of ruminal pH and ammonia-N were significantly ($P \leq 0.05$) higher; while value of total VFA's was significantly ($P \leq 0.05$) lower for animals fed fungal treated LSH than those fed untreated LSH. Values of blood total protein, albumin, globulin, glucose and urea-N were not significantly different between the two groups, except Alanine Transaminase and Aspartate Transaminase values were significantly ($P \leq 0.05$) higher for animals fed treated LSH.

Effects of feeding zinc sources on biochemical parameters, concentration of some hormones and zinc in serum of Holstein dairy cows

Sobhanirad, S.[1], Valizadeh, R.[2], Moghimi, A.[3] and Naserian, A.[2], [1]Azad university of Mashhad, Animal science, Golbahar - Mashhad, 9195648499, Iran, [2]Ferdowsi University of Mashhad, Animal science, Azadi sq., 917751163, Iran, [3]Ferdowsi University of Mashhad, Biology, Azadi sq., 917751163, Iran; ssobhani2002@yahoo.com

Fifty-four lactating dairy cows randomly allocated to one of three groups. Ttreatment groups received the basal diet with no supplemental Zn (control), basal diet plus 500 mg of Zn/kg of DM as $ZnSO_4.H_2O$ (ZnS) and basal diet plus 500 mg of Zn/kg of DM as Zinc methionine (ZnM). This experiment was started in the first phase of lactation (35 ± 3 days after parturition). The data were analysed using the mixed procedure of SAS (9.1) for a block randomized design with repeated measures. Plasma and serum were taken from samples of blood on 0, 2, 4, 6, 8 and 10 weeks of experiment. The number of red blood cell, haemoglobin, MCHC, number of platelet and serum zinc in the ZnM and ZnS groups were significantly higher than control group ($P < 0.05$). Mentioned factors were also higher ($P < 0.05$) in dairy cattle that received ZnM than ZnS. There were no differences for WBC, PCV, MCH and PLT. There were not significant differences among different sources of Zn for ALP, but LDH and SOD in ZnM group were significantly higher than ZnS. In the current study, lactating dairy cattles fed ZnM had higher concentration of Zn in blood serum, some of haematological parameters and enzyme activity of LDH and SOD which is consisted with previous studies. In this study, despite the diet was fortified at levels exceeding the NRC requirements, improvements in blood parameters and enzyme activities of LDH and SOD were observed when organic Zn sources were supplemented.

The productive performance of Egyptian dairy buffaloes receiving biosynthetic bovine somatotropin (rbST) with or without monensin

Helal, F.I.S. and Lasheen, M.A., National Research Center Cairo, Egypt, Animal Production, 10 sabel el kazendar st. Daher - ABBASSYA, 1111, Egypt; mazenfhelal_310@yahoo.com

20 lactating buffaloes divided into 4 groups (5 animals each) were used to evaluate the singular and combined effect of bovine somatotropin (rbST) and monensin (M) on the productive performance of Egyptian dairy buffaloes. Treatments were T1: control group fed a 70% concentrate feed mixture (CFM), 15% rice straw and 15% berseem fodder (on dry – matter basis); T2: control + injection subcutaneously of 500 mg rbST per animal per 14 days; T3: control + 400 mg M per animal per day added on the top of the CFM; T4: rbST + M in combination during the last 14 days before expected calving up to 120 days after parturition. Dry matter intake (DMI) was higher by 7.9, 3.7, and 5.3% for T2, T3 and T4 respectively compared to T1. Although DMI was increased by rbST and M treatments, digestibility of DM, OM, CF, EE and NFE were not affected by treatments, except CP digestibility which was decreased by M treatment. Milk yield and 4% fat corrected milk yield were significantly ($P<0.01$) higher for animals treated with rbST than for animals in other groups (milk yield of 11.2, 16.1, 12.6 and 14.4 kg/day for T1, T2, T3 and T4 respectively and the same trend for the fat corrected milk). The effect of T3 on milk yield was less pronounced than for T2, whit an increase of 13.3% on milk yield was observed in buffaloes received (M) in their diet. Milk fat, total solids (TS), protein (TP) &ash contents were not significantly ($P>0.05$) changed by treatments, however, milk lactose content was significantly ($P<0.01$) increased by treatments The results of the present study suggest that rbST is efficacious in increasing milk yield without effect on milk composition and without any adverse effects on lactating buffaloes.

Effect of dietary inorganic chromium supplementation on rumen microbial fermentation rate using *in vitro* gas production

Ghiasi, S.E., Valizadeh, R., Naserian, A.A. and Tahmasbi, A.M., Ferdowsi university of Mashhad, Department of animal science, Mashhad azadi square, 91775-1163, Iran; rvalizadh@yahoo.com

Insulin resistance (IR) is a problem in periparturient dairy cows. Chromium has the potential for lowering plasma free fatty acids and cholesterol concentration and IR. The role of chromium is probably associated with increasing the insulin internalisation and amplifying insulin signaling through activation of cellular insulin receptors. Chromium supplementation in the diet may have effects on microbial fermentation in the rumen. The aim of this study was to evaluate the effects of dietary chromium sulphate supplementation on microbial fermentation rate using an *in vitro* gas production method. Two fistulated beluchi sheep were sampled for rumen fluid. The basal diet for the gas production test was mixed with 0, 50, 100 and 200 ppm of chromium as chromium supplement. Feed samples were milled (1-mm screen). Gas production was recorded at various times up to 72 h of incubation with four replicates. The degradable parameters of DM were determined using the equation of $P=b(1-e^{-ct})$ with P = accumulative gas production during time t; b = produced gas from fermentable fraction; c = gas production constant rate and t = time of incubation. There were no significant differences among treatments for c. The gas produced from fermentable fraction (b) of the 200 ppm Cr group was significantly ($P<0.05$) less than other treatments except 100 ppm group (0: 64.49±1.72[a], 50: 65.83±1.49[a], 100: 60.38±1.48[ab], 200: 55.91±2.97[b]). This depression in gas production for concentration of 200 ppm may be the chromium toxicity in microbial environment. The results indicate that, based on *in vitro* measurements, chromium supplementation to a level of 100 ppm per diet has no reverse effects on microbial fermentation in the rumen.

Effects of an amylase inhibitor (acarbose) on ruminal fermentation and animal metabolism in lactating cows fed a high-carbohydrate ration

Blanch, M.[1], Calsamiglia, S.[1], Devant, M.[2] and Bach, A.[2], [1]UAB (Universitat Autònoma De Barcelona), Edifici V Campus UAB, 08193 Bellaterra, Spain, [2]IRTA (Institut De Recerca I Tecnologia Agroalimentàries), Torre Marimon, 08140 Caldes De Montbui, Spain; marta.blanch@irta.cat

The objective of this study was to evaluate the effects of an amylase inhibitor (acarbose, Pfizer Limited, Corby, UK) in dairy cows in a 2x2 (2x28d) cross-over experiment. Eight Holstein cows fitted with a rumen cannula (milk yield = 24.3± kg/d, BW = 622±54 kg, DIM = 183±67, 5 multiparous and 3 primiparous) were used to study animal metabolism and ruminal fermentation. Throughout the study, animals were fed a high non-fibre carbohydrates (NFC) partial mixed ration (PMR, 17.6% CP, 28.3% NDF, and 46.5% NFC, based DM) and concentrate during milking. Treatments were: control (no additive, CTR) and amylase and glucosidase inhibitor (0.75 g acarbose-premix/cow/d, AMI). Animals were blood sampled to determine blood glucose, insulin, and urea within the first hour after the morning feeding in two separate days in each period. Samples of ruminal contents were collected during three days in each period at 0h, 4h and 8h post feeding to determine volatile fatty acid and ammonia-N concentrations, quantification of protozoa, Streptococcus bovis and Megasphaera elsdenii. Rumen pH was recorded electronically at 22-min intervals during six days in each period. Data were analysed using a mixed-effects model. Differences were declared at $P<0.05$. Cows on AMI treatment spent less hours below pH = 5.6 compared with CTR group (3.74 and 6.52±0.704 h/d, respectively) and AMI had greater daily average pH compared with CTR (6.05 and 5.92±0.042, respectively). AMI animals tended ($P=0.09$) to have lower S. bovis to M. elsdenii ratio than CTR (4.09 and 26.8±12.0, respectively). These results suggest that supplementing diets with acarbose to dairy cattle fed high-yielding rations may be effective in reducing subacute ruminal acidosis in lactating cows with no negative effects on ruminal fermentation and animal metabolism.

The effect of cinnamaldehyde, eugenol and a garlic oil standardized in propyl propyl thiosulfonate content on rumen fermentation and on methane production measured with gas production technique

Cavini, S.[1], Calsamiglia, S.[1], Bravo, D.[2], Rodriguez, M.[1], Schroeder, G.[3] and Ferret, A.[1], [1]Univeristat Autonoma de Barcelona, Facultat de Veterinaria, 08193-Bellaterra, Spain, [2]Pancosma, Voie-des-Traz 6, C.P. 143, CH-1218 Le Grand-Saconnex, Switzerland, [3]Cargill, Elk River, MN, USA; cavinisara@yahoo.it

A mixture of cinnamaldehyde and eugenol (CIE) and garlic oil has been shown to decrease the acetate to propionate ratio, the ammonia-N concentration and the methane to volatile fatty acid (VFA) ratio in rumen fluid. The aim of this experiment was to study the effect of increasing doses (20, 40, 80, 120 and 160 mg/l) of a mixture of CIN+EUG (38 and 68%, respectively) and a garlic extract standardised for propyl propyl thiosulfonate (PPT) content on rumen fermentation and methane production. The test was conducted using an in vitro gas production technique. A negative control (C) with nonessential oil and a positive control with 500 mg/l of monensin (MON) were included in the trial. A 0.25 g DM of a 60:40 forage:concentrate diet was incubated with 30 ml of a 1:4 ruminal fluid-to-buffer solution and incubated for 72h at 39 °C. Methane, ammonia-N and VFA's were analysed 24 h after the beginning of the incubation, and gas production was determined at 0:45, 1:15, 2, 4, 8, 12, 24, 48 and 72 h. Each treatment was tested in duplicate and in two replicated periods. Results were analysed with PROC MIXED of SAS and significance declared at $P<0.05$. Treatments had no effect on total gas production and only reduced total VFA concentration in 160-CIE and 160-PPT, and 160-PPT also reduced methane production. Ammonia-N was reduced by 80-CIE, 120-CIE, 20-PPT, 40-PPT and 160-CIE. Acetate proportion was lower in 120-PPT and 160-PPT, and higher in 120-CIE. Propionate was higher in 160-PPT. Acetate to propionate ratio was lower in 120-PPT and 160-PPT. Branched-VFA's were higher in 40-CIE and 160-PPT but lower in 80-CIE, 120-CIE and 160-PPT. Results suggest garlic oil standardised in PPT may improve the efficiency of rumen fermentation.

Effect of medicinal herbs or spices on *in vitro* ruminal nutrients disappearance of alfalfa hay
Jahani-Azizabadi, H., Danesh Mesgaran, M., Vakili, A.R. and Heravi Moussavi, A.R., Ferdowsi university of Mashhad, Dept. of Animal Science, Faculty of Agriculture, P.O. Box 91775-1163, Mashhad, Baranb, Iran; vakili_ar@yahoo.com

The objective of the present study was to evaluate the *in vitro* effect of medicinal herbs or spices on ruminal disappearance of various nutrients of alfalfa hay. Treatments were incubated in a medium containing cell free rumen fluid and buffer as 0.4: 0.6. Approximately 400 mg of dried alfalfa hay (as control) [Neutral detergent fiber (NDF) = 537 and crude protein (CP) = 150 g/kg DM] or plus 16 mg of garlic (GA), cinnamon (CI), cumin (CU), nutmeg (NU) or tumeric (TU) were placed in a 100 ml bottle (n=4) containing 45 ml of medium and 5 ml of mixed rumen microbes and incubated for 24 h at 38.5 °C. Then, bottle content was filtered (48 μm pore size) and unfiltered residual was dried at 60 °C for 48 h. Dry matter, CP and NDF concentrations of the residues were determined. Data were statistically analyzed using SAS (V. 9/1) and Duncan test was used to compare the means ($P<0.05$). Results indicated that all medicinal herbs or spices reduced significantly ($P<0.05$) *in vitro* DM (levels were 543, 494, 516, 511, 503 and 509, respectively for control, GA, CI, CU, NU and TU), CP (levels were 610.8, 578.1, 572.9, 560.5, 590.7 and 540.1, respectively for control, GA, CI, CU, NU and TU) and NDF (levels were 364.9, 269.8, 294.1, 305.5, 365.3 and 310.3, respectively for control, GA, CI, CU, NU and TU) disappearance. These results indicate that medicinal herbs and spices decrease the ruminal disappearance of alfalfa hay nutrients.

Effect of cinnamaldehyde, eugenol and capsicum on methane production from DDGS
Bravo, D.[1], Calsamiglia, S.[2], Doane, P.H.[3] and Pyatt, N.A.[3], [1]Pancosma, Voie-des-Traz 6, C.P. 143, CH-1218 Le Grand-Saconnex, Switzerland, [2]Universitat Autonoma de Barcelona, Facultat de Veterinaria, 08193-Bellaterra, Spain, [3]ADM, 1000 North 30th Street, Quincy, IL 62301, USA; David.BRAVO@pancosma.ch

Interactions between a mixture of eugenol and cinnamaldehyde (CIE) and capsicum oleoresin (CAP) at two different fermentation environments, dairy (rumen fluid from dairy cattle and pH 7.0) vs beef (rumen fluid from beef cattle and pH 5.0) were investigated to determine the effect on rumen microbial fermentation *in vitro*. Each treatment was tested in triplicate and in 2 periods. A 50-ml of a 1:1 rumen fluid to buffer solution preparation was introduced into polypropylene tubes supplied with 0.5 g of DM of DDGS, and incubated for 24 h at 39 °C. Samples were collected for ammonia N and volatile fatty acid concentrations. Results were analysed within type of fermentation environment using SAS, and differences were declared at $P<0.05$. The beef-type fermentation resulted in lower total volatile fatty acid concentrations (162.8 vs 176.2 mM), acetate (41.7 vs 58.7 mol/100mol) and butyrate (8.4 vs 9.5 mol/100mol) proportions, acetate to propionate ratio (1.02 vs 2.47), and ammonia-N (16.0 vs 18.9 mgN/dl) and methane (17.4 vs 29.5) concentrations; and higher propionate proportion (41.3 vs 24.4 mol/100mol), as expected. In the beef-type fermentation, supplementation with CIE increased total volatile fatty acids (+10%) and the proportion of propionate (+5%), and decreased methane (-19%) concentration, while capsicum had minimal effects. In the dairy-type environment, CIE increased propionate proportion (+17%) and decreased acetate proportion (-6%), the acetate to propionate ratio (-23%), and methane production (-18%). The effects of CAP were small, which was expected because its main effect on ruminants has been reported on increasing dry matter intake. The mixture CIE reduced methane production by rumen bacteria *in vitro*, and this effect was larger in a dairy-type environment without addition of CAP, compared with the beef-type environment.

Lactobacillus crispatus as an effective probiotic candidate among lactic acid bacteria for chickens

Taheri, H.R.[1], Moravej, H.[1], Tabandeh, F.[2], Zaghari, M.[1] and Shivazad, M.[1], [1]University of Tehran, Department of Animal Sciences, University College of Agricultural and Natural Resources, Karaj, Iran, [2]National Institute of Genetic Engineering and Biotechnology (NIGEB), Industrial and Environmental Biotechnology Department, Pajouhesh Blv., Tehran, Iran; taherihr@gmail.com

The objective of this research was to isolate, characterize and select the most effective strain in terms of adherence ability, antibacterial effects and enzymatic activities from crop, ileum and cecum contents of broilers for potential application as a chicken probiotic. These tests have not been performed so far in broiler chickens. Thus the final selected strain might be different from those which have been isolated in other studies. Therefore its probiotic effects may be more applicable for chick's health and digestion. A total of 332 lactic acid bacteria of broiler gut were investigated as putative probiotic candidates. The used tests for screening were aggregation, antibacterial and enzymatic activities, cell surface hydrophobicity, co-aggregation and tolerance to bile salts and acidic conditions, respectively. *Lactobacillus crispatus* was selected as a source of chicken probiotic because of its predominant characteristics in comparison with other strains isolated from the gastrointestinal tract. There were only noticeable differences in aggregation, amylase and phytase tests among lactic acid bacteria and the selected strain had high aggregation and amylase activity. Also this strain had phytase and protease activities, cell surface hydrophobicity (92%), and high resistance to pH 2 and bile salts (0.15% in medium culture). This strain did not show any co-aggregation and lipase activity as well as other strains. This study showed that the screening methods are of vital importance which can affect the final selection. It also indicates that screening procedures must be comprehensive and include more features in order to obtain a more applicable strain as a probiotic supplement.

The effect of feed composition and delayed access to feed in the neonatal period on broiler chick's performance

Tabeidian, S.A.[1], Samei, A.[2], Pourreza, J.[2] and Sadeghi, G.H.[3], [1]Islamic Azad University, Khorasgan Branch, Department of Animal Science, College of Agriculture, Isfahan, Iran, [2]Isfahan University of Technology, Department of Animal Science, College of Agriculture, Isfahan, Iran, [3]University of Kurdistan, Department of Animal Science, College of Agriculture, 416-Sanadaj, Iran; tabeidian@yahoo.com

Broiler performance in the end of the rearing period may be influenced by the post-hatch feeding program. Five hundred and forty Ross 308 broiler chicks aged 6h after hatching were allotted according a completely randomised design (CRD) scheduled of 9 treatments and 4 replicates of each 15 chicks. Treatments included a corn-soybean meal diet as base, fasting for 24 and 48 hours, feeding base diet supplemented with 15% egg powder for 24 and 48 hours after hatching, feeding base diet supplemented with 20% corn syrup for 24 and 48 hours after hatching and feeding base diet supplemented with both corn syrup and egg powder for 24 and 48 hours after hatching. Fasting for 48 hours had a negative effect on weight gain of broilers at 21 and 42 days of age and feed conversion ratio at 21 days of age. Feeding a diet supplemented with corn syrup for 24 hours after hatching resulted in higher ($P<0.05$) weight gain and lower feed conversion ratios at 7 and 42 days of age. The higher ($P<0.05$) weight gain at 21 days of age was recorded for chickens fed with a diet supplemented by both corn syrup and egg powder. Daily feed intake was not affected by treatments in the first 21 days of the experiment, however the groups receiving the base diet and the diet supplemented with corn syrup for 48 hours had lower ($P<0.05$) feed intake at 42 days of age. The results suggest that long term fasting after hatching could negatively affect broiler productivity at market age and offering a diet supplemented with corn syrup or corn syrup plus egg powder for 24 hours after hatching could improve broiler performance.

Effects of *E. coli* phytase supplementation to diet on the performance and bone mineralization in broilers

Han, Y.K.[1], Lee, W.I.[1] and Shen, S.Y.[2], [1]Sungkyunkwan University, Food Science and Biotechnology, 300 Chunchun-dong, Jangan-gu, 440-746, Suwon, Korea, South, [2]JBS United, Technical Management, P.O. Box 108, 4310 W.State Road 38, Sheridan, In 46069, USA; swisshan@paran.com

Two trials were conducted to evaluate different sources of phosphorus [dihydrated dicalcium phosphate(DCP), tricalcium phosphate(TCP) from an inositol by-product (TCPI), tricalcium phosphate from deflorinated phosphate(TCPD)] and *E.coli* phytase (expressed in yeast, Optiphos TM, JBS United, Inc.)] addition on growth performance, nutrient digestibility, blood composition and bone mineralisation in broilers fed corn-soybean meal based diets. Each trial utilised male chicks (Ross) from d 0 to d 33. Trial 1 used five pens of ten chicks and Trial 2 used six pens of five chicks. For trials 1, and 2 the experimental diet was a corn-soybean meal based diet that was analysed to contain 0.55% total P (0.20% estimated available P, as fed basis). Trial 1 was done to evaluate the growth performance responses of broilers fed diets supplemented with DCP (1.59%), TCPI (1.74%), or TCPD (1.57%). There were improved growth performance from d 0 to 7 for chickens fed TCPD supplemented diets compared with the TCPI ($P<0.05$). There was no difference in weight gain, feed intake and feed conversion from d7 to 33, and blood composition between broilers fed diets DCP, TCPI or TCPD. The digestibility of crude protein and phosphorus was significantly ($P<0.05$) lower for broilers fed the TCPI supplemented diet than for broilers fed the DCP diet. Trial 2 compared the effects of three graded levels of phytase (0, 250, 500, 1000 FTU/kg) in TCPI diet on growth performance, nutrient digestibility, plasma Ca and P, and bone mineralisation. Digestibility of DM, CP, energy, Cys, Leu, Met, and Phe responded linearly ($P<0.05$) to dietary phytase. Bone ca and bone breaking strength at d 33 responses increased linearly ($P<0.05$) in response to phytase.

The effects of Natuzyme supplementation on performance of broilers fed a threonine deficient diet

Khalaji, S., Zaghari, M. and Shivazad, M., University of Tehran, Animal science, Karaj Tehran, Iran; mzaghari@ut.ac.ir

Day-old male broiler chicks were randomly assigned to 24 battery cages pens with 7 birds per cage in a completely randomized block design with 4 replicates per treatment and grown to 6 wk of age. Birds were fed dietary treatments from 1 to 42 d of age. The 6 treatments were basal diet contained 0.56 and 0.46% standardized ileal threonine (SID Thr) in starter and grower periods, treatments 2 to 6 were basal diet supplemented with 0.1, 0.2, 0.3, 0.4 and 0.5 gr/kg Natuzyme. Body weight and feed intake of birds were recorded at 28 and 42 d of age. *In vivo* lymphoproliferation against Phytohemagglutinin-P (PHA-P) was measured at 36 d of age. Data analyzed using GLM procedure of SAS software. Means were compared by Duncan's multiple range test. Supplementing diets with Natuzyme significantly ($P<0.001$) improved feed conversion ratio (FCR) at 28 and 42 d of age. The effect of Natuzyme on body weight was significant ($P<0.05$) at 28 d of age but it had no effect on body weight at 42 d of age. *In vivo* lymphoproliferation against Phytohemagglutinin-P (PHA-P) was significant ($P<0.05$) at 36 d of age. There were no significant differences in breast, thigh and abdominal fat at 42 d of age. These Results showed that supplementing diets with each level of Natuzyme improved efficiency of broiler fed low threonine diet at starter period.

Effects of yeast *Saccharomyces cerevisiae* on rumen fermentation and blood parameters in Ghizel sheep

Ansari, A. and Taghizadeh, A., University of Tabriz, Department of Animal Science, 51664, Iran; ataghius2000@yahoo.com

Interest in the use of live yeast Saccharomyces cerevisiae as a feed additive for ruminants has increased in recent years. Addition of Saccharomyces cerevisiae to diets has been shown to decrease lactic acid content and redox potential of the rumen fluid and increase ruminal pH and VFA production. The aim of this study was to evaluate effects of Saccharomyces cerevisiae on rumen and blood parameters. Sixteen Ghizel sheep were randomly divided into 4 treatments and were fed with alfalfa as ad-libitum. The treatments were without yeast culture (YC) and with 2.5, 5.0 and 7.5 g/d YC, respectively. At the end of experimental period (day 21), rumen fluid samples were collected from all rams 2 h after morning feeding (by a stomach tube) for analysis of pH, total volatile fatty acids (TVFA), NH3-N, methylene blue reduction time (MBRT) and sedimentation and flotation time (SFT). Blood samples were collected from the jugular vein of each ram and transferred into collection tubes and plasma was stored at -20 °C for determination of blood glucose, urea, triglyceride, cholesterol, total protein and albumin. Data were analysed as a completely randomized design by using the GLM procedure of SAS software. Ruminal pH and TVFA in YC consuming sheep was higher ($P<0.05$) than for sheep in the control treatment. NH3-N, MBRT and SFT decreased by YC and a significant difference ($P<0.05$) was observed between treatments. There was no significant difference ($P<0.05$) in blood metabolites of sheep offered the diets with different levels of YC. The obtained data showed an improvement in the rumen fermentation yield, when YC was used. The increase of ruminal pH in YC containing treatments could be due to the stimulation of the activity of lactate utilising bacteria such as Selenomonas ruminantium. It was concluded that YC had positive effects on rumen parameters and Saccharomyces cerevisiae had sufficient potential to manipulate the rumen ecosystem.

An *in vitro* study of antimicrobial activity of clove and garlic essential oils and neomycin on rumen fluid

Ghanbari, F.[1], Ghoorchi, T.[1], Ebrahimi, T.[2] and Khomeiri, M.[2], [1]Gorgan University of Agricultural Science and Natural Resources, Animal Science, Beheshti Street, 4913815739, Iran, [2]Gorgan University of Agricultural Science and Natural Resources, Food Science and Technology, Beheshti Street, Gorgan, 4913815739, Iran; farzadghanbari@yahoo.com

In order to compare antimicrobial activity of garlic and clove essential oils with neomycin on rumen fluid, an experiment was conducted in Complete Randomized Design. Garlic and clove essential oils were infused into sterile 6 mm filter paper discs in 5 levels (0.5, 1, 2, 4 and 8µg). Neomycin was also infused into discs at a level of 50 µg. Then discs were put on Muller Hinton Agar which was cultured with rumen fluid. After 24 hours, inhibition zone diameters were measured with caliper. The Inhibition zone diameter was significantly different among treatments ($P<0.05$). Clove essential oil with 8 µg disk potency had the highest Inhibition zone diameter compared to other treatments(15.78 mm), while garlic essential oil with 0.5 µg disk potency had the lowest (5.3 mm). The results showed that clove essential oil had higher antimicrobial activity than neomycin. But more studies are needed to investigate the special effects of these compounds on the rumen microorganisms for replacing neomycin by clove essential oil.

Effects of dietary probiotic, prebiotic and butyric acid glycerides on Coccidiosis in broiler chickens
Taherpour, K.[1], Moravej, H.[1], Shivazad, M.[1], Adibmoradi, M.[2] and Yakhchali, B.[3], [1]University of Tehran, Department of Animal Sciences, Karaj, Iran, [2]University of Tehran, Faculty of veterinary medicine, Tehran, Iran, [3]National Institute of Genetic Engineering and Biotechnology, Industrial and Environmental Biotechnology Department, Tehran, Iran; hmoraveg@ut.ac.ir

Coccidiosis is one of the major parasitic diseases of poultry and is caused by the apicomplexan parasites *Eimeria*. Drugs and live vaccines are two main control measures of the disease; however, due to increasing concerns with prophylactic drug use and the high cost of vaccines, alternative control methods are needed. This study was conducted to evaluate the effects of probiotic (Primalac), prebiotic (Fermacto) and butyric acid glycerides (Baby C_4) on the development of coccidiosis on Ross 308 male broiler chickens. Seven hundred and four day-old broilers were randomly distributed in a $2\times2\times2$ factorial arrangement with two levels of probiotic, prebiotic and butyric acid glycerides. Eight treatments with four replicates and twenty two birds per replicate were used. At 28 d of age, two birds of each replicate were infected with *E. Acervulina*, *E. Maxima*, and *E. Tenella* and were placed in the separate cages. At this time, control was divided in two groups (with and without supplementation of salinomycin). The birds of different treatments fed diets that contained without supplement (control), salinomaycin, probiotic, prebiotic, butyric acid glycerides, probiotic+prebiotic, probiotic+butyric acid glycerides, prebiotic+butyric acid glycerides and probiotic+prebiotic+butyric acid glycerides. Lesion scores (LS) and oocyst counts (OC) were performed 7 d post-infection. The birds fed diets that contained salinomycin, probiotic + prebiotic + butyric acid glycerides and probiotic + butyric acid glycerides had lowest LS and OC than other groups ($P<0.05$). In conclusion, these results demonstrate that there are potential feed additives that, used in combination, can partially protect broilers against coccidiosis.

Effect of a mannanoligosaccharide on the growth performance and immune status of rainbow trout (Oncorhynchus mykiss)
Staykov, Y.[1], Spring, P.[2], Denev, S.[3] and Sweetman, J.[4], [1]Trakia University, Aquaculture, Student Campus, 6000 Stara Zagora, Bulgaria, [2]Swiss College of Agriculture, Länggasse 85, CH-3052 Zollikofen, Switzerland, [3]Trakia University, Biochemistry & Microbiology, Student Campus, 6000 Stara Zagora, Bulgaria, [4]Alltech Inc., Lixouri, 28200 Livadi, Cephalonia, Greece; stefandenev@hotmail.com

The objective of these experimental trials was to determine the effect of a mannan oligosaccharide (MOS) derived from the outer cell wall of Saccharomyces cerevisiae strain 1,026 on the growth performance and immune status of rainbow trout. Two experiments were conducted, one with eight net cages and the other with eight raceways. The net cage experiment (42 days) involved 14,400 fish with an initial average weight of 30 g. The raceway experiment (90 days) was conducted with 40,000 fish with an initial average weight of 101 g. Both experiments compared a commercial extruded diet with and without 2,000 ppm MOS supplementation. The calculated daily feed was supplied in six equal rations. Body weight, feed intake, and mortality were recorded and samples were taken for analysis of indicators of immune status. Significantly improved performance and immune status were observed in the net cage trial — improved weight gain of 13.7% ($P<0.01$), reduced feed conversion ratio (FCR) ($P<0.05$), reduced mortality ($P<0.01$), and improved indicators of immune status ($P<0.01$) for fish fed the MOS supplement compared with controls. Similar significantly improved performance was observed for the MOS-treated groups in the raceway trial — 9.97% improved weight gain ($P<0.01$), lower FCR ($P<0.01$), and reduced mortality compared with the control treatment. In the raceway trials, however, only the indicators of immune status lysozyme concentration, APCA, and CPCA were significantly improved by MOS treatment ($P<0.05$). These experimental trials demonstrated the ability of MOS to improve the growth performance, survival, and immune status of rainbow trout produced in net cages or raceways.

Long-term effects of fertilizer applications on natural grasslands within the aspect of environmentally sustainable grazing systems in highlands of the Eastern Turkey

Bozkurt, Y.[1], Basayigit, L.[2] and Kaya, I.[3], [1]Suleyman Demirel University, Faculty of Agriculture, Department of Animal Science, Isparta, 32260, Turkey, [2]Suleyman Demirel University, Faculty of Agriculture, Department of Soil Science, Isparta, 32260, Turkey, [3]Kafkas University, Faculty of Veterinary Medicine, Department of Animal Science, Kars, 36600, Turkey; ybozkurt@ziraat.sdu.edu.tr

The aim of this study was to evaluate long term effects of artificial fertiliser applications on natural grasslands in highlands of the Eastern part of Turkey with altitude of above 2,000 m. For this purpose an experiment was conducted to compare two grazing areas; one with naturally growing (NG) and the other one with artificial fertilizer applied (FG). Two grassland areas were chosen next to each other and FG area was fertilised with CAN fertilizer (Calcium Ammonium Nitrate) and TSP (Triple Super-Phosphate) at the rate of 180 and 190 kg/ha respectively. The experiment lasted during the years 2005 and 2007. In order to monitor biomas and chemical composition of grass, 6 sub-plots (16 m^2) were fenced within both areas to collect the soil and grass samples from non-grazed areas every two weeks. The results showed that there was no statistical differences ($P>0.05$) in botanical composition between NG and FG areas. Mean biomass dry matter per quadrat and sward height in NG and FG was statistically significant ($P<0.05$) only for the first year. While there were no significant differences in crude fibre and dry matter content of grass samples in all years, there were significant differences in N, Mg, K and P content of both areas. However, there was no statistical significant differences ($P>0.05$) in organic matter and mineral content of the soil between NG and FG during the experimental period. Therefore, it was concluded that there was no subsequent effect of artificial fertiliser application to improve grassland conditions in the highlands of the Eastern part of Turkey, which should be reconsidered in terms of environmentally sustainable natural grasslands.

Long-term performance of beef cattle herds on Mediterranean rangeland

Ungar, E.D., Agricultural Research Organization – the Volcani Center, Department of Agronomy and Natural Resources, Institute of Plant Sciences, POB 6, Bet Dagan 50250, Israel; eugene@volcani.agri.gov.il

The production efficiency of beef cattle herds on Mediterranean rangeland appears to be far from its potential, and many producers suspect that it is in decline. Since the cow can be productive for as long as 20 years, it is important to evaluate the technical performance of a herd in a long-term perspective. There are a few herds in Israel that have maintained records for decades, and data from two of them – Keshet and Har Zion – were analyzed using the Boker Tov software program. The program facilitates analysis according to four axes: chronological time, calving month, age of cow, and year of entry of cow into the herd. The dataset for Keshet spans the period 1975-2000 and contains data on 2,204 cows (15,487 cow-years) and 10,694 calves. The dataset for Har Zion spans the period 1960-2005 and contains data on 879 cows (8,139 cow-years) and 5,901 calves. The gross production efficiency of the Keshet and Har Zion herds was 0.69 and 0.73 calves per cow-year, respectively. The proportion of calves that contributed to production was 91.1 and 94.6%, respectively. The net production efficiency was 0.63 and 0.69 calves per cow-year, respectively. Mean weaning weight was 252 and 240 kg, respectively. Production efficiency in terms of meat production was 159 and 165 kg weaned meat per cow-year, respectively. At both herds there has been a steady decline in the number of calves weaned per cow-year since the mid-1980s, partly accounted for by a steady increase in calf losses. The present analysis lends weight to an existing suspicion that there is a sector-wide problem. Various possible causes for this decline have been suggested, including inappropriate breeding strategies, inadequate supplementary feeding, animal health problems, and rangeland degradation. However, a major impediment to progress in this sector lies in basic data recording, collection and analysis at the individual animal level.

Effects of the reduction of feeding frequency in dry beef cows on animal performance and labour costs
Casasús, I.[1], Blanco, M.[1], Alvarez-Rodríguez, J.[1], Sanz, A.[1] and Revilla, R.[2], [1]CITA, Av. Montañana 930, 50059 Zaragoza, Spain, [2]CTA, Av. Movera, 50194 Zaragoza, Spain; icasasus@aragon.es

The increasing size of cattle herds implies that more efficient use of labour force has to be implemented. High labour consuming tasks such as feeding should be simplified if possible, particularly in non-producing animals such as dry cows. Therefore, this study analysed the effect of reducing feeding frequency on beef cow performance and labour costs. Thirty-eight dry cows were either kept indoors during the winter (68 d) and fed once daily 10 kg of a total mixed ration (1D) or kept on forest pastures and fed once weekly 70 kg of the same diet (1W). In the spring, both groups grazed together on forest pastures with no supplement for 79 d. Cows were weighed fortnightly through the winter and at the end of the spring. At 2-week intervals during the winter blood samples were taken for metabolite analyses and faecal samples collected for determination of N content. Time consumed by feeding distribution was registered, and remaining forage in the stalls was annotated daily in the morning and afternoon. Cows on 1W treatment consumed 89% of the offer on the first 4 days (13.9±4.3 kg DM/d), and their lower faecal N compared to 1D indicates that they could also have grazed on the low quality pasture available. Blood metabolites indicate a lower nutritional level in 1W cows than 1D cows, except for urea. Response was quicker in NEFA, different from the second sample, and more delayed in b-OH-butyrate and triglycerides, only different at the end of the winter. Cows on 1W had minor BW and BCS losses during the winter, while 1D cows gained both BW and BCS in this period, and afterwards gains were similar in the spring. The total time devoted to feed and bed straw distribution was almost three-fold higher in 1D than in 1W treatment. Therefore, consequences on performance were minor but the reduction of feeding frequency allowed for a large reduction in labour costs.

Management strategies in hill pastures of Central Italy grazed by rotational-stocked cattle
D'ottavio, P.[1], Trombetta, M.F.[2] and Santilocchi, R.[1], [1]Marche Polytechnic University, SAPROV, via Brecce Bianche, 60100 Ancona, Italy, [2]Marche Polytechnic University, SAIFET, via Brecce Bianche, 60100 Ancona, Italy; p.dottavio@univpm.it

The aim of the paper is to discuss various management aspects for beef cattle breeding in a multi-paddock rotational stocking unit. The experiment was performed on clay soils characterised by different morphology and slope located at a mean altitude of about 500 m a.s.l. The climate of the study area is characterised by a mean annual temperature of 12.6 °C and a mean annual precipitation of 945 mm. The pasture surface was divided into paddocks by electric fences, with 2 mobile water-points and rotationally grazed by a herd of about 30 cows and 10 calves of local Marchigiana breed. In each paddock botanical composition, DM yield and forage quality were assessed before, during and after the utilisation throughout the grazing period 2005-2008. The main results concern with the following aspects: grazing management, forage yield and quality, mean daily intake and forage balance.

Cattle physiology and behavior in a Mediterranean oak woodland

Schoenbaum, I.[1], Henkin, Z.[2], Brosh, A.[3], Kigel, J.[1] and Ungar, E.D.[2], [1] Faculty of Agricultural, Food &Environment. Hebrew University of Jerusalem, Institute for Plant Sciences and Genetics in Agriculture, Rehovot, 76100, Israel, [2]ARO -Volcani Center, Agronomy and Natural Resources, Bet Dagan, 50250, Israel, [3]ARO-Newe Yaar Research Center, Beef Cattle Section, Ramat Yishay, 30095, Israel; isi_shin@yahoo.com

Sustainable utilization of woodlands by cattle is a serious challenge world-wide. The domination by woody vegetation of poor nutritional quality and the low yield of herbaceous vegetation limit cattle performance. Shortages of rangelands for grazing cattle in Israel necessitate finding an optimal grazing management for the woodlands that, on the one hand, will improve cattle performance and, on the other hand, will create an open-parkland landscape and decrease fire hazards. In this research the behavior and physiology of cattle in woodland habitats was studied. The experiment was conducted in patchy scrub-oak woodland in the Western Galilee, Israel. The area was classified into six vegetation types according to its formation. Spatial distribution, activity and heart rate of the cattle were monitored during winter, spring and summer. Lotek GPS collars with activity sensors and Polar heart rate sensors were deployed on six cows at each period. It was found that vegetation structure and seasonal conditions determined the spatial distribution and activity dynamics of the cattle in this habitat. No seasonal differences were found in the overall time devoted to each activity (rest, graze or walk). However, there were seasonal variations in the diurnal pattern of activity. Although cattle are naturally grazers, they browsed the woody vegetation in the summer when the quality of the herbaceous vegetation was low. Changes in activity over the course of the day were reflected in the heart rate pattern. Daily energy expenditure varied according to reproductive state and the availability and quality of the forage. Therefore it appears that cattle grazing may constitute a sustainable management tool for conservation in Mediterranean oak woodland.

Structure and perspective of the dairy cattle industry in Greece

Georgoudis, A.[1], Tsiokos, D.[1], Kotroni, E.[2] and Daskalopoulou, E.[2], [1]Aristotle University of Thessaloniki, Dept. of Animal Production, Faculty of Agriculture, Thessaloniki, 54006, Greece, [2]Holstein Association of Greece, P.O. Box 51603, Lagadas, 56403, Thessaloniki, Greece; andgeorg@agro.auth.gr

Dairy cattle in Greece are kept in environments, which range from the upper medium to high level of inputs. The dairy farms belong to a 65% to production system, in which food tends to consist predominantly of maize silage offered to cows with reasonable high levels of concentrates. During the last few years significant changes have taken place concerning the size of farms and their performance. Dairying has been concentrated on fewer, larger farms resulting in a corresponding decrease of the total number of farms employed in the sector and more importantly increasing the abandonment of small sized holdings. Intensive dairy cow units are now found in many areas especially in good farming land and in sub-urban regions, in proximity to dairy processing plants. Dairy farms are not necessarily associated directly with farmland and usually do not have enough land to produce their own fodder needed to support the herd. Very few farmers have available grassland for complementary grazing for 3-4 months. Thus, grazing options are from limited to zero. Forage self-sufficiency varies from zero to 50%. In 2008 the population of dairy cows including the replacement heifers was 203,461 heads from which 99.0% are Holstein-Friesians. The total number of cattle units in 2007-2008 is 5,630 with a total milk production of 716,318 tons, while the corresponding numbers in 2000-01 were 12,402 and 774,471 respectively. The milk of the most of the units is channeled in a network of pasteurization and milk processing factories, which is extended in the various regions of the country with the 70% of them located in Macedonia and Thrace. The minimum dairy herd size today in Greece which could 'guarantee' an acceptable farm income is considered to be 80 to 100 cows.

Evaluation of German Holstein Friesian cattle at Goht Al-Sultan farm, Libya II- Genotype environment interaction on production traits of German Holsteins Friesian in Libya

Mohamed, S.[1], Almasli, I.[1], Gargoum, R.[1] and Abusneina, A.[1,2], [1]University Of Garyounis, Department Of Zoology, P.O. Box 5035, Benghazi, Libyan Arab Jamahiriya, [2]Laurentian University, 935 Ramsey Lake RD MSR # 107, P3E 2C6, Canada; ax_abusneina@laurentian.ca

Genotype x Environment interaction and genetic correlations were studied through daughters of sires from Germany, Holland and Libya. Estimated transmitting ability (ETA) of sires was computed using LSM and REML method. Sires were used as a random effect. Estimates for ETA were compared to those under European and North American environment. Heritability estimates for a total milk yield, 305 daily milk mature equivalent M.E. or a total milk yield, and total milk production 305 day milk yield 2x M.E. were: medium (0.18), low (0.09), and medium (0.30 and 0.19), respectively. In addition, heritability for dry period, lactation period and reproductive traits were low and ranged between 0.01 and 0.03. Milk yield traits except dry period had a higher ETA during the Libyan management. Age at first calving and reproductive traits had also a higher ETA but, during the Dutch management. Correlations between ETA for milk yield ranged between 0.16 and 0.56 during Libyan, European and North American conditions. ETA of both North American and European sires were medium, low and negatively ranked under Libyan environment. ETA of milk yield traits showed higher variability under the Libyan environment. Sires with low ETA (> 500) revealed negative ranking under the Libyan conditions. These differences in heritability estimates due to the management system and or to environmental factors reflect Genotype x Environment interaction.

Cadmium, lead and arsenic in liver, kidney and meat in calves from extensive breedin in Castilla y León (Spain)

Escudero-Población, A.[1], Gonzalez-Montaña, J.R.[1], Prieto-Montaña, F.R.[1], Gutierrez-Chavez, A.J.[1] and Benedito-Castellote, J.L.[2], [1]University of León, Campus de Vegazana, 24071, Spain, [2]University of Santiago de Compostela, Campus de Lugo, 27002, Spain; andres.escudero@unileon.es

The objective of this study have been to determine the concentrations of three toxic elements (arsenic, cadmium and lead) in the livers, kidneys and the muscle of beefs reared in Castilla y León, which is one of the most important region in the production of beef cattle in Spain. A total of 62 cattles, between 6 months to 2 years old, from 4 different farms were collected from one slaughterhouse during 2006. Information on each calf was obtained from farm documentation that accompanied the calves to slaughter. These samples came to of the liver's caudal lobule, of the right kidney and of the diaphragms to avoid disturbances. Samples were acid digested and toxic metals concentrations were determined by ICP-MS/OES. The arithmetic mean concentration in all animals was for the liver 3.58; 22.06; and 7.24 ug kg^{-1} w.wt^{-1}. for arsenic cadmium and lead respectively. In the kidney concentrations were 17; 109.83 and 25,97 ug kg^{-1} w.wt^{-1}. for arsenic, cadmium and lead respectively. And for the diaphragm were 6.101; 1.285; and 3.7 ug kg^{-1} w.wt-1 for arsenic, cadmium and lead. All result has been lower than the limit established for the European Commission but all of them could be compared with the results of other authors including correlations between these metals metals. Knowledge of these concentrations is very important to the consumers because they are aware whit the Food Safety nowadays.

Possibility for increasing of nucleus herd size of simmental cattle in Serbia
Stjelja, S., Bogdanovic, V., Djedovic, R. and Perisic, P., Faculty of Agriculture, University of Belgrade, Nemanjina 6, 11080 Zemun-Belgrade, Serbia; vlbogd@agrif.bg.ac.rs

The best part of one cattle population, regarding milk traits, fertility, and body development traits, should be represented by nucleus herd, in which genetic improvement is realized. Nucleus herd of Simmental cattle in Serbia is consisting of only 1% of the best cows. In order to increase selection intensity, that has a great influence on potential genetic improvement, it is necessary to increasing the number of animals in nucleus herd. Establishing of nucleus herd in Simmental cattle population in Serbia might be done by above average cows exhibited on annual cattle shows, given that they are controlled and have detailed records keeping. The results of the analysis of milk yield from Simmental cows exhibited on cattle shows in Serbia, held from 2001 to 2007 are presented. Data set included 1,143 Simmental cows with average milk yield higher than of 5,500 kg in standard lactation. Possibility for increasing of size of Simmental nucleus herd in Serbia and further sustainable breeding is discussed.

Optimal age at first calving for improved milk yield and long productive life in Holsteins under tropical conditions
Ben Gara, A.[1], Rekik, B.[1], Hammami, H.[2] and Hamdene, M.[3], [1]ESAM, Départemet des productions animales, 7030, Mateur, Tunisia, [2]FSAGX, 2, Passage des Déportés, B-5030, Gembloux, Belgium, [3]OEP, Direction Amélioration Génétique, 2020, Sidi Thabet, Tunisia; bengara.abderrahmen@iresa.agrinet.tn

The effects of age at first calving (AFC) on milk production and true herd life (THL) were studied in Tunisian Holsteins. There were 33,407 first lactation records of cows born between 1987 and 2001 in 166 herds. Firstly, AFC was analysed using an animal model that included herd, calving month and year, herd-calving year interaction and age of dam as fixed effects and the random animal additive genetic value. Secondly, variations of first lactation and productive life milk yields and THL were explained by AFC in addition to herd, month and year at first calving and herd- year interaction. A cow produced on the average 5,669.8 kg milk during a 305-d first lactation period and 19,496 kg during her lifetime in the herd estimated at 3.3 lactations. All factors in milk yield and THL models were significant. Coefficients of determination ranged from 14% for THL to 64% for first lactation milk yield. The mean of THL was 38.6 months and the mean AFC was 28.7 months. Posterior mean of heritability of AFC derived by a Markov Chain Monte Carlo Bayesian method via a Gibbs sampling algorithm was 0.08. The reduction of AFC to around 24 months may result in improved 305-d and lifetime milk yields and a long THL in Tunisian Holsteins.

Performance of Friesian cattle under two environmental conditions

Badran, A.E., Aziz, M.A. and Sharaby, M.A., Faculty of Agriculture, Alexandria University, Animal Production Department, Aflaton St. elshatby, alexandria, 1256, Egypt; me6hat@yahoo.com

Recording of 200 Friesian cow were used in this study. Cows were imported from United Kingdom and located at the experimental station of university of Alexandria. The main object was to determine performance of Friesian cows under the Egyptian conditions compared to performance of their ancestor made under British condition. Cows had significantly lower average for milk yield (4,447.20 kg) than dams, granddames and dams of their sires. Fat % was higher in cows (4.07%) than that of dam and granddames (3.93% and 3.85%) but it was lower than that of dam of sire (4.14%). simple regression models were used to determine the relationship between cows and dams (I), dams and granddames (II) and cows and dams of sires (III) generally, the regression coefficients (b1) were proved to be significant ($P<0.01$) and ranged between -0.012 and 0.299 for fat% however, the coefficient of determination were extremely low and did not exceed limit of 0.1. residuals resulting from the above mentioned models were kept and were assumed to be representative of the environmental conditions under which the animal were raised. Regression coefficients (-0.008 for milk yield and -0.0009 for fat %) however, coefficients on residuals of II of residuals of III (0.909 for milk yield and 0.924 for fat %) were significant Than the Egyptian environment. Therefore, performance of cows was better under the British environmental.

Estimation of genetic parameters and breeding values from single- and multiple-trait animal models analysis for Friesian cattle

Oudah, E.Z.M., Department of Animal Production, Faculty of Agriculture., Mansoura University, 35516, Mansoura, Egypt; saidauda@yahoo.com

This study was carried out to compare the estimates of genetic parameters and breeding values (BV) using single- and multi-trait animal model analyses for a governmental herd of Friesian cattle in Egypt. The farm located at the northern part of the Nile Delta. Genetic parameters and breeding values for productive and reproductive traits were estimated from 1011 first lactation records. Single-trait analysis of total milk yield (TMY), 305-day milk yield (305-dMY), lactation period (LP), birth weight (BW), weaning weight (WW), number of service per conception (NOS), and age at first calving (AFC) was investigated as a method to genetic parameters and breeding values for individual traits. Multiple-trait analysis (four traits) including 305-dMY, LP, AFC and NOS; and multiple-trait analysis (two traits) included BW and WW were performed. Data were analyzed using Multi-trait Derivative Free Restricted Maximum Likelihood to calculate the genetic parameters and BV. Fixed effects of month and year of calving, and different covariates were added to the statistical models to remove its environmental effects on different studied traits. Heritability estimates for single-traits analysis were 0.07, 0.27, 0.12, 0.19, 0.06, 0.04 and 0.25 for TMY, 305-dMY, LP, BW, WW, NOS, and AFC, respectively. Heritability estimates for multiple-traits analysis were 0.28, 0.03, 0.31, 0.04, 0.04 and 0.10 for 305-dMY, LP, AFC, NOS, BW, WW, respectively. Genetic correlations of 305-dMY and each of fertility traits (AFC and NOS) were unfavourable (0.39 and 0.49 respectively) and between BW and WW was 0.71. However, there were differences between estimated breeding values for the same trait from single versus multiple-trait analyses. Breeding values for all animals regarding 305-dMY ranged from -3.43 to 5.62 and -4.38 to 7.67 kg for single versus multiple-trait analyses, respectively. The corresponding values for AFC were -4.01 to7.94 and -3.92 to 6.94 months, respectively.

Udder health traits as related to economic milk losses in Friesian cattle

El-Awady, H.G.[1] and Oudah, E.Z.M.[2], [1]Animal Production Department, Faculty of Agriculture, Kafr El Sheikh University, Kafr El Sheikh, Egypt, [2]Animal Production Department, Faculty of Agriculture, Mansoura University, 35516, Mansoura, Egypt; saidauda@yahoo.com

A total number of 4,752 lactation records of Friesian cows from 2000 to 2005 were used to determine the relationship between somatic cell count (SCC), udder health traits (UHS) and economic losses in milk production. Studied traits were milk yield traits {i.e., 305-milk yield, (MY), fat yield (FY) and protein yield (PY)} and udder health traits {i.e. SCC, mastitis (MAST) and udder quarter infection (UDQI)}. Least square analysis was used to estimate the fixed effects of month and year of calving, parity and stage of lactation on different studied traits. Data were analyzed using Multi-trait Derivative Free Restricted Maximum Likelihood to estimate the genetic parameters. The effects of SCC, MAST and UDQI on milk traits were also studied. Unadjusted means of MY, FY, PY and SCC were 3936, 121, 90 kg and 453,000 cells/ml, respectively. All fixed effects were significantly ($P<0.01$) affect all traits except the effects of month of calving on both FY and PY was not significant. The SCC, MAST, UHS and UQI increased during winter and summer than spring and autumn. Additionally, SCC and MAST noticeably increased with advancing in parities. Increasing SCC from 300,000 to 3,000,000 cells/ml increased UDQI from 5.5 to 23.2%. Losses in monthly and lactationally milk yields per cow ranged from 14 to 89 and from 105 to 921 kg, respectively. Losses in monthly and lactationally milk yields per cow ranged from 14 to 89 and from 105 to 921 kg, respectively. Losses in monthly and lactationally milk yields return per cow at the same level of SCC ranged from 24.5 to 155.75 LE and from 50.8 to 1612 LE, respectively. Heritability estimates of MY, FY, PY, SCC, MAST, UHS, UDQI were 0.31, 0.33, 0.35, 0.14, 0.23, 0.13, and 0.09, respectively. All milk production traits phenotypically and genetically correlated negatively with SCC, MAST and UDQI. The SCC can be used as a perfect tool for UHS and milk quality.

Evaluation of German Holstein Friesian cattle at Goht Al-Sultan farm, Libya I. genetics and environmental factors affecting milk yield and reproductive traits

Mohamed, S.[1], Almasli, I.[1], Gargoum, R.[1] and Abusneina, A.[1,2], [1]Garyounis University, Zoology, P.O. Box 5035, Benghazi, Libyan Arab Jamahiriya, [2]Laurentian University, Chemistry-Biochemistry, 935 RAMSEY LAKE RD., MSR 107, P3E 2C6, SUDBURY, ON, Canada; ax_abusneina@laurentian.ca

In the present study, the productive and reproductive performance of German Holstein Friesian dairy cows and subsequent generations born in Libya were evaluated and factors affecting productive and reproductive performance were assessed, based on data records collected from Ghot Al-Sultan station in Benghazi / Libya. A total of 694 pregnant heifers were imported from Germany. The number of cows that had the first lactation record and included in the study was 2,094. Data were analyzed using REML to estimate the environmental and genetic factors affecting milk yield and reproductive traits. Total milk yield was generally high (8409 Liters) and attained along with reproductive efficiency (110 days open) during Dutch management, and at the expense of the reproductive traits (161 days open) during Libyan Management. Management, calving year, calving month, age at first calving, lactation period, generation, origin of sire and sire all had a significant effect on both milk yield and reproductive traits. The performance of North American, European and Libyan sires was different for milk yield and reproductive traits. Heritability estimate was 0.18 for total milk yield indicating genetic differences among sires. Genetic correlations were found to be high among milk yield traits. Genetic correlations between age at first calving, dry period and reproductive traits were positive and low whereas correlations between calving intervals and days open were also high. Genetic correlation between milk yield and reproductive traits were low and negative.

Main aspects to promote organic goat production in mountains areas of Andalusia, Spain

Mena, Y.[1], Ruiz, F.A.[2], Castel, J.M.[1], González, O.[3] and Nahed, J.[4], [1]Sevilla University, Ctra de Utrera km 1, 41013 Seville, Spain, [2]IFAPA, Agrarian Economy, Ctra. Alcalá del Rio, 41200 Seville, Spain, [3]Asociación de Criadores de la Raza Caprina Payoya, Arco 123, 11680 Algodonales, Spain, [4]ECOSUR, Ctra Panamericana, 29290 San Cristobal de las Casas, Mexico; yomena@us.es

Organic goat production is poorly developed in Spain. In Andalusia (southern Spain) there is an important number of grazing dairy goat farms located in mountains areas, which are very close to the organic model. During 2006 and 2007, 13 grazing goats farms (4 of which certificated as organic) was monthly monitoring. Information about different aspects of the production system was generated. This information was used for researchers, technicians and farmers to make a diagnosis about the proximity to the organic model of dairy goat systems in mountainous areas. This diagnosis was made using the SWOT (Strengths, Weaknesses, Opportunities, and Threats) method. Later, weaknesses and threats were ordered by priority, using the structural method. As results, both, weaknesses and threats, were classified into four groups: power, conflict, independent and exit. The first two groups are the most important because they affect most of the others. The main difficulties for the transformation to organic model are: the little feed self-sufficiency of farm; the scarcity of organic feed in the market; the lack of training of technicians and farmers in grazing management; the difficulties for marketing of organic products (milk and meat); the delay in the payment of economical aids; the limited demand of organic goat products by consumers. Some of the proposals for the transformation to organic model are: to improve the management of grazing; to minimize the consumption of concentrate, although this implies a slight decrease in milk produced; to promote partnership between stock farmers and farmers to obtain food for the animals; to promote also partnership between stock farmers and handmade cheese factories to improve the commercialization of organic products.

Economic profitability of organic dairy goat farms in south of Spain: preliminary results

Angón, E.[1], Perea, J.[1], Santos, M.V.F.[2], García, A.[1] and Acero, R.[1], [1]University of Cordoba, Animal Production, Edificio Produccion Animal, Campus Rabanales, 14071, Spain, [2]University Federal Rural of Pernambuco, Campus Pernambuco, 52171-900, Brazil; pa2pemuj@uco.es

The aim of this study was to evaluate the economic profitability of organic dairy goat farms located in south of Spain. The area of study includes two Spanish regions (Andalucía, and Castilla la Mancha) that concentrate the 44% of the organic dairy goat farms registered in Spain. The sample of farms comprised 15 organic dairy farms (100% of the official census in these regions). The farms generate an average income of 49,348 € which comes from three main groups: the sale of milk (54.87%), sales of kids (17.71%) and subsidies (21.32%). The milk is sold through two channels: conventional or organic. Only 13% of the milk marketed by the organic channel (0.95 €/l), whereas the conventional channel absorbs 87% at a lower price (0 49 €/l). The costs amounting to 46,910 €, and include food (30.7%), labor costs (30.4%) and amortizations (9.86%), which representing over 70% of total costs. The remaining costs were distributed as follows: financing costs (0.17%), insurance premiums (0.14%), independent professional services (1.68%), supplies (1.73%), grazing leases (6.28%), taxes (4.64%) and other costs (4.24%). The costs can be grouped into fixed (62%) or variable (38%), and as farm size increases the variable costs gain more weight. The high labor cost reflects the lack of technology of many farms, which is replaced by workforce (145.7 goats/WU). The high outlay involved in the repair and maintenance (6.26%) compared to depreciation reiterated in sectorial technologic deficit. The amortization comes as follows: 51% animals, 25% machinery, 20% facilities and 4% constructions. The unitary cost is 0.87 €/l while the unitary income reached 1.1 €/l, so is sufficient to generate profits. The average level of production necessary to reach a balance between incomes and costs is 41,370 l, while the real production reach 54,300 l, so profits are generated.

Technical characterisation of organic dairy goat farms in south of Spain: preliminary results

Angón, E., Garcia, A., Perea, J. and Acero, R., Universitiy of Cordoba, Animal Production, Edificio Produccion Animal, Campus Rabanales, 14071, Spain; pa2pemuj@uco.es

The aim of this study was to characterise organic dairy goat farms through structural, technical and productive aspects located in south of Spain. The area of study includes two Spanish regions (Andalucía, and Castilla la Mancha) that concentrate the 44% of the organic dairy goat farms registered in Spain. The sample of farms comprised 15 organic dairy farms (100% of the official census in these regions). The production of organic dairy goat farms is developed in farms following a semi-extensive production system and located in economically and demographically depressed areas of the south. The farms are small-medium size, about 200 goats of mean in an average area of 253.8 ha. The breeds used are native biotypes with high rusticity and adaptation to the environment, emphasizing the Murciano-Granadina (33%), Malagueña (22%) and Florida (20%). The food is based on the use of fodder resources from farm, the average stocking rate is 0.2 LU/ha, and the concentrate is used only for goats in production. The average consumption of concentrate per animal is 277 kg. It carries out management with lots and two seasons, one in February-March and another in September-October. With the male effect and food control to achieve adequate results in most of the farms (fertility rate 1.26, replacement rate 10.4%). The average productivity per goat is 321 l, the rate of commercial kids amounts to 1.3 and the mortality rate is 8.8%, lower than those obtained in conventional intensified. The average investment per farm is 387,00 € (without taking into account the land investment), although it is generally outdated and inadequate to production system. The lack of technology is replace with workforce, thereby reducing the number of goats per worker to 145.7 goats/WU.

Sociological approach to organic dairy goat farms in south of Spain: preliminary results

García, A.[1], Angón, E.[1], Perea, J.[1], Acero, R.[1] and Santos, M.V.F.[2], [1]University of Cordoba, Animal Production, Edificio Produccion Animal,Campus Rabanales, 14071, Spain, [2]University Federal Rural of Pernambuco, Campus Pernambuco, 52171-900, Brazil; pa2accrr@uco.es

The new Common Agricultural Policy increases the importance of denominated non-productive functions, like the role of farms in environmental conservation or in the rural development. The aim of this study was to analyze the organic dairy goat farms located in south of Spain trough a sociological perspective. The area of study includes two Spanish regions (Andalucía, and Castilla la Mancha) that concentrate the 44% of the organic dairy goat farms registered in Spain. The sample of farms comprised 15 organic dairy farms (100% of the census). The farms are located in economically and demographically depressed and rugged areas of south peninsular. The goat activity is the only or main source of income for 91% of farmers, which indicates the little significant economic activities in these areas. The farms are highly specialized and are family (99% of the workforce), which generate a mean of 1.25 stable jobs and an average of 3.8 people are economically dependent of each farm. Under these conditions, the farm can be considered an engine of development: set rural people in unfavorable areas and ensure the sustenance of the family unit. The farmers are young (42.94 years) with 14.42 years of experience, which often come from farming families of great roots and goat tradition. Likewise, the 80% of the farmers have basic studies or training, and 10% university studies. This profile supports the continued of the farm in the medium term and allows the farmer faces the future sectorial challenges. The level of association is low: 85% of farmers belong to any Sanitary Defense Association, while only 30% belong to any association racial or any kind of cooperative. Finally note that the woman has an active role in 46% of farms. The formation of women exceeds that of their spouses and generally develop an active role in management of the farm. It also develops supplementary works in milking and raising kids.

The effect of dietary ionophores on feedlot performance and carcass characteristics of lambs

Greyling, J.P.C., Price, M.M., De Witt, F.H., Einkamerer, O.B. and Fair, M.D., University of the Free State, Animal, Wildlife & Grassland Sciences, P.O. Box 339, 9300, Bloemfontein, South Africa; greylijp.sci@ufs.ac.za

This study was conducted to evaluate the effect of different rumen fermentation modifiers (ionophores) in feedlot finisher diets on the production performance and various carcass characteristics of S.A. Mutton Merino whether lambs. Monensin (16.4 mg/kg), Lasalocid (33.0 mg/kg) or Salinomycin (17.5 mg/kg) was incorporated into a commercial high-protein (398 g CP/kg DM) concentrate. Treatment diets consisted of maize meal (650 g/kg), lucerne hay (150 g/kg) and a protein concentrate (200 g/kg) (containing the specific ionophore or not) to supply in an isonitrogenous (177 g CP/kg DM) and isocaloric (18.2 MJ GE/kg DM) total mixed diet used during the experimental period (60 days). Sixty lambs (initial BW 29.7±2.5 kg) were randomly allocated to the respective treatment groups (n=15/treatment) and each treatment was further subdivided into 5 replicates (n=3/replicate). Individual body weight and average feed intake per replicate was recorded weekly and used to calculate the feed conversion ratio (FCR) and average daily gain (ADG). All animals were slaughtered at termination of the study and cold (4°C) carcass weights as well as carcass characteristics (carcass length, shoulder- and buttock circumference, dressing percentage and back fat thickness) were recorded 24h post slaughter. Dietary ionophore treatment had no effect on feedlot performance parameters (mean±s.d.) (feed intake: 1476.5±84.5 g DM/day; ADG: 320.3±18.3 g/day; FCR: 4.6±0.0 g DM intake/kg live weight gained) or carcass characteristics (dressing percentage: 50.1±0.4% carcass weight: 24.4±0.7 kg; carcass length: 57.4±0.6 cm; shoulder circumference: 78.5±0.7 cm; buttock circumference: 67.1±0.5 cm; back fat thickness: 3.8±0.5 mm) during the study. Results suggest the efficiency of the different rumen fermentation modifiers to be similar, postulating that financial implications and/or animal preference could influence ionophore usage in finisher diets of lambs.

Optimization of protein requirement for dairy goats during rearing period with local protein feed stuff from bioethanol production and best roughage

Ringdorfer, F., LFZ Raumberg-Gumpenstein, Sheep and goats, Raumberg 38, 8952 Irdning, Austria; ferdinand.ringdorfer@raumberg-gumpenstein.at

Optimal diets for dairy goats are very important for high performance and healthy animals. To get a high milk yield the animals must be fed according their requirement during rearing period. In this time the use of concentrate is necessary. Because of high cost of concentrate the amount in the ration must be considered well. Also the protein component in concentrate is a question. In Austria mostly soybeans are used, but this must be imported from other countries. With the building of a factory for ethanol production from corn dried distillers grains (DDG) are available for feeding ruminants. The use DDG in the diet for goats a feeding experiment was carried out. The aim was to replace soybeans as protein component in the concentrate with DDG. Three different concentrates was offered: group K-0 with 11.8% soybeans and no DDG, group K-50 with 5.9% soybeans and 9.4% DDG and group K-100 without soybeans and 18.7% DDG. The protein content was the same in all 3 groups, 15.6%, the energy content was 12.26, 12.15 and 12.04 MJ ME/kgDM. Additionally to the concentrate the animals had ad libitum access to good quality hay. The experiment was carried out with 36 female Saanen goats with a body weight of 21 kg at the begin and a final body weight of 51 kg. Animals where housed individual in small pens on straw. Water was available for free intake. Body weight was measured once a week. Daily hay and concentrate intake was recorded. The results showed no significant differences in average daily gains (191, 201 and 198 g/d), in daily feed intake (1.21, 1.19 and 1.21 g DM/d) and in feed conversion (6.71, 6.36 and 6.55 gDM/kg gain). In conclusion DDG is an internal feedstuff and as a protein component in the diet for rearing kids well useable. Additionally, these data indicate that soybeans can be replaced by 100% in the concentrate by dried distillers grains.

Drinking behaviour and water intake of Boer goats and German Blackhead Mutton sheep

Al-Ramamneh, D., Riek, A. and Gerken, M., Department of Animal Science, Albrecht-Thaer Weg 3, 37075 Goettinegn, Germany; dal-ram@gwdg.de

The aim of this investigation was to study differences in average daily water intakes between small ruminant species differing in their adaptation to climatic conditions. Boar goats were chosen as arid adapted species and compared to German Black Head mutton originating from temperate climates. Sixteen non-lactating female animals (8 Boer goats, BW of 64.7±3.7 kg and 8 German Black Head mutton ewes, BW: 67.5±8.8 kg; mean±SD) were kept under controlled stable conditions (room temperature: 13.6±0.4 °C; light schedule: 10 h dark: 14 h light). Animals had access to hay and water ad libitum. Diurnal drinking behaviour was recorded by video. Individual water intake was estimated from water kinetics using the deuterium dilution technique during 2 wks. Simultaneously, water intake was directly measured by weighing water buckets every 24 h. additionally; individual hay intake was measured daily. The average daily water intakes (l) differed significantly (P=0.01) between the two species, with higher intakes in sheep (4.68±1.54 l) than in goats (2.34±0.86 l); these significant differences were maintained when relating water intake to metabolic body weight resulting in 195±60 (sheep) vs. 104±39 (goat) g/kg BW0.75. Daily hay intake differed significantly between sheep and goats, whether expressed as kg per day (1.64±0.50 vs.1.29±0.50 kg DM; P=0.011) or as g per metabolic weight (68±20 vs. 57±23 g/kg BW0.75; P=0.040). The higher amount of water intake in sheep was also reflected by the drinking behaviour: sheep spent approximately 2% of 24 h (31±19 min/d) drinking while Boer goats spent only 0.7% (10±9 min/d). It is suggested that the lower water intake in Boer goats seems to be an adaptive mechanism to arid climates.

Effect Lactobacilli probiotic supplementation during late pregnancy and early lactation on dairy goats

Galina, M., FES-Cuautitlan Universidad Nacional Autónoma de México, Ciencias Pecuarias, Km 2.5 Carretera Cuautitlan teoloyucan San Sebastian Xhala, Cuautitlan Izcalli Estado de México 54714, Mexico; miguelgalina@correo.unam.mx

An experiment on 50 Alpine goats was performed to determine the influence of a Lactobacilli probiotic (LAB) blend either in late pregnancy diet (PD) or on in lactation diet (LD), adding a slow intake urea supplement (SIUS). Two diets, pre and postpartum were fed to two groups of 25 animals each. First starting 21 d before kidding, were subdivided in two experimental groups. 1) PD without supplement; 2) PD with LAB and SIUS. Secondly, 50 post kidding dairy (25 animals each) were fed a lactation diet (LD) 3) with LAB/SIUS or 4) LD without LAB/SIUS supplementation. LAB Probiotic blend contained approximately 4 x 10^7 cfu of lactic bacteria spread on a liquid 150 g supplement/goat per day during both pre partum and post partum period. SIUS composition was: 17% corn, 17% molasses, 16% poultry litter, 14% rice polishing, 8% cottonseed meal, 5% animal lard, 4% fish meal, 4% salt, 4% urea, 3.2% calcium carbonate, 3% orthophosphate, 2.2% ammonium sulfate, 1.6% cement kiln dust, 1% mineral salt. DMI, milk yield, and milk protein content were higher for goats receiving the LAB/SIUS probitoic compared with non LAB/SIUS diet. Blood glucose and insulin levels were higher and plasma nonesterified fatty acids (NEFA) levels were lower for goat's receiving LAB/SIUS during the postpartum period. Supplementation increased DMI and milk production post partum in the LAB/SIUS dairy goats group. Blood metabolite information suggested that this response was associated with more glucose being made available and less fatty acids mobilized from lipid stores in dairy animals. Augmented DMI in LAB/SIUS probably resulted from higher cell wall utilization and development of bacterial protein from non protein nitrogen.

Conjugated linoleic acids effects on adipogenesis gene expression in ovine preadipocytes

Soret, B., Martinez, P., Alfonso, L., Encio, I. and Arana, A., Universidad Publica de Navarra, Produccion Agraria, Campus Arrosadia, 31006, Spain; soret@unavarra.es

Conjugated linoleic acids (CLA) have wide interest regarding their condition of dietary bioactives with potential ability to modify the adipogenesis, although their effects are contradictory and not well understood. In order to explore the effects of these compounds in ovine adipose tissue, we analysed the effect of isomers trans10, cis12 and cis9, trans11 on the differentiation of ovine preadipocytes. Preadipocytes obtained from lamb subcutaneous (SC) and omental (OM) tissues were cultured and induced to differentiate by 1.6 microg/ml insulin, 2nM tri-iodothyronine, 10 nM dexamethasone, 10 microM rosiglitazone and one of the two isomers (50 microM) or a mixture of the two (25 microM each). mRNA expression levels of the transcription factors PRARg, ADD1 and C/EBPa and the lipogenic enzymes lipoprotein lipase (LPL) and acetyl CoA carboxylase (ACC) were analyzed through the differentiation period (7 days) by quantitative real time RT-PCR. The delta delta Ct method was used to calculate the relative gene expression. Data were analysed by Anova. All markers increased their expression during the differentiation process but addition of CLA had little effect on expression levels of the genes analysed as only the levels of expression of ADD1 on the 7th day of differentiation of SC preadipocytes were different from the control treatment ($P>0.05$). These, and previous results showing a tendency to increase the number of cells when CLA were added, suggest that CLA may act through a mechanism that could stimulate lipid filling without involvement of the adipogenesis cascade.

The use of ultrasounds to predict lean meat proportion of lamb carcasses

Cadavez, V.A.P., CIMO - Escola Superior Agrária de Bragança, Deparment of Animal Science, Campus de Santa Apolónia, Apartado 1172, 5301-854 Bragança, Portugal; vcadavez@ipb.pt

The objective of this work was to study the use of ultrasounds to predict lean meat proportion of lamb carcasses. One hundred and twenty lambs (80 males and 40 females) of Churra Bragançana Portuguese local breed with a mean live weight of 23.0±6.9 kg were used. Lambs were scanned using an ALOKA SSD-500V ultrasound machine, equipped with one probe of 7.5 MHz, at lumbar and sternal regions. The images were analysed in order to measure the longissimus muscle depth (LD), subcutaneous fat thickness (SF), between the 12th and 13th vertebrae, and breast bone tissue thickness at 3rd (BT3) sternebrae. Lambs were slaughtered after 24 h fasting and carcasses were cooled at 4 °C for 24 hours. Carcass left side was dissected into muscle, subcutaneous fat, intermuscular fat and bone and remainder (major blood vessels, ligaments, tendons, and thick connective tissue sheets associated with some muscles). The LD, SF, BT3 and hot carcass weight were fitted as independent variables to predict lean meat proportion. of carcasses. Hot carcass weight explained 18,9% of lean meat proportion with a mean square error of 3.5%. Model including tissues measurements explained 57.2% of lean meat proportion with a root mean square error of 2.6%. These results indicate that carcass tissues measured by ultrasounds can be used to predict the lean meat proportion of Churra Bragançana lambs.

Statistic-genetical analysis of auction price of Texel, Suffolk and German white-headed mutton rams
Maxa, J.[1], Borchers, N.[2], Thomsen, H.[3], Simianer, H.[1], Gauly, M.[1] and Sharifi, A.R.[1], [1]University of Göttingen, Institute of Animal Breeding and Genetics, Albrecht-Thaer-Weg 3, 37075 Göttingen, Germany, [2]Chamber of Agriculture of Schleswig-Holstein, Section of Animal Husbandry and Breeding, Futterkamp, 24327 Blekendorf, Germany, [3]University of Aarhus, Research Centre Foulum, Department of Genetics and Biotechnology, P.O. Box 50, 8830 Tjele, Denmark; jmaxa@gwdg.de

The impact of performance traits recorded at licensing as well as various factors that affect auction price were determined for Texel, Suffolk and German white-headed mutton rams in Schleswig-Holstein. Furthermore, genetic parameters for the performance traits and for the auction price were estimated. Data used in this study were collected from 1988 to 2007 by the Sheep Breeding Organisation in Schleswig-Holstein, Germany. The auction prices in the range from 150 to 1,000 € of rams of mentioned breeds between 170 and 270 days of age were included in the analysis. General Linear Models and Multiple Regression Analysis were used to identify the impact of traits and effects on the auction price. Estimation of (co)variance components was carried out using multivariate animal model. From the traits recorded at licensing, live weight, followed by type traits of conformation and muscle mass, had the highest impact on the auction price. The results indicated that ultrasound measurements of muscle and fat depth had a minimum impact on the final auction price. The effect of breed, Prion Protein genotype and owner of the ram attached the importance of customers and affected significantly the auction price of the rams. The estimates of genetic parameters showed favourable genetic correlations between auction price and other performance traits.

Influence of breed, farm, age at first lambing and number of lambing on length of productive life in sheep
Kern, G., Traulsen, I., Kemper, N. and Krieter, J., Institute of Animal Breeding and Husbandry, Hermann-Rodewald-Str. 6, 24098 Kiel, Germany; gkern@tierzucht.uni-kiel.de

The life performance of female sheep is one of the most important economic traits in sheep husbandry. The objective of this study was to perform a survival analysis to estimate the effects on length of productive life in sheep. Therefore, a dataset with 16,972 female sheep from a breeding association of northern Germany was analysed. The data included information about the breed, lambing dates, number of lambs each lambing, date of birth, date of death or culling, farm and breeder, total number of lambings in lifetime associated with the total number of born lambs alive or dead in lifetime. About 25% of the records were right-censored. A proportional hazard model assuming a Weibull distribution for the baseline hazards function was applied (Survival Kit V3.12). The final model included the effects of the breed (n=7) and the age at first lambing in classes (n=4) as fixed effects, the farm (n=319) as random effect and the lambing number (n=10) as a time-dependent effect. All effects showed a significant effect on length of productive life ($P<0.001$). The German Blackheaded Mutton breed showed the lowest risk ratios being culled (RR=0.55) compared to the Texel breed (RR=1.00). The highest culling risk within the effect of the breed was found for the Suffolk breed (RR=1.19). By increasing number of lambings the hazard slightly decreased from first (RR=1.00) until seventh lambing (RR=0.09). Thereafter, the risk ratio increased until the tenth lambing and higher. The hazard rate of the effect age at first lambing showed an unexpected trend. If ewes were older than two years at first lambing the risk ratio decreased (RR=0.74) compared to the age at first lambing from 395 till 455 days (RR=1.00).

Variance component estimation of station tested German Blackface sheep

Baulain, U.[1], Brandt, H.[2], Schön, A.[3] and Brade, W.[3], [1]Institute of Farm Animal Genetics Mariensee, Friedrich-Loeffler-Institute, Hoeltystr. 10, 31535 Neustadt, Germany, [2]Institute of Animal Breeding and Genetics, Justus-Liebig-University, Ludwigstr. 21b, 35390 Gießen, Germany, [3]Chamber of Agriculture Lower Saxony, Johannssenstr. 10, 30159 Hannover, Germany; ulrich.baulain@fli.bund.de

Data of 1,300 station tested male German Blackface lambs descending from 174 rams tested from 1993 to 2007 were analysed to estimate variance components. According to the progeny testing instructions lambs were tested between 20 and 42 kg of live weight. All animals were slaughtered in the experimental slaughterhouse in Mariensee. In the observed time period average daily gain increased from 440 to 490 g and nutrient energy ultilisation (measured as group average) decreased from 38 to 33 MJ per kg weight gain. Carcass quality was predominantly described by subjective conformation scores. These were 5.7 and 7.5 for loin and leg muscle, and 7.0 and 6.4 for fat coverage and kidney fat, respectively. The area of the loin eye muscle measured between 5th and 6th thoracic vertebra was 8.2 cm^2 on average. Variance components were estimated by an animal model using VCE 6 including the year as fixed effect and beside the animal effect the farm of origin as additional random effect. For daily gain a relatively high heritability of 0.58 was estimated. The estimates for conformation scores were lower with 0.33 and 0.21 for loin and leg muscle, respectively. Both, for fat coverage and kidney fat heritabilities of 0.36 were found. For the loin eye area a heritability of 0.35 was estimated.

Nutritional and neuroendocrinological involvement in the control of luteinizing hormone secretion of Mediterranean female goats during the onset of the breeding season

Zarazaga, L.A.[1], Celi, I.[1], Guzmán, J.L.[1] and Malpaux, B.[2], [1]University of Huelva, E.P.S. La Rábida, 21819, Palos de la Frontera, Spain, [2]aPhysiologie de la Reproduction et des Comportements, INRA, 37380 Nouzilly, France; zarazaga@uhu.es

The role of nutrition and different neuroendocrinological systems (opioidergic, dopaminergic and serotonergic systems) on LH secretion was investigated in Mediterranean goats during the onset of the breeding season. At the onset of the experiment, the animals were allocated to two experimental groups differing on the level of nutrition. The control nutrition group (C, n=9) received their maintenance requirements and the low nutrition group (L, n=9) received 0.7 times their maintenance requirements. Both groups were balanced for live weight (LW) and body condition score (BCS) at the beginning of the study. All animals were ovariectomized, treated with an oestradiol implant, and subjected to a series of 3 months of short (8L:16D) and long days (16L:8D) to stimulate or inhibit their LH secretion. The effects of intravenous injections of the opiate receptor antagonist naloxone (2 mg/kg), the dopaminergic D_2 receptor antagonist pimozide (0.75 mg/kg) and the 5-hydroxytryptamine receptor antagonist cyproheptadine (0.75 mg/kg) on LH pulsatility were assessed during the onset of the breeding season. Blood samplings were done from 3 h before treatment until 3 h after treatment at 10 min intervals. A clear effect of level of nutrition was observed on mean LH concentrations before injection of the different antagonist (0.95±0.04 ng/ml vs 0.57±0.03 ng/ml, respectively, $P<0.001$). In comparison with the pre-injection period, naloxone significantly increased the mean LH concentrations in C group. These results provide evidence that dopaminergic or serotoninergic systems seems to be not involved in the inhibition of LH secretion at the onset of the breeding season, however the ability of naloxone to increase LH concentrations at this period could be enhanced by a higher plane of nutrition in Mediterranean goat females.

Nutritional and neuroendocrinological involvement in the control of luteinizing hormone secretion of Mediterranean female goats during the seasonal anoestrous

Zarazaga, L.A.[1], Celi, I.[1], Guzmán, J.L.[1] and Malpaux, B.[2], [1]University of Huelva, E.P.S. La Rábida, 21819, Palos de la Frontera, Spain, [2]Physiologie de la Reproduction et des Comportements, INRA, 37380 Nouzilly, France; zarazaga@uhu.es

The role of nutrition and different neuroendocrinological systems (opioidergic, dopaminergic and serotonergic systems) on LH secretion was investigated in Mediterranean goats during the seasonal anoestrous. At the onset of the experiment, the animals were allocated to two experimental groups differing on the level of nutrition. The control nutrition group (C, n=10) received their maintenance requirements and the low nutrition group (L, n=10) received 0.7 times their maintenance requirements. Both groups were balanced for live weight (LW) and body condition score (BCS) at the beginning of the study. All animals were ovariectomized, treated with an oestradiol implant, and subjected to a series of 3 months of short (8L:16D) and long days (16L:8D) to stimulate or inhibit their LH secretion. The effects of intravenous injections of the opiate receptor antagonist naloxone (2 mg/kg), the dopaminergic D_2 receptor antagonist pimozide (0.75 mg/kg) and the 5-hydroxytryptamine receptor antagonist cyproheptadine (0.75 mg/kg) on LH pulsatility were assessed during the seasonal anoestrous. Blood samplings were done from 3 h before treatment until 3 h after treatment at 10 min intervals. A clear effect of level of nutrition was observed on mean LH concentrations before injection of the different antagonist (0.23 ± 0.02 ng/ml vs 0.09 ± 0.01 ng/ml, for C and L group, respectively, $P<0.001$). In comparison with the pre-injection period, pimozide significantly increased the mean LH concentrations in C group. These results provide evidence that endogenous opioid or the dopaminergic system do not seem to be involved in the inhibition of LH secretion in late anoestrous in Mediterranean goat females. That ability of pimozide to increase LH concentrations in deep anoestrous could be enhanced by a higher plane of nutrition in Mediterranean goat females.

Nutritional and neuroendocrinological involvement in the control of luteinizing hormone secretion of Mediterranean female goats during the onset of the seasonal anoestrous season

Zarazaga, L.A.[1], Celi, I.[1], Guzmán, J.L.[1] and Malpaux, B.[2], [1]University of Huelva, E.P.S. La Rábida, 21819, Palos de la Frontera, Spain, [2]aPhysiologie de la Reproduction et des Comportements, INRA, 37380, Nouzilly, France; zarazaga@uhu.es

The role of nutrition and different neuroendocrinological systems (opioidergic, dopaminergic and serotonergic systems) on LH secretion was investigated in Mediterranean goats during the onset of the seasonal anoestrous. At the onset of the experiment, the animals were allocated to two experimental groups differing on the level of nutrition. The control nutrition group (C, n=10) received their maintenance requirements and the low nutrition group (L, n=10) received 0.7 times their maintenance requirements. Both groups were balanced for live weight (LW) and body condition score (BCS) at the beginning of the study. All animals were ovariectomized, treated with an oestradiol implant, and subjected to a series of 3 months of short (8L:16D) and long days (16L:8D) to stimulate or inhibit their LH secretion. The effects of intravenous injections of the opiate receptor antagonist naloxone (2 mg/kg), the dopaminergic D_2 receptor antagonist pimozide (0.75 mg/kg) and the 5-hydroxytryptamine receptor antagonist cyproheptadine (0.75 mg/kg) on LH pulsatility were assessed during the onset of the seasonal anoestrous. Blood samplings were done from 3 h before treatment until 3 h after treatment at 10 min intervals. A clear effect of level of nutrition was observed on mean LH concentrations before injection of the different antagonist (0.33 ± 0.03 ng/ml vs 0.16 ± 0.02 ng/ml, respectively, $P<0.001$). In comparison with the pre-injection period, pimozide significantly increased the mean LH concentrations in C group. These results provide evidence that opioidergic or serotoninergic systems seems to be not involved in the inhibition of LH secretion at the onset of the seasonal anoestrous. However, the ability of pimozide to increase LH concentrations could be enhanced by a higher plane of nutrition in Mediterranean goat females.

Utilisation de la pulpe de caroube (Ceratonia siliqua) pour l'alimentation des ovins

Gasmi-Boubaker, A.[1], Abdouli, H.[1], Mosquera-Losada, R.[2], Khaldi, A.[3], Boulbaba, R.[1] and Soula, M.[1], [1]Institut National Agronomique, 43 avenue Charles Nicolle, 1082 Tunis, Tunisia, [2]Ecole polytechnique supérieure, université de Santiago de Compostela, 27002 Lugo, Spain, [3]Institut National des Recherches en Génie Rural Eaux et Forêts, rue Hedi Karray, 1082 Tunis, Tunisia; azizaboubaker@ymail.com

La hausse du prix des matières premières utilisées pour la fabrication des aliments concentrés relance l'intérêt de valoriser les sous produits agro-industriels. Le caroubier (Ceratonia siliqua) présente un intérêt socioéconomique et des potentialités favorables au développement rural. La pulpe, obtenue après extraction des graines, peut être valorisée par les ruminants. Comparée à l'orge, elle est riche en sucres (40%), en cellulose brute (10.5 à 13.45%), en calcium (0.48 à 0.63%) et en potassium (1.10 à 1.20%) mais pauvre matières azotées totales (5.72 à 6.36%). Sa valeur énergétique est estimée à 0.9 UFL/kg de matière sèche. L'incorporation de la pulpe de caroube jusqu'à 50% dans l'aliment concentré distribué à des agneaux nourris à base de foin d'avoine, n'affecte ni l'ingestion du fourrage grossier ni celle de la ration totale. Toutefois, on note une diminution de la digestibilité *in vivo* de la matière organique et celle des matières azotées de la ration totale. Incorporée à raison de 25% dans l'aliment concentré, la pulpe de caroube permet aux agneaux de réaliser des croissances journalières équivalentes à celles obtenues avec le concentré commercial (à base d'orge) ce qui permettrait de réduire le coût d'alimentation.

Effect of supplementary feeding on growth performance and carcass traits in male and female Merghoze suckling kids

Souri, M., Moeini, M. and Abdolmaleki, Z., Razi University, Animal Science, Kermanshah, 67155, Iran; msouri@yahoo.com

A study was conducted to investigate the effect of supplementary feeding using concentrate on growth performance and carcass traits in male and female suckling kids of the Merghoze goat. Sixteen male and female kids were used in a 2×2 factorial arrangement based on sex and weight and allocated into two feeding groups 1) MAC: Milk plus Alfalfa leave and concentrate 2) MA: Milk plus Alfalfa. The rations were standard and balanced for protein and energy. Dry matter intake (DMI), average daily gain (ADG), live weight and feed conversion ratio (FCR) were recorded. Kids were slaughtered after 126 days and carcass traits were measured. The results indicated that there was no difference between MA and MAC treatment groups. Male kids had heavier birth weight, final body weight, greater body gain weight and ADG than female ($P<0.05$). The results of carcass measurements show that all carcass traits were not affected by treatment except for head and empty body weight (EBW). EBW and cold carcass weight (CCW) were higher than female ($P<0.01$). The result of carcass conformation show that except for hind quarter perimeter, pelvic limb compactness and ccw, other measurement were not difference significantly ($P>0.05$). In conclusion the result of this study indicated that there was no difference between MA and MAC and the supplementary feeding of concentrate had no effect on growth and carcass traits.

Effect of yeast preparations Saccharomyces cerevisiae on meat performance traits and hematological indices in sucking lambs

Milewski, S., Brzostowski, H., Tański, Z., Zaleska, B., Ząbek, K. and Kosińska, K., University of Warmia and Mazury, Department of Sheep and Goat Breeding, Olsztyn, Oczapowskiego 5, 10-719, Poland

The aim of the study was to establish the influence of dried yeast Saccharomyces cerevisiae and an extract containing β-glucanand mannan-oligosaccharides (MOS), on meat performance traits and hematological indices in lambs. The study was conducted on 48 sucking lambs, the offspring of Kamieniec ewes, divided into three equal groups: I - control, II and III - experimental. Throughout a 70-day rearing period, experimental group lambs were fed diets supplemented with dried brewer's yeast Saccharomyces cerevisiae (group II) or Biolex-MB40 - a yeast extract containing 25-30% β-1,3/1,6-D-glucan and 20-25% MOS (group III). Meat performance traits, i.e. body weight, daily gains, growth rate, cross-section dimensions of musculus longissimus dorsi (ultrasound examination) and fat thickness over the loin 'eye', as well as hematological indices were determined. It was found that both dried brewer's yeast Saccharomyces cerevisiae and Biolex-MB40 added to concentrated feed had a beneficial influence on the meat performance traits of sucking lambs, including body weight, daily gains, growth rate and m. longissimus dorsi dimensions indicating muscle tissue development. It was also demonstrated that the above supplements caused a significant increase in the values of hematological indices: WBC, RBC and HBG and in the number of lymphocytes in the leukogram, suggesting immune system stimulation. The effect of both yeast supplements was comparable.

Effect of dried yeast Saccharomyces cerevisiae on meat performance and non-specific humoral immunity mechanism

Milewski, S.[1], Wójcik, R.[2], Zaleska, B.[1], Małaczewska, J.[2], Tański, Z.[1], Brzostowski, H.[1] and Siwicki, A.[2], [1]University of Warmia and Mazury, Department of Sheep and Goat Breeding, Olsztyn, Oczapowskiego 5, 10-719, Poland, [2]University of Warmia and Mazury, Department of Microbiology and Clinical Immunology, Olsztyn, Oczapowskiego 13, 10-719, Poland

The aim of this study was to establish the influence of dried brewer's yeastSaccharomyces cerevisiae, given to lambs for a 60-day-period, on their meat performance and non-specific humoral immunity indices. The experiment was performed on 32 lambs of Kamieniec sheep, divided into two equal groups: I - control and II - experimental. Lambs from group II were receiving concentrate with dried yeast supplement, in a proportion of 50 g per one kg concentrate. Meat performance rating contained: body weight, daily gains, growth rate and ultrasound cross-section dimensions of musculus longissimus dorsi and fat thickness over the loin 'eye' examination. Non-specific humoral immunity analysis included: lysozyme and ceruloplasmin activity in blood plasma and total protein and gammaglobulin content in blood serum. It was found that dried yeast added to the concentrate feed for lambs had a beneficial influence on their meat performance and a muscle tissue development. Simultaneously, a significant increase of lysozyme and ceruloplasmin activity, held constant in the next examination period, and a gammaglobulin content increase on the 30th day of the experiment were recorded. No changes in the protein content were noticed. That indicates lambs immunity increase, which could have a positive effect on their meat performance.

Performance of ewes and lambs submitted to different levels of energy during gestation and lactation

Ribeiro, E.L.A., Castro, F.A.B., Mizubuti, I.Y., Silva, L.D.F. and Barbosa, M.A.A., Universidade Estadual de Londrina, Zootecnia, Caixa Postal 6001, 86051-990, Londrina, Brazil; elar@uel.br

The objective of this work was to evaluate the influence of feed energy levels during the last third of gestation and lactation on productive and behavioral parameters of ewes and its lambs. Thirty crossbred or Santa Ines ewes were used with 105 d of gestation, average live weight of 57.6 kg and average body condition score of 3.4 at the beginning of the experiment. The animals were confined in collective corrals, distributed randomly in three treatments: 2.4 (T24); 2.2 (T22) and 2.0 (T20) Mcal of ME/kg of DM. The rations were supplied ad libitum. Behavioral parameters (vocalization, the act of smell and lick the lamb, the position of the ewe standing up, the latency of the lamb to stand up, to suck for the first time and total suckling time) were evaluated during the first two hours after lambing; they were not affected by treatment ($P>0.05$). Weights and body condition scores were similar ($P>0.05$) at lambing. Lambs from T20 were lighter (4.7 kg) at birth than the lambs from the other two treatments (5.8 and 6.1 kg) ($P<0.05$). The effects of treatments were realized up to weaning (70 d), where ewes from T22 and T20 were lighter (53.1 and 47.4 kg) and presented worse body condition scores (2.3 and 1.6) than ewes from T24 (61.2 kg and 3.8 points). The average daily milk production was also affected ($P<0.05$); being 0.69, 0.47 and 0.19 liters for T24, T22 and T20, respectively. Lambs average daily gains from birth to weaning (0.30, 0.23 and 0.14 kg) and weights at weaning (26.8, 22.2 and 14.4 kg) were also affected ($P<0.05$) by treatments (T24, T22 and T20, respectively). It can be concluded that although the levels of energy tested had no effect on ewe traits evaluated at lambing, only the highest level allowed good weights and body condition scores to ewes at weaning, allowing them to enter immediately in the next breeding season; and at the same time, the lambs were ready to be slaughtered.

Influence of yeast Saccharomyces cerevisiae and β-1,3/1,6-D-Glucan on protein level and its fractions in sheep's milk

Ząbek, K.M.[1], Milewski, S.[1], Zaleska, B.[1], Antoszkiewicz, Z.[2] and Wielgosz-Groth, Z.[3], [1]University of Warmia and Mazury, Department of Sheep and Goat Breeding, Oczapowskiego 5, 10-719 Olsztyn, Poland, [2]University of Warmia and Mazury, Department of Animal Nutrition and Feed Managment, Oczapowskiego 5, 10-719 Olsztyn, Poland, [3]University of Warmia and Mazury, Department of Cattle Breeding and Milk Quality Evaluation, Oczapowskiego 5, 10-719 Olsztyn, Poland; khoik@uwm.edu.pl

The aim of the study was to determine the influence of dried yeast Saccharomyces cerevisiae and β-1,3/1,6-D-glucan, extracted from yeast cells, on protein level and its fractions in sheep's milk. The experiment was performed on a conservation herd of Kamieniec sheep. The experimental materials comprised 39 ewes. The animals were divided into three groups: I - control, II and III – experimental. Experimental ewes from group II were receiving concentrate with dried yeast Saccharomyces cerevisiae Inter Yeast S® supplement (50g/kg concentrate), group III concentrate was supplemented with Biolex®-Beta S (3 g/kg concenterate), inclusive about 70% β-1,3/1,6-D-glucan isolated from yeast Saccharomyces cerevisiae cell walls. Dried brewer's yeast Saccharomyces cerevisiae specimen didn't affected the protein level, nor in 28th neither in 70th day of lactation. In group fed with β-1,3/1,6-D-glucan protein level in milk in 28th day was comparable, however in the terminal phase was significantly higher than in control group ($P\leq0.05$). Casein level in experimental sheep's milk was significantly higher in comparison to the control group in 70th day of lactation. Adopted specimens affected casein fraction ratio. Dried brewer's yeast Saccharomyces cerevisiae specimen had an influence on their formation. Lowering b-casein level can contribute to lowering milk antigenicity, but increasing k-casein amount can cause better technological usage of milk.

Chitotriosidase activity in goat kids, diet effects

Moreno-Indias, I.[1], Castro, N.[1], Morales-Delanuez, A.[1], Capote, J.[2] and Argüello, A.[1], [1]Las Palmas de Gran Canaria University, Animal Science Unit, Fac. Veterinaria, Transmontaña s/n, 35413, Arucas, Spain, [2]Instituto Canario de Investigaciones Agrarias, Carretera el Boquerón s/n, La laguna, Tenerife, Spain; aarguello@dpat.ulpgc.es

The aim of the present study was to evaluate the effect of the diet on plasma Chitotriosidase (CHT) activity in goat kids. Recently CHT has been reported as part of innate immune system on small ruminants and no information about diet effects are available. 30 Majorera goat male kids were distributed in three groups which were fed with different diets. During the first two days of life, all goat kids received colostrum and after that, until day 60, they were fed with goat milk (group GM), milk replacer (group MR) or milk replacer plus 20 g/kg DM CLA (group CLA). Blood samples were collected at 1, 2, 3, 4, 5, 10, 20, 30, 40, 50, 60 days of life. After centrifuge, plasma was obtained and frozen until analysis. CHT activity was measure using a synthetic substrate in a fluorescence assay. CHT activity increased during the first 60 days of life in all groups, starting at 1,127 and finishing at 2,524 nmol/ml/h. At day 10, 20, 30, 40, 50, and 60 CHT activity was higher in GM and CLA than in MR group. Milk replacer formula must to be remake due to results observed in the present study and previous ones where goat kid diet affected complement system.

The effect of weight loss on productive characteristics of Damara, Dorper and Australian Merino

Almeida, A.M.[1], Kilminster, T.[2], Scanlon, T.[2], Greef, J.[2], Milton, J.[3] and Oldham, C.[2], [1]IICT & CIISA, CVZ, FMV Av. Univ Tecnica, 1300-477 Lisboa, Portugal, [2]Dep Agriculture and Food Western Australia, 3 Baron-Hay Court, South Perth WA 6151, Australia, [3]University of Western Australia, 35 Stirling Highway, Crawley WA 6009, Australia; aalmeida@fmv.utl.pt

Alternative sheep breeds, especially those showing tolerance to seasonal weight loss are considered of particular relevance in the context of the development of small ruminant production in dry areas. Two South African breeds, the Dorper and the Damara (fat tailed sheep) have recently been introduced to Australia. Their production characteristics in this country are however still largely unknown. In this trial we aimed to compare the effect of weight loss on productive performances (growth and carcass traits) of the three breeds when subjected to weight loss. A total of 72 male animals were used (24 per breed) and each breed was divided into two weight matched groups: control and restricted feeding. Animals were euthanized at the end of 30 days of trial. Damara and Dorper had significantly higher dressing percentages than Merino, although they had similar weights at slaughter. The color of Merino and Dorper meat were similar, although less dark than Damara meat.

Growth, meatiness and fattiness *in vivo* in lambs of chosen breeds and crossbreeds

Kuchtik, J.[1], Petr, R.[1], Dobes, I.[1] and Hegedusova, Z.[2], [1]Mendel University of Agriculture and Forestry, MSM 2B06108, Zemedelska 1, 61300 Brno, Czech Republic, [2]Research Institute for Cattle Breeding, MSMT LA 330, Vyzkumniku 267, 78813 Vikyrovice, Czech Republic; kuchtik@mendelu.cz

The aim of the study was the evaluation of the growth, meatiness and fattiness *in vivo* in lambs (n=122) of chosen breeds (Ch: Charollais and T: Texel) and crossbreeds (Ch x Sf (Suffolk), Ch x T, EF (East Friesian) x Ch and EF x T. Within the frame of the study were also evaluated the effects of sex (S), litter size (LS), age of dams (AD) and year of the observation (YO) on aforementioned indicators. The study was carried out on the farm in Kunčice during the years 2004 and 2005. Both years of the study, all lambs were kept with their mothers on the pasture. The daily feeding ration of lambs consisted of ad libitum grazing on the permanent pasture, mother´s milk (*ad libitum*) and meadow hay (*ad libitum*). Within the evaluation of meatiness and fattiness *in vivo* in lambs the following ultrasound measurement were carried out: area (Am.l.l.t., in cm^2) and depth (Dm.l.l.t., in mm) of m. *longissimus lumborum et thoracis* (m.l.l.t.) and fat thickness (FT, in mm). Genotype had a significant effect on body weight (BW) at birth and on daily gains (DGs) from 30 to 100 days of age and from birth to 100 days of age. Sex had a significant effect on most of BWs under study and on DGs from birth to 30 days of age, from 30 to 70 days of age and from birth to 100 days of age. The LS, AD and YO had a significant effect on most of growth indicators under study. Genotype had a significant effect on Am.l.l.t. at the age of 70 days, Dm.l.l.t. and FT at the age of 100 days. The S had a significant effect on Dm.l.l.t. and Am.l.l.t. at the age of 70 days and on Am.l.l.t. at the age of 100 days. The LS and YO had a significant effect on most of indicators of meatiness and fattiness *in vivo*. On the other hand only the AD had not significant effect on all parameters of meatiness and fattiness.

Result of crossing native Awassi sheep breed with exotic mutton and prolific ram breeds on growth and meat performance of lamb in Jordan

Momani Shaker, M.[1], Malinová, M.[1], Kridli, R.[2], Šáda, I.[1] and Abdullah, A.Y.[2], [1]Czech University of Life Sciences Prague, Department of Animal Science and Food Processing in Tropics and Subtropics, Kamýcká 129, Suchdol, 165 21 Prague 6, Czech Republic, [2]Jordan University of Science and Technology-IRBID, Faculty of Agriculture, IRBID, 3030, Jordan; Momani@its.czu.cz

The aim of the present research is to evaluate the effect of crossbreeding Awassi sheep with two exotic ram breeds (Charollais and Romanov) on growth, slaughter value and quality of lamb meat. 575 male and female lambs of different genotypes [Awassi (A) n=184; Awassi x Charollais (ACH) n=156; Awassi x Romanov (AR) n=235] were investigated from birth until weaning for a period of two years (2000 and 20067) in Jordan. Live birth weight, weaning weight and weight gain from birth to weaning were evaluated by using mathematical program SAS.STAT. Average birth weight, average weaning weight and average weight gain from birth to weaning of lambs was influenced by the genotype of lambs, sex, litter size and year of birth. The average live weight of lambs at birth was (4.14 kg), at weaning (18.83 kg) and average weight gain from birth to weaning (244 g). The carcass evaluation showed that leg percentage of carcass weight in ACH ram-lambs (34.85%) was significantly different from AR ram-lambs (33.54%). Crossbreeding Romanov and Charollais rams with Awassi sheep breed in Jordan had a positive influence on breeding performance.

The effects of internal and external factors on lamb growth of Romanov sheep under field conditions in Czech Republic

Momani Shaker, M., Žáčková, M. and Malinová, M., Czech University of Life Sciences Prague, Department of Animal Science and Food Processing in Tropics and Subtropics, Kamýcká 129, Suchdol, 165 21 Prague 6, Czech Republic; Momani@its.czu.cz

The objective of this study was to evaluate the effect of sirson body weight and growth ability of Romanov sheep lambs from birth to 100 days, including the effect of litter size, sex, ewe age, herd, birth month and year of lambing. In the years 1990 and 2007 the live weight was determined in 325 lambs at birth, 30 and 100 days of age by weighing on a digital scale with accuracy of 0.1 kg. Average live weight of lambs at birth was 2.935 ± 2.534 kg and at the age of 30 and 100 days 8.760 ± 23.245 kg and 23.030 ± 162.119 kg, respectively. ADG of lambs from birth until 100 days of lambs age was 0.201 ± 0.014 kg. Sex of lambs, litter size, ewe age and year of lambing affected ADG, live weight of lambs at birth, 30 and 100 days significantly ($P\leq0.05$-0.001). Investigation of the effect of sex on live weight of lambs at birth and at 100 days showed that the differences between males and females were statistically significant($P\leq0.001$). Differences in ADG and live weight at 30 and 100 days according to dam age were significant ($P\leq0.001$). The effect of litter size on live weight of lambs at birth, ADG until 100 days was highly significant ($P\leq0.001$). The effect of sires on ADG, live weight of lambs at birth, at 30 and 100 days of age was not confirmed. Likewise, live weight of lambs at birth, ADG until weaning and live weight at 30 and 100 days of age were affected by the year of lambing ($P\leq0.01$-0.001). Romanov sheep; lamb; growth ability; internal and external factors.

Effect of breed, sex and weight at slaughter, on carcass quality of hair sheep and wool sheep in Canary Islands

Camacho, A.[1], Capote, J.[2], Mata, J.[1], Argüello, A.[3] and Bermejo, L.A.[1], [1]University of La Laguna, Department of Agrarian Engineering, Production and Economy, Carretera General de Geneto, La Laguna, Spain, [2]Instituto Canario de Investigaciones Agrarias, Carretera del Boquerón s/n, La Laguna, Tenerife, Spain, [3]Las Palmas de Gran Canaria University, Animal Science Unit, Fac. Veterinaria, Transmontaña s/n, 35413, Arucas, Spain; jcapote@icia.es

Hair sheep breed (Canary Hair Sheep) has reached a significant development in the Canary Islands due to its adaptation to the subtropical and semiarid Canarian environments and its easy management. Wool sheep (Canary Wool Sheep) are the traditional breed in the Canaries and they are a dairy sheep. The main aim of this study has been the initial description of meat and production traits of these breeds in order to establish the basis to characterize the production of local breeds. The effect of breed and slaughter weight (10, 16 and 25 kg) on several characteristics of carcass were studied (pH, meat and fat color, conformation, carcass measure and yield, etc.). Canonical discriminate analysis was used with breed by slaughter weight interaction as grouping variable (6 groups). The distances among breeds increased as long as the slaughter weight increased. Hence the highest statistical distance was among breeds at heaviest lambs. The fat distribution, fat thickness, dressing percentage and conformation were the most discriminate variables, mainly due to a remarkable increase of fat variables in relationship with growth in wool sheep and an increase of carcass yield and conformation in hair sheep.

Relative growth of body fat depots and carcass composition in Churra da Terra Quente ewes with different body condition score
Silva, S.R., Guedes, C., Santos, V., Mena, E., Lourenço, A., Gomes, M., Azevedo, J. and Santos, A., CECAV-UTAD, POBox 1013, 5001-801 Vila Real, Portugal; assantos@utad.pt

The autochthonous Churra da Terra Quente (CTQ) sheep breed is reared in Northeast of Portugal under extensive production systems. These systems frequently involve relevant body composition changes due to the storing and mobilization of body reserves. However, information regarding the range of tissue changes is lacking. The objectives of this study were: examine the carcass composition variation and understand the fat depots distribution in CTQ ewes with different body condition score (BCS). Thirty seven non-lactating and non-pregnant CTQ ewes with 42.3±7.5 kg body weight (BW) and a BCS range from 1.5 to 4.5 were used. The BCS was obtained using the system based on a scale of 1 to 5 scores with a middle scale point. After slaughter, internal fat depots (omental- OF, mesenteric- MF, perirenal- PF and pelvic) were obtained and the carcasses were entirely dissected into muscle, fat (subcutaneous fat-SF and intermuscular fat-IF) and bone. Carcass composition and fat depots variations were analyzed by ANOVA. Allometric coefficients of the fat depots in relation to total fat were established using the Huxley equation with a log transformation. Carcass composition and internal fat varied considerably with BCS from 1.5 to 4.5 points. Muscle weight varied between 6.8 and 13.4 kg ($P<0.05$) and fat weight varied between 1.6 and 10.9 kg ($P<0.05$). Internal fat also exhibit a large variation (0.7 to 9.3 kg) ($P<0.05$). These findings reflect the variation of BW and clearly show that fat is the most variable tissue. In relation to total fat, OF and PF depots grew more rapidly (b=1.25 and 1.39, b>1, $P<0.05$, respectively), SF and IF grew at the same rate (b close to 1) and MF and pelvic fat depots grew more slowly (allometric growth coefficients <1).

The use of ultrasound measurements for monitoring subcutaneous fat and muscle depths of Churra da Terra Quente ewes submitted to a long-term feed restriction
Silva, S.R., Guedes, C., Santos, V., Lourenço, A., Mena, E. and Azevedo, J., CECAV-UTAD, POBox 1013, 5001-801, Portugal; eligomes@utad.pt

Throughout the Northeast of Portugal most sheep are raised under extensive production systems with seasonality of feed resource availability. Depending on the severity and length of the scarcity, the ewes use their body reserves, which results in significant variations of body weight (BW) and body condition. Objective techniques, such as real time ultrasonography (RTU), are able to monitor body reserves during food shortage periods. Our aim was to monitor subcutaneous fat (SF) and Longissimus thoracis et lumborum muscle (LM) depletion using RTU. Eighteen non-lactating and non-pregnant Churra da Terra Quente ewes were fed restricted diets to lose BW (49±5 to 37±1kg) during 42 weeks and were scanned biweekly using an AlokaSSD500V with a 7.5 MHz probe. The probe was placed over the 13th thoracic (13T) and between the 3rd and the 4th (34L) vertebrae. At these sites SF depth (SFD) and LM depth (LMD) were measured by image analysis using ImageJ software on digitised ultrasound images. Relationships between BW and RTU measurements and the relative growth of SFD and LMD were analysed by regression. The SFD growth relative to BW was obtained after log transformation of Huxley allometric equation. Both SFD and LMD were depleted (SFD: 7 to 3mm; LMD: 22 to 14mm). The decrease of 1mm on SFD and LMD represented 2.4 and 1.4kg of BW loss, respectively. The relative growth of SFD (SFD13T: b=1.73; SFD34L: b=1.62, both b>1, $P<0.01$) confirmed that subcutaneous fat varies highly with body condition. Results suggest that RTU can be used to monitor SFD and LMD depletion. It can also be concluded that RTU is useful to evaluate the relative growth of subcutaneous fat in different body regions.

Carcass quality of F1 crossings of Ile de France rams with local Romanian sheep

Pascal, C.[1], Ivancia, M.[1], Gilca, I.[1], Nacu, G.H.[1] and Iftimie, N.[2], [1]University of Agricultural Sciences and Veterinary Medicine, Faculty of Animal Science Iasi, 8 Mihail Sadoveanu Alley, 700489, iasi, Romania, [2]SCDOC Secuieni, Bacau, 607570, Romania; pascalc@univagro-iasi.ro

The carried out research is part of a wider complex of activities run in Romania, in order to identify the best methods and technologies to improve sheep meat production. The used biological material consisted of lambs of local breeds, Tigaie and Turcana as well as F1 crossbreds between Ile de France rams and ewes of the previously specified indigenous breeds. The purpose of the investigations was to verify the growth intensity and the way how the paternal breed may contribute to achieve higher meat production of better quality. The lambs had the same age (80 days) at the beginning of fattening in order to ensure groups' uniformity and to achieve some conclusive results. Check of the fattening ability has been done by applying an intensive fattening technology, with three phases (adaptation, growing and finishing) during a period of 85 days. Based on the achieved results, it was found that, during fattening, the growing intensity was higher in the crossbred groups. While the Ile de France x Tigaie group had a live weight 15.54% higher than the group composed of pure Tigaie breed individuals, the relative difference of weight between Ile de France x Turcana group and Turcana control group was slightly over 20%, proving that the use of industrial crossings is a solution to improve fattening performance. Concerning the average carcass weight, significant differences occurred between the investigated groups ($P>0.01$). In Ile de France x Tigaie crossbreds, the slaughtering efficiency was over 3.5% improved, the meat yield representing 70.5% of carcass weight, while the bone: meat ratio reached 1:4.29. All results confirmed that Ile de France breed was a good improver for carcass quality and conformation.

Carcass yield and slaughter traits of kids from Serbian fawn goat population

Ćinkulov, M.[1], Krajinović, M.[1], Pihler, I.[1], Josipović, S.[2], Ivačković, J.[1] and Žujović, M.[2], [1]Faculty of Africulture, Department of Animal Science, Trg Dositeja Obradovića 8, 21000 Novi Sad, Serbia, [2]Institute for Animal Husbandry, Autoput 16, 11080 Zemun, Serbia; mircink@polj.ns.ac.yu

Fro, 1950`s to mid-1980`s, in former Yugoslavia, of which Serbia was a part, goat keeping was forbidden. As a result, in Serbia in the 1980`s there was a small population of the Balkan goat breed. Nowadays, in Central part of Serbia, breeders keep white goats (different crosses of Balkan and Saanen goats), while in Vojvodina, south part of Serbia, fawn goats are dominating – Alpine, German Fawn and their crosses. The main aim of goat breeding in Serbia is milk production, but due to the increasing of number of goats the number of kids for slaughtering also increases. Only small number of kids are slaughtered in slaughterhouses. As a consequence, in Serbia there are no standards for carcass classification and meat quality. In our paper, we present results of carcass yield and percentage proportion of eatable and non eatable entrails of 30 male kids from Serbian fawn population of goats. Kids were suckling for two months, and after that they were fed in the barns for the next two months. They were slaughtered when they were approximately 4 months old. The average body weight of kids at slaughtering was 23.06 kg. Hot carcass weight with entrails was 11.28kg (48.87%), while the average hot carcass weight without entrails was 10.38 kg (44.94%). The average percentage proportion of entrails, legs and skin were as fallows: liver: 1.84%; lungs: 1.32%; spleen: 1.1%; hart: 0.07%; peritoneum: 0.26%; testes: 0.88%; rumen: 2.4% small intestine: 2.51%; rennet: 0.58%; mesentery: 0.74%; legs: 2.68%; skin: 7.58%; waste: 7.61%. The presented results are part of the national project in which further measurements are planned to find out the influence of the sex, age and system of feeding on carcass characteristics of the fawn goat population.

Prediction of the chemical body composition of suckling goat kids protected by the PGI 'Cabrito de Barroso' from ultrasound measurements

Silva, S.R., Gomes, M.J., Silva, M., Guedes, C., Lourenço, A., Mena, E., Santos, V. and Azevedo, J.,; eligomes@utad.pt

In the Barroso region (Portugal) there is traditional consumption of light goat kids (Cabrito do Barroso, CB), belonging to the local breeds Serrana and Bravia. Accurate measurements of changes in body chemical composition (BCC) of live animals are important to understand the response to the production system. Among others the real time ultrasonography (RTU) associated with image analysis is able to predict BCC. The present study was undertaken to determine the best combination of RTU measurement and live weight (LW) to predict the BCC in PGI kids. Data from 43 kids (10±2 kg LW) were scanned with an Aloka SSD500V real time scanner using a linear probe of 7.5 MHz, which was placed over the 8th, 11th, 13th thoracic and over the 4th lumbar vertebrae. At these points the subcutaneous fat depth (SF) and *Longissimus thoracis et lumborum* muscle depth (MD) were measured. The probe was also positioned over the 3rd sternebra of the sternum and over the 11th rib at the middle of thoracic cage, the SF and the tissue depth (TD) being recorded. The RTU measurements were obtained after image analysis using the ImageJ software. Carcasses and all non-carcass body components were ground. Two samples were obtained and analysed for moisture, protein, fat and ash. Stepwise regression analyses were established between chemical body components and RTU measurements and LW as independent variables. The best fitting regression equations were evaluated by the coefficients of determination (r2), residual standard deviation (rsd) and Mallows statistic (Cp). For moisture and fat the best fit was achieved with 3 RTU measurements and LW (r^2=0.948; RSD =181g; Cp=1.98 and r^2=0.924; RSD=130g; Cp=0.95; respectively). For protein the best fit was obtained with LW and one RTU measurement (r^2=0.794; RSD=129g; Cp=7.1). These results showed the usefulness of *in vivo* RTU in assessing body composition of kids protected by the PGI Cabrito de Barroso.

Estimating body weight in Turkish Hair goats using body measurements

Cam, M.A., Olfaz, M. and Soydan, E., Ondokuz Mayis University, Agricultural Faculty,Department of Animal Science, Samsun, 55139, Turkey; makifcam@omu.edu.tr

Body weight estimation could be calculated more accurately by combination of two or more measurements. This study was carried out to estimate the body weight from different body measurements of Turkish Hairy Goat (Kilkeçi) reared under rural conditions. Relationships between body weight and wither height, heart girth, body length, heart width, rump height and body length were studied using data from four different Turkish Hair Goat (Kılkeçi) farms with 175 observations. Body weight was regressed on the body measurements. The correlation coefficient between body weight and body measurement was positive and strong ($P<0.01$). The highest determination coefficients (R) of body weight were found on heart depth and heart girth (0.775 and 0.847, respectively).It was concluded that the liveweight could be estimated by the equation of Y=-47.8+1.12 HG; R2= 0.717 in Turkish Hair Goat.

Slaughter value of goat kids as related to breed

Kawęcka, A. and Sikora, J., National Research Institute of Animal Production, St. Krakowska 1, 32-083 Balice, Poland; akawecka@izoo.krakow.pl

The growing interest in goat breeding, which has recently been observed in Poland, is due to the increased demand for goat milk products and is also related to the production of young kids for slaughter. Poland has no tradition of eating young goat meat, which is largely used by breeders for the purposes of agritourism farms among others. Despite its high nutritive and taste value, goat meat is undervalued by consumers. It contains more protein than lamb meat and veal, has low-cholesterol fat and contains more unsaturated acids and L-carnitine than beef, which determines the heath-promoting properties of this product. The aim of the present study was to evaluate the slaughter value of kids representing three dairy breeds: Saanen, Alpine and White Improved. Goat kids were slaughtered at the age of 3 and 6 months. Kid origin and slaughter age resulted in significant differences in dressing percentage, which was the lowest in White Improved (43.56 and 41.52 at 3 and 6 months, respectively) and the highest in Saanen goats (48.34 and 46.26, respectively). In Alpine goats, the respective values were 47.2 and 43.83. The weight of valuable cuts (best end of neck, saddle, shoulder and leg) was higher in kids slaughtered at 6 months of age. The mean content of valuable cuts in half-carcasses was lower in kids slaughtered at 6 months compared to those slaughtered at 3 months of age.

In vivo Serrana goat kid carcass composition prediction by ultrasound measurements

Monteiro, A.[1], Azevedo, J.[2], Lourenço, A.[2], Silva, S.[2] and Teixeira, A.[3], [1]IPV, ESAV, 3500-606 Viseu, Portugal, [2]CECAV, POBox 1013, 5001-801 Vila Real, Portugal, [3]CIMO, POBox 1172, 3501-855 Bragança, Portugal; teixeira@ipb.pt

The use of real time ultrasound (RTU) to predict carcass composition was widely used for cattle, swine and sheep. However, for goat and particularly for light goat kids, this technique was less investigated. Thus the aim of this work was to *in vivo* predict carcass composition of goat kids using RTU measurements. Forty two goat kids of the Serrana breed (13.4±5.2 kg live weight) were utilized. The *in vivo* RTU images were made with an ALOKA 500V scanner equipped with a 5 MHz probe. The probe was placed over the 9th, 11th thoracic vertebrae and over the 1st, 3rd and 5th lumbar vertebrae. Images between 3-4ª sternebrae were also captured. All RTU images were analysed using the ImageJ software. With the images obtained on thoracic and lumbar the depth, width, perimeter and area of Longissimus dorsi muscle (LM) and the subcutaneous fat thickness above this muscle (SFL) were determined. At sternum, the subcutaneous fat depth (SFS) was measured. After slaughter the carcasses were stored at 4 °C for 24 h. After this period the carcasses were divided and the left half was entirely dissected into muscle, dissected fat (subcutaneous fat plus intermuscular fat) and bone. Prior to the dissection measurements equivalent to those obtained *in vivo* with RTU were recorded. Using the Statistica 5, correlation and regression analyses were performed. The correlation between RTU and carcass measurements were significant (r >0.58, $P<0.01$) for all muscle measurements. For fat measurements only the RTU SFS was significantly correlated with carcass measurement (r=0.96, $P<0.01$). The RTU measurements can explain the kid goat carcass composition variation (r2 between 0.40-0.89; 0.24-0.58 and 0.31-0.83, $P<0.01$, for muscle, dissected fat and bone respectively). This research shows that RTU is able to *in vivo* measure LM but not the SFL, due to its small amount. It can also be concluded that RTU measurements can explain kid goat carcass tissue variation.

Body measurements reflect body weights and carcass yields in Karayaka sheep
Cam, M.A., Olfaz, M. and Soydan, E., Ondokuz Mayis University, Agricultural Faculty Department of Animal Science, Samsun, 55139, Turkey; molfaz@omu.edu.tr

The accurate estimation of body weight and meat yield is very important in extensive animal production systems. Body weight can be estimated by using body measurements. This study was designed to investigate the relationships between live weights, body measurements and meat yields of Karayaka sheep. Animals were brought from different regions to slaughterhouse. Approximately 8 to 18 months-aged of Karayaka male (n=67) and female (n=55) sheep were used to investigate the relationships between body weights and body measurements such as heart girth (HG), height at withers (HW), chest depth (CD), width at withers (WW), body length (BL), height at rump (HR), circumference of hind leg (CHL) and circumference of canon (CC). There were significant correlations between live weight and height at withers, chest depth ($P<0.001$). Consequently, live body weight could be estimated by using the equation of Y=-25.8+2.11 CD; R^2=0.773.

Fitting regression models for fat-tailed and carcass weight with PCA analysis method in Makoei sheep
Moradi Shahrbabak, M.[1], Moradi Shahrbabak, H.[1], Khaltabadi Farahani, K.H.[2] and Atashi, H.[1], [1]University College of Agriculture & Natural Resources, University of Tehran, Animal Science, faculty of Agronomy and Aanimal Science, P.O.Box:4111,KARAJ, 31587-11167, Iran, [2]Faculty of Agriculture, University of Arak, Animal Science, Arak, 31587-11167, Iran; Moradim@ut.ac.ir

This Research conducted for first time in Iran.samples were collected from a total 576 Makoei fat-Tailed lamb of both sexes. the lamb were used to predict *in vivo* carcass composition(fat-tail, abdominal fat weight), some body and Fat-tailed measurements.lamb were 6-7 months old and weighed about 15.1-41.4 kg. Measurements were included body weight, upper, middle and lower tail depth and width, upper, middle and lower tail and neck circumference, right, middle and left tail length, body length, whither height, abdominal and heart girth. Having slaughtered each animal, omental mesenteric and tail fat separated and weighed. weight of fat-tailed and sum of omental and mesenteric contents varied from 50-2,290gr and from 8.1-536 gr respectively and had relatively high correlation with all fat-tail and body measurements. the highest correlations were observed between fat-tailed weight and lower circumference and live body weight measures(68% and 67% respectively).the fitting regression model for fat tailed weight is: Fat-tailed weight = 1003/54 + 123/7(prin1) + 79/4(prin2) - 124/52(prin3) the fitting regression model foe carcass weight is: carcass weight = 11/83 + 0/62(prin1) + 0/19(prin2) + 0.44(prin3) - 0/5(prin4).

Contact between livestock and wildlife in free-ranging systems: overview of shared and emerging infectious diseases in western Europe

Bastian, S.[1] and Hars, J.[2], [1]National Veterinary School, Department of Farm Animals and Public Health, BP 40706, 44307 Nantes Cedex 03, France, [2]Office National de la Chasse et de la Faune Sauvage (ONCFS), Unité sanitaire de la faune, 5 Allée de Bethléem, 38610 Gières, France; bastian@vet-nantes.fr

Free-ranging animal production systems are traditionnally important in western Europe, particularly in areas with Protected Geographical Indication productions, or natural areas like mountain environments or marshlands. In these systems, as well as in the less extensive outdoor systems, contact may be frequent with wild animal species taxonomically related to domestic livestock, like wild ruminants (cervids or bovids), suids or birds, but also with rodents and carnivores. Several diseases of the notifiable disease list of the OIE (World Organisation for Animal Health) can be enzootic in wildlife, jeopardizing efforts to control them in livestock. Tuberculosis, Brucellosis, Classical Swine Fever, Aujeszky's disease or Avian Influenza for example are prevalent or accidentally present in certain areas. Disease transmission can happen by close contacts in shared pastures or drenching points, animal intrusion in fenced facilities or even cross-reproduction between wild and domestic species, as seen frequently in suids. Most emerging diseases issue from the wild fauna or include it in their epidemiological cycles. Several factors of global change influence the distribution and nature of vectors and pathogens. Epidemiosurveillance systems in Europe try to adapt to these challenges, but have important limitations in terms of cost and feasibility. Risk management possibilities are diverse, ranging from active protection by fencing to acting on prevalence in the wild, either by vaccination or culling. Strategies have to be adapted for every disease and local situation, but when the disease cannot be controlled in wildlife, strict biosecurity measures still are necessary.

Biosecurity challenges in outdoor husbandry systems

Arné, P., Ecole Nationale Veterinaire Alfort, Zootechnie, 7 avenue du général de gaulle, 94700 Maisons-Alfort, France; parne@vet-alfort.fr

Infectious diseases outbreaks impose major constraints on the productivity and profitability of the livestock industry. Current strategies to control this crucial hazard involve sanitation and biosecurity practices to prevent the introduction and spread of infectious agent in the herd. Biosecurity plans generally imply restricted access to the farm (animals, visitors, vehicles, feedstuffs and equipment control) and internal measures to limit pest proliferation or infection spread. Beside these basic procedures, each farm should benefit from an individual risk assessment taking into account the probability of targeted infections occurrence, the susceptibility of the herd to the identified pathogens and the economic impact of infections. In the case of domestic animals kept outdoors or in open-facilities, exposure to contaminated wildlife, coming in close contact especially around the feeding, drinking or bathing areas must be carefully evaluated. Birds for example can fly from farm to farm and might be efficient carriers of pathogens after having been visiting infected herds. Considering potential threats and management tools (vaccination, sanitary measures, confinement) in such farm systems, consistent risk abatement procedures should be carefully identified and prioritized implementing protocols according to need and cost, evaluating modifying biosecurity procedures as risk changes.

Biosecurity issues for the open-air pig production in Spain, as related to contact with wildlife and wildlife management

López-Olvera, J.R.[1], Vicente, J.[2], Boadella, M.[2] and Gortázar, C.[2], [1]Servicio de Ecopatología de Fauna Salvaje, (http://www.uab.es/sefas), Facultat de Veterinària, Universitat Autònoma de Barcelona, Bellaterra, E-08193, Barcelona, Spain, [2]Instituto de Investigación en Recursos Cinegéticos, (http://www.uclm.es/irec), Ronda de Toledo s/n, E-13005, Ciudad Real, Spain; Jordi.Lopez.Olvera@uab.cat

Open-air pig production of both autochthonous and white domestic pig is increasing in the Iberian Peninsula, whereas wild boar is being increasingly managed, including fencing, feeding, and translocations, but also extensive or even intensive farming. This artificial management of wild boar is linked with increased disease transmission risks due to higher densities and aggregation. Biosecurity concerns therefore arise in the interface between wild boars and extensively bred domestic pigs. Several diseases, such as Aujeszky's disease and bovine tuberculosis (TB) are relevant among the Spanish wild boars, management having an effect on disease prevalence. Wild boar acts as a true wildlife reservoir for TB in the ecological, epidemiological, and management situation in Southern Spain. However, whereas pig industry is severely sanitary controlled, wild boars are less sanitarily controlled. Hence, wild boars suppose a disease re-introduction risk for sympatric domestic pigs. Several actions intend to assess this transmission risk and to suggest control measures. First, applied research on disease transmission at the wildlife-livestock interface is taking place. Second, Spain has a draft national wildlife disease surveillance scheme. Third, experimental research aimed at developing new diagnostic and prevention (including vaccination) tools to improve wildlife disease surveillance and control. The results of this research may influence legislation, increasing sanitary control on wildlife, including an upcoming decree to control sanitary issues in wildlife translocations.

Characterization of biosecurity measures on swine farms located in the Valencian community region

Martinez, M.[1], Torres, A.[2] and Lainez, M.[3], [1]TRAGSEGA, Centro de Tecnología Animal (CITA-IVIA) Poligono la Esperanza, 100 apartado 187, 12400 Segorbe (Castellón), Spain, [2]Instituto de Ciencia y Tecnología Animal, Universidad Politécnica de Valencia, 46021 Valencia, Spain, [3]Consellería de Agricultura, Pesca y Alimentación. Generalitat Valenciana, Amadeo de Saboya, 2, 46010 Valencia, Spain; mmarti36@tragsa.es

The aim of this study was to asses biosecurity measures in swine farms located in the Valencian Community region (Spain). During three years (2005-2008), data from 264 swine producers were collected in a questionnaire. Each questionnaire had information related to farm management practices, equipment, location, preventive hygiene and biosecurity measures. Answers were analyzed with regard to type of production system (owner-operated and integrated farms), livestock type (sows units, nursery farms, farrow to finish, and fattening pigs), and province (Castellón, Valencia and Alicante). Two statistical procedures were used in this study: univariant analysis and multiple-correspondence analysis (two-step clustering procedure). Univariant analysis showed most farms had good biosecurity level. Nevertheless, biosecurity level was higher in integrated than owner-operated farms. Around 93.5% of the visited farms presented a perimetral fence. Moreover, 77.1% of farms presented bird-proof nets in windows. Water tanks, pipes and drinkers were cleaned and disinfected as part of a regular routine in the 68.5% of farms. Cleaning and disinfection procedures were carried out in two phases. Firstly, water was applied with high pressures, although only 1% of farmers used detergent during cleaning. Secondly, disinfection was done. Multiple-correspondence analysis showed three well differentiated clusters according to biosecurity measures and farm age. The first cluster presented new buildings and high level of biosecurity. The second cluster presented old buildings and the worst level of biosecurity measures. The third cluster included middle-age buildings and medium level of biosecurity.

Prevalence and burden of helminthes in local free range laying hens

Kaufmann, F. and Gauly, M., Department of Animal Science, Albrecht Thaer Weg 3, 37075 Göttingen, Germany; fkaufma@gwdg.de

Chickens in free range production systems are sometimes exposed to unfavorable hygienic conditions. The aim of the present study was to estimate the prevalence and burden of helminthes in free range layers in Germany. Therefore 144 laying hens of 5 genotypes were collected at different ages from 11 farms in Germany. Animals were slaughtered, the gastrointestinal tract and trachea were removed and examined for the presence of helminthes. Faecal samples were taken to estimate faecal egg counts (EpG) and faecal oozyst counts (OpG), respectively. The prevalence (P) and mean worm burden (WB) of individual helminth species and correlations between WB and EpG were calculated using SAS (1999). Genotype differences (Tetra Brown, Lohmann Brown, Isa Brown) were analyzed with ANOVA. 92.4% of the birds were parasite positive with a total mean worm burden of 147.7 (SD±158.3). The following nematodes were found (P; WB): Heterakis gallinarum (84%; 97.6), Capillaria spp. (71.5%, 45.7), Ascaridia galli (66.6%; 16), Acuaria hamulosa (1.4%; 1). The following cestodes were found: Raillietina cesticillus (24.3%; 41.3), Choanotaenia infundibulum (2.8%; 26.8) and Hymenolepis cantaniana (2.1%; 11.3). 7% of the faecal samples were positive for coccidial oozysts. There was a significant correlation between number of nematodes and EpG (r=0.48; $P<0.0001$). TETRA hens had significant higher worm burden when compared with ISA and LB hens (221.1a; 123.9b; 85. 4b; $P<0.0001$). The high prevalence of helminthes indicates that the majority of the chickens are subclinically infected, which may cause economic losses. Genotype differences in parasite resistance were described earlier. This have propably to be taken into account when birds are selected for such systems.

Disease transmission risks between wild ruminants and free-ranging livestock

López-Olvera, J.R., Marco, I. and Lavín, S., Servicio de Ecopatología de Fauna Salvaje, (http://www.uab. es/sefas), Facultat de Veterinària, Universitat Autònoma de Barcelona, Bellaterra, E-08193, Barcelona, Spain; Jordi.Lopez.Olvera@uab.cat

Wild ruminants are increasing in number in Spain, due to protection and management measures. Therefore, there is a higher probability of contact with domestic animals, particularly with those extensively bred, thus becoming a concern for the management of diseases in the wildlife-livestock interface. Wild ruminants share several diseases with their domestic relatives. Transmission of infectious diseases has been reported to happen in both directions, i.e. from wildlife to livestock and from livestock to wildlife. Different epidemiological situations have been reported in the wildlife-livestock interface: the disease spills-over from domestic livestock to wild ruminants, which act as dead-end host, therefore not participating in the epidemiology of the disease; the disease spreads from domestic livestock to wild ruminants, which population maintains the disease. Then, wildlife can act as a reservoir, being a source of re-infection for domestic ruminants, or the disease may establish two apparently independent cycles, one for domestic and another for wild ruminants, with no disease transmission between them; the disease circulates freely among domestic and wild ruminants in a single cycle. Dynamics of disease may differ widely between domestic and wild ruminants, due to differences in host susceptibility and survival, population density, and prevention, control and treatment measures feasible for each species. Consequently, the epidemiological situation of a disease depends on the factors of the animal population (either wild or domestic) affected, like density, immune status, vaccination programs, and others. When implementing sanitary control measures in domestic animals, the possibility of wild species acting as a reservoir, which may spread or maintain the disease, must be considered, both for protection of wild species and for control and/or eradication of disease in domestic livestock.

Shigatoxin producing *Escherichia coli* in dairy cattle in Northern Germany: prevalence and influences on shedding patterns
Kemper, N.[1], Menrath, A.[1] and Wieler, L.H.[2], [1]Institute of Animal Breeding and Husbandry, CAU Kiel, Olshausenstraße 40, 24098 Kiel, Germany, [2]Institute of Microbiology and Epizootics, FU Berlin, Philippstraße 13, 10115 Berlin, Germany; nkemper@tierzucht.uni-kiel.de

Since their first mention in 1977, shigatoxin producing *Escherichia coli* (STEC) have been an emerging issue for veterinary public health. These zoonotic strains can cause gastroenteritides and severe disease patterns in children. Cattle and cattle derived foods are the primary reservoir for STEC. In this longitudinal study selected dairy cows were monitored over a period of twelve months. Their faecal excretion patterns of STEC were surveyed with special emphasis on the shedding in the individual animal and in the herd. On six dairy farms in Northern Germany, 1,626 faecal samples from 140 cows in different lactation numbers were examined once per month in the period between 02/2007 and 01/2008. After cultural isolation, STEC were characterized by biochemical reaction, PCR and colony-hybridization. The data was analyzed using the SAS-procedure 'logistic'. In 24.8% of all samples, STEC were detected. Within the herds, prevalence varied between 11.1% and 32.3%. Over the whole sampling period, only 13.6% of the cows were detected as constantly negative. Most of the animals excreted STEC at least once (45.5%) and 17.1% were detected as positive in more than 50% of their samples. Cows with equal or more than four consecutive positive samplings (10.0%) were classified as persistently infected shedders. Significant influences for the excretion of STEC were month of sampling, lactation number, days in milk and the presence of a persistent shedder in the herd. Congruent to other literature, prevalence was highest in the late summer months. Even though many studies on other risk factors for the excretion of STEC in cattle exist, the main influencing factors still remain unknown. However, persistent shedders represent a high risk with respect to contamination in the food chain and the maintenance of STEC and the infection cycle in the herd.

Prevention of *Coxiella burnetii* shedding by dairy cows in infected herds using an inactivated monovalent phase I *C. burnetii* vaccine
Guatteo, R.[1], Fourichon, C.[1], Joly, A.[2], Seegers, H.[1] and Beaudeau, F.[1], [1]Veterinary School - INRA, UMR1300 BioEpAR, BP40706, 44307 Nantes, France, [2]GDS, BP110, 56003 Vannes, France; fourichon@vet-nantes.fr

Q fever is an endemic worldwide zoonosis caused by *Coxiella burnetii* (Cb). Ruminants, which shed Cb in milk, vaginal mucus and faeces are the main sources for human infection. Therefore, any measure leading to a significant decrease of shedding by dairy cows would be of interest in order to control the spread of this bacterium within a herd and therefore to limit the zoonotic risk. The main objective of this study was to assess the efficacy of a vaccine containing phase I Cb to prevent shedding in susceptible dairy cows within infected herds in comparison to a placebo. A total of 336 dairy cows and heifers, from 6 spontaneously infected herds, were followed over a one-year period. The allocation of a treatment was performed randomly within pregnant and non pregnant animals. After treatment (D0), the animals were subject to systematic sampling (milk, vaginal mucus and faeces) on D90, D180, D270 and D360 to detect putative shedding (using real time PCR). In addition, the same samples were taken within 10 days after calving. The effect of the treatment on the probability for an initially susceptible animal of becoming shedder was assessed using survival analysis (Cox regression model). Almost all heifers were detected as susceptible before treatment. When vaccinated while not pregnant, an animal had a 5 times lower probability of becoming a shedder than an animal receiving placebo. An animal which was vaccinated while pregnant had a similar probability of becoming shedder as an animal receiving the placebo. These results highlight the value of implementing vaccination, if possible, in non infected herds. In infected herds, the vaccination should be implemented in quite all presumably susceptible animals, i.e. at least the heifers. The vaccination of the dairy cows should be performed when the infection has not spread widely (ie low within-herd seroprevalence).

Risk factors for Salmonella in fattening pigs in Northern Germany

Hotes, S. and Krieter, J., Christian-Albrechts-University, Institute of Animal Breeding and Husbandry, Olshausenstraße 40, 24098 Kiel, Germany; shotes@tierzucht.uni-kiel.de

Human illnesses caused by Salmonella are one of the most important foodborne diseases worldwide. Even in countries with a high hygienic standard it is still a problem. One reason for infection is the consumption of contaminated pork. The most important risk factors for the prevalence of Salmonella should be revealed by analysing blood samples of 1836 fattening pigs housed in 59 stables of 32 farms. Therefore, data about husbandry, management and hygiene have been collected by a questionnaire. The analysis of the blood samples showed that almost 14% of the pigs had to be classified as Salmonella-positive (cut off at optical density percentage of 40%). These pigs were located in 35 of the 59 stables and just five farms had no positive blood sample. For further analysis a logistic regression model was used. Out of 20 variables a set of seven was chosen by using stepwise selection. All of them had a significant impact on the detection of Salmonella. The application of antibiotics increased the Odds Ratio (OR) by a factor of 5.21 compared to untreated pigs. Furthermore, proximity to other swine herds (max. 2 km) increased the prevalence of infected pigs significantly (OR 3.76). The data showed also that hygienic aspects such as overalls for visitors, fully slatted floors and cleaning of the feed tube (ORs <0.55) should be realised with regard to Salmonella. In contrast to our expectations, the results suggested that a partition which enabled contact from pen to pen reduced the chance of a positive blood sample (OR 0.56) as well as the occurrence of rodents and birds in the barn (OR 0.33). The former might be explained by the differentiation of pen partitions. Even so called 'closed' pen partitions enabled the pigs to have contact with their neighbours at least by snout. The effect of the rodents and birds were difficult to interpret. A possible explanation is that the small number of farms biased the data set.

Extraintestinal pathogenic *Escherichia coli* in pets and livestock: a rising concern over zoonotic infections?

Ewers, C.[1], Diehl, I.[1], Philipp, H.C.[2], Sharifi, A.R.[3] and Wieler, L.H.[1], [1]Microbiology and Epizootics, Philippstr. 13, 10115 Berlin, Germany, [2]Lohmann Tierzucht GmbH, Am Seedeich 9-11, 27472 Cuxhaven, Germany, [3]Animal Breeding and Genetics, Albrecht-Thaer-Weg 3, 37075 Goettingen, Germany; ewers. christa@vetmed.fu-berlin.de

Extraintestinal pathogenic *E. coli* (ExPEC) constitute a group of strains including uropathogenic, avian pathogenic and neonatal meningitic E. coli. This term has been proposed to reflect their shared ability to cause disease at multiple anatomical sites outside the intestine in various hosts. ExPEC are linked with systemic infections in poultry, urogenital tract infections and septicemia in companion and livestock animals, and the MMA syndrome in swine, while in humans they are predominantly associated with urinary tract infections and new born meningitis. Although in general ExPEC exhibit considerable diversity regarding serotypes, virulence gene pattern and phylogenetic background, profound molecular typing tools revealed that a substantive number of isolates, be it of human or animal origin, have identical phylogenetic backgrounds and overlapping virulence features. Likewise, human ExPEC are capable of causing severe systemic infections in birds, as demonstrated in a chicken model. We have new evidence suggesting a particular region in the *E. coli* genome being more predictive for the virulence of a strain than its clinical or host origin. This region promises to be a diagnostic marker not only to determine highly virulent strains but also to unravel the role of the animal gut as a reservoir for putative virulent strains, thereby contributing to the development of intestinal intervention strategies in the future. There is currently no evidence for the presence of any factor conveying host specificity to ExPEC strains, nor do our results of genotyping, phylogenetic typing and animal experiments account for that. It might therefore be concluded that a considerable number of animal ExPEC strains are capable of causing severe infections in humans, further leading to the aspect of an increased risk of zoonotic infections.

Proteomics evaluation of molecular mechanisms involved in pathogenesis of *Salmonella* spp

Roncada, P.[1], Deriu, F.[2], Soggiu, A.[2], Gaviraghi, A.[2] and Bonizzi, L.[2], [1]Istituto Sperimentale Italiano L. Spallanzani, Laboratorio di Proteomica, c/o Università degli Studi di Milano, Facoltà di Medicina Veterinaria, 20133 Milano, Italy, [2]Università degli Studi di Milano, DIPAV, sezione di diagnostica sperimentale e di laboratorio, via Celoria 10, 20133 Milano, Italy; paola.roncada@unimi.it

Salmonella species are an important group of enteric pathogens which cause severe disease both in human and animals. Many foods, particularly of animal origin, but also fruit and vegetables have been identified as vehicles for these pathogens due to faecal contamination. Spread of these pathogens may occur in the food processing through cross contamination from raw food or food handlers. The molecular bases for *Salmonella* adherence and invasion of epithelial cells are different and complex. A large number of *Salmonella* genes are required for entry into epithelial cells and causing disease. *Salmonella* serotypes are closely related genetically but they are significantly different in pathogenic activity. Deep inside the molecular mechanisms of pathogenesis may be a key to control intestinal salmonella infections. Proteomics of *S. Enteriditis* and *S. Thyphimurium* highlighted some significant differences both in metabolic enzymes, and in proteins as elongation factors, heat shock proteins, membrane proteins or membrane proteins precursors. Particularly last ones should be correlated with pathogenesis and virulence mechanisms. About 37 proteins had shown differential expression patterns (anova $P<0.05$). S. thyphimurium had increased levels of structural protein involved in folding (GRO EL) and synthesis (50S ribosomal sub prt), in motility of bacteria (FliC) and ion transport (OmpD) respect to S. enteriditis. The final purposes is to find biomarkers to better understanding pathogenesis and to identify new molecular targets to improve diagnosis and therapies. This work was supported by Ministero Salute,(2007) Ricerca Finalizzata U.O. prof Luigi Bonizzi.

Fermented liquid feed to pigs: microbial and nutritional characteristics and effects on digestive diseases

Canibe, N. and Jensen, B.B., University of Aarhus, Faculty of Agricultural Sciences, Dept. of Animal Health, Welfare and Nutrition, Blichers Allé 20, P.O. Box 50, 8830 Tjele, Denmark; Nuria.Canibe@agrsci.dk

The use of liquid feed in pig nutrition has gained interest due to the political wish of decreasing the use of antibiotics in pig production. Further, fluctuations in feed prices make liquid feed, with the possibility of using cheap liquid ingredients, an interesting feeding strategy. The effect of fermented liquid feed (FLF), a product obtained by deliberate and controlled fermentation of liquid feed, on gastrointestinal health and growth performance of pigs is one of the subjects being investigated in the last years. In order to obtain FLF of good microbial quality, that is, biosafe, fresh feed and water are mixed with material from a previous successful fermentation, which acts as inoculum for the new mixture. The nutritional characteristics of the mixture can be positively or negatively affected by fermentation. For example, the concentration of some compounds that can be considered as antinutrional compounds are reduced e.g., α-galactosides, but degradation of free lysine can also occur. Several factors affect the characteristics of the final product and therefore knowledge on the influence of these factors on the nutritional and microbial quality of the mixture is crucial. Incubation temperature, addition of starter cultures, and feed processing are among the parameters that can affect the quality of FLF. Feeding FLF of good quality results in reduction of the number of enteric pathogens, like coliforms and Salmonella, along the gastrointestinal tract of the animals; and the few published studies on its effect on important pig diseases like porcine proliferative enteropathy and swine dysentery also indicate reduction of the incidence of these diseases in pigs fed FLF.

Water-, feed intake and eating rank of pregnant sows

Kruse, S., Traulsen, I. and Krieter, J., Christian-Albrechts-University, Institute of Animal Breeding and Husbandry, Olshausenstr. 40, D-24098 Kiel, Germany; skruse@tierzucht.uni-kiel.de

The aim of the present study was to investigate the water- (WI), feed intake (FI) and daily eating rank (RE) of pregnant sows. Data recording was performed on the research farm of the University of Kiel from April 2007 to June 2008. A water flow meter was installed in the dry sow house to measure the individual daily water intake of 90 sows. Daily feed was given by an electronic sow feeder, which recorded the sow number, the beginning and the ending of feed intake as well as the amount of feed. From the 90 gestating sows the RE was defined according to their visiting at the feeding station. WI, FI and RE were analysed by applying linear mixed Fixed Regression models. The fixed effects were parity class (3 parity classes: 1, 2, ≥ 3 parity) and test day. For WI and FI a polynomial of second degree for day of pregnancy within parity class was considered and for RE the day within the sow group and parity class was taken into account. For all models a random sow effect was included. In order to account for repeated measurement the residual term was modelled with the spatial (exponential) covariance structure. The results showed that first parity class sows drank less water than second (2.7 l/d, $P<0.001$) or third parity class sows (4.2 l/d, $P<0.001$). WI remained constant during the pregnancy with a small increase at the end of the pregnancy. Furthermore, the variation of WI between and within sows was high. The curve of the feed intake was similar but the variation within and between sows was smaller. First and second parity sows had a tendency to higher feed intake than third parity class sows (0.1 kg/d, 0.06 kg/d, respectively). The regression of day within the sow group and parity class had the major influence on the RE. With longer stay within the dynamic sow group the RE increased.

Effect of crude protein and phosphorus level in a phytase supplemented grower finisher pig diet on phosphorus and nitrogen metabolism

Varley, P.F., Callan, J.J. and O'doherty, J.V., University College Dublin, Animal Nutrition, Lyons Research Farm, Newcastle, Co. Dublin, Ireland; patrick.varley@ucd.ie

The objective of this study was to investigate the effects of dietary crude protein (CP) level on phosphorus (P) metabolism in a low and high P diet supplemented with phytase. A 2×2 factorial experiment was conducted to investigate the effect of CP inclusion (130 vs 200g/kg) and dietary P inclusion (3.9 vs 5.8g/kg) on P metabolism in pigs. The dietary treatments contained 500 FYT/kg of phytase. Sixteen entire male pigs (live-weight 78kg) were randomly allocated to the diets in a digestibility and mineral balance study. Pigs offered high P diets had higher faecal dry matter (DM) ($P<0.05$) and faecal DM output ($P<0.05$) than pigs offered low P diets. Pigs offered low P diets had higher gross energy digestibility ($P<0.05$) than pigs offered high P diets. There was an interaction between CP and P level on P ($P<0.05$) retention. Pigs offered low CP-high P diets had higher P retention compared with pigs offered high CP-high P diets however there was no effect of P level at the low P diets on P retention. There was an interaction between CP and P level on P digestibility ($P<0.05$). Pigs offered high CP-low P diets had higher P digestibility compared with pigs offered high CP-high P diets however there was no effect of P level at the low CP diets on P digestibility. Pigs offered low P diets had higher nitrogen absorption ($P<0.01$) than those offered high P diets. In conclusion pigs offered a high CP-high P diet had reduced P digestibility.

Influence of soya bean meal and synthetics amino acids prices in the cost of nutritional Best Available Technique in Spain

Pineiro, C.[1], Montalvo, G.[2], Garcia, M.A.[2] and Bigeriego, M.[3], [1]PigCHAMP Pro Europa, Segovianos, 40195 Segovia, Spain, [2]Tragsega, Julian Camarillo, 28037 Madrid, Spain, [3]Spnaish Ministry of the Environmental and Rural and Marine Affairs, Afonso II, 28071 Madrid, Spain; carlos.pineiro@pigchamp-pro.com

The Directive 96/61/EC concerning integrated pollution, prevention and control (IPPC) define the Best Available Techniques (BAT) as the most effective techniques to reduce emissions. BAT shall mean those developed under viable conditions, taking into consideration the costs. Therefore, it is necessary to have common methodology to calculate costs of abatement techniques to decide if a technique is BAT or not. Spanish Ministry of the Environment and Rural and Marine Affairs developed a calculation on cost of every BAT, because define the most cost-effective methods for reducing ammonia emissions from Spanish farms. The calculation was carried out according to the methodology set out in the EU Reference Document on BAT for Intensive Rearing of Pigs. Unit used for assessing costs were € per place per year for feed and housing techniques. All these costs have been expressed also as € per kg pig produced, because in the pig sector it is more easily understood. Since being protein more expensive (up 280 €/t in 2007-2008 period), formulation tend to use the minimum amount of soya to achieve the desired feed, what meant that the use of low protein diets (16.5% CP for pigs between 20-60 kg and 13.5% CP for pigs between 60-100 kg body weigh) an average savings of 0.77 € per place per year or 0.26 € per ton of pig produced were achieved because of the use of this technique and market prices. When soya price decreased, low protein diets technique could be not so economically favorable, therefore cataloguing low protein diet as BAT will depend on soya bean meal and synthetics amino acids prices.

Best available techniques in french pig production

Guingand, N., IFIP Institut du Porc, La Motte au Vicomte, 35651 Le Rheu, France; nadine.guingand@ifip.asso.fr

Ammonia is one of the main gaseous compound emitted by pig units. Since 2002, European regulation impose to pig breeders to declare the whole quantity of ammonia produced by their farms. The NEC directive in 2001 combined to the IPPC directive adopted in 1996 fixed the level of emission and proposed technical tools with Best Available Techniques (BAT) to reduce ammonia emitted by pig farms. Intensive livestock concerned are installations for the intensive rearing of pigs with more than 2,000 places for production pigs (over 30 kg) or 750 places for sows. BAT are not only applied in order to reduce ammonia. Water and energy consumptions are concerned. The BAT's list is presented in a technical synthesis – the BREF document – dedicated to intensive rearing poultry and pigs. Because ammonia can be emitted by building, the storage units and during the slurry spreading, BAT concerned all those aspects. In France, more than 3,000 installations for the intensive rearing of pigs and poultry are concerned by the IPPC directive and directly by the application of BAT. Most of the BAT proposed on the storage and the landspreading of manure are already commonly applied in intensive pig farms in relation with specific French regulation. At the opposite, the application of BAT dedicated to the reduction of ammonia emission from pig housing could be more difficult. Actually, in France, more than 75% of grower-finisher pigs are housed on fully-slatted floor with underlying deep collection pit. Most of the BAT identified in the BREF document are based on the reduction of the surface of the collection pit with partially-slatted floor and with frequent evacuation of manure. The aim of this article is to analyse the environmental impact including ammoniac, odours, water and energy of those short-term modifications on French pig production and to propose alternative techniques which are not yet BAT. As often as possible, economic and welfare aspects will be added to environmental data.

Ammonia and nitrous oxide emissions following land application of high and low nitrogen pig manures to winter wheat at three growth stages

Meade, G.[1], Pierce, K.[1], O'Doherty, J.V.[1], Mueller, C.[1], Lanigan, G.[2] and Mc Cabe, T.[1], [1]University College Dublin, School of Agriculture, Food and Veterinary Medicine, Belfield, Dublin 4, Ireland, [2]Teagasc, Johnstown Castle, Wexford, Co. Wexford, Ireland; grainne.meade@ucdconnect.ie

Pig manure can be a valuable nutrient source in cereal crop production but significant nitrogen (N) losses may occur through nitrous oxide (N_2O) and ammonia (NH_3) emissions. Agriculture in Ireland contributes to a 27.7% of GHG emissions and 98% of NH_3 emissions so strategies to reduce losses must be investigated. In a field trial study on a winter wheat crop (cv. Alchemy) at UCD Lyons Farm, gaseous emissions of NH_3 and N_2O and manure N uptake (CNU) by the crop after land application of pig manures were investigated. In the study grower-finisher pigs were assigned to one of two diets; a high CP diet (230 g/kg) and a low CP diet (160 g/kg) to produce high and low N manures (HN/LN). In a 2 x 3 factorial experimental design the manure products were applied to a winter wheat crop at three timings; at growth stage (G.S) 25, G.S 30-31 and G.S 37-39, at a rate of 30,000 l/ha using a 6 meter band spread applicator to plots 30m x 24m. NH_3 emissions were measured for 7 days after each manure application using the micrometeorological mass balance technique with passive flux samplers placed 5 heights above ground level. N_2O fluxes were measured using the closed chamber technique at 3 day intervals for 6 weeks after manure application. Pig manure application resulted in increased grain yield and CNU relative to control plots ($P<0.0001$). Higher N_2O emissions were measured from the HN manure treatments ($P<0.02$). Manure spreading ($P<0.005$) and sampling dates ($P<0.0001$) also affected N_2O emissions. NH_3 emissions were also higher from the HN treatment ($P<0.024$). Spread date influenced NH_3 emissions with the first spread date having the highest total emissions ($P<0.0005$). Ammonia concentrations were highest 1 hour post spreading ($P<0.0001$), with 95% of emissions recorded within the first 24 hours after application.

Effects of glycinin or β-conglycinin on the growth performance and apparent crude protein digestibility of pigs at different growth phases

Zhao, Y., Qin, G., Sun, Z., Wang, T., Wang, B. and Zhang, B., Jilin Agricultural University, College of Animal Science and Technology, No.2888 Xincheng Street, 130118,Changchun, China; guixin@public.cc.jl.cn

Glycinin and β-conglycinin have been identified as major food/feed allergens. The present study was conducted to investigate the effects of glycinin or β-conglycinin on growth performance and protein digestibility in the digestive tract of piglets, growing pigs and finishing pigs in three experiments. In experiment 1,2, and 3, 15 weanling (7.06±0.18 kg), 15 growing (44.54±9.04 kg), and 15 finishing (78.93±7.37 kg) General No.1 barrows, 28 days weaned, were used, respectively. The pigs were randomly allotted to three (A, B, C) treatments with five replicates. The pigs in the A group (control group) were fed diets without ingredients originating from leguminous products, while the pigs in the B or C groups were fed the diets containing purified glycinin or β-conglycinin, which replaced protein in group A by 4%. After experimental periods of 7 days, the digesta in stomach, duodenum, middle jejunum, ileum, and caecum were collected to determine the apparent digestibility of dry matter and crude protein. The results indicated that glycinin or β-conglycinin has negative effect on growth performance and apparent digestibility of dry matter and crude protein for piglets and growers, but not for finishers. The ileum apparent digestibility of piglets and growers fed with Glycinin or β-conglycinin were lower 3-6% than the control. Only in stomach, the opposite decreased extent of digestibility of crude protein for piglets were larger than growers (Glycinin, $P=0.005$; β-Conglycinin, P =0.001), but there were no difference at other sites of digestive tract between piglets and growers ($P>0.05$). This study is essential criteria to explore the soybean-induced hypersensitivity.

Influence of varying Duroc gene proportion on fattening, carcass and meat characteristics in organic pig production

Baulain, U.[1], Lapp, J.[2], Brandt, H.[2], Brade, W.[3], Fischer, K.[4] and Weißmann, F.[5], [1]Institute of Farm Animal Genetics Mariensee, Friedrich-Loeffler-Institute, Hoeltystr. 10, 31535 Neustadt, Germany, [2]Institute of Animal Breeding and Genetics, Justus-Liebig-University, Ludwigstr. 21b, 35390 Gießen, Germany, [3]Chamber of Agriculture Lower Saxony, Johannssenstr. 10, 30159 Hannover, Germany, [4]Department of Safety and Quality of Meat, Max Rubner-Institute, E.-C.-Baumann-Straße 20, 95326 Kulmbach, Germany, [5]Institute of Organic Farming, Johann Heinrich von Thuenen-Institute, Trenthorst 32, 23847 Westerau, Germany; ulrich.baulain@fli.bund.de

A total of 194 pigs of varying proportion of Duroc genes (0, 25, 50, and 75%) were housed and fed under the regulations of organic farming to deduce the optimal Duroc gene percentage for this branch of production. Pigs were fed ad libitum in two phases (13.3 and 12.5 MJ ME/kg, 0.87 and 0.64 g Lysine/MJ ME)and slaughtered at a live weight of approx. 115 kg (carcass weight 93 kg). Due to the applied feeding system high average daily gains from 935 to 975 g were achieved. Feed intake of Du-75 was significantly higher than in the other breeds, but this group was inferior in feed conversion. In carcass quality Du-0 and Du-50 were superior. Lean meat content (FOM) was 55% and 54%, respectively. Significant differences between experimental groups were observed in meat quality traits. In Du-0 pigs a higher electrical conductivity in the loin muscle 24 h p.m. was observed than in groups with Duroc genes. In agreement, the highest drip loss was found in the Du-0 control which is due to the Piétrain sires in this group. The intramuscular fat content was significantly influenced by Duroc gene percentage and increased nearly linear from 1.5% in Du-0 to 2.7% in Du-75. In a carcass quality based marketing system Duroc gene percentage should not exceed 50%, whereas already 25% Duroc genes significantly promote meat quality. Only for marketing systems very strictly based on meat quality Duroc gene portion should be 75% due to a significant increase of intramuscular fat content.

Evaluation of resource based pig feeding and management system in households of North Eastern India

Mohan, N.H.[1], Tamuli, M.K.[1], Das, A.[1] and Bujarbaruah, K.M.[2], [1]National Research Centre on Pig, Indian Council of Agricultural Research, Rani (near airport), Guwahati, 781131, Assam, India, [2]Indian Council of Agricultural Research, Krishi Bhavan, Rajendra Prasad Road, New Delhi, 110114., India; mohannh. icar@nic.in

Pig rearing is one of the most important occupations of rural society especially the tribal masses of India. The North Eastern India (NEI) comprising of 8 states and holds about 28% of the country's pig population. The NEI region is characterized by inaccessibility, cultural heterogeneity, ethnicity and rich biodiversity. In NEI, survey spanning 325 pig faming households, mostly tribals, was conducted to document local swine husbandry practices. The survey was necessitated by availability of limited information of pig husbandry practices followed in this region and also for development of strategies for sustainable pork production system. The results of the survey indicate that the farmers follow a low input low output pig production system, where feed mostly consists of locally available feed ingredients. The pig feed consists of cooked locally available greens, tubers, kitchen waste etc with salt. The tribal families were also found to use waste of traditional grain beer which they prepare as regular pig feed. The nutrient availability in the system has been indirectly calculated and seems to be at the lower plane than the standard recommendations. The present study also describes housing, general and reproductive management, marketing of pork, local pork products and economics of pork production in the existing system. The study also assessed the health practices and availability of germplasm. It was observed that the limited supply of good quality germplasm limits the pork production system in this region. Since the farmers follow jhum based cultivation, the pork production system can be considered more or less organic. Based on the study it was concluded that the pig farming can be made more economically sustainable by farmer adaptable interventions with respect to mainly feeding and breeding.

Impact of the technological level in the economic variables and productive efficiency in pork farms of complete cycle of the States of Guanajuato, Jalisco, Sonora and Yucatan

Trueta, R. and Nava, J., Universidad Nacional Autónoma de México, Economia, Administración y Desarrollo rural, Avenida Universidad 3000. Coyoacán DF, 04510 Mèxico DF, Mexico; trueta@servidor.unam.mx

The objective of the investigation was the technological classification and the determination of its impact in the productive efficiency, and in the main economic indicators; of a random sample of complete cycle pig farms of the 4 states with the greater production of Mexico. The 31.1% of the sample were located in the highest technical level in the semitechnified level the 29.5% of the farms, and the rest 39.3% were located like low technical level. The Kruskall-Wallis non-parametric statistical analysis showed significant evidence that the technified level offers a lower production cost: $9.45 against $11,20 and $11,84 of the semi and low technified respectively (P=0.009). In the productive efficiency, the technified companies produced and sent a greater number of pigs to the market by sow per year: 18.60, in comparison to 11.67 and 15.97 pigs of the semitechnified one and lower level of technification respectively (P=0.0001).

Effect of feed-water ratio in diet on water intake, fattening performance and behaviour of growing-finishing pigs depending on environmental conditions

Paschma, J.M., National Research Institute of Animal Production, Dept. Technology, Ecology and Economics of Animal Production, Balice n. Krakow, 32-083, Poland; mtyra@izoo.krakow.pl

The aim of this study was to determine the effect of feed-water ratio in diet on water intake, fattening performance and behaviour of pigs. Two growth experiments were carried out in summer and winter seasons on total of 192 Polish Landrace pigs. In each trial, starting from 25 kg of body weight, pigs were assigned to four groups, each of 24 animals, that differed in feed-water ratio in diets. Group I (control) received only a dry complete feed with free access to automatic drinkers. The experimental groups (II, III and IV) were fed a wet feed in which the complete feed to water ratio were: 1:1.5; 1:2.5 and 1:4.0, respectively. During the final week of each experiment, continuous of 24-hours video observation of pigs behaviour were made. The experimental fattening to approximately 105 kg body weight ended with slaughter and dissection of carcasses. The amount of water intake, behaviour, average daily gains, feed conversion and carcass traits were determined. Total water intake per pig varied according to the group, relatively less (0.672 m^3) in the control group and the most in IV group (1.053 m^3). A beneficial effect on daily gains, feed conversion and on some traits of carcass meatiness were achieved by pigs fed a wet feed in which feed to water ratio was 1:2.5. The results of behavioural observations indicated lower physical activities with a reduced frequency of aggressive behaviour and greater resting activity in pigs of experimental groups II and III.

Prediction of protein supply from bio-ethanol co-products by the National Research Council 2001 Model and Dutch System (DVE/OEB)

Nuez, W. and Yu, P., University of Saskatchewan, Department of Animal and Poultry Science, 51 Campus Drive, S7N5A8, Saskatoon, SK, Canada; wgn887@mail.usask.ca

The objective was to compare the supply of protein from dry distillers grains with solubles (DDGS) to dairy cows: 1) among different types of DDGS, 2) between DDGS and their corresponding feedstock grains, and 3) between different origins within the same type of DDGS. The prediction models used were the NRC 2001 model and the DVE/OEB system. Comparisons were made in terms of: 1) ruminally synthesized microbial protein (AMCP), 2) rumen undegraded feed protein (ARUP); 3) truly absorbed protein in the small intestine (MP), and 4) degraded protein balance (DPB). Corn DDGS, wheat DDGS, blend DDGS (70% wheat:30% corn), and wheat and corn feedstock samples from different batches were obtained during 2007 in Canada. The results showed significant differences ($P<0.05$) among the three different types of DDGS, between DDGS and their corresponding feedstock grains, and between different origins within the same type of DDGS. The DVE/OEB system showed AMCP values of 76, 54, 56, 51, and 53 g/kg DM for wheat, corn, wheat DDGS, corn DDGS and blend DDGS respectively; ARUP values were 36, 56, 200, 204, and 235 g/kg DM; MP values were 107, 108, 249, 251, and 281 g/kg DM; and DPB values were 3, -44, 72, 11 and 55 g/kg DM. Similar trend was predicted by the NRC 2001 model. The results from both models showed that AMCP was generally higher in original wheat grain than in wheat DDGS. However, MP and DPB were higher in DDGS ($P<0.05$). Among DDGS samples, the highest MP was supplied by blend DDGS ($P<0.05$), while the closest to zero DPB was observed in corn DDGS ($P<0.05$). The plant effect was also observed in MP predicted from both models ($P<0.05$). These differences are due to differences in the protein characteristics of original grain and in the heating applied during the drying process. According to both prediction models, blend DDGS is a superior source of metabolizable protein.

Effects of feeding protected linseed or fish oil on the lipid composition and characteristics of Suffolk ram semen

Estuty, N., Chikunya, S. and Scaife, J., Writtle College, School of Equine and Animal Science, Lordship Road, Chelmsford, Essex, CM1 3RR, United Kingdom; Naser.Estuty@writtle.ac.uk

Previously we reported that C22:6n-3 (DHA) levels in ram semen appear to be insensitive to increased dietary supply of DHA. In other studies in the literature, testicular cells have been demonstrated to synthesise DHA from a-linolenic acid (C18:3n-3) *in vitro*. This study aimed to investigate the effect of feeding sources of C18:3n-3 (linseed) or DHA (fish-oil) on the PUFA content and characteristics of ram semen. Suffolk rams (15) were fed a 70:30 basal diet of haylage and concentrate. Five rams were randomly allocated to each of three protected fatty acid (FA) sources; linseed (LIN), fish oil (FO), and a 50:50 mix of the two sources (LINFO). After 9 weeks of feeding, semen and blood samples were collected and analysed. None of the fat sources affected sperm concentration, motility, viability or morphology. The linseed supplemented diets increased the levels of C18:3n-3 in plasma (7.3, 1.2 and 3.0 g/100g FA, $P<0.001$) and in semen (0.2, 0.1 and 0.1 g/100g FA $P<0.001$) for LIN, FO and LINFO respectively. On the other hand, fish oil diets tended to elevate EPA (C20:5n-3) levels relative to linseed, rams fed LIN, FO and LINFO, had concentrations of 2.2, 20.7 and 11.1 g EPA/100g FA ($P<0.001$) respectively in plasma, whilst in semen the concentrations were 0.2, 0.5 and 0.5 g EPA/100g FA ($P<0.001$). DHA levels in plasma (0.7, 5.9 and 3.5 g /100g FA for LIN, FO and LINFO, $P<0.001$) were enhanced by fish oil containing diets, but levels in semen ejaculates were not and DHA concentration averaged 40.1 g /100g FA across all diets. These results illustrate that protected linseed induced a six fold increase in plasma C18:3n-3 compared to fish oil, and only marginal C18:3n-3 increases in the ejaculate. These slight increases in C18:3n-3 following linseed supplementation were not sufficiently high to elicit significant deposition of DHA into semen lipids via elongation and desaturation of C18:3n-3.

Comparing the *in vitro* proteolysis of feed proteins described with an exponential model using three different enzymes

Guedes, C.M.[1], Lourenço, A.[1], Rodrigues, M.[1], Cone, J.W.[2], Silva, S.R.[1] and Dias-Da-Silva, A.[1], [1]CECAV-UTAD, POBox 1013, 5001-8001 Vila Real, Portugal, [2]Animal Nutrition Group, Department of Animal Sciences, POBox 338, NL-6700 Wageningen, Netherlands; analou@utad.pt

This experiment was conducted to compare the suitability of three enzymes for *in vitro* proteolysis of feed proteins. Fifteen protein supplements (ground at 1 mm) were incubated with proteases from Streptomyces griseus (SG), bromelain and ficin for 1, 2, 4, 6, 8, 24 and 48h and the residues were analysed for nitrogen. The data on proteolysis over time were fitted using an exponential model p=a+b[1-exp(-ct)]. All data on proteolysis over time and the constants derived from the model were statistically analysed. The exponential model fitted well to the data for all the enzymes. Differences ($P<0.05$) on proteolysis between enzymes were observed at most incubation times. Although the protein solubility (constant a) of SG and ficin was not different ($P>0.05$) and was lower than with bromelain ($P<0.05$), the extent of proteolysis (a+b) with ficin was higher ($P<0.05$) than with SG or bromelain. The proteolysis with ficin was ($P<0.05$) faster (c=0.34/h) than with SG (c=0.20/h) or bromelain (c=0.24/h). These three enzyme-based methods described *in vitro* proteolysis of feed proteins according to the exponential model. However, significant different constants, derived from the model, were observed between enzymes.

Use of principal component analysis to illustrate the relationships between rumen protein degradation and the composition of nitrogen fraction

Guedes, C.M.[1], Lourenço, A.[1], Rodrigues, M.[1], Cone, J.W.[2], Silva, S.R.[1] and Dias-Da-Silva, A.[1], [1]CECAV-UTAD, POBox 1013, 5001-801 Vila Real, Portugal, [2]Animal Nutrition Group, Department of Animal Sciences, POBox 338, NL-6700 Wageningen, Netherlands

The aim of this study was to investigate the relationships between rumen protein degradation and the composition of the nitrogen fraction using principal component analysis (PCA). Protein supplements (n=21) were incubated in the rumen of 3 fistulated rams for 2, 4, 8, 16, 24, 48, 72 and 96h. The degradation constants were estimated from the equation p=a−b[1-exp(-ct)] and the effective degradability was calculated as a+[(bc)/(c+k)] with a passage rate (k) of 0.02, 0.05 and 0.07/h. Samples were analysed for nitrogen bound fibre, water soluble nitrogen, soluble proteins (albumins, globulins, prolamins, glutelins) and protein nitrogen. The PCA analysis was based on the correlation matrix of 16 variables for rumen degradation and 10 variables for composition of nitrogen fraction. Loading plots of PCA appear to offer an interesting approach to illustrate the relationship between protein degradation and composition of nitrogen fraction. Using this tool, rumen undegradable protein was identified to be dependent on nitrogen bound to fibre, specially for low passage rates (0.02/h). Moreover, this study revealed that degradation rate (c) was related to soluble proteins (globulins, glutelins) and that the slowly degradable protein (b) was related to insoluble nitrogen not bound to fibre.

Effects of pressure toasting on *in situ* degradability and intestinal protein and protein-free organic matter digestibility of rapeseed
Azarfar, A.[1], Ferreira, C.S.[2], Goelema, J.O.[2] and Van Der Poel, A.F.B.[2], [1]University of Lorestan, Animal Sciences, Khorramabad, P.O. Box 465, Iran, [2]Wageningen University, Animal sciences, P. O. Box 338, 6700 AA Wageningen, Netherlands; Arash.Azarfar@gmail.com

Rapeseed is a protein supplement that contains up to 40% crude protein (CP) on a dry matter (DM) basis, but a large part of its protein can be easily degraded in the rumen. Therefore, before inclusion in ruminant's diet, the extent of its protein degradation in the rumen must be reduced without altering its intestinal digestibility. A study was conducted to investigate the effects of pressure toasting (T, 130 °C) at two residence times (1.5 and 10 min) alone or in combination with soaking in water (ST, 4 h) on ruminal degradability and intestinal digestibility of CP and protein-free organic matter (PFOM) in whole fullfat rapeseed. Regardless of the processing time (1.5 or 10 min), T significantly ($P<0.05$) increased the fraction of undegraded intake protein (UIP) compared to the untreated rapeseed samples. Soaking prior to further toasting did not improve the rumen degradation characteristics of rapeseed CP. Compared to the untreated rapeseed samples, both T and ST significantly ($P<0.0001$) the true protein digested in the small intestine (DVE) and degraded protein balance (OEB), effects that were more evident in samples heated for 10 min. Soaking prior to pressure toasting, however, did not further improve the DVE and OEB in the rapeseed samples in comparison with T treatment. It was concluded that ruminal protein degradability of rapeseed decreased after pressure toasting, without seriously affecting its intestinal digestibility.

***In vitro* gas production as a technique to characterise the degradative behaviour of feedstuffs**
Azarfar, A.[1] and Tamminga, S.[2], [1]University of Lorestan, Animal Sciences, Khorramabad, P.O. Box 465, Iran, [2]Wageningen University, Animal Nutrition Group, P. O. Box 338, 6700 AA Wageningen, Netherlands; Arash.Azarfar@gmail.com

Introducing the automated gas production system (APES) enables us to fit the gas production profiles with an enormous number of data points, to a multi-phasic model as described by Groot *et al.* (1996) in order to characterise the degradative behaviour of feedstuffs. More importantly, when combined with a fractionation method that fractionates feeds into its inherent fractions (non-washable, insoluble washable and soluble washable fraction), the gas production technique enables us to study the kinetics of gas production of the washable fraction, whose degradative behaviour can not be measured with the routine *in situ* method. However, the extent to which a feed is effectively degraded in the rumen not only depends on the rate of degradation but also on the rate of its outflow from the rumen. France *et al.* (2000) proposed a mathematical approach to estimate the extent of degradation (E) using the data of the *in vitro* gas production technique. Using this approach the E was estimated for the the non-washable fraction (NWF) and insoluble washable fraction (ISWF) in barley, maize, milo, lupins, peas and faba beans. The E value for the NWF was estimated assuming a kp of 0.06 h^{-1} (accepted fractional passage rate of particulates in the DVE/OEB system) whilst for the insoluble washable fraction (ISWF) it was estimated at both kp of 0.06 and 0.08 h^{-1} (the latter being the suggested value for the kp of fine particles). At the kp of 0.06 h^{-1}, the E were 45 and 40% in ISWF and NWF, respectively. However, at the fractional passage rate of 0.08 h^{-1} for ISWF, the E was 17.5% lower in ISWF than in NWF (33 vs. 40%). Therefore, it seems that a considerable quantity of small particles that are usually rich in starch (except in lupins) can escape form ruminal degradation. This escape may affect the proportion of rumen degradable nitrogen to rumen degradable carbohydrates and as a consequence the supply of nutrients for the animal.

Liquid feed fermented with a *Lactobacillus* strain with probiotic properties, reduces susceptibility of broiler chickens to *Salmonella enterica typhimurium* Sal 1344 nalr

Savvidou, S.[1], Beal, J.D.[1], La Ragione, R.M.[2] and Brooks, P.H.[1], [1]University of Plymouth, Biological Sciences, Drake Circus, PL4 8AA Plymouth, United Kingdom, [2]Veterinary Laboratory Agency, New Haw, KT15 3NB Surrey, United Kingdom; soumela.savvidou@plymouth.ac.uk

This study investigated the potential of fermented liquid feed (FLF) to reduce Salmonella carriage in broiler chickens. In this experiment, a strain identified as *Lactobacillus salivarius* NCIMB 41606 (Lb salivarius) that had been isolated from chicken gut and had been selected for its fermentation and potential probiotic properties, was assessed for its efficacy in reducing the shedding of *Salmonella enterica typhimurium* Sal 1344 nal[r] (S. *typhimurium*) in broilers. A total of 68 hatchlings were randomly divided into four groups. One group was provided with a daily dose of 10^7 cfu/ml of Lb *salivarius*, delivered via drinking water from day one of age. The second group was provided with 10^9 cfu/gr of Lb salivarius delivered in fermented liquid feed (FLF). The third group (ALF) was provided with feed acidified with 30.3 ml of lactic acid/ kg of wet feed from day one of age and the last group was the control without any addition of Lb salivarius or acid. Birds fed FLF were free of Salmonella on 51% of days compared with 23% for birds fed ALF ($P<0.05$) and 19% for birds provided with Lb *salivarius* via water, and 8% of Salmonella negative days in CON birds. FLF, produced using Lb *salivarius*, is suggested as an effective means of controlling S. *typhimurium* infection to poultry.

In situ degradation of fresh and frozen-thawed 15N-labelled alfalfa and ryegrass

Tahmasbi, R.[1], Nolan, J.V.[1] and Dobos, R.C.[2], [1]University of New England, Animal Science, Armidale, NSW, 2351, Australia, [2]NSW Department of Primary Industries, Armidale, NSW, 2351, Australia; reza.tahmasbi@gmail.com

Alfalfa and perennial ryegrass were labeled with 15N during growth in a glasshouse, harvested at similar growth phases and samples (either fresh (F) or frozen and thawed (FT)) were incubated *in situ* in the rumen. The results were fitted to a model describing the degradation of DM and total N with time. There was no difference between forage species for the instantly soluble fraction (a) and the insoluble but potentially degradable fraction (b) estimates for DM and N disappearance over time but alfalfa 15N values for a and b were significantly ($P<0.05$) higher and lower, respectively, than for ryegrass. Forage species had significant ($P<0.05$) effect on the degradation rate of b (c) and potential degradability on DM, N and 15N, however, alfalfa had higher degradation rates of b for DM, N and 15N than ryegrass. Samples prepared as FT had higher a estimates than for F preparations, while estimates for b were higher for F than FT in both DM and N. There was no significant effect due to sample preparation in estimates of degradation rate (c) for DM and 15N but FT had a significantly ($P<0.05$) higher estimate than F for N degradation rate. Potential degradability was significantly ($P<0.05$) higher for FT than F for DM and N but not for 15N. There were significant ($P<0.05$) interactions between forage species and sample preparation method for DM degradation rate and potential degradability because FT increased the rate of degradation and potential degradability in alfalfa but decreased them in ryegrass. A significant ($P<0.05$) interaction was also found between forage species and sample preparation method for quickly degradable protein. Because of the differences in prediction of DM and N degradation parameters among sample preparation methods, it is recommended that for *in situ* degradation determination utilize the same sample preparation method throughout.

An evaluation of the factors affecting the rate and extent of NDF digestion and a mathematical procedure for defining rates of digestion
Raffrenato, E.[1,2], Van Soest, P.J.[2] and Van Amburgh, M.E.[2], [1]CoRFiLaC, Regione Siciliana, 97100 Ragusa, Italy, [2]Cornell University, Ithaca, NY 14853, USA; er53@cornell.edu

Digestion in ruminants can be empirically and mechanistically described by models of varying complexity. In models like the Cornell Net Carbohydrate and Protein System (CNCPS), rate of NDF digestion (k_d) is an input variable in the feed library. Routinely estimation of k_d has not been achieved, in part, because of lengthy analyses and statistical interpretation. Most NDF k_ds have been calculated with multiple time-points and nonlinear models, assuming completion of digestion at 96 hr. We have developed a simple mathematical procedure for calculating NDF k_d using only one or two time points, assuming a specific fraction of indigestible NDF (iNDF) is correctly computed. Further, we are developing forage specific prediction equations for iNDF based on improved recoveries of acid detergent lignin (ADL). Although highly correlated, ADL and Klason lignin suggest the presence of acid soluble lignin that potentially dilutes the NDF solubles and might impact the apparent energy content of the forage. Further, measurements of ADL do not always account for variability in digestibility, especially at fermentation times representing rumen residence times of forages.The chemistry of ADL cross-linkages with cell wall polysaccharides rather than amount of ADL has been suggested as a better predictor of NDF digestibility (NDFD). The content of ester and ether linked p-coumaric (PCA) and ferulic (FA) acids in NDF and ADF residues showed negative correlations with 24hr NDFD, -016 to -0.87 for NDF and -0.71 to -0.95 for ADF. The ester linked PCA content of ADF explained 92% of the variation in 24hr NDFD. Correlations decreased for longer fermentation times, demonstrating that PCA and FA limit only rate and not extent of in-vitro digestion. Analyses from an in-vivo study confirmed the in-vitro results, demonstrating the highest total tract NDFD (70%) for the silage with the lowest ester linked PCA content in ADF.

Effect of using different rumen fluid to buffer ratios on the *in vitro* degradability and fermentation profile of forages
Khan, M.M.H. and Chaudhry, A.S., Newcastle University, School of Agriculture, Food and Rural Development, Newcastle upon Tyne, NE4 6PL, United Kingdom; m.m.h.khan@ncl.ac.uk

The objective of this study was to investigate the effect of different rumen fluid (RF) to buffer (B) ratios (RFB) in an inoculum on the *in vitro* degradability (IVD), pH, ammonia and methane production from different forages. A 7x2 factorial arrangement in duplicate for each of the two different times of 48h and 96h was used to assess the IVD and fermentation profiles of seven forages (rice straw, wheat straw, hay, silage, sugarcane bagasse, rape seed plant and rye grass) with two RF:B ratios (1:2 and 1:4). This 7x2 factorial arrangement in duplicate was also used to methane production for 96 h only. While the main effect of forage was significant ($P<0.001$) for IVD, and *in vitro* organic matter degradability (IVOMD) for both incubation times, the main effect of RFB on IVOMD was insignificant ($P>0.05$) for 48 h but significant at 96 h ($P<0.05$). The IVD and IVOMD were highest for rye grass and lowest for sugarcane bagasse at both 48 and 96 h of incubations. The pH of RF was highest in sugarcane bagasse at 48 and 96h and lowest for hay at 48h and silage at 96 h. The pH significantly differed ($P<0.001$) for different RFB where it was higher at 1:4 than 1:2 RFB for both incubation times. Main effectsof forages, RFB and their interaction were significant for rumen ammonia level (AL) for each incubation time ($P<0.05$). AL was higher for rye grass and rape seed plant compare to other forages and lower in wheat straw. AL value was higher at 1:2 than 1:4 RFB. Methane production was highest in rape seed plant at 1:2 and in rye grass at 1:4 RFB. Methane production was lowest in sugarcane bagasse compared with other forages at both RFB.

The performance of Iberian piglets growing from 10 to 25 kg BW as affected by the protein: energy ratio in the diet and the level of feeding

Conde-Aguilera, J.A., Aguinaga, M.A., Aguilera, J.F. and Nieto, R., Instituto de Nutrición Animal (IFNA), Estación Experimental del Zaidín (CSIC), Cno del Jueves s/n, 18100 Armilla, Granada, Spain; jose.aguilera@eez.csic.es

Forty-eight weaned Iberian piglets, castrated males, were randomly assigned at 10 kg BW to one of eight dietary treatments. These consisted in four isoenergetic diets (12.58±0.061 MJ ME)) differing in protein content (PC; 10.87, 9.20, 7.86 and 5.96 g digestible ideal protein (DP) / MJ ME, for diets A, B, C and D, respectively), offered at two feeding levels (FL), 0.70 x and 0.95 x ad libitum intake. A feeding experiment, with six piglets per dietary treatment, was performed up to the piglets attained 25 kg BW. It included a digestibility and N-balance trial placed approximately at the middle of the feeding period. The daily ration was weekly adjusted according to treatment and body weight. The average ME intakes (MEI) were 5,438±174 and 8,280±179 kJ/day according to the feeding level imposed, with no significant differences among diets. Performance parameters were significantly affected by both treatment factors, PC and FL ($P<0.001$). The maximum value for the average daily gain (ADG) was obtained when either diet A or diet B were given at the highest FL (412±10.9 g). The average gain: MEI ratio (g/MJ) was significantly higher ($P<0.05$) for diets A and B (39.5 and 39.8) vs. diets C and D (34.3 and 33.0). The N-balance measurements revealed that both factors, PC and FL highly influence N retention (NR; $P<0.001$). Nevertheless, differences in absolute NR failed to attain significance among the three diets with the highest PC. ADG and NR were related by the equation ADG, g = 141±17.8 + 28.5±2.61 x NR, g/day; $P<0.001$; se=38.9; R^2=72.1. The N-balance measurements suggest that maximum protein deposition, achieved when these diets were offered at 0.95 x *ad libitum*, was 55±2.8 g/day, a value well below figures reported in piglets from lean genotypes. It is concluded that optimum growth and N retention of the Iberian piglets can be achieved with the diet providing 9.20 g DP/MJ ME.

Nutritional composition of microalgal biomass for biodiesel production as animal feed

Hurtado, A.[1], Vázquez, P.[1], Alonso, I.[1], Guarnizo, P.[2], Hidasi, N.[2], Núñez, N.[3], Lázaro, R.[3] and García-Rebollar, P.[3], [1]CESFAC, Diego de León, 54, 28006 Madrid, Spain, [2]Biotecnología de Microalgas S.L. (Grupo AURANTIA), Velázquez, 19, 28001 Madrid, Spain, [3]Departamento Producción Animal, UPM, Ciudad Universitaria s/n, 28040 Madrid, Spain; paloma.grebollar@upm.es

Microalgae seem to be the most promising oil source for renewable biodiesel without adversely affecting supply of food and other crop products. On the scale needed for biofuel production, large amounts of dried biomass could be also available for the compound feed industry in the next years. The proximate composition of two microalgal biomass classes, the marine eustigmatophyte Nannochloropsis spp. (NCP) and the continental clorophyte Nannochloris spp. (NC), was determined to assess their usefulness for animal feeding. Microalgae were grown in outdoor photobioreactors, harvested by tangential flow filtration, and dried by refractance window technology. Gross composition (on dry matter basis) was similar for NCP and NC: moisture (6.1 and 6.3%), ash (8.1 and 7.7%), protein (51.8 and 52.0%) and non protein nitrogen (7.5 and 6.1%), respectively. Total carbohydrates ranged from 8.9% (NCP) to 15.6% (NC), and soluble fiber (<1.0%), hemicelluloses (4.1 and 2.2%), cellulose (2.0 and 0.8%) and lignin (<0.5%) were low for NCP and NC, respectively. Ether extract contents differ between classes of microalgae (20.8 vs 14.1% for NCP vs NC), but the same high proportion (40%) of non elutable material was obtained. The fatty acid (FA) profiles (% of total FAs) were also different. The main FAs were eicosapentanoic (40.5%), palmitic (16.8%) and palmitoleic (14.4%) acids for NCP and palmitic (30.7%), linoleic (21.1%) and linolenic (19.7%) acids for NC. Similar concentrations (%) of Ca (0.30), but higher of P, Mg and Na (1.41; 0.35; 0.97 vs 1.10; 0.28; 0.58) and lower of K (1.02 vs 1.62%) were determined for NCP vs NC, respectively. Based on their composition, the two microalgal biomasses seem valuable ingredients for monogastric diets. Feeding trials will be needed to confirm this.

Effect of crude protein in milk replacer on heat production due to feed intake in veal calves

Labussiere, E.[1,2], Van Milgen, J.[2], Dubois, S.[2], Bertrand, G.[1] and Noblet, J.[2], [1]Institut Elevage, Monvoisin BP85225, 35652 Le Rheu, France, [2]INRA, UMR SENAH, Domaine de la Prise, 35590 Saint-Gilles, France; etienne.labussiere@rennes.inra.fr

The use of milk replacers containing casein is known to favour clotting of protein and fat in the abomasum of veal calves but it would not affect the flow of carbohydrates. The clot formation, dependent on the dietary balance between protein and fat would alter digestion and heat production (HP) kinetics. These effects on HP have been assessed during 2 experiments. In the first experiment, veal calves received isocaloric diets with 4 levels of crude protein (CP) at 3 stages of fattening (mean body weight: 72, 136 and 212 kg). They were placed in an open-circuit respiration chamber for measuring their HP, feed intake and physical activity during 6 days. Their protein fractional degradation rate (FDR) was estimated by measuring urinary 3-methylhistidine excretion. Their HP kinetic was partitioned between components due to physical activity, feed intake and metabolic utilization of nutrients (thermic effect of feeding (TEF), short term component) and resting metabolism. During an additional fasting day, their fasting HP was estimated and the difference with HP due to resting metabolism was assumed as a long term component of TEF. During the first stage, but not later, short term TEF increased when dietary CP decreased, resulting in a negative long term TEF for the lowest CP levels at the first stage. These differences can be explained by differences in digestive kinetic and/or in metabolic use of nutrients since FDR increased when dietary CP increased. Both hypotheses have been tested in a second experiment where 2 batches of 4 calves received the 4 dietary CP levels during successive periods of 2 days. This design may not affect digestive kinetic of nutrients but prevent metabolic adaption to CP level. As short term TEF was no more affected by dietary CP level, the differences observed in the first experiment may be due to metabolic adaptation to dietary CP level.

Antibody absorption by Santa Ines lambs fed bovine colostrum from Holstein cows or sheep colostrum from Santa Ines ewes

Moretti, D.B., Pauletti, P., Kindlein, L. and Machado-Neto, R., University of São Paulo ESALQ/USP, Animal Science - Zootecnia, Av. Pádua Dias, 11, 13418900, Brazil; raul.machado@esalq.usp.br

The objective of this study was to investigate antibody acquisition in newborn lambs fed bovine or ovine colostrum obtained from an intensive management of production. Pools of bovine and ovine colostrum were constitute from Holstein cows and Santa Ines ewes. Pools of bovine and ovine colostrum, analyzed by radial immunodiffusion (RID) to quantify immunoglobulin G (IgG), showed 115.69±35.40 and 48.12±13.19 mg/ml, respectively. Twenty four Santa Ines lambs received 250 mL of bovine colostrum (BC group) or 250 ml of ovine colostrum (OC group), at 0 and 6 hrs of life. Blood samples were analyzed for quantification of IgG by RID and serum total protein (TP) by the biuret method. The IgG serum concentration at 6, 24 and 72 hrs were significantly higher ($P<0.05$) for the BC group, 16.32±6.19, 33.80±5.68 and 27.95±5.46 mg/mL, compared with the OC group, 11.31±6.08, 21.02±6.53 and 19.88±7.31 mg/ml, respectively. Considering the group means, BC showed serum IgG concentrations 56% higher than OC. The BC group also showed higher ($P<0.05$) serum TP values at 24 and 72 hrs (7.29±0.87 and 6.89±0.30 g/dl, respectively) in relation to the OC group (5.73±1.35 and 5.69±0.57 g/dl, respectively). Serum IgG represented about 34.76% of serum TP and the correlation between TP and IgG was r=0.81 ($P<0.05$), confirming that the pattern of serum TP fluctuation in this critical period of life, depends significantly on colostrum IgG. The results indicate that the bovine colostrum from Holstein cows can be used as an alternative source of IgG for newborn Santa Ines lambs, and with advantage, since bovine colostrum is naturally richer in immunoglobulins than colostrum from ovine. Supported by FAPESP – The State of São Paulo Research Foundation.

Evaluation of three wheat treatments for dairy cattle: rolled wheat, ensiled ground wheat and ensiled whole wheat in brewers' grains

De Campeneere, S., Vanacker, J. and De Brabander, D.L., Institute for Agricultural and Fisheries Research, Animal Sciences, Scheldeweg 68, 9090 Melle, Belgium; sam.decampeneere@ilvo.vlaanderen.be

To evaluate 3 wheat treatments, three diets were compared using 18 lactating Holstein cows in a Latin square design with 3 periods of 4 weeks. Dry wheat was either rolled, ground and ensiled or ensiled as whole grain after mixing with brewers' grains (proportion: 46/54 DM whole wheat/brewers' grains).The basal diet consisted of maize silage and prewilted grass silage (55/45 on DM base) fed ad lib and completed with 3.2 kg of wheat and 11.8 kg brewers' grains (separately for rolled and ground wheat, mixed for whole wheat). At the start of the trial soybean meal and concentrate level was calculated individually to supply 105% of the net energy (NE) and digestible protein requirements based on expected milk yield and composition. The results indicated that by ensiling whole wheat in brewers' grain, the risk for acidosis was strongly reduced. The treatments with ensiled ground and rolled wheat showed clear indications of a lack of physical structure (reduced DM intake and strongly reduced milk fat content), while the treatment with ensiled whole wheat did not. However, from that, it is too simple to conclude that the ensiled whole grain is preferable, as analyses on samples of the faeces indicated that the starch content in the faeces of the animals fed the ensiled whole grain was higher (115 g versus 26 and 21 g/kg DM faeces) than for the ground and rolled wheat, respectively. Obviously, the whole grain was not weakened enough for a total digestion in the intestinal tract leaving important amounts of starch in the faeces. This also resulted in a lower NE value for that form of wheat. However, as a consequence of the presumed lack of physical structure for the treatments with ground and rolled wheat, the FPCM production for these treatments was significantly reduced as compared to the ensiled whole grain treatment. The ensiled ground wheat conserved well, without a relevant fermentation.

Supplementation of high and low crude protein diets with chitosan promotes Enterobacteria in the caecum and colon and subsequently increases manure odour emissions from finisher pigs

O'shea, C.J., Lynch, M.B., Callan, J.J. and O'doherty, J.V., University College Dublin, Lyons Research Farm, Newcastle, Co. Dublin, Ireland; cormacoshea@gmail.com

The hypothesis of this study is that supplementation of both high and low crude protein (CP), wheat-based diets with chitosan may increase protein-fermenting bacteria in the large intestine at the expense of predominately carbohydrate-fermenting bacteria, thus indicating the role of the former in subsequent manure odour emissions. A 2x2 factorial experiment was conducted to investigate the effect of dietary chitosan inclusion (0 vs 20g/kg) and CP concentration (200 vs 150g/kg) on intestinal microflora, volatile fatty acids concentrations (VFA), ammonia and subsequent manure odour emissions from finisher boars (n=7, 60.3kg). The inclusion of chitosan decreased *Lactobacilli* spp. and increased Enterobacteria in the caecum ($P<0.05$) and colon ($P<0.001$) compared with pigs offered unsupplemented diets. Dietary chitosan decreased the molar proportion of butyric acid and increased valeric acid in the caecum ($P<0.05$) and colon ($P<0.001$) compared with unsupplemented diets. Dietary chitosan increased caecal ammonia ($P<0.05$) compared with unsupplemented diets. Dietary chitosan increased manure odour emissions ($P<0.05$) at 72h post excretion. In conclusion, dietary chitosan decreased *Lactobacilli* spp. and increased Enterobacteria in the hind gut and subsequently increased manure odour emissions.

Nutritional quality of gamma and electron beam-irradiated canola meal

Taghinejad, M., Islamic Azad University, Tabriz Branch, Department of Animal science, P.O. Box 51589-1655, Tabriz, Iran; taghinejad_mehdi@yahoo.com

The growing demand by humans for vegetable oils has provided canola meal (CM) for use in ruminant's diets because it has an excellent balance of AA. However, canola meal is not an effective source of AA for high-producing dairy cows because of its high effective ruminal protein degradabilities which ranged from 44.3 to 74%. It also contains antinutritional factors such as glucosinolates and phytic acid. Radiation processing (exposing a product to ionizing radiations such as gamma rays or electron beam) has been recognized as a reliable and safe method for improving the nutritional value and inactivation or removal of certain antinutritional factors in foods/feeds. The major aim of the present study was to ascertain the impact of gamma and electron beam irradiation on the nutritional and antinutritional components of canola meal. We studied the effects of gamma (γ) and electron beam (EB) irradiation at doses of 15, 30 and 45 kGy on antinutritional contents, ruminal Crude Protein (CP) degradability and *in vitro* CP digestibility of CM. Results showed that total glucosinolate and phytic acid contents in γ and EB irradiated CM were decreased ($P<0.01$) as irradiation dose increased. At irradiation doses equal/more than 30 kGy, washout fractions and degradation rate of the b fraction of CP were decreased and hence effective degradibility (ED) of CP were also decreased compared to untreated sample ($P<0.05$). On the contrary digestibility of ruminally undegraded CP of irradiated CM at those doses was improved ($P<0.05$). In conclusion, irradiation processing of CM at doses equal or higher than 30 kGy could be successfully utilized to improve its nutritional quality. Results showed that EB irradiation had similar effect to γ-irradiation. However, EB irradiation is more practical than γ irradiation and has a lower cost but the cost of 30 kGy irradiation might prevent this processing method is recommended for feed industry, now.

Effect of sorghum silage on rumen microbial counts in buffalo

Chiariotti, A., Puppo, S., Grandoni, F., Barile, V.L. and Antonelli, S., CRA-PCM, Via Salaria, 31 Monterotondo (Roma), 00015, Italy; antonella.chiariotti@entecra.it

Sorghum silage diets were tested to reduce water consumption and nitrogen environmental input due to cultivation of corn. Four cannulated buffalo milking cows were fed ad libitum three diets for 180 days, according to a Latin square design. Diet M1(%): corn silage 64.4, alpha-alpha hay 13.9, concentrate 21.7; diet S1 and S2: sorghum silage 54.9, alpha-alpha hay 17.7, concentrate 27.4. The diets had the same energy (0.90 milk FU/kg DM) and protein content (CP 155.0 g/kg DM), differed only for sorghum variety (S1: BMR 333-SIS; S2: Nicol-PIONEER plus Trudan 8-NK). Animal growth rate together with differences in microbiological counts (total viable (TV), cellulolytic (CB) and xilanolytic bacteria (XB), fungi (F)) and pH of rumen samples were determined. Animal daily weight gain was similar (M1:0.7 vs. S1/2:0.6 kg/d) such as growing rate (M1:16,1% vs. S1/2:16,0%). Microbial counts (expressed as n° cell/ g dry rumen) of XB were similar between the two diets (S1vs.S2) (1.22x109 vs. 1.33 x109), and pH values (6.6 vs. 6.7), neither versus M1 (1.2x109 and 6,5 respectively). On the contrary F values difference (S2: 4.12x105a vs. S1:2.63x105a, M1: 1,73E+05b) was significant at $P>0.05$ as the CB values (S2:4.44x109a vs. M1:1.97 x109 b) but showed the tendency of being higher in S2 diet when compared to S1 (3.18x109 $P>0.10$). Differences resulted also in TV (S1:6.52x1011 a, S2: 3.17x1012b, M1:1.91x1011c) ($P>0.10$). Even if there are differences between the two sorghum silages probably due to vegetative habit of varieties nevertheless their effect seemed not relevant. The difference between M1 and S1/S2 diets was probably due to the more digestible fiber in diet Sorghum respect to corn (NDF 32.4 vs. 36.8; ADF 15.6 vs. 22.4 respectively). As the authors previously found buffalo had higher numbers of cellulolytic bacteria when the animals were fed digestible fiber. Considering that the two diets had the same energy and protein content it seems that sorghum silage is a valid substitute to corn silage.

Assessment of nutritional and chemical changes in rice straw during drying

Coelho, M.[1] and Robinson, P.H.[2], [1]Instituto Superior de Agronomia, Producao animal, Tapada da ajuda, 1349-017 Lisboa, Portugal, [2]UC Davis, animal science, Shields avenue, 95616 california, USA; mariojmcoelho@hotmail.com

The aim of this study was to evaluate the impact of post harvest dry down of M-206 rice straw on its *in vitro* gas production as well as the Si distribution in detergent fibre fractions. Rice plants (n=6 groups) grown in a greenhouse were analyzed fresh (i.e., chopped to <1mm lengths), and after drying the same material at ambient temperature, for dry matter (DM), ash, acid detergent fibre (ADF) and acid detergent/neutral detergent fibre (ND/ADF - residues from AD extraction sequentially extracted with ND), Si in ADF and ND/ADF as well as *in vitro* gas production at 4, 24 and 72 h. The rice straws contained 5.48% Si (DM). Drying sharply decreased gas production at 4 h of incubation (26.4 to 20.2 ml/g organic matter (OM)) as well as 24 h (130.9 to 91.2 ml/g OM) and 72 h (228.5 to 173.4 ml/g OM), but increased ($P<0.01$) the ADF and AD/NDF values from 38.2 to 47.4 and 33.4 to 39.4% (DM) respectively. Drying reduced ($P<0.01$) the percentage of total Si recovered in ADF from 53.9 to 48.5%, but markedly increased ($P<0.01$) the percentage of Si recovered in AD/NDF from 19.6 to 34.1%. As the Si in rice straw can be separated into three fractions in this scheme (i.e., that soluble in AD, that insoluble in AD but soluble in ND and that insoluble in both AD and ND), it may be that it is the increase in the latter fraction with field drying which is responsible for the reduction in *in vitro* gas production of rice straw. While the role of Si in digestibility of rice straw remains unclear, these results demonstrate that the depression in *in vitro* gas production caused by drying is associated with changes in the locations of the Si in the plant.

Determination of apparent metabolizable energy of sunflower seed meal in broiler chickens

Nassiri Moghaddam, H., Salari, S., Arshami, J. and Golian, A., Ferdowsi university of Mashhad, Agricultural faculty, Excellent center of animal science, Azadi square, 0098-Mashhad, Iran; hnassirim@gmail.com

This experiment was conducted to determine the AME_n value of SFSM with a multilevel assay including 3 dietary inclusion levels (7, 14, and 21%) of SFSM that is incorporated to the basal diet of broiler (2 to 3 wk of age) and contained 0.3% chromium oxide as an indigestible marker. One-day old male chicks fed a standard broiler diet for 2 wk. On d 10, 80 birds were placed at random in 16 cages for 4 replicates per dietary treatments. On d 15, the birds were starved for 4 hours and then received the experimental diets from 15 to 21 d of age. During the last 3 d, excreta samples from each cage were collected and stored at -20 °C. After being thawed, excreta were homogenized, dried, and ground through a 1-mm screen. Diets and excreta were analysed for dry matter, CP, chromium oxide, and gross energy. Apparent metabolizable energy was calculated as follows: ME(kcal/kg)=dietary gross energy×[1-(dietCr$_2$O$_3$/excretaCr$_2$O$_3$)×(excreta gross energy/diet gross energy)]. The correction of AME to zero nitrogen retention (AME_n) was based on a factor of 8.22 kcal/g of retained N. The AME_n value of SFSM was calculated using the following equation: AME_n=(AME_nT−α×AME_nB)/b, where T is the test diet, α is the proportion of the basal diet in the test diets, B is the basal diet, and b is the proportion of SFSM in the test diets. Results showed that increasing inclusion rate of SFSM decreased the AME_n of the diets significantly. The AME_n (kcal/kg) of SFSM, calculated by difference ranged from 95 to 1233 kcal/kg. The AME_n values obtained for the diets were regressed against the level of SFSM in the basal diet to estimate the AME_n content in SFSM. The equation derived by fitting a linear model was the following: y = 2957-1.735x; R^2=0.736. An estimate of the AME_n of SFSM, obtained by extrapolation of this equation gave a value of 1222 kcal/kg.

The effect of supplemented diet by sucrose and/or starch on Fibrobacter succinogenes population in ruminal fluid of Holstein steers determined by real-time PCR

Danesh Mesgaran, M., Rezaii, F., Vakili, A.R. and Heravi Moussavi, A., Ferdowsi University of Mashhad, Dept. of Animal Science, Excellence Center for Animal Science, Faculty of Agriculture, 91775-1163, Iran; danesh@um.ac.ir

The objective of the present study was to investigate the effect of diets containing different types of non-fiber carbohydrates (NFC) on runminal Fibrobacter succinogenes (FBS) population determined by real-time polymerase chain reaction (RT-PCR). Four Holstein steers were used in a 4×4 Latin square design (21 days each period). A basal diet (BD) was formulated to be contained of alfalfa hay, barley grain, soybean meal and sugar beet pulp (400, 290, 190 and 50 g/kg, respectively). Sucrose (Su) or starch (St) or a 1:1 mixture of sucrose and starch (SuSt) was added to the BD at the rate of 70 g/kg DM. Diets were offered as 2-2.5 times of maintenance requirements (7 kg DM/d). Rumen fluid samples were collected before and 4 h after the morning feeding at the last day of each period. Samples were analyzed for FBS quantitation by qPCR. DNA was extracted from the samples using the QIAamp® DNA stool mini kit (Qiagen Ltd, Crawley, West Sussex, UK). Fibrobacter succinogenes rDNA concentration was measured by RT-PCR relative to total bacteria amplification ($\Delta\Delta$Ct). The 16s rRNA gene-targeted primer sets used in the present study were forward: GTTCGGAATTACTGGGCGTAAA and reverse: CGCCTGCCCCTGAACTATC. Cycling conditions were 95 °C for 5 min, forty cycles of 95 °C for 15 s, 60 °C for 15 s and 72 °C for 30 s. Mixed procedure of SAS (2003; Y = Mean + Treatment + Animal + Period + residual) was applied to analyzed the data, and the means compared by Tukey test ($P<0.05$). The results indicated that different types of non-fiber carbohydrates did not have any effect on FBS population relative to total bacterial population ($\Delta\Delta$Ct) in ruminal fluid before and 4 h after the morning feeding (BD = 0.193 and 0.179, Su = 0.179 and 0.230, St = 0.123 and 0.144, SuSt = 0.0911 and 0.189, SEM = 0.0629 and 0.054, respectively).

Effects of combination of ethylene di amine tetra acetic acid and microbial phytase on the serum concentration and digestibility of some minerals in broiler chicks

Ebrahim Nezhad, Y.[1], Eshrat Khah, B.[1], Hatefi Nezhad, K.[1] and Karkoodi, K.[2], [1]Islamic Azad University, Shabestar Branch, Department of Animal Science, Faculty Agriculture, Shabestar, East Azerbaijan, 5381637181, Iran, [2]Islamic Azad University, Saveh Branch, Department of Animal Science, Faculty Agriculture, Saveh, 39187/366, Iran; ebrahimnezhad@gmail.com

This experiment was conducted to evaluate the combined effects of ethylene di amine tetra acetic acid (EDTA) and microbial phytase (MP) on the serum concentration and digestibility of some minerals in broiler chicks. This experiment was conducted using 360 Ross-308 broiler chicks. In a completely randomized design with a 3×2 factorial arrangement (0, 0.1 and 0.2% EDTA and 0 and 500 IU MP). Four replicate of 15 chicks per each were fed dietary treatments including 1) P-deficient basal diet [0.2% available phosphorus (aP)] (NC); 2) NC + 500 IU MP per kilogram of diet; 3) NC + 0.1% EDTA per kilogram of diet; 4) NC + 0.1% EDTA + 500 IU MP per kilogram of diet; (v) NC + 0.2% EDTA per kilogram; and (vi) NC + 0.2% EDTA + 500 IU MP per kilogram of diet. The concentration of zinc, cupper and manganese of serum and their digestibility and also digestibility of apparent metabolizable energy (AMEn) was evaluated. The results showed that phytase supplementation of P-deficient diets significantly increased zinc concentration of serum ($P<0.05$). Interaction effect of EDTA×MP on serum concentration of copper and manganese and also digestibility of zinc was significant ($P<0.05$). EDTA supplementation of P-deficient diets significantly increased manganese digestibility in broiler chicks ($P<0.01$).

Effect of biological treatment on chemical composition, nitrogen fraction and protein solubility of linseed straw

Sobhy, H.M.[1], Mikhail, W.A.[2], Abo-Donia, F.M.[3], Abedo, A.A.[4], El-Gamal, K.M.[5] and Fadel, M.[6], [1]Instit. of African Research, Natural Research Dep., Cairo Univ., Giza, Egypt, [2]Instit. of African Research, Natural Research Dep., Cairo Univ., Gza, Egypt, [3]Anim. Prod. Instit., By-Product Dep., Dokki, Giza, Egypt, [4]National Research Center, Animal Production Dep., Dokki, Giza, Egypt, [5]Ministry of Agric., Technical Office, Dokki, Giza, Egypt, [6]National Research Center, Microbial Chem. Dep., Dokki, Giza, Egypt; abedoaa@hotmail.com

This work aimed to improve the nutritive value of linseed straw (LSS) using the biological treatment. In the first experiment LSS was treated with T. viride, T. harzianum and T. reesei for 7 days. In the second experiment LSS was treated with the proper fungi (T. reesei) for 1 to 10 days. Results of the first experiment indicated that T. reesei was the appropriate fungi, whereas the highest (6.33) CP was achieved compared with 3.12, 2.79 and 2.22% for treated LSS with T. viride, T. harzianum and untreated LSS, respectively. Results of the second experiment showed that 6 days was the appropriate period to treatment, whereas the highest CP content (6.56%) were recorded and the CF content was decreased to 58.64 compared with 2.22 and 66.72%, respectively for untreated LSS. Total nitrogen (TN) and the true protein N content were increased with advancing the treatment period, from 0.95 to 1.05 and from 37.89 to 53.33% as a percentage of TN in the first day and after 6 days of treatment, respectively. Protein solubility in sodium chloride and in pepsin was decreased from 63.69 to 44.61 and from 65.54 to 55.69% after 1 and 6 days of treatment, respectively.

In vitro fermentation profile of *Solanum lycocarpum* St hil. volatile fatty acid response

Lima Neto, H., Chaudhry, A.S. and Khan, M.M.H., Newcastle University, Kings Road, Agriculture Building, NE1 7RU, Newcastle upon Tyne, United Kingdom; helio.limaneto@gmail.com

A triplicate 5x2x4 factorial experiment aimed to settle volatile fatty acid (VFA) *in vitro* fermentation profile of five dried meal-like fractions of Solanum lycocarpum (SL) (Flower=Fl, Fruit=Fr, Leaf=Lf, Stem= St and Root=Rt) at two levels (0.2 and 0.4g of DM) at 4 distinct incubation times (08, 24, 48 and 96h). Samples of about 0.2 and 0.4g dried ground SL were weighed into 50 ml centrifuge tubes to which 40 ml of the inoculum was added. Rumenfluid (RF) was obtained from 2 fistulated sheep, strained through cheese cloth into pre-warmed flasks It was then mixed in 50 ml polypropylene centrifuge tubes (PCT) with the pre-warmed buffer at 1:4 ratio. CO_2 was added and samples incubated in water bath (WB) at 39 °C. After digestion times 2 ml liquid solution of PCT were added to 0.5 ml deproteinasing solution. Mixture of rumen fluid was centrifuged at 3,000 rpm for 10 minutes. 0.5 ml of supernatant was transferred into a Gas Chromatography (GC) vial. Computing integrator was used to peak acetic, propionic, isobutyric, butyric, isovaleric and valeric productions. Data were analyzed by using GLM to settle the main effects of SL fractions, levels and interactions. Main effects of SL fractions, concentration and incubation times were significant for all VFAs ($P<0.001$) and that all fractions are fermentable at distinct rates ($P<0.001$). Total VFA was 40.79, 31.12, 26.49, 19.06, 27.37 mM, /100 mol for Fr, Fl, Lf, St and Rt respectively Highest partial productions were observed for Fr and Fl reaching 20.52 and 18.41 mM of Acetic Acid respectively ($P<0.001$). Effective production rates adjusted by regression model of Acetic, Propionic and Butyric were 0.09, 0.045 and 0.0015 mM per hour for the top fermented (Fr) showing a pattern similar to high fermentable forages ($P<0.05$). Further studies are being conduced to assess additional information regarding the fermentation characteristics of this potential animal feedstock.

Nutritive value assessment for some Acacia species grown in Saudi Arabia

Al-Soqeer, A.A., Qassim University, Department of Plant Production and Protection, Faculty of Agriculture and Veterinary Medicine, Buriedah-51452, Saudi Arabia; wasmy84@hotmail.com

Fourteen Acacia species namely *A. coriacea, A. cuthbertsonii, A. ineguilatera, A. iteaphylla, A. kempeana, A. ligulata, A. microbotrya, A. nilotica, A. oswaldii, A. pruinocarpa, A. saligna, A. sclerosperma, A. seyal* and *A. victoria* were collected from Prince Sultan Research Center for Environment, Water and Desert of King Saud University (Al-Riyadh- Saudi Arabia). Chemical composition and *In vitro* gas production technique were used to assess the nutritive values of *Acacia* spp. and alfalfa hay. The crude protein content among tested *Acacia* spp ranged from 8.0 to 16.7%, which was comparable in *A. iteaphylla* with that of alfalfa hay (i.e. 17.1%). Amongst tested Acacia spp, the *A. ineguilatera* had the highest condensed tannin (CT), the CT value ranged from 10.4 to 77.0 mg/ g DM. No significant differences ($P>0.05$) for degradable fractions were found between alfalfa hay in one hand and *A. kempeana, A. nilotica* and *A. seyal* in the other hand. Calculated metabolizable energy (ME, MJ/ kg DM) for the tested species of acacia. ranged from 4.35 to 6.69 MJ/ kg DM which could supply the animals with the 53-84% of the ME in comparison with that of alfalfa hay.

Effect of replacement alfalfa with tanniniferous feedstuff (dried grape by-product) on rumen protozoa population

Taghizadeh, A. and Besharati, M., University of Tabriz, Animal science, Collage of Agriculture, 51664, Iran; ataghius2000@yahoo.com

The aim of the present study was to evaluate effects of alfalfa replacement by multiple level of dried grape by-product on ciliate protozoa population. 16 mature Ghizel wether sheep of live weight 34 kg (±1.5) were used. The animals were allocated individually in boxes with free-access to salt block and water. Four diets were used, one as basal diet (alfalfa) and the rest as mixed diets (DGB with alfalfa). The sheep were fed twice daily, at 09:00 and 17:00. For the first week, sheep received alfalfa for ad libitum intake. The amounts of consumed and refused feeds for every sheep were recorded. For the second and third week, DGB replaced 0, 15, 30, or 45% of the alfalfa DM. On the last day at 2 h after feeding, the digesta samples collected and were bulked for counting of rumen ciliate protozoa. The population of protozoa counted using Dehority (1978) method. Data obtained from study was subjected to ANOVA as a completely randomized design with 4 replicates by the GLM procedure (SAS Inst. Inc., Cary, NC), and treatment means were compared by the Duncan test. Average total protozoal concentration increased when sheep were fed alfalfa plus dried grape by-product, but it was not significantly differ ($P>0.05$). Average total protozoal concentration in sheep fed only alfalfa was lower but not different from the other test diets. Within the given diets, there were no differences among sheep in total ruminal protozoa concentrations. The concentration of Diplodinium was observed to increase when sheep were fed with dried grape by-product. No differences were found in either the concentration of *Entodinium* spp., *Holotricha* or *Opharyoscolex* species between the treatments. The concentrations of *Entodinium* spp., *Diplodinium* spp., *Holotricha* and *Opharyoscolex* species were higher when 15% dried grape by-product was included in the diet. It was concluded that replacement alfalfa with dried grape by-product (DGB) in diets had not significant effects on rumen ciliate protozoa population.

Voluntary feed intake, nutrient digestibility and rumen fermentation characteristics of Iranian Balouchi sheep fed *Kochia scoparia*

Riasi, A.[1], Danesh Mesgaran, M.[2] and Fathi, M.H.[1], [1]University of Birjand, Department of Animal Science, Faculty of Agriculture, 97175/331, Birjand, Iran, [2]Ferdowsi University of Mashhad, Department of Animal Science, Faculty of Agriculture, 91775/1163, Mashhad, Iran; riasi2008@gmail.com

Kochia spp. are halophytic plants that typically growth in salty land and may have good potential as forages in ruminant feeding. Consumable parts of *Kochia scoparia* (leaves and stems) were harvested at mid bloom stage from Salinity Research Centre of Birjand University (with alkaline soil and maximum air temperature 40 °C). Ten cannulated Balouchi ewes (48±2 kg) were transferred to metabolism cages and randomly allocated to two dietary treatments (100% kochia or 100% alfalfa). The forage samples were chopped and composite before air-drying. Dried samples were analyzed for total N, NDF, ash, Na, Cl and K. Animals had *ad libitum* access to feed and water. The results showed the voluntary dry matter intake of kochia (579.8 g/d) was lower than alfalfa (1,052.1 g/d) ($P<0.01$). Kochia had lower apparent digestibility of DM, OM, CP, and NDF compared with alfalfa (452, 412, 490 and 310 vs 620, 629, 682, and 515, respectively) ($P<0.05$). At different times after feeding (0, 0.5, 1, 2, 3, 4, 6, 8 hours) mean ruminal pH were similar for two treatments. However, the ammonia-N concentration was higher ($P<0.05$) in ruminal fluid of ewes fed alfalfa ($P<0.05$). The linear relationship for the sampling time was significant ($P<0.05$).

Effects of ascorbic acid, α-tocopherol and red chicory on *in vitro* hind-gut fermentation of two pig feeds

Cattani, M., Tagliapietra, F. and Schiavon, S., Università degli Studi di Padova, Dipartimento Scienze Animali, Viale dell'Università, 16, 35020, Legnaro, PD, Italy; mirko.cattani@unipd.it

Possible effects on hind-gut fermentation pattern due to an addition of ascorbic acid (AA), α-tocopherol (α-T) or red chicory water extract (RCWE) were studied with an *in vitro* automatic batch system, equipped with a gas pressure detector. A pig commercial feed (CF) and a corn meal (CM) were incubated for 48 h at 37 °C with buffered caecal fluid collected from 3 pigs at the slaughterhouse. Each feed sample (1 g), milled at 1 mm, was tested without additive (C) or with the addition of AA, α-T or RCWE (1 mg/g of feed) in 4 replications. Four blanks without feeds were also included. Data of gas production (GP) were fitted with a monophasic model. At the end of incubation the residual fluid was analysed for ammonia (NH_3) and volatile fatty acids (VFA) contents. Data were subjected to ANOVA. After 48 h of incubation significant differences between CF and CM were observed for GP at 48 h (217.4 vs. 278.7 ml; $P<0.01$), NH_3 (11.1 vs. 8.0 mmol/l; $P<0.01$) and VFA (3.06 vs. 3.27 mg/ml; $P<0.01$). With respect to C, all the 3 additives significantly increased the asymptotic GP (254.4 vs. 263.7, 264.0 and 264.2 ml for C vs. AA, α-T and RCWE, respectively; $P<0.01$), but did not influence the measured GP at 48 h and the others GP kinetic parameters. With respect to C, the 3 additives significantly decreased the fluid content of propionate (0.93 vs. 0.89, 0.87 and 0.90 mg/ml, for C vs. AA, α-T and RCWE, respectively; $P<0.05$) and, consequently reduced total VFA. With respect to C, RCWE also significantly reduced NH_3 (9.8 vs. 9.3 mmol/l; $P<0.05$), while the other additives had no effects. In conclusion, all the 3 additives showed to exert effects on the *in vitro* hind-gut fermentation.

Fermentation characteristics and ruminal microbial population in buffaloes fed low, medium or high concentrate based diets

Mirza Aghazadeh, A.[1], Gholipour, G.[1] and Mansouri, H.[2], [1]Urmia university, Animal science, Urmia, 57153-165, Iran, [2]Animal science and nutural resource research center, Animal science, Tehran, 57153-165, Iran; a.aghazadeh@urmia.ac.ir

Water Buffaloe is an integral part of sustainable small farming systems in Iran. Fermentation characteristics were measured and numbers of microbial population were determined in ruminal liquid samples collected from three rumen fistulated adult male (450 ± 20 kg)buffaloes fed on diets containing 35% (low concentrate), 45% (medium concentrate) or 55% (high concentrate). The animals were kept on each diet for a period of 15 days for microbial adaptation. The rumen liquor samples were collected before feeding (zero hour) and subsequently at 3, 6, 9 and 12 h post feeding. The results showed that the highest rumen pH (6.97) before feeding and the lowest (6.42) 6 h post feeding. The least TVFA (92.53 mmol/l) were found before feeding and the highest (120.2 mmol/l) after 3h post feeding on diet based on 55% concentrate. The bacterial counts were significantly ($P<0.01$) higher in animals on 55% concentrate diet, protozoa count were significantly ($P<0.01$) higher in animals on 45% concentrate diet. Rumen fungi were significantly ($P<0.01$) higher in animals on 35% concentrate diet.

The effect of replacing corn silage with triticale silage on the performance of feedlot Zel lambs

Hajilari, D.[1], Yazdani, A.R.[1], Fazaeli, H.[2], Zerehdaran, S.[1] and Mohajer, M.[2], [1]Gorgan University of Agricultural sciences and natural resources, Animal science, Shahid Beheshti Ave., 49138-15739, Gorgan, Iran, [2]Faculty member of animal science research institute, Heydar Abad, Karaj, Iran; aryazdani@yahoo.com

An experiment was conducted with a completely randomize design with 4 treatments and 10 replications. Treatments consisted of 0 (T_1), 33.3 (T_2), 66.6 (T_3) and 100% (T_4) triticale silage replaced by corn silage. The experiment lasted for 84 days. Chemical features and Aerobic stability of triticale and corn silage compared. Weight gain, feed intake, feed conversion efficiency, carcass characteristics and blood nitrogen urea and glucose concentration (3 stages) were determined. Results showed that there were significant differences in pH, dry matter intake, crude protein, ADF and NDF among treatments ($P<0.05$). Time of peak temperature differed between triticale and corn silages but treatments didn't show significant difference in temperature peak ($P>0.05$). Dry matter intake in T_4 was lower than other treatment groups ($P<0.05$). There were no significant differences in body weight gain, carcass characteristics, blood nitrogen urea and glucose content among different treatments ($P>0.05$). Ruminal pH and ammonia didn't affect by treatment at any stages ($P>0.05$). Apparent digestibility of dry matter intake, organic matter and ADF were altered among treatment groups but treatment had no differences in crude protein, NDF and ash digestibility ($P>0.05$). Results of current experiment showed that complete replacing of corn silage with triticale silage had no negative effect on performance of feedlot male Zel lambs.

Chemical composition, antinutritional factors and *in vitro* fermentability of two chickpea (*Cicer arietinum* L.) varieties

Ronchi, B.[1], Danieli, P.P.[1], Primi, R.[1], Bernabucci, U.[1] and Bani, P.[2], [1]Università della Tuscia, Dipartimento di Produzioni Animali, Via S. C. de Lellis snc, 01100 Viterbo, Italy, [2]Università Cattolica del Sacro Cuore, Istituto di Zootecnica, Via E. Parmense, 84, 29100 Piacenza, Italy; ronchi@unitus.it

Nutritional assessment of 'traditional' legumes to be used in animal feeding, should be regarded as a priority to provide a suitable alternative to the imported industrial protein sources, in primis soybean products. Therefore, the present study was aimed to evaluate some nutritional parameters and antinutritional factors (ANFs) of seeds of two varieties (Sultano and Pascià) of chickpea (*Cicer arietinum* L.) obtained by agronomic trials at different time and density of sowing. Chickpea seeds were assessed for crude protein (CP), total starch, crude fiber, NDF, ADF, ADL, total lipids, ash, trypsin inhibition activity (TIA), tannins, amino acidic profile and *in vitro* ruminal fermentability. Nutritional characterization showed a good CP level (18.3-24.9% DM) and an excellent starch content (36.6-40.8% DM). Amino acidic profile of chickpea proteins showed a slightly lower content of sulfur, aromatic and essential amino acids (18.31, 6.19, 25.82 g/16 gN, respectively) compared with values reported in literature for soybean. Among the ANFs, TIA was lower compared with soybean (12-18 vs 43-84 TIU/mg DM) and moderate levels of condensed and total tannins (50-150 mg/100 g and 170-240 mg/100 g on DM basis) were found. *In vitro* gas production (GP) pointed out a rapid fermentability of chickpea, with almost half of the final (48 h) GP expressed in the first 8 h of fermentation. Sultano vs Pascià produced more gas and had a lower final pH, although differences were numerically small. DM degradation was almost complete (98.26% on average) without significant differences between varieties. Chemical parameters and *in vitro* tests showed that chickpea may be an alternative to soy products, both under nutritional point of view and in terms of integration of livestock farming with sustainable cropping systems.

Effect of biological treatments on recovery, *in situ* disappearance, chemical composition and nutritive value of some desert by-products

El-Bordeny, N.E.[1], Abo-Eid, H.A.[2], Salem, M.F.[3], El-Ashry, M.A.[1] and Khorshed, M.M.[1], [1]Ain shams university faculty of Agriculture, Animal production, Ain shams univ. fac. of Agric, Animal production Dept, 11241, Egypt, [2]Minufiya University,Environmental Studies & Research, Minufiya University,Environmental Studies & Research, 11111, Egypt, [3]Minufiya Unive,Genetic Engineering and Biotechnology Research Institute, Minufiya Unive,Genetic Engineering and Biotechnology Research Institute, 11111, Egypt; nasr_elbordeny@yahoo.com

The present study aimed to evaluate the effects of biological treatment (Trichoderma reesei application) to some by-products desert namely, pinnae of date palm tree (PDP), olive pulp (OP), date seeds (DS), barley straw (BS) and peanut straw (PS) at different moisture levels and incubation periods on recovery rate, *In situ* DM and OM disappearance and chemical composition. Recovery rate of OP was higher ($P<0.01$) than the other by-products while lowest recovery rate was recorded for PS. Recovery rate after two weeks incubation period was higher ($P<0.01$) than the other incubation periods (4 or 6 weeks). Significant increases was observed in CP percent for the treated materials compared to the control, which CP% increased from 8.13 to 16.07-16.56% and from 7.09 to 14.41-14.65% for OP and DS, respectively. The crude fiber decreased significantly ($P<0.01$) by fungal treatments from 33.99 to 29.38-28.92% and from 10.98 to 10.46-10.27% for RS and CS, respectively. While, the NFE decreased significantly ($P<0.01$) by fungal treatments from 73.06 to 66.70-66.38% for DS. Overall means of DM and OM *in situ* disappearance for different incubation periods (6, 12 and 24 hrs) in the rumen increased for treated residuals compared to untreated residuals, this may have been related to the lower CF and higher CP contents of these treatments than untreated for all roughages used. The best level of moisture for the desert by-products was 40% and the best incubation period for the treatment was 2 weeks.

Reaplacing triticale silage treated with microbial additive and molasses with corn silage on performance of fattening male calves
Kocheh Loghmani, M.[1], Forughi, A.R.[1] and Tahmasebi, A.M.[2], [1]Hasheminejhad high ejucational center, animal science, shahid kalantary street mashhad, 9178695398, Iran, [2]ferdosi univercity, Animal science, shahid kalantary street mashhad, 9178695398, Iran; afroghi@yahoo.com

This research was conducted to study the effect of replacing corn silage treated with microbial additive and molasses on performance of fattening male calves, In complete randomize design, twenty Brown Swiss male calves were used to examine the effects of replacing corn silage with triticale silage treated with microbial additive and molasses on their performances in 84 days. Cows fed TMR with forge to concentrate ratio of 30:70. Four experimental diets were included: 1) control, 2) triticale silage without additive, 3) triticale silage treated with molasses, 4) triticale silage treated with microbial additives. Comparing to control using triticale silage treated with molasses and microbial additive had not increased ADG, blood metabolite and feed intake significantly ($P>0/05$). Triticale silage treated with molasses improved the apparent dry matter digestibility significantly ($P<0/05$). Ruminal pH in diet 4 was significantly higher than others ($P<0/05$) Comparing to control using triticale silage treated with molasses and microbial additive had not increased feed conversion ratio and Body weight significantly ($P>0/05$). The time spent rumination and total chewing activity not affected by treatment, but time eating was influenced significantly by treatments ($P<0/05$). The results of this experiment indicated that replacing com silage with triticale silage treated with microbial additive and molasses had not effect on performance of fattening male caves and triticale silage treated with molasses improved the apparent dry matter digestibility.

Effects of extruded linen seed incorporation on lamb's growth, carcass and meat quality and fatty acids profile
Saidi, C., Atti, N., Mehouachi, M. and Methlouthi, N., Institut National de Recherche Agronomique de Tunis, Zootechnie, rue Hédi Karray, 2049 Ariana, Tunisia; S.cherifa@gmail.com

Effects of extruded linseed incorporation on growth, carcass characteristics, meat quality and intramuscular fat fatty acids (FA) composition were studied on 36 lambs from Tunisian local breed (Queue Fine de l'Ouest). All lambs were male with an average body weight of 21.3 kg at the beginning of the experiment. Animals were randomly divided into three groups of 12 lambs each. Sheep in each group received individually oat hay ad-libitum and one of three concentrates containing 0 (Control), 15 (L15), and 30 (L30) percent of extruded linseed in dry matter. All concentrates were isocaloric and isonitrogenic. At the end of the experiment which lasted 85 days, all the lambs were cut down. Samples of longissimus dorsi muscle were taken for meat quality measurements and FA determination. Dry matter intake, sheep growth and carcass characteristics were not significantly affected by diet treatment. Also, meat chemical composition, colour parameters, the cooking loss and meat pH were similar for all regimens. However, the intramuscular FA composition was affected by linseed presence ($P<0.01$). Particularly, saturated fatty acid percent was lower for L15 and L30 groups than for control one. C18: 3 content and the content of all poly-unsaturated fatty acids were higher for L15 and L30 groups than for control one ($P<0.05$). The omega 3 contents were 0.52, 2.86 and 2.41 for Control, L15 and L3, respectively; the omega 6 to omega 3 ratio was 7.1, 0.9 and 2.2 for Control, L15 and L3, respectively. In conclusion, the incorporation of extrude linseed as 15 percent of concentrate improves the dietetic quality of lamb meat.

The effect of feeding triticale, corn and wheat on the performance of crossbred feedlot cattle

Yazdani, A.R. and Hajilari, D., Gorgan University of Agricultural sciences and natural resources, Animal science, Shahid Beheshti Ave., 49138-15739, Gorgan, Iran; aryazdani@yahoo.com

An investigation was conducted to evaluate the management effect of feeding triticale on the performance of crossbred feedlot cattle. At the end of 83 day finishing period, measurements were made of daily feed intake, average daily gain, total gain and feed / gain ratio. The animals were allocated by randomized block design on the basis of live weight to one of three dietary high concentrates rations of triticale, corn and wheat grain for crossbred feedlot cattle. Average daily gains of steers fed triticale, corn and wheat were1.14, 1.39 and 1.28 kg, respectively, and were significantly different ($P<0.05$).Daily feed intake was significantly ($P<0.05$) higher for corn fed steers (8.38 kg) than the values for crossbred steers fed triticale (7.22 kg) or wheat (7.19 kg).Feed gain/ratio even though not significant, but tended to be the best on wheat and poorest on triticale.It is therefore concluded that corn was slightly superior to triticale and wheat as a finishing ration and triticale is potentially an important feed grain in beef finishing rations.

Ascorbic acid, α-tocopherol acetate and wine marc extract incubated *in vitro* with rumen fluid: gas production and fermentation products

Tagliapietra, F., Cattani, M., Bailoni, L. and Schiavon, S., Università degli Studi di Padova, Dipartimento Scienze Animali, Viale della Università, 16, 35020, Legnaro, Italy; franco.tagliapietra@unipd.it

Effects on rumen fermentation processes of high doses of ascorbic acid (AA), α-tocopherol acetate (α-T), and a purified wine marc extract rich in phenols (WME) were studied with a *in vitro* gas production (GP) technique. Meadow hay (MH) and corn meal (CM) were incubated for 48 h at 39 °C with rumen fluid (25 ml) and buffer (50 ml). Each feed (0.5 g) was tested without additive (C) or with the addition of 0.01 g of AA, α-T or WME in 4 replications. A 2nd incubation was stopped at t½, the time on which half of the asymptotic GP of each feed occurred (20 and 11 h for MH and CM, respectively). The fluid of 2nd incubation, treated with methyl-cellulose, was centrifuged: the solid fraction was analysed for total N content and the liquid for total N, NH_3-N and volatile fatty acids (VFA). At t½, the correlations between GP and GP estimated, by stoichiometry, from VFA yields were R^2=0.91 and 0.93, for MH and CM respectively. With respect to C, AA slightly increased the GP at 24 and 48 h ($P<0.05$) but did not affect the amount and the proportions of VFA on both feeds. With respect to C, α-T increased NH_3-N concentration (5.6 vs. 7.7 mg; $P<0.01$) and reduced the amount of N potentially used by the microbes (11.3 vs. 9.2 mg, $P<0.05$) on MH, but not in CM, however GP and VFA were not significantly changed. With respect to C, WME incubated with MH reduced GP (GP at 24 h: 39 vs. 27 ml, $P<0.05$; GP at 48 h: 56 vs. 34 ml, $P<0.05$), while with CM, WME decreased GP at 24 h (120 vs. 106 ml, $P<0.05$) and not at 48 h, consequently t½ was significantly increased (11 vs. 17 h; $P<0.05$). WME significantly reduced VFA yields, too. In conclusion high dosages of AA had no effects on the *in vitro* fermentation processes, while for α-T and WME some substrate-dependent effects on microbial fermentation were observed.

In vitro fermentation of diets incorporating different levels of carob pulp

Gasmi-Boubaker, A.[1], Bergaoui, R.[1], Khaldi, A.[2] and Mosquera-Losada, R.[3], [1]Institut National Agronomique de Tunisie, 43, avenue Charles Nicolle, 1082 Tunis, Tunisia, [2]Institut National des Recherches en Génie Rural, Eaux et Forêts, Rue Hédi Karray, 1082 Tunis, Tunisia, [3]Escuela Polytecnica Superior, Universidade de Santiago de Compostela, Lugo, Lugo, Spain; azizaboubaker@ymail.com

Fermentation characteristics of diets incorporating 0% (D1), 10% (D2), 20% (D3) and 100% (D4) dry matter (DM) of carob pulp were determined in an *in vitro* experiment using cecal contents collected from 4 rabbits. Samples of diets were incubated in glass syringe for 72 h and various fermentation variables were determined. Diets varied in their crude protein and fermentable carbohydrates. Potential gas production ranged from 123 (D3) to 179 (carob pulp) ml/g DM and was similar ($P > 0.05$) for D1, D2 and D3. Low value of pH after 72 h fermentation was observed in D4 (carob pulp: 6.47), and the highest was in diet incorporating 10% of carob pulp (6.66). Of all diets, carob pulp was fermented the most rapidly and had higher ($P < 0.05$) organic matter digestibility (64.3%) than D1 (62%), D2 (60.3%) and D3 (58.6%). *In vivo* experiments are necessary to confirm these results.

The effect of heat treatment on protein quality of soybean

Sudzinová, J., Chrenková, M., Čerešňáková, Z., Mlyneková, Z. and Mihina, Š., Animal Production Research Centre, Department of Nutrition, Hlohovecká 2, 951 41 Lužianky, Slovakia (Slovak Republic); sudzinova@scpv.sk

The main purpose was to determine the quality of proteins in native and heat treated soybean by chemical and biological tests in animals. In the tested feeds we assessed the content of nutrients according to the Decree of the Ministry of Agriculture of the Slovak Republic No. 1497/4/1997-100. Amino acids were determined by ion-exchange chromatography after acid hydrolysis 6-M HCl. Methionine and Cysteine were determined after oxidizing hydrolysis. Effective degradation and degradation parameters were determined by in sacco method on the three rumen fistulated cows with outflow rate of $0.06.h^{-1}$. Intestinal digestibility of by pass protein was determined by method mobile bags on three cows with duodenal T-cannula. Results were evaluated by statistical programme Statistics ($P < 0.01$, $P < 0.05$). Heat treatment had no significant effect in chemical and amino acid composition of soybean. We determined the crude protein solubility of untreated soybean about 50%. The heat treatment caused decrease of crude protein solubility by flaked soybean (to 16.7%), roasted soybean (140 °C) (to 15%) and also by toasted soybean (to 17.6%). Heat treatment (all used methods) reduce effective CP degradability ($P < 0.05$). The most significant decrease we found by toasted soybean (from 83.4% to 66.7%). Heat treatment had significant effect on decrease of soluble fraction 'a' and on increase of insoluble fraction 'b' ($P < 0.05$) by all treated feeds. Intestinal digestibility of by pass protein was kept high (97.6-98.4%).

Effect of folic acid and panthothenic acid supplementation on ruminal metabolism and nutrient flow at the duodenum

Ragaller, V., Lebzien, P., Huether, L. and Flachowsky, G., Friedrich-Loeffler-Institute, Federal Research Institute for Animal Health, Institute of Animal Nutrition, Bundesallee 50, 38116 Braunschweig, Germany; peter.lebzien@fli.bund.de

Folic acid (FA) and pantothenic acid (PA) are both B-complex vitamins. The active form of FA plays an essential role in methionine and DNA metabolism. As a part of Coenzyme A and acyl carrier protein, PA is required in nearly every cell. It seems that FA and PA are the only B-vitamins for which microbial synthesis does not meet the estimated requirements. The objectives of the presented experiments were to study the influence of supplements of FA and PA on ruminal metabolism, with two diets differing in concentrate (C) to forage (F) ratio in cows fistulated at the rumen and the proximal duodenum. In the first experiment, a supplementation of 1 g FA per cow and day, and in the second experiment, a supplementation of 1 g PA per cow and day, were added to diets consisting either of $\frac{2}{3}$ C and $\frac{1}{3}$ F, or $\frac{1}{3}$ C and $\frac{2}{3}$ F. It seems that both vitamins have no influence on ruminal pH and NH_3 concentration and, only minor influences on some short chain fatty acids at both rations. FA at $\frac{1}{3}$ C ration had no influence on fermented organic matter (FOM), apparent NDF digestibility, microbial protein synthesis, protein degradation and uCP. However, at $\frac{2}{3}$ C-ration, FA tended to decrease the proportion of FOM to organic matter intake. Additionally, the amount of microbial protein arriving at the duodenum as well as the efficiency of the microbial protein synthesis significantly decreased due to FA at $\frac{2}{3}$ C-ration. Thus, FA at $\frac{2}{3}$ C-ration may result in less available energy for microbial protein synthesis. Apart from an increased FOM at $\frac{1}{3}$ C-ration and a decreased microbial protein synthesis at $\frac{2}{3}$ C-ration, no parameters were significantly influenced due to PA supplementation. Therefore, it seems that both vitamins had varying effects on ruminal metabolism at the different rations.

The effect of various mechanical processing on *in situ* degradation of sugar beet pulp

Mojtahedi, M., Danesh Mesgaran, M., Heravi Moussavi, A. and Tahmasbi, A., Ferdowsi University of Mashhad, Department of Animal Science, Exellence center for Animal Sicence, P.O. Box: 91775-1163, Mashhad, Iran; danesh@um.ac.ir

The aim of this study was to evaluate the effect of the mechanical processing on *in situ* rumen dry matter (DM) degradation of sugar beet pulp (SBP). Samples were provided as unmolasses wet shred (UWS), molasses dried shred (MDS), molasses pelleted (MPL) and molasses blocked (MBL). Four rumen fistulated Holstein steers (400±12 kg, body weight) were used. Samples were milled (2-mm screen) and weighed (5 g, DM) into bags (12×19 cm) made of polyester cloth with 52 μm pore size (n=4). Bags were incubated in the rumen for 2, 4, 8, 16, 24, 48 and 72 h. A part of bags was washed with cold tap water to estimate the wash-out at zero time. After each rumen incubation, bags were hand-washed, then, dried in a forced-air oven (60 °C, 48 h). The degradable parameters of DM were determined using the equation of $P = a + b (1 - e^{-ct})$; P: potential degradability, a: rapidly degradable fraction, b: slowly degradable fraction, c: fractional degradation rate constant (h^{-1}). Results showed that the *in situ* degradation parameters of DM of SBP were influenced by the processing ($P<0.05$). The fraction of (a) was the highest for MBL sample (UWS = 0.03±0.02, MPL = 0.41±0.02, MDS = 0.42±0.02 and MBL = 0.58±0.01). While, slowly degradable fraction of UWS was higher than the others (UWS = 0.97±0.03, MPL = 0.54±0.02, MDS = 0.54±0.03 and MBL = 0.39±0.02). Molasses blocked and MPL had the highest fraction of (c) of DM ($P<0.05$) compared with MDS and UWS (0.080±0.011, 0.071±0.008, 0.055±0.009 and 0.052±0.009, respectively).

In situ dry matter and crud protein degradability of raw and steamed Iranian *Lathyrus sativus* seed
Golizadeh, M., Riasi, A., Fathi, M.H. and Allahrasani, A., University of Birjand, Department of Animal Science, Faculty of Agriculture, 97175/331, Birjand, Iran; Riasi2008@gmail.com

Lathyrus sativus is a plant that typically grown in tropical and subtropical regions.Grass pea seeds have high protein content and its starch can supply abundant energy.An experiment was conducted to test the *in situ* dry matter (DM) and crude protein (CP) degradability of raw and steamed *Lathyrus sativus* seed (10, 20, and 30 min). Raw and steamed samples were ground topass through a 2-mm screen and weighed (1.5g DM) into the polyester bags (6*10 cm, 50 µm pore size).The bags were incubated in the rumen of fistulated cows (450±11 kg) for 0, 2, 4, 8, 16, 24 and 48 hours. After removal from the rumen, the bags were washed using cold tab water and then dried in an oven (70 °C for 48 h). The equation of $p = a + b(1 - e^{-ct})$ was used for determination of degradability parameters. The results indicated that dry matter and crude protein slowly degradable fraction of raw samples (0.545±0.028 and 0.695±0.025, respectively) was higher than those the steamed samples, however 20 min steamed samples had the highest slowly degradable fraction (b) of DM and CP (0.471±0.127 and 0.466±0.18, respectively) and its degradation constant rate (c) was lower than the other treatments. It was concluded that the steaming for 20 min may have beneficial effects on DM and CP degradability of *Lathyrus sativus* seed.

Effect of fungal treatment of linseed hulls on chemical composition, nitrogen fraction and protein solubility
Abedo, A.A.[1], Sobhy, H.M.[2], Mikhail, W.A.[3], Abo-Donia, F.M.[4], Fadel, M.[5] and El-Gamal, K.M.[6], [1]National Research Center, Animal Production Dep., Dokki, Giza, Egypt, [2]Institute of African Research and Students, Natural Research Dep., Cairo Univ., Giza, Egypt, [3]Institute of African Research and Students, Natural Research Dep., Cairo Univ., Giza, Egypt, [4]Animal Production Research Institute, By-Product Utilization Dep., Dokki, Giza, Egypt, [5]National Research Center, Microbial Chem.Dep, Dokki, Giza, Egypt, [6]Ministry of Agric., Technical Office, Dokki, Giza, Egypt; abedoaa@hotmai.com

Linseed hulls (LSH) in the first trial was treated with three species of *Trichoderma* (T); *T. viride, T. harzianum* and *T. reesei*. In the second trial LSH was fermented with *T. harzianum* for 10 dads. The results of the first trial showed that *T. harzianum* was the proper fungi, wherease the CP content was increased to 11.67 compared with 8.84, 9.16 and 7.98%, while the CF was decreased to 32.05% compared with 36.08, 33.43 and 43.40% for treated LSH with *T. viride, T. reesei* and untreated LSH. The results of the second trial indicated that 9 days was the suitable treatment period, since the highest value of CP (12.59%) and the lowest value of CF (33.13%) were achieved compared with 7.98 and 43.40% for untreated LSH, respectively. Total N content was gradually increased (from 1.33 to 2.01%); also true protein N was increased from 58.65 to 80.10% as a percentage of total N in the first day and in the ninth day of treatment. While Protein solubility in sodium chloride and in pepsin was gradually decreased from 65.78 to 24.37 and from 84.44 to 40.03 in the first day and in the ninth day of treatment. It could be concluded that *T. harzianum* was the proper fungi for fungal treatment of LSH and 9 days was the suitable fermentation period.

Comparison of grass species influence on dry matter degradability and its prediction using chemical composition

Jančík, F., Homolka, P. and Koukolová, V., Institute of Animal Science, Nutrition and feeding of farm animals, Přátelství 815, 104 00, Praha Uhříněves, Czech Republic; jancik.filip@vuzv.cz

The objectives of this experiment was to compare the most widely used grass species (*Dactylis glomerata* L., *Phleum pratense* L., *Lolium perene* L., *Festuca arundinacea* L. and hybrid Felina) conserved by ensiling process. These silages were compared according to dry matter rumen degradability parameters. The regression equations for prediction of effective dry matter rumen degradability (ED_{DM}) of grass silages were based on chemical composition of estimated samples. Tested silages made from different grass species were compared among each other according to dry matter (DM) rumen degradability parameters (a = portion of DM solubilized at initiation of incubation, b = fraction of DM potentially degradable in the rumen, c = rate constant of disappearance of fraction b and ED_{DM} = effective degradability of DM, estimated for each ingredient assuming rumen solid outflow rates of 0.02 (ED_{DM2}), 0.05 (ED_{DM5}) and 0.08 (ED_{DM8}) h^{-1}). The best values of ED_{DM} were determined for Lolium perenne (ED_{DM2} = 753.2, ED_{DM5} = 631.1 and ED_{DM8} = 567.7 g/kg DM). The best predictor was NDF (R^2-values of 0.757 (ED_{DM2}), 0.863 (ED_{DM5}) and 0.906 (ED_{DM8})). Using two predictors was increased accuracy level. Combination of CF and NDF gave R^2-values 0.892, 0.920 and 0.929 for ED_{DM2}, ED_{DM5} and ED_{DM8}, respectively. The regression equations based on the most important grass species harvested at different vegetation period seems to be a useful tool for practical use. There was not found significant ($P<0.05$) impact of ensiling process in relation to dry matter rumen degradability parameters. This research was supported by the Ministry of Agriculture (grants MZE 0002701403 and MZE 0002701404).

Yeast cultures in ruminant nutrition

Denev, S.A.[1], Peeva, T.Z.[2], Radulova, P.[3], Stancheva, N.[2], Staykova, G.[2], Beev, G.[1], Todorova, P.[3] and Tchobanova, S.[1], [1]Trakia University, Agricultural Faculty, Student Campus, 6000 Stara Zagora, Bulgaria, [2]Agricultural Institute, Large & Small Ruminants, 3 Simeon Veliki Str., 9700 Shumen, Bulgaria, [3]Research Institute of Mountain Stockbreeding and Agriculture, Animal Nutrition, 283 Vasil Levski Str., 5600 Troyan, Bulgaria; stefandenev@hotmail.com

Interest in the use of fungal direct-fed microbials in ruminant nutrition is considerable. The ban of antibiotic growth promoters in feed for production of animal foods has increased interest in evaluating the effect of yeast cultures (YC) on the gastrointestinal ecosystem (GIE), rumen microbial populations and function. The effects of specific YC preparations on the rumen environment and performance of ruminants have been well documented, and has generated considerable scientific interest over the last two decades. The precise mode of action by which YC,which are mostly derived from Saccharomyces cerevisiae,improve livestock performance has attracted the attention of a number of researchers in the world. It is clear from these research efforts that YC supplements can beneficially modify microbial activities, fermentative and digestive functions in the rumen. The research has demonstrated that viable YC preparations can stimulate specific groups of beneficial bacteria in the rumen, and has provided mechanistic models that can explain their effects on animal performance. The effects of YC on animal productivity are strain-dependant. So, all YC preparations are not equivalent in efficiency. This aspect opens a new field of research for new strains, each being more specialized in its use. The goal of many of these research activities has been to define the application and production strategies that can optimize animal responses to YC. Continuous research with live YC has clearly established scientifically-proven strategies for modifying and optimizing microbial activities in the GIE and techniques for improving performance and health of ruminants. This article reviews the current status of the use of live yeast cultures in ruminant nutrition.

Effect of a dietary sulfur amino acid deficiency on the amino acid composition of body proteins in piglets

Conde-Aguilera, J.A.[1], Barea, R.[1] and Van Milgen, J.[2], [1]Institute of Animal Nutrition, Estación Experimental del Zaidín (CSIC), Cno. del Jueves s/n, 18100, Armilla, Granada, Spain, [2]INRA, UMR 1079, Domaine de la Prise, 35590, Saint-Gilles, France; roberto.barea@eez.csic.es

The sulfur containing amino acids (AA), methionine (Met) and cysteine (Cys) are often considered to be the second or third limiting AA in nursery pigs. The concept of ideal protein relies on the assumption of a constant AA composition of protein gain. The objective of this study was to test the response of piglets to diets differing in total sulfur AA supply on the AA composition of body protein and the characterization and quantification of the main muscular proteins in piglets. Two diets (based on wheat, peas, soybean meal and corn starch) were formulated differing in Met supply (0.20 and 0.41% Met on a standardized ileal digestible basis) and total sulfur AA (0.45 and 0.70% Met+Cys). A total of 18 piglets (6 blocks of 3 pigs each) at 42 d of age were used in a comparative slaughter study. After slaughter, the whole animal was divided into five body components (carcass, blood, intestines, carcass, liver, and remaining parts). Samples of *Longissimus dorsi* (LD) were analyzed for collagen, actin and myosin contents. A deficient supply of sulfur AA resulted in decrease in the Met content of proteins in the whole body, blood, LD muscle, and remaining parts. The Cys content was lower in blood protein and tended to be lower in liver protein. A deficient supply of sulfur AA also reduced the Met content in protein gain in the whole body and LD muscle and tended to decrease the Met content of carcass protein gain. The myosin and actin content (per 16 g N) was reduced with age but was not affected by the sulfur AA content in the diet. The myosin to actin ratio was not affected either. In conclusion, the AA supply appears to affect the AA composition of body protein and different body proteins are affected to a different extent. This questions the use of a constant ideal amino profile, but also illustrates the plasticity of the animal to cope with nutritional challenges.

Effect of raw material for distillers grains production on protein quality for ruminants

Chrenková, M., Čerešňáková, Z., Sudzinová, J., Fľak, P., Poláčiková, M. and Mlyneková, Z., Animal Production Research Centre, Nutrition, Hlohovecká 2, 951 41 Lužianky, Slovakia (Slovak Republic); chrenkova@scpv.sk

The main attention regarding the use of distillers grains (DG) as animal feedstuffs dedicate on stability in nutritional characteristics and quality of the product. Our experiment was focused on quality of DDGS produced from corn (CDDGS,n=8), wheat (W,n=6) triticale (T,n=1), and wet corn DGS (CWDG,n=7) for ruminants. The products were tested for CP, NDF, ADF, NNDF, NADF, AA profile, CP degradability and intestinal digestibility (ID) of by-pass CP. DG are a relatively high-protein feed with average content of CP (g/kgDM) in CDDGS 289.9, WDDGS 356.9, TDDGS 332.4 and CWDGS 302.0 ($P<0.01$). We found the highest content of essential AA in CWDGS (133.23 g/kgDM or 440.1 g/160gN, Lys 29.33 g/160gN and Met 20.47 g/160gN) followed with CDDGS (122.5 g/kgDM or 413.61 g/160gN, Lys 21.59 g/160gN, Met 19.07 g/160gN). The most variable was the content of Lys in WDDGS (from 13.69 to 22.79 g/160gN). In all samples of DDGS and WDGS was very high and different ($P<0.01$). Content of NDF and ADF in g/kg DM was very high (average of 383.4 and 179.8 in corn DDGS, 542.5 and 211.8 corn WDGS, 367.7 and 223.7 WDDGS, 386.2 and 268.2 TDDGS, respectively). We found for wheat DDGS from 16.8 to 36.2% of CP bound in ADF. The lowest share of N was in the ADF in CWDGS (from 2.8 to 6.6% of total N). With high percentage of N bound in ADF is affected ruminally degradable CP and ID of by-pass protein in W and TDDGS that are the lowest from all tested types of DG. Effective CP degradability of WDDGS with the highest NADF (36.2%) was only 50.2% and ID of by –pass protein 82.2%, total essential AA 81.2%, Lys 77.69% and Met 81.4%. The ruminally degradable CP fraction was in the range of 52.6 to 70.2% and the rate of CP degradation from 0.031 to 0.054/h for CDDGS. ID of total and essential by-pass AA was very high 96.2% and 96.1%, Lysine 90.9% and Methionine 96.7%. From these data it can be concluded that DDGS from grains are a good source of protein for ruminants.

Impact of drying on *in vitro* gas production of rice straw
Santos, M.[1,2], Nader, G.[1], Robinson, P.H.[1], Gomes, M.J.[2] and Juchem, S.O.[1], [1]UCCE, University of California, Davis, 95616 Davis, USA, [2]CECAV-UTAD, Vila Real, 5001-801, Portugal; marta.bl.santos@gmail.com

In vitro gas production was measured to determine effects of field drying on chemical composition and fermentation characteristics of rice straw.Rice plants were sampled from the field 14, 10, 6 and 2 days before(stage pre-harvest,PRE) and at harvest, the seeds and senesced flag leaves were removed. Plants collected at harvest were stored under cover in simulated windrows in a naturally ventilated area and sampled on days 1, 2, 3, 4 (stage dry down, DD), 6, 8, 12, 19 and 33 (stage dry, D) relative to harvest. Seeds and senesced flag leaves were removed. Cumulative gas production was recorded at 1, 2, 3, 4, 5, 6, 12, 18, 19, 20, 22, 24, 26, 28, 50 and 72 h of incubation and kinetics of gas production (Y; ml) was described as $Y = b(1-e^{-k * t})$, where b is potential gas production (ml) and k is the rate (/h) at which gas is produced. There was a small decrease ($P<0.05$) on crude protein content during dry down (43.0 to 39.0 g/kg of DM), whereas fibre fractions (i.e., NDF, ADF and lignin(sa)) were not affected by drying. Gas production was higher ($P<0.05$) before harvest, with 6.8, 15.6 and 8.9% more gas relative to D stage at 4, 24 and 72 h of fermentation respectively. Potential gas production also declined from PRE to DD stage by 5.7%, but the rate of gas production was not affected. These findings clearly show thatfield drying has a negative impact on rate and extent of fermentation of rice straw. Since the fibre content of rice straw was not affected by stage relative to harvest,it appears that straw undergoes changes during drying that are not identified by traditional chemical assays, which reduce its fermentability.More research is needed to understand these changes, and to define strategies to prevent this loss in nutritive value.

Energy expenditure of splanchnic tissues relative to the total in Iberian growing gilts
Lachica, M., Rodríguez-López, J.M., González-Valero, L. and Fernández-Fígares, I., CSIC, Animal Nutrition, Profesor Albareda 1, 18008, Spain; manuel.lachica@eez.csic.es

The aim of this work was to determine splanchnic tissues heat production (HP) relative to total HP of Iberian growing gilts. Two trials were carried out with six female gilts (29.3 kg average BW). Three catheters were placed in each pig: in carotid artery and portal vein for blood sampling, and in mesenteric vein for para-aminohippuric acid (PAH) infusion to measure blood flow. Pigs were on metabolic cages and fed at 85% *ad libitum* (160 g/kg CP and 14-14.5 MJ EM/kg DM). First trial began when animals were recovered of surgery. Forty five min prior blood sampling a 15 ml pulse dose of PAH (2%) was infused into mesenteric vein, followed by continuous infusion of 0.8 ml/min. A 4.5 ml blood sample was anaerobically taken simultaneously from carotid artery and portal vein 0.5, 1, 1.5, 2, 2.5, 3, 3.5, 4, 5 and 6 h after feeding 25% of total daily ration, into an heparinized tube for instantaneous measurement of O_2 concentration and saturation of haemoglobin (by an hemoximeter). The rest of blood was centrifuged for PCV and plasma harvest and stored at -20 °C until PAH analysis. Whole-blood flow and O_2 consumption rates were based on Fick principle. The energy equivalent for O_2 was 20.4 kJ/l. For the second trial, pigs were individually moved on its own metabolic cage to a respirometry chamber. After 18 h of adaptation, total HP was measured over 6 h following the same protocol of first trial by physic principles and from O_2 consumption and CO_2 production. The HP of splanchnic tissues and whole body was 194.0 ± 5.8 and 785.8 ± 31.4 kJ/kg$^{-0.75}$ per day, respectively; this represents 25% of the total HP. In conclusion, the use of arterio-venous preparations is a suitable technique for Iberian pigs and permit to evaluate the relative importance of splanchnic tissues on total animal energy expenditure and could explain differences in digestive and absorptive capacities relative to modern breeds.

An individual double choice test to study preferences for nutrients and other tastants in weaned pigs

Tedo, G.[1], Ruiz-De La Torre, J.L.[2], Manteca, X.[2], Colom, C.[1] and Roura, E.[1], [1]Lucta SA, R&D Feed Additives Division, Ctra.Masnou a Granollers, 08170 Montornès del Vallès, Spain, [2]Universitat Autònoma de Barcelona, Animal Science, Veterinary Faculty, 08193 Cerdanyola del Vallès, Spain; gemma.tedo@lucta.es

Multiple choice models to assess preferences in drinking water have been used to evaluate palatability of nutrients and other tastants in many animal species. Pig models include training sessions of 3-5 days before conducting experimental tests where animals had been previously housed individually during long periods of time to get them adapted to isolation. The aim of this study was to develop a model for double choice testing in piglets avoiding the isolation time prior to training. Eighty two pigs (12 ± 2 kg BW, mixed sexes) were maintained in groups and housed individually only during the training and test sessions. During 4 consecutive days, animals were under an operant conditioning scheme of two 10-minute training sessions (9 am and 12 pm) where the double choice test was between water and a 500mM sucrose solution (250g each). On the last training day, 40 animals showing no attempts to escape, a latency time to try both liquids of less than 30s, no vocalizations and 100g of minimum consumption were selected for the 2-minute experimental double-choices between a test solution and water. Test solutions were water (negative control) or 500mM sucrose (positive control) or 500mM Monosodium Glutamate (MSG). Preferences (in % -P-) and consumption rates (in g/s –CR-) were recorded. Both P and CR for MSG were higher than the negative control ($P<0.05$) and similar to the positive control ($P>0.05$). No differences were detected between water and the negative control (water). Results showed that avoiding an adaptation period to social isolation might be adequate if animals are previously selected based on their behaviour and consumption. It is concluded that a 2-minute double-choice model following a 4-day training period might be effective in detecting differences among nutrients or other tastants.

Comparison of splanchnic tissues mass of two sheep breeds offered two diets differing in roughage level

Gomes, M.J., Silva, S., Mena, E., Azevedo, J.M.T., Lourenço, A. and Dias-Da-Silva, A., UTAD-CECAV, P.O. Box 1013, 5000-801 Vila Real, Portugal; analou@utad.pt

It is a general believe that the local sheep breed Churra da Terra Quente (CTQ), reared in the Northeast of Portugal is well adapted to the conditions of its production system. However, there are large gaps on our knowledge to allow understanding of such claimed adaptation. Splanchnic tissues (ST) can account for 35-50% of body energy expenditure, this high demand for energy being positively related with the protein and cell turnover, which in turn allows adaptability and flexibility so that animals can respond to multiple physiological and environmental challenges. The objective of the present study was to compare ST weights of lambs from CTQ and from an exotic breed (Ile de France, IF) at the same mature body weight (MBW) and fed two diets differing in roughage level. The study was carried out with 15 female lambs from each of the two breeds. Initial body weight was 20.7 ± 1.3 and 34.0 ± 0.7 kg, for CTQ and IF respectively, corresponding to ca 45% of their MBW. Animals were assigned to be slaughtered as an initial group or when they reached ca 65% of their MBW, after being fed ad libitum a high (HR, 70% of meadow hay) or a low-roughage diet (LR, 25% of meadow hay), according to a 2x2 factorial design. ST were removed and weighed. Per kg empty BW, CTQ had higher weights of small intestine, large intestine, liver and total ST ($P<0.05$), the weights of rumen-reticulum and omasum being higher only at 65% of MBW ($P<0.01$). Weights of the rumen-reticulum, omasum, large intestine and total ST were higher ($P<0.001$) in sheep fed HR diet irrespective of breed, but no differences were observed for the weight of abomasum, small intestine and liver ($P<0.05$). This study confirms the existence of differences on relative weights of ST between sheep breeds at the same level of maturity and between animals fed diets with different roughage levels. In conclusion, the high ST weights of the local breed may lead to enhanced adaptation ability to its environment.

Ruminal dry matter degradation of sodium hydroxide treated cottonseed hulls using *in situ* technique
Faramarzi-Garmroodi, A., Danesh Mesgaran, M., Jahani-Azizabadi, H., Vakili, A.R., Tahmasbi, A. and Heravi Moussavi, A.R., Ferdowsi University of Mashhad, Dept. Animal Science (Excellence Center for Animal Science), P.O. Box 91775-1163, Mashhad, Iran; Danesh@um.ac.ir

The objective of this study was to determine the effect of alkali treating on ruminal dry matter (DM) degradation of cottonseed hulls using *in situ* technique. Cottonseed hulls were treated with NaOH as 20 or 40 g/kg DM [a 20% or 40% solution of NaOH was sprayed on CSH and kept for 0.5 h (CSH2Na0.5 and CSH4Na0.5, respectively), or 48 h (CSH2Na48 and CSH4Na48, respectively) at room temperature. Then, samples were dried using air-forced oven (60 °C). Four sheep (44±5 kg body weight) fitted with rumen fistulae were used. Bags (17×12 cm) were made of polyester cloth with pore size of 52 μm. About 5 g DM of each sample was placed in each bag, then incubated (n=4) for each time (2, 4, 6, 8, 12, 16, 24, 48, 72, 96 and 120 h). For zero time, bags were washed using cold tap water. The equation of $P = a + b(1-e-ct)$ was applied to determine the degradation coefficients (a= quickly degradable fraction, b= slowly degradable fraction, c= constant rate of fractional degradation). Dry matter degradation parameters (a, b and c) of the samples were: CSH2Na0.5=0.03±0.01, 0.50±0.06, 0.018±0.005; CSH2Na48=0.08±0.014, 0.50±0.08, 0.013±0.004; CSH4Na0.5=0.057±0.015, 0.50±0.05, 0.018±0.004; CSH4Na48=0.09±0.011, 0.49±0.06, 0.013±0.004, respectively. It was concluded that the spray-treated of CSH with NaOH solution for 48 h caused to increase the (a) fraction. However, there was no significant effect of the spraying time on b and c fractions.

Protein digestibility of different varieties of white lupine
Homolka, P., Jančík, F. and Koukolová, V., Institute of Animal Science, Department of Nutrition and Feeding of Farm Animals, Přátelství 815, 104 00 Prague, Czech Republic; homolka.petr@vuzv.cz

The objective of this study was to determine the intestinal digestibility (DSI) of rumen undergraded protein in different varieties of white lupine (Amiga, Butan, Dieta). The digestibility profiles of protein and individual amino acids (AA) were evaluated using mobile bag technique in three rumen and T-piece duodenal cannulated cows. The daily diet per cow was based on 4 kg of alfalfa hay, 1 kg of barley meal with 100 g of vitamin and mineral supplement, and water was available ad libitum. Concentration of dry matter, crude protein, ether extract, crude fibre, neutral detergent fibre, acid detergent fibre, ash and individual essential AA (arginine, histidine, isoleucine, leucine, lysine, methionine, phenylalanine, threonine, valine) and non-essential AA (alanine, aspartate, cysteine, glutamate, glycine, proline, serine, tyrosine) in lupines was determined. DSI of rumen undergraded protein was 71, 84 and 76% for Amiga, Butan and Dieta, respectively. DSI of essential and nonessential AA was found to be 86% and 83% (Amiga), 87% and 85% (Butan), and 84% and 82% (Dieta), respectively. The significant differences among the estimated varieties of lupines ($P < 0.05$) were declared for DSI of protein (Butan vs Dieta, Amiga vs Butan) and subsequently for glutamate (Butan vs Dieta) and proline (Amiga vs Butan). The study was supported by the Ministry of Agriculture of the Czech Republic (NAZV QG 60142).

Impact of heat treatment on rumen degradability of barley grain

Nemati, Z. and Tagizadeh, A., University of Tabriz, Animal Science, Tabriz, 56166, Iran; taghius@yahoo.com

Impact of heat treatment on rumen degradability of barley grain This study carried out to evaluate the effects of Autoclaving treatment and treatment time on *in situ* rumen degradation of protein in barley' grains. The experimental feedstuff was barley grain. The feedstuff samples were ground on hammer mill through a 2 mm screen and a sample was taken as untreated Barley grain. After milled other samples heated using autoclave and treatment named Untreated Barley Grain (UBG), treated barley grain at 120 °C, (5') (TBG1) and 20' (TBG2), treated barley grain at 100 °C (5') (TBG3) and 20' (TBG4). In all samples, nitrogen was determined as described by AOAC. Ruminal degradation measurements were carried out using *in situ* methods mainly as described by Madsen *et al.*, and Prestl¢kken (1999). Three fistulated Gizel sheep (37±3.5 kg) were used. The sheep fed diet content 40% alfalfa: 60% concentrate containing 2.9 Mcal kg-1 DM and 14% CP. The nylon bags containing 5 g of samples were incubated into rumen at, 0, 4, 8, 16, 24 and 48 h. The crude protein data were fitted to the exponential equation $Y(t) = a + b - 1 - e^{-ct}$), where $Y(t)$ is degraded proportion at time t, a the water soluble and momentary degradable fraction, b not water soluble, but potentially rumen degradable fraction, and c is (/h) the fractional rate of degradation of fraction b. The crude protein disappearance data were analyzed using statistical Analysis System (2001). Rumen crude protein degradation of treatments (%), UBG, TBG1, TBG2, TBG3 and TBG4 at 48 h after incubation were 59.5, 40.9, 51.6, 46.5 and 55.5 respectively. Treatment of barley grain by heat decreased CP degradability resulting increased escaped CP into lower digestion tract in sheep. Effects of processing barley grain have been extensively evaluated for cattle, but little is known for the effects on productivity of sheep's and this warrants further investigation.

Kinetics of fermentation of straw from different varieties of chick peas

Kafilzadeh, F. and Maleki, E., Razi University, Animal Science, School of Agriculture, Kermanshah, 6719685416, Iran; kafilzadeh@razi.ac.ir

Crop residues provide much of the feed resources for ruminants in developing countries. The feed value of crop residues has not received enough attention in the development of crop varieties. Gas production technique was used to investigate the extent of diversity of degradation profile of straw from different varieties of chickpeas. Straw from five different varieties of Kabuli chickpeas cultivated under similar agronomic condition were used. Varieties were ILC482 (V_1), Arman (V_2), Hashem (V_3), FLIP9393 (V_4) and Bivanij (V_5). The mean values for CP, NDF and ADF across varieties were 31.6±3.4, 620±21.5 and 465.5±3.8 respectively. There were significant ($P<0.05$) differences between potential gas production (b) of straws from different chickpea varieties. The lowest b constant was in V_5 (27.20 ml/200mg) and the highest b was found in V_4 (37.85 ml/200 mg). Rate of gas production (c) was not affected by variety. However the effective degradability of straws estimated at 4% outflow rate was significantly lower in V_5 compared with the other varieties.

Removal of tannins can cause improvements in ruminal cell wall degradation and total tract apparent NDF digestibility of sainfoin (*Onobrychis vicifolia*)

Rezayazdi, K., Khalilvandi-Behroozyar, H. and Dehghan-Banadaki, M., University of Tehran, Animal science department, Karaj, Tehran, 3158777871, Iran; khalilvandi@ut.ac.ir

Sainfoin is temperate legume forage, with medium to high concentrations of condensed tannins (from 25 to 100 g/kg dry matter). The objective of this study was to examine the effectiveness of tannin deactivation on the availability of cell wall. Second cut forage were chopped and exposed to 5% solution of polyethylene glycol (PEG 6000 MW) with v/w ratio of 1:1 and soaked with tap water (v/w ratio of 4:1). Condensed tannin determined with butanol-HCl reagent. Degradability determined using three ruminally fistulated Holstein cows. Animal were fed balanced diets formulated using CNCPS V5 to meet 110% of maintenance requirements with forage: concentrate ratios of 60:40. Samples were ground to pass 2 mm screen and 5 g were weighed into nylon bags with 50 μm pore size to create sample size: surface area of 12.5 mg/cm^2. Duplicates were incubated for 4, 8, 12, 24, 48, 72 and 96h in the ventral rumen. The effective degradability (ED) was calculated using NEWAY computer package. Animals and forages, as described above, were used in an *in vivo* digestibility trial, using 3*3 change over design. Each period consisted of 10 days for adaptation and 7 days for sample collection. Forages were fed as the sole diet. Acid insoluble ash used as digestibility marker. CRD design, SAS 9.1, GLM procedure and Duncan Test option was used for data analysis. The condensed tannin concentration of the control forage was 21.3±0.4 g/kg DM. Tannin deactivation led to over 90% reduction in CT content and an increased 'a' fraction compared with control. Differences between treatments were not significant in the case of 'b' fraction and rate of degradation of it. Effective degradability in K=0.05 were 27.10^c±2.6, 40.93^a±0.6, 36.47^b±2.6, and total tract NDF digestibility were 49.25^b±3.1, 60.18^a±5.8, 58.44^{ab}±3.2, for Control, PEG and Water soaking, respectively.

Effects of zinc supplementation on growth performance, blood metabolites and lameness in Holstein male calves

Fagari-Nobijari, H.[1], Amanlo, H.[1] and Dehghan-Banadaky, M.[2], [1]Zanjan university, Animal Science, Zanjan, 1362 AR, Iran, [2]University of Tehran, Animal Science, Karaj, 1360 AR, Iran; hadifakari@gmail.com

106 Holstein male calves (initial body weight, 376.7±18.3 kg) were randomly allocated in two treatments in a completely randomized design for 60 days. Treatments consisted of 1) basal diet with no supplemental Zn (control), 2) basal diet plus 150 mg of Zn/kg of DM as $ZnSO_4$. Calves received fresh total mixed ration (TMR) (13.2% DM CP and 29.25% DM NDF) at 900, 1200, 1500 as ad libitum feeding allowing 10% orts. Group dry matter intakes were measured daily and individual body weight changes were recorded monthly. Blood samples were taken on day 56 (2 h after morning meal). Average daily gain (ADG) ($P<0.01$) and feed efficiency (F:G) ($P<0.01$) were decreased by supplementation of $ZnSO_4$. However, dry matter intake was not affected ($P=0.2938$). Plasma total protein ($P<0.01$), urea nitrogen ($P<0.01$), albumin ($P<0.01$) and Zn serum concentration were higher in calves fed $ZnSO_4$. Cholesterol serum concentration was higher in control group ($P<0.01$). The prevalence of lameness was higher in control group than $ZnSo_4$ supplemented treatment (28% vs. 12.5%; odds ratio = 2.7). These data suggest that high levels of supplemental Zn negatively impact ADG, F:G, however, it can decrease prevalence of lameness in male calves.

Determination of nutritive value of Iran fruit and vegetable residues in autumn using *in vitro* and *in vivo* techniques

Karkoodi, K.[1], Fazaeli, F.[2] and Ebrahimnezhad, Y.[3], [1]Islamic Azad University, Saveh Branch, Department of Animal Science, 39187/366 Saveh, Iran, [2]State Animal Science Research Institute, 3195683491 Karaj, Iran, [3]Islamic Azad University, Shabestar Branch, Department of Animal Science, 5381637181 shabestar, Iran; kkarkoodi@yahoo.com

This study was conducted to evaluate nutritive value of Iran fruit and vegetable residues in autumn. Samples of fruit and vegetables residues were collected during 3 months of autumn every other week per month. After separation of exogenous materials, all samples were dried and prepared for chemical analysis and *in vitro* assay. Data were analyzed using ANOVA in a CRD based statistical design in 3 replicates and mean values were tested using Duncan's least significant range test. The means of DM, CP, Ash, CF, EE, NFE, NFC, NDF, ADF, ADL, Ca and P were 11.11, 13.3, 26.05, 12.3, 1.76, 46.57, 25.97, 32.90, 21.33, 3.66, 1.81 and 0.32 percent respectively and the means of Cu, Zn, Pb were 15.49, 91.41 and 14.34 mg/kg respectively and the average of GE was 3152.38 kcal/kg. Except for DM, ADL, Cu and Pb, significant differences were observed for the chemical composition among the months of sampling. The means of *in vitro* digestibility for DMD, OMD and DOMD were 68.73, 72.61 and 53.67 percent respectively that were not significantly different between the months. The means of *in vivo* digestibility for DM, OM, CP, CF, EE, NFE, TDN, and ME were 58.9, 73.63, 67.29, 59.2, 53.21, 80.08, 55.16% and 1941.32 kcal/kg respectively. It is obvious that nutritive value of dried fruit and vegetable residues in autumn may be comparable with medium quality alfalfa, except for the higher ash content in fruit and vegetable residues.

Performance enhancement by using acidifiers in poultry diets

Urbaityte, R., Roth, N. and Acosta Aragon, Y., Biomin, Industriestrasse 21, 3130 Herzogenburg, Austria; yunior.acostaaragon@biomin.net

Performance improvements through supplementation of diets with organic acids, as well as their benefits in assisting in disease prevention have been observed. The extent to which organic acids effect growth performance of animal depends on the type and inclusion level of acids used, as well as diet, animal and environment factors. In several studies the growth promoting and antimicrobial efficacy of formic (FO), propionic (PR) acids blend in a ration 1:1 or FO, PR and lactic (LAC) acids blend in a ratio 1:1:0.3 were tested. The enhancement of animal growth performance is achieved by inhibition of microbial growth and acidification of the feed following by acidification of the stomach and the upper intestinal tract. The present study showed that a blend consisting of FO-PR-LAC acids at inclusion levels of 0.3 and 0.5% reduced the pH-value by 3.4 and 5.8%, respectively, and the buffering capacity (BC) by 20.6 and 31.7%, respectively, in commercial poultry grower diets. The pH and BC reducing effect in feed was directly related with the inclusion level of acidifiers and the diet composition. Many authors showed the inhibition activity of FO, PR and LAC acids and their blends against Enterobacteriaceae. In the present study the dietary supplementation with FO-PR acids blend at an inclusion of 0.3% reduced *E. coli* counts by 10% in fresh broiler ceacal samples compared to those in the control group. The FO-PR-LAC acids blend at an inclusion level 0.2% in drinking water reduced the coliforms and *E. coli* counts in broiler faecal samples by 41 and 21%, respectively, compared to those in the control group. In several *in vivo* studies acidifiers consisting of FO, PR and LAC acids on average improved feed conversion rate and decreased mortality up to 8 and 50%, respectively, compared to broilers of the control group. The results of the various studies showed that acidification of poultry diets with FO-PR or FO-PR-LAC acids blends improves feed stability due to decreased pH, buffer capacity and microbial loads in the diets, thus enhances growth performance in poultry.

Effect of tannin deactivation on *in vitro* digestibility of grass pea (*Lathyrus sativus*)
Vahdani, N., Moravej, H. and Dehgan-Banadaki, M., University of Tehran, Animal science, Karaj, Tehran, 3413966733, Iran; hmoraveg@ut.ac.ir

Legumes such as *Lathyrus* sp. contain a variety of anti-nutritional factors (ANFs) which hinder free nutritional utilization in animals. Condensed tannin (CT) is one of these ANFs which can affect on digestibility. So tannin deactivation by some treatments can improve the nutritive value of this forage. Detanning process applied, involved 0.03 M solution of $KMnO_4$, 0.05 M solution of NaOH and 0.1 M solution of $NaHCO_3$, Wood Ash (180 g/lit) and water soaking (The ratio of forage to reagent volume was kept at 1:4 (W/V)). The amount of polyethylene glycol (PEG) (50g/1 kg of DM) was based on a kg of DM. In Urea treatment (20g/ 100 ml/1 kg of DM) sample was stored in a plastic bag to create anaerobic conditions for one week. *In vitro* digestibility of grass pea was determined in 2 steps. Samples were incubated in triplicate with three jars containing only rumen liquor. DM and OM digestibility were corrected for the blank. 0.5 mg of sample (milled through 1 mm sieve), was incubated with rumen fluid and pepsin in 2 steps. After 96 h, digestion residue was filtered and then dried at 105 °C overnight and igniting in muffle furnace at 525 °C for 4.5 h, for determined DM and OM digestibility, respectively. The present study results have shown that, in NaOH treatment, in spite of low reduction in CT content (7.32), dry matter digestibility (DMD) was highest (87.48) in comparision with others. DMD has improved by urea (86.42) and $NaHCO_3$ (85.48), compared to control. The highest amount of organic matter digestibility (OMD) was achieved in $NaHCO_3$ treatment. Tannin deactivation (%) in these treatments were, 85.15 and 98.8 (%) for urea and $NaHCO_3$, respectively. In conclusion, small amount of sample in *in vitro* digestibility trial, can not show the effect of CT on digestibility, in forages containing ANFs. So the estimated digestibility by this method must be compare to another methods to suppose one of them.

Comparision of *in situ* crude protein degradability and gas production of grass pea (*Lathyrus sativus*)
Vahdani, N., Rezayazdi, K. and Moravej, H., University of Tehran, Animal Science, Karaj, 3413966733 Tehran, Iran; hmoraveg@ut.ac.ir

Grass pea is a forage legume that contains several antinuritonal factors that decrease nutritive value of grass pea.The present study aims at development of treatment of this compounds. Chopped forage was treated with different chemical solutions. The ratio of forage to reagent volume was kept at 1:4 (W/V) for 0.05 M solution of NaOH and 0.1 M solution of NaHCO3, Wood Ash (180 g/l) and water soaking. The amount of polyethylene glycol (PEG) (50g/1 kg of DM) was based on a kg of DM. The treated forages dried for 48 h at 40 °C in a forced air Oven. Three ruminally cannulated sheep were used to determine the degradability of samples by nylon bag techniques.Dacron bags with 40-45 μm pore size were filled with 3 g of sample, in duplicate, and incubated for 4,8,12,24,48,72 and 96 h in the rumen of each sheep. Four rumen fistulated Taleshi cows were used to obtain rumen fluid for *in vitro* gas production trial.Rumen fluid was incubated alone as a blank, or with 200 mg dry weight of sample, in triplicate. Rate and extent of gas production was determined by reading gas volumes before incubation at 2, 4, 6, 8, 12, 24, 48, 72 and 96 h of incubation time. Results showed that water soaking and PEG treated forage improved digradability of crude protein, but it was not significantly in all of incubation times.Water soaking and PEG have showed highest rapidly soluble fraction (a) (75.93%) and potentialy degradable fraction (b) (25.69%), respectively. Effective digradability in different rumen dilution rates were high for water soaking treated grass pea.PEG has showed highest gas production in all of incubation times and (b) (49.33), compared to others. In rich protein forages like grass pea (23.24%),because of accumulation of gas in syringe, the *in vitro* gas production is not a soutable method. Low correlation coefficients of relationships between *in situ* CP disappearance and gas production (-0.38), confirm this oponion.

Evaluation of the effects of four types of enzyme based xylanase and β-glucanase on the performance, wet litter and jejunal digesta viscosity of broilers fed wheat/barley-based diet

Shirzadi, H., Moravej, H. and Shivazad, M., Tehran University, Animal Science, Faculty of Agriculture, Karaj, 3158777871 Tehran, Iran; hmoraveg@ut.ac.ir

The objective of the study was compare to effects of four enzyme preparations containing xylanase and β-glucanase activities on performance of broiler chicks fed wheat/barley-based diet with and without enzyme and compare to those that fed corn-based diet without enzyme too. 234 day-old male broiler chicks (Ross 308) were randomly allocated to 6 treatment groups, with 3 replicates and 13 birds per replicate in floor pen. All data was analyzed through the General Linear Model procedure of SAS for a randomized complete block design. The six dietary treatments consisted of a 60% corn-based ration without enzymes and five other rations containing wheat & barley (30 & 30%) supplemented with and without enzyme (A, B, C, and D added over the top to diets). Measured traits were body weight (BW), feed intake (FI), feed conversion ratio (FCR), wet Litter, and jejunal digesta viscosity. All of them were measured at 42d, except viscosity (at 28d). BW was increased by addition all enzymes ($P<0.05$). However, FI was not significantly affected by enzyme supplementation. FCR were lower in diets containing enzymes compare to the barley-based diet without enzyme ($P<0.05$), however there were no significant differences between diets containing enzyme and the corn-based diet. Moisture of litter was not affected by enzyme, and these results were similar to results of corn-based diet ($P<0.05$). The results demonstrated that the viscosity of jejunal contents was ($P<0.05$) reduced by only enzyme A and D compared to barley-based diet. Our results led to the conclusion that there were similar improvements in performance of birds fed diets with enzyme supplementation, and choice preference of enzymes should be based on cost of them.

Effect size and density of rumen papillae in the dependence on composition of diet

Hucko, B.[1], Kodes, A.[1], Mudrik, Z.[1] and Christodoulou, V.[2], [1]Czech University of Life Sciences, Microbiology, Nutrition and Dietetics, Kamycka 129, 165 21, Prague 6, Czech Republic, [2]Animal Research Institute, National Agricultural Research Foundation (NAGREF), Animal Nutrition, Giannitsa, Greece, 58100, Greece; hucko@af.czu.cz

In a series of attempts to breed Holstein calves we monitored the presence of hay dose in the rations of calves to weaning, for the development of ruminal papillae and cellulolytic activity of microorganisms in the rumen fluid. Action of hay, we compared with effects of starters, contened mainly of cereals, included in the ration of calves as early as the third day after birth. Presence of hay at a dose of calves retarded the development of rumen papillae. The addition of concentrade feed with grossly modified cereals or rolled cereal components in the mixture, a positive effect on the size of papillae, their length, width and number per unit area. Similarly, significant findings were noted in the evaluation of ruminal cellulolytic activity of microorganisms. Ration for calves with hay, braked the development of microorganisms which digested net cellulose - fiber. Group with hay have papillae length 2588,2 μm, width 1113,3 μm and density 59,73, digestibility of cellulose 16,48%. Group with starter have papillae length 3086,8 μm, width 1405,4 μm and density 74,19, digestibility of cellulose 19,00%.

Influences of ensiling wet barley distillers' grain with sugar beet pulp on lactating performance of dairy cows
Tahmasbi, A.M., Kazemi, M., Valizadeh, R., Danesh Mesgran, M. and Gholami Hossein Abad, F., Department of Animal Science, Excellence Center of Animal Science, Ferdowsi University of Mashhad, P.O. Box 917794897, Iran; a.tahmasbi@lycos.com

An experiment was conducted to evaluate the effect of ensiling wet barley distillers' grain with different levels of sugar beet pulp on dairy cattle performance was examined. Eighteen Holstein dairy cows (86 ± 10 day's postpartum 600 ± 20 kg BW), were penned individually and randomly allocated to three treatments in a balanced completely randomized design (repeated measures). Three diets were formulated to compose 7.5% of TMR. Treatments were an ensiled mixture of 60% barley distillers' grain (BDG)with 40% beet pulp (BP) (BDGBP 40%), ensiled mixture of 80% BDG with 20% BP (BDGBP 20%), and BDG 100% ensiled without 0.0%BP (BDGBP 0%). Experimental diets contained, 40% forage (60:40, corn silage: alfalfa hay) and 60% concentrate. Cows were fed a total mixed ration and milked three times daily. Diet concentrations of NDF, ADF and CP were 33.2, 19.6 and 16.4% DM (for BGBP 40%), 33.7, 20 and 16.8% DM (for BGBP 20%) and 33.8, 20.5 and 16.8% DM (BGBP 0%), respectively. The feed intakes, daily milk yield and milk composition were not significantly different between treatments. There was no significant effect of treatments on rumen pH and NH_3-N. Also there was no significant effect on blood plasma metabolites. It is concluded that partial substitution BG ensiled with or without BP for corn silage did not have any negative effect on the performance of Holstein dairy cows.

Assessment of nutritional values of Caraway seed pulp (CSP) by *in situ* and *in vitro* technique
Tahmasbi, A.M., Moheghi, M.M., Nasserian, A., Aslaminejad, A.G. and Kazemi, M., Department of Animal Science, Excelence Center for Animal Science, College of Agriculture, Ferdowsi University of Mashhad, Iran; a.tahmasbi@lycos.com

Factors obstruct animal production in Afghanistan, as well as other countries are numerous and serious. Among of them unstable price and shortage of some of the conventional protein sources. Caraway seed pulp (CSP) is a major agro-industrial by-product from extract herb factory. The CSP are high in protein and fat with reported ether extract (EE) of 6%, CP of 15.2%, ADF of 51% and NDF of 55%. Because of the nutrient content, CSPappear to be a desirable feed ingredient in ruminant nutrition. The ruminal degradability of dry matter in CSP were estimated up to 48 h in three rumen fistulaed Holstein steer using the mobile nylon bag method. Treated Samples of CSP with PEG and urea were also fermented using Menk and Steingass gas production technique up to 96 h. Samples were run in duplicate in two separate runs. The estimated parameters of degradability's of DM indicated that CSP had high potential of degradability. The quickly degradable DN fraction (a) was 25.41 and potentially degradable DM fraction (b) of CSP was 55.15. Cumulative gas production after 96 h incubation had no affected by PEG or urea although amount of gas produced in CSP containing PEG was 18% higher that control group.

Correlation between *in vitro* gas production and *in situ* technique for evaluation of various sources of dried citrus pulp

Bayat Koohsar, J., Safari, R., Tahmasebi, A.M., Nasserian, A.A. and Valizadeh, R., Ferdowsi University, Mashhad, 51667, Iran; javad_bayat@yahoo.com

This study was aimed to evaluate correlation between *in vitro* gas production and *in situ* nylon bag technique for various sources of Dried Citrus Pulp (DCP). *In vivo* experiments to determine organic matter degradability (OMD) values of forages are expensive and laborious, impair animal welfare and are not suited for routine analysis. Therefore alternative *in vitro* techniques to predict OMD were developed in the past 40 years, such as the *in vitro* gas production technique. The method of sample preparation to measure DM degradability with the *in situ* nylon bag method was used as described by Dulphy *et al.* (1999). Samples were incubated in the rumen of 3 fistulated steers. The bags were removed following 0, 2, 4, 8, 12, 24, 36, 48, 72 and 96 h after incubation. *In vitro* gas production for the same components, with rumen liquor of same animals at same times, was measure. Regression model was made among the data of *in situ* and *in vitro* measurements for DM with the best r^2 calculated by SAS 9.1 program. Correlations for DM degradability measured by two methods were 0.94, 0.94, 0.92, 0.88, 0.95, 0.94, 0.96, 0.94, 0.96 and 0.91 for C. *Sinensis* cul. *Tamson*, C. *Sinensis* cul.*siahvaras*, C. *Sinensis* cul. *Brohen*, *Citrus unchiu*, C.*reticulata* cul. Page, C.*reticulata* cul. *Celemantin*, C.*reticulata* cul.*younesi*, *Citrus aurantifolia*, *Citrus aurantium* and *Citrus paradisi* respectively. R^2 between gas production and *in situ* DM degradability are high and indicated that both techniques can apply for determination of fermentation potential in the rumen. Observed DM values of degradability quite agree with gas production data during different time of incubation. According to the regression models, *in vitro* gas production for the DM contents of various sources of DCP could be used instead of the *in situ* technique at lower cost in shorter time.

Dry matte degradation of various sources of citrus pulp measured by mobile nylon bag technique

Bayat Koohsar, J., Safari, R., Tahmasebi, A.M., Nasserian, A.A. and Valizadeh, R., Ferdowsi University, Mashhad, 51667, Iran; rashid_safari@yahoo.com

This study was aimed to evaluate the degradability fractions of dry matter of different variety of Dried Citrus pulp (DCP) using *in situ* technique. DCP as a by-product of juices extracting factory has a desirable potential for ruminates. *In situ* DM degradability were estimated for different sources of citrus pulp; orange (*Citrus sinensis*), tangerine (*Citrus reticulate*), lemon (*Citrus auranifolia*) and grapefruit (*Citrus paradisi*) using the modified *in situ* polyester bag technique described by Ørskov (1999). Dried sample were ground to pass through a 2-mm screen and weighed (5 g DM) into the polyester bags (3 bags for each sample at each time). The bags were incubated in the rumen of three fistulated steers for 0, 2, 4, 8, 12, 24, 36, 48, 72 and 96 h. Data were subjected to analysis of variances using SAS program. Fraction 'a' was higher in *Citrus paradisi* (63%) and lower for *Citrus aurantifolia* (32%). However, *Citrus paradisi* and *Citrus aurantifolia* had lowest and highest fraction of 'b' (36% vs. 66% respectively). C. *Sinensis* cul.siahvaras had a high rate of degradability among other DCP (0.01, 0.15, 0.14, 0.10, 0.01, 0.01, 0.10, 0.07, 0.09 and 0.14 ml/g DM /h for C. *Sinensis* cul. Tamson, C. *Sinensis* cul.siahvaras, C. *Sinensis* cul. Brohen, *Citrus unchiu*, C. *reticulata* cul. Page, C. *reticulata* cul. Celemantin, C. *reticulata* cul.younesi, *Citrus aurantifolia*, *Citrus aurantium* and *Citrus paradisi* respectively). Potential of degradability of Citrus unchiu was higher than other DCP (99, 98, 99, 99, 99, 95, 98, 97, 96 and 98% for C. *Sinensis* cul. Tamson, C. *Sinensis* cul.siahvaras, C. *Sinensis* cul. Brohen, *Citrus unchiu*, C. *reticulata* cul. Page, C. *reticulata* cul. Celemantin, C. *reticulata* cul.younesi, *Citrus aurantifolia*, *Citrus aurantium* and *Citrus paradisi* respectively). Obtained results from this experiment indicated that DCP has a high potential as a feed sources in livestock and their fraction of degradability in rumen are desirable.

Effects of protected fat supplements in relation to parity on production of early lactation Holstein Cows

Ganjkhanlou, M.[1], Ghorbani, G.R.[1], Raza Yazdi, K.[1], Dehghan Banadaki, M.[1], Morraveg, H.[1] and Yang, W.Z.[2], [1]tehran university, animal science, karaj, 45841, Iran, [2]Lethbridge Research Centre, cLethbridge Research Centre, Agriculture and Agri-Food Canada, P.O. Box 3000 Lethbridge Alberta, T1J 4B1 Canada, 3000, Canada; ganjkhanloum@yahoo.com

This study was conducted to evaluate production response of early lactating cows to rumen protected fat. Twelve (nine multiparous and three primiparous) Holstein cows (36±4 day in milk) were used in a replicated 3 × 3 Latin square design with 21-d experimental period and three treatments: control (no fat supplementation), and supplemented with 30 g/kg prilled protected fat (Energizer-10) or 35 g/kg Ca salt of protected fat (Magnapac). Cows were fed ad libitum a total mixed ration consisting of 200 g/kg corn silage, 200 g/kg alfalfa hay and 600 g/kg concentrate mix. Each period had 14 days of adaptation and 7 days for sampling. Intakes of dry matter (DM), organic matter (OM) and neutral detergent fibre (NDF) were decreased by 7% with supplementation of rumen protected fat in multiparous cows. However, intakes of nutrients were not affected with fat supplementation in primiparous cows. Production of milk and 3.5% fat corrected milk (FCM), composition and yield of milk fat, protein and lactose were not affected by fat supplements in both primiparous and multiparous cows. As a result, milk efficiency (3.5% FCM/DM intake) was improved by 9.1 and 8.4% with supplementation of Energizer-10 and Magnapac, respectively, compared with control diet in multiparous cows. Feeding fat supplements increased ruminating time for primiparous cows but not for multiparous cows. These results indicate that supplementation of early lactating diet with rumen protected fat decreased feed intake but without altering milk production, milk composition and body weight, thus improved milk efficiency.

The potential of feeding goats sun dried rumen contents with or without bacterial inoculums from slaughtered animals as replacement for berseem clover and the effects on lactating goats productive performance

Khattab, H.M.[1], Kholif, A.E.[2], Gado, H.M.[1], Mansour, A.M.[1] and Kholif, A.M.[2], [1]Faculty of Agriculture, Ain Shams University, Animal Production, Hadaeq Shubra, 11241, Cairo, Egypt, [2]National Research Center, Dairy Science, Bohos street, 12622, Dokki, Egypt; ae_kholif@hotmail.com

Twelve lactating Baladi goats weighed 26±0.5 kg in the first week of lactation were randomly assigned among four experimental treatments (n=3) using 4x4 Latin square design to study the nutritional evaluation of replacing berseem clover (BC) with sun dried rumen contents (DRC) untreated or treated with anaerobic bacteria and enzyme in lactating goats ration. The period of this trial divided into four experimental periods each of 30 days. Animals were fed the following treatments: 60% concentrate feed mixture (CFM) + 40% BC (Control); 60% CFM + 20% BC + 20% DRC (T1); 60% CFM + 20% BC + 20% DRC treated with biological compound ZAD (T2); 60% CFM + 20% BC + 20% DRC treated with ZAD compound + 20g biological compound ZADO /head/d fed directly before feeding (T3). Results showed that T3 and T2 groups recorded higher values of digestibility coefficients compared with control and T1 groups. Groups contained DRC recorded higher values ($P>0.05$) of ruminal pH than the control group. The biologically treated groups (T2 and T3) showed higher ($P<0.05$) values for rumen liquor ammonia and total volatile fatty acids (TVFA's) ($P>0.05$) compared with T1 group. Biological treated groups (T3 and T2) increased daily milk yield, fat, total solids (TS), solids not fat (SNF), total protein (TP), lactose compared with T1 group. Control ration increased ($P>0.05$) milk acidity compared with ration contained DRC. It could be concluded that feeding animals on rations containing DRC treated with ZADO and/or ZAD compounds as a partial substitute of BC in lactating goats ration improved the performance without any adverse effect on animals' health.

The comparison of different treatment of chicken pea in broiler chicken

Moeini, M.[1], Heidari, M.[1] and Sanjabi, M.R.[2], [1]Razi University, Animal Science, kermanshah, 67155, Iran, [2]irost, animal science, 02126144128 Tehran, Iran; msanjabii@gmail.com

There are huge amount of chicken and split pea harvested in west of Iran which could be used as a cheap protein source for poultry nutrition. A study was conducted to determine the optimum intake of treated pea in broilers feed. Total 180 day old chicks broilers (Cobb breed) were chosen in a completely randomized design in six treatment groups; T1) 35% raw pea T2) 35% soaked pea in water for 72 hours T3) 35% soaked pea in water for 48 hours T4) 35 toasted pea for 20 minutes and T5) 35% toasted pea for 30 minutes and C) control (without pea). Each treatment had three replications with 12 observations in each treatment. The similar standard diets were fed and ration was iso caloric and iso nitrogenous. Feed intake, weight gain and feed conversion ratio (FCR) were measured weekly (from day 7 to day 49). At the end of experiment from each treatment 12 chicks selected randomly for the percentage of leg, chest and fat percentage of ventricle area. Statistical analyses were performed using Duncan's mean test. The results showed that the mean daily gain and FCR had no significant difference among treatments but the mean weight gain was lower in T1 compared with other treatments ($P<0.05$). There was no significant difference between treatments for ventricle area, the percentage of leg, fat and chest. It can be concluded that treated chicken pea up to 35% could be used as a good protein and energy source in broiler feed.

Impact of tolerance thresholds for unapproved GM soy on the EU feed sector

Aramyan, L., Van Wagenberg, C., Backus, G. and Valeeva, N., LEI Wageningen UR, Postbox 35, 6700 AA Wageningen, Netherlands; Lusine.Aramyan@wur.nl

The EU is a major importer of protein rich feedstuffs as soy, which is increasingly produced with GM varieties. Only GM varieties approved in a lengthy approval procedure can be imported in the EU. During the procedure the variety can already been grown and marketed. The EU zero tolerance policy for unapproved GM means that only soy batches without traces of EU-unapproved GM varieties can be imported. The coming years new GM soy varieties are expected. Cross contamination and mixing of EU-unapproved GM varieties with non GM and EU-approved GM varieties combined with zero tolerance, can lead to difficulties for the EU to import sufficient soy. The problem is especially severe for soy as it cannot be sufficiently produced in the EU, so insufficient import results in a substantial decline in feed and animal production. Relaxing the zero tolerance policy might prevent the import decline and consequential problems. To evaluate alternative tolerance thresholds for EU-unapproved GM soy using a computer based model of the soy supply. The analysis uses a supply chain of soy producers in USA, Brazil and Argentina, EU importers and feed producers. The time horizon is 2009-2012. We distinguish non-GM soy, EU-approved GM soy and EU-unapproved GM soy. The model has 2 steps. Step 1 adds primary production, processing, transport, and market costs to calculate the expected price of a non GM and EU-approved GM soy batch. Market costs are the price premium if soy availability is restricted. Relating soy price to feed composition determines EU demand for non-GM and approved GM soy. Expert estimation of the probability of contamination with EU-unapproved GM soy determines the available amount of non GM and EU-approved GM soy for each tolerance threshold. Step 2 is a partial equilibrium model that uses EU-demand and the available non GM and EU-approved GM soy to determine market costs. The model calculates feed prices, for various threshold levels.

Simultaneous validation of variance component estimation and BLUP software

Wensch-Dorendorf, M.[1], Wensch, J.[2] and Swalve, H.H.[1], [1]Institute of Agricultural and Nutritional Science, University of Halle, Adam-Kuckhoff-Str. 35, 06099 Halle, Germany, [2]Institute of Scientific Computing, Technical University Dresden, Fachrichtung Mathematik, Zellescher Weg 12-14, 01062 Dresden, Germany; monika.dorendorf@landw.uni-halle.de

The prediction of breeding values depends on reliable estimation of variance components. With knowledge of variance components the breeding value prediction via mixed-model equations (BLUP) requires the solution of linear systems, only. Variance component estimation, on the other hand, is a more complex task leading to nonlinear minimization problems that have to be solved by iterative numerical algorithms. In order to evaluate the reliability of these algorithms benchmark problems have to be constructed where the exact solution is a priori known. We develop techniques to construct such benchmark problems for mixed models including fixed and random effects, 1-way and 2-way classification, ML and REML predictors. Besides the construction of artificial data that produce the desired variance components we describe a projection method to construct benchmark data from simulated data.

Genetic heterogeneity of environmental variance: estimation of variance components using double hierarchical generalized linear models

Ronnegard, L.[1,2], Felleki, M.-B.[1], Fikse, W.F.[1] and Strandberg, E.[1], [1]SLU, Dept. Animal Breeding and Genetics, Box 7023, 75007 Uppsala, Sweden, [2]Dalarna University, Rodav. 3, 78170 Borlange, Sweden; lrn@du.se

Previous studies have shown that environmental sensitivity (i.e. the capability of an animal to adapt to changes in the environment) may be under genetic control, which is essential to take into account if we wish to breed robust farm animals. Linear mixed models including a genetic effect explaining heterogeneity of the environmental variance have previously been used and parameters estimated using EM and MCMC algorithms. We propose the use of double hierarchical generalized linear models (DHGLM), where the squared residuals are assumed to be gamma distributed and the residual variance is fitted using a generalized linear model (GLM). The algorithm iterates between two sets of mixed model equations (MME), one on the level of observations and one on the level of variances. We show by means of simulations that the variance of the random genetic effects in the model for residual variances can be accurately estimated using DHGLM in a population of 10,000 individuals (10 sires with each 1,000 offspring). For each animal, an observation was generated as the sum of a fixed effect (2 levels), a random genetic effect (u) and a random residual. The residual effect was sampled from N(0,phi), where log(phi) was generated as the sum of a fixed effect (2 levels) and a random genetic effect (g). Both genetic effects (u and g) were negatively correlated and sampled from a multivariate normal distribution. We replicated the simulation 20 times and obtained estimates of variance components using DHGLM. The estimated variance components were empirically unbiased (albeit with quite large SE due to the low number of sires). An advantage of DHGLM is that the calculations are based on MME and sparse matrix techniques developed for other MME-based algorithms are possible to utilize. GLM theory provides model checking tools for DHGLM and the h-likelihood gives model selection criteria.

Estimating the covariance structure for maternal temporary environmental effects in weaning weights of beef cattle

Cantet, R.J.C. and Birchmeier, A.N., Universidad de Buenos Aires, Facultad de Agronomía -CONICET, Departamento de Producción Animal, Av San Martín 4453, 1417 Buenos Aires, Argentina; rcantet@agro. uba.ar

The negative value of the additive covariance between direct and maternal effects (CADM) for weaning weight in beef cattle remains to be a controversial issue. For example, a large breed association set CADM in its genetic evaluation to zero to avoid any conflict between direct and maternal effects. It was postulated, many years ago, that the presence of an environment covariance between direct and maternal effects (CEDM) would render CADM less negative, thus decreasing the antagonistic relationship between direct and maternal breeding values. Fitting CEDM in the phenotypic covariance structure presents a formidable computing problem. We present here an animal model with maternal effects including random additive direct and maternal effects, and random temporary environmental effects (TE) on the individual records that display non-independence between offspring and dam due to the CEDM parameter. Error terms are independent and identically distributed. The covariance structure of TE for the progeny of the same dam is a function of the time elapsed between the birth dates of any pair of maternal-half-sib (or full-sib) calves. A Bayesian method using conjugate prior densities and Gibbs sampling is used for estimating all parameters. Records are grouped by lines of dam, i.e. those records of an ancestor dam, their dam descendants and all their progeny, raised naturally (not by embryo transfer). Within any line of dam TE effects are correlated but are independent with those from the other lines. The maximum number of animals involved in any line was as large as 20 in a data set of 1,943 Brangus records and 118 in a Hereford data set of 6,860 records. The proposed model was fitted to both data sets, as well as the regular maternal animal model with permanent environmental maternal effects. For both breeds estimates of CADM from the model including CEDM were less negative than in the usual model.

Analysis of genetically structured variance heterogeneity and the Box-Cox transformation

Yang, Y., Christensen, O.F. and Sorensen, D., Faculty of Agricultural Sciences, Department of Genetics and Biotechnology, Aarhus University, 8830 Tjele, Denmark; ye.yang@agrsci.dk

The classical model of quantitative genetics assumes that genotypes affect the mean of a trait but that the variance of phenotype, given genotype (environmental variance) is the same for all genotypes. An extension (HET) postulates that both mean and variability differ between genotypes. Over recent years, statistical support for a genetic component at the level of the environmental variance has come from fitting the HET model to field or experimental data in various species. Since the skewness of the marginal distribution of the data under the HET model is directly proportional to the coefficient of correlation between genes affecting mean and variance, there is the concern that statistical support for the HET model may be an artifact of the scale of measurement. One may pose the question: Is there still support for the HET model when the data are analysed in the 'correct' scale? This was investigated by extending a previously developed Bayesian McMC-based model to accommodate the family of Box and Cox transformations. Litter size data in rabbits that had previously been analysed in the untransformed scale were reanalysed in a scale equal to the posterior mode of the Box-Cox parameter. The posterior means (95% posterior intervals, in brackets) of the correlation between genes affecting mean and variance and of the additive variance in environmental variance, in the untransformed scale, were -0.73 (-0.85; -0.50) and 0.13 (0.06; 0.20). The posterior distribution of the coefficient of correlation is shifted a long way from zero and so is the additive genetic variance in variance. However in the transformed scale, the respective figures were 0.29 (-0.24; 0.79) and 0.06 (0.03; 0.10) indicating a markedly weaker support for a genetically structured variance heterogeneity. The study confirms that inferences on variances can be strongly affected by the presence of asymmetry in the distribution of the data. The results are obtained through the EC-funded FP6 Project 'SABRETRAIN'.

Comparisons of three models for canalising selection or genetic robustness

Garcia, M.[1], David, I.[1], Garreau, H.[1], Ibañez-Escriche, N.[2], Mallard, J.[3], Masson, J.P.[3], Pommeret, D.[4], Robert-Granié, C.[1] and Bodin, L.[1], [1]INRA, UMR631, SAGA, 31326 Castanet-Tolosan, France, [2]IRTA, Genètica i millora Animal, Lleida, Spain, [3]ENSAR, Agrocampus, Rennes, France, [4]IML, Univ. Méd., Marseille, France; milagros.garcia@toulouse.inra.fr

Genetic robustness can be defined as the genetic ability to maintain a production level in changing environment or a low production variability in different environmental conditions. The aim of this paper is the comparison of three different statistical models for robustness. The three models are extensions of the infinitesimal model and aimed at estimating genetic and environmental effects which contribute to the mean and to the environmental variability of a trait. The first model considers an exponential link between predictive parameters and the environmental variance of each observation. The second model considers a square root function, and the third model deals with a linear link function on the environmental standard deviation. In this study we define the equivalence between the three models and we compare their ability to fit the data under several simulated scenarios. Further, real data of rabbit birth weight are used to compare the predictive ability of these models. The implementation is based on Bayesian theory and uses MCMC (Markov chain of Monte Carlo) algorithms.

Comparison of Monte Carlo EM REML and Bayesian estimation by Gibbs sampling in estimation of genetic parameters for a test day model

Matilainen, K.[1], Lidauer, M.[1], Strandén, I.[1], Thompson, R.[2] and Mäntysaari, E.A.[1], [1]MTT Agrifood Research Finland, Biotechnology and Food Research, Biometrical Genetics, FI-31600 Jokioinen, Finland, [2]Rothamsted Research, Biomathematics and Bioinformatics, Harpenden, AL5 2JQ, United Kingdom; esa.mantysaari@mtt.fi

Random regression test day (TD) models have multiplied the number of equations needed in variance component (VC) estimation. Computation of REML estimates using analytical calculation of likelihood is limited to only small data sets. Thus, many studies have applied Bayesian estimation of VC which can be done effectively via Gibbs sampling (GS). We compared results from EM REML algorithm based on Monte Carlo (MC) estimation of prediction error variances, to the estimates of same parameters using GS. The data consisted of 185,007 TD records of milk, protein and fat for 19,709 Finnish Ayrshire first lactation cows. Pedigree had 31,255 animals. MCEM took 1219 REML rounds to reach convergence criterion of 1.0e-9, which was proved to be sufficient in earlier studies. GS was done 200,000 rounds with the first 50,000 as burn-in period and a sampling interval of 20. Heritabilities for milk, protein and fat by the EM (GS in brackets) were 0.38 (0.34), 0.32 (0.31) and 0.34 (0.33), respectively. Genetic correlations by the EM (GS) were 0.85 (0.86), 0.67 (0.69) and 0.79 (0.78) between milk and protein, milk and fat, and protein and fat, respectively. Both methods gave phenotypic correlations of 0.93, 0.80 and 0.84 between milk and protein, milk and fat, and protein and fat, respectively. To address the difference in heritability for milk, analyses were remade with simplified model and a data set small enough to use analytical EM REML. Solutions from this analysis confirmed differences seen with full data and model. Differences may be due to approach used; joint mode of REML likelihood vs means of marginal posterior densities. The MCEM analysis was found reliable with favourable properties such as small memory need and relatively short computing time.

Breeding programs for genetic improvement of traits affected by social interactions among individuals
Bijma, P. and Ellen, E.D., Wageningen University, Animal Breeding and Genomics Centre, Marijkeweg 40, 6709PF Wageningen, Netherlands; piter.bijma@wur.nl

Evidence is growing that many traits important in livestock are affected by social interactions among individuals. Examples are growth rate and feed intake in pigs, and mortality due to cannibalism in laying hens. Traits affected by social interactions respond differently to selection, because the social effect of an individual on the trait value of another contains a heritable component. Thus efficient improvement of socially affected traits may require modification of breeding programs. With interactions among n individuals, an individual's total breeding value equals $TBV_i = A_{D,i} + (n-1)A_{S,i}$, where A_D and A_S indicate direct and social breeding values. Heritable variance in a trait equals $Var(TBV) = Var(A_D) + 2(n-1)Cov(A_D,A_S) + (n-1)^2Var(A_S)$, and response to selection equals $R = i\ r_{IH}\ sigma(TBV)$, where r_{IH} represents the accuracy, i.e., the correlation between the selection criterion and the TBV of an individual. Thus response in socially affected traits can be expressed in terms familiar to animal breeders. Key factors determining accuracy are relatedness among group members and relative emphasis on group vs. individual performance in the selection criterion. Accuracy increases strongly with relatedness among group members, irrespective of the selection method (e.g., mass selection versus selection on BLUP-EBV). When genetic parameters for social effects are unknown, which is common, breeding schemes relying on information from groups composed of family members are robust, and yield ~80% or more of the theoretical maximum response in most cases. The dependency of the TBV on group size potentially creates genotype by group size interaction, which may reduce efficiency of nucleus breeding for production environments with large groups. Our first results for growth in pigs, however, suggest that social breeding values decrease almost linearly with (n-1), making the TBV nearly independent of n, so that genotype by group size interaction is practically absent.

Genetic evaluation considering phenotypic data and limited molecular information using a novel equivalent model: case study using effect of the mh locus on milk production in the dual-purpose Belgian Blue breed
Colinet, F.G.[1] and Gengler, N.[1,2], [1]Gembloux Agricultural University, Passage des Déportés 2, 5030 Gembloux, Belgium, [2]National Fund for Scientific Research, Rue Egmont 5, 1000 Brussels, Belgium; colinet.f@fsagx.ac.be

The introduction of molecular information into genetic evaluation systems is currently under research. Based on an equivalent method, we developed from existing theory a new alternative strategy for the prediction of gene effects and especially their smooth integration into genetic evaluations. Underlying hypothesis were based on the idea that knowledge of genotypes will not affect overall additive genetic variance but only change expected values of genetic effects for animals with known genotypes. However, all animals could not be genotyped. Thus, the developed equations were modified to allow the integration of the known genotype for a portion of the population. This strategy was tested for the mh locus (responsible for the double-muscling phenotype) in dual-purpose Belgian Blue cattle. The genotype was determined for 123 bulls and 1,940 cows (+/+ 19.5, mh/+ 39.3 and mh/mh 41.2%). These animals had 11,150 daughters with test-day (TD) records. The genotypes were incorporated into a modified genetic evaluation based on the current routine multi-trait multi-breed test-day model used in the Walloon Region (Belgium). Data used included 12,829,309 TD records for 689,057 dairy cows in production. The pedigree file contained 1,606,024 animals (cows with TD records and ancestors). Computation of the modified mixed model equations was done solving iteratively two systems of equations, one for the polygenic effects and one for the gene effect until the relative differences in the gene solutions were below 10^{-5}. A linear extrapolation was also used to speed up the convergence of gene effects. As expected, the mh locus exerts negative effects on milk production traits. For the first three lactations, the average estimated allelic substitution effects were -158.7 kg milk, -8.93 kg fat and -5.64 kg protein per lactation (305 days).

Detecting imprinted QTL in general pedigrees: a cautionary tale

Rowe, S.J.[1], Pong-Wong, R.[1], Haley, C.S.[2], Knott, S.A.[3] and De Koning, D.J.[1], [1]Roslin & R(D)SVS University of Edinburgh, Roslin, Midlothian, EH259PS, United Kingdom, [2]Medical Research Council, Human Genetics Unit, Crewe Road, Edinburgh EH4 2XU, United Kingdom, [3]University of Edinburgh, Institute of evolutionary biology, Kings Buildings, Edinburgh, EH9 3JT, United Kingdom; suzanne.rowe@roslin.ed.ac.uk

Genomic imprinting is the preferential expression of genes depending on sex of the parent from which they were inherited. Examples include Callipyge in sheep, and Igf2 in pigs. Recent Genome scans incorporating parent of origin effects have highlighted their importance in livestock. Simulation was used to compare power of the variance component approach to detect imprinted QTL in poultry, pig, and humans. Effect of population structure on estimation of variance components and distribution of test statistics were evaluated over a range of additive, dominant and imprinted QTL effects. An additive QTL model was compared with modelling maternal and paternal QTL effects separately (ie. removed the assumption that each explains exactly 50% of the variance at the test position). For imprinted QTL power was greatest under a model incorporating separate parental components, and could be used to search for additive QTL with little loss of power. Overall power was high; >95% to detect imprinted effects of 0.2 explaining >4% of phenotypic variance, however, type 1 error rates were very high, particularly with large additive and dominance effects. There was spurious imprinting of 10-70% with moderate – large additive QTL effects (2-20% phen variance), and 70-80% with over-dominance with highest rates observed in pigs. Estimates of variance components were also dependent on pedigree. Paternal QTL were underestimated in the pig and the chicken populations. Distribution of the test statistic differed markedly depending on genetic background and population structure, indicating that the test should be used with caution. We are exploring new methods for permutation analysis. Acknowledgements Work was funded by SABRE, BBSRC, RCUK, and Genesis-Faraday.

Power and robustness of three whole genome association mapping approaches in selected populations

Erbe, M., Ytournel, F., Pimentel, E. and Simianer, H., University of Goettingen, Department of Animal Sciences - Animal Breeding and Genetics Group, Albrecht-Thaer-Weg 3, 37075 Goettingen, Germany; merbe@gwdg.de

Selection is known to influence the linkage disequilibrium (LD) pattern in livestock populations. This may lead to an increased rate of false positive associations in whole genome association mapping. We compared power and robustness of three different approaches: single marker regression (SMR) and two two-step approaches, where in the first step a mixed linear model was fitted to the data including a random polygenic component. In the first two-step method (GRAMMAR), residuals estimated in step one were analysed with a single marker regression. In the second method (MTDT), Mendelian sampling terms were derived from estimated breeding values and were analysed with a quantitative transmission disequilibrium test. The three approaches were compared in a simulation study. We simulated the evolution of a population over 1,000 generations to generate the LD structure. 50 randomly distributed QTL were generated afterwards following a gamma-distribution. After ten generations of either random mating or selection, a five generation pedigree comprising 2,500 phenotyped individuals was obtained. These 15 generations were replicated 100 times for both scenarios. The genomic data were composed of 10,000 unevenly distributed SNPs on 10 chromosomes of each 1 Morgan length. Without selection, the number of detected correct (false) associations was 2.79 (1.35) with GRAMMAR, 7.96 (16.15) with MTDT and 9.74 (45.48) with SMR on a genome wide 1% error level. In the selected populations, the number of detected correct associations was reduced with MTDT (6.70) and SMR (9.11), while it was hardly affected (2.85) with GRAMMAR. Regarding the detected false associations the number almost doubled with GRAMMAR (2.49) and decreased with MTDT (13.15) and SMR (38.26).

A dynamic system to manage subdivided populations using molecular markers

Fernández, J.[1], Toro, M.A.[2] and Caballero, A.[3], [1]INIA, Ctra. Coruña Km 7.5, 28040 Madrid, Spain, [2]ETSIA, UPM, Ciudad Universitaria, 28040 Madrid, Spain, [3]Facultad Biología, Universidad de Vigo, 36310 Vigo, Spain; miguel.toro@upm.es

Within the context of a conservation program the management of subdivided populations implies a compromise between the control of the global genetic diversity, the avoidance of high inbreeding levels, and, sometimes, the maintenance of a certain degree of differentiation between subpopulations. Previously, we have presented a dynamic and flexible methodology for attaining these goals, based on genealogical information. Here we extend the method to the situation where only molecular information on markers is available. The objective is the maximization of the genetic diversity measured through the global population molecular coancestry (expected heterozygosity) in captive subdivided populations while controlling/restricting the levels of molecular inbreeding (observed heterozygosity). The method is able to implement specific restrictions on the desired relative levels of molecular coancestry between and within subpopulations. By accounting for the particular genetic population structure, the method determines the optimal contributions (i.e., number of offspring) of each individual, the number of migrants, and the particular subpopulations involved in the exchange of individuals. Computer simulations are used to illustrate the procedure and its performance in a range of reasonable scenarios and for different number of markers and alleles/marker.

Prioritising breeds for conservation using allelic diversity

Caballero, A. and Rodriguez-Ramilo, S.T., Departamento de Bioquímica, Genética e Inmunología, Facultad de Biología, Universidad de Vigo, 36310, Spain; silviat@uvigo.es

A new method is proposed for the analysis of allelic diversity (number of segregating alleles) in the context of subdivided populations. The definition of an allelic distance between breeds allows for the partition of total allelic diversity into within- and between-breed components, in a way analogous to the classical partition of gene diversity. A new definition of allelic differentiation between breeds results from this partition, and is contrasted with a previous proposal. The partition of allelic diversity makes it possible to establish the relative contribution of each breed to within and between- breeds components of diversity with implications in management and priorisation for conservation. The difference between this new method and a previous one is illustrated with simulated and empirical data.

QTLMAP, a software for the detection of QTL in full and half sib families

Elsen, J.M.[1], Filangi, O.[2], Gilbert, H.[3], Legarra, A.[1], Le Roy, P.[2] and Moreno-Romieux, C.[1], [1]INRA, SAGA, BP52627, 31326 Castanet, France, [2]INRA, GARen, ENSAR, 35000 Rennes, France, [3]INRA, GABI, Jouy en Josas, 78352, France; elsen@toulouse.inra.fr

QTLMAP is a software developed for the detection of QTL controlling traits in outbred populations comprising mixture of large full sib and half sib families. The underlying methods are based on Linkage Analysis. The main features of QTLMAP are the possibilities of varying the genetic model: one or two l inked QTL, single trait or multi-trait analysis, non normal observations including discrete traits and survival data, non mendelian inheritance. It allows the possibility of searching expressionQTL. It deals with high numbers of markers (SNPs). The statistical model may include fixed nuisance effects or covariates, and the polygenic relationships between the animals may be included using an animal model. The test statistic is an approximated likelihood ratio test. The distribution of the quantitative phenotype is modelled as a mixture of sub-distributions corresponding to each QTL genotype, the proportion of which being the dam phases probabilities. The test statistic can be simplified to its first order, corresponding to the Haley - Knott regression approach. QTLMAP includes a choice of simulation procedures aiming at calculating the rejection thresholds by description of the distribution of the test statistic under the null distribution and at estimating the power of experimental designs. QTLMAP is available on the web (http://gqp.jouy.inra.fr/) and its source is available on request. In the future it should incorporate additional options, in particular a modelling of epistatis and a Linkage Disequilibrium Linkage Analysis proposed by Legarra and Fernando. These results are obtained through the EC-funded FP6 Project 'SABRE'.

The effect of founder haplotype information content on likelihood profile in fine mapping experiments

Baes, C.F.[1], Mayer, M.[2], Bennewitz, J.[1] and Reinsch, N.[2], [1]Universität Hohenheim, Animal Breeding and Biotechnology, Garbenstraße 17, 70593 Stuttgart, Germany, [2]Research Institute for the Biology of Farm Animals, Genetics and Biometry, Wilhelm-Stahl-Allee 2, 18196 Dummerstorf, Germany; baes@uni-hoheneheim.de

Founder haplotypes in linkage analysis (e.g. maternal haplotypes of sires and maternal and paternal haplotypes of grandsires in a granddaughter design) are considered unrelated and their pair-wise probabilities of being identical by descent (IBD) are assumed to be zero. For fine mapping experiments, historical recombination is frequently used as a means of exploiting the non-random allelic association between QTL and closely linked markers in the population in addition to linkage information (combined linkage / linkage disequilibrium mapping, LELD). The information content of founder haplotypes is, however, not necessarily equal across all putative QTL positions. The information content of founder haplotypes can be calculated as 1 minus the haplotype block entropy, where entropy is the deviation from equilibrium for all possible haplotypes at a given locus. The likelihood profiles of three independent QTL regions previously identified on bos taurus autosomes (BTA) 2, 18 and 27 in the German Holstein population were examined. We show that likelihood profiles are not only dependant on haplotype definition (i.e. chromosomal length, number of markers or SNP) and linkage disequilibrium (the relationship between marker haplotype and QTL), but can also be highly sensitive to the information content of founder haplotypes. Our results show that information content of founder haplotypes must be considered when interpreting fine mapping results.

Modelling censored discrete time data with a view towards applications in animal quantitative genetics
Labouriau, R. and Madsen, P., Aarhus University, Faculty of Agricultural Sciences, Genetics and Biotechnology. P.O. Box 50, DK-8830, Tjele, Denmark; rodrigo.labouriau@agrsci.dk

Two practical examples involving Danish Holstein cows are considered: the study of the number of inseminations until conception and the number of lactations until death or slaughtering. In both examples, the data contains some right censored observations, i.e. observations for which we know that the number of inseminations or number lactations was larger than the observed. Moreover, the type of applications we have in mind, quantitative genetics, involves very large data sets and requires the use of a complex structure of random effects representing among other factors the pedigree information (under classical genetic models) and some natural grouping of the observations (e.g. representation of herds or AI technicians as random effects). A suitable multivariate model for discrete time data allowing right censoring will be presented. These multivariate models can also include traits of different nature, as for example classical Gaussian traits as yield, and traits following other distributions in the framework of generalized linear mixed model. These models are indeed based on suitable extensions of multivariate generalized linear mixed models and were fit using the statistical software DMU. After illustrating the use of this technique in the two concrete examples above, we will also briefly present some simulation results showing that the procedures proposed produce reasonable results under a typical scenario of animal quantitative genetics.

Application of the grouped data model to the study of genetic and environmental factors influencing calving difficulty of beef cows
Tarrés, J., Fina, M. and Piedrafita, J., Universitat Autònoma de Barcelona, G2R, Ciència Animal i dels Aliments, Campus de Bellaterra, 08193-Cerdanyola del Vallès (Barcelona), Spain; joaquim.tarres@uab.cat

Calving difficulties are usually scored by farmers from 1 (no assistance) to 5 (caesarean). The discrete nature of the variable is usually taken into account for the genetic evaluation by using threshold linear models. However, the grouped data model, currently used in survival analysis, can also deal with discrete variables. The most attractive reasons for using this model are that 1) it is a semi-parametric model that allows working with unknown and extremely skewed distributions like the calving difficulty one and 2) there is a unique assumption about proportionality of hazards that can be relaxed when necessary by considering that some effects change along the dependent variable. The aim of this study was to use this model for the analysis of calving difficulty in Bruna dels Pirineus beef cows. Its observed distribution was 82.0% calvings without assistance, 10.4% slightly assisted by the farmer, 5.7% strongly assisted by the farmer, 0.9% assisted by the veterinary practitioner, and 1.0% caesarean. Calving difficulty was very different in each herd, in part for the different subjective way of scoring of each farmer. Calving difficulty associated with male calvings was higher than that with female calvings ($P<0.001$; 1.8% of caesarean versus 0.5%). Primiparous cows calved with higher difficulty ($P<0.001$), i.e. 62.1% calvings without assistance and 2.9% needing a caesarean. Calving difficulty increased in autumn, from October to December, ($P<0.001$), associated to the higher birth weight of the calves born in these months. Only 65.4% of calves with very large birth weights were born without assistance. Calving difficulty had also an important genetic component. The heritability estimates of direct and maternal effects were 0.18 and 0.07, respectively. Overall, these results suggest that the grouped data model could be a powerful tool for the genetic evaluation of calving difficulty in beef cattle.

Separation of additive-genomic imprinting variance from additive-genetic maternal variance

Neugebauer, N.[1], Luther, H.[2] and Reinsch, N.[1], [1]Forschungsinstitut für die Biologie landwirtschaftlicher Nutztiere (FBN), Wilhelm-Stahl-Allee 2, 18196 Dummerstorf, Germany, [2]SUISAG, AG für Dienstleistung in der Schweineproduktion, Allmend, 6204 Sempach, Switzerland; neugebauer@fbn-dummerstorf.de

Imprinted genes are involved in many aspects of development in mammals and may play a role in growth and carcass composition of slaughter animals. In the presence of genomic imprinting the expression and, consequently, the effect on the phenotype of maternal and paternal alleles is different. Genomic imprinting can be accounted for by incorporating two additive genetic effects per animal in the genetic evaluation. The first corresponds to a paternal and the second to a maternal expression pattern of imprinted genes. This model fits whatever the mode of imprinting may be: paternal or maternal, full or partial, or any combination thereof. In the livestock industry slaughter and carcass traits are a class of economically important phenotypes. These traits are however only available from non-parents. It can be shown, that for these traits a probably existing maternal genetic variance component cannot be separated from the additive-genetic imprinting variance. On the other hand there are traits, which can be recorded on dams and progeny, like e.g. litter size or milk production traits. For the latter class of traits we try to separate the genetic variances due to genomic imprinting and the heritable maternal effect. Comprising information on litter size from an analysis of a large data set from a commercial pig population will be presented and genomic imprinting seems to be responsible for notable differences in the number of piglets born alive.

Evidence of a major gene for tick- and worm resistance in Tropical Beef cattle via complex segregation analyses

Kadarmideen, H.N.[1], De Klerk, B.[2] and Prayaga, K.C.[1], [1]CSIRO, Livestock Industries, Ibis av, Rockhampton 4701, Australia, [2]WUR, Animal Science, Postbus 9101, Wageningen, Netherlands; britt.deklerk@wur.nl

The main objective of this study was to conduct complex segregation analysis (CSA) on data collected from a crossbreeding experiment near Rockhampton, Australia. Phenotypic measurements on tick counts (TICK) and faecal worm egg counts (EPG) were made on Bos taurus derived tropically adapted British (BB), Bos indicus derived Zebu (ZZ) and Sanga derived (SS) breeds. Hence there were 6 different datasets (n=147 to 307) with different pedigree files (n=409 to 613). TICK and EPG traits were log-and cube root-transformed, respectively. On the transformed scale, phenotypic means (SD) for different traits were as follows; TICK-BB = 1.29 (0.48), TICK-SS = 1.41 (0.50), TICK-ZZ = 1.04 (0.54), EPG-BB = 8.17 (2.78), EPG-SS = 8.22 (2.97), and EPG-ZZ = 6.12 (2.33). A general linear model fitted to traits had fixed effects; sex, previous lactation stage, contemporary group (year and season of birth and dam age). In general, all effects were significant at $P<0.001$ to $P<0.01$ for both traits. In the second step, CSA fitting a polygenic effect and a bi-allelic major locus, were performed on each one of the datasets to investigate the mode of inheritance of these two traits. TICK-SS and TICK-ZZ datasets led to illogical estimates. Mendelian transmission probabilities (MTP) of A allele given AA, Aa and aa genotypes for TICK and EPG were 0.75, 0.45 and 0.19 and 0.81, 0.54 and 0.25, respectively, close to expectation. There were significant additive genetic variances at the major gene; 0.43, 2.22, 2.96 and 3.90 for TICK-BB, EPG-BB, EPG-SS and EPG-ZZ, respectively. Therefore this study reports evidence of a major gene for tick and intestinal worm resistance in tropically adapted beef cattle. The results reported here will be useful in breeding programs aimed at improving tick and worm resistance and provide a basis for more molecular investigations such as whole genome association studies.

A Bayesian change-point recursive model: an application on litter size and number of stillborn piglets

Ibañez-Escriche, N.[1], López De Maturana, E.[2], Noguera, J.L.[1] and Varona, L.[3], [1]IRTA, Gènetica i Millora Animal, Av. Rovira Roure 191, 25198 Lleida, Spain, [2]INIA, Mejora Genética Animal, Crta, de la Coruña, km 7.5, 28040 Madrid, Spain, [3]Universidad de Zaragoza, Anatomía, Embriología y Genética Animal, Miguel Servet 177, 50013 Zaragoza, Spain; noelia.ibanez@irta.es

The purpose of this study was to develop a change-point recursive model for the investigation of the relationships between litter size (LS) and number of stillborn piglets (NSB). This approach allows to estimate the change point in the analysis of a multiple segment modelling of non linear relationships between phenotypes, and to consider the continuity between the change points. Field data were provided by a Large White selection nucleus from a commercial breeding company. The data file contained LS and NSB of 4462 farrows. After the analysis, 2 change points were located ($T_{1LS} \sim 16$ and $T_{2LS} \sim 20$). Different structural (regression) coefficients between the change points were also obtained, since their interval highest posterior densities at 95% did not included zero and they were not overlapped. However, posterior distributions of correlations were similar across groups of LS (between change points), except for those between residuals. Posterior means of the heritabilities were low and similar to those obtained in a previous study using standard mixed models. The posterior means of the structural coefficients showed negative effect of LS on NSB. The NSB would increase by 0.13, 0.16 and 0.20 piglets for each additional born piglet, respectively. These results confirm the existence of a non linear relationship between LS and NSB and it would support the adequacy of a change-point recursive model. Nevertheless, a suitable model comparison with different number of change points would be convenient.

MolabIS: effective management of data in molecular farm animal biodiversity studies

Cong, T.V.C., Duchev, Z. and Groeneveld, E., Institute of Farm Animal Genetics, Mariensee, FLI, Department of Breeding and Genetic Resources, Höltystr 10, Mariensee, 31535 Neustadt, Germany; eildert.groeneveld@fli.bund.de

Recent advances of biodiversity studies in farm animals have rapidly increased the amount of processed and stored molecular genetics data. In different labs, the methodologies and procedures of data handling are also different. Besides, complex experiment workflows and heterogeneous data formats make the management of genetic data more challenging. To address these problems, we have developed an Web-based integrated information system (MolabIS) to effectively collect and manage farm animal genetic data. Our formalized data model meets the most common demands of various molecular genetics labs. The application allows samples and experimental results to be captured and tracked easily at each step in the workflow from sample collection to DNA sequencing and microsatellite genotyping. Lab data might be searched, updated and reported quickly at different levels. Under the Open Source GNU public licence, MolabIS will be released as a free application.

Statistical tools to detect a sex dimorphism in piglet birth weight
Wittenburg, D., Teuscher, F. and Reinsch, N., Forschungsinstitut für die Biologie landwirtschaftlicher Nutztiere (FBN), Genetik und Biometrie, Wilhelm-Stahl-Allee 2, 18196 Dummerstorf, Germany; wittenburg@ fbn-dummerstorf.de

The uniformity of birth weight is an important factor for piglet survival. Uniformity of body weight is desirable not only at birth but also in later ages, since it allows for slaughtering at nearly the same age and makes management of fattening pigs easier. Thus, breeding to reduce the variation in birth weight should be aimed. Breeding would be most efficient if birth weight was equal among sexes, but this trait is sex-specifically expressed. Such a sex dimorphism may negatively influence the genetic gain. We studied whether the difference in birth weight of male and female piglets is partly under genetic control. For that purpose we set up a linear mixed model for birth weight, which included additive genetic components differing between sexes. We defined a hypothesis testing problem to detect whether the breeding values significantly differ between sexes. In a second step, we studied the effect of sex-linked genes explicitly. Concerning this, the additive genetic effect was partitioned into autosomal and gonosomal effects. We set up a relationship matrix accounting for the X-chromosomal inheritance. The Y-chromosomal effect was traced back to the male founder animals. The proposed tools were applied to practical datasets of two pig lines. In both lines, a fixed effect accounting for sex was significant, but only in one line the breeding values significantly differed between sexes. Using the second approach, the difference was ascribed to the Y-chromosome.

GSEVM v.3: MCMC software to analyze genetically structured environmental variance models
Garcia, M.[1] and Ibañez-Escriche, N.[2], [1]INRA, UMR631, SAGA, 31326 Castanet-Tolosan, France, [2]IRTA, Genètica i millora Animal, Lleida, Spain; milagros.garcia@toulouse.inra.fr

One of the main problems in animal breeding to analyze genetically the variance of phenotypic traits is the lack of available software.ÿThe purpose of this paper is to present a new software that implement Bayesian-MCMC methods to fit genetically structured variance models. The program GSEVM v.3 (genetically structured environmental variance model) implements three different structural mixed linear models considering:1) an exponential link between predictive parameters and the environmental variance of each observation; 2) a square root function on the environmental variance; 3) a simpler linear link function on the environmental standard deviation. Programming is in FORTRAN90 and the executable file runs in DOS-Windows and UNIX operating systems. The GSEVM v.3 software is friendly, easy to run, and it is driven by a parameter file that defines the input files, attributes, models, and the parameters of variance priors. The input files (data and pedigree) used by the program should be prepared as an ASCII file with separated columns. The output files provide elementary statistics of estimated parameters (min, max, mean, std), Monte Carlo estimates of marginal posterior distributions of parameters of interest, the Kolmogorov normality test of the posterior distributions, autocorrelation test and the deviance information criterion (DIC), a measure of goodness fit a model. The program is quite flexible, allowing the user to fit a variety of models at the level of the mean and the variance.

Estimation of variance components for binary threshold models
Fikse, W.F.[1] and Rönnegård, L.[1,2], [1]SLU, Dept. Animal Breeding and Genetics, Box 7023, 750 07 Uppsala, Sweden, [2]Dalarna University, Rödavägen 3, 78170 Borlänge, Sweden; Freddy.Fikse@hgen.slu.se

The statistical literature has reported severe underestimation of variance components by a common estimation technique (penalized quasi-likelihood, PQL) for binary threshold models when the number of random effects is large compared to the number of observations. Several general purpose software packages often used in the context of animal breeding are based on PQL (e.g. ASReml, DMU). The objective of this study is to evaluate the potential bias in variance components for a dairy cattle and a horse data structure. For the dairy cattle scenario, observations (0/1) were simulated for about 90,000 first-lactation cows after 760 sires. For the horse scenario, observations (0/1) were simulated for 1,250 offspring of 370 stallions. The model contained a random effect of sire (and herd-year for the dairy cattle case) in addition to several fixed effects. Five different incidence levels were simulated: 2.5%, 5%, 10%, 25% or 50%. Results indicated that software packages using the PQL technique severely underestimated variance components (up to 50%) at low incidence levels. The bias was largest for low to intermediate heritabilities (0.05 and 0.15) and reduced to about 10% when a heritability of 0.25 was simulated. The bias in sire variance was largest when the herd-year effect explained a large proportion of the phenotypic variance. MCMC based methods yielded in general unbiased estimates of variance components, with the exception of a few scenarios with very low simulated variances and incidence levels.

Combining co-variance components estimated by different models using iterative summing of expanded part matrices approach
Negussie, E.[1], Lidauer, M.[1], Stranden, I.[1], Aamand, G.P.[2], Nielsen, U.S.[2], Pösö, J.[3], Johansson, K.[4], Eriksson, J.-Å.[4] and Mäntysaari, E.A.[1], [1]MTT Agrifood Research Finland, BGE, 31600 Jokioinen, Finland, [2]Nordic Cattle Genetic Evaluation, Udkærsvej 15, Skejby, DK-8200 Århus N, Denmark, [3]Faba Breeding, Box 40, 01301 Vantaa, Finland, [4]Svensk Mjölk, Box 210, S-101 24 Stockholm, Sweden; enyew.negussie@mtt.fi

In this study estimation of co-variance components for a BLUP meta-model which has both longitudinal and non-longitudinal traits were considered. The co-variance components for these traits can be estimated directly fitting the meta-model. Alternatively, they can be estimated separately and later combined using the iterative summing of expanded part matrices approach. These two methods were compared in analysis of the joint Nordic dairy cattle udder health model fitting a multi-trait random regression. The meta-model had 5 traits: test-day somatic cell score (TDSCS), two clinical mastitis (CM) traits (CM1: -15 to 50 and CM2: 51 to 300 DIM) and two udder type traits (fore udder attachment (UA) and udder depth (UD)). Data from about 34,000 first-lactation Danish Holstein cows were used. Co-variance components were estimated using the AI-REML procedure. Heritability estimates using the combining approach were 0.05, 0.03, 0.27 and 0.41 for CM1, CM2, UA and UD, respectively, but ranged from 0.08 to 0.15 for TD SCS. Genetic correlations between TDSCS and CM1, CM2, UA and UD ranged from 0.45 to 0.57, 0.55 to 0.68,-0.17 to -0.23 and -0.32 to -0.40, respectively at different stages of lactation. Heritability estimates from the direct meta-model analyses were 0.06, 0.03, 0.21 and 0.39, for CM1, CM2, UA and UD, respectively, while they ranged from 0.08 to 0.14 for TD SCS. Genetic correlations between TDSCS and CM1, CM2, UA and UD ranged from 0.45 to 0.57, 0.55 to 0.70,-0.16 to -0.22 and -0.30 to -0.36, respectively at different stages of lactation. Estimates from both methods were close. However, estimation using the direct meta-model analysis was slow and had poor convergence characteristics.

Bioinformatics and network analysis of reproductive proteins

Arefnezhad, B.[1], Sharifi, A.[1] and Farahmand, H.[2], [1]Tehran University, Animal Science, Agricultural Faculty, Karaj-Tehran, 81578, Iran, [2]Tehran University, Department of Aquaculture, Agricultural Faculty, Karaj-Tehran, 81578, Iran; babaref@yahoo.com

Reproduction is a fundamental biological process in eukaryotes. There is an adaptive diversification in reproductive protein evolution. Like other biological processes, reproduction is a robust and redundant system that many factors control the robustness of the system. DNA and Protein sequence of the reproductive proteins in the paracrine and endocrine communication were implemented from the NCBI mammalian ResNet databases. Comparative sequence analysis was performed by BLOSUM62 matrix and for pathway reconstruction we used Pathway Studio Software. In this study we have shown that rapidly evolved proteins are more redundant and make less robust module. According to text mining the previous literature, we reconstruct the hub of network of the reproductive proteins. More rapidly evolved proteins made the subordinate factors in the networks.

Identification of selection signatures across dairy cattle genome based on the Illumina 54,000 SNP panel

Ajmone Marsan, P.[1], Marino, R.[1], Perini, D.[1], Negrini, R.[1], Nicolazzi, E.[1], Pariset, L.[2], Valentini, A.[2], Vicario, D.[3], Santus, E.[4], Blasi, M.[5], Fontanesi, L.[6], Schiavini, F.[7], Bagnato, A.[7], Russo, V.[6], Maciotta, N.[8] and Nardone, A.[2], [1]UNICATT, Via E. Parmense, 29100 Piacenza, Italy, [2]UNITUS, Via C. de Lellis, 01100 Viterbo, Italy, [3]ANAPRI, I. Nievo, 33100 Udine, Italy, [4]ANARB, Loc. Ferlina, 37012 Bussolengo (VR), Italy, [5]LGS, Via Bergamo, 26100 Cremona, Italy, [6]DIPROVAL, Via F.lli Rosselli, 017, 42100 Reggio Emilia, Italy, [7]VSA, Via Celoria, 20133 Milano, Italy, [8]UNISS, Via E. de Nicola, 07100 Sassari, Italy; paolo.ajmone@unicatt.it

Advances in marker technologies have allowed the development of large scale SNP panels in livestock species. These could be exploited in a number of applications, spanning from the characterization of genetic diversity to genomic selection. We have assayed 493 Pezzata Rossa (aka Simmenthal) and 775 Bruna Italiana (aka Brown Swiss) bulls with the BovineSNP50 BeadChip (Illumina, USA). Observed and expected heterozogosities were calculated at genome-wide scale and by chromosome. In both breeds observed heterozygosities was consistently lower than expected ones. Loss of heterozygosity was use to calculate FIS values along adjacent 1Mb genome regions. Eight regions with $P<0.05$ significance at chromosome-wide level, 5 in Bruna Italiana and 3 in Pezzata Rossa were identified. These figures indicate a high directional selection pressure in these regions. Since Brown selection indices include k-casein genotype, we expected to find high Fis values in chromosome 6 where caseins are coded. Low Fis values were observed around the casein loci around 88 Mbp. Further investigations are needed to better explain this behaviour since mechanisms that reduce Fis, like disassortative mating and balancing selection do not seem likely. Poor linkage to the casein superlocus of the markers used could also be ruled out since the markers are quite dense, alleles at k-casein are still segregating and selection on these is very recent, therefore we expect to find an extended linkage disequilibrium around these loci.

Variance component method for QTL mapping in F2 populations

Zimmer, D., Mayer, M. and Reinsch, N., Research Institute for the Biology of Farm Animals (FBN), Research unit Genetics and Biometry, Wilhelm-Stahl-Allee 2, 18196 Dummerstorf, Germany; zimmer@ fbn-dummerstorf.de

The variance component method (VCM) is often used for mapping of quantitative trait loci (QTL) using flanking marker information. We have developed an efficient approach to set up the additive genetic, dominance and pairwise epistatic relationship matrices in an F2 population from a cross of two different parental inbred lines, when marker and QTL positions do not coincide. We determine conditional QTL genotype probabilities and fundamental QTL relationship matrices containing the IBD probabilities, when QTL genotypes are assumed to be known. In contrast to previous studies we consider possible dependencies between QTL genotypes. In our long approach we consider conditional genotypic effects for each individual, but for multiple QTL the VCM becomes computationally demanding with increasing progeny number. Therefore, we have developed a short computation method for the relationship matrices which are independent of experimental size. In this approach the F2 individuals are grouped depending on their marker genotypes and an average genotypic effect is estimated conditional on the marker genotype. The number of mixed model equations is reduced and essentially less computing time is required. The estimated QTL positions with short and long VCM were compared with multiple interval mapping (MIM; Kao *et al.*, 1999). First results from simulations indicate that our VCM approach is competitive with MIM in terms of the accuracy of the estimated QTL positions.

Developing the method of estimating genetic similarity between populations

Nilforooshan, M.A., Swedish University of Agricultural Sciences, Department of Animal Breeding and Genetics, P.O. Box 7023, 750 07, Uppsala, Sweden; Mohammad.Nilforooshan@hgen.slu.se

An existing method for measuring genetic similarity between populations was further developed to include information from ancestors. Genetic similarity is the extent of genetic exchange or the amount of the genetic material that is shared between the gene pools of two populations. Genetic correlations and the number of exchanged individuals have been used traditionally as indicators of genetic similarity. However, genetic correlations are trait-dependent and under the influence of genotype by environment interaction and the number of exchanged individuals do not show the extent of spreading genes from exchanged individuals into two populations. The present method uses the number of progenies from the shared individuals in the two populations without using the parental information of the shared individuals. With the new method, most of the pedigree information is capture without the need to use a relationship matrix. Therefore, it is easy to implement and has a low computational demand. This new method can be used in studies of biodiversity or screening of populations for which genetic correlations are going to be estimated.

Selective genotyping in commercial populations

Duijvesteijn, N.[1] and De Koning, D.J.[2], [1]IPG, Institute for Pig Genetics, P.O. Box 43, 6640 AA Beuningen, Netherlands, [2]Roslin Institute and R(D)SVS, University of Edinburgh, Roslin, United Kingdom; naomi. duijvesteijn@ipg.nl

Genotyping large numbers of pigs with a very large number of markers is now possible, but still costly. When working within the constraints of a commercial breeding programme, it is crucial to optimize the design of the association study. For the actual pedigree structure of an IPG commercial line, phenotypes and genotypes were simulated using the software programmme MORGAN. A chromosome with ten markers and 1 QTL was simulated as well as a second chromosome with ten markers, but without a QTL to determine the false-positive rate and the empirical threshold. After this simulation, different selection methods were applied to select 1,000 animals from 1,920 (description of the real situation). Four selection methods were compared: random selection (rand), selecting large families (>35 half sibs; fam), 500 high and 500 low based on phenotypic values (hilo), and high and low phenotypes within families (hilo-fam). ANOVA was used to analyze each marker against the phenotype and F-test values for each simulated SNP were saved from each replicate (1,000 replicates) using R, a free software environment. The 95th percentiles of the F-test values of the simulated chromosome without a QTL were computed and used as thresholds for the analysis of the chromosome with one QTL. Random selection and selecting large families resulted in a significantly lower power compared with hilo, hilo-fam and no selective genotyping (i.e. typing all 1,920 individuals; $P<0.01$). Hilo-fam showed a trend to get a higher power of analysis than just hilo across all families, but was not significant ($P>0.05$). Genotyping all animals would result in similar power as selective genotyping using hilo or hilo-fam. The possibility of using your own pedigrees makes it a valuable method for commercial companies to apply for selective genotyping and saving genotyping costs for any trait and pedigree structure. These results are obtained through the EC-funded FP6 Project 'SABRE'.

Gametic gene flow method accounts for genomic imprinting and inbreeding

Börner, V. and Reinsch, N., Forschungsinstitut für die Biologie landwirtschaftlicher Nutztiere (FBN), Genetik und Biometrie, Wilhelm-Stahl-Allee 2, 18196 Dummerstorf, Germany; boerner@fbn-dummerstorf.de

Findings within the last fifteen years emphasise the possible role of genomic imprinting for trait expression in livestock species. In genetic evaluation, genomically imprinted traits can be treated by models with two different breeding values per animal; one accounts for the paternal and the other for the maternal expression pattern. Relative weighting factors for these breeding values were derived by a generalised version of the discounted gene flow method, which was extended to a gametic level to account for parent-of-origin effects. The gametic approach proved also useful for calculating the expected increase in inbreeding induced by one round of selection and its dynamics over time. The gametic gene flow method was applied to a hypothetical pig breeding programme. Relative weighting factors were higher for the paternally inherited genetic effect even in female selection paths, but heavily depend on the breeding scheme. The maximum increase in inbreeding due to selection exceeded the long term increase in a range of 20% to 100%.

Including copy number variation in association studies

Calus, M.P.L.[1], De Koning, D.J.[2] and Haley, C.S.[2,3], [1]Animal Sciences Group, Wageningen University and Research Centre, Animal Breeding and Genomics Centre, P.O. Box 65, 8200 AB Lelystad, Netherlands, [2]Roslin Institute and Royal (Dick) School of Veterinary Studies, University of Edinburgh, Division of Genetics and Genomics, EH25 9PS, Roslin, United Kingdom, [3]Western General Hospital, MRC Human Genetics Unit, Crewe Road, Edinburgh, EH4 2XU, United Kingdom; mario.calus@wur.nl

Genome-wide association studies are typically performed using markers such as microsatellites and single nucleotide polymorphisms (SNPs) that represent a sample of the variation in the genome. Another source of structural genomic variation are Copy Number Polymorphisms (CNP). Considering that CNPs may be directly associated with phenotypic variation, an important question is whether this phenotypic variance can also be captured using a dense SNP map, or whether CNPs should be genotyped and included in GWA studies. Deriving CNP genotypes from raw hybridizations may however be difficult, especially if more than two alleles are segregating. Therefore, the objective of this study was to investigate the ability to explain genetic variation resulting from a CNP by including the CNP, either by its genotype or by its raw hybridization, alone or together with a nearby SNP in the model. Stochastic simulation and derivation of deterministic formulas were used to investigate this objective. Under the assumption that x copies at a CNP locus lead to the effect of x times the effect of 1 copy, including the raw hybridizations of a CNP locus in the model together with the genotype of a nearby SNP increased power to explain variation at the CNP locus, even when the raw hybridization explained only 25% of the variation at the CNP locus. These results are obtained through the EC-funded FP6 Project 'SABRE'.

Breeding value estimation combining QTL and polygenic information

Mulder, H.A., Calus, M.P.L. and Veerkamp, R.F., Animal Breeding and Genomics Center, Animal Sciences Group, Wageningen UR, P.O. Box 65, 8200 AB Lelystad, Netherlands; herman.mulder@wur.nl

Optimal selection is based on breeding values that combine QTL information with polygenic breeding values estimated simultaneously from phenotypic records. Generally a large proportion of animals is not genotyped however. The aim of this study was to compare two methods to deal with ungenotyped animals: 1) predicting their marker haplotypes and include them in the marker-assisted breeding value estimation (ma-blup) or 2) perform conventional breeding value estimation and include estimated haplotype effects for genotyped animals in a total EBV by using selection index weights (ma-index). MIXBLUP-software was used for breeding value estimation. To compare both methods, a population was simulated with one additive QTL and an additive polygenic genetic effect (heritability was 0.30; QTL explained 15% of genetic variance). Using the same estimates of haplotype effects and optimal selection index weights calculated from the true variances and accuracies, both methods resulted in very similar accuracy of total EBV, as expected. However, in practice accuracies need to be approximated and therefore, index weights are not optimal. For genotyped juveniles without phenotype this resulted in an average accuracy that was 0.01-0.03 lower with ma-index in comparison to ma-blup. Furthermore, in some replicates (18%) the accuracy for ma-index was even lower than for conventional breeding value estimation without QTL information. Even with ma-blup and equal weights on haplotype effects and polygenic effects, 12% of the replicates showed lower accuracy than with conventional breeding value estimation. With optimal weights, however, the accuracy was always equal to or higher than that of conventional breeding value estimation. Therefore, it can be concluded that marker-assisted breeding value estimation is better than using a selection index to blend haplotype effects and conventional breeding values. These results were obtained through the EC-funded FP6 Project 'SABRE'.

Aggregated phenotypes for molecular genetic analyses

Edel, C., Emmerling, R. and Goetz, K.-U., Bavarian State Research Center for Agriculture, Institute of Animal Breeding, Prof.-Duerrwaechter-Platz 1, 85586 Poing-Grub, Germany; Christian.Edel@LfL.bayern. de

Molecular genetic analyses and related applications like marker-assisted prediction of breeding values (MA-BLUP) or genomic selection rely on the availability of aggregated phenotypes and weighing factors for genotyped animals. In QTL-mapping an approach based on daughter yield deviations (DYD) of bulls and effective daughter contributions (EDC) as weighing factors might be sufficient. In MA-BLUP, ignoring the phenotypes of genotyped bulldams leads to a substantial loss of information. For our implementation of MA-BLUP in German Fleckvieh we developed two approaches of combining DYD of bulls and cows and yield deviations (YD) of cows. As weighing factors we used approximate multivariate reliabilities after transforming them to equivalent numbers of own performances (EOP). Since our MA-BLUP evaluation was developed as univariate, aggregation was not only within one trait but also over traits defining a biological trait or are combined to a selection criterion (e.g. first, second and higher lactation milk yield). To validate our approaches, we correlated the resulting ebvs from a data-set of genotyped animals including parents (7090 records) to the solutions of the routine evaluation (November 2008). From approach 1, that was developed for multivariate random-regression test-day models (milk production traits), we found correlations of 0.99, 0.94 and 0.97 for proven bulls, bulldams and candidates respectively (milk yield). Lactation YD of cows based on 'best prediction' were superior compared to calculations using YD functions. Approach 2 was developed for standard multivariate breeding value estimation with many missing values. Here, DYD and YD are transformed to breeding values using phenotype specific selection indices. These breeding values were then deregressed and combined to the final phenotype. Correlations of 0.99 for bulls, bull dams and candidates (somatic cell count) show that this approach works successfully even under difficult conditions.

What do artificial neural networks tell us about the genetic structure of populations? The example of European pig populations

Nikolic, N.[1], Park, Y.S.[2], San Cristobal, M.[1], Lek, S.[3] and Chevalet, C.[1], [1]INRA, animal genetics, Laboratoire de génétique cellulaire, BP52627, 31326 Castanet Tolosan cedex, France, [2]Kyung Hee University, Department of Biology, Dongdaemun-gu, Seoul 130-701, Korea, South, [3]CNRS, Laboratoire Evolution de la Diversité Biologique, 118 route de Narbonne, 31400 Toulouse cedex 4, France; magali.san-cristobal@ toulouse.inra.fr

General and genetic statistical methods are commonly used to deal with microsatellite data (highly variable neutral genetic markers). In this work, Self-Organizing Maps (SOM) that belong to the unsupervised Artificial Neural Networks (ANNs) were applied to analyse the structure of 58 European and 2 Chinese pig populations (Sus scrofa) including commercial lines, local breeds and cosmopolitan breeds. Results were compared to other unsupervised classification methods (Factorial Correspondance Analysis, hierarchical clustering from an Allele Sharing distance, Bayesian genetic model) and to supervised approaches (Principal Components Analysis and Neighbor joining from genetic distances). Like other methods, SOM were able to classify individuals according to their breed origin and to visualise similarities between breeds. They provided additional information on the between- and within-populations diversity, allowed differences between similar populations to be highlighted and helped differentiate different groups of populations.

Investigations on fat protein ratio of milk and daily energy balance in Holstein Friesians

Buttchereit, N.[1], Stamer, E.[2], Junge, W.[1] and Thaller, G.[1], [1]CAU, Institute of Animal Breeding and Husbandry, Hermann-Rodewald-Str.6, 24098 Kiel, Germany, [2]TiDa GmbH, Bosseer Str.4c, 24259 Westensee, Germany; nbuttchereit@tierzucht.uni-kiel.de

The fat to protein ratio in milk (FPR) could serve as a measure of energy balance status and could be used as a selection criterion to improve metabolic stability. Therefore, the fit of several fixed and random regression models, describing FPR and daily energy balance (EB), was tested to establish models appropriate for genetic evaluations. Data were collected on the Karkendamm dairy research farm running a bull-dam performance test. EB was calculated using milk yield, feed intake per day and live weight. Weekly FPR measurements were available. Three data sets were created containing records of 577 heifers with observations from lactation day 11 to 180 as well as records of 613 heifers and 76 cows with observations from lactation day 11 to 305. Five parametric functions of days in milk (Ali and Schaeffer, Guo and Swalve, Wilmink, Legendre polynomials of third and fourth degree) were used to model both fixed and random regression coefficients. Evaluation of goodness of fit was based upon different information criteria, correlation between the real observations and estimated values, and on residuals plotted against days in milk. The random regression models showed superior fit to the data. In general, the Ali and Schaeffer model performed best. Thus, this model was chosen to analyse the relationship between FPR and EB for different lactation stages. FPR is highest in the initial lactation period when energy deficit is most severe. EB stabilizes at the same time as FPR stops decreasing. The mirror-inverted patterns point to a causal relationship between these traits. A similar pattern was also observed for repeatability of both traits, with repeatability being highest at the beginning. Correlations between cow effects were highest in the initial lactation (r=-0.43). Results support the idea that FPR serves as a suitable indicator for energy status, especially during the most critical period of metabolic stress.

The effects of malate supplementation on productive, metabolic and acid-base balance parameters in calves fed a corn-based high-grain diet

Pereira, V.[1], Castillo, C.[1], Hernández, J.[1], Vázquez, P.[2], Vilariño, O.[1], Méndez, J.[3] and Benedito, J.L.[1], [1]Facultad de Veterinaria de Lugo - USC, Patología Animal, Campus Universitario, 27002 Lugo, Spain, [2]CESFAC - I+D+i Research Department, Diego de León 54, 28006 Madrid, Spain, [3]Coren SCL, Juan XXIII, 32003 Orense, Spain; victor.pereira@usc.es

This study investigated the effects of malate supplementation on blood acid-base balance, serum L-lactate levels and final productive performance in bull calves during an entire productive cycle, considering both short- and longer-term effects on the parameters considered. A 137-day feedlot metabolic study was conducted using 26 Belgian Blue bull calves. Animals were allotted randomly to one of the two experimental groups: 1) control group (no supplementation; n=10), and 2) supplementation with 2.8 g of disodium malate-calcium malate (Rumalato®) per kg (dry matter basis) ([n=16). Blood pH, pCO_2, HCO_3^- and base excess (BE) were determined in whole blood, using a hand-held portable analyser. L-lactate was determined in serum. The malate-supplemented calves showed slightly lower mean average daily gains (ADG) and feed intake than the control animals, but the difference in ADG was not statistically significant. Feed-to-gain ratio was similar in the two groups. In respect of productivity parameters, malate supplementation of a corn-based diet for feedlot beef cattle appears to have no beneficial effects. Furthermore, the observed effects on internal acid-base balance suggest a need for more research on malate effects on internal balance, for although supplemented animals showed more stable pH values than controls, their blood buffer bases were in several moments lower than non supplemented animals. Finally, despite the well-known *in vitro* effects of malate on lactate levels, in the present study, serum L-lactate remained higher in malate-supplemented animals than in controls, suggesting that the effects of malate can be dependent on the characteristics of the diet being fed.

Milk urea content as an estimator of nutritive balance on grazing or silage conditions

Roca Fernández, A.I., González Rodríguez, A. and Vázquez Yáñez, O.P., Agrarian Research Centre of Mabegondo, Animal Production, Ctra. AC-542 Betanzos-Santiago km 7,5, 15080, Spain; anairf@ciam.es

A proper balance between rumen degradable protein and rapidly fermentable carbohydrate allows the animal to make the best use of protein. However, it is difficult to make recommendations for the crude protein (CP) content in the diet as it depends on milk yield, milk CP, growth rate, body weight (BW), energy content and type, as well as amino acid composition and degradability of dietary protein. The aim of this study was to devise an index for diagnosis of nutritive balance under grazing or silage feeding conditions, and to study whether milk urea (MU) content could be used as a management aid to balance the ration. Milk yield and composition were analyzed from March to August in three herds of cows (n=92) at different stages of lactation, two under grazing (G) - spring-calving (S) and autumn-calving (A) - and one indoors (I) and spring-calving with supplementation. Data analysis was performed using the statistical program SPSS 15.0. There were no significant differences in milk production between GS and IS (24.3 and 25.6 kg day^{-1}, respectively) compared to 18.4 kg day^{-1} for GA. However, MU was significantly higher in IS than in GS (231 vs 192 g kg^{-1}). Protein deficiencies were detected by a decrease in the MU content for IS. Estimates of MU in the grazing treatments indicated that there was a balance between protein and carbohydrates with high grass quality and high total dry matter intake. Milk protein was significantly higher in GA (31.5 g kg^{-1}) than in the spring treatments. There were significant differences between treatments in BW, but not in body condition score (average value of 3). The IS significantly increased BW through a significant increase in silage and concentrate intakes. Using the MU content provides an opportunity to increase milk production, reduce feed costs and improve profitability of the herd with less environmental impact from N in manure.

Limits to prediction of energy balance from milk composition measures at individual cow level

Løvendahl, P.[1], Ridder, C.[1,2] and Friggens, N.C.[1], [1]Aarhus University, Faculty of Agricultural Sciences, Research Centre Foulum, P.O. Box 50, DK 8830 Tjele, Denmark, [2]Lattec I/S, Slangerupgade 69, DK 3400 Hillerød, Denmark; Peter.Lovendahl@agrsci.dk

Monitoring of individual cow status has been shown to provide real benefits in terms of early identification of cows with health and reproductive problems. Prediction of energy balance from milk composition measures has recently been shown to be very accurate on a group basis through lactation, but with a poorer fit when used at the level of individual cows. The purpose of the present study was to characterize the between cow variation in prediction of energy balance, when prediction was based on milk composition (EBalMilk) vs predictions based on body condition score (BCS) and live weight (EBalBody). The study was based on 623 lactations from 299 cows of Red Dane, Holstein and Jersey breeds, in their first 3 parities. Records of milk composition were available on a weekly basis and BCS and weight were available fortnightly. Assessment of EbalBody was obtained during three lactation stages, A=0-28, B=49-70, and C=112-301 days in milk (DIM), each with distinct mobilization or deposition characteristics. During the same periods, EbalMilk was estimated using a partial least squares model including milk fat content, fat:protein ratio, together with 3 first derivative variables, which are the current minus the previous value of the milk measure in question namely milk yield, fat:protein ratio, and protein yield. Both traits were analyzed with a mixed model, having cow within parity and period as random. The repeatability for EbalBody was always larger than for EbalMilk (0.93, 0.91 and 0.86 vs 0.53, 0.41 and 0.43 in periods A, B and C, respectively). The within period individual level correlation between EbalMilk and EbalBody was low and non-significant in all three periods. We conclude that Ebal estimates based on milk composition are far less reliable than estimates based on BCS and weight.

Predicting bovine milk fat composition of winter and summer milk using infrared spectroscopy

Rutten, M.J.M.[1], Bovenhuis, H.[1], Hettinga, K.A.[2], Van Valenberg, H.J.F.[2] and Van Arendonk, J.A.M.[1], [1]Wageningen University, Animal Breeding and Genomics Centre, P.O. Box 338, 6700 AH Wageningen, Netherlands, [2]Wageningen University, Dairy Science and Technology Group, P.O. Box 8129, 6700 EV Wageningen, Netherlands; marc.rutten@wur.nl

Fat percentage measurement of bovine milk is part of routine milk recording and is determined by infrared spectroscopy. At present, no information on detailed milk fat composition is routinely collected whereas this might be of importance for producing specialized dairy products. Recently, it has been shown that spectra can be used for prediction of detailed fat composition. In this study we constructed prediction equations for milk fat composition based on 3,631 milk samples. The average validation r-square was 0.88 for saturated fatty acids C4:0-C18:0, 0.48 for unsaturated C18, and 0.88 for the ratio of saturated to unsaturated fatty acids (g/100g fat). We investigated the sensitivity of the prediction equations for the effect of season by distinguishing between winter and summer milk samples as significant differences in milk fat composition exist between winter and summer. For models calibrated on winter data and validated on summer rather than winter data (CwVs), average r-square dropped from 0.71 to 0.55, from 0.46 to 0.33, and from 0.83 to 0.82 for the listed fatty acids (g/100g fat). For models calibrated on summer data and validated on winter rather than summer data (CsVw), these values dropped from 0.67 to 0.54, from 0.45 to 0.29 and from 0.91 to 0.77 for the listed fatty acids (g/100g fat). Prediction bias expressed as a percentage of the mean for the model CwVs went up from on average 0.89% to 3.72% (g/100g fat) and for the model CsVw went up from on average 0.69 to 1.16% (g/100g fat). We conclude that a representative sample including observations collected in various seasons, or other sources of variation with respect to milk fat composition, is absolutely critical for model calibration and subsequent unbiased prediction of milk fat composition.

Adding value to test-day data by using modified best prediction method

Gillon, A.[1], Abras, S.[2], Mayeres, P.[2], Bertozzi, C.[2] and Gengler, N.[1,3], [1]Gembloux Agricultural University, Passage des Déportés 2, 5030 Gembloux, Belgium, [2]Walloon Breeding Association, Rue des Champs Elysées 4, 5590 Ciney, Belgium, [3]National Fund for Scientific Research, Rue Egmont 5, 1000 Brussels, Belgium; gillon.a@fsagx.ac.be

Computation of lactation yields from test-day yield has lost much of its importance for genetic evaluations as the use of test-day models is rather widespread. At the same time its importance for intra-farm management increases at farms as a base for advanced management tools. The most common official method to compute lactation yield is the Test Interval Method (TIM). Alternative methods for computing cumulated productions were developed. These methods can be considered as improvements of TIM as the interpolation method, or completely different methods as multiple-trait prediction (MTP) and best prediction (BP). Research in this field has shown the potential to compute lactation parameters (e.g., cumulated production) with test-day models. The aim of this study was to develop a new method which takes into account advantages and disadvantages of existing methods, and to test its potential to provide useful tools to help farmers to make management decisions. The second objective was to compare the accuracy and the robustness of this method with those of BP and TIM. Because of its similarities with BP, the method developed here was called mBP, for modified-BP. The main difference from BP is the definition of the standard lactation curve. To minimize bias, components of standard lactation curves proper to each herd are computed jointly with random individual effects. Recently a new version of mBP was tested that puts expectations of constant animal effects to observed average values using Bayesian prediction, a feature also used by MTP.

Using Fuzzy-Logic to model airborne spread of foot and mouth disease

Traulsen, I. and Krieter, J., Christian-Albrechts-University, Institute of Animal Breeding and Husbandry, Olshausenstraße 40, 24098 Kiel, Germany; itraulsen@tierzucht.uni-kiel.de

A Fuzzy-Logic model (FLM) was developed to model the airborne spread of foot and mouth disease virus (FMDV). According to the Gaussian Dispersion model as reference, wind speed (u), stability class (k), amount of virus emitted at the origin farm (q) as well as x- and y-coordinates of the receiving farm were used as input parameters. A binomial output parameter was defined: 0 if the FMDV concentration arriving at the receiving farm was below the threshold for one animal (cattle: 0.06, sheep: 1.11, pigs: 7.7 tissue culture infective dose, $TCID_{50}/m^3$), 1 if it was above. Based on real weather data from Vienna, two data sets with 10,000 observations each were generated. For both data sets, a virus emission rate of $4*10^{10}$ $TCID_{50}/m^3*day$ was assumed (matched 1000 infected pigs). In the first data set, x (10-10,000 m) and y (10-3,000 m) were varied, while u (4 m/s), and k (4) remained constant. In the second data set, u was varied also (1-12 m/s). From each data set 90% of the values were used to estimate the FLM (training data) and the remaining 10% for the validation (testing data). Sensitivity (SE in %), specificity (SP in %) and error rate (ER in %) described the goodness-of-fit between the two models. Varying only the x and y coordinates resulted in high SEs for training and testing data (80.1, 81.1) and high SPs (93.0, 94.1), if the cattle threshold concentration was assumed. The ERs were relatively small (10.6, 10.0). With increasing threshold concentrations, the SE and SP increased (90.8-100.0, 99.9-100.0) but the ER increased also (33.3-52.4). Varying u, decreased SE and SP and increased ER. Lowest ERs were found for the cattle threshold (37.3, 39.3) with accordant SE of 75.5 and 75.3, and SP of 91.6 and 92.1. For the sheep and pigs, threshold SE and SP were larger (min. 73.9,) but also the ERs increased (min. 75.0). In conclusion, the FLM was adequate to estimate if the threshold concentration on a certain farm was reached. Best results were obtained for cattle, the most susceptible species to FMDV.

A survey on the selenium status in cattle herds in Wallonia

Robaye, V., Dotreppe, O., Hornick, J.L., Istasse, L., Dufrasne, I. and Knapp, E., Liege University, Nutrition Unit, Bd de Colonster 20, 4000 Liege, Belgium; eknapp@ulg.ac.be

Selenium (Se) is a trace element of importance for animals as it is implicated in many organs and functions by specific selenoproteins or metabolites such as the metylselenol. A survey was carried on 166 beef and/or dairy farms located in the 4 major geographical areas of Wallonia, the Southern part of Belgium. The Se status was assessed by glutathion peroxydase activity in the red blood cells and expressed as µg Se/l. The overall average Se concentration in 743 samples was 45.2±27.4 µg Se/l. This value was considerably lower than the 70 µg Se/l considered normal. There were 26 animals with Se concentrations below 10 µg/l and 37 with over 100 µg/l. The Se concentration was significantly higher in the dairy herds than in the beef herds (52.0 vs. 40.2 µg Se/l, $P<0.001$) but there were no differences between cows and heifers (45.6 vs. 46.6 µg Se/l). In dairy herds, large amounts of compound feedstuffs along with mineral mixtures are included in the diet. Both of these are normally supplemented with Se. By contrast, in beef herds, locally produced forages are the main dietary ingredients. They are low in Se owing to the low Se content in the soil. There were also large differences between geographical areas. Concentrations were highest in the grassland areas at 55.8 µg Se/l and they were lowest in Ardennes at 39.4 µg Se/l. The differences between areas may be ascribed mainly to the type of cattle herds - milk production in the grassland areas and mainly beef production in Ardennes. The geographical effect was thus confounded with the animal type effect. Owing to the large extent of the deficiency in Se status, Se intake should be increased by use of supplements high in mineral or organic Se, or by use of feedstuffs grown with Se enriched fertilizers. The use of such fertilizers could be of interest in the management of suckling beef herds based only on feedstuffs produced on the farm.

Serum metabolite and enzyme activities as biomarkers of high-grain diet consumption in finishing bull calves

Castillo, C.[1], Hernandez, J.[1], Pereira, V.[1], Mendez, J.[2], Vazquez, P.[3], Miranda, M.[4] and Benedito, J.L.[1], [1]Veterinary Faculty, USC, Animal Pathology, Universitary Campus, 27002 LUGO, Spain, [2]Coren, Research Department, Ourense, 32003, Ourense, Spain, [3]CESFAC, I+D+I, Diego de León, 28006, Madrid, Spain, [4]Veterinary Faculty, USC, Veterinary Clinical Sciences, Universitary Campus, 27002 Lugo, Spain; cristina.castillo@usc.es

This study evaluated the effects of three high-grain diets fed to growing/finishing feedlot cattle in Galicia (NW Spain) on serum parameters, seeking metabolic indicators of nutritional status of the animals. A 80-day feedlot study used 30 Belgian Blue bull calves allotted randomly to one of three experimental groups of 10 each, defined by the cereal grain in their diet, i.e. predominantly maize (group M), predominantly barley (group B), or predominantly a mixture of maize and barley (group MB). The serum parameters determined were glucose, non esterified fatty acids (NEFA), total protein, albumin, serum urea N (SUN), creatinine, L-lactate, aspartate amino transferase (AST) and gamma glutamyl transpeptidase (GGT). All the parameters measured fell within the physiological ranges for beef. This finding, in addition to the lack of clinical symptoms of ruminal disturbances, suggests that none of the diets were detrimental to health, due to the protein content of the ration and the forage fibre source. In respect of the metabolic parameters, we found that values of NEFA, SUN, creatinine, albumin and GGT cannot be considered useful biomarkers, as they exhibited significant time×treatment interactions. Only serum L-lactate and AST were directly influenced by diet, with significant differences among groups. Particularly interesting is the finding that blood glucose levels were not affected by type of diet, possibly because of genotype-dependent metabolic characteristics for this double-muscled breed. In conclusion, animals fed a high-grain diet with equal proportions of maize and barley did not have significantly better metabolic indicators than those fed a high-grain diet composed mainly by barley.

Use of dairy herd test-day effects stemming from genetic evaluations for herd management purposes

Leclerc, H.[1] and Ducrocq, V.[2], [1]Institut de l Elevage, Département Génétique, Batiment 211, 78352 Jouy-en-Josas, France, [2]INRA, UMR 1313 GABI, Batiment 211, 78352 Jouy-en-Josas, France; helene.leclerc@jouy.inra.fr

The development of genetic evaluations based on individual test-day records instead of classical 305-day lactation offers numerous perspectives for herd management. The main interests of the test-day model are its ability to account for environmental effects occurring on the day of milk recording through the herd test-day effect (HTD) and to account for the effect of days in milk. It also gives the opportunity to model individual differences in the shape of lactation curves. As all effects are estimated simultaneously, it allows an independent interpretation of each one. As HTD is related to short-term environmental effect, it can be used as an indicator of the herd management efficiency, and its accurate prediction is a major challenge. Three approaches were compared to forecast HTD. The first one is based on the decomposition of HTD into its predictable elements: a within-herd moving average and a within herd month average. The second is based on the mixed model methodology with fixed herd effects and random herd-year effects. The third one uses the Holt-Winters methodology based on time series analysis. Comparison of the three methods was made for milk, fat and protein yields and contents. Correlations between predicted effects and those estimated a posteriori were between 0.68 and 0.85 depending to the trait analysed and the prediction method used, with a small advantage to the mixed model methodology. Two dynamic applications can be developed from predicted HTD: a prospective tool to forecast herd production for the next month and a monitoring tool to assist technicians and/or farmers in detection and identification of herd management problems through comparison of predicted HTD with real ones estimated a posteriori. Expressed as a deviation from a regional mean level, HTD can be used to evaluate the technical level of the herd and to track down its strong and weak points in order to concentrate future effort on the latter ones.

Effects of suckling restriction and parity on metabolic and reproductive function of autumn-calving beef cows

Álvarez-Rodríguez, J.[1], Palacio, J.[2] and Sanz, A.[1], [1]Centro de Investigación y Tecnología Agroalimentaria, Gobierno de Aragón, Tecnología en Producción Animal, Av. Montañana, 930, 50059 Zaragoza, Spain, [2]Universidad de Zaragoza, Departamento de Patología Animal, c/Miguel Servet, 177, 50013 Zaragoza, Spain; jalvarezr@aragon.es

This experiment evaluated the effect of suckling restriction and parity on productive, metabolic and reproductive function of beef cows. Autumn-calving Parda de Montaña cows (n=46) were assigned to three nursing frequencies from the day after calving: Once-daily nursing (RESTR1), twice-daily nursing (RESTR2) and ad libitum nursing (ADLIB). Heifers (n=18) were maintained with free access to their calves as in the ADLIB cow's group. Cow daily gains were greater while milk yield and calf daily gains were lower in RESTR1 than in RESTR2 and ADLIB cows ($P<0.05$). Peripheral cholesterol and IGF-I did not differ across suckling systems ($P>0.10$), but their mean concentration was lower in ADLIB-cows than in ADLIB-heifers ($P<0.05$). Serum NEFA was lower in RESTR1-cows than in their RESTR2 and ADLIB counterparts on weeks 7 and 9 of lactation ($P<0.05$) whereas both RESTR1 and RESTR2 treatments showed lower serum NEFA than ADLIB-cows on week 11 post-partum ($P<0.05$). Serum NEFA were higher in ADLIB-cows than in ADLIB-heifers on week 1 and after week 7 of lactation ($P<0.05$). Serum β-hydroxybutyrate was lower in RESTR1 and RESTR2 than in ADLIB-cows ($P<0.05$). Calf management did not affect significantly the interval to first post-partum ovulation or oestrus in multiparous cows ($P>0.10$) but ADLIB-cows had shorter post-partum intervals to first ovulation than ADLIB-heifers ($P<0.05$). The different productive and metabolic function due to suckling restriction did not trigger remarkable differences in cow reproductive parameters. Adult animals had different metabolic traits compared to heifers, but they were unrelated to the observed delay in the onset of ovarian cyclicity of primiparous dams.

Introducing ethics in science higher education

Marie, M.[1] and Rollet, L.[2], [1]Nancy-Université, ENSAIA-INPL, B.P. 172, 54505 Vandoeuvre, France, [2]Nancy-Université, ENSGSI-INPL, B.P. 647, 54010 Nancy, France; Michel.Marie@ensaia.inpl-nancy.fr

In our multi-cultural societies, in a context of rapid evolution of sciences and techniques, the need of ethics is growing. As well as in other fields, the responsibility of the professionals involved in livestock production (scientists, technicians, operators) is engaged towards the society by the consequences of their action on life and environment, and by the growing public concerns. This situation requires the development of the awareness and competency of professionals in this field, either during their education or through further training. A difficulty arises in the necessary multidisciplinarity of such formations, combining both scientific and philosophical expertises. After a rapid presentation of the educational offer in Europe, we present a module proposed in a Master of science. This course, named Bioethics, Science and Society, is offered to students specializing in the domains of biology, food, nutrition, or forestry, agronomy and environment. Animal ethics represent a significant part of the issues, as these students may have to deal with animals either in experiments, or in management of livestock or wildlife. But the scope of the course is broader and aims at giving the philosophical bases as well as tools necessary to handle ethical issues in general. After a presentation of the main concepts and elements of moral philosophy, case studies are developed in small exercises or with more complex methods such as the ethical matrix or the method of reflexive equilibrium. A home work based on a report of an ethical committee offers the opportunity to practice such analyses. The second part of the module is devoted to science and society (responsibility of the scientist, ethics and research, intellectual property, ethical committees and guidelines, or citizen conferences), with the use of problem-based learning and role playing. The evaluation of the module shows that most of the students discover ethics and gain awareness and know-how after completing this course.

Animal welfare, environment and food quality interaction studies in Central and South-eastern Europe in virtual environment

Szücs, E.[1,2], Bozkurt, Z.[3], Gaál, K.[2], Sossidou, E.N.[4], Venglovsky, J.[5], Peneva, M.[6], Konrád, S.[2] and Cziszter, L.T.[7], [1]Szent István University, Páter Károly u. 1, 2103 Gödöllö, Hungary, [2]University of West Hungaryí, Vár 2, 9200 Mosonmagyaróvár, Hungary, [3]Afyon Kocatepe University, Ahmet Necder Sezer Campus, 03200 Afyonkarahisar, Turkey, [4]National Agricultural Research Foundation, P.O.Box 376, 57008, Ionia-Thessaloniki, Greece, [5]University of Veterinary Medicine, Komenskeho, 73 041 81 Kosice, Slovakia (Slovak Republic), [6]University of National and World Economy, Studentski grad, Hristo Botev, 1700 Sofia, Bulgaria, [7]Banat University of Agricultural Sciences and Veterinary Medicine, Calea Aradului 119, 300645 Timisoara, Romania; Szucs.Endre@mkk.szie.hu

In a LdV Pilot Project 'Promoting quality assurance in animal welfare – environment – food quality interaction studies through upgraded e-Learning' (Welfood) coordinated by HU new methods of advanced and multilingual and multilevel vocational training programs were developed in the field of (1) animal welfare, (2) environmental impacts on and of animals, and (3) food quality and safety interaction studies with focus on ethical, issues in animal production change. Partner countries were BE, EE, GR, and PL. The sustainability of results has been ensured by expressions of interest from further countries (BG, RO, SK TR) in a Transfer of Innovation project 'A new approach on different aspects of welfare, environment and food interactions in Central and South-Eastern Europe with use of ICT' (Welanimal) coordinated by the Turkey. The aim is to transfer and adapt the products in further regions with different socio-cultural, religious, regional and historical environment for better understanding of animal welfare concepts. The target groups are students, teachers, sectored enterprise workers and family members. The expected impact in the short term is to train target groups within their cultural perspective in the topic of animal welfare with courses and printed products, as well as encouragement to use ICT techniques.

Fundamental moral attitudes and their role in judgement

Stassen, E.N.[1], Brom, F.W.A.[2] and Cohen, N.[1], [1]Wageningen University, Animal Sciences, Marijkeweg 40, 6709 PG Wageningen, Netherlands, [2]Rathenau Institute, Anna van Saksenlaan 51, 2593 HW The Hague, Netherlands; elsbeth.stassen@wur.nl

An empirical model to describe fundamental moral attitudes to animals and their role in judgement on animal issues has been developed. The theoretical framework of the model will be presented. Our aim was to develop a model to describe the diversity of people's fundamental moral attitudes (FMAs) to animals. Furthermore, we aimed to clarify the role of these FMAs in the public debate about the culling of healthy animals in an animal disease epidemic. In the model we used criteria from philosophical animal ethics to describe and understand the moral basis of FMAs and the dynamics of FMAs in debates. Moreover, the criteria provides us with a moral language for communication between philosophical animal ethics, FMAs and public debates. The results of a survey performed in the Netherlands will be presented. Two dominant FMAs were identified among the respondents. More FMA1 respondents were men, were older, lived in smaller towns or in the country, with a higher education, and with less contact with animals than FMA2 respondents. The FMAs also differed with respect to their views on the hierarchical position of animals, in the valuation of convictions, and in judgement. The FMA1 group considered humans to be superior to animals, while the FMA2 group considered both to be equal. The FMA2 group valued convictions about animals higher and more were opposed to the culling. The model appeared useful in discussions on various animal issues among international biomedical students. It structured the discussion because the differences in convictions and their relevance for the topic became clear.

Drawing a transparent line between acceptable and unacceptable welfare in livestock production

Jensen, K.K., Danish Centre for Bioethics and Risk Assessment, Rolighedsvej 25, DK-1958 Frederiksberg C, Denmark; kkje@life.ku.dk

This paper reports preliminary considerations on how to draw the line between acceptable and unacceptable welfare in livestock production. These considerations belong to a larger interdisciplinary research project, initiated in 2009 in Denmark, called 'On-farm animal welfare assessment for farmers and authorities'. The project will test three hypotheses: (1) It is possible to make a valid and accurate risk-based identification of farms with a high level of welfare problems based on limited information (mostly from central data registers), (2) It is possible, by combining reflection on animal ethics, economics, legal requirements and measurement theory, to set up transparent models of how to draw the line between acceptable and unacceptable welfare, and (3) it is possible to design training courses for farmers with unacceptable welfare leading to increased understanding and more far-reaching and lasting improvements than legal orders from the authorities. The present paper addresses hypothesis (2). Animal welfare is not a singular, directly measurable parameter. Assessment of welfare involves interpretation of several indicators, and the interpretation is dependent of the underlying understanding of welfare. Three theories of welfare are known: Hedonism (welfare is the degree of positive quality of mental states); Preference satisfaction theory (one state involves higher welfare than another, if it is preferred); and perfectionism (welfare is the realisation of the potential specific for the species). Aggregation of welfare on farm level involves comparisons of different welfare states and raises the question of whether, and to which degree, trade-offs should be allowed. Finally, the determination of the line between acceptable and unacceptable is dependent of the weight of other relevant concerns, such as costs of improvements and the justice of legal regulation. The paper will address these questions by making underlying value assumptions transparent.

Development of a tool for the overall assessment of animal welfare at farm level

Botreau, R.[1], Perny, P.[2], Champciaux, P.[1], Brun, J.-P.[1], Lamadon, A.[1], Capdeville, J.[3] and Veissier, I.[1], [1]INRA, UR1213 Herbivores, Site de Theix, 63122 Saint-Genès-Champanelle, France, [2]Université Paris 6, LIP6, 104 avenue du Président Kennedy, 75016 Paris, France, [3]Institut de l Elevage, BP18, 31321 Castanet-Tolosan, France; raphaelle.botreau@clermont.inra.fr

Taking into account the increasing societal concern for animal welfare, the European research project Welfare Quality® aimed at designing systems to monitor the welfare of cattle, pigs and poultry on farms. An assessment tool was developed with a view of helping farmers identify welfare problems and monitor progresses, as well as providing information to consumers about the animals from which they buy products. Repeatable and feasible measures have been defined to cover all dimensions of welfare. The data collected on a given farm are combined to produce welfare scores for 12 welfare criteria to be fulfilled to ensure welfare (absence of hunger, absence of thirst, comfort around resting…). The scores are aggregated into four principle-scores corresponding to main welfare dimensions: Good feeding, Good housing, Good health, Appropriate behaviour. Welfare scores are expressed on a 0-100 value scale, where 0 is for worst situations and 100 for best ones. An overall welfare assessment of the animal unit is finally produced by comparing the four principle-scores to reference profiles delimiting four welfare categories. These categories (Excellent, Enhanced, Acceptable, Not classified) were defined with stakeholders according to the potential uses they would make of a welfare assessment tool. A software chain is proposed to ease the collection of data on farms, to store them on a web database, to calculate scores, to synthesise the results for end-users (producers, retailers, consumers…), and to simulate welfare improvements. The tool offers a standardised way to check animal welfare and performs all necessary calculations. It should thus largely facilitate the implementation of welfare programs.

Automation systems for farm animals: potential impacts on the human animal relationship and on animal welfare

Cornou, C., Faculty of Life Sciences, University of Copenhagen, Department of Large Animal Sciences, Groennegaardsvej 2, 1870 Frederiksberg C. Copenhagen, Denmark; cec@life.ku.dk

The use of automation systems in animal farming raises ethical issues. These systems automatically collect various kinds of information about an animal and allow the farmer to monitor it remotely. It is argued that the relationship between the farmer and the individual animal is becoming increasingly distant and impoverished. Although this may protect the animal from some negative interactions, it is less clear whether use of these systems will lead to an increase in positive interactions of the kind beneficial for animal welfare. A better monitoring of the individual animal seems a priori beneficial for animal welfare. Examples of potential positive applications are an earlier and better detection of diseases, or a more homogeneous manner to perform routine tasks (as with a milking robot). However, it is suggested that increasing automation may result in a growing objectification of the animal. As automation systems replace traditional tasks, the role of the farmer is changing drastically. This may lead to deskilling in the farmer, which in turn may affect the animal welfare. Farmers need to be aware of the risks, both of overlooking problems that are not detected by the sensors, and that this remote monitoring may risk impairing their sensitivity towards their livestock, and make them less capable of taking action to treat the individual animal. The value of automation systems in increasing productivity is clear; however, it is questioned the extent to which these systems can be used to enhance animal welfare. It is argued that ethically acceptable development of automation systems for farm animals can only be achieved if these systems prove to be beneficial in respect of animal welfare.

An ethical analysis of pig castration and the alternatives

Edwards, S.A.[1], Von Borell, E.[2], Fredriksen, B.[3], Lundstrom, K.[4], Oliver, M.A.[5], De Roest, K.[6] and Bonneau, M.[7], [1]Newcastle University, Newcastle upon Tyne, NE1 7RU, United Kingdom, [2]Martin Luther University, Halle, Germany, [3]Animalia, Oslo, Norway, [4]SLU, Uppsala, Sweden, [5]IRTA, Monells, Spain, [6]CRPA, Reggio Emilia, Italy, [7]INRA, St Gilles, France; sandra.edwards@ncl.ac.uk

The EU PIGCAS project carried out a review of the available research and other information on surgical castration without pain relief and its alternatives: surgical castration with pain relief (anaesthesia, analgesia), alternative castration methods (e.g. chemical castration, immunocastration), meat production from entire male pigs, future technologies (e.g. sperm sorting, genetic modification). These reviews were subsequently discussed at an expert workshop, and 38 experts from academia and industry then scored each option according to five different aspects: public attitudes, practicality, animal welfare, pigmeat quality, resource efficiency and economy. These scores were used to carry out an ethical appraisal of the options using the framework of the Ethical Matrix, giving consideration of the principles of wellbeing, autonomy (choice) and justice (fairness) in relation to each of the stakeholder groups: farmers, consumers, animals and the environment. This highlighted key ethical dilemmas, with options which improve animal welfare relative to the current situation often having significant detrimental aspects for farmers, in terms of workload, safety and income, or to consumers, in terms of the cost and quality of pigmeat.

Pig welfare model development: the animals' perspective

Averos, X.[1], Brossard, L.[1], Edwards, S.A.[2], Edge, H.L.[2], De Greef, K.H.[3], Dourmad, J.-Y.[1] and Meunier-Salaün, M.C.[1], [1]INRA, UMR1079 SENAH, 35000 Rennes, France, [2]University of Newcastle, NE1 7RU, Newcastle upon Tyne, United Kingdom, [3]Animal Sciences Group of Wageningen UR, P.O. Box 65, 8200 AB Lelystadt, Netherlands; xavier.averos@rennes.inra.fr

The objective of the EU QPorkChains project is to improve the quality of pork taking into account different dimensions, including animal welfare. In this context, Workpackage VI.3 aims to integrate the existing knowledge on animal welfare, combining both the scientific and the social views, through a modelling approach to predict the effect of changes in the production systems. Three key interest-perspectives are considered: animal, farmer, and citizen-consumer. An animal model is being developed using a multi-agent approach, which takes into account the variability between pigs. Individuals are characterized by attributes including genotypic and phenotypic traits (breed, sex, age, body weight …), and welfare criteria (behaviour, health and performance). The model also considers the two main factors implicated in the evolution of the system, i.e. housing and feeding conditions, and their impact on the welfare criteria. Relationships between factors and animal attributes are extracted from literature. For instance the link between space allowance, described using an allometric basis, and the lying behaviour is assessed using a broken-stick analysis. Relationships with other factors such as group size, floor type, ambient temperature, enrichment, feeding level and access will also be evaluated. An overall model will be further developed linking the animal model with the farmer and the citizen-consumers models. This overall model expects to provide information on how the choice of farming practices, consumer purchases, and ethical positions, can impact on pigs' welfare and production systems.

Pig welfare model development: the farmers' perspective

Edge, H.L.[1], Averos, X.[2], Brossard, L.[2], De Greef, K.H.[3], Dourmad, J.-Y.[2], Meunier-Salaun, M.C.[2] and Edwards, S.A.[1], [1]University of Newcastle, Newcastle upon Tyne, NE1 7RU, United Kingdom, [2]INRA, UMR1079 SENAH, F-35000, Rennes, France, [3]Wageningen UR, Animal Sciences Group, P.O. Box 65, 8200 AB Lelystadt, Netherlands; h.l.edge@ncl.ac.uk

The Q Porkchains project has identified four key interest perspectives with regards to animal welfare (the animal, the farmer, the consumer and the citizen) and aims to integrate existing knowledge from all four perspectives through a modelling based approach. The farmer can influence pig welfare both directly through the day-to-day interactions with the animals, and indirectly by the business decisions which are taken. These decisions have an impact on the uptake of new technologies, the systems in which animals are produced and consumers perceptions of animal agriculture. Whilst some management decisions can be made on a purely financial basis, those relating to animal welfare are usually the result of a trade off between the farmers desire to run a successful business and their moral views as a member of society. The model of factors impacting on farmers' decisions with regards to animal welfare identifies the main drivers as personal, economic and legal. It identifies how the individual personality characteristics and socio-demographics of the farmer, along with their personal views with regards to welfare and the wider social framework in which they farm, interact with various external constraints (economic, legal, human, geographical) to determine management choices.

Pig welfare model development: consumers and citizens

De Greef, K.H.[1], Averos, X.[2], Brossard, L.[2], Dourmad, J.-Y.[2], Edge, H.L.[3], Edwards, S.A.[3], Meunier-Salaün, M.C.[2] and Ursinus, W.[1], [1]Animal Sciences Group of Wageningen UR, Lelystad, Netherlands, [2]INRA, UMR1079 SENAH, Rennes, France, [3]University of Newcastle, Newcastle upon Tyne, United Kingdom; karel.degreef@wur.nl

In the societal search for acceptable and economic pork production systems, the Farmer, the Animal, the Consumer and the Citizen can be seen as agents that have different views on what is 'a good life' or 'good welfare'. In the EU-project Quality Pork Chains, these four agents are modelled separately, to be combined later into one multi-agent model that depicts possible effects of system changes. The non-farmer human actor can be seen as being comprised of two mixed agents: consumer and citizen. The agent Consumer is defined as the human being in its meat purchasing role; the agent Citizen as the human being taking or feeling social responsibility. Empirical studies report both substantial social interest in animal welfare and a considerable willingness to pay for additional welfare claims on meat. Nevertheless, the pork system in NW-Europe is dominated by conventional pork without explicit welfare claim. To model the seemingly discordant behaviour, the classic distinction between Knowledge, Attitude and Behaviour is used as the core concept. This approach is derived from Fishbein & Ajzen (1975), elaborated in the Theory of Planned Behaviour. Consumer behaviour is modelled on basis of the food choice process. The role of influencing factors on public view and acceptance such as the media need to be explored. But first, a useable concept for public acceptance is sought. Until date, some connections between the consumer role and the citizen role are represented as political consumerism and donation behaviour. Within the project, is has been realised that it is a challenge to bridge gaps between the technical (animal) sciences and the relevant social sciences. Still, it is believed that viable welfare arrangements can only be built on expertise that understands both the animal and the humans involved (farmers, consumers and citizens).

Authors index

A

Aamand, G.P.	608
Aarts, H.F.M.	23
Abadjieva, D.	370
Abbasi, H.	262
Abbasi, M.A.	259, 259
Abd El-Khalek, A.E.	272
Abd Elsamee, L.D.	45
Abd-Allah, S.A.E.	377
Abdel Hakeem, A.	462
Abdel-Magid, S.	380
Abdelhadi, O.M.A.	61
Abdi-Benemar, H.	346
Abdolmaleki, Z.	540
Abdouli, H.	540
Abdullah, A.Y.	544
Abe, T.	142
Abecia, J.A.	275
Abedo, A.A.	517, 573, 582
Abel, H.J.	342
Abeni, F.	97
Abi Said, M.	284
Abilleira, E.	49, 53
Abo El-Nor, S.A.H.	356
Abo-Donia, F.M.	517, 573, 582
Abo-Eid, H.A.	577
Abou-Fandoud, E.I.	462
Abramson, M.	410
Abras, S.	616
Abreu, G.C.G.	287
Abu Siam, M.	74
Abusneina, A.	528, 531
Acciaro, M.	348
Acero, R.	237, 237, 238, 532, 533, 533
Acosta Aragón, Y.	374, 416, 590
Adibmoradi, M.	524
Aferri, G.	408
Afonso, F.P.	13
Agabriel, C.	311
Agabriel, J.	238
Aghaziarati, N.	380
Agrícola, R.	307, 420
Aguado, D.	472
Aguilar, I.	298, 301
Aguilera, J.F.	369, 567
Aguinaga, M.A.	369, 567
Aguzzi, J.	118
Agüera, E.	427
Ahadi, A.H.	21, 418
Ahmadi, A.B.	326
Ahola, V.	205
Aida, H.	249
Aihara, M.	201

Ait-Saidi, A.	490, 493
Ajmone Marsan, P.	17, 149, 186, 609
Akhlaghi, A.	397, 397, 398
Aksoy, Y.	10, 98
Aksu, S.	176
Aktoprakligil, D.	176
Al Aïn, S.	134
Al Baqain, A.	74
Al Yacoub, A.N.	277, 277
Al-Ramamneh, D.	535
Al-Soqeer, A.A.	574
Alabart, J.L.	82, 263, 275, 278
Albanell, E.	362, 469, 470
Albaqain, A.	73
Albaqain, R.	73
Albar, J.	230
Albera, A.	145, 148, 149
Albers, G.	211
Albertsdóttir, E.	217
Albertí, P.	38
Albini, S.	339
Albisu, M.	49, 53
Alborali, G.L.	452
Albrecht, A.K.	438
Albrecht, C.	349
Alcalde, M.J.	36, 52, 53, 115, 464
Alcazar, E.	213
Alegría, D.	393
Aletru, M.	491
Alexandre, G.	255, 258
Alfonso, L.	178, 434, 536
Alipanah, M.	360
Alipoor, K.	63
Allahrasani, A.	358, 582
Allain, D.	257, 257, 314
Allais, S.	132
Allen, P.	418
Allen, W.R.	303
Allipour, F.	274
Almasli, I.	528, 531
Almeida, A.M.	99, 543
Alonso, I.	567
Aloulou, R.	187
Altarriba, J.	140
Althaus, R.L.	409
Álvarez, F.	19
Alvarez, I.	175, 265, 267
Alvarez, M.J.	402
Alvarez, S.	80, 402
Álvarez-Rodríguez, J.	82, 87, 253, 286, 390, 391, 526, 619
Alves, E.	161
Alves, S.P.	36

Amador, C.	323	Ariño, L.	454
Amanlo, H.	589	Arkoudelos, J.	48, 85
Amanloo, H.	362	Árnason, T.	217
Amanlou, H.	62, 364, 365, 380	Arnau, J.	34
Ambord, S.	392	Arnold, E.T.	388
Ambrosiadis, J.	48, 85	Arnould, V.M.-R.	203
Amigues, Y.	181, 460	Arné, P.	551
Amills, M.	157, 158, 267, 268	Aro, J.	205
Aminafshar, M.	151, 269	Arquet, R.	126, 258, 460
Amirinia, C.	151	Arranz, J.	49, 53, 73
Amirinia, S.	151	Arranz, J.J.	261
Ammon, C.	344	Arsenos, G.	84
Amores, G.	49	Arshami, J.	357, 571
Ampuero Kragten, S.	35	Arvelius, P.	65
Anastasiou, I.	84, 469	Asadzadeh, N.	91, 92
Anastasopoulos, V.	346	Ashabi, S.M.	378
Andanson, L.	311	Ashmawy, T.A.	272, 273
Andersen, B.H.	75	Ashworth, C.J.	295
Anderson, L.	505	Aslaminejad, A.G.	593
Anderson, T.J.C.	505	Aslan, O.	176
Andersson, G.	67	Asmini, E.	387
Andersson, L.	14	Asquini, E.	247
Andrade, P.L.	91, 147	Astolfi, A.	136
Andrejsová, L.	69, 70, 176, 431	Astruc, J.M.	3, 507
Andrews, S.	437	Atashi, H.	550
Andrieu, S.	34, 43	Atti, N.	578
Angelozzi, G.	87, 370	Autran, P.	257, 257
Angiolillo, A.	268	Avendano, S.	212
Angón, E.	238, 532, 533, 533	Averos, X.	623, 623, 624
Anguita, M.	513	Avgeris, I.	469
Anjos, M.A.	373	Avilés-López, K.	154, 309, 391
Ansari, A.	378, 523	Avon, L.	5
Antkowiak, I.	204	Awadalla, I.	380
Anton, I.	190	Awata, T.	155
Antonelli, S.	570	Ayadi, M.	60, 368, 470, 473
Antonini, M.	316	Azarfar, A.	564, 564
Antoszkiewicz, Z.	542	Azevedo, J.M.T.	546, 546, 548, 549, 586
Antunovic, Z.	477	Aziz, M.A.	530
Apaza Castillo, N.	317	Azor, P.J.	267, 421
Arab, A.	358	Azzi, R.	407
Aragni, C.	459	Azzini, I.	496
Aramyan, L.	596		
Arana, A.	178, 536	**B**	
Arandia, A.	292	Baars, T.	20
Arat, S.	176	Babaei, M.	91, 92, 386, 387
Arata, S.	159	Babiker, S.A.	61
Arav, A.	408	Babilônia, J.L.	373
Arce, C.	505, 508	Babo, H.	86
Archibald, A.L.	135, 505	Babot, D.	458, 483
Ardalan, M.	516	Bacci, M.L.	288, 431, 432, 452
Ardiyanti, A.	142	Bacciu, N.	169
Arefnezhad, B.	609	Bach, A.	519
Ares, J.L.	268, 268	Bach, R.	86
Arfsten, M.	171	Backus, G.	332, 596
Argente, M.J.	171	Badaoui, B.	268
Argüello, A.	468, 474, 543, 545	Badiola, J.J.	391

Badran, A.E.	530	Baumgartner, J.	437
Baena, F.	268, 268	Baumont, R.	424
Baes, C.F.	510, 603	Baumung, R.	4, 109, 191, 409
Bagis, H.	176	Bay, E.	77
Bagnato, A.	609	Bayat Koohsar, J.	383, 384, 594, 594
Bahelka, I.	443, 453	Bazin, C.	285
Bailoni, L.	311, 579	Beal, J.D.	565
Bain, M.M.	169	Beattie, E.M.	185
Balasch, S.	116	Beaudeau, F.	554
Balcells, I.	156	Beaumont, C.	138
Balcioglu, K.	176	Bébin, D.	232
Baldi, F.	241, 400	Becchetti, T.A.	327
Baldi, M.	117	Bed'hom, B.	169
Baldo, A.	486, 494	Bednarczyk, M.	168
Balenović, M.	15	Beduin, J.M.	357
Balenović, T.	15	Bee, G.	35, 35
Balieiro, J.C.C.	14, 18, 189	Beerda, B.	181
Balvay, B.	490	Beev, G.	583
Bambou, J.C.	460	Begley, N.	206
Bampidis, V.A.	47, 48, 49, 85, 85, 364	Beja, F.	429
Banabazi, M.H.	91, 92	Bellec, T.	209
Banchero, G.	241, 400	Beltran, J.A.	51, 54
Bani, P.	577	Beltrán, M.C.	409
Bannink, A.	25	Beltrán De Heredia, I.	49, 53, 73, 76
Barac, Z.	477	Ben Gara, A.	529
Barandiarán, M.	19	Ben Hamouda, M.	187
Barbosa, E.	185	Ben M'Rad, M.	368, 470
Barbosa, M.A.A.	542	Ben Salem, I.	272
Barbour, E.K.	284	Ben-Noon, I.	410
Barcaccia, G.	288	Bendixen, C.	156, 211
Barcelos, B.	367	Benedettini, A.	422
Barea, R.	584	Benedito, J.L.	234, 411, 413, 528, 614, 618
Barile, V.L.	152, 389, 570	Benhajali, H.	283
Barillet, F.	507	Benito, J.M.	11
Barmat, A.	473	Benne, F.	192
Barnett, J.L.	437	Bennett, L.H.	488
Baro, J.A.	123, 142, 398	Bennett, L.N.	326
Barragán, C.	137, 161, 213	Bennewitz, J.	6, 320, 603
Barrefors, P.	196	Benoit, M.	292
Barrey, E.	226	Benradi, Z.	268
Barrón, L.J.R.	49, 53	Beretta, E.	486, 494
Bartiaux-Thill, N.	233, 337, 357, 501	Berg, P.	108, 123, 130, 319
Bartol, F.F.	456	Berg, W.	344
Bartolomé, E.	8, 315, 423	Bergaoui, R.	375, 580
Bartolomé, J.	235, 334, 334, 499	Bergero, D.	426
Barton, L.	146	Bergsma, R.	134, 162
Basayigit, L.	403, 525	Béri, B.	16, 103, 399, 415
Bassoul, C.	427	Bermejo, L.A.	545
Bastiaans, J.A.H.P.	348	Bermingham, M.L.	183
Bastiaansen, J.	296, 300, 301	Bernabucci, U.	117, 577
Bastian, S.	551	Bernard, L.	184
Bastin, C.	106, 106, 205	Bernard-Capel, C.	132
Batellier, F.	304	Bernardini, C.	452
Bathrachalam, C.	318	Bernués, A.	76, 290
Bauchart, D.	61	Berri, C.M.	138
Baulain, U.	538, 560	Berruga, M.I.	419, 467

Berry, D.P.	129, 141, 145, 148, 183, 400	Boichard, D.	2
Berry, S.	185	Boisdon, I.	311
Bertechini, A.G.	56	Boissy, A.	257
Bertin, G.	427	Boisteanu, P.C.	328, 333
Bertozzi, C.	106, 357, 616	Boivin, X.	283
Bertrand, G.	568	Bojanovsky, J.	381
Bertschinger, H.U.	505	Bokaian, J.	500
Besharati, M.	574	Bokkers, E.A.M.	111
Besle, J.M.	313	Bolger, T.	350
Bessa, R.J.B.	36, 40, 40, 56, 59, 443	Bömcke, E.	298, 302
Bevilacqua, C.	184	Bonavitacola, F.	252
Bewley, J.M.	103	Bonde, M.K.	111, 113
Bezirtzoglou, E.	84, 469	Bonet, J.	39
Biagini, D.	404, 404	Bongiorni, S.	17, 149
Bibé, B.	3, 257, 257, 260, 507	Bonin, M.N.	20
Bidanel, J.P.	134, 159, 164, 209, 285, 434	Bonizzi, L.	9, 556
Bielfeldt, J.C.	163	Bonneau, M.	332, 497, 622
Biffani, S.	186	Bonnefont, C.	507
Bigeriego, M.	327, 328, 439, 439, 558	Bonnet, A.	192
Bijma, P.	134, 321, 600	Bonnot, A.	507
Billon, Y.	134, 159, 282, 285	Bontempo, V.	515
Binnendijk, G.P.	116	Boogaard, B.K.	496
Birchmeier, A.N.	598	Boraie, M.A.	377
Birgele, E.	59	Borba, A.E.S.	40, 450
Biscarini, F.	504	Borba, A.R.	56, 382
Bishop, J.	483	Borchers, N.	163, 537
Bishop, S.C.	459, 460, 503	Börner, V.	611
Bispo, S.V.	361, 361	Borrás, M.	409
Bittante, G.	145, 148, 350	Borys, A.	37
Bízková, Z.	363	Borys, B.	37, 47
Björnerfeldt, S.	67	Boscher, M.Y.	507
Blair, H.	395	Bostad, E.	331
Blanch, M.	519	Bősze, Z.	265
Blanco, I.	413	Botega, L.M.G.	56, 88, 89
Blanco, M.	98, 526	Botreau, R.	621
Blanco-Penedo, I.	234	Botti, S.	133
Blasco, A.	173, 173	Bouabidi, M.A.	368
Blasco, I.	87	Bouche, R.	459
Blasco, M.E.	278	Bouchel, D.	507
Blasi, D.A.	484	Boudry, C.	456, 457
Blasi, M.	609	Bouffartigue, B.	507
Bloettner, S.	32	Bouffaud, M.	285
Blouin, C.	220	Bouix, J.	257, 257, 260, 270, 507
Bo, N.	28	Boulanger, L.	181
Boadella, M.	552	Boulbaba, R.	540
Bobić, T.	412	Boulesteix, P.	283
Bochu, J.L.	232	Bouquet, A.	122
Bocquier, F.	485	Boushaba, N.	264
Bodas, R.	64	Bouvier, F.	260
Bodin, L.	192, 260, 270, 271, 276, 491, 599	Bouwman, A.C.	134
Bodó, I.	16, 227	Boué, P.	270
Boer, H.M.T.	187	Bovenhuis, H.	140, 157, 169, 183, 296, 300, 504, 616
Boer, M.	226		
Bogdanović, V.	61, 412, 529	Bowman, P.	294
Bogner, P.	166	Boyce, R.	417
Bohte-Wilhelmus, D.	6	Bozkurt, Y.	79, 403, 525

Carlini, G.	69	Cerra, Y.	50
Carlström, C.	107	Cerri, D.	471
Carmona, K.	393	Cervantes, I.	8, 8, 198, 225, 227
Carnier, P.	145, 148	Chalkias, H.	436
Carné, S.	489, 490, 492, 493, 493, 494	Chamani, M.	340
Caroli, A.	190	Champciaux, P.	621
Carolino, I.	19	Chapaux, P.	241
Carolino, M.I.	12, 198	Charpigny, G.	248
Carolino, N.	12, 13, 198, 224	Charvatova, V.	62
Carrasco, A.	213	Chatzipanagiotou, A.	364
Carretta, A.	224	Chatziplis, D.	32
Carriedo, J.A.	258	Chaudhry, A.S.	347, 566, 573
Carriquiry, M.	241, 400	Chavatte-Palmer, P.	181, 305
Carrizosa, J.	267, 268, 268	Chaveiro, A.	307, 420
Carrión, D.	139	Chedid, M.	284
Carvajal, A.	508	Cherr, C.	327
Carvajal, J.A.	154, 391	Chesnais, J.	31
Carvalhais, I.	420	Chessa, S.	189
Casabianca, F.	75, 81	Chevalet, C.	613
Casals, R.	60, 368, 469, 470	Chiariotti, A.	570
Casao, A.	275	Chibon, J.	507
Casaponsa, J.	458	Chikunya, S.	562
Casasús, I.	98, 286, 526	Chomón Gallo, N.	22
Casavola, V.	152, 394	Choroszy, Z.	77
Casellas, J.	79, 86, 264	Chrenková, M.	580, 584
Cassandro, M.	78	Christensen, O.F.	211, 299, 598
Castejon, R.	512	Christodoulou, V.	47, 48, 49, 85, 85, 364, 592
Castel, J.M.	532	Churcher, C.	505
Castellana, E.	152, 394	Ciampolini, R.	69
Castellano, R.	369	Ciani, E.	69, 152, 394
Castellanos Moncho, M.	421	Ciani, F.	422
Castiglioni, B.	189	Çiftçioğlu, G.	206, 496
Castillo, C.	234, 411, 413, 614, 618	Cilev, G.	385
Castillo, V.	469, 470	Čílová, D.	69, 70
Castro, F.A.B.	542	Činkulov, M.	547
Castro, N.	468, 474, 543	Ciocîrlie, N.	331
Castro, P.	38	Cirera, S.	505
Castro, T.	64	Citek, J.	440, 441, 442, 442, 446, 446
Catillo, G.	224	Claudi-Magnussen,, C.	75
Cattani, M.	575, 579	Clausen, A.	228
Caubet, C.	507	Cloete, S.W.P.	263
Cavalieri, A.	117, 118	Closter, A.M.	140
Cavini, S.	519	Cocero, M.J.	275
Cazaux, J.G.	230	Coelho, A.V.	99
Cazemier, C.H.	20	Coelho, M.	571
Ceacero, F.	329	Coenen, M.	513
Cecchi, F.	69	Cohen, N.	620
Cecchinato, A.	148	Coleman, J.	129
Cecchinato, R.	311	Colinet, F.G.	12, 600
Celaya, R.	38	Colitti, M.	247
Celi, I.	538, 539, 539	Collado-Romero, M.	505, 508
Celorrio, I.	195	Collares-Pereira, M.J.	308
Čerešňáková, Z.	580, 584	Colli, L.	186
Cerina, S.	41, 55	Colom, C.	586
Cerino, S.	288, 426, 429, 431, 432	Colombo, M.	136
Cerisuelo, A.	39, 381	Coma, J.	39

Comella, M.	166	Dall'olio, S.	164
Commun, L.	118	Dalmau, A.	112, 254, 287
Conde-Aguilera, J.A.	567, 584	Dammann, M.	70, 71
Cone, J.W.	312, 563, 563	Dämmgen, U.	402
Cong, T.V.C.	606	Danchin-Burge, C.	68, 71, 121
Conte, G.	189	D'Andrea, M.	17, 133, 149
Conti, R.M.C.	367	D'Andrea, S.	117, 118
Contiero, B.	306	Danesh Mesgaran, M.	313, 371, 373, 379, 383,
Contò, G.	365		520, 572, 575, 581, 587, 593
Cornou, C.	622	Daniel, J.	93
Cornu, A.	313	Danieli, P.P.	577
Correa, J.A.	254	Danvy, S.	222
Correia, M.J.	429, 432	Darabi, S.	274
Corte, R.R.P.S.	408	Dardenne, P.	202, 203
Cortés, O.	153	Daridan, D.	130
Coster, A.	157, 296, 300, 301	Darnhofer, I.	289
Cothran, E.G.	175	Das, A.	246, 560
Coueron, E.	336	Dashab, G.H.	360, 360
Coughlan, F.	350	Daskalopoulou, E.	527
Cournut, S.	290	Dastar, B.	57, 353
Couvreur, S.	324	David, I.	260, 599
Cozzi, G.	332	David, V.	250
Crenshaw, J.	449	Davis, S.R.	185
Crepaldi, P.	186, 317	Davis, T.A.	94
Crepon, K.	230	Davoli, R.	133, 164, 166
Crespo, D.G.	516	Davy, J.	327
Crespo, I.	124	Dawson, K.	244
Crespo, J.P.	516	Daza, A.	246
Crews Jr., D.H.	148	Daza, J.	315
Croiseau, P.	2, 294	Daß, G.	342
Cromie, A.R.	141, 183	De Argüello Díaz, S.	22
Crompton, L.A.	25	De Boer, I.J.M.	111
Cronin, G.M.	437	De Boever, J.L.	354, 514
Crooijmans, R.P.	140	De Brabander, D.L.	96, 354, 569
Crook, B.J.	150	De Campeneere, S.	354, 569
Crowley, J.J.	148	De Freitas, M.A.R.	152
Cruz, V.	115	De Greef, K.H.	1, 125, 480, 497, 623, 623, 624
Csapó, J.	472	De Haas, Y.	6
Cucco, D.C.	189	De Klerk, B.	605
Cue, R.I.	180, 195	De Koning, D.J.	135, 138, 169, 178, 601, 611, 612
Curran, J.	310	De La Chevrotière, C.	256, 460
Cutullic, E.	104	De La Fuente, J.	11
Cyrino, J.E.P.	356	De La Fuente, L.F.	185, 258
Cziszter, L.T.	620	De Marchi, M.	148
		De Montera, B.	181
D		De Ondiz, A.	154, 405
D'Abbadie, F.	169	De Pedro, E.	213
Dabiri, N.	45	De Renobales, M.	49, 53
Daetwyler, H.D.	293	De Roest, K.	622
Daftarian, P.M.	177	De Roos, A.P.W.	29
Daga, C.	146	De Smet, K.	514
Dal Maso, M.	350	De Vries, M.	111
D'Alessandro, A.G.	84, 84, 90	De Wit, A.A.C.	181
Dalin, G.	215	De Witt, F.H.	534
Daliri, M.	397, 398	D'Eath, R.B.	281, 282
Dallan, E.M.	494	Debus, N.	485

Farant, A.	126, 256	Fike, K.E.	484
Farhangfar, H.	358, 358, 379	Fikse, W.F.	65, 141, 196, 597, 608
Faria, P.B.	88, 89, 91, 373	Fikselová, M.	376
Farias, I.	361, 361	Filangi, O.	322, 603
Farid, A.	177	Filik, G.	41
Farkas, V.	190	Filipcik, R.	410, 471
Farruggia, A.	313	Filippini, F.	17
Fatehi, J.	205	Fina, M.	79, 604
Fatet, A.	270	Finocchiaro, R.	186
Fathi Nasri, M.H.	358, 358, 379, 575, 582	Fiore, G.	250, 252, 487, 496
Faucitano, L.	254	Fiorotto, M.L.	94
Faucon, F.	184	Fischer, K.	436, 560
Fayazi, J.	45	Fitie, A.	43
Faye, B.	61	Flachowsky, G.	243, 341, 581
Fazaeli, F.	44, 590	Flak, P.	584
Fazaeli, H.	576	Flamarique, F.	434
Fazekas, G.	15	Fleurance, G.	424
Fekete, Z.S.	412	Flisikowski, K.	194, 510
Felicetti, M.	226, 288	Foisnet, A.	393
Felleki, M.-B.	597	Folch, J.	82, 263, 275, 278
Ferm, K.	67	Foltys, V.	43
Fernandes, R.	432	Fonseca, A.J.M.	56
Fernandes, R.H.R.	367, 407	Font I Furnols, M.	52
Fernández, A.	11, 156, 213	Fontanesi, L.	136, 164, 609
Fernández, A.I.	137, 156, 246, 509	Fontes, C.M.G.A.	516
Fernández, A.M.	329	Forabosco, F.	31
Fernández, C.	381	Forbes, J.M.	24
Fernandez, I.	175, 265, 267	Forcada, F.	275
Fernández, J.	119, 120, 323, 602	Formigoni, A.	370
Fernández, J.A.	7, 88	Forni, M.	452
Fernàndez, X.	235	Foroughi, A.R.	371, 372
Fernández-Cabanás, V.	267	Forughi, A.R.	353, 578
Fernández Casado, J.A.	355	Forzale, F.	471
Fernández Irizar, J.	22	Fotou, K.	84, 469
Fernández-Fígares, I.	585	Foucras, G.	507
Fernández-Rodríguez, A.	135, 156	Fouilloux, M.N.	122
Fernando, R.L.	303	Foulquie, D.	257, 257
Ferns, L.E.	177	Fourichon, C.	338, 554
Ferrão, S.B.P.	91	Foury, A.	279, 281, 282, 285
Ferrarini, A.	247	Fradinho, M.J.	429, 432
Ferraz, J.B.S.	13, 14, 18, 20, 100, 189	Fraisse, D.	313
Ferreira, C.S.	564	Francàs, C.	52
Ferreira, G.	450	France, J.	25
Ferreira, L.M.A.	516	François, D.	257, 257, 260, 276, 507
Ferreira, M.A.	361, 361	Frappat, B.	338
Ferreira, R.L.C.	312	Fratini, F.	471
Ferreira-Dias, G.	429, 432	Fredholm, M.	208, 505
Ferrer, J.	330	Fredriksen, B.	622
Ferret, A.	519	Frelich, J.	78, 405
Ferretti, L.	186	French, P.	350
Ferri, N.	341, 496	Friedrich, M.	278
Fésüs, L.	190	Fries, R.	510
Ficco, A.	92	Friggens, N.C.	615
Fiems, L.O.	96	Frijters, A.C.J.	181
Fievez, V.	43	Fritz, S.	2
Fife, M.	503	Froidmont, E.	337, 357

Gigli, S.	420	Grabherr, H.	341
Gilain-Galliot, C.	490	Grabow, M.	243
Gilbert, H.	134, 159, 322, 603	Gracio, V.	429
Gilca, I.	547	Graebner, M.	99
Gillon, A.	106, 616	Grandoni, F.	570
Giorgetti, A.	422	Granier, R.	440
Giovagnoli, G.	288, 431, 432	Graulet, B.	313
Giovanetti, V.	348, 514	Gravel, C.	394
Girard, N.	3	Gredler, B.	182, 191, 409, 415
Gispert, M.	52, 76, 116, 235	Greef, J.	543
Giuffra, E.	133	Grenier, B.	245
Giza, E.	469	Greppi, G.F.	9
Glasser, T.A.	74	Greyling, J.P.C.	534
Glick, G.	192	Grgas, A.	15
Głowacz, K.	460	Grigorova, S.	370
Goddard, M.E.	294	Grimard, B.	248
Goddard, P.	112	Grodzycki, M.	437
Godino, R.F.	509	Groenen, M.A.M.	140
Goelema, J.O.	564	Groeneveld, E.	14, 20, 125, 129, 606
Goering, H.H.H.	505	Groenewald, I.B.	236
Goetz, K.-U.	613	Grosso, G.	341
Gogué, J.M.	285	Große Beilage, E.	438
Golian, A.G.	357, 571	Große-Brinkhaus, C.	161
Golik, M.	192	Grullon, L.	154
Golizadeh, M.	358, 582	Grzeskiewicz, S.	37
Golzar-Adabi, S.	102, 352	Grzeskowiak, G.	37
Gomes, F.A.	56	Guarcini, R.	17
Gomes, M.J.	546, 548, 585, 586	Guàrdia, M.D.	116
Gomes, R.C.	100, 100	Guarnizo, P.	567
Gómez, G.	237, 237, 238, 312	Guatteo, R.	554
Gómez, M.D.	8, 217, 223, 225, 427	Guedes, C.M.	407, 546, 546, 548, 563, 563
Gonyou, H.	254	Guéniot, F.	459
Gonzales Castillo, M.L.	317	Guérin, G.	215
Gonzalez, A.	18	Guerra, L.	152, 394
Gonzalez, J.	76	Guerreiro, C.I.P.D.	516
González, O.	532	Guéry, L.	164
González, P.	267	Guidi, L.	425
González, R.	329, 472	Guillaume, F.	2
González López, F.	332	Guillouet, P.	134
Gonzalez Lopez, V.	163	Guingand, N.	438, 558
Gonzalez-Montaña, J.R.	528	Gunn, G.	114
González-Martín, S.	493	Gunnarsson, E.	270
González-Recio, O.	180, 295	Guo, G.	296
González Rodríguez, A.	310, 310, 355, 615	Guridi, M.	178
González-Valero, L.	585	Gürtler, P.	349
Gonçalves, T.M.	147	Gutiérrez, J.P.	8, 198, 227, 265, 315
Good, M.	183	Gutierrez-Adan, A.	154
Goorchi, T.	102	Gutierrez-Chavez, A.J.	528
Gootwine, E.	73, 74, 510	Gutierrez-Estrada, J.C.	18
Gorjanc, G.	461	Gutiérrez-Gil, B.	149, 261
Gortázar, C.	552	Guy, D.R.	503
Götz, K.-U.	5	Guy, J.H.	116
Goulas, P.	84	Guzmán, J.L.	538, 539, 539
Gourdine, J.L.	1, 125, 126		
Goyache, F.	8, 175, 227, 265, 267	**H**	
Graber, M.	388	Habibi, A.	420

Habier, D.	163, 297
Haedari, M.	48
Hagiya, K.	201
Haider, A.	273
Haile, A.	122
Hajda, Z.	42
Hajilari, D.	355, 576, 579
Halachmi, I.	417
Haley, C.S.	135, 138, 178, 503, 601, 612
Hamadeh, S.K.	284
Hamann, H.	419
Hamasaki, Y.	416
Hamdene, M.	529
Hamed, A.	462
Hameister, T.	286
Hamilton, A.	503
Hammadi, M.	473
Hammami, H.	529
Hamouda, L.	264
Han, Y.K.	515, 522
Hanenberg, E.	231
Hanigan, M.D.	25
Hanna, N.	284
Hanoglu, H.	462
Hansen, H.H.	498
Hansen Axelsson, H.	107
Hanusová, E.	443, 453
Hanzen, C.	344, 357
Haque, F.	326, 488
Harangi, S.	15, 103, 399
Haresign, W.	459
Harlizius, B.	137
Haro, A.	369
Harper, E.	251, 251
Harper, J.H.	325
Harper, J.M.	327
Hars, J.	551
Hartl, K.	70
Hasani, S.	366
Haščík, P.	376
Hashish, S.M.	45
Hasler, H.	131
Hassani, S.	102, 352, 354
Hatefi Nezhad, K.	572
Haubitz, M.	212
Hausberger, M.	426
Hayashi, T.	155
Haydari, K.	45
Hayes, B.	294
Heard, C.	252
Hedayati, M.	276
Hedegaard, J.	280
Hedhammar, Å.	67
Hegedusova, Z.	50, 544
Heidari, M.	596
Heidinger, B.	191
Heindl, J.	62
Heine, A.	71
Heinrich, I.	413
Helal, F.I.S.	518
Hellbrügge, B.	450
Hellemans, B.	178
Help, H.	447
Hemsworth, P.H.	437
Hendriks, W.H.,	312
Henke, S.	249
Henkin, Z.	527
Hennessy, D.P.	437
Henning, M.D.	11
Henryon, M.	123
Heravi Moussavi, A.R.	313, 371, 373, 572, 520, 581, 587
Heringstad, B.	179
Hermansen, J.E.	75
Hernández, J.	234, 411, 413, 614, 618
Hernández, P.	173, 335
Hernández-Jover, M.	458, 483, 487, 495
Herold, P.	4, 73, 127
Herrera, M.	18
Herrero, M.	327, 328
Hettinga, K.A.	616
Heuven, H.C.M.	157
Hidalgo, C.O.	11
Hidasi, N.	567
Hiemstra, S.J.	5, 6, 77, 120, 321
Higgins, I.M.	183
Higuera, M.A.	139, 433
Hiraga, A.	249
Hocquette, J.F.	61, 96, 132, 319, 403
Hodate, K.	390
Hodžić, A.	188
Hoekman, A.J.W.	33
Hoelker, M.	182
Hofer, A.	208
Hoffmann, C.	340
Hofherr, J.	250, 252, 487
Hofmanová, B.	229, 430
Hogewerf, P.H.	484
Holcvart, M.	415
Holló, G.	80, 81
Holló, I.	80, 81
Holm, B.	132, 158
Holm, L.E.	156
Holmgren, N.	480
Holroyd, S.	185
Holt, T.	256
Holtz, J.	490
Holtz, W.	271, 277, 277, 278
Holyoake, P.K.	495
Homola, M.	50
Homolka, P.	363, 583, 587
Honarvar, M.	193

Jensen, B.B.	556	Kamada, H.	389
Jensen, K.K.	621	Kamalalavi, M.	358
Jensen, S.K.	231	Kamali, M.A.	102, 151, 352
Jerónimo, E.	36, 59, 86, 443	Kaminski, S.	447
Ježková, A.	406	Kanetani, T.	444, 447, 449
Jiang, H.Z.	316	Kanis, E.	162
Jiang, L.	506, 508	Kanitz, E.	99, 286
Jiménez-Marín, A.	505, 509	Kantanen, J.	126
Joerg, H.	303	Kantas, D.	84
Johansson, K.	107, 107, 608	Kanyar, R.	42
Joller, D.	505	Kanz, C.	4
Joly, A.	554	Kapell, D.N.R.G.	295
Jonas, E.	161	Kaptan, C.	462
Jones, J.H.	249	Karacaören, B.	135
Jonkus, D.	197, 414, 414	Karadjole, I.	15, 492
Jonmundsson, J.V.	270	Karagiannidou, A.	346
Jönsson, L.	214, 215	Karalazos, A.	57
Jordana, J.	267, 268, 268	Karalazos, V.	57
Jørgensen, C.B.	208, 505	Karamzadeh Omrani, H.	396
Jori, F.	336	Karbo, N.	498
Josipović, S.	547	Karimi, D.	262
Jouhet, E.	507	Karkabounas, S.	84
Jouneau, L.	181	Karkinen, K.T.	115, 500
Journaux, L.	132	Karkoodi, K.	44, 572, 590
Jovanovic, S.J.	235, 502	Karlskov-Mortensen, P.	505
Jovellar, L.C.	402	Karlsson, Å.	67
Joy, M.	82, 87, 98, 253, 262, 390	Karoui, S.	195
Juchem, S.O.	585	Kasuya, E.	390
Juga, J.	66, 66, 128	Katanos, I.	85
Jullien, E.	507	Katila, T.	219
Junge, W.	203, 614	Kato, K.	143
Jungerius, A.	504	Katoh, K.	142
Jüngst, H.	161	Kauffold, J.	456
Juniper, D.	427	Kaufmann, F.	342, 553
Jurado, J.J.	263, 278	Kaufmann, T.	388
Jurie, C.	61, 403	Kawęcka, A.	464, 465, 465, 466, 473, 549
Juska, R.	457	Kaya, I.	403, 525
Juskiene, V.	457	Kayan, A.	42
Juste, M.C.	468	Kaygısız, F.	206
Juszczuk-Kubiak, E.	194	Kazemi, M.	383, 593, 593
		Keane, M.G.	141, 400, 418
K		Kearney, F.	77
Kaart, T.	204	Kebreab, E.	25
Kababya, D.	74	Keidane, D.	59
Kačániová, M.	376	Keikha-Saber, M.	360, 360
Kaczor, U.	47	Kemp, B.	101
Kadarmideen, H.N.	605	Kempe, R.	368
Kadokawa, H.	389	Kemper, N.	450, 479, 537, 554
Kadowaki, H.	155	Kennedy, M.	252
Kafilzadeh, F.	62, 588	Kenny, D.A.	148, 400
Kaim, M.	406, 410	Kent, M.	300
Kairiša, D.	414, 414	Kern, G.	537
Kaiser, P.	503	Kernaleguen, L.	209
Kalbe, C.	95	Ketoja, E.	386, 394
Kalita, D.	246	Kettlewell, P.	252
Kalm, E.	160	Kettlewell, P.J.	251, 251

Kristensen, N.B.	24, 369, 372, 401	Larson, S.R.	325, 327
Kristensen, T.	231, 233	Larzul, C.	159, 434
Kritas, S.	515	Lasheen, M.A.	518
Kronberg, S.L.	37	Lassen, J.	108
Kruijt, L.	33, 34, 279	Latorre, M.A.	454, 455
Kruse, S.	557	Laugé, V.	497
Kubesova, M.	405	Lautrou, Y.	324
Kučević, D.	491	Lauvie, A.	75
Kuchida, K.	416	Lavaf, A.	21
Kuchtik, J.	410, 471, 544	Lavon, Y.	406
Kukovics, S.	265, 472, 475	Lavín, S.	553
Kunej, T.	197	Lawlor, P.G.	451
Kunz, P.L.	324	Lawrence, A.B.	281
Kuran, M.	10, 98, 392	Lazar, C.	466
Kurt, E.	281, 282	Lázaro, R.	567
Kushibiki, S.	390	Lazzari, M.	486, 494
Kusza, S.Z.	15, 228, 265	Lazzaroni, C.	404, 404
Kutlu, H.R.	41	Le Bihan-Duval, E.	138, 169
Kvapilík, J.	458	Le Cozler, Y.	238
Kyser, G.B.	327	Le Roy, P.	159, 169, 322, 603
		Leão, M.I.	361, 361
L		Lebboroni, G.	316, 318
La Manna, V.	317, 317, 318, 428, 428	Lebedová, L.	69
La Ragione, R.M.	565	Leboeuf, B.	270
La Terza, A.	315, 318	Lebzien, P.	243, 341, 581
Labatut, J.	3	Lecerf, F.	271
Laborda, P.	173	Leclerc, H.	196, 618
Labouriau, R.	287, 604	Lecomte, P.	232
Labrinea, E.	48, 85	Ledvinka, Z.	363
Labussiere, E.	568	Lee, W.I.	522
Laca, E.A.	325, 327	Leeb, T.	182
Lacetera, N.	117	Leenhouwers, J.I.	231
Lachica, M.	585	Legarra, A.	216, 298, 301, 603
Ladeira, M.M.	147	Lehel, L.	42
Laforest, J.P.	254	Lehmann, S.	308
Laga, V.	85	Leifert, C.	234
Lagant, H.	134	Leikus, R.	457
Lagriffoul, G.	270	Lek, S.	613
Lahoz, B.	275, 278	Leme, P.R.	100, 100, 408
Lahučký, R.	453	Lendelova, J.	336
Lainez, M.	39, 330, 552	León, M.	36
Laitat, M.	233	Lepetit, J.	132
Laitinen, M.	386	Leroux, C.	184
Lajudie, P.	283	Leroy, G.	68, 71
Laloux, L.	106	Leslie, E.E.C.	495
Lamadon, A.	621	Leury, B.J.	94
Lambe, N.R.	459	Levéziel, H.	132
Lambert-Derkimba, A.	75, 81	Lewczuk, D.	221
Lambertini, L.	87, 253	Lewis, F.	114
Landau, S.	74	Leymarie, C.	507
Landete-Castillejos, T.	329	Li, M.-H.	126
Langlois, B.	220	Li Destri Nicosia, D.	288, 432
Lanigan, G.	559	Licón, C.	467
Lanini, M.	117	Lidauer, M.	599, 608
Lapp, J.	560	Lien, S.	300
Larroque, H.	184	Liermann, T.	351

Marchi, E.	341, 496	Maté Caballero, J.	485
Marco, I.	553	Maupertuis, F.	230
Maretto, F.	145, 219	Maurice-Van Eijndhoven, M.H.T.	6, 7
Marguin, L.	490	Maxa, J.	537
Marguš, D.	15	Maximini, L.	409
Mari, F.	133	Mayer, M.	603, 610
Marie, M.	501, 619	Mayeres, P.	616
Marino, R.	609	Mayor, P.	335, 336, 396
Marković, D.	15	Mazzi, M.	389
Marlin, D.	248, 252	Mazzone, G.	87, 253
Marmaryan, G.	57	Mc Cabe, T.	559
Marnet, P.G.	326	Mc Cartney, E.	515
Marot, G.	248	Mcandrew, B.J.	503
Marqués, M.	154	Mccauley, I.	437
Marriott, D.	443	Mcdade, K.	169
Marsalek, M.	405	Mcgee, M.	148
Martemucci, G.	90	Mcgourty, G.T.	325
Martí, J.I.	82, 275	Mclean, K.A.	459
Martin, C.	118	Meade, G.	559
Martin, G.	293	Medel, P.	515
Martin, O.	291	Medina, C.	225
Martin, P.	184, 270	Mehouachi, M.	578
Martin, P.G.P.	245	Mehrabani-Yeganeh, H.	46, 269
Martín-Collado, D.	77	Mehri, M.	360, 360
Martín-Palomino, P.	246	Meirelles, F.V.	13, 18
Martínez, A.	38	Meldrum, K.	252
Martínez, G.	312	Mele, M.	189
Martínez, M.	381, 552	Melis, J.	185
Martinez, P.	536	Mello, A.C.L.	417
Martínez-Giner, M.	135	Melo, C.	158
Martínez-Royo, A.	262, 263, 278	Meléndez, A.J.	36
Martinez Villamor, V.	398	Mena, E.	407, 546, 546, 548, 586
Martini, A.	422	Mena, Y.	532
Martins Da Silva, A.	152	Menard, O.	184
Martyniuk, E.	83	Menčik, S.	15, 492
Marubashi, T.	515	Méndez, J.	413, 614, 618
Marzouk, K.	273	Menéndez-Buxadera, A.	142, 223, 256
Masamitsu, T.	109	Menesatti, P.	117, 118
Maschio, M.	389	Mengi, A.	206
Masri, A.	459	Menrath, A.	554
Massaoudi, I.	375	Mereu, A.	348, 514
Massault, C.	178	Merino, M.J.	11
Masson, J.P.	599	Merks, J.W.M.	231
Mata, J.	545	Mesías, F.J.	497
Matas, C.	154	Mészáros, G.	133
Mateus, L.	432	Metges, C.C.	99
Mathlouthi, N.	375	Methlouthi, N.	578
Mathur, P.	207	Metsios, A.	84, 84
Matika, O.	459	Meunier-Salaün, M.C.	623, 623, 624
Matilainen, K.	599	Meuwissen, T.H.E.	6, 120, 120, 300, 320
Maton, C.	485	Meyer, H.H.D.	349
Matson, T.	307	Meyer, U.	243
Matthews, A.	27	M'hamdi, N.	187
Mattos, E.C.	14, 18, 189	Mialon, M.M.	118
Mattsson, B.	480	Michas, V.	57
Matás, C.	309, 391	Michel, G.	104

Micol, D.	118, 403	Moioli, B.	224
Miculis, J.	41	Mojtahedi, M.	581
Migdał, W.	454	Moladoost, K.	262
Miglior, F.	31, 104, 106, 127, 205	Molina, A.	8, 8, 217, 218, 222, 223, 225, 227, 315,
Mihailova, G.S.	367		423, 430, 467
Mihina, Š.	336, 580	Molina, E.	76
Mihók, S.	72, 227, 228	Molina, M.P.	409, 419, 467
Mijić, P.	412	Moll, J.	303
Mikawa, S.	155	Molla Salehi, M.R.	397
Mikhail, W.A.	517, 573, 582	Molle, G.	514
Mikko, S.	14	Molnár, A.	472
Mikkola, M.	394	Momani Shaker, M.	544, 545
Mikulec, Ž.	492	Mömke, S.	184
Milan, D.	159	Momm, H.	127
Milanesi, E.	186, 317	Monazami, H.R.	500
Milerski, M.	458	Monget, P.	192
Milewski, S.	475, 476, 476, 541, 541, 542	Monleón, E.	391
Milisits, G.	166, 167	Monniaux, D.	192
Mills, A.A.	228, 307, 309	Montalvo, G.	327, 328, 439, 439, 558
Milne, C.	112	Monteiro, A.	549
Milton, J.	543	Montossi, F.	335
Milán, M.J.	329, 472, 489, 490, 499	Monzón, M.	391
Minogue, D.	350	Moors, E.	477
Mioc, B.	477	Moradi Shahrbabak, H.	269, 550
Miotello, S.	311	Moradi Shahrbabak, M.	128, 269, 550
Mirabito, L.	250, 332	Morais, R.	407
Miraei-Ashtiani, S.R.	193, 202	Morais, S.B.	373
Miraglia, N.	425, 426	Morales-Delanuez, A.	468, 543
Miranda, G.	184	Moravej, H.	46, 379, 521, 524, 591, 591, 592
Miranda, M.	234, 618	More, S.J.	183
Mirhosseini, S.Z.	340	Moreira, O.C.	443
Mirkena, T.	122	Moreira Da Silva, F.	307, 420
Mirza Aghazadeh, A.	93, 375, 576	Morek-Kopec, M.	194
Mirzaei-Aghsaghali, A.	63	Morel, C.	388
Misztal, I.	263, 298, 301	Moreno, A.	16, 17
Mitchell, M.A.	251, 251	Moreno, C.	140, 322, 460, 507
Mitterwallner, I.	190	Moreno-Indias, I.	474, 543
Miyagi, A.	51	Moreno-Romieux, C.	603
Mizeli, C.	77	Moreno-Sánchez, N.	141
Mizubuti, I.Y.	542	Moretti, D.B.	568
Mlyneková, Z.	580, 584	Mori, H.	444, 447, 449
Moalem, U.	410	Moriya, N.	390
Möckel, P.	351	Mormède, P.	279, 281, 282, 282, 284, 285
Modesto, E.C.	417	Morraveg, H.	595
Moeini, M.M.	21, 48, 101, 274, 274, 418, 540, 596	Morravege, H.	385
Moevi, I.	403	Morris, D.G.	240
Mogensen, L.	231	Morsy, A.	273
Moghadam, A.A.	274	Moset, V.	330
Moghimi, A.	517	Moslemipur, F.	102, 102, 351, 352, 352, 353, 354
Mohajer, M.	576	Mosquera-Losada, R.	540, 580
Mohamed, M.I.	377	Mota De Azevedo, S.	420
Mohamed, S.	528, 531	Mota-Velasco, J.	503
Mohammadian, B.	387	Mougios, V.	57
Mohan, N.H.	246, 560	Mounaix, B.	250
Moheghi, M.M.	593	Mourão, G.B.	13, 20, 189
Moigneau, C.	282	Mourão, J.	330

Moya, J.	330	Nassiry, M.	144, 168, 266
Mozafari, N.	151	Natale, F.	250, 252, 487
Mudrik, Z.	377, 381, 592	Nauta, W.J.	20
Muela, E.	50, 51, 54, 92	Nava, J.	561
Muelas, R.	171	Nava, S.	486, 494
Mueller, C.	559	Navajas, E.A.	401
Mueller, U.	200	Naves, M.	126, 256, 258
Muižniece, I.	414	Naya, H.	295
Muklada, H.	74	Nedelec, Y.	501
Mulder, H.A.	181, 321, 612	Negrini, R.	186, 609
Mulindwa, H.	291	Negussie, E.	179, 608
Müller, A.	388	Nejati Javaremi, A.	128, 193, 397, 397, 398
Müller, A.B.	344	Nejsum, P.	505
Mulsant, P.H.	192, 271	Nemati, I.	259
Münch, C.	424	Nemati, Z.	588
Munim, T.	109, 444, 447, 449	Németh, T.	265, 472, 475
Munksgaard, L.	283, 349	Neser, F.W.C.	150
Muñoz, G.	213	Nettier, B.	289
Muñoz, M.	161	Neuenschwander, S.	212
Mura, M.C.	146	Neuenschwander, T.F.-O.	104
Murai, M.	389	Neugebauer, N.	605
Murani, E.	247, 281, 282	Newbold, C.J.	371
Murasawa, N.	416	Newman, S.	131
Murphy, E.J.	37	Nicks, B.	233
Murphy, P.	350	Nicolazzi, E.	609
Murray, B.B.	482	Nicoloso, L.	186, 317
Muzzachi, S.	394	Nielsen, B.	208, 211
Mwai, A.O.	291	Nielsen, T.R.	111
Myllymäki, H.	205	Nielsen, U.S.	608
		Nieto, R.	369, 567
		Niidome, K.	376
N		Nikkhah, A.	346
Nachreiner, R.	67	Nikkonen, T.	66
Nacu, G.H.	547	Nikolić, D.	214
Nadaf, J.	138	Nikolic, N.	613
Nadaf, S.	308	Nilforooshan, M.A.	610
Nader, G.	585	Nili, N.	399
Naderfrad, H.R.	21	Nistor, E.	47
Naeemipoor, H.	358	Nita, S.	57
Nafarrate, L.	292	Nitas, D.	47, 57, 85
Nagy, I.	265	Niżnikowski, R.	460
Nahed, J.	532	Noblet, J.	134, 568
Najar, T.	368, 470	Noé, G.	181
Nájera, A.I.	49, 53	Nogueira Filho, J.C.M.	100, 408
Nakahashi, Y.	416	Noguera, J.L.	135, 157, 165, 606
Nakajima, H.	142	Nolan, J.V.	565
Nakamura, M.	389	Nollet, L.	34, 43
Nanni Costa, L.	164	Nooriyan Sarvar, E.	101
Narahara, H.	177	Norman, H.D.	242
Nardone, A.	17, 117, 149, 609	Norouzian, M.A.	60
Naserian, A.A.	63, 371, 372, 373, 517, 518	Norviliene, J.	457
Näsholm, A.	14, 141, 215, 216	Nováková, I.	376
Naskar, S.	246	Nováková, K.	70
Näslund, J.	196	Nuchchanart, W.	247
Nasr-Esfahani, M.H.	399	Nuernberg, G.	95
Nasserian, A.A.	384, 593, 594, 594	Nuernberg, K.	81
Nassiri Moghaddam, H.	357, 399, 571		

Pavão, A.L.	198	Philipsson, J.	31, 107, 107, 214, 215, 216
Paya, I.	206	Phillips, K.	112
Payeras, L.	175	Phocas, F.	122, 130, 283
Pazzola, M.	146, 261	Phongpiachan, P.	42
Pecaud, D.	338	Piacère, A.	270
Pécsi, A.	15	Picard, B.	61
Pedersen, L.D.	319	Piccand, V.	324
Pediconi, D.	315	Picron, P.	337
Pedonese, F.	471	Piedrafita, J.	79, 86, 264, 604
Pedrosa, V.B.	14	Pierce, K.M.	129, 199, 206, 559
Peeters, K.	221	Pierni, E.	425
Peeva, T.Z.	583	Pieta, M.	229
Peinado, J.	434	Pigozzi, G.	306
Peippo, J.	182, 205, 386, 394	Pihler, I.	547
Pellegrini, P.	283	Pijet, B.	95
Pellicer-Rubio, M.T.	270	Pijet, M.	95
Pellikaan, W.F.	422, 423	Pikuła, R.J.	225
Pelmus, R.	466	Piles, M.	9, 172
Peltonen, A.	433	Pilla, F.	17, 149
Peña, F.	315, 427	Pillet, E.	304
Pena, R.N.	135, 153, 154, 156, 157, 165	Pimentel, E.C.G.	302, 320, 601
Penasa, M.	78	Pinard-Van Der Laan, M.-H.	169, 502
Penazzi, P.	452	Pinato, T.	483
Peneva, M.	620	Pineiro, C.	327, 328, 439, 439, 558
Pentelescu, O.N.	332	Pinheiro, M.	124
Perea, J.	237, 237, 238, 312, 532, 533, 533	Pinheiro, V.	330
Pereira, A.A.	88, 89, 373	Pinto, M.	292
Pereira, A.S.C.	408	Pinton, P.	245
Pereira, E.	461	Piquer, O.	381
Pereira, V.	234, 411, 413, 614, 618	Pirany, N.	266
Perestrello, F.	429	Piras, G.	261
Perevolotsky, A.	74	Piraux, E.	357
Pérez, P.	50	Pirlo, G.	97
Perez-Almero, J.L.	115, 464	Pitel, F.	169
Pérez-Cabal, M.A.	198	Pizzi, F.	5
Pérez-Elortondo, F.J.	49, 53	Plachy, V.L.	377, 381
Perez-Pardal, L.	175, 265, 267	Plaixats, J.	334
Pérez-Quintero, G.	9	Planchon, V.	357
Perini, D.	609	Plante, Y.	267
Perisic, P.	529	Plavšić, M.	491
Perny, P.	621	Pőcze, O.	167
Peškovičová, D.	417, 443, 453, 468	Pogorzelska, A.	95
Pessoa, R.A.S.	361, 361	Pogran, S.	336
Petersson, K.-J.	107	Poigner, J.	133
Pethick, D.W.	403	Poirel, D.	282
Petkov, G.S.	367	Poix, C.	290
Petkova, M.	370	Pokorna, M.	471
Petr, R.	544	Pol, A.	334
Petridou, A.	57	Poláčiková, M.	584
Pettersson, G.	107	Polák, P.	417
Peura, J.	368	Polgár, J.P.	190
Pezzi, P.	370	Polo, J.	449, 479
Pfeiffer, A.-M.	351	Połoszynowicz, J.	194
Phatsara, C.	161, 182, 497	Polvillo, O.	53, 267
Philipp, H.C.	555	Pommeret, D.	599
Philippe, F.X.	233	Ponce De Leon, F.A.	267

Rekik, B.	529	Rodrigues, A.P.O.	356
Rekik, M.	272	Rodrigues, E.C.	56, 88, 89, 91
Remondet, M.	130	Rodrigues, M.	563, 563
Renand, G.	122, 132, 242	Rodrigues, M.C.O.	373
Renard, J.-P.	181	Rodrigues, S.	461
Renieri, C.	315, 316, 317, 317, 318, 318	Rodríguez, A.	11
Repa, I.	166	Rodríguez, C.	246, 479
Rérat, M.	339	Rodriguez, M.	519
Revilla, R.	526	Rodríguez, M.C.	137, 156, 161, 213
Reynaud, A.	313	Rodriguez, P.	112, 287
Reynolds, C.K.	24	Rodríguez Latorre, A.	343
Reza Yazdi, K.	385	Rodríguez-López, J.M.	585
Rezaeian, M.	378	Rodriguez-Navarro, A.	169
Rezaeipour, V.	366	Rodriguez-Ramilo, S.T.	602
Rezaii, F.	572	Rodríguez-Sánchez, J.A.	454, 455
Rezayazdi, K.	346, 382, 516, 589, 591	Roehe, R.	160, 295, 401
Rezende, F.M.	13, 18, 20	Roepstorff, A.	505
Riasi, A.	358, 358, 575, 582	Roepstorff, L.	215, 216
Ribeca, C.	145	Roessler, R.	127
Ribeiro, E.L.A.	542	Rogel-Gaillard, C.	245
Ribikauskiene, D.	138	Rognon, X.	68, 71, 264
Ribo, O.	325	Rohde, H.	105
Ricard, A.	215, 216, 220, 222	Rojas-Olivares, M.A.	489, 492, 493
Ricard, E.	491	Røjen, B.A.	369
Ricardo, C.F.	147	Rollet, L.	619
Richardson, R.I.	149, 401	Romera, A.J.	326
Rickard, B.A.	484	Ron, M.	192
Ridder, C.	615	Roncada, P.	9, 556
Ridolfo, E.	288	Roncancio-Peña, C.	513
Riek, A.	535	Ronchi, B.	577
Riha, J.A.N.	50, 147	Rondia, P.	501
Rinaldi III, A.E.	495	Rönnegård, L.	597, 608
Rinaldi III, A.R.	482	Røntved, C.M.	506
Ringdorfer, F.	534	Rosa, A.	429
Rings, F.	182	Rosa, G.J.M.	295
Rinnhofer, B.	285	Rosa, H.J.D.	40, 40, 56, 450
Rios, Á.F.L.	152	Rose-Meierhöfer, S.	344
Ripoll, G.	52, 290, 454, 455	Rosell, R.	506
Ripoll, R.	76, 87	Rösler, H.J.	345
Riquet, J.	159, 209	Ross, D.W.	401
Ritvos, O.	386	Rossato, L.V.	56, 88, 89
Rius-Vilarrasa, E.	119, 459	Rossi, R.O.D.S.	147
Robalo Silva, J.	307	Roth, M.	342
Robaye, V.	33, 347, 425, 617	Roth, N.	374, 590
Robert-Granié, C.	184, 192, 507, 599	Rotz, C.A.	230
Robertson, M.W.	94	Roughsedge, T.	119
Robinson, P.H.	359, 571, 585	Roura, E.	586
Robles, J.	50	Roussel, S.	326
Roca, M.I.	409	Rousset, S.	132
Roca Fernández, A.I.	310, 310, 355, 615	Rowe, S.J.	601
Roche, A.	275, 278	Royer, E.	230, 440
Roche, S.	308	Royo, L.J.	175, 265, 267
Rodenburg, J.	339, 482	Rózsa Várszegi, Z.S.	72
Rodero, A.	427	Różycki, M.	445, 445
Rodero, E.	18, 115, 464	Rubio, R.	467
Rodrigañez, J.	246	Rucinski, M.	284

Rudolphi, B.	239	San Cristobal, M.	192, 613
Rueda, J.	141	San Primitivo, F.	185, 258
Rufí, J.	86	Sánchez, A.	161, 268, 487, 506
Ruiz, F.A.	532	Sánchez, J.	515
Ruiz, R.	49, 53, 73, 76, 290	Sánchez, J.L.	213
Ruiz, S.	154, 405	Sánchez, J.P.	185, 258
Ruiz De Gordoa, J.C.	49, 53	Sanchez, L.	16, 17
Ruiz-De La Torre, J.L.	586	Sanchez, M.D.	115
Rumbles, I.	482	Sánchez, M.J.	315
Ruotolo, E.	250	Sanchez, M.P.	159
Rupp, R.	507	Sánchez, P.	275
Russell, L.	449	Sánchez-Macías, D.	468, 474
Russo, V.	136, 164, 166, 609	Sánchez Recio, J.M.	80, 402
Rutten, M.J.M.	7, 616	Sanchez-Vazquez, M.	114
Ryan, S.E.	484	Sanjabi, M.R.	21, 274, 418, 596
Ryan, T.P.	451	Santacreu, M.A.	173
Rydhmer, L.	1, 107, 125, 163, 281, 497	Santamaria, P.	292
		Santamarina, C.	458, 483
S		Santana, M.T.	91
Saa, C.	489	Santibañez, A.	470
Saarinen, K.	115	Santilocchi, R.	526
Sabioni, S.	288, 432	Santos, A.S.	422, 423, 546
Šáda, I.	544	Santos, I.	19
Sadeghi, A.A.	340	Santos, M.	585
Sadeghi, B.	144, 168, 266	Santos, M.V.F.	237, 312, 417, 532, 533
Sadeghi, G.H.	360, 360, 521	Santos, R.	218, 222, 423, 430
Sadeghipanah, H.	91, 92, 386, 387	Santos, V.	407, 546, 546, 548
Saez, J.L.	83	Santos Fadista, J.P.	156
Safari, R.	383, 384, 594, 594	Santos-Silva, J.	36, 59, 86
Safari, S.	500	Santos-Silva, M.F.	12, 19
Safarzadeh-Torghabeh, H.	354	Santus, E.	186, 609
Šafus, P.	200, 201	Sañudo, C.	38, 50, 51, 52, 53, 54, 92, 335
Sahana, G.	180, 506	Sanz, A.	82, 87, 253, 286, 390, 391, 526, 619
Saïd, B.	55	Sanz, F.	439, 439
Saidi, C.	578	Sanz, M.J.	327, 439, 439
Saïdi-Mehtar, N.	264	Sapa, J.	283
Saintilan, R.	242	Saraiva, V.	86
Sainz, R.D.	100	Saran Netto, A.S.	367, 407
Sairanen, J.	219	Saravanaperumal, S.A.	315
Sakai, E.	177	Saremi, B.	63, 371, 372
Salajeanu, A.	193	Sargentini, C.	422
Salama, A.A.K.	462, 489, 490, 492, 493, 493, 494	Sargolzaei, M.	127
Salama, R.	377	Sarhadi, F.	91, 92
Salangoudis, A.	364	Sarmah, B.C.	246
Salari, S.	357, 571	Sartin, J.L.	93
Saleh, M.	271	Sasaki, O.	159, 201
Salehi, S.	386, 387	Sasaki, S.	160
Salem, M.F.	577	Sassi, T.	375
Salemi, M.	374	Sato, K.	376
Salilew-Wondim, D.	182, 205	Sato, M.	144
Salimei, E.	428, 428	Sato, S.	155
Sallam, A.A.	273	Satoh, M.	159, 160, 201
Sallam, M.	273	Sauerwein, H.	98
Salmi, B.	434	Sauvant, D.	291
Samei, A.	521	Savic, M.	235, 502
Samiei, R.	353	Savvidou, S.	565

Printed in the United States
by Baker & Taylor Publisher Services